AF868914

Geotechnical Engineering

Principles and Practices of Soil Mechanics and Foundation Engineering

V. N. S. Murthy
Consulting Geotechnical Engineer
Bangalore, India

MARCEL DEKKER, INC. NEW YORK • BASEL

Civil and Environmental Engineering

A Series of Reference Books and Textbooks

Editor

Michael D. Meyer
Department of Civil and Environmental Engineering
Georgia Institute of Technology
Atlanta, Georgia

1. **Preliminary Design of Bridges for Architects and Engineers**
   Michele Melaragno
2. **Concrete Formwork Systems**
   Awad S. Hanna
3. **Multilayered Aquifer Systems: Fundamentals and Applications**
   Alexander H.-D. Cheng
4. **Matrix Analysis of Structural Dynamics: Applications and Earthquake Engineering**
   Franklin Y. Cheng
5. **Hazardous Gases Underground: Applications to Tunnel Engineering**
   Barry R. Doyle
6. **Cold-Formed Steel Structures to the AISI Specification**
   Gregory J. Hancock, Thomas M. Murray, Duane S. Ellifritt
7. **Fundamentals of Infrastructure Engineering: Civil Engineering Systems: Second Edition, Revised and Expanded**
   Patrick H. McDonald
8. **Handbook of Pollution Control and Waste Minimization**
   Abbas Ghassemi
9. **Introduction to Approximate Solution Techniques, Numerical Modeling, and Finite Element Methods**
   Victor N. Kaliakin
10. **Geotechnical Engineering: Principles and Practices of Soil Mechanics and Foundation Engineering**
    V. N. S. Murthy

Additional Volumes in Production

Chemical Grouting and Soil Stabilization: Third Edition, Revised and Expanded
Reuben H. Karol
Estimating Building Costs
Calin M. Popescu, Kan Phaobunjong, Nuntapong Ovararin

Geotechnical Engineering

Principles and Practices of Soil Mechanics and Foundation Engineering

V. N. S. Murthy
Consulting Geotechnical Engineer
Bangalore, India

MARCEL DEKKER
MARCEL DEKKER, INC. NEW YORK • BASEL

Library of Congress Cataloging-in-Publication Data

Murthy, V.N.S.
Geotechnical engineering : principles and practices of soil mechanics and foundation engineering / V.N.S. Murthy.
p. cm. -- (Civil and environmental engineering ; 10)
ISBN 0-8247-0873-3 (alk. Paper)
1. Engineering geology. 2. Soil mechanics. 3. Foundations. I. Title. II. Series.

TA705 .M85 2002
624.1'51--dc21

2002073993

Typesetting and illustrations by: Aicra Info Mates Pvt Ltd, Hyderabad, India.

This book is printed on acid-free paper.

Headquarters
Marcel Dekker, Inc.
270 Madison Avenue, New York, NY 10016
tel: 212-696-9000; fax: 212-685-4540

Eastern Hemisphere Distribution
Marcel Dekker AG
Hutgasse 4, Postfach 812, CH-4001 Basel, Switzerland
tel: 41-61-260-6300; fax: 41-61-260-6333

World Wide Web
http://www.dekker.com

The publisher offers discounts on this book when ordered in bulk quantities. For more information, write to Special Sales/Professional Marketing at the headquarters address above.

Copyright © 2003 by Marcel Dekker, Inc. All Rights Reserved.

Neither this book nor any part may be reproduced or transmitted in any form or by any means, electronic or mechanical, including photocopying, microfilming, and recording, or by any information storage and retrieval system, without permission in writing from the publisher.

Current printing (last digit):
10 9 8 7 6 5 4

PRINTED IN THE UNITED STATES OF AMERICA

Dedicated to the Cause of Students

FOREWORD

Geotechnical Engineering: Principles and Practices of Soil Mechanics and Foundation Engineering is a long title befitting a major work. I am pleased to introduce this superb volume destined for a readership of students, professors, and consultants. What makes this text different from other books on these subjects that appear each year and why am I recommending it to you? I have been working and teaching in the area of geotechnical engineering for 25 years. I have read and used scores of textbooks in my classes and practice. Dr. Murthy's text is by far the most comprehensive text I have found. You will find that his organization of the subject matter follows a logical progression. His example problems are numerous and, like the text, start from fundamental principles and progressively develop into more challenging material. They are the best set of example problems I have seen in a textbook. Dr. Murthy has included ample homework problems with a range of difficulty meant to help the student new to the subject to develop his/her confidence and to assist the experienced engineer in his/her review of the subject and in professional development.

As the technical editor I have read the entire manuscript three times. I have been impressed by the coverage, the clarity of the presentation, and the insights into the hows and whys of soil and foundation behavior. Often I have been astonished at Dr. Murthy's near-conversational approach to sharing helpful insights. You get the impression he's right there with you guiding you along, anticipating your questions, and providing instruction and necessary information as the next steps in the learning process. I believe you will enjoy this book and that it will receive a warm welcome wherever it is used.

I thank Dr. Murthy for his commitment to write this textbook and for sharing his professional experience with us. I thank him for his patience in making corrections and considering suggestions. I thank Mr. B. J. Clark, Senior Acquisitions Editor at Marcel Dekker Inc., for the opportunity to be associated with such a good book. I likewise express my appreciation to Professor Pierre Foray of l'Ecole Nationale Superieure d'Hydraulique et de Mecanique de Grenoble, Institut National Polytechnique de Grenoble, France for his enthusiastic and unflagging support while I edited the manuscript.

Mark T. Bowers, Ph.D., P. E.
Associate Professor of Civil Engineering
University of Cincinnati

FOREWORD

It gives me great pleasure to write a foreword for *Geotechnical Engineering: Principles and Practices of Soil Mechanics and Foundation Engineering*. This comprehensive, pertinent and up-to-date volume is well suited for use as a textbook for undergraduate students as well as a reference book for consulting geotechnical engineers and contractors. This book is well written with numerous examples on applications of basic principles to solve practical problems.

The early history of geotechnical engineering and the pioneering work of Karl Terzaghi in the beginning of the last century are described in Chapter 1. Chapters 2 and 3 discuss methods of classification of soil and rock, the chemical and the mechanical weathering of rock, and soil phase relationships and consistency limits for clays and silts. Numerous examples illustrate the relationship between the different parameters. Soil permeability and seepage are investigated in Chapter 4. The construction of flow nets and methods to determine the permeability in the laboratory and in the field are also explained.

The concept of effective stress and the effect of pore water pressure on effective stress are discussed in Chapter 5. Chapter 6 is concerned with stress increase in soil caused by surface load and methods to calculate stress increase caused by spread footings, rafts, and pile groups. Several examples are given in Chapter 6. Consolidation of soils and the evaluation of compressibility in the laboratory by oedometer tests are investigated in Chapter 7. Determination of drained and undrained shear strength by unconfined compression, direct shear or triaxial tests is treated in Chapter 8.

The important subject of soil exploration is discussed in Chapter 9, including the use of penetration tests such as SPT and CPT in different countries. The stability of slopes is investigated in Chapter 10. Methods using plain and circular slip surfaces to evaluate stability are described such as the methods proposed by Bishop, Fellenius, Morgenstern, and Spencer. Chapter 11 discusses methods to determine active and passive earth pressures acting on retaining and sheet pile walls.

Bearing capacity and settlement of foundation and the evaluation of compressibility in the laboratory by oedometer tests are discussed in Chapters 12, 13, and 14. The effect of inclination and eccentricity of the load on bearing capacity is also examined. Chapter 15 describes different pile types, the concept of critical depth, methods to evaluate the bearing capacity of piles in cohesive and cohesionless soils, and pile-driving formulae. The behavior of laterally loaded piles is investigated in Chapter 16 for piles in sand and in clay. The behavior of drilled pier foundations and the effect of the installation method on bearing capacity and uplift are analyzed in Chapter 17. Foundations on swelling and collapsible soils are treated in Chapter 18 as are methods that can be used to reduce heave. This is an important subject, seldom treated in textbooks. The design of retaining walls is covered in Chapter 19, as well as the different factors that affect active and passive earth pressures. Different applications of geotextiles are covered in this chapter as well as the topic of reinforced earth. Cantilever, anchored, and strutted sheet pile walls are investigated in Chapter 20, as are methods to evaluate stability and the moment distribution. Different soil improvement methods, such as compaction of granular soils, sand compaction piles, vibroflotation, preloading, and stone columns, are described in Chapter 21. The chapter also discusses lime and cement stabilization. Appendix A provides a list of SI units, and Appendix B compares methods that have been proposed.

This textbook by Prof. V. N. S. Murthy is highly recommended for students specializing in geotechnical engineering and for practicing civil engineers in the United States and Europe. The book includes recent developments such as soil improvement and stabilization methods and applications of geotextiles to control settlements and lateral earth pressure. Numerous graphs and examples illustrate the most important concepts in geotechnical engineering. This textbook should serve as a valuable reference book for many years to come.

Bengt B. Broms, Ph. D.
Nanyang Technical University, Singapore (retired).

PREFACE

This book has the following objectives:

1. To explain the fundamentals of the subject from theory to practice in a logical way
2. To be comprehensive and meet the requirements of undergraduate students
3. To serve as a foundation course for graduate students pursuing advanced knowledge in the subject

There are 21 chapters in this book. The first chapter traces the historical background of the subject and the second deals with the formation and mineralogical composition of soils. Chapter 3 covers the index properties and classification of soil. Chapters 4 and 5 explain soil permeability, seepage, and the effect of water on stress conditions in soil. Stresses developed in soil due to imposed surface loads, compressibility and consolidation characteristics, and shear strength characteristics of soil are dealt with in Chapters 6,7, and 8 respectively. The first eight chapters develop the necessary tools for computing compressibility and strength characteristics of soils.

Chapter 9 deals with methods for obtainig soil samples in the field for laboratory tests and for determining soil parameters directly by use of field tests. Chapters 10 to 20 deal with stability problems pertaining to earth embankments, retaining walls, and foundations. Chapter 21 explains the various methods by which soil in situ can be improved. Many geotechnical engineers have not appreciated the importance of this subject. No amount of sophistication in the development of theories will help the designers if the soil parameters used in the theory are not properly evaluated to simulate field conditions. Professors who teach this subject should stress this topic.

The chapters in this book are arranged in a logical way for the development of the subject matter. There is a smooth transition from one chapter to the next and the continuity of the material is maintained. Each chapter starts with an introduction to the subject matter, develops the theory, and explains its application to practical problems. Sufficient examples are worked out to help students understand the significance of the theories. Many homework problems are given at the end of each chapter.

The subject matter dealt with in each chapter is restricted to the requirements of undergraduate students. Half-baked theories and unconfirmed test results are not developed in this book. Chapters are up-to-date as per engineering standards. The information provided in Chapter 17 on drilled pier foundations is the latest available at the time of this writing. The design of mechanically stabilized earth retaining walls is also current. A new method for predicting the nonlinear behavior of laterally loaded vertical and batter piles is described in Chapter 16.

The book is comprehensive, rational, and pertinent to the requirements of undergraduate students. It serves as a foundation course for graduate students, and is useful as a reference book for designers and contractors in the field of geotechnical engineering.

ACKNOWLEDGEMENTS

It is my pleasure to thank Marcel Dekker, Inc., for accepting me as a single author for the publication of my book. The man who was responsible for this was Mr. B.J. Clark, the Executive Acquisition Editor. It was my pleasure to work under his guidance. Mr. Clark is a refined gentleman personified, polished, and clear sighted. I thank him cordially for the courtesies and help extended to me during the course of writing the manuscript. I remain ever grateful to him.

Writing a book for American Universities by a nonresident of America is not an easy task. I needed an American professor to edit my manuscript and guide me with regards to the requirements of undergraduate students in America. Dr. Mark T. Bowers, Associate Professor of Civil Engineering, University of Cincinnati, accepted to become my consultant and chief editor. Dr. Bowers is a man of honesty and integrity. He is dedicated to the cause of his profession. He worked hard for over a year in editing my book and helped me to streamline to make it acceptable to the undergraduate students of American Universities. I thank Dr. Bowers for the help extended to me.

There are many in India who helped me during the course of writing this book. Some provided me useful suggestions and others with references. I acknowledge their services with thanks. The members are:

Mr. S. Pranesh	Managing Director Prism Books Pvt Ltd Bangalore
Dr. K.S.Subba Rao	Professor of Civil Engineering Indian Institute of Science Bangalore
Dr. T.S. Nagaraj	Professor of Civil Engineering (Emeritus), Indian Institute of Science, Bangalore
Dr. C. Subba Rao	Professor of Civil Engineering Indian Institute of Technology Kharagpur

Chaitanya Graphics, Bangalore, provided the artwork for the book. I thank Mr S.K. Vijayasimha, the designer, for the excellent job done by him.

My son Prakash was associated with the book since its inception. He carried on correspondence with the publishers on my behalf and sent reference books as needed. My wife Sharadamani was mainly responsible for keeping my spirit high during the years I spent in writing the book. I remain grateful to my son and my wife for all they did.

I sincerely thank Mr. Brian Black for his continuous efforts in the production of this book. I immensely thank Mr. Janardhan and Mr. Rajeshwar, computer engineers of Aicra Info Mates Pvt Ltd., Hyderabad, for their excellent typesetting work on this book.

V.N.S. Murthy

CONTENTS

CHAPTER 1

INTRODUCTION

1.1 GENERAL REMARKS

Karl Terzaghi writing in 1951 (Bjerrum, et. al., 1960), on 'The Influence of Modern Soil Studies on the Design and Construction of Foundations' commented on foundations as follows:

> *Foundations can appropriately be described as a necessary evil. If a building is to be constructed on an outcrop of sound rock, no foundation is required. Hence, in contrast to the building itself which satisfies specific needs, appeals to the aesthetic sense, and fills its matters with pride, the foundations merely serve as a remedy for the deficiencies of whatever whimsical nature has provided for the support of the structure at the site which has been selected. On account of the fact that there is no glory attached to the foundations, and that the sources of success or failures are hidden deep in the ground, building foundations have always been treated as step children; and their acts of revenge for the lack of attention can be very embarrassing.*

The comments made by Terzaghi are very significant and should be taken note of by all practicing Architects and Engineers. Architects or Engineers who do not wish to make use of the growing knowledge of foundation design are not rendering true service to their profession. Since substructures are as important as superstructures, persons who are well qualified in the design of substructures should always be consulted and the old proverb that a 'stitch in time saves nine' should always be kept in mind.

The design of foundations is a branch of Civil Engineering. Experience has shown that most of these branches have passed in succession through two stages, the empirical and the scientific, before they reached the present one which may be called the stage of maturity.

The stage of scientific reasoning in the design of foundations started with the publication of the book *Erdbaumechanik* (means Soil Mechanics) by Karl Terzaghi in 1925. This book represents the first attempt to treat Soil Mechanics on the basis of the physical properties of soils. Terzaghi's contribution for the development of Soil Mechanics and Foundation Engineering is so vast that he may truly be called the *Father of Soil Mechanics.* His activity extended over a period of about 50 years starting from the year 1913. He was born on October 2, 1883 in Prague and died on October 25, 1963 in Winchester, Massachusetts, USA. His amazing career is well documented in the book *'From Theory to Practice in Soil Mechanics'* (Bjerrum, L., et. al., 1960).

Many investigators in the field of Soil Mechanics were inspired by Terzaghi. Some of the notable personalities who followed his footsteps are Ralph B. Peck, Arthur Casagrande, A. W. Skempton, etc. Because of the unceasing efforts of these and other innumerable investigators, Soil Mechanics and Foundation Engineering has come to stay as a very important part of the Civil Engineering profession.

The transition of foundation engineering from the empirical stage to that of the scientific stage started almost at the commencement of the 20th century. The design of foundations during the empirical stage was based mostly on intuition and experience. There used to be many failures since the procedure of design was only by trial and error.

However, in the present scientific age, the design of foundations based on scientific analysis has received a much impetus. Theories have been developed based on fundamental properties of soils. Still one can witness unsatisfactory performance of structures constructed even on scientific principles. The reasons for such poor performance are many. The soil mass on which a structure is to be built is heterogeneous in character and no theory can simulate field conditions. The fundamental properties of soil which we determine in laboratories may not reflect truly the properties of the soil *in-situ.* A judicial combination of theory and experience is essential for successful performance of any structure built on earth. Another method that is gaining popularity is the *observational approach.* This procedure consists in making appropriate observations soon enough during construction to detect signs of departure of the real conditions from those assumed by the designer and in modifying either the design or the method of construction in accordance with the findings.

1.2 A BRIEF HISTORICAL DEVELOPMENT

Many structures that were built centuries ago are monuments of curiosity even today. Egyptian temples built three or four thousand years ago still exist though the design of the foundations were not based on any presently known principles. Romans built notable engineering structures such as harbors, breakwaters, aqueducts, bridges, large public buildings and a vast network of durable and excellent roads. The *leaning tower of Pisa* in Italy completed during the 14th century is still a center of tourist attraction. Many bridges were also built during the 15th to 17th centuries. Timber piles were used for many of the foundations.

Another marvel of engineering achievement is the construction of the famed mausoleum Taj Mahal outside the city of Agra. This was constructed in the 17th century by the Mogul Emperor of Delhi, Shahjahan, to commemorate his favorite wife Mumtaz Mahal. The mausoleum is built on the bank of the river Jamuna. The proximity of the river required special attention in the building of the foundations. It is reported that masonry cylindrical wells have been used for the foundations. It goes to the credit of the engineers who designed and constructed this grand structure which is still quite sound even after a lapse of about three centuries.

The first rational approach for the computation of earth pressures on retaining walls was formulated by Coulomb (1776), a famous French scientist. He proposed a theory in 1776 called the "Classical Earth Pressure Theory". Poncelet (1840) extended Coulomb's theory by giving an elegant graphical method for finding the magnitude of earth pressure on walls. Later, Culmann (1875) gave the Coulomb-Poncelet theory a geometrical formulation, thus supplying the method with a broad scientific basis. Rankine (1857) a Professor of Civil Engineering in the University of

Glasgow, proposed a new earth pressure theory, which is also called a *Classical Earth Pressure Theory*.

Darcy (1856), on the basis of his experiments on filter sands, proposed a law for the flow of water in permeable materials and in the same year Stokes (1856) gave an equation for determining the terminal velocity of solid particles falling in liquids. The rupture theory of Mohr (1900) Stress Circles are extensively used in the study of shear strength of soils. One of the most important contributions to engineering science was made by Boussinesq (1885) who proposed a theory for determining stress distribution under loaded areas in a semi-infinite, elastic, homogeneous, and isotropic medium.

Atterberg (1911), a Swedish scientist, proposed simple tests for determining the consistency limits of cohesive soils. Fellenius (1927) headed a Swedish Geotechnical Commission for determining the causes of failure of many railway and canal embankments. The so-called *Swedish Circle method* or otherwise termed as the *Slip Circle method* was the outcome of his investigation which was published in 1927.

The development of the science of Soil Mechanics and Foundation Engineering from the year 1925 onwards was phenomenal. Terzaghi laid down definite procedures in his book published in 1925 for determining properties and the strength characteristics of soils. The modern soil mechanics was born in 1925. The present stage of knowledge in Soil Mechanics and the design procedures of foundations are mostly due to the works of Terzaghi and his band of devoted collaborators.

1.3 SOIL MECHANICS AND FOUNDATION ENGINEERING

Terzaghi defined Soil Mechanics as follows:

> *Soil Mechanics is the application of the laws of mechanics and hydraulics to engineering problems dealing with sediments and other unconsolidated accumulations of solid particles produced by the mechanical and chemical disintegration of rocks regardless of whether or not they contain an admixture of organic constituents.*

The term *Soil Mechanics* is now accepted quite generally to designate that discipline of engineering science which deals with the properties and behavior of soil as a structural material.

All structures have to be built on soils. Our main objective in the study of soil mechanics is to lay down certain principles, theories and procedures for the design of a safe and sound structure. The subject of *Foundation Engineering* deals with the design of various types of substructures under different soil and environmental conditions.

During the design, the designer has to make use of the properties of soils, the theories pertaining to the design and his own practical experience to adjust the design to suit field conditions. He has to deal with natural soil deposits which perform the engineering function of supporting the foundation and the superstructure above it. Soil deposits in nature exist in an extremely erratic manner producing thereby an infinite variety of possible combinations which would affect the choice and design of foundations. The foundation engineer must have the ability to interpret the principles of soil mechanics to suit the field conditions. The success or failure of his design depends upon how much in tune he is with Nature.

CHAPTER 2

SOIL FORMATION AND CHARACTERIZATION

2.1 INTRODUCTION

The word 'soil' has different meanings for different professions. To the agriculturist, soil is the top thin layer of earth within which organic forces are predominant and which is responsible for the support of plant life. To the geologist, soil is the material in the top thin zone within which roots occur. From the point of view of an engineer, soil includes all earth materials, organic and inorganic, occurring in the zone overlying the rock crust.

The behavior of a structure depends upon the properties of the soil materials on which the structure rests. The properties of the soil materials depend upon the properties of the rocks from which they are derived. A brief discussion of the parent rocks is, therefore, quite essential in order to understand the properties of soil materials.

2.2 ROCK CLASSIFICATION

Rock can be defined as a compact, semi-hard to hard mass of natural material composed of one or more minerals. The rocks that are encountered at the surface of the earth or beneath, are commonly classified into three groups according to their modes of origin. They are igneous, sedimentary and metamorphic rocks.

Igneous rocks are considered to be the primary rocks formed by the cooling of molten magmas, or by the recrystallization of older rocks under heat and pressure great enough to render them fluid. They have been formed on or at various depths below the earth surface. There are two main classes of igneous rocks. They are:

1. Extrusive (poured out at the surface), and
2. Intrusive (large rock masses which have not been formed in contact with the atmosphere).

Initially both classes of rocks were in a molten state. Their present state results directly from the way in which they solidified. Due to violent volcanic eruptions in the past, some of the molten materials were emitted into the atmosphere with gaseous extrusions. These cooled quickly and eventually fell on the earth's surface as volcanic ash and dust. Extrusive rocks are distinguished, in general, by their glass-like structure.

Intrusive rocks, cooling and solidifying at great depths and under pressure containing entrapped gases, are wholly crystalline in texture. Such rocks occur in masses of great extent, often going to unknown depths. Some of the important rocks that belong to the igneous group are *granite* and *basalt*. Granite is primarily composed of feldspar, quartz and mica and possesses a massive structure. Basalt is a dark-colored fine-grained rock. It is characterized by the predominance of plagioclase, the presence of considerable amounts of pyroxene and some olivine and the absence of quartz. The color varies from dark-grey to black. Both granite and basalt are used as building stones.

When the products of the disintegration and decomposition of any rock type are transported, redeposited, and partly or fully consolidated or cemented into a new rock type, the resulting material is classified as a *sedimentary rock*. The sedimentary rocks generally are formed in quite definitely arranged beds, or strata, which can be seen to have been horizontal at one time although sometimes displaced through angles up to 90 degrees. Sedimentary rocks are generally classified on the basis of grain size, texture and structure. From an engineering point of view, the most important rocks that belong to the group are *sandstones*, *limestones,* and *shales.*

Rocks formed by the complete or incomplete recrystallization of igneous or sedimentary rocks by high temperatures, high pressures, and/or high shearing stresses are *metamorphic rocks.* The rocks so produced may display features varying from complete and distinct foliation of a crystalline structure to a fine fragmentary partially crystalline state caused by direct compressive stress, including also the cementation of sediment particles by siliceous matter. Metamorphic rocks formed without intense shear action have a massive structure. Some of the important rocks that belong to this group are *gneiss, schist, slate* and *marble.* The characteristic feature of gneiss is its structure, the mineral grains are elongated, or platy, and banding prevails. Generally gneiss is a good engineering material. Schist is a finely foliated rock containing a high percentage of mica. Depending upon the amount of pressure applied by the metamorphic forces, schist may be a very good building material. Slate is a dark colored, platy rock with extremely fine texture and easy cleavage. Because of this easy cleavage, slate is split into very thin sheets and used as a roofing material. Marble is the end product of the metamorphism of limestone and other sedimentary rocks composed of calcium or magnesium carbonate. It is very dense and exhibits a wide variety of colors. In construction, marble is used for facing concrete or masonry exterior and interior walls and floors.

Rock Minerals

It is essential to examine the properties of the rock forming minerals since all soils are derived through the disintegration or decomposition of some parent rock. A *'mineral'* is a natural inorganic substance of a definite structure and chemical composition. Some of the very important physical properties of minerals are crystal form, color, hardness, cleavage, luster, fracture, and specific gravity. Out of these only two, specific gravity and hardness, are of foundation engineering interest. The specific gravity of the minerals affects the specific gravity of soils derived from them. The specific gravity of most rock and soil forming minerals varies from 2.50 (some feldspars) and 2.65 (quartz) to 3.5 (augite or olivine). Gypsum has a smaller value of 2.3 and salt (NaCl) has 2.1. Some iron minerals may have higher values, for instance, magnetite has 5.2.

It is reported that about 95 percent of the known part of the lithosphere consists of igneous rocks and only 5 percent of sedimentary rocks. Soil formation is mostly due to the disintegration of igneous rock which may be termed as a parent rock.

Table 2.1 Mineral composition of igneous rocks

Mineral	Percent
Quartz	12–20
Feldspar	50–60
Ca, Fe and Mg, Silicates	14–17
Micas	4–8
Others	7–8

The average mineral composition of igneous rocks is given in Table 2.1. Feldspars are the most common rock minerals, which account for the abundance of clays derived from the feldspars on the earth's surface. Quartz comes next in order of frequency. Most sands are composed of quartz.

2.3 FORMATION OF SOILS

Soil is defined as a natural aggregate of mineral grains, with or without organic constituents, that can be separated by gentle mechanical means such as agitation in water. By contrast rock is considered to be a natural aggregate of mineral grains connected by strong and permanent cohesive forces. The process of weathering of the rock decreases the cohesive forces binding the mineral grains and leads to the disintegration of bigger masses to smaller and smaller particles. Soils are formed by the process of weathering of the parent rock. The weathering of the rocks might be by mechanical disintegration, and/or chemical decomposition.

Mechanical Weathering

Mechanical weathering of rocks to smaller particles is due to the action of such agents as the expansive forces of freezing water in fissures, due to sudden changes of temperature or due to the abrasion of rock by moving water or glaciers. Temperature changes of sufficient amplitude and frequency bring about changes in the volume of the rocks in the superficial layers of the earth's crust in terms of expansion and contraction. Such a volume change sets up tensile and shear stresses in the rock ultimately leading to the fracture of even large rocks. This type of rock weathering takes place in a very significant manner in arid climates where free, extreme atmospheric radiation brings about considerable variation in temperature at sunrise and sunset.

Erosion by wind and rain is a very important factor and a continuing event. Cracking forces by growing plants and roots in voids and crevasses of rock can force fragments apart.

Chemical Weathering

Chemical weathering (decomposition) can transform hard rock minerals into soft, easily erodable matter. The principal types of decomposition are *hydration, oxidation, carbonation, desilication* and *leaching*. Oxygen and carbon dioxide which are always present in the air readily combine with the elements of rock in the presence of water.

2.4 GENERAL TYPES OF SOILS

It has been discussed earlier that soil is formed by the process of physical and chemical weathering. The individual size of the constituent parts of even the weathered rock might range from the smallest state (colloidal) to the largest possible (boulders). This implies that all the weathered constituents of a parent rock cannot be termed soil. According to their grain size, soil particles are classified as cobbles, gravel, sand, silt and clay. Grains having diameters in the range of 4.75 to 76.2 mm are called gravel. If the grains are visible to the naked eye, but are less than about 4.75 mm in size the soil is described as sand. The lower limit of visibility of grains for the naked eyes is about 0.075 mm. Soil grains ranging from 0.075 to 0.002 mm are termed as silt and those that are finer than 0.002 mm as clay. This classification is purely based on size which does not indicate the properties of fine grained materials.

Residual and Transported Soils

On the basis of origin of their constituents, soils can be divided into two large groups:

1. Residual soils, and
2. Transported soils.

Residual soils are those that remain at the place of their formation as a result of the weathering of parent rocks. The depth of residual soils depends primarily on climatic conditions and the time of exposure. In some areas, this depth might be considerable. In temperate zones residual soils are commonly stiff and stable. An important characteristic of residual soil is that the sizes of grains are indefinite. For example, when a residual sample is sieved, the amount passing any given sieve size depends greatly on the time and energy expended in shaking, because of the partially disintegrated condition.

Transported soils are soils that are found at locations far removed from their place of formation. The transporting agencies of such soils are glaciers, wind and water. The soils are named according to the mode of transportation. *Alluvial* soils are those that have been transported by running water. The soils that have been deposited in quiet lakes, are *lacustrine* soils. *Marine soils* are those deposited in sea water. The soils transported and deposited by wind are *aeolian* soils. Those deposited primarily through the action of gravitational force, as in land slides, are *colluvial* soils. *Glacial* soils are those deposited by glaciers. Many of these transported soils are loose and soft to a depth of several hundred feet. Therefore, difficulties with foundations and other types of construction are generally associated with transported soils.

Organic and Inorganic Soils

Soils in general are further classified as *organic* or *inorganic*. Soils of organic origin are chiefly formed either by growth and subsequent decay of plants such as peat, or by the accumulation of fragments of the inorganic skeletons or shells of organisms. Hence a soil of organic origin can be either organic or inorganic. The term organic soil ordinarily refers to a transported soil consisting of the products of rock weathering with a more or less conspicuous admixture of decayed vegetable matter.

Names of Some Soils that are Generally Used in Practice

Bentonite is a clay formed by the decomposition of volcanic ash with a high content of montmorillonite. It exhibits the properties of clay to an extreme degree.

Varved Clays consist of thin alternating layers of silt and fat clays of glacial origin. They possess the undesirable properties of both silt and clay. The constituents of varved clays were transported into fresh water lakes by the melted ice at the close of the ice age.

Kaolin, China Clay are very pure forms of white clay used in the ceramic industry.

Boulder Clay is a mixture of an unstratified sedimented deposit of glacial clay, containing unsorted rock fragments of all sizes ranging from boulders, cobbles, and gravel to finely pulverized clay material.

Calcareous Soil is a soil containing calcium carbonate. Such soil effervesces when tested with weak hydrochloric acid.

Marl consists of a mixture of calcareous sands, clays, or loam.

Hardpan is a relatively hard, densely cemented soil layer, like rock which does not soften when wet. Boulder clays or glacial till is also sometimes named as hardpan.

Caliche is an admixture of clay, sand, and gravel cemented by calcium carbonate deposited from ground water.

Peat is a fibrous aggregate of finer fragments of decayed vegetable matter. Peat is very compressible and one should be cautious when using it for supporting foundations of structures.

Loam is a mixture of sand, silt and clay.

Loess is a fine-grained, air-borne deposit characterized by a very uniform grain size, and high void ratio. The size of particles ranges between about 0.01 to 0.05 mm. The soil can stand deep vertical cuts because of slight cementation between particles. It is formed in dry continental regions and its color is yellowish light brown.

Shale is a material in the state of transition from clay to slate. Shale itself is sometimes considered a rock but, when it is exposed to the air or has a chance to take in water it may rapidly decompose.

2.5 SOIL PARTICLE SIZE AND SHAPE

The size of particles as explained earlier, may range from gravel to the finest size possible. Their characteristics vary with the size. Soil particles coarser than 0.075 mm are visible to the naked eye or may be examined by means of a hand lens. They constitute the coarser fractions of the soils. Grains finer than 0.075 mm constitute the finer fractions of soils. It is possible to distinguish the grains lying between 0.075 mm and 2 μ (1 μ = 1 micron = 0.001 mm) under a microscope. Grains having a size between 2 μ and 0.1 μ can be observed under a microscope but their shapes cannot be made out. The shape of grains smaller than 1 μ can be determined by means of an electron microscope. The molecular structure of particles can be investigated by means of X-ray analysis.

The coarser fractions of soils consist of gravel and sand. The individual particles of gravel, which are nothing but fragments of rock, are composed of one or more minerals, whereas sand grains contain mostly one mineral which is quartz. The individual grains of gravel and sand may be angular, subangular, sub-rounded, rounded or well-rounded as shown in Fig. 2.1. Gravel may contain grains which may be flat. Some sands contain a fairly high percentage of mica flakes that give them the property of elasticity.

Silt and clay constitute the finer fractions of the soil. Any one grain of this fraction generally consists of only one mineral. The particles may be angular, flake-shaped or sometimes needle-like.

Table 2.2 gives the particle size classification systems as adopted by some of the organizations in the USA. The Unified Soil Classification System is now almost universally accepted and has been adopted by the American Society for Testing and Materials (ASTM).

Specific Surface

Soil is essentially a particulate system, that is, a system in which the particles are in a fine state of subdivision or dispersion. In soils, the dispersed or the solid phase predominates and the dispersion medium, soil water, only helps to fill the pores between the solid particles. The significance of the concept of dispersion becomes more apparent when the relationship of surface to particle size is considered. In the case of silt, sand and larger size particles the ratio of the area of surface of the particles to the volume of the sample is relatively small. This ratio becomes increasingly large as

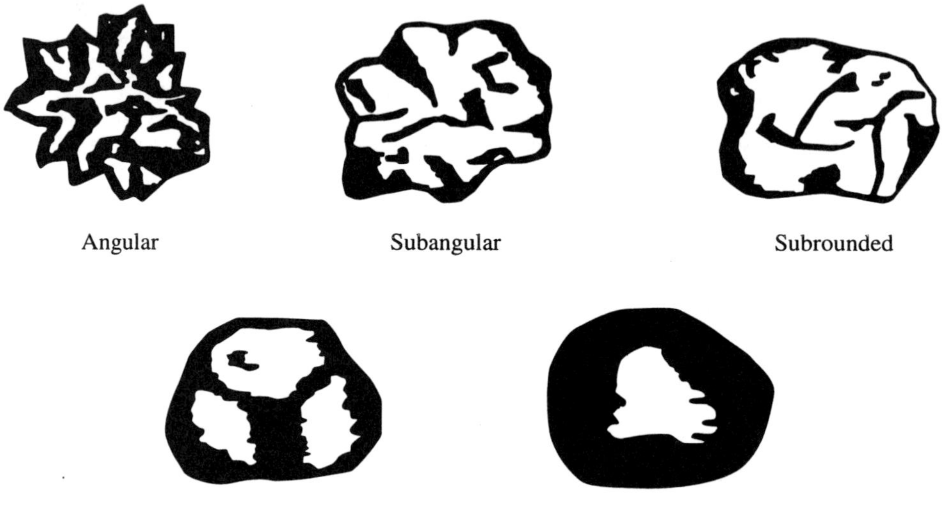

Figure 2.1 Shapes of coarser fractions of soils

size decreases from 2 μ which is the upper limit for clay-sized particles. A useful index of relative importance of surface effects is the *specific surface* of grain. The specific surface is defined as the total area of the surface of the grains expressed in square centimeters per gram or per cubic centimeter of the dispersed phase.

The shape of the clay particles is an important property from a physical point of view. The amount of surface per unit mass or volume varies with the shape of the particles. Moreover, the amount of contact area per unit surface changes with shape. It is a fact that a sphere has the smallest surface area per unit volume whereas a plate exhibits the maximum. Ostwald (1919) has emphasized the importance of shape in determining the specific surface of colloidal systems. Since disc-shaped particles can be brought more in intimate contact with each other, this shape has a pronounced effect upon the mechanical properties of the system. The interparticle forces between the surfaces of particles have a significant effect on the properties of the soil mass if the particles in the media belong to the clay fraction. The surface activity depends not only on the specific surface but also on the chemical and mineralogical composition of the solid particles. Since clay particles

Table 2.2 Particle size classification by various systems

Name of the organization	Particle size (mm)			
	Gravel	Sand	Silt	Clay
Massachusetts Institute of Technology (MIT)	> 2	2 to 0.06	0.06 to 0.002	< 0.002
US Department of Agriculture (USDA)	> 2	2 to 0.05	0.05 to 0.002	< 0.002
American Association of State Highway and Transportation Officials (AASHTO)	76.2 to 2	2 to 0.075	0.075 to 0.002	< 0.002
Unified Soil Classification System, US Bureau of Reclamation, US Army Corps of Engineers and American Society for Testing and Materials	76.2 to 4.75	4.75 to 0.075	Fines (silts and clays) < 0.075	

are the active portions of a soil because of their high specific surface and their chemical constitution, a discussion on the chemical composition and structure of minerals is essential.

2.6 COMPOSITION OF CLAY MINERALS

The word 'clay' is generally understood to refer to a material composed of a mass of small mineral particles which, in association with certain quantities of water, exhibits the property of plasticity. According to the clay mineral concept, clay materials are essentially composed of extremely small crystalline particles of one or more members of a small group of minerals that are commonly known as clay minerals. These minerals are essentially hydrous aluminum silicates, with magnesium or iron replacing wholly or in part for the aluminum, in some minerals. Many clay materials may contain organic material and water-soluble salts. Organic materials occur either as discrete particles of wood, leaf matter, spores, etc., or they may be present as organic molecules adsorbed on the surface of the clay mineral particles. The water-soluble salts that are present in clay materials must have been entrapped in the clay at the time of accumulation or may have developed subsequently as a consequence of ground water movement and weathering or alteration processes.

Clays can be divided into three general groups on the basis of their crystalline arrangement and it is observed that roughly similar engineering properties are connected with all the clay minerals belonging to the same group. An initial study of the crystal structure of clay minerals leads to a better understanding of the behavior of clays under different conditions of loading. Table 2.3 gives the groups of minerals and some of the important minerals under each group.

2.7 STRUCTURE OF CLAY MINERALS

Clay minerals are essentially crystalline in nature though some clay minerals do contain material which is non-crystalline (for example allophane). Two fundamental building blocks are involved in the formation of clay mineral structures. They are:

1. Tetrahedral unit.
2. Octahedral unit.

The tetrahedral unit consists of four oxygen atoms (or hydroxyls, if needed to balance the structure) placed at the apices of a tetrahedron enclosing a silicon atom which combines together to form a shell-like structure with all the tips pointing in the same direction. The oxygen at the bases of all the units lie in a common plane.

Each of the oxygen ions at the base is common to two units. The arrangement is shown in Fig. 2.2. The oxygen atoms are negatively charged with two negative charges each and the silicon with four positive charges. Each of the three oxygen ions at the base shares its charges with the

Table 2.3 Clay minerals

	Name of mineral	Structural formula
I.	Kaolin group	
	1. Kaolinite	$Al_4Si_4O_{10}(OH)_8$
	2. Halloysite	$Al_4Si_4O_6(OH)_{16}$
II.	Montmorillonite group	
	Montmorillonite	$Al_4Si_8O_{20}(OH)_4nH_2O$
III.	Illite group	
	Illite	$K_y(Al_4Fe_2.Mg_4.Mg_6)Si_{8-y}$ $Al_y(OH)_4O_{20}$

adjacent tetrahedral unit. The sharing of charges leaves three negative charges at the base per tetrahedral unit and this along with two negative charges at the apex makes a total of 5 negative charges to balance the 4 positive charges of the silicon ion. The process of sharing the oxygen ions at the base with neighboring units leaves a net charge of –1 per unit.

The second building block is an octahedral unit with six hydroxyl ions at apices of an octahedral enclosing an aluminum ion at the center. Iron or magnesium ions may replace aluminum ions in some units. These octahedral units are bound together in a sheet structure with each hydroxyl ion common to three octahedral units. This sheet is sometimes called as *gibbsite* sheet. The Al ion has 3 positive charges and each hydroxyl ion divides its –1 charge with two other neighboring units. This sharing of negative charge with other units leaves a total of 2 negative charges per unit [(1/3) × 6]. The net charge of a unit with an aluminum ion at the center is +1. Fig. 2.3 gives the structural arrangements of the units. Sometimes, magnesium replaces the aluminum atoms in the octahedral units in this case, the octahedral sheet is called a *brucite sheet*.

Formation of Minerals

The combination of two sheets of silica and gibbsite in different arrangements and conditions lead to the formation of different clay minerals as given in Table 2.3. In the actual formation of the sheet silicate minerals, the phenomenon of *isomorphous substitution* frequently occurs. Isomorphous (meaning same form) substitution consists of the substitution of one kind of atom for another.

Kaolinite Mineral

This is the most common mineral of the kaolin group. The building blocks of gibbsite and silica sheets are arranged as shown in Fig. 2.4 to give the structure of the kaolinite layer. The structure is composed of a single tetrahedral sheet and a single alumina octahedral sheet combined in units so that the tips of the silica tetrahedrons and one of the layers of the octahedral sheet form a common layer. All the tips of the silica tetrahedrons point in the same direction and towards the center of the unit made of the silica and octahedral sheets. This gives rise to strong ionic bonds between the silica and gibbsite sheets. The thickness of the layer is about 7 Å (one angstrom = 10^{-8} cm) thick. The kaolinite mineral is formed by stacking the layers one above the other with the base of the silica sheet bonding to hydroxyls of the gibbsite sheet by hydrogen bonding. Since hydrogen bonds are comparatively strong, the kaolinite

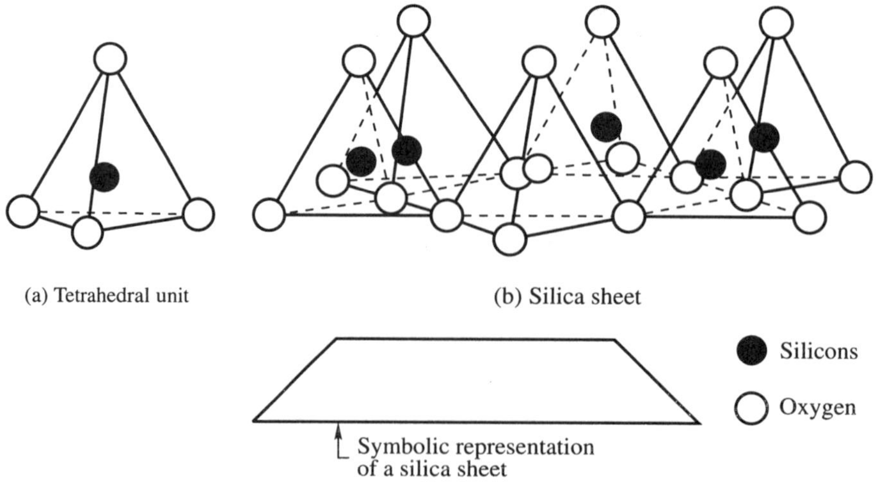

Figure 2.2 Basic structural units in the silicon sheet (Grim, 1959)

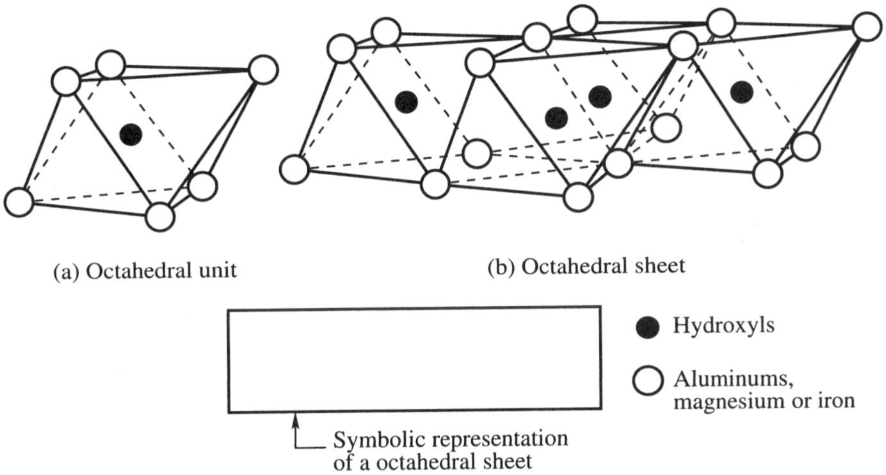

Figure 2.3 Basic structural units in octahedral sheet (Grim, 1959)

crystals consist of many sheet stackings that are difficult to dislodge. The mineral is therefore, stable, and water cannot enter between the sheets to expand the unit cells. The lateral dimensions of kaolinite particles range from 1000 to 20,000 Å and the thickness varies from 100 to 1000 Å. In the kaolinite mineral there is a very small amount of isomorphous substitution.

Halloysite Mineral

Halloysite minerals are made up of successive layers with the same structural composition as those composing kaolinite. In this case, however, the successive units are randomly packed and may be separated by a single molecular layer of water. The dehydration of the interlayers by the removal of the water molecules leads to changes in the properties of the mineral. An important structural feature of halloysite is that the particles appear to take tubular forms as opposed to the platy shape of kaolinite.

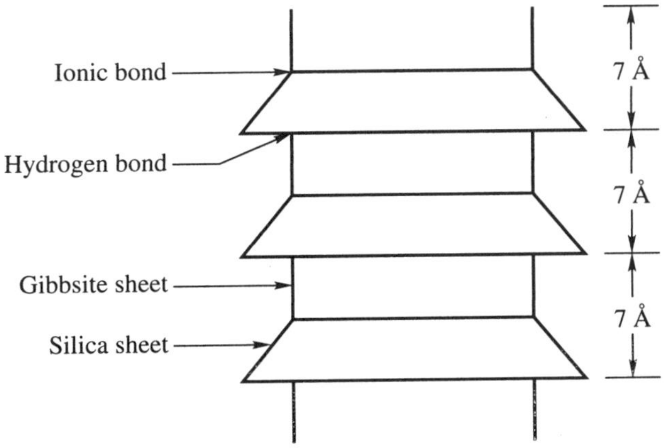

Figure 2.4 Structure of kaolinite layer

Montmorillonite Mineral

Montmorillonite is the most common mineral of the montmorillonite group. The structural arrangement of this mineral is composed of two silica tetrahedral sheets with a central alumina octahedral sheet. All the tips of the tetrahedra point in the same direction and toward the center of the unit. The silica and gibbsite sheets are combined in such a way that the tips of the tetrahedrons of each silica sheet and one of the hydroxyl layers of the octahedral sheet form a common layer. The atoms common to both the silica and gibbsite layer become oxygen instead of hydroxyls. The thickness of the silica-gibbsite-silica unit is about 10 Å (Fig. 2.5). In stacking these combined units one above the other, oxygen layers of each unit are adjacent to oxygen of the neighboring units with a consequence that there is a very weak bond and an excellent cleavage between them. Water can enter between the sheets, causing them to expand significantly and thus the structure can break into 10 Å thick structural units. Soils containing a considerable amount of montmorillonite minerals will exhibit high swelling and shrinkage characteristics. The lateral dimensions of montmorillonite particles range from 1000 to 5000 Å with thickness varying from 10 to 50 Å. Bentonite clay belongs to the montmorillonite group. In montmorillonite, there is isomorphous substitution of magnesium and iron for aluminum.

Illite

The basic structural unit of illite is similar to that of montmorillonite except that some of the silicons are always replaced by aluminum atoms and the resultant charge deficiency is balanced by potassium ions. The potassium ions occur between unit layers. The bonds with the nonexchangeable K^+ ions are weaker than the hydrogen bonds, but stronger than the water bond of montmorillonite. Illite, therefore, does not swell as much in the presence of water as does montmorillonite. The lateral dimensions of illite clay particles are about the same as those of montmorillonite, 1000 to 5000 Å, but the thickness of illite particles is greater than that of montmorillonite particles, 50 to 500 Å. The arrangement of silica and gibbsite sheets are as shown in Fig. 2.6.

2.8 CLAY PARTICLE-WATER RELATIONS

The behavior of a soil mass depends upon the behavior of the discrete particles composing the mass and the pattern of particle arrangement. In all these cases water plays an important part. The

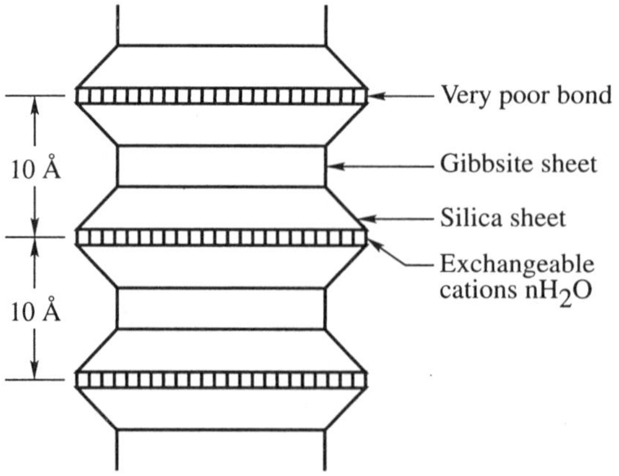

Figure 2.5 Structure of montmorillonite layer

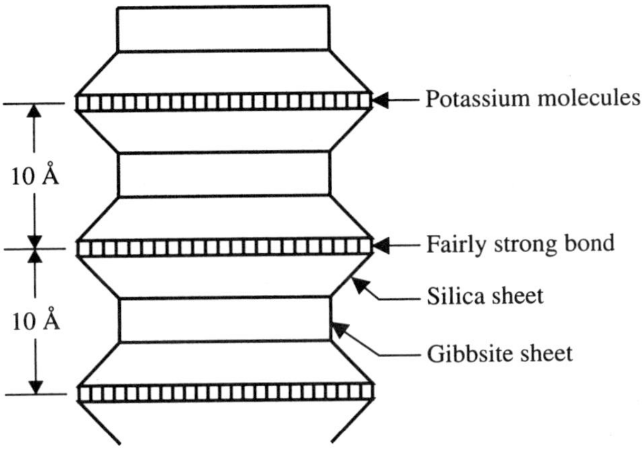

Figure 2.6 Structure of illite layer

behavior of the soil mass is profoundly influenced by the inter-particle-water relationships, the ability of the soil particles to adsorb exchangeable cations and the amount of water present.

Adsorbed Water

The clay particles carry a net negative charge on their surface. This is the result of both isomorphous substitution and of a break in the continuity of the structure at its edges. The intensity of the charge depends to a considerable extent on the mineralogical character of the particle. The physical and chemical manifestations of the surface charge constitute the surface activity of the mineral. Minerals are said to have high or low surface activity, depending on the intensity of the surface charge. As pointed out earlier, the surface activity depends not only on the specific surface but also on the chemical and mineralogical composition of the solid particle. The surface activity of sand, therefore, will not acquire all the properties of a true clay, even if it is ground to a fine powder. The presence of water does not alter its properties of coarser fractions considerably excepting changing its unit weight. However, the behavior of a saturated soil mass consisting of fine sand might change under dynamic loadings. This aspect of the problem is not considered here. This article deals only with clay particle-water relations.

In nature every soil particle is surrounded by water. Since the centers of positive and negative charges of water molecules do not coincide, the molecules behave like dipoles. The negative charge on the surface of the soil particle, therefore, attracts the positive (hydrogen) end of the water molecules. The water molecules are arranged in a definite pattern in the immediate vicinity of the boundary between solid and water. More than one layer of water molecules sticks on the surface with considerable force and this attractive force decreases with the increase in the distance of the water molecule from the surface. The electrically attracted water that surrounds the clay particle is known as the *diffused double-layer of water*. The water located within the zone of influence is known as the *adsorbed layer* as shown in Fig. 2.7. Within the zone of influence the physical properties of the water are very different from those of free or normal water at the same temperature. Near the surface of the particle the water has the property of a solid. At the middle of the layer it resembles a very viscous liquid and beyond the zone of influence, the properties of the water become normal. The adsorbed water affects the behavior of clay particles when subjected to external stresses, since it comes between the particle surfaces. To drive off the adsorbed water, the clay particle must be heated to more than 200 °C, which would indicate that the bond between the water molecules and the surface is considerably greater than that between normal water molecules.

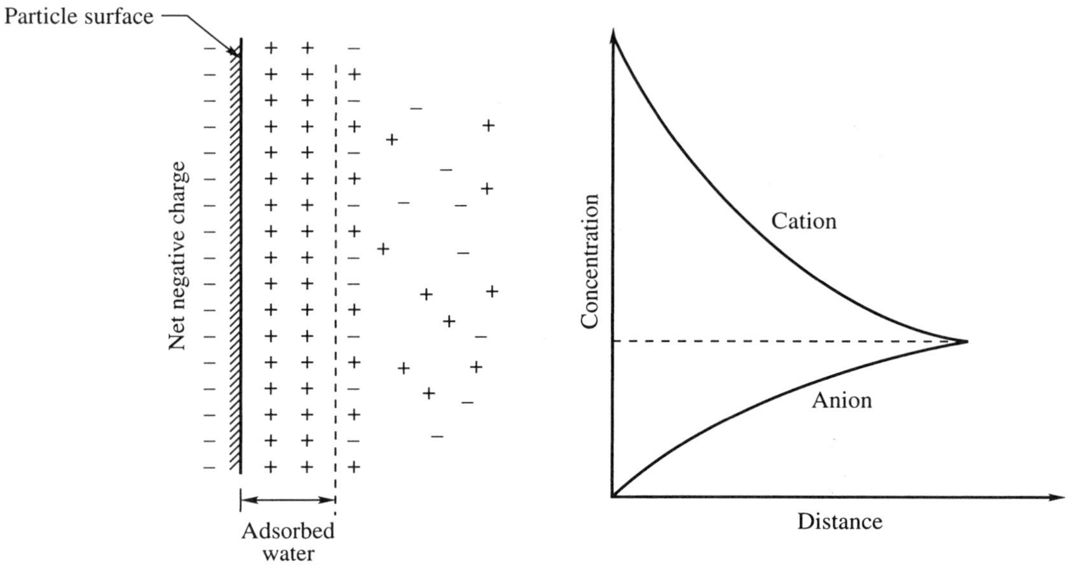

Figure 2.7 Adsorbed water layer surrounding a soil particle

The adsorbed film of water on coarse particles is thin in comparison with the diameter of the particles. In fine grained soils, however, this layer of adsorbed water is relatively much thicker and might even exceed the size of the grain. The forces associated with the adsorbed layers therefore play an important part in determining the physical properties of the very fine-grained soils, but have little effect on the coarser soils.

Soils in which the adsorbed film is thick compared to the grain size have properties quite different from other soils having the same grain sizes but smaller adsorbed films. The most pronounced characteristic of the former is their ability to deform plastically without cracking when mixed with varying amounts of water. This is due to the grains moving across one another supported by the viscous interlayers of the films. Such soils are called *cohesive soils*, for they do not disintegrate with pressure but can be rolled into threads with ease. Here the cohesion is not due to direct molecular interaction between soil particles at the points of contact but to the shearing strength of the adsorbed layers that separate the grains at these points.

Base Exchange

Electrolytes dissociate when dissolved in water into positively charged cations and negatively charged anions. Acids break up into cations of hydrogen and anions such as Cl or SO_4. Salts and bases split into metallic cations such as Na, K or Mg, and nonmetallic anions. Even water itself is an electrolyte, because a very small fraction of its molecules always dissociates into hydrogen ions H^+ and hydroxyl ions OH^-. These positively charged H^+ ions migrate to the surface of the negatively charged particles and form what is known as the adsorbed layer. These H^+ ions can be replaced by other cations such as Na, K or Mg. These cations enter the adsorbed layers and constitute what is termed as an *adsorption complex*. The process of replacing cations of one kind by those of another in an adsorption complex is known as *base exchange*. By base exchange is meant the capacity of

Table 2.4 Exchange capacity of some clay minerals

Mineral group	Exchange capacity (meq per 100 g)
Kaolinites	3.8
Illites	40
Montmorillonites	80

Table 2.5 Cations arranged in the order of decreasing shear strength of clay

$NH_4^+ > H^+ > K^+ > Fe^{+++} > Al^{+++} > Mg^+ > Ba^{++} > Ca^{++} > Na^+ > Li^+$

colloidal particles to change the cations adsorbed on their surface. Thus a hydrogen clay (colloid with adsorbed H cations) can be changed to sodium clay (colloid with adsorbed Na cations) by a constant percolation of water containing dissolved Na salts. Such changes can be used to decrease the permeability of a soil. Not all adsorbed cations are exchangeable. The quantity of exchangeable cations in a soil is termed *exchange capacity.*

The base exchange capacity is generally defined in terms of the mass of a cation which may be held on the surface of 100 gm dry mass of mineral. It is generally more convenient to employ a definition of base exchange capacity in milli-equivalents (meq) per 100 gm dry soil. One meq is one milligram of hydrogen or the portion of any ion which will combine with or displace 1 milligram of hydrogen.

The relative exchange capacity of some of the clay minerals is given in Table 2.4.

If one element, such as H, Ca, or Na prevails over the other in the adsorption complex of a clay, the clay is sometimes given the name of this element, for example H-clay or Ca-clay. The thickness and the physical properties of the adsorbed film surrounding a given particle, depend to a large extent on the character of the adsorption complex. These films are relatively thick in the case of strongly water-adsorbent cations such as Li^+ and Na^+ cations but very thin for H^+. The films of other cations have intermediate values. Soils with adsorbed Li^+ and Na^+ cations are relatively more plastic at low water contents and possess smaller shear strength because the particles are separated by a thicker viscous film. The cations in Table 2.5 are arranged in the order of decreasing shear strength of clay.

Sodium clays in nature are a product either of the deposition of clays in sea water or of their saturation by saltwater flooding or capillary action. Calcium clays are formed essentially by fresh water sediments. Hydrogen clays are a result of prolonged leaching of a clay by pure or acidic water, with the resulting removal of all other exchangeable bases.

2.9 SOIL MASS STRUCTURE

The orientation of particles in a mass depends on the size and shape of the grains as well as upon the minerals of which the grains are formed. The structure of soils that is formed by natural deposition can be altered by external forces. Figure 2.8 gives the various types of structures of soil. Fig. 2.8(a) is a *single grained structure* which is formed by the settlement of coarse grained soils in suspension in water. Fig. 2.8(b) is a *flocculent structure* formed by the deposition of the fine soil fraction in water. Fig. 2.8(c) is a *honeycomb structure* which is formed by the disintegration of a flocculent structure under a superimposed load. The particles oriented in a flocculent structure will have edge-to-face contact as shown in Fig. 2.8(d) whereas in a

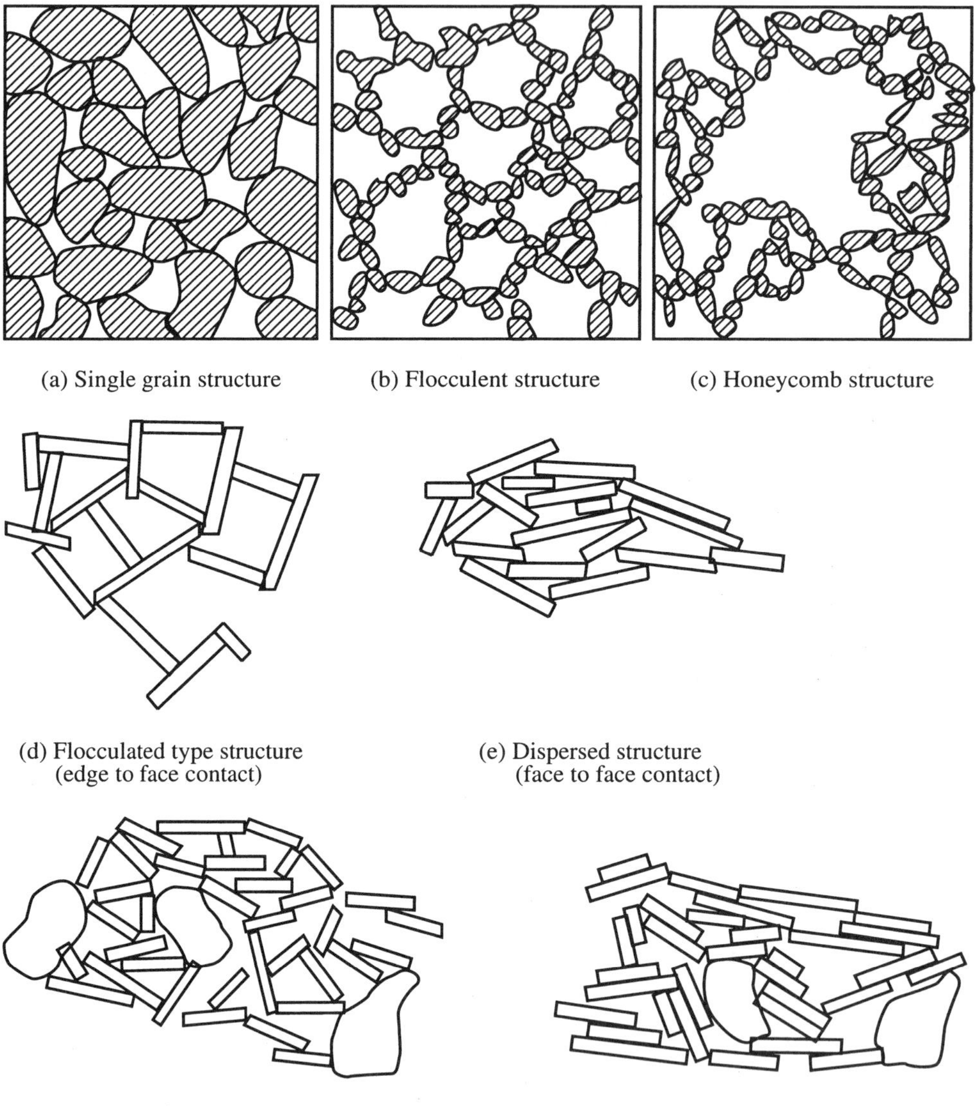

(a) Single grain structure (b) Flocculent structure (c) Honeycomb structure

(d) Flocculated type structure (edge to face contact) (e) Dispersed structure (face to face contact)

(f) Undisturbed salt water deposit (g) Undisturbed fresh water deposit

Figure 2.8 Schematic diagrams of various types of structures (Lambe, 1958a)

honeycomb structure, the particles will have face-to-face contact as shown in Fig. 2.8(e). Natural clay sediments will have more or less flocculated particle orientations. Marine clays generally have a more open structure than fresh water clays. Figs. 2.8(f) and (g) show the schematic views of salt water and fresh water deposits.

CHAPTER 3

SOIL PHASE RELATIONSHIPS, INDEX PROPERTIES AND CLASSIFICATION

3.1 SOIL PHASE RELATIONSHIPS

Soil mass is generally a three phase system. It consists of solid particles, liquid and gas. For all practical purposes, the liquid may be considered to be water (although in some cases, the water may contain some dissolved salts) and the gas as air. The phase system may be expressed in SI units either in terms of mass-volume or weight-volume relationships. The inter relationships of the different phases are important since they help to define the condition or the physical make-up of the soil.

Mass-Volume Relationship

In SI units, the mass M, is normally expressed in kg and the density ρ in kg/m^3. Sometimes, the mass and densities are also expressed in g and g/cm^3 or Mg and Mg/m^3 respectively. The density of water ρ_o at 4 °C is exactly 1.00 g/cm^3 (= 1000 kg/m^3 = 1 Mg/m^3). Since the variation in density is relatively small over the range of temperatures encountered in ordinary engineering practice, the density of water ρ_w at other temperatures may be taken the same as that at 4 °C. The volume is expressed either in cm^3 or m^3.

Weight-Volume Relationship

Unit weight or weight per unit volume is still the common measurement in geotechnical engineering practice. The density ρ, may be converted to unit weight, γ by using the relationship

$$\gamma = \rho g \tag{3.1a}$$

The 'standard' value of g is 9.807 m/s^2 (= 9.81 m/s^2 for all practical purposes).

Conversion of Density of Water ρ_w to Unit Weight γ_w

From Eq. (3.1a)

$$\gamma_w = \rho_w g \tag{3.1b}$$

Substituting ρ_w = 1000 kg/m^3 and g = 9.81 m/s^2, we have

$$\gamma_w = 1000\frac{\text{kg}}{\text{m}^3}\left(9.81\frac{\text{m}}{\text{s}^2}\right) = 9810\frac{\text{kg.m}}{\text{m}^3\text{s}^2} \tag{3.1c}$$

Since 1N (newton) = $\frac{1\text{ kg}\cdot\text{m}}{\text{s}^2}$, we have,

$$\gamma_w = 9810\frac{\text{N}}{\text{m}^3} = 9.81\frac{\text{kN}}{\text{m}^3} \tag{3.1d}$$

or $$\gamma_w = \left(1\times\frac{\text{g}}{\text{cm}^3}\right)\times 9.81 = 9.81\text{ kN/m}^3 \tag{3.1e}$$

In general, the unit weight of a soil mass may be obtained from the equation

$$\gamma = 9.81\rho\text{ kN/m}^3 \tag{3.1f}$$

where in Eq. (3.1f), ρ is in g/cm^3. For example, if a soil mass has a dry density, ρ_d = 1.7 g/cm^3, the dry unit weight of the soil is

$$\gamma_d = 9.81 \times 1.7 = 16.68\text{ kN/m}^3 \tag{3.1g}$$

3.2 MASS-VOLUME RELATIONSHIPS

The phase-relationships in terms of mass-volume and weight-volume for a soil mass are shown by a block diagram in Fig. 3.1. A block of unit sectional area is considered. The volumes of the different constituents are shown on the right side and the corresponding mass/weights on the right and left sides of the block. The mass/weight of air may be assumed as zero.

Volumetric Ratios

There are three volumetric ratios that are very useful in geotechnical engineering and these can be determined directly from the phase diagram, Fig. 3.1.

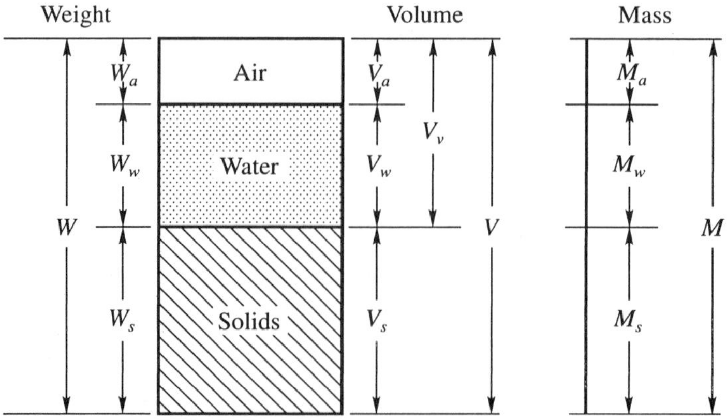

Figure 3.1 Block diagram—three phases of a soil element

1. The *void ratio*, e, is defined as

$$e = \frac{V_v}{V_s} \tag{3.2}$$

where, V_v = volume of voids, and V_s = volume of the solids.
The void ratio e is always expressed as a *decimal*.

2. The *porosity* n is defined as

$$n = \frac{V_v}{V} \times 100\% \tag{3.3}$$

where, V = total volume of the soil sample.
The *porosity* n is always expressed as a *percentage*.

3. The *degree of saturation* S is defined as

$$S = \frac{V_w}{V_v} \times 100\% \tag{3.4}$$

where, V_w = volume of water
It is always expressed as a *percentage*. When $S = 0\%$, the soil is completely dry, and when $S = 100\%$, the soil is fully saturated.

Mass-Volume Relationships

The other aspects of the phase diagram connected with mass or weight can be explained with reference to Fig. 3.1.

Water Content, *w*

The water content, w, of a soil mass is defined as the ratio of the mass of water, M_w, in the voids to the mass of solids, M_s, as

$$w = \frac{M_w}{M_s} \times 100 \tag{3.5}$$

The water content, which is usually expressed as a percentage, can range from zero (dry soil) to several hundred percent. The natural water content for most soils is well under 100%, but for the soils of volcanic origin (for example bentonite) it can range up to 500% or more.

Density

Another very useful concept in geotechnical engineering is *density* (or, unit weight) which is expressed as mass per unit volume. There are several commonly used densities. These may be defined as the total (or bulk), or moist density, ρ_t; the dry density, ρ_d; the saturated density, ρ_{sat}; the density of the particles, solid density, ρ_s; and density of water ρ_w. Each of these densities is defined as follows with respect to Fig. 3.1.

Total density, $$\rho_t = \frac{M}{V} \tag{3.6}$$

Dry density, $$\rho_d = \frac{M_s}{V} \tag{3.7}$$

Saturated density, $$\rho_{sat} = \frac{M}{V} \tag{3.8}$$

for $S = 100\%$

Density of solids, $$\rho_s = \frac{M_s}{V_s} \tag{3.9}$$

Density of water, $$\rho_w = \frac{M_w}{V_w} \tag{3.10}$$

Specific Gravity

The specific gravity of a substance is defined as the ratio of its mass in air to the mass of an equal volume of water at reference temperature, 4 °C. The specific gravity of a mass of soil (including air, water and solids) is termed as bulk specific gravity G_m. It is expressed as

$$G_m = \frac{\rho_t}{\rho_w} = \frac{M}{V\rho_w} \tag{3.11}$$

The specific gravity of solids, G_s, (excluding air and water) is expressed by

$$G_s = \frac{\rho_s}{\rho_w} = \frac{M_s}{V_s\rho_w} \tag{3.12}$$

Interrelationships of Different Parameters

We can establish relationships between the different parameters defined by equations from (3.2) through (3.12). In order to develop the relationships, the block diagram Fig. 3.2 is made use of. Since the sectional area perpendicular to the plane of the paper is assumed as unity, the heights of the blocks will represent the volumes. The volume of solids may be represented as $V_s = 1$. When the soil is fully saturated, the voids are completely filled with water.

Relationship Between *e* and *n* (Fig. 3.2)

$$e = \frac{V_v}{V_s} = \frac{V_v}{1} = V_v, \quad n = \frac{V_v}{V} = \frac{e}{1+e}$$

$$\text{or } e = \frac{n}{1-n} \tag{3.13}$$

Relationship Between *e*, G_s and *S*

Case 1: When partially saturated ($S < 100\%$)

$$S = \frac{V_w}{V_v} = \frac{V_w}{e}; \text{ but } V_w = \frac{M_w}{\rho_w} = \frac{wM_s}{\rho_w} = \frac{wG_sV_s\rho_w}{\rho_w} = wG_s$$

Therefore, $S = \frac{wG_s}{e}$ or $e = \frac{wG_s}{S}$ (3.14a)

Case 2: When saturated (S = 100%)

From Eq. (3.14a), we have (for S = 1)

$$e = wG_s \tag{3.14b}$$

Relationships Between Density ρ and Other Parameters

The density of soil can be expressed in terms of other parameters for cases of soil (1) partially saturated (S < 100%); (2) fully saturated (S = 100%); (3) Fully dry (S = 0); and (4) submerged.

Case 1: For S < 100%

$$\rho_t = \frac{M}{V} = \frac{M_s(1+w)}{1+e} = \frac{G_s\rho_w(1+w)}{1+e} \tag{3.15}$$

From Eq. (3.14a) $w = eS/G_s$; substituting for w in Eq. (3.15), we have

$$\rho_t = \frac{\rho_w(G_s + eS)}{1+e} \tag{3.16}$$

Case 2: For S = 100%

From Eq. (3.16)

$$\rho_t = \rho_{sat} = \frac{\rho_w(G_s + e)}{1+e} \tag{3.17}$$

Case 3: For S = 0%

From Eq. (3.16)

$$\rho_t = \rho_d = \frac{\rho_w G_s}{1+e} \tag{3.18}$$

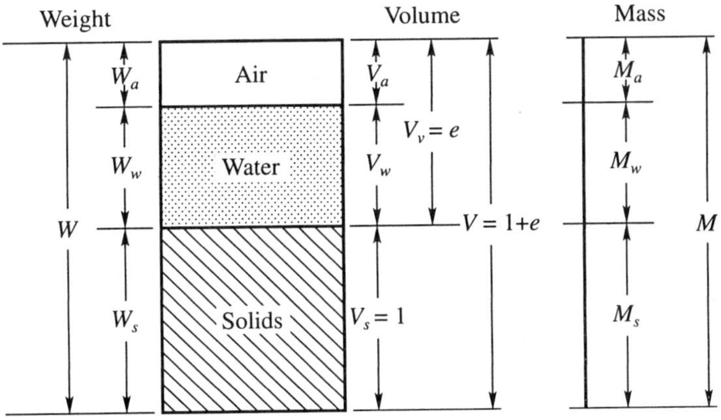

Figure 3.2 Block diagram—three phases of a soil element

Case 4: When the soil is submerged

If the soil is submerged, the density of the submerged soil ρ_b, is equal to the density of the saturated soil reduced by the density of water, that is

$$\rho_b = (\rho_{sat} - \rho_w) = \frac{\rho_w (G_s + e)}{1+e} - \rho_w = \frac{\rho_w (G_s - 1)}{1+e} \tag{3.19}$$

Relative Density

The looseness or denseness of sandy soils can be expressed numerically by relative density D_r, defined by the equation

$$D_r = \frac{e_{max} - e}{e_{max} - e_{min}} \times 100 \tag{3.20}$$

in which

e_{max} = void ratio of sand in its loosest state having a dry density of ρ_{dm}

e_{min} = void ratio in its densest state having a dry density of ρ_{dM}

e = void ratio under in-situ condition having a dry density of ρ_d

From Eq. (3.18), a general equation for e may be written as

$$e = \frac{\rho_w G_s - \rho_d}{\rho_d}$$

Now substituting the corresponding dry densities for e_{max}, e_{min} and e in Eq. (3.20) and simplifying, we have

$$D_r = \frac{\rho_{dM}}{\rho_d} \times \frac{\rho_d - \rho_{dm}}{\rho_{dM} - \rho_{dm}} \times 100 \tag{3.21}$$

The loosest state for a granular material can usually be created by allowing the dry material to fall into a container from a funnel held in such a way that the free fall is about one centimeter. The densest state can be established by a combination of static pressure and vibration of soil packed in a container.

ASTM Test Designation D-2049 (1991) provides a procedure for determining the minimum and maximum dry unit weights (or densities) of granular soils. This procedure can be used for determining D_r in Eq. (3.21).

3.3 WEIGHT-VOLUME RELATIONSHIPS

The weight-volume relationships can be established from the earlier equations by substituting γ for ρ and W for M. The various equations are tabulated below.

1. Water content $$w = \frac{W_w}{W_s} \times 100 \tag{3.5a}$$

2. Total unit weight $$\gamma_t = \frac{W}{V} \tag{3.6a}$$

3. Dry unit weight $$\gamma_d = \frac{W_s}{V} \tag{3.7a}$$

4. Saturated unit weight $$\gamma_{sat} = \frac{W}{V} \tag{3.8a}$$

5. Unit weight of solids $$\gamma_s = \frac{W_s}{V_s} \tag{3.9a}$$

6. Unit weight of water $$\gamma_w = \frac{W_w}{V_w} \tag{3.10a}$$

7. Mass specific gravity $$G_m = \frac{W}{V\gamma_w} \tag{3.11a}$$

8. Specific gravity of solids $$G_s = \frac{W_s}{V_s\gamma_w} \tag{3.12a}$$

9. Total unit weight for $S < 100$ $$\gamma_t = \frac{G_s\gamma_w(1+w)}{1+e} \tag{3.15a}$$

or $$\gamma_t = \frac{\gamma_w(G_s+eS)}{1+e} \tag{3.16a}$$

10. Saturated unit weight $$\gamma_{sat} = \frac{\gamma_w(G_s+e)}{1+e} \tag{3.17a}$$

11. Dry unit weight $$\gamma_d = \frac{\gamma_w G_s}{1+e} \tag{3.18a}$$

12. Submerged unit weight $$\gamma_b = \frac{\gamma_w(G_s-1)}{1+e} \tag{3.19a}$$

13. Relative density $$D_r = \frac{\gamma_{dM}}{\gamma_d} \times \frac{\gamma_d - \gamma_{dm}}{\gamma_{dM} - \gamma_{dm}} \tag{3.21a}$$

3.4 COMMENTS ON SOIL PHASE RELATIONSHIPS

The void ratios of natural sand deposits depend upon the shape of the grains, the uniformity of grain size, and the conditions of sedimentation. The void ratios of clay soils range from less than unity to 5 or more. The soils with higher void ratios have a loose structure and generally belong to the montmorillonite group. The specific gravity of solid particles of most soils varies from 2.5 to 2.9. For most of the calculations, G_s can be assumed as 2.65 for cohesionless soils and 2.70 for clay soils. The dry unit weights (γ_d) of granular soils range from 14 to 18 kN/m^3, whereas, the saturated unit weights of fine grained soils can range from 12.5 to 22.7 kN/m^3. Table 3.1 gives typical values of porosity, void ratio, water content (when saturated) and unit weights of various types of soils.

Table 3.1 Porosity, void ratio, water content, and unit weights of typical soils in natural state

Soil no.	Description of soil	Porosity n %	Void ratio e	Water content w%	Unit weight kN/m^3 γ_d	γ_{sat}
1	2	3	4	5	6	7
1	Uniform sand, loose	46	0.85	32	14.0	18.5
2	Uniform sand, loose	34	0.51	19	17.0	20.5
3	Mixed-grained sand, loose	40	0.67	25	15.6	19.5
4	Mixed-grained sand, dense	30	0.43	16	18.2	21.2
5	Glacial till, mixed grained	20	0.25	9	20.8	22.7
6	Soft glacial clay	55	1.20	45	11.9	17.3
7	Soft glacial clay	37	0.60	22	16.7	20.3
8	Soft slightly organic clay	66	1.90	70	9.1	15.5
9	Soft highly organic clay	75	3.00	110	6.8	14.0
10	Soft bentonite	84	5.20	194	4.2	12.4

Example 3.1

A sample of wet silty clay soil has a mass of 126 kg. The following data were obtained from laboratory tests on the sample: Wet density, $\rho_t = 2.1$ g/cm^3, $G = 2.7$, water content, $w = 15\%$.

Determine (i) dry density, ρ_d, (ii) porosity, (iii) void ratio, and (iv) degree of saturation.

Solution

Mass of sample $M = 126$ kg.

$$\text{Volume} \quad V = \frac{126}{2.1 \times 10^3} = 0.06 \ \text{m}^3$$

$$\text{Now,} \ M_s + M_w = M, \ \text{or} \ M_s + wM_s = M_s(1+w) = M$$

$$\text{Therefore,} \quad M_s = \frac{M}{1+w} = \frac{126}{1.15} = 109.57 \ \text{kg}; \quad M_w = M - M_s = 16.43 \ \text{kg}$$

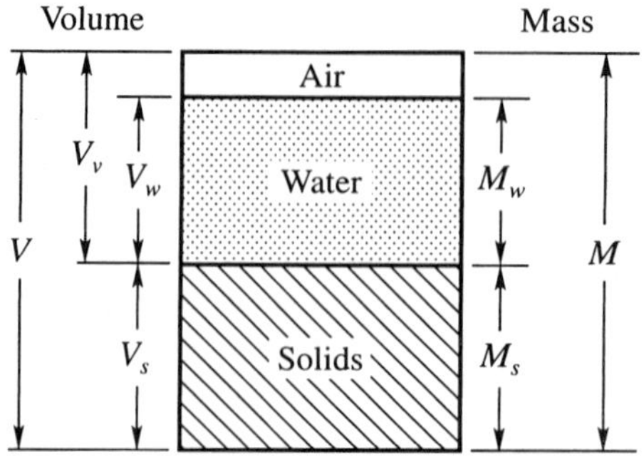

Figure Ex. 3.1

Now, $V_w = \frac{M_w}{\rho_w} = \frac{16.43}{1000} = 0.01643 \ \text{m}^3;$

$$V_s = \frac{M_s}{G_s \rho_w} = \frac{109.57}{2.7 \times 1000} = 0.04058 \ \text{m}^3;$$

$$V_v = V - V_s = 0.06000 - 0.04058 = 0.01942 \ \text{m}^3.$$

(i) Dry density, $\rho_d = \frac{M_s}{V} = \frac{109.57}{0.06} = 1826.2 \ \text{kg/m}^3$

(ii) Porosity, $n = \frac{V_v}{V} \times 100 = \frac{0.01942 \times 100}{0.06} = 32.37\%$

(iii) Void ratio, $e = \frac{V_v}{V_s} = \frac{0.01942}{0.04058} = 0.4786$

(iv) Degree of saturation, $S = \frac{V_w}{V_v} \times 100 = \frac{0.01643}{0.01942} \times 100 = 84.6\%$

Example 3.2

Earth is required to be excavated from borrow pits for building an embankment. The wet unit weight of undisturbed soil is 18 kN/m^3 and its water content is 8%. In order to build a 4 m high embankment with top width 2 m and side slopes 1 : 1, estimate the quantity of earth required to be excavated per meter length of embankment. The dry unit weight required in the embankment is 15 kN/m^3 with a moisture content of 10%. Assume the specific gravity of solids as 2.67. Also determine the void ratios and the degree of saturation of the soil in both the undisturbed and remolded states.

Solution

The dry unit weight of soil in the borrow pit is

$$\gamma_d = \frac{\gamma_t}{1+w} = \frac{18}{1.08} = 16.7 \quad \text{kN/m}^3$$

Volume of embankment per meter length V_e

$$V_e = \frac{2+10}{2} \times 4 = 24 \ \text{m}^3$$

The dry unit weight of soil in the embankment is 15 kN/m^3

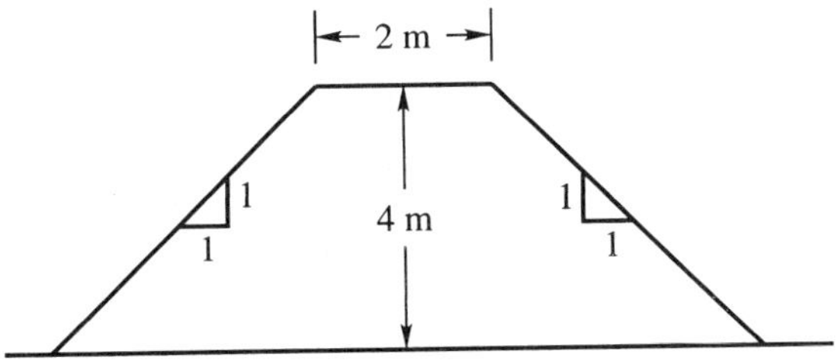

Figure Ex. 3.2

Volume of earth required to be excavated V_{ex} per meter

$$V_{ex} = 24 \times \frac{15}{16.7} = 21.55 \text{ m}^3$$

Undisturbed state

$$V_s = \frac{\gamma_d}{G_s \gamma_w} = \frac{16.7}{2.67} \times \frac{1}{9.81} = 0.64 \text{ m}^3; \quad V_v = 1 - 0.64 = 0.36 \text{ m}^3$$

$$e = \frac{0.36}{0.64} = 0.56, \quad W_w = 18.0 - 16.7 = 1.3 \text{ kN}$$

Degree of saturation, $S = \frac{V_w}{V_v} \times 100$, where

$$V_w = \frac{W_w}{\gamma_w} = \frac{1.3}{9.81} = 0.133 \text{ m}^3$$

Now, $S = \frac{0.133}{0.36} \times 100 = 36.9\%$

Remolded state

$$V_s = \frac{\gamma_d}{G\gamma_w} = \frac{15}{2.67 \times 9.81} = 0.57 \text{ m}^3$$

$$V_v = 1 - 0.57 = 0.43 \text{ m}^3$$

$$e = \frac{0.43}{0.57} = 0.75; \quad \gamma_t = \gamma_d(1+w) = 15 \times 1.1 = 16.5 \text{ kN/m}^3$$

Therefore, $W_w = 16.5 - 15.0 = 1.5 \text{ kN}$

$$V_w = \frac{1.5}{9.81} = 0.153 \text{ m}^3$$

$$S = \frac{0.153}{0.43} \times 100 = 35.6\%$$

Example 3.3

The moisture content of an undisturbed sample of clay belonging to a volcanic region is 265% under 100% saturation. The specific gravity of the solids is 2.5. The dry unit weight is 21 lb/ft^3. Determine (i) the saturated unit weight, (ii) the submerged unit weight, and (iii) void ratio.

Solution

(i) Saturated unit weight, $\gamma_{sat} = \gamma_t$

$$W = W_w + W_s = wW_s + W_s = W_s(1+w)$$

From Fig. Ex. 3.3, $\gamma_t = \frac{W}{V} = \frac{W}{1} = W$. Hence

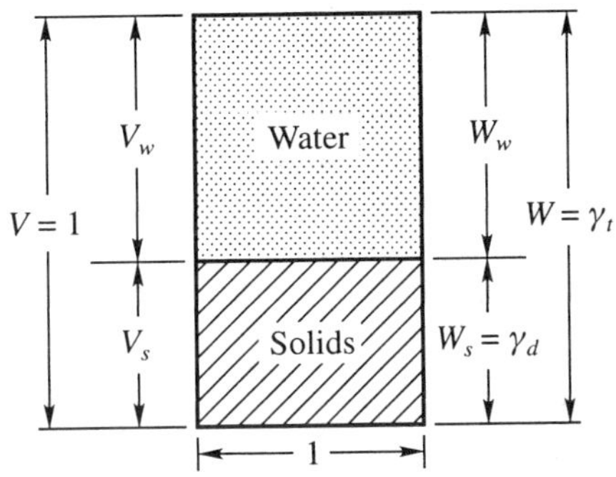

Figure Ex. 3.3

$$\gamma_t = 21(1 + 2.65) = 21 \times 3.65 = 76.65 \text{ lb/ft}^3$$

(ii) Submerged unit weight, γ_b

$$\gamma_b = \gamma_{sat} - \gamma_w = 76.65 - 62.4 = 14.25 \text{ lb/ft}^3$$

(iii) Void ratio, e

$$V_s = \frac{\gamma_d}{G_s \gamma_w} = \frac{21}{2.5 \times 62.4} = 0.135 \text{ ft}^3$$

Since $S = 100\%$

$$V_v = V_w = \frac{w \times W_s}{\gamma_w} = 2.65 \times \frac{21}{62.4} = 0.89 \text{ ft}^3$$

$$e = \frac{V_v}{V_s} = \frac{0.89}{0.135} = 6.59$$

Example 3.4

A sample of saturated clay from a consolidometer test has a total weight of 3.36 lb and a dry weight of 2.32 lb: the specific gravity of the solid particles is 2.7. For this sample, determine the water content, void ratio, porosity and total unit weight.

Solution

$$w = \frac{W_w}{W_s} \times 100\% = \frac{3.36 - 2.32}{2.32} = 44.9\% \approx 45\%$$

$$e = \frac{wG_s}{S} = \frac{0.45 \times 2.7}{1} = 1.215$$

$$n = \frac{e}{1+e} = \frac{1.215}{1 + 1.215} = 0.548 \text{ or } 54.8\%$$

$$\gamma_t = \frac{\gamma_w (G_s + e)}{1+e} = \frac{62.4(2.7 + 1.215)}{1 + 1.215} = 110.3 \text{ lb/ft}^3$$

Example 3.5

A sample of silty clay has a volume of 14.88 cm³, a total mass of 28.81 g, a dry mass of 24.83 g, and a specific gravity of solids 2.7. Determine the void ratio and the degree of saturation.

Solution

Void ratio

$$V_s = \frac{M_s}{G_s \rho_w} = \frac{24.83}{2.7(1)} = 9.2 \text{ cm}^3$$

$$V_v = V - V_s = 14.88 - 9.2 = 5.68 \text{ cm}^3$$

$$e = \frac{V_v}{V_s} = \frac{5.68}{9.2} = 0.618$$

Degree of saturation

$$w = \frac{M_w}{M_s} = \frac{28.81 - 24.83}{24.83} = 0.16$$

$$S = \frac{G_s w}{e} = \frac{2.7(0.16)}{0.618} = 0.70 \text{ or } 70\%$$

Example 3.6

A soil sample in its natural state has a weight of 5.05 lb and a volume of 0.041 ft³. In an oven-dried state, the dry weight of the sample is 4.49 lb. The specific gravity of the solids is 2.68. Determine the total unit weight, water content, void ratio, porosity, and degree of saturation.

Solution

$$\gamma_t = \frac{W}{V} = \frac{5.05}{0.041} = 123.2 \text{ lb/ft}^3$$

$$w = \frac{W_w}{W_s} = \frac{5.05 - 4.49}{4.49} = 0.125 \text{ or } 12.5\%$$

$$e = \frac{V_v}{V_s}, \quad V_s = \frac{W_s}{G_s \gamma_w} = \frac{4.49}{2.68 \times 62.4} = 0.0268 \text{ ft}^3$$

$$V_v = V - V_s = 0.041 - 0.0268 = 0.0142 \text{ ft}^3$$

$$e = \frac{0.0142}{0.0268} = 0.53$$

$$n = \frac{e}{1+e} = \frac{0.53}{1+0.53} = 0.3464 \text{ or } 34.64\%$$

$$S = \frac{wG_s}{e} = \frac{0.125 \times 2.68}{0.53} = 0.632 = 63.2\%$$

Example 3.7

A soil sample has a total unit weight of 16.97 kN/m^3 and a void ratio of 0.84. The specific gravity of solids is 2.70. Determine the moisture content, dry unit weight and degree of saturation of the sample.

Solution

Degree of saturation [from Eq. (3.16a)]

$$\gamma_t = \frac{\gamma_w (G_s + eS)}{1+e} \quad \text{or} \quad 16.97 = \frac{(9.81)\,(2.7 + 0.84S)}{1+0.84} \quad \text{or} \quad S = 58\%$$

Dry unit weight (Eq. 3.18a)

$$\gamma_d = \frac{\gamma_w G_s}{1+e} = \frac{(9.81)\,2.7}{1+0.84} = 14.4 \ \text{kN/m}^3$$

Water content (Eq. 3.14a)

$$w = \frac{Se}{G_s} = \frac{0.58 \times 0.84}{2.7} = 0.18 \ \text{or} \ 18\%$$

Example 3.8

A soil sample in its natural state has, when fully saturated, a water content of 32.5%. Determine the void ratio, dry and total unit weights. Calculate the total weight of water required to saturate a soil mass of volume 10 m^3. Assume $G_s = 2.69$.

Solution

Void ratio (Eq. 3.14a)

$$e = \frac{wG_s}{S} = \frac{32.5 \times 2.69}{(1) \times 100} = 0.874$$

Total unit weight (Eq. 3.15a)

$$\gamma_t = \frac{G_s \gamma_w (1+w)}{1+e} = \frac{2.69\,(9.81)\,(1+0.325)}{1+0.874} = 18.7 \ \text{kN/m}^3$$

Dry unit weight (Eq. 3.18a)

$$\gamma_d = \frac{\gamma_w G_s}{1+e} = \frac{2.69 \times 9.81}{1+0.874} = 14.08 \ \text{kN/m}^3$$

From Eq. (3.6a), $W = \gamma_t V = 18.66 \times 10 = 186.6$ kN

From Eq. (3.7a), $W_s = \gamma_d V = 14.08 \times 10 = 140.8$ kN

Weight of water $= W - W_s = 186.6 - 140.8 = 45.8$ kN

3.5 INDEX PROPERTIES OF SOILS

The various properties of soils which would be considered as index properties are:

1. The size and shape of particles.
2. The relative density or consistency of soil.

The index properties of soils can be studied in a general way under two classes. They are:

1. Soil grain properties.
2. Soil aggregate properties.

The principal soil grain properties are the size and shape of grains and the mineralogical character of the finer fractions (applied to clay soils). The most significant aggregate property of cohesionless soils is the relative density, whereas that of cohesive soils is the consistency. Water content can also be studied as an aggregate property as applied to cohesive soils. The strength and compressibility characteristics of cohesive soils are functions of water content. As such water content is an important factor in understanding the aggregate behavior of cohesive soils. By contrast, water content does not alter the properties of a cohesionless soil significantly except when the mass is submerged, in which case only its unit weight is reduced.

3.6 THE SHAPE AND SIZE OF PARTICLES

The shapes of particles as conceived by visual inspection give only a qualitative idea of the behavior of a soil mass composed of such particles. Since particles finer than 0.075 mm diameter cannot be seen by the naked eye, one can visualize the nature of the coarse grained particles only. Coarser fractions composed of angular grains are capable of supporting heavier static loads and can be compacted to a dense mass by vibration. The influence of the shape of the particles on the compressibility characteristics of soils are:

1. Reduction in the volume of mass upon the application of pressure.
2. A small mixture of mica to sand will result in a large increase in its compressibility.

The classification according to size divides the soils broadly into two distinctive groups, namely, coarse grained and fine grained. Since the properties of coarse grained soils are, to a considerable extent, based on grain size distribution, classification of coarse grained soils according to size would therefore be helpful. Fine grained soils are so much affected by structure, shape of grain, geological origin, and other factors that their grain size distribution alone tells little about their physical properties. However, one can assess the nature of a mixed soil on the basis of the percentage of fine grained soil present in it. It is, therefore, essential to classify the soil according to grain size.

The classification of soils as gravel, sand, silt and clay as per the different systems of classification is given in Table 2.2. Soil particles which are coarser than 0.075 mm are generally termed as *coarse grained* and the finer ones as silt, clay and peat (organic soil) are considered *fine grained*. From an engineering point of view, these two types of soils have distinctive characteristics. In coarse grained soils, gravitational forces determine the engineering characteristics. Interparticle forces are predominant in fine grained soils. The dependence of the behavior of a soil mass on the size of particles has led investigators to classify soils according to their size.

The physical separation of a sample of soil by any method into two or more fractions, each containing only particles of certain sizes, is termed *fractionation*. The determination of the mass of material in fractions containing only particles of certain sizes is termed Mechanical Analysis. Mechanical analysis is one of the oldest and most common forms of soil analysis. It provides the basic information for revealing the uniformity or gradation of the materials within established size ranges and for textural classifications. The results of a mechanical analysis are not equally valuable in different branches of engineering. The size of the soil grains is of importance in such cases as construction of earth dams or railroad and highway embankments, where earth is used as a material that should satisfy definite specifications. In foundations of structures, data from mechanical analyses are generally illustrative; other properties such as compressibility and shearing resistance are of more importance. The normal method adopted for separation of particles in a fine grained soil mass is the hydrometer analysis and for the coarse grained soils the sieve analysis. These two methods are described in the following sections.

3.7 SIEVE ANALYSIS

Sieve analysis is carried out by using a set of standard sieves. Sieves are made by weaving two sets of wires at right angles to one another. The square holes thus formed between the wires provide the limit which determines the size of the particles retained on a particular sieve. The sieve sizes are given in terms of the number of openings per inch. The number of openings per inch varies according to different standards. Thus, an ASTM 60 sieve has 60 openings per inch width with each opening of 0.250 mm. Table 3.2 gives a set of ASTM Standard Sieves (same as US standard sieves).

The usual procedure is to use a set of sieves which will yield equal grain size intervals on a logarithmic scale. A good spacing of soil particle diameters on the grain size distribution curve will be obtained if a nest of sieves is used in which each sieve has an opening approximately one-half of the coarser sieve above it in the nest. If the soil contains gravel, the coarsest sieve that can be used to separate out gravel from sand is the No. 4 Sieve (4.75 mm opening). To separate out the silt-clay fractions from the sand fractions, No. 200 sieve may be used. The intermediate sieves between the coarsest and the finest may be selected on the basis of the principle explained earlier. The nest of sieves consists of Nos 4 (4.75 mm), 8 (2.36 mm), 16 (1.18 mm) 30 (600 μm), 50 (300 μm), 100 (150 μm), and 200 (75 μm).

The sieve analysis is carried out by sieving a known dry mass of sample through the nest of sieves placed one below the other so that the openings decrease in size from the top sieve downwards, with a pan at the bottom of the stack as shown in Fig. 3.3. The whole nest of sieves is given a horizontal shaking for about 10 minutes (if required, more) till the mass of soil remaining on each sieve reaches a constant value (the shaking can be done by hand or using a mechanical shaker, if available). The amount of shaking required depends on the shape and number of particles. If a sizable portion of soil is retained on the No. 200 sieve, it should be washed. This is done by placing the sieve with a pan at the bottom and pouring clean water on the screen. A spoon may be used to stir the slurry. The soil which is washed through is recovered, dried and weighed. The mass of soil recovered is subtracted from the mass retained on the No. 200 sieve before washing and added to the soil that has passed through the No. 200 sieve by dry sieving. The mass of soil required for sieve analysis is of oven-dried soil with all

Table 3.2 US Standard sieves

Designation	Opening mm	Designation	Opening mm
2 in	50.80	35	0.50
1½ in	38.10	40	0.425
¾ in	19.00	50	0.355
3/8 in	9.51	60	0.250
4	4.75	70	0.212
8	2.36	80	0.180
10	2.00	100	0.150
14	1.40	120	0.125
16	1.18	170	0.090
18	1.00	200	0.075
30	0.60	270	0.053

Table 3.3 Sample size for sieve analysis

Max particle size	Min. sample size in g
3 in	6000
2 in	4000
1 in	2000
1/2 in	1000
No. 4	200
No. 10	100

the particles separated out by some means. The minimum size of sample to be used depends upon the maximum particle size as given in Table 3.3 (US Army Corps of Engineers). By determining the mass of soil sample left on each sieve, the following calculations can be made.

1. Percentage retained on any sieve $= \dfrac{\text{mass of soil retained}}{\text{total soil mass}} \times 100$

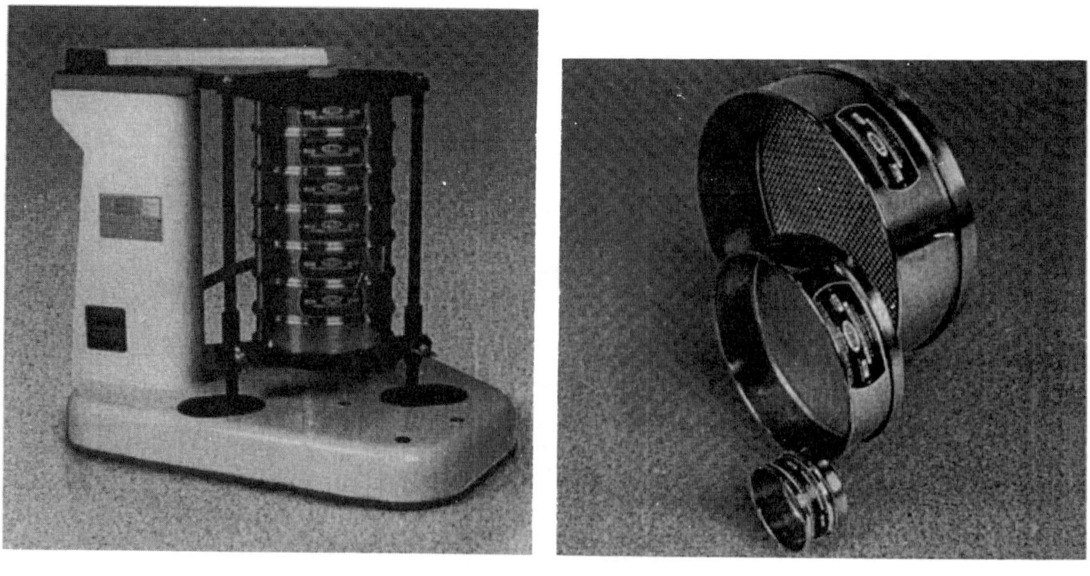

(a) (b)

Figure 3.3 (a) Sieve shaker and (b) a set of sieves for a test in the laboratory (Courtesy: Soiltest, USA)

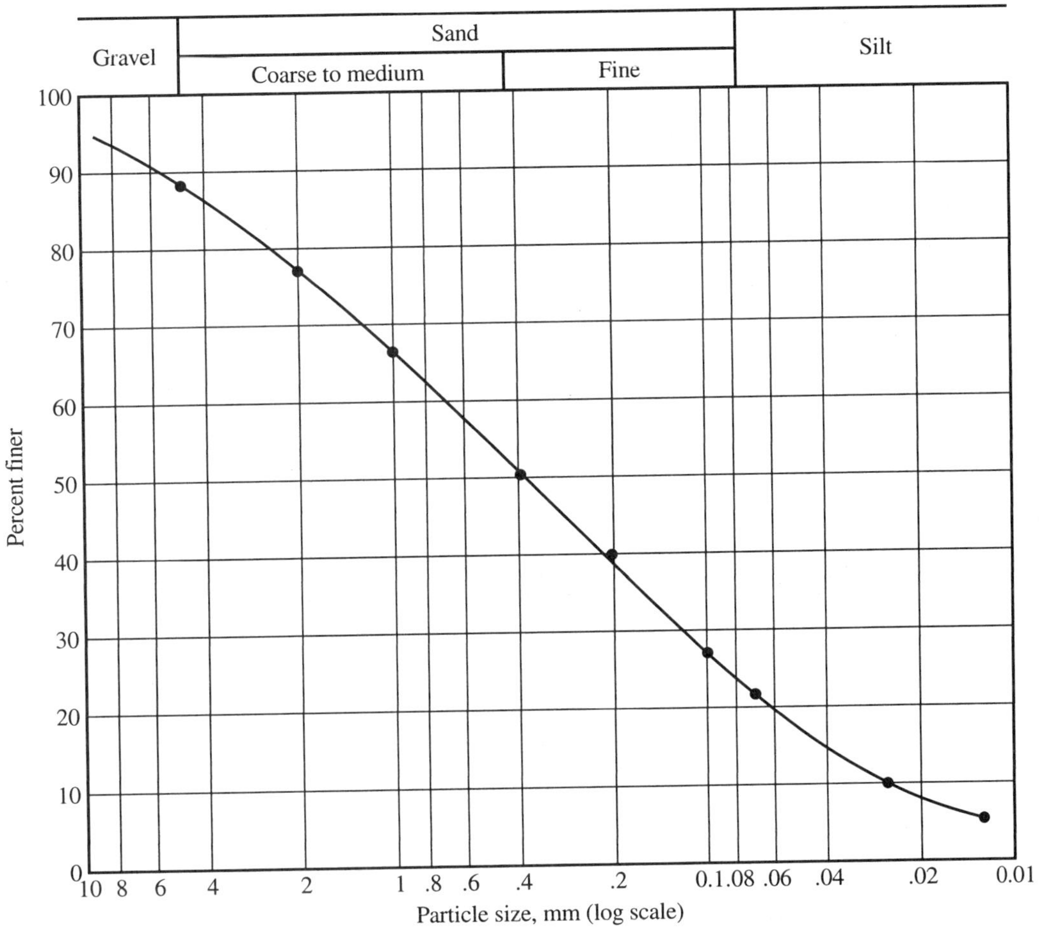

Figure 3.4 Particle-size distribution curve

2. Cumulative percentage retained on any sieve = Sum of percentages retained on all coarser sieves.
3. Percentage finer than any sieve size, *P* = 100 per cent minus cumulative percentage retained.

The results may be plotted in the form of a graph on semi-log paper with the percentage finer on the arithmetic scale and the particle diameter on the log scale as shown in Fig. 3.4.

3.8 THE HYDROMETER METHOD OF ANALYSIS

The hydrometer method was originally proposed in 1926 by Prof. Bouyoucos of Michigan Agricultural College, and later modified by Casagrande (1931). This method depends upon variations in the density of a soil suspension contained in a 1000 mL graduated cylinder. The density of the suspension is measured with a hydrometer at determined time intervals; then the coarsest diameter of particles in suspension at a given time and the percentage of particles finer than that coarsest (suspended) diameter are computed. These computations are based on Stokes' formula which is described below.

Stokes' Law

Stokes (1856), an English physicist, proposed an equation for determining the terminal velocity of a falling sphere in a liquid. If a single sphere is allowed to fall through a liquid of indefinite extent, the terminal velocity, v can be expressed as,

$$v = \frac{\gamma_s - \gamma_w}{18\mu} D^2 \tag{3.22}$$

in which,

v = terminal velocity of fall of a sphere through a liquid $= \frac{\text{distance}}{\text{time}} = \frac{L}{t}$

γ_s = unit weight of solid sphere

γ_w = unit weight of liquid

μ = absolute viscosity of liquid

D = diameter of sphere.

From Eq. (3.22), after substituting for v, we have

$$D = \sqrt{\frac{18\mu}{(G_s - 1)\gamma_w}} \sqrt{\frac{L}{t}} \tag{3.23}$$

in which $\gamma_s = G_s \gamma_w$

If L is in cm, t is in min, γ_w in g/cm^3, μ in (g-sec)/cm^2 and D in mm, then Eq. (3.23) may be written as

$$\frac{D(\text{mm})}{10} = \sqrt{\frac{18\mu}{(G_s - 1)\gamma_w}} \sqrt{\frac{L}{t \times 60}}$$

or $$D = \sqrt{\frac{30\mu}{(G_s - 1)\gamma_w}} \sqrt{\frac{L}{t}} = K\sqrt{\frac{L}{t}} \tag{3.24}$$

where, $$K = \sqrt{\frac{30\mu}{(G_s - 1)}} \tag{3.25}$$

by assuming $\gamma_w \approx 1 \text{g/cm}^3$

It may be noted here that the factor K is a function of temperature T, specific gravity G_s of particles and viscosity of water. Table 3.4a gives the values of K for the various values of G_s at different temperatures T. If it is necessary to calculate D without the use of Table 3.4a we can use Eq. (3.24) directly. The variation of μ with temperature is required which is given in Table 3.4b.

Assumptions of Stokes Law and its Validity

Stokes' law assumes spherical particles falling in a liquid of infinite extent, and all the particles have the same unit weight γ_s. The particles reach constant terminal velocity within a few seconds after they are allowed to fall.

Since particles are not spherical, the concept of an equivalent diameter has been introduced. A particle is said to have an equivalent diameter D_e, if a sphere of diameter D having the same unit weight as the particle, has the same velocity of fall as the particle. For bulky grains $D_e \approx D$, whereas for flaky particles $D/D_e = 4$ or more.

Table 3.4a Values of *K* for use in Eq. (3.24) for several specific gravity of solids and temperature combinations

	G_s of Soil Solids							
Temp °C	2.50	2.55	2.60	2.65	2.70	2.75	2.80	2.85
16	0.0151	0.0148	0.0146	0.0144	0.0141	0.0139	0.0139	0.0136
17	0.0149	0.0146	0.0144	0.0142	0.0140	0.0138	0.0136	0.0134
18	0.0148	0.0144	0.0142	0.0140	0.0138	0.0136	0.0134	0.0132
19	0.0145	0.0143	0.0140	0.0138	0.0136	0.0134	0.0132	0.0131
20	0.0143	0.0141	0.0139	0.0137	0.0134	0.0133	0.0131	0.0129
21	0.0141	0.0139	0.0137	0.0135	0.0133	0.0131	0.0129	0.0127
22	0.0140	0.0137	0.0135	0.0133	0.0131	0.0129	0.0128	0.0126
23	0.0138	0.0136	0.0134	0.0132	0.0130	0.0128	0.0126	0.0124
24	0.0137	0.0134	0.0132	0.0130	0.0128	0.0126	0.0125	0.0123
25	0.0135	0.0133	0.0131	0.0129	0.0127	0.0125	0.0123	0.0122
26	0.0133	0.0131	0.0129	0.0127	0.0125	0.0124	0.0122	0.0120
27	0.0132	0.0130	0.0128	0.0126	0.0124	0.0122	0.0120	0.0119
28	0.0130	0.0128	0.0126	0.0124	0.0123	0.0121	0.0119	0.0117
29	0.0129	0.0127	0.0125	0.0123	0.0121	0.0120	0.0118	0.0116
30	0.0128	0.0126	0.0124	0.0122	0.0120	0.0118	0.0117	0.0115

Table 3.4b Properties of distilled water (μ = absolute viscosity)

Temp °C	Unit weight of water, g/cm^3	Viscosity of water, poise
4	1.00000	0.01567
16	0.99897	0.01111
17	0.99880	0.0108
18	0.99862	0.0105
19	0.99844	0.01030
20	0.99823	0.01005
21	0.99802	0.00981
22	0.99780	0.00958
23	0.99757	0.00936
24	0.99733	0.00914
25	0.99708	0.00894
26	0.99682	0.00874
27	0.99655	0.00855
28	0.99627	0.00836
29	0.99598	0.00818
30	0.99568	0.00801

The effect of influence of one particle over the other is minimized by limiting the mass of soil for sedimentation analysis to 60 g in a sedimentation jar of 10^3 cm^3 capacity.

Hydrometer Analysis

Figure 3.5 shows a streamlined hydrometer of the type ASTM 152 H used for hydrometer analysis. The hydrometer possesses a long stem and a bulb. The hydrometer is used for the determination of unit weight of suspensions at different depths and particular intervals of time. A unit volume of soil suspension at a depth L and at any time t contains particles finer than a particular diameter D. The value of this diameter is determined by applying Stokes' law whereas the percentage finer than this diameter is determined by the use of the hydrometer. The principle of the method is that the reading of the hydrometer gives the unit weight of the suspension at the center of volume of the hydrometer. The first step in the presentation of this method is to calibrate the hydrometer.

Let the sedimentation jar contain a suspension of volume V with total mass of solids M_s. Let the jar be kept vertically on a table after the solids are thoroughly mixed. The initial density ρ_i of the suspension at any depth z from the surface at time $t = 0$ may be expressed as

$$\rho_i = \frac{M_s}{V} + \left(1 - \frac{M_s}{G_s V \rho_o}\right)\rho_w = \frac{M_s}{V} + \left(1 - \frac{M_s}{G_s V}\right) \qquad (3.26a)$$

where ρ_o = density of water at 4°C and ρ_w density of water at test temperature T, and G_s = specific gravity of the solids. For all practical purposes $\rho_o = \rho_w = 1$ g/cm^3.

After a lapse of time t, a unit volume of suspension at a depth z contains only particles finer than a particular diameter D, since particles coarser than this diameter have fallen a distance greater than z as per Stokes'law. The coarsest diameter of the particle in a unit volume of the suspension at depth z and time t is given by Eq. (3.24) where $z = L$. Let M_d be the mass of all particles finer than D in the sample taken for analysis. The density of the suspension ρ_f after an elapsed time t may be expressed as

$$\rho_f = \frac{M_D}{V} + \left(1 - \frac{M_D}{G_s V}\right) \qquad (3.26b)$$

where $\frac{M_D}{V}$ = Mass of particles of diameter smaller than diameter D in the unit volume of suspension at depth z at an elapsed time t.

From Eq. (3.26b) we may write

$$M_D = \frac{G_s}{(G_s - 1)} V\,(\rho_f - 1) \qquad (3.26c)$$

The ASTM 152 H type hydrometer, normally used for the analysis, is calibrated to read from 0 to 60 g of soil in a 1000 mL soil-water mixture with the limitation that the soil particles have a specific gravity $G_s = 2.65$. The reading is directly related to the specific gravity of the suspension. In Eq. (3.26c) the mass of the solids M_D in the suspension varies from 0 to 60 grams. The reading R on the stem of the hydrometer (corrected for meniscus) may be expressed as

$$R = M_D = \frac{G_s}{(G_s - 1)} V(\rho_f - 1) = 1.6061 \times 10^3 (\rho_f - 1) \qquad (3.26d)$$

where,

$G_s = 2.65$, and $V = 1000$ mL

ρ_f = density of suspension per unit volume = specific gravity of the suspension.

From Eq. (3.26d), it is clear that the ASTM 152H hydrometer is calibrated in such a way that the reading on the stem will be

$R = 0$ when $\rho_f = 1$, and $R = 60$ when $\rho_f = 1.0374$

The ASTM 152 H hydrometer gives the distance of any reading R on the stem to the center of volume and is designated as L as shown in Fig. 3.5. The distance L varies linearly with the reading R. An expression for L may be written as follows for any reading R for the ASTM 152 H hydrometer (Fig. 3.5).

$$L = L_1 + \frac{L_2}{2} \tag{3.27}$$

where L_1 = distance from reading R to the top of the bulb
L_2 = length of hydrometer bulb = 14 cm for ASTM 152 H hydrometer

When the hydrometer is inserted into the suspension, the surface of the suspension rises as shown in Fig. 3.6. The distance L in Fig. 3.6 is the actual distance through which a particle of diameter D has fallen. The point at level A_1 at depth L occupies the position A_2 (which coincides with the center of volume of the hydrometer) in the figure after the immersion of the hydrometer and correspondingly the surface of suspension rises from B_1 to B_2. The depth L' is therefore greater than L through which the particle of diameter D has fallen. The effective value of L can be obtained from the equation

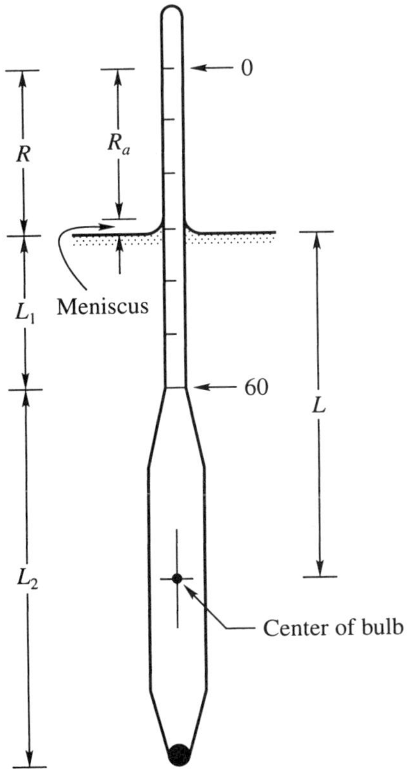

Figure 3.5 ASTM 152 H type hydrometer

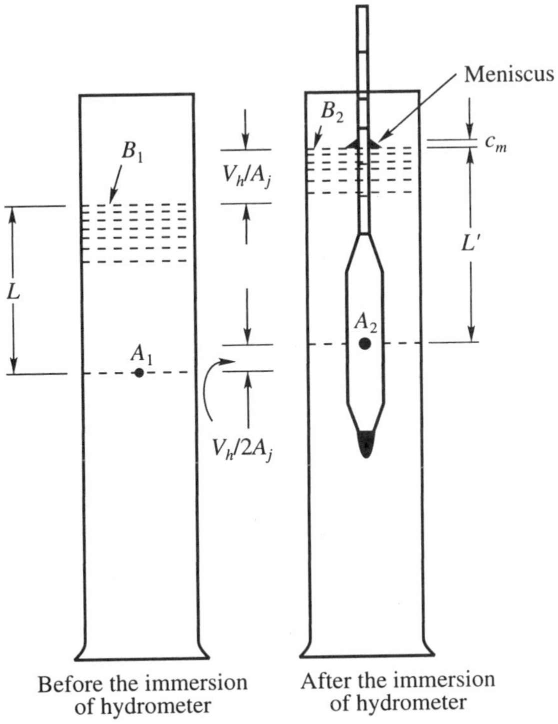

Figure 3.6 Immersion correction

Table 3.5 Values of *L* (effective depth) for use in Stokes' formula for diameters of particles for ASTM soil hydrometer 152H

Original hydrometer reading (corrected for meniscus only)	Effective depth *L* cm	Original hydrometer reading (corrected for meniscus only)	Effective depth *L* cm	Original hydrometer reading (corrected for meniscus only)	Effective depth *L* cm
0	16.3	21	12.9	42	9.4
1	16.1	22	12.7	43	9.2
2	16.0	23	12.5	44	9.1
3	15.8	24	12.4	45	8.9
4	15.6	25	12.2	46	8.8
5	15.5	26	12.0	47	8.6
6	15.3	27	11.9	48	8.4
7	15.2	28	11.7	49	8.3
8	15.0	29	11.5	50	8.1
9	14.8	30	11.4	51	7.9
10	14.7	31	11.2	52	7.8
11	14.5	32	11.1	53	7.6
12	14.3	33	10.9	54	7.4
13	14.2	34	10.7	55	7.3
14	14.0	35	10.5	56	7.1
15	13.8	36	10.4	57	7.0
16	13.7	37	10.2	58	6.8
17	13.5	38	10.1	59	6.6
18	13.3	39	9.9	60	6.5
19	13.2	40	9.7		
20	13.0	41	9.6		

$$L = L' - \frac{V_h}{2A_j} = L_1 + \frac{1}{2}\left(L_2 - \frac{V_h}{A_j}\right) \tag{3.28}$$

where V_h = volume of hydrometer (152 H) = 67 cm^3; A_j = cross-sectional area of the sedimentation cylinder = 27.8 cm^2 for 1000 mL graduated cylinder .

For an ASTM 152 H hydrometer, the value of L for any reading R (corrected for meniscus) may be obtained from

$$L = 16.3 - 0.1641R \tag{3.29}$$

Table 3.5 gives the values of L for various hydrometer readings of R for the 152 H hydrometer.

Determination of Percent Finer

The ASTM 152 H hydrometer is calibrated to read from 0 to 60 g of soil in a 1000 mL suspension with the limitation that the soil has a specific gravity G_s = 2.65. The reading is, of course, directly related to the specific gravity of the suspension. The hydrometer gives readings pertaining to the specific gravity of the soil-water suspension at the center of the bulb. Any soil particles larger than those still in suspension in the zone shown as L (Fig 3.5) have fallen below the center of volume, and this constantly decreases the specific gravity of the suspension at the center of volume of the hydrometer. Lesser the specific gravity of the suspension, the deeper the hydrometer will sink into the suspension. It must also be remembered here, that the specific gravity of water decreases as the temperature rises from 4° C. This will also cause the hydrometer to sink deeper into the suspension.

The readings of the hydrometer are affected by the rise in temperature during the test. The temperature correction is a constant. The use of a dispersing agent also affects the hydrometer reading. Corrections for this can be obtained by using a sedimentation cylinder of water from the same source and with the same quantity of dispersing agent as that used in the soil-water suspension to obtain a *zero correction*. This jar of water should be at the same temperature as that of the soil water suspension.

A reading of less than zero in the standard jar of water is recorded as a (–) correction value; a reading between 0 and 60 is recorded as a (+) value. All the readings are taken to the top of the meniscus in both the standard jar (clear water) and soil suspension.

If the temperature during the test is quite high, the density of water will be equally less and hydrometer will sink too deep. One can use a temperature correction for the soil-water suspension. Table 3.6 gives the values of temperature correlation C_T. The zero correction C_o can be (±) and the temperature correction also has (±) sign.

The actual hydrometer reading R_a has to be corrected as follows

1. correction for meniscus C_m only for use in Eq. (3.24)
2. zero correction C_o and temperature correction C_T for obtaining percent finer.

Reading for use in Eq. (3.24)

$$R = R_a + C_m \tag{3.30a}$$

Reading for obtaining percent finer

$$R_c = R_a - C_o + C_T \tag{3.30b}$$

Percent Finer

The 152 H hydrometer is calibrated for a suspension with a specific gravity of solids $G_s = 2.65$. If the specific gravity of solids used in the suspension is different from 2.65, the percent finer has to be corrected by the factor C_{sg} expressed as

Table 3.6 Temperature correction factors C_T

Temp °C	C_T	Temp °C	C_T
15	–1.10	23	+0.70
16	–0.90	24	+1.00
17	–0.70	25	+1.30
18	–0.50	26	+1.65
19	–0.30	27	+2.00
20	0.00	28	+2.50
21	+0.20	29	+3.05
22	+0.40	30	+3.80

$$C_{sg} = \frac{1.65G_s}{2.65(G_s - 1)} \tag{3.31}$$

Typical values of C_{sg} are given in Table 3.7.

Now the percent finer with the correction factor C_{sg} may be expressed as

$$\text{Percent finer,} \quad P' = \frac{C_{sg} R_c}{M_s} \times 100 \tag{3.32}$$

where R_c = grams of soil in suspension at some elapsed time t [corrected hydrometer reading from Eq. (3.30b)]

M_s = mass of soil used in the suspension in gms (not more than 60 gm for 152 H hydrometer)

Eq. (3.32) gives the percentage of particles finer than a particle diameter D in the mass of soil M_s used in the suspension. If M_p is the mass of soil particles passing through 75 micron sieve (greater than M_s) and M the total mass taken for the combined sieve and hydrometer analysis, the percent finer for the entire sample may be expressed as

$$\text{Percent finer(combined),} \quad P = P'\% \times \frac{M_p}{M} \tag{3.33}$$

Now Eq. (3.33) with Eq. (3.24) gives points for plotting a grain size distribution curve.

Test procedure

The suggested procedure for conducting the hydrometer test is as follows:

1. Take 60 g or less dry sample from the soil passing through the No. 200 sieve
2. Mix this sample with 125 mL of a 4% of $NaPO_3$ solution in a small evaporating dish
3. Allow the soil mixture to stand for about 1 hour. At the end of the soaking period transfer the mixture to a dispersion cup and add distilled water until the cup is about two-thirds full. Mix for about 2 min.
4. After mixing, transfer all the contents of the dispersion cup to the sedimentation cylinder, being careful not to lose any material Now add temperature-stabilized water to fill the cylinder to the 1000 mL mark.
5. Mix the suspension well by placing the palm of the hand over the open end and turning the cylinder upside down and back for a period of 1 min. Set the cylinder down on a table.
6. Start the timer immediately after setting the cylinder. Insert the hydrometer into the suspension just about 20 seconds before the elapsed time of 2 min. and take the first reading at 2 min. Take the temperature reading. Remove the hydrometer and the thermometer and place both of them in the control jar.
7. The control jar contains 1000 mL of temperature-stabilized distilled water mixed with 125 mL of the same 4% solution of $NaPO_3$.

Table 3.7 Correction factors C_{sg} for unit weight of solids

G_s of soil solids	Correction factor C_{sg}	G_s of soil solids	Correction factor C_{sg}
2.85	0.96	2.65	1.00
2.80	0.97	2.60	1.01
2.75	0.98	2.55	1.02
2.70	0.99	2.50	1.04

8. The hydrometer readings are taken at the top level of the meniscus in both the sedimentation and control jars.
9. Steps 6 through 8 are repeated by taking hydrometer and temperature readings at elapsed times of 4, 8, 16, 30, 60 min. and 2, 4, 8, 16, 32, 64 and 96 hr.
10. Necessary computations can be made with the data collected to obtain the grain-distribution curve.

3.9 GRAIN SIZE DISTRIBUTION CURVES

A typical set of grain size distribution curves is given in Fig. 3.7 with the grain size D as the abscissa on the logarithmic scale and the percent finer P as the ordinate on the arithmetic scale. On the curve C_1 the section AB represents the portion obtained by sieve analysis and the section $B'C'$ by hydrometer analysis. Since the hydrometer analysis gives equivalent diameters which are generally less than the actual sizes, the section $B'C'$ will not be a continuation of AB and would occupy a position shown by the dotted curve. If we assume that the curve BC is the actual curve obtained by sketching it parallel to $B'C'$, then at any percentage finer, say 20 per cent, the diameters D_a and D_e represent the actual and equivalent diameters respectively. The ratio of D_a to D_e can be quite high for flaky grains.

The shapes of the curves indicate the nature of the soil tested. On the basis of the shapes we can classify soils as:

1. Uniformly graded or poorly graded.
2. Well graded.
3. Gap graded.

Uniformly graded soils are represented by nearly vertical lines as shown by curve C_2 in Fig. 3.7. Such soils possess particles of almost the same diameter. A well graded soil, represented by curve C_1, possesses a wide range of particle sizes ranging from gravel to clay size particles. A gap graded soil, as shown by curve C_3 has some of the sizes of particles missing. On this curve the soil particles falling within the range of XY are missing.

The grain distribution curves as shown in Fig. 3.7 can be used to understand certain grain size characteristics of soils. Hazen (1893) has shown that the permeability of clean filter sands in a loose state can be correlated with numerical values designated D_{10}, the effective grain size. The effective grain size corresponds to 10 per cent finer particles. Hazen found that the sizes smaller than the effective size affected the functioning of filters more than did the remaining 90 per cent of the sizes. To determine whether a material is uniformly graded or well graded, Hazen proposed the following equation:

$$C_u = \frac{D_{60}}{D_{10}} \tag{3.34}$$

where D_{60} is the diameter of the particle at 60 per cent finer on the grain size distribution curve. The *uniformity coefficient*, C_u, is about one if the grain size distribution curve is almost vertical, and the value increases with gradation. For all practical purposes we can consider the following values for granular soils.

$C_u > 4$ for well graded gravel
$C_u > 6$ for well graded sand
$C_u < 4$ for uniformly graded soil containing particles of the same size

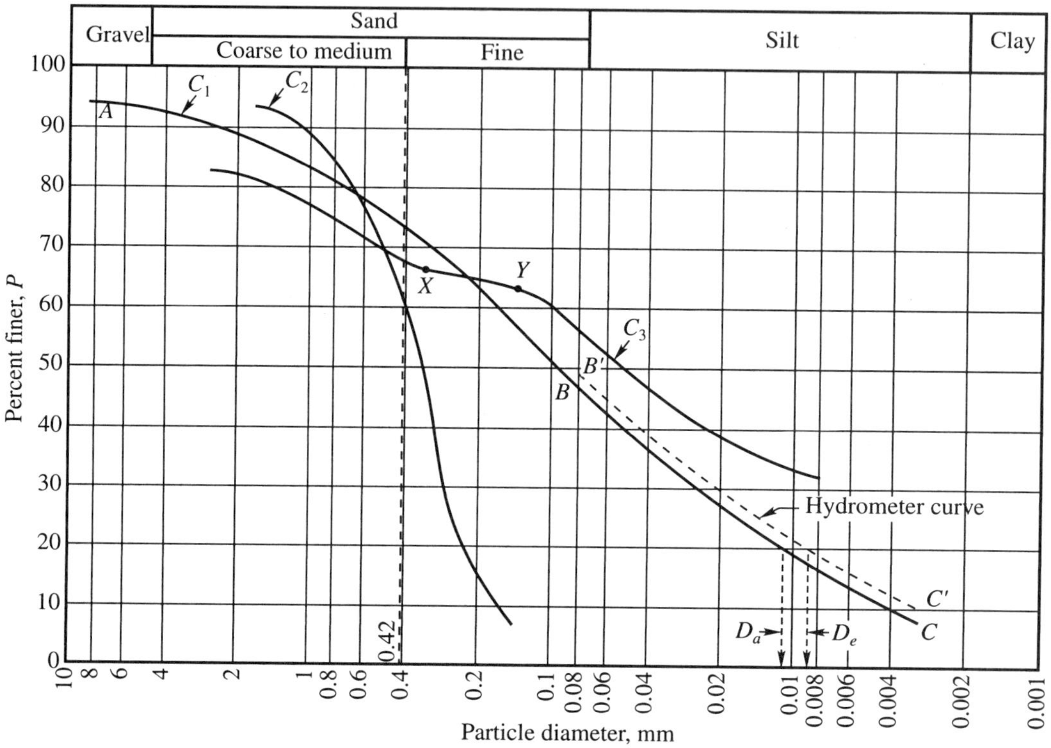

Figure 3.7 Grain size distribution curves

There is another step in the procedure to determine the gradation of particles. This is based on the term called the *coefficient of curvature* which is expressed as

$$C_c = \frac{D_{30}^2}{D_{10} \times D_{60}} \tag{3.35}$$

wherein D_{30} is the size of particle at 30 percent finer on the gradation curve. The soil is said to be well graded if C_c lies between 1 and 3 for gravels and sands.

Two samples of soils are said to be similarly graded if their grain size distribution curves are almost parallel to each other on a semilogarithmic plot. When the curves are almost parallel to each other the ratios of their diameters at any percentage finer approximately remain constant. Such curves are useful in the design of filter materials around drainage pipes.

3.10 RELATIVE DENSITY OF COHESIONLESS SOILS

The density of granular soils varies with the shape and size of grains, the gradation and the manner in which the mass is compacted. If all the grains are assumed to be spheres of uniform size and packed as shown in Fig. 3.8(a), the void ratio of such a mass amounts to about 0.90. However, if the grains are packed as shown in Fig. 3.8(b), the void ratio of the mass is about 0.35. The soil corresponding to the higher void ratio is called loose and that corresponding to the lower void ratio is called dense. If the soil grains are not uniform, then smaller grains fill in the space between the bigger ones and the void ratios of such soils are reduced to as low as 0.25 in the densest state. If the grains are angular, they tend to form looser structures than rounded grains

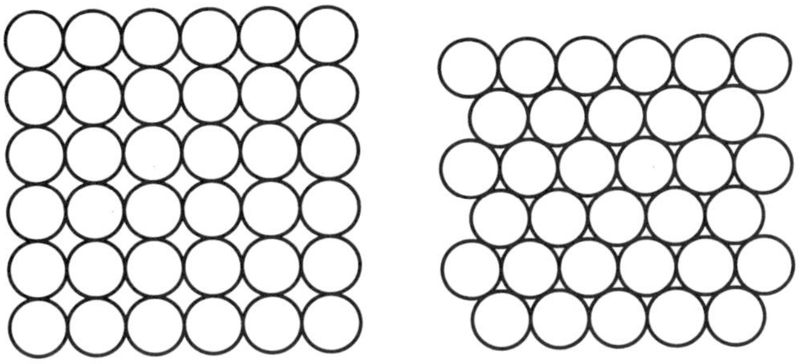

(a) Loosest state (b) Densest state

Figure 3.8 Packing of grains of uniform size

Table 3.8 Classification of sandy soils

Relative density, D_r, %	Type of soil
0–15	Very loose
15–50	Loose
50–70	Medium dense
70–85	Dense
85–100	Very dense

because their sharp edges and points hold the grains further apart. If the mass with angular grains is compacted by vibration, it forms a dense structure. Static load alone will not alter the density of grains significantly but if it is accompanied by vibration, there will be considerable change in the density. The water present in voids may act as a lubricant to a certain extent for an increase in the density under vibration. The change in void ratio would change the density and this in turn changes the strength characteristics of granular soils. Void ratio or the unit weight of soil can be used to compare the strength characteristics of samples of granular soils of the same origin. The term used to indicate the strength characteristics in a qualitative manner is termed as relative density which is already expressed by Eq. (3.20). On the basis of relative density, we can classify sandy soils as loose, medium or dense as in Table 3.8.

3.11 CONSISTENCY OF CLAY SOIL

Consistency is a term used to indicate the degree of firmness of cohesive soils. The consistency of natural cohesive soil deposits is expressed qualitatively by such terms as very soft, soft ,stiff, very stiff and hard. The physical properties of clays greatly differ at different water contents. A soil which is very soft at a higher percentage of water content becomes very hard with a decrease in water content. However, it has been found that at the same water content, two samples of clay of different origins may possess different consistency. One clay may be relatively soft while the other may be hard. Further, a decrease in water content may have little effect on one sample of clay but may transform the other sample from almost a liquid to a very firm condition. Water content alone, therefore, is not an adequate index of consistency for engineering and many other purposes. Consistency of a soil can be expressed in terms of:

1. Atterberg limits of soils
2. Unconfined compressive strengths of soils.

Atterberg Limits

Atterberg, a Swedish scientist, considered the consistency of soils in 1911, and proposed a series of tests for defining the properties of cohesive soils. These tests indicate the range of the plastic state (plasticity is defined as the property of cohesive soils which possess the ability to undergo changes of shape without rupture) and other states. He showed that if the water content of a thick suspension of clay is gradually reduced, the clay water mixture undergoes changes from a liquid state through a plastic state and finally into a solid state. The different states through which the soil sample passes with the decrease in the moisture content are depicted in Fig. 3.9. The water contents corresponding to the transition from one state to another are termed as *Atterberg Limits* and the tests required to determine the limits are the *Atterberg Limit Tests*. The testing procedures of Atterberg were subsequently improved by A. Casagrande (1932).

The transition state from the liquid state to a plastic state is called the *liquid limit*, w_l. At this stage all soils possess a certain small shear strength. This arbitrarily chosen shear strength is probably the smallest value that is feasible to measure in a standardized procedure. The transition from the plastic state to the semisolid state is termed the *plastic limit*, w_p. At this state the soil rolled into threads of about 3 mm diameter just crumbles. Further decrease of the water contents of the same will lead finally to the point where the sample can decrease in volume no further. At this point the sample begins to dry at the surface, saturation is no longer complete, and further decrease in water in the voids occurs without change in the void volume. The color of the soil begins to change from dark to light. This water content is called the *shrinkage limit*, w_s. The limits expressed above are all expressed by their percentages of water contents. The range of water content between the liquid and plastic limits, which is an important measure of plastic behavior, is called the *plasticity index*, I_p, i.e.,

$$I_p = w_l - w_p \tag{3.36}$$

Figure 3.10 depicts the changes in volume from the liquid limit to the shrinkage limit graphically. The soil remains saturated down to the shrinkage limit and when once this limit is crossed, the soil becomes partially saturated. The air takes the place of the moisture that is lost due to evaporation. At about 105° to 110°C, there will not be any normal water left in the pores and soil at this temperature is said to be *oven-dry*. A soil sample of volume V_o and water content w_o is represented by point *A* in the figure.

As the soil loses moisture content there is a corresponding change in the volume of soils. The volume change of soil is equal to the volume of moisture lost. The straight line, *AE*, therefore, gives the volume of the soil at different water contents. Points *C* and *D* represent the transition stages of soil sample at liquid and plastic limits respectively. As the moisture content is reduced further beyond the point *D*, the decrease in volume of the soil sample will not be linear

States	Limit	Consistency	Volume change
Liquid		Very soft	↑
.... w_l	Liquid limit	Soft	
Plastic		Stiff	Decrease in volume
.... w_p	Plastic limit	Very stiff	
Semi solid			
.... w_s	Shrinkage limit	Extremely stiff	↓
Solid		Hard	Constant volume

Figure 3.9 Different states and consistency of soils with Atterberg limits

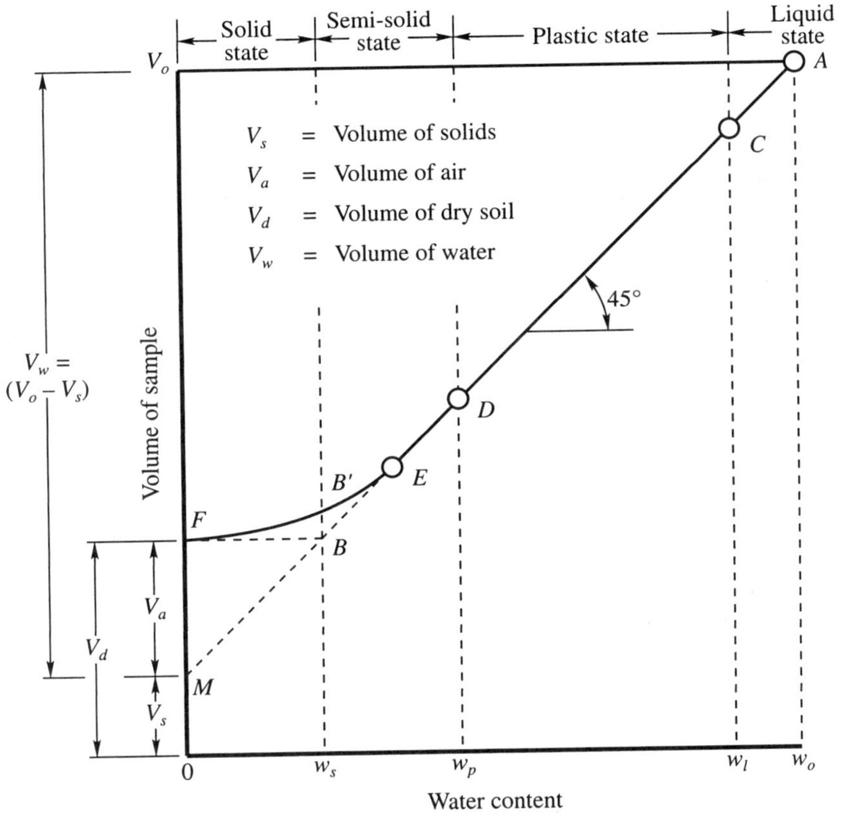

Figure 3.10 Curve showing transition stages from the liquid to solid state

with the decrease in moisture beyond a point E due to many causes. One possible cause is that air might start entering into the voids of the soil. This can happen only when the normal water between the particles is removed. If the normal water between some particles is removed, the soil particles surrounded by absorbed water will come in contact with each other. Greater pressure is required if these particles are to be brought still closer. As such the change in volume is less than the change in moisture content. Therefore, the curve *DEB'F* depicts the transition from plastic limit to the dry condition of soil represented by point *F*. However, for all practical purposes, the abscissa of the point of intersection *B* of the tangents *FB* and *EB* may be taken as the shrinkage limit, w_s. The straight line *AB* when extended meets the ordinate at point *M*. The ordinate of *M* gives the volume of the solid particles V_s. Since the ordinate of *F* is the dry volume, V_d, of the sample, the volume of air V_a, is given by $(V_d - V_s)$.

3.12 DETERMINATION OF ATTERBERG LIMITS

Liquid Limit

The apparatus shown in Fig. 3.11 is the Casagrande Liquid Limit Device used for determining the liquid limits of soils. Figure 3.12 shows a hand-operated liquid limit device. The device contains a brass cup which can be raised and allowed to fall on a hard rubber base by turning the handle. The cup is raised by one cm. The limits are determined on that portion of soil finer than a No. 40 sieve (ASTM Test Designation D-4318). About 100 g of soil is mixed thoroughly with distilled water into a uniform paste. A portion of the paste is placed in the cup and leveled to a maximum depth of 10 mm. A channel of the dimensions of 11 mm width and 8 mm depth is cut through the sample along the

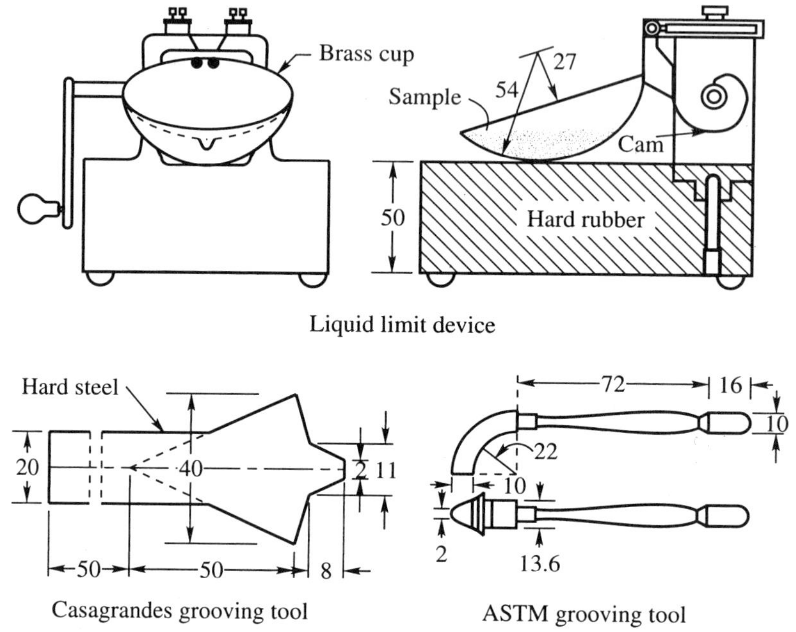

Figure 3.11 Casagrande's liquid limit apparatus

symmetrical axis of the cup. The grooving tool should always be held normal to the cup at the point of contact. The handle is turned at a rate of about two revolutions per second and the number of blows necessary to close the groove along the bottom for a distance of 12.5 mm is counted. The groove should be closed by a flow of the soil and not by slippage between the soil and the cup. The water content of the soil in the cup is altered and the tests repeated. At least four tests should be carried out by adjusting the water contents in such a way that the number of blows required to close the groove may fall within the range of 5 to 40. A plot of water content against the log of blows is made as shown in Fig. 3.13. Within the range of 5 to 40 blows, the plotted points lie almost on a straight line. The curve so obtained is known as a '*flow curve*'. The water content corresponding to 25 blows is termed the *liquid limit*. The equation of the flow curve can be written as

$$w = -I_f \log N + C \tag{3.37}$$

where,
w = water content
I_f = slope of the flow curve, termed as flow index
N = number of blows
C = a constant.

Liquid Limit by One-Point Method

The determination of liquid limit as explained earlier requires a considerable amount of time and labor. We can use what is termed the 'one-point method' if an approximate value of the limit is required. The formula used for this purpose is

$$w_l = w\left(\frac{N}{25}\right)^n \tag{3.38}$$

Figure 3.12 Hand-operated liquid limit device (Courtesy: Soiltest, USA)

where w is the water content corresponding to the number of blows N, and n, an index whose value has been found to vary from 0.068 to 0.121. An average value of 0.104 may be useful for all practical purposes. It is, however, a good practice to check this method with the conventional method as and when possible.

Liquid Limit by the Use of Fall Cone Penetrometer

Figure 3.14 shows the arrangement of the apparatus. The soil whose liquid limit is to be determined is mixed well into a soft consistency and pressed into the cylindrical mold of 5 cm diameter and 5 cm high. The cone which has a central angle of 31° and a total mass of 148 g will be kept free on the surface of the soil. The depth of penetration y of the cone is measured in mm on the graduated scale after 30 sec of penetration. The liquid limit w_l may be computed by using the formula,

$$w_l = w_y + 0.01(25 - y)(w_y + 15) \qquad (3.39)$$

where w_y is the water content corresponding to the penetration y.

The procedure is based on the assumption that the penetration lies between 20 and 30 mm. Even this method has to be used with caution.

Plastic Limit

About 15 g of soil, passing through a No. 40 sieve, is mixed thoroughly. The soil is rolled on a glass plate with the hand, until it is about 3 mm in diameter. This procedure of mixing and rolling is repeated till the soil shows signs of crumbling. The water content of the crumbled portion of the thread is determined. This is called the plastic limit.

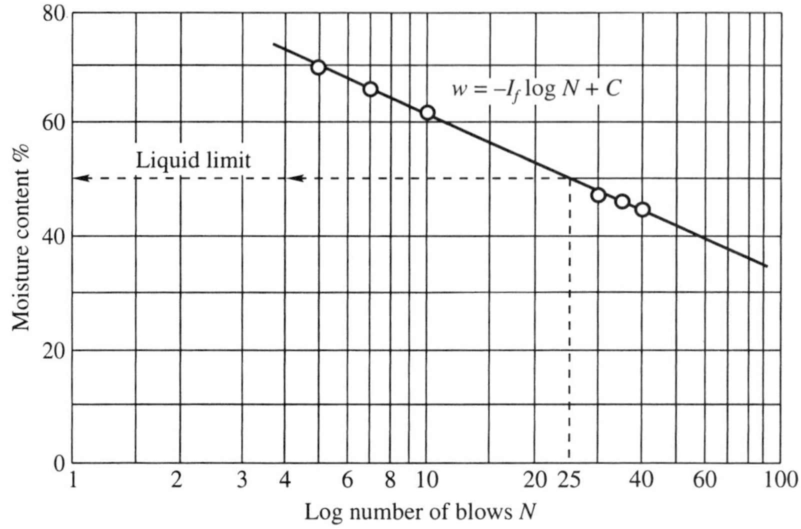

Figure 3.13 Determination of liquid limit

Shrinkage Limit

The shrinkage limit of a soil can be determined by either of the following methods:

1. Determination of w_s, when the specific gravity of the solids G_s is unknown.
2. Determination of w_s, when the specific gravity of the solids, G_s is known.

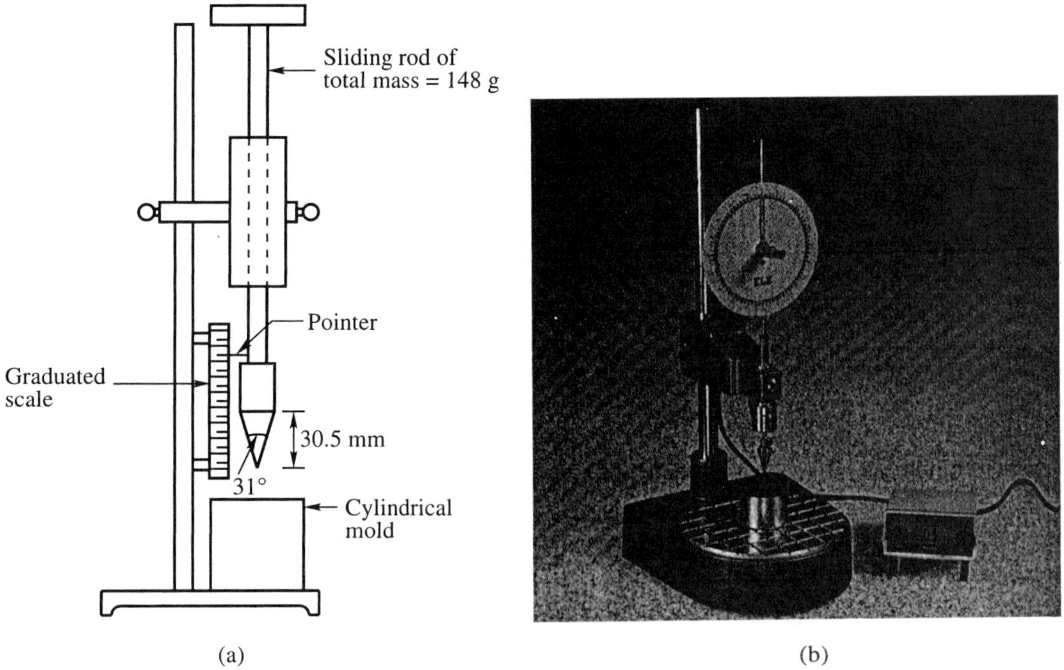

Figure 3.14 Liquid limit by the use of the fall cone penetrometer: (a) a schematic diagram, and (b) a photograph (Courtesy: Soiltest, USA)

Method I When G_s is Unknown

Three block diagrams of a sample of soil having the same mass of solids M_s, are given in Fig. 3.15. Block diagram (a) represents a specimen in the plastic state, which just fills a container of known volume, V_o. The mass of the specimen is M_o. The specimen is then dried gradually, and as it reaches the shrinkage limit, the specimen is represented by block diagram (b). The specimen remains saturated up to this limit but reaches a constant volume V_d. When the specimen is completely dried, its mass will be M_s whereas its volume remains as V_d.

These different states are represented in Fig. 3.10. The shrinkage limit can be written as

$$w_s = \frac{M_w}{M_s} \times 100 \tag{3.40}$$

where, $M_w = M_o - M_s - (V_o - V_d)\,\rho_w$

$$\text{Therefore } w_s = \frac{(M_o - M_s) - (V_o - V_d)\rho_w}{M_s} \times 100\% \tag{3.41}$$

The volume of the dry specimen can be determined either by the displacement of mercury method or wax method. Many prefer the wax method because wax is non-toxic. The wax method is particularly recommended in an academic environment.

Determination of Dry Volume V_d of Sample by Displacement in Mercury

Place a small dish filled with mercury up to the top in a big dish. Cover the dish with a glass plate containing three metal prongs in such a way that the plate is entrapped. Remove the mercury spilt over into the big dish and take out the cover plate from the small dish. Place the soil sample on the mercury. Submerge the sample with the pronged glass plate and make the glass plate flush with the top of the dish. Weigh the mercury that is spilt over due to displacement. The volume of the sample is obtained by dividing the weight of the mercury by its specific gravity which may be taken as 13.6. Figure 3.16 shows the apparatus used for the determination of dry volume.

Method II When G_s is Known

$$w_s = \frac{M_w}{M_s} \times 100 \text{ where, } M_w = (V_d - V_s)\rho_w = \left(V_d - \frac{M_s}{G_s \rho_w}\right)\rho_w$$

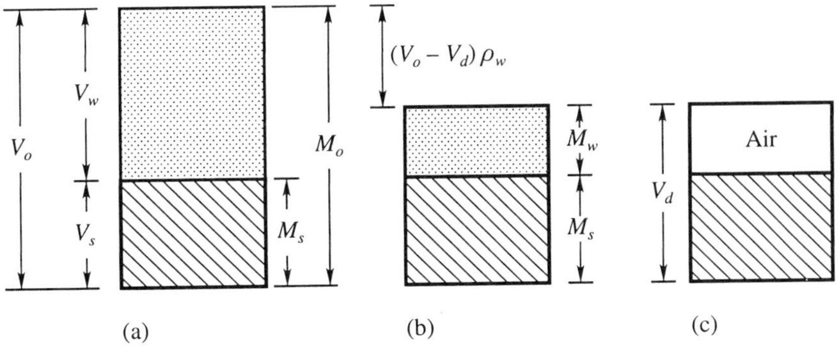

Figure 3.15 Determination of shrinkage limit

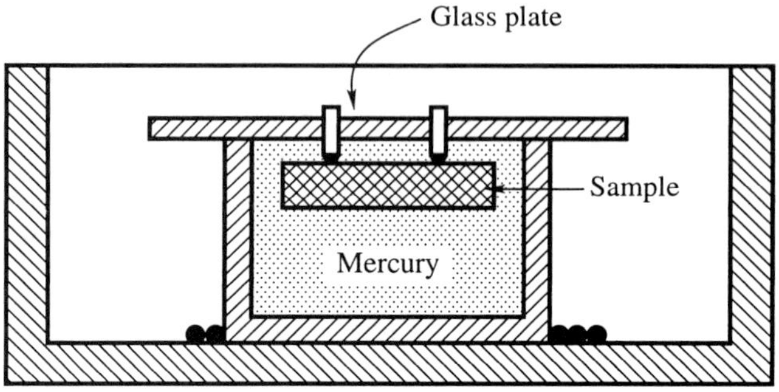

Figure 3.16 Determination of dry volume by mercury displacement method

$$\text{Therefore, } w_s = \frac{\left(V_d - \dfrac{M_s}{G_s \rho_w}\right)\rho_w}{M_s} \times 100 = \left[\frac{V_d \rho_w}{M_s} - \frac{\rho_w}{G_s \rho_w}\right] \times 100 \tag{3.42}$$

or

$$w_s = \left[\frac{V_d}{M_s} - \frac{1}{G_s}\right] \times 100 \tag{3.43}$$

where, $\rho_w = 1$ for all practical purposes.

3.13 DISCUSSION ON LIMITS AND INDICES

Plasticity index and liquid limit are the important factors that help an engineer to understand the consistency or plasticity of a clay. Shearing strength, though constant at liquid limits, varies at plastic limits for all clays. A highly plastic clay (sometimes called a fat clay) has higher shearing strength at the plastic limit and the threads at this limit are rather hard to roll whereas a lean clay can be rolled easily at the plastic limit and thereby possesses low shearing strength.

There are some fine grained soils that appear similar to clays but they cannot be rolled into threads so easily. Such materials are not really plastic. They may be just at the border line between plastic and non-plastic soils. In such soils, one finds the liquid limit practically identical with the plastic limit and $I_p = 0$.

Two soils may differ in their physical properties even though they have the same liquid and plastic limits or the same plasticity index. Such soils possess different flow indices. For example in Fig. 3.17 are shown two flow curves C_1 and C_2 of two samples of soils. C_1 is flatter than C_2. It may be assumed for the sake of explanation that both the curves are straight lines even when the moisture content in the soil is nearer the plastic limit and that the same liquid limit device is used to determine the number of blows required to close the groove at lower moisture contents. The plasticity index I_p is taken to be the same for both the soils. It can be seen from the figure that the sample of flow curve C_1 has liquid and plastic limits of 100 and 80 percent respectively, giving thereby a plasticity index I_p of 20 per cent. The sample of flow curve C_2 has liquid and plastic limits of 54 and 34 percent giving thereby the same plasticity index value of 20 percent. Though the plasticity indices of the two samples remain the same, the resistance offered by the two samples for slippage at their plastic limits is different. Sample one takes 90 blows for slippage whereas the second one takes only 40 blows. This indicates that at the plastic limits, the cohesive strength of sample 1 with a lower flow index is larger than that of sample 2 with a higher flow index.

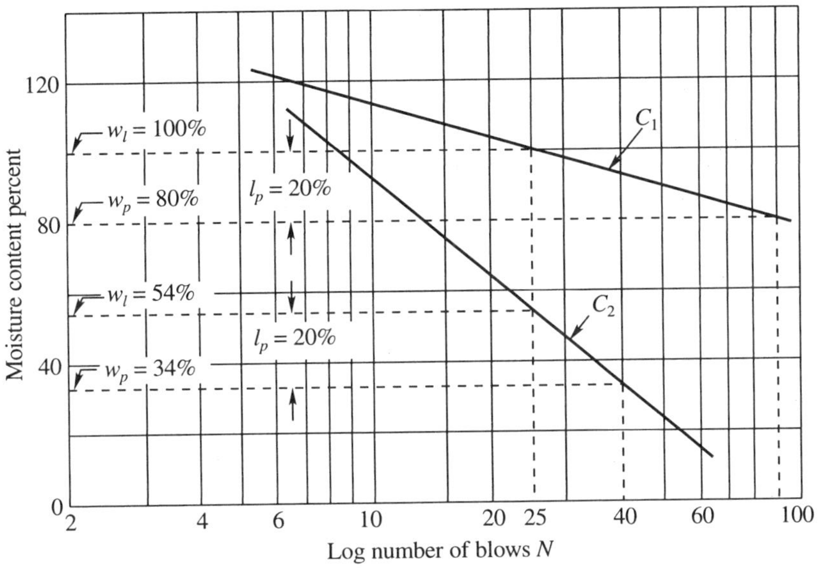

Figure 3.17 Two samples of soils with different flow indices

Plasticity Index I_p

Plasticity index I_p indicates the degree of plasticity of a soil. The greater the difference between liquid and plastic limits, the greater is the plasticity of the soil. A cohesionless soil has zero plasticity index. Such soils are termed non-plastic. Fat clays are highly plastic and possess a high plasticity index. Soils possessing large values of w_1 and I_p are said to be highly plastic or fat. Those with low values are described as slightly plastic or lean. Atterberg classifies the soils according to their plasticity indices as in Table 3.9.

A liquid limit greater than 100 is uncommon for inorganic clays of non-volcanic origin. However, for clays containing considerable quantities of organic matter and clays of volcanic origin, the liquid limit may considerably exceed 100. Bentonite, a material consisting of chemically disintegrated volcanic ash, has a liquid limit ranging from 400 to 600. It contains approximately 70 percent of scale-like particles of colloidal size as compared with about 30 per cent for ordinary highly plastic clays. Kaolin and mica powder consist partially or entirely of scale like particles of relatively coarse size in comparison with highly colloidal particles in plastic clays. They therefore possess less plasticity than ordinary clays. Organic clays possess liquid limits greater than 50. The plastic limits of such soils are equally higher. Therefore soils with organic content have low plasticity indices corresponding to comparatively high liquid limits.

Table 3.9 Soil classifications according to Plasticity Index

Plasticity index	Plasticity
0	Non-plastic
<7	Low plastic
7–17	Medium plastic
>17	Highly plastic

Toughness Index, I_t

The shearing strength of a clay at the plastic limit is a measure of its toughness. Two clays having the same plasticity index possess toughness which is inversely proportional to the flow indices. An approximate numerical value for the toughness can be derived as follows.

Let s_l = shearing strength corresponding to the liquid limit, w_l, which is assumed to be constant for all plastic clays.

s_p = shearing strength at the plastic limit, which can be used as a measure of toughness of a clay.

Now $w_l = -I_f \log N_l + C,\ w_p = -I_f \log N_p + C$

where N_l and N_p are the number of blows at the liquid and plastic limits respectively. The flow curve is assumed to be a straight line extending into the plastic range as shown in Fig. 3.17.

Let, $N_l = ms_l$, $N_p = ms_p$, where m is a constant.

We can write

$$w_l = -I_f \log ms_l + C,\ w_p = -I_f \log ms_p + C$$

Therefore $$I_p = w_l - w_p = I_f(\log ms_p - \log ms_l) = I_f \log \frac{s_p}{s_l}$$

or $$I_t = \frac{I_p}{I_f} = \log \frac{s_p}{s_l} \tag{3.44}$$

Since we are interested only in a relative measure of toughness, I_t can be obtained from Eq. (3.44) as the ratio of plasticity index and flow index. The value of I_t generally falls between 0 and 3 for most clay soils. When I_t is less than one, the soil is friable at the plastic limit. I_t is quite a useful index to distinguish soils of different physical properties.

Liquidity Index I_l

The Atterberg limits are found for remolded soil samples. These limits as such do not indicate the consistency of undisturbed soils. The index that is used to indicate the consistency of undisturbed soils is called the *liquidity index*. The liquidity index is expressed as

$$I_l = \frac{w_n - w_p}{I_p} \tag{3.45}$$

where, w_n is the natural moisture content of the soil in the undisturbed state. The liquidity index of undisturbed soil can vary from less than zero to greater than 1. The value of I_l varies according to the consistency of the soil as in Table 3.10.

The liquidity index indicates the state of the soil in the field. If the natural moisture content of the soil is closer to the liquid limit the soil can be considered as soft, and the soil is stiff if the natural moisture content is closer to the plastic limit. There are some soils whose natural moisture contents are higher than the liquid limits. Such soils generally belong to the montmorillonite group and possess a brittle structure. A soil of this type when disturbed by vibration flows like a liquid. The liquidity index values of such soils are greater than unity. One has to be cautious in using such soils for foundations of structures.

Table 3.10 Values of I_l and I_c according to consistency of soil

Consistency	I_l	I_c
Semisolid or solid state	Negative	>1
Very stiff state ($w_n = w_p$)	0	1
Very soft state ($w_n = w_l$)	1	0
Liquid state (when disturbed)	>1	Negative

Consistency Index, I_c

The consistency index may be defined as

$$I_c = \frac{w_l - w_n}{I_p} \tag{3.46}$$

The index I_c reflects the state of the clay soil condition in the field in an undisturbed state just in the same way as I_l described earlier. The values of I_c for different states of consistency are given in Table 3.10 along with the values I_l. It may be seen that values of I_c and I_c are opposite to each other for the same consistency of soil.

From Eqs (3.45) and (3.46) we have

$$I_l + I_c = \frac{w_l - w_p}{I_p} = 1 \tag{3.47}$$

Effect of Drying on Plasticity

Drying produces an invariable change in the colloidal characteristics of the organic matter in a soil. The distinction between organic and inorganic soils can be made by performing two liquid limit tests on the same material. One test is made on an air-dried sample and the other on an oven-dried one. If the liquid limit of the oven-dried sample is less than about 0.75 times that for the air-dried sample, the soils may be classed as organic. Oven-drying also lowers the plastic limits of organic soils, but the drop in plastic limit is less than that for the liquid limit.

Shrinking and Swelling of Soils

If a moist cohesive soil is subjected to drying, it loses moisture and shrinks. The *degree of shrinkage*, S_r, is expressed as

$$S_r = \frac{V_o - V_d}{V_o} \times 100 \tag{3.48a}$$

where,

V_o = original volume of a soil sample at saturated state

V_d = final volume of the sample at shrinkage limit

On the basis of the degree of shrinkage, Scheidig (1934) classified soils as in Table 3.11.

Shrinkage Ratio *SR*

Shrinkage ratio is defined as the ratio of a volume change expressed as a percentage of dry volume to the corresponding change in water content above the shrinkage limit.

Table 3.11 Soil classification according to degree of shrinkage S_r

S_r%	Quality of soil
< 5	Good
5–10	Medium good
10–15	Poor
> 15	Very poor

$$SR = \frac{(V_o - V_d)/V_d}{w_o - w_s} \times 100 \quad (3.48b)$$

where

V_o = initial volume of a saturated soil sample at water content w_o
V_d = the final volume of the soil sample at shrinkage limit w_s
$(w_o - w_s)$ = change in the water content

$$= \frac{(V_o - V_d)\,\rho_w}{M_d}$$

M_d = mass of dry volume, V_d, of the sample
Substituting for $(w_o - w_s)$ in Eq (3.48b) and simplifying, we have

$$SR = \frac{M_d}{\rho_w V_d} = \frac{\rho_d}{\rho_w} = \frac{\gamma_d}{\gamma_w} = G_m \quad (3.48c)$$

Thus the shrinkage ratio of a soil mass is equal to the mass specific gravity of the soil in its dry state.

Volumetric Shrinkage S_v

The volumetric shrinkage or volumetric change is defined as the decrease in volume of a soil mass, expressed as a percentage of the dry volume of the soil mass when the water content is reduced from the initial w_o to the final w_s at the shrinkage limit.

$$S_v = \frac{V_o - V_d}{V_d} \times 100 = (w_o - w_s)\,SR \quad (3.49)$$

Linear shrinkage can be computed from the volumetric change by the following equation

$$LS = \left[1 - \left(\frac{1.0}{S_v + 1.0}\right)^{1/3}\right] \times 100 \text{ percent} \quad (3.50)$$

The volumetric shrinkage S_v is used as a decimal quantity in Eq. (3.50). This equation assumes that the reduction in volume is both linear and uniform in all directions.

Linear shrinkage can be directly determined by a test [this test has not yet been standardized in the United States (Bowles, 1992)]. The British Standard BS 1377 used a half-cylinder of mold of diameter 12.5 mm and length L_o = 140 mm. The wet sample filled into the mold is dried and the final length L_f is obtained. From this, the linear shrinkage LS is computed as

$$LS = \frac{L_o - L_f}{L_o} \tag{3.51}$$

Activity

Skempton (1953) considers that the significant change in the volume of a clay soil during shrinking or swelling is a function of plasticity index and the quantity of colloidal clay particles present in soil. The clay soil can be classified *inactive*, *normal* or *active* (after Skempton, 1953). The *activity* of clay is expressed as

$$\text{Activity } A = \frac{\text{Plasticity index, } I_p}{\text{Percent finer than 2 micron}} \tag{3.52}$$

Table 3.12 gives the type of soil according to the value of A. The clay soil which has an activity value greater than 1.4 can be considered as belonging to the swelling type. The relationship between plasticity index and clay fraction is shown in Fig. 3.18(a).

Figure 3.18(b) shows results of some tests obtained on prepared mixtures of various percentage of particles less than and greater than 2 μ. Several natural soils were separated into fractions greater and less than 2 μ and then the two fractions were combined as desired. Fig 3.18(c) shows the results obtained on clay minerals mixed with quartz sand.

Table 3.12 Soil classification according to activity

A	Soil type
<0.75	Inactive
0.75–1.40	Normal
>1.40	Active

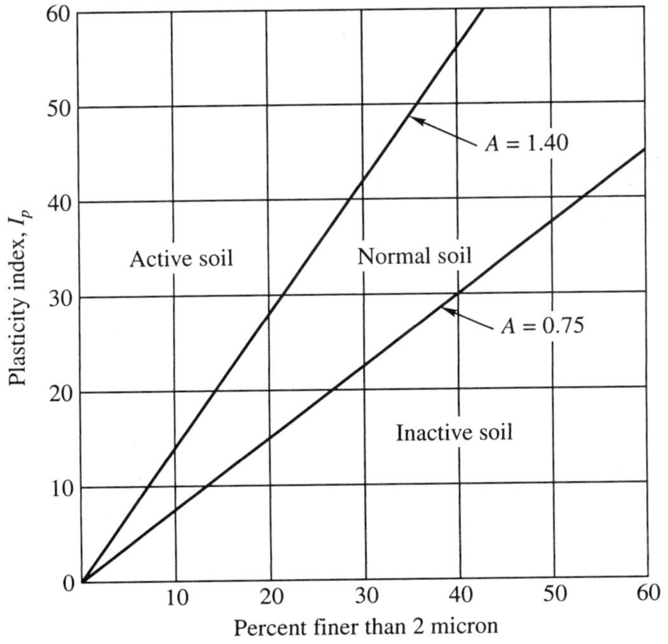

Figure 3.18(a) Classification of soil according to activity

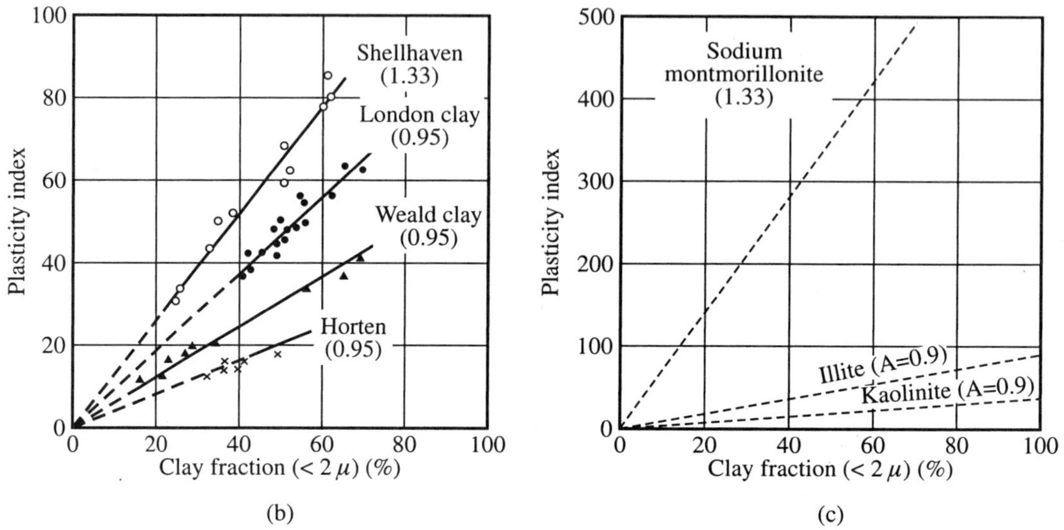

Figure 3.18(b, c) Relation between plasticity index and clay fraction. Figures in parentheses are the *activities* of the clays (after Skempton, 1953)

Consistency of Soils as per the Unconfined Compressive Strength

The consistency of a natural soil is different from that of a remolded soil at the same water content. Remolding destroys the structure of the soil and the particle orientation. The liquidity index value which is an indirect measure of consistency is only qualitative. The consistency of undisturbed soil varies quantitatively on the basis of its unconfined compressive strength. The unconfined compressive strength, q_u, is defined as the ultimate load per unit cross sectional area that a cylindrical specimen of soil (with height to diameter ratio of 2 to 2.5) can take under compression without any lateral pressure. Water content of the soil is assumed to remain constant during the duration of the test which generally takes only a few minutes. Table 3.13 indicates the relationship between consistency and q_u.

As explained earlier, remolding of an undisturbed sample of clay at the same water content alters its consistency, because of the destruction of its original structure. The degree of disturbance of undisturbed clay sample due to remolding can be expressed as

Table 3.13 Relationship between consistency of clays and q_u

Consistency	q_u, kN/m^2	Consistency	q_u, kN/m^2
Very soft	<25	Stiff	100–200
Soft	25–50	Very stiff	200–400
Medium	50–100	Hard	>400

Table 3.14 Soil classification on the basis of sensitivity (after Skempton and Northey, 1954)

S_t	Nature of clay	S_t	Nature of clay
1	Insensitive clays	4–8	Sensitive clays
1–2	Low-sensitive clays	8–16	Extra-sensitive clays
2–4	Medium sensitive clays	>16	Quick clays

$$\text{Sensitivity, } S_t = \frac{q_u, \text{ undisturbed}}{q'_u, \text{ remolded}} \tag{3.53}$$

where q'_u is the unconfined compressive strength of remolded clay at the same water content as that of the undisturbed clay.

When q'_u is very low as compared to q_u the clay is highly sensitive. When $q_u = q'_u$ the clay is said to be insensitive to remolding. On the basis of the values of S_t clays can be classified as in Table 3.14.

The clays that have sensitivity greater than 8 should be treated with care during construction operations because disturbance tends to transform them, at least temporarily, into viscous fluids. Such clays belong to the montmorillonite group and possess flocculent structure.

Thixotropy

If a remolded clay sample with sensitivity greater than one is allowed to stand without further disturbance and change in water content, it may regain at least part of its original strength and stiffness. This increase in strength is due to the gradual reorientation of the absorbed molecules of water, and is known as *thixotropy* (from the Greek *thix,* meaning '*touch*' and *tropein,* meaning '*to change*'). The regaining of a part of the strength after remolding has important applications in connection with pile-driving operations, and other types of construction in which disturbance of natural clay formations is inevitable.

3.14 PLASTICITY CHART

Many properties of clays and silts such as their dry strength, compressibility and their consistency near the plastic limit can be related with the Atterberg limits by means of a *plasticity chart* as shown is Fig. 3.19. In this chart the ordinates represent the plasticity index I_p and the abscissas the

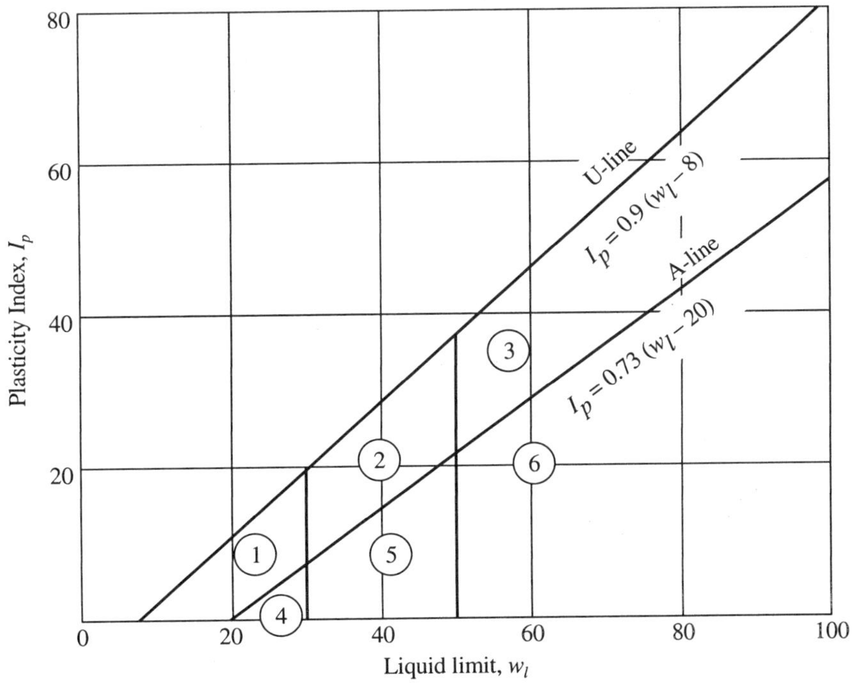

Figure 3.19 Plasticity chart

corresponding liquid limit w_l. The chart is divided into six regions, three above and three below line A. The equation to the line A is

$$I_p = 0.73\,(w_l - 20) \tag{3.51}$$

If a soil is known to be inorganic its group affiliation can be ascertained on the basis of the values of I_p and w_l alone. However, points representing organic clays are usually located within the same region as those representing inorganic silts of high compressibility, and points representing organic silts in the region assigned to inorganic silts of medium compressibility.

Casagrande (1932) studied the consistency limits of a number of soil and proposed the plasticity chart shown in Fig. 3.19. The distribution of soils according to the regions are given below.

Region	Liquid limit w_l	Type of soil
Above A-line		
1	Less than 30	Inorganic clays of low plasticity and cohesionless soils
2	$30 < w_l < 50$	Inorganic clays of medium plasticity
3	$w_l > 50$	Inorganic clays of high plasticity
Below A-line		
4	$w_l < 30$	Inorganic silts of low compressibility
5	$30 < w_l < 50$	Inorganic silts of medium compressibility and organic silts
6	$w_l > 50$	Inorganic silts of high compressibility and organic clay

The upper limit of the relationship between plasticity index and liquid limit is provided by another line called the U-line whose equation is

$$I_p = 0.9(w_l - 8) \tag{3.52}$$

Example 3.9

Determine the times (t) required for particles of diameters 0.2, 0.02, 0.01 and 0.005 mm to fall a depth of 10 cm from the surface in water.

Given: $\mu = 8.15 \times 10^{-3}$ poises, $G = 2.65$. (Note: 1 poise = 10^{-3} gm-sec/cm^2)

Solution

$\mu = 8.15 \times 10^{-3} \times 10^{-3} = 8.15 \times 10^{-6}$ gm-sec/cm^2.

Use Eq. (3.24)

$$t = \frac{30\mu}{(G_s - 1)}\,\frac{L}{D^2} = \frac{30 \times 8.15 \times 10^{-6}}{(2.65-1)} \times \frac{10}{D^2} = \frac{1.482 \times 10^{-3}}{D^2}\ \text{min}$$

The times required for the various values of D are as given below.

D (mm)	t
0.2	2.22 sec
0.02	3.71 min
0.01	14.82 min
0.005	59.28 min

Example 3.10

A sedimentation analysis by the hydrometer method (152 H) was conducted with 50 g (= M_s)of oven dried soil. The volume of soil suspension is $V = 10^3$ cm^3. The hydrometer reading $R_a = 19.50$ after a lapse of 60 minutes after the commencement of the test.

Given: C_m (meniscus) = 0.52, L (effective) = 14.0 cm, C_o (zero correction) = +2.50, $G_s = 2.70$ and $\mu = 0.01$ poise.

Calculate the smallest particle size, which would have settled a depth of 14.0 cm and the percentage finer than this size. Temperature of test = 25° C.

Solution

From Eq. (3.24)

$$D(\text{mm}) = \sqrt{\frac{30\mu}{(G_s - 1)}} \times \sqrt{\frac{L}{t}}$$

where $\mu = 0.01 \times 10^{-3}$ (gm-sec)/cm^2.
Substituting

$$D = \sqrt{\frac{30 \times 0.01 \times 10^{-3}}{(2.7-1)}} \times \sqrt{\frac{14}{60}} = 0.0064 \text{ mm.}$$

From Eq. (3.31)

$$R_c = R_a - C_o + C_t$$

From Table 3.6 for $T = 25$ °C, $C_T = +1.3$. Therefore,

$$R_c = 19.5 - 2.5 + 1.3 = 18.3$$

From Eqs (3.32) and (3.31), we have

$$P'\% = \frac{C_{sg} R_c}{M_s} \times 100, \quad C_{sg} = \frac{1.65 G_s}{2.65(G_s - 1)}$$

$$C_{sg} = \frac{1.65 \times 2.7}{2.65(2.7-1)} = 0.99, \quad P'\% = \frac{0.99 \times 18.3}{50} \times 100 = 36.23\%$$

Example 3.11

A 500 g sample of dry soil was used for a combined sieve and hydrometer analysis (152 H type hydrometer). The soil mass passing through the 75 μ sieve = 120 g. Hydrometer analysis was carried out on a mass of 40 g that passed through the 75 μ sieve. The average temperature recorded during the test was 30°C.

Given: $G_s = 2.55$, C_m (meniscus) = 0.50, $C_o = +2.5$, $\mu = 8.15 \times 10^{-3}$ poises.

The actual hydrometer reading $R_a = 15.00$ after a lapse of 120 min after the start of the test. Determine the particle size D and percent finer $P'\%$ and $P\%$.

Solution

From Eq. (3.29)

$$L = 16.3 - 0.1641 R$$

where, $R = R_a + C_m = 15.0 + 0.5 = 15.5$

$$L = 16.3 - 0.1641 \times 15.5 = 13.757$$

From Eq. (3.24)

$$D = \sqrt{\frac{30 \times 8.15 \times 10^{-6}}{2.55 - 1}} \times \sqrt{\frac{13.757}{120}} = 0.0043 \text{ mm}$$

From Eq. (3.32)

$$\text{Percent finer, } P'\% = \frac{C_{sg} R_c}{M_s} \times 100$$

From Table 3.7, $C_{sg} = 1.02$ for $G_s = 2.55$

From Table 3.6, $C_T = +3.8$ for $T = 30$ °C

$$\text{Now, } R_c = R_a - C_o + C_T = 15 - 2.5 + 3.8 = 16.3$$

$$\text{Now, } P' = \frac{1.02 \times 16.3}{40} \times 100 = 41.57\%$$

$$P\% = 41.57 \times \frac{120}{500} \approx 10\%$$

Example 3.12

500 g of dry soil was used for a sieve analysis. The masses of soil retained on each sieve are given below:

US standard sieve	Mass in g	US standard sieve	Mass in g
2.00 mm	10	500 μ	135
1.40 mm	18	250 μ	145
1.00 mm	60	125 μ	56
		75 μ	45

Plot a grain size distribution curve and compute the following:

(a) Percentages of gravel, coarse sand, medium sand, fine sand and silt, as per the Unified Soil Classification System, (b) uniformity coefficient (c) coefficient of curvature.

Comment on the type of soil.

Solution

Computation of percent finer

US standard sieve	Diameter, *D* of grains in mm	Mass retained in g	% retained	Cumulative % retained	% finer *P*
2.00 mm	2.00	10	2.0	2.0	98.0
1.40 mm	1.40	18	3.6	5.6	94.4
1.00 mm	1.00	60	12.0	17.6	82.4
500 μ	0.500	135	27.0	44.6	55.4
250 μ	0.25	145	29.0	73.6	26.4
125 μ	0.125	56	11.2	84.8	15.2
75 μ	0.075	45	9.0	93.8	6.2

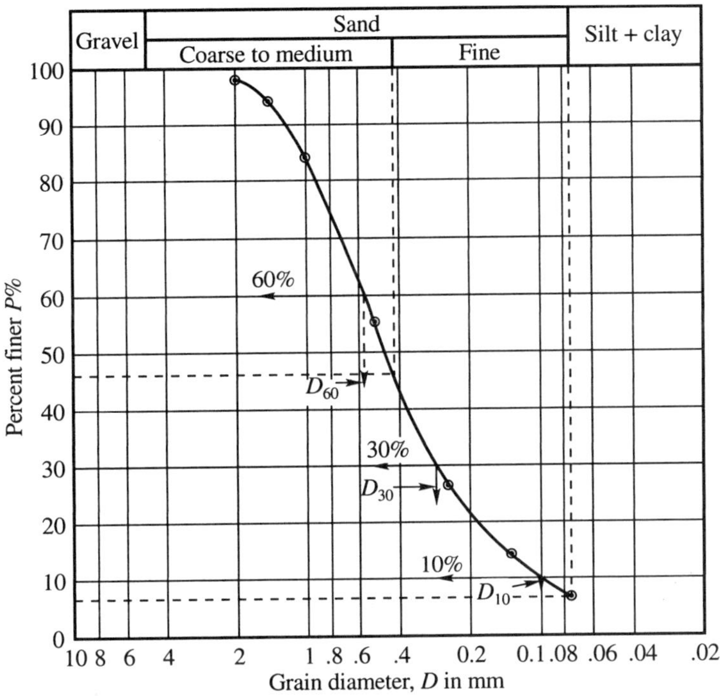

Figure Ex. 3.12

(a) Percentage coarse to medium sand = 98 – 48 = 50 percent

Percentage fine sand = 48 – 6.2 = 41.8 percent

Percentage silt and clay = 6.2 percent.

(b) Uniformity coefficient $C_u = \dfrac{D_{60}}{D_{10}} = \dfrac{0.58}{0.098} = 5.92$

(c) Coefficient of curvature $C_c = \dfrac{(D_{30})^2}{D_{10} \times D_{60}} = \dfrac{(0.28)^2}{0.098 \times 0.58} = 1.38$

The soil is just on the border line of well graded sand.

Example 3.13

Liquid limit tests on a given sample of clay were carried out. The data obtained are as given below.

Test No.	1	2	3	4
Water content, %	70	64	47	44
Number of blows, N	5	8	30	45

Draw the flow curve on semi-log paper and determine the liquid limit and flow index of the soil.

Solution

Figure Ex. 3.13 gives the flow curve for the given sample of clay soil. As per the curve,

Liquid limit, $w_l = 50\%$

Flow index, $I_f = 29$

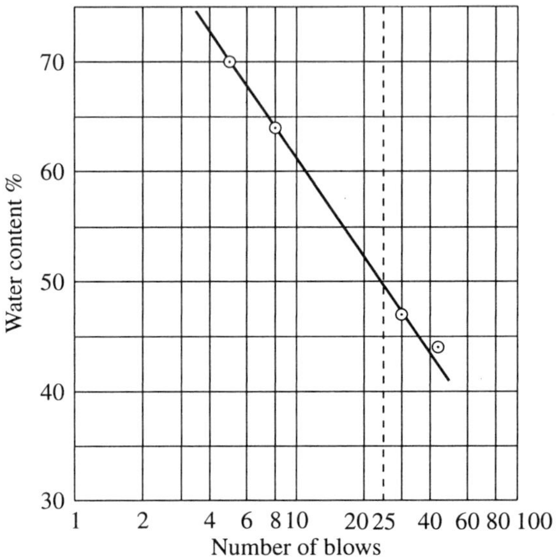

Figure Ex. 3.13

Example 3.14

The laboratory tests on a sample of soil gave the following results:

$w_n = 24\%$, $w_l = 62\%$, $w_p = 28\%$, percentage of particles less than 2 $\mu = 23\%$

Determine: (a) The liquidity index, (b) activity (c) consistency and nature of soil.

Solution

(a) Plasticity index, $I_p = w_l - w_p = 62 - 28 = 34\%$

$$\text{Liquidity index, } I_l = \frac{w_n - w_p}{I_p} = \frac{24 - 28}{34} = -0.12.$$

$$\text{(b) Activity, } A = \frac{I_p}{\%\text{ of particles } < 2\mu} = \frac{34}{23} = 1.48.$$

(c) Comments:

(i) Since I_l is negative, the consistency of the soil is very stiff to extremely stiff (semisolid state).

(ii) Since I_p is greater than 17% the soil is highly plastic.

(iii) Since A is greater than 1.40, the soil is active and is subject to significant volume change (shrinkage and swelling).

Example 3.15

Two soil samples tested in a soil mechanics laboratory gave the following results:

	Sample no. 1	Sample no. 2
Liquid limit	50%	40%
Plastic limit	30%	20%
Flow indices, I_f	27	17

(a) Determine the toughness indices and
(b) comment on the types of soils.

Solution

(a) $I_t = \dfrac{w_l - w_p}{I_f}$

Sample 1, $I_t = \dfrac{50-30}{27} = \dfrac{20}{27} = 0.74$; Sample 2, $I_t = \dfrac{40-20}{17} = \dfrac{20}{17} = 1.18$

(b)
(i) Both the soils are clay soils as their toughness indices lie between 0 and 3.
(ii) Soil one is friable at the plastic limit since its I_t value is less than one.
(iii) Soil two is stiffer than soil one at the plastic limit since the I_t value of the latter is higher.

Example 3.16

The natural moisture content of an excavated soil is 32%. Its liquid limit is 60% and plastic limit is 27%. Determine the plasticity index of the soil and comment about the nature of the soil.

Solution

Plasticity index, $I_p = w_l - w_p = 60 - 27 = 33\%$

The nature of the soil can be judged by determining its liquidity index, I_l from Eq. (3.45)

$$I_l = \frac{w_n - w_p}{I_p} = \frac{32-27}{33} = +0.15$$

since the value of I_l is very close to 0, the nature of the soil as per Table 3.10 is very stiff.

Example 3.17

A soil with a liquidity index of –0.20 has a liquid limit of 56% and a plasticity index of 20%. What is its natural water content? What is the nature of this soil?

Solution

As per Eq. (3.45)

Liquidity index, $I_l = \dfrac{w_n - w_p}{I_p}$

$w_p = w_l - I_p = 56 - 20 = 36,$

$w_n = I_l I_p + w_p = -0.20 \times 20 + 36 = 32.$

Since I_l is negative, the soil is in a semisolid or solid state as per Table 3.10.

Example 3.18

Four different types of soils were encountered in a large project. The liquid limits (w_l), plastic limits (w_p), and the natural moisture contents (w_n) of the soils are given below

Soil type	w_l%	w_p%	w_n%
1	120	40	150
2	80	35	70
3	60	30	30
4	65	32	25

Determine: (a) the liquidity indices I_l of the soils, (b) the consistency of the natural soils and (c) the possible behavior of the soils under vibrating loads.

Solution

(a) $I_l = \frac{w_n - w_p}{I_p}$

By substituting the appropriate values in this equation, we have

Type	I_l
1	1.375
2	0.778
3	0
4	–0.21

(b) From Table 3.10, Type 1 is in a liquid state, Type 2 in a very soft state, Type 3 in very stiff state, and Type 4 in a semisolid state.

(c) Soil types 3 and 4 are not much affected by vibrating loads. Type 1 is very sensitive even for small disturbance and as such is not suitable for any foundation. Type 2 is also very soft, with greater settlement of the foundation or failure of the foundation due to development of pore pressure under saturated condition taking place due to any sudden application of loads.

Example 3.19

A shrinkage limit test on a clay soil gave the following data. Compute the shrinkage limit. Assuming that the total volume of dry soil cake is equal to its total volume at the shrinkage limit, what is the degree of shrinkage? Comment on the nature of soil

Mass of shrinkage dish and saturated soil	M_1	=	38.78 g
Mass of shrinkage dish and oven dry soil	M_2	=	30.46 g
Mass of shrinkage dish	M_3	=	10.65 g
Volume of shrinkage dish	V_o	=	16.29 cm^3
Total volume of oven dry soil cake	V_d	=	10.00 cm^3

Solution

Refer to Fig. 3.15

The equation for shrinkage limit $w_s = \frac{M_w}{M_s}$

where M_w = mass of water in the voids at the shrinkage limit.

M_o = mass of sample at the plastic state = $M_1 - M_3$ = 38.78 – 10.65 = 28.13 g

Volume of water lost from the plastic state to the shrinkage limit $\Delta V = (V_o - V_d)$

or $\Delta V = 16.29 - 10.00 = 6.29\ \text{cm}^3$

Mass of dry soil $= M_s = M_2 - M_3 = 30.46 - 10.65 = 19.81$ g

Now, $M_w = M_o - M_s - (V_o - V_d)\,\rho_w = 28.13 - 19.81 - (6.29)\,(1) = 2.03$ g

From Eq. (3.41), $$w_s = \frac{(M_o - M_s) - (V_o - V_d)\rho_w}{M_s} = \frac{M_w}{M_s} = \frac{2.03}{19.81} = 0.102 = 10.2\%$$

As per Eq. (3.48a), the degree of shrinkage, S_r is

$$S_r = \frac{V_o - V_d}{V_o} \times 100 = \frac{(16.29 - 10.0) \times 100}{16.29} = 38.6\%$$

From Table 3.11 the soil is of very poor quality.

3.15 GENERAL CONSIDERATIONS FOR CLASSIFICATION OF SOILS

It has been stated earlier that soil can be described as gravel, sand, silt and clay according to grain size. Most of the natural soils consist of a mixture of organic material in the partly or fully decomposed state. The proportions of the constituents in a mixture vary considerably and there is no generally recognized definition concerning the percentage of, for instance, clay particles that a soil must have to be classified as clay, etc.

When a soil consists of the various constituents in different proportions, the mixture is then given the name of the constituents that appear to have significant influence on its behavior, and then other constituents are indicated by adjectives. Thus a sandy clay has most of the properties of a clay but contains a significant amount of sand.

The individual constituents of a soil mixture can be separated and identified as gravel, sand, silt and clay on the basis of mechanical analysis. The clay mineral that is present in a clay soil is sometimes a matter of engineering importance. According to the mineral present, the clay soil can be classified as kaolinite, montmorillonite or illite. The minerals present in a clay can be identified by either X-ray diffraction or differential thermal analysis. A description of these methods is beyond the scope of this book.

Buildings, bridges, dams etc. are built on natural soils (undisturbed soils), whereas earthen dams for reservoirs, embankments for roads and railway lines, foundation bases for pavements of roads and airports are made out of remolded soils. Sites for structures on natural soils for embankments, etc, will have to be chosen first on the basis of preliminary examinations of the soil that can be carried out in the field. An engineer should therefore be conversant with the field tests that would identify the various constituents of a soil mixture.

The behavior of a soil mass under load depends upon many factors such as the properties of the various constituents present in the mass, the density, the degree of saturation, the environmental conditions etc. If soils are grouped on the basis of certain definite principles and rated according to their performance, the properties of a given soil can be understood to a certain extent, on the basis of some simple tests. The objectives of the following sections of this chapter are to discuss the following:

1. Field identification of soils.
2. Classification of soils.

3.16 FIELD IDENTIFICATION OF SOILS

The methods of field identification of soils can conveniently be discussed under the headings of coarse-grained and fine-grained soil materials.

Coarse-Grained Soil Materials

The coarse-grained soil materials are mineral fragments that may be identified primarily on the basis of grain size. The different constituents of coarse-grained materials are sand and gravel. As described in the earlier sections, the size of sand varies from 0.075 mm to 4.75 mm and that of gravel from 4.75 mm to 80 mm. Sand can further be classified as coarse, medium and fine. The engineer should have an idea of the relative sizes of the grains in order to identify the various fractions. The description of sand and gravel should include an estimate of the quantity of material in the different size ranges as well as a statement of the shape and mineralogical composition of the grains. The mineral grains can be rounded, subrounded, angular or subangular. The presence of mica or a weak material such as shale affects the durability or compressibility of the deposit. A small magnifying glass can be used to identify the small fragments of shale or mica. The properties of a coarse grained material mass depend also on the uniformity of the sizes of the grains. A well-graded sand is more stable for a foundation base as compared to a uniform or poorly graded material.

Fine-Grained Soil Materials

Inorganic Soils: The constituent parts of fine-grained materials are the silt and clay fractions. Since both these materials are microscopic in size, physical properties other than grain size must be used as criteria for field identification. The classification tests used in the field for preliminary identification are

1. Dry strength test
2. Shaking test
3. Plasticity test
4. Dispersion test

Dry strength: The strength of a soil in a dry state is an indication of its cohesion and hence of its nature. It can be estimated by crushing a 3 mm size dried fragment between thumb and forefinger. A clay fragment can be broken only with great effort, whereas a silt fragment crushes easily.

Shaking test: The shaking test is also called as dilatancy test. It helps to distinguish silt from clay since silt is more permeable than clay. In this test a part of soil mixed with water to a very soft consistency is placed in the palm of the hand. The surface of the soil is smoothed out with a knife and the soil pat is shaken by tapping the back of the hand. If the soil is silt, water will rise quickly to the surface and give it a shiny glistening appearance. If the pat is deformed either by squeezing or by stretching, the water will flow back into the soil and leave the surface with a dull appearance. Since clay soils contain much smaller voids than silts and are much less permeable, the appearance of the surface of the pat does not change during the shaking test. An estimate of the relative proportions of silt and clay in an unknown soil mixture can be made by noting whether the reaction is rapid, slow or nonexistent.

Plasticity test: If a sample of moist soil can be manipulated between the palms of the hands and fingers and rolled into a long thread of about 3 mm diameter, the soil then contains a significant amount of clay. Silt cannot be rolled into a thread of 3 mm diameter without severe cracking.

Dispersion test: This test is useful for making a rough estimate of sand, silt and clay present in a material. The procedure consists in dispersing a small quantity of the soil in water taken in a glass cylinder and allowing the particles to settle. The coarser particles settle first followed by finer ones. Ordinarily sand particles settle within 30 seconds if the depth of water is about 10 cm. Silt particles settle in about 1/2 to 240 minutes, whereas particles of clay size remain in suspension for at least several hours and sometimes several days.

Organic soils: Surface soils and many underlying formations may contain significant amounts of solid matter derived from organisms. While shell fragments and similar solid matter are found at some locations, organic material in soil is usually derived from plant or root growth and consists of almost completely disintegrated matter, such as muck or more fibrous material, such as peat. The soils with organic matter are weaker and more compressible than soils having the same mineral composition but lacking in organic matter. The presence of an appreciable quantity of organic material can usually be recognized by the dark-grey to black color and the odor of decaying vegetation which it lends to the soil.

Organic silt: It is a fine grained more or less plastic soil containing mineral particles of silt size and finely divided particles of organic matter. Shells and visible fragments of partly decayed vegetative matter may also be present.

Organic clay: It is a clay soil which owes some of its significant physical properties to the presence of finely divided organic matter. Highly organic soil deposits such as peat or muck may be distinguished by a dark-brown to black color, and by the presence of fibrous particles of vegetable matter in varying states of decay. The organic odor is a distinguishing characteristic of the soil. The organic odor can sometimes be distinguished by a slight amount of heat.

3.17 CLASSIFICATION OF SOILS

Soils in nature rarely exist separately as gravel, sand, silt, clay or organic matter, but are usually found as mixtures with varying proportions of these components. Grouping of soils on the basis of certain definite principles would help the engineer to rate the performance of a given soil either as a sub-base material for roads and airfield pavements, foundations of structures, etc. The classification or grouping of soils is mainly based on one or two index properties of soil which are described in detail in earlier sections. The methods that are used for classifying soils are based on one or the other of the following two broad systems:

1. A textural system which is based only on grain size distribution.
2. The systems that are based on grain size distribution and limits of soil.

Many systems are in use that are based on grain size distribution and limits of soil. The systems that are quite popular amongst engineers are the AASHTO Soil Classification System and the Unified Soil Classification System.

3.18 TEXTURAL SOIL CLASSIFICATION

U.S. Department of Agriculture System (USDA)

The boundaries between the various soil fractions of this systems are given in Table 3.15.

By making use of the grain size limits mentioned in the table for sand, silt and clay, a triangular classification chart has been developed as shown in Fig. 3.20 for classifying mixed soils. The first step in the classification of soil is to determine the percentages of sand, silt and clay-size materials in a given sample by mechanical analysis. With the given relative percentages of the sand, silt and clay, a point is located on the triangular chart as shown in Fig. 3.20. The designation given on the chart for the area in which the point falls is used as the classification of the sample. This method of classification does not reveal any properties of the soil other than grain-size distribution. Because of its simplicity, it is widely used by workers in the field of agriculture. One significant

Table 3.15 Soil Fractions as per U.S. Department of Agriculture

Soil fraction	Diameter in mm
Gravel	>2.00
Sand	2–0.05
Silt	0.05–0.002
Clay	<0.002

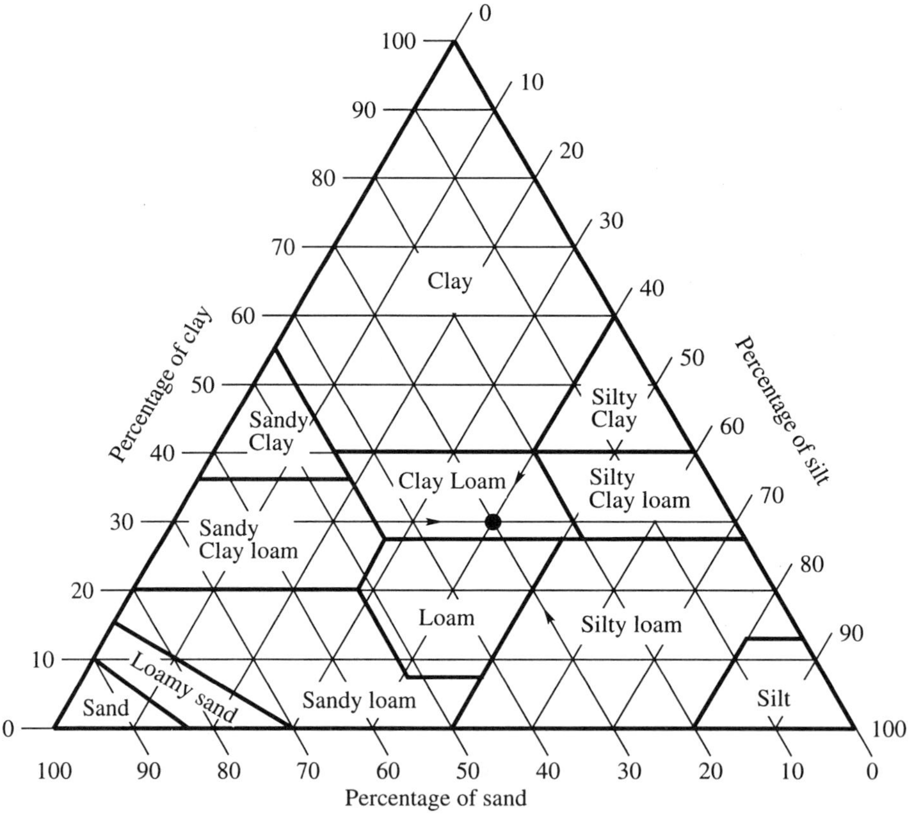

Figure 3.20 U.S. Department of Agriculture textural classification

disadvantage of this method is that the textural name as derived from the chart does not always correctly express the physical characteristics of the soil. For example, since some clay size particles are much less active than others, a soil described as clay on the basis of this system may have physical properties more typical of silt.

3.19 AASHTO SOIL CLASSIFICATION SYSTEM

This system was originally proposed in 1928 by the U.S. Bureau of Public Roads for use by highway engineers. A Committee of highway engineers for the Highway Research Board, met in 1945 and made an extensive revision of the PRA System. This system is known as the AASHTO (American Association of State Highway and Transportation Officials) System (ASTM D-3242, AASHTO

Method M 145). The revised system comprises seven groups of inorganic soils, A-1 to A-7 with 12 subgroups in all. The system is based on the following three soil properties:

1. Particle-size distribution
2. Liquid Limit
3. Plasticity Index.

A Group Index is introduced to further differentiate soils containing appreciable fine-grained materials. The characteristics of various groups are defined in Table 3.16. The Group Index may be determined from the equation.

$$\text{Group Index } (GI) = 0.2a + 0.005ac + 0.01bd \tag{3.56a}$$

in which,

a = that portion of percentage of soil particles passing No. 200 (ASTM) sieve greater than 35 = (F –35).

b = that portion of percentage of soil particles passing No. 200 sieve, greater than 15 = (F –15).

c = that portion of the liquid limit greater than 40 = (w_l –40).

d = that portion of the plasticity index greater than 10 = (I_p –10).

F = percent passing No. 200 sieve. If $F < 35$, use $(F - 35) = 0$

It may be noted here that if $GI < 0$, use $GI = 0$. There is no upper limit for GI. When calculating the GI for soils that belong to groups A-2-6 and A-2-7, use the partial group index (PGI) only, that is (From Eq. 3.56a)

$$PGI = 0.01\text{bd} = 0.01(F - 15)(I_p - 10) \tag{3.56b}$$

Figure 3.21 provides a rapid means of using the liquid and plastic limits (and plasticity index I_p) to make determination of the A-2 subgroups and the A-4 through A-7 classifications. Figure 3.21 is based on the percent passing the No. 200 sieve (whether greater or less than 35 percent)

The group index is a means of rating the value of a soil as a subgrade material within its own group. It is not used in order to place a soil in a particular group, that is done directly from the results of sieve analysis, the liquid limit and plasticity index. The higher the value of the group index, the poorer is the quality of the material. The group index is a function of the amount of material passing the No. 200 sieve, the liquid limit and the plasticity index.

If the pertinent index value for a soil falls below the minimum limit associated with a, b, c or d, the value of the corresponding term is zero, and the term drops out of the group index equation. The group index value should be shown in parenthesis after a group symbol such as A-6(12) where 12 is the group index.

Classification procedure*:* With the required data in mind, proceed from left to right in the chart. The correct group will be found by a process of elimination. The first group from the left consistent with the test data is the correct classification. The *A*-7 group is subdivided into *A*-7-5 or *A*-7-6 depending on the plasticity index, I_p.

For A-7-5, $I_p \leq w_l - 30$

For A-7-6, $I_p \geq w_l - 30$

Table 3.16 AASHTO soil classification

General classification	Granular Materials (35 percent or less of total sample passing No. 200)							Silt-clay Materials (More than 35 percent of total sample passing No. 200)			
Group classification	A-1		A-3	A-2				A-4	A-5	A-6	A-7
	A-1-*a*	*A*-1-*b*		A-2-4	A-2-5	A-2-6	A-2-7				A-7-5 A-7-6
Sieve analysis percent passing											
No. 10	50 max										
No. 40	30 max	50 max	51 min								
No. 200	15 max	25 max	10 max	35 max	35 max	35 max	35 max	36 min	36 min	36 min	36 min
Characteristics of fraction passing No. 40											
Liquid limit				40 max	41 min	40 max	41 min	40 max	41 min	40 max	41 min
Plasticity Index	6 max		N.P.	10 max	10 max	11 min	11 max	10 max	10 max	11 min	11 min
Usual types of significant constituent materials	Stone fragments—gravel and sand		Fine sand	Silty or clayey gravel and sand				Silty soils		Clayey soils	
General rating as subgrade	Excellent to good					Fair to poor					

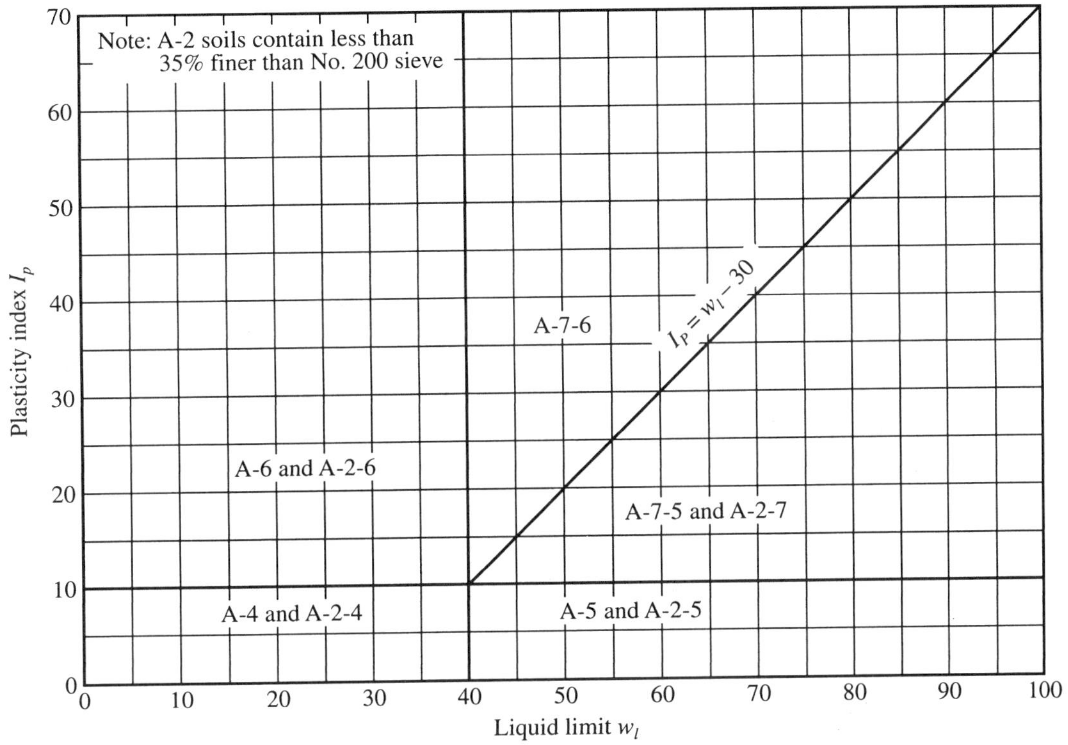

Figure 3.21 Chart for use in AASHTO soil classification system

3.20 UNIFIED SOIL CLASSIFICATION SYSTEM (USCS)

The Unified Soil Classification System is based on the recognition of the type and predominance of the constituents considering grain-size, gradation, plasticity and compressibility. It divides soil into three major divisions: coarse-grained soils, fine grained soils, and highly organic (peaty) soils. In the field, identification is accomplished by visual examination for the coarse-grained soils and a few simple hand tests for the fine-grained soils. In the laboratory, the grain-size curve and the Atterberg limits can be used. The peaty soils are readily identified by color, odor, spongy feel and fibrous texture.

The Unified Soil Classification System is a modified version of A. Casagrande's Airfield Classification (AC) System developed in 1942 for the Corps of Engineers. Since 1942 the original classification has been expanded and revised in cooperation with the Bureau of Reclamation, so that it applies not only to airfields but also to embankments, foundations, and other engineering features. This system was adopted in 1952. In 1969 the American Society for Testing and Materials (ASTM) adopted the Unified System as a standard method for classification for engineering purposes (ASTM Test Designation D-2487).

Table 3.17 presents the primary factors to consider in classifying a soil according to the Unified Soil Classification system.

The following subdivisions are considered in the classification:

1. Gravels and sands are GW, GP, SW, or SP

   if less than 5 percent of the material passes the No. 200 sieve; G = gravel; S = sand; W = well-graded; P = poorly-graded. The well- or poorly-graded designations depend on C_u and C_c as defined in section 3.9 and numerical values shown in Table 3.16

Table 3.17 The Unified Soil Classification System (Source: Bowles, 1992)

Major divisions			Group symbol	Typical names	Classification criteria for coarse-grained soils	
Coarse-grained soils (more than half of material is larger than No. 200)	Gravels (more than half of coarse fraction is larger than No. 4 sieve size)	Clean gravels (little or no fines)	GW	Well-graded gravels, gravel-sand mixtures, little or no fines	$C_u \geq 4$ $1 \leq C_c \leq 3$	
			GP	Poorly graded gravels, gravel-sand mixtures, little or no fines	Not meeting all gradation requirements for GW ($C_u < 4$ or $1 > C_c > 3$)	
		Gravels with fines (appreciable amount of fines)	GM d/u	Silty gravels, gravel-sand-silt mixture	Atterberg limits below *A* line or $I_p < 4$	Above A line with $4 < I_p < 7$ are borderline cases requiring use of dual symbols
			GC	Clayey gravels, gravel-sand-clay mixture	Atterberg limits above *A* line with $I_p > 7$	
	Sands (more than half of coarse fraction is smaller than No. 4 sieve size)	Clean sands (little or no fines)	SW	Well-graded sands, gravelly sands, little or no fines	$C_u \geq 6$ $1 \leq C_c \leq 3$	
			SP	Poorly graded sands, gravelly sands, little or no fines	Not meeting all gradation requirements for SW ($C_u < 6$ or $1 > C_c > 3$)	
		Sands with fines (appreciable amount of fines)	SM d/u	Silty sands, sand-silt mixture	Atterberg limits below *A* line or $I_p < 4$	Above A line with $4 \leq I_p \leq 7$ are borderline cases requiring use of dual symbols
			SC	Clayey sands, sand-silt mixture	Atterberg limits above *A* line with $I_p > 7$	
Fine-grained soils (more than half of material is smaller than No. 200)	Silts and clays (liquid limit < 50)		ML	Inorganic silts and very fine sands, rock flour, silty or clayey fine sands, or clayey silts with slight plasticity	1. Determine percentages of sand and gravel from grain-size curve. 2. Depending on percentages of fines (fraction smaller than 200 sieve size), coarse-grained soils are classified as follows: Less than 5%–GW, GP, SW, SP More than 12%–GM, GC, SM, SC 5 to 12%–Borderline cases requiring dual symbols	
			CL	Inorganic clays of very low to medium plasticity, gravelly clays, sandy clays, silty clays, lean clays		
			OL	Organic silts and organic silty clays of low plasticity		
	Silts and clays (liquid limit > 50)		MH	Inorganic silts, micaceous or diatomaceous fine sandy or silty soils, elastic silts	$C_u = \frac{D_{60}}{D_{10}}$ $C_c = \frac{D_{30}^2}{D_{10}D_{60}}$	
			CH	Inorganic clays or high plasticity, fat clays		
			OH	Organic clays of medium to high plasticity, organic silts		
	Highly organic soils		Pt	Peat and other highly organic soils		

2. Gravels and sands are GM, GC, SM, or SC

   if more than 12 percent passes the No. 200 sieve; M = silt; C = clay. The silt or clay designation is determined by performing the liquid and plastic limit tests on the (–) No. 40 fraction and using the plasticity chart of Fig. 3.22. This chart is also a Casagrande contribution to the USC system, and the A line shown on this chart is sometimes called Casagrande's A line.

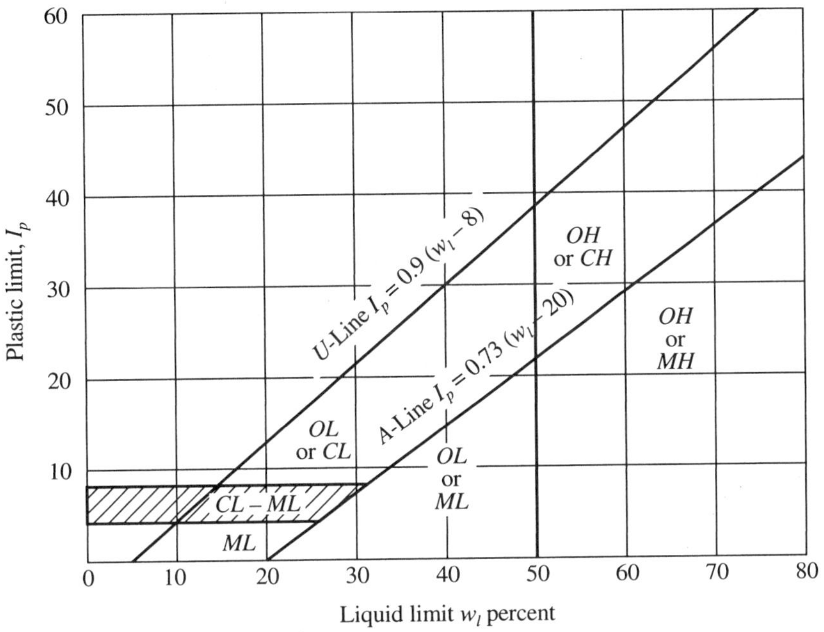

Figure 3.22 Plasticity chart for fine-grained soils

The chart as presented here has been slightly modified based on the Corps of Engineers findings that no soil has so far been found with coordinates that lie above the "upper limit" or *U* line shown. This chart and lines are part of the ASTM D 2487 standard.

3. Gravels and sands are (note using dual symbols)

   GW-GC SW-SC GP-GC SP-SC, or GW-GM SW-SM GP-GM SP-SM

   if between 5 and 12 percent of the material passes the No. 200 sieve. It may be noted that the M or C designation is derived from performing plastic limit tests and using Casagrande's plasticity chart.

4. Fine-grained soils (more than 50 percent passes the No. 200 sieve) are:

   ML, OL, or CL

   if the liquid limits are < 50 percent; M = silt; O = organic soils; C = clay. L = Less than 50 percent for w_l

5. Fine grained soils are

   MH, OH, or CH

   if the liquid limits are ≥ 50 percent; H = Higher than 50 percent. Whether a soil is a Clay (C), Silt (M), or Organic (O) depends on whether the soil coordinates plot above or below the A line on Fig. 3.22.

   The organic (O) designation also depends on visual appearance and odor in the USC method. In the ASTM method the O designation is more specifically defined by using a comparison of the air-dry liquid limit w_l and the oven-dried w'_l. If the oven dried value is

   $w_l' < 0.75 w_l$

   and the appearance and odor indicates "organic" then classify the soil as O.

Table 3.18 Unified Soil Classification System—fine-grained soils (more than half of material is larger than No. 200 sieve size)

Soil	Major divisions	Group symbols	Identification procedures on fraction smaller than No. 40 sieve size		
			Dry strength	Dilatancy	Toughness
Silt and clays	Liquid limit less than 50	*ML*	None to slight	Quick to slow	None
		CL	Medium to high	None to very slow	Medium
		OL	Slight to medium	Slow	Slight
	Liquid limit more than 50	*MH*	Slight to medium	Slow to none	Slight to medium
		CH	High to very high	None	High
		OH	Medium to high	None to very slow	Slight to medium
Highly organic soils		*Pt*	Readily identified by color, odor, spongy feel and frequently by fibrous texture		

The liquid and plastic limits are performed on the (–) No. 40 sieve fraction of all of the soils, including gravels, sands, and the fine-grained soils. Plasticity limit tests are not required for soils where the percent passing the No. 200 sieve ≤ 5 percent. The identification procedure of fine grained soils are given in Table 3.18.

A visual description of the soil should accompany the letter classification. The ASTM standard includes some description in terms of sandy or gravelly, but color is also very important. Certain areas are underlain with soil deposits having a distinctive color (e.g., Boston blue clay, Chicago blue clay) which may be red, green, blue, grey, black, and so on. Geotechnical engineers should become familiar with the characteristics of this material so the color identification is of considerable aid in augmenting the data base on the soil.

3.21 COMMENTS ON THE SYSTEMS OF SOIL CLASSIFICATION

The various classification systems described earlier are based on:

1. The properties of soil grains.
2. The properties applicable to remolded soils.

The systems do not take into account the properties of intact materials as found in nature. Since the foundation materials of most engineering structures are undisturbed, the properties of intact materials only determine the soil behavior during and after construction. The classification of a soil according to any of the accepted systems does not in itself enable detailed studies of soils to be dispensed with altogether. Solving flow, compression and stability problems merely on the basis of soil classification can lead to disastrous results. However, soil classification has been found to be a valuable tool to the engineer. It helps the engineer by giving general guidance through making available in an empirical manner the results of field experience.

Example 3.20

A sample of inorganic soil has the following grain size characteristics

Size (mm)	Percent passing
2.0 (No. 10)	95
0.075 (No. 200)	75

The liquid limit is 56 percent, and the plasticity index 25 percent. Classify the soil according to the AASHTO classification system.

Solution

Percent of fine grained soil = 75

Computation of Group Index [Eq. (3.56a)]:

$a = 75 - 35 = 40$

$b = 75 - 15 = 60$

$c = 56 - 40 = 16$, $d = 25 - 10 = 15$

Group Index, $GI = 0.2 \times 40 + 0.005 \times 40 \times 16 + 0.01 \times 60 \times 15 = 20.2$

On the basis of percent of fine-grained soils, liquid limit and plasticity index values, the soil is either A-7-5 or A-7-6. Since $(w_l - 30) = 56 - 30 = 26 > I_p\ (25)$, *the soil classification is A-7-5(20).*

Example 3.21

Mechanical analysis on four different samples designated as *A*, *B*, *C* and *D* were carried out in a soil laboratory. The results of tests are given below. Hydrometer analysis was carried out on sample *D*. The soil is non-plastic.

Sample *D*: liquid limit = 42, plastic limit = 24, plasticity index = 18

Classify the soils per the Unified Soil Classification System.

Samples	*A*	*B*	*C*	*D*
ASTM Sieve Designation	Percentage finer than			
63.0 mm	100		93	
20.0 mm	64		76	
6.3	39	100	65	
2.0 mm	24	98	59	
600 μ	12	90	54	
212 μ	5	9	47	100
63 μ	1	2	34	95
20 μ			23	69
6 μ			7	46
2 μ			4	31

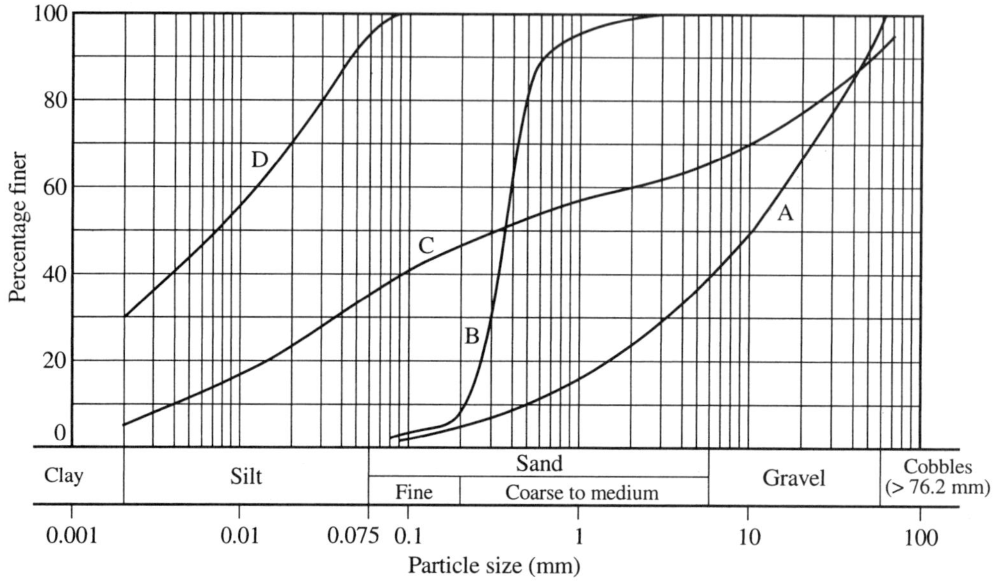

Figure Ex. 3.21

Solution

Grain size distribution curves of samples *A, B, C* and *D* are given in Fig. Ex. 3.21. The values of C_u and C_c are obtained from the curves as given below.

Sample	D_{10}	D_{30}	D_{60}	C_u	C_c
A	0.47	3.5	16.00	34.0	1.60
B	0.23	0.30	0.41	1.8	0.95
C	0.004	0.036	2.40	600.0	0.135

Sample *A*: Gravel size particles more than 50%, fine grained soil less than 5%. C_u, greater than 4, and C_c lies between 1 and 3. Well graded sandy gravel classified as *GW*.

Sample *B*: 96% of particles are sand size. Finer fraction less than 5%. $C_u = 1.8$, C_c is not between 1 and 3. Poorly-graded sand, classified as *SP*.

Sample *C*: Coarse grained fraction greater than 66% and fine grained fraction less than 34%. The soil is non-plastic. C_u is very high but C_c is only 0.135. Gravel-sand-silt mixture, classified as *GM*.

Sample *D*: Finer fraction 95% with clay size particles 31%. The point plots just above the *A*-line in the *CL* zone on the plasticity chart. Silty-clay of low plasticity, classified as *CL*.

Example 3.22

The following data refers to a silty clay that was assumed to be saturated in the undisturbed condition. On the basis of these data determine the liquidity index, sensitivity, and void ratio of the saturated soil. Classify the soil according to the Unified and AASHTO systems. Assume $G_s = 2.7$.

Index property	Undisturbed	Remolded
Unconfined compressive strength, q_u kN/m^2	244 kN/m^2	144 kN/m^2
Water content, %	22	22
Liquid limit, %		45
Plastic limit, %		20
Shrinkage limit, %		12
% passing no. 200 sieve		90

Solution

Liquidity Index, $I_l = \dfrac{w_n - w_p}{w_l - w_p} = \dfrac{22-20}{45-20} = 0.08$

Sensitivity, $S = \dfrac{q_u \text{ undisturbed}}{q'_u \text{ disturbed}} = \dfrac{244}{144} = 1.7$

Void ratio, $e = \dfrac{V}{V_s}$

For $S = 1$, $e = wG_s = 0.22 \times 2.7 = 0.594$.

Unified Soil Classification

Use the plasticity chart Fig. 3.22. $w_l = 45$, $I_p = 25$. The point falls above the A-line in the CL-zone, that is the soil is inorganic clay of low to medium plasticity.

AASHTO System

Group Index $GI = 0.2a + 0.005ac + 0.01bd$

$a = 90 - 35 = 55$

$b = 90 - 15 = 75$

$c = 45 - 40 = 5$

$d = 25 - 10 = 15$ (here $I_p = w_l - w_p = 45 - 20 = 25$)

Group index $GI = 0.2 \times 55 + 0.005 \times 55 \times 5 + 0.01 \times 75 \times 15$

$= 11 + 1.315 + 11.25 = 23.63$ or say 24

Enter Table 3.15 with the following data

% passing 200 sieve = 90%

Liquid limit = 45%

Plasticity index = 25%

With this, the soil is either A-7-5 or A-7-6. Since $(w_l - 30) = (45 - 30) = 15 < 25$ (I_p) the soil is classified as A-7-6. According to this system the soil is clay, A-7-6 (24).

3.22 PROBLEMS

3.1 A soil mass in its natural state is partially saturated having a water content of 17.5% and a void ratio of 0.87. Determine the degree of saturation, total unit weight, and dry unit weight. What is the weight of water required to saturate a mass of 10 m^3 volume? Assume $G_s = 2.69$.

3.2 The void ratio of a clay sample is 0.5 and the degree of saturation is 70%. Compute the water content, dry and wet unit weights of the soil. Assume $G_s = 2.7$.

3.3 A sample of soil compacted according to a standard Proctor test has a unit weight of 130.9 lb/ft^3 at 100% compaction and at optimum water content of 14%. What is the dry unit weight? If the voids become filled with water what would be the saturated unit weight? Assume $G_s = 2.67$.

3.4 A sample of sand above the water table was found to have a natural moisture content of 15% and a unit weight of 18.84 kN/m^3. Laboratory tests on a dried sample indicated values of $e_{min} = 0.50$ and $e_{max} = 0.85$ for the densest and loosest states respectively. Compute the degree of saturation and the relative density. Assume $G_s = 2.65$.

3.5 How many cubic meters of fill can be constructed at a void ratio of 0.7 from 119,000 m^3 of borrow material that has a void ratio of 1.2?

3.6 The natural water content of a sample taken from a soil deposit was found to be 11.5%. It has been calculated that the maximum density for the soil will be obtained when the water content reaches 21.5%. Compute how much water must be added to 22,500 lb of soil (in its natural state) in order to increase the water content to 21.5%. Assume that the degree of saturation in its natural state was 40% and $G_s = 2.7$.

3.7 In an oil well drilling project, drilling mud was used to retain the sides of the borewell. In one liter of suspension in water, the drilling mud fluid consists of the following material:

Material	Mass (g)	Sp. gr
Clay	410	2.81
Sand	75	2.69
Iron filings	320	7.13

Find the density of the drilling fluid of uniform suspension.

3.8 In a field exploration, a soil sample was collected in a sampling tube of internal diameter 5.0 cm below the ground water table. The length of the extracted sample was 10.2 cm and its mass was 387 g. If $G_s = 2.7$, and the mass of the dried sample is 313 g, find the porosity, void ratio, degree of saturation, and the dry density of the sample.

3.9 A saturated sample of undisturbed clay has a volume of 19.2 cm^3 and weighs 32.5 g. After oven drying, the weight reduces to 20.2 g. Determine the following:

(a) water content, (b) specific gravity, (c) void ratio, and (d) saturated density of the clay sample.

3.10 The natural total unit weight of a sandy stratum is 117.7 lb/ft^3 and has a water content of 8%. For determining of relative density, dried sand from the stratum was filled loosely into a 1.06 ft^3 mold and vibrated to give a maximum density. The loose weight of the sample in the mold was 105.8 lb, and the dense weight was 136.7 lb. If $G_s = 2.66$, find the relative density of the sand in its natural state.

3.11 An earth embankment is to be compacted to a density of 120.9 lb/ft^3 at a moisture content of 14 percent. The in-situ total unit weight and water content of the borrow pit are

114.5 lb/ft^3 and 8% respectively. How much excavation should be carried out from the borrow pit for each ft^3 of the embankment? Assume $G_s = 2.68$.

3.12 An undisturbed sample of soil has a volume of 29 cm^3 and weighs 48 g. The dry weight of the sample is 32 g. The value of $G_s = 2.66$. Determine the (a) natural water content, (b) in-situ void ratio, (c) degree of saturation, and (d) saturated unit weight of the soil.

3.13 A mass of soil coated with a thin layer of paraffin weighs 0.982 lb. When immersed in water it displaces 0.011302 ft^3 of water. The paraffin is peeled off and found to weigh 0.0398 lb. The specific gravity of the soil particles is 2.7 and that of paraffin is 0.9. Determine the void ratio of the soil if its water content is 10%.

3.14 225 g of oven dried soil was placed in a specific gravity bottle and then filled with water to a constant volume mark made on the bottle. The mass of the bottle with water and soil is 1650 g. The specific gravity bottle was filled with water alone to the constant volume mark and weighed. Its mass was found to be 1510 g. Determine the specific gravity of the soil.

3.15 It is required to determine the water content of a wet sample of silty sand weighing 400 g. This mass of soil was placed in a pycnometer and water filled to the top of the conical cup and weighed (M_3). Its mass was found to be 2350 g. The pycnometer was next filled with clean water and weighed and its mass was found to be 2200 g (M_4). Assuming $G_s = 2.67$, determine the water content of the soil sample.

3.16 A clay sample is found to have a mass of 423.53 g in its natural state. It is then dried in an oven at 105 °C. The dried mass is found to be 337.65 g. The specific gravity of the solids is 2.70 and the density of the soil mass in its natural state is 1700 kg/m^3. Determine the water content, degree of saturation and the dry density of the mass in its natural state.

3.17 A sample of sand in its natural state has a relative density of 65 percent. The dry unit weights of the sample at its densest and loosest states are respectively 114.5 and 89.1 lb/ft^3. Assuming the specific gravity of the solids as 2.64, determine (i) its dry unit weight, (ii) wet unit weight when fully saturated, and (iii) submerged unit weight.

3.18 The mass of wet sample of soil in a drying dish is 462 g. The sample and the dish have a mass of 364 g after drying in an oven at 110 °C overnight. The mass of the dish alone is 39 g. Determine the water content of the soil.

3.19 A sample of sand above the water table was found to have a natural moisture content of 10% and a unit weight of 120 lb/ft^3. Laboratory tests on a dried sample indicated values $e_{min} = 0.45$, and $e_{max} = 0.90$ for the densest and loosest states respectively. Compute the degree of saturation, S, and the relative density, D_r. Assume $G_s = 2.65$.

3.20 A 50 cm^3 sample of moist clay was obtained by pushing a sharpened hollow cylinder into the wall of a test pit. The extruded sample had a mass of 85 g, and after oven drying a mass of 60 g. Compute w, e, S, and ρ_d. $G_s = 2.7$.

3.21 A pit sample of moist quartz sand was obtained from a pit by the sand cone method. The volume of the sample obtained was 150 cm^3 and its total mass was found to be 250 g. In the laboratory the dry mass of the sand alone was found to be 240 g. Tests on the dry sand indicated $e_{max} = 0.80$ and $e_{min} = 0.48$. Estimate ρ_s, w, e, S, ρ_d and D_r of the sand in the field. Given $G_s = 2.67$.

3.22 An earthen embankment under construction has a total unit weight of 99.9 lb/ft^3 and a moisture content of 10 percent. Compute the quantity of water required to be added per 100 ft^3 of earth to raise its moisture content to 14 percent at the same void ratio.

3.23 The wet unit weight of a glacial outwash soil is 122 lb/ft^3, the specific gravity of the solids is $G_s = 2.67$, and the moisture content of the soil is $w = 12\%$ by dry weight. Calculate (a) dry unit weight, (b) porosity, (c) void ratio, and (d) degree of saturation.

3.24 Derive the equation $e = wG_s$ which expresses the relationship between the void ratio e, the specific gravity G_s and the moisture content w for full saturation of voids.

3.25 In a sieve analysis of a given sample of sand the following data were obtained. Effective grain size = 0.25 mm, uniformity coefficient 6.0, coefficient of curvature = 1.0.

Sketch the curve on semilog paper.

3.26 A sieve analysis of a given sample of sand was carried out by making use of US standard sieves. The total weight of sand used for the analysis was 522 g. The following data were obtained.

Sieve size in mm	4.750	2.000	1.000	0.500	0.355	0.180	0.125	0.075
Weight retained in g	25.75	61.75	67.00	126.0	57.75	78.75	36.75	36.75
Pan								31.5

Plot the grain size distribution curve on semi-log paper and compute the following:

(i) Percent gravel

(ii) Percent of coarse, medium and fine sand

(iii) Percent of silt and clay

(iv) Uniformity coefficient

(v) Coefficient of curvature

3.27 Combined mechanical analysis of a given sample of soil was carried out. The total weight of soil used in the analysis was 350 g. The sample was divided into coarser and finer fractions by washing it through a 75 microns sieve The finer traction was 125 g. The coarser fraction was used for the sieve analysis and 50 g of the finer fraction was used for the hydrometer analysis. The test results were as given below:

Sieve analysis:

Particle size	Mass retained g	Particle size	Mass retained g
4.75 mm	9.0	355 μ	24.5
2.00 mm	15.5	180 μ	49.0
1.40 mm	10.5	125 μ	28.0
1.00 mm	10.5	75 μ	43.0
500 μ	35.0		

A hydrometer (152 H type) was inserted into the suspension just a few seconds before the readings were taken. It was next removed and introduced just before each of the subsequent readings. Temperature of suspension = 25°C.

Hydrometer analysis: Readings in suspension

Time, min	Reading, R_a	Time, min	Reading, R_a
1/4	28.00	30	5.10
1/2	24.00	60	4.25
1	20.50	120	3.10
2	17.20	240	2.30
4	12.00	480	1.30
8	8.50	1440	0.70
15	6.21		

Meniscus correction $C_m = +0.4$, zero correction $C_o = +1.5$, $G_s = 2.75$

(i) Show (step by step) all the computations required for the combined analysis.

(ii) Plot the grain size distribution curve on semi-log paper

(iii) Determine the percentages of gravel, sand, and fine fractions present in the sample

(iv) Compute the uniformity coefficient and the coefficient of curvature

(v) Comment on the basis of the test results whether the soil is well graded or not

3.28 Liquid limit tests were carried out on two given samples of clay. The test data are as given below.

Test Nos	1	2	3	4
Sample no. 1				
Water content %	120	114	98	92
Number of blows, N	7	10	30	40
Sample no. 2				
Water content %	96	74	45	30
Number of blows, N	9	15	32	46

The plastic limit of Sample No. 1 is 40 percent and that of Sample No. 2 is 32 percent.

Required:

(i) The flow indices of the two samples

(ii) The toughness indices of the samples

(iii) Comment on the type of soils on the basis of the toughness index values

3.29 Four different types of soils were encountered in a large project. Their liquid limits (w_l), plastic limits (w_p) and their natural moisture contents (w_n) were as given below:

Soil type	w_l%	w_p%	w_n%
1	120	40	150
2	80	35	70
3	60	30	30
4	65	32	25

Required:

(i) The liquidity indices of the soils, (ii) the consistency of the natural soils (i.e., whether soft, stiff, etc.)

(ii) and the possible behavior of the soils under vibrating loads

3.30 The soil types as given in Problem 3.29 contained soil particles finer than 2 microns as given below:

Soil type	1	2	3	4
Percent finer than 2 micron	50	55	45	50

Classify the soils according to their activity values.

3.31 A sample of clay has a water content of 40 percent at full saturation. Its shrinkage limit is 15 percent. Assuming G_s = 2.70, determine its degree of shrinkage. Comment on the quality of the soil.

3.32 A sample of clay soil has a liquid limit of 62% and its plasticity index is 32 percent.

(i) What is the state of consistency of the soil if the soil in its natural state has a water content of 34 percent?

(ii) Calculate the shrinkage limit if the void ratio of the sample at the shrinkage limit is 0.70

Assume G_s = 2.70.

3.33 A soil with a liquidity index of –0.20 has a liquid limit of 56 percent and a plasticity index of 20 percent. What is its natural water content?

3.34 A sample of soil weighing 50 g is dispersed in 1000 mL of water. How long after the commencement of sedimentation should the hydrometer reading be taken in order to estimate the percentage of particles less than 0.002 mm effective diameter, if the center of the hydrometer is 150 mm below the surface of the water?

Assume: G_s = 2.7; $\mu = 8.15 \times 10^{-6}$ g-sec/cm^2.

3.35 The results of a sieve analysis of a soil were as follows:

Sieve size (mm)	Mass retained (g)	Sieve size (mm)	Mass retained (g)
20	0	2	3.5
12	1.7	1.4	1.1
10	2.3	0.5	30.5
6.3	8.4	0.355	45.3
4.75	5.7	0.180	25.4
2.8	12.9	0.075	7.4

The total mass of the sample was 147.2 g.

(a) Plot the particle-size distribution curve and describe the soil. Comment on the flat part of the curve

(b) State the effective grain size

3.36 A liquid limit test carried out on a sample of inorganic soil taken from below the water table gave the following results:

Fall cone penetration y (mm)	15.5	18.2	21.4	23.6
Moisture content, w_y%	34.6	40.8	48.2	53.4

A plastic limit test gave a value of 33%. Determine the average liquid limit and plasticity index of this soil and give its classification.

3.37 The oven dry mass of a sample of clay was 11.26 g. The volume of the dry sample was determined by immersing it in mercury and the mass of the displaced liquid was 80.29 g. Determine the shrinkage limit, w_s, of the clay assuming G_s = 2.70.

3.38 Particles of five different sizes are mixed in the proportion shown below and enough water is added to make 1000 cm^3 of the suspension.

Particle size (mm)	Mass (g)	
0.050	6	
0.020	20	
0.010	15	
0.005	5	
0.001	4	Total 50 g

It is ensured that the suspension is thoroughly mixed so as to have a uniform distribution of particles. All particles have specific gravity of 2.7.

(a) What is the largest particle size present at a depth of 6 cm after 5 mins from the start of sedimentation?

(b) What is the density of the suspension at a depth of 6 cm after 5 mins from the start of sedimentation?

(c) How long should sedimentation be allowed so that all the particles have settled below 6 cm? Assume $\mu = 0.9 \times 10^{-6}$ kN-s/m^2

3.39 A sample of clayey silt is mixed at its liquid limit of 40%. It is placed carefully in a small porcelain dish with a volume of 19.3 cm^3 and weighs 34.67 g. After oven drying, the soil pat displaced 216.8 g of mercury.

(a) Determine the shrinkage limit, w_s, of the soil sample

(b) Estimate the dry unit weight of the soil

3.40 During the determination of the shrinkage limit of a sandy clay, the following laboratory data was obtained:

Wet wt. of soil + dish	=	87.85 g
Dry wt. of soil + dish	=	76.91 g
Wt of dish	=	52.70 g
The volumetric determination of the soil pat:		
Wt. of dish + mercury	=	430.8 g
Wt. of dish	=	244.62 g
Calculate the shrinkage limit, assuming G_s	=	2.65

3.41 A sedimentation analysis by a hydrometer (152 H type) was conducted with 50 g of oven dried soil sample. The hydrometer reading in a 1000 cm^3 soil suspension 60 mins after the commencement of sedimentation is 19.5. The meniscus correction is 0.5. Assuming $G_s = 2.70$ and $\mu = 1 \times 10^{-6}$ kN-s/m^2 for water, calculate the smallest particle size which would have settled during the time of 60 mins and percentage of particles finer than this size. Assume: $C_0 = +2.0$, and $C_T = 1.2$

3.42 Classify the soil given below using the Unified Soil Classification System.

Percentage passing No. 4 sieve	72
Percentage passing No. 200 sieve	33
Liquid limit	35
Plastic limit	14

3.43 Soil samples collected from the field gave the following laboratory test results:

Percentage passing No. 4 sieve	100
Percentage passing No. 200 sieve	76
Liquid limit	65
Plastic limit	30

Classify the soil using the Unified Soil Classification System.

3.44 For a large project, a soil investigation was carried out. Grain size analysis carried out on the samples gave the following average test results.

Sieve No.	Percent finer
4	96
10	60
20	18
40	12
60	7
100	4
200	2

Classify the soil by using the Unified Soil Classification System assuming the soil is non-plastic.

3.45 The sieve analysis of a given sample of soil gave 57 percent of the particles passing through 75 micron sieve. The liquid and plastic limits of the soil were 62 and 28 percent respectively. Classify the soil per the AASHTO and the Unified Soil Classification Systems.

CHAPTER 4

SOIL PERMEABILITY AND SEEPAGE

4.1 SOIL PERMEABILITY

A material is permeable if it contains continuous voids. All materials such as rocks, concrete, soils etc. are permeable. The flow of water through all of them obeys approximately the same laws. Hence, the difference between the flow of water through rock or concrete is one of degree. The permeability of soils has a decisive effect on the stability of foundations, seepage loss through embankments of reservoirs, drainage of subgrades, excavation of open cuts in water bearing sand, rate of flow of water into wells and many others.

Hydraulic Gradient

When water flows through a saturated soil mass there is certain resistance for the flow because of the presence of solid matter. However, the laws of *fluid mechanics* which are applicable for the flow of fluids through pipes are also applicable to flow of water through soils. As per *Bernoulli's equation,* the total head at any point in water under steady flow condition may be expressed as

Total head = pressure head + velocity head + elevation head

This principle can be understood with regards to the flow of water through a sample of soil of length L and cross-sectional area A as shown in Fig. 4.1(a). The heads of water at points A and B as the water flows from A to B are given as follows (with respect to a datum)

Total head at A, $$H_A = Z_A + \frac{p_A}{\gamma_w} + \frac{V_A^2}{2g}$$

Total head at B, $$H_B = Z_B + \frac{p_B}{\gamma_w} + \frac{V_B^2}{2g}$$

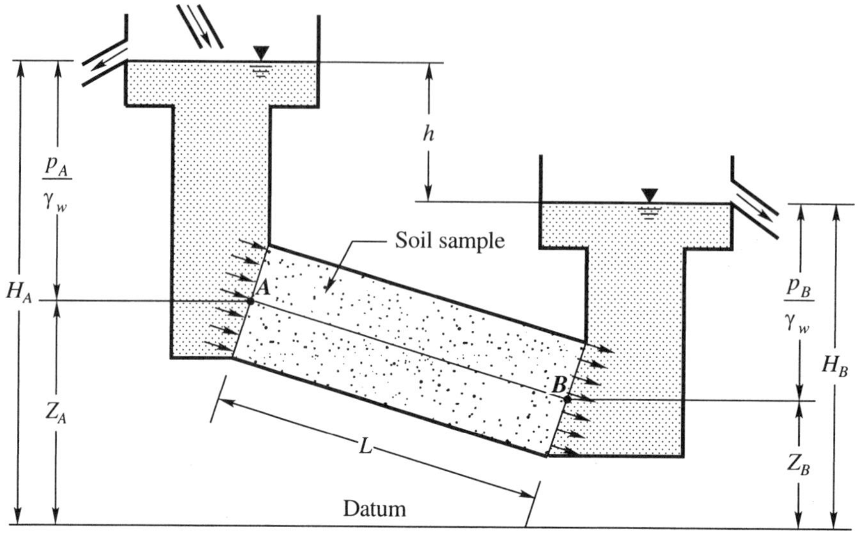

Figure 4.1(a) Flow of water through a sample of soil

As the water flows from A to B, there is an energy loss which is represented by the difference in the total heads H_A and H_B

$$\text{or} \quad H_A - H_B = \left(Z_A + \frac{p_A}{\gamma_w} + \frac{V_A^2}{2g}\right) - \left(Z_B + \frac{p_B}{\gamma_w} + \frac{V_B^2}{2g}\right) = h$$

where, p_A and p_B = pressure heads, V_A and V_B = velocity, g = acceleration due to gravity, γ_w = unit weight of water, h = loss of head.

For all practical purposes the velocity head is a small quantity and may be neglected. The loss of head of h units is effected as the water flows from A to B. The loss of head per unit length of flow may be expressed as

$$i = \frac{h}{L} \tag{4.1}$$

where i is called the *hydraulic gradient.*

Laminar and Turbulent Flow

Problems relating to the flow of fluids in general may be divided into two main classes:

1. Those in which the flow is laminar.
2. Those in which the flow is turbulent.

There is a certain velocity, v_c, below which for a given diameter of a straight tube and for a given fluid at a particular temperature, the flow will always remain laminar. Likewise there is a higher velocity, v_T, above which the flow will always be turbulent. The lower bound velocity, v_T, of turbulent flow is about 6.5 times the upper bound velocity v_c of laminar flow as shown in Fig. 4.1(b). The upper bound velocity of laminar flow is called the *lower critical velocity.* The fundamental laws that determine the state existing for any given case were determined by Reynolds (1883). He found the lower critical velocity is inversely proportional to the diameter of

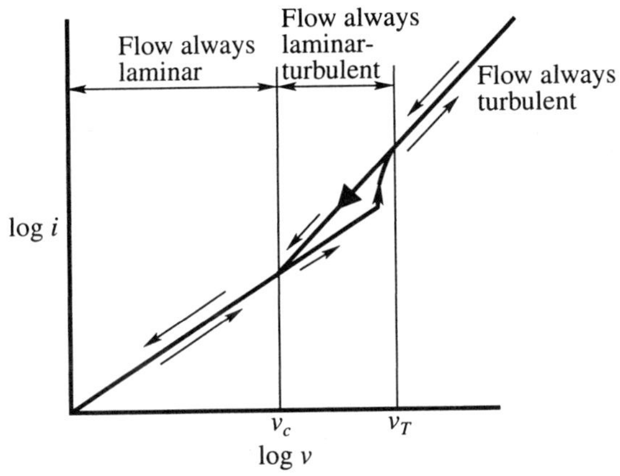

Figure 4.1(b) Relationship between velocity of flow and hydraulic gradient for flow of liquids in a pipe

the pipe and gave the following general expression applicable for any fluid and for any system of units.

$$N_R = \frac{v_c D \gamma_0}{\mu g} = 2000$$

where, N_R = Reynolds Number taken as 2000 as the maximum value for the flow to remain always laminar, D = diameter of pipe, v_c = critical velocity below which the flow always remains laminar, γ_0 = unit weight of fluid at 4 °C, μ = viscosity of fluid, g = acceleration due to gravity.

The principal difference between laminar flow and turbulent flow is that in the former case the velocity is proportional to the first power of the hydraulic gradient, i, whereas in the latter case it is 4/7 the power of i. According to Hagen-Poiseuille's' Law the flow through a capillary tube may be expressed as

$$q = \frac{\gamma_w R^2 a i}{8\mu} \tag{4.2a}$$

or $$v = \frac{q}{a} = \frac{\gamma_w R^2 i}{8\mu} \tag{4.2b}$$

where, R = radius of a capillary tube of sectional area a, q = discharge through the tube, v = average velocity through the tube, μ = coefficient of viscosity.

4.2 DARCY'S LAW

Darcy in 1856 derived an empirical formula for the behavior of flow through saturated soils. He found that the quantity of water q per sec flowing through a cross-sectional area of soil under hydraulic gradient i can be expressed by the formula

$$q = kiA \tag{4.3}$$

or the velocity of flow can be written as

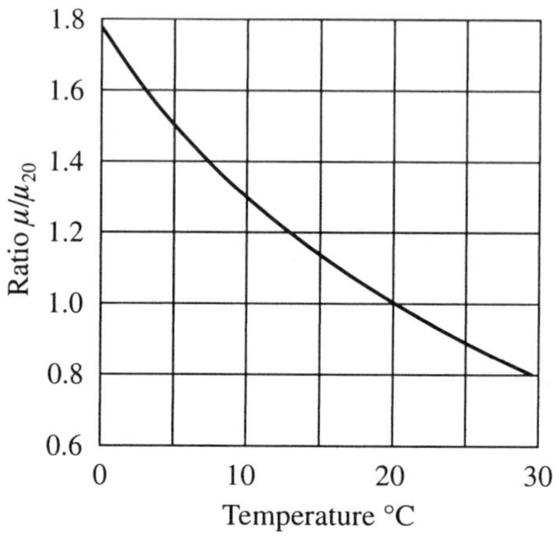

Figure 4.2 Relation between temperature and viscosity of water

$$v = \frac{q}{A} = ki \tag{4.4}$$

where k is termed the *hydraulic conductivity* (or coefficient of permeability)with units of velocity. A in Eq. (4.4) is the cross-sectional area of soil normal to the direction of flow which includes the area of the solids and the voids, whereas the area a in Eq. (4.2) is the area of a capillary tube. The essential point in Eq. (4.3) is that the flow through the soils is also proportional to the first power of the hydraulic gradient i as propounded by Poiseuille's Law. From this, we are justified in concluding that the flow of water through the pores of a soil is laminar. It is found that, on the basis of extensive investigations made since Darcy introduced his law in 1856, this law is valid strictly for fine grained types of soils.

The hydraulic conductivity is a measure of the ease with which water flows through permeable materials. It is inversely proportional to the viscosity of water which decreases with increasing temperature as shown in Fig. 4.2. Therefore, permeability measurements at laboratory temperatures should be corrected with the aid of Fig. 4.2 before application to field temperature conditions by means of the equation

$$k_f = \frac{\mu_T}{\mu_f} k_T \tag{4.5}$$

where k_f and k_T are the hydraulic conductivity values corresponding to the field and test temperatures respectively and μ_f and μ_T are the corresponding viscosities. It is customary to report the values of k_T at a standard temperature of 20°C. The equation is

$$k_{20} = \frac{\mu_T}{\mu_{20}} k_T \tag{4.6}$$

4.3 DISCHARGE AND SEEPAGE VELOCITIES

Figure 4.3 shows a soil sample of length L and cross-sectional area A. The sample is placed in a cylindrical horizontal tube between screens. The tube is connected to two reservoirs R_1 and R_2 in which the water levels are maintained constant. The difference in head between R_1 and R_2 is h. This difference in head is responsible for the flow of water. Since Darcy's law assumes no change in the

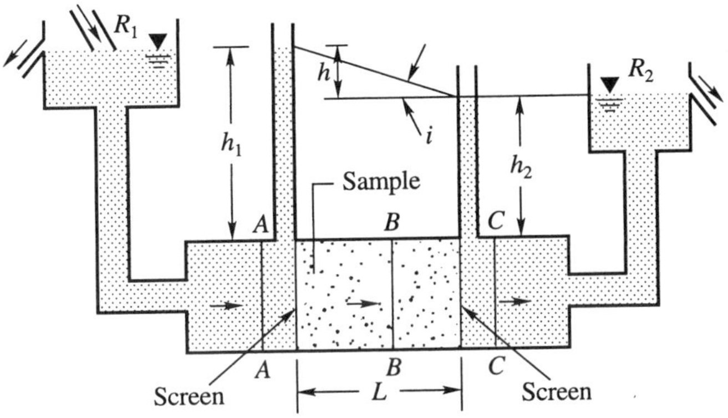

Figure 4.3 Flow of water through a sample of soil

volume of voids and the soil is saturated, the quantity of flow past sections *AA*, *BB* and *CC* should remain the same for steady flow conditions. We may express the equation of continuity as follows

$$q_{aa} = q_{bb} = q_{cc}$$

If the soil be represented as divided into solid matter and void space, then the area available for the passage of water is only A_v. If v_s is the velocity of flow in the voids, and v, the average velocity across the section then, we have

$$A_v v_s = Av \quad \text{or} \quad v_s = \frac{A}{A_v} v$$

$$\text{Since,} \quad \frac{A}{A_v} = \frac{1}{n} = \frac{1+e}{e}, \quad v_s = \frac{v}{n} = \left(\frac{1+e}{e}\right) v \tag{4.7}$$

Since $(1 + e)/e$ is always greater than unity, v_s is always greater than v. Here, v_s is called the *seepage velocity* and v the *discharge velocity*.

4.4 METHODS OF DETERMINATION OF HYDRAULIC CONDUCTIVITY OF SOILS

Methods that are in common use for determining the coefficient of permeability k can be classified under laboratory and field methods.

Laboratory methods:
1. Constant head permeability method
2. Falling head permeability method

Field methods:
1. Pumping tests
2. Bore hole tests

Indirect Method: Empirical correlations

The various types of apparatus which are used in soil laboratories for determining the permeability of soils are called *permeameters*. The apparatus used for the constant head permeability test is called a *constant head permeameter* and the one used for the falling head test is a *falling head permeameter*. The soil samples used in laboratory methods are either undisturbed or disturbed. Since it is not possible to obtain undisturbed samples of cohesionless soils, laboratory tests on cohesionless materials are always conducted on samples which are reconstructed to the same density as they exist in nature. The results of tests on such reconstructed soils are often misleading since it is impracticable to obtain representative samples and place them in the test apparatus to give exactly the same density and structural arrangement of particles. Direct testing of soils in place is generally preferred in cases where it is not possible to procure undisturbed samples. Since this method is quite costly, it is generally carried out in connection with major projects such as foundation investigation for dams and large bridges or building foundation jobs where lowering of the water table is involved. In place of pumping tests, bore hole tests as proposed by the U.S. Bureau of Reclamation are quite inexpensive as these tests eliminate the use of observation wells.

Empirical correlations have been developed relating grain size and void ratio to hydraulic conductivity and will be discussed later on.

4.5 CONSTANT HEAD PERMEABILITY TEST

Figure 4.4(a) shows a constant head permeameter which consists of a vertical tube of lucite (or any other material) containing a soil sample which is reconstructed or undisturbed as the case may be. The diameter and height of the tube can be of any convenient dimensions. The head and tail water levels are kept constant by overflows. The sample of length L and cross-sectional area A is subjected to a head h which is constant during the progress of a test. A test is performed by allowing water to flow through the sample and measuring the quantity of discharge Q in time t.

The value of k can be computed directly from Darcy's law expressed as follows

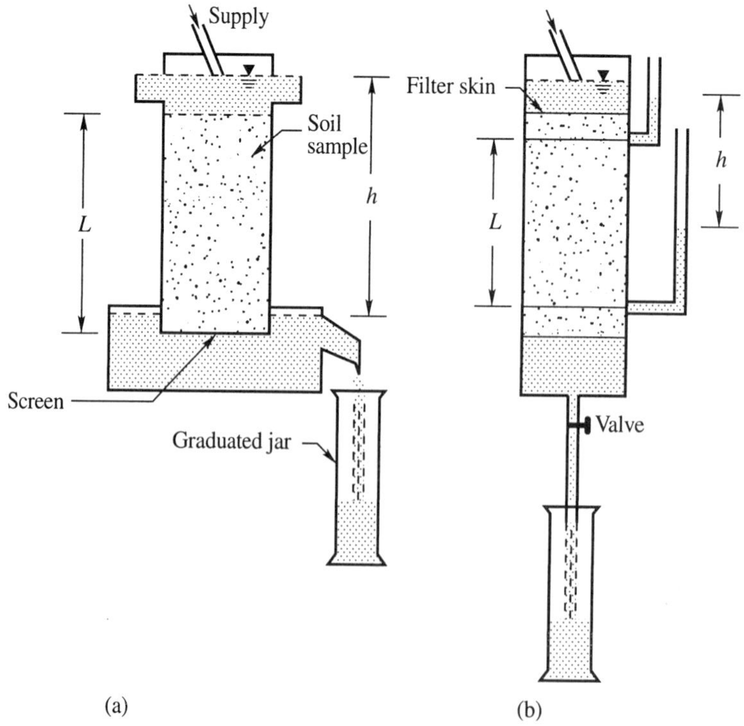

Figure 4.4 Constant head permeability test

$$Q = k\frac{h}{L}At \tag{4.8}$$

$$\text{or} \quad k = \frac{QL}{hAt} \tag{4.9}$$

The constant head permeameter test is more suited for coarse grained soils such as gravelly sand and coarse and medium sand. Permeability tests in the laboratory are generally subjected to various types of experimental errors. One of the most important of these arises from the formation of a filter skin of fine materials on the surface of the sample. The constant head permeameter of the type shown in Fig. 4.4(b) can eliminate the effect of the surface skin. In this apparatus the loss of head is measured through a distance in the interior of the sample, and the drop in head across the filter skin has no effect on the results.

4.6 FALLING HEAD PERMEABILITY TEST

A falling head permeameter is shown in Fig. 4.5(a). The soil sample is kept in a vertical cylinder of cross-sectional area A. A transparent stand pipe of cross sectional area, a, is attached to the test cylinder. The test cylinder is kept in a container filled with water, the level of which is kept constant by overflows. Before the commencement of the test the soil sample is saturated by allowing the water to flow continuously through the sample from the stand pipe. After saturation is complete, the stand pipe is filled with water up to a height of h_0 and a stop watch is started. Let the initial time be t_0. The time t_1 when the water level drops from h_0 to h_1 is noted. The hydraulic conductivity k can be determined on the basis of the drop in head $(h_0 - h_1)$ and the elapsed time $(t_1 - t_0)$ required for the drop as explained below.

Let h be the head of water at any time t. Let the head drop by an amount dh in time dt. The quantity of water flowing through the sample in time dt from Darcy's law is

$$dQ = kiA\,dt = k\frac{h}{L}A\,dt \tag{4.10}$$

where, $i = h/L$ the hydraulic gradient.

The quantity of discharge dQ can be expressed as

$$dQ = -a\,dh \tag{4.11}$$

Since the head decreases as time increases, dh is a negative quantity in Eq. (4.11). Eq. (4.10) can be equated to Eq. (4.11)

$$-a\,dh = k\frac{h}{L}A\,dt \tag{4.12}$$

The discharge Q in time $(t_1 - t_0)$ can be obtained by integrating Eq. (4.10) or (4.11). Therefore, Eq. (4.12) can be rearranged and integrated as follows

$$-a\int_{h_0}^{h_1}\frac{dh}{h} = \frac{kA}{L}\int dt \quad \text{or} \quad a\log_e\frac{h_0}{h_1} = \frac{kA}{L}(t_1 - t_0)$$

The general expression for k is

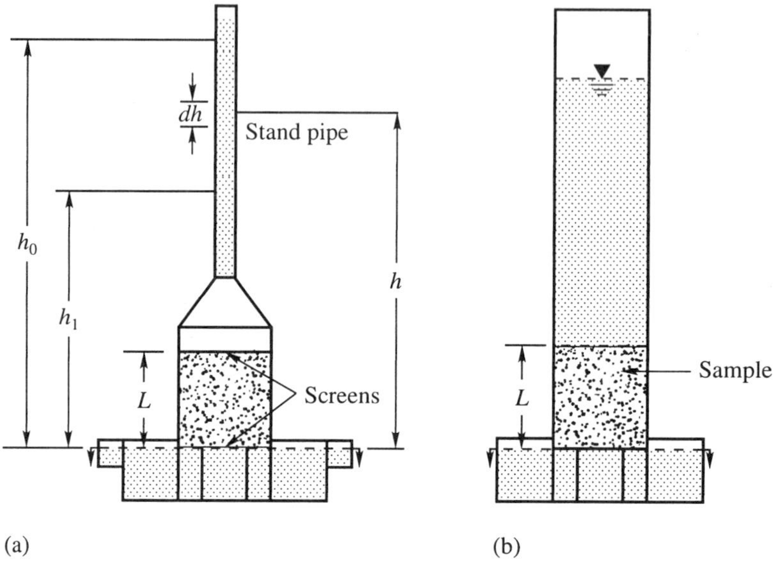

Figure 4.5 Falling head permeability test

$$k = \frac{aL}{A(t_1 - t_0)} \log_e \frac{h_0}{h_1} \quad \text{or} \quad k = \frac{2.3aL}{A(t_1 - t_0)} \log_{10} \frac{h_0}{h_1} \tag{4.13}$$

The setup shown in Fig. 4.5(a) is generally used for comparatively fine materials such as fine sand and silt where the time required for the drop in head from h_0 to h_1 is neither unduly too long nor too short for accurate recordings. If the time is too long evaporation of water from the surface of the water might take place and also temperature variations might affect the volume of the sample. These would introduce serious errors in the results. The set up is suitable for soils having permeabilities ranging from 10^{-3} to 10^{-6} cm per sec. Sometimes, falling head permeameters are used for coarse grained soils also. For such soils, the cross sectional area of the stand pipe is made the same as the test cylinder so that the drop in head can conveniently be measured. Fig. 4.5(b) shows the test set up for coarse grained soils. When $a = A$, Eq. (4.13) is reduced to

$$k = \frac{2.3L}{(t_1 - t_0)} \log_{10} \frac{h_0}{h_1} \tag{4.14}$$

Example 4.1

A constant head permeability test was carried out on a cylindrical sample of sand 4 in. in diameter and 6 in. in height. 10 in^3 of water was collected in 1.75 min, under a head of 12 in. Compute the hydraulic conductivity in ft/year and the velocity of flow in ft/sec.

Solution

The formula for determining k is

$$k = \frac{Q}{Ait}$$

$$Q = 10 \text{ in}^3, \; A = 3.14 \times \frac{4^2}{4} = 12.56 \text{ in.}^2$$

$$i = \frac{h}{L} = \frac{12}{6} = 2, \quad t = 105 \text{ sec.}$$

$$\text{Therefore} \quad k = \frac{10}{12.56 \times 2 \times 105} = 3.79 \times 10^{-3} \text{ in./sec} = 31.58 \times 10^{-5} \text{ ft/sec} = 9960 \text{ ft/year}$$

$$\text{Velocity of flow} = ki = 31.58 \times 10^{-5} \times 2 = 6.316 \times 10^{-4} \text{ ft/sec}$$

Example 4.2

A sand sample of 35 cm^2 cross sectional area and 20 cm long was tested in a constant head permeameter. Under a head of 60 cm, the discharge was 120 ml in 6 min. The dry weight of sand used for the test was 1120 g, and $G_s = 2.68$. Determine (a) the hydraulic conductivity in cm/sec, (b) the discharge velocity, and (c) the seepage velocity.

Solution

$$\text{Use Eq. (4.9),} \quad k = \frac{QL}{hAt}$$

where Q = 120 ml, t = 6 min, A = 35 cm^2, L = 20 cm, and h = 60 cm. Substituting, we have

$$k = \frac{120 \times 20}{60 \times 35 \times 6 \times 60} = 3.174 \times 10^{-3} \text{ cm/sec}$$

$$\text{Discharge velocity,} \quad v = ki = 3.174 \times 10^{-3} \times \frac{60}{20} = 9.52 \times 10^{-3} \text{ cm/sec}$$

Seepage velocity v_s

$$\gamma_d = \frac{W_s}{V} = \frac{1120}{35 \times 20} = 1.6 \text{ g/cm}^3$$

$$\text{From Eq. (3.18a),} \quad \gamma_d = \frac{\gamma_w G_s}{1+e} \quad \text{or} \quad e = \frac{G_s}{\gamma_d} - 1 \quad \text{since} \quad \gamma_w = 1 \text{ g/cm}^3$$

$$\text{Substituting,} \qquad e = \frac{2.68}{1.6} - 1 = 0.675$$

$$n = \frac{e}{1+e} = \frac{0.675}{1+0.675} = 0.403$$

$$\text{Now,} \qquad v_s = \frac{v}{n} = \frac{9.52 \times 10^{-3}}{0.403} = 2.36 \times 10^{-2} \text{ cm/sec}$$

Example 4.3

Calculate the value of k of a sample of 2.36 in. height and 7.75 in^2 cross-sectional area, if a quantity of water of 26.33 in^3 flows down in 10 min under an effective constant head of 15.75 in. On oven drying, the test specimen weighed 1.1 lb. Assuming $G_s = 2.65$, calculate the seepage velocity of water during the test.

Solution

$$\text{From Eq. (4.9),}\quad k = \frac{QL}{hAt} = \frac{26.33 \times 2.36}{15.75 \times 7.75 \times 10 \times 60} = 0.8 \times 10^{-3} \text{ in./sec}$$

$$\text{Discharge velocity,}\quad v = ki = k\frac{h}{L} = 0.8 \times 10^{-3} \times \frac{15.75}{2.36} = 5.34 \times 10^{-3} \text{ in./sec}$$

$$\gamma_d = \frac{W_s}{V} = \frac{1.1}{7.75 \times 2.36} = 0.0601 \text{ lb/in}^3 = 103.9 \text{ lb/ft}^3$$

$$\text{From Eq. (3.18a),}\quad e = \frac{\gamma_w G_s}{\gamma_d} - 1$$

$$\text{or } e = \frac{62.4 \times 2.65}{103.9} - 1 = 0.5915$$

$$n = \frac{e}{1+e} = \frac{0.5915}{1+0.5915} = 0.372$$

$$\text{Seepage velocity,}\quad v_s = \frac{v}{n} = \frac{5.34 \times 10^{-3}}{0.372} = 14.35 \times 10^{-3} \text{ in./sec}$$

Example 4.4

The hydraulic conductivity of a soil sample was determined in a soil mechanics laboratory by making use of a falling head permeameter. The data used and the test results obtained were as follows: diameter of sample = 2.36 in, height of sample = 5.91 in, diameter of stand pipe = 0.79 in, initial head h_0 = 17.72 in. final head h_1 = 11.81 in. Time elapsed = 1 min 45 sec. Determine the hydraulic conductivity in ft/day.

Solution

The formula for determining k is [Eq. (4.13)]

$$k = \frac{2.3aL}{At}\log_{10}\frac{h_0}{h_1} \quad \text{where } t \text{ is the elapsed time.}$$

$$\text{Area of stand pipe,}\quad a = \frac{3.14 \times 0.79 \times 0.79}{4 \times 12 \times 12} = 34 \times 10^{-4} \text{ ft}^2$$

$$\text{Area of sample,}\quad A = \frac{3.14 \times 2.36 \times 2.36}{4 \times 12 \times 12} = 304 \times 10^{-4} \text{ ft}^2$$

$$\text{Height of sample,}\quad L = \frac{(17.72 - 11.81)}{12} = 0.4925 \text{ ft}$$

$$\text{Head,}\quad h_0 = \frac{17.72}{12} = 1.477 \text{ ft},\quad h_1 = \frac{11.81}{12} = 0.984 \text{ ft}$$

Elapsed time, $t = 105 \text{ sec} = \dfrac{105}{60 \times 60 \times 24} = 12.15 \times 10^{-4}$ days

$$k = \frac{2.3 \times 34 \times 10^{-4} \times 0.4925}{304 \times 10^{-4} \times 12.15 \times 10^{-4}} \times \log \frac{1.477}{0.984} = 18 \text{ ft/day}$$

4.7 DIRECT DETERMINATION OF *k* OF SOILS IN PLACE BY PUMPING TEST

The most reliable information concerning the permeability of a deposit of coarse grained material below the water table can usually be obtained by conducting pumping tests in the field. Although such tests have their most extensive application in connection with dam foundations, they may also prove advisable on large bridge or building foundation jobs where the water table must be lowered.

The arrangement consists of a test well and a series of observation wells. The test well is sunk through the permeable stratum up to the impermeable layer. A well sunk into a water bearing stratum, termed an *aquifer*, and tapping free flowing ground water having a free ground water table under atmospheric pressure, is termed a *gravity* or *unconfined well*. A well sunk into an aquifer where the ground water flow is confined between two impermeable soil layers, and is under pressure greater than atmospheric, is termed as *artesian* or *confined well*. Observation wells are drilled at various distances from the test or pumping well along two straight lines, one oriented approximately in the direction of ground water flow and the other at right angles to it. A minimum of two observation wells and their distances from the test well are needed. These wells are to be provided on one side of the test well in the direction of the ground water flow.

The test consists of pumping out water continuously at a uniform rate from the test well until the water levels in the test and observation wells remain stationary. When this condition is achieved the water pumped out of the well is equal to the inflow into the well from the surrounding strata. The water levels in the observation wells and the rate of water pumped out of the well would provide the necessary additional data for the determination of k.

As the water from the test well is pumped out, a steady state will be attained when the water pumped out will be equal to the inflow into the well. At this stage the depth of water in the well will remain constant. The drawdown resulting due to pumping is called the *cone of depression*. The maximum drawdown D_0 is in the test well. It decreases with the increase in the distance from the test well. The depression dies out gradually and forms theoretically, a circle around the test well called the *circle of influence*. The radius of this circle, R_i, is called the *radius of influence* of the depression cone.

Equation for *k* for an Unconfined Aquifer

Figure 4.6 gives the arrangement of test and observation wells for an unconfined aquifer. Only two observation wells at radial distances of r_1 and r_2 from the test well are shown. When the inflow of water into the test well is steady, the depths of water in these observation wells are h_1 and h_2 respectively.

Let h be the depth of water at radial distance r. The area of the vertical cylindrical surface of radius r and depth h through which water flows is

$$A = 2\pi rh$$

The hydraulic gradient is $i = \dfrac{dh}{dr}$

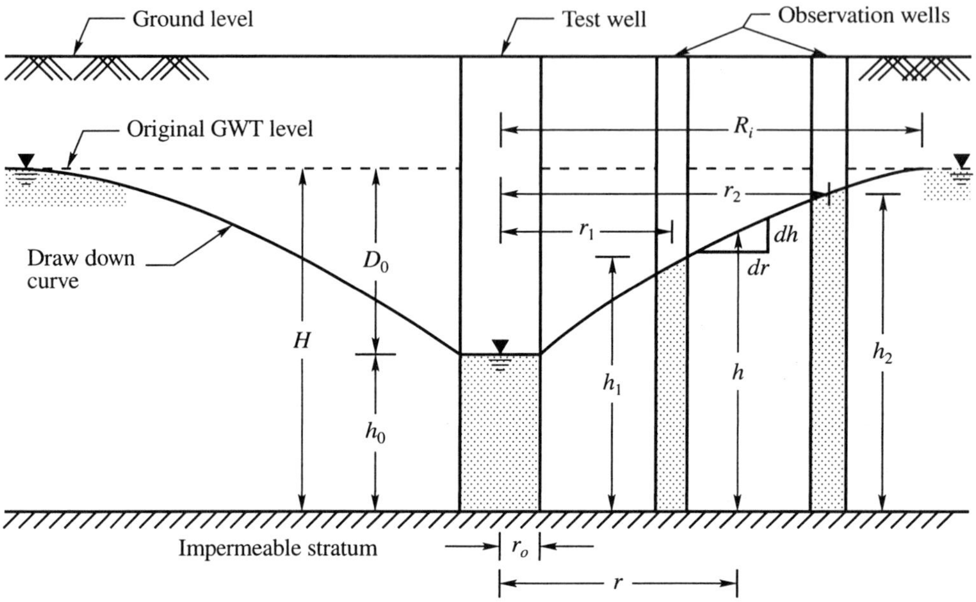

Figure 4.6 Pumping test in an unconfined aquifer

As per Darcy's law the rate of inflow into the well when the water levels in the wells remain stationary is

$$q = kiA$$

Substituting for A and i the rate of inflow across the cylindrical surface is

$$q = k\frac{dh}{dr}2\pi rh$$

Rearranging the terms, we have

$$\frac{dr}{r} = \frac{2\pi khdh}{q}$$

The integral of the equation within the boundary limits is

$$\int_{r_1}^{r_2}\frac{dr}{r} = \frac{2\pi k}{q}\int_{h_1}^{h_2} h\,dh \tag{4.15}$$

The equation for k after integration and rearranging is

$$k = \frac{2.3q}{\pi(h_2^2 - h_1^2)}\log\frac{r_2}{r_1} \tag{4.16}$$

Proceeding in the same way as before another equation for k in terms of r_o, h_o and R_i can be established as (referring to Fig. 4.6)

$$k = \frac{2.3q}{\pi(H^2 - h_0^2)} \log \frac{R_i}{r_0} \tag{4.17}$$

If we write $h_0 = (H - D_0)$ in Eq. (4.17), where D_0 is the depth of maximum drawdown in the test well, we have

$$k = \frac{2.3q}{\pi D_0 (2H - D_0)} \log \frac{R_i}{r_0} \tag{4.18}$$

Now from Eq. (4.18), the maximum yield from the well may be written as

$$q = \frac{\pi D_0 k (2H - D_0)}{2.3} \frac{1}{\log(R_i / r_0)} \tag{4.19}$$

Radius of Influence R_i: Based on experience, Sichardt (1930) gave an equation for estimating the radius of influence for the stabilized flow condition as

$$R_i = 3000 D_0 \sqrt{k} \text{ meters} \tag{4.20}$$

where D_0 = maximum drawdown in meters
k = hydraulic conductivity in m/sec

Equation for *k* in a Confined Aquifer

Figure 4.7 shows a confined aquifer with the test and observation wells. The water in the observation wells rises above the top of the aquifer due to artesian pressure. When pumping from such an artesian well two cases might arise. They are:

Case 1. The water level in the test well might remain above the roof level of the aquifer at steady flow condition.

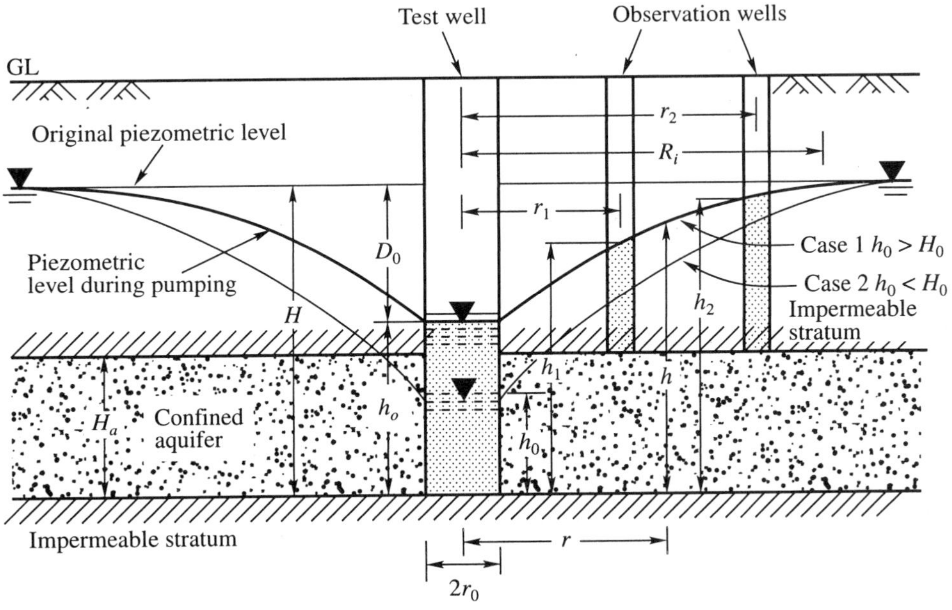

Figure 4.7 Pumping test in confined aquifer

Case 2. The water level in the test well might fall below the roof level of the aquifer at steady flow condition.

If H_0 is the thickness of the confined aquifer and h_0 is the depth of water in the test well at the steady flow condition Case 1 and Case 2 may be stated as—

Case 1. When $h_0 > H_0$. **Case 2.** When $h_0 < H_0$.

Case 1. When $h_0 > H_0$

In this case, the area of a vertical cylindrical surface of any radius *r* does not change, since the depth of the water bearing strata is limited to the thickness H_0. Therefore, the discharge surface area is

$$A = 2\pi r H_0 \tag{4.21}$$

Again writing $i = \frac{dh}{dr}$, the flow equation as per Darcy's law is

$$q = kiA = k\frac{dh}{dr}2\pi r H_0$$

The integration of the equation after rearranging the terms yields

$$\int_{h_1}^{h_2} dh = \frac{q}{2\pi k H_0}\int_{r_1}^{r_2}\frac{dr}{r} \quad \text{or} \quad (h_2 - h_1) = \frac{q}{2\pi H_0}\log_e\frac{r_2}{r_1} \tag{4.22}$$

The equation for *k* is

$$k = \frac{2.3q}{2\pi H_0 (h_2 - h_1)}\log\frac{r_2}{r_1} \tag{4.23}$$

Alternate Equations

As before we can write the following equation for determining *k*

$$k = \frac{2.3q}{2\pi H_0 (h_1 - h_0)}\log\frac{r_1}{r_0} \tag{4.24a}$$

$$\text{or} \quad k = \frac{2.3q}{2\pi H_0 (H - h_0)}\log\frac{R_i}{r_0} \tag{4.24b}$$

$$\text{or} \quad k = \frac{2.3q}{2\pi H_0 D_0}\log\frac{R_i}{r_0} \tag{4.24c}$$

Case 2. When $h_0 < H_0$

Under the condition when h_0 is less than H_0, the flow pattern close to the well is similar to that of an unconfined aquifer whereas at distances farther from the well the flow is artesian. Muskat (1946) developed an equation to determine the hydraulic conductivity. The equation is

$$k = \frac{2.3q}{\pi(2HH_0 - H_0^2 - h_0^2)}\log\frac{R_i}{r_0} \tag{4.25}$$

4.8 BOREHOLE PERMEABILITY TESTS

Two types of tests may be carried out in auger holes for determining k. They are

(a) Falling water level method
(b) Rising water level method

Falling Water Level Method (cased hole and soil flush with bottom)

In this test auger holes are made in the field that extend below the water table level. Casing is provided down to the bottom of the hole (Fig. 4.8(a)). The casing is filled with water which is then allowed to seep into the soil. The rate of drop of the water level in the casing is observed by measuring the depth of the water surface below the top of the casing at 1, 2 and 5 minutes after the start of the test and at 5 minutes intervals thereafter. These observations are made until the rate of drop becomes negligible or until sufficient readings have been obtained. The coefficient of permeability is computed as [Fig. 4.8(a)]

$$k = \frac{2.3\,\pi r_0}{5.5(t_2 - t_1)} \log \frac{H_1}{H_2} \tag{4.26}$$

where, H_1 = piezometric head at $t = t_1$, H_2 = piezometric head at $t = t_2$.

Rising Water Level Method (cased hole and soil flush with bottom)

This method, most commonly referred to as the time-lag method, consists of bailing the water out of the casing and observing the rate of rise of the water level in the casing at intervals until the rise in water level becomes negligible. The rate is observed by measuring the elapsed time and the depth of the water surface below the top of the casing. The intervals at which the readings are required will vary somewhat with the permeability of the soil. Eq. (4.26) is applicable for this case, [Fig. 4.8(b)]. A rising water level test should always be followed by sounding the bottom of the holes to determine whether the test created a quick condition.

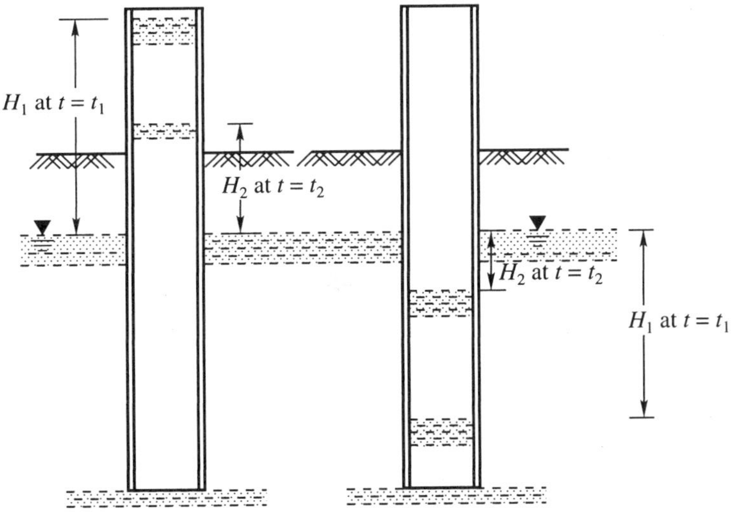

(a) Falling water head method (b) Rising water head method

Figure 4.8 Falling and rising water method of determining *k*

4.9 APPROXIMATE VALUES OF THE HYDRAULIC CONDUCTIVITY OF SOILS

The coefficients of permeability of soils vary according to their type, textural composition, structure, void ratio and other factors. Therefore, no single value can be assigned to a soil purely on the basis of soil type. The possible coefficients of permeability of some soils are given in Table 4.1

4.10 HYDRAULIC CONDUCTIVITY IN STRATIFIED LAYERS OF SOILS

Hydraulic conductivity of a disturbed sample may be different from that of the undisturbed sample even though the void ratio is the same. This may be due to a change in the structure or due to the stratification of the undisturbed soil or a combination of both of these factors. In nature we may find fine grained soils having either flocculated or dispersed structures. Two fine-grained soils at the same void ratio, one dispersed and the other flocculated, will exhibit different permeabilities.

Soils may be stratified by the deposition of different materials in layers which possess different permeability characteristics. In such stratified soils engineers desire to have the average permeability either in the horizontal or vertical directions. The average permeability can be computed if the permeabilities of each layer are determined in the laboratory. The procedure is as follows:

$k_1, k_2, \ldots, k_n$ = hydraulic conductivities of individual strata of soil either in the vertical or horizontal direction.

$z_1, z_2 \ldots z_n$ = thickness of the corresponding strata.

$$Z = z_1 + z_2 + \cdots + z_n$$

k_h = average hydraulic conductivity parallel to the bedding planes (usually horizontal).

k_v = average hydraulic conductivity perpendicular to the bedding planes (usually vertical).

Flow in the Horizontal Direction (Fig. 4.9)

When the flow is in the horizontal direction the hydraulic gradient i remains the same for all the layers. Let $v_1, v_2, \ldots, v_n$ be the discharge velocities in the corresponding strata. Then

$$Q = kiZ = (v_1 z_1 + v_2 z_2 + \cdots + v_n z_n) \quad = (k_1 i z_1 + k_2 i z_2 + \cdots + k_n i z_n)$$

Therefore,

$$k_h = \frac{1}{Z}(k_1 z_1 + k_2 z_2 + \cdots + k_n z_n) \tag{4.27}$$

Table 4.1 Hydraulic conductivity of some soils (after Casagrande and Fadum, 1939)

k (cm/sec)	Soils type	Drainage conditions
10^1 to 10^2	Clean gravels	Good
10^1	Clean sand	Good
10^{-1} to 10^{-4}	Clean sand and gravel mixtures	Good
10^{-5}	Very fine sand	Poor
10^{-6}	Silt	Poor
10^{-7} to 10^{-9}	Clay soils	Practically impervious

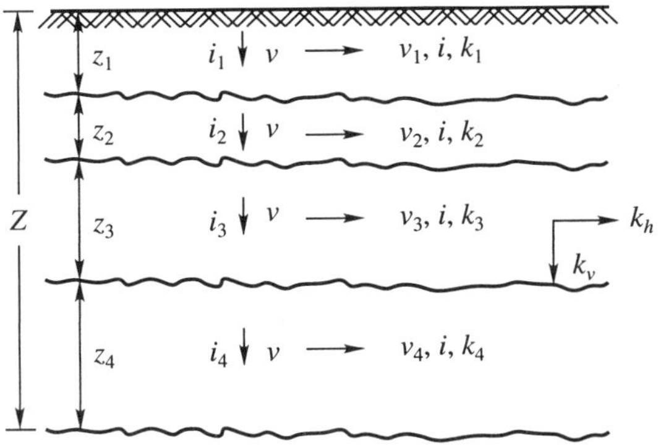

Figure 4.9 Flow through stratified layers of soil

Flow in the Vertical Direction

When flow is in the vertical direction, the hydraulic gradients for each of the layers are different. Let these be denoted by $i_1, i_2, \ldots, i_n$. Let h be the total loss of head as the water flows from the top layer to the bottom through a distance of Z. The average hydraulic gradient is h/Z. The principle of continuity of flow requires that the downward velocity be the same in each layer. Therefore,

$$v = k_v \frac{h}{Z} = k_1 i_1 = k_2 i_2 = \cdots = k_n i_n$$

If $h_1, h_2, \ldots, h_n$, are the head losses in each of the layers, we have

$$h = h_1 + h_2 + \cdots + h_n$$

or $h = z_1 i_1 + z_2 i_2 + \cdots + z_n i_n$

Solving the above equations we have

$$k_v = \frac{Z}{\dfrac{z_1}{k_1} + \dfrac{z_2}{k_2} + \cdots + \dfrac{z_n}{k_n}} \tag{4.28}$$

It should be noted that in all stratified layers of soils the horizontal permeability is generally greater than the vertical permeability. Varved clay soils exhibit the characteristics of a layered system. However, loess deposits possess vertical permeability greater than the horizontal permeability.

4.11 EMPIRICAL CORRELATIONS FOR HYDRAULIC CONDUCTIVITY

Granular Soils

Some of the factors that affect the permeability are interrelated such as grain size, void ratio, etc. The smaller the grain size, the smaller the voids which leads to the reduced size of flow channels and lower permeability.

The average velocity of flow in a pore channel from Eq. (4.2b) is

$$v_a = \frac{q}{a} = \frac{\gamma_w R_i^2}{8\mu} = \frac{\gamma_w i d^2}{32\mu} \tag{4.29}$$

where d is the average diameter of a pore channel equal to $2R$.

Eq. (4.29) expresses for a given hydraulic gradient i, the velocity of water in a circular pore channel is proportional to the square of the diameter of the pore channel. The average diameter of the voids in a soil at a given porosity increases practically in proportion to the grain size, D

Extensive investigations of filter sands by Hazen (1892) led to the equation

$$k(\text{m/s}) = CD_e^{\,2} \tag{4.30}$$

where D_e is a characteristic effective grain size which was determined to be equal to D_{10} (10% size). Fig. 4.10 gives a relationship between k and effective grain size D_{10} of granular soil which validates Eq. (4.30). The permeability data approximates a straight line with a slope equal to 2 consistent with Eq. (4.30). These data indicate an average value of $C = 10^{-2}$ where k is expressed in m/s and D_{10} in mm. According to the data in Fig. 4.10, Eq. (4.30) may underestimate or overestimate the permeability of granular soils by a factor of about 2.

Further investigations on filter sands were carried out by Kenney et al., (1984). They found the effective grain size D_5 would be a better choice compared to D_{10}. Fig. 4.11 gives relationships between D_5 and k. The sand they used in the investigation had a uniformity coefficient ranging from 1.04 to 12.

Hydraulic Conductivity as a Function of Void Ratio for Granular Soils

Further analysis of hydraulic conductivity in granular soils based on Hagen-Poiseuille's Eq. (4.2b) leads to interesting relationships between k and void ratio e. Three types of relationships may be expressed as follows.

It can be shown that the hydraulic conductivity k can be expressed as

$$k = \bar{k}F(e) \tag{4.31}$$

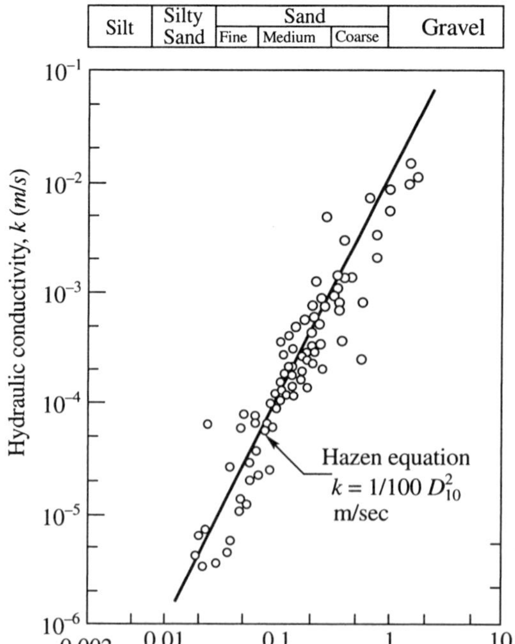

Figure 4.10 Hazen equation and data relating hydraulic conductivity and D_{10} of granular soils (after Louden, 1952)

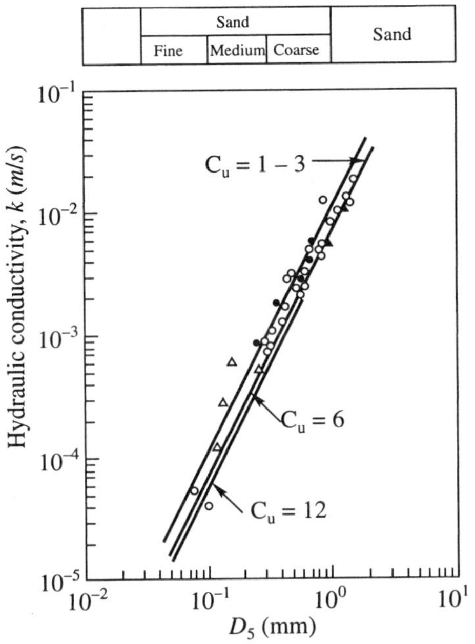

Figure 4.11 Influence of gradation on permeability on granular soils (after Kenney et al., 1984)

where $\bar{k}$ = a soil constant depending on temperature and void ratio e.

$F(e)$ may be expressed as

$$F(e) = \frac{2e^{2(1+x)}}{1+e} \tag{4.32}$$

When $e = 1$, $F(e) = 1$. Therefore $\bar{k}$ represents the hydraulic conductivity corresponding to void ratio $e = 1$. Since $\bar{k}$ is assumed as a constant, k is a function of e only.

By substituting in $F(e)$, the limiting values, $x = 0$, $x = 0.25$, and $x = 0.5$, we get

For $x = 0$,
$$F_1(e) = \frac{2e^2}{1+e} \tag{4.33}$$

$x = 0.25$,
$$F_2(e) = \frac{2e^{5/2}}{1+e} \tag{4.34}$$

$x = 0.50$
$$F_3(e) = \frac{2e^3}{1+e} \tag{4.35}$$

F_2(e) represents the geometric mean of $F_1(e)$ and $F_3(e)$

$$F_2(e) = \sqrt{F_1(e)F_3(e)}$$

The arithmetic mean of the functions $F_1(e)$ and $F_3(e)$ is

$$F_4(e) = \frac{e^2(1+e)}{1+e} = e^2 \tag{4.36}$$

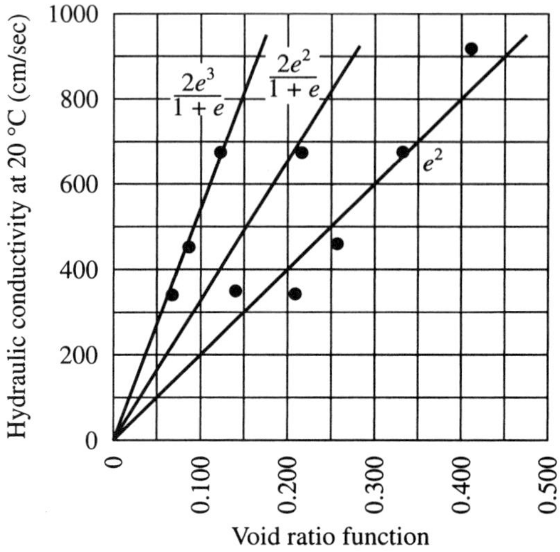

Figure 4.12 Relationship between void ratio and permeability for coarse grained soils

Best Value for *x* for Coarse Grained Soils

From laboratory tests determine k for various void ratios e of the sample. Then plot curves k versus $2e^{2(1+x)}/(1+e)$ for values of $x = 0, 0.25, 0.5$ and k versus e^2. The plot that fits well gives the best value of x. It has been found from experimental results that the function

$$F_3(e) = \frac{2e^3}{1+e} \tag{4.37}$$

gives better agreement than the other functions. However, the function $F_4(e) = e^2$ is sometimes preferred because of its simplicity and its fair degree of agreement with the experimental data. Fig. 4.12 present experimental data in the form of k versus functions of e.

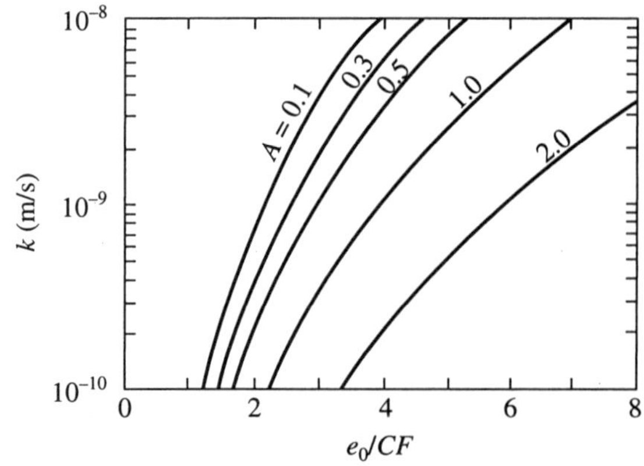

Figure 4.13 *In situ* permeability of soft clays in relation to initial void ratio, e_o; clay fraction; *CF*; and activity *A* (After Mesri et al., 1994)

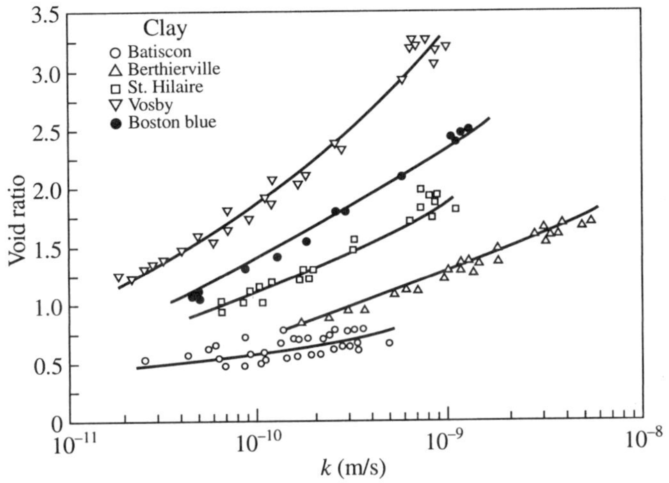

Figure 4.14 Results of falling-head and constant-head permeability tests on undisturbed samples of soft clays (Terzaghi, Peck and Mesri, 1996)

Fine Grained Soils

Laboratory experiments have shown that hydraulic conductivity of very fine grained soils are not strictly a function of void ratio since there is a rapid decrease in the value of k for clays below the plastic limit. This is mostly due to the much higher viscosity of water in the normal channels which results from the fact that a considerable portion of water is exposed to large molecular attractions by the closely adjacent solid matter. It also depends upon the fabric of clays especially those of marine origin which are often flocculated. Fig. 4.13 shows that the hydraulic conductivity in the vertical direction, at *in situ* void ratio e_0, is correlated with clay fraction (CF) finer than 0.002 mm and with the activity A ($= I_p/CF$).

Consolidation of soft clays may involve a significant decrease in void ratio and therefore of permeability. The relationships between e and k (log-scale) for a number of soft clays are shown in Fig. 4.14 (Terzaghi, Peck, and Mesri 1996).

Example 4.5

A pumping test was carried out for determining the hydraulic conductivity of soil in place. A well of diameter 40 cm was drilled down to an impermeable stratum. The depth of water above the bearing stratum was 8 m. The yield from the well was 4 m^3/min at a steady drawdown of 4.5 m. Determine the hydraulic conductivity of the soil in m/day if the observed radius of influence was 150 m.

Solution

The formula for determining k is [Eq. (4.18)]

$$k = \frac{2.3q}{\pi D_0 (2H - D_0)} \log \frac{R_i}{r_0}$$

$$q = 4 \text{ m}^3/\text{min} = 4 \times 60 \times 24 \text{ m}^3/\text{day}$$

$$D_0 = 4.5 \text{ m}, \quad H = 8 \text{ m}, \quad R_i = 150 \text{ m}, \quad r_0 = 0.2 \text{ m}$$

$$k=\frac{2.3\times 4\times 60\times 24}{3.14\times 4.5(2\times 8-4.5)}\log\frac{150}{0.2}=234.4 \text{ m/day}$$

Example 4.6

A pumping test was made in pervious gravels and sands extending to a depth of 50 ft, where a bed of clay was encountered. The normal ground water level was at the ground surface. Observation wells were located at distances of 10 and 25 ft from the pumping well. At a discharge of 761 ft^3 per minute from the pumping well, a steady state was attained in about 24 hr. The draw-down at a distance of 10 ft was 5.5 ft and at 25 ft was 1.21 ft. Compute the hydraulic conductivity in ft/sec.

Solution

Use Eq. (4.16) where

$$k=\frac{2.3q}{\pi\,(h_2^2-h_1^2)}\log\frac{r_2}{r_1}$$

where $q=\dfrac{761}{60}=12.683 \text{ ft}^3/\text{sec}$

$r_1=10$ ft, $r_2=25$ ft, $h_2=50-1.21=48.79$ ft, $h_1=50-5.5=44.5$ ft

$$k=\frac{2.3\times 12.683}{3.14(48.79^2-44.5^2)}\log\frac{25}{10}=9.2\times 10^{-3} \text{ ft/sec.}$$

Example 4.7

A field pumping test was conducted from an aquifer of sandy soil of 4 m thickness confined between two impervious strata. When equilibrium was established, 90 liters of water was pumped

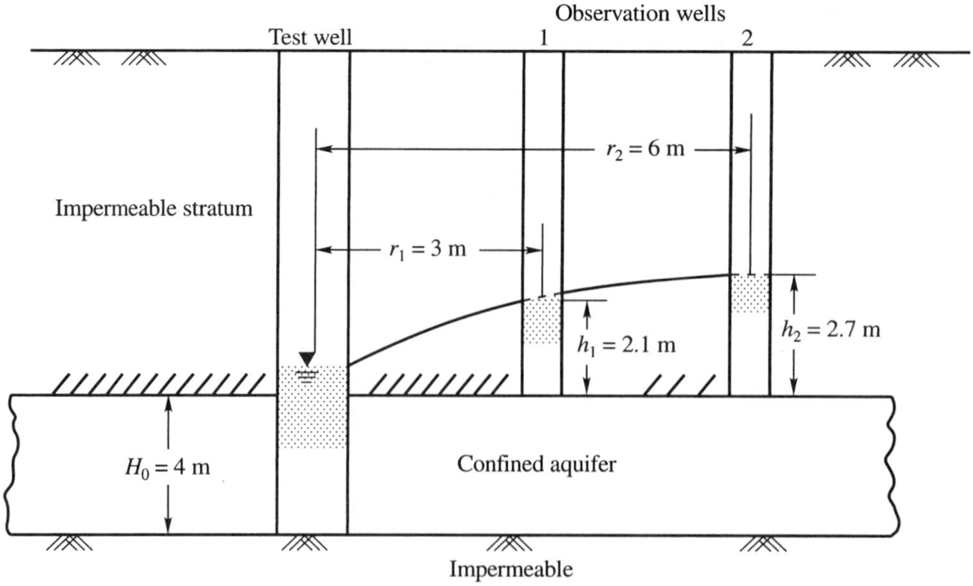

Figure Ex. 4.7

out per hour. The water elevation in an observation well 3.0 m away from the test well was 2.1 m and another 6.0 m away was 2.7 m from the roof level of the impervious stratum of the aquifer. Find the value of k of the soil in m/sec. (Fig. Ex. 4.7)

Solution

Use Eq. (4.24a)

$$k = \frac{2.3q}{2\pi H_0 (h_2 - h_1)} \log \frac{r_2}{r_1}$$

$$q = 90 \times 10^3 \text{ cm}^3/\text{hr} = 25 \times 10^{-6} \text{ m}^3/\text{sec}$$

$$k = \frac{2.3 \times 25 \times 10^{-6}}{2 \times 3.14 \times 4(2.7 - 2.1)} \log \frac{6}{3} = 1.148 \times 10^{-6} \text{ m/sec}$$

Example 4.8

Calculate the yield per hour from a well driven into a confined aquifer. The following data are available:

height of original piezometric level from the bed of the aquifer, $H = 29.53$ ft,
thickness of aquifer, $H_a = 16.41$ ft,
the depth of water in the well at steady state, $h_0 = 18.05$ ft,
hydraulic conductivity of soil = 0.079 ft/min,
radius of well, $r_0 = 3.94$ in. (0.3283 ft), radius of influence, $R_i = 574.2$ ft.

Solution

Since h_0 is greater than H_a the equation for q (refer to Fig 4.7) is Eq. (4.24b)

$$q = \frac{2\pi k H_0 (H - h_0)}{2.3 \log(R_i / r_0)}$$

where $k = 0.079$ ft/min = 4.74 ft/hr

$$\text{Now } q = \frac{2 \times 3.14 \times 4.74 \times 16.41(29.53 - 18.05)}{2.3 \log(574.2 / 0.3283)} = 751.87 \text{ ft}^3/\text{hour} \approx 752 \text{ ft}^3/\text{hour}$$

Example 4.9

A sand deposit contains three distinct horizontal layers of equal thickness (Fig. 4.9). The hydraulic conductivity of the upper and lower layers is 10^{-3} cm/sec and that of the middle is 10^{-2} cm/sec. What are the equivalent values of the horizontal and vertical hydraulic conductivities of the three layers, and what is their ratio?

Solution

Horizontal flow

$$k_h = \frac{1}{Z}(k_1 z_1 + k_2 z_2 + k_3 z_3) = \frac{1}{3}(k_1 + k_2 + k_3) \quad \text{since} \quad z_1 = z_2 = z_3$$

$$k_h = \frac{1}{3}(10^{-3} + 10^{-2} + 10^{-3}) = \frac{1}{3}(2 \times 10^{-3} + 10^{-2}) = 4 \times 10^{-3} \text{ cm/sec}$$

Vertical flow

$$k_v = \frac{Z}{\frac{z_1}{k_1} + \frac{z_2}{k_2} + \frac{z_3}{k_3}} = \frac{3}{\frac{1}{k_1} + \frac{1}{k_2} + \frac{1}{k_3}} = \frac{3}{\frac{2}{k_1} + \frac{1}{k_2}}$$

$$= \frac{3k_1k_2}{2k_2 + k_1} = \frac{3 \times 10^{-3} \times 10^{-2}}{2 \times 10^{-2} + 10^{-3}} = 1.43 \times 10^{-3} \text{ cm/sec}$$

$$\frac{k_h}{k_v} = \frac{4 \times 10^{-3}}{1.43 \times 10^{-3}} = 2.8$$

Example 4.10

The following details refer to a test to determine the value of k of a soil sample: sample thickness = 2.5 cm, diameter of soil sample = 7.5 cm, diameter of stand pipe = 10 mm, initial head of water in the stand pipe = 100 cm, water level in the stand pipe after 3 h 20 min = 80 cm. Determine the value of k if e = 0.75. What is the value of k of the same soil at a void ratio e = 0.90?

Solution

Use Eq. (4.13) where, $k = \dfrac{2.3aL}{A(t_1 - t_0)} \log \dfrac{h_0}{h_1}$

$$a = \frac{3.14}{4}(1)^2 = 0.785 \text{ cm}^2$$

$$A = \frac{3.14}{4}(7.5)^2 = 44.16 \text{ cm}^2$$

t = 12000 sec

By substituting the value of k for e_1 = 0.75

$$k = k_1 = \frac{2.3 \times 0.785 \times 2.5}{44.16 \times 12000} \times \log \frac{100}{80} = 0.826 \times 10^{-6} \text{ cm/sec}$$

For determining k at any other void ratio, use Eq. (4.35)

$$\frac{k_1}{k_2} = \frac{e_1^3}{1+e_1} \bigg/ \frac{e_2^3}{1+e_2} = \frac{1+e_2}{1+e_1} \times \left(\frac{e_1}{e_2}\right)^3$$

Now, $k_2 = \dfrac{1+e_1}{1+e_2} \times \left(\dfrac{e_2}{e_1}\right)^3 \times k_1$

For e_2 = 0.90

$$k_2 = \frac{1.75}{1.90} \times \left(\frac{0.9}{0.75}\right)^3 \times 0.826 \times 10^{-6} = 1.3146 \times 10^{-6} \text{ cm/sec}$$

Example 4.11

In a falling head permeameter, the sample used is 20 cm long having a cross-sectional area of 24 cm^2. Calculate the time required for a drop of head from 25 to 12 cm if the cross-sectional area of the stand pipe is 2 cm^2. The sample of soil is made of three layers. The thickness of the first layer from the top is 8 cm and has a value of $k_1 = 2 \times 10^{-4}$ cm/sec, the second layer of thickness 8 cm has $k_2 = 5 \times 10^{-4}$ cm/sec and the bottom layer of thickness 4 cm has $k_3 = 7 \times 10^{-4}$ cm/sec. Assume that the flow is taking place perpendicular to the layers (Fig. Ex. 4.11).

Solution

Use Eq. (4.28)

$$k_v = \frac{Z}{\frac{z_1}{k_1} + \frac{z_2}{k_2} + \frac{z_3}{k_3}} = \frac{20}{\frac{8}{2 \times 10^{-4}} + \frac{8}{5 \times 10^{-4}} + \frac{4}{7 \times 10^{-4}}} = 3.24 \times 10^{-4} \text{ cm/sec}$$

Now from Eq. (4.13), $k = \dfrac{2.3aL}{A(t_1 - t_0)} \log \dfrac{h_0}{h_1}$

$$\text{or} \quad (t_1 - t_0) = t = \frac{2.3aL}{Ak} \log \frac{h_0}{h_1} = \frac{2.3 \times 2 \times 20}{24 \times 3.24 \times 10^{-4}} \log \frac{25}{12}$$

$$= 3771 \text{ sec} = 62.9 \text{ minutes}$$

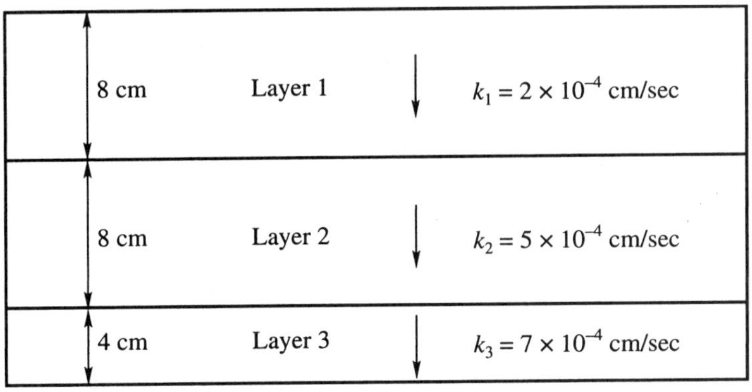

Figure Ex. 4.11

Example 4.12

The data given below relate to two falling head permeameter tests performed on two different soil samples:

(a) stand pipe area = 4 cm^2, (b) sample area = 28 cm^2, (c) sample height = 5 cm, (d) initial head in the stand pipe = 100 cm, (e) final head = 20 cm, (f) time required for the fall of water level in test 1, t = 500 sec, (g) for test 2, t = 15 sec.

Determine the values of k for each of the samples. If these two types of soils form adjacent layers in a natural state with flow (a) in the horizontal direction, and (b) flow in the vertical direction, determine the equivalent permeability for both the cases by assuming that the thickness of each layer is equal to 150 cm.

Solution

Use Eq. (4.13)

$$k = \frac{2.3aL}{At}\log\frac{h_1}{h_2}$$

For test 1

$$k_1 = \frac{2.3 \times 4 \times 5}{28 \times 500}\log\frac{100}{20} = 2.3 \times 10^{-3} \text{ cm/sec}$$

For test 2

$$k_2 = \frac{2.3 \times 4 \times 5}{28 \times 15}\log\frac{100}{20} = 76.7 \times 10^{-3} \text{ cm/sec}$$

Flow in the horizontal direction

Use Eq. (4.27)

$$k_h = \frac{1}{Z}(k_1 z_1 + k_2 z_2) = \frac{1}{300}(2.3 \times 150 + 76.7 \times 150) \times 10^{-3} = 39.5 \times 10^{-3} \text{ cm/sec}$$

Flow in the vertical direction

Use Eq. (4.28)

$$k_v = \frac{Z}{\dfrac{z_1}{k_1} + \dfrac{z_2}{k_2}} = \frac{300}{\dfrac{150}{2.3 \times 10^{-3}} + \dfrac{150}{76.7 \times 10^{-3}}} = 4.46 \times 10^{-3} \text{ cm/sec}$$

4.12 HYDRAULIC CONDUCTIVITY OF ROCKS BY PACKER METHOD

Packers are primarily used in bore holes for testing the permeability of rocks under applied pressures. The apparatus used for the pressure test is comprised of a water pump, a manually adjusted automatic pressure relief valve, pressure gage, a water meter and a packer assembly. The packer assembly consists of a system of piping to which two expandable cylindrical rubber sleeves, called packers, are attached. The packers which provide a means of sealing a limited section of bore hole for testing, should have a length five times the diameter of the hole. They may be of the pneumatically or mechanically expandable type. The former are preferred since they adapt to an oversized hole whereas the latter may not. However, when pneumatic packers are used, the test apparatus must also include an air or water supply connected, through a pressure gage, to the packers by means of a higher pressure hose. The piping of a packer assembly is designed to permit testing of either the portion of the hole between the packers or the portion below the lower packer. The packers are usually set 50, 150 or 300 cm apart. The wider spacings are used for rock which is more uniform. The short spacing is used to test individual joints which may be the cause of high water loss in otherwise tight strata.

Two types of packer methods are used for testing of permeability. They are:

1. Single packer method.
2. Double packer method.

The single packer method is useful where the full length of the bore hole cannot stand uncased/ungrouted in soft rocks, such as soft sand stone, clay shale or due to the highly fractured and sheared nature of the rocks, or where it is considered necessary to have permeability values side by side with drilling. Where the rocks are sound and the full length of the hole can stand without casing/grouting, the double packer method may be adopted. The disadvantage of the double packer method is that leakage through the lower packer can go unnoticed and lead to overestimation of water loss.

Single Packer Method

The method used for performing water percolation tests in a section of a drilled hole using a single packer is shown in Fig. 4.15a. In this method the hole should be drilled to a particular depth desirable for the test. The core barrel should then be removed and the hole cleaned with water. The packer should be fixed at the desired level above the bottom of the hole and the test performed. Water should be pumped into the section under pressure. Each pressure should be maintained until the readings of water intake at intervals of 5 min show a nearly constant reading of water intake for one particular pressure. The constant rate of water intake should be noted. After performing the test the entire assembly should be removed. The drilling should then proceed for the next test section.

Double Packer Method

In this method the hole is first drilled to the final depth and cleaned. The packer assembly may be fixed at any desired test section as shown in Fig. 4.15b.

Both packers are then expanded and water under pressure is introduced into the hole between the packers. The tests are conducted as before.

Regardless of which procedure is used, a minimum of three pressures should be used for each section tested. The magnitude of these pressures are commonly 100, 200 and 300 kPa. (1, 2 and 3 kg/cm^2) above the natural piezometric level. However in no case should the excess pressure be greater than about 20 kPa per meter of soil and rock overburden above the upper packer. The limitation is imposed to insure against possible heavy damage to the foundation.

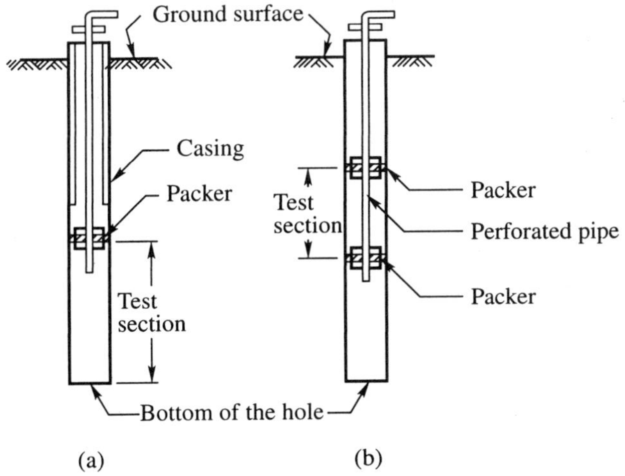

Figure 4.15 Test sections for single and double packer methods

The formulae used to compute the permeability from pressure test data are (from US Bureau of Reclamation, 1968)

$$k = \frac{q}{2\pi LH}\log_e \frac{L}{r_0} \quad \text{for } L \geq 10r_0 \tag{4.38a}$$

$$k = \frac{q}{2\pi LH}\sinh^{-1}\frac{L}{2r_0} \quad \text{for } 10r_0 > L \geq r_0 \tag{4.38b}$$

where

k = hydraulic conductivity

q = constant rate of flow into the hole

L = length of the test section

H = differential head on the test section

r_0 = radius of the bore hole

4.13 SEEPAGE

The interaction between soils and percolating water has an important influence on:

1. The design of foundations and earth slopes,
2. The quantity of water that will be lost by percolation through a dam or its subsoil.

Foundation failures due to '*piping*' are quite common. Piping is a phenomenon by which the soil on the downstream sides of some hydraulic structures get lifted up due to excess pressure of water. The pressure that is exerted on the soil due to the seepage of water is called the *seepage force or pressure*. In the stability of slopes, the seepage force is a very important factor. Shear strengths of soils are reduced due to the development of neutral stress or pore pressures. A detailed understanding of the hydraulic conditions is therefore essential for a satisfactory design of structures.

The computation of seepage loss under or through a dam, the uplift pressures caused by the water on the base of a concrete dam and the effect of seepage on the stability of earth slopes can be studied by constructing flow nets.

Flow Net

A flow net for an isometric medium is a network of flow lines and equipotential lines intersecting at right angles to each other.

The path which a particle of water follows in its course of seepage through a saturated soil mass is called a *flow line.*

Equipotential lines are lines that intersect the flow lines at right angles. At all points along an equipotential line, the water would rise in piezometric tubes to the same elevation known as the *piezometric head* . Fig. 4.16 gives a typical example of a flow net for the flow below a sheet pile wall. The head of water on the upstream side of the sheet pile is h_t and on the downstream side h_d. The head lost as the water flows from the upstream to the downstream side is h.

4.14 LAPLACE EQUATION

Figure 4.16(a) illustrates the flow of water along curved lines which are parallel to the section shown. The figure represents a section through an impermeable diaphragm extending to a depth D below the horizontal surface of a homogeneous stratum of soil of depth H.

It is assumed that the difference h between the water levels on the two sides of the diaphragm is constant. The water enters the soil on the upstream side of the diaphragm, flows in a downward direction and rises on the downstream side towards the surface.

Consider a prismatic element P shown shaded in Fig. 4.16(a) which is shown on a larger scale in (b). The element is a parallelopiped with sides dx, dy and dz. The x and z directions are as shown in the figure and the y direction is normal to the section. The velocity v of water which is tangential to the stream line can be resolved into components v_x and v_z in the x and z directions respectively.

Let,

$$i_x = -\frac{\partial h}{\partial x}, \text{ the hydraulic gradient in the horizontal direction.}$$

$$i_z = -\frac{\partial h}{\partial z}, \text{ the hydraulic gradient in the vertical direction.}$$

k_x = hydraulic conductivity in the horizontal direction.

k_z = hydraulic conductivity in the vertical direction.

If we assume that the water and soil are perfectly incompressible, and the flow is steady, then the quantity of water that enters the element must be equal to the quantity that leaves it.

The quantity of water that enters the side $ab = v_x dzdy$

The quantity of water that leaves the side $cd = \left(v_x + \frac{\partial v_x}{\partial x}dx\right)dzdy$

The quantity of water that enters the side $bc = v_z dxdy$

The quantity of water that leaves the side $ad = \left(v_z + \frac{\partial v_z}{\partial z}dz\right)dxdy$

Therefore, we have the equation,

$$v_x dzdy + v_z dxdy = \left(v_x + \frac{\partial v_x}{\partial x}dx\right)dzdy + \left(v_z + \frac{\partial v_z}{\partial z}dz\right)dxdy$$

After simplifying, we obtain,

$$\frac{\partial v_x}{\partial x} + \frac{\partial v_z}{\partial z} = 0 \qquad (4.39)$$

Equation (4.39) expresses the necessary condition for continuity of flow. According to Darcy's Law we may write,

$$v_x = -k_x \frac{\partial h}{\partial x}, \quad v_z = -k_z \frac{\partial h}{\partial z}$$

Substituting for v_x and v_z we obtain,

$$\frac{\partial}{\partial x}\left(-k_x \frac{\partial h}{\partial x}\right) + \frac{\partial}{\partial z}\left(-k_z \frac{\partial h}{\partial z}\right) = 0$$

$$\text{or } k_x \frac{\partial^2 h}{\partial x^2} + k_z \frac{\partial^2 h}{\partial z^2} = 0 \qquad (4.40)$$

When $k_z = k_x$, i.e., when the permeability is the same in all directions, Eq. (4.40) reduces to

$$\frac{\partial^2 h}{\partial x^2} + \frac{\partial^2 h}{\partial z^2} = 0 \tag{4.41}$$

Equation (4.41) is the *Laplace Equation* for homogeneous soil. It says that the change of gradient in the x-direction plus the change of gradient in the z-direction is zero. The solution of this equation gives a family of curves meeting at right angles to each other. One family of these curves represents flow lines and the other equipotential lines. For the simple case shown in Fig. 4.16, the flow lines represent a family of semi-ellipses and the equipotential lines semi-hyperbolas.

Anisotropic Soil

Soils in nature do possess permeabilities which are different in the horizontal and vertical directions. The permeability in the horizontal direction is greater than in the vertical direction in sedimentary deposits and in most earth embankments. In loess deposits the vertical permeability is greater than the horizontal permeability. The study of flow nets would be of little value if this variation in the permeability is not taken into account. Eq. (4.40) applies for a soil mass where anisotropy exists. This equation may be written in the form

$$\frac{\partial^2 h}{\left(\frac{k_z}{k_x}\right)\partial x^2} + \frac{\partial^2 h}{\partial z^2} = 0 \tag{4.42}$$

If we consider a new coordinate variable x_c measured in the same direction as x multiplied by a constant, expressed by

$$x_c = x_x\sqrt{\frac{k_z}{k_x}} \tag{4.43}$$

Eq. (4.42) may be written as

$$\frac{\partial^2 h}{\partial x_c^2} + \frac{\partial^2 h}{\partial z^2} = 0 \tag{4.44}$$

Now Eq. (4.44) is a *Laplace equation* in the coordinates x_c and z. This equation indicates that a cross-section through an anisotropic soil can be transformed to an imaginary section which possesses the same permeability in all directions. The transformation of the section can be effected as per Eq. (4.43) by multiplying the x-coordinates by $\sqrt{k_z/k_x}$ and keeping the z-coordinates at the natural scale. The flow net can be sketched on this transformed section. The permeability to be used with the transformed section is

$$k_e = \sqrt{k_x k_z} \tag{4.45}$$

4.15 FLOW NET CONSTRUCTION

Properties of a Flow Net

The properties of a flow net can be expressed as given below:

1. Flow and equipotential lines are smooth curves.
2. Flow lines and equipotential lines meet at right angles to each other.

3. No two flow lines cross each other.
4. No two flow or equipotential lines start from the same point.

Boundary Conditions

Flow of water through earth masses is in general three dimensional. Since the analysis of three-dimensional flow is too complicated, the flow problems are solved on the assumption that the flow is two-dimensional. All flow lines in such a case are parallel to the plane of the figure, and the condition is therefore known as two-dimensional flow. All flow studies dealt with herein are for the steady state case. The expression for boundary conditions consists of statements of head or flow conditions at all boundary points. The boundary conditions are generally four in number though there are only three in some cases. The boundary conditions for the case shown in Fig. 4.16 are as follows:

1. Line *ab* is a boundary equipotential line along which the head is h_t
2. The line along the sheet pile wall is a flow boundary
3. The line *xy* is a boundary equipotential line along which the head is equal to h_d
4. The line *m n* is a flow boundary (at depth *H* below bed level).

If we consider any flow line, say, $p_1\ p_2\ p_3$ in Fig. 4.16, the potential head at p_1 is h_t and at p_3 is h_d. The total head lost as the water flows along the line is *h* which is the difference between the upstream and downstream heads of water. The head lost as the water flows from p_1 to equipotential line *k* is Δh which is the difference between the heads shown by the piezometers. This loss of head Δh is a fraction of the total head lost.

Flow Net Construction

Flow nets are constructed in such a way as to keep the ratio of the sides of each block bounded by two flow lines and two equipotential lines a constant. If all the sides of one such block are equal, then the flow net must consist of squares. The square block referred to here does not constitute a square according to the strict meaning of the word, it only means that the average width of the square blocks are equal. For example, in Fig. 4.16, the width a_1 of block 1 is equal to its length b_1.

The area bounded by any two neighboring flow lines is called a *flow channel*. If the flow net is constructed in such a way that the ratio *a/b* remains the same for all blocks, then it can be shown that there is the same quantity of seepage in each flow channel. In order to show this consider two blocks 1 and 2 in one flow channel and another block 3 in another flow channel as shown in Fig. 4.16. Block 3 is chosen in such a way that it lies within the same equipotential lines that bound the block 2. Darcy's law for the discharge through any block such as 1 per unit length of the section may be written as

$$\Delta q = kia = \frac{\Delta h}{b} a = k\Delta h \frac{a}{b}$$

where Δh represents the head loss in crossing the block. The expressions in this form for each of the three blocks under consideration are

$$\Delta q_1 = k\Delta h \frac{a_1}{b_1}, \qquad \Delta q_2 = k\Delta h_2 \frac{a_2}{b_2}$$

In the above equation the value of hydraulic conductivity *k* remains the same for all the blocks. If the blocks are all squares then

$$\frac{a_1}{b_1} = \frac{a_2}{b_2} = \frac{a_3}{b_3} = 1$$

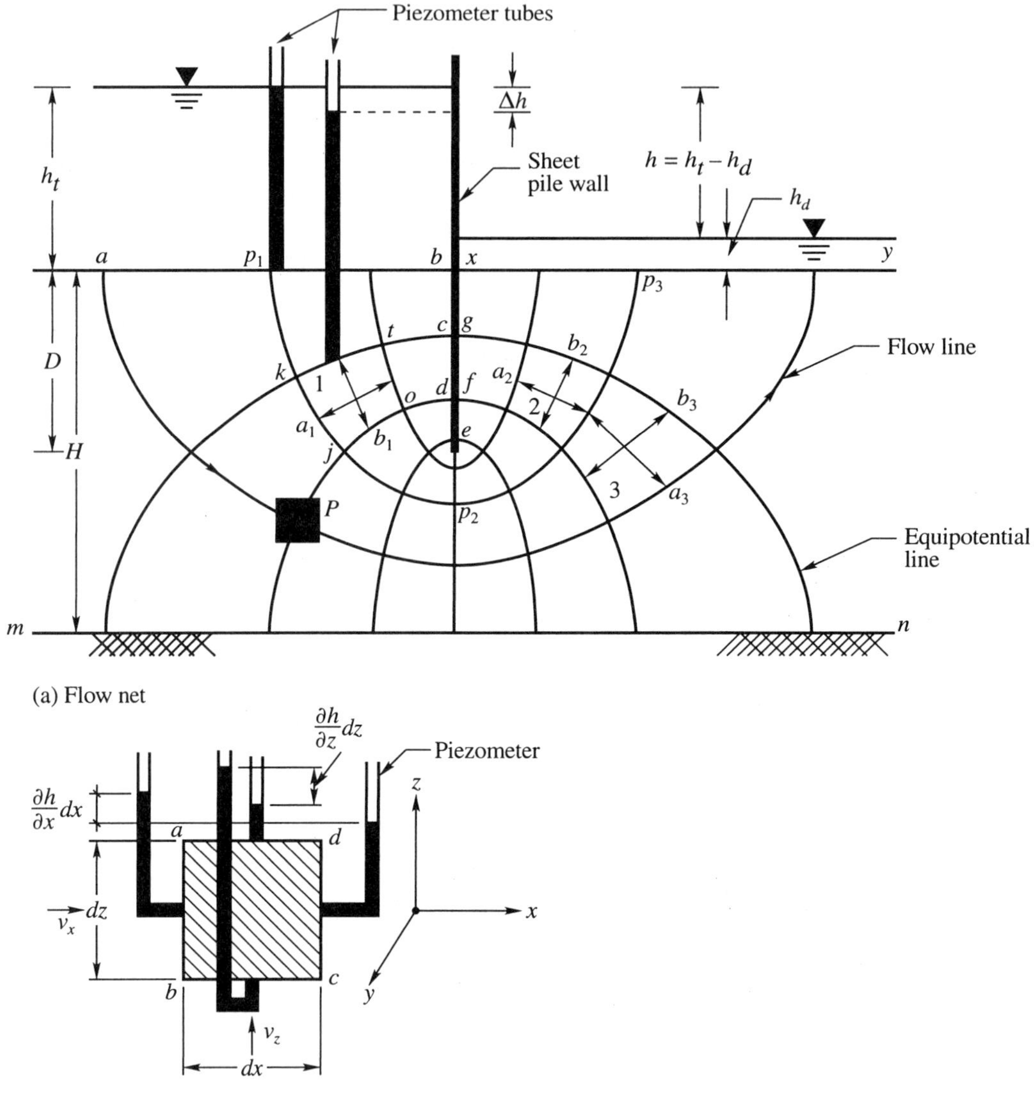

(a) Flow net

(b) Flow through a prismatic element

Figure 4.16 Flow through a homogeneous stratum of soil

Since blocks 1 and 2 are in the same flow channel, we have $\Delta q_1 = \Delta q_2$. Since blocks 2 and 3 are within the same equipotential lines we have $\Delta h_2 = \Delta h_3$. If these equations are inserted we obtain the following relationship:

$$\Delta q_1 = \Delta q_2 \text{ and } \Delta h_1 = \Delta h_2$$

This proves that the same quantity flows through each block and there is the same head drop in crossing each block if all the blocks are squares or possess the same ratio *a*/*b*. Flow nets are constructed by keeping the ratio *a*/*b* the same in all figures. Square flow nets are generally used in practice as this is easier to construct.

There are many methods that are in use for the construction of flow nets. Some of the important methods are

1. Analytical method,
2. Electrical analog method,
3. Scaled model method,
4. Graphical method.

The analytical method, based on the Laplace equation although rigorously precise, is not universally applicable in all cases because of the complexity of the problem involved. The mathematics involved even in some elementary cases is beyond the comprehension of many design engineers. Although this approach is sometimes useful in the checking of other methods, it is largely of academic interest.

The electrical analogy method has been extensively made use of in many important design problems. However, in most of the cases in the field of soil mechanics where the estimation of seepage flows and pressures are generally required, a more simple method such as the graphical method is preferred.

Scaled models are very useful to solve seepage flow problems. Soil models can be constructed to depict flow of water below concrete dams or through earth dams. These models are very useful to demonstrate the fundamentals of fluid flow, but their use in other respects is limited because of the large amount of time and effort required to construct such models.

The graphical method developed by Forchheimer (1930) has been found to be very useful in solving complicated flow problems. A. Casagrande (1937) improved this method by incorporating many suggestions. The main drawback of this method is that a good deal of practice and aptitude are essential to produce a satisfactory flow net. In spite of these drawbacks, the graphical method is quite popular among engineers.

Graphical Method

The usual procedure for obtaining flow nets is a graphical, trial sketching method, sometimes called the Forchheimer Solution. This method of obtaining flow nets is the quickest and the most practical of all the available methods. A. Casagrande (1937) has offered many suggestions to the beginner who is interested in flow net construction. Some of his suggestions are summarized below:

1. Study carefully the flow net pattern of well-constructed flow nets.
2. Try to reproduce the same flow nets without seeing them.
3. As a first trial, use not more than four to five flow channels. Too many flow channels would confuse the issue.
4. Follow the principle of 'whole to part', i.e., one has to watch the appearance of the entire flow net and when once the whole net is found approximately correct, finishing touches can be given to the details.
5. All flow and equipotential lines should be smooth and there should not be any sharp transitions between straight and curved lines.

The above suggestions, though quite useful for drawing flow nets, are not sufficient for a beginner. In order to overcome this problem, Taylor (1948) proposed a procedure known as the procedure by explicit trials. Some of the salient features of this procedure are given below:

1. As a first step in the explicit trial method, one trial flow line or one trial equipotential line is sketched adjacent to a boundary flow line or boundary equipotential.
2. After choosing the first trial line (say it is a flow line), the flow path between the line and the boundary flow line is divided into a number of squares by drawing equipotential lines.

These equipotential lines are extended to meet the bottom flow line at right angles keeping in view that the lines drawn should be smooth without any abrupt transitions.

3. The remaining flow lines are next drawn, adhering rigorously to square figures.
4. If the first trial is chosen property, the net drawn satisfies all the necessary conditions. Otherwise, the last drawn flow line will cross the bottom boundary flow line, indicating that the trial line chosen is incorrect and needs modification.
5. In such a case, a second trial line should be chosen and the procedure repeated.

A typical example of a flow net under a sheet pile wall is given in Fig. 4.16. It should be understood that the number of flow channels will be an integer only by chance. That means, the bottom flow line sketched might not produce full squares with the bottom boundary flow line. In such a case the bottom flow channel will be a fraction of a full flow channel. It should also be noted that the figure formed by the first sketched flow line with the last equipotential line in the region is of irregular form. This figure is called a *singular square*. The basic requirement for such squares, as for all the other squares, is that continuous sub-division of the figures give an approach to true squares. Such singular squares are formed at the tips of sheet pile walls also. Squares must be thought of as valid only where the Laplace equation applies. The Laplace equation applies to soils which are homogeneous and isotropic. When the soil is anisotropic, the flow net should be sketched as before on the transformed section. The transformed section can be obtained from the natural section explained earlier.

4.16 DETERMINATION OF QUANTITY OF SEEPAGE

Flow nets are useful for determining the quantity of seepage through a section. The quantity of seepage q is calculated per unit length of the section. The flow through any square can be written as

$$\Delta q = k\Delta h \tag{4.46}$$

Let the number of flow channel and equipotential drops in a section be N_f and N_d, respectively. Since all drops are equal, we can write

$$\Delta h = \frac{h}{N_d}$$

Since the discharge in each flow channel is the same we can write,

$$q = N_f \Delta q$$

Substituting for Δq and Δh, we have

$$q = kh\frac{N_f}{N_d} \tag{4.47}$$

Eq. (4.47) can also be used to compute the seepage through anisotropic sections by writing k_e in place of k. As per Eq. (4.45), k_e is equal to $\sqrt{k_x k_z}$, where k_x and k_z are the hydraulic conductivities in the x and z directions, respectively. The validity of this relationship can be proved as follows. Consider a figure bounded by flow and equipotential lines in which the flow is parallel to the x direction. In Fig. 4.17 the figure in question is drawn to a transformed scale in (b) and the same to the natural scale in (a).

In Fig. 4.17(b) the permeability has the effective value k_e in both the x and z directions and the flow through the square according to Eq. (4.46), is

$$\Delta q = k_e \frac{\Delta h a}{a} = k_e \Delta h \tag{4.48}$$

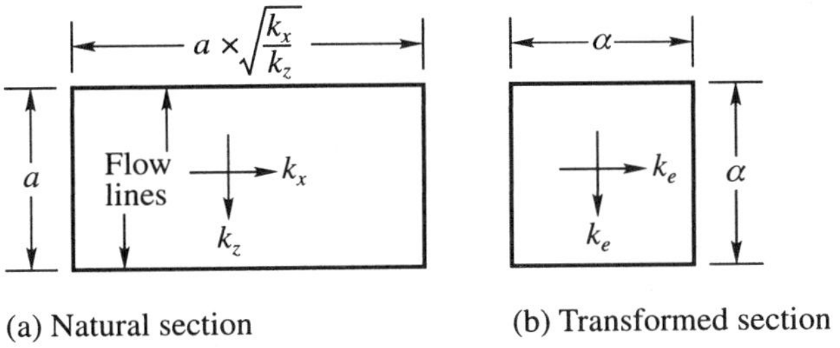

Figure 4.17 Flow through anisotropic soil

In Fig. 4.17(a) the hydraulic conductivity k_x in the horizontal section must apply because the flow is horizontal and the sketch is to the natural scale. The flow equation is, therefore,

$$\Delta q = k_x iA = k_x \frac{\Delta ha}{a \times \sqrt{\frac{k_x}{k_z}}} \tag{4.49}$$

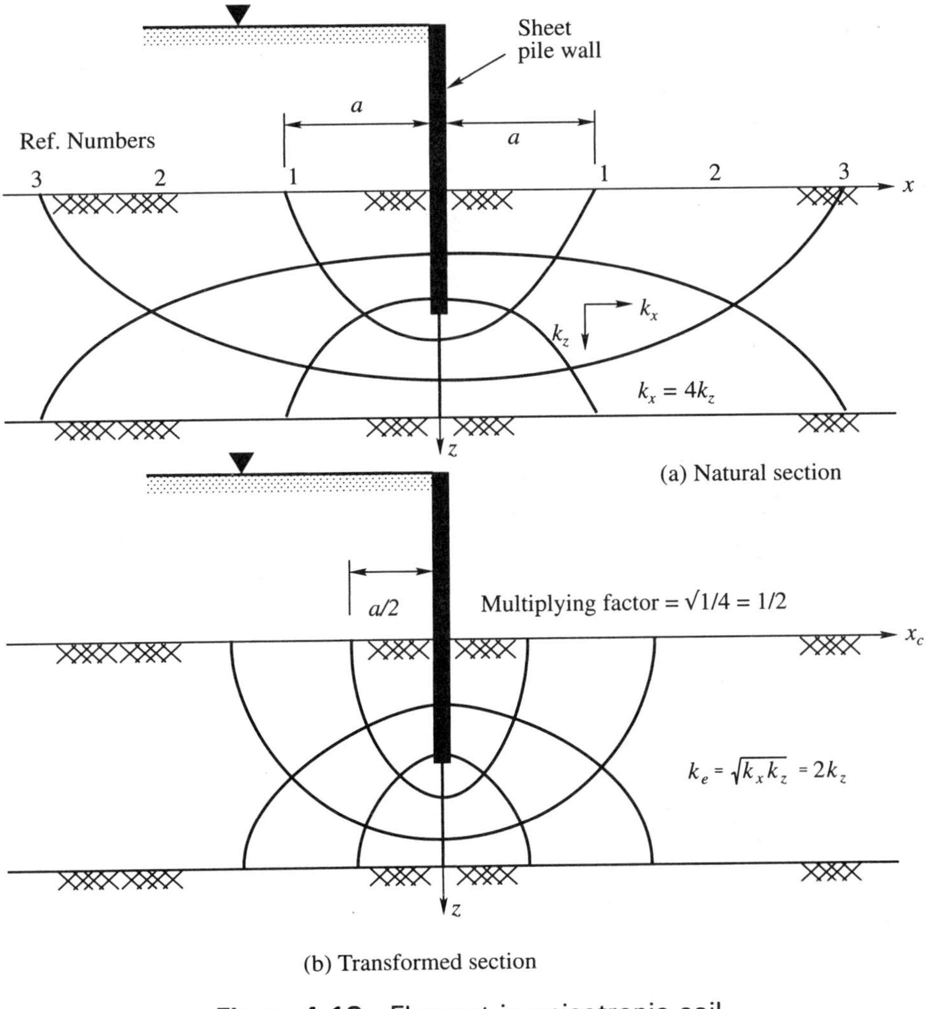

Figure 4.18 Flownet in anisotropic soil

Equating Eq. (4.48) and (4.49), we obtain

$$k_e = \sqrt{k_x k_z} \tag{4.50}$$

Flow Net in Anisotropic Soils

To obtain a flow net for anisotropic soil conditions, the natural cross-section has to be redrawn to satisfy the condition of Laplace Eq. (4.41).

The transformed section may be obtained by multiplying either the natural horizontal distances by $\sqrt{k_z / k_x}$ or the vertical distances by $\sqrt{k_x / k_z}$ keeping the other dimension unaltered. Normally the vertical dimensions are kept as they are but the horizontal dimensions are multiplied by $\sqrt{k_z / k_x}$. The natural section gets shortened or lengthened in the x- direction in accordance with the condition that k_x is greater or less than k_z.

Fig. 4.18(a) is a natural section with flow taking place around a sheet pile wall. The horizontal permeability is assumed to be 4 times that of the vertical permeability. Fig. 4.18(b) is transformed section with the horizontal dimensions multiplied by a factor equal to $\sqrt{k_z / k_x} = \sqrt{1/4} = 1/2$. This section is now assumed to possess the same permeability of $k_e = \sqrt{4k_z^2} = 2k$ in all directions. The flow nets are constructed on this section in the usual way. The same flow net is transferred to the natural section in (a) of Fig. 4.18, by multiplying the x-coordinates of points on the flow and equipotential lines by the factor 2. On the natural cross-section the flow net will not be composed of squares but of rectangles elongated in the direction of greater permeability.

4.17 DETERMINATION OF SEEPAGE PRESSURE

Flow nets are useful in the determination of the seepage pressure at any point along the flow path. Consider the cubical element 1 in Fig. 4.16(a) with all the sides equal to a. Let h_1 be the piezometric head acting on the face kt and h_2 on face jo.

The total force on face $kt = P_1 = a^2\gamma_w h_1$

The total force on face $jo = P_2 = a^2 \gamma_w h_2$

The differential force acting on the element is

$$P_1 - P_2 = P_3 = a^2\gamma_w(h_1 - h_2)$$

Since $(h_1 - h_2)$ is the head drop Δh, we can write

$$P_3 = a^2\gamma_w \Delta h = a^3 \frac{\Delta h}{a} \gamma_w = a^3 i \gamma_w$$

where a^3 is the volume of the element. The force per unit volume of the element is, therefore,

$$p_s = i\gamma_w \tag{4.51}$$

This force exerts a drag on the element known as the *seepage pressure*. It has the dimension of unit weight, and at any point its line of action is tangent to the flow line. The seepage pressure is a very important factor in the stability analysis of earth slopes. If the line of action of the seepage force acts in the vertical direction upward as on an element adjacent to point x in Fig. 4.16(a), the force that is acting downward to keep the element stable is the buoyant unit weight of the element. When these two forces balance, the soil will just be at the point of being lifted up, and there will be effectively no grain-to-grain pressures. The gradient at which this occurs can be computed from the balance of forces given by Eqs. (3.19a) and (4.51). Therefore we can write

$$i_c \gamma_w = \frac{(G_s - 1)\gamma_w}{1+e} \quad \text{or } i_c = \frac{G_s - 1}{1+e} \tag{4.52}$$

The soil will be in *quick condition* at this gradient, which is therefore called i_c, the *critical hydraulic gradient*.

4.18 DETERMINATION OF UPLIFT PRESSURES

Water that seeps below masonry dams or weirs founded on permeable soils exerts pressures on the bases of structures. These pressures are called *uplift pressures*. Uplift pressures reduce the effective weight of the structure and thereby cause instability. It is therefore very essential to determine the uplift pressures on the base of dams or weirs accurately. Accurate flow nets should be constructed in cases where uplift pressures are required to be determined. The method of determining the uplift pressures can be explained as follows.

Consider a concrete dam Fig. 4.19a founded on a permeable foundation at a depth D below the ground surface. The thickness of the permeable strata is H. The depth of water on the upstream side is h_t and on the downstream side is zero. Water flows from the upstream to the downstream

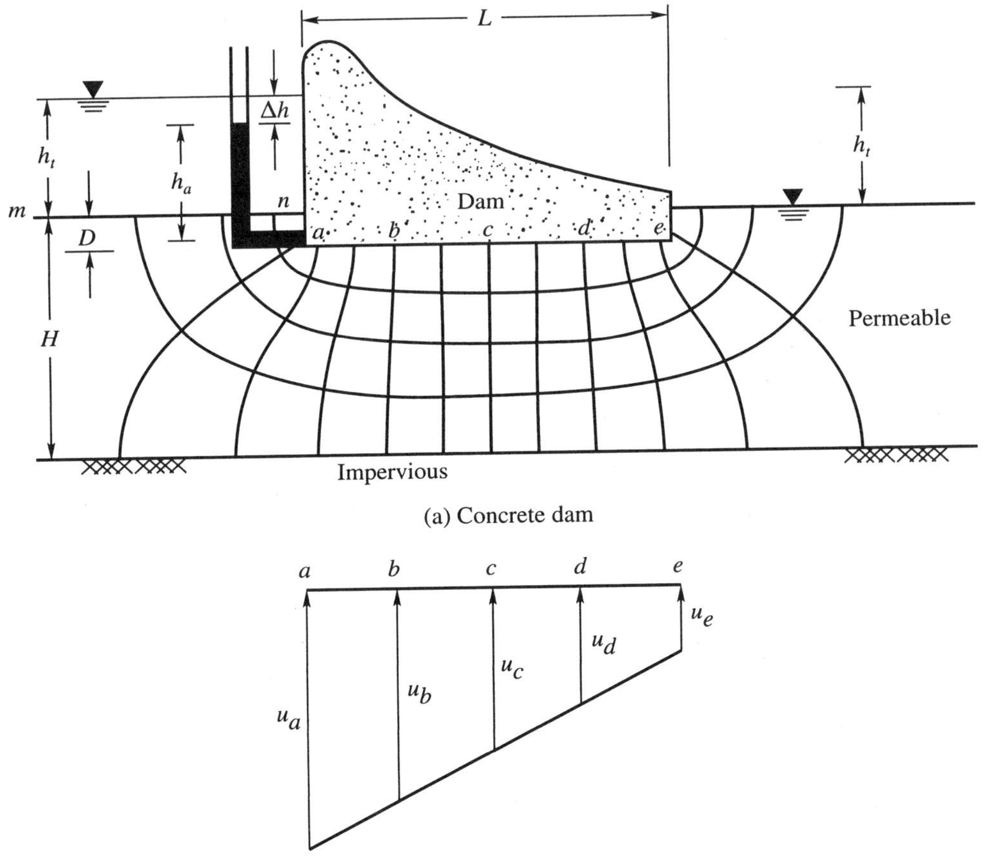

(a) Concrete dam

(b) Uplift-pressure distribution

Figure 4.19 Uplift pressure on the base of a concrete dam side. It is necessary to determine the uplift pressure on the base of the dam by means of flow nets as shown in the figure.

The difference in head between the upstream and downstream water levels is h_t. Let the number of equipotential drops be N_d. The head lost per drop be Δh (= h_t/N_d). As the water flows along the side and base of the dam, there will be equal drops of head between the equipotential lines that meet the dam as shown in the figure. A piezometer tube at point a (coinciding with the corner of the dam in the figure) gives a pressure head h_a. Now the uplift pressure at point a may be expressed as

$$u_a = h_a \gamma_w = (h_t + D - \Delta h)\,\gamma_w \tag{4.53a}$$

Similarly, the uplift pressure at any other point, say e (see the figure), may be estimated from the expression

$$u_e = (h_t + D - n_d \Delta h)\,\gamma_w \tag{4.53b}$$

where n_d = the number of equipotential drops to the point e.

Fig. 4.19b shows the distribution of uplift pressure on the base of the dam.

Example 4.13

In order to compute the seepage loss through the foundation of a cofferdam, flownets were constructed. The result of the flownet study gave $N_f = 6$, $N_d = 16$. The head of water lost during seepage was 19.68 ft. If the hydraulic conductivity of the soil is $k = 13.12 \times 10^{-5}$ ft/min, compute the seepage loss per foot length of dam per day.

Solution

The equation for seepage loss is

$$q = kh\frac{N_f}{N_d}$$

Substituting the given values,

$$q = 13.12\times 10^{-5} \times 19.68 \times \frac{6}{16} = 9.683\times 10^{-4}\ \text{ft}^3/\text{min} = 1.39\ \text{ft}^3/\text{day} \text{ per ft length of dam.}$$

Example 4.14

Two lines of sheet piles were driven in a river bed as shown in Fig. Ex. 4.14. The depth of water over the river bed is 8.20 ft. The trench level within the sheet piles is 6.6 ft below the river bed. The water level within the sheet piles is kept at trench level by resorting to pumping. If a quantity of water flowing into the trench from outside is 3.23 ft^3/hour per foot length of sheet pile, what is the hydraulic conductivity of the sand? What is the hydraulic gradient immediately below the trench bed?

Solution

Fig. Ex. 4.14 gives the flow net and other details. The differential head between the bottom of trench and the water level in the river is 14.8 ft.

Number of channels = 6

Number of equipotential drops = 10

$$q = kh\frac{N_f}{N_d} \quad \text{or} \quad 3.23 = 14.8 \times \frac{6}{10} \times k$$

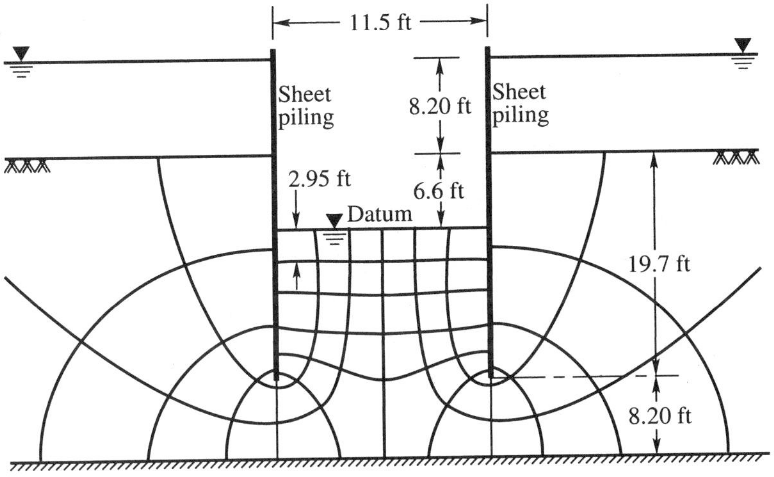

Figure Ex. 4.14

$$\text{or } k = \frac{3.23 \times 10}{14.8 \times 6} \times \frac{1}{60 \times 60} = 1 \times 10^{-4} \text{ ft / sec}$$

The distance between the last two equipotentials given is 2.95 ft. The calculated hydraulic gradient is

$$i = \frac{\Delta h}{\Delta s} = \frac{14.8}{10 \times 2.95} = 0.50$$

$$F_s = \frac{i_{crit}}{i} = \frac{1.0}{0.5} = 2 < 5 \text{ to } 6 \text{ which is normally required for sand.}$$

Example 4.15

A concrete dam (Fig 4.19) is constructed across a river over a permeable stratum of soil of limited thickness. The water heads are upstream side 16 m and 2 m on the downstream side. The flow net constructed under the dam gives $N_f = 4$ and $N_d = 12$. Calculate the seepage loss through the subsoil if the average value of the hydraulic conductivity is 6×10^{-3} cm/sec horizontally and 3×10^{-4} cm/sec vertically. Calculate the exit gradient if the average length of the last field is 0.9 m. Assuming $e = 0.56$, and $G_s = 2.65$, determine the critical gradient. Comment on the stability of the river bed on the downstream side.

Solution

Upstream side $h_1 = 16$ m and downstream side $h_2 = 2$ m, therefore $h = 16 - 2, = 14$ m

$N_f = 4$, $N_d = 12$

$k_h = 6 \times 10^{-3}$ cm/sec, $k_v = 3 \times 10^{-4}$ cm/sec

$$k_e = \sqrt{k_h k_v} = \sqrt{(6 \times 10^{-3}) \times (3 \times 10^{-4})} = 1.34 \times 10^{-3} \text{ cm / sec}$$

$$q = k_e h \frac{N_f}{N_d} = (1.34 \times 10^{-3}) \times (14 \times 100) \times \frac{4}{12} = 0.626 \text{ cm}^3 \text{ / sec}$$

The head loss per potential drop $= \dfrac{h}{N_d} = \dfrac{14}{12} = 1.17$ m

The exit gradient $i = \dfrac{\Delta h}{l} = \dfrac{1.17}{0.9} = 1.30$

As per Eq. (4.52), the critical gradient i_e is

$$i_e = \frac{G_s - 1}{1+e} = \frac{2.65-1}{1+0.56} = 1.06$$

Since the exit gradient is greater than the critical gradient, the river bed on the down stream side will be subjected to a quick condition. One solution would be to provide a sheet pile wall on the upstream side below the dam to prevent this condition.

4.19 SEEPAGE FLOW THROUGH HOMOGENEOUS EARTH DAMS

In almost all problems concerning seepage beneath a sheet pile wall or through the foundation of a concrete dam all boundary conditions are known. However, in the case of seepage through an earth dam the upper boundary or the uppermost flow line is not known. This upper boundary is a free water surface and will be referred to as the *line of seepage or phreatic line*. The seepage line may therefore be defined as the line above which there is no hydrostatic pressure and below which there is hydrostatic pressure. In the design of all earth dams, the following factors are very important.

1. The seepage line should not cut the downstream slope.
2. The seepage loss through the dam should be the minimum possible.

The two important problems that are required to be studied in the design of earth dams are:

1. The prediction of the position of the line of seepage in the cross-section.
2. The computation of the seepage loss.

If the line of seepage is allowed to intersect the downstream face much above the toe, more or less serious sloughing may take place and ultimate failure may result. This mishap can be prevented by providing suitable drainage arrangements on the downstream side of the dam.

The section of an earth dam may be homogeneous or non-homogeneous. A homogeneous dam contains the same material over the whole section and only one coefficient of permeability may be assumed to hold for the entire section. In the non homogeneous or the composite section, two or more permeability coefficients may have to be used according to the materials used in the section. When a number of soils of different permeabilities occur in a cross-section, the prediction

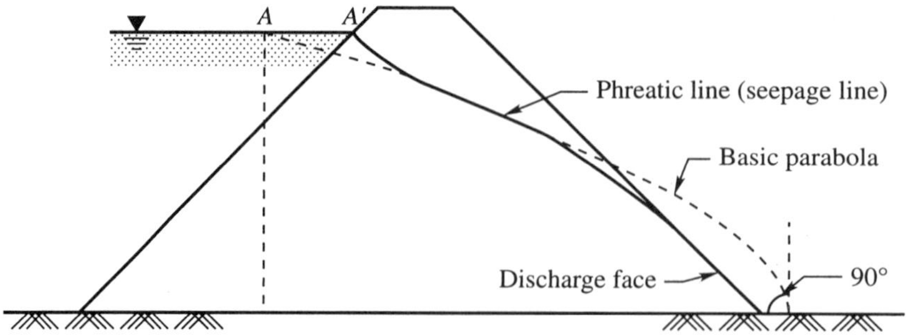

Figure 4.20 Basic parabola and the phreatic line for a homogeneous earth dam

of the position of the line of seepage and the computation of the seepage loss become quite complicated.

It has been noticed from experiments on homogeneous earth dam models that the line of seepage assumes more or less the shape of a parabola as illustrated in Fig. 4.20. In some sections a little divergence from a regular parabola is required at the surfaces of entry and discharge of the line of seepage. In some ideal sections where conditions are favorable the entire seepage line may be considered as a parabola. When the entire seepage line is a parabola, all the other flow lines will be confocal parabolas. The equipotential lines for this ideal case will be conjugate confocal parabolas as shown in Fig. 4.21. As a first step it is necessary to study the ideal case where the entire flow net consists of conjugate confocal parabolas.

4.20 FLOW NET CONSISTING OF CONJUGATE CONFOCAL PARABOLAS

As a prelude to the study of an ideal flow net comprising of parabolas as flow and equipotential lines, it is necessary to understand the properties of a single parabola. The parabola ACV illustrated in Fig. 4.21, is defined as the curve whose every point is equidistant from a point F called the *focus* and a line DG called the *directrix*. If we consider any point, say, A, on the curve, we can write $FA = AG$, where the line AG is normal to the directrix. If F is the origin of coordinates, and the coordinates of point A are (x, y), we can write

$$AF = \sqrt{x^2 + y^2} = AG = x + y_0$$

$$\text{or } x = \frac{y^2 - y_0^2}{2y_0} \tag{4.54}$$

where, $y_0 = FD$

Eq. (4.54) is the equation of the basic parabola. If the parabola intersects the y-axis at C, we can write

$$FC = CE = y_0$$

Similarly for the vertex point V, the focal distance a_0 is

$$FV = VD = a_0 = y_0/2 \tag{4.55}$$

Figure 4.21 illustrates the ideal flow net consisting of conjugate confocal parabolas. All the parabolas have a common focus F.

The boundary lines of such an ideal flow net are:

1. The upstream face AB, an equipotential line, is a parabola.
2. The downstream discharge face FV, an equipotential line, is horizontal.
3. ACV, the phreatic line, is a parabola.
4. BF, the bottom flow line, is horizontal.

The known boundary conditions are only three in number. They are, the two equipotential lines AB and FV, and the bottom flow line BF. The top flow line ACV is the one that is unknown. The theoretical investigation of Kozeny (1931) revealed that the flow net for such an ideal condition mentioned above with a horizontal discharge face FV consists of two families of confocal parabolas with a common focus F. Since the conjugate confocal parabolas should intersect at right angles to each other, all the parabolas crossing the vertical line FC should have their intersection points lie on this line.

Since the seepage line is a line of atmospheric pressure only, the only type of head that can exist along it is the elevation head. Therefore, there must be constant drops in elevation between the points at which successive equipotentials meet the top flow line, as shown in Fig. 4.21.

In all seepage problems connected with flow through earth dams, the focus *F* of the basic parabola is assumed to lie at the intersection of the downstream discharge face *FV* and the bottom flow line *BF* as shown in Fig. 4.21. The point *F* is therefore known. The point *A*, which is the intersection point of the top flow line of the basic parabola and the upstream water level, is also supposed to be known. When the point *A* is known, its coordinates (*d*, *h*) with respect to the origin *F* can be determined. With these two known points, the basic parabola can be constructed as explained below. We may write

$$FA = AG = \sqrt{d^2 + h^2}$$

$$FD = 2a_0 = \sqrt{d^2 + h^2} - d$$

$$a_0 = \frac{1}{2}\left(\sqrt{d^2 + h^2} - d\right) \tag{4.56}$$

Seepage Loss Through the Dam

The seepage flow *q* across any section can be expressed according to Darcy's law as

$$q = kiA \tag{4.57}$$

Considering the section *FC* in Fig. 4.21, where the sectional area *A* is equal to y_0, the hydraulic gradient *i* can be determined analytically as follows:

From Eq. (4.54), the equation of the parabola can be expressed as

$$y = \sqrt{2xy_0 + y_0^2} \tag{4.58}$$

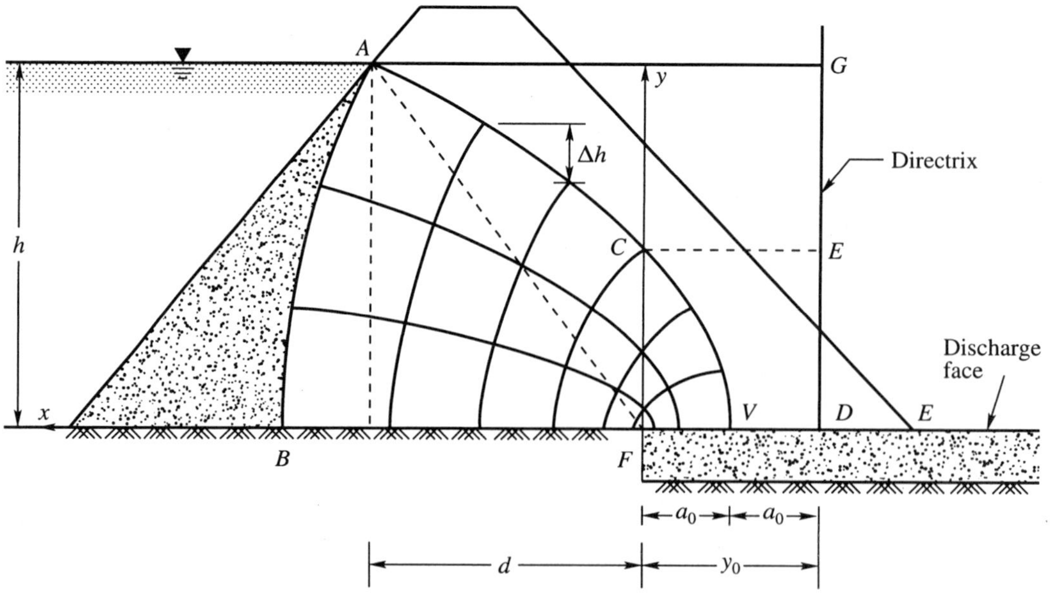

Figure 4.21 Ideal flownet consisting of conjugate confocal parabolas

The hydraulic gradient i at any point on the seepage line in Fig. 4.21 can be expressed as

$$\frac{dy}{dx} = \frac{y_0}{\sqrt{2xy_0 + y_0^2}} \tag{4.59}$$

For the point C which has coordinates $(0, y_0)$, the hydraulic gradient from Eq. (4.59) is

$$\frac{dy}{dx} = \frac{y_0}{\sqrt{y_0^2}} = 1$$

Therefore, the seepage quantity across section FC is

$$q = k\frac{dy}{dx}y_0 = ky_0 \tag{4.60}$$

Seepage Through Homogeneous and Isotropic Earth Dams

Types of Entry and Exit of Seepage lines

The flow net consisting of conjugate confocal parabolas is an ideal case which is not generally met in practice. Though the top flow line resembles a parabola for most of its length, the departure from the basic parabola takes place at the faces of entry and discharge of the flow line. The departure from the basic parabola depends upon the conditions prevailing at the points of entrance and discharge of the flow line as illustrated in Fig. 4.22 from (a) to (e).

The seepage line should be normal to the equipotential line at the point of entry as shown in Fig. 4.22(a). However, this condition is violated in Fig. 4.22(b), where the angle made by the upstream face AB with the horizontal is less than 90°. It can be assumed in this case the coarse material used to support the face AB is highly permeable and does not offer any resistance for flow. In such cases AB taken as the upstream equipotential line. The top flow line cannot therefore be

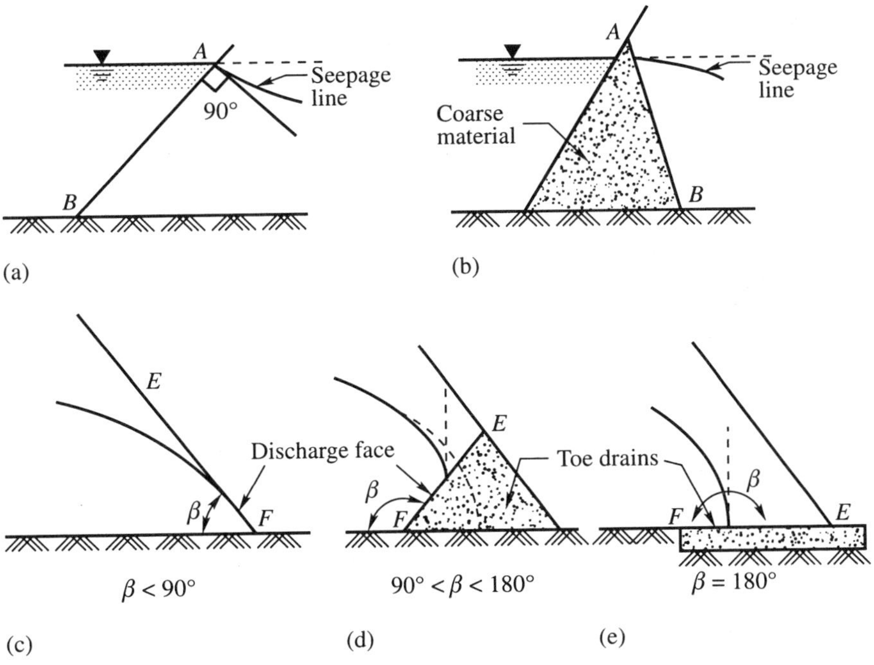

Figure 4.22 Types of entry and exit of seepage lines

normal to the equipotential line. However, this line possesses zero gradient and velocity at the point of entry. This zero condition relieves the apparent inconsistency of deviation from a normal intersection.

The conditions prevailing at the downstream toe of the dam affect the type of exit of the flow line at the discharge face. In Fig. 4.22(c) the material at the toe is the same as in the other parts of the dam whereas in (d) and (e) rock toe drains are provided. This variation in the soil condition at the toe affects the exit pattern of the flow line. The flow line will meet the discharge face *FE* tangentially in 4.22(c). This has to be so because the particles of water as they emerge from the pores at the discharge face have to conform as nearly as possible to the direction of gravity. But in cases where rock toe drains are provided, the top flow line becomes tangential to the vertical line drawn at the point of exit on the discharge face as shown in (d) and (e) of Fig. 4.22.

Method of Locating Seepage Line

The general method of locating the seepage line in any homogeneous dam resting on an impervious foundation may be explained with reference to Fig. 4.23(a). As explained earlier, the focus *F* of the basic parabola is taken as the intersection point of the bottom flow line *BF* and the discharge face *EF*. In this case the focus coincides with the toe of the dam. One more point is required to construct the basic parabola. Analysis of the location of seepage lines by A. Casagrande has revealed that the basic parabola with focus *F* intersects the upstream water surface at A such that *AA′* = 0.3 m, where *m* is the projected length of the upstream equipotential line *A′B* on the water surface. Point *A* is called the corrected entrance point. The parabola *APSV* may now be constructed as per Eq. (4.54). The divergence of the seepage line from the basic parabola is shown as *A′P* and *SD* in Fig. 4.23(a). For dams with flat slopes, the divergences may be sketched by eye keeping in view the boundary requirements. The error involved in sketching by eye, the divergence on the downstream side, might be considerable if the slopes are steeper.

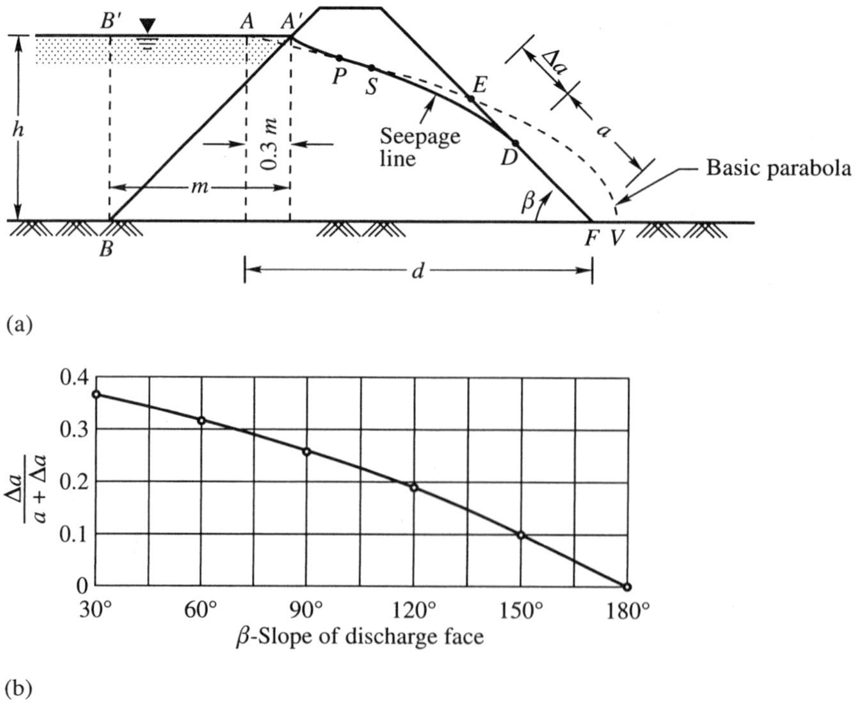

Figure 4.23 Construction of seepage line

Procedures have therefore been developed to sketch the downstream divergence as explained below. As shown in Fig. 4.23(a), E is the point at which the basic parabola intersects the discharge face. Let the distance ED be designated as Δa and the distance DF as a. The values of Δa and $a + \Delta a$ vary with the angle, β, made by the discharge face with the horizontal measured clockwise. The angle may vary from 30° to 180°. The discharge face is horizontal as shown in Fig. 4.22(e). Casagrande (1937) determined the ratios of $\Delta a / (a + \Delta a)$ for a number of discharge slopes varying from 30° to 180° and the relationship is shown in a graphical form Fig. 4.23(b).

The distance $(a + \Delta a)$ can be determined by constructing the basic parabola with F as the focus. With the known $(a + \Delta a)$ and the discharge face angle β, Δa can be determined from Fig. 4.23(b). The point D may therefore be marked out at a distance of Δa from E. With the point D known, the divergence DS may be sketched by eye.

It should be noted that the discharge length a, is neither an equipotential nor a flow line, since it is at atmospheric pressure. It is a boundary along which the head at any point is equal to the elevation.

Analytical Solutions for Determining *a* and *q*

Casagrande (1937) proposed the following equation for determining a for $\beta < 30°$

$$a = \frac{d}{\cos\beta} - \sqrt{\frac{d^2}{\cos^2\beta} - \frac{h^2}{\sin^2\beta}} \tag{4.61}$$

L. Casagrande (1932) gave the following equation for a when β lies between 30° and 90°.

$$a = \sqrt{h^2 + d^2} - \sqrt{d^2 - h^2 \cot^2\beta} \tag{4.62}$$

The discharge q per unit length through any cross-section of the dam may be expressed as follows:

For $\beta < 30°$, $\quad q = ka \sin\beta \tan\beta$ (4.63)

For $30° < \beta < 90°$, $\quad q = ka \sin^2\beta$ (4.64).

4.21 PIPING FAILURE

Piping failures caused by heave can be expected to occur on the downstream side of a hydraulic structure when the uplift forces of seepage exceed the downward forces due to the submerged weight of the soil.

The mechanics of failure due to seepage was first presented by Terzaghi. The principle of this method may be explained with respect to seepage flow below a sheet pile wall. Fig. 4.24(a) is a sheet pile wall with the flow net drawn. The uplift pressures acting on a horizontal plane ox can be determined as explained in Sect. 4.18. The ordinates of curve C in Fig. 4.24(b) represent the uplift pressure at any point on the line ox. It is seen that the uplift pressure is greatest close to the wall and gradually becomes less with an increase in the distance from the wall. When the upward forces of seepage on a portion of ox near the wall become equal to the downward forces exerted by the submerged soil, the surface of the soil rises as shown in Fig. 4.24(a). This heave occurs simultaneously with an expansion of the volume of the soil, which causes its permeability to increase. Additional seepage causes the sand to boil, which accelerates the flow of water and leads to complete failure. Terzaghi determined from model tests that heave occurs within a distance of about $D/2$ (where D is the depth of penetration of the pile) from the sheet pile and the critical section ox passes through the lower edge of the sheet pile.

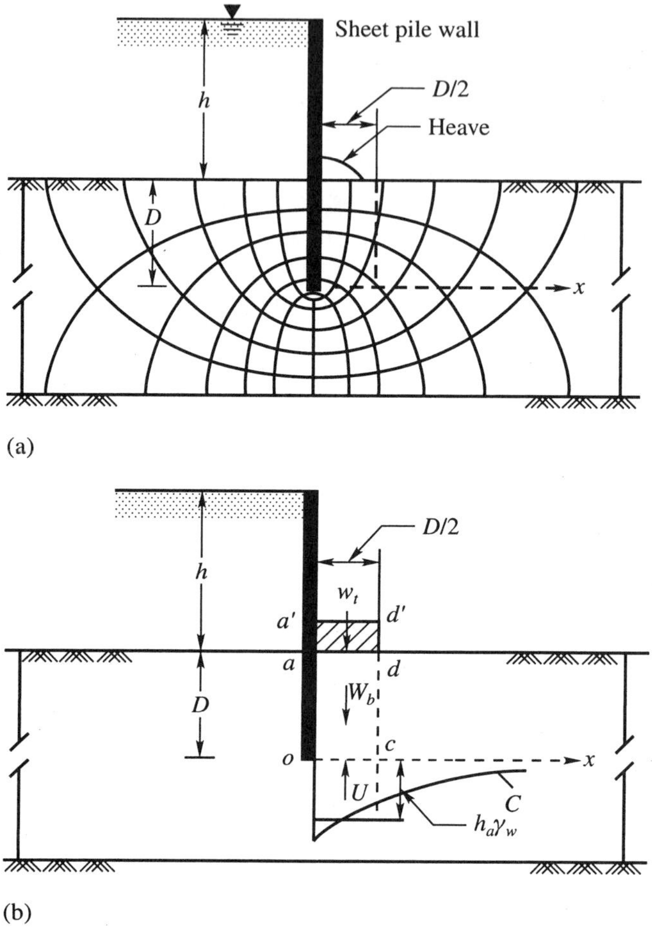

Figure 4.24 Piping failure

Factor of Safety Against Heave

The prism *aocd* in Fig. 4.24(b) subjected to the possible uplift has a depth of D and width $D/2$.

The average uplift pressure on the base of prism is equal to $\gamma_w h_a$. The total uplift force per unit length of wall is

$$U = \frac{1}{2}\gamma_w h_a D \tag{4.65}$$

The submerged weight of the prism *aocd* is

$$W_b = \frac{1}{2}\gamma_b D^2$$

where γ_b is the submerged unit weight of the material. The factor of safety with respect to piping can therefore be expressed as

$$F_s = \frac{W_b}{U} = \frac{D\gamma_b}{h_a \gamma_w} \tag{4.66}$$

If it is not economical to drive the sheet piles deeply enough to prevent heave, the factor of safety can be increased by placing a weighted filter over the prism *aocd* as shown by the prism *aa'd'd*. If the weight of such a filter is W_t, the new factor of safety can be written as

$$F_s = \frac{W_b + W_t}{U} \tag{4.67}$$

Filter Requirements to Control Piping

Filter drains are required on the downstream sides of hydraulic structures and around drainage pipes. A properly graded filter prevents the erosion of soil in contact with it due to seepage forces. To prevent the movement of erodible soils into or through filters, the pore spaces between the filter particles should be small enough to hold some of the protected materials in place. Taylor (1948) shows that if three perfect spheres have diameters greater than 6.5 times the diameter of a small sphere, the small spheres can move through the larger as shown in Fig. 4.25(a). Soils and aggregates are always composed of ranges of particle sizes, and if pore spaces in filters are small enough to hold the 85 per cent size (D_{85}) of the protected soil in place, the finer particles will also be held in place as exhibited schematically in Fig. 4.25(b).

The requirements of a filter to keep the protected soil particles from invading the filter significantly are based on particle size. These requirements were developed from tests by Terzaghi which were later extended by the U.S. Army Corps of Engineers (1953). The resulting filter specifications relate the grading of the protective filter to that of the soil being protected by the following;

$$\frac{D_{15\,\text{filter}}}{D_{85\,\text{soil}}} \leq 4, \quad 4 < \frac{D_{15\,\text{filter}}}{D_{15\,\text{soil}}} \leq 20, \quad \frac{D_{50\,\text{filter}}}{D_{50\,\text{soil}}} \leq 25 \tag{4.68}$$

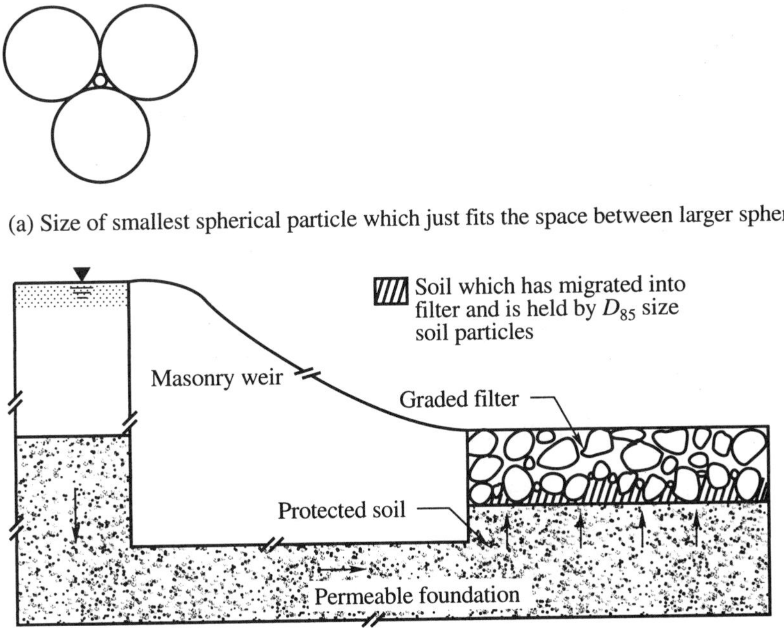

(a) Size of smallest spherical particle which just fits the space between larger spheres

(b) Condition of the boundary between protected soil and the filter material

Figure 4.25 Requirements of a filter

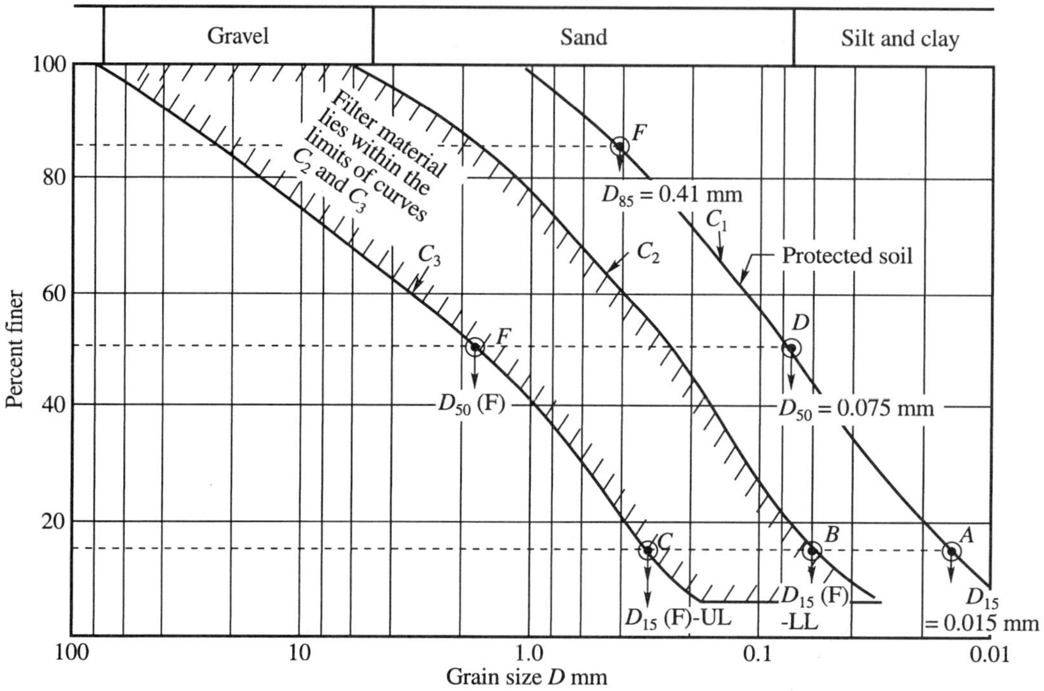

Figure 4.26 Grain size distribution curves for graded filter and protected materials

The criteria may be explained as follows:

1. The 15 per cent size (D_{15}) of filter material must be less than 4 times the 85 per cent size (D_{85}) of a protected soil. The ratio of D_{15} of a filter to D_{85} of a soil is called the *piping ratio*.
2. The 15 per cent size (D_{15}) of a filter material should be at least 4 times the 15 per cent size (D_{15}) of a protected soil but not more than 20 times of the latter.
3. The 50 per cent size (D_{50}) of filter material should be less than 25 times the 50 per cent size (D_{50}) of protected soil.

Experience indicates that if the basic filter criteria mentioned above are satisfied in every part of a filter, piping cannot occur under even extremely severe conditions.

A typical grain size distribution curve of a protected soil and the limiting sizes of filter materials for constructing a graded filter is given in Fig. 4.26. The size of filter materials must fall within the two curves C_2 and C_3 to satisfy the requirements.

Example 4.16

Fig. Ex. 4.16 gives the section of a homogeneous dam with a hydraulic conductivity $k = 7.874 \times 10^{-5}$ in/sec. Draw the phreatic line and compute the seepage loss per foot length of the dam.

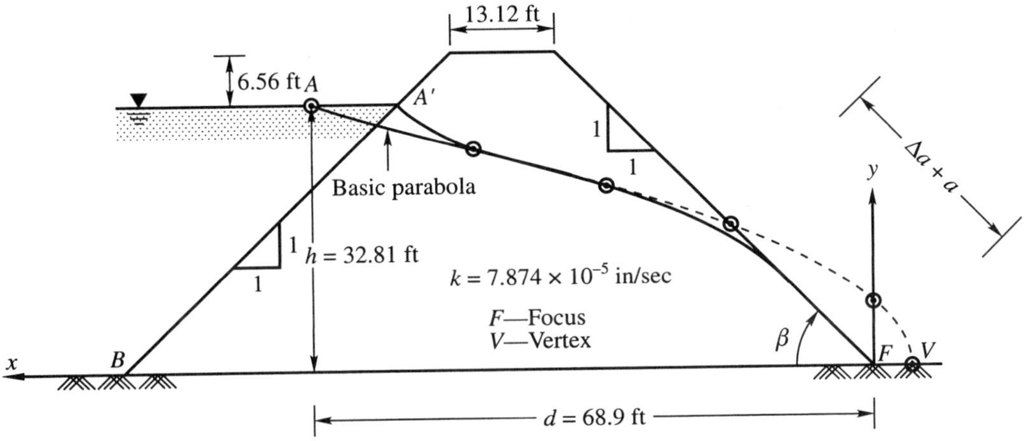

Figure Ex. 4.16

Solution

The depth h of water on upstream side = 32.81 ft.

The projected length of slope $A'B$ on the water surface = 32.81 ft.

The point A on the water level is a point on the basic parabola. Therefore

$AA' = 0.3 \times 32.81 = 9.84$ ft.

F is the focus of the parabola. The distance of the directrix from the focus F is

$$y_0 = \sqrt{d^2 + h^2} - d$$

where $d =$ 68.9 ft, $h = 32.81$ ft. Therefore

$$y_0 = \sqrt{(68.9)^2 + (32.81)^2} - 68.9 = 7.413 \text{ ft}$$

The distance of the vertex of the parabola from F is

$$FV = a_0 = \frac{y_0}{2} = \frac{7.413}{2} = 3.706 \text{ ft}$$

The (x, y) coordinates of the basic parabola may be obtained from Eq. (4.58) as

$$x = \frac{y^2 - y_0^2}{2y_0} = \frac{y^2 - (7.413)^2}{2 \times 7.413} = \frac{y^2 - 55}{14.83}$$

or $y = \sqrt{14.83x + 55}$

Given below are values of y for various values of x

x (ft)	0	15	30	45	68.9
y (ft)	7.416	16.65	22.36	26.88	32.81

The parabola has been constructed with the above coordinates as shown in Fig. Ex. 4.16.

From Fig. Ex. 4.16 $\Delta a + a = 24.6$ ft

From Fig. 4.23, for a slope angle $\beta = 45°$

$$\frac{\Delta a}{a+\Delta a} = 0.35$$

or $\Delta a = 0.35\ (a + \Delta a) = 0.35 \times 24.6 = 8.61$ ft

From Eq. (4.60)

$q = ky_0$

where $k = 7.874 \times 10^{-5}$ in/sec or 6.56×10^{-6} ft/sec and $y_0 = 7.413$ ft

$q = 6.56 \times 10^{-6} \times 7.413 = 48.63 \times 10^{-6}$ ft³/sec per ft length of dam.

Example 4.17

An earth dam which is anisotropic is given in Fig. Ex. 4.17(a). The hydraulic conductivities k_x and k_z in the horizontal and vertical directions are respectively 4.5×10^{-8} m/s and 1.6×10^{-8} m/s. Construct the flow net and determine the quantity of seepage through the dam. What is the pore pressure at point P?

Solution

The transformed section is obtained by multiplying the horizontal distances by $\sqrt{k_z / k_x}$ and by keeping the vertical dimensions unaltered. Fig. Ex. 4.17(a) is a natural section of the dam. The scale factor for transformation in the horizontal direction is

$$\text{Scale factor} = \sqrt{\frac{k_z}{k_x}} = \sqrt{\frac{1.6 \times 10^{-8}}{4.5 \times 10^{-8}}} = 0.6$$

The transformed section of the dam is given in Fig. Ex. 4.17(b). The isotropic equivalent coefficient of permeability is

$$k_e = \sqrt{k_x k_z} = \left(\sqrt{4.5 \times 1.6}\right) \times 10^{-8} = 2.7 \times 10^{-8} \text{ m/s}$$

Confocal parabolas can be constructed with the focus of the parabola at A. The basic parabola passes through point G such that

$GC = 0.3\ \overline{HC} = 0.3 \times 27 = 8.10$ m

The coordinates of G are:

$x = +40.80$ m, $z = +18.0$ m

As per Eq. (4.58) $$x = \frac{z^2 - 4a_0^2}{4a_0} \qquad \text{(a)}$$

Substituting for x and z, we get, $40.80 = \dfrac{18^2 - 4a_0^2}{4a_0}$

Simplifying we have, $4a_0^2 + 163.2a_0 - 324 = 0$

Solving, $a_0 = 1.9$ m

Substituting for a_0 in Eq. (a) above, we can write

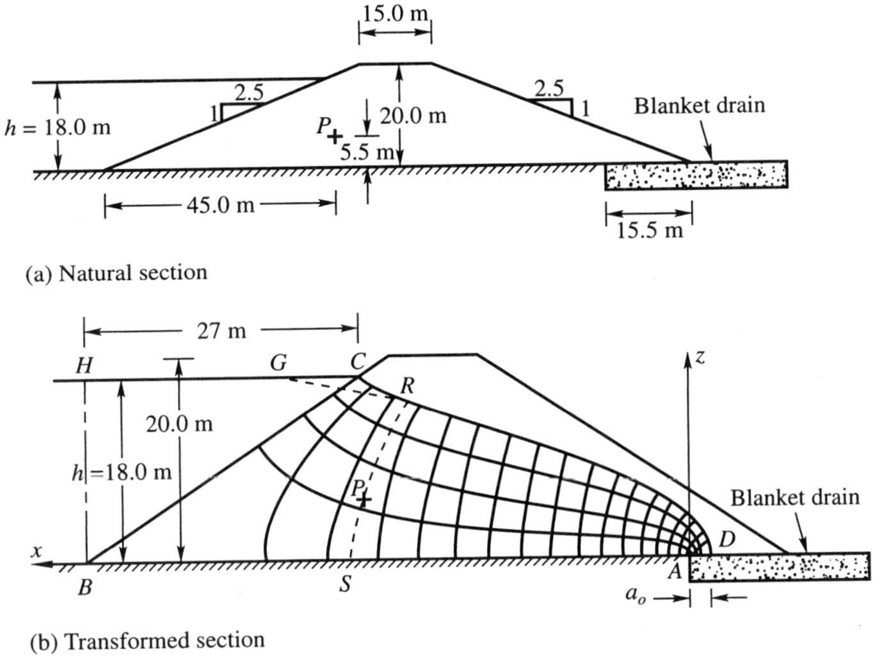

Figure Ex. 4.17

$$x = \frac{z^2 - 14.4}{7.6} \tag{b}$$

By using Eq. (b), the coordinates of a number of points on the basic parabola may be calculated.

x(m)	−1.9	0.0	5.0	10.0	20.0	30.0
z(m)	0.0	3.8	7.24	9.51	12.9	15.57

The basic parabola is shown in Fig. Ex. 4.17(b).

The flownet is completed by making the entry corrections by ensuring that the potential drops are equal between the successive equipotential lines at the top seepage line level.

As per Fig. Ex. 4.17(b), there are 3.8 flow channels and 18 equipotential drops. The seepage per unit length of dam is

$$q = k_e h \frac{N_f}{N_d} = (2.7 \times 10^{-8}) \times 18 \times \frac{3.8}{18} = 1 \times 10^{-7} \text{ m}^3/\text{s}$$

The quantity of seepage across section Az can also be calculated without the flownet by using Eq. (4.60)

$$q = k_e y_0 = 2k_e a_0 = 2 \times 2.7 \times 10^{-8} \times 1.9 \approx 1 \times 10^{-7} \text{ m}^3/\text{sec per meter}$$

Pore pressure at *P*

Let *RS* be the equipotential line passing through *P*. The number of equipotential drops up to point *P* equals 2.4

Total head loss = h = 18 m, number of drops = 18

Head loss per drop $= \frac{18}{18} = 1\text{ m}$.

Therefore the head at point $P = 18 - 2.4(\Delta h) = 18 - 2.4(1) = 15.6$ m

Assuming the base of the dam as the datum, the elevation head of point $P = 5.50$ m.

Therefore the pressure head at $P = 15.6 - 5.5 = 10.1$ m.

The pore pressure at P is, therefore, $u_w = 10.1 \times 9.81 = 99$ kN/m^2

Example 4.18

A sheet pile wall was driven across a river to a depth of 6 m below the river bed. It retains a head of water of 12.0 m. The soil below the river bed is silty sand and extends up to a depth of 12.0 m where it meets an impermeable stratum of clay. Flow net analysis gave $N_f = 6$ and $N_d = 12$. The hydraulic conductivity of the sub-soil is $k = 8 \times 10^{-5}$ m /min. The average uplift pressure head h_a at thc bottom of the pile is 3.5 m. The saturated unit weight of the soil $\gamma_{sat} = 19.5$ kN/m^3. Determine:

(a) The seepage less per meter length of pile per day.

(b) The factor of safety against heave on the downstream side of the pile.

Solution

(a) Seepage loss,

The loss of head $h = 12$ m

$$q = kh\frac{N_f}{N_d} = (8\times10^{-5})\times12\times\frac{6}{12} = 48\times10^{-5}\ \text{m}^3/\text{min} = 69.12\times10^{-2}\ \text{m}^3/\text{day}$$

(b) The F_s as per Eq. (4.67) is (Ref. Fig 4.24)

$$F_s = \frac{W_b + W_t}{U} = \frac{D\gamma_b}{h_a\gamma_w}$$

$$h_a = 3.5\text{ m}$$

$$\gamma_b = \gamma_{sat} - \gamma_w = 19.5 - 9.81 = 9.69\text{ kN/m}^3$$

$$\text{Therefore, } F_s = \frac{6\times9.69}{3.5\times9.81} = 1.69$$

4.22 PROBLEMS

4.1 A constant head permeability test was carried out on a cylindrical sample of sand of 10 cm diameter and 15 cm height. 200 cm^3 of water was collected in 2.25 min under a head of 30 cm. Compute the hydraulic conductivity in m/sec.

4.2 Calculate the hydraulic conductivity of a soil sample 6 cm in height and 50 cm^2 cross-sectional area if 430 mL of water was collected in 10 min under a constant head of 40 cm.

On oven-drying the test specimen had a mass of 498 g. Assuming $G_s = 2.65$, calculate the seepage velocity.

4.3 A constant head permeability test was carried out on a sample of sand. The diameter and the length of the sample were 10 and 20 cm. respectively. The head of water was maintained at 35 cm. If 110 cm^3 of water is collected in 80 seconds, compute the hydraulic conductivity of the sand.

4.4 A falling head permeability test was performed on a sample of silly sand. The time required for the head to fall in the stand pipe from 60 cm to the 30 cm mark was 70 min. The cross sectional area of the stand pipe was 1.25 cm^2. If the height and diameter of the sample were respectively 10 and 9 cm, determine the value k in cm/min.

4.5 In a falling head permeability test, the time taken for the head to fall from h_1 to h_2 is t. If the test is repeated with the same initial head h_1, what would be the final head in a time interval of $t/2$?

4.6 In a falling head permeameter test the initial head at t = 0 is 40 cm. The head drops by 5 cm in 10 minutes. Determine the time required to run the test for the final head to be at 20 cm. Given: Height of sample = 6 cm; cross sectional areas of sample = 50 cm^2 and stand pipe = 0.5 cm^2 Determine the hydraulic conductivity in cm/sec.

4.7 The hydraulic conductivity of a soil sample at a temperature of 30°C was 8×10^{-5} cm/sec. Determine its permeability at 20°C.
Given: Viscosity of water at (a) 30 °C $\mu_T = 8.0 \times 10^{-7}$ kN-sec/m^2, and (b) 20°C, $\mu_{20} = 10.09 \times 10^{-7}$ kN-sec/m^2.

4.8 Fig. Prob. 4.8 gives a test well with observation wells for conducting a pumping tests. The following data are available.
Maximum $D_o = 0.5$ m, $r_o = 20$ cm, H = 8 m, $k = 8 \times 10^{-4}$m/sec.
Determine the maximum yield in m^3/hour.

4.9 Refer to Fig. Prob. 4.8.
Given: $H = 52$ ft, $h_1 = 47$ ft, $h_2 = 50.75$ ft, discharge q under steady condition = 80 ft^3/min, $r_1 = 10$ ft, and $r_2 = 20$ ft.
Required: The hydraulic conductivity in ft/year.

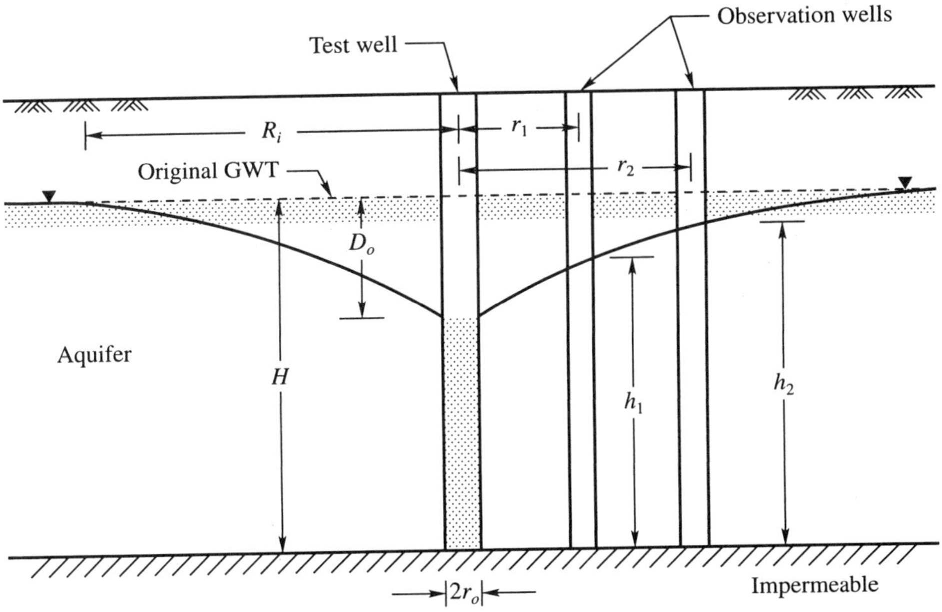

Figure Prob. 4.8

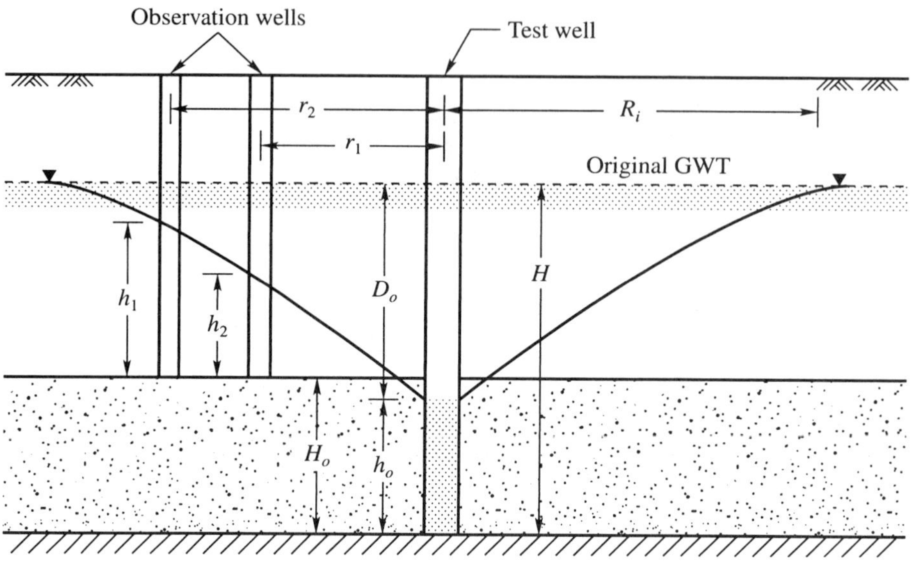

Figure Prob. 4.12

4.10 Refer to Fig. Prob. .8. Determine the hydraulic conductivity of the aquifer in m/hr under a steady state discharge of 240 m^3/hr with the following data:

$H = 30.5$m, $h_1 = 26.5$ m, $h_2 = 29.8$ m, $r_1 = 10$m, $r_2 = 50$ m. Diameter of the test well = 20 cm.

4.11 Refer to Prob. 4.10. For a maximum $D_0 = 4.9$ m, and radius of influence $R_i = 30$m, calculate the value of k.

4.12 Fig. Prob. 4.12 gives the sectional profile of a confined aquifer.

Given: $H_0 = 5$m, D_0 (max) = 4.5m, $R_i = 100$m, radius of test well $r_o = 10$ cm. and $H = 10$ m. Determine the hydraulic conductivity in cm/sec assuming q = 1.5 m^3/min under steady state conditions.

4.13 For the Prob 4.12, if D_o (max) = 5.5m, determine k. All the other data remain the same.

4.14 Calculate the yield in ft^3 per hour from a well driven into a confined aquifer (Fig. Prob. 4.12).

Given: H = 35 ft, H_0 = 15ft, h_0 = 18 ft, k = 0.09 ft/min, r_0 = 4in., R_i = 600 ft.

14.15 The soil investigation at a site revealed three distinct layers of sandy soil (Fig. 4.9). The data available are:

Layer No	Thickness (m)	k (cm/sec)
1	2	8×10^{-3}
2	3	6×10^{-2}
3	4	5×10^{-3}

Determine the equivalent values of k both in the horizontal and vertical directions.

4.16 Laboratory tests on a sample of undisturbed silty sand gave the following data:

void ratio = 0.62; $k = 4 \times 10^{-2}$ cm/sec.

Estimate the value of k of another similar sample whose void ratio is 1.05.

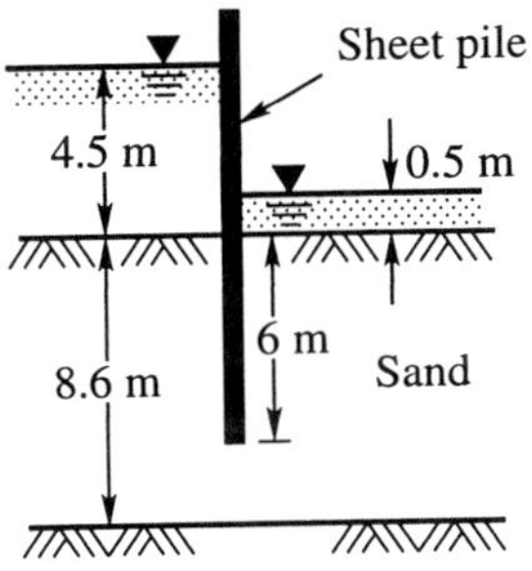

Figure Prob. 4.17

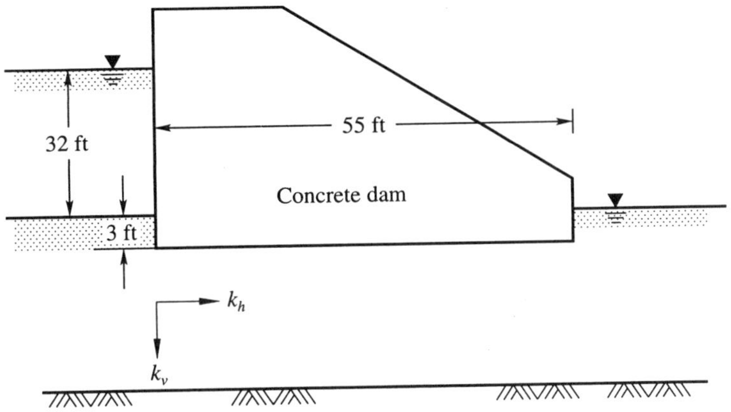

Figure Prob. 4.18

4.17 Figure Prob. 4.17 shows sheet piles driven into a permeable stratum. Construct the flow net and determine the quantity of seepage in m^3/hour per meter length of piling. Assume $k = 8 \times 10^{-4}$ cm/sec.

4.18 Fig. Prob. 4.18 gives a cross section of a concrete dam. The subsoil is anisotropic and has permeabilities $k_h = 0.8 \times 10^{-6}$ in./sec and $k_v = 2.0 \times 10^{-7}$ in./sec. Find the rate of flow beneath the dam per foot length of the dam. Assume $N_f = 4$, and $N_d = 8$.

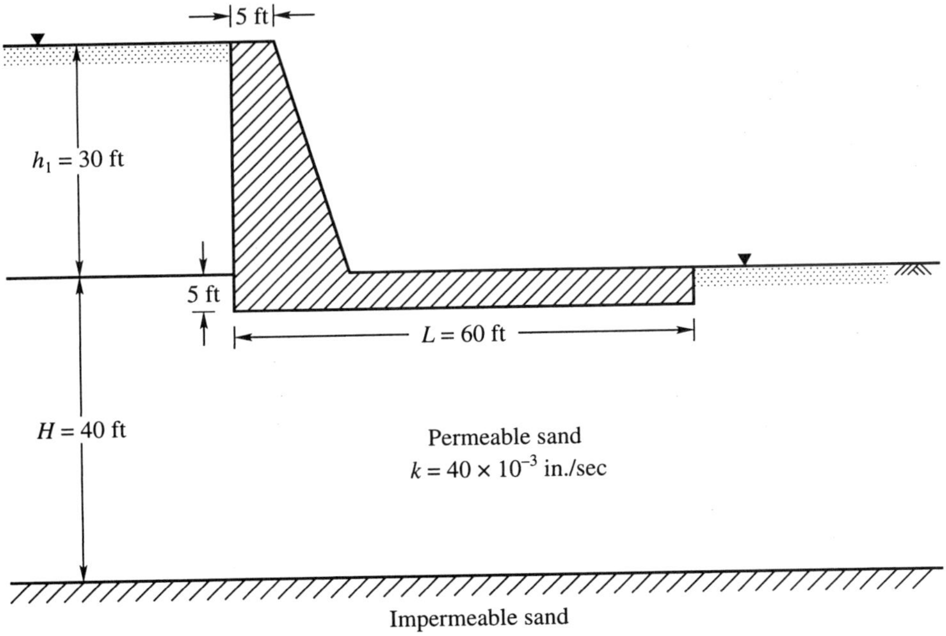

Figure Prob. 4.19

4.19 Construct a flow net in Fig. Prob. 4.19 and estimate the seepage loss in ft^3 per hour per foot length of weir.

4.20 A homogeneous earth dam is shown in Fig. Prob. 4.20. Sketch the phreatic line and estimate the quantity of seepage.

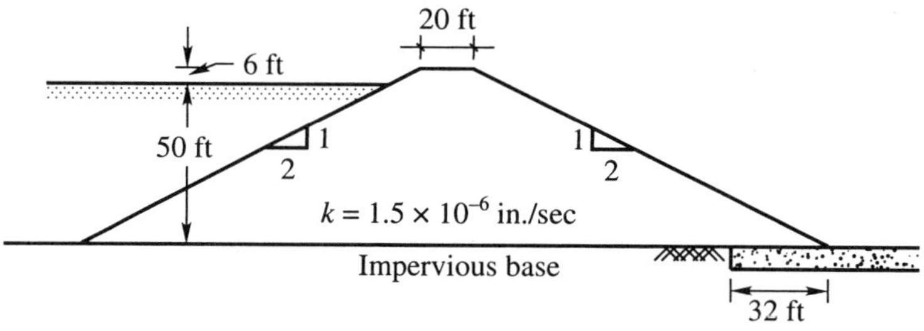

Figure Prob. 4.20

CHAPTER 5

EFFECTIVE STRESS AND PORE WATER PRESSURE

5.1 INTRODUCTION

The pressure transmitted through grain to grain at the contact points through a soil mass is termed as *intergranular* or *effective pressure*. It is known as effective pressure since this pressure is responsible for the decrease in the void ratio or increase in the frictional resistance of a soil mass.

If the pores of a soil mass are filled with water and if a pressure induced into the pore water, tries to separate the grains, this pressure is termed as *pore water pressure* or *neutral stress*. The effect of this pressure is to increase the volume or decrease the frictional resistance of the soil mass.

The effects of the intergranular and pore water pressures on a soil mass can be illustrated by means of simple practical examples.

Consider a rigid cylindrical mold, Fig. 5.1(a), in which dry sand is placed. Assume that there is no side friction. Load Q is applied at the surface of the soil through a piston. The load applied at the surface is transferred to the soil grains in the mold through their points of contact. If the load is quite considerable, it would result in the compression of the soil mass in the mold. The compression might be partly due to the elastic compression of the grains at their points of contact and partly due to relative sliding between particles. If the sectional area of the cylinder is A, the average stress at any level XY may be written as

$$\sigma_a = \frac{Q}{A} \tag{5.1}$$

The stress σ_a is the average stress and not the actual stress prevailing at the grain to grain contacts which is generally very high. Any plane such as XY will not pass through all the points of contact and many of the grains are cut by the plane as shown in Fig. 5.1(b). The actual points of contact exhibit a wavy form. However, for all practical purposes the average stress is considered. Since this stress is responsible for the deformation of the soil mass, it is termed the *intergranular* or *effective stress*. We may therefore write,

$$\sigma_a = \sigma' \tag{5.2}$$

where σ' is the *effective stress*.

Consider now another experiment. Let the soil in the mold be fully saturated and made completely watertight. If the same load Q is placed on the piston, this load will not be transmitted to the soil grains as in the earlier case. If we assume that water is incompressible, the external load Q will be transmitted to the water in the pores. This pressure that is developed in the water is called the *pore water* or *neutral stress* u_w as shown schematically in Fig. 5.1(c). This pore water pressure u_w prevents the compression of the soil mass. The value of this pressure is

$$u_w = \frac{Q}{A} \tag{5.3}$$

If the valve V provided in the piston is opened, immediately there will be expulsion of water through the hole in the piston. The flow of water continues for some time and then stops.

The expulsion of water from the pores decreases the pore water pressure and correspondingly increases the intergranular pressure. At any stage the total pressure Q/A is divided between water and the points of contact of grains. A new equation may therefore be written as

$$\text{Total pressure } \sigma_t = \frac{Q}{A} = \text{Intergranular pressure} + \text{pore water pressure}$$

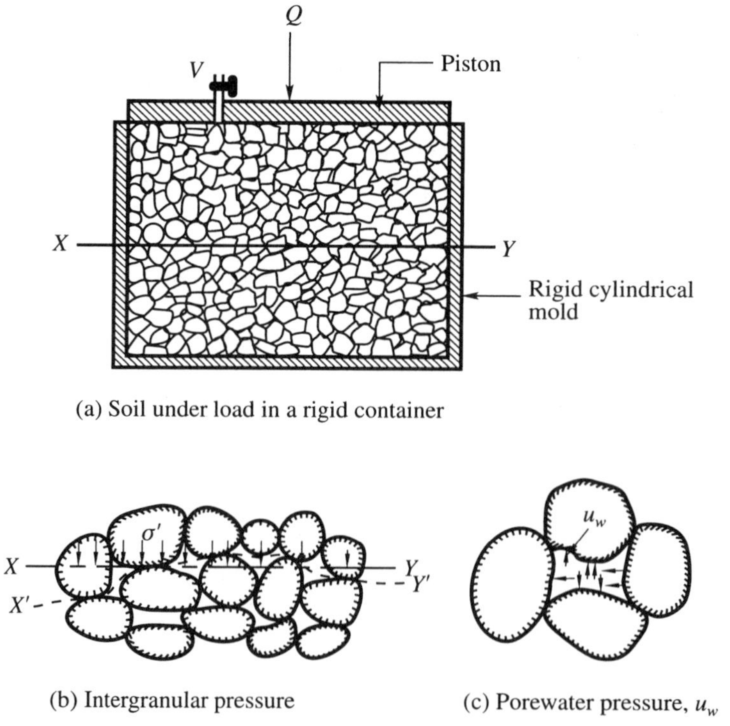

Figure 5.1 Effective and pore water pressures

or $\sigma_t = \sigma' + u_w$ (5.4)

Final equilibrium will be reached when there is no expulsion of water. At this stage the pore water pressure $u_w = 0$. All the pressure will be carried by the soil grains. Therefore, we can write,

$$\sigma_t = \sigma' \tag{5.5}$$

The pore water pressure u_w can be induced in the pores of a soil mass by a head of water over it. When there is no flow of water through the pores of the mass, the intergranular pressure remains constant at any level. But if there is flow, the intergranular pressure increases or decreases according to the direction of flow. In partially saturated soils part of the void space is occupied by water and part by air. The pore water pressure u_w must always be less than the pore air pressure (u_a). Bishop (1955) proposed an equation for computing the effective pressure in partially saturated soils. This equation contains a parameter which cannot be determined easily. Since this equation is only of academic interest, no further discussion is necessary here.

5.2 STRESSES WHEN NO FLOW TAKES PLACE THROUGH THE SATURATED SOIL MASS

In Fig. 5.2 the container A is filled with sand to a depth z_1 and water to a depth z_2 above the sand surface. A flexible tube connects the bottom of the container *A* to another container *B*. The water levels are kept constant in these two containers.

The water surfaces in both the containers in Fig. 5.2(a) are kept at the same level. Under this condition, no flow takes place from one container to another.

Consider two points *M* and *N* as shown in the figure on a horizontal plane. The water pressure at *M* should be equal to the pressure at *N* according to the laws of hydraulics. Therefore,

the water pressure at $N = u_z = (z + z_2)\gamma_w$ (5.6)

The pressure u_z is termed as the pore water pressure acting on the grains at depth z from the surface of the sample. However, the total pressure at point *N* is due to the water head plus the weight of the submerged soil above *N*. If γ_b is the submerged unit weight of the soil, the total pressure at *N* is

$$\sigma_z = z\gamma_b + (z + z_2)\gamma_w \tag{5.7}$$

The intergranular or effective pressure at the point *N* is the difference between the total and the pore water pressures. Therefore, the effective pressure σ_z' is

$$\sigma_z' = \sigma_z - u_z = z\gamma_b + (z + z_2)\gamma_w - (z + z_2)\gamma_w = z\gamma_b \tag{5.8a}$$

Equation (5.8a) clearly demonstrates that the effective pressure σ_z' is independent of the depth of water z_2 above the submerged soil surface. The total pore water and effective pressures at the bottom of the soil sample are as follows

Total pressure $\sigma_t = \sigma_C = (z_1 + z_2)\gamma_w + z_1\gamma_b$ (5.8b)

Pore water pressure $u_C = (z_1 + z_2)\gamma_w$ (5.8c)

Effective pressure $\sigma_C' = (\sigma_C - u_C) = z_1\gamma_b$ (5.8d)

The stress diagrams are shown in Fig. 5.2(b).

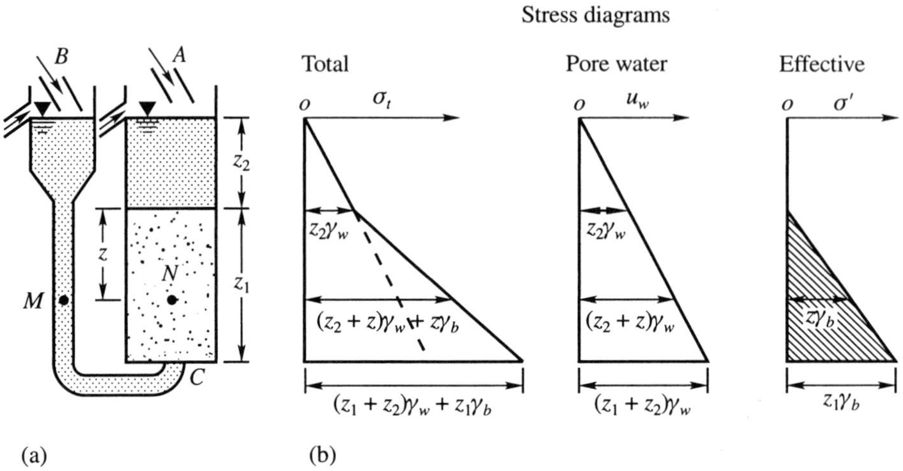

Figure 5.2 Stresses when no flow takes place

5.3 STRESSES WHEN FLOW TAKES PLACE THROUGH THE SOIL FROM TOP TO BOTTOM

In Fig. 5.3(a) the water surface in container *B* is kept at *h* units below the surface in *A*. This difference in head permits water to flow from container *A* to *B*.

Since container *B* with the flexible tube can be considered as a piezometer tube freely communicating with the bottom of container *A*, the piezometric head or the pore water pressure head at the bottom of container *A* is $(z_1 + z_2 - h)$. Therefore, the pore water pressure u_C at the bottom level is

$$u_C = (z_1 + z_2 - h)\gamma_w \tag{5.9}$$

As per Fig. 5.3(a), the pore water pressure at the bottom of container *A* when no flow takes place through the soil sample is

$$u_C = (z_1 + z_2)\gamma_w \tag{5.10}$$

It is clear from Eq. (5.9) and (5.10) that there is a decrease in pore water pressure to the extent of $h\gamma_w$ when water flows through the soil sample from top to bottom. It may be understood that this decrease in pore water pressure is not due to velocity of the flowing water. The value of the velocity head $V^2/2g$ is a negligible quantity even if we take the highest velocity of flow that is encountered in natural soil deposits. As in Fig. 5.2(a), the total pressure σ_C at the bottom of the container in this case also remains the same. Therefore,

$$\sigma_C = (z_1 + z_2)\gamma_w + z_1\gamma_b \tag{5.11}$$

The effective pressure σ_C' at the bottom of the container is

$$\sigma_C' = \sigma_C - u_C = (z_1 + z_2)\gamma_w + z_1\gamma_b - (z_1 + z_2 - h)\gamma_w = z_1\gamma_b + h\gamma_w \tag{5.12}$$

Equation (5.12) indicates that in this case there is an increase in the effective pressure by $h\gamma_w$ at the bottom of the container *A* as compared to the earlier case. The effective pressure at the top surface of the sample is zero as before. Therefore, the effective pressure σ_z' at any depth *z* can be written as

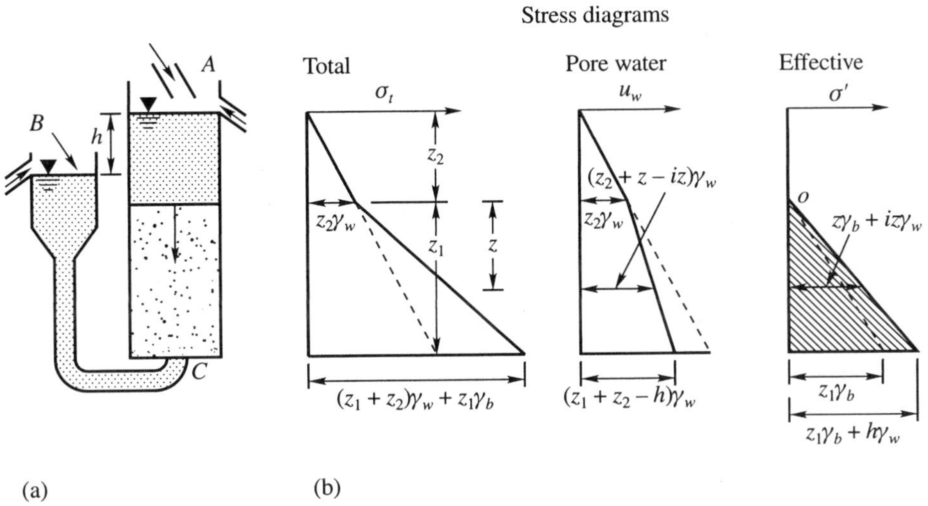

Figure 5.3 Stresses when flow takes place from top to bottom

$$\sigma'_z = \sigma'_C \frac{z}{z_1} = (z_1\gamma_b + h\gamma_w)\frac{z}{z_1} = z\gamma_b + \frac{hz\gamma_w}{z_1} \tag{5.13}$$

Equation (5.13) indicates that $hz\gamma_w/z_1$ is the increase in the effective pressure as the water flows from the surface to a depth z. This increase in effective pressure due to the flow of water through the pores of the soil is known as *seepage pressure*. It may be noted that h is the total loss of head as the water flows from the top surface of the sample to a depth z_1.

The corresponding loss of head at depth z is $(z/z_1)h$. Since $(h/z_1) = i$, the *hydraulic gradient*, the loss of head at depth z can be expressed as iz. Therefore the seepage pressure at any depth may be expressed as $iz\gamma_w$. The effective pressure at depth z can be written as

$$\sigma'_z = z\gamma_b + iz\gamma_w \tag{5.14}$$

The distribution of pore water and effective pressures are shown in Fig. 5.3(b). In normal soil deposits when flow takes place in the direction of gravity there will be an increase in the effective pressure.

5.4 STRESSES WHEN FLOW TAKES PLACE THROUGH THE SOIL FROM BOTTOM TO TOP

In Fig. 5.4(a), the water surface in container B is kept above that of A by h units. This arrangement permits water to flow upwards through the sample in container A. The total piezometric or the pore water head at the bottom of the sample is given by

$$(z_1 + z_2 + h)$$

Therefore, the pore water pressure u_C at the bottom of the sample is

$$u_C = (z_1 + z_2 + h)\gamma_w \tag{5.15}$$

As before the total pressure head σ_C at the bottom of the sample is

$$\sigma_C = (z_1 + z_2)\gamma_w + z_1\gamma_b \tag{5.16}$$

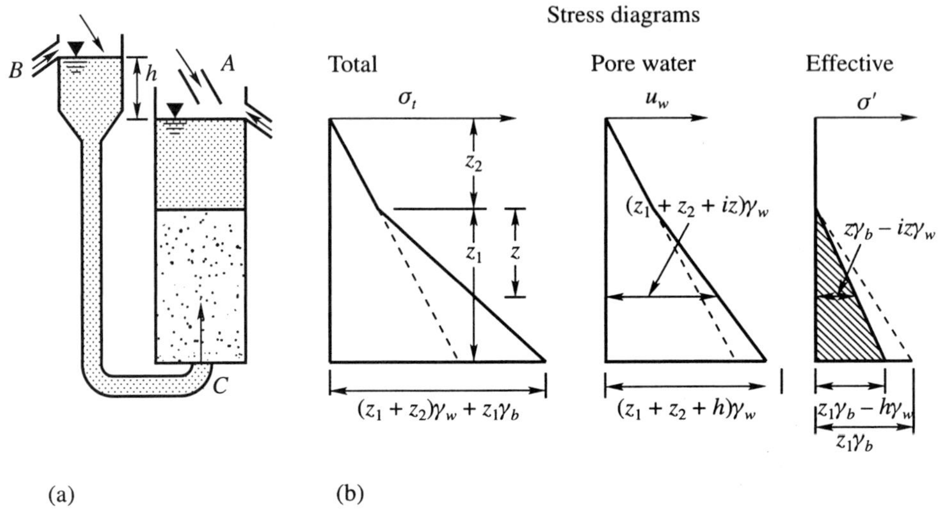

Figure 5.4 Stresses when flow takes place from bottom to top

The effective pressure σ_C' at the bottom of sample is, therefore,

$$\sigma_C' = \sigma_C - u_C = (z_1 + z_2)\gamma_w + z_1\gamma_b - (z_1 + z_2 + h)\gamma_w = z_1\gamma_b - h\gamma_w \tag{5.17}$$

As in Eq. (5.14) the effective pressure at any depth z can be written as

$$\sigma_z' = z\gamma_b - iz\gamma_w \tag{5.18}$$

Equation (5.18) indicates that there is a decrease in the effective pressure due to upward flow of water. At any depth z, $z\gamma_b$ is the pressure of the submerged soil acting downward and $iz\gamma_b$ is the seepage pressure acting upward. The effective pressure σ_z' reduces to zero when these two pressures balance. This happens when

$$\sigma_z' = z\gamma_b - iz\gamma_w = 0 \quad \text{or} \quad i = i_c = \frac{\gamma_b}{\gamma_w} \tag{5.19}$$

Equation (5.19) indicates that the effective pressure reduces to zero when the hydraulic gradient attains a maximum value which is equal to the ratio of the submerged unit weight of soil and the unit weight of water. This gradient is known as the *critical hydraulic gradient* i_c. In such cases, cohesionless soils lose all of their shear strength and bearing capacity and a visible agitation of soil grains is observed. This phenomenon is known as *boiling* or a *quick sand condition*. By substituting in Eq. (5.19) for γ_b

$$\gamma_b = \frac{\gamma_w(G_s - 1)}{1+e}, \quad \text{we have}$$

$$i_c = \frac{G_s - 1}{1+e} \tag{5.20}$$

The critical gradient of natural granular soil deposits can be calculated if the void ratios of the deposits are known. For all practical purposes the specific gravity of granular materials can be assumed as equal to 2.65. Table 5.1 gives the critical gradients of granular soils at different void ratios ranging from 0.5 to 1.0.

Table 5.1 Critical hydraulic gradients of granular soils

Soil No.	Void ratio	i_c
1	0.5	1.10
2	0.6	1.03
3	0.7	0.97
4	0.8	0.92
5	1.0	0.83

It can be seen from Table 5.1 that the critical gradient decreases from 1.10 by about 25 percent only as the void ratio increases by 100 percent from an initial value of 0.5 to 1.0. The void ratio of granular deposits generally lies within the range of 0.6 to 0.7 and as such a critical gradient of unity can justifiably be assumed for all practical purposes. It should be remembered that a quick condition does not occur in clay deposits since the cohesive forces between the grains prevent the soil from boiling.

Quick conditions are common in excavations below the ground water table. This can be prevented by lowering the ground water elevation by pumping before excavation. Quick conditions occur most often in fine sands or silts and cannot occur in coarse soils. The larger the particle size, the greater is the porosity. To maintain a critical gradient of unity, the velocity at which water must be supplied at the point of inflow varies as the permeability. Therefore a quick condition cannot occur in a coarse soil unless a large quantity of water can be supplied.

5.5 EFFECTIVE PRESSURE DUE TO CAPILLARY WATER RISE IN SOIL

The term water level, water table and phreatic surface designate the locus of the levels to which water rises in observation wells in free communication with the voids of the soil at a site. The water table can also be defined as the surface at which the neutral stress u_w in the soil is equal to zero.

If the water contained in the soil were subjected to no force other than gravity, the soil above the water table would be perfectly dry. In reality, every soil in the field is completely saturated above this level up to a certain height. The water that occupies the voids of the soil located above the water table constitutes *soil moisture*.

If the lower part of the mass of dry soil comes into contact with water, the water rises in the voids to a certain height above the free water surface. The upward flow into the voids of the soil is attributed to the *surface tension* of the water. The height to which water rises above the water table against the force of gravity is called *capillary rise*. The height of capillary rise is greatest for very fine grained soil materials. The water that rises above the water table attains the maximum height h_c only in the smaller voids. A few large voids may effectively stop capillary rise in certain parts of the soil mass. As a consequence, only a portion of the capillary zone above the free water surface remains fully saturated and the remainder is partially saturated.

The seat of the surface tension is located at the boundary between air and water. Within the boundary zone the water is in a state of tension comparable to that in a stretched rubber membrane attached to the walls of the voids of a soil. However, in contrast to the tension in a stretched membrane, the surface tension in the boundary film of water is entirely unaffected by either the contraction or stretching of the film. The water held in the pores of soil above the free water surface is retained in a state of reduced pressure. This reduced pressure is called *capillary pressure* or *soil moisture suction pressure*.

The existence of surface tension can be demonstrated as follows:

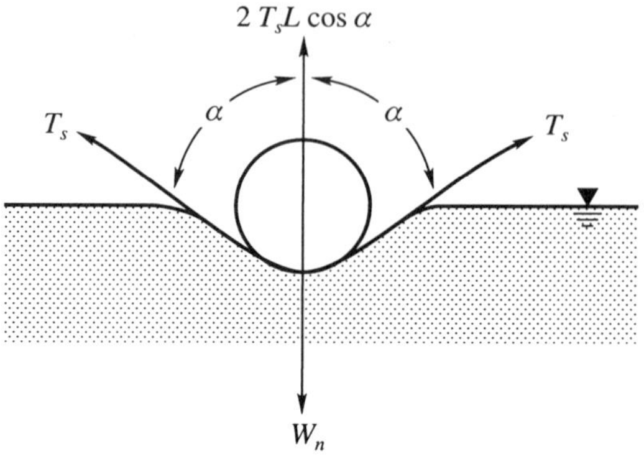

Figure 5.5 Needle smeared with grease floating on water

A greased sewing needle, Fig. 5.5, can be made to float on water because water has no affinity to grease, and, therefore, the water surface curves down under the needle until the upward component of the surface tension is large enough to support the weight of the needle. In Fig. 5.5, T_s is the surface tension per unit length of the needle and W_n the weight of the needle. The upward vertical force due to surface tension is $2T_sL\cos\alpha$, where L is the length of the needle. The needle floats when this vertical force is greater than the weight of the needle W_n acting downwards.

Rise of Water in Capillary Tubes

The phenomenon of capillary rise can be demonstrated by immersing the lower end of a very small diameter glass tube into water. Such a tube is known as capillary tube. As soon as the lower end of the tube comes into contact with water, the attraction between the glass and the water molecules combined with the surface tension of the water pulls the water up into the tube to a height h_c above the water level as shown in Fig. 5.6(a). The height h_c is known as the height of capillary rise. The upper surface of water assumes the shape of a cup, called the '*meniscus*' that joins the walls of the tube at an angle α known as the *contact angle*.

On the other hand, if the tube is dipped into mercury a depression of the surface develops in the tube below the surface of the mercury, with the formation of a convex meniscus as shown in Fig. 5.6(b). The reason for the difference between the behavior of water and mercury resides in the different affinity between the molecules of the solid and water or mercury. If there is a strong affinity between the molecules of the solid and the liquid, the surface of the liquid will climb up on the wall of the solid until a definite contact angle α is established. The contact angle between a clean moist glass surface and water is zero, that is, the water surface touches the glass surface tangentially. For the case of a dry glass surface and water, α is not a constant. It may be as high as 45° at first then gradually reducing to much smaller values. Probably the inevitable contamination of surfaces cleaned by ordinary methods, and the humidity of air are responsible for such variations. Fig. 5.6(c) shows the contact angles between water and the surfaces under different conditions.

Surface Tension

Surface tension is a force that exists at the surface of the meniscus. Along the line of contact between the meniscus in a tube and the walls of the tube itself, the surface tension, T_s, is expressed as the force per unit length acting in the direction of the tangent as shown in Fig. 5.7(a). The components of this force along the wall and perpendicular to the wall are

Along the wall = $T_s\cos\alpha$ per unit length of wall

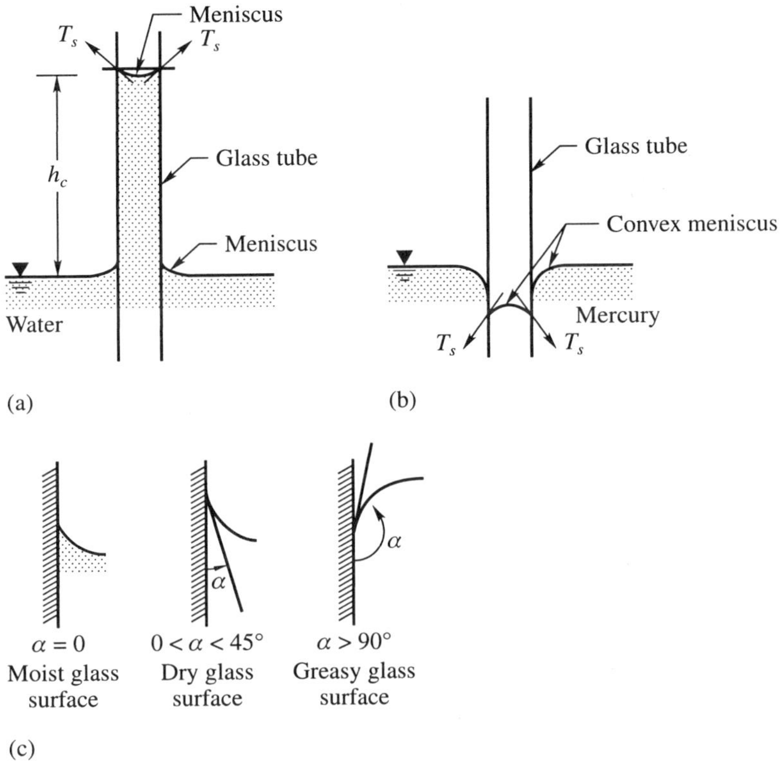

Figure 5.6 Capillary rise and meniscus

Normal to the wall = $T_s \sin \alpha$ per unit length of wall.

The force normal to the wall tries to pull the walls of the tube together and the one along the wall produces a compressive force in the tube below the line of contact.

The meniscus can be visualized as a suspension bridge in three dimensions which is supported on the walls of the tube. The column of water of height h_c below the meniscus is suspended from this bridge by means of the molecular attraction of the water molecules. If the meniscus has stopped moving upward in the tube, then there must be equilibrium between the weight of the column of water suspended from the meniscus and the force with which the meniscus is clinging to the wall of the tube. We can write the following equation of equilibrium

$$\pi d T_s \cos \alpha = \frac{\pi d^2 h_c \gamma_w}{4} \quad \text{or} \quad h_c = \frac{4T_s \cos \alpha}{d\gamma_w} \tag{5.21}$$

The surface tension T_s for water at 20 °C can be taken as equal to 75 × 10^{-8} kN per cm. The surface tensions of some of the common liquids are given in Table 5.2.

Equation (5.21) can be simplified by assuming $\alpha = 0$ for moist glass and by substituting for T_s. Therefore, for the case of water, the capillary height h_c can be written as

$$h_c = \frac{4T_s}{d\gamma_w} = \frac{4 \times 75 \times 10^{-8} \times 10^6}{d \times 9.81} = \frac{0.3}{d} \tag{5.22}$$

In Eq. (5.22) h_c and d are expressed in cm, and, $\gamma_w = 9.81$ kN/m^3.

Table 5.2 Surface tension of some liquids at 20 °C

Liquids	T_s kN/cm × 10^{-8}
Ethyl Alcohol	22.03
Benzene	28.90
Carbon Tetra Chloride	26.80
Mercury	573.00
Petroleum	26.00
Water	75.00

Stress Distribution in Water Below the Meniscus

Figure 5.7(b) shows a capillary tube with its bottom end immersed in water. The pressure is atmospheric at points *A* and *B*. Since point *C* is at the same level as *A*, according to the laws of hydraulics, the pressure at *C* is also atmospheric. Since the point *D* which is just below the meniscus is higher than point *C* by the head h_c, the pressure at *D* must be less than atmospheric by the amount $h_c\gamma_w$. Therefore, the pressure at any point in water between *C* and *D* is less than atmospheric. That means, the water above point *C* is in tension if we refer to atmospheric pressure as zero pressure. The tension in water at any height *h* above *C* is given by $h\gamma_w$. By contrast, the pressure in the water below the free surface *A* is above atmospheric and therefore is in compression. The stress distribution in water is given in Fig. 5.7(b).

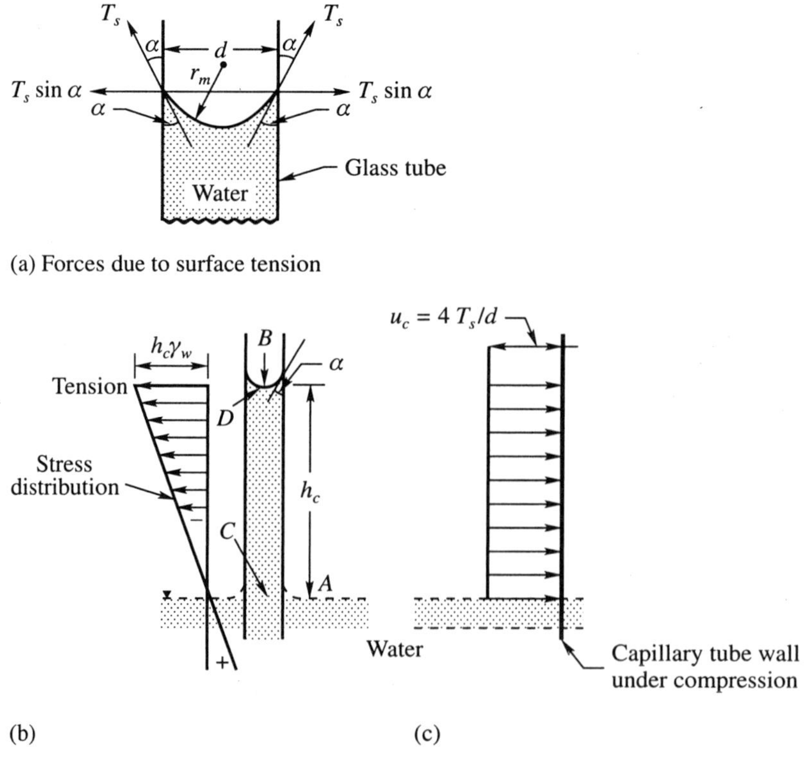

Figure 5.7 Capillary pressure

Thus the tension u_w in water immediately below the meniscus is given by

$$u_w = -h_c \gamma_w = -\frac{4T_s \cos\alpha}{d} \tag{5.23}$$

If r_m is the radius of the meniscus, Fig. 5.7(a), we can write,

$$r_m = \frac{d}{2\cos\alpha} \quad \text{or} \quad d = 2r_m \cos\alpha$$

Substituting for d in Eq. (5.23), we have

$$u_w = -\frac{4T_s \cos\alpha}{2r_m \cos\alpha} = -\frac{2T_s}{r_m} \tag{5.24}$$

It may be noted here that at the level of the meniscus the magnitude of the capillary pressure u_c that compresses the wall of the tube is also equal to the capillary tension in the water just below the meniscus. The magnitude of the capillary pressure u_c remains constant with depth as shown in Fig. 5.7(c) whereas the capillary tension, u_w, in water varies from a maximum of $h_c\gamma_w$ at the meniscus level to zero at the free water surface level as shown in Fig. 5.7(b).

Capillary Rise of Water in Soils

In contrast to capillary tubes the continuous voids in soils have a variable width. They communicate with each other in all directions and constitute an intricate network of voids. When water rises into the network from below, the lower part of the network becomes completely saturated. In the upper part, however, the water occupies only the narrowest voids and the wider areas remain filled with air.

Fig. 5.8(a) shows a glass tube filled with fine sand. Sand would remain fully saturated only up to a height h'_c which is considerably smaller than h_c. A few large voids may effectively stop capillary rise in certain parts. The water would rise, therefore, to a height of h_c only in the smaller voids. The zone between the depths $(h_c - h'_c)$ will remain partially saturated.

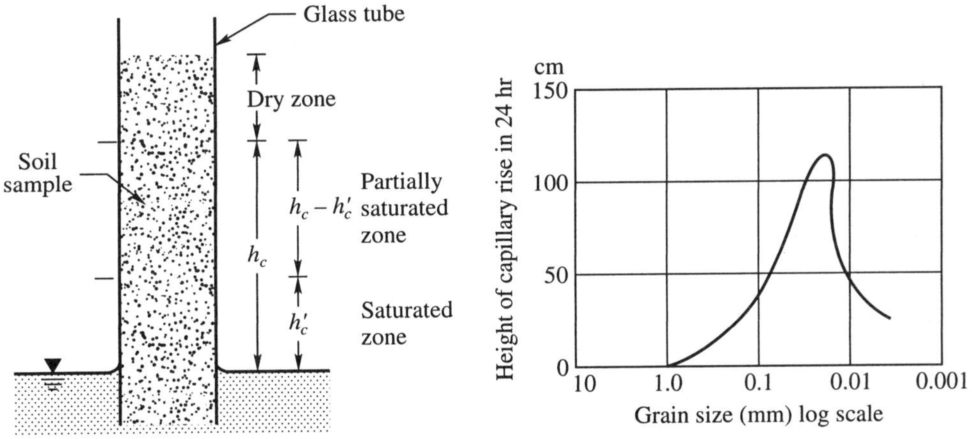

(a) Height of capillary rise

(b) Rate of capillary rise in soil consisting of uniform quartz powder

Figure 5.8 Capillary rise in soils

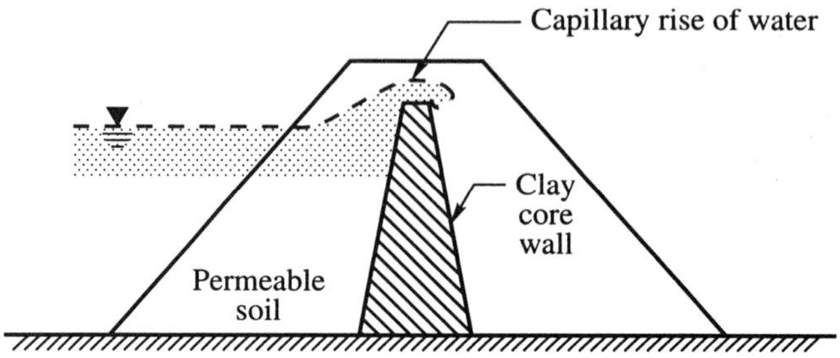

Figure 5.9 Capillary siphoning

The height of the capillary rise is greatest for very fine grained soils materials, but the rate of rise in such materials is slow because of their low permeability. Fig. 5.8(b) shows the relationship between the height of capillary rise in 24 hours and the grain size of a uniform quartz powder. This clearly shows that the rise is a maximum for materials falling in the category of silts and fine sands.

As the effective grain size decreases, the size of the voids also decreases, and the height of capillary rise increases. A rough estimation of the height of capillary rise can be determined from the equation,

$$h_c = \frac{C}{eD_{10}} \tag{5.25}$$

in which e is the void ratio, D_{10} is *Hazen's effective diameter* in centimeters, and C is an empirical constant which can have a value between 0.1 and 0.5 sq. cm.

Capillary Siphoning

Capillary forces are able to raise water against the force of gravity not only into capillary tubes or the voids in columns of dry soil, but also into narrow open channels or V-shaped grooves. If the highest point of the groove is located below the level to which the surface tension can lift the water, the capillary forces will pull the water into the descending part of the groove and will slowly empty the vessel. This process is known as *capillary siphoning*. The same process may also occur in the voids of soil. For example, water may flow over the crest of an impermeable core in a dam in spite of the fact that the elevation of the free water surface is below the crest of the core as shown in Fig. 5.9.

Capillary Pressure in Soils

The tension u_w in water just below the meniscus is given by Eq. (5.23) as

$$u_w = -\frac{4T_s \cos\alpha}{d}$$

Since this pressure is below atmospheric pressure, it draws the grains of soils closer to each other at all points where the menisci touch the soil grains. Intergranular pressure of this type is called *capillary pressure*. The effective or intergranular pressure at any point in a soil mass can be expressed by

$$\sigma' = \sigma_t - u_w \tag{5.26}$$

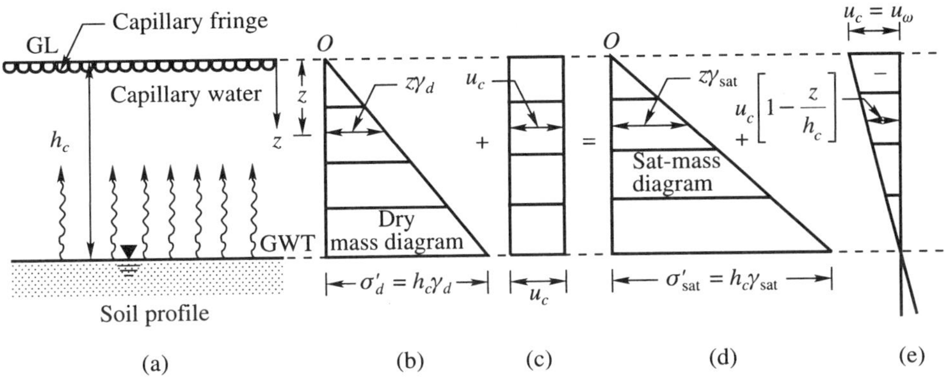

Figure 5.10 Effect of capillary pressure u_c on soil vertical stress diagram

where σ_t is the total pressure, σ' is the effective or the intergranular pressure and u_w is the pore water pressure. When the water is in compression u_w is positive, and when it is in tension u_w is negative. Since u_w is negative in the capillary zone, the intergranular pressure is increased by u_w. The equation, therefore, can be written as

$$\sigma' = \sigma_t - (-u_w) = \sigma_t + u_w \tag{5.27}$$

The increase in the intergranular pressure due to capillary pressure acting on the grains leads to greater strength of the soil mass.

Stress Condition in Soil due to Surface Tension Forces

It is to be assumed here that the soil above the ground water table remains dry prior to the rise of capillary water. The stress condition in the dry soil mass changes due to the rise of capillary water.

Now consider the soil profile given in Fig. 5.10(a). When a dry soil mass above the GWT comes in contact with water, water rises by capillary action. Let the height of rise be h_c and assume that the soil within this zone becomes saturated due to capillary water. Assume that the menisci formed at height h_c coincide with the ground surface. The plane of the menisci is called the *capillary fringe*.

The vertical stress distribution of the dry soil mass is shown in Fig 5.10(b). The vertical stress distribution of the saturated mass of soil is given in Fig 5.10(d). The tension in the water is maximum at the menisci level, say equal to u_w and zero at the GWT level as shown in Fig. 5.10(e).

Prior to capillary rise the maximum pressure of the dry mass, σ'_d, at the GWT level is

$$\sigma'_d = \gamma_d h_c$$

where, γ_d = dry unit weight of soil.

After the capillary rise, the maximum pressure of the saturated weight of soil at the GWT level is

$$\sigma'_{sat} = \gamma_{sat} h_c$$

Since the pore water pressure at the GWT level is zero, it is obvious that the difference between the two pressures σ'_{sat} and σ'_d represents the increase in pressure due to capillary rise which is actually the capillary pressure, which may be expressed as

$$u_c = h_c(\gamma_{sat} - \gamma_d) \tag{a}$$

By substituting for

$$\gamma_{sat} = \frac{(G_s + e)\gamma_w}{1+e}, \text{ and } \gamma_d = \frac{G_s\gamma_w}{1+e}$$

in Eq. (a), we have, after simplifying

$$u_c = \frac{e}{1+e} h_c \gamma_w = n h_c \gamma_w \tag{5.28}$$

where, e = void ratio,

n = porosity

It is clear from Eq. (5.28) that the capillary pressure for soil is directly proportional to the porosity of the soil and this pressure is very much less than $h_c\gamma_w$ which is used only for a fine bore and uniform diameter capillary tube.

The distribution of capillary pressure u_c (constant with depth) is given in Fig. 5.10(c). The following equation for the pressure at any depth z may be written as per Fig. 5.10

$$z\gamma_d + u_c = z\gamma_{sat} + u_c\left[1 - \frac{z}{h_c}\right]. \tag{5.29}$$

Example 5.1

The depth of water in a well is 3 m. Below the bottom of the well lies a layer of sand 5 meters thick overlying a clay deposit. The specific gravity of the solids of sand and clay are respectively 2.64 and 2.70. Their water contents are respectively 25 and 20 percent. Compute the total, intergranular and pore water pressures at points A and B shown in Fig. Ex. 5.1.

Solution

The formula for the submerged unit weight is

$$\gamma_b = \frac{\gamma_w(G_s - 1)}{1+e}$$

Since the soil is saturated,

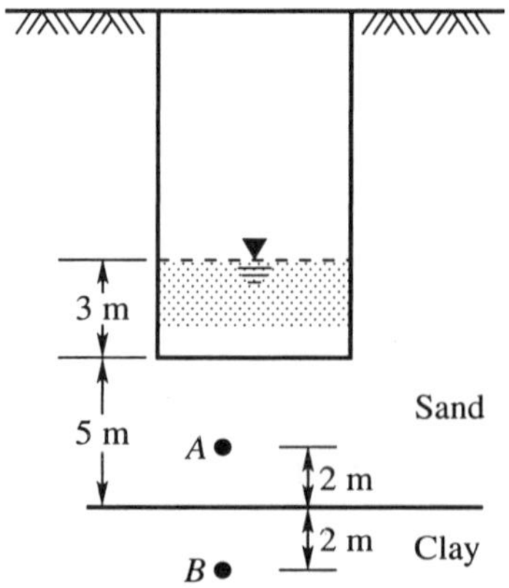

Figure Ex. 5.1

$$e = wG_s, \quad \gamma_b = \frac{\gamma_w (G_s - 1)}{1 + wG}$$

$$\text{For sand,} \quad \gamma_b = \frac{9.81(2.64 - 1)}{1 + 0.25 \times 2.64} = 9.7 \text{ kN/m}^3$$

$$\text{For clay,} \quad \gamma_b = \frac{9.81(2.70 - 1)}{1 + 0.20 \times 2.70} = 10.83 \text{ kN/m}^3$$

Pressure at point A

(i) Total pressure = 3 × 9.7 (sand) + 6 × 9.81 = 29.1 + 58.9 = 88 kN/m^2

(ii) Effective pressure = 3 × 9.7 = 29.1 kN/m^2

(iii) Pore water pressure = 6 × 9.81 = 58.9 kN/m^2

Pressure at point B

(i) Total pressure = 5 × 9.7 + 2 × 10.83 + 10 × 9.81 = 168.3 kN/m^2

(ii) Intergranular pressure = 5 × 9.7 + 2 × 10.83 = 70.2 kN/m^2

(iii) Pore water pressure = 10 × 9.81 = 98.1 kN/m^2

Example 5.2

If water in the well in example 5.1 is pumped out up to the bottom of the well, estimate the change in the pressures at points *A* and *B* given in Fig. Ex. 5.1.

Solution

Change in pressure at points *A* and *B*

(i) Change in total pressure	=	decrease in water pressure due to pumping
	=	3 × 9.81 = 29.43 kN/m^2
(ii) Change in effective pressure	=	0
(iii) Change in pore water pressure	=	decrease in water pressure due to pumping
	=	3 × 9.81 = 29.43 kN/m^2

Example 5.3

A trench is excavated in fine sand for a building foundation, up to a depth of 13 ft. The excavation was carried out by providing the necessary side supports for pumping water. The water levels at the sides and the bottom of the trench are as given Fig. Ex. 5.3. Examine whether the bottom of the trench is subjected to a quick condition if $G_s = 2.64$ and $e = 0.7$. If so, what is the remedy?

Solution

As per Fig. Ex. 5.3 the depth of the water table above the bottom of the trench = 10 ft. The sheeting is taken 6.5 ft below the bottom of the trench to increase the seepage path.

The equation for the critical gradient is $i_c = \dfrac{G_s - 1}{1 + e}$

If the trench is to be stable, the hydraulic gradient, *i*, prevailing at the bottom should be less than i_c. The hydraulic gradient *i* is

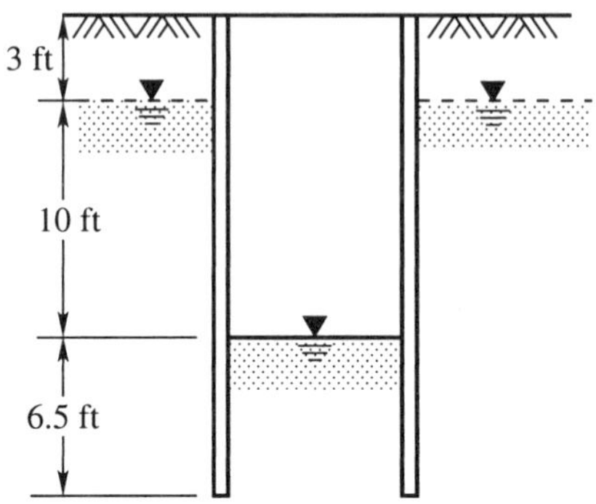

Figure Ex. 5.3

$$i = \frac{h}{L},$$

There will be no quick condition if,

$$\frac{h}{L} < \frac{G_s - 1}{1 + e}$$

From the given data

$$i_c = \frac{2.64 - 1}{1 + 0.7} = \frac{1.64}{1.7} = 0.96$$

$$\frac{h}{L} = \frac{10}{6.5} = 1.54$$

It is obvious that $h/L > i_c$. There will be quick condition.

Remedy:

(i) Increase L to at least a 13 ft depth below the bottom of trench so that $h/L = 0.77$ which gives a margin of factor of safety.

or (ii) Keep the water table outside the trench at a low level by pumping out water. This reduces the head h.

or (iii) Do not pump water up to the bottom level of the trench. Arrange the work in such a way that the work may be carried out with some water in the trench.

Any suggestion given above should be considered by keeping in view the site conditions and other practical considerations.

Example 5.4

A clay layer 3.66 m thick rests beneath a deposit of submerged sand 7.92 m thick. The top of the sand is located 3.05 m below the surface of a lake. The saturated unit weight of the sand is 19.62 kN/m^3 and of the clay is 18.36 kN/m^3.

Compute (a) the total vertical pressure, (b) the pore water pressure, and (c) the effective vertical pressure at mid height of the clay layer (Refer to Fig. Ex. 5.4).

Solution

(a) Total pressure

The total pressure σ_t over the midpoint of the clay is due to the saturated weights of clay and sand layers plus the weight of water over the bed of sand, that is

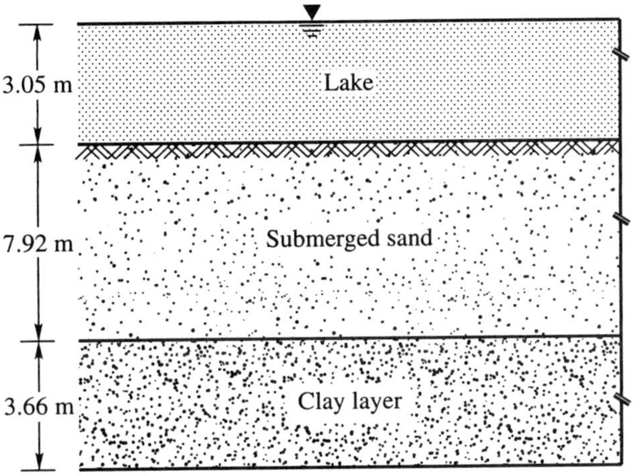

Figure Ex. 5.4

$$\sigma_t = \frac{3.66}{2} \times 18.36 + 7.92 \times 19.62 + 3.05 \times 9.81 = 33.6 + 155.4 + 29.9 = 218.9 \text{ kN/m}^2$$

(b) Pore water pressure is due to the total water column above the midpoint.
That is

$$u_w = \frac{3.66}{2} \times 9.81 + 7.92 \times 9.81 + 3.05 \times 9.81 = 125.6 \text{ kN/m}^2$$

(c) Effective vertical pressure

$$\sigma_t - u_w = \sigma' = 218.9 - 125.6 = 93.3 \text{ kN/m}^2$$

Example 5.5

The surface of a saturated clay deposit is located permanently below a body of water as shown in Fig. Ex. 5.5. Laboratory tests have indicated that the average natural water content of the clay is 47% and that the specific gravity of the solid matter is 2.74. What is the vertical effective pressure at a depth of 37 ft below the top of the clay.

Solution

To find the effective pressure, we have to find first the submerged unit weight of soil expressed as

$$\gamma_b = \frac{(G_s - 1)\gamma_w}{1 + e}$$

Now from Eq. (3.14a), $e = \frac{wG_s}{S} = wG_s$ since $S = 1$

$$\text{or} \quad e = 0.47 \times 2.74 = 1.29$$

Therefore,

$$\gamma_b = \frac{(2.74 - 1.00) \times 62.4}{1 + 1.29} = 47.41 \text{ lb/ ft}^3$$

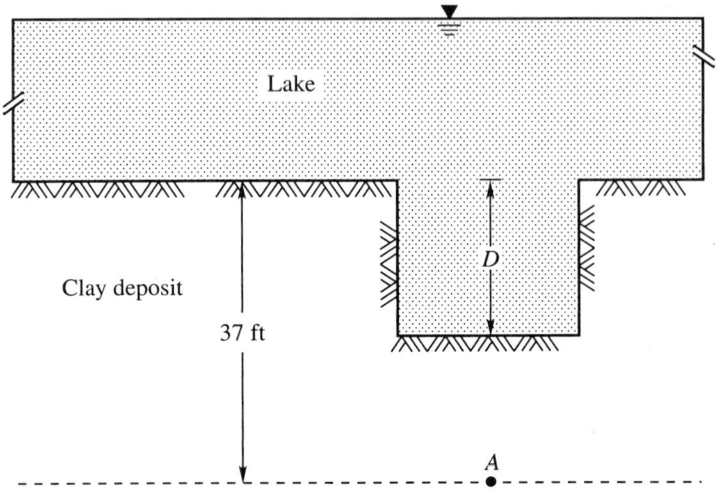

Figure Ex. 5.5

Effective pressure, $\sigma' = 37 \times 47.41 = 1754$ lb/ft^2

Example 5.6

If the water level in Ex. 5.5 remains unchanged and an excavation is made by dredging, what depth of clay must be removed to reduce the effective pressure at point *A* at a depth of 37 ft by 1000 lb/ft^2? (Fig. Ex. 5.5)

Solution

As in Ex. 5.5, $\gamma_b = 47.41$ lb/ft^3, let the depth of excavation be *D*. The effective depth over the point *A* is (37 – *D*) ft. The depth of *D* must be such which gives an effective pressure of (1754 – 1000) lb/ft^3 = 754 lb/ft^2

or $(37 - D) \times 47.41 = 754$

or $D = \dfrac{37 \times 47.41 - 754}{47.41} = 21.1$ ft

Example 5.7

The water table is lowered from a depth of 10 ft to a depth of 20 ft in a deposit of silt. All the silt is saturated even after the water table is lowered. Its water content is 26%. Estimate the increase in the effective pressure at a depth of 34 ft on account of lowering the water table. Assume $G_s = 2.7$.

Solution

Effective pressure before lowering the water table.

The water table is at a depth of 10 ft and the soil above this depth remains saturated but not submerged. The soil from 10 ft to 20 ft remains submerged. Therefore, the effective pressure at 34 ft depth is

$$\sigma_1' = 10\gamma_{sat} + (34 - 10)\gamma_b$$

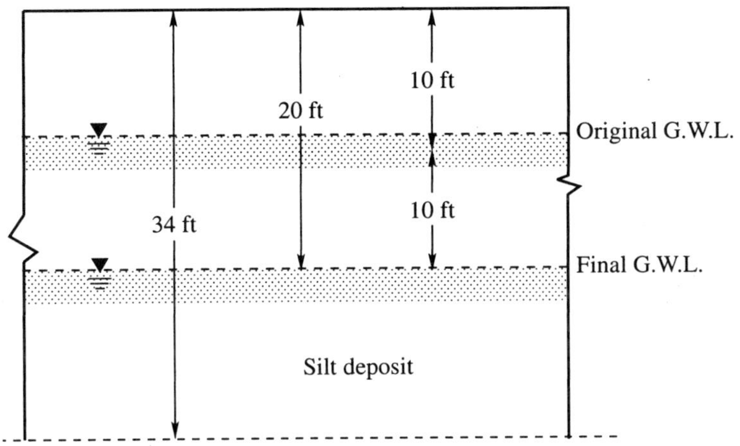

Figure Ex. 5.7

Now, $\gamma_{sat} = \frac{\gamma_w(G_s + e)}{1+e}$, $\gamma_b = \frac{\gamma_w(G_s - 1)}{1+e}$, $\gamma_w = 62.4$ lb/ft^3, $e = wG_s$ for $S = 1$

Therefore, $e = 0.26 \times 2.7 = 0.70$

$$\gamma_{sat} = \frac{62.4(2.7+0.7)}{1+0.7} = 124.8 \text{ lb/ft}^3$$

$$\gamma_b = \frac{62.4(2.7-1)}{1+0.7} = 62.4 \text{ lb/ft}^3$$

$$\sigma_1' = 10 \times 124.8 + 24 \times 62.4 = 2745.6 \text{ lb/ft}^2$$

Effective pressure after lowering of water table

After lowering the water table to a depth of 20 ft, the soil above this level remains saturated but effective and below this submerged. Therefore, the altered effective pressure is

$$\sigma_2' = 20\gamma_{sat} + (34-20)\gamma_b = 20 \times 124.8 + 14 \times 62.4 = 3369.6 \text{ lb/ft}^2$$

The increase in the effective pressure is

$$\sigma_2' - \sigma_1' = \Delta\sigma' = 3369.6 - 2745.6 = 624.0 \text{ lb/ft}^2$$

Example 5.8

Compute the critical hydraulic gradients for the following materials: (a) Coarse gravel, $k = 10$ cm/sec, $G_s = 2.67$, $e = 0.65$ (b) sandy silt, $k = 10^{-6}$ cm/sec, $G_s = 2.67$, $e = 0.80$

Solution

As per Eq. (5.20), the critical gradient i_c may be expressed as

$$i_c = \frac{G_s - 1}{1+e}$$

(a) Coarse gravel

$$i_c = \frac{2.67-1}{1+0.65} = 1.01$$

(b) Sandy silt

$$i_e = \frac{2.67-1}{1+0.80} = 0.93$$

Example 5.9

A large excavation is made in a stiff clay whose saturated unit weight is 109.8 lb/ft^3. When the depth of excavation reaches 24.6 ft, cracks appear and water begins to flow upward to bring sand to the surface. Subsequent borings indicate that the clay is underlain by sand at a depth of 36.1 ft below the original ground surface.

What is the depth of the water table outside the excavation below the original ground level?

Solution

Making an excavation in the clay creates a hydraulic gradient between the top of the sand layer and the bottom of the excavation. As a consequence, water starts seeping in an upward direction from the sand layer towards the excavated floor. Because the clay has a very low permeability, flow equilibrium can only be reached after a long period of time. The solution must be considered over a short time interval.

The floor of the excavation at depth d is stable only if the water pressure σ_w at the top of the sand layer at a depth of 36.1 ft is counterbalanced by the saturated weight σ_c per unit area of the clay above it disregarding the shear strength of the clay.

Let H = total thickness of clay layer = 36.1 ft, d = depth of excavation in clay = 24.6 ft, h = depth of water table from ground surface, γ_{sat} = saturated unit weight of the clay.

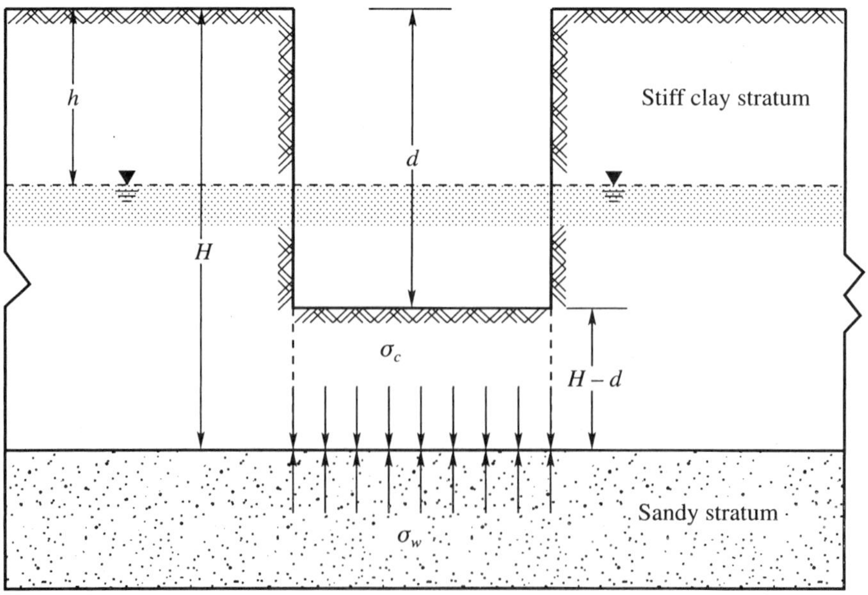

Figure Ex. 5.9

$(H - d) = 36.1 - 24.6 = 11.5$ ft, the thickness of clay strata below the bottom of the trench.

$$\sigma_c = \gamma_{sat}(H - d) = 109.8 \times 11.5 = 1263 \text{ lb/ft}^2$$

$$\sigma_w = \gamma_w (H - h) = 62.4 \times (36.1 - h) \text{ lb/ft}^2$$

cracks may develop when $\sigma_c = \sigma_w$

$$\text{or} \quad 1263 = 62.4(36.1 - h), \quad \text{or} \quad h = 36.1 - \frac{1263}{62.4} = 15.86 \text{ ft}$$

Example 5.10

The water table is located at a depth of 3.0 m below the ground surface in a deposit of sand 11.0 m thick (Fig. Ex. 5.10). The sand is saturated above the water table. The total unit weight of the sand is 20 kN/m^3. Calculate the (a) the total pressure, (b) the pore water pressure and (c) the effective pressure at depths 0, 3.0, 7.0, and 11.0 m from the ground surface, and draw the pressure distribution diagram.

Solution

$\gamma_{sat} = 20$ kN/m^3, $\gamma_b = 20 - 9.81 = 10.19$ kN/m^3

Depth (m)	Total pressure σ_t (kN/m^2)	Pore water pressure u_w(kN/m^2)	Effective pressure σ' (kN/m^2)
0	0	0	0
3	3 × 20 = 60	0	60
7	7 × 20 = 140.00	4 × 9.81 = 39.24	100.76
11	11 × 20 = 220.00	8 × 9.81 = 78.48	141.52

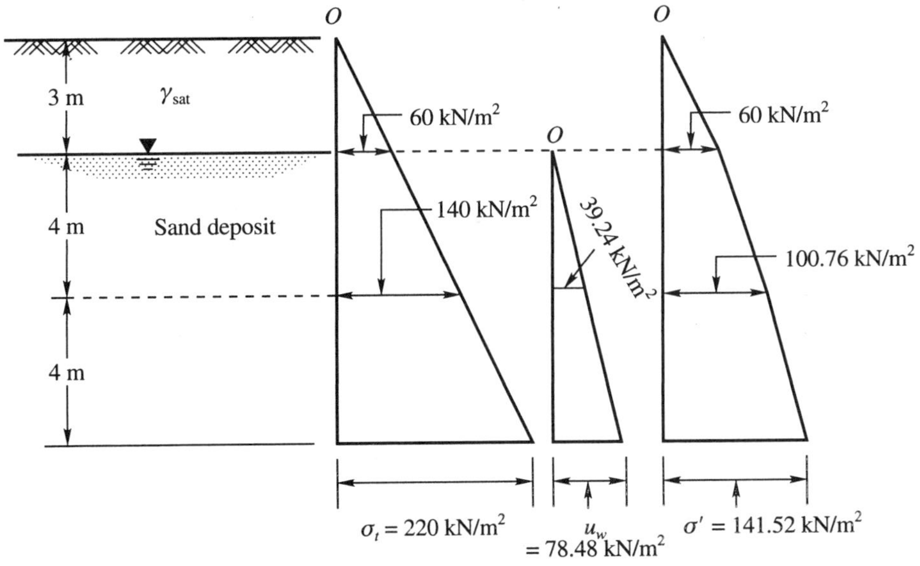

Figure Ex. 5.10

The pressure distribution diagrams of σ_t, u_w and σ' are given in Fig. Ex. 5.10.

Example 5.11

A clay stratum 8.0 m thick is located at a depth of 6 m from the ground surface. The natural moisture content of the clay is 56% and $G_s = 2.75$. The soil stratum between the ground surface and the clay consists of fine sand. The water table is located at a depth of 2 m below the ground surface. The submerged unit weight of fine sand is 10.5 kN/m^3, and its moist unit weight above the water table is 18.68 kN/m^3. Calculate the effective stress at the center of the clay layer.

Solution

Fine sand:

Above water table: $\gamma_t = 18.68$ kN/m^3

Below WT: $\gamma_b = 10.5$ kN/m^3

$\gamma_{sat} = 10.5 + 9.81 = 20.31$ kN/m^3

Clay stratum:

For $S = 1.0$,

$$e = wG_s = 0.56 \times 2.75 = 1.54$$

$$\gamma_{sat} = \frac{\gamma_w (G_s + e)}{1+e} = \frac{9.81(2.75 + 1.54)}{1 + 1.54} = 16.57 \text{ kN/m}^3$$

$$\gamma_b = 16.57 - 9.81 = 6.76 \text{ kN/m}^3$$

At a depth 10.0 m from GL, that is, at the center of the clay layer,

$$\sigma_t = 2 \times 18.68 + 4 \times 20.31 + 4 \times 16.57$$

$$= 37.36 + 81.24 + 66.28 = 184.88 \text{ kN/m}^2$$

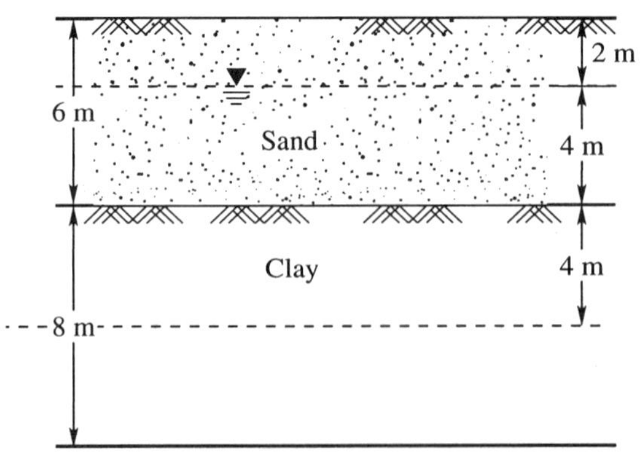

Figure Ex. 5.11

$$u_w = 4 \times 9.81 + 4 \times 9.81 = 39.24 + 39.24 = 78.48 \text{ kN/m}^2$$

Effective stress, $\sigma' = \sigma_t - u_w = 184.88 - 78.48 = 106.40 \text{ kN/m}^2$

Example 5.12

A 39.4 ft thick layer of relatively impervious saturated clay lies over a gravel aquifer. Piezometer tubes introduced to the gravel layer show an artesian pressure condition with the water level standing in the tubes 9.8 ft above the top surface of the clay stratum. The properties of the clay are $e = 1.2$, $G_s = 2.7$ and $\gamma_{sat} = 110.62$ lb/ft^3.

Determine (a) the effective stress at the top of the gravel stratum layer, and (b) the depth of excavation that can be made in the clay stratum without bottom heave.

Solution

(a) At the top of the gravel stratum

$$\sigma_c = 39.4 \times 110.62 = 4358.43 \text{ lb/ft}^2$$

The pore water pressure at the top of the gravel is

$$u_w = 62.4 \times 49.2 = 3070 \text{ lb/ft}^2$$

The effective stress at the top of the gravel is

$$\sigma' = \sigma_c - u_w = 4358.43 - 3070 = 1288.43 \text{ lb/ft}^2$$

(b) If an excavation is made into the clay stratum as shown in Fig. Ex. 5.12, the depth must be such that

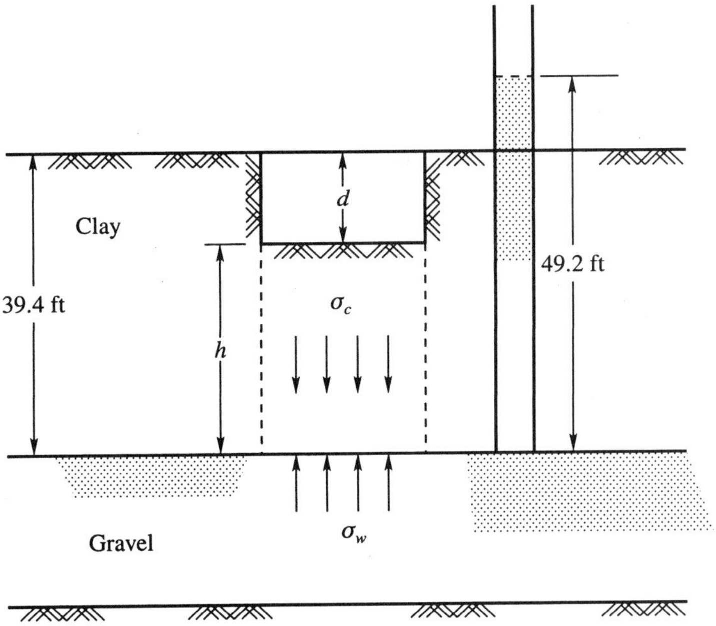

Figure Ex. 5.12

$$\sigma_c \le u_w$$

Let the bottom of the excavation be h ft above the top of gravel layer. Now the downward pressure acting at the top of the gravel layer is

$$\sigma_c = \gamma_t h = 110.62h \text{ lb/ft}^2$$

$$u_w = 3070 \text{ lb/ft}^2$$

Now, $110.62h = 3070$ or $h = \dfrac{3070}{110.62} = 27.75$ ft

Depth of excavation, $d = 39.4 - 27.75 = 11.65$ ft

This is just the depth of excavation with a factor of safety $F_s = 1.0$. If we assume a minimum $F_s = 1.10$

$$h = \frac{3070 \times 1.1}{110.62} = 30.52 \text{ ft}$$

Depth of excavation $= 39.4 - 30.52 = 8.88$ ft

Example 5.13

The diameter of a clean capillary tube is 0.08 mm. Determine the expected rise of water in the tube.

Solution

Per Eq. (5.22), the expected rise, h_c, in the capillary tube is

$$h_c = \frac{0.3}{d} = \frac{0.3}{0.008} = 37.5 \text{ cm}$$

where, d is in centimeters

Example 5.14

The water table is at a depth of 10 m in a silty soil mass. The sieve analysis indicates the effective diameter D_{10} of the soil mass is 0.05 mm. Determine the capillary rise of water above the water table and the maximum capillary pressure (a) by using Eq. (5.23) and (b) by using Eq. (5.28). Assume the void ratio $e = 0.51$.

Solution

Using Eq. (5.25) and assuming $C = 0.5$, the capillary rise of water is

$$h_c = \frac{C}{eD_{10}} = \frac{0.5}{0.51 \times 0.005} = 196 \text{ cm}$$

(a) Per Eq. (5.23)

the capillary pressure is $u_w = -h_c\gamma_w = -1.96 \times 9.81 = -19.2 \text{ kN/m}^2$

(b) Per Eq. (5.28)

$$\text{Porosity, } n = \frac{e}{1+e} = \frac{0.51}{1+0.51} = 0.338$$

$$u_w = u_c = -n\, h_c \gamma_w = -0.338 \times 19.2 = 6.49 \text{ kN/m}^2$$

Example 5.15

A layer of silty soil of thickness 5 m lies below the ground surface at a particular site and below the silt layer lies a clay stratum. The ground water table is at a depth of 4 m below the ground surface. The following data are available for both the silt and clay layers of soil.

Silt layer: $D_{10} = 0.018$ mm, $e = 0.7$, and $G_s = 2.7$

Clay layer: $e = 0.8$ and $G_s = 2.75$

Required: (a) Height of capillary rise, (b) capillary pressure, (c) the effective pressure at the ground surface, at GWT level, at the bottom of the silt layer and at a depth of $H = 6$ m below ground level, and (d) at a depth 2 m below ground level.

Solution

For the silty soil:

$$\gamma_d = \frac{G_s \gamma_w}{1+e} = \frac{2.7 \times 9.81}{1.7} = 15.6 \text{ kN/m}^3$$

$$\gamma_{sat} = \frac{(G_s + e)\gamma_w}{1+e} = \frac{(2.7+0.7)9.81}{1.7} = 19.62 \text{ kN/m}^3$$

$$\gamma_b = \gamma_{sat} - \gamma_w = 19.62 - 9.81 = 9.81 \text{ kN/m}^3$$

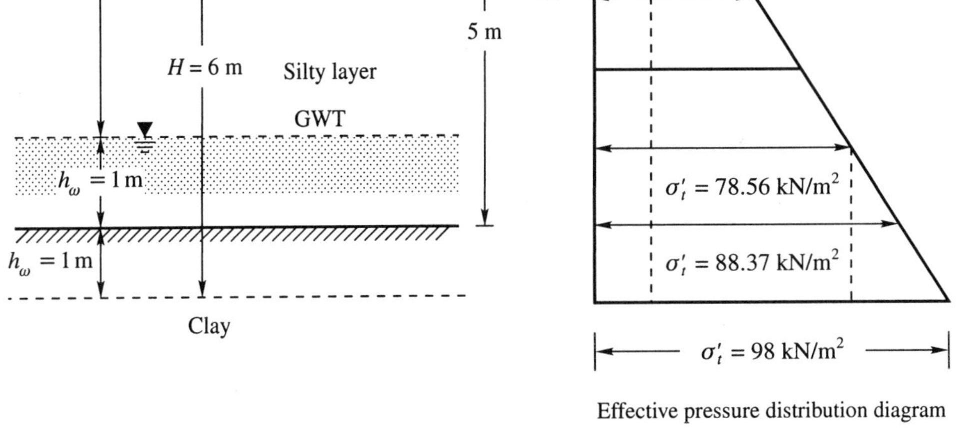

Figure Ex. 5.15

In the clay stratum:

$$\gamma_{sat} = \frac{(2.75 + 0.8)9.81}{1.8} = 19.35 \text{ kN/m}^3$$

$$\gamma_b = 19.35 - 9.81 = 9.54 \text{ kN/m}^3$$

(a) Height of capillary rise

$$h_c = \frac{C}{eD_{10}} \text{ per Eq. (9.5)}$$

Assume $C = 0.5$ sq. cm.

We have $h_c = \dfrac{0.5}{0.7 \times 0.0018} = 397$ cm or say 4.0 m

It is clear from h_c that the plane of menisci formed by the capillary water coincides with the ground surface as the water table is also at a depth of 4 m from ground level.

(b) *Capillary pressure* u_c

Per Eq. (9.28), $u_c = nh_c\gamma_w = \dfrac{e}{1+e}h_c\gamma_w$

or $u_c = \dfrac{0.7}{1.7} \times 4 \times 9.81 = 16.16 \text{ kN/m}^2$

(c) *The effective pressure at GL*

Since the plane of menisci coincides with the ground surface, the effective pressure at GL is equal to the capillary pressure u_c

Total effective pressure at GWT level, σ'_{sat}

Per Fig. Ex. 5.15

$$\sigma'_{sat} = \sigma'_d + u_c = \gamma_d h_c + u_c$$

$$\sigma'_{sat} = 15.6 \times 4 + 16.16 = 78.56 \text{ kN/m}^2$$

Total effective pressure at the bottom of the silt layer

The bottom of the silt layer is at a depth of 1 m below GWT level. The effective pressure due to this depth is

$\sigma' = \gamma_b h_w = 9.81 \times 1 = 9.81 \text{ kN/m}^2$

Total effective pressure, $\sigma'_t = \sigma'_{sat} + \sigma' = 78.56 + 9.81 = 88.37 \text{ kN/m}^2$

Total effective pressure at a depth of 6 m below GL

This point lies in the clay stratum at a depth of 1 m below the bottom of the silty layer. The increase in effective pressure at this depth is

$\sigma' = \gamma_b h_w = 9.54 \times 1 = 9.54 \text{ kN/m}^2$

The total effective pressure $\sigma'_t = 88.37 + 9.54 = 97.91 \text{ kN/m}^2 \approx 98 \text{ kN/m}^2$

(d) σ'_z at 2 m below GL

$\sigma'_z = u_c + z\gamma_d = 16.16 + 2 \times 15.6 = 47.36 \text{ kN/m}^2$

The pressure distribution diagram is given in Fig. Ex. 5.15.

Example 5.16

At a particular site lies a layer of fine sand 8 m thick below the ground surface and having a void ratio of 0.7. The GWT is at a depth of 4 m below the ground surface. The average degree of saturation of the sand above the capillary fringe is 50%. The soil is saturated due to capillary action to a height of 2.0 m above the GWT level. Assuming $G_s = 2.65$, calculate the total effective pressures at depths of 6 m and 3 m below the ground surface.

Solution

$$\gamma_d = \frac{G_s \gamma_w}{1+e} = \frac{2.65 \times 9.81}{1.7} = 15.29 \text{ kN / m}^3$$

$$\gamma_{sat} = \frac{(e+G_s)\gamma_w}{1+e} = \frac{(0.7+2.65)\times 9.81}{1.7} = 19.33 \text{ kN / m}^3$$

$$\gamma_b = \gamma_{sat} - \gamma_w = 19.33 - 9.81 = 9.52 \text{ kN / m}^3$$

The moist unit weight of soil above the capillary fringe is

$$\gamma_m = \frac{(G_s + eS)\gamma_w}{1+e} = \frac{(2.65 + 0.7 \times 0.5)\times 9.81}{1.7} = 17.31 \text{ kN / m}^3$$

Capillary pressure,

$$u_c = nh_c\gamma_w = \frac{e}{1+e}h_c\gamma_w = \frac{0.7}{1.7}\times 2 \times 9.81 = 8.08 \text{ kN / m}^2$$

Effective stresses at different levels

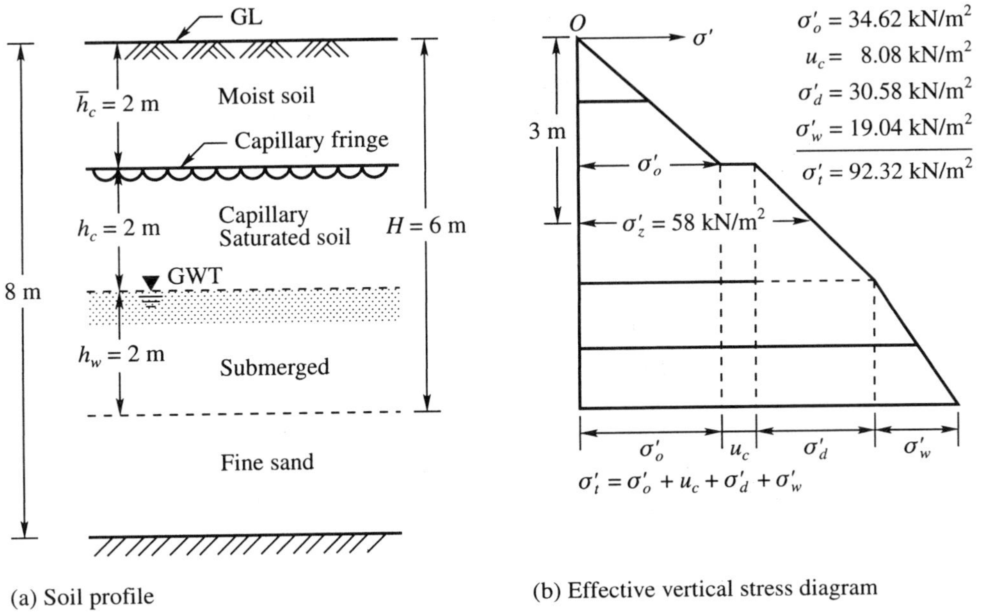

Figure Ex. 5.16

(a) At ground level $\sigma' = 0$

(b) Overburden pressure at fringe level = $\sigma'_o = \bar{h}_c \gamma_m = 2 \times 17.31 = 34.62$ kN/m^2

(c) Effective pressure at fringe level = $\sigma'_c = \sigma'_o + u_c = 34.62 + 8.08 = 42.70$ kN/m^2

(d) Effective pressure at GWT level = $\sigma'_{sat} = \sigma'_c + \sigma'_d = 42.70 + 2 \times 15.29$

$= 42.70 + 30.58 = 73.28$ kN/m^2

(e) Effective pressure at 6 m below GL

$\sigma'_t = \sigma'_{sat} + h_w \gamma_b = 73.28 + 2 \times 9.52 = 73.28 + 19.04 = 92.32$ kN/m^2

Effective stress at a depth 3 m below GL

Refer Fig. Ex. 5.16.

$\sigma'_z = \sigma'_o + u_c + (z - h_c)\gamma_d = 34.62 + 8.08 + (3 - 2) \times 15.29 \approx 58$ kN/m^2

5.6 PROBLEMS

5.1 The depth of water in a lake is 3 m. The soil properties as obtained from soil exploration below the bed of the lake are as given below.

Depth from bed of lake (m)	Type of soil	Void ratio e	Sp. gr. G_s
0–4	Clay	0.9	2.70
4–9	Sand	0.75	2.64
9–15	Clay	0.60	2.70

Calculate the following pressures at a depth of 12 m below the bed level of the lake.
(i) The total pressure, (ii) the pore pressure and (iii) the intergranular pressure.

5.2 The water table in a certain deposit of soil is at a depth of 6.5 ft below the ground surface. The soil consists of clay up to a depth of 13 ft from the ground and below which lies sand. The clay stratum is saturated above the water table.

Given: Clay stratum: $w = 30$ percent, $G_s = 2.72$; Sandy stratum: $w = 26$ percent, $G_s = 2.64$.

Required:

(i) The total pressure, pore pressure and effective pressure at a depth of 26 ft below the ground surface.

(ii) The change in the effective pressure if the water table is brought down to a level of 13 ft below the ground surface by pumping.

5.3 Water flows from container B to A as shown in Fig. 5.4. The piezometric head at the bottom of container A is 2.5 m and the depth of water above the sand deposit is 0.25 m. Assuming the depth of the sand deposit is 1.40 m, compute the effective pressure at the middle of the sand deposit. Assume $e = 0.65$ and $G_s = 2.64$ for the sand.

5.4 In order to excavate a trench for the foundation of a structure, the water table level was lowered from a depth of 4 ft to a depth of 15 ft in a silty sand deposit. Assuming that the soil above the water table remained saturated at a moisture content of 28 percent, estimate the increase in effective stress at a depth of 16 ft. Given $G_s = 2.68$

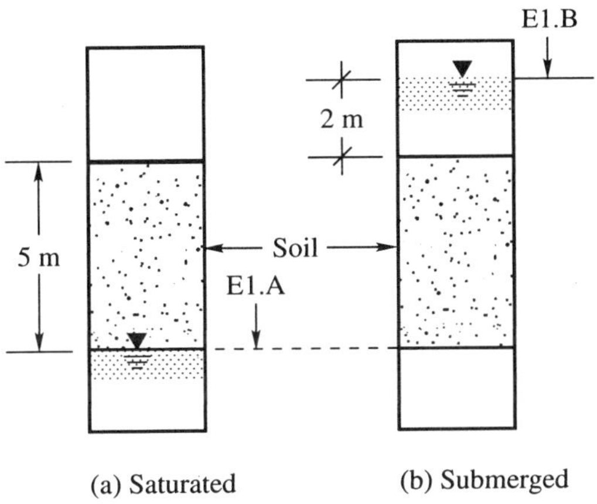

Figure Prob. 5.5

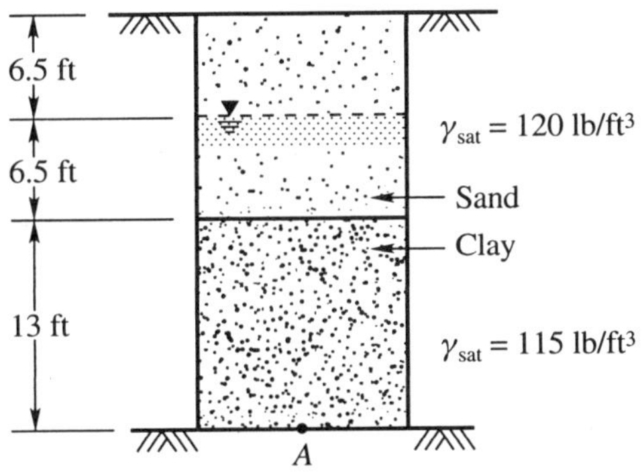

Figure Prob. 5.6

5.5 Soil is placed in the containers shown in Fig. Prob. 5.5. The saturated unit weight of soil is 20 kN/m^3. Calculate the pore pressure, and the effective stress at elevation *A*, when (a) the water table is at elevation *A*, and (b) when the water table rises to E1.B.

5.6 Figure Prob. 5.6 gives a soil profile. Calculate the total and effective stresses at point *A*. Assume that the soil above the water table remains saturated.

5.7 For the soil profile given in Fig. Prob. 5.6, determine the effective stress at point A for the following conditions: (a) water table at ground level, (b) water table at E1.A. (assume the soil above this level remains saturated), and (c) water table 6.5 ft above ground level.

5.8 A glass tube, opened at both ends, has an internal diameter of 0.002 mm. The tube is held vertically and water is added from the top end. What is the maximum height *h* of the column of water that will be supported?

5.9 Calculate (a) the theoretical capillary height and pressure h_c, and (b) the capillary pressure, u_c, in a silty soil with D_{10} = 0.04 mm. Assume the void ratio is equal to 0.50.

5.10 Calculate the height to which water will rise in a soil deposit consisting of uniform fine silt. The depth of water below the ground surface is 20 m. Assume the surface tension is 75×10^{-8} kN/cm and the contact angle is zero. The average size of the pores is 0.004 mm.

CHAPTER 6

STRESS DISTRIBUTION IN SOILS DUE TO SURFACE LOADS

6.1 INTRODUCTION

Estimation of vertical stresses at any point in a soil-mass due to external vertical loadings are of great significance in the prediction of settlements of buildings, bridges, embankments and many other structures. Equations have been developed to compute stresses at any point in a soil mass on the basis of the theory of elasticity. According to elastic theory, constant ratios exist between stresses and strains. For the theory to be applicable, the real requirement is not that the material necessarily be elastic, but there must be constant ratios between stresses and the corresponding strains. Therefore, in non-elastic soil masses, the elastic theory may be assumed to hold so long as the stresses induced in the soil mass are relatively small. Since the stresses in the subsoil of a structure having adequate factor of safety against shear failure are relatively small in comparison with the ultimate strength of the material, the soil may be assumed to behave elastically under such stresses.

When a load is applied to the soil surface, it increases the vertical stresses within the soil mass. The increased stresses are greatest directly under the loaded area, but extend indefinitely in all directions. Many formulas based on the theory of elasticity have been used to compute stresses in soils. They are all similar and differ only in the assumptions made to represent the elastic conditions of the soil mass. The formulas that are most widely used are the Boussinesq and Westergaard formulas. These formulas were first developed for point loads acting at the surface. These formulas have been integrated to give stresses below uniform strip loads and rectangular loads.

The extent of the elastic layer below the surface loadings may be any one of the following:

1. Infinite in the vertical and horizontal directions.
2. Limited thickness in the vertical direction underlain with a rough rigid base such as a rocky bed.

The loads at the surface may act on flexible or rigid footings. The stress conditions in the elastic layer below vary according to the rigidity of the footings and the thickness of the elastic layer. All the external loads considered in this book are vertical loads only as the vertical loads are of practical importance for computing settlements of foundations.

6.2 BOUSSINESQ'S FORMULA FOR POINT LOADS

Figure 6.1 shows a load Q acting at a point O on the surface of a semi-infinite solid. A semi-infinite solid is the one bounded on one side by a horizontal surface, here the surface of the earth, and infinite in all the other directions. The problem of determining stresses at any point P at a depth z as a result of a surface point laod was solved by Boussinesq (1885) on the following assumptions.

1. The soil mass is elastic, isotropic, homogeneous and semi-infinite.
2. The soil is weightless.
3. The load is a point load acting on the surface.

The soil is said to be isotropic if there are identical elastic properties throughout the mass and in every direction through any point of it. The soil is said to be homogeneous if there are identical elastic properties at every point of the mass in identical directions.

The expression obtained by Boussinesq for computing vertical stress σ_z at point P (Fig. 6.1) due to a point load Q is

$$\sigma_z = \frac{3Q}{2\pi z^2}\frac{1}{[1+(r/z)^2]^{5/2}} = \frac{Q}{z^2}I_B \tag{6.1}$$

where, r = the horizontal distance between an arbitrary point P below the surface and the vertical axis through the point load Q.

z = the vertical depth of the point P from the surface.

$$I_B = \text{Boussinesq stress coefficient} = \frac{3}{2\pi}\frac{1}{[1+(r/z)^2]^{5/2}} \tag{6.1a}$$

The values of the Boussinesq coefficient I_B can be determined for a number of values of r/z. The variation of I_B with r/z in a graphical form is given in Fig. 6.2. It can be seen from this figure

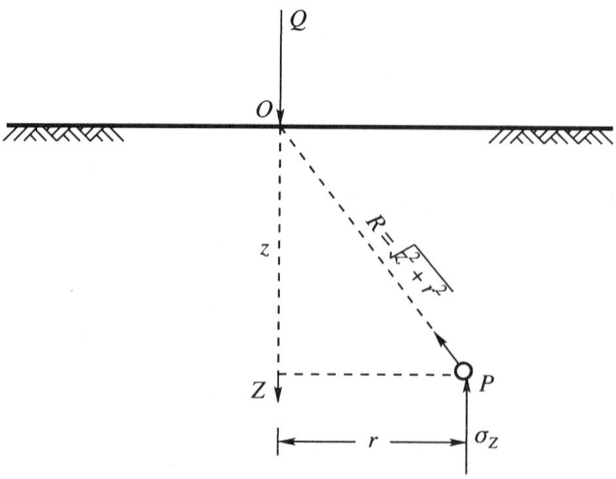

Figure 6.1 Vertical pressure within an earth mass

that I_B has a maximum value of 0.48 at $r/z = 0$, i.e., indicating thereby that the stress is a maximum below the point load.

6.3 WESTERGAARD'S FORMULA FOR POINT LOADS

Boussinesq assumed that the soil is elastic, isotropic and homogeneous for the development of a point load formula. However, the soil is neither isotropic nor homogeneous. The most common type of soils that are met in nature are the water deposited sedimentary soils. When the soil particles are deposited in water, typical clay strata usually have their lenses of coarser materials within them. The soils of this type can be assumed as laterally reinforced by numerous, closely spaced, horizontal sheets of negligible thickness but of infinite rigidity, which prevent the mass as a whole from undergoing lateral movement of soil grains. Westergaard, a British Scientist, proposed (1938) a formula for the computation of vertical stress σ_z by a point load, Q, at the surface as

$$\sigma_z = \frac{Q}{2\pi z^2} \frac{\sqrt{(1-2\mu)/(2-2\mu)}}{[(1-2\mu)/(2-\mu)+(r/z)^2]^{3/2}} = \frac{Q}{z^2} I_w \tag{6.2}$$

in which μ is Poisson's ratio. If μ is taken as zero for all practical purposes, Eq. (6.2) simplifies to

$$\sigma_z = \frac{Q}{\pi z^2} \frac{1}{[1+2(r/z)^2]^{3/2}} = \frac{Q}{z^2} I_w \tag{6.3}$$

where $I_w = \dfrac{(1/\pi)}{[1+2(r/z)^2]^{3/2}}$ is the Westergaard stress coefficient. The variation of I_w with the ratios of (r/z) is shown graphically in Fig. 6.2 along with the Boussinesq's coefficient I_B. The value of I_w at $r/z = 0$ is 0.32 which is less than that of I_B by 33 per cent.

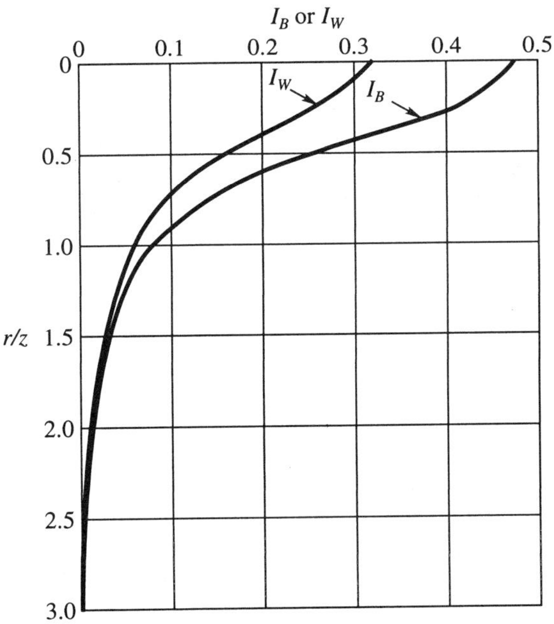

Figure 6.2 Values of I_B or I_W for use in the Boussinesq or Westergaard formula

Geotechnical engineers prefer to use Boussinesq's solution as this gives conservative results. Further discussions are therefore limited to Boussinesq's method in this chapter.

Example 6.1

A concentrated load of 1000 kN is applied at the ground surface. Compute the vertical pressure (i) at a depth of 4 m below the load, (ii) at a distance of 3 m at the same depth. Use Boussinesq's equation.

Solution

The equation is

$$\sigma_z = \frac{Q}{z^2} I_B, \text{ where } I_B = \frac{3/2\pi}{\left[1+(r/z)^2\right]^{5/2}}$$

(i) When $r/z = 0$, $I_B = 3/2\ \pi = 0.48$, $\sigma_z = 0.48\frac{Q}{z^2} = 0.48 \times \frac{1000}{4 \times 4} = 30 \text{ kN/m}^2$

(ii) When $r/z = 3/4 = 0.75$

$$I_B = \frac{3/2\pi}{\left[1+(0.75)^2\right]^{5/2}} = 0.156, \quad \sigma_z = \frac{0.156 \times 1000}{4 \times 4} = 9.8 \text{ kN/m}^2$$

Example 6.2

A concentrated load of 45000 lb acts at foundation level at a depth of 6.56 ft below ground surface. Find the vertical stress along the axis of the load at a depth of 32.8 ft and at a radial distance of 16.4 ft at the same depth by (a) Boussinesq, and (b) Westergaard formulae for $\mu = 0$. Neglect the depth of the foundation.

Solution

(a) Boussinesq Eq. (6.1a)

$$\sigma_z = \frac{Q}{z^2} I_B, \quad I_B = \frac{3}{2\pi}\left[\frac{1}{1+(r/z)^2}\right]^{5/2}$$

Substituting the known values, and simplifying

$I_B = 0.2733$ for $r/z = 0.5$

$$\sigma_z = \frac{45000}{(32.8)^2} \times 0.2733 = 11.43 \text{ lb/ft}^2$$

(b) Westergaard (Eq. 6.3)

$$\sigma_z = \frac{Q}{z^2} I_w, \quad I_w = \frac{1}{\pi}\left[\frac{1}{1+2(r/z)^2}\right]^{3/2}$$

Substituting the known values and simplifying, we have,

$I_w = 0.1733$ for $r/z = 0.5$

therefore,

$$\sigma_z = \frac{45000}{(32.8)^2} \times 0.1733 = 7.25 \text{ lb/ft}^2$$

Example 6.3

A rectangular raft of size 30 × 12 m founded at a depth of 2.5 m below the ground surface is subjected to a uniform pressure of 150 kPa. Assume the center of the area is the origin of coordinates (0, 0), and the corners have coordinates (6, 15). Calculate stresses at a depth of 20 m below the foundation level by the methods of (a) Boussinesq, and (b) Westergaard at coordinates of (0, 0), (0, 15), (6, 0) (6, 15) and (10, 25). Also determine the ratios of the stresses as obtained by the two methods. Neglect the effect of foundation depth on the stresses (Fig. Ex. 6.3).

Solution

Equations (a) Boussinesq: $\sigma_z = \frac{Q}{z^2} I_B,\ I_B = \frac{0.48}{\left[1+(r/z)^2\right]^{5/2}}$

(b) Westergaard: $\sigma_z = \frac{Q}{z^2} I_w,\ I_w = \frac{0.32}{\left[1+2(r/z)^2\right]^{3/2}}$

The ratios of r/z at the given locations for $z = 20$ m are as follows:

Location	r/z	Location	r/z
(0, 0)	0	(6, 15)	$\left(\sqrt{6^2+15^2}\right)/20 = 0.81$
(6, 0)	6/20 = 0.3	(10, 25)	$\left(\sqrt{10^2+25^2}\right)/20 = 1.35$
(0, 15)	15/20 = 0.75		

The stresses at the various locations at z = 20 m may be calculated by using the equations given above. The results are tabulated below for the given total load $Q = qBL = 150 \times 12 \times 30 = 54000$ kN acting at (0, 0) coordinate. $Q/z^2 = 135$.

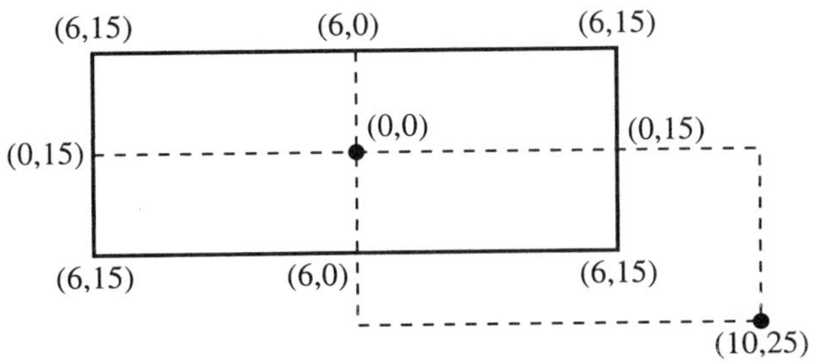

Figure Ex. 6.3

Location	r/z	Boussinesq		Westergaard		σ_B/σ_W
		I_B	σ_B(kPa)	I_W	σ_W(kPa)	
(0, 0)	0	0.48	65	0.32	43	1.51
(6, 0)	0.3	0.39	53	0.25	34	1.56
(0, 15)	0.75	0.16	22	0.10	14	1.57
(6,15)	0.81	0.14	19	0.09	12	1.58
(10, 25)	1.35	0.036	5	0.03	4	1.25

6.4 LINE LOADS

The basic equation used for computing σ_z, at any point P in an elastic semi-infinite mass is Eq. (6.1) of Boussinesq. By applying the principle of his theory, the stresses at any point in the mass due to a line load of infinite extent acting at the surface may be obtained. The state of stress encountered in this case is that of a plane strain condition. The strain at any point P in the Y-direction parallel to the line load is assumed equal to zero. The stress σ_y normal to the XZ-plane (Fig. 6.3) is the same at all sections and the shear stresses on these sections are zero. By applying the theory of elasticity, stresses at any point P (Fig. 6.3) may be obtained either in polar coordinates or in rectangular coordinates. The vertical stress σ_z at point P may be written in rectangular coordinates as

$$\sigma_z = \frac{q}{z}\frac{2/\pi}{[1+(x/z)^2]^2} = \frac{q}{z}I_z \tag{6.4}$$

where, I_z is the influence factor equal to 0.637 at $x/z = 0$.

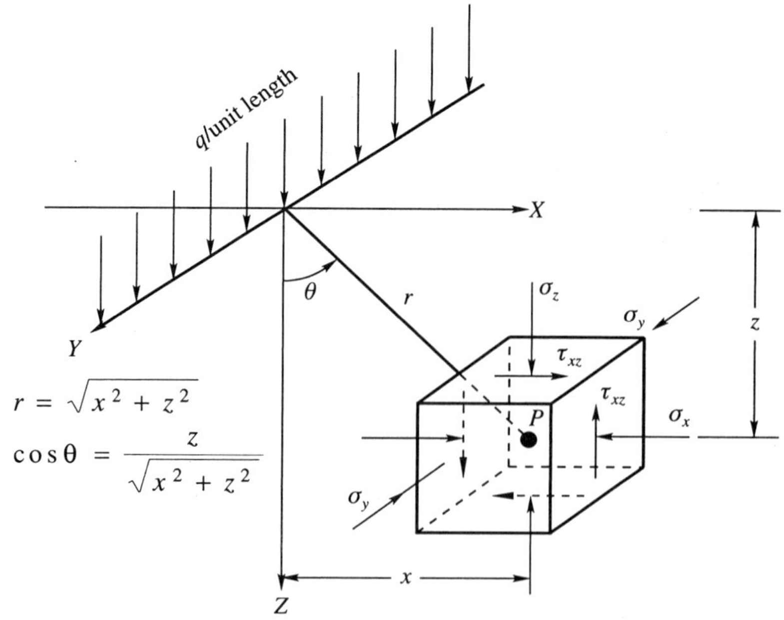

Figure 6.3 Stresses due to vertical line load in rectangular coordinates

6.5 STRIP LOADS

The state of stress encountered in this case also is that of a plane strain condition. Such conditions are found for structures extended very much in one direction, such as strip and wall foundations, foundations of retaining walls, embankments, dams and the like. For such structures the distribution of stresses in any section (except for the end portions of 2 to 3 times the widths of the structures from its end) will be the same as in the neighboring sections, provided that the load does not change in directions perpendicular to the plane considered.

Fig. 6.4(a) shows a load q per unit area acting on a strip of infinite length and of constant width B. The vertical stress at any arbitrary point P due to a line load of $q d\bar{x}$ acting at $x = \bar{x}$ can be written from Eq. (6.4) as

$$d\sigma_z = \frac{2q}{\pi} \frac{z^3}{[(x-\bar{x})^2 + z^2]^2} \tag{6.5}$$

Applying the principle of superposition, the total stress σ_z at point P due to a strip load distributed over a width $B(= 2b)$ may be written as

$$\sigma_z = \frac{2q}{\pi} \int_{-b}^{+b} \frac{z^3}{[(x-\bar{x})^2 + z^2]^2} dx$$

or

$$\sigma_z = \frac{q}{\pi}\left[\tan^{-1}\frac{z}{x-b} - \tan^{-1}\frac{z}{x+b} - \frac{2bz(x^2-b^2-z^2)}{(x^2-b^2+z^2)^2 + 4b^2z^2}\right] \tag{6.6}$$

The non-dimensional values of σ_z/q are given graphically in Fig. 6.5. Eq. (6.6) can be expressed in a more convenient form as

$$\sigma_z = \frac{q}{\pi}[\beta + \sin\beta\cos(\beta + 2\delta)] \tag{6.7}$$

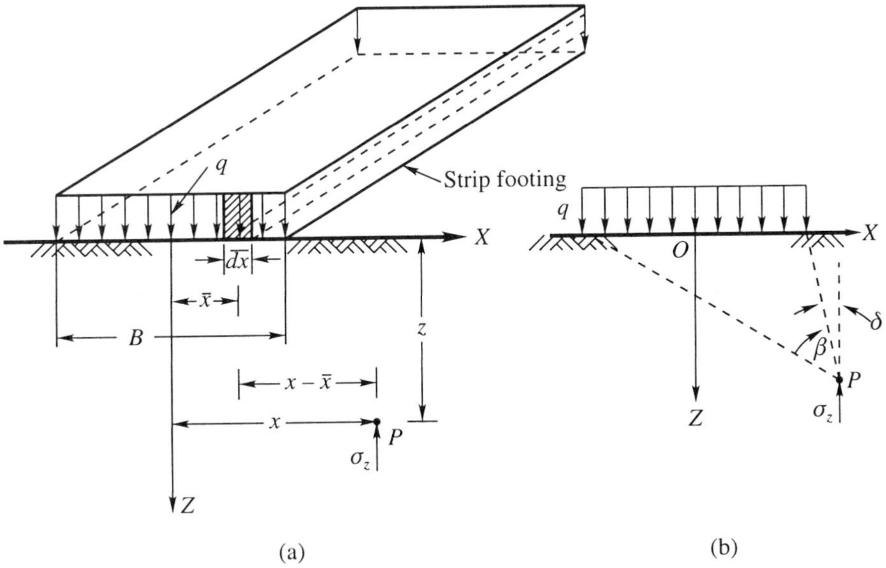

Figure 6.4 Strip load

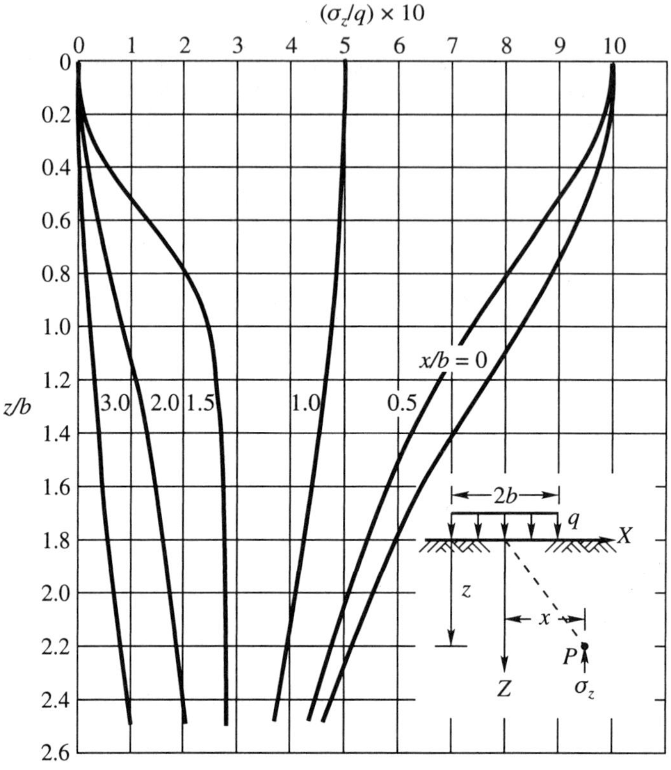

Figure 6.5 Non-dimensional values of σ_z/q for strip load

where β and δ are the angles as shown in Fig. 6.4(b). Equation (6.7) is very convenient for computing σ_z, since the angles β and δ can be obtained graphically for any point P. The principal stresses σ_1 and σ_3 at any point P may be obtained from the equations.

$$\sigma_1 = \frac{q}{\pi}(\beta + \sin\beta) \tag{6.8}$$

$$\sigma_3 = \frac{q}{\pi}(\beta - \sin\beta) \tag{6.9}$$

Example 6.4

Three parallel strip footings 3 m wide each and 5 m apart center to center transmit contact pressures of 200, 150 and 100 kN/m^2 respectively. Calculate the vertical stress due to the combined loads beneath the centers of each footing at a depth of 3 m below the base. Assume the footings are placed at a depth of 2 m below the ground surface. Use Boussinesq's method for line loads.

Solution

From Eq. (6.4), we have

$$\sigma_z = \frac{q}{z}\frac{2/\pi}{\left[1+(x/z)^2\right]^2} = \frac{q}{z}I_z$$

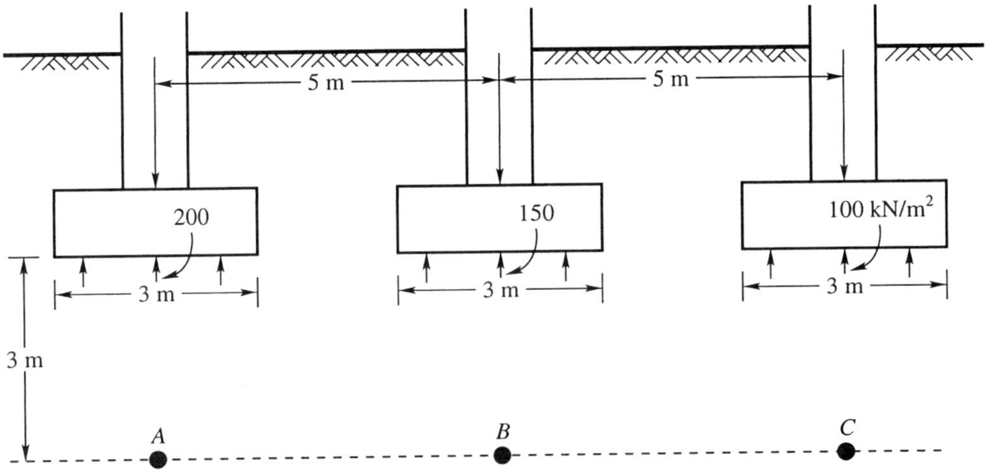

Figure Ex. 6.4 Three parallel footings

The stress at *A* (Fig. Ex. 6.4) is

$$\left(\sigma_z\right)_A = \frac{2\times 200}{3.14\times 3}\left[\frac{1}{1+(0/3)^2}\right]^2 + \frac{2\times 150}{3.14\times 3}\left[\frac{1}{1+(5/3)^2}\right]^2$$

$$+\frac{2\times 100}{3.14\times 3}\left[\frac{1}{1+(10/3)^2}\right]^2 = 45 \text{ kN/m}^2$$

The stress at *B*

$$\left(\sigma_z\right)_B = \frac{2\times 200}{3\pi}\left[\frac{1}{1+(5/3)^2}\right]^2 + \frac{2\times 150}{3\pi}\left[\frac{1}{1+(0/3)^2}\right]^2$$

$$+\frac{2\times 100}{3\pi}\left[\frac{1}{1+(5/3)^2}\right]^2 = 36.3 \text{ kN/m}^2$$

The stress at *C*

$$\left(\sigma_z\right)_C = \frac{2\times 200}{3\pi}\left[\frac{1}{1+(10/3)^2}\right]^2 + \frac{2\times 150}{3\pi}\left[\frac{1}{1+(5/3)^2}\right]^2 + \frac{2\times 100}{3\pi} = 23.74 \text{ kN/m}^2$$

6.6 STRESSES BENEATH THE CORNER OF A RECTANGULAR FOUNDATION

Consider an infinitely small unit of area of size $db \times dl$, shown in Fig. 6.6. The pressure acting on the small area may be replaced by a concentrated load dQ applied to the center of the area.

Hence

$$dQ = q\, db\, .\, dl \tag{6.10}$$

The increase of the vertical stress σ_z due to the load dQ can be expressed per Eq. (6.11) as

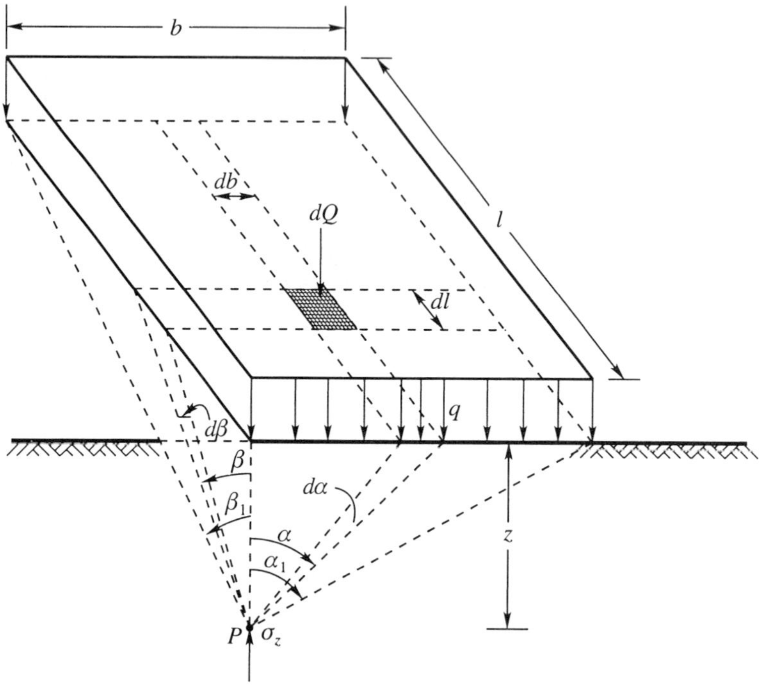

Figure 6.6 Vertical stress under the corner of a rectangular foundation

$$d\sigma_z = \frac{dQ}{2\pi} \frac{3z^3}{(z^2 + r^2)^{5/2}} \tag{6.11}$$

The stress produced by the pressure q over the entire rectangle $b \times l$ can then be obtained by expressing dl, db and r in terms of the angles α and β, and integrating

$$\sigma_z = \int_{\alpha=0}^{\alpha=\alpha_1} \int_{\beta=0}^{\beta=\beta_1} d\sigma_z \tag{6.12}$$

There are several forms of solution for Eq. (6.12). The one that is normally used is of the following form

$$\sigma_z = q\frac{1}{4\pi}\left[\frac{2mn(m^2+n^2+1)^{1/2}}{m^2+n^2+m^2n^2+1}\left(\frac{m^2+n^2+2}{m^2+n^2+1}\right) + \tan^{-1}\frac{2mn(m^2+n^2+1)^{1/2}}{m^2+n^2-m^2n^2+1}\right] \tag{6.13}$$

or $$\sigma_z = qI \tag{6.14}$$

wherein, $m = b/z$, $n = l/z$, are pure numbers. I is a dimensionless factor and represents the influence of a surcharge covering a rectangular area on the vertical stress at a point located at a depth z below one of its corners.

Eq. (6.14) is presented in graphical form in Fig. 6.7. This chart helps to compute pressures beneath loaded rectangular areas. The chart also shows that the vertical pressure is not materially altered if the length of the rectangle is greater than ten times its width. Fig. 6.8 may also be used for computing the influence value I based on the values of m and n and may also be used to determine stresses below points that lie either inside or outside the loaded areas as follows.

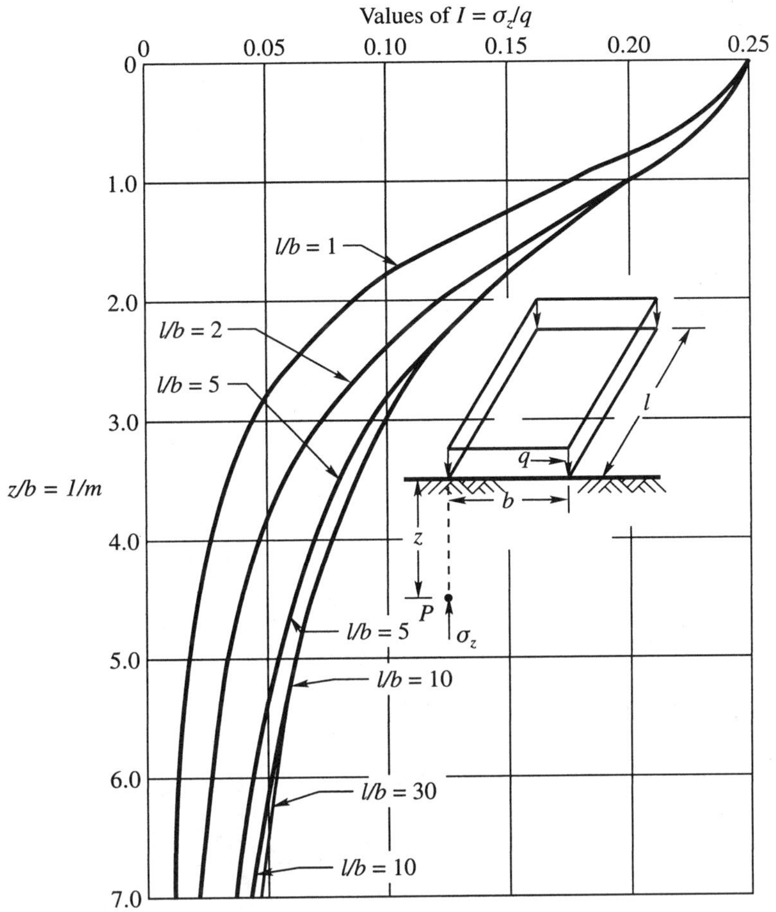

Figure 6.7 Chart for computing σ_z below the corner of a rectangular foundation (after Steinbrenner, 1934)

When the Point is Inside

Let O be an interior point of a rectangular loaded area $ABCD$ shown in Fig. 6.9(a). It is required to compute the vertical stress σ_z below this point O at a depth z from the surface. For this purpose, divide the rectangle $ABCD$ into four rectangles marked 1 to 4 in the Fig. 6.9(a) by drawing lines through O. For each of these rectangles, compute the ratios z/b. The influence value I may be obtained from Fig. 6.7 or 6.8 for each of these ratios and the total stress at P is therefore

$$\sigma_z = q\,(I_1 + I_2 + I_3 + I_4) \tag{6.15}$$

When the Point is Outside

Let O be an exterior point of loaded rectangular area $ABCD$ shown in Fig. 6.9(b). It is required to compute the vertical stress σ_z below point O at a depth z from the surface.

Construct rectangles as shown in the figure. The point O is the corner point of the rectangle OB_1CD_1. From the figure it can be seen that

$$\text{Area } ABCD = OB_1CD_1 - OB_1BD_2 - OD_1DA_1 + OA_1AD_2 \tag{6.16}$$

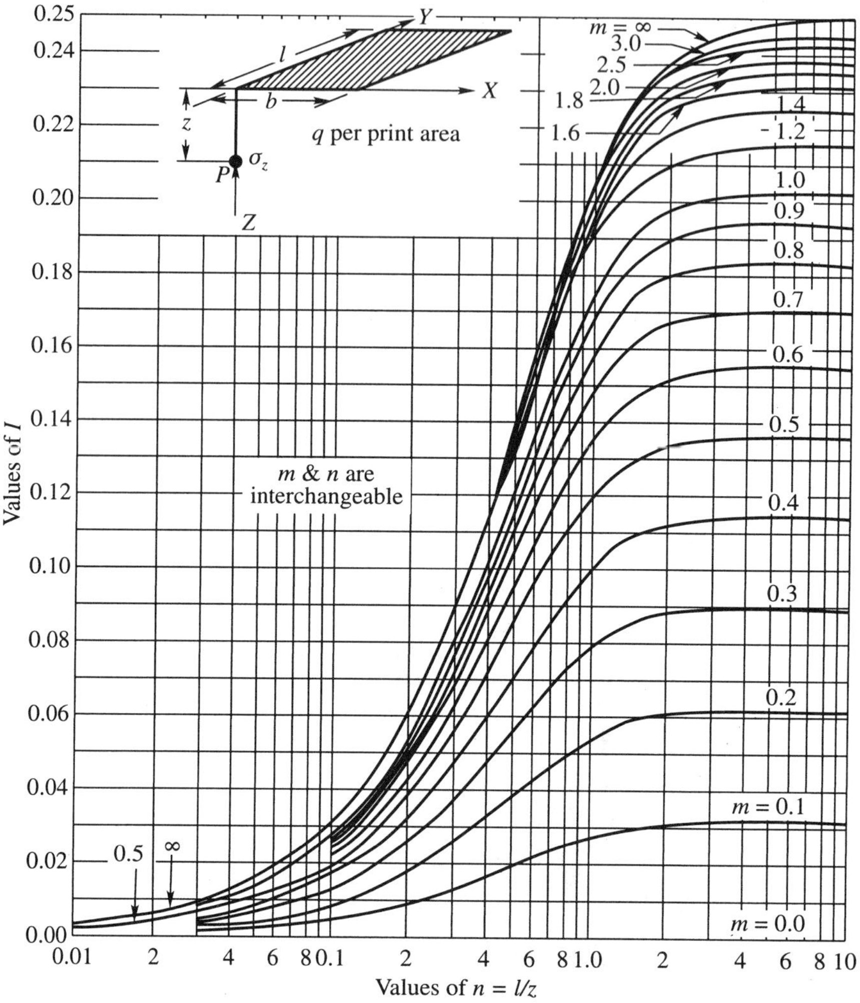

Figure 6.8 Graph for determining influence value for vertical normal stress σ_z at point P located beneath one corner of a uniformly loaded rectangular area. (After Fadum, 1948)

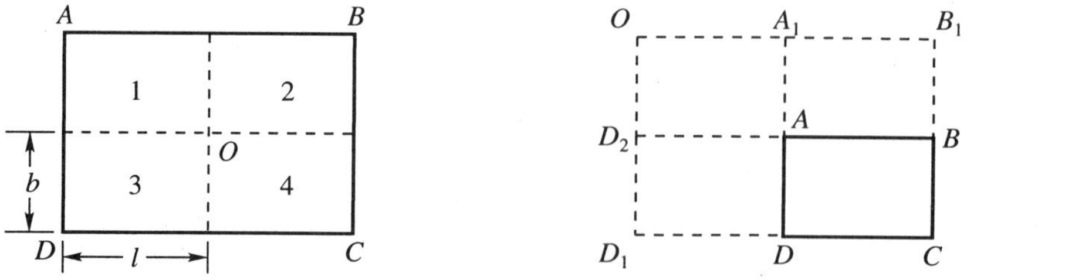

(a) When the point 'O' is within the rectangle (b) When the point 'O' is outside the rectangle

Figure 6.9 Computation of vertical stress below a point

The vertical stress at point P located at a depth z below point O due to a surcharge q per unit area of $ABCD$ is equal to the algebraic sum of the vertical stresses produced by loading each one of the areas listed on the right hand side of the Eq. (6.16) with q per unit of area. If I_1 to I_4 are the influence factors of each of these areas, the total vertical stress is

$$\sigma_z = q\,(I_1 - I_2 - I_3 + I_4) \qquad (6.17)$$

Example 6.5

$ABCD$ is a raft foundation of a multi-story building [Fig. 6. 9(b)] wherein AB = 65.6 ft, and BC = 39.6 ft. The uniformly distributed load q over the raft is 7310 lb/ft^2. Determine σ_z at a depth of 19.7 ft below point O [Fig. 6.9(b)] wherein AA_1 = 13.12 ft and A_1O = 19.68 ft. Use Fig. 6.8.

Solution

Rectangles are constructed as shown in [Fig. 6.9(b)].

$$\text{Area } ABCD = OB_1CD_1 - OB_1BD_2 - OD_1DA_1 + OA_1AD_2$$

Rectangle	l (ft)	b (ft)	m	n	I
OB_1CD_1	85.28	52.72	2.67	4.33	0.245
OB_1BD_2	85.28	13.12	0.67	4.33	0.168
OD_1DA_1	52.72	19.68	1.00	2.67	0.194
OA_1AD_2	19.68	13.12	0.67	1.00	0.145

Per Eq. (6.17)

$$\sigma_z = q\,(I_1 - I_2 - I_3 + I_4) = 7310\,(0.245 - 0.168 - 0.194 + 0.145) = 204.67 \text{ lb/ft}^2$$

The same value can be obtained using Fig. 6.7.

Example 6.6

A rectangular raft of size 30 × 12 m founded on the ground surface is subjected to a uniform pressure of 150 kN/m^2. Assume the center of the area as the origin of coordinates (0, 0), and corners with coordinates (6, 15). Calculate the induced stress at a depth of 20 m by the exact method at location (0, 0).

Solution

Divide the rectangle 12 × 30 m into four equal parts of size 6 × 15 m.

The stress below the corner of each footing may be calculated by using charts given in Fig. 6.7 or Fig. 6.8. Here Fig. 6.7 is used.

For a rectangle 6 × 15 m, $z/b = 20/6 = 3.34$, $l/b = 15/6 = 2.5$.

For $z/b = 3.34$, $l/b = 2.5$, $\bar{\sigma}_z/q = 0.07$

Therefore, $\sigma_z = 4\bar{\sigma}_z = 4 \times 0.07q = 4 \times 0.07 \times 150 = 42 \text{ kN/m}^2$.

6.7 STRESSES UNDER UNIFORMLY LOADED CIRCULAR FOOTING

Stresses Along the Vertical Axis of Symmetry

Figure 6.10 shows a plan and section of the loaded circular footing. The stress required to be determined at any point *P* along the axis is the vertical stress σ_z.

Let *dA* be an elementary area considered as shown in Fig. 6.10. *dQ* may be considered as the point load acting on this area which is equal to *q dA*. We may write

$$dQ = q\,dA = qr\,d\theta\,dr \tag{6.18}$$

The vertical stress $d\sigma$ at point *P* due to point load *dQ* may be expressed [Eq. (6.1a)] as

$$d\sigma_z = \frac{3q}{2\pi}\frac{z^3 r\,d\theta\,dr}{(r^2+z^2)^{5/2}} \tag{6.19}$$

The integral form of the equation for the entire circular area may be written as

$$\sigma_z = \int_{\theta=0}^{\theta=2\pi}\int_{r=0}^{r=R_0} d\sigma_z = \frac{3qz^3}{2\pi}\int_{\theta=0}^{\theta=2\pi}\int_{r=0}^{r=R_0}\frac{r\,d\theta\,dr}{(r^2+z^2)^{5/2}}$$

On integration we have, $$\sigma_z = q\left[1-\frac{z^3}{(R_0^2+z^2)^{3/2}}\right] \tag{6.20}$$

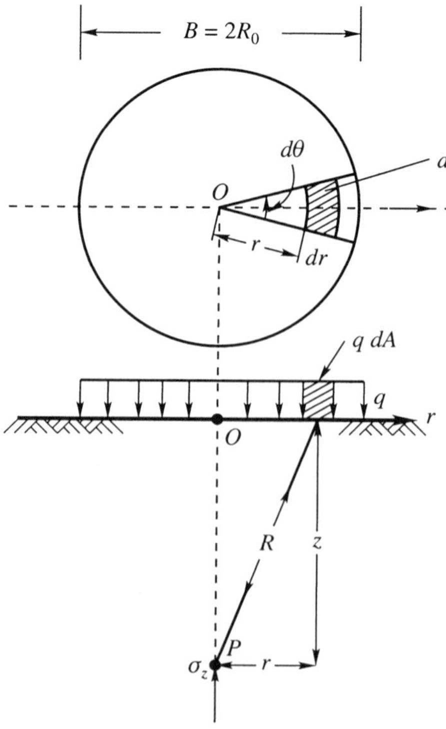

Figure 6.10 Vertical stress under uniformly loaded circular footing

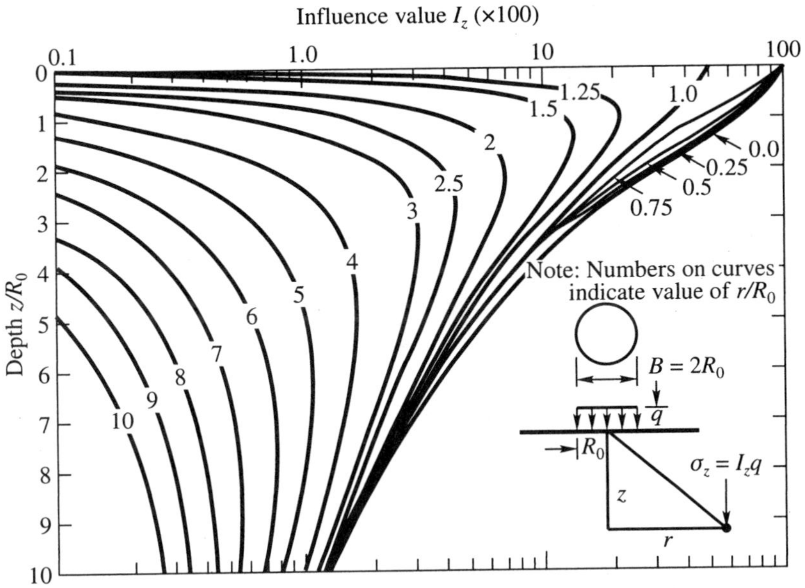

Figure 6.11 Influence diagram for vertical normal stress at various points within an elastic half-space under a uniformly loaded circular area. (After Foster and Ahlvin, 1954)

or

$$\sigma_z / q = 1 - \left[\frac{1}{1 + (R_0 / z)^2} \right]^{3/2} = I_z \tag{6.21}$$

where, I_z *is the Influence coefficient*. The stress at any point P on the axis of symmetry of a circular loaded area may be calculated by the use of Eq. (6.21) Vertical stresses σ_z may be calculated by using the influence coefficient diagram given in Fig. 6.11.

Example 6.7

A water tank is required to be constructed with a circular foundation having a diameter of 16 m founded at a depth of 2 m below the ground surface. The estimated distributed load on the foundation is 325 kN/m². Assuming that the subsoil extends to a great depth and is isotropic and homogeneous, determine the stresses σ_z at points (i) $z = 8$ m, $r = 0$, (ii) $z = 8$ m, $r = 8$ m, (iii) $z = 16$ m, $r = 0$ and (iv) $z = 16$ m, $r = 8$ m, where r is the radial distance from the central axis. Neglect the effect of the depth of the foundation on the stresses. (Use Fig. 6.11)

Solution

$q = 325$ kN/m², $R_0 = 8$ m. The results are given in a tabular form as follows:

	Point	z/R_0	r/R_0	I	σ_z kN/m²
(i)	(8, 0)	1	0	0.7	227.5
(ii)	(8, 8)	1	1.0	0.33	107.25
(iii)	(16, 0)	2	0	0.3	97.5
(iv)	(16, 8)	2	1.0	0.2	65

Example 6.8

For a raft of size 98.4 x 39.36 ft, compute the stress at 65.6 ft depth below the center of the raft by assuming that the rectangle can be represented by an equivalent circle. The load intensity on the raft is 3133 lb/ft^2.

Solution

The radius of a fictitious circular footing of area equal to the rectangular footing of size 98.4 x 39.36 ft is

$$\pi R_0^2 = 98.4 \times 39.36 = 3873 \text{ sq. ft} \quad \text{or } R_0 = \sqrt{\frac{3873}{\pi}} = 35.12 \text{ ft}$$

Use Eq. (6.21) for computing σ_z at 35.6 ft depth

Now, $z/R_0 = \dfrac{65.6}{35.12} = 1.9$, and $r/R_0 = 0$. From Fig. 6.11, $I_z = 0.3$

Therefore, $\sigma_z = 0.3\, q = 0.3 \times 3133 = 940$ lb/ft^2.

6.8 VERTICAL STRESS BENEATH LOADED AREAS OF IRREGULAR SHAPE

Newmark's Influence Chart

When the foundation consists of a large number of footings or when the loaded mats or rafts are not regular in shape, a chart developed by Newmark (1942) is more practical than the methods explained before. It is based on the following procedure. The vertical stress σ_z below the center of a circular area of radius R which carries uniformly distributed load q is determined per Eq. (6.21).

It may be seen from Eq. (6.21) that when $R/z = \infty$, $\sigma_z/q = 1$, that is $\sigma_z = q$. This indicates that if the loaded area extends to infinity, the vertical stress in the semi-infinite solid at any depth z is the same as unit load q at the surface. If the loaded area is limited to any given radius R_1 it is possible to determine from Eq. (6.21) the ratios R/z for which the ratio of σ_z/q may have any specified value, say 0.8 or 0.6. Table 6.1 gives the ratios of R/z for different values of σ_z/q.

Table 6.1 may be used for the computation of vertical stress σ_z at any depth z below the center of a circular loaded area of radius R. For example, at any depth z, the vertical stress $\sigma_z = 0.8\, q$ if the radius of the loaded area at the surface is $R = 1.387\, z$. At the same depth, the vertical stress is $\sigma_z = 0.7\, q$ if $R = 1.110\, z$. If instead of loading the whole area, if only the annular space between the circles of radii $1.387\, z$ and $1.110\, z$ are loaded, the vertical stress at z at the center of the circle is $\Delta\sigma_z = 0.8\, q - 0.7\, q = 0.1\, q$. Similarly if the annular space between circles of radii $1.110\, z$ and $0.917\, z$ are loaded, the vertical stress at the same depth z is $\Delta\sigma_z = 0.7\, q - 0.6\, q = 0.1\, q$. We may therefore draw a series of concentric circles on the surface of the ground in such a way that when the annular space between any two consecutive circles is loaded with a load q per unit area, the vertical stress $\Delta\sigma_z$ produced at any depth z below the center remains a constant fraction of q. We may write, therefore,

$$\Delta\sigma_z = Cq \tag{6.22}$$

where C is constant. If an annular space between any two consecutive concentric circles is divided into n equal blocks and if any one such block is loaded with a distributed load q, the vertical stress produced at the center is, therefore,

Table 6.1 Values of R/z for different values of σ_z/q

σ_z/q	R/z	σ_z/q	R/z
0.00	0.000	0.80	1.387
0.10	0.270	0.90	1.908
0.20	0.401	0.92	2.094
0.30	0.518	0.94	2.351
0.40	0.637	0.96	2.748
0.50	0.766	0.98	3.546
0.60	0.917	1.00	∞
0.70	1.110	–	–

$$\frac{\Delta\sigma_z}{n} = \frac{C}{n}q = C_i q \tag{6.23}$$

$$\frac{\Delta\sigma_z}{n} = C_i \text{ when } q = 1.$$

A load $q = 1$ covering one of the blocks will produce a vertical stress C_i. In other words, the '*influence value*' of each loaded block is C_i. If the number of loaded blocks is N, and if the intensity of load is q per unit area, the total vertical stress at depth z below the center of the circle is

$$\sigma_z = C_i N q \tag{6.24}$$

The graphical procedure for computing the vertical stress σ_z due to any surface loading is as follows.

Select some definite scale to represent depth z. For instance a suitable length AB in cm as shown in Fig. 6.12 to represent depth z in meters. In such a case, the scale is 1 cm = z/AB meters. The length of the radius $R_{0.8}$ which corresponds to $\sigma_z/q = 0.8$ is then equal to $1.387 \times AB$ cm, and a circle of that radius may be drawn. This procedure may be repeated for other ratios of σ_z/q, for instance, for $\sigma_z/q = 0.7$, 0. 5 etc. shown in Fig. 6.12.

The annular space between the circles may be divided into n equal blocks, and in this case n = 20. The influence value C_i is therefore equal to 0.1/20 = 0.005. A plan of the foundation is drawn on a tracing paper to a scale such that the distance AB on the chart corresponds to the depth z at which the stress σ_z is to be computed. For example, if the vertical stress at a depth of 9 m is required, and if the length AB chosen is 3 cm, the foundation plan is drawn to a scale of 1 cm = 9/3 = 3 m. In case the vertical stress at a depth 12 m is required, a new foundation plan on a separate tracing paper is required. The scale for this plan is 1 cm = 12/AB = 12/3 = 4 m.

This means that a different tracing has to be made for each different depth whereas the chart remains the same for all. Fig. 6.12(b) gives a foundation plan, which is loaded with a uniformly distributed load q per unit area. It is now required to determine the vertical stress σ_z at depth vertically below point O shown in the figure. In order to determine σ_z, the foundation plan is laid over the chart in such a way that the surface point O coincides with the center O' of the chart as shown in Fig. 6.12. The number of small blocks covered by the foundation plan is then counted. Let this number be N. Then the value of σ_z at depth z below O is

$\sigma_z = C_i Nq$, which is the same as Eq. (6.24).

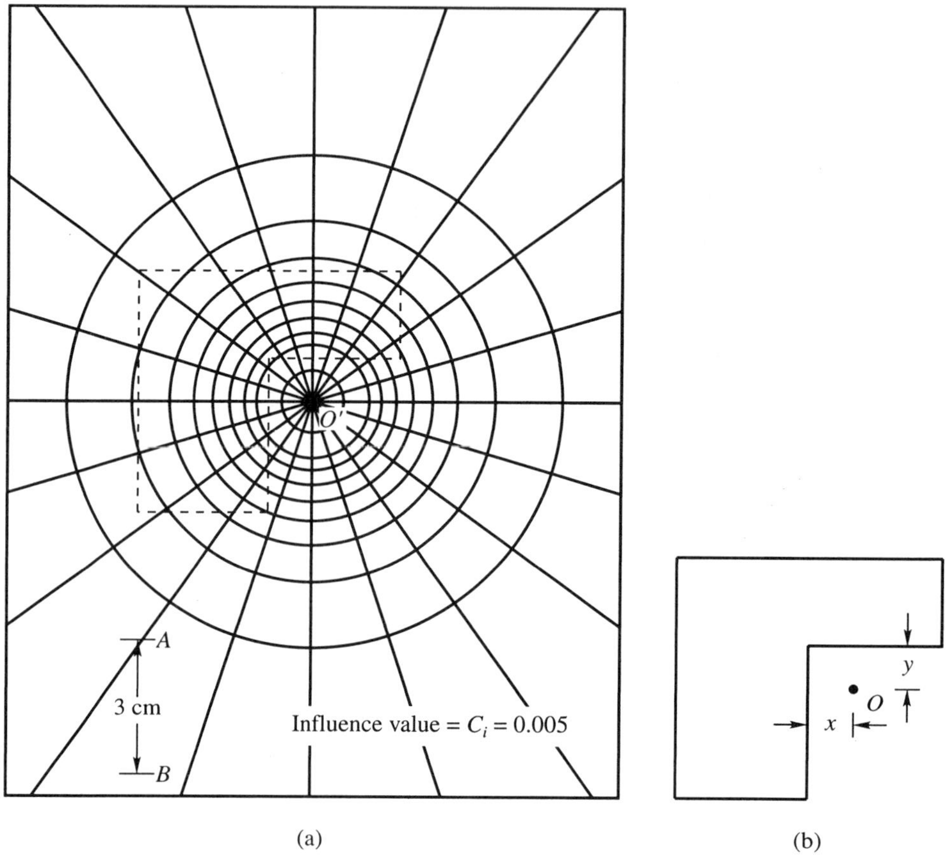

Figure 6.12 Newmark's influence chart

Example 6.9

A ring footing of external diameter 8 m and internal diameter 4 m rests at a depth 2 m below the ground surface. It carries a load intensity of 150 kN/m^2. Find the vertical stress at depths of 2, 4 and 8 m along the axis of the footing below the footing base. Neglect the effect of the excavation on the stress.

Solution

From Eq. (6.21) we have,

$$\frac{\sigma_z}{q} = 1 - \left[\frac{1}{1+(R_0/z)^2}\right]^{3/2} = I_z$$

where q = contact pressure 150 kN/m^2, I_z = Influence coefficient.

The stress σ_z at any depth z on the axis of the ring is expressed as

$$\sigma_z = \sigma_{z_1} - \sigma_{z_2} = q(I_{z_1} - I_{z_2})$$

where σ_{z_1} = stress due to the circular footing of diameter 8 m, and $I_z = I_{z_1}$ and $R_0/z = (R_1/z)$
σ_{z_2} = stress due to the footing of diameter 4 m, $I_z = I_{z_2}$ and $R_0/z = (R_2/z)$.

The values of I_z may be obtained from Table 6.1 for various values of R_0/z. The stress σ_z at depths 2, 4 and 8 m are given below:

Depth (m)	R_1/z	I_{z_1}	R_2/z	I_{z_2}	$(I_{z_1} - I_{z_2})q = \sigma_z$ kN/m²
2	2	0.911	1.0	0.697	39.6
4	1.0	0.647	0.5	0.285	54.3
8	0.5	0.285	0.25	0.087	29.7

Example 6.10

A raft foundation of the size given in Fig. Ex. 6.10 carries a uniformly distributed load of 300 kN/m². Estimate the vertical pressure at a depth 9 m below the point O marked in the figure.

Solution

The depth at which σ_z required is 9 m.

Using Fig. 6.12, the scale of the foundation plan is AB = 3 cm = 9 m or 1 cm = 3 m. The foundation plan is required to be made to a scale of 1 cm = 3 m on tracing paper. This plan is superimposed on Fig. 6.12 with O coinciding with the center of the chart. The plan is shown in dotted lines in Fig. 6.12.

Number of loaded blocks occupied by the plan, $N = 62$

Influence value, $C_i = 0.005$, $q = 300$ kN/m²

The vertical stress, $\sigma_z = C_i Nq = 0.005 \times 62 \times 300 = 93$ kN/m².

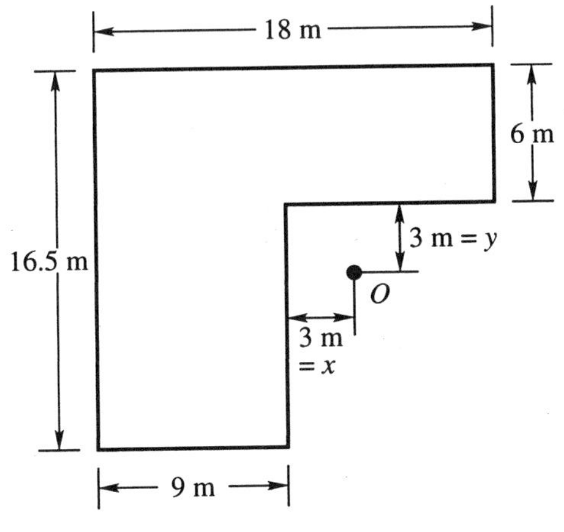

Figure Ex. 6.10

6.9 EMBANKMENT LOADINGS

Long earth embankments with sloping sides represent trapezoidal loads. When the top width of the embankment reduces to zero, the load becomes a triangular strip load. The basic problem is to determine stresses due to a linearly increasing vertical loading on the surface.

Linearly Increasing Vertical Loading

Fig. 6.13(a) shows a linearly increasing vertical loading starting from zero at *A* to a finite value *q* per unit length at *B*. Consider an elementary strip of width *db* at a distance *b* from *A*. The load per unit length may be written as

$$dq = (q/a)\, b\, db$$

If dq is considered as a line load on the surface, the vertical stress $d\sigma_z$ at *P* [Fig. 6.13(a)] due to dq may be written from Eq. (6.4) as

$$d\sigma_z = \left(\frac{1}{a}\right)\left(\frac{2q}{\pi}\right)\frac{z^3 b\, db}{\left[(x-b)^2 + z^2\right]^2}$$

Therefore,

$$\sigma_z = \int d\sigma_z = \left(\frac{1}{a}\right)\left(\frac{2q}{\pi}\right)\int_{b=0}^{b=a}\frac{z^3 b\, db}{\left[(x-b)^2 + z^2\right]^2}$$

$$\text{on integration,}\quad \sigma_z = \frac{q}{2\pi}\left(\frac{2x}{a}\alpha - \sin 2\beta\right) = qI_z \tag{6.25}$$

where I_z is non-dimensional coefficient whose values for various values of x/a and z/a are given in Table 6.2.

If the point *P* lies in the plane *BC* [Fig. 6.13(a)], then $\beta = 0$ at $x = a$. Eq. (6.25) reduces to

$$\sigma_z = \frac{q}{\pi}(\alpha) \tag{6.26}$$

Figs. 6.13(b) and (c) show the distribution of stress σ_z on vertical and horizontal sections under the action of a triangular loading as a function of *q*. The maximum vertical stress occurs below the center of gravity of the triangular load as shown in Fig. 6.13(c).

Vertical Stress Due to Embankment Loading

Many times it may be necessary to determine the vertical stress σ_z beneath road and railway embankments, and also beneath earth dams. The vertical stress beneath embankments may be

Table 6.2 I_z for triangular load (Eq. 6.25)

x/a	z/a						
	0.00	0.5	1.0	1.5	2	4	6
−1.500	0.00	0.002	0.014	0.020	0.033	0.051	0.041
−1.00	0.00	0.003	0.025	0.048	0.061	0.060	0.041
0.00	0.00	0.127	0.159	0.145	0.127	0.075	0.051
0.50	0.50	0.410	0.275	0.200	0.155	0.085	0.053
0.75	0.75	0.477	0.279	0.202	0.163	0.082	0.053
1.00	0.50	0.353	0.241	0.185	0.153	0.075	0.053
1.50	0.00	0.056	0.129	0.124	0.108	0.073	0.050
2.00	0.00	0.017	0.045	0.062	0.069	0.060	0.050
2.50	0.00	0.003	0.013	0.041	0.050	0.049	0.045

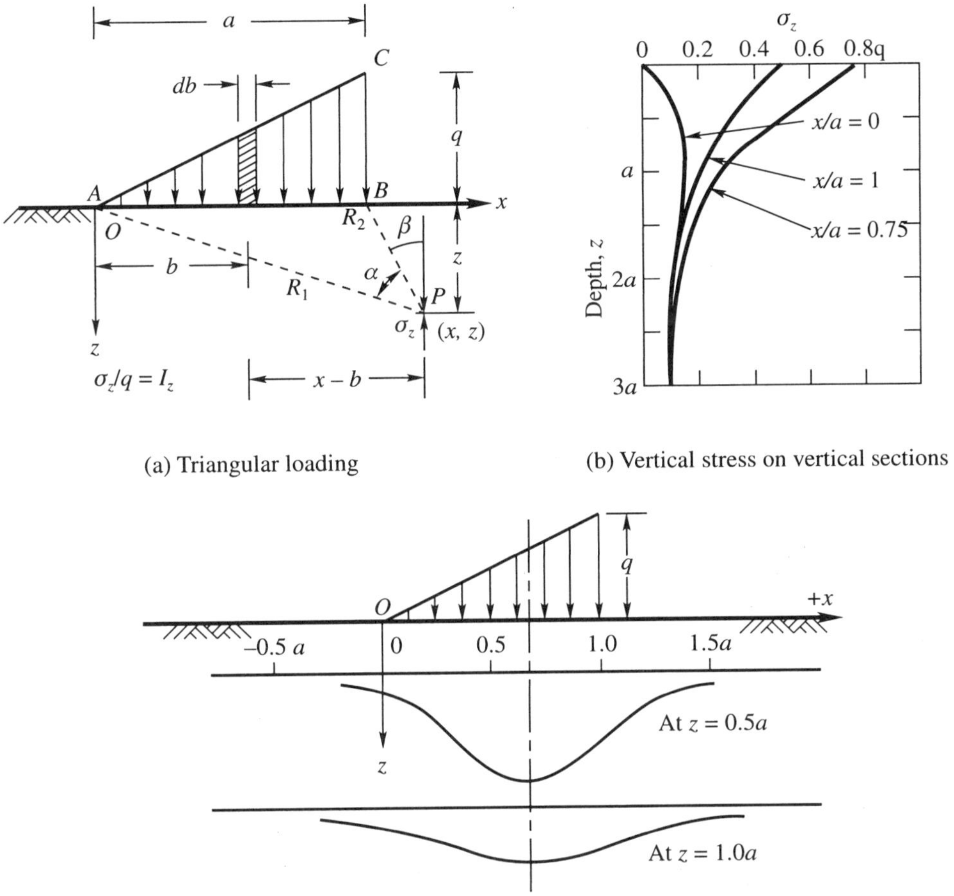

(a) Triangular loading

(b) Vertical stress on vertical sections

(c) Vertical stress on horizontal sections

Figure 6.13 Stresses in a semi-infinite mass due to triangular loading on the surface

determined either by the method of superposition by making use of Eq. (6.26) or by making use of a single formula which can be developed from first principles.

σ_z by Method of Superposition

Consider an embankment given in Fig. 6.14. σ_z at P may be calculated as follows:

The trapezoidal section of embankment *ABCD*, may be divided into triangular sections by drawing a vertical line through point P as shown in Fig. 6.14. We may write

$$ABCD = AGE + FGB - EDJ - FJC \tag{6.27}$$

If σ_{z1}, σ_{z2}, σ_{z3}, and σ_{z4} are the vertical stresses at point P due to the loadings of figures *AGE*, *FGB*, *EDJ* and *FJC* respectively, the vertical stress σ_z due to the loading of figure *ABCD* may be written as

$$\sigma_z = \sigma_{z_1} + \sigma_{z_2} - \sigma_{z_3} - \sigma_{z_4} \tag{6.28}$$

By applying the principle of superposition for each of the triangles by making use of Eq. (6.26), we obtain

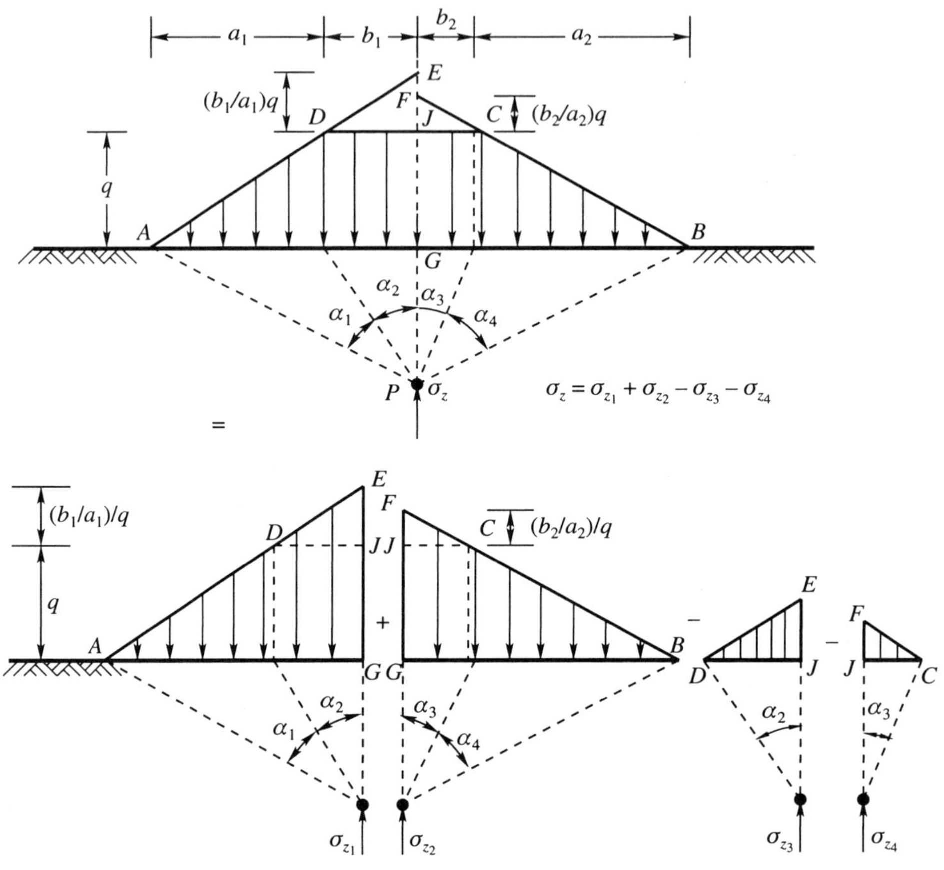

Figure 6.14 Vertical stress due to embankment

$$\sigma_z = \frac{q}{\pi}\left[(\alpha_1 + \alpha_2 + \alpha_3 + \alpha_4) + (b_1 / a_1)\alpha_1 + (b_2 / a_2)\alpha_4\right] \tag{6.29}$$

$$\sigma_z = qI_z = \frac{q}{\pi} f(a / z, b / z) \tag{6.30}$$

where I_z is the influence factor for a trapezoidal load which is a function of a/z and b/z.

The values of I_z for various values of a/z and b/z are given in Fig. 6.15. (After Osterberg, 1957)

σ_z from a Single Formula for Asymmetrical Trapezoidal Loading

A single formula can be developed for trapezoidal loading for computing σ_z at a point P (Fig. 6.16) by applying Eq. (6.26). The origin of coordinates is as shown in the figure. The final equation may be expressed as

$$\sigma_z = \frac{q}{\pi}\left[(\alpha_1 + \alpha_2 + \alpha_3) + \frac{b}{a_1}(\alpha_1 + R\alpha_3) + \frac{x}{a_1}(\alpha_1 - R\alpha_3)\right] \tag{6.31}$$

where α_1, α_2, and α_3 are the angles subtended at the point P in the supporting medium by the loading and $R = a_1/a_2$. When $R = 1$, the stresses are due to that of a symmetrical trapezoidal loading.

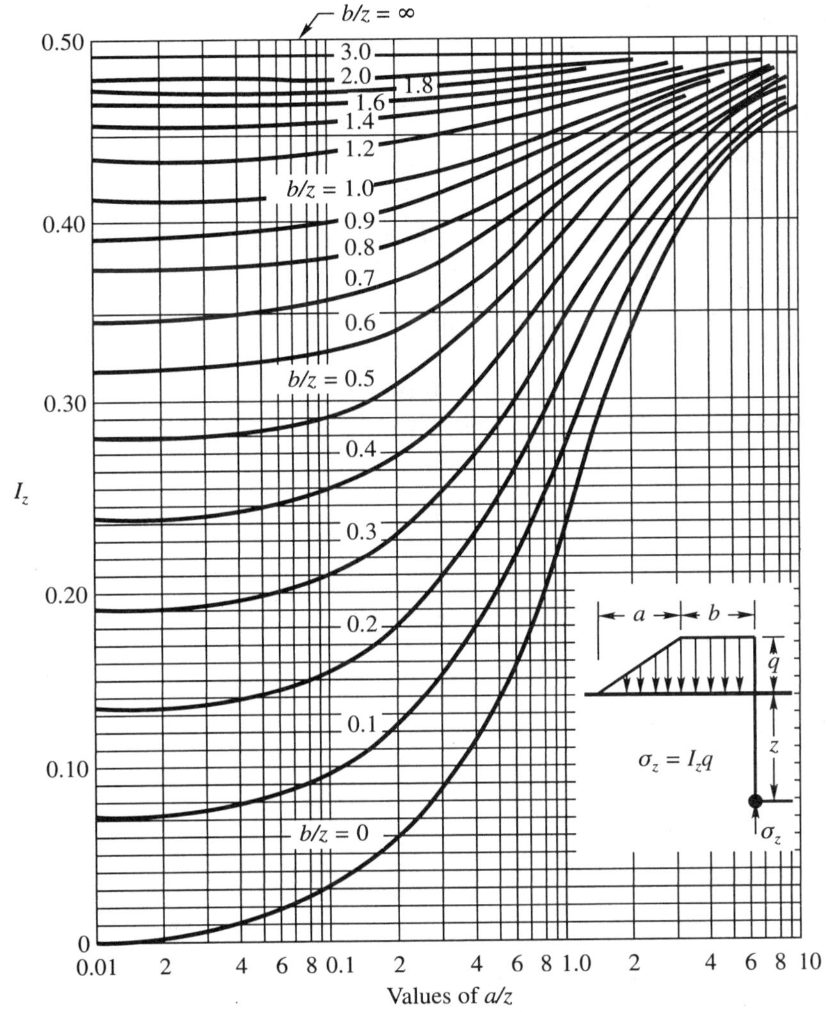

Figure 6.15 A graph to determine compressive stresses from a load varying by straight line law (After Osterberg, 1957)

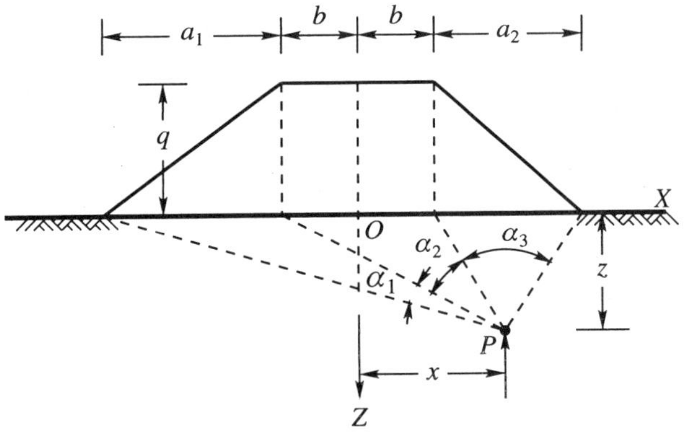

Figure 6.16 Trapezoidal loads

When the top width is zero, i.e, when $b = 0$, $\alpha_2 = 0$, the vertical stress σ_z will be due to a triangular loading. The expression for triangular loading is

$$\sigma_z = \frac{q}{\pi}\left[(\alpha_1 + \alpha_3) + \frac{x}{a_1}(\alpha_1 - R\alpha_3)\right] \tag{6.32}$$

Eq. (6.31) and Eq. (6.32) can be used to compute σ_z at any point in the supporting medium. The angles α_1, α_2, and α_3 may conveniently be obtained by a graphical procedure where these angles are expressed as radians in the equations.

Example 6.11

A 3 m high embankment is to be constructed as shown in Fig. Ex. 6. 11. If the unit weight of soil used in the embankment is 19. 0 kN/m^3, calculate the vertical stress due to the embankment loading at points P_1, P_2, and P_3.

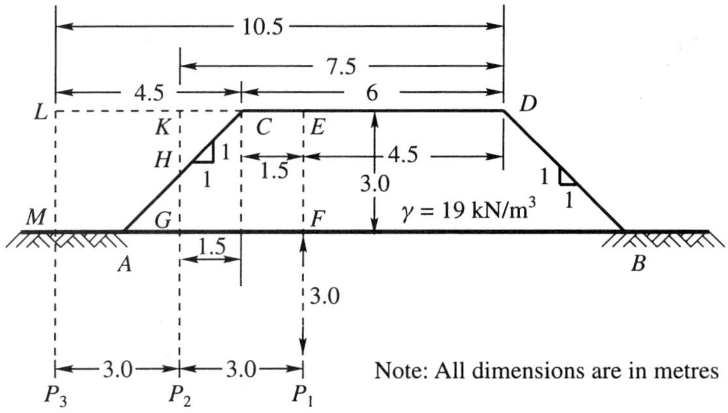

Figure Ex. 6.11 Vertical stresses at P_1, P_2 & P_3

Solution

$q = \gamma H = 19 \times 3 = 57$ kN/m^2, $z = 3$ m

The embankment is divided into blocks as shown in Fig. Ex. 6.11 for making use of the graph given in Fig. 6. 15. The calculations are arranged as follows:

Point	Block	b (m)	a (m)	b/z	a/z	I
P_1	*ACEF*	1.5	3	0.5	1	0.39
	EDBF	4.5	3	1.5	1	0.477
P_2	*AGH*	0	1.5	0	0.5	0.15
	GKDB	7.5	3	2.5	1.0	0.493
	HKC	0	1.5	0	0.5	0.15
P_3	*MLDB*	10.5	3.0	3.5	1.0	0.498
	MACL	1.5	3.0	0.5	1.0	0.39

Vertical stress σ_z

At point P_1, σ_z = $(0.39 + 0.477) \times 57 = 49.4$ kN/m^2

At point P_2, σ_z = $0.15 \times (57/2) + 0.493 \times 57 - 0.15 \times (57/2) = 28.1$ kN/m^2

At point P_3, σ_z = $(0.498 - 0.39)\ 57 = 6.2$ kN/m^2

6.10 APPROXIMATE METHODS FOR COMPUTING σ_z

Two approximate methods are generally used for computing stresses in a soil mass below loaded areas. They are

1. Use of the point load formulas such as Boussinesq's equation.
2. 2 : 1 method which gives an average vertical stress σ_z at any depth z. This method assumes that the stresses distribute from the loaded edge points at an angle of 2 (vertical) to 1 (horizontal)

The first method if properly applied gives the point stress at any depth which compares fairly well with exact methods, whereas the second does not give any point stress but only gives an average stress σ_z at any depth. The average stress computed by the second method has been found to be in error depending upon the depth at which the stress is required.

Point Load Method

Eq. (6.1) may be used for the computation of stresses in a soil mass due to point loads acting at the surface. Since loads occupy finite areas, the point load formula may still be used if the footings are divided into smaller rectangles or squares and a series of concentrated loads of value $q\ dA$ are assumed to act at the center of each square or rectangle. Here dA is the area of the smaller blocks and q the pressure per unit area. The only principle to be followed in dividing a bigger area into smaller blocks is that the width of the smaller block should be less than one-third the depth z of the point at which the stress is required to be computed. The loads acting at the centers of each smaller area may be considered as point loads and Boussinesq's formula may then be applied. The difference between the point load method and the exact method explained earlier is clear from

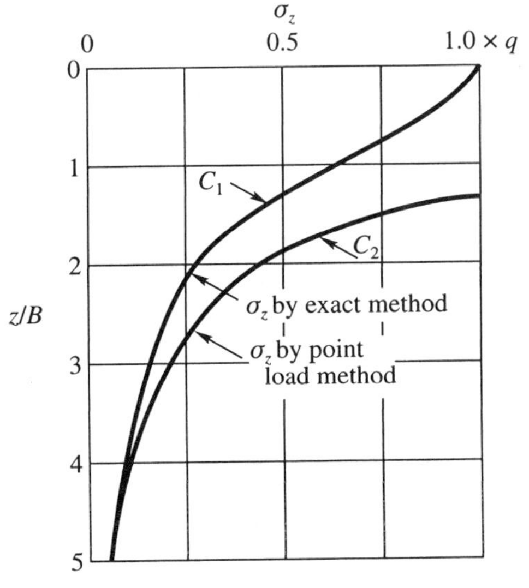

Figure 6.17 σ_z by point load method

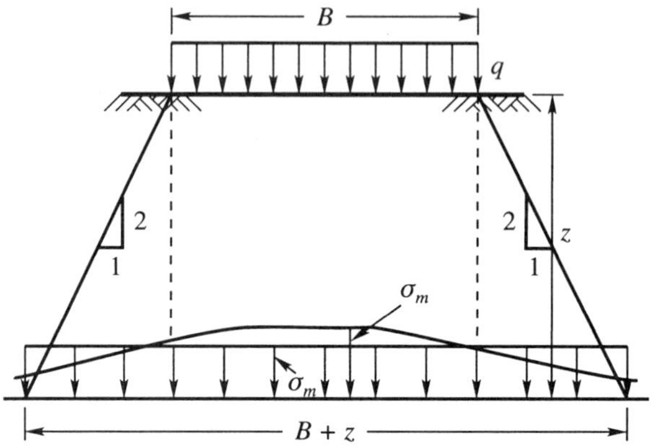

Figure 6.18 σ_m 2 : 1 method

Fig. 6.17. In this figure the abscissa of the curve C_1 represents the vertical stress σ_z at different depths z below the center of a square area $B \times B$ which carries a surcharge q per unit area or a total surcharge load of B^2q. This curve is obtained by the exact method explained under Sect. 6.6. The abscissa of the curve C_2 represents the corresponding stresses due to a concentrated load $Q = B^2q$ acting at the center of the square area. The figure shows that the difference between the two curves becomes very small for values of z/B in excess of three. Hence in a computation of the vertical stress σ_z at a depth z below an area, the area should be divided into convenient squares or rectangles such that the least width of any block is not greater than $z/3$.

2 : 1 Method

In this method, the stress is assumed to be distributed uniformly over areas lying below the foundation. The size of the area at any depth is obtained by assuming that the stresses spread out at an angle of 2 (vertical) to 1 (horizontal) from the edges of the loaded areas shown in Fig. 6.18. The average stress at any depth z is

$$\sigma_a = \frac{Q}{(B+z)(L+z)} \tag{6.33}$$

The maximum stress σ_m by an exact method below the loaded area is different from the average stress σ_a at the same depth. The value of σ_m/σ_a reaches a maximum of about 1.6 at z/b = 0-5, where b = half width.

6.11 PRESSURE ISOBARS

Definition

An *isobar* is a line which connects all points of equal stress below the ground surface. In other words, an isobar is a stress contour. We may draw any number of isobars as shown in Fig. 6.19 for any given load system. Each isobar represents a fraction of the load applied at the surface. Since these isobars form closed figures and resemble the form of a bulb, they are also termed *bulb of pressure* or simply the *pressure bulb*. Normally isobars are drawn for vertical, horizontal and shear stresses. The one that is most important in the calculation of settlements of footings is the vertical pressure isobar.

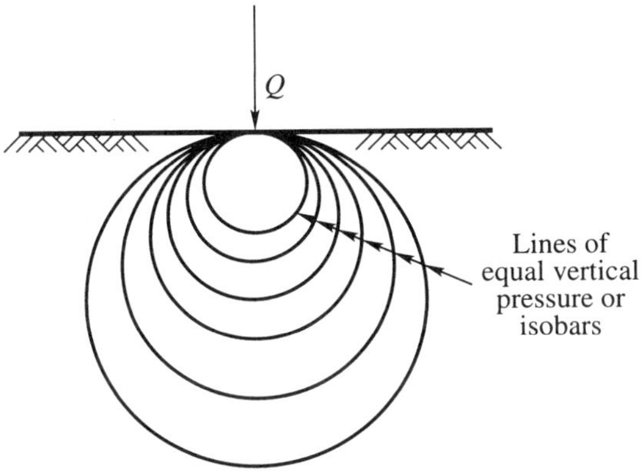

Figure 6.19 Bulb of pressure

Significant Depth

In his opening discussion on settlement of structures at the First International Conference on Soil Mechanics and Foundation Engineering (held in 1936 at Harvard University in Cambridge, Mass, USA), Terzaghi stressed the importance of the bulb of pressure and its relationship with the seat of settlement. As stated earlier we may draw any number of isobars for any given load system, but the one that is of practical significance is the one which encloses a soil mass which is responsible for the settlement of the structure. The depth of this stressed zone may be termed as the *significant depth* D_s which is responsible for the settlement of the structure. Terzaghi recommended that for all practical purposes one can take a *stress contour* which represents 20 per cent of the foundation contact pressure q, i.e, equal to $0.2q$. The depth of such an isobar can be taken as the *significant depth* D_s which represents the seat of settlement for the foundation. Terzaghi's recommendation was based on his observation that direct stresses are considered of negligible magnitude when they are smaller than 20 per cent of the intensity of the applied stress from structural loading, and that most of the settlement, approximately 80 per cent of the total, takes place at a depth less than D_s. The depth D_s is approximately equal to 1.5 times the width of square or circular footings [Fig. 6.20(a)].

If several loaded footings are spaced closely enough, the individual isobars of each footing in question would combine and merge into one large isobar of the intensity as shown in [Fig. 6.20(b)]. The combined significant depth D_s is equal to about 1.5 $\bar{B}$.

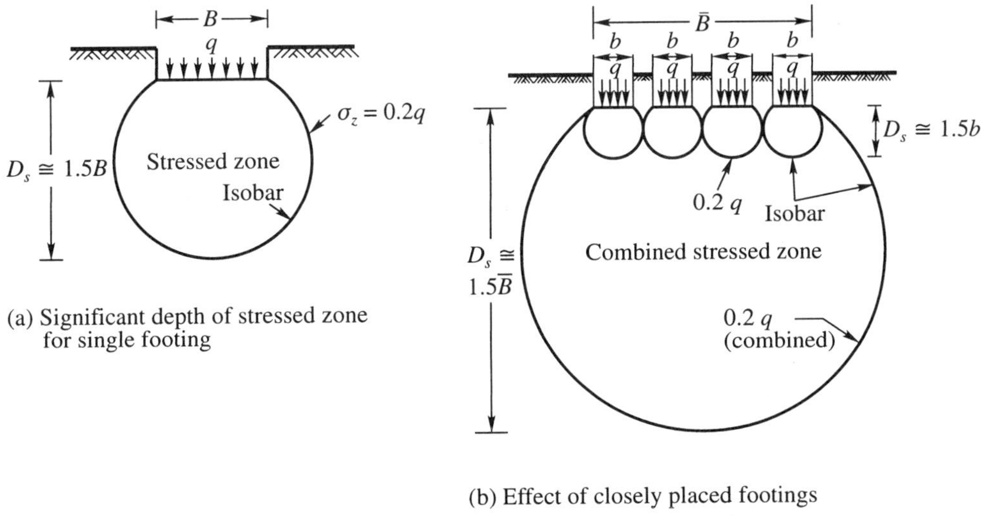

(a) Significant depth of stressed zone for single footing

(b) Effect of closely placed footings

Figure 6.20 Significant depth of stressed zone

Pressure Isobars for Footings

Pressure isobars of square, rectangular and circular footings may conveniently be used for determining vertical pressure, σ_z, at any depth, z, below the base of the footings. The depths z from the ground surface, and the distance r (or x) from the center of the footing are expressed as a function of the width of the footing B. In the case of circular footing B represents the diameter.

The following pressure isobars are given based on either Boussinesq or Westergaard's equations

1. Boussinesq isobars for square and continuous footings, Fig. 6.21.
2. Boussinesq isobar for circular footings, Fig. 6.22.
3. Westergaard isobars for square and continuous footings, Fig. 6.23.

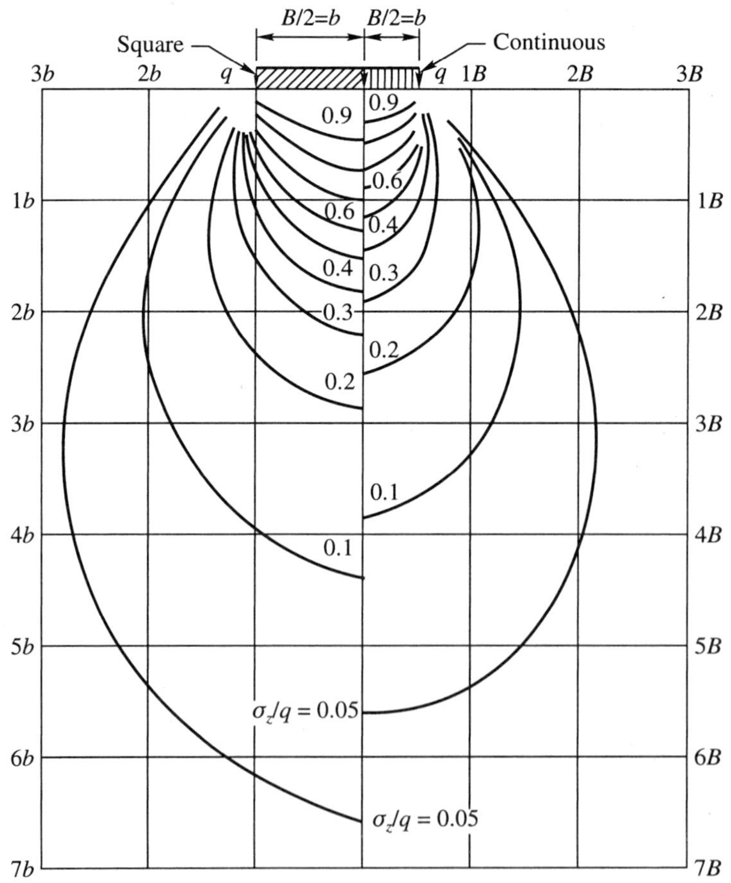

Figure 6.21 Pressure isobars based on Boussinesq equation for square and continuous footings

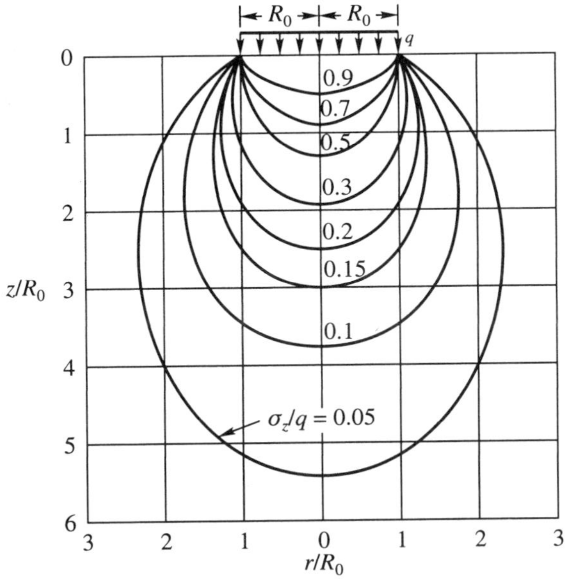

Figure 6.22 Pressure isobars based on Boussinesq equation for uniformly loaded circular footings

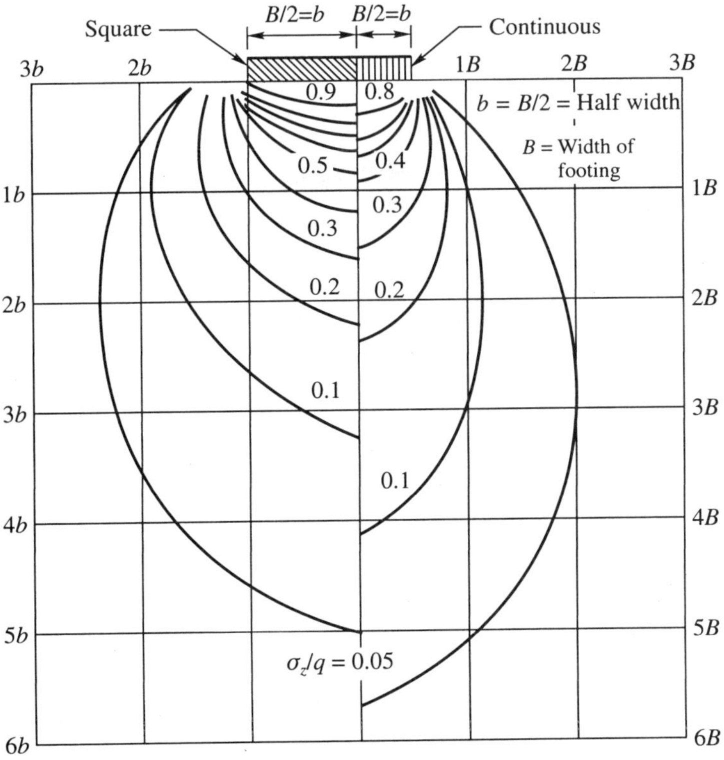

Figure 6.23 Pressure isobars based on Westergaard equation for square and continuous footing

Example 6.12

A single concentrated load of 1000 kN acts at the ground surface. Construct an isobar for $\sigma_z = 40$ kN/m^2 by making use of the Boussinesq equation.

Solution

From Eq. (6.1a) we have

$$\sigma_z = \frac{3Q}{2\pi z^2}\left[\frac{1}{1+(r/z)^2}\right]^{5/2}$$

We may now write by rearranging an equation for the radial distance r as

$$r = \sqrt{z}\sqrt{\left(\frac{3Q}{2\pi z^2 \sigma_z}\right)^{2/5} - 1}$$

Now for $Q = 1000$ kN, $\sigma_z = 40$ kN/m^2, we obtain the values of r_1, r_2, r_3, etc. for different depths z_1, z_2, z_3, etc. The values so obtained are

z (m)	r (m)
0.25	1.34
0.50	1.36
1.0	1.30
2.0	1.04
3.0	0.60
3.455	0.00

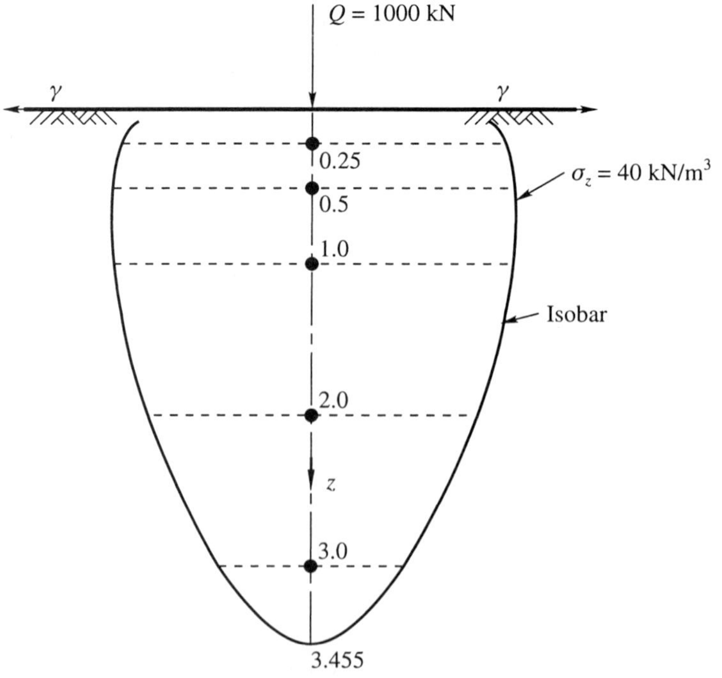

Figure Ex. 6.12

The isobar for σ_z = 40 kN/m^2 may be obtained by plotting z against r as shown in Fig. Ex. 6.12.

6.12 PROBLEMS

6.1 A column of a building transfers a concentrated load of 225 kips to the soil in contact with the footing. Estimate the vertical pressure at the following points by making use of the Boussinesq and Westergaard equations.

(i) Vertically below the column load at depths of 5, 10, and 15 ft.

(ii) At radial distances of 5, 10 and 20 ft and at a depth of 10 ft.

6.2 Three footings are placed at locations forming an equilateral triangle of 13 ft sides. Each of the footings carries a vertical load of 112.4 kips. Estimate the vertical pressures by means of the Boussinesq equation at a depth of 9 ft at the following locations :

(i) Vertically below the centers of the footings.

(ii) Below the center of the triangle.

6.3 A reinforced concrete water tank of size 25 ft × 25 ft and resting on the ground surface carries a uniformly distributed load of 5.25 kips/ft^2. Estimate the maximum vertical pressures at depths of 37.5 and 60 ft by point load approximation below the center of the tank.

6.4 Two footings of sizes 13 × 13 ft and 10 × 10 ft are placed 30 ft center to center apart at the same level and carry concentrated loads of 337 and 281 kips respectively. Compute the vertical pressure at depth 13 ft below point C midway between the centers of the footings.

6.5 A and B are two footings of size 1.5 × 1.5 m each placed in position as shown in Fig. Prob. 6.5. Each of the footings carries a column load of 400 kN. Determine by the

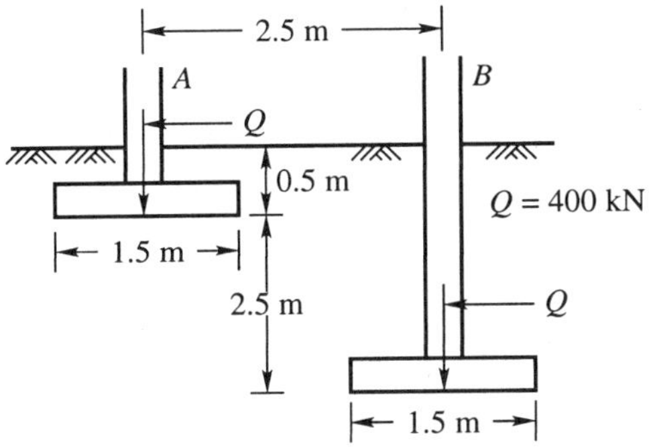

Figure Prob. 6.5

Boussinesq method, the excess load footing B carries due to the effect of the load on A. Assume the loads at the centers of footings act as point loads.

6.6 If both footings A and B in Fig. Prob. 6.5 are at the same level at a depth of 0.5 m below the ground surface, compute the stress σ_z midway between the footings at a depth of 3 m from the ground surface. Neglect the effect of the size for point load method.

6.7 Three concentrated loads Q_1 = 255 kips, Q_2 = 450 kips and Q_3 = 675 kips act in one vertical plane and they are placed in the order Q_1–Q_2–Q_3. Their spacings are 13 ft–10 ft. Determine the vertical pressure at a depth of 5 ft along the center line of footings using Boussinesq's point load formula.

6.8 A square footing of 13 × 13 ft is founded at a depth of 5 ft below the ground level. The imposed pressure at the base is 8732 lb/ft^2. Determine the vertical pressure at a depth of 24 ft below the ground surface on the center line of the footing.

6.9 A long masonry wall footing carries a uniformly distributed load of 200 kN/m ^{2}. If the width of the footing is 4 m, determine the vertical pressures at a depth of 3 m below the (i) center, and (ii) edge of the footing.

6.10 A long foundation 0.6 m wide carries a line load of 100 kN/m. Calculate the vertical stress σ_z at a point *P*, the coordinates of which are x = 2.75 m, and z = 1.5 m, where the x-coordinate is normal to the line load from the central line of the footing.

6.11 A strip footing 10 ft wide is loaded on the ground surface with a pressure equal to 4177 lb/ft^2. Calculate vertical stresses at depths of 3, 6, and 12 ft under the center of the footing.

6.12 A rectangular footing of size 25 × 40 ft carries a uniformly distributed load of 5200 lb/ft^2. Determine the vertical pressure 20 ft below a point *O* which is located at a distance of 35 ft from the center of the footing on its longitudinal axis by making use of the curves in Fig. 6.8.

6.13 The center of a rectangular area at the ground surface has cartesian coordinate (0,0) and the corners have coordinates (6,15). All dimensions are in foot units. The area carries a uniform pressure of 3000 lb/ft^2. Estimate the stresses at a depth of 30 ft below ground surface at each of the following locations: (0,0), (0,15), (6,0).

6.14 Calculate the vertical stress at a depth of 50 ft below a point 10 ft oubide the corner (along the longer side) of a rectangular loaded area 30 × 80 ft carrying a uniform load of 2500 lb/ft^2.

6.15 A rectangular footing 6 × 3 m carries a uniform pressure of 300 kN/m^2 on the surface of a soil mass. Determine the vertical stress at a depth of 4.5 m below the surface on the center line 1.0 m inside the long edge of the foundation.

6.16 A circular ring foundation for an overhead tank transmits a contact pressure of 300 kN/m^2. Its internal diameter is 6 m and external diameter 10 m. Compute the vertical stress on the center line of the footing due to the imposed load at a depth of 6.5 m below the ground level. The footing is founded at a depth of 2.5 m.

6.17 In Prob. 6.16, if the foundation for the tank is a raft of diameter 10 m, determine the vertical stress at 6.5 m depth on the center line of the footing. All the other data remain the same.

6.18 How far apart must two 20 m diameter tanks be placed such that their combined stress overlap is not greater than 10% of the surface contact stress at a depth of 10 m?

6.19 A water tower is founded on a circular ring type foundation. The width of the ring is 4 m and its internal radius is 8 m. Assuming the distributed load per unit area as 300 kN/m^2, determine the vertical pressure at a depth of 6 m below the center of the foundation.

6.20 An embankment for road traffic is required to be constructed with the following dimensions :

Top width = 8 m, height = 4 m, side slopes= 1 V : 1.5 Hor

The unit weight of soil under the worst condition is 21 kN/m^3. The surcharge load on the road surface may be taken as 50 kN/m^2. Compute the vertical pressure at a depth of 6 m below the ground surface at the following locations:

(i) On the central longitudinal plane of the embankment.

(ii) Below the toes of the embankment.

6.21 If the top width of the road given in Prob. 6.20 is reduced to zero, what would be the change in the vertical pressure at the same points?

6.22 A square footing of size 13 × 13 ft founded on the surface carries a distributed load of 2089 lb/ft^2. Determine the increase in pressure at a depth of 10 ft by the 2:1 method

6.23 A load of 337 kips is imposed on a foundation 10 ft square at a shallow depth in a soil mass. Determine the vertical stress at a point 16 ft below the center of the foundation (a) assuming the load is uniformly distributed over the foundation, and (b) assuming the load acts as a point load at the center of the foundation.

6.24 A total load of 900 kN is uniformly distributed over a rectangular footing of size 3 × 2 m. Determine the vertical stress at a depth of 2.5 m below the footing at point *C* (Fig. Prob. 6.24), under one corner and *D* under the center. If another footing of size 3 × 1 m with a total load of 450 kN is constructed adjoining the previous footing, what is the additional stress at the point *C* at the same depth due to the construction of the second footing?

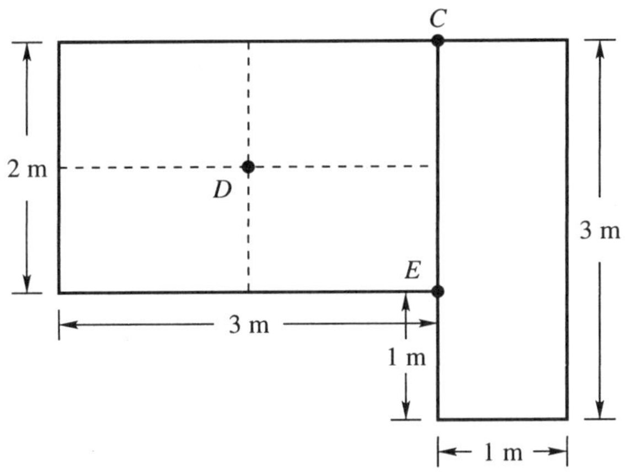

Figure Prob. 6.24

6.25 Refer to Prob. 6.24. Determine the vertical stress at a depth of 2.5 m below point *E* in Fig. Prob. 6.24. All the other data given in Prob. 6.24 remain the same.

CHAPTER 7

COMPRESSIBILITY AND CONSOLIDATION

7.1 INTRODUCTION

Structures are built on soils. They transfer loads to the subsoil through the foundations. The effect of the loads is felt by the soil normally up to a depth of about two to three times the width of the foundation. The soil within this depth gets compressed due to the imposed stresses. The compression of the soil mass leads to the decrease in the volume of the mass which results in the settlement of the structure.

The displacements that develop at any given boundary of the soil mass can be determined on a rational basis by summing up the displacements of small elements of the mass resulting from the strains produced by a change in the stress system. The compression of the soil mass due to the imposed stresses may be almost immediate or time dependent according to the permeability characteristics of the soil. Cohesionless soils which are highly permeable are compressed in a relatively short period of time as compared to cohesive soils which are less permeable. The compressibility characteristics of a soil mass might be due to any or a combination of the following factors:

1. Compression of the solid matter.
2. Compression of water and air within the voids.
3. Escape of water and air from the voids.

It is quite reasonable and rational to assume that the solid matter and the pore water are relatively incompressible under the loads usually encountered in soil masses. The change in volume of a mass under imposed stresses must be due to the escape of water if the soil is saturated. But if the soil is partially saturated, the change in volume of the mass is partly due to the compression and escape of air from the voids and partly due to the dissolution of air in the pore water.

The compressibility of a soil mass is mostly dependent on the rigidity of the soil skeleton. The rigidity, in turn, is dependent on the structural arrangement of particles and, in fine grained soils, on the degree to which adjacent particles are bonded together. Soils which possess a honeycombed structure possess high porosity and as such are more compressible. A soil composed predominantly of flat grains is more compressible than one containing mostly spherical grains. A soil in an undisturbed state is less compressible than the same soil in a remolded state.

Soils are neither truly elastic nor plastic. When a soil mass is under compression, the volume change is predominantly due to the slipping of grains one relative to another . The grains do not spring back to their original positions upon removal of the stress. However, a small elastic rebound under low pressures could be attributed to the elastic compression of the adsorbed water surrounding the grains.

Soil engineering problems are of two types. The first type includes all cases wherein there is no possibility of the stress being sufficiently large to exceed the shear strength of the soil, but wherein the strains lead to what may be a serious magnitude of displacement of individual grains leading to settlements within the soil mass. Chapter 7 deals with this type of problem. The second type includes cases in which there is danger of shearing stresses exceeding the shear strength of the soil. Problems of this type are called *Stability Problems* which are dealt with under the chapters of earth pressure, stability of slopes, and foundations.

Soil in nature may be found in any of the following states

1. Dry state.
2. Partially saturated state.
3. Saturated state.

Settlements of structures built on granular soils are generally considered only under two states, that is, either dry or saturated. The stress-strain characteristics of dry sand, depend primarily on the relative density of the sand, and to a much smaller degree on the shape and size of grains. Saturation does not alter the relationship significantly provided the water content of the sand can change freely. However, in very fine-grained or silty sands the water content may remain almost unchanged during a rapid change in stress. Under this condition, the compression is time-dependent. Suitable hypotheses relating displacement and stress changes in granular soils have not yet been formulated. However, the settlements may be determined by semi-empirical methods (Terzaghi, Peck and Mesri, 1996).

In the case of cohesive soils, the dry state of the soils is not considered as this state is only of a temporary nature. When the soil becomes saturated during the rainy season, the soil becomes more compressible under the same imposed load. Settlement characteristics of cohesive soils are, therefore, considered only under completely saturated conditions. It is quite possible that there are situations where the cohesive soils may remain partially saturated due to the confinement of air bubbles, gases etc. Current knowledge on the behavior of partially saturated cohesive soils under external loads is not sufficient to evolve a workable theory to estimate settlements of structures built on such soils.

7.2 CONSOLIDATION

When a saturated clay-water system is subjected to an external pressure, the pressure applied is initially taken by the water in the pores resulting thereby in an excess pore water pressure. If drainage is permitted, the resulting hydraulic gradients initiate a flow of water out of the clay mass and the mass begins to compress. A portion of the applied stress is transferred to the soil skeleton, which in turn causes a reduction in the excess pore pressure. This process, involving a gradual compression occurring simultaneously with a flow of water out of the mass and with a gradual transfer of the applied pressure from the pore water to the mineral skeleton is called *consolidation*. The process opposite to consolidation is called *swelling*, which involves an increase in the water content due to an increase in the volume of the voids.

Consolidation may be due to one or more of the following factors:

1. External static loads from structures.
2. Self-weight of the soil such as recently placed fills.
3. Lowering of the ground water table.
4. Desiccation.

The total compression of a saturated clay strata under excess effective pressure may be considered as the sum of

1. Immediate compression,
2. Primary consolidation, and
3. Secondary compression.

The portion of the settlement of a structure which occurs more or less simultaneously with the applied loads is referred to as the *initial* or *immediate settlement*. This settlement is due to the immediate compression of the soil layer under undrained condition and is calculated by assuming the soil mass to behave as an elastic soil.

If the rate of compression of the soil layer is controlled solely by the resistance of the flow of water under the induced hydraulic gradients, the process is referred to as *primary consolidation*. The portion of the settlement that is due to the primary consolidation is called *primary consolidation settlement* or *compression*. At the present time the only theory of practical value for estimating time-dependent settlement due to volume changes, that is under primary consolidation is *the one-dimensional theory*.

The third part of the settlement is due to secondary consolidation or compression of the clay layer. This compression is supposed to start after the primary consolidation ceases, that is after the excess pore water pressure approaches zero. It is often assumed that secondary compression proceeds linearly with the logarithm of time. However, a satisfactory treatment of this phenomenon has not been formulated for computing settlement under this category.

The Process of Consolidation

The process of consolidation of a clay-soil-water system may be explained with the help of a mechanical model as described by Terzaghi and Frohlich (1936).

The model consists of a cylinder with a frictionless piston as shown in Fig. 7.1. The piston is supported on one or more helical metallic springs. The space underneath the piston is completely filled with water. The springs represent the mineral skeleton in the actual soil mass and the water below the piston is the pore water under saturated conditions in the soil mass. When a load of p is placed on the piston, this stress is fully transferred to the water (as water is assumed to be incompressible) and the water pressure increases. The pressure in the water is

$$u = p$$

This is analogous to pore water pressure, u, that would be developed in a clay-water system under external pressures. If the whole model is leakproof without any holes in the piston, there is no chance for the water to escape. Such a condition represents a highly impermeable clay-water system in which there is a very high resistance for the flow of water. It has been found in the case of compact plastic clays that the minimum initial gradient required to cause flow may be as high as 20 to 30.

If a few holes are made in the piston, the water will immediately escape through the holes. With the escape of water through the holes a part of the load carried by the water is transferred to the springs. This process of transference of load from water to spring goes on until the flow stops

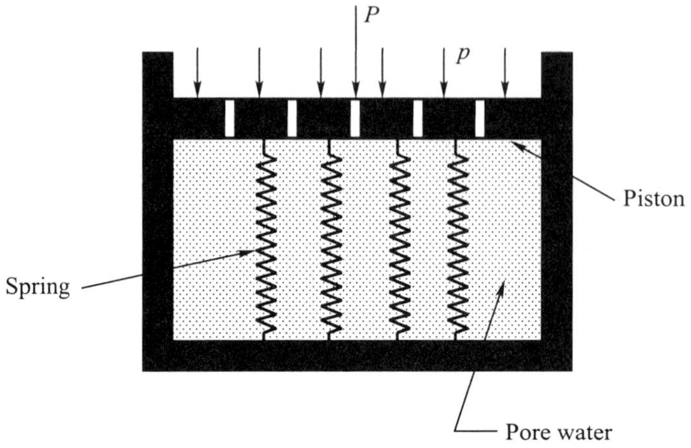

Figure 7.1 Mechanical model to explain the process of consolidation

when all the load will be carried by the spring and none by the water. The time required to attain this condition depends upon the number and size of the holes made in the piston. A few small holes represents a clay soil with poor drainage characteristics.

When the spring-water system attains equilibrium condition under the imposed load, the settlement of the piston is analogous to the compression of the clay-water system under external pressures.

One-Dimensional Consolidation

In many instances the settlement of a structure is due to the presence of one or more layers of soft clay located between layers of sand or stiffer clay as shown in Fig. 7.2A. The adhesion between the soft and stiff layers almost completely prevents the lateral movement of the soft layers. The theory that was developed by Terzaghi (1925) on the basis of this assumption is called the *one-dimensional consolidation theory.* In the laboratory this condition is simulated most closely by the *confined compression* or *consolidation test.*

The process of consolidation as explained with reference to a mechanical model may now be applied to a saturated clay layer in the field. If the clay strata shown in Fig 7.2 B(a) is subjected to an excess pressure Δp due to a uniformly distributed load p on the surface, the clay layer is compressed over

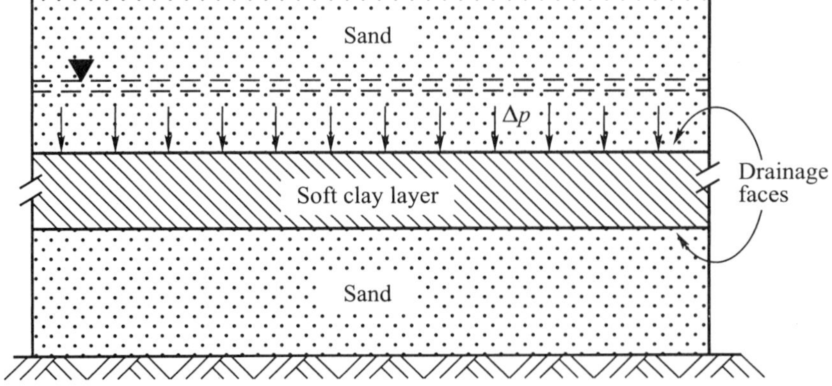

Figure 7.2A Clay layer sandwiched between sand layers

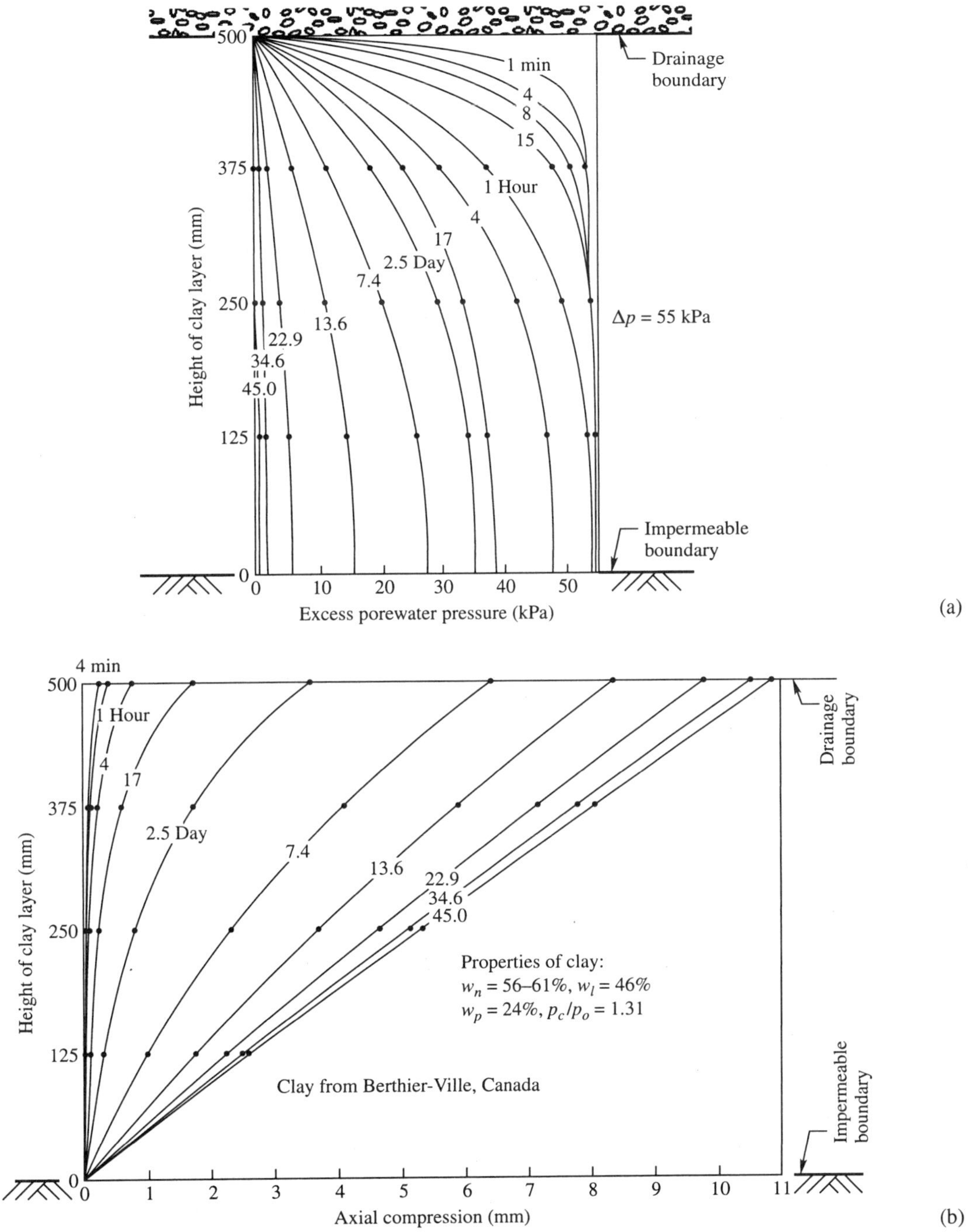

Figure 7.2B (a) Observed distribution of excess pore water pressure during consolidation of a soft clay layer; (b) observed distribution of vertical compression during consolidation of a soft clay layer (after Mesri and Choi, 1985, Mesri and Feng, 1986)

time and excess pore water drains out of it to the sandy layer. This constitutes the process of *consolidation*. At the instant of application of the excess load Δp, the load is carried entirely by water in the voids of the soil. As time goes on the excess pore water pressure decreases, and the effective vertical pressure in the layer correspondingly increases. At any point within the consolidating layer, the value u of the excess pore water pressure at a given time may be determined from

$$u = u_i - \Delta p_z$$

where, u = excess pore water pressure at depth z at any time t

u_i = initial total pore water pressure at time $t = 0$

Δp_z = effective pressure transferred to the soil grains at depth z and time t

At the end of primary consolidation, the excess pore water pressure u becomes equal to zero. This happens when $u = 0$ at all depths.

The time taken for full consolidation depends upon the drainage conditions, the thickness of the clay strata, the excess load at the top of the clay strata etc. Fig. 7.2B (a) gives a typical example of an observed distribution of excess pore water pressure during the consolidation of a soft clay layer 50 cm thick resting on an impermeable stratum with drainage at the top. Figure 7.2B(b) shows the compression of the strata with the dissipation of pore water pressure. It is clear from the figure that the time taken for the dissipation of pore water pressure may be quite long, say a year or more.

7.3 CONSOLIDOMETER

The compressibility of a saturated, clay-water system is determined by means of the apparatus shown diagrammatically in Fig. 7.3(a). This apparatus is also known as an *oedometer*. Figure 7.3(b) shows a table top consolidation apparatus.

The consolidation test is usually performed at room temperature, in floating or fixed rings of diameter from 5 to 11 cm and from 2 to 4 cm in height. Fig. 7.3(a) is a fixed ring type. In a floating ring type, the ring is free to move in the vertical direction.

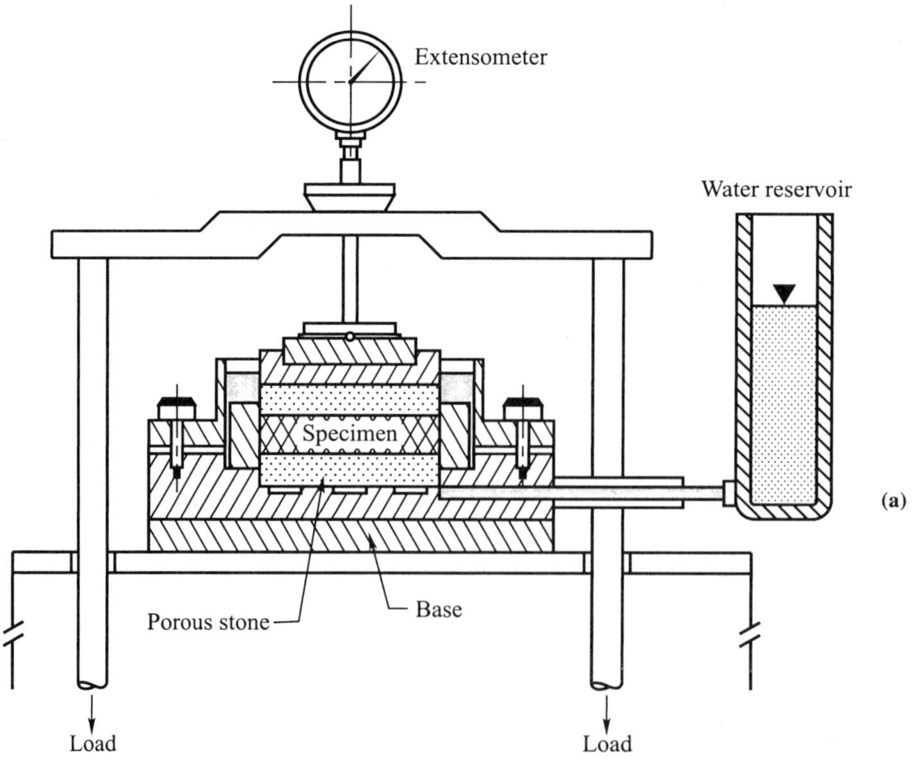

Figure 7.3 (a) A schematic diagram of a consolidometer

Figure 7.3 (b) Table top consolidation apparatus (Courtesy: Soiltest, USA)

The soil sample is contained in the brass ring between two porous stones about 1.25 cm thick. By means of the porous stones water has free access to and from both surfaces of the specimen. The compressive load is applied to the specimen through a piston, either by means of a hanger and dead weights or by a system of levers. The compression is measured on a dial gauge.

At the bottom of the soil sample the water expelled from the soil flows through the filter stone into the water container. At the top, a well-jacket filled with water is placed around the stone in order to prevent excessive evaporation from the sample during the test. Water from the sample also flows into the jacket through the upper filter stone. The soil sample is kept submerged in a saturated condition during the test.

7.4 THE STANDARD ONE-DIMENSIONAL CONSOLIDATION TEST

The main purpose of the consolidation test on soil samples is to obtain the necessary information about the compressibility properties of a saturated soil for use in determining the magnitude and rate of settlement of structures. The following test procedure is applied to any type of soil in the standard consolidation test.

Loads are applied in steps in such a way that the successive load intensity, *p*, is twice the preceding one. The load intensities commonly used being 1/4, 1/2, 1, 2, 4, 8, and 16 tons/ft^2 (25, 50, 100, 200, 400, 800 and 1600 kN/m^2). Each load is allowed to stand until compression has practically ceased (no longer than 24 hours). The dial readings are taken at elapsed times of 1/4, 1/2, 1, 2, 4, 8, 15, 30, 60, 120, 240, 480 and 1440 minutes from the time the new increment of load is put on the sample (or at elpased times as per requirements). Sandy samples are compressed in a relatively short time as compared to clay samples and the use of one day duration is common for the latter.

After the greatest load required for the test has been applied to the soil sample, the load is removed in decrements to provide data for plotting the expansion curve of the soil in order to learn its elastic properties and magnitudes of plastic or permanent deformations. The following data should also be obtained:

1. Moisture content and weight of the soil sample before the commencement of the test.
2. Moisture content and weight of the sample after completion of the test.
3. The specific gravity of the solids.
4. The temperature of the room where the test is conducted.

7.5 PRESSURE-VOID RATIO CURVES

The pressure-void ratio curve can be obtained if the void ratio of the sample at the end of each increment of load is determined. Accurate determinations of void ratio are essential and may be computed from the following data:

1. The cross-sectional area of the sample A, which is the same as that of the brass ring.
2. The specific gravity, G_s, of the solids.
3. The dry weight, W_s, of the soil sample.
4. The sample thickness, h, at any stage of the test.

Let V_s = volume of the solids in the sample

where

$$V_s = \frac{W}{G_s \gamma_w}$$

where γ_w = unit weight of water

We can also write

$$V_s = h_s A \quad \text{or} \quad h_s = \frac{V_s}{A}$$

where, h_s = thickness of solid matter.

If e is the void ratio of the sample, then

$$e = \frac{Ah - Ah_s}{Ah_s} = \frac{h - h_s}{h_s} \tag{7.1}$$

In Eq. (7.1) h_s is a constant and only h is a variable which decreases with increment load. If the thickness h of the sample is known at any stage of the test, the void ratio at all the stages of the test may be determined.

The equilibrium void ratio at the end of any load increment may be determined by the change of void ratio method as follows:

Change of Void-Ratio Method

In one-dimensional compression the change in height Δh per unit of original height h equals the change in volume ΔV per unit of original volume V.

$$\frac{\Delta h}{h} = \frac{\Delta V}{V} \tag{7.2}$$

V may now be expressed in terms of void ratio e.

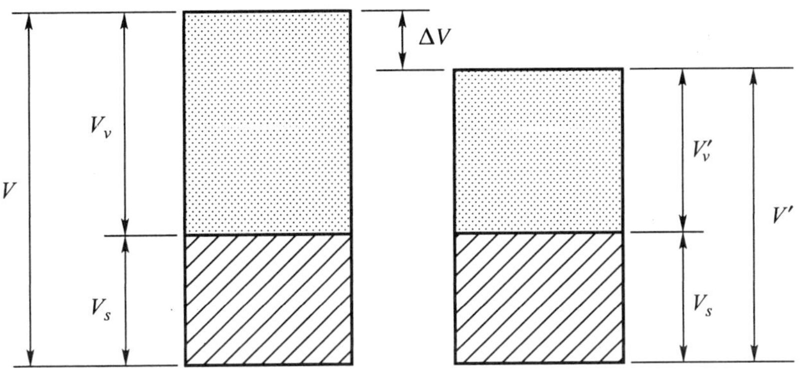

Figure 7.4 Change of void ratio

We may write (Fig. 7.4),

$$V_v = eV_s, \quad V = V_s(1 + e), \quad V_v' = e'V_s$$

$$V' = V_s(1+e')$$

$$\frac{\Delta V}{V} = \frac{V - V'}{V} = \frac{V_s(1+e) - V_s(1+e')}{V} = \frac{e-e'}{1+e} = \frac{\Delta e}{1+e}$$

Therefore,

$$\frac{\Delta h}{h} = \frac{\Delta e}{1+e}$$

or

$$\Delta e = \frac{1+e}{h}\Delta h \tag{7.3}$$

wherein, Δe = change in void ratio under a load, h = initial height of sample, e = initial void ratio of sample, e' = void ratio after compression under a load, Δh = compression of sample under the load which may be obtained from dial gauge readings.

Typical pressure-void ratio curves for an undisturbed clay sample are shown in Fig. 7.5, plotted both on arithmetic and on semilog scales. The curve on the log scale indicates clearly two branches, a fairly horizontal initial portion and a nearly straight inclined portion. The coordinates of point A in the figure represent the void ratio e_0 and effective overburden pressure p_0 corresponding to a state of the clay in the field as shown in the inset of the figure. When a sample is extracted by means of the best of techniques, the water content of the clay does not change significantly. Hence, the void ratio e_0 at the start of the test is practically identical with that of the clay in the ground. When the pressure on the sample in the consolidometer reaches p_0, the e-log p curve should pass through the point A unless the test conditions differ in some manner from those in the field. In reality the curve always passes below point A, because even the best sample is at least slightly disturbed.

The curve that passes through point A is generally termed as a *field curve* or *virgin curve*. In settlement calculations, the field curve is to be used.

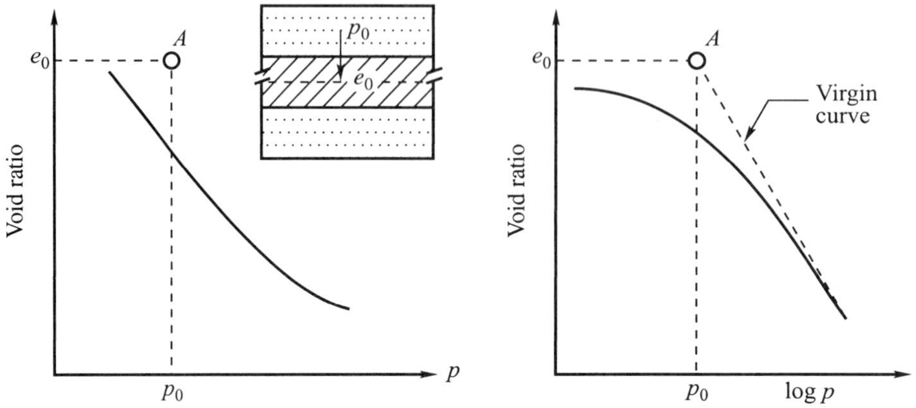

Figure 7.5 Pressure-void ratio curves

Pressure-Void Ratio Curves for Sand

Normally, no consolidation tests are conducted on samples of sand as the compression of sand under external load is almost instantaneous as can be seen in Fig. 7.6(a) which gives a typical curve showing the time versus the compression caused by an increment of load.

In this sample more than 90 per cent of the compression has taken place within a period of less than 2 minutes. The time lag is largely of a frictional nature. The compression is about the same whether the sand is dry or saturated. The shape of typical *e-p* curves for loose and dense sands are shown in Fig. 7.6(b). The amount of compression even under a high load intensity is not significant as can be seen from the curves.

Pressure-Void Ratio Curves for Clays

The compressibility characteristics of clays depend on many factors.

The most important factors are

1. Whether the clay is normally consolidated or overconsolidated
2. Whether the clay is sensitive or insensitive.

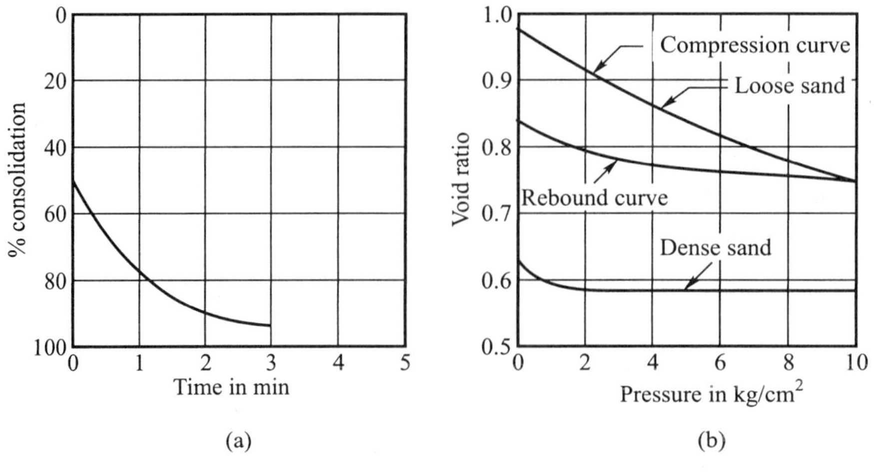

Figure 7.6 Pressure-void ratio curves for sand

Normally Consolidated and Overconsolidated Clays

A clay is said to be normally consolidated if the present effective overburden pressure p_0 is the maximum pressure to which the layer has ever been subjected at any time in its history, whereas a clay layer is said to be overconsolidated if the layer was subjected at one time in its history to a greater effective overburden pressure, p_c, than the present pressure, p_0. The ratio p_c / p_0 is called the *overconsolidation ratio* (OCR).

Overconsolidation of a clay stratum may have been caused due to some of the following factors

1. Weight of an overburden of soil which has eroded
2. Weight of a continental ice sheet that melted
3. Desiccation of layers close to the surface.

Experience indicates that the natural moisture content, w_n, is commonly close to the liquid limit, w_l, for normally consolidated clay soil whereas for the overconsolidated clay, w_n is close to plastic limit w_p.

Fig. 7.7 illustrates schematically the difference between a normally consolidated clay strata such as *B* on the left side of Section *CC* and the overconsolidated portion of the same layer *B* on the right side of section *CC*. Layer *A* is overconsolidated due to desiccation.

All of the strata located above bed rock were deposited in a lake at a time when the water level was located above the level of the present high ground when parts of the strata were removed by erosion, the water content in the clay stratum *B* on the right hand side of section *CC* increased slightly, whereas that of the left side of section *CC* decreased considerably because of the lowering of the water table level from position D_0D_0 to *DD*. Nevertheless, with respect to the present overburden, the clay stratum *B* on the right hand side of section *CC* is overconsolidated clay, and that on the left hand side is normally consolidated clay.

While the water table descended from its original to its final position below the floor of the eroded valley, the sand strata above and below the clay layer *A* became drained. As a consequence, layer *A* gradually dried out due to exposure to outside heat. Layer *A* is therefore said to be overconsolidated by desiccation.

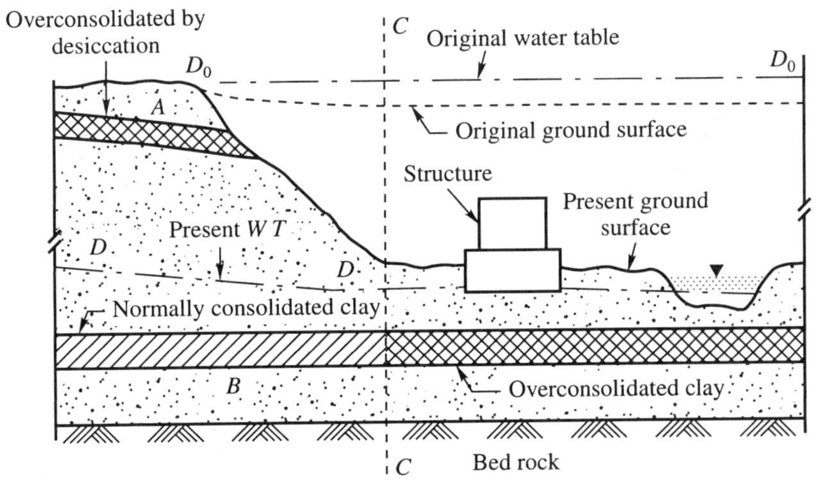

Figure 7.7 Diagram illustrating the geological process leading to overconsolidation of clays (After Terzaghi and Peck, 1967)

7.6 DETERMINATION OF PRECONSOLIDATION PRESSURE

Several methods have been proposed for determining the value of the maximum consolidation pressure. They fall under the following categories. They are

1. Field method,
2. Graphical procedure based on consolidation test results.

Field Method

The field method is based on geological evidence. The geology and physiography of the site may help to locate the original ground level. The overburden pressure in the clay structure with respect to the original ground level may be taken as the preconsolidation pressure p_c. Usually the geological estimate of the maximum consolidation pressure is very uncertain. In such instances, the only remaining procedure for obtaining an approximate value of p_c is to make an estimate based on the results of laboratory tests or on some relationships established between p_c and other soil parameters.

Graphical Procedure

There are a few graphical methods for determining the preconsolidation pressure based on laboratory test data. No suitable criteria exists for appraising the relative merits of the various methods.

The earliest and the most widely used method was the one proposed by Casagrande (1936). The method involves locating the point of maximum curvature, *B*, on the laboratory *e*-log *p* curve of an undisturbed sample as shown in Fig. 7.8. From *B*, a tangent is drawn to the curve and a horizontal line is also constructed. The angle between these two lines is then bisected. The abscissa of the point of intersection of this bisector with the upward extension of the inclined straight part corresponds to the preconsolidation pressure p_c.

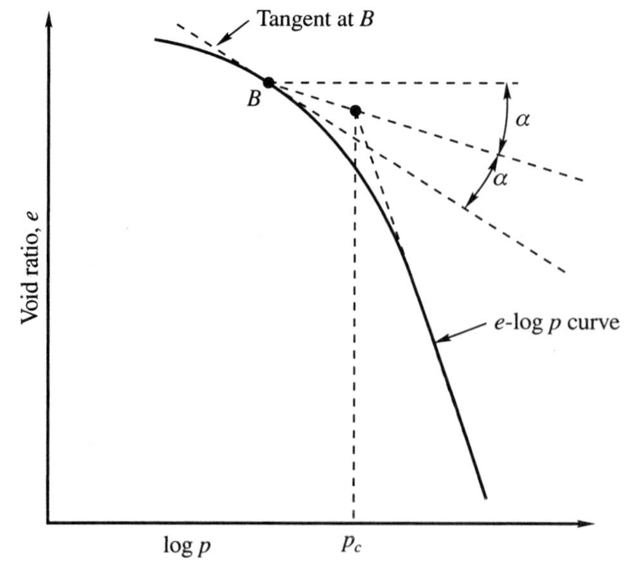

Figure 7.8 Method of determining p_c by Casagrande method

7.7 *e-log p* FIELD CURVES FOR NORMALLY CONSOLIDATED AND OVERCONSOLIDATED CLAYS OF LOW TO MEDIUM SENSITIVITY

It has been explained earlier with reference to Fig. 7.5, that the laboratory *e*-log *p* curve of an undisturbed sample does not pass through point *A* and always passes below the point. It has been found from investigation that the inclined straight portion of *e*-log *p* curves of undisturbed or remolded samples of clay soil intersect at one point at a low void ratio and corresponds to $0.4e_0$ shown as point *C* in Fig. 7.9 (Schmertmann, 1955). It is logical to assume the field curve labelled as K_f should also pass through this point. The field curve can be drawn from point *A*, having coordinates (e_0, p_0), which corresponds to the *in-situ* condition of the soil. The straight line *AC* in Fig. 7.9(a) gives the field curve K_f for normally consolidated clay soil of low sensitivity.

The field curve for overconsolidated clay soil consists of two straight lines, represented by *AB* and *BC* in Fig. 7.9(b). Schmertmann (1955) has shown that the initial section *AB* of the field curve is parallel to the mean slope *MN* of the rebound laboratory curve. Point *B* is the intersection point of the vertical line passing through the preconsolidation pressure p_c on the abscissa and the sloping line *AB*. Since point *C* is the intersection of the laboratory compression curve and the horizontal line at void ratio $0.4e_0$, line *BC* can be drawn. The slope of line *MN* which is the slope of the rebound curve is called the *swell index* C_s.

Clay of High Sensitivity

If the sensitivity S_t is greater than about 8 [sensitivity is defined as the ratio of unconfined compressive strengths of undisturbed and remolded soil samples refer to Eq. (3.50)], then the clay is said to be highly sensitive. The natural water contents of such clay are more than the liquid limits. The *e*-log *p* curve K_u for an undisturbed sample of such a clay will have the initial branch almost flat as shown in Fig. 7.9(c), and after this it drops abruptly into a steep segment indicating there by a structural breakdown of the clay such that a slight increase of pressure leads to a large decrease in void ratio. The curve then passes through a point of inflection at *d* and its slope decreases. If a tangent is drawn at the point of inflection *d*, it intersects the line e_0A at *b*. The pressure corresponding to *b* (p_b) is approximately equal to that at which the structural breakdown takes place. In areas underlain by soft highly sensitive clays, the excess pressure Δp over the layer should be limited to a fraction of the difference of pressure $(p_b - p_0)$. Soil of this type belongs mostly to volcanic regions.

7.8 COMPUTATION OF CONSOLIDATION SETTLEMENT

Settlement Equations for Normally Consolidated Clays

For computing the ultimate settlement of a structure founded on clay the following data are required

1. The thickness of the clay stratum, H
2. The initial void ratio, e_0
3. The consolidation pressure p_0 or p_c
4. The field consolidation curve K_f

The slope of the field curve K_f on a semilogarithmic diagram is designated as the *compression index* C_c (Fig. 7.9)

The equation for C_c may be written as

$$C_c = \frac{e_0 - e}{\log p - \log p_0} = \frac{e_0 - e}{\log p/p_0} = \frac{\Delta e}{\log p/p_0} \tag{7.4}$$

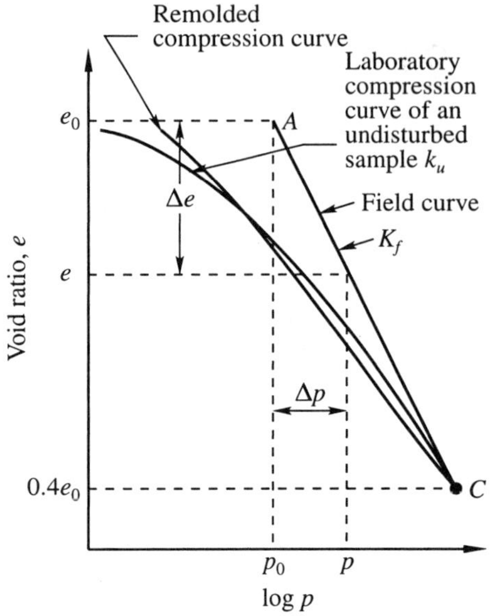

(a) Normally consolidated clay soil

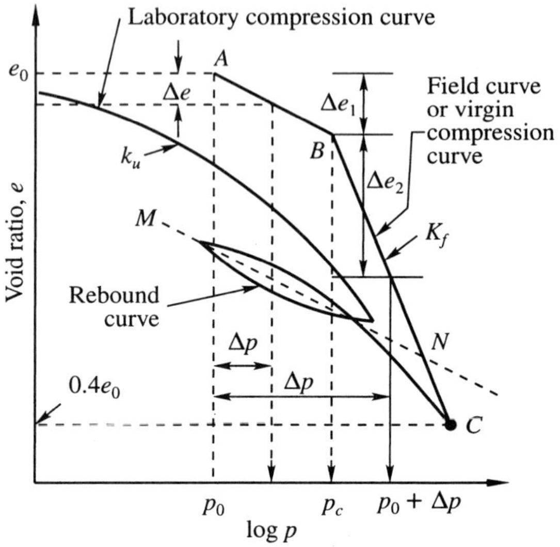

(b) Preconsolidated clay soil

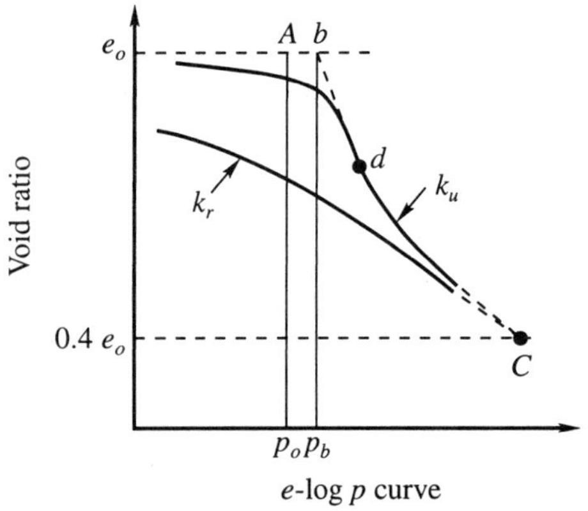

(c) Typical e-log p curve for an undisturbed sample of clay of high sensitivity (Peck et al., 1974)

Figure 7.9 Field *e*-log *p* curves

In one-dimensional compression, as per Eq. (7.2), the change in height ΔH per unit of original H may be written as equal to the change in volume ΔV per unit of original volume V (Fig. 7.10).

$$\frac{\Delta H}{H} = \frac{\Delta V}{V} \tag{7.5}$$

Considering a unit sectional area of the clay stratum, we may write

$$V = H, \quad V_1 = H_1$$

$$\Delta V = (H - H_1) = H_s(1 + e_0) - H_s(1 + e_1) = H_s(e_0 - e_1)$$

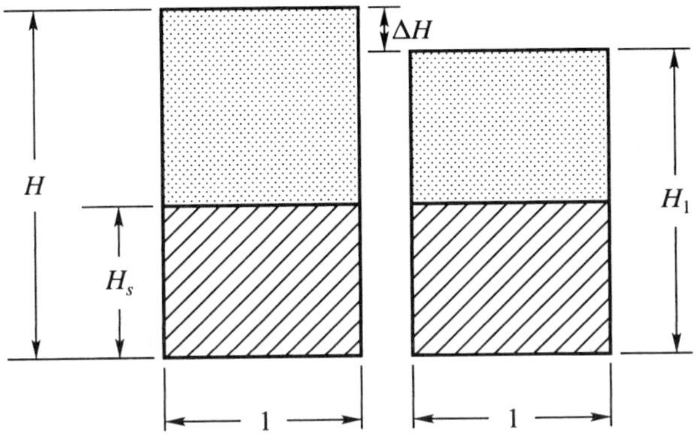

Figure 7.10 Change of height due to one-dimensional compression

Therefore,

$$\frac{\Delta V}{V} = \frac{H_s(e_0 - e_1)}{H_s(1+e_0)} = \frac{e_0 - e_1}{(1+e_0)} = \frac{\Delta e}{1+e_0} \tag{7.6}$$

Substituting for $\Delta V/V$ in Eq. (7.5)

$$\Delta H = H\frac{\Delta e}{1+e_0} \tag{7.7}$$

If we designate the compression ΔH of the clay layer as the total settlement S_t of the structure built on it, we have

$$\Delta H = S_t = H\frac{\Delta e}{1+e_0} \tag{7.8}$$

Settlement Calculation from *e*-log *p* Curves

Substituting for Δe in Eq. (7.8) we have

$$S_t = \frac{C_c}{1+e_0} H \log\frac{p}{p_0} \tag{7.9}$$

or

$$S_t = \frac{C_c}{1+e_0} H \log\frac{p_0 + \Delta p}{p_0} \tag{7.10}$$

The net change in pressure Δp produced by the structure at the middle of a clay stratum is calculated from the Boussinesq or Westergaard theories as explained in Chapter 6.

If the thickness of the clay stratum is too large, the stratum may be divided into layers of smaller thickness not exceeding 3 m. The net change in pressure Δp at the middle of each layer will have to be calculated. Consolidation tests will have to be completed on samples taken from the middle of each of the strata and the corresponding compression indices will have to be determined. The equation for the total consolidation settlement may be written as

$$S_t = \quad H_i \frac{C_c}{1+e_0} \log\frac{p_0 + \Delta p}{p_0} \tag{7.11}$$

where the subscript i refers to each layer in the subdivision. If there is a series of clay strata of thickness H_1, H_2, etc., separated by granular materials, the same Eq. (7.10) may be used for calculating the total settlement.

Settlement Calculation from *e-p* Curves

We can plot the field *e-p* curves from the laboratory test data and the field *e*-log *p* curves. The weight of a structure or of a fill increases the pressure on the clay stratum from the overburden pressure p_0 to the value $p_0 + \Delta p$ (Fig. 7.11). The corresponding void ratio decreases from e_0 to e. Hence, for the range in pressure from p_0 to $(p_0 + \Delta p)$, we may write

$$e_0 - e = \Delta e = a_v \Delta p$$

or

$$a_v (\text{cm}^2 / \text{gm}) = \frac{\Delta e}{p(\text{cm}^2 / \text{gm})} \tag{7.12}$$

where a_v is called the *coefficient of compressibility.*

For a given difference in pressure, the value of the coefficient of compressibility decreases as the pressure increases. Now substituting for Δe in Eq. (7.8) from Eq. (7.12), we have the equation for settlement

$$S_t = \frac{a_v H}{1+e_0} \Delta p = m_v H \ \Delta p \tag{7.13}$$

where $m_v = a_v/(1 + e_0)$ is known as the *coefficient of volume compressibility.*

It represents the compression of the clay per unit of original thickness due to a unit increase of the pressure.

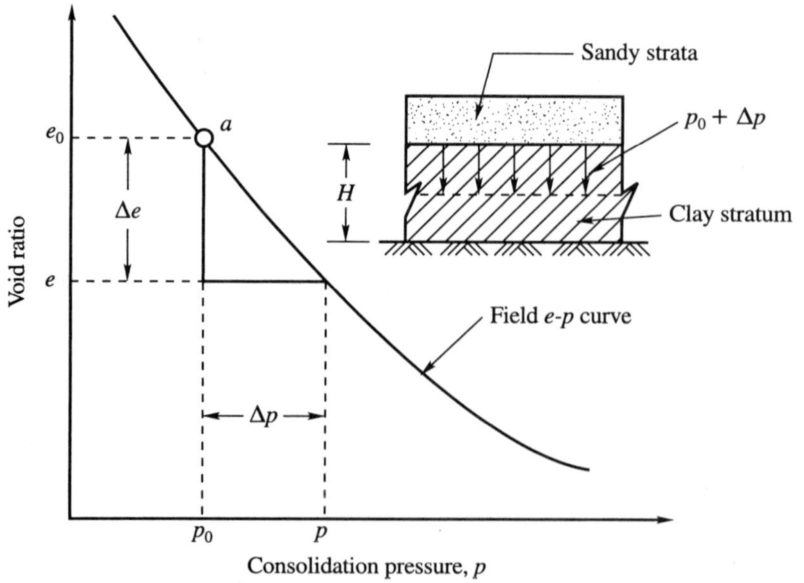

Figure 7.11 Settlement calculation from *e-p* curve

Settlement Calculation from *e*-log *p* Curve for Overconsolidated Clay Soil

Fig. 7.9(b) gives the field curve K_f for preconsolidated clay soil. The settlement calculation depends upon the excess foundation pressure Δp over and above the existing overburden pressure p_0.

Settlement Computation, if $p_0 + \Delta p \le p_c$ (Fig. 7.9(b))

In such a case, use the sloping line *AB*. If C_s = slope of this line (also called the swell index), we have

$$C_s = \frac{\Delta e}{\log \frac{(p_0 + \Delta p)}{p_0}} \quad (7.14a)$$

$$\text{or } \Delta e = C_s \log \frac{p_0 + \Delta p}{p_0} \quad (7.14b)$$

By substituting for Δe in Eq. (7.8), we have

$$S_t = \frac{C_s H}{1 + e_0} \log \frac{p_0 + \Delta p}{p_0} \quad (7.15a)$$

Settlement Computation, if $p_0 < p_c < p_0 + \Delta p$

We may write from Fig. 7.9(b)

$$\Delta e = \Delta e_1 + \Delta e_2 = C_s \log \frac{p_c}{p_0} + C_c \log \frac{p_0 + \Delta p}{p_c} \quad (7.15b)$$

In this case the slope of both the lines *AB* and *BC* in Fig. 7.9(b) are required to be considered. Now the equation for S_t may be written as [from Eq. (7.8) and Eq. (7.15b)]

$$S_t = \frac{C_s H}{1 + e_0} \log \frac{p_c}{p_0} + \frac{C_c H}{1 + e_0} \log \frac{p_0 + \Delta p}{p_c} \quad (7.15c)$$

The swell index $C_s \approx$ 1/5 to 1/10 C_c can be used as a check.

Nagaraj and Murthy (1985) have proposed the following equation for C_s as

$$C_s = 0.0463 \left(\frac{w_l}{100} \right) G_s$$

where w_l = liquid limit, G_s = specific gravity of solids.

Compression Index C_c—Empirical Relationships

Research workers in different parts of the world have established empirical relationships between the *compression index* C_c and other soil parameters. A few of the important relationships are given below.

Skempton's Formula

Skempton (1944) established a relationship between C_c and liquid limits for remolded clays as

$$C_c = 0.007\,(w_l - 10) \quad (7.16)$$

where w_l is in percent.

Terzaghi and Peck Formula

Based on the work of Skempton and others, Terzaghi and Peck (1948) modified Eq. (7.16) applicable to normally consolidated clays of low to moderate sensitivity as

$$C_c = 0.009\,(w_l - 10) \tag{7.17}$$

Azzouz et al., Formula

Azzouz et al., (1976) proposed a number of correlations based on the statistical analysis of a number of soils. The one of the many which is reported to have 86 percent reliability is

$$C_c = 0.37\,(e_0 + 0.003\,w_l + 0.0004\,w_n - 0.34) \tag{7.18}$$

where e_0 = *in-situ* void ratio, w_l and w_n are in per cent. For organic soil they proposed

$$C_c = 0.115\,w_n \tag{7.19}$$

Hough's Formula

Hough (1957), on the basis of experiments on precompressed soils, has given the following equation

$$C_c = 0.3\,(e_0 - 0.27) \tag{7. 20}$$

Nagaraj and Srinivasa Murthy Formula

Nagaraj and Srinivasa Murthy (1985) have developed equations based on their investigation as follows

$$C_c = 0.2343\,e_l \tag{7.21}$$

$$C_c = 0.39\,e_0 \tag{7.22}$$

where e_l is the void ratio at the liquid limit, and e_0 is the *in-situ* void ratio.

In the absence of consolidation test data, one of the formulae given above may be used for computing C_c according to the judgment of the engineer.

7.9 SETTLEMENT DUE TO SECONDARY COMPRESSION

In certain types of clays the secondary time effects are very pronounced to the extent that in some cases the entire time-compression curve has the shape of an almost straight sloping line when plotted on a semilogarithmic scale, instead of the typical inverted S-shape with pronounced primary consolidation effects. These so called secondary time effects are a phenomenon somewhat analogous to the creep of other overstressed material in a plastic state. A delayed progressive slippage of grain upon grain as the particles adjust themselves to a more dense condition, appears to be responsible for the secondary effects. When the rate of plastic deformations of the individual soil particles or of their slippage on each other is slower than the rate of decreasing volume of voids between the particles, then secondary effects predominate and this is reflected by the shape of the time compression curve. The factors which affect the rate of the secondary compression of soils are not yet fully understood, and no satisfactory method has yet been developed for a rigorous and reliable analysis and forecast of the magnitude of these effects. Highly organic soils are normally subjected to considerable secondary consolidation.

The rate of secondary consolidation may be expressed by the *coefficient of secondary compression,* $\overline{C}_\alpha$ as

$$\overline{C}_\alpha = \frac{\Delta e}{1+e_0}\frac{1}{\log (t_2/t_1)} = \frac{C_\alpha}{1+e_0} \text{ or } \Delta e = C_\alpha \log \frac{t_2}{t_1} \quad (7.23)$$

where C_a, the slope of the straight-line portion of the e-log t curve, is known as the *secondary compression index*. Numerically C_a is equal to the value of Δe for a single cycle of time on the curve (Fig. 7.12(a)). Compression is expressed in terms of decrease in void ratio and time has been normalized with respect to the duration t_p of the primary consolidation stage. A general expression for settlement due to secondary compression under the final stage of pressure p_f may be expressed as

$$S_s = \frac{\Delta e}{1+e_0}H \quad (7.24)$$

The value of Δe from $t/t_p = 1$ to any time t may be determined from the e versus t/t_p curve corresponding to the final pressure p_f.

Eq. (7.23) may now be expressed as

$$\Delta e = C_\alpha \log\frac{t}{t_p} \quad (7.25)$$

For a constant value C_α between t_p and t, Equation (7.24) may be expressed as

$$S_s = \frac{C_\alpha}{1+e_0}H\log\frac{t}{t_p} \quad (7.26)$$

where, e_0 = initial void ratio

H = thickness of the clay stratum.

The value of $\overline{C}_\alpha$ for normally loaded compressible soils increases in a general way with the compressibility and hence, with the natural water content, in the manner shown in Fig. 7.12(b) (Mesri, 1973). Although the range in values for a given water content is extremely large, the relation gives a conception of the upper limit of the rate of secondary settlement that may be anticipated if the deposit is normally loaded or if the stress added by the proposed construction will appreciably exceed the preconsolidation stress. The rate is likely to be much less if the clay is strongly preloaded or if the stress after the addition of the load is small compared to the existing overburden pressure. The rate is also influenced by the length of time the preload may have acted,

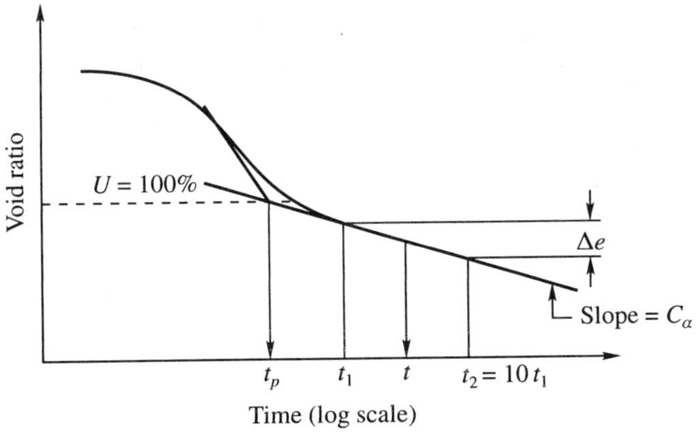

Figure 7.12(a) e-log p time curve representing secondary compression

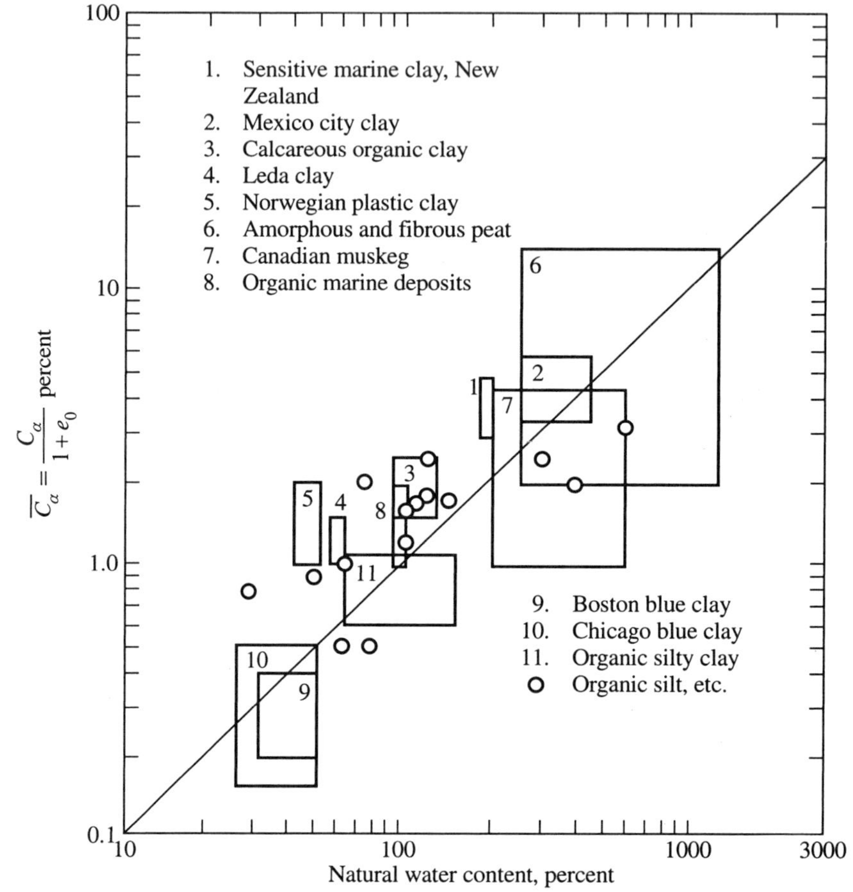

Figure 7.12(b) Relationship between coefficient of secondary consolidation and natural water content of normally loaded deposits of clays and various compressible organic soils (after Mesri, 1973)

by the existence of shearing stresses and by the degree of disturbance of the samples. The effects of these various factors have not yet been evaluated. Secondary compression is high in plastic clays and organic soils. Table 7.1 provides a classification of soil based on secondary compressibility. If 'young, normally loaded clay', having an effective overburden pressure of p_0 is left undisturbed for thousands of years, there will be creep or secondary consolidation. This will reduce the void ratio and consequently increase the preconsolidation pressure which will be much greater than the existing effective overburden pressure p_0. Such a clay may be called an *aged, normally consolidated clay*.

Mesri and Godlewski (1977) report that for any soil the ratio C_α/C_c is a constant (where C_c is the compression index). This is illustrated in Fig. 7.13 for undisturbed specimens of brown Mexico City clay with natural water content w_n = 313 to 340%, w_l = 361%, w_p = 91% and p_c/p_o = 1.4

Table 7.2 gives values of C_α/C_c for some geotechnical materials (Terzaghi, et al., 1996).

It is reported (Terzaghi et al., 1996) that for all geotechnical materials C_α/C_c ranges from 0.01 to 0.07. The value 0.04 is the most common value for inorganic clays and silts.

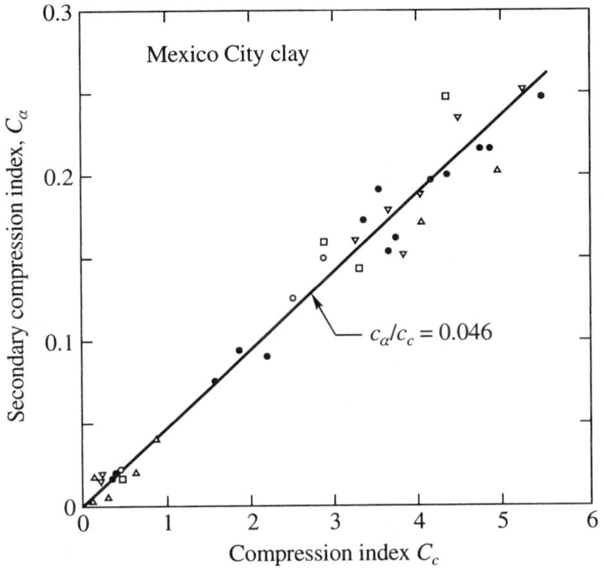

Figure 7.13 An example of the relation between C_α and C_c (after Mesri and Godlewski, 1977)

Table 7.1 Classification of soil based on secondary compressibility (Terzaghi, et al., 1996)

C_α	Secondary compressibility
< 0.002	Very low
0.004	Low
0.008	Medium
0.016	High
0.032	Very high
0.064	Extremely high

Table 7.2 Values of C_α/C_c for geotechnical materials (Terzaghi, et al., 1996)

Material	C_α/C_c
Granular soils including rockfill	0.02 ± 0.01
Shale and mudstone	0.03 ± 0.01
Inorganic clay and silts	0.04 ± 0.01
Organic clays and silts	0.05 ± 0.01
Peat and muskeg	0.06 ± 0.01

Example 7.1

During a consolidation test, a sample of fully saturated clay 3 cm thick (= h_0) is consolidated under a pressure increment of 200 kN/m^2. When equilibrium is reached, the sample thickness is reduced to 2.60 cm. The pressure is then removed and the sample is allowed to expand and absorb water. The final thickness is observed as 2.8 cm (h_f) and the final moisture content is determined as 24.9%.

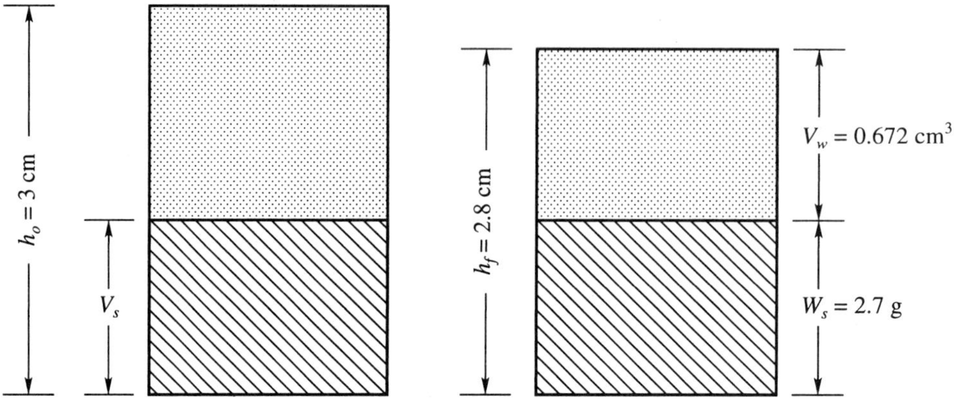

Figure Ex. 7.1

If the specific gravity of the soil solids is 2.70, find the void ratio of the sample before and after consolidation.

Solution

Use equation (7.3)

$$\Delta e = \frac{1+e}{h}\Delta h$$

1. *Determination of* e_f

   Weight of solids $= W_s = V_s G_s \gamma_\omega = 1 \times 2.70 \times 1 = 2.70$ g.

   $$\frac{W_w}{W_s} = 0.249 \text{ or } W_w = 0.249 \times 2.70 = 0.672 \text{ gm}, \; e_f = V_w = 0.672.$$

2. *Changes in thickness from final stage to equilibrium stage with load on*

   $$\Delta h = 2.80 - 2.60 = 0.20 \text{ cm}, \qquad \Delta e = \frac{(1+0.672)\; 0.20}{2.80} = 0.119.$$

   Void ratio after consolidation $= e_f - \Delta e = 0.672 - 0.119 = 0.553$.

3. *Change in void ratio from the commencement to the end of consolidation*

   $$\Delta e = \frac{1+0.553}{2.6}(3.00 - 2.60) = \frac{1.553}{2.6} \times 0.40 = 0.239.$$

   Void ratio at the start of consolidation $= 0.553 + 0.239 = 0.792$

Example 7.2

A recently completed fill was 32.8 ft thick and its initial average void ratio was 1.0. The fill was loaded on the surface by constructing an embankment covering a large area of the fill. Some months after the embankment was constructed, measurements of the fill indicated an average void ratio of 0.8. Estimate the compression of the fill.

Solution

Per Eq. (7.7), the compression of the fill may be calculated as

$$\Delta H = \frac{\Delta e}{1+e_0} H_0$$

where ΔH = the compression, Δe = change in void ratio, e_0 = initial void ratio, H_0 = thickness of fill.

Substituting, $\Delta H = \frac{1.0-0.8}{1+1.0} \times 32.8 = 3.28$ ft.

Example 7.3

A stratum of normally consolidated clay 7 m thick is located at a depth 12 m below ground level. The natural moisture content of the clay is 40.5 per cent and its liquid limit is 48 per cent. The specific gravity of the solid particles is 2.76. The water table is located at a depth 5 m below ground surface. The soil is sand above the clay stratum. The submerged unit weight of the sand is 11 kN/m^3 and the same weighs 18 kN/m^3 above the water table. The average increase in pressure at the center of the clay stratum is 120 kN/m^2 due to the weight of a building that will be constructed on the sand above the clay stratum. Estimate the expected settlement of the structure.

Solution

1. Determination of e and γ_b for the clay [Fig. Ex. 7.3]

$$\frac{W_w}{W_s} = w_n, \quad W_s = V_s G_s \gamma_w = 1 \times 2.76 \times 1 = 2.76 \text{ g}$$

$$W_w = \frac{40.5}{100} \times 2.76 = 1.118 \text{ g}$$

$$V_w = \frac{W_w}{\gamma_w} = \frac{1.118}{1} = 1.118 \text{ cm}^3$$

$$e_0 = \frac{V_w}{V_s} = \frac{1.118}{1} = 1.118$$

$$W = W_w + W_s = 1.118 + 2.76 = 3.878 \text{ g}$$

$$\gamma_t = \frac{W}{1+e_0} = \frac{3.88}{2.118} = 1.83 \text{ g/cm}^3$$

$$\gamma_b = (1.83 - 1) = 0.83 \text{ g/cm}^3.$$

2. Determination of overburden pressure p_0

$$p_0 = \gamma_1 h_1 + \gamma_2 h_2 + \gamma_3 h_3 \text{ or}$$

$$p_0 = 0.83 \times 9.81 \times 3.5 + 11 \times 7 + 18 \times 5 = 195.5 \text{ kN/m}^2$$

3. Compression index [Eq. 11.17]

$$C_c = 0.009(w_l - 10) = 0.009 \times (48 - 10) = 0.34$$

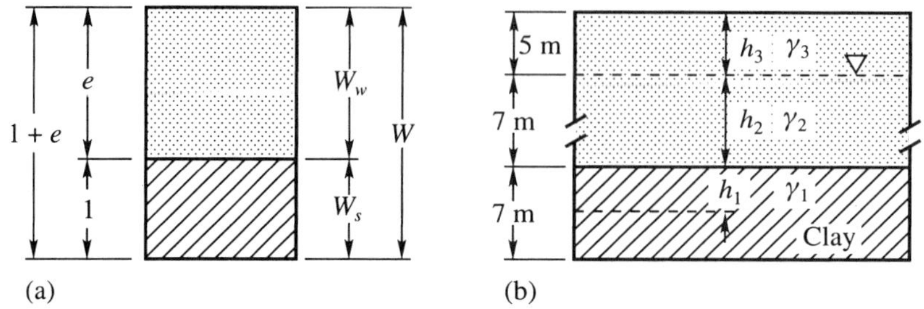

Fig. Ex. 7.3

4. Excess pressure

$$\Delta p = 120 \ \text{kN/m}^2$$

5. Total Settlement

$$S_t = \frac{C_c}{1+e_0} H \log \frac{p_0 + \Delta p}{p_0}$$

$$= \frac{0.34}{2.118} \times 700 \log \frac{195.5+120}{195.5} = 23.3 \ \text{cm}$$

Estimated settlement = 23.3 cm.

Example 7.4

A column of a building carries a load of 1000 kips. The load is transferred to sub soil through a square footing of size 16 × 16 ft founded at a depth of 6.5 ft below ground level. The soil below the footing is fine sand up to a depth of 16.5 ft and below this is a soft compressible clay of thickness 16 ft. The water table is found at a depth of 6.5 ft below the base of the footing. The specific gravities of the solid particles of sand and clay are 2.64 and 2.72 and their natural moisture contents are 25 and 40 percent respectively. The sand above the water table may be assumed to remain saturated. If the plastic limit and the plasticity index of the clay are 30 and 40 percent respectively, estimate the probable settlement of the footing (see Fig. Ex. 7.4)

Solution

1. Required Δp at the middle of the clay layer using the Boussinesq equation

$$\frac{z}{B} = \frac{24.5}{16} = 1.53 < 3.0$$

Divide the footing into 4 equal parts so that $Z/B > 3$

The concentrated load at the center of each part = 250 kips

Radial distance, $r = 5.66$ ft

By the Boussinesq equation the excess pressure Δp at depth 24.5 ft is ($I_B = 0.41$)

$$\Delta p = 4 \times \frac{Q}{z^2} I_B = \frac{4 \times 250}{24.5^2} \times 0.41 = 0.683 \ \text{k/ft}^2$$

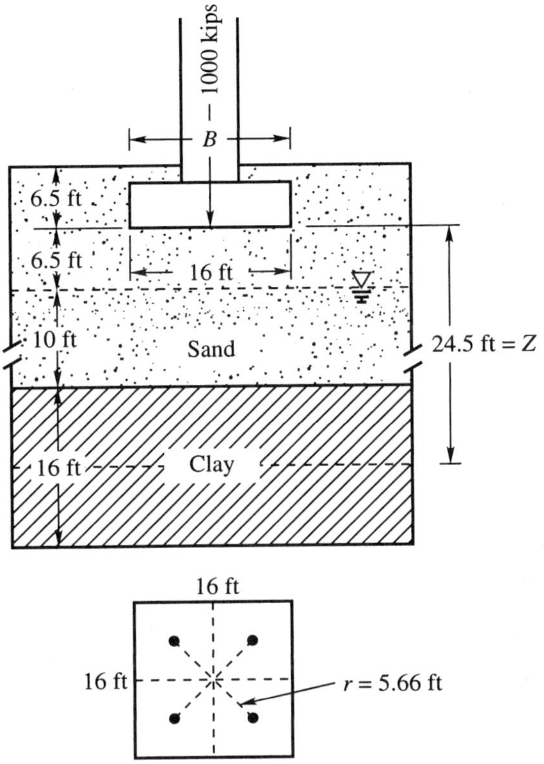

Figure Ex. 7.4

2. Void ratio and unit weights

Per the procedure explained in Ex. 7.3

For sand $\gamma_t = 124$ lb/ft^3 $\gamma_b = 61.6$ lb/ft^3

For clay $\gamma_b = 51.4$ lb/ft^3 $e_0 = 1.09$

3. Overburden pressure p_0

$$p_0 = 8 \times 51.4 + 10 \times 62 + 13 \times 124 = 2639 \text{ lb/ft}^2$$

4. Compression index

$w_l = I_p + w_p = 40 + 30 = 70\%$, $C_c = 0.009\,(70 - 10) = 0.54$

Settlement $$S_t = \frac{0.54}{1+1.09} \times 16 \times \log \frac{2639+683}{2639} = 0.413 \text{ ft} = 4.96 \text{ in.}$$

Example 7.5

Soil investigation at a site gave the following information. Fine sand exists to a depth of 10.6 m and below this lies a soft clay layer 7.60 m thick. The water table is at 4.60 m below the ground surface. The submerged unit weight of sand γ_b is 10.4 kN/m^3, and the wet unit weight above the water table is 17.6 kN/m^3. The water content of the normally consolidated clay $w_n = 40\%$, its liquid limit $w_l = 45\%$, and the specific gravity of the solid particles is 2.78. The proposed construction will transmit a net stress of 120 kN/m^2 at the center of the clay layer. Find the average settlement of the clay layer.

Solution

For calculating settlement [Eq. (7.15a)]

$$S_t = \frac{C_c}{1+e_0} H \log \frac{p_0 + \Delta p}{p_0} \qquad \text{where } \Delta p = 120 \text{ kN/m}^2$$

From Eq. (7.17), $C_c = 0.009\,(w_l - 10) = 0.009(45 - 10) = 0.32$

From Eq. (3.14a), $e_0 = \frac{wG}{S} = wG = 0.40 \times 2.78 = 1.11$ since $S = 1$

γ_b, the submerged unit weight of clay, is found as follows

$$\gamma_{sat} = \frac{\gamma_w (G_s + e_0)}{1+e_0} = \frac{9.81(2.78+1.11)}{1+1.11} = 18.1 \text{ kN/m}^3$$

$$\gamma_b = \gamma_{sat} - \gamma_w = 18.1 - 9.81 = 8.28 \text{ kN/m}^3$$

The effective vertical stress p_0 at the mid height of the clay layer is

$$p_0 = 4.60 \times 17.6 + 6 \times 10.4 + \frac{7.60}{2} \times 8.28 = 174.8 \text{ kN/m}^2$$

Now $$S_t = \frac{0.32 \times 7.60}{1+1.11} \log \frac{174.8+120}{174.8} = 0.26 \text{ m} = 26 \text{ cm}$$

Average settlement = 26 cm.

Example 7.6

A soil sample has a compression index of 0.3. If the void ratio e at a stress of 2940 lb/ft^2 is 0.5, compute (i) the void ratio if the stress is increased to 4200 lb/ft^2, and (ii) the settlement of a soil stratum 13 ft thick.

Solution

Given: $C_c = 0.3$, $e_1 = 0.50$, $p_1 = 2940$ lb/ft^2, $p_2 = 4200$ lb/ft^2.

(i) Now from Eq. (7.4),

$$C_c = \frac{e_1 - e_2}{\log p_2 - \log p_1} = \frac{e_1 - e_2}{\log p_2/p_1}$$

or $e_2 = e_1 - C_c \log p_2/p_1$

substituting the known values, we have,

$$e_2 = 0.5 - 0.3 \log \frac{4200}{2940} = 0.454$$

(ii) The settlement per Eq. (7.10) is

$$S = \frac{C_c}{1+e_1} H \log \frac{p_2}{p_1} = \frac{0.3 \times 13 \times 12}{1.5} \log \frac{4200}{2940} = 4.83 \text{ in.}$$

Example 7.7

Two points on a curve for a normally consolidated clay have the following coordinates.

Point 1: $e_1 = 0.7$, $p_1 = 2089$ lb/ft^2
Point 2: $e_2 = 0.6$, $p_2 = 6266$ lb/ft^2

If the average overburden pressure on a 20 ft thick clay layer is 3133 lb/ft^2, how much settlement will the clay layer experience due to an induced stress of 3340 lb/ft^2 at its middepth.

Solution

From Eq. (7.4) we have

$$C_c = \frac{e_1 - e_2}{\log p_2/p_1} = \frac{0.7-0.6}{\log(6266/2089)} = 0.21$$

We need the initial void ratio e_0 at an overburden pressure of 3133 lb/ft^2.

$$C_c = \frac{e_0 - e_2}{\log p_2/p_0} = 0.21$$

or $(e_0 - 0.6) = 0.21 \log (6266/3133) = 0.063$

or $e_0 = 0.6 + 0.063 = 0.663.$

$$\text{Settlement, } S = \frac{C_c}{1+e_0} H \log \frac{p_0 + \Delta p}{p_0}$$

Substituting the known values, with $\Delta p = 3340$ lb/ft^2

$$S = \frac{0.21 \times 20 \times 12}{1.663} \log \frac{3133+3340}{3133} = 9.55 \text{ in}$$

7.10 RATE OF ONE-DIMENSIONAL CONSOLIDATION THEORY OF TERZAGHI

One dimensional consolidation theory as proposed by Terzaghi is generally applicable in all cases that arise in practice where

1. Secondary compression is not very significant,
2. The clay stratum is drained on one or both the surfaces,
3. The clay stratum is deeply buried, and
4. The clay stratum is thin compared with the size of the loaded areas.

The following assumptions are made in the development of the theory:

1. The voids of the soil are completely filled with water,
2. Both water and solid constituents are incompressible,
3. Darcy's law is strictly valid,
4. The coefficient of permeability is a constant,
5. The time lag of consolidation is due entirely to the low permeability of the soil, and
6. The clay is laterally confined.

Differential Equation for One-Dimensional Flow

Consider a stratum of soil infinite in extent in the horizontal direction (Fig. 7.14) but of such thickness H, that the pressures created by the weight of the soil itself may be neglected in comparison to the applied pressure.

Assume that drainage takes place only at the top and further assume that the stratum has been subjected to a uniform pressure of p_0 for such a long time that it is completely consolidated under that pressure and that there is a hydraulic equilibrium prevailing, i.e., the water level in the piezometric tube at any section XY in the clay stratum stands at the level of the water table (piezometer tube in Fig. 7.14).

Let an increment of pressure Δp be applied. The total pressure to which the stratum is subjected is

$$p_1 = p_0 + \Delta p \tag{7.27}$$

Immediately after the increment of load is applied the water in the pore space throughout the entire height, H, will carry the additional load and there will be set up an excess hydrostatic pressure u_i throughout the pore water equal to Δp as indicated in Fig. 7.14.

After an elapsed time $t = t_1$, some of the pore water will have escaped at the top surface and as a consequence, the excess hydrostatic pressure will have been decreased and a part of the load transferred to the soil structure. The distribution of the pressure between the soil and the pore water, p and u respectively at any time t, may be represented by the curve as shown in the figure. It is evident that

$$p_1 = p + u \tag{7.28}$$

at any elapsed time t and at any depth z, and u is equal to zero at the top. The pore pressure u, at any depth, is therefore a function of z and t and may be written as

$$u = f(z, t) \tag{7.29}$$

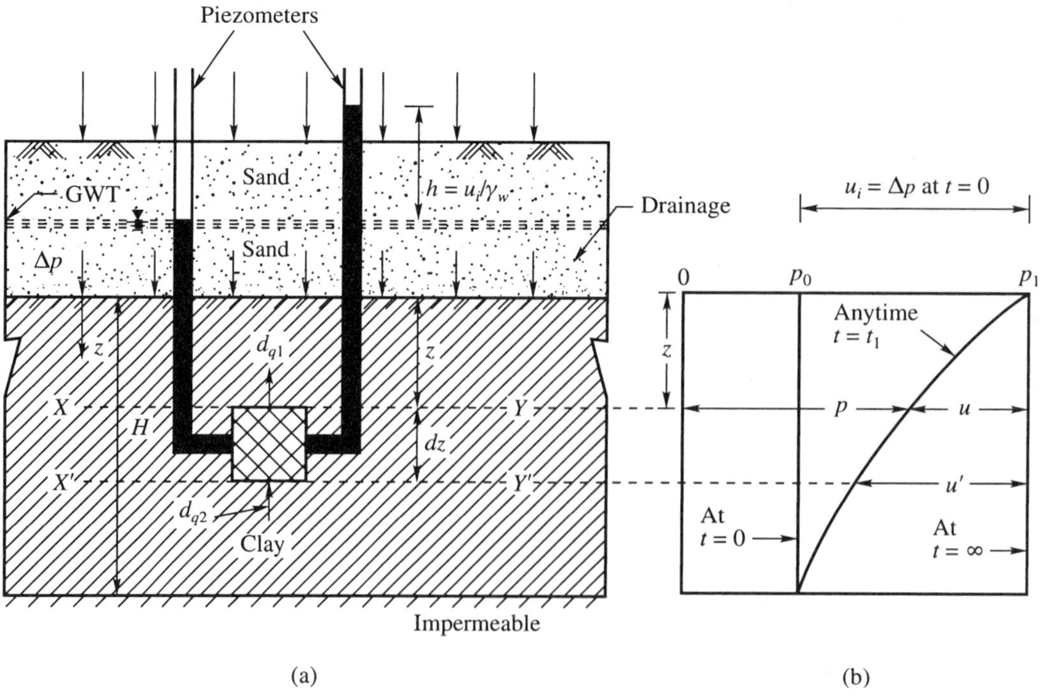

Figure 7.14 One-dimensional consolidation

Consider an element of volume of the stratum at a depth z, and thickness dz (Fig. 7.14). Let the bottom and top surfaces of this element have unit area.

The consolidation phenomenon is essentially a problem of non-steady flow of water through a porous mass. The difference between the quantity of water that enters the lower surface at level $X'Y'$ and the quantity of water which escapes the upper surface at level XY in time element dt must equal the volume change of the material which has taken place in this element of time. The quantity of water is dependent on the hydraulic gradient which is proportional to the slope of the curve t.

The hydraulic gradients at levels XY and $X'Y'$ of the element are

$$i = \frac{1}{\gamma_w}\frac{\partial u}{\partial z}$$

$$i' = \frac{1}{\gamma_w}\frac{\partial}{\partial z}\left(u + \frac{\partial u}{\partial z}dz\right) = \frac{1}{\gamma_w}\frac{\partial u}{\partial z} + \frac{1}{\gamma_w}\frac{\partial^2 u}{\partial z^2}dz \tag{7.30}$$

If k is the hydraulic conductivity the outflow from the element at level XY in time dt is

$$dq_1 = ikdt = \frac{k}{\gamma_w}\frac{\partial u}{\partial z}dt \tag{7.31}$$

The inflow at level $X'Y'$ is

$$dq_2 = ikdt = \frac{k}{\gamma_w}\frac{\partial u}{\partial z}dt + \frac{\partial^2 u}{\partial z^2}dz\,dt \tag{7.32}$$

The difference in flow is therefore

$$dq = dq_1 - dq_2 = -\frac{k}{\gamma_w}\frac{\partial^2 u}{\partial z^2}dz\,dt \tag{7.33}$$

From the consolidation test performed in the laboratory, it is possible to obtain the relationship between the void ratios corresponding to various pressures to which a soil is subjected. This relationship is expressed in the form of a pressure-void ratio curve which gives the relationship as expressed in Eq. (7.12)

$$de = a_v dp \tag{7.34}$$

The change in volume Δdv of the element given in Fig. 7.14 may be written as per Eq. (7.7).

$$\Delta dv = \Delta dz = \frac{de}{1+e}dz \tag{7.35}$$

Substituting for de, we have

$$\Delta dv = \frac{a_v}{1+e}dp\,dz \tag{7.36}$$

Here dp is the change in effective pressure at depth z during the time element dt. The increase in effective pressure dp is equal to the decrease in the pore pressure, du.

Therefore, $$dp = -du = \frac{\partial u}{\partial t}dt \tag{7.37}$$

Hence, $\Delta dv = -\dfrac{a_v}{1+e}\dfrac{\partial u}{\partial t}dt\,dz = -m_v\dfrac{\partial u}{\partial t}dt\,dz$ (7.38)

Since the soil is completely saturated, the volume change Δdv of the element of thickness dz in time dt is equal to the change in volume of water dq in the same element in time dt.

Therefore, $dq = \Delta dv$ (7.39)

or $$-\frac{k}{\gamma_w}\frac{\partial^2 u}{\partial z^2}dz\,dt = -\frac{a_v}{1+e}\frac{\partial u}{\partial t}dz\,dt$$

or $$\frac{k(1+e)}{\gamma_w a_v}\frac{\partial^2 u}{\partial z^2} = \frac{\partial u}{\partial t} = c_v\frac{\partial^2 u}{\partial z^2} \tag{7.40}$$

where $$c_v = \frac{k(1+e)}{\gamma_w a_v} = \frac{k}{\gamma_w m_v} \tag{7.41}$$

is defined as the *coefficient of consolidation.*

Eq. (7.40) is the differential equation for one-dimensional flow. The differential equation for three-dimensional flow may be developed in the same way. The equation may be written as

$$\frac{\partial u}{\partial t} = \frac{1+e}{\gamma_w a_v}\left(k_x\frac{\partial^2 u}{\partial x^2} + k_y\frac{\partial^2 u}{\partial y^2} + k_z\frac{\partial^2 u}{\partial z^2}\right) \tag{7.42}$$

where k_x, k_y and k_z are the coefficients of permeability (hydraulic conductivity) in the coordinate directions of x, y and z respectively.

As consolidation proceeds, the values of k, e and a_v all decrease with time but the ratio expressed by Eq. (7.41) may remain approximately constant.

Mathematical Solution for the One-Dimensional Consolidation Equation

To solve the consolidation Eq. (7.40) it is necessary to set up the proper boundary conditions. For this purpose, consider a layer of soil having a total thickness $2H$ and having drainage facilities at both the top and bottom faces as shown in Fig. 7.15. Under this condition no flow will take place across the center line at depth H. The center line can therefore be considered as an impervious barrier. The boundary conditions for solving Eq. (7.40) may be written as

1. $u = 0$ when $z = 0$
2. $u = 0$ when $z = 2H$
3. $u = \Delta p$ for all depths at time $t = 0$

On the basis of the above conditions, the solution of the differential Eq. (7.40) can be accomplished by means of Fourier Series.

The solution is

$$u = \sum_{N=0}^{N=\infty}\frac{2\Delta p}{m}\sin\left[\frac{mz}{H}\right]\varepsilon^{-m^2T} \tag{7.43}$$

where $m = \dfrac{(2N+1)\pi}{2}$, $T = \dfrac{c_v t}{H^2}$ = a non-dimensional time factor.

Eq. (7.43) can be expressed in a general form as

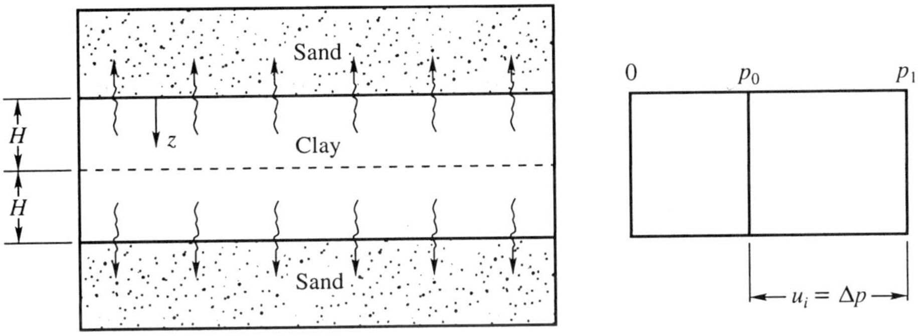

Figure 7.15 Boundary conditions

$$\frac{u}{\Delta p} = f\left(\frac{z}{H}, T\right) \tag{7.44}$$

Equation (7.44) can be solved by assuming T constant for various values of z/H. Curves corresponding to different values of the *time factor* T may be obtained as given in Fig. 7.16. It is of interest to determine how far the consolidation process under the increment of load Δp has progressed at a time t corresponding to the *time factor* T at a given depth z. The term U_z is used to express this relationship. It is defined as the ratio of the amount of consolidation which has already taken place to the total amount which is to take place under the load increment.

The curves in Fig. 7.16 shows the distribution of the pressure Δp between solid and liquid phases at various depths. At a particular depth, say $z/H = 0.5$, the stress in the soil skeleton is represented by AC and the stress in water by CB. AB represents the original excess hydrostatic pressure $u_i = \Delta p$. The degree of consolidation U_z percent at this particular depth is then

$$U_z\% = 100 \times \frac{AC}{AB} = \frac{\Delta p - u}{\Delta p} = 100\left(1 - \frac{u}{\Delta p}\right) \tag{7.45}$$

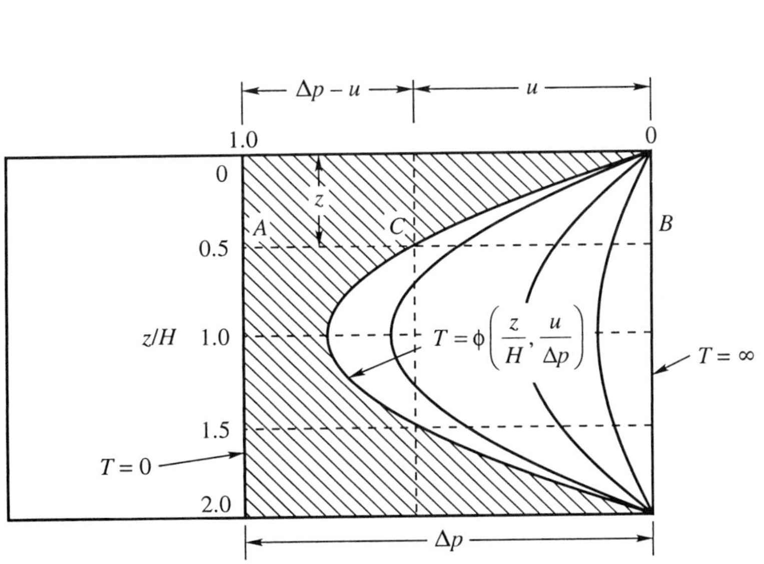

Figure 7.16 Consolidation of clay layer as a function T

Following a similar reasoning, the average degree of consolidation $U\%$ for the entire layer at a time factor T is equal to the ratio of the shaded portion (Fig. 7.16) of the diagram to the entire area which is equal to $2H\,\Delta p$.

Therefore

$$U\% = \frac{\int_0^{2H} (\Delta p - u)dz}{2H\Delta p} \times 100$$

or
$$U\% = \frac{100}{2H}\left[2H - \frac{1}{\Delta p}\int_0^{2H} u dz\right] \tag{7.46}$$

Hence, Eq. (7.46) after integration reduces to

$$U\% = 100\left[1 - \sum_{N=0}^{N=\infty} \frac{2}{m^2}\varepsilon^{-m^2 T}\right] \tag{7.47}$$

It can be seen from Eq. (7.47) that the degree of consolidation is a function of the time factor T only which is a dimensionless ratio. The relationship between T and $U\%$ may therefore be established once and for all by solving Eq. (7.47) for various values of T. Values thus obtained are given in Table 7.3 and also plotted on a semilog plot as shown in Fig. 7.17.

For values of $U\%$ between 0 and 60%, the curve in Fig. 7.17 can be represented almost exactly by the equation

$$T = \frac{\pi}{4}\left(\frac{U\%}{100}\right)^2 \tag{7.48}$$

which is the equation of a parabola. Substituting for T, Eq. (7.48) may be written as

$$\frac{U\%}{100} = \sqrt{\frac{4c_v}{\pi H^2}}\sqrt{t} \tag{7.49}$$

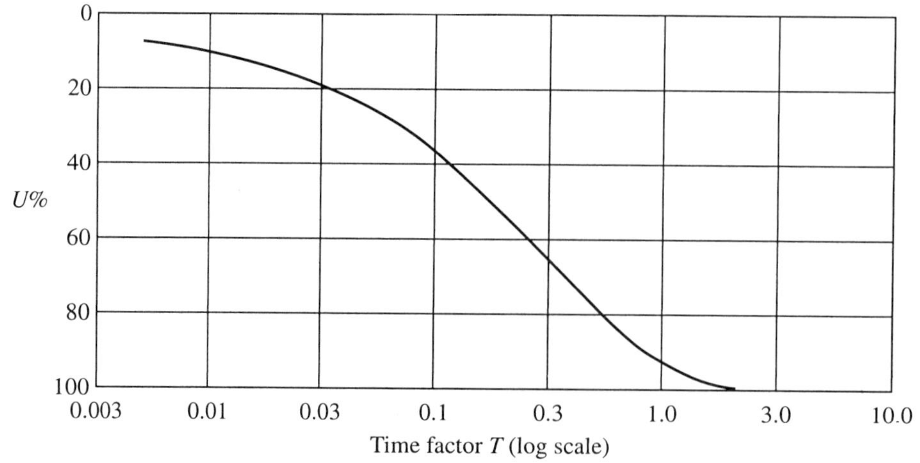

Figure 7.17 *U* versus *T*

Table 7.3 Relationship between U and T

$U\%$	T	$U\%$	T	$U\%$	T
0	0	40	0.126	75	0.477
10	0.008	45	0.159	80	0.565
15	0.018	50	0.197	85	0.684
20	0.031	55	0.238	90	0.848
25	0.049	60	0.287	95	1.127
30	0.071	65	0.342	100	∞
35	0.096	70	0.405		

In Eq. (7.49), the values of c_v and H are constants. One can determine the time required to attain a given degree of consolidation by using this equation. It should be noted that H represents half the thickness of the clay stratum when the layer is drained on both sides, and it is the full thickness when drained on one side only.

TABLE 7.4 Relation between U% and T (Special Cases)

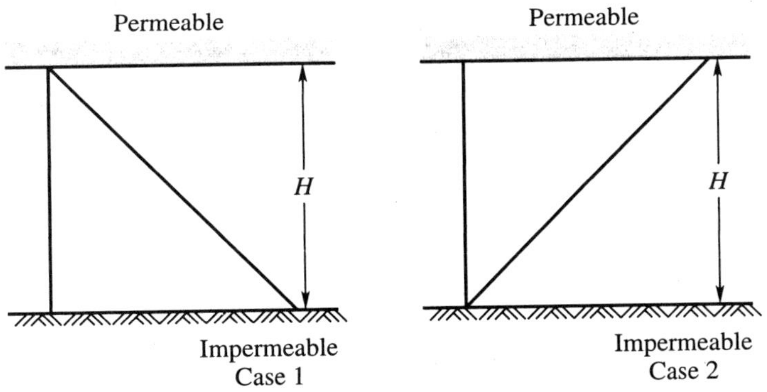

Time Factors, T

$U\%$	Consolidation pressure increase with depth	Consolidation pressure decreases with depth
00	0	0
10	0.047	0.003
20	0.100	0.009
30	0.158	0.024
40	0.221	0.048
50	0.294	0.092
60	0.383	0.160
70	0.500	0.271
80	0.665	0.44
90	0.94	0.72
95	1	0.8
100	∞	∞

For values of $U\%$ greater than 60%, the curve in Fig. 7.17 may be represented by the equation

$$T = 1.781 - 0.933 \log (100 - U\%) \tag{7.50}$$

Effect of Boundary Conditions on Consolidation

A layer of clay which permits drainage through both surfaces is called an *open layer*. The thickness of such a layer is always represented by the symbol $2H$, in contrast to the symbol H used for the thickness of half-closed layers which can discharge their excess water only through one surface.

The relationship expressed between T and U given in Table 7.3 applies to the following cases:

1. Where the clay stratum is drained on both sides and the initial consolidation pressure distribution is uniform or linearly increasing or decreasing with depth.
2. Where the clay stratum is drained on one side but the consolidation pressure is uniform with depth.

Separate relationships between T and U are required for half closed layers with thickness H where the consolidation pressures increase or decrease with depth. Such cases are exceptional and as such not dealt with in detail here. However, the relations between $U\%$ and T for these two cases are given in Table 7.4.

7.11 DETERMINATION OF THE COEFFICIENT OF CONSOLIDATION

The coefficient of consolidation c_v can be evaluated by means of laboratory tests by fitting the experimental curve with the theoretical.

There are two laboratory methods that are in common use for the determination of c_v. They are

1. Casagrande Logarithm of Time Fitting Method.
2. Taylor Square Root of Time Fitting Method.

Logarithm of Time Fitting Method

This method was proposed by Casagrande and Fadum (1940).

Figure 7.18 is a plot showing the relationship between compression dial reading and the logarithm of time of a consolidation test. The theoretical consolidation curve using the log scale for the time factor is also shown. There is a similarity of shape between the two curves. On the laboratory curve, the intersection formed by the final straight line produced backward and the tangent to the curve at the point of inflection is accepted as the 100 per cent primary consolidation point and the dial reading is designated as R_{100}. The time-compression relationship in the early stages is also parabolic just as the theoretical curve. The dial reading at zero primary consolidation R_0 can be obtained by selecting any two points on the parabolic portion of the curve where times are in the ratio of 1 : 4. The difference in dial readings between these two points is then equal to the difference between the first point and the dial reading corresponding to zero primary consolidation. For example, two points A and B whose times 10 and 2.5 minutes respectively, are marked on the curve. Let z_1 be the ordinate difference between the two points. A point C is marked vertically over B such that $BC = z_1$. Then the point C corresponds to zero primary consolidation. The procedure is repeated with several points. An average horizontal line is drawn through these points to represent the theoretical zero percent consolidation line.

The interval between 0 and 100% consolidation is divided into equal intervals of percent consolidation. Since it has been found that the laboratory and the theoretical curves have better

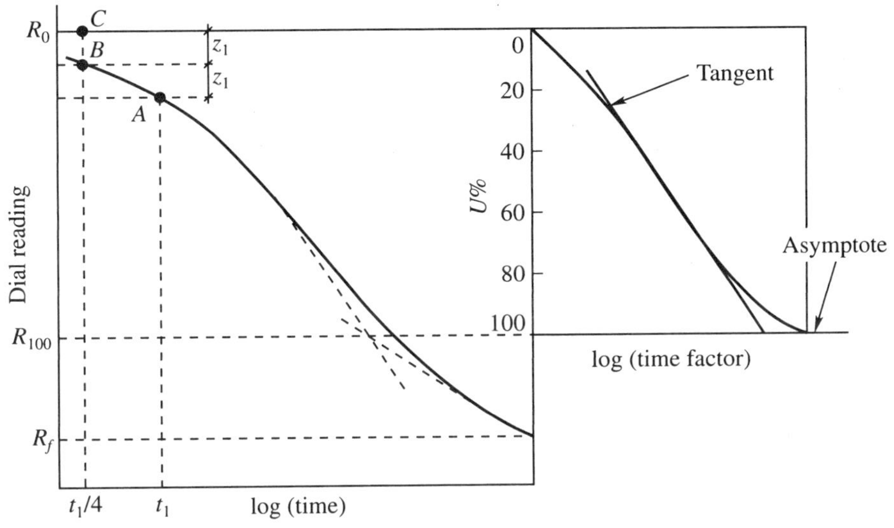

(a) Experimental curve (b) Theoretical curve

Figure 7.18 Log of time fitting method

correspondence at the central portion, the value of c_v is computed by taking the time t and time factor T at 50 percent consolidation. The equation to be used is

$$T_{50} = \frac{c_v t_{50}}{H_{dr}^2} \quad \text{or} \quad c_v = \frac{T_{50}}{t_{50}} H_{dr}^2 \tag{7.51}$$

where H_{dr} = drainage path

From Table 7.3, we have at $U = 50\%$, $T = 0.197$. From the initial height H_i of specimen and compression dial reading at 50% consolidation, H_{dr} for double drainage is

$$H_{dr} = \frac{H_i - \Delta H}{2} \tag{7.52}$$

where ΔH= Compression of sample up to 50% consolidation.

Now the equation for c_v may be written as

$$c_v = 0.197 \frac{H_{dr}^2}{t_{50}} \tag{7.53}$$

Square Root of Time Fitting Method

This method was devised by Taylor (1948). In this method, the dial readings are plotted against the square root of time as given in Fig. 7.19(a). The theoretical curve U versus $\sqrt{T}$ is also plotted and shown in Fig. 7.19(b). On the theoretical curve a straight line exists up to 60 percent consolidation while at 90 percent consolidation the abscissa of the curve is 1.15 times the abscissa of the straight line produced.

The fitting method consists of first drawing the straight line which best fits the early portion of the laboratory curve. Next a straight line is drawn which at all points has abscissa 1.15 times as great as those of the first line. The intersection of this line and the laboratory curve is taken as the 90 percent (R_{90}) consolidation point. Its value may be read and is designated as t_{90}.

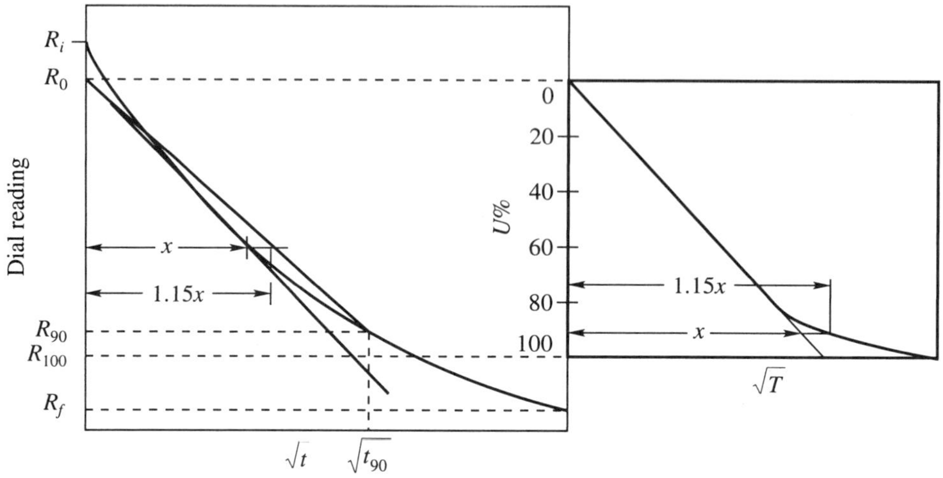

(a) Experimental curve (b) Theoretical curve

Figure 7.19 Square root of time fitting method

Usually the straight line through the early portion of the laboratory curve intersects the zero time line at a point (R_o) differing somewhat from the initial point (R_i). This intersection point is called the *corrected zero point*. If one-ninth of the vertical distance between the corrected zero point and the 90 per cent point is set off below the 90 percent point, the point obtained is called the "*100 percent primary compression point*" (R_{100}). The compression between zero and 100 per cent point is called "*primary compression*".

At the point of 90 percent consolidation, the value of $T = 0.848$. The equation of c_v may now be written as

$$c_v = 0.848 \frac{H_{dr}^2}{t_{90}} \tag{7.54}$$

where H_{dr} = drainage path (average)

7.12 RATE OF SETTLEMENT DUE TO CONSOLIDATION

It has been explained that the ultimate settlement S_t of a clay layer due to consolidation may be computed by using either Eq. (7.10) or Eq. (7.13). If S is the settlement at any time t after the imposition of load on the clay layer, the degree of consolidation of the layer in time t may be expressed as

$$U\% = \frac{S}{S_t} \times 100 \text{ percent} \tag{7.55}$$

Since U is a function of the time factor T, we may write

$$U\% = 100 f(T) = \frac{S}{S_t} \times 100 \tag{7.56}$$

The rate of settlement curve of a structure built on a clay layer may be obtained by the following procedure:

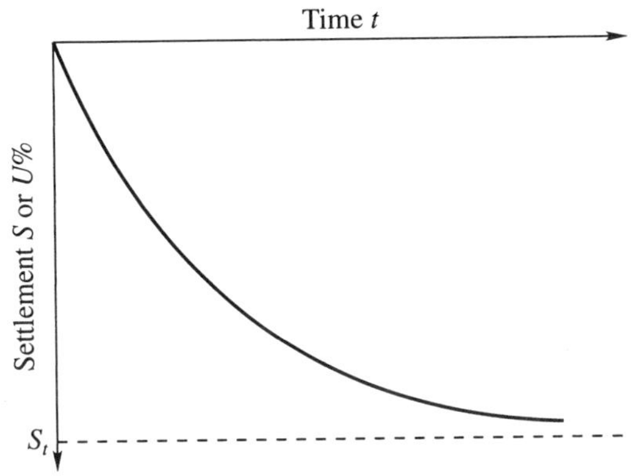

Figure 7.20 Time-settlement curve

1. From consolidation test data, compute m_v and c_v.
2. Compute the total settlement S_t that the clay stratum would experience with the increment of load Δp.
3. From the theoretical curve giving the relation between U and T, find T for different degrees of consolidation, say 5, 10, 20, 30 percent etc.
4. Compute from equation $t = \frac{TH_{dr}^2}{c_v}$ the values of t for different values of T. It may be noted here that for drainage on both sides H_{dr} is equal to half the thickness of the clay layer.
5. Now a curve can be plotted giving the relation between t and $U\%$ or t and S as shown in Fig. 7.20.

7.13 TWO- AND THREE-DIMENSIONAL CONSOLIDATION PROBLEMS

When the thickness of a clay stratum is great compared with the width of the loaded area, the consolidation of the stratum is three-dimensional. In a three-dimensional process of consolidation the flow occurs either in radial planes or else the water particles travel along flow lines which do not lie in planes. The problem of this type is complicated though a general theory of three-dimensional consolidation exists (Biot, et al., 1941). A simple example of three-dimensional consolidation is the consolidation of a stratum of soft clay or silt by providing sand drains and surcharge for accelerating consolidation.

The most important example of two dimensional consolidation in engineering practice is the consolidation of the case of a hydraulic fill dam. In two-dimensional flow, the excess water drains out of the clay in parallel planes. Gilboy (1934) has analyzed the two dimensional consolidation of a hydraulic fill dam.

Example 7.8

A 2.5 cm thick sample of clay was taken from the field for predicting the time of settlement for a proposed building which exerts a uniform pressure of 100 kN/m^2 over the clay stratum. The sample was loaded to 100 kN/m^2 and proper drainage was allowed from top and bottom. It was seen that 50 percent of the total settlement occurred in 3 minutes. Find the time required for 50 percent of the total settlement of the building, if it is to be constructed on a 6 m thick layer of clay which extends from the ground surface and is underlain by sand.

Solution

T for 50% consolidation = 0.197.

The lab sample is drained on both sides. The coefficient of consolidation c_v is found from

$$c_v = \frac{TH_{dr}^2}{t} = 0.197 \times \frac{(2.5)^2}{4} \times \frac{1}{3} = 10.25 \times 10^{-2} \text{ cm}^2 / \text{min.}$$

The time t for 50% consolidation in the field will be found as follows.

$$t = \frac{0.197 \times 300 \times 300 \times 100}{10.25 \times 60 \times 24} = 120 \text{ days}.$$

Example 7.9

The void ratio of a clay sample A decreased from 0.572 to 0.505 under a change in pressure from 122 to 180 kN/m^2. The void ratio of another sample *B* decreased from 0.61 to 0.557 under the same increment of pressure. The thickness of sample *A* was 1.5 times that of *B*. Nevertheless the time taken for 50% consolidation was 3 times larger for sample *B* than for *A*. What is the ratio of coefficient of permeability of sample *A* to that of *B*?

Solution

Let H_a = thickness of sample *A*, H_b = thickness of sample *B*, m_{va} = coefficient of volume compressibility of sample *A*, m_{vb} = coefficient of volume compressibility of sample *B*, c_{va} = coefficient of consolidation for sample *A*, c_{vb} = coefficient of consolidation for sample *B*, Δp_a = increment of load for sample *A*, Δp_b = increment of load for sample *B*, k_a = coefficient of permeability for sample *A*, and k_b = coefficient of permeability of sample *B*.

We may write the following relationship

$$m_{va} = \frac{\Delta e_a}{1+e_a} \frac{1}{\Delta p_a}, \quad m_{vb} = \frac{\Delta e_b}{1+e_b} \frac{1}{\Delta p_b}$$

where e_a is the void ratio of sample *A* at the commencement of the test and Δe_a is the change in void ratio. Similarly e_b and Δe_b apply to sample *B*.

$$\frac{m_{va}}{m_{vb}} = \frac{\Delta p_b}{\Delta p_a} \frac{\Delta e_a}{\Delta e_b} \frac{1+e_b}{1+e_a}, \quad \text{and } T_a = \frac{c_{va} t_a}{H_a^2}, \quad T_b = \frac{c_{vb} t_b}{H_b^2}$$

wherein T_a, t_a, T_b and t_b correspond to samples *A* and *B* respectively. We may write

$$\frac{c_{va}}{c_{vb}} = \frac{T_a}{T_b} \frac{H_a^2}{H_b^2} \frac{t_b}{t_a}, \qquad k_a = c_{va} m_{va} \gamma_w, \qquad k_b = c_{vb} m_{vb} \gamma_w$$

Therefore, $\dfrac{k_a}{k_b} = \dfrac{c_{va}}{c_{vb}} \dfrac{m_{va}}{m_{vb}}$

Given $e_a = 0.572$, and $e_b = 0.61$

$$\Delta e_a = 0.572 - 0.505 = 0.067, \ \Delta e_b = 0.610 - 0.557 = 0.053$$

$$\Delta p_a = \Delta p_b = 180 - 122 = 58 \text{ kN/m}^2, \qquad H_a = 1.5\, H_b$$

But $t_b = 3t_a$

We have,
$$\frac{m_{va}}{m_{vb}} = \frac{0.067}{0.053} \times \frac{1+0.61}{1+0.572} = 1.29$$

$$\frac{c_{va}}{c_{vb}} = 1.5^2 \times 3 = 6.75$$

Therefore,
$$\frac{k_a}{k_b} = 6.75 \times 1.29 = 8.7$$

The ratio is 8.7 : 1.

Example 7.10

A strata of normally consolidated clay of thickness 10 ft is drained on one side only. It has a hydraulic conductivity of $k = 1.863 \times 10^{-8}$ in/sec and a coefficient of volume compressibility $m_v = 8.6 \times 10^{-4}$ in²/lb. Determine the ultimate value of the compression of the stratum by assuming a uniformly distributed load of 5250 lb/ft² and also determine the time required for 20 percent and 80 percent consolidation.

Solution

Total compression,

$$S_t = m_v H \Delta p = 8.6 \times 10^{-4} \times 10 \times 12 \times 5250 \times \frac{1}{144} = 3.763 \text{ in}.$$

For determining the relationship between $U\%$ and T for 20% consolidation use the equation

$$T = \frac{\pi}{4}\left(\frac{U\%}{100}\right)^2 \text{ or } T = \frac{3.14}{4} \times \left(\frac{20}{100}\right)^2 = 0.0314$$

For 80% consolidation use the equation

$$T = 1.781 - 0.933 \log (100 - U\%)$$

Therefore $T = 1.781 - 0.933 \log_{10} (100 - 80) = 0.567.$

The coefficient of consolidation is

$$c_v = \frac{k}{\gamma_w m_v} = \frac{1.863 \times 10^{-8}}{3.61 \times 10^{-2} \times 8.6 \times 10^{-4}} = 6 \times 10^{-4} \text{ in}^2/\text{sec}$$

The times required for 20% and 80% consolidation are

$$t_{20} = \frac{H_{dr}^2 T}{c_v} = \frac{(10 \times 12)^2 \times 0.0314}{6 \times 10^{-4} \times 60 \times 60 \times 24} = 8.72 \text{ days}$$

$$t_{80} = \frac{H_{dr}^2 T}{c_v} = \frac{(10 \times 12)^2 \times 0.567}{6 \times 10^{-4} \times 60 \times 60 \times 24} = 157.5 \text{ days}$$

Example 7.11

The loading period for a new building extended from May 1995 to May 1997. In May 2000, the average measured settlement was found to be 11.43 cm. It is known that the ultimate settlement will be about 35.56 cm. Estimate the settlement in May 2005. Assume double drainage to occur.

Solution

For the majority of practical cases in which loading is applied over a period, acceptable accuracy is obtained when calculating time-settlement relationships by assuming the time datum to be midway through the loading or construction period.

S_t = 11.43 cm when t = 4 years and S = 35.56 cm.

The settlement is required for t = 9 years, that is, up to May2005. Assuming as a starting point that at t = 9 years, the degree of consolidation will be = 0.60. Under these conditions per Eq. (7.48), $U = 1.13\sqrt{T}$.

If S_{t_1} = settlement at time t_1, S_{t_2} = settlement at time t_2

$$\frac{S_{t_1}}{S_{t_2}} = \frac{U_1}{U_2} = \sqrt{\frac{T_1}{T_2}} = \sqrt{\frac{t_1}{t_2}} \quad \text{since} \quad T = \frac{c_v t}{H_{dr}^2}$$

where $\frac{c_v}{H_{dr}^2}$ is a constant. Therefore $\frac{11.43}{S_{t_2}} = \sqrt{\frac{4}{9}}$ or $S_{t_2} = 17.15\,\text{cm}$

Therefore at t = 9 years, $U = \frac{17.5}{35.56} = 0.48$

Since the value of U is less than 0.60 the assumption is valid. Therefore the estimated settlement is 17.15 cm. In the event of the degree of consolidation exceeding 0.60, equation (7.50) has to be used to obtain the relationship between T and U.

Example 7.12

An oedometer test is performed on a 2 cm thick clay sample. After 5 minutes, 50% consolidation is reached. After how long a time would the same degree of consolidation be achieved in the field where the clay layer is 3.70 m thick? Assume the sample and the clay layer have the same drainage boundary conditions (double drainage).

Solution

The time factor T is defined as $T = \frac{c_v t}{H_{dr}^2}$

where H_{dr} = half the thickness of the clay for double drainage.

Here, the time factor T and coefficient of consolidation are the same for both the sample and the field clay layer. The parameter that changes is the time t. Let t_1 and t_2 be the times required to reach 50% consolidation both in the oedometer and field respectively. t_1 = 5 min

Therefore $\frac{c_v t_1}{H_{dr(1)}^2} = \frac{c_v t_2}{H_{dr(2)}^2}$

Now $t_2 = \left(\frac{H_{dr(2)}}{H_{dr(1)}}\right)^2 t_1 = \left(\frac{370}{2}\right)^2 \times 5 \times \frac{1}{60} \times \frac{1}{24}\,\text{days} \approx 119\,\text{days}$

Example 7.13

A laboratory sample of clay 2 cm thick took 15 min to attain 60 percent consolidation under a double drainage condition. What time will be required to attain the same degree of consolidation for a clay layer 3 m thick under the foundation of a building for a similar loading and drainage condition?

Solution

Use Eq. (7.50) for $U > 60\%$ for determining T

$$T = 1.781 - 0.933 \log (100 - U\%)$$

$$= 1.781 - 0.933 \log (100 - 60) = 0.286.$$

From Eq. (7.51) the coefficient of consolidation, c_v is

$$c_v = \frac{TH_{dr}^2}{t} = \frac{0.286 \times (1)^2}{15} = 1.91 \times 10^{-2} \text{ cm}^2/\text{min}.$$

The value of c_v remains constant for both the laboratory and field conditions. As such, we may write,

$$\left(\frac{TH_{dr}^2}{t}\right)_{lab} = \left(\frac{TH_{dr}^2}{t}\right)_{field}$$

where H_{dr} = half the thickness = 1 cm for the lab sample and 150 cm for field stratum, and t_{lab} = 15 min.

Therefore,

$$\left(\frac{T(1)^2}{0.25}\right)_{lab} = \left(\frac{T(150)^2}{t}\right)_{field}$$

or $t_f = (150)^2 \times 0.25 = 5625$ hr or 234 days (approx).
for the field stratum to attain the same degree of consolidation.

7.14 PROBLEMS

7.1 A bed of sand 10 m thick is underlain by a compressible of clay 3 m thick under which lies sand. The water table is at a depth of 4 m below the ground surface. The total unit weights of sand below and above the water table are 20.5 and 17.7 kN/m^3 respectively. The clay has a natural water content of 42%, liquid limit 46% and specific gravity 2.76. Assuming the clay to be normally consolidated, estimate the probable final settlement under an average excess pressure of 100 kN/m^2.

7.2 The effective overburden pressure at the middle of a saturated clay layer 12 ft thick is 2100 lb/ft^2 and is drained on both sides. The overburden pressure at the middle of the clay stratum is expected to be increased by 3150 lb/ft^2 due to the load from a structure at the ground surface. An undisturbed sample of clay 20 mm thick is tested in a consolidometer. The total change in thickness of the specimen is 0.80 mm when the applied pressure is 2100 lb/ft^2. The final water content of the sample is 24 percent and the specific gravity of the solids is 2.72. Estimate the probable final settlement of the proposed structure.

7.3 The following observations refer to a standard laboratory consolidation test on an undisturbed sample of clay.

Pressure kN/m^2	Final Dial Gauge Reading $\times 10^{-2}$ mm	Pressure kN/m^2	Final Dial Gauge Reading $\times 10^{-2}$ mm
0	0	400	520
50	180	100	470
100	250	0	355
200	360		

The sample was 75 mm in diameter and had an initial thickness of 18 mm. The moisture content at the end of the test was 45.5%; the specific gravity of solids was 2.53.

Compute the void ratio at the end of each loading increment and also determine whether the soil was overconsolidated or not. If it was overconsolidated, what was the overconsolidation ratio if the effective overburden pressure at the time of sampling was 60 kN/m^2?

7.4 The following points are coordinates on a pressure-void ratio curve for an undisturbed clay.

p	0.5	1	2	4	8	16	$kips/ft^2$
e	1.202	1.16	1.06	0.94	0.78	0.58	

Determine (i) C_c, and (ii) the magnitude of compression in a 10 ft thick layer of this clay for a load increment of 4 $kips/ft^2$. Assume $e_0 = 1.320$, and $p_0 = 1.5$ $kips/ft^2$

7.5 The thickness of a compressible layer, prior to placing of a fill covering a large area, is 30 ft. Its original void ratio was 1.0. Sometime after the fill was constructed measurements indicated that the average void ratio was 0.8. Determine the compression of the soil layer.

7.6 The water content of a soft clay is 54.2% and the liquid limit is 57.3%. Estimate the compression index, by equations (7.17) and (7.18). Given $e_0 = 0.85$

7.7 A layer of normally consolidated clay is 20 ft thick and lies under a recently constructed building. The pressure of sand overlying the clay layer is 6300 lb/ft^2, and the new construction increases the overburden pressure at the middle of the clay layer by 2100 lb/ft^2. If the compression index is 0.5, compute the final settlement assuming $w_n = 45\%$, $G_s = 2.70$, and the clay is submerged with the water table at the top of the clay stratum.

7.8 A consolidation test was made on a sample of saturated marine clay. The diameter and thickness of the sample were 5.5 cm and 3.75 cm respectively. The sample weighed 650 g at the start of the test and 480 g in the dry state after the test. The specific gravity of solids was 2.72. The dial readings corresponding to the final equilibrium condition under each load are given below.

Pressure, kN/m^2	*DR* cm $\times 10^{-4}$	Pressure, kN/m^2	*DR* cm $\times 10^{-4}$
0	0	106	1880
6.7	175	213	3340
11.3	275	426	5000
26.6	540	852	6600
53.3	965		

(a) Compute the void ratios and plot the e-log p curve.
(b) Estimate the maximum preconsolidation pressure by the Casagrande method.
(c) Draw the field curve and determine the compression index.

7.9 The results of a consolidation test on a soil sample for a load increased from 200 to 400 kN/m^2 are given below:

Time in Min.	Dial reading division	Time in Min.	Dial reading division
0	1255	16	1603
0.10	1337	25	1632
0.25	1345	36	1651
0.50	1355	49	1661
1.00	1384	64	1670
2.25	1423	81	1677
4.00	1480	100	1682
9.00	1557	121	1687

The thickness of the sample corresponding to the dial reading 1255 is 1.561 cm. Determine the value of the coefficient of consolidation using the square root of time fitting method in cm^2/min. One division of dial gauge corresponds to 2.5×10^{-4} cm. The sample is drained on both faces.

7.10 A 2.5 cm thick sample was tested in a consolidometer under saturated conditions with drainage on both sides. 30 percent consolidation was reached under a load in 15 minutes. For the same conditions of stress but with only one way drainage, estimate the time in days it would take for a 2 m thick layer of the same soil to consolidate in the field to attain the same degree of consolidation.

7.11 The dial readings recorded during a consolidation test at a certain load increment are given below.

Time min	Dial Reading cm $\times 10^{-4}$	Time min	Dial Reading cm $\times 10^{-4}$
0	240	15	622
0.10	318	30	738
0.25	340	60	842
0.50	360	120	930
1.00	385	240	975
2.00	415	1200	1070
4.00	464		
8.00	530	-	-

Determine c_v by both the square root of time and log of time fitting methods. The thickness of the sample at DR 240 = 2 cm and the sample is drained both sides.

7.12 In a laboratory consolidation test a sample of clay with a thickness of 1 in. reached 50% consolidation in 8 minutes. The sample was drained top and bottom. The clay layer from which the sample was taken is 25 ft thick. It is covered by a layer of sand through which water can escape and is underlain by a practically impervious bed of intact shale. How long will the clay layer require to reach 50 per cent consolidation?

7.13 The following data were obtained from a consolidation test performed on an undisturbed clay sample 3 cm in thickness:

(i) $p_1 = 3.5$ kips/ft^2, $e_1 = 0.895$

(ii) $p_2 = 6.5$ kips/ft^2, $e_2 = 0.782$

By utilizing the known theoretical relationship between percent consolidation and time factor, compute and plot the decrease in thickness with time for a 10 ft thick layer of this clay, which is drained on the upper surface only. Given : $e_0 = 0.92$ $p_0 = 4.5$ kips/ft^2, $\Delta p = 1.5$ kips/ft^2, $c_v = 4.2 \times 10^{-5}$ ft^2/min.

7.14 A structure built on a layer of clay settled 5 cm in 60 days after it was built. If this settlement corresponds to 20 percent average consolidation of the clay layer, plot the time settlement curve of the structure for a period of 3 years from the time it was built. Given : Thickness of clay layer = 3m and drained on one side

7.15 A 30 ft thick clay layer with single drainage settles 3.5 in. in 3.5 yr. The coefficient consolidation for this clay was found to be 8.43×10^{-4} in.^{2}/sec. Compute the ultimate consolidation settlement and determine how long it will take to settle to 90% of this amount.

7.16 The time factor T for a clay layer undergoing consolidation is 0.2. What is the average degree of consolidation (consolidation ratio) for the layer?

7.17 If the final consolidation settlement for the clay layer in Prob. 7.16 is expected to be 1.0 m, how much settlement has occurred when the time factor is (a) 0.2 and (b) 0.7?

7.18 A certain compressible layer has a thickness of 12 ft. After 1 yr when the clay is 50% consolidated, 3 in. of settlement has occurred. For similar clay and loading conditions, how much settlement would occur at the end of 1 yr and 4 yr, if the thickness of this new layer were 20 ft?

7.19 A layer of normally consolidated clay 14 ft thick has an average void ratio of 1.3. Its compression index is 0.6. When the induced vertical pressure on the clay layer is doubled, what change in thickness of the clay layer will result? Assume: $p_0 = 1200$ lb/ft^2 and $\Delta p = 600$ lb/ft^2.

7.20 Settlement analysis for a proposed structure indicates that 2.4 in. of settlement will occur in 4 yr and that the ultimate total settlement will be 9.8 in. The analysis is based on the assumption that the compressible clay layer is drained on both sides. However, it is suspected that there may not be drainage at the bottom surface. For the case of single drainage, estimate the time required for 2.4 in. of settlement.

7.21 The time to reach 60% consolidation is 32.5 sec for a sample 1.27 cm thick tested in a laboratory under conditions of double drainage. How long will the corresponding layer in nature require to reach the same degree of consolidation if it is 4.57 m thick and drained on one side only?

7.22 A certain clay layer 30 ft thick is expected to have an ultimate settlement of 16 in. If the settlement was 4 in. after four years, how much longer will it take to obtain a settlement of 6 in?

7.23 If the coefficient of consolidation of a 3 m thick layer of clay is 0.0003 cm^2/sec, what is the average consolidation of that layer of clay (a) in one year with two-way drainage, and (b) the same as above for one-way drainage.

7.24 The average natural moisture content of a deposit is 40%; the specific gravity of the solid matter is 2.8, and the compression index C_c is 0.36. If the clay deposit is 6.1 m thick drained on both sides, calculate the final consolidation settlement S_t. Given: $p_0 = 60$ kN/m^2 and $\Delta p = 30$ kN/m^2

7.25 A rigid foundation block, circular in plan and 6 m in diameter rests on a bed of compact sand 6 m deep. Below the sand is a 1.6 m thick layer of clay overlying on impervious bed rock. Ground water level is 1.5 m below the surface of the sand. The unit weight of sand above water table is 19.2 kN/m^3, the saturated unit weight of sand is 20.80 kN/m^3, and the saturated unit weight of the clay is 19.90 kN/m^3.

A laboratory consolidation test on an undisturbed sample of the clay, 20 mm thick and drained top and bottom, gave the following results:

Pressure (kN/m^2)	50	100	200	400	800
Void ratio	0.73	0.68	0.625	0.54	0.41

If the contact pressure at the base of the foundation is 200 kN/m^2, and e_0 = 0.80, calculate the final average settlement of the foundation assuming 2:1 method for the spread of the load.

7.26 A stratum of clay is 2 m thick and has an initial overburden pressure of 50 kN/m^2 at the middle of the clay layer. The clay is overconsolidated with a preconsolidation pressure of 75 kN/m^2. The values of the coefficients of recompression and compression indices are 0.05 and 0.25 respectively. Assume the initial void ratio e_0 = 1.40. Determine the final settlement due to an increase of pressure of 40 kN/m^2 at the middle of the clay layer.

7.27 A clay stratum 5 m thick has the initial void ration of 1.50 and an effective overburden pressure of 120 kN/m^2. When the sample is subjected to an increase of pressure of 120 kN/m^2, the void ratio reduces to 1.44. Determine the coefficient of volume compressibility and the final settlement of the stratum.

7.28 A 3 m thick clay layer beneath a building is overlain by a permeable stratum and is underlain by an impervious rock. The coefficient of consolidation of the clay was found to be 0.025 cm^2/min. The final expected settlement for the layer is 8 cm. Determine (a) how much time will it take for 80 percent of the total settlement, (b) the required time for a settlement of 2.5 cm to occur, and (c) the settlement that would occur in one year.

7.29 An area is underlain by a stratum of clay layer 6 m thick. The layer is doubly drained and has a coefficient of consolidation of 0.3 m^2/month. Determine the time required for a surcharge load to cause a settlement of 40 cm if the same load cause a final settlement of 60 cm.

7.30 In an oedometer test, a clay specimen initially 25 mm thick attains 90% consolidation in 10 minutes. In the field, the clay stratum from which the specimen was obtained has a thickness of 6 m and is sandwiched between two sand layers. A structure constructed on this clay experienced an ultimate settlement of 200 mm. Estimate the settlement at the end of 100 days after construction.

CHAPTER 8

SHEAR STRENGTH OF SOIL

8.1 INTRODUCTION

One of the most important and the most controversial engineering properties of soil is its shear strength or ability to resist sliding along internal surfaces within a mass. The stability of a cut, the slope of an earth dam, the foundations of structures, the natural slopes of hillsides and other structures built on soil depend upon the shearing resistance offered by the soil along the probable surfaces of slippage. There is hardly a problem in the field of engineering which does not involve the shear properties of the soil in some manner or the other.

8.2 BASIC CONCEPT OF SHEARING RESISTANCE AND SHEARING STRENGTH

The basic concept of shearing resistance and shearing strength can be made clear by studying first the basic principles of friction between solid bodies. Consider a prismatic block *B* resting on a plane surface *MN* as shown in Fig. 8.1. Block *B* is subjected to the force P_n which acts at right angles to the surface *MN*, and the force F_a that acts tangentially to the plane. The normal force P_n remains constant whereas F_a gradually increases from zero to a value which will produce sliding. If the tangential force F_a is relatively small, block *B* will remain at rest, and the applied horizontal force will be balanced by an equal and opposite force F_r on the plane of contact. This resisting force is developed as a result of roughness characteristics of the bottom of block *B* and plane surface *MN*. The angle δ formed by the resultant *R* of the two forces F_r and P_n with the normal to the plane *MN* is known as the *angle of obliquity*.

If the applied horizontal force F_a is gradually increased, the resisting force F_r will likewise increase, always being equal in magnitude and opposite in direction to the applied force. Block *B* will start sliding along the plane when the force F_a reaches a value which will increase the angle of obliquity to a certain maximum value δ_m. If block *B* and plane surface *MN* are made of the same

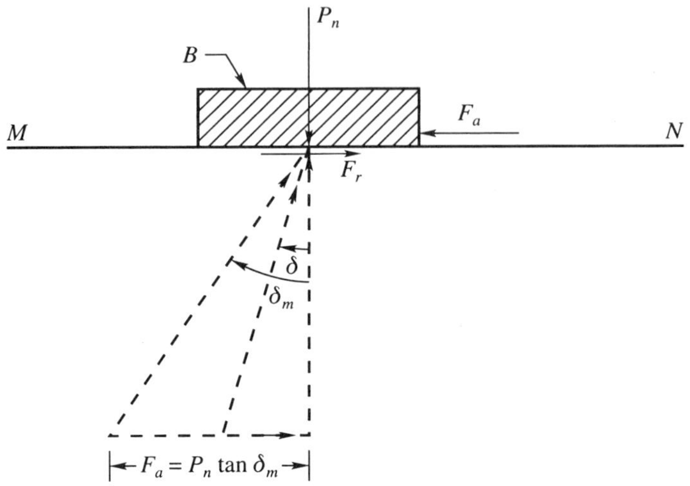

Figure 8.1 Basic concept of shearing resistance and strength.

material, the angle δ_m is equal to ϕ which is termed the *angle of friction*, and the value $\tan \phi$ is termed the *coefficient of friction*. If block B and plane surface MN are made of dissimilar materials, the angle δ is termed the *angle of wall friction*. The applied horizontal force F_a on block B is a shearing force and the developed force is friction or *shearing resistance*. The maximum shearing resistance which the materials are capable of developing is called the *shearing strength*.

If another experiment is conducted on the same block with a higher normal load P_n the shearing force F_a will correspondingly be greater. A series of such experiments would show that the shearing force F_a is proportional to the normal load P_n, that is

$$F_a = P_n \tan \phi \tag{8.1}$$

If A is the overall contact area of block B on plane surface MN, the relationship may be written as

$$\text{shear strength, } s = \frac{F_a}{A} = \frac{P_n}{A} \tan \phi$$

$$\text{or} \qquad s = \sigma \tan \phi \tag{8.2}$$

8.3 THE COULOMB EQUATION

The basic concept of friction as explained in Sect. 8.2 applies to soils which are purely granular in character. Soils which are not purely granular exhibit an additional strength which is due to the cohesion between the particles. It is, therefore, still customary to separate the shearing strength s of such soils into two components, one due to the cohesion between the soil particles and the other due to the friction between them. The fundamental shear strength equation proposed by the French engineer Coulomb (1776) is

$$s = c + \sigma \tan \phi \tag{8.3}$$

This equation expresses the assumption that the cohesion c is independent of the normal pressure σ acting on the plane of failure. At zero normal pressure, the shear strength of the soil is expressed as

$$s = c \tag{8.4}$$

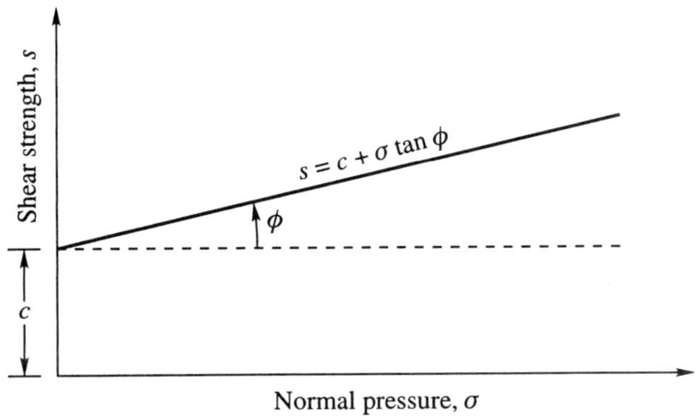

Figure 8.2 Coulomb's law

According to Eq. (8.4), the cohesion of a soil is defined as the shearing strength at zero normal pressure on the plane of rupture.

In Coulomb's equation c and ϕ are empirical parameters, the values of which for any soil depend upon several factors; the most important of these are :

1. The past history of the soil.
2. The initial state of the soil, i.e., whether it is saturated or unsaturated.
3. The permeability characteristics of the soil.
4. The conditions of drainage allowed to take place during the test.

Since c and ϕ in Coulomb's Eq. (8.3) depend upon many factors, c is termed as *apparent cohesion* and ϕ the angle of shearing resistance. For cohesionless soil $c = 0$, then Coulomb's equation becomes

$$s = \sigma \tan \phi \tag{8.5}$$

The relationship between the various parameters of Coulomb's equation is shown diagrammatically in Fig. 8.2.

8.4 METHODS OF DETERMINING SHEAR STRENGTH PARAMETERS

Methods

The shear strength parameters c and ϕ of soils either in the undisturbed or remolded states may be determined by any of the following methods:

1. *Laboratory methods*
   (a) Direct or box shear test
   (b) Triaxial compression test
2. *Field method:* Vane shear test or by any other indirect methods

Shear Parameters of Soils in-situ

The laboratory or the field method that has to be chosen in a particular case depends upon the type of soil and the accuracy required. Wherever the strength characteristics of the soil in-situ are required, laboratory tests may be used provided undisturbed samples can be extracted from the stratum. However, soils are subject to disturbance either during sampling or extraction from the sampling tubes in the laboratory even though soil particles possess cohesion. It is practically impossible to obtain undisturbed samples of cohesionless soils and highly pre-consolidated clay soils. Soft sensitive clays are nearly always remolded during sampling. Laboratory methods may, therefore, be used only in such cases where fairly good undisturbed samples can be obtained. Where it is not possible to extract undisturbed samples from the natural soil stratum, any one of the following methods may have to be used according to convenience and judgment :

1. Laboratory tests on remolded samples which could at best simulate field conditions of the soil.
2. Any suitable field test.

The present trend is to rely more on field tests as these tests have been found to be more reliable than even the more sophisticated laboratory methods.

Shear Strength Parameters of Compacted Fills

The strength characteristics of fills which are to be constructed, such as earth embankments, are generally found in a laboratory. Remolded samples simulating the proposed density and water content of the fill materials are made in the laboratory and tested. However, the strength characteristics of existing fills may have to be determined either by laboratory or field methods keeping in view the limitations of each method.

8.5 SHEAR TEST APPARATUS

Direct Shear Test

The original form of apparatus for the direct application of shear force is the shear box. The box shear test, though simple in principle, has certain shortcomings which will be discussed later on. The apparatus consists of a square brass box split horizontally at the level of the center of the soil sample, which is held between metal grilles and porous stones as shown in Fig. 8.3(a). Vertical load is applied to the sample as shown in the figure and is held constant during a test. A gradually increasing horizontal load is applied to the lower part of the box until the sample fails in shear. The shear load at failure is divided by the cross-sectional area of the sample to give the ultimate shearing strength. The vertical load divided by the area of the sample gives the applied vertical stress σ. The test may be repeated with a few more samples having the same initial conditions as the first sample. Each sample is tested with a different vertical load.

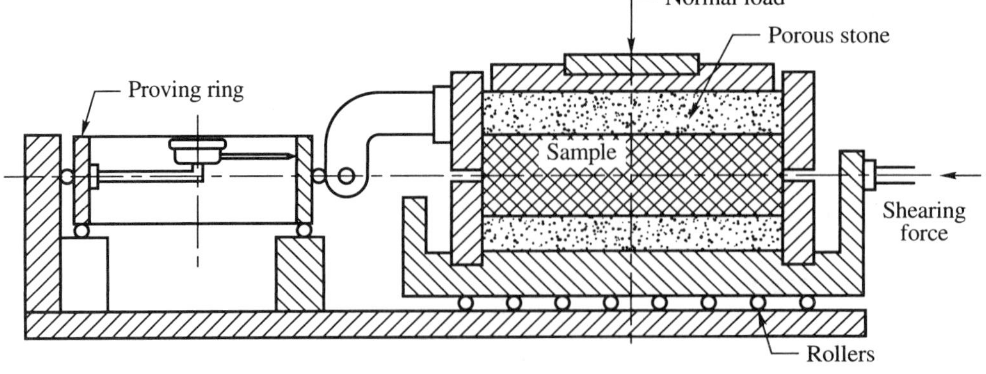

Figure 8.3(a) Constant rate of strain shear box

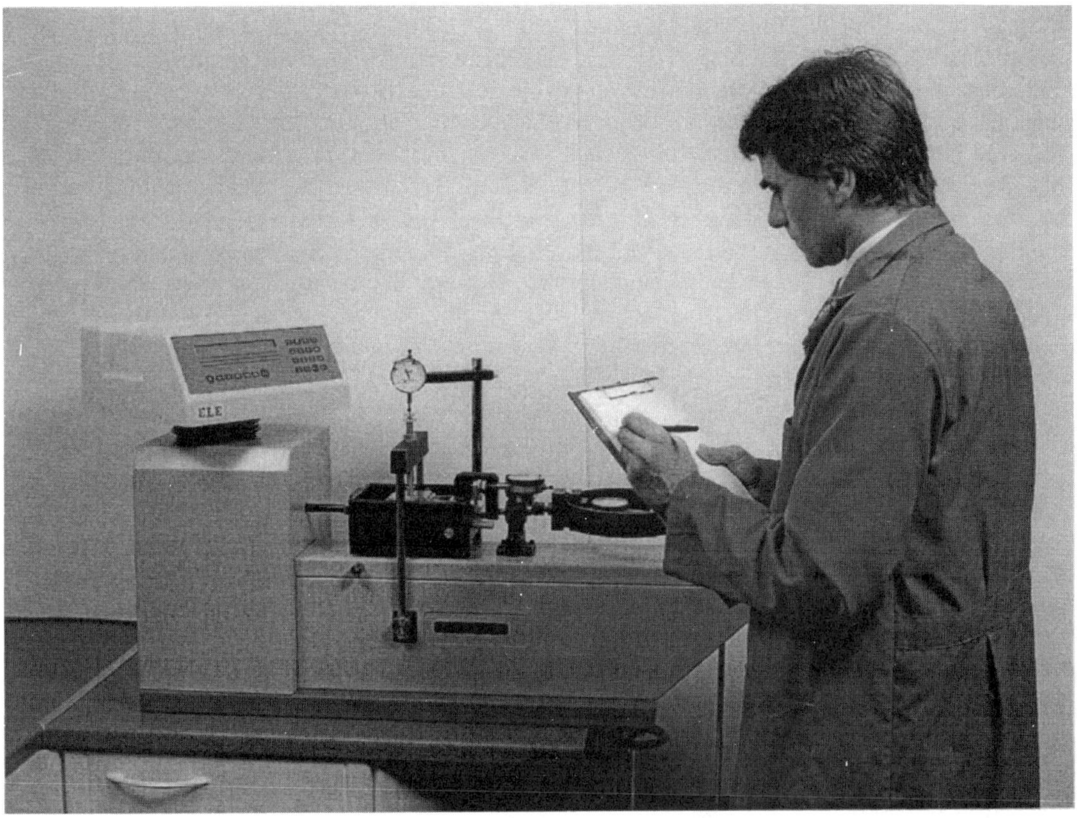

Figure 8.3(b) Strain controlled direct shear apparatus (Courtesy: Soiltest)

The horizontal load is applied at a constant rate of strain. The lower half of the box is mounted on rollers and is pushed forward at a uniform rate by a motorized gearing arrangement. The upper half of the box bears against a steel proving ring, the deformation of which is shown on the dial gauge indicating the shearing force. To measure the volume change during consolidation and during the shearing process another dial gauge is mounted to show the vertical movement of the top platen. The horizontal displacement of the bottom of the box may also be measured by another dial gauge which is not shown in the figure. Figure 8.3(b) shows a photograph of strain controlled direct shear test apparatus.

Procedure for Determining Shearing Strength of Soil

In the direct shear test, a sample of soil is placed into the shear box. The size of the box normally used for clays and sands is 6 × 6 cm and the sample is 2 cm thick. A large box of size 30 × 30 cm with sample thickness of 15 cm is sometimes used for gravelly soils.

The soils used for the test are either undisturbed samples or remolded. If undisturbed, the specimen has to be carefully trimmed and fitted into the box. If remolded samples are required, the soil is placed into the box in layers at the required initial water content and tamped to the required dry density.

After the specimen is placed in the box, and all the other necessary adjustments are made, a known normal load is applied. Then a shearing force is applied. The normal load is held constant throughout the test but the shearing force is applied at a constant rate of strain (which will be explained later on). The shearing displacement is recorded by a dial gauge.

Dividing the normal load and the maximum applied shearing force by the cross-sectional area of the specimen at the shear plane gives respectively the unit normal pressure σ and the shearing strength s at failure of the sample. These results may be plotted on a shearing diagram where σ is the abscissa and s the ordinate. The result of a single test establishes one point on the graph representing the Coulomb formula for shearing strength. In order to obtain sufficient points to draw the Coulomb graph, additional tests must be performed on other specimens which are exact duplicates of the first. The procedure in these additional tests is the same as in the first, except that a different normal stress is applied each time. Normally, the plotted points of normal and shearing stresses at failure of the various specimens will approximate a straight line. But in the case of saturated, highly cohesive clay soils in the undrained test, the graph of the relationship between the normal stress and shearing strength is usually a curved line, especially at low values of normal stress. However, it is the usual practice to draw the best straight line through the test points to establish the Coulomb Law. The slope of the line gives the angle of shearing resistance and the intercept on the ordinate gives the apparent cohesion (See. Fig. 8.2).

Triaxial Compression Test

A diagrammatic layout of a triaxial test apparatus is shown in Fig. 8.4(a). In the triaxial compression test, three or more identical samples of soil are subjected to uniformly distributed fluid pressure around the cylindrical surface. The sample is sealed in a watertight rubber membrane. Then axial load is applied to the soil sample until it fails. Although only compressive load is applied to the soil sample, it fails by shear on internal faces. It is possible to determine the shear strength of the soil from the applied loads at failure. Figure 8.4(b) gives a photograph of a triaxial test apparatus.

Advantages and Disadvantages of Direct and Triaxial Shear Tests

Direct shear tests are generally suitable for cohesionless soils except fine sand and silt whereas the triaxial test is suitable for all types of soils and tests. Undrained and consolidated undrained tests on clay samples can be made with the box-shear apparatus. The advantages of the triaxial over the direct shear test are:

1. The stress distribution across the soil sample is more uniform in a triaxial test as compared to a direct shear test.
2. The measurement of volume changes is more accurate in the triaxial test.
3. The complete state of stress is known at all stages during the triaxial test, whereas only the stresses at failure are known in the direct shear test.
4. In the case of triaxial shear, the sample fails along a plane on which the combination of normal stress and the shear stress gives the maximum angle of obliquity of the resultant with the normal, whereas in the case of direct shear, the sample is sheared only on one plane which is the horizontal plane which need not be the plane of actual failure.
5. Pore water pressures can be measured in the case of triaxial shear tests whereas it is not possible in direct shear tests.
6. The triaxial machine is more adaptable.

Advantages of Direct Shear Tests

1. The direct shear machine is simple and fast to operate.
2. A thinner soil sample is used in the direct shear test thus facilitating drainage of the pore water quickly from a saturated specimen.
3. Direct shear requirement is much less expensive as compared to triaxial equipment.

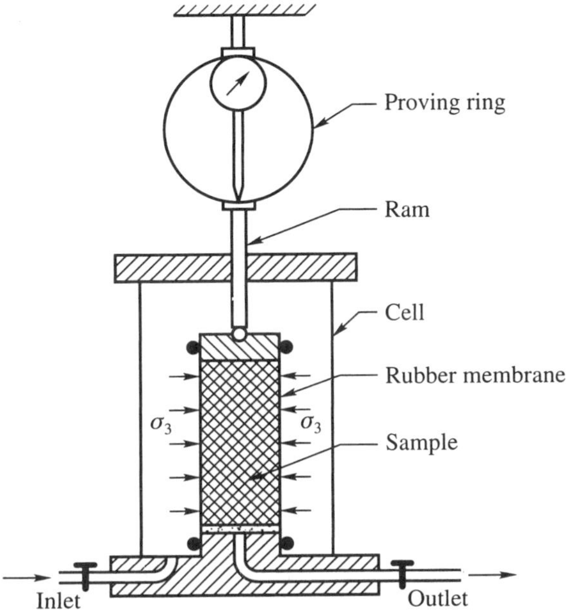

(a) Diagrammatic layout

(b) Multiplex 50-E load frame triaxial test apparatus (Courtesy: Soiltest USA)

Figure 8.4 Triaxial test apparatus

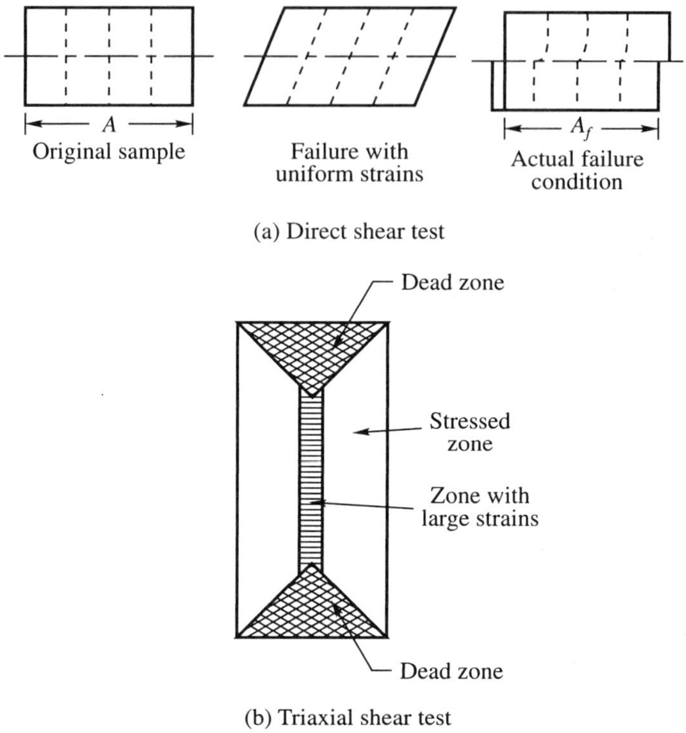

Figure 8.5 Condition of sample during shearing in direct and triaxial shear tests

The stress conditions across the soil sample in the direct shear test are very complex because of the change in the shear area with the increase in shear displacement as the test progresses, causing unequal distribution of shear stresses and normal stresses over the potential surface of sliding. Fig. 8.5(a) shows the sample condition before and after shearing in a direct shear box. The final sheared area A_f is less than the original area A.

Fig. 8.5(b) shows the stressed condition in a triaxial specimen. Because of the end restraints, dead zones (non-stressed zones) triangular in section are formed at the ends whereas the stress distribution across the sample midway between the dead zones may be taken as approximately uniform.

8.6 STRESS CONDITION AT A POINT IN A SOIL MASS

Through every point in a stressed body there are three planes at right angles to each other which are unique as compared to all the other planes passing through the point, because they are subjected only to normal stresses with no accompanying shearing stresses acting on the planes. These three planes are called *principal planes*, and the normal stresses acting on these planes are *principal stresses*. Ordinarily the three principal stresses at a point differ in magnitude. They may be designated as the major principal stress σ_1, the intermediate principal stress σ_2, and the minor principal stress σ_3. Principal stresses at a point in a stressed body are important because, once they are evaluated, the stresses on any other plane through the point can be determined. Many problems in foundation engineering can be approximated by considering only two-dimensional stress conditions. The influence of the intermediate principal stress σ_2 on failure may be considered as not very significant.

A Two-Dimensional Demonstration of the Existence of Principal Planes

Consider the body (Fig. 8.6(a)) is subjected to a system of forces such as F_1, F_2, F_3 and F_4, whose magnitudes and lines of action are known.

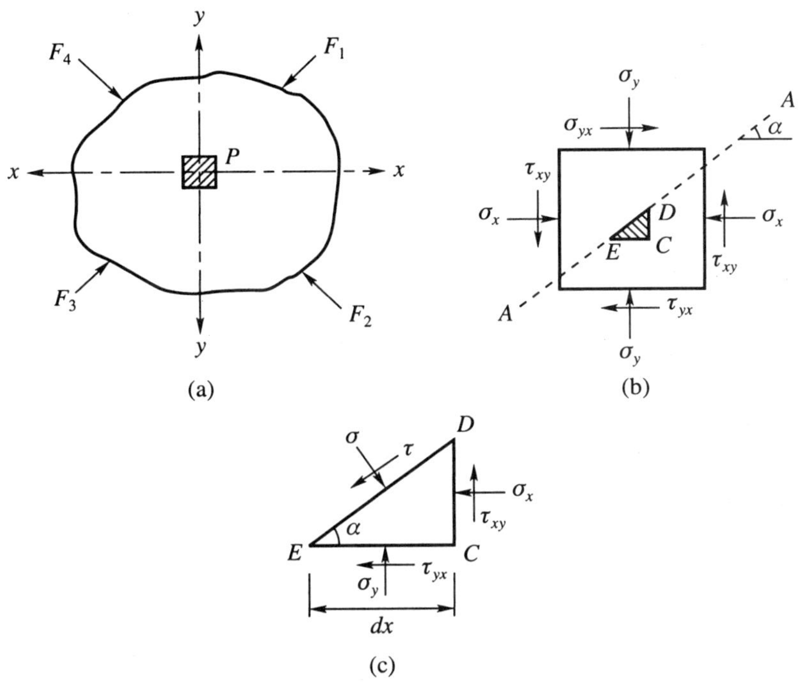

Figure 8.6 Stress at a point in a body in two dimensional space

Consider a small prismatic element *P*. The stresses acting on this element in the directions parallel to the arbitrarily chosen axes *x* and *y* are shown in Fig. 8.6(b).

Consider a plane *AA* through the element, making an angle α with the *x*-axis. The equilibrium condition of the element may be analyzed by considering the stresses acting on the faces of the triangle *ECD* (shaded) which is shown to an enlarged scale in Fig. 8.6(c). The normal and shearing stresses on the faces of the triangle are also shown.

The unit stress in compression and in shear on the face *ED* are designated as σ and τ respectively.

Expressions for σ and τ may be obtained by applying the principles of statics for the equilibrium condition of the body. The sum of all the forces in the *x*-direction is

$$\sigma_x dx \tan\alpha + \tau_{xy} dx + \tau\, dx \sec\alpha \cos\alpha - \sigma dx \sec\alpha \sin\alpha = 0 \tag{8.6}$$

The sum of all the forces in the *y*-direction is

$$\sigma_y dx + \tau_{xy} dx \tan\alpha - \tau\, dx \sec\alpha \sin\alpha - \sigma dx \sec\alpha \cos\alpha = 0 \tag{8.7}$$

Solving Eqs. (8.6) and (8.7) for σ and τ, we have

$$\sigma = \frac{\sigma_y + \sigma_x}{2} + \frac{\sigma_y - \sigma_x}{2} \cos 2\alpha + \tau_{xy} \sin 2\alpha \tag{8.8}$$

$$\tau = \frac{1}{2}\left(\sigma_y - \sigma_x\right) \sin 2\alpha - \tau_{xy} \cos 2\alpha \tag{8.9}$$

By definition, a principal plane is one on which the shearing stress is equal to zero. Therefore, when τ is made equal to zero in Eq. (8.9), the orientation of the principal planes is defined by the relationship

$$\tan 2\alpha = \frac{2\tau_{xy}}{\sigma_y - \sigma_x} \tag{8.10}$$

Equation (8.10) indicates that there are two principal planes through the point *P* in Fig. 8.6(a) and that they are at right angles to each other. By differentiating Eq. (8.8) with respect to α, and equating to zero, we have

$$\frac{d\sigma}{d\alpha} = -\sigma_y \sin 2\alpha + \sigma_x \sin 2\alpha + 2\tau_{xy} \cos 2\alpha = 0$$

or $$\tan 2\alpha = \frac{2\tau_{xy}}{\sigma_y - \sigma_x} \tag{8.11}$$

Equation (8.11) indicates the orientation of the planes on which the normal stresses σ are maximum and minimum. This orientation coincides with Eq. (8.10). Therefore, it follows that the principal planes are also planes on which the normal stresses are maximum and minimum.

8.7 STRESS CONDITIONS IN SOIL DURING TRIAXIAL COMPRESSION TEST

In triaxial compression test a cylindrical specimen is subjected to a constant all-round fluid pressure which is the minor principal stress σ_3 since the shear stress on the surface is zero. The two ends are subjected to axial stress which is the major principal stress σ_1. The stress condition in the specimen goes on changing with the increase of the major principal stress σ_1. It is of interest to analyze the state of stress along inclined sections passing through the sample at any stress level σ_1 since failure occurs along inclined surfaces.

Consider the cylindrical specimen of soil in Fig. 8.7(a) which is subjected to principal stresses σ_1 and σ_3 ($\sigma_2 = \sigma_3$).

Now *CD*, a horizontal plane, is called a principal plane since it is normal to the principal stress σ_1 and the shear stress is zero on this plane. *EF* is the other principal plane on which the principal stress σ_3 acts. *AA* is the inclined section on which the state of stress is required to be analyzed.

Consider as before a small prism of soil shown shaded in Fig. 8.7(a) and the same to an enlarged scale in Fig. 8.7(b). All the stresses acting on the prism are shown. The equilibrium of the prism requires

$$\Sigma \text{ Horizontal forces} = \sigma_3 \sin \alpha\, dl - \sigma \sin \alpha\, dl + \tau \cos \alpha\, dl = 0 \tag{8.12}$$

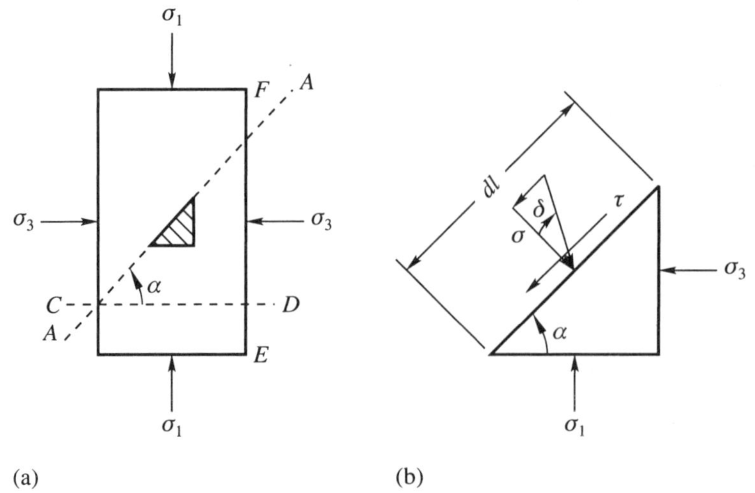

Figure 8.7 Stress condition in a triaxial compression test specimen

$$\Sigma \text{ Vertical forces} = \sigma_1 \cos \alpha \, dl - \sigma \cos \alpha \, dl - \tau \sin \alpha \, dl = 0 \tag{8.13}$$

Solving Eqs. (8.12) and (8.13) we have

$$\sigma = \frac{\sigma_1 + \sigma_3}{2} + \frac{\sigma_1 - \sigma_3}{2} \cos 2\alpha \tag{8.14}$$

$$\tau = \frac{1}{2}(\sigma_1 - \sigma_3) \sin 2\alpha \tag{8.15}$$

Let the resultant of σ and τ make an angle δ with the normal to the inclined plane. One should remember that when α is less than 90°, the shear stress τ is positive, and the angle δ is also positive.

Eqs. (8.14) and (8.15) may be obtained directly from the general Eqs. (8.8) and (8.9) respectively by substituting the following:

$$\sigma_y = \sigma_1, \sigma_x = \sigma_3 \text{ and } \tau_{xy} = 0$$

8.8 RELATIONSHIP BETWEEN THE PRINCIPAL STRESSES AND COHESION *c*

If the shearing resistance *s* of a soil depends on both friction and cohesion, sliding failure occurs in accordance with the Coulomb Eq. (8.3), that is, when

$$\tau = s = c + \sigma \tan \phi \tag{8.16}$$

Substituting for the values of σ and τ from Eqs. (8.14) and (8.15) into Eqs. (8.16) and solving for σ_1 we obtain

$$\sigma_1 = \sigma_3 + \frac{c + \sigma_3 \tan \phi}{\sin \alpha \cos \alpha - \cos^2 \alpha \tan \phi} \tag{8.17}$$

The plane with the least resistance to shearing along it will correspond to the minimum value of σ_1 which can produce failure in accordance with Eq. (8.17). σ_1 will be at a minimum when the denominator in the second member of the equation is at a maximum, that is, when

$$\frac{d}{d\alpha}(\sin \alpha \cos \alpha - \cos^2 \alpha \tan \phi) = 0$$

Differentiating, and simplifying, we obtain (writing $\alpha = \alpha_c$)

$$\alpha_c = 45° + \phi/2 \tag{8.18}$$

Substituting for α in Eq. (8.17) and simplifying, we have

$$\sigma_1 = \sigma_3 \tan^2 (45° + \phi/2) + 2c \tan (45° + \phi/2) \tag{8.19}$$

or

$$\sigma_1 = \sigma_3 N_\phi + 2c\sqrt{N_\phi} \tag{8.20}$$

where $N_\phi = \tan^2 (45° + \phi/2)$ is called the *flow value*.

If the cohesion $c = 0$, we have

$$\sigma_1 = \sigma_3 N_\phi \tag{8.21}$$

If $\phi = 0$, we have

$$\sigma_1 = \sigma_3 + 2c \tag{8.22}$$

If the sides of the cylindrical specimen are not acted on by the horizontal pressure σ_3, the load required to cause failure is called the unconfined compressive strength q_u. It is obvious that an unconfined compression test can be performed only on a cohesive soil. According to Eq. (8.20), the unconfined compressive strength q_u is equal to

$$\sigma_1 = q_u = 2c\sqrt{N_\phi} \tag{8.23}$$

If $\phi = 0$, then $q_u = 2c$ (8.24a)

or the shear strength

$$s = c = \frac{q_u}{2} \tag{8.24b}$$

Eq. (8.24b) shows one of the simplest ways of determining the shear strength of cohesive soils.

8.9 MOHR CIRCLE OF STRESS

Squaring Eqs. (8.8) and (8.9) and adding, we have

$$\left[\sigma - \frac{\sigma_y + \sigma_x}{2}\right]^2 + \tau^2 = \left(\frac{\sigma_y - \sigma_x}{2}\right)^2 + \tau_{xy}^2 \tag{8.25}$$

Now, Eq. (8.25) is the equation of a circle whose center has coordinates

$$\left[\frac{1}{2}(\sigma_y + \sigma_x), 0\right] \text{ and whose radius is } \frac{1}{2}\sqrt{(\sigma_y - \sigma_x)^2 - 4\tau_{xy}^2}$$

The coordinates of points on the circle represent the normal and shearing stresses on inclined planes at a given point. The circle is called the *Mohr circle of stress*, after Mohr (1900), who first recognized this useful relationship. Mohr's method provides a convenient graphical method for determining

1. The normal and shearing stress on any plane through a point in a stressed body.
2. The orientation of the principal planes if the normal and shear stresses on the surface of the prismatic element (Fig. 8.6) are known. The relationships are valid regardless of the mechanical properties of the materials since only the considerations of equilibrium are involved.

If the surfaces of the element are themselves principal planes, the equation for the Mohr circle of stress may be written as

$$\tau^2 + \left[\sigma_y - \frac{\sigma_1 + \sigma_3}{2}\right]^2 = \left(\frac{\sigma_1 - \sigma_3}{2}\right)^2 \tag{8.26}$$

The center of the circle has coordinates $\tau = 0$, and $\sigma = (\sigma_1 + \sigma_3)/2$, and its radius is $(\sigma_1 - \sigma_3)/2$. Again from Mohr's diagram, the normal and shearing stresses on any plane passing through a point in a stressed body (Fig. 8.7) may be determined if the principal stresses σ_1 and σ_3 are known. Since σ_1 and σ_3 are always known in a cylindrical compression test, Mohr's diagram is a very useful tool to analyze stresses on failure planes.

8.10 MOHR CIRCLE OF STRESS WHEN A PRISMATIC ELEMENT IS SUBJECTED TO NORMAL AND SHEAR STRESSES

Consider first the case of a prismatic element subjected to normal and shear stresses as in Fig. 8.8(a).

Sign Convention

1. Compressive stresses are positive and tensile stresses are negative.
2. Shear stresses are considered as positive if they give a clockwise moment about a point above the stressed plane as shown in Fig. 8.8(b), otherwise negative.

The normal stresses are taken as abscissa and the shear stresses as ordinates. It is assumed the normal stresses σ_x , σ_y and the shear stress τ_{xy} ($\tau_{xy} = \tau_{yx}$) acting on the surface of the element are known. Two points P_1 and P_2 may now be plotted in Fig. 8.8(b), whose coordinates are

$$P_1 = (\sigma_x , \tau_{xy})$$
$$P_2 = (\sigma_y , - \tau_{xy})$$

If the points P_1 and P_2 are joined, the line intersects the abscissa at point C whose coordinates are $[(\sigma_x + \sigma_y)/2, 0]$.

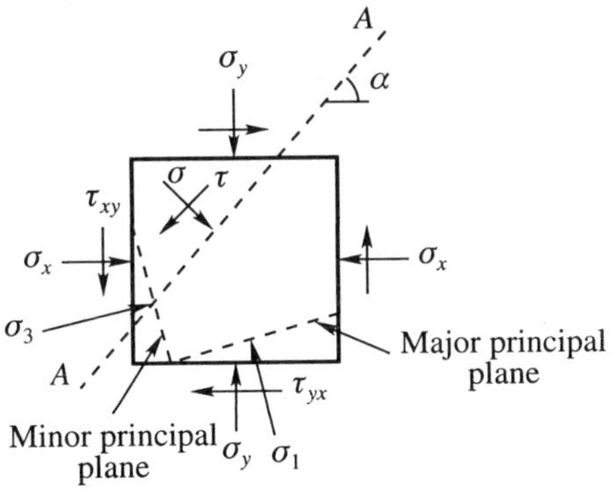

(a) A prismatic element subjected to normal and shear stresses

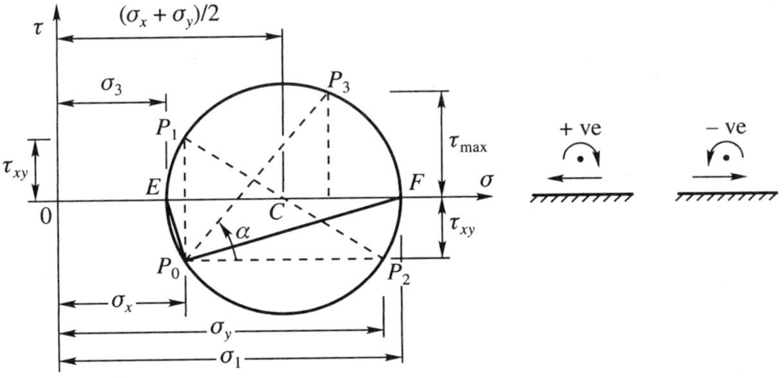

(b) Mohr circle of stress

Figure 8.8 Mohr stress circle for a general case

Point O is the origin of coordinates for the center of the Mohr circle of stress. With center C a circle may now be constructed with radius

$$\frac{1}{2}\sqrt{(\sigma_y-\sigma_x)^2+4\tau_{xy}^2}$$

This circle which passes through points P_1 and P_2 is called the *Mohr circle of stress*. The Mohr circle intersects the abscissa at two points E *and* F. The major and minor principal stresses are σ_1 (= OF) and σ_3 (= OE) respectively.

Determination of Normal and Shear Stresses on Plane AA [Fig. 8.8(a)]

Point P_1 on the circle of stress in Fig. 8. 8(b) represents the state of stress on the vertical plane of the prismatic element; similarly point P_2 represents the state of stress on the horizontal plane of the element. If from point P_1 a line is drawn parallel to the vertical plane, it intersects the circle at point P_0 and if from the point P_2 on the circle, a line is drawn parallel to the horizontal plane, this line also intersects the circle at point P_0. The point P_0 so obtained is called the *origin of planes* or the *pole*. If from the pole P_0 a line is drawn parallel to the plane AA in Fig. 8.8(a) to intersect the circle at point P_3 (Fig. 8.8(b)) then the coordinates of the point give the normal stress σ and the shear stress τ on plane AA as expressed by equations 8.8 and 8.9 respectively. This indicates that a line drawn from the pole P_0 at any angle α to the σ-axis intersects the circle at coordinates that represent the normal and shear stresses on the plane inclined at the same angle to the abscissa.

Major and Minor Principal Planes

The orientations of the principal planes may be obtained by joining point P_0 to the points E and F in Fig 8.8(b). P_0F is the direction of the major principal plane on which the major principal stress σ_1 acts; similarly P_0E is the direction of the minor principal plane on which the minor principal stress σ_3 acts. It is clear from the Mohr diagram that the two planes P_0E and P_0F intersect at a right angle, i.e., angle EP_0 F = 90°.

8.11 MOHR CIRCLE OF STRESS FOR A CYLINDRICAL SPECIMEN COMPRESSION TEST

Consider the case of a cylindrical specimen of soil subjected to normal stresses σ_1 and σ_3 which are the major and minor principal stresses respectively (Fig. 8.9)

From Eqs. (8.14) and (8.15), we may write

$$\tau^2+\left[\sigma-\left(\frac{\sigma_1+\sigma_3}{2}\right)\right]^2=\left(\frac{\sigma_1-\sigma_3}{2}\right)^2 \tag{8.27}$$

Again Eq. (8.27) is the equation of a circle whose center has coordinates

$$\sigma=\frac{\sigma_1+\sigma_3}{2} \text{ and } \tau=0 \text{ and whose radius is } \frac{\sigma_1-\sigma_3}{2}$$

A circle with radius $(\sigma_1 - \sigma_3)/2$ with its center C on the abscissa at a distance of $(\sigma_1 + \sigma_3)/2$ may be constructed as shown in Fig. 8.9. This is the Mohr circle of stress. The major and minor principal stresses are shown in the figure wherein $\sigma_1 = OF$ and $\sigma_3 = OE$.

From Fig. 8.8, we can write equations for σ_1 and σ_3 and τ_{max} as follows

$$\sigma_1,\ \sigma_3=\frac{\sigma_x+\sigma_y}{2}\pm\frac{1}{2}\sqrt{(\sigma_y-\sigma_x)^2+4\tau_{xy}^2} \tag{8.28}$$

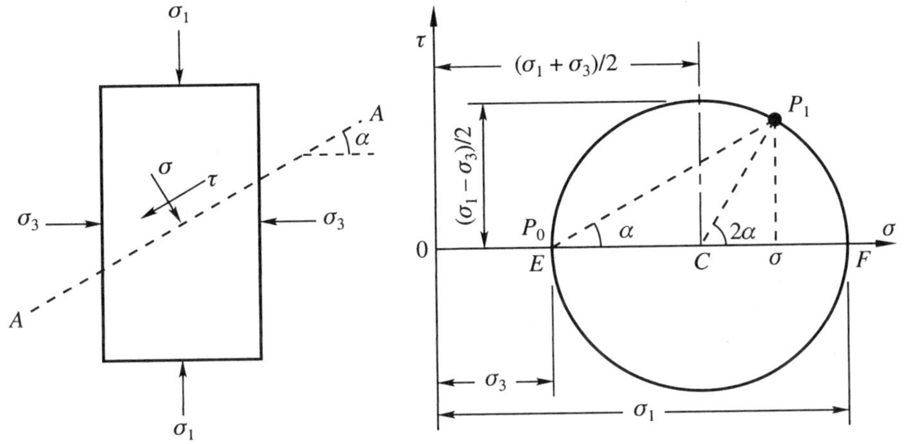

Figure 8.9 Mohr stress circle for a cylindrical specimen

$$\tau_{max} = \frac{1}{2}\sqrt{\left(\sigma_y - \sigma_x\right)^2 + 4\tau_{xy}^2} \tag{8.29}$$

where τ_{max} is the maximum shear stress equal to the radius of the Mohr circle.

The origin of planes or the pole P_0 (Fig. 8.9) may be obtained as before by drawing lines from points E and F parallel to planes on which the minor and major principal stresses act. In this case, the pole P_0 lies on the abscissa and coincides with the point E.

The normal stress σ and shear stress τ on any arbitrary plane AA making an angle α with the major principal plane may be determined as follows.

From the pole P_0 draw a line $P_0 P_1$ parallel to the plane AA (Fig. 8.9). The coordinates of the point P_1 give the stresses σ and τ. From the stress circle we may write

$$\angle P_1CF = 2\alpha$$

$$\sigma = \frac{\sigma_1 + \sigma_3}{2} + \frac{\sigma_1 - \sigma_3}{2}\cos 2\alpha \tag{8.30}$$

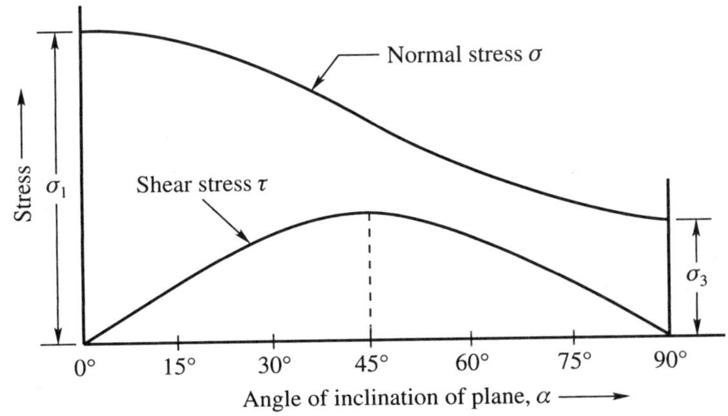

Figure 8.10 Variation of σ and τ with α

$$\tau = \frac{\sigma_1 - \sigma_3}{2} \sin 2\alpha \tag{8.31}$$

Equations (8.30) and (8.31) are the same as Eqs. (8.14) and (8.15) respectively.

It is of interest to study the variation of the magnitudes of normal and shear stresses with the inclination of the plane.

Eqs. (8.30) and (8.31) are plotted with α as the abscissa shown in Fig. 8.10. The following facts are clear from these curves:

1. The greatest and least principal stresses are respectively the maximum and minimum normal stresses on any plane through the point in question.
2. The maximum shear stress occurs on planes at 45° to the principal planes.

8.12 MOHR-COULOMB FAILURE THEORY

Various theories relating to the stress condition in engineering materials at the time of failure are available in the engineering literature. Each of these theories may explain satisfactorily the actions of certain kinds of materials at the time they fail, but no one of them is applicable to all materials. The failure of a soil mass is more nearly in accordance with the tenets of the Mohr theory of failure than with those of any other theory and the interpretation of the triaxial compression test depends to a large extent on this fact. The Mohr theory is based on the postulate that a material will fail when the shearing stress on the plane along which the failure is presumed to occur is a unique function of the normal stress acting on that plane. The material fails along the plane only when the angle between the resultant of the shearing stress and the normal stress is a maximum, that is, where the combination of normal and shearing stresses produces the maximum obliquity angle δ.

According to Coulomb's Law, the condition of failure is that the shear stress

$$\tau \leqq c + \sigma \tan \phi \tag{8.32}$$

In Fig 8.11(b) M_0N and M_0N_1 are the lines that satisfy Coulomb's condition of failure. If the stress at a given point within a cylindrical specimen under triaxial compression is represented by Mohr circle 1, it may be noted that every plane through this point has a shearing stress which is smaller than the shearing strength.

For example, if the plane *AA* in Fig. 8.11(a) is the assumed failure plane, the normal and shear stresses on this plane at any intermediate stage of loading are represented by point *b* on Mohr circle 1 where the line P_0b is parallel to the plane *AA*. The shearing stress on this plane is *ab* which is less than the shearing strength *ac* at the same normal stress *Oa*. Under this stress condition there is no possibility of failure. On the other hand it would not be possible to apply the stress condition represented by Mohr stress circle 2 to this sample because it is not possible for shearing stresses to be greater than the shearing strength. At the normal stress *Of*, the shearing stress on plane *AA* is shown to be *fh* which is greater than the shear strength of the materials *fg* which is not possible. Mohr circle 3 in the figure is tangent to the shear strength line M_0N and M_0N_l at points *e* and e_1 respectively. On the same plane *AA* at normal stress *Od*, the shearing stress *de* is the same as the shearing strength *de*. Failure is therefore imminent on plane *AA* at the normal stress *Od* and shearing stress *de*. The equation for the shearing stress *de* is

$$s = de = de' + e'e = c + \sigma \tan \phi \tag{8.33}$$

where ϕ is the slope of the line M_0N which is the maximum angle of obliquity on the failure plane. The value of the obliquity angle can never exceed $\delta_m = \phi$, the angle of shearing resistance, without the occurrence of failure. The shear strength line M_0N which is tangent to Mohr circle 3 is called the

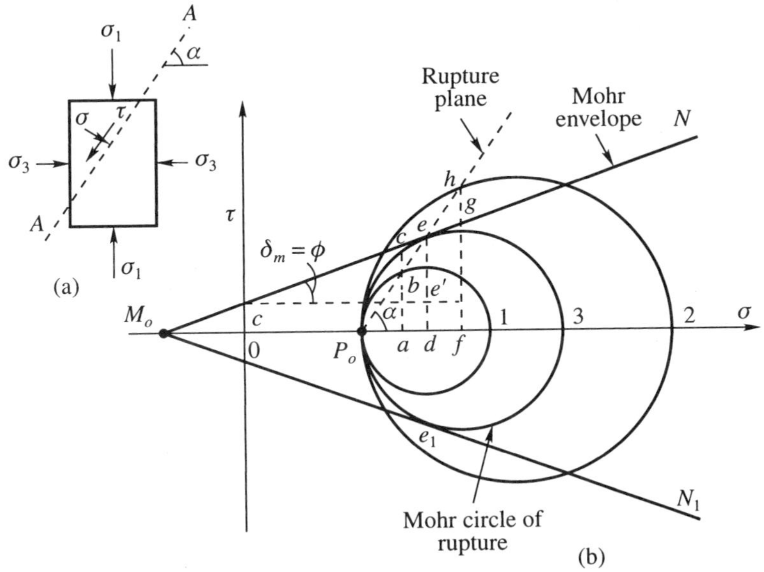

Figure 8.11 Diagram presenting Mohr's theory of rupture

Mohr envelope or *line of rupture*. The Mohr envelope may be assumed as a straight line although it is curved under certain conditions. The Mohr circle which is tangential to the shear strength line is called the *Mohr circle of rupture*. Thus the Mohr envelope constitutes a shear diagram and is a graph of the Coulomb equation for shearing stress. This is called the *Mohr-Coulomb Failure Theory*. The principal objective of a triaxial compression test is to establish the Mohr envelope for the soil being tested. The cohesion and the angle of shearing resistance can be determined from this envelope. When the cohesion of the soil is zero, that is, when the soil is cohesionless, the Mohr envelope passes through the origin.

8.13 MOHR DIAGRAM FOR TRIAXIAL COMPRESSION TEST AT FAILURE

Consider a cylindrical specimen of soil possessing both cohesion and friction is subjected to a conventional triaxial compression test. In the conventional test the lateral pressure σ_3 is held constant and the vertical pressure σ_1 is increased at a constant rate of stress or strain until the sample fails. If σ_1 is the peak value of the vertical pressure at which the sample fails, the two principal stresses that are to be used for plotting the Mohr circle of rupture are σ_3 and σ_1. In Fig. 8.12 the values of σ_1 and σ_3 are plotted on the σ-axis and a circle is drawn with $(\sigma_1 - \sigma_3)$ as diameter. The center of the circle lies at a distance of $(\sigma_1 + \sigma_3)/2$ from the origin. As per Eq. (8.18), the soil fails along a plane which makes an angle $\alpha = 45° + \phi/2$ with the major principal plane. In Fig. 8.12 the two lines P_0P_1 and P_0P_2 (where P_0 is the origin of planes) are the conjugate rupture planes. The two lines M_0N and M_0N_1 drawn tangential to the rupture circle at points P_1 and P_2 are called Mohr envelopes. If the Mohr envelope can be drawn by some other means, the orientation of the failure planes may be determined.

The results of analysis of triaxial compression tests as explained in Sect. 8.8 are now presented in a graphical form in Fig. 8.12. The various information that can be obtained from the figure includes

1. The angle of shearing resistance ϕ = the slope of the Mohr envelope.

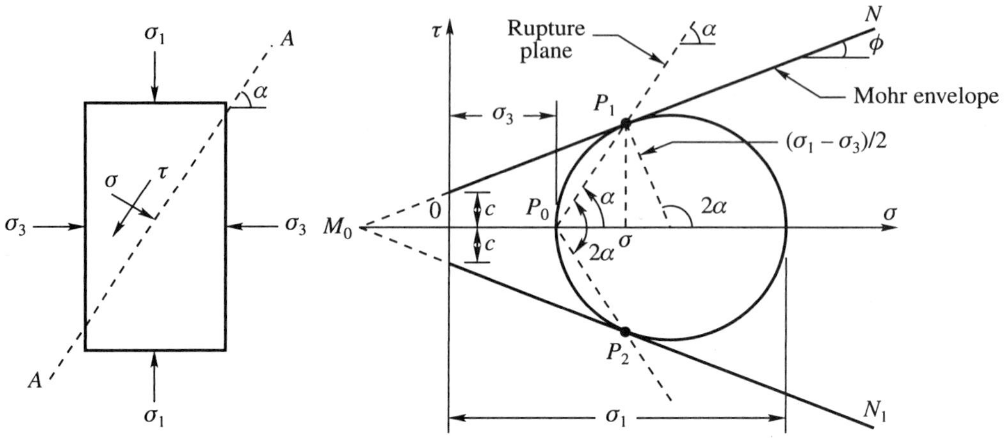

Figure 8.12 Mohr diagram for triaxial test at failure for c-ϕ soil

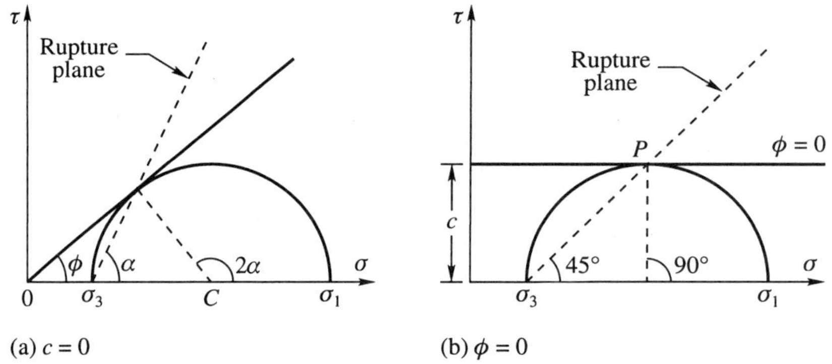

Figure 8.13 Mohr diagram for soils with $c = 0$ and $\phi = 0$

2. The apparent cohesion c = the intercept of the Mohr envelope on the τ-axis.
3. The inclination of the rupture plane = α.
4. The angle between the conjugate planes = 2α.

If the soil is cohesionless with $c = 0$ the Mohr envelopes pass through the origin, and if the soil is purely cohesive with $\phi = 0$ the Mohr envelope is parallel to the abscissa. The Mohr envelopes for these two types of soils are shown in Fig. 8.13.

8.14 MOHR DIAGRAM FOR A DIRECT SHEAR TEST AT FAILURE

In a direct shear test the sample is sheared along a horizontal plane. This indicates that the failure plane is horizontal. The normal stress σ on this plane is the external vertical load divided by the area of the sample. The shear stress at failure is the external lateral load divided by the area of the sample.

Point P_1 on the stress diagram in Fig. 8.14 represents the stress condition on the failure plane. The coordinates of the point are

normal stress = σ, shear stress $\tau = s$.

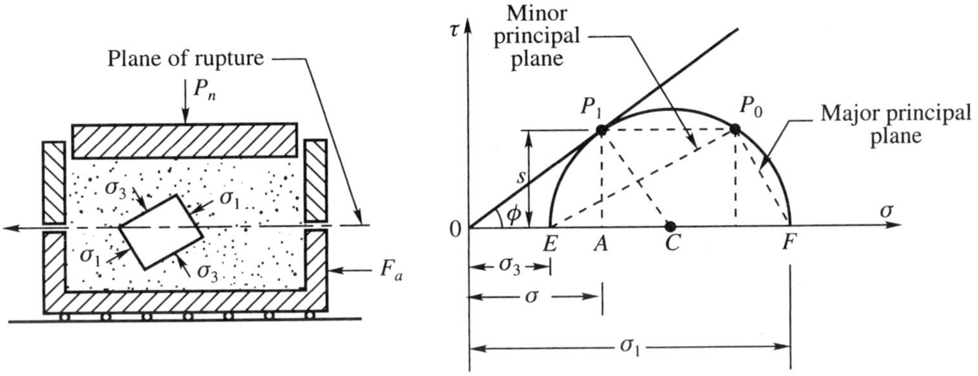

Figure 8.14 Mohr diagram for a direct shear test at failure

If it is assumed that the Mohr envelope is a straight line passing through the origin (for cohesionless soil or normally consolidated clays), it follows that the maximum obliquity δ_m occurs on the failure plane and $\delta_m = \phi$. Therefore the line OP_1 must be tangent to the Mohr circle, and the circle may be constructed as follows:

Draw P_1C normal to OP_1. Point C which is the intersection point of the normal with the abscissa is the center of the circle. CP_1 is the radius of the circle. The Mohr circle may now be constructed which gives the major and minor principal stresses σ_1 and σ_3 respectively.

Since the failure is on the horizontal plane, the origin of planes P_0 may be obtained by drawing a horizontal line through P_1 giving P_0. P_0F and P_0E give the directions of the major and minor principal planes respectively.

Example 8.1

What is the shearing strength of soil along a horizontal plane at a depth of 4 m in a deposit of sand having the following properties:

Angle of internal friction, $\phi = 35°$

Dry unit weight, $\gamma_d = 17$ kN/m^3

Specific gravity, $G_s = 2.7$.

Assume the ground water table is at a depth of 2.5 m from the ground surface. Also find the change in shear strength when the water table rises to the ground surface.

Solution

The effective vertical stress at the plane of interest is

$$\sigma' = 2.50 \times \gamma_d + 1.50 \times \gamma_b$$

Given $\gamma_d = 17$ kN/m^3 and $G_s = 2.7$

$$\text{We have } \gamma_d = 17 = \frac{G_s}{1+e}\gamma_w = \frac{2.7}{1+e} \times 9.81$$

$$\text{or} \quad 17e = 26.5 - 17 = 9.49 \quad \text{or} \quad e = \frac{9.49}{17} = 0.56$$

$$\text{Therefore, } \gamma_b = \frac{G_s - 1}{1+e}\gamma_w = \frac{2.7-1.0}{1+0.56} \times 9.81 = 10.7 \text{ kN/m}^3$$

Hence $\sigma' = 2.5 \times 17 + 1.5 \times 10.7 = 58.55$ kN/m^2

Hence, the shearing strength of the sand is

$s = \sigma' \tan \phi = 58.55 \times \tan 35° = 41$ kN/m^2

If the water table rises to the ground surface i.e., by a height of 2.5 m, the change in the effective stress will be,

$\Delta\sigma' = \gamma_d \times 2.5 - \gamma_b \times 2.5 = 17 \times 2.5 - 10.7 \times 2.5 = 15.75$ kN/m^2 (negative)

Hence the decrease in shear strength will be,

$= \Delta\sigma' \tan 35° = 15.75 \times 0.70 = 11$ kN/m^2

Example 8.2

Direct shear tests were conducted on a dry sand. The size of the samples used for the tests was 2 in. × 2 in. × 0.75 in. The test results obtained are given below:

Test No.	Normal load (lb)	Normal stress σ (lb/ft^2)	Shear force at failure (lb)	Shear stress (lb/ft^2)
1	15	540	12	432
2	20	720	18	648
3	30	1080	23	828
4	60	2160	47	1692
5	120	4320	93	3348

Determine the shear strength parameters c and ϕ.

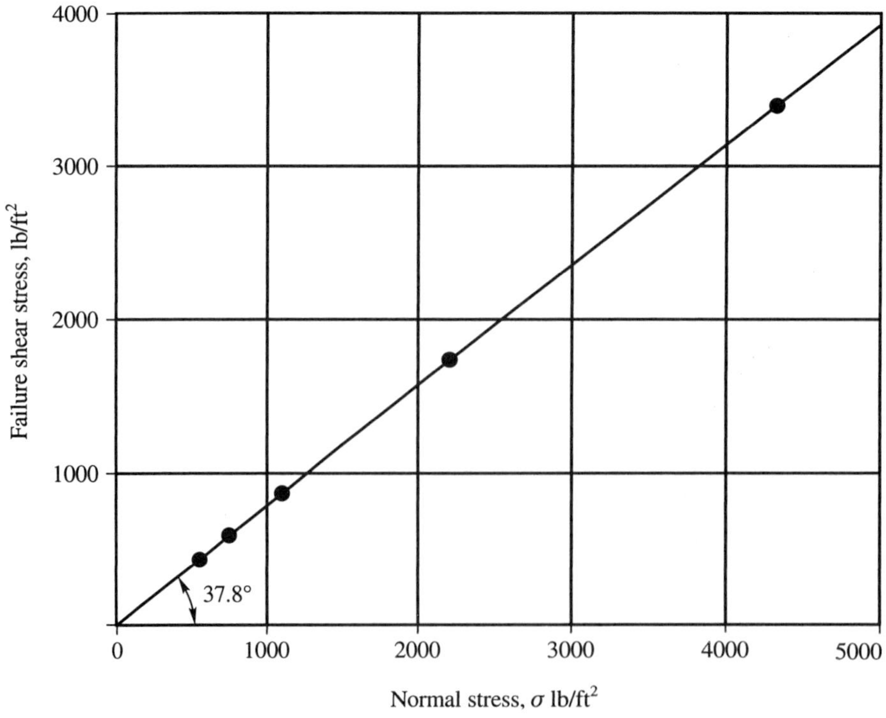

Figure Ex. 8.2

Solution

The failure shear stresses τ_f, as obtained from the tests are plotted against the normal stresses σ, in Figure Ex 8.2. The shear parameters from the graph are: $c = 0$, $\phi = 37.8°$.

Example 8.3

A direct shear test, when conducted on a remolded sample of sand, gave the following observations at the time of failure: Normal load = 288 N; shear load = 173 *N*. The cross sectional area of the sample = 36 cm^2.

Determine: (i) the angle of internal friction, (ii) the magnitude and direction of the principal stresses in the zone of failure.

Solution

Such problems can be solved in two ways, namely graphically and analytically. The analytical solution has been left as an exercise for the students.

Graphical Solution

(i) Shear stress $\tau = \frac{173}{36} = 4.8 \text{ N/cm}^2 = 48 \text{ kN/m}^2$

Normal stress $\sigma = \frac{288}{36} = 8.0 \text{ N/cm}^2 = 80 \text{ kN/m}^2$

We know one point on the Mohr envelope. Plot point *A* (Fig. Ex. 8.3) with coordinates τ = 48 kN/m^2, and σ = 80 kN/m^2. Since cohesion $c = 0$ for sand, the Mohr envelope *OM* passes through the origin. The slope of *OM* gives the angle of internal friction $\phi = 31°$.

(ii) In Fig. Ex. 8.3, draw line *AC* normal to the envelope *OM* cutting the abscissa at point *C*. With *C* as center, and *AC* as radius, draw Mohr circle C_1 which cuts the abscissa at points *B* and *D*, which gives

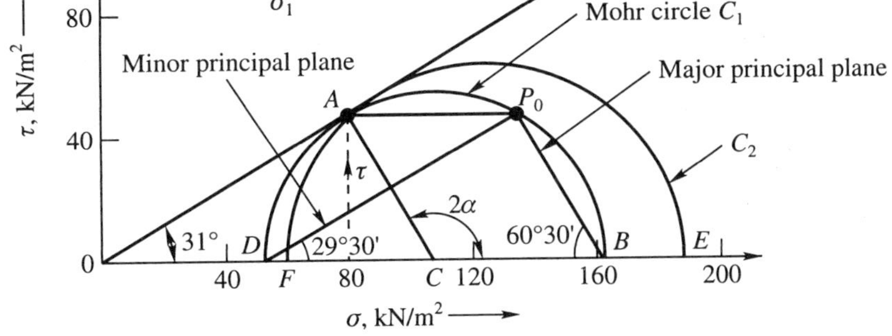

Figure Ex. 8.3

major principal stress = $OB = \sigma_1 = 163.5\ \text{kN/m}^2$

minor principal stress = $OD = \sigma_3 = 53.5\ \text{kN/m}^2$

Now, $\angle ACB = 2\alpha$ = twice the angle between the failure plane and the major principal plane. Measurement gives

$2\alpha = 121°$ or $\alpha = 60.5°$

Since in a direct shear test the failure plane is horizontal, the angle made by the major principal plane with the horizontal will be 60.5°. The minor principal plane should be drawn at a right angle to the major principal plane.

The directions of the principal planes may also be found by locating the pole P_o. P_o is obtained by drawing a horizontal line from point A which is parallel to the failure plane in the direct shear test. Now P_oB and P_oD give the directions of the major and minor principal planes respectively.

8.15 EFFECTIVE STRESSES

So far, the discussion has been based on consideration of total stresses. It is to be noted that the strength and deformation characteristics of a soil can be understood better by visualizing it as a compressible skeleton of solid particles enclosing voids. The voids may completely be filled with water or partly with water and air. Shear stresses are to be carried only by the skeleton of solid particles. However, the total normal stresses on any plane are, in general, the sum of two components.

Total normal stress = component of stress carried by solid particles
+ pressure in the fluid in the void space.

This visualization of the distribution of stresses between solid and fluid has two important consequences:

1. When a specimen of soil is subjected to external pressure, the volume change of the specimen is not due to the total normal stress but due to the difference between the total normal stress and the pressure of the fluid in the void space. The pressure in the fluid is the pore pressure u. The difference which is called the effective stress σ' may now be expressed as

$$\sigma' = \sigma - u \tag{8.34}$$

2. The shear strength of soils, as of all granular materials, is largely determined by the frictional forces arising during slip at the contacts between the soil particles. These are clearly a function of the component of normal stress carried by the solid skeleton rather than of the total normal stress. For practical purposes the shear strength equation of Coulomb is given by the expression

$$s = c' + (\sigma - u)\tan\phi' = c' + \sigma'\tan\phi' \tag{8.35}$$

where c' = apparent cohesion in terms of effective stresses
ϕ' = angle of shearing resistance in terms of effective stresses
σ = total normal pressure to the plane considered
u = pore pressure.

The effective stress parameters c' and ϕ' of a given sample of soil may be determined provided the pore pressure u developed during the shear test is measured. The pore pressure u is developed when the testing of the soil is done under undrained conditions. However, if free drainage takes place during testing, there will not be any development of pore pressure. In such cases, the total stresses themselves are effective stresses.

8.16 SHEAR STRENGTH EQUATION IN TERMS OF EFFECTIVE PRINCIPAL STRESSES

The principal stresses may be expressed either as total stresses or as effective stresses if the values of pore pressure are known.

If u is the pore pressure developed during a triaxial test, we may write as before

$$\begin{aligned} \sigma_1' &= \sigma_1 - u \\ \sigma_3' &= \sigma_3 - u \end{aligned} \tag{8.36}$$

where σ_1' and σ_3' are the effective principal stresses. The equation for shear strength in terms of effective stresses is

$$s = \frac{\sigma_1' - \sigma_3'}{2} \sin 2\alpha = \frac{\sigma_1 - \sigma_3}{2} \sin 2\alpha = \frac{\sigma_1 - \sigma_3}{2} \cos \phi' \tag{8.37}$$

where $2\alpha = 90° + \phi'$

Coulomb's equation in terms of effective stresses is

$s = c' + (\sigma - u) \tan \phi'$

Therefore, $\dfrac{\sigma_1 - \sigma_3}{2} \cos \phi' = c' + (\sigma - u) \tan \phi'$

Since, $\sigma = \dfrac{\sigma_1 + \sigma_3}{2} + \dfrac{\sigma_1 - \sigma_3}{2} \cos(90 + \phi')$

we have $\dfrac{\sigma_1 - \sigma_3}{2} \cos \phi' = c' + \dfrac{\sigma_1 + \sigma_3}{2} \tan \phi'$

$$+ \frac{\sigma_1 - \sigma_3}{2} \cos(90 + \phi') \tan \phi' - u \tan \phi'$$

Simplifying

$$\frac{\sigma_1 - \sigma_3}{2} \cos^2 \phi' = c' \cos \phi' + \frac{\sigma_1 + \sigma_3}{2} \sin \phi' - \frac{\sigma_1 - \sigma_3}{2} \sin^2 \phi' - u \sin \phi'$$

or $$\frac{1}{2}(\sigma_1 - \sigma_3)_f = \frac{c' \cos \phi' + (\sigma_3 - u) \sin \phi'}{1 - \sin \phi'} \tag{8.38}$$

where $(\sigma_1 - \sigma_3)_f$ indicates the maximum deviator stress at failure. Eq (8.38) may also be expressed in a different form as follows by considering effective principal stresses

$$\frac{1}{2}(\sigma_1' - \sigma_3')_f = \frac{c' \cos \phi' + \sigma_3' \sin \phi'}{1 - \sin \phi'}$$

or $$\frac{1}{2}(\sigma_1' - \sigma_3')_f (1 - \sin \phi') = c' \cos \phi' + \sigma_3' \sin \phi'$$

Simplifying, we have

$$(\sigma_1' - \sigma_3')_f = (\sigma_1' + \sigma_3')\sin\phi' + 2c'\cos\phi' \tag{8.39}$$

8.17 STRESS-CONTROLLED AND STRAIN-CONTROLLED TESTS

Direct shear tests or triaxial compression tests may be carried out by applying stresses or strains at a particularly known rate. When the stress is applied at a constant rate it is called a *stress-controlled test* and when the strain is applied at a constant rate it is called a *strain-controlled test.* The difference between the two types of tests may be explained with respect to box shear tests.

In the stress-controlled test [Fig. 8.15(a)] the lateral load F_a which induces shear is gradually increased until complete failure occurs. This can be done by placing weights on a hanger or by filling a counterweighted bucket of original weight W at a constant rate. The shearing displacements are measured by means of a dial gauge G as a function of the increasing load F_a. The shearing stress at any shearing displacement, is

$$\tau = \frac{F_a}{A}$$

where A is the cross sectional area of the sample. A typical shape of a stress-strain curve of the stress-controlled test is shown in Fig. 8.15(a).

A typical arrangement of a box-shear test apparatus for the strain-controlled test is shown in Fig. 8.15(b). The shearing displacements are induced and controlled in such a manner that they occur at a constant fixed rate. This can be achieved by turning the wheel either by hand or by means of any electrically operated motor so that horizontal motion is induced through the worm gear B. The dial gauge G gives the desired constant rate of displacement. The bottom of box C is mounted on frictionless rollers D. The shearing resistance offered to this displacement by the soil sample is measured by the proving ring E. The stress-strain curves for this type of test have the shape shown in Fig. 8.15(b).

Both stress-controlled and strain-controlled types of test are used in connection with all the direct triaxial and unconfined soil shear tests. Strain-controlled tests are easier to perform and have the advantage of readily giving not only the peak resistance as in Fig. 8.15 (b) but also the ultimate resistance which is lower than the peak such as point b in the same figure, whereas the stress controlled gives only the peak values but not the smaller values after the peak is achieved. The stress-controlled test is preferred only in some special problems connected with research.

8.18 TYPES OF LABORATORY TESTS

The laboratory tests on soils may be on

1. Undisturbed samples, or
2. Remolded samples.

Further, the tests may be conducted on soils that are :

1. Fully saturated, or
2. Partially saturated.

The type of test to be adopted depends upon how best we can simulate the field conditions. Generally speaking, the various shear tests for soils may be classified as follows:

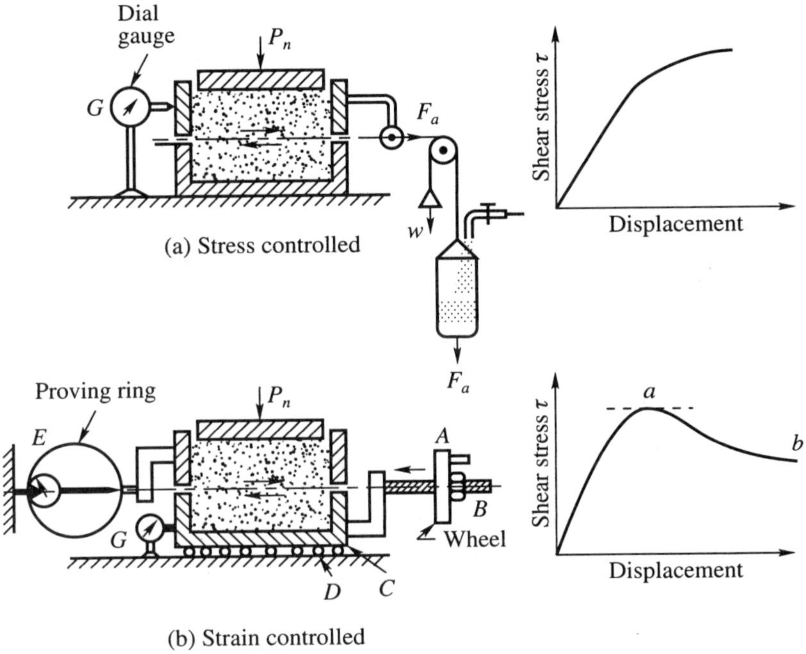

Figure 8.15 Stress and strain controlled box shear tests

1. Unconsolidated-Undrained Tests (UU)

The samples are subjected to an applied pressure under conditions in which drainage is prevented, and then sheared under conditions of no drainage.

2. Consolidated-Undrained or Quick Tests (CU)

The samples are allowed to consolidate under an applied pressure and then sheared under conditions of no drainage.

3. Consolidated-Drained or Slow Tests (CD)

The samples are consolidated as in the previous test, but the shearing is carried out slowly under conditions of no excess pressure in the pore space.

The drainage condition of a sample is generally the deciding factor in choosing a particular type of test in the laboratory. The purpose of carrying out a particular test is to simulate field conditions as far as possible. Because of the high permeability of sand, consolidation occurs relatively rapidly and is usually completed during the application of the load. Tests on sand are therefore generally carried out under drained conditions (drained or slow test).

For soils other than sands the choice of test conditions depends upon the purpose for which the shear strength is required. The guiding principle is that drainage conditions of the test should conform as closely as possible to the conditions under which the soils will be stressed in the field.

Undrained or quick tests are generally used for foundations on clay soils, since during the period of construction only a small amount of consolidation will have taken place and consequently the moisture content will have undergone little change. For clay slopes or cuts undrained tests are used both for design and for the investigation of failures.

Consolidated-undrained tests are used where changes in moisture content are expected to take place due to consolidation before the soil is fully loaded. An important example is the condition known as "sudden drawdown" such as that occurs in an earth dam behind which the water level is lowered at a faster rate than at which the material of the dam can consolidate. In the consolidated-undrained tests used in this type of problem, the consolidation pressures are chosen to represent the initial conditions of the soil, and the shearing loads correspond to the stresses called into play by the action of sudden drawdown.

As already stated, drained tests are always used in problems relating to sandy soils. In clay soils drained tests are sometimes used in investigating the stability of an earth dam, an embankment or a retaining wall after a considerable interval of time has passed.

Very fine sand, silts and silty sands also have poor drainage qualities. Saturated soils of these categories are likely to fail in the field under conditions similar to those under which consolidated quick tests are made.

Shearing Test Apparatus for the Various Types of Tests

The various types of shear tests mentioned earlier may be carried out either by the box shear test or the triaxial compression test apparatus. Tests that may be made by the two types of apparatus are:

Box Shear Test Apparatus

1. Undrained and consolidated- undrained tests on clay samples only.
2. Drained or Slow tests on any soil.

The box shear test apparatus is not suited for undrained or consolidated-undrained tests on samples other than clay samples, because the other soils are so permeable that even a rapid increase of the stresses in the sample may cause at least a noticeable change of the water content.

Triaxial Compression Test Apparatus

All types of tests can conveniently be carried out in this apparatus.

8.19 SHEARING STRENGTH TESTS ON SAND

Shear tests on sand may be made when the sand is either in a dry state or in a saturated state. No test shall be made when the soil is in a moist state as this state exists only due to apparent cohesion between particles which would be destroyed when it is saturated. The results of shear tests on saturated samples are almost identical with those on the same sand at equal relative density in a dry state except that the angle ϕ is likely to be 1 or 2 degrees smaller for the saturated sand.

The usual type of test used for coarse to medium sand is the slow shear test. However, consolidated undrained tests may be conducted on fine sands, sandy silts etc. which do not allow free drainage under changed stress conditions. If the equilibrium of a large body of saturated fine sand in an embankment is disturbed by rapid drawdown of the surface of an adjoining body of water, the change in water content of the fill lags behind the change in stress.

In all the shearing tests on sand, only the remolded samples are used as it is not practicable to obtain undisturbed samples. The soil samples are to be made approximately to the same dry density as it exists in-situ and tested either by direct shear or triaxial compression tests.

Tests on soils are generally carried out by the strain-controlled type apparatus. The principal advantage of this type of test on dense sand is that its peak-shear resistance, as well as the shear resistances smaller than the peak, can be observed and plotted.

Direct Shear Test

Only the drained or the slow shear tests on sand may be carried out by using the box shear test apparatus. The box is filled with sand to the required density. The sample is sheared at a constant vertical pressure σ. The shear stresses are calculated at various displacements of the shear box. The test is repeated with different pressures σ.

If the sample consists of loose sand, the shearing stress increases with increasing displacement until failure occurs. If the sand is dense, the shear failure of the sample is preceded by a decrease of the shearing stress from a peak value to an ultimate value (also known as residual value) lower than the peak value.

Typical stress-strain curves for loose and dense sands are shown in Fig. 8.16(a).

The shear stress of a dense sand increases from 0 to a peak value represented by point *a*, and then gradually decreases and reaches an ultimate value represented by point *b*. The sample of sand in a dense state is closely packed and the number of contact points between the particles are more than in the loose state. The soil grains are in an interlocked state. As the sample is subjected to shear stress, the stress has to overcome the resistance offered by the interlocked arrangement of the particles. Experimental evidence indicates that a significant percent of the peak strength is due to the interlocking of the grains. In the process of shearing one grain tries to slide over the other and the void ratio of the sample which is the lowest at the commencement of the test reaches the maximum value at point *a*, in the Fig 8.16(a). The shear stress also reaches the maximum value at this level. Any further increase of strain beyond this point is associated with a progressive disintegration of the structure of the sand resulting in a decrease in the shear stress. Experience shows that the change in void ratio due to shear depends on both the vertical load and the relative density of the sand. At very low vertical pressure, the void ratio at failure is larger and at very high pressure it is smaller than the initial void ratio, whatever the relative density of the sand may be. At

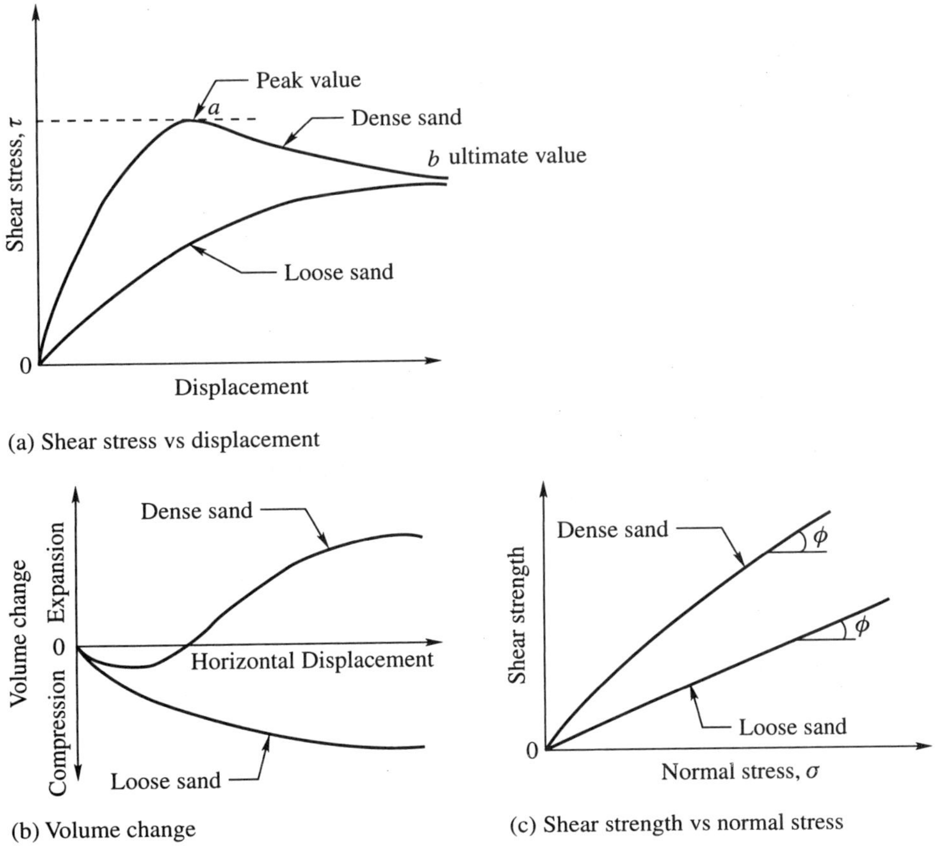

Figure 8.16 Direct shear test on sand

Table 8.1 Typical values of ϕ and $\bar{\phi}_u$ for granular soils

Types of soil	ϕ deg	$\bar{\phi}_u$ deg
Sand: rounded grains		
Loose	28 to 30	
Medium	30 to 35	26 to 30
Dense	35 to 38	
Sand: angular grains		
Loose	30 to 35	
Medium	35 to 40	30 to 35
Dense	40 to 45	
Sandy gravel	34 to 48	33 to 36

intermediate values of pressure, the shearing force causes a decrease in the void ratio of loose sand and an increase in the void ratio of dense sand. Fig 8.16(b) shows how the volume of dense sand decreases up to a certain value of horizontal displacement and with further displacement the volume increases, whereas in the case of loose sand the volume continues to decrease with an increase in the displacement. In saturated sand a decrease of the void ratio is associated with an expulsion of pore water, and an increase with an absorption of water. The expansion of a soil due to shear at a constant value of vertical pressure is called *dilatancy*. At some intermediate state or degree of density in the process of shear, the shear displacement does not bring about any change in volume, that is, density. The density of sand at which no change in volume is brought about upon the application of shear strains is called the *critical density*. The porosity and void ratio corresponding to the critical density are called the *critical porosity* and the *critical void ratio* respectively.

By plotting the shear strengths corresponding to the state of failure in the different shear tests against the normal pressure a straight line is obtained for loose sand and a slightly curved line for dense sand [Fig. 8.16(c)]. However, for all practical purposes, the curvature for the dense sand can be disregarded and an average line may be drawn. The slopes of the lines give the corresponding angles of friction ϕ of the sand. The general equation for the lines may be written as

$$s = \sigma \tan \phi$$

For a given sand, the angle ϕ increases with increasing relative density. For loose sand it is roughly equal to the *angle of repose*, defined as the angle between the horizontal and the slope of a heap produced by pouring clean dry sand from a small height. The angle of friction varies with the shape of the grains. Sand samples containing well graded angular grains give higher values of ϕ as compared to uniformly graded sand with rounded grains. The angle of friction ϕ for dense sand at peak shear stress is higher than that at ultimate shear stress. Table 8.1 gives some typical values of ϕ (at peak) and $\bar{\phi}_u$ (at ultimate).

Triaxial Compression Test

Reconstructed sand samples at the required density are used for the tests. The procedure of making samples should be studied separately (refer to any book on Soil Testing). Tests on sand may be conducted either in a saturated state or in a dry state. Slow or consolidated undrained tests may be carried out as required.

Drained or Slow Tests

At least three identical samples having the same initial conditions are to be used. For slow tests under saturated conditions the drainage valve should always be kept open. Each sample should be

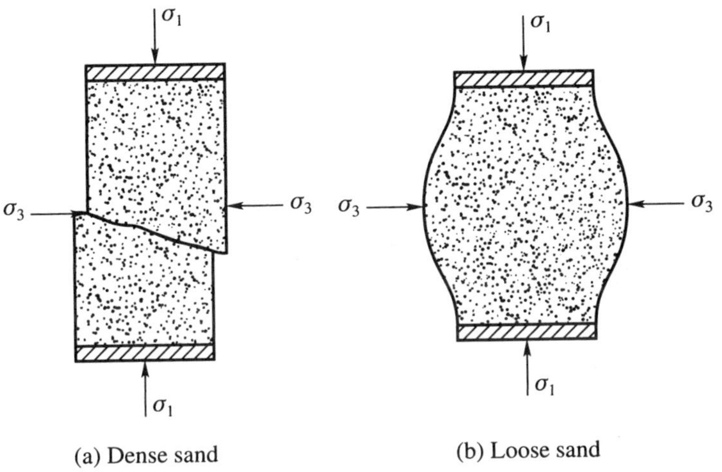

Figure 8.17 Typical shapes of dense and loose sands at failure

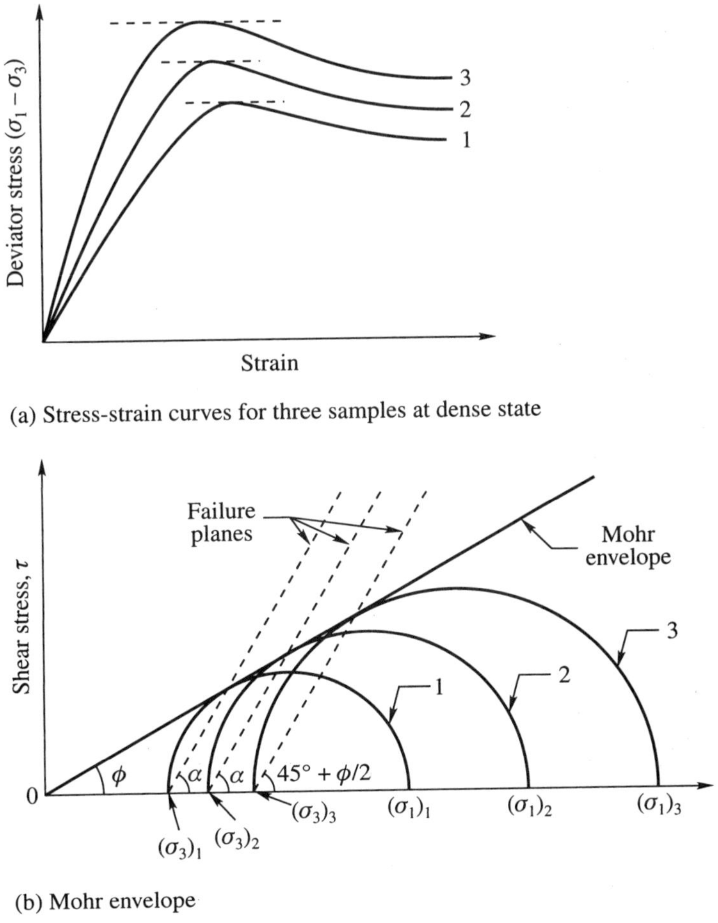

Figure 8.18 Mohr envelope for dense sand

tested under different constant all-round pressures for example, 1, 2 and 3 kg/cm^2. Each sample is sheared to failure by increasing the vertical load at a sufficiently slow rate to prevent any build up of excess pore pressures.

At any stage of loading the major principal stress is the all-round pressure σ_3 plus the intensity of deviator stress $(\sigma_1 - \sigma_3)$. The actually applied stresses are the effective stresses in a slow test, that is $\sigma_1 = \sigma'_1$ and $\sigma_3 = \sigma'_3$, Dense samples fail along a clearly defined rupture plane whereas loose sand samples fail along many planes which result in a symmetrical bulging of the sample. The compressive strength of a sample is defined as the difference between the major and minor principal stresses at failure $(\sigma_1 - \sigma_3)_f$. Typical shapes of dense and loose sand samples at failure are shown in Fig. 8.17.

Typical stress-strain curves for three samples in a dense state and the Mohr circles for these samples at peak strength are shown in Fig. 8.18.

If the experiment is properly carried out there will be one common tangent to all these three circles and this will pass through the origin. This indicates that the Mohr envelope is a straight line for sand and the sand has no cohesion. The angle made by the envelope with the σ-axis is called the angle of internal friction. The failure planes for each of these samples are shown in Fig. 8.18(b). Each of them make an angle α with the horizontal which is approximately equal to

$$\alpha = 45° + \phi/2$$

From Fig. 8.18(b) an expression for the angle of internal friction may be written as

$$\sin\phi = \frac{\sigma_1 - \sigma_3}{\sigma_1 + \sigma_3} = \frac{\sigma_1 / \sigma_3 - 1}{\sigma_1 / \sigma_3 + 1} \tag{8.40}$$

Example 8.4

Determine the magnitude of the deviator stress if a sample of the same sand with the same void ratio as given in Ex. 8.3 was tested in a triaxial apparatus with a confining pressure of 60 kN/m^2.

Solution

In the case of a triaxial test on an identical sample of sand as given in Ex. 8.3, use the same Mohr envelope *OM* (Fig. Ex. 8.3). Now the point *F* on the abscissa gives the confining pressure $\sigma_3 = 60$ kN/m^2. A Mohr circle C_2 may now be drawn passing through point *F* and tangential to the Mohr envelope *OM*. The point *E* gives the major principal stress σ_1 for the triaxial test.

Now $\sigma_1 = OE = 188$ kN/m^2, $\sigma_3 = 60$ kN/m^2

Therefore $\sigma_1 - \sigma_3 = 188 - 60 = 128$ kN/m^2 = deviator stress

Example 8.5

A consolidated drained triaxial test was conducted on a granular soil. At failure $\sigma'_1/\sigma'_3 = 4.0$. The effective minor principal stress at failure was 100 kN/m^2. Compute ϕ' and the principal stress difference at failure.

Solution

$$\sin\phi' = \frac{\sigma'_1/\sigma'_3 - 1}{\sigma'_1/\sigma'_3 + 1} = \frac{4-1}{4+1} = 0.6 \quad \text{or} \quad \phi' = 37°$$

The principal stress difference at failure is

$$(\sigma_1' - \sigma_3') = \sigma_3' \left(\frac{\sigma_{1f}'}{\sigma_{3f}'} - 1 \right) = 100(4-1) = 300 \text{ kN/m}^2$$

Example 8.6

A drained triaxial test on sand with $\sigma_3' = 3150$ lb/ft^2 gave $(\sigma_1'/\sigma_3')_f = 3.7$. Compute (a) σ_{1f}' (b) $(\sigma_1 - \sigma_3)_f$, and (c) ϕ'.

Solution

(a) $\dfrac{\sigma_1'}{\sigma_3'} = 3.7$

Therefore, $\sigma_1' = 3.7\sigma_3' = 3.7 \times 3150 = 11{,}655$ lb/ft^2

(b) $(\sigma_1 - \sigma_3)_f = (\sigma_1' - \sigma_3')_f = 11{,}655 - 3150 = 8505$ lb/ft^2

(c) $\sin\phi' = \dfrac{\sigma_1'/\sigma_3' - 1}{\sigma_1'/\sigma_3' + 1} = \dfrac{3.7-1}{3.7+1} = 0.574$ or $\phi' = 35°$

Example 8.7

Assume the test specimen in Ex. 8.6 was sheared undrained at the same total cell pressure of 3150 lb/ft^2. The induced excess pore water pressure at failure u_f was equal to 1470 lb/ft^2. Compute:

(a) σ_{1f}'
(b) $(\sigma_1 - \sigma_3)_f$
(c) ϕ in terms of total stress,
(d) the angle of the failure plane α_f.

Solution

(a) and (b): Since the void ratio after consolidation would be the same for this test as for Ex. 8.6, assume ϕ' is the same.

As before $(\sigma_1 - \sigma_3)_f = \sigma_{3f}' \left(\dfrac{\sigma_1'}{\sigma_3'} - 1 \right)$

$$\sigma_{3f}' = \sigma_{3f} - u_f = 3150 - 1470 = 1680 \text{ lb/ft}^2$$

So $(\sigma_1 - \sigma_3)_f = 1680(3.7-1) = 4536$ lb/ft^2

$$\sigma_{1f}' = (\sigma_1 - \sigma_3)_f + \sigma_{3f}' = 4536 + 1680 = 6216 \text{ lb/ft}^2$$

(c) $\sin\phi_{\text{total}} = \dfrac{\sigma_1 - \sigma_3}{\sigma_1 + \sigma_3} = \dfrac{4536}{6216+1470} = 0.59$ or $\phi_{\text{total}} = 36.17°$

(d) From Eq. (8.18)

$$\alpha_f = 45° + \frac{\phi'}{2} = 45° + \frac{35}{2} = 62.5°$$

$$\alpha_f = 45^\circ + \frac{\phi'}{2} = 45^\circ + \frac{35}{2} = 62.5^\circ$$

where ϕ' is taken from Ex. 8.6.

Example 8.8

A saturated specimen of cohesionless sand was tested under drained conditions in a triaxial compression test apparatus and the sample failed at a deviator stress of 482 kN/m^2 and the plane of failure made an angle of 60° with the horizontal. Find the magnitudes of the principal stresses. What would be the magnitudes of the deviator stress and the major principal stress at failure for another identical specimen of sand if it is tested under a cell pressure of 200 kN/m^2?

Solution

Per Eq. (8.18), the angle of the failure plane α is expressed as equal to

$$\alpha = 45^\circ + \frac{\phi}{2}$$

Since $\alpha = 60°$, we have $\phi = 30°$.

From Eq. (8.40), $\quad \sin\phi = \dfrac{\sigma_1 - \sigma_3}{\sigma_1 + \sigma_3}$

with $\phi = 30°$, and $\sigma_1 - \sigma_3 = 482$ kN/m^2. Substituting we have

$$\sigma_1 + \sigma_3 = \frac{\sigma_1 - \sigma_3}{\sin\phi} = \frac{482}{\sin 30^\circ} = 964 \text{ kN/m}^2 \tag{a}$$

$$\sigma_1 - \sigma_3 = 482 \text{ kN/m}^2 \tag{b}$$

solving (a) and (b) we have

$$\sigma_1 = 723 \text{ kN/m}^2, \quad \text{and} \quad \sigma_3 = 241 \text{ kN/m}^2$$

For the identical sample

$$\phi = 30^\circ, \quad \sigma_3 = 200 \text{ kN/m}^2$$

From Eq. (8.40), we have

$$\sin 30^\circ = \frac{\sigma_1 - 200}{\sigma_1 + 200}$$

Solving for σ_1 we have $\sigma_1 = 600$ kN/m^2 and $(\sigma_1 - \sigma_3) = 400$ kN/m^2

8.20 UNCONSOLIDATED-UNDRAINED TEST

Saturated Clay

Tests on saturated clay may be carried out either on undisturbed or on remolded soil samples. The procedure of the test is the same in both cases. A series of samples (at least a minimum of three) having the same initial conditions are tested under undrained conditions. With σ_3, the all-round pressure, acting on a sample under conditions of no drainage, the axial pressure is increased until failure occurs at a deviator stress $(\sigma_1 - \sigma_3)$. From the deviator stress, the major principal stress σ_1 is determined. If the other samples are tested in the same way but with different values of σ_3, it is found that for all types of saturated clay, the deviator stress at failure (compressive strength) is entirely independent of the magnitude of σ_3 as shown in Fig. 8.19. The diameters of all the Mohr circles are equal and the Mohr envelope is parallel to the σ-axis indicating that the angle of shearing resistance $\phi_u = 0$. The symbol ϕ_u represents the angle of shearing resistance under undrained conditions. Thus saturated clays behave as purely cohesive materials with the following properties:

$$\phi_u = 0, \text{ and } c_u = \frac{1}{2}(\sigma_1 - \sigma_3) \tag{8.41}$$

where c_u is the symbol used for cohesion under undrained conditions. Eq. (8.41) holds true for the particular case of an unconfined compression test in which $\sigma_3 = 0$. Since this test requires a very simple apparatus, it is often used, especially for field work, as a ready means of measuring the shearing strength of saturated clay, in this case

$$c_u = \frac{q_u}{2}, \quad \text{where} \quad q_u = (\sigma_1 - \sigma_3)_f = (\sigma_1)_f \tag{8.42}$$

Effective Stresses

If during the test, pore-pressures are measured, the effective principal stresses may be written as

$$\sigma_1' = \sigma_1 - u$$

$$\sigma_3' = \sigma_3 - u \tag{8.43}$$

where u is the pore water pressure measured during the test. The effective deviator stress at failure may be written as

$$(\sigma_1' - \sigma_3')_f = (\sigma_1 - u)_f - (\sigma_3 - u)_f = (\sigma_1 - \sigma_3)_f \tag{8.44}$$

Eq. (8.44) shows that the deviator stress is not affected by the pore water pressure. As such the effective stress circle is only shifted from the position of the total stress circle as shown in Fig. 8.19.

Partially Saturated Clay

Tests on partially saturated clay may be carried out either on undisturbed or on remolded soil samples. All the samples shall have the same initial conditions before the test, i.e., they should possess the same water content and dry density. The tests are conducted in the same way as for saturated samples. Each sample is tested under undrained conditions with different all-round pressures σ_3.

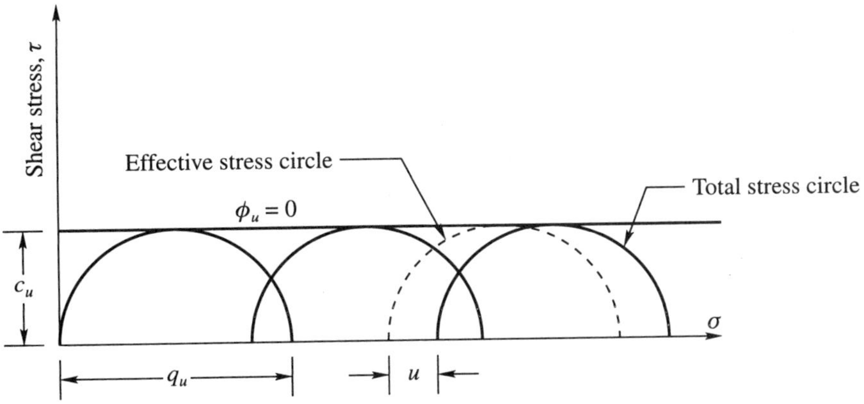

Figure 8.19 Mohr circle for undrained shear test on saturated clay

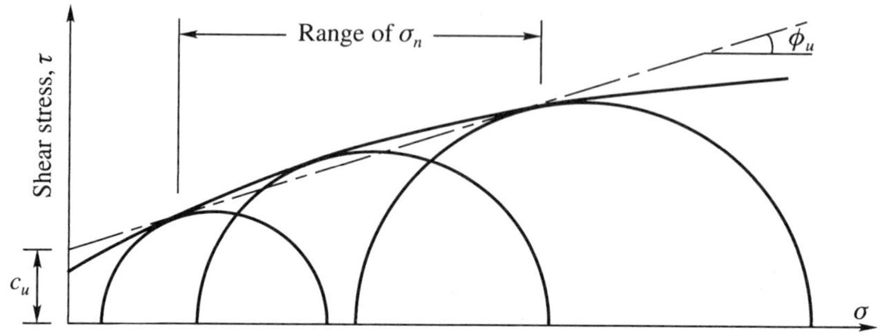

Figure 8.20 Mohr circle for undrained shear tests on partially saturated clay soils

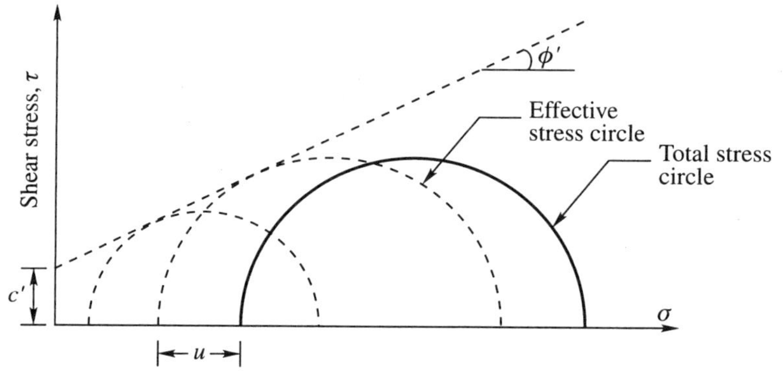

Figure 8.21 Effective stress circles for undrained shear tests on partially saturated clay soils

Mohr circles for three soil samples and the Mohr envelope are shown in Fig. 8.20. Though all the samples had the same initial conditions, the deviator stress increases with the increase in the all-round pressure σ_3 as shown in the figure. This indicates that the strength of the soil increases with increasing values of σ_3. The degree of saturation also increases with the increase in σ_3. The Mohr envelope which is curved at lower values of σ_3 becomes almost parallel to the σ-axis as full saturation is reached. Thus it is not strictly possible to quote single values for the parameters c_u and ϕ_u for partially saturated clays, but over any range of normal pressure σ_n encountered in a practical example, the envelope can be approximated by a straight line and the approximate values of c_u and ϕ_u can be used in the analysis.

Effective Stresses

If the pore pressures are measured during the test, the effective circles can be plotted as shown in Fig. 8.21 and the parameters c' and ϕ' obtained. The envelope to the Mohr circles, when plotted in terms of effective stresses, is linear.

Typical undrained shear strength parameters for partially saturated compacted samples are shown in Table 8.2.

8.21 UNCONFINED COMPRESSION TESTS

The unconfined compression test is a special case of a triaxial compression test in which the all-round pressure $\sigma_3 = 0$ (Fig. 8.22). The tests are carried out only on saturated samples which can stand without any lateral support. The test, is, therefore, applicable to cohesive soils only. The test

Table 8.2 Probable undrained shear strength parameters for partially saturated soils

Types of soil	c_u (tsf)	ϕ_u	c' (tsf)	ϕ'
Sand with clay binder	0.80	23°	0.70	40°
Lean silty clay	0.87	13°	0.45	31°
Clay, moderate plasticity	0.93	9°	0.60	28°
Clay, very plastic	0.87	8°	0.67	22°

is an undrained test and is based on the assumption that there is no moisture loss during the test. The unconfined compression test is one of the simplest and quickest tests used for the determination of the shear strength of cohesive soils. These tests can also be performed in the field by making use of simple loading equipment.

Figure 8.22 Unconfined compression test equipment (Courtesy: Soiltest)

Any compression testing apparatus with arrangement for strain control may be used for testing the samples . The axial load σ_1 may be applied mechanically or pneumatically.

Specimens of height to diameter ratio of 2 are normally used for the tests. The sample fails either by shearing on an inclined plane (if the soil is of brittle type) or by bulging. The vertical stress at any stage of loading is obtained by dividing the total vertical load by the cross-sectional area. The cross-sectional area of the sample increases with the increase in compression. The cross-sectional area A at any stage of loading of the sample may be computed on the basic assumption that the total volume of the sample remains the same. That is

$$A_0 h_0 = Ah$$

where A_0, h_0 = initial cross-sectional area and height of sample respectively.
A, h = cross-sectional area and height respectively at any stage of loading

If Δh is the compression of the sample, the strain is

$$\varepsilon = \frac{\Delta h}{h_0} \quad \text{since} \quad \Delta h = h_0 - h, \text{ we may write}$$

$$A_0 h_0 = A(h_0 - \Delta h)$$

Therefore,
$$A = \frac{A_0 h_0}{h_0 - \Delta h} = \frac{A_0}{1 - \Delta h / h_0} = \frac{A_0}{1 - \varepsilon} \tag{8.45}$$

The average vertical stress at any stage of loading may be written as

$$\sigma_1 = \frac{P}{A} = \frac{P(1-\varepsilon)}{A_0} \tag{8.46}$$

where P is the vertical load at the strain ε.

Using the relationship given by Eq. (8.46) stress-strain curves may be plotted. The peak value is taken as the unconfined compressive strength q_u, that is

$$(\sigma_1)_f = q_u \tag{8.47}$$

The unconfined compression test (UC) is a special case of the unconsolidated-undrained (UU) triaxial compression test (TX-AC). The only difference between the UC test and UU test is that a total confining pressure under which no drainage was permitted was applied in the latter test. Because of the absence of any confining pressure in the UC test, a premature failure through a weak zone may terminate an unconfined compression test. For typical soft clays, premature failure is not likely to decrease the undrained shear strength by more than 5%. Fig 8.23 shows a comparison of undrained shear strength values from unconfined compression tests and from triaxial compression tests on soft-Natsushima clay from Tokyo Bay. The properties of the soil are:

Natural moisture content w_n = 80 to 90%
Liquid limit w_n= 100 to 110%
Plasticity index I_p = 60%

There is a unique relationship between remolded undrained shear strength and the liquidity index, I_p, as shown in Fig. 8.24 (after Terzaghi et al., 1996). This plot includes soft clay soil and silt deposits obtained from different parts of the world.

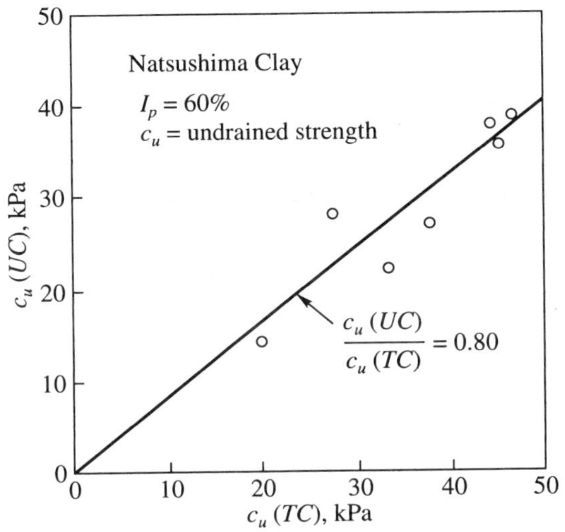

Figure 8.23 Relation between undrained shear strengths from unconfined compression and triaxial compression tests on Natsushima clay (data from Hanzawa and Kishida, 1982)

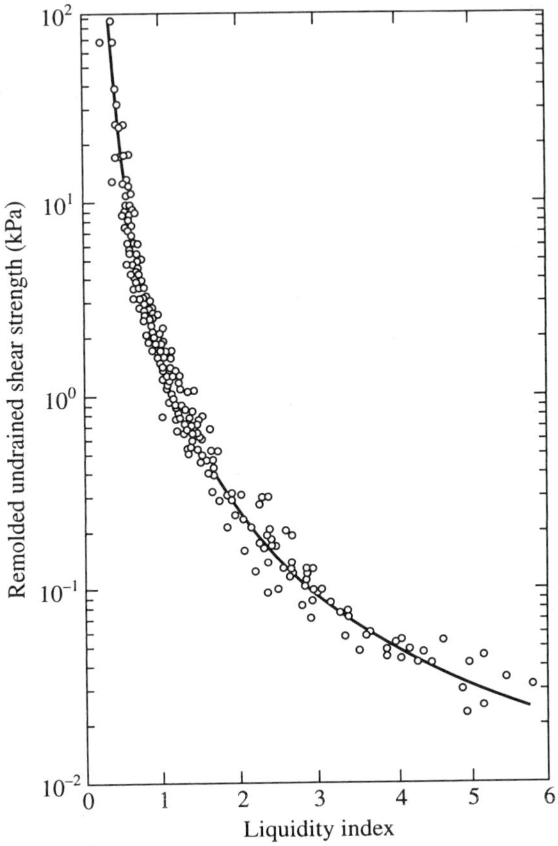

Figure 8.24 Relation between undrained shear strength and liquidity index of clays from around the world (after Terzaghi et al., 1996)

Example 8.9

Boreholes reveal that a thin layer of alluvial silt exists at a depth of 50 ft below the surface of the ground. The soil above this level has an average dry unit weight of 96 lb/ft^3 and an average water content of 30%. The water table is approximately at the surface. Tests on undisturbed samples give the following data: c_u = 1008 lb/ft^2, ϕ_u = 13°, c_d = 861 lb/ft^2, ϕ_d = 23°. Estimate the shearing resistance of the silt on a horizontal plane (a) when the shear stress builds up rapidly, and (b) when the shear stress builds up very slowly.

Solution

Bulk unit weight $\gamma_t = \gamma_d(1+w) = 96 \times 1.3 = 124.8$ lb/ft^3

Submerged uint weight $\gamma_b = 124.8 - 62.4 = 62.4$ lb/ft^3

Total normal pressure at 50 ft depth = 50 × 124.8 = 6240 lb/ft^2

Effective pressure at 50 ft depth = 50 × 62.4 = 3120 lb/ft^2

(a) For rapid build-up, use the properties of the undrained state and total pressure.

At a total pressure of 6240 lb/ft^2

shear strength, $s = c + \sigma \tan \phi$ = 1008 + 6240 tan 13° = 2449 lb/ft^2

(b) For slow build-up, use effective stress properties

At an effective stress of 3120 lb/ft^2,

shear strength = 861 + 3120 tan 23° = 2185 lb/ft^2

Example 8.10

When an undrained triaxial compression test was conducted on specimens of clayey silt, the following results were obtained:

Specimen No.	1	2	3
σ_3 (kN/m^2)	17	44	56
σ_1 (kN/m^2)	157	204	225
u (kN/m^2)	12	20	22

Determine the values of shear parameters considering (a) total stresses and (b) effective stresses.

Solution

(a) Total stresses

For a solution with total stresses, draw Mohr circles C_1, C_2 and C_3 for each of the specimens using the corresponding principal stresses σ_1 and σ_3.

Draw a Mohr envelope tangent to these circles as shown in Fig. Ex. 8.10. Now from the figure

$c = 48$ kN/m^2, $\phi = 15°$

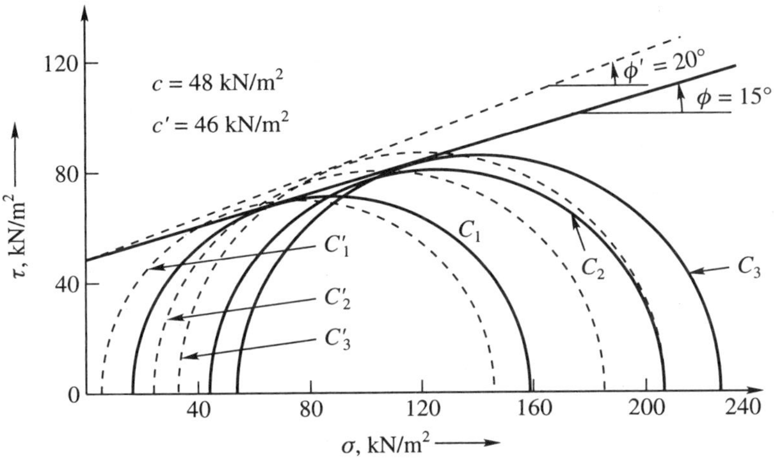

Figure Ex. 8.10

(b) With effective stresses

The effective principal stresses may be found by subtracting the pore pressures u from the total principal stresses as given below.

Specimen No.	1	2	3
$\sigma'_3 = (\sigma_3 - u)$ kN/m²	5	24	34
$\sigma'_1 = (\sigma_1 - u)$ kN/m²	145	184	203

As before draw Mohr circles C'_1, C'_2 and C'_3 for each of the specimens as shown in Fig. Ex. 8.10. Now from the figure

$c' = 46$ kN/m², $\phi' = 20°$

Example 8.11

A soil has an unconfined compressive strength of 120 kN/m². In a triaxial compression test a specimen of the same soil when subjected to a chamber pressure of 40 kN/m² failed at an additional stress of 160 kN/m². Determine:

(i) The shear strength parameters of the soil, (ii) the angle made by the failure plane with the axial stress in the triaxial test.

Solution

There is one unconfined compression test result and one triaxial compression test result. Hence two Mohr circles, C_1, and C_2 may be drawn as shown in Fig. Ex. 8.11. For Mohr circle C_1, $\sigma_3 = 0$ and $\sigma_1 = 120$ kN/m², and for Mohr circle C_2, $\sigma_3 = 40$ kN/m² and $\sigma_1 = (40 + 160) = 200$ kN/m². A common tangent to these two circles is the Mohr envelope which gives

(i) $c = 43$ kN/m² and $\phi = 19°$

(ii) For the triaxial test specimen, A is the point of tangency for Mohr circle C_2 and C is the center of circle C_2. The angle made by AC with the abscissa is equal to twice the angle between the failure plane and the axis of the sample = 2θ. From Fig. Ex. 8.11, $2\theta = 71°$ and $\theta = 35.5°$. The angle made by the failure plane with the σ-axis is $\alpha = 90° - 35.5° = 54.5°$.

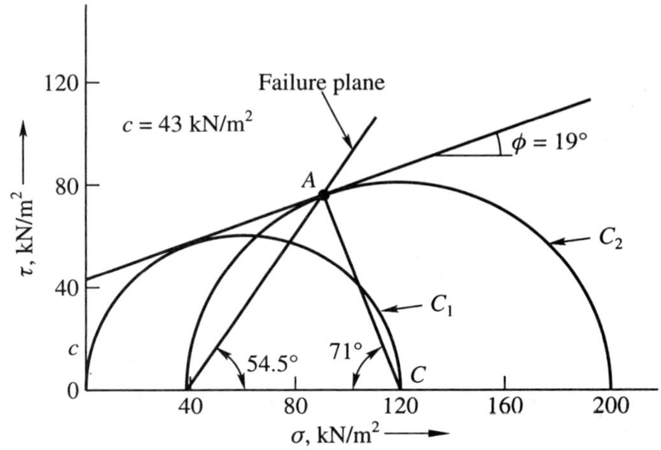

Figure Ex. 8.11

Example 8.12

A cylindrical sample of saturated clay 4 cm in diameter and 8 cm high was tested in an unconfined compression apparatus. Find the unconfined compression strength, if the specimen failed at an axial load of 360 N, when the axial deformation was 8 mm. Find the shear strength parameters if the angle made by the failure plane with the horizontal plane was recorded as 50°.

Solution

Per Eq. (8.46), the unconfined compression strength of the soil is given by

$$\sigma_1 = \frac{P(1-\varepsilon)}{A_o}, \quad \text{where } P = 360\ N$$

$$A_o = \frac{3.14}{4} \times (4)^2 = 12.56 \text{ cm}^2, \quad \varepsilon = \frac{0.8}{8} = 0.1$$

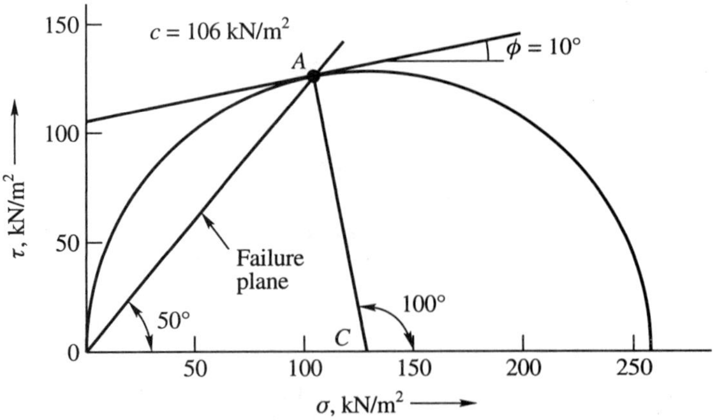

Figure Ex. 8.12

Therefore $\sigma_1 = \dfrac{360(1-0.1)}{12.56} = 25.8\ \mathrm{N/cm^2} = 258\ \mathrm{kN/m^2}$

Now $\phi = 2\alpha - 90°$ (Refer to Fig. 8.12) where $\alpha = 50°$. Therefore $\phi = 2 \times 50 - 90° = 10°$.

Draw the Mohr circle as shown in Fig. Ex. 8.12 ($\sigma_3 = 0$ and $\sigma_1 = 258$ kN/m^2) and from the center C of the circle, draw CA at $2\alpha = 100°$. At point A, draw a tangent to the circle. The tangent is the Mohr envelope which gives

$c = 106$ kN/m^2, and $\phi = 10°$

Example 8.13

An unconfined cylindrical specimen of clay fails under an axial stress of 5040 lb/ft^2. The failure plane was inclined at an angle of 55° to the horizontal. Determine the shear strength parameters of the soil.

Solution

From Eq. (8.20),

$$\sigma_1 = \sigma_3 N_\phi + 2c\sqrt{N_\phi}, \quad \text{where} \quad N_\phi = \tan^2\left(45° + \frac{\phi}{2}\right)$$

since $\sigma_3 = 0$, we have

$$\sigma_1 = 2c\sqrt{N_\phi} = 2c\tan\left(45° + \frac{\phi}{2}\right), \quad \text{where } \sigma_1 = 5040\ \text{lb/ft}^2 \qquad \text{(a)}$$

From Eq. (8.18), the failure angle α is

$$\alpha = 45° + \frac{\phi}{2}, \quad \text{since } \alpha = 55°, \quad \text{we have}$$

$$\phi = (55 - 45) \times 2 = 20°$$

From Eq. (a),

$$c = \frac{\sigma_1}{2\tan\left(45° + \dfrac{\phi}{2}\right)} = \frac{5040}{2\tan 55°} = 1765\ \text{lb/ft}^2$$

Example 8.14

A cylindrical sample of soil having a cohesion of 80 kN/m^2 and an angle of internal friction of 20° is subjected to a cell pressure of 100 kN/m^2.

Determine: (i) the maximum deviator stress ($\sigma_1 - \sigma_3$) at which the sample will fail, and (ii) the angle made by the failure plane with the axis of the sample.

Graphical solution

$\sigma_3 = 100$ kN/m^2, $\phi = 20°$, and $c = 80$ kN/m^2.

A Mohr circle and the Mohr envelope can be drawn as shown in Fig. Ex. 8.14(a). The circle cuts the σ-axis at B (= σ_3), and at E (= σ_1). Now $\sigma_1 = 433$ kN/m^2, and $\sigma_3 = 100$ kN/m^2.

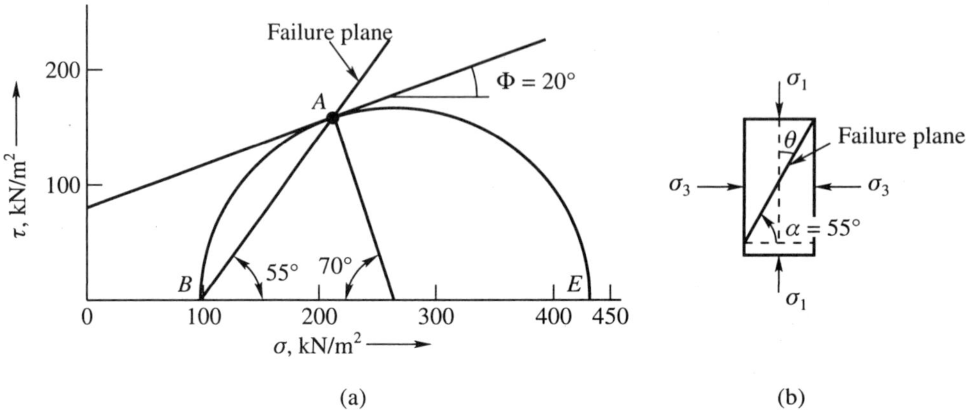

Figure Ex. 8.14

$(\sigma_1 - \sigma_3) = 433 - 100 = 333 \text{ kN/m}^2$.

Analytical solution

Per Eq. (8.20)

$$\sigma_1 = \sigma_3 \tan^2\left(45° + \frac{\phi}{2}\right) + 2c\tan\left(45° + \frac{\phi}{2}\right)$$

Substituting the known values, we have

$\tan(45° + \phi/2) = \tan(45° + 10) = \tan 55° = 1.428$

$\tan^2(45° + \phi/2) = 2.04$.

Therefore,

$\sigma_1 = 100 \times 2.04 + 2 \times 80 \times 1.428 \approx 433 \text{ kN/m}^2$

$(\sigma_1 - \sigma_3) = (433 - 100) = 333 \text{ kN/m}^2$

If θ = angle made by failure planes with the axis of the sample, (Fig. Ex. 8.14(b))

$2\theta = 90 - \phi = 90 - 20 = 70°$ or $\theta = 35°$.

Therefore, the angle made by the failure plane with the σ-axis is

$\alpha = 90 - 35 = 55°$

8.22 CONSOLIDATED-UNDRAINED TEST ON SATURATED CLAY

Normally Consolidated Saturated Clay

If two clay samples 1 and 2 are consolidated under ambient pressures of p_1 and p_2 and are then subjected to undrained triaxial tests without further change in cell pressure, the results may be expressed by the two Mohr circles C_1 and C_2 respectively as shown in Fig. 8.25(b). The failure envelope tangential to these circles passes through the origin and its slope is defined by ϕ_{cu}, the angle of shearing resistance in consolidated undrained tests. If the pore pressures are measured the effective stress Mohr circles C'_1 and C'_2 can also be plotted and the slope of this envelope is ϕ'_{cu}. The effective principal stresses are:

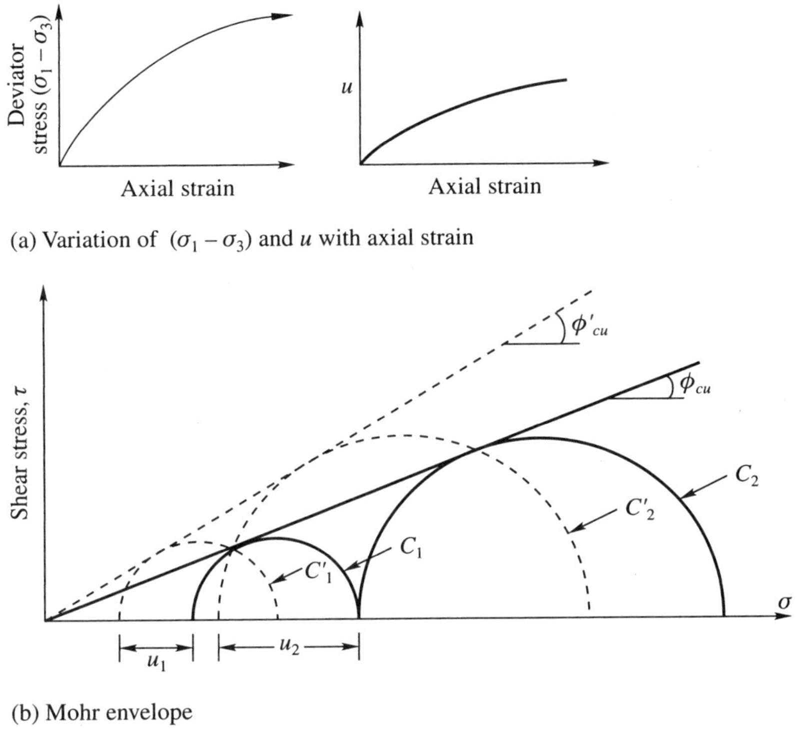

Figure 8.25 Normally consolidated clay under undrained triaxial test

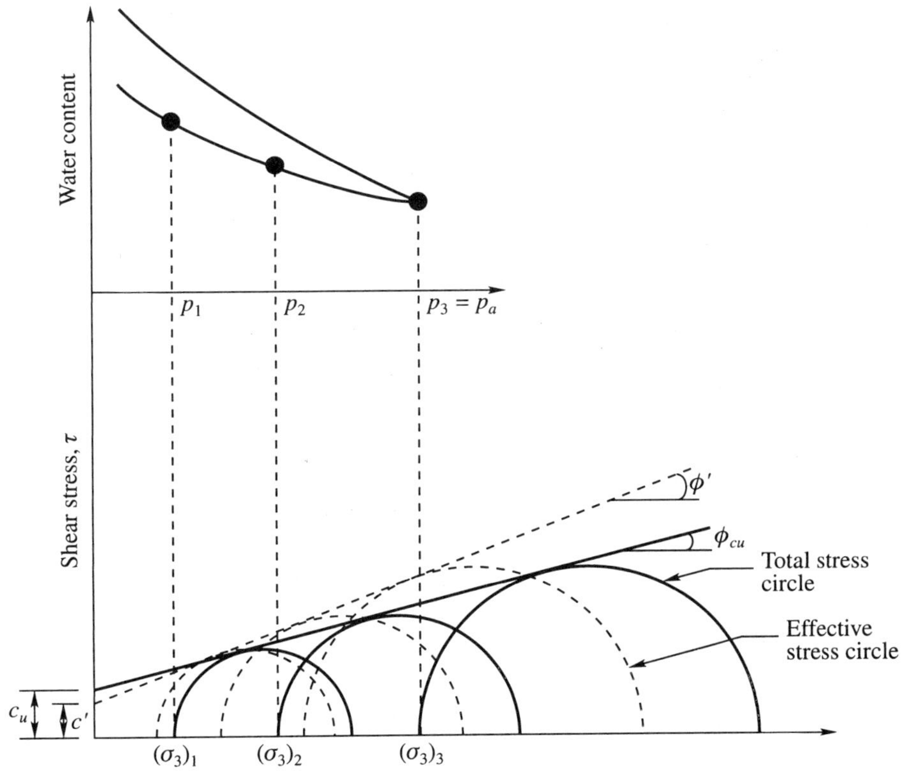

Figure 8.26 Consolidated-undrained tests on saturated overconsolidated clay

$$\sigma'_{11} = \sigma_{11} - u_1;\ \sigma'_{12} = \sigma_{12} - u_2$$

$$\sigma'_{31} = \sigma_{31} - u_1;\ \sigma'_{32} = \sigma_{32} - u_2$$

where u_1 and u_2 are the pore water pressures for the samples 1 and 2 respectively.

It is an experimental fact that the envelopes to the total and effective stress circles are linear. Fig. 8.25(a) shows the nature of the variation on the deviator stress ($\sigma_1 - \sigma_3$) and the pore water pressure u in the specimen during the test with the axial strain. The pore water pressure builds up during shearing with a corresponding decrease in the volume of the sample.

Overconsolidated Clay

Let a saturated sample 1 be consolidated under an ambient pressure p_a and then allowed to swell under the pressure p_1. An undrained triaxial test is carried out on this sample under the all-round pressure $p_1(=\sigma_{31})$. Another sample 2 is also consolidated under the same ambient pressure p_a and allowed to swell under the pressure $p_2(=\sigma_{32})$. An undrained triaxial test is carried out on this sample under the same all-round pressure p_2. The two Mohr circles are plotted and the Mohr envelope tangential to the circles is drawn as shown in Fig. 8.26. The shear strength parameters are c_u and ϕ_{cu}. If pore water pressure is measured, effective stress Mohr circles may be plotted as shown in the figure. The strength parameters for effective stresses are represented by c' and ϕ'.

8.23 CONSOLIDATED-DRAINED SHEAR STRENGTH TEST

In drained triaxial tests the soil is first consolidated under an ambient pressure p_a and then subjected to an increasing deviator stress until failure occurs, the rate of strain being controlled in such a way that at no time is there any appreciable pore-pressure in the soil. Thus at all times the applied stresses are effective, and when the stresses at failure are plotted in the usual manner, the failure envelope is directly expressed in terms of effective stresses. For normally consolidated clays and for sands the envelope is linear for normal working stresses and passes through the origin as shown in Fig. 8.27. The failure criterion for such soils is therefore the angle of shearing resistance in the drained condition ϕ_d.

The drained strength is

$$\frac{1}{2}(\sigma_1 - \sigma_3)_f = \frac{p \sin\phi_d}{1 - \sin\phi_d} \tag{8.48}$$

Eq. (8.48) is obtained from Eq. (8.38)

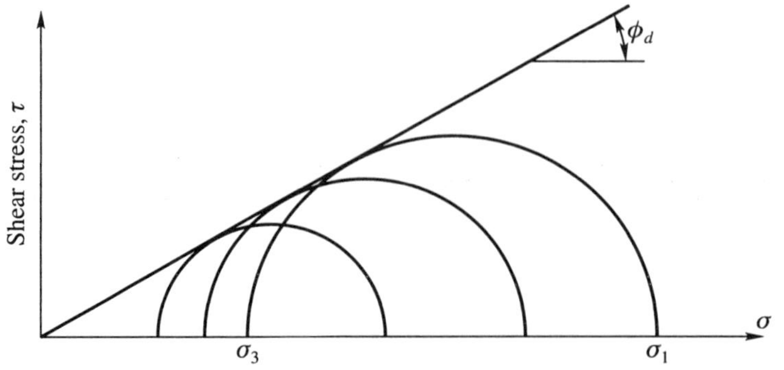

Figure 8.27 Drained tests on normally consolidated clay samples

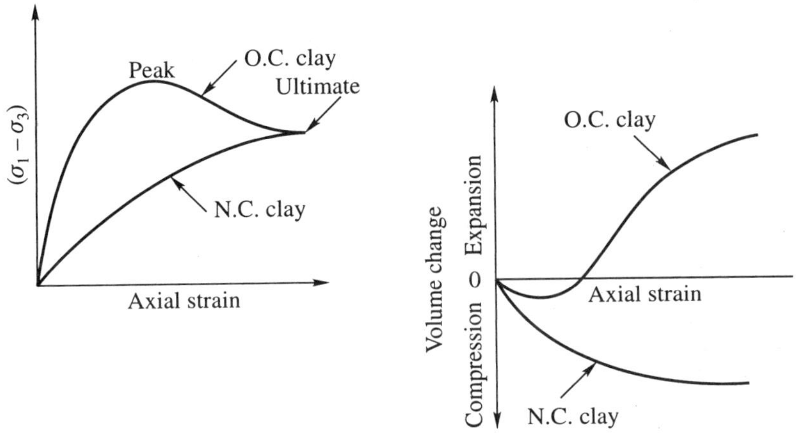

(a) Variation of $(\sigma_1 - \sigma_3)$ with axial strain

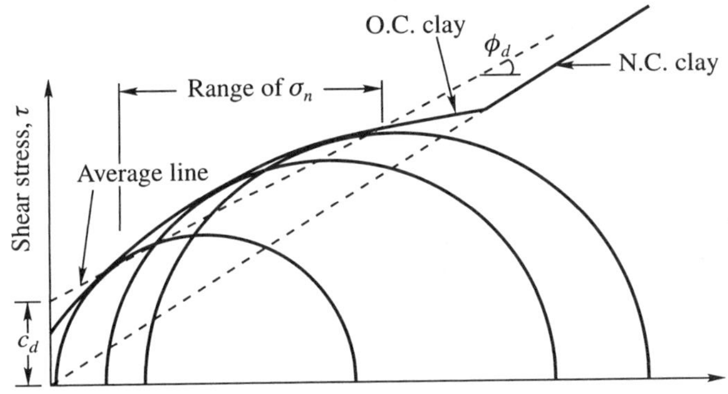

(b) Mohr envelope

Figure 8.28 Drained tests on overconsolidated clays

For overconsolidated clays, the envelope intersects the axis of zero pressure at a value c_d. The apparent cohesion in the drained test and the strength are given by the expression.

$$\frac{1}{2}(\sigma_1 - \sigma_3)_f = \frac{c_d \cos\phi_d + p\sin\phi_d}{1 - \sin\phi_d} \tag{8.49}$$

The Mohr envelope for overconsolidated clays is not linear as may be seen in Fig. 8.28(b). An average line is to be drawn within the range of normal pressure σ_n. The shear strength parameters c_d and ϕ_d are referred to this line.

Since the stresses in a drained test are effective, it might be expected that a given ϕ_d would be equal to ϕ' as obtained from undrained tests with pore-pressure measurement. In normally consolidated clays and in loose sands the two angles of shearing resistance are in fact closely equal since the rate of volume change in such materials at failure in the drained test is approximately zero and there is no volume change throughout an undrained test on saturated soils. But in dense sands and heavily overconsolidated clays there is typically a considerable rate of positive volume change at failure in drained tests, and work has to be done not only in overcoming the shearing resistance of the soils, but also in increasing the volume of the specimen against the ambient pressure. Yet in undrained tests on the same soils, the volume change is zero and consequently ϕ_d for dense sands and heavily overconsolidated clays is greater than ϕ'. Fig. 8.28(a) shows the nature of variation of the deviator stress with axial strain. During the application of the deviator stress, the volume of the specimen gradually reduces for normally consolidated clays. However, overconsolidated clays go through some reduction of volume initially but then expand.

8.24 PORE PRESSURE PARAMETERS UNDER UNDRAINED LOADING

Soils in nature are at equilibrium under their overburden pressure. If the same soil is subjected to an instantaneous additional loading, there will be development of pore pressure if drainage is delayed under the loading. The magnitude of the pore pressure depends upon the permeability of the soil, the manner of application of load, the stress history of the soil, and possibly many other factors. If a load is applied slowly and drainage takes place with the application of load, there will practically be no increase of pore pressure. However, if the hydraulic conductivity of the soil is quite low, and if the loading is relatively rapid, there will not be sufficient time for drainage to take place. In such cases, there will be an increase in the pore pressure in excess of the existing hydrostatic pressure. It is therefore necessary many times to determine or estimate the excess pore pressure for the various types of loading conditions. Pore pressure parameters are used to express the response of pore pressure to changes in total stress *under undrained conditions*. Values of the parameters may be determined in the laboratory and can be used to predict pore pressures in the field under similar stress conditions.

Pore Pressure Parameters Under Triaxial Test Conditions

A typical stress application on a cylindrical element of soil under triaxial test conditions is shown in Fig. 8.29 ($\Delta\sigma_1 > \Delta\sigma_3$). Δu is the increase in the pore pressure without drainage. From Fig. 8.29, we may write

$\Delta u_3 = B\Delta\sigma_3$, $\Delta u_1 = AB(\Delta\sigma_1 - \Delta\sigma_3)$, therefore,

$$\Delta u = \Delta u_1 + \Delta u_3 = B[\Delta\sigma_3 + A(\Delta\sigma_1 - \Delta\sigma_3)] \tag{8.50}$$

$$\text{or } \Delta u = B\Delta\sigma_3 + \bar{A}(\Delta\sigma_1 - \Delta\sigma_3) \tag{8.51}$$

where, $\bar{A} = AB$

for saturated soils $B = 1$, so

$$\Delta u = \Delta\sigma_3 + A(\Delta\sigma_1 - \Delta\sigma_3) \tag{8.52}$$

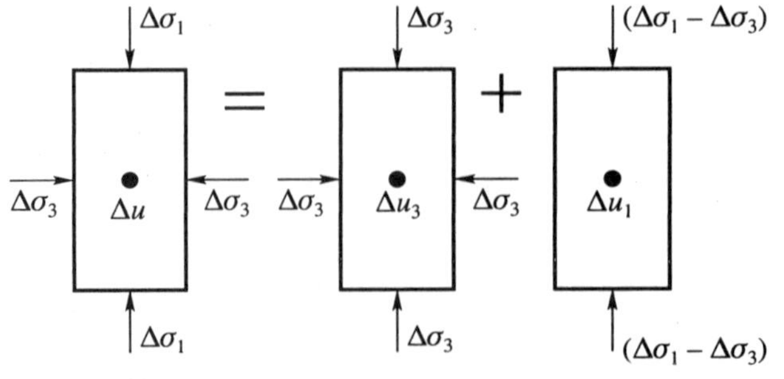

Figure 8.29 Excess water pressure under triaxial test conditions

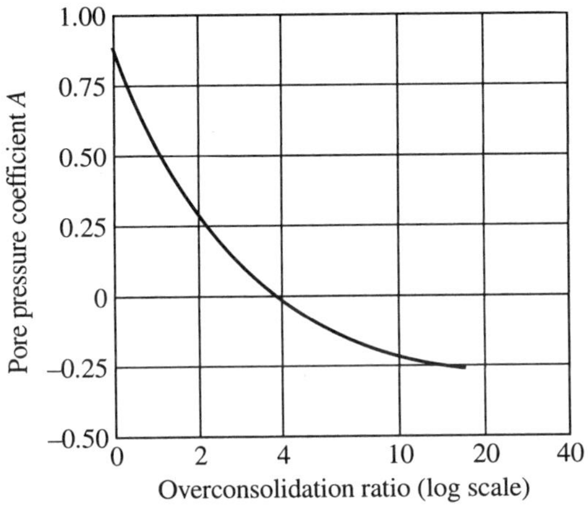

Figure 8.30 Relationship between overconsolidation ratio and pore pressure coefficient *A*

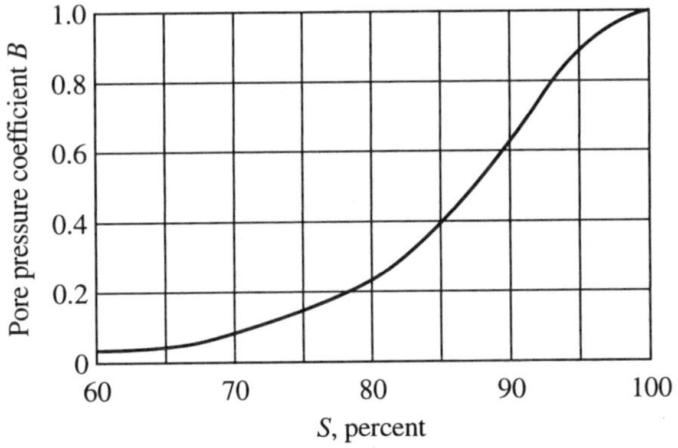

Figure 8.31 Typical relationship between *B* and degree of saturation *S*.

where *A* and *B* are called pore pressure parameters. The variation of *A* under a failure condition (A_f) with the overconsolidation ratio, O_{CR}, is given in Fig. 8.30. Some typical values of A_f are given in Table 8.3. The value of *B* varies with the degree of saturation as shown in Fig. 8.31.

Table 8.3 Typical values of A_f

Type of Soil	Volume change	A_f
Highly sensitive clay	large contraction	+ 0.75 to + 1.5
Normally consolidated clay	contraction	+ 0.5 to + 1.0
Compacted sandy clay	slight contraction	+ 0.25 to + 0.75
Lightly overconsolidated clay	none	+ 0.00 to + 0.5
Compacted clay gravel	expansion	– 0.25 to + 0.25
Heavily overconsolidated clay	expansion	– 0.5 to 0

8.25 VANE SHEAR TESTS

From experience it has been found that the vane test can be used as a reliable in-situ test for determining the shear strength of soft-sensitive clays. It is in deep beds of such material that the vane test is most valuable, for the simple reason that there is at present no other method known by which the shear strength of these clays can be measured. Repeated attempts, particularly in Sweden, have failed to obtain undisturbed samples from depths of more than about 10 meters in normally consolidated clays of high sensitivity even using the most modern form of thin-walled piston samplers. In these soils the vane is indispensable. The vane should be regarded as a method to be used under the following conditions:

1. The clay is normally consolidated and sensitive.
2. Only the undrained shear strength is required.

It has been determined that the vane gives results similar to those obtained from unconfined compression tests on undisturbed samples.

The soil mass should be in a saturated condition if the vane test is to be applied. The vane test cannot be applied to partially saturated soils to which the angle of shearing resistance is not zero.

Description of the Vane

The vane consists of a steel rod having at one end four small projecting blades or vanes parallel to its axis, and situated at 90° intervals around the rod. A post hole borer is first employed to bore a hole up to a point just above the required depth. The rod is pushed or driven carefully until the vanes are embedded at the required depth. At the other end of the rod above the surface of the ground a torsion head is used to apply a horizontal torque and this is applied at a uniform speed of about 0.1° per sec until the soil fails, thus generating a cylinder of soil. The area consists of the peripheral surface of the cylinder and the two round ends. The first moment of these areas divided by the applied moment gives the unit shear value of the soil. Fig. 8.32(*a*) gives a diagrammatic sketch of a field vane.

Determination of Cohesion or Shear Strength of Soil

Consider the cylinder of soil generated by the blades of the vane when they are inserted into the undisturbed soil in-situ and gradually turned or rotated about the axis of the shaft or vane axis. The turning moment applied at the torsion head above the ground is equal to the force multiplied by the eccentricity.

Let the force applied = P eccentricity (lever arm) = x units

Turning moment = Px

The surface resisting the turning is the cylindrical surface of the soil and the two end faces of the cylinder.

Therefore,

resisting moment = $(2\pi r \times L \times c_u \times r + 2\pi r^2 \times c_u \times 0.67r) = 2\pi r^2 c_u(L + 0.67r)$

where r = radius of the cylinder and c_u the undrained shear strength.

At failure the resisting moment of the cylinder of soil is equal to the turning moment applied at the torsion head.

Therefore, $Px = 2\pi r^2 c_u(L + 0.67r)$

$$c_u = \frac{Px}{2\pi r^2(L+0.67r)} \tag{8.53}$$

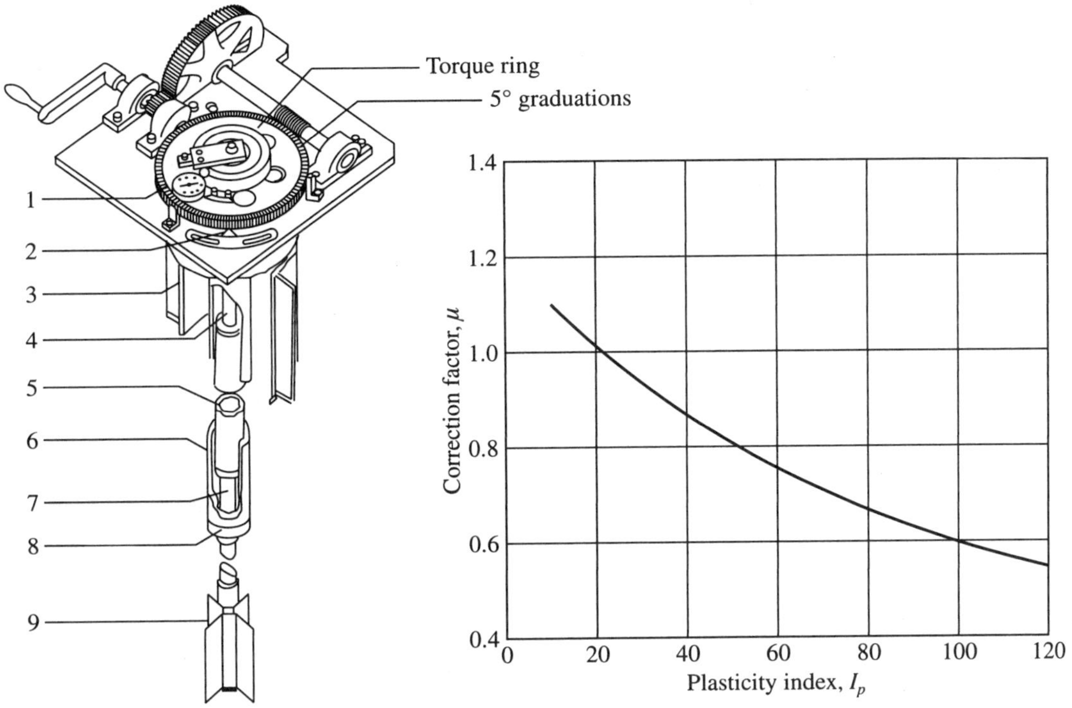

1. Straingauge for reading torque
2. Rotation indicator
3. 8-in casing with side fins for anchoring torque assembly
4. Torque rod
5. A-rod for applying torque to vane. Made up in 5-ft lengths
6. BX casing for housing torque rod and A rod
7. Vane rod
8. BX-casing-point containing bearing and water seals for vane rod
9. Vane varying sizes
   2 in dia by 4 in
   3 in dia by 6 in
   4 in dia by 8 in

(a) (b)

Figure 8.32 Vane shear test (a) diagrammatic sketch of a field vane, (b) correction factor μ (Bjerrum, 1973)

The standard dimensions of field vanes as recommended by ASTM (1994) are given in Table 8.4.

Some investigators believe that vane shear tests in cohesive soil gives a values of the shear strength about 15 per cent greater than in unconfined compression tests. There are others who believe that vane tests give lower values.

Table 8.4 Recommended dimensions of field vanes (ASTM, 1994)

Casing size	Height, mm (L)	Diameter, mm (d)	Blade thickness mm	Diameter of rod mm
AX	76.2	38.1	1.6	12.7
BX	101.6	50.8	1.6	12.7
NX	127.0	63.5	3.2	12.7

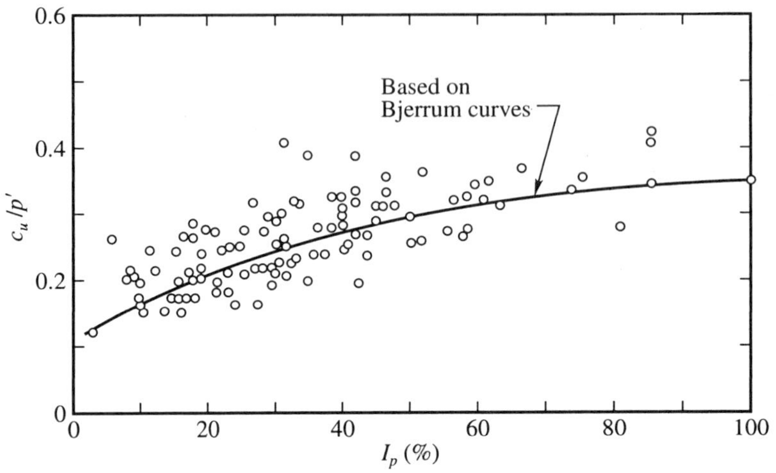

Figure 8.33 Undrained shear strengths from field vane tests on inorganic soft clays and silts (after Tavenas and Leroueil, 1987)

Bjerrum (1973) back computed a number of embankment failures on soft clay and concluded that the vane shear strength tended to be too high. Fig. 8.32(b) gives correction factors for the field vane test as a function of plasticity index, I_p (Ladd et al., 1977). We may write

$$c_u \text{ (field)} = \mu c_u \text{ (vane)} \tag{8.54}$$

where μ is the correction factor (Fig. 8.32b).

Fig. 8.33 give relationships between plasticity index I_p and c_u/p' where c_u is the undrained shear strength obtained by field vane and p' the effective overburden pressure. This plot is based on comprehensive test data compiled of Tavenas and Leroueil (1987). Necessary correction factors have been applied to the data as per Fig. 8.32 (b) before plotting.

8.26 OTHER METHODS FOR DETERMINING UNDRAINED SHEAR STRENGTH OF COHESIVE SOILS

We have discussed in earlier sections three methods for determining the undrained shear strength of cohesive soils. They are

1. Unconfined compression test
2. UU triaxial test
3. Vane shear test

In this section two more methods are discussed. The instruments used for this purpose are

1. Torvane (TV)
2. Pocket penetrometer (PP)

Torvane

Torvane, a modification of the vane, is convenient for investigating the strength of clays in the walls of test pits in the field or for rapid scanning of the strength of tube or split spoon samples. Fig 8.34(a) gives a diagrammatic sketch of the instrument. Figure 8.34(b) gives a photograph of the same. The vanes are pressed to their full depth into the clay below a flat surface, whereupon a torque is applied through a calibrated spring until the clay fails along the cylindrical surface

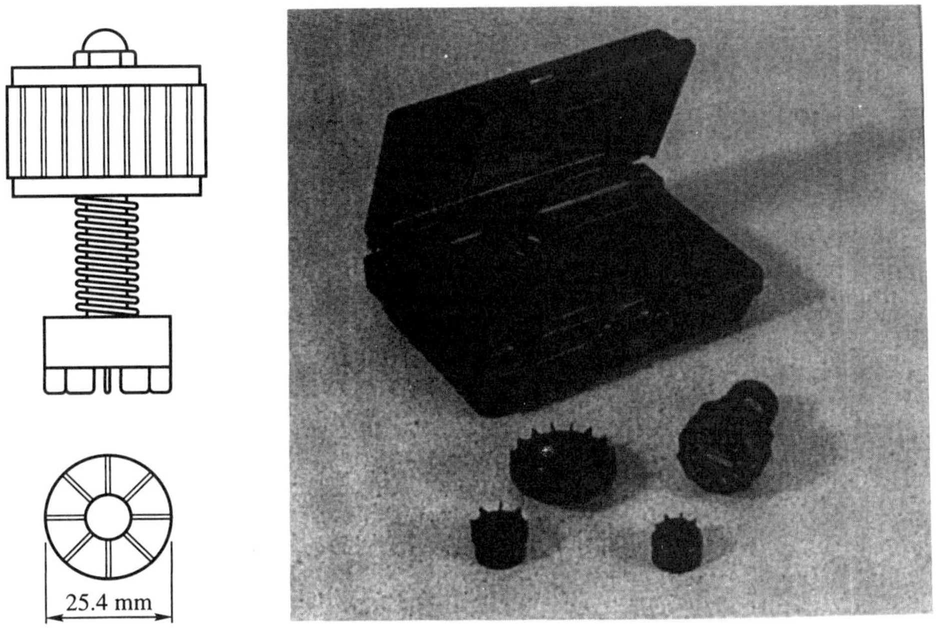

(a) (b)

Figure 8.34 Torvane shear device (a) a diagrammatic sketch, and (b) a photograph (Courtesy: Soiltest)

circumscribing the vanes and simultaneously along the circular surface constituting the base of the cylinder. The value of the shear strength is read directly from the indicator on the calibrated spring.

Specification for three sizes of vanes are given below (Holtz et al., 1981)

Diameter (mm)	Height of vane (mm)	Maximum shear strength (kPa)
19	3	250
25	5	100 (standard)
48	5	20

Pocket Penetrometer

Figure 8.35 shows a pocket penetrometer (Holtz et al., 1981) which can be used to determine undrained shear strength of clay soils both in the laboratory and in the field. The procedure consists in pushing the penetrometer directly into the soil and noting the strength marked on the calibrated spring.

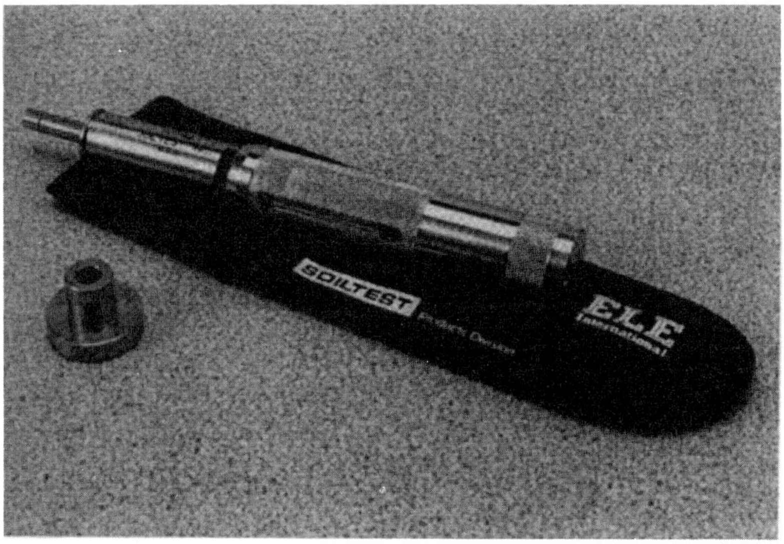

Figure 8.35 Pocket penetrometer (PP), a hand-held device which indicates unconfined compressive strength (Courtesy: Soiltest, USA)

8.27 THE RELATIONSHIP BETWEEN UNDRAINED SHEAR STRENGTH AND EFFECTIVE OVERBURDEN PRESSURE

It has been discussed in previous sections that the shear strength is a function of effective consolidation pressure. If a relationship between undrained shear strength, c_u, and effective consolidation pressure p' can be established, we can determine c_u if p' is known and vice versa. If a soil stratum in nature is normally consolidated the existing effective overburden pressure p_0' can be determined from the known relationship. But in overconsolidated natural clay deposits, the preconsolidation pressure p_c' is unknown which has to be estimated by any one of the available methods. If there is a relationship between p_c' and c_u, c_u can be determined from the known value of p_c'. Alternatively, if c_u is known, p_c' can be determined. Some of the relationships between c_u and p' are presented below. A typical variation of undrained shear strength with depth is shown in Fig. 8.36 for both normally consolidated and heavily overconsolidated clays. The higher shear strength as shown in Fig. 8.36(a) for normally consolidated clays close to the ground surface is due to desiccation of the top layer of soil.

Skempton (1957) established a relationship which may be expressed as

$$\frac{c_u}{p'} = 0.10 + 0.004\, I_p \tag{8.55}$$

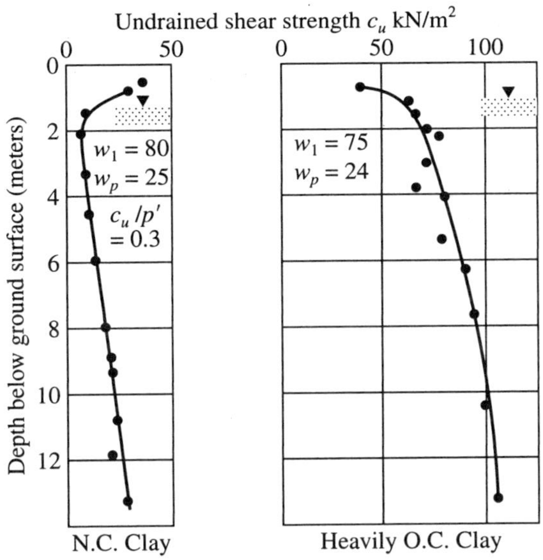

Figure 8.36 Typical variations of undrained shear strength with depth (After Bishop and Henkel, 1962)

He found a close correlation between c_u/p' and I_p as illustrated in Fig. 8.37. Though the Eq. (8.55) was originally meant for normally consolidated clays, it has been used for overconsolidated clays also. p' may be replaced by p_0' as the existing effective overburden pressure for normally consolidated clays, and by p_c' for overconsolidated clays. Peck et al., (1974) has extensively used this relationship for determining preconsolidation pressure p_c'. Eq. (8.55) may also be used for determining p_c' indirectly. If p_c' can be determined independently, the value of the undrained shear strength c_u for overconsolidated clays can be obtained from Eq. (8.55). The values of c_u so obtained may be checked with the values determined in the laboratory on undisturbed samples of clay.

Bjerrum and Simons (1960) proposed a relationship between c_u/p' and plasticity index I_p *as*

$$\frac{c_u}{p'} = 0.45\,(I_p)^{\frac{1}{2}} \text{ for } I_p > 5\% \tag{8.56}$$

The scatter is expected to be of the order of ± 25 percent of the computed value.

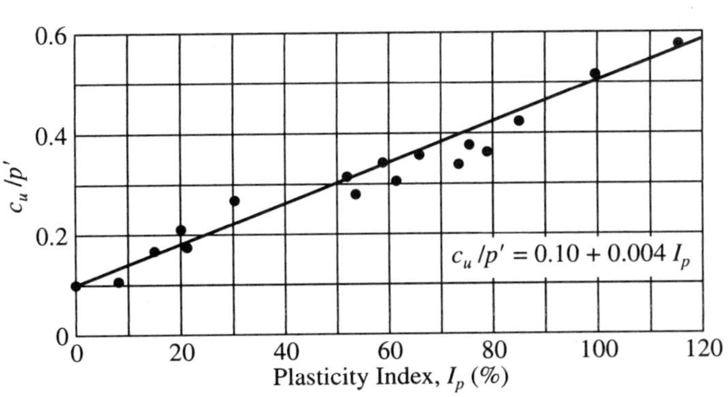

Figure 8.37 Relation between c_u/p' and plasticity index

Another relationship expressed by them is

$$\frac{c_u}{p'} = 0.18\,(I_l)^{-\frac{1}{2}} \quad \text{for } I_l > 0.5 \tag{8.57}$$

where I_l is the liquidity index. The scatter is found to be of the order of ± 30 percent.

Karlsson and Viberg (1967) proposed a relationship as

$$\frac{c_u}{p'} = 0.005 w_l \quad \text{for } w_l > 20 \text{ percent} \tag{8.58}$$

where w_l is the liquid limit in percent. The scatter is of the order of ± 30 percent.

The engineer has to use judgment while selecting any one of the forms of relationships mentioned above.

c_u/p' Ratio Related to Overconsolidation Ratio p_c'/p_0'

Ladd and Foott (1974) presented a non-dimensional plot (Fig. 8.38) giving a relationship between a non-dimensional factor N_f and overconsolidation ratio *OCR*. Figure 8.38 is based on direct simple shear tests carried out on five clays from different origins. The plot gives out a trend but requires further investigation.

The non-dimensional factor N_f is defined as

$$N_f = \frac{(c_u/p_0')_{NC}}{(c_u/p_0')_{OC}} \tag{8.59}$$

where p_0' = existing overburden pressure

OC = overconsolidated

NC = normally consolidated

From the plot in Fig. 8.38 the shear strength c_u of overconsolidated clay can be determined if p_0' and $(c_u/p_0')_{NC}$ are known.

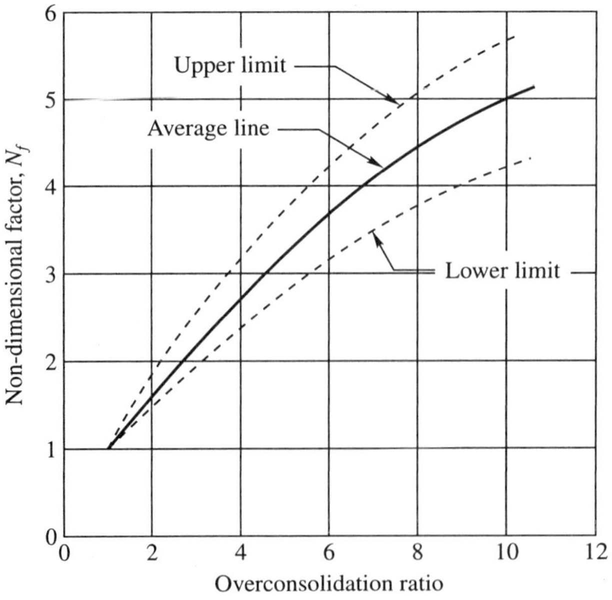

Figure 8.38 Relationship between N_f and overconsolidation ratio *OCR* (Ladd and Foott, 1974)

Example 8.15

A normally consolidated clay was consolidated under a stress of 3150 lb/ft^2, then sheared undrained in axial compression. The principal stress difference at failure was 2100 lb/ft^2, and the induced pore pressure at failure was 1848 lb/ft^2. Determine (a) the Mohr-Coulomb strength parameters, in terms of both total and effective stresses analytically. (b) compute $(\sigma_1/\sigma_3)_f$ and $(\sigma'_1/\sigma'_3)_f$, and (c) determine the theoretical angle of the failure plane in the specimen.

Solution

The parameters required are: effective parameters c' and ϕ', and total parameters c and ϕ.

(a) Given $\sigma_{3f} = 3150$ lb/ft^2, and $(\sigma_1 - \sigma_3)_f = 2100$ lb/ft^2. The total principal stress at failure σ_{1f} is obtained from

$\sigma_{1f} = (\sigma_1 - \sigma_3)_f + \sigma_{3f} = 2100 + 3150 = 5250$ lb/ft^2

Effective $\sigma'_{1f} = \sigma_{1f} - u_f = 5250 - 1848 = 3402$ lb/ft^2

$\sigma'_{3f} = \sigma_{3f} - u_f = 3150 - 1848 = 1302$ lb/ft^2

Now $\sigma_1 = \sigma_3 \tan^2(45° + \phi/2) + 2c\tan(45° + \phi/2)$

Since the soil is normally consolidated, $c = 0$. As such

$$\frac{\sigma_1}{\sigma_3} = \tan^2(45° + \phi/2) = \frac{1+\sin\phi}{1-\sin\phi}, \quad \text{or } \sin\phi = \frac{\sigma_1 - \sigma_3}{\sigma_1 + \sigma_3}$$

$$\text{Total } \phi = \sin^{-1}\frac{2100}{5250+3150} = \sin^{-1}\frac{2100}{8400} = 14.5°$$

$$\text{Effective } \phi' = \sin^{-1}\frac{2100}{3402+1302} = \sin^{-1}\frac{2100}{4704} = 26.5°$$

(b) The stress ratios at failure are

$$\frac{\sigma_1}{\sigma_3} = \frac{5250}{3150} = 1.67, \quad \frac{\sigma'_1}{\sigma'_3} = \frac{3402}{1302} = 2.61$$

(c) From Eq. (8.18)

$$\alpha_f = 45° + \frac{\phi'}{2} = 45° + \frac{26.5}{2} = 58.25°$$

The above problem can be solved graphically by constructing a Mohr-Coulomb envelope.

Example 8.16

The following results were obtained at failure in a series of consolidated-undrained tests, with pore pressure measurement, on specimens of saturated clay. Determine the values of the effective stress parameters c' and ϕ' by drawing Mohr circles.

σ_3 kN/m^2	$\sigma_1 - \sigma_3$ kN/m^2	u_w kN/m^2
150	192	80
300	341	154
450	504	222

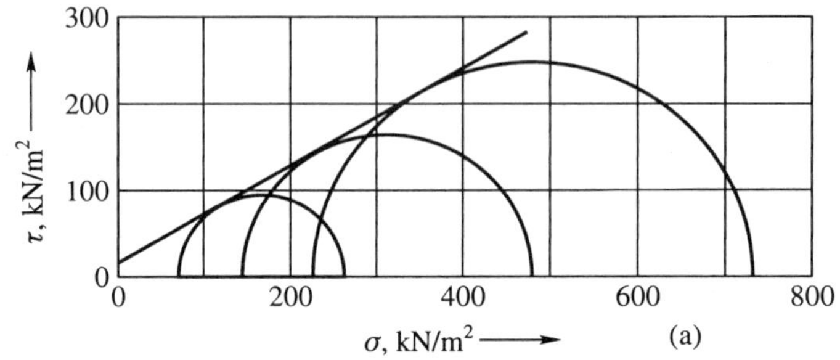

Figure Ex. 8.16

Solution

The values of the effective principal stresses σ'_1 and σ'_3 at failure are tabulated below

σ_3 kN/m²	σ_1 kN/m²	σ'_3 kN/m²	σ'_1 kN/m²
150	342	70	262
300	641	146	487
450	954	228	732

The Mohr circles in terms of effective stresses and the failure envelope are drawn in Fig. Ex. 8.16. The shear strength parameters as measured are :

$c' = 16$ kN/m²; $\phi' = 29°$

Example 8.17

The following results were obtained at failure in a series of triaxial tests on specimens of a saturated clay initially 38 mm in diameter and 76 mm long. Determine the values of the shear strength parameters with respect to (a) total stress, and (b) effective stress.

Type of test	σ_3 kN/m²	Axial load (N)	Axial compression (mm)	Volume change (cm³)
(a) Undrained	200	222	9.83	–
	400	215	10.06	–
	600	226	10.28	–
(b) Drained	200	467	10.81	6.6
	400	848	12.26	8.2
	600	1265	14.17	9.5

Solution

The principal stress difference at failure in each test is obtained by dividing the axial load by the cross-sectional area of the specimen at failure. The corrected cross-sectional area is calculated from Eq. (8.45). There is, of course, no volume change during an undrained test on a saturated clay. The initial values of length, area and volume for each specimen are $h_0 = 76$ mm, $A_0 = 11.35$ cm²; $V_0 = 86.0$ cm³ respectively.

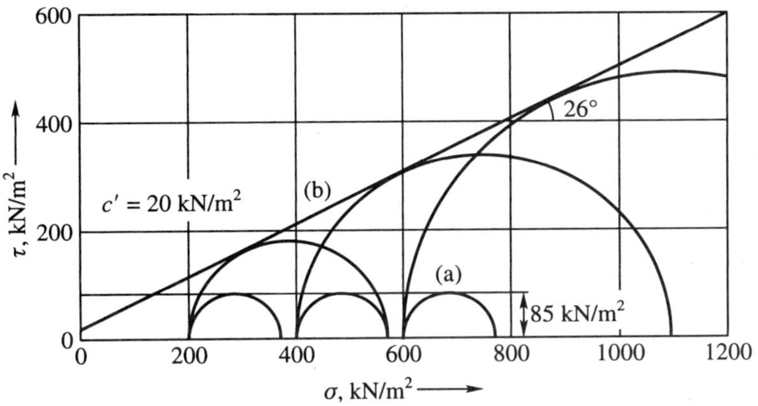

Figure Ex. 8.17

The Mohr circles at failure and the corresponding failure envelopes for both series of tests are shown in Fig. Ex. 8.17. In both cases the failure envelope is the line nearest to the common tangent to the Mohr circles. The total stress parameters representing the undrained strength of the clay are:

$c_u = 85$ kN/m^2; $\phi_u = 0$

The effective stress parameters, representing the drained strength of the clay, are:

$c' = 20$ kN/m^2; $\phi = 26°$

	σ_3 kN/m^2	$\Delta h/h_0$	$\Delta V/V_0$	Area (corrected) cm^2	$\sigma_1 - \sigma_3$ kN/m^2	σ_1 kN/m^2
a	200	0.129		13.04	170	370
	400	0.132		13.09	160	564
	600	0.135	-	13.12	172	772
b	200	0.142	0.077	12.22	382	582
	400	0.161	0.095	12.25	691	1091
	600	0.186	0.110	12.40	1020	1620

Example 8.18

An embankment is being constructed of soil whose properties are $c' = 1071$ lb/ft^2, $\phi' = 21°$ (all effective stresses), and $\gamma = 99.85$ lb/ft^3. The pore pressure parameters as determined from triaxial tests are $A = 0.5$, and $B = 0.9$. Find the shear strength of the soil at the base of the embankment just after the height of fill has been raised from 10 ft to 20 ft. Assume that the dissipation of pore pressure during this stage of construction is negligible, and that the lateral pressure at any point is one-half of the vertical pressure.

Solution

The equation for pore pressure is [Eq. (8.51)]

$$\Delta u = B\left[\Delta\sigma_3 + A(\Delta\sigma_1 - \Delta\sigma_3)\right]$$

$\Delta\sigma_1$ = Vertical pressure due to 10 ft of fill = 10 × 99.85 = 998.5 lb/ft^2

$$\Delta\sigma_3 = \frac{998.5}{2} = 499.25 \text{ lb/ft}^2$$

Therefore, $\Delta u = 0.9[499.25 + 0.5 \times 499.25] = 674 \text{ lb/ft}^2$

Original pressure, $\sigma_1 = 10 \times 99.85 = 998.5 \text{ lb/ft}^2$

Therefore, $\sigma' = \sigma_1 + \Delta\sigma_1 - \Delta u$

$$= 998.5 + 998.5 - 674 = 1323 \text{ lb/ft}^2$$

Shear strength, $s = c' + \sigma' \tan\phi' = 1071 + 1323 \tan 21° = 1579 \text{ lb/ft}^2$

Example 8.19

At a depth of 6 m below the ground surface at a site, a vane shear test gave a torque value of 6040 N-cm. The vane was 10 cm high and 7 cm across the blades. Estimate the shear strength of the soil.

Solution

Per Eq. (8.53)

$$c_u = \frac{\text{Torque } (T)}{2\pi r^2 \left[L + (2/3)r\right]}$$

where $T = 6040$ N-cm, $L = 10$ cm, $r = 3.5$ cm.
substituting,

$$c_u = \frac{6040}{2 \times 3.14 \times 3.5^2 (10 + 0.67 \times 3.5)} = 6.4 \text{ N/cm}^2 = 64 \text{ kN/m}^2$$

Example 8.20

A vane 11.25 cm long, and 7.5 cm in diameter was pressed into soft clay at the bottom of a borehole. Torque was applied to cause failure of soil. The shear strength of clay was found to be 37 kN/m^2. Determine the torque that was applied.

Solution

From Eq. (8.53),

Torque, $T = c_u [2\pi r^2 (L + 0.67r)]$ where $c_u = 37 \text{ kN/m}^2 = 3.7 \text{ N/cm}^2$

$$= 3.7[2 \times 3.14 \times (3.75)^2 (11.25 + 0.67 \times 3.75)] = 4500 \text{ N-cm}$$

8.28 GENERAL COMMENTS

One of the most important and the most controversial engineering properties of soil is its shear strength. The behavior of soil under external load depends on many factors such as arrangement of particles in the soil mass, its mineralogical composition, water content, stress history and many others. The types of laboratory tests to be performed on a soil sample depends upon the type of soil and its drainage condition during the application of external loads in the field. It is practically very difficult (if not impossible) to obtain undisturbed samples of granular soils from the field for laboratory tests. In such cases laboratory tests on remolded samples are mostly of academic interest. The angle of shearing resistance of granular soils is normally determined by the relationships established between ϕ and penetration resistance in the field. The accuracy of this method is sufficient for all practical purposes. The penetrometer used may be standard penetration equipment or static cone penetrometer. Shear strength tests on cohesive soils depend mostly on the accuracy with which undisturbed samples can be obtained from the field.

Undisturbed samples are extracted in sampling tubes of diameter 75 or 100 mm. Samples for triaxial tests are extracted in the laboratory from the samples in the sampling tubes by using sample extractors. Samples may be disturbed at both the stages of extraction. If we are dealing with a highly overconsolidated clay the disturbance is greater at both the stages. Besides there is another major disturbance which affects the test results very significantly. A highly overconsolidated clay is at equilibrium in its *in-situ* conditions. The overconsolidation pressures of such soils could be of the order 1000 kPa (10 tsf) or more. The standard penetration value N in such deposits could be 100 or more. The shear strength of such a soil under the *in-situ* condition could be in the order of 600 kPa or more. But if an undisturbed sample of such a soil is tested in standard triaxial equipment, the shear strength under undrained conditions could be very low. This is mostly due to the cracks that develop on the surface of the samples due to the relief of the *in-situ* overburden pressure on the samples. Possibly the only way of obtaining the *in-situ* strength in the laboratory is to bring back the state of the sample to its original field condition by applying all-around pressures on the samples equal to the estimated overconsolidation pressures. This may not be possible in standard triaxial equipment due to its limitations. The present practice is therefore to relate the *in-situ* shear strength to some of the field tests such as standard penetration tests, static cone penetration tests or pressuremeter tests.

8.29 QUESTIONS AND PROBLEMS

8.1 Explain Coulomb's equation for shear strength of a soil. Discuss the factors that affect the shear strength parameters of soil.

8.2 Explain the method of drawing a Mohr circle for a cylindrical sample in a triaxial test. Establish the geometrical relationships between the stresses on the failure plane and externally applied principal stresses.

8.3 Classify the shear tests based on drainage conditions. Explain how the pore pressure variation and volume change take place during these tests. Enumerate the field conditions which necessitate each of these tests.

8.4 What are the advantages and disadvantages of a triaxial compression test in comparison with a direct shear test?

8.5 For what types of soils, will the unconfined compression test give reliable results? Draw a Mohr circle for this test. How do you consider the change in the area of the specimen which takes place during the test in the final results?

8.6 What types of field tests are necessary for determining the shear strength parameters of sensitive clays? Derive the relationships that are useful for analyzing the observations of this test.

8.7 For loose and dense sands, draw the following typical diagrams:

(i) deviator stress vs. linear strain, and

(ii) volumetric strain vs. linear strain. Discuss them.

8.8 Discuss the effects of drainage conditions on the shear strength parameters of clay soil.

8.9 A direct shear test on specimens of fine sand gave the following results:

Normal stress (lb/ft^2)	2100	3700	4500
Shearing stress (lb/ft^2)	970	1700	2080

Determine :

(i) the angle of internal friction of the soil, and

(ii) shear strength of the soil at a depth of 15 ft from the ground surface.

The specific gravity of solids is 2.65, void ratio 0.7 and the ground water table is at a depth of 5 ft from the ground surface. Assume the soil above ground watar table is saturated.

8.10 A specimen of clean sand when subjected to a direct shear test failed at a stress of 2520 lb/ft^2 when the normal stress was 3360 lb/ft^2.

Determine:

(i) the angle of internal friction, and

(ii) the deviator stress at which the failure will take place, if a cylindrical sample of the same sand is subjected to a triaxial test with a cell pressure of 2000 lb/ft^2. Find the angle made by the failure planc with the horizontal.

8.11 A specimen of fine sand, when subjected to a drained triaxial compression test, failed at a deviator stress of 8400 lb/ft^2. It was observed that the failure plane made an angle of 30° with the axis of the sample. Estimate the value of the cell pressure to which this specimen would have been subjected.

8.12 A specimen of sandy silt, when subjected to a drained triaxial test failed at major and minor principal stresses of 120 kN/m^2 and 50 kN/m^2 respectively. At what value of deviator stress would another sample of the same soil fail,if it were subjected to a confining pressure of 75 kN/m^2?

8.13 A sand is hydrostatically consolidated in a triaxial test apparatus to 8820 lb/ft^2 and then sheared with the drainage valves open. At failure, $(\sigma_1 - \sigma_3)$ is 22 $kips/ft^2$. Determine the major and minor principal stresses at failure and the angle of shearing resistance.

8.14 The same sand as in Prob. 8.13 is tested in a direct shear apparatus under a normal pressure of 8820 lb/ft^2. The sample fails when a shear stress of 5880 lb/ft^2 is reached. Determine the major and minor principal stresses at failure and the angle of shearing resistance. Plot the Mohr diagram.

8.15 A sample of dense sand tested in a triaxial *CD* test failed along a well defined failure plane at an angle of 66° with the horizontal. Find the effective confining pressure of the test if the principal stress difference at failure was 100 kPa.

8.16 A drained triaxial test is performed on a sand with σ'_{3f} = 10.5 $kips/ft^2$. At failure $\sigma'_1/\sigma'_3 = 4$. Find σ'_{1f}, $(\sigma_1 - \sigma_3)_f$ and ϕ'.

8.17 If the test of Prob. 8.16 had been conducted undrained, determine $(\sigma_1 - \sigma_3)_f$, ϕ', ϕ_{total} and the angle of the failure plane in the specimen. The pore water pressure u = 2100 lb/ft^2.

8.18 If the test of Prob. 8.16 is conducted at an initial confining pressure of 21 $kips/ft^2$, estimate the principal stress difference and the induced pore pressure at failure.

8.19 A sample of silty sand was tested consolidated drained in a triaxial cell where σ_3 = 475 kPa. If the total axial stress at failure was 1600 kPa while σ_3 = 475 kPa, compute the angle of shearing resistance and the theoretical orientation of the failure plane with respect to the horizontal.

8.20 A drained triaxial test is to be performed on a uniform dense sand with rounded grains. The confining pressure is 4200 lb/ft^2. At what vertical pressure will the sample fail?

8.21 Compute the shearing resistance along a horizontal plane at a depth of 6.1 m in a deposit of sand. The water table is at a depth of 2.13 m. The unit weight of moist sand above the water table is 18.54 kN/m^3 and the saturated weight of submerged sand is 20.11 kN/m^3. Assume that the sand is drained freely and ϕ_d for the wet sand is 32°.

8.22 A sample of dry sand was tested in a direct shear device under a vertical pressure of 137.9 kN/m^2. Compute the angle of internal friction of the sand. Assume shearing resistance = 96.56 kN/m^2.

8.23 The sand in a deep natural deposit has an angle of internal friction of 40° in the dry state and dry unit weight of 17.28 kN/m^3. If the water table is at a depth of 6.1 m, what is the shearing resistance of the material along a horizontal plane at a depth of 3.05 m? Assume: G_s = 2.68.

8.24 Compute the shearing resistance under the conditions specified in Prob. 8.23, if the water table is at the ground surface.

8.25 A drained triaxial test was conducted on dense sand with a confining pressure of 3000 lb/ft^2. The sample failed at an added vertical pressure of 11,000 lb/ft^2. Compute the angle of internal friction ϕ and the angle of inclination α of the failure planes on the assumption that Coulomb's law is valid.

8.26 A saturated sample of dense sand was consolidated in a triaxial test apparatus at a confining pressure of 143.6 kN/m^2. Further drainage was prevented. During the addition of vertical load, the pore pressure in the sample was measured. At the instant of failure, it amounted to 115 kN/m^2. The added vertical pressure at this time was 138.85 kN/m^2. What was the value of ϕ for the sand?

8.27 An undrained triaxial test was carried out on a sample of saturated clay with a confining pressure of 2000 lb/ft^2. The unconfined compressive strength obtained was 7300 lb/ft^2. Determine the excess vertical pressure in addition to the all-round pressure required to make the sample fail.

8.28 A series of undrained triaxial tests on samples of saturated soil gave the following results

σ_3, kN/m^2	100	200	300
u, kN/m^2	20	70	136
$(\sigma_1 - \sigma_3)$, kN/m^2	290	400	534

Find the values of the parameters c and ϕ

(a) with respect to total stress, and (b) with respect to effective stress.

8.29 When an unconfined compression test was conducted on a specimen of silty clay, it showed a strength of 3150 lb/ft^2 . Determine the shear strength parameters of the soil if the angle made by the failure plane with the axis of the specimen was 35°.

8.30 A normally consolidated clay was consolidated under a stress of 150 kPa, then sheared undrained in axial compression. The principal stress difference at failure was 100 kPa and the induced pore pressure at failure was 88 kPa. Determine analytically (a) the slopes of the total and effective Mohr stress envelopes, and (b) the theoretical angle of the failure plane.

8.31 A normally consolidated clay sample was consolidated in a triaxial shear apparatus at a confining pressure of 21 kips/ft^2 and then sheared under undrained condition. The $(\sigma_1 - \sigma_3)$ at failure was 21 kips/ft^2 . Determine ϕ_{cu} and α.

8.32 A *CD* axial compression triaxial test on a normally consolidated clay failed along a clearly defined failure plane of 57°. The cell pressure during the test was 4200 lb/ft^2. Estimate ϕ', the maximum σ'/σ'_3, and the principal stress difference at failure.

8.33 Two identical samples of soft saturated normally consolidated clay were consolidated to 150 kPa in a triaxial apparatus. One specimen was sheared under drained conditions, and the principal stress difference at failure was 300 kPa. The other specimen was sheared undrained, and the principal stress difference at failure was 200 kPa. Determine ϕ_d and ϕ_{cu}.

8.34 When a triaxial compression test was conducted on a soil specimen, it failed at an axial pressure of 7350 lb/ft^2. If the soil has a cohesion of 1050 lb/ft^2 and an angle of internal friction of 24°, what was the cell pressure of the test? Also find the angle made by the failure plane with the direction of σ_3.

8.35 Given the following triaxial test data, plot the results in a Mohr diagram and determine ϕ.

σ_3 kN/m^2	Peak σ_1 kN/m^2	σ_3 kN/m^2	Peak σ_1 kN/m^2
69	190	276	759
138	376	345	959
207	580	414	1143

8.36 Two sets of triaxial tests were carried out on two samples of glacial silt. The results are (a) σ_{11} = 400 kN/m^2, σ_{31} = 100 kN/m^2 (b) σ_{12} = 680 kN/m^2, σ_{32} = 200 kN/m^2.
The angle of the failure plane in both tests was measured to be 59°. Determine the magnitudes of ϕ and c.

8.37 A triaxial compression test on a cylindrical cohesive sample gave the following effective stresses:
(a) Major principal stress, σ'_1 = 46,000 lb/ft^2
(b) Minor principal stress, σ'_3 = 14,500 lb/ft^2
(c) The angle of inclination of the rupture plane = 60° with the horizontal.
Determine analytically the (i) normal stress, (ii) the shear stress, (iii) the resultant stress on the rupture plane through a point, and (iv) the angle of obliquity of the resultant stress with the shear plane.

8.38 Given the results of two sets of triaxial shear tests:
σ_{11} = 1800 kN/m^2; σ_{31} = 1000 kN/m^2
σ_{12} = 2800 kN/m^2; σ_{32} = 2000 kN/m^2
Compute ϕ and c.

8.39 What is the shear strength in terms of effective stress on a plane within a saturated soil mass at a point where the normal stress is 295 kN/m^2 and the pore water pressure 120 kN/m^2 ? The effective stress parameters for the soil are c' = 12 kN/m^2, and ϕ' = 30°.

8.40 The effective stress parameters for a fully saturated clay are known to be c' = 315 lb/ft^2 and ϕ' = 29°. In an unconsolidated-undrained triaxial test on a sample of the same clay the confining pressure was 5250 lb/ft^2 and the principal stress difference at failure was 2841 lb/ft^2. What was the value of the pore water pressure in the sample at failure?

8.41 It is believed that the shear strength of a soil under certain conditions in the field will be governed by Coulomb's law, wherein c = 402 lb/ft^2, and ϕ = 22°. What minimum lateral pressure would be required to prevent failure of the soil at a given point if the vertical pressure were 9000 lb/ft^2?

8.42 The following data refer to three triaxial tests performed on representative undisturbed samples:

Test	Cell pressure kN/m^2	Axial dial reading (division) at failure
1	50	66
2	150	106
3	250	147

The load dial calibration factor is 1.4 N per division. Each sample is 75 mm long and 37.5 mm diameter. Find by graphical means, the value of the apparent cohesion and the angle of internal friction for this soil.

8.43 In a triaxial test a soil specimen was consolidated under an allround pressure of 16 kips/ft^2 and a back pressure of 8 kips/ft^2 . Thereafter, under undrained conditions, the allround pressure was raised to 19 kips/ft^2, resulting in a pore water pressure of 10.4 kips/ft^2, then (with the confining pressure remaining at 19 kips/ft^2) axial load was applied to give a principal stress difference of 12.3 kips/ft^2 and a pore water pressure of 13.8 kips/ft^2. Calculate the values of the pore pressure coefficients A and B.

8.44 In an *in-situ* vane test on a saturated clay a torque of 35 N-m is required to shear the soil. The vane is 50 mm in diameter and 100 mm long. What is the undrained strength of the clay?

8.45 In a vane test a torque of 46 N-m is required to cause failure of the vane in a clay soil. The vane is 150 mm long and has a diameter of 60 mm. Calculate the apparent shear strength of the soil from this test when a vane of 200 mm long and 90 mm in diameter is used in the same soil and the torque at failure was 138 N-m. Calculate the ratio of the shear strength of the clay in a vertical direction to that in the horizontal direction.

8.46 A vane of 80 mm diameter and 160 mm height has been pushed into a soft clay stratum at the bottom of a bone hole. The torque required to rotate the vane was 76 N-m. Determine the undrained shear strength of the clay. After the test the vane was rotated several times and the ultimate torque was found to be 50 N-m. Estimate the sensitivity of the clay.

8.47 A normally consolidated deposit of undisturbed clay extends to a depth of 15 m from the ground surface with the ground water level at 5 m depth from ground surface. Laboratory test on the clay gives a plasticity index of 68%, saturated and dry unit weights of 19.2 kN/m^3 and 14.5 kN/m^3 respectively. An undisturbed specimen for unconfined compressive strength is taken at 10 m depth. Determine the unconfined compressive strength of the clay.

8.48 A triaxial sample was subjected to an ambient pressure of 200 kN/m^2, and the pore pressure recorded was 50 kN/m^2 at a fully saturated state. Then the cell pressure was raised to 300 kN/m^2. What would be the value of pore pressure? At this stage a deviator stress of 150 kN/m^2 was applied to the sample. Determine the pore pressure assuming pore pressure parameter $A = 0.50$.

8.49 In a triaxial test on a saturated clay, the sample was consolidated under a cell pressure of 160 kN/m^2. After consolidation, the cell pressure was increased to 350 kN/m^2, and the sample was then failed under undrained condition. If the shear strength parameters of the soil are $c' = 15.2$ kN/m^2, $\phi' = 26°$, $B = 1$, and $A_f = 0.27$, determine the effective major and minor principal stresses at the time of failure of the sample.

8.50 A thin layer of silt exists at a depth of 18 m below the surface of the ground. The soli above this level has an average dry unit weight of 15.1 kN/m^3 and an average water content of 36%. The water table is almost at the ground surface level. Tests on undisturbed samples of the silt indicate the following values:

$c_u = 45$ kN/m^2, $\phi_u = 18°$, $c' = 36$ kN/m^2 and $\phi' = 27°$.

Estimate the shearing resistance of the silt on a horizontal plane when (a) the shear stress builds up rapidly, and (b) the shear stress builds up slowly.

CHAPTER 9

SOIL EXPLORATION

9.1 INTRODUCTION

The stability of the foundation of a building, a bridge, an embankment or any other structure built on soil depends on the strength and compressibility characteristics of the subsoil. The field and laboratory investigations required to obtain the essential information on the subsoil is called *Soil Exploration or Soil Investigation.* Soil exploration happens to be one of the most important parts of Foundation Engineering and at the same time the most neglected part of it. Terzaghi in 1951 (Bjerrum, et al., 1960) had rightly remarked, that "*Building foundations have always been treated as step children*". His remarks are relevant even today. The success or failure of a foundation depends essentially on the reliability of the various soil parameters obtained from the field investigation and laboratory testing, and used as an input into the design of foundations. *Sophisticated theories alone will not give a safe and sound design.*

Soil exploration is a *must* in the present age for the design of foundations of any project. The extent of the exploration depends upon the magnitude and importance of the project. Projects such as buildings, power plants, fertilizer plants, bridges etc., are localized in areal extent. The area occupied by such projects may vary from a few square meters to many square kilometers. Transmission lines, railway lines, roads and other such projects extend along a narrow path. The length of such projects may be several kilometers. Each project has to be treated as per its requirements. The principle of soil exploration remains the same for all the projects but the program and methodology may vary from project to project.

The elements of soil exploration depend mostly on the importance and magnitude of the project, but generally should provide the following:

1. Information to determine the type of foundation required such as a shallow or deep foundation.
2. Necessary information with regards to the strength and compressibility characteristics of the subsoil to allow the Design Consultant to make recommendations on the safe bearing pressure or pile load capacity.

Soil exploration involves broadly the following:

1. Planning of a program for soil exploration.
2. Collection of disturbed and undisturbed soil or rock samples from the holes drilled in the field. The number and depths of holes depend upon the project.
3. Conducting all the necessary *in-situ* tests for obtaining the strength and compressibility characteristics of the soil or rock directly or indirectly.
4. Study of ground-water conditions and collection of water samples for chemical analysis.
5. Geophysical exploration, if required.
6. Conducting all the necessary tests on the samples of soil /rock and water collected.
7. Preparation of drawings, charts, etc.
8. Analysis of the data collected.
9. Preparation of report.

9.2 BORING OF HOLES

Auger Method

Hand Operated Augers

Auger boring is the simplest of the methods. Hand operated or power driven augers may be used. Two types of hand operated augers are in use as shown in Fig. 9.1

The depths of the holes are normally limited to a maximum of 10 m by this method. These augers are generally suitable for all types of soil above the water table but suitable only in clayey soil below the water table (except for the limitations given below). A string of drill rods is used for advancing the boring. The diameters of the holes normally vary from 10 to 20 cm. Hand operated augers are not suitable in very stiff to hard clay nor in granular soils below the water table. Hand augering is not practicable in dense sand nor in sand mixed with gravel even if the strata lies above the water table.

Power Driven Augers

In many countries the use of power driven continuous flight augers is the most popular method of soil exploration for boring holes. The flights act as a screw conveyor to bring the soil to the surface.

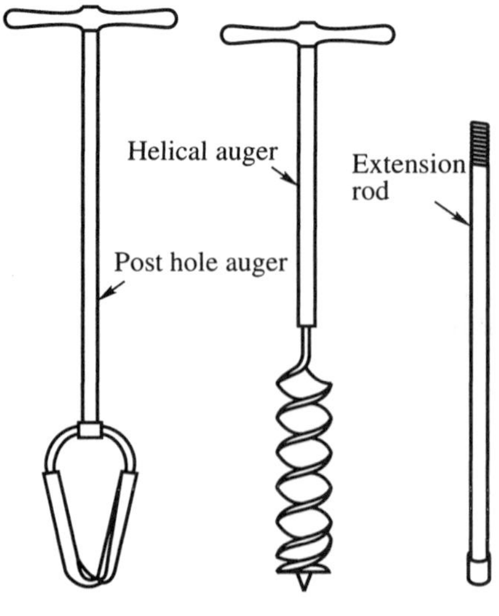

Figure 9.1 Hand augers

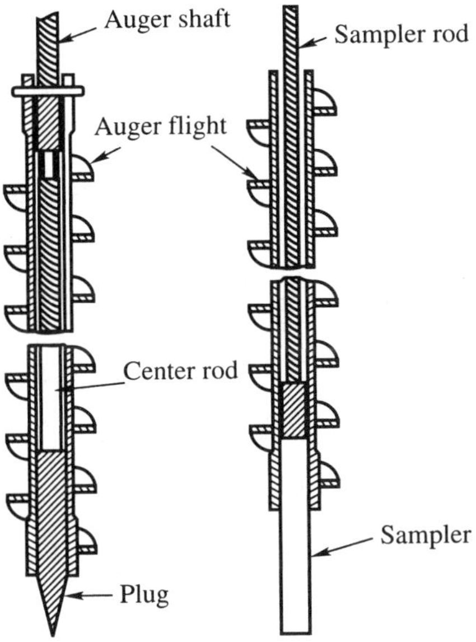

Figure 9.2 Hollow-stem auger
(a) Plugged while advancing the auger, and (b) plug removed and sampler inserted to sample soil below auger

This method may be used in all types of soil including sandy soils below the water table but is not suitable if the soil is mixed with gravel, cobbles etc. The central stem of the auger flight may be hollow or solid. A hollow stem is sometimes preferred since standard penetration tests or sampling may be done through the stem without lifting the auger from its position in the hole. Besides, the flight of augers serves the purpose of casing the hole. The hollow stem can be plugged while advancing the bore and the plug can be removed while taking samples or conducting standard penetration tests (to be described) as shown in Fig. 9.2. The drilling rig can be mounted on a truck or a tractor. Holes may be drilled by this method rapidly to depths of 60 m or more.

Wash Boring

Wash boring is commonly used for boring holes. Soil exploration below the ground water table is usually very difficult to perform by means of pits or auger-holes. Wash boring in such cases is a very convenient method provided the soil is either sand, silt or clay. The method is not suitable if the soil is mixed with gravel or boulders.

Figure 9.3 shows the assembly for a wash boring. To start with, the hole is advanced a short depth by auger and then a casing pipe is pushed to prevent the sides from caving in. The hole is then continued by the use of a chopping bit fixed at the end of a string of hollow drill rods. A stream of water under pressure is forced through the rod and the bit into the hole, which loosens the soil as the water flows up around the pipe. The loosened soil in suspension in water is discharged into a tub. The soil in suspension settles down in the tub and the clean water flows into a sump which is reused for circulation. The motive power for a wash boring is either mechanical or man power. The bit which is hollow is screwed to a string of hollow drill rods supported on a tripod by a rope or steel cable passing over a pulley and operated by a winch fixed on one of the legs of the tripod.

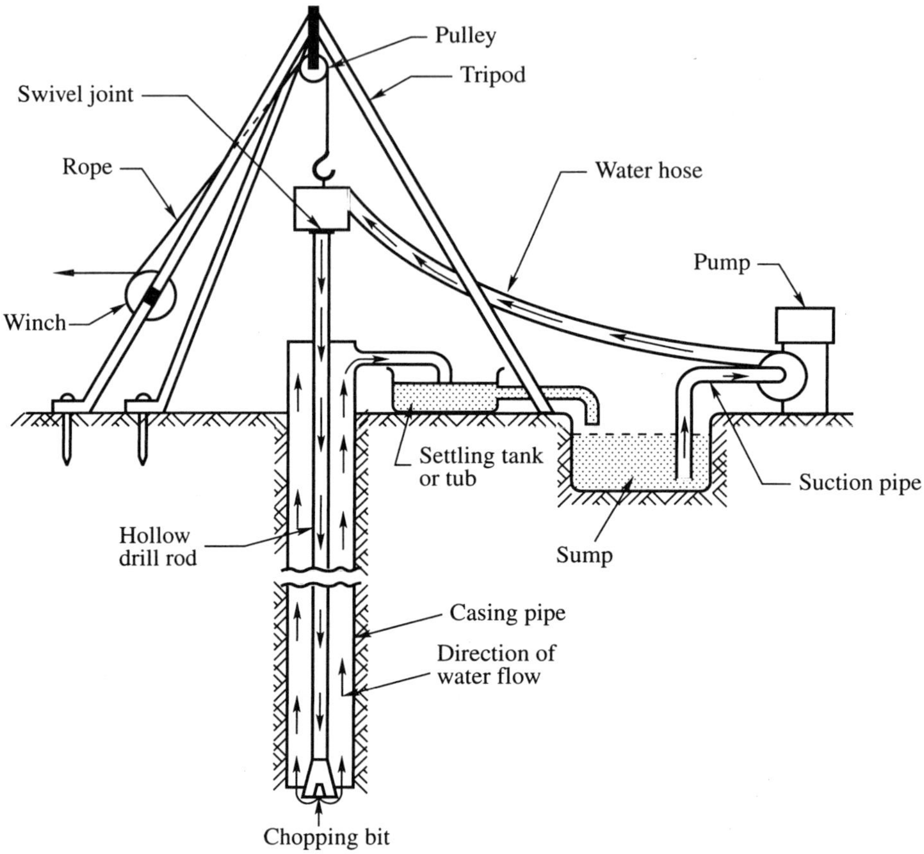

Figure 9.3 Wash boring

The purpose of wash boring is to drill holes only and not to make use of the disturbed washed materials for analysis. Whenever an undisturbed sample is required at a particular depth, the boring is stopped, and the chopping bit is replaced by a sampler. The sampler is pushed into the soil at the bottom of the hole and the sample is withdrawn.

Rotary Drilling

In the rotary drilling method a cutter bit or a core barrel with a coring bit attached to the end of a string of drill rods is rotated by a power rig. The rotation of the cutting bit shears or chips the material penetrated and the material is washed out of the hole by a stream of water just as in the case of a wash boring. Rotary drilling is used primarily for penetrating the overburden between the levels of which samples are required. Coring bits, on the other hand, cut an annular hole around an intact core which enters the barrel and is retrieved. Thus the core barrel is used primarily in rocky strata to get rock samples.

As the rods with the attached bit or barrel are rotated, a downward pressure is applied to the drill string to obtain penetration, and drilling fluid under pressure is introduced into the bottom of the hole through the hollow drill rods and the passages in the bit or barrel. The drilling fluid serves the dual function of cooling the bit as it enters the hole and removing the cuttings from the bottom of the hole as it returns to the surface in the annular space between the drill rods and the walls of the hole. In an uncased hole, the drilling fluid also serves to support the walls of the hole. When boring

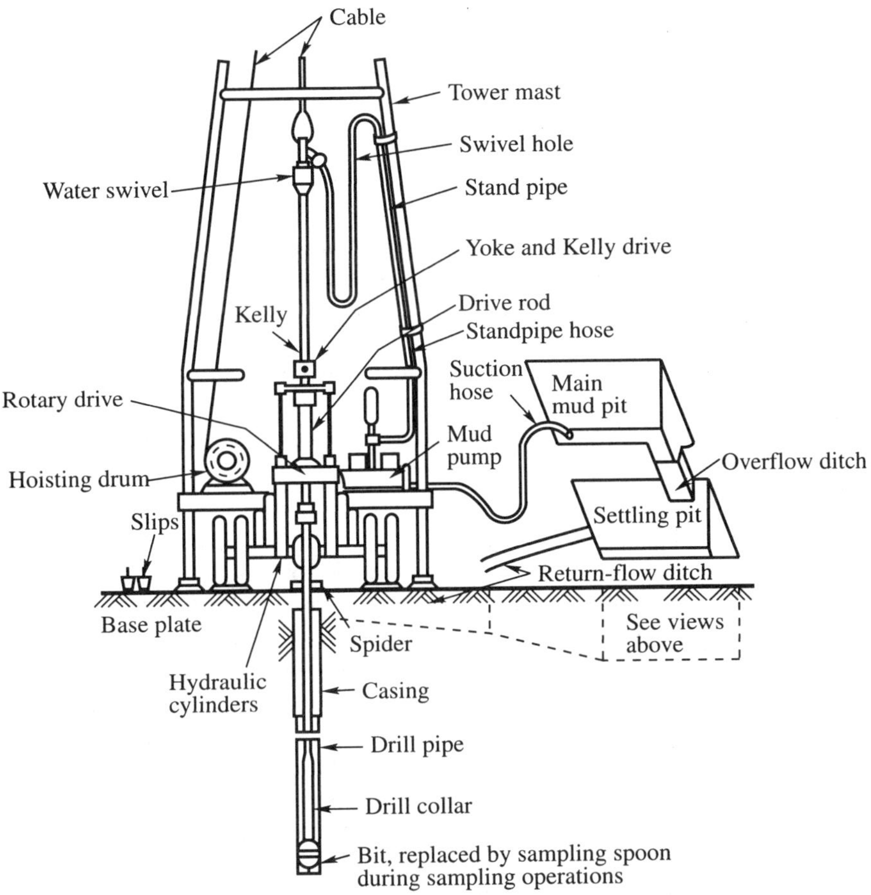

Figure 9.4 Rotary drilling rig (After Hvorslev, 1949)

in soil, the drilling bit is removed and replaced by a sampler when sampling is required, but in rocky strata the coring bit is used to obtain continuous rock samples. The rotary drilling rig of the type given in Fig. 9.4 can also be used for wash boring and auger boring.

Coring Bits

Three basic categories of bits are in use. They are diamond, carbide insert, and saw tooth. Diamond coring bits may be of the surface set or diamond impregnated type. Diamond coring bits are the most versatile of all the coring bits since they produce high quality cores in rock materials ranging from soft to extremely hard. Carbide insert bits use tungsten carbide in lieu of diamonds. Bits of this type are used to core soft to medium hard rock. They are less expensive than diamond bits but the rate of drilling is slower than with diamond bits. In saw-tooth bits, the cutting edge comprises a series of teeth. The teeth are faced and tipped with a hard metal alloy such as tungsten carbide to provide wear resistance and thereby increase the life of the bit. These bits are less expensive but normally used to core overburden soil and very soft rocks only.

Percussion Drilling

Percussion drilling is another method of drilling holes. Possibly this is the only method for drilling in river deposits mixed with hard boulders of the quartzitic type. In this method a heavy drilling bit is alternatively raised and dropped in such a manner that it powders the underlying materials which form a slurry with water and are removed as the boring advances.

9.3 SAMPLING IN SOIL

Soils met in nature are heterogeneous in character with a mixture of sand, silt and clay in different proportions. In water deposits, there are distinct layers of sand, silt and clay of varying thicknesses and alternating with depth. We can bring all the deposits of soil under two distinct groups for the purpose of study, namely, coarse grained and fine grained soils. Soils with particles of size coarser than 0.075 mm are brought under the category of coarse grained and those finer than 0.075 mm under fine grained soils. Sandy soil falls in the group of coarse grained, and silt and clay soils in the fine grained group. A satisfactory design of a foundation depends upon the accuracy with which the various soil parameters required for the design are obtained. The accuracy of the soil parameters depends upon the accuracy with which representative soil samples are obtained from the field.

Disturbed Samples

Auger samples may be used to identify soil strata and for field classification tests, but are not useful for laboratory tests. The cuttings or chopping from wash borings are of little value except for indicating changes in stratification to the boring supervisor. The material brought up with the drilling mud is contaminated and usually unsuitable even for identification.

For proper identification and classification of a soil, representative samples are required at frequent intervals along the bore hole. Representative samples can usually be obtained by driving or pushing into the strata in a bore hole an open-ended sampling spoon called a split spoon sampler (Fig. 9.5) which is used for conducting standard penetration tests (refer Sect. 9.5). It is made up of a driving shoe and a barrel. The barrel is split longitudinally into two halves with a coupling at the upper end for connection to the drill rods. The dimensions of the split spoon are given in Fig. 9.5. In a test the sampler is driven into the soil a measured distance. After a sample is taken, the cutting shoe and the coupling are unscrewed and the two halves of the barrel separated to expose the material. Experience indicates that samples recovered by this device are likely to be highly disturbed and as such can only be used as disturbed samples. The samples so obtained are stored in glass or plastic jars or bags, referenced and sent to the laboratory for testing. If spoon samples are to be transported to the laboratory without examination in the field, the barrel is often cored out to hold a cylindrical thin-walled tube known as a *liner*. After a sample has been obtained, the liner and the sample it contains are removed from the spoon and the ends are sealed with caps or with metal discs and wax. Samples of cohesionless soils below the water table cannot be retained in conventional sampling spoons without the addition of a *spring core catcher*.

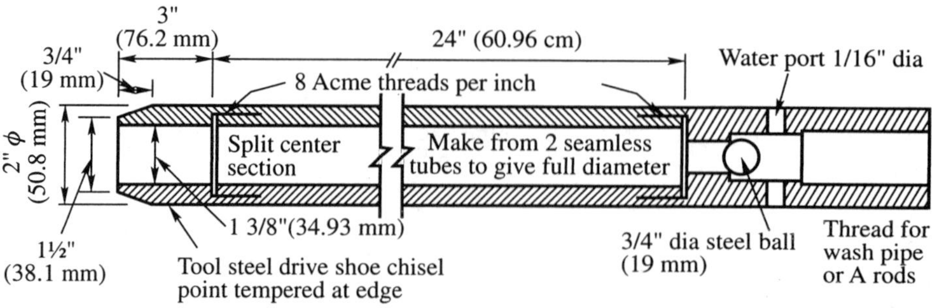

Figure 9.5 Split barrel sampler for standard penetration test

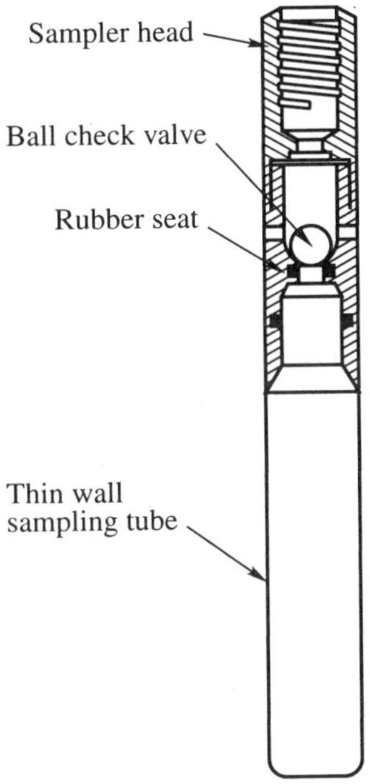

Figure 9.6 Thin wall Shelby tube sampler

Sampler

Many types of samplers are in use for extracting the so called undisturbed samples. Only two types of samplers are described here. They are,

1. Open drive sampler,
2. Piston sampler.

Open Drive Sampler

The wall thickness of the open drive sampler used for sampling may be thin or thick according to the soil conditions met in the field. The samplers are made of seamless steel pipes. A thin-walled tube sampler is called as shelby tube sampler (Fig. 9.6), consists of a thin wall metal tube connected to a sampler head. The sampler head contains a ball check valve and ports which permit the escape of water or air from the sample tube as the sample enters it. The thin wall tube, which is normally formed from 1/16 to 1/8 inch metal, is drawn in at the lower end and is reamed so that the inside diameter of the cutting edge is 0.5 to 1.5 percent less than that of the inside diameter of the tube. The exact percentage is governed by the size and wall thickness of the tube. The wall thickness is governed by the *area ratio*, A_r, which is defined as

$$A_r = \frac{d_o^2 - d_i^2}{d_i^2} \times 100 \text{ percent}, \tag{9.1}$$

where, d_i = inside diameter,

d_o = outside diameter.

A_r is a measure of the volume of the soil displacement to the volume of the collected sample. Well-designed sampling tubes have an area ratio of about 10 percent. However, the area ratio may have to be much more than 10 percent when samples are to be taken in very stiff to hard clay soils mixed with stones to prevent the edges of the sampling tubes from distortion during sampling.

Sample Extraction

The thin-wall tube sampler is primarily used for sampling in soft to medium stiff cohesive soils. The wall thickness has to be increased if sampling is to be done in very stiff to hard strata. For best results it is better to push the sampler statically into the strata. Samplers are driven into the strata where pushing is not possible or practicable. The procedure of sampling involves attaching a string of drill rods to the sampler tube adapter and lowering the sampler to rest on the bottom of the bore hole which was cleaned of loose materials in advance. The sampler is then pushed or driven into the soil. Over driving or pushing should be avoided. After the sampler is pushed to the required depth, the soil at the bottom of the sampler is sheared off by giving a twist to the drill rod at the top. The sampling tube is taken out of the bore hole and the tube is separated from the sampler head. The top and bottom of the sample are either sealed with molten wax or capped to prevent evaporation of moisture. The sampling tubes are suitably referenced for later identification.

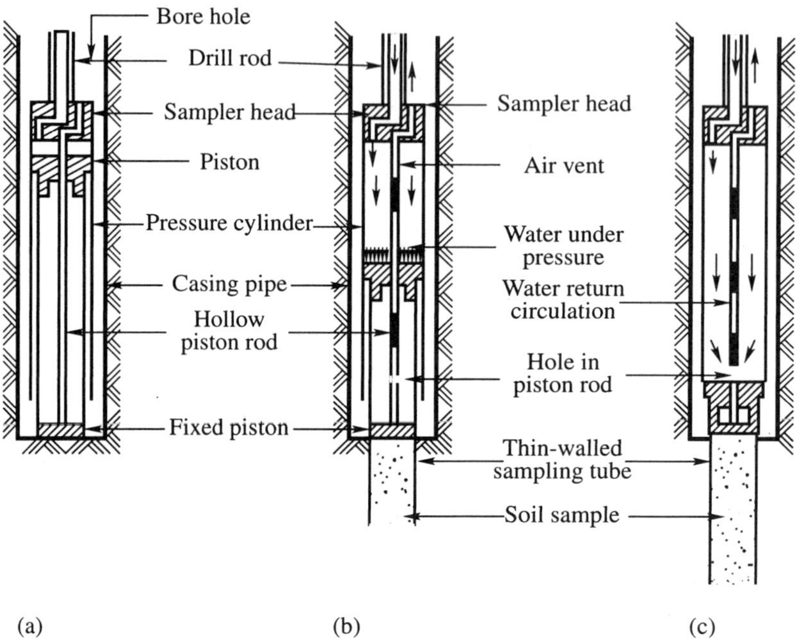

Figure 9.7 Osterberg Piston Sampler (a) Sampler is set in drilled hole. (b) Sample tube is pushed hydraulically into the soil. (c) Pressure is released through hole in piston rod.

Piston Sampler (After Osterberg 1952)

To improve the quality of samples and to increase the recovery of soft or slightly cohesive soils, a *piston sampler* is normally used. Such a sampler consists of a thin walled tube fitted with a piston that closes the end of the sampling tube until the apparatus is lowered to the bottom of the bore hole (Fig. 9.7(a)). The sampling tube is pushed into the soil hydraulically by keeping the piston stationary (Fig. 9.7(b)). The presence of the piston prevents the soft soils from squeezing rapidly into the tube and thus eliminates most of the distortion of the sample. The piston also helps to increase the length of sample that can be recovered by creating a slight vacuum that tends to retain the sample if the top of the column of soil begins to separate from the piston. During the withdrawal of the sampler, the piston also prevents water pressure from acting on the top of the sample and thus increases the chances of recovery. The design of piston samplers has been refined to the extent that it is sometimes possible to take undisturbed samples of sand from below the water table. However, piston sampling is relatively a costly procedure and may be adopted only where its use is justified.

Example 9.1

The following dimensions are given for a shelby tube sampler:

External diameter = 51 mm
Internal diameter = 48 mm
Determine the area ratio

Solution

Per Eq (9.1) the area ratio A_r is

$$A_r = \frac{d_o^2 - d_i^2}{d_i^2} = \frac{51^2 - 48^2}{48^2} = 0.129 = 12.9\%$$

Example 9.2

75 mm is the external diameter of a sampling tube. If the area ratio required is 20%, determine the thickness of the sampling tube. In what type of clay would such a high area ratio be required?

Solution

$$A_r = \frac{75^2 - d_i^2}{d_i^2} = 0.20$$

Solving $d_i = 68.465$ mm.

The wall thickness $= \frac{75.0 - 68.465}{2} = 3.267$ mm

When samples are to be taken in very stiff to hard clay soils mixed with stones, sampling tubes with high area ratios are required.

9.4 ROCK CORE SAMPLING

Rock coring is the process in which a sampler consisting of a tube (core barrel) with a cutting bit at its lower end cuts an annular hole in a rock mass, thereby creating a cylinder or core of rock which is recovered in the core barrel. Rock cores are normally obtained by rotary drilling.

The primary purpose of core drilling is to obtain intact samples. The behavior of a rock mass is affected by the presence of fractures in the rock. The size and spacing of fractures, the degree of weathering of fractures, and the presence of soil within the fractures are critical items. Figure 9.8 gives a schematic diagram of core barrels with coring bits at the bottom. As discussed earlier, the cutting element may consist of diamonds, tungsten carbide inserts or chilled shot. The core barrel may consist of a single tube or a double tube. Samples taken in a single tube barrel are likely to experience

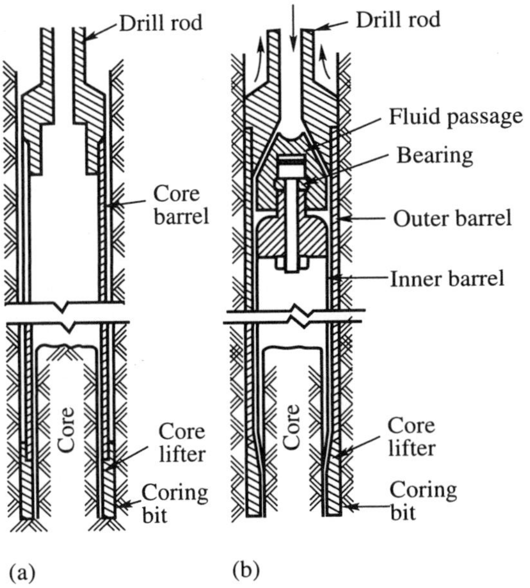

Figure 9.8 Schematic diagram of core barrels (a) Single tube. (b) Double tube.

considerable disturbance due to torsion, swelling and contamination by the drilling fluid, but these disadvantages are not there if the coring is conducted in hard, intact, rocky strata. However, if a double tube barrel is used, the core is protected from the circulating fluid. Most core barrels are capable of retaining cores up to a length of 2 m. Single barrel is used in Calyx drilling. Standard rock cores range from about 1¼ inches to nearly 6 inches in diameter. The more common sizes are given in Table 9.1.

The *recovery ratio* R_r, defined as the percentage ratio between the length of the core recovered and the length of the core drilled on a given run, is related to the quality of rock encountered in boring, but it is also influenced by the drilling technique and the type and size of core barrel used. Generally the use of a double tube barrel results in higher recovery ratios than can be obtained with single tube barrels. A better estimate of in-situ rock quality is obtained by a modified core recovery ratio known as the *Rock Quality Designation* (RQD) which is expressed as

$$RQD = \frac{\overline{L}_a}{L_t} \tag{9.2}$$

where, $\overline{L}_a$ = total length of intact hard and sound pieces of core of length greater than 4 in. arranged in its proper position,

L_t = total length of drilling.

Breaks obviously caused by drilling are ignored. The diameter of the core should preferably be not less than $2^1/_8$ inches. Table 9.2 gives the rock quality description as related to RQD.

Table 9.1 Standard sizes of core barrels, drill rods, and compatible casing (Peck et al., 1974)

Core Barrel			Drill Rod		Casing		
Symbol	Hole dia (in)	Core dia (in)	Symbol	Outside dia (in)	Symbol	Outside dia (in)	Inside dia (in)
EWX, EWM	1½	$1^3/_{16}$	E	$1^{15}/_{16}$	–	–	–
AWX, AWM	$1^{15}/_{16}$	$1^3/_{16}$	A	$1^5/_8$	EX	$1^{13}/_{16}$	1½
BWX, BWM	$2^3/_8$	$1^5/_8$	B	$1^7/_8$	AX	$2^1/_4$	$1^{29}/_{32}$
NWX, NWM	3	$2^1/_8$	N	$2^3/_8$	BX	$2^7/_8$	$2^3/_8$
$2^3/_4 \times 3^7/_8$	$3^7/_8$	$2^{11}/_{16}$	–	–	NX	$3^1/_2$	3

Note: Symbol X indicates single barrel, M indicates double barrel.

Table 9.2 Relation of RQD and *in-situ* Rock Quality (Peck et al., 1974)

RQD %	Rock Quality
90-100	Excellent
75-90	Good
50-75	Fair
25-50	Poor
0-25	Very Poor

9.5 STANDARD PENETRATION TEST

The SPT is the most commonly used *in situ* test in a bore hole in the USA. The test is made by making use of a split spoon sampler shown in Fig. 9.7. The method has been standardized as ASTM D-1586 (1997) with periodic revision since 1958. The method of carrying out this test is as follows:

1. The split spoon sampler is connected to a string of drill rods and is lowered into the bottom of the bore hole which was drilled and cleaned in advance.
2. The sampler is driven into the soil strata to a maximum depth of 18 in by making use of a 140 lb weight falling freely from a height of 30 in on to an anvil fixed on the top of drill rod. The weight is guided to fall along a guide rod. The weight is raised and allowed to fall by means of a manila rope, one end tied to the weight and the other end passing over a pulley on to a hand operated winch or a motor driven cathead.
3. The number of blows required to penetrate each of the successive 6 in depths is counted to produce a total penetration of 18 in.
4. To avoid seating errors, the blows required for the first 6 in of penetration are not taken into account; those required to increase the penetration from 6 in to 18 in constitute the *N*-value.

As per some codes of practice if the *N*-value exceeds 100, it is termed as refusal, and the test is stopped even if the total penetration falls short of the last 300 mm depth of penetration. Standardization of refusal at 100 blows allows all the drilling organizations to standardize costs so that higher blows if required may be eliminated to prevent the excessive wear and tear of the equipment. The SPT is conducted normally at 2.5 to 5 ft intervals. The intervals may be increased at greater depths if necessary.

Standardization of SPT

The validity of the SPT has been the subject of study and research by many authors for the last many years. The basic conclusion is that the best results are difficult to reproduce. Some of the important factors that affect reproducibility are

1. Variation in the height of fall of the drop weight (hammer) during the test
2. The number of turns of rope around the cathead, and the condition of the manila rope
3. Length and diameter of drill rod
5. Diameter of bore hole
6. Overburden pressure

There are many more factors that hamper reproducibility of results. Normally corrections used to be applied for a quick condition in the hole bottom due to rapid withdrawal of the auger. ASTM 1586 has stipulated standards to avoid such a quick condition. Discrepancies in the input driving energy and its dissipation around the sampler into the surrounding soil are the principal factors for the wide range in *N* values. The theoretical input energy may be expressed as

$$E_{in} = Wh \tag{9.3}$$

where W = weight or mass of the hammer

h = height of fall

Investigation has revealed (Kovacs and Salomone, 1982) that the actual energy transferred to the driving head and then to the sampler ranged from about 30 to 80 percent. It has been suggested that the SPT be standardized to some energy ratio R_e keeping in mind the data collected so far from the existing SPT. Bowles (1996) suggests that the observed SPT value N be reduced to a standard blow count corresponding to 70 percent of standard energy. Terzaghi, et al., (1996) suggest 60 percent. The standard energy ratio may be expressed as

$$R_{es} = \frac{\text{Actual hammer energy to sampler}, E_a}{\text{Input energy}, E_{in}} \tag{9.4}$$

Corrections to the Observed SPT Value

Three types of corrections are normally applied to the observed N values. They are:

1. Hammer efficiency correction
2. Drillrod, sampler and borehole corrections
3. Correction due to overburden pressure

1. Hammer Efficiency Correction, E_h

Different types of hammers are in use for driving the drill rods. Two types are normally used in USA. They are (Bowles, 1996)

1. Donut with two turns of manila rope on the cathead with a hammer efficiency $E_h = 0.45$.
2. Safety with two turns of manila rope on the cathead with a hammer efficiency as follows:

   Rope-pulley or cathead = 0.7 to 0.8;
   Trip or automatic hammer = 0.8 to 1.0.

2. Drill Rod, Sampler and Borehole Corrections

Correction factors are used for correcting the effects of length of drill rods, use of split spoon sampler with or without liner, and size of bore holes. The various correction factors are (Bowles, 1996).

a) Drill rod length correction factor C_d

Length (m)	Correction factor (C_d)
> 10 m	1.0
4–10 m	0.85–0.95
< 4.0 m	0.75

b) Sampler correction factor, C_s

Without liner $C_s = 1.00$
With liner,
Dense sand, clay = 0.80
Loose sand = 0.90

c) Bore hole diameter correction factor, C_b

Bore hole diameter	Correction factor, C_b
60-120 mm	1.0
150 mm	1.05
200 mm	1.15

3. Correction Factor for Overburden Pressure in Granular Soils, C_N

The C_N as per Liao and Whitman (1986) is

$$C_N = \left[\frac{95.76}{\rho'_o}\right]^{1/2} \tag{9.5}$$

where, ρ'_o =effective overburden pressure in kN/m^2

There are a number of empirical relations proposed for C_N. However, the most commonly used relationship is the one given by Eq. (9.5).

N_{cor} may be expressed as

$$N_{cor} = C_N N E_h C_d C_s C_b \tag{9.6}$$

N_{cor} is related to the standard energy ratio used by the designer. N_{cor} may be expressed as N_{70} or N_{60} according to the designer's choice.

In Eq (9.6) $C_N N$ is the corrected value for overburden pressure only. The value of C_N as per Eq. (9.5) is applicable for granular soils only, whereas $C_N = 1$ for cohesive soils for all depths.

Example 9.3

The observed standard penetration test value in a deposit of fully submerged sand was 45 at a depth of 6.5 m. The average effective unit weight of the soil is 9.69 kN/m^3. The other data given are (a) hammer efficiency = 0.8, (b) drill rod length correction factor = 0.9, and (c) borehole correction factor = 1.05. Determine the corrected SPT value for standard energy (a) R_{es} = 60 percent, and (b) R_{es} = 70%.

Solution

Per Eq (9.6), the equation for N_{60} may be written as

(i) $N_{60} = C_N N E_h C_d C_s C_b$

where N = observed SPT value

C_N = overburden correction

Per Eq (9.5) we have

$$C_N = \left[\frac{95.76}{p'_o}\right]^{1/2}$$

where p'_0 = effective overburden pressure

$= 6.5 \times 9.69 = 63$ kN/m^2

Substituting for p'_0,

$$C_N = \left(\frac{95.76}{60}\right)^{1/2} = 1.233$$

Substituting the known values, the corrected N_{60} is

$N_{60} = 1.233 \times 45 \times 0.8 \times 0.9 \times 1.05 = 42$

For 70 percent standard energy

$$N_{70} = 42 \times \frac{0.6}{0.7} = 36$$

9.6 SPT VALUES RELATED TO RELATIVE DENSITY OF COHESIONLESS SOILS

Although the SPT is not considered as a refined and completely reliable method of investigation, the N_{cor} values give useful information with regard to consistency of cohesive soils and relative density of cohesionless soils. The correlation between N_{cor} values and relative density of granular soils suggested by Peck, et al., (1974) is given in Table 9.3.

Before using Table 9.3 the observed N value has to be corrected for standard energy and overburden pressure. The correlations given in Table 9.3 are just a guide and may vary according to the fineness of the sand.

Meyerhof (1956) suggested the following approximate equations for computing the angle of friction ϕ from the known value of D_r.

For granular soil with fine sand and more than 5 percent silt,

$$\phi^\circ = 25 + 0.15D_r \tag{9.7}$$

For granular soils with fine sand and less than 5 percent silt,

$$\phi^\circ = 30 + 0.15D_r \tag{9.8}$$

where D_r is expressed in percent.

9.7 SPT VALUES RELATED TO CONSISTENCY OF CLAY SOIL

Peck et al., (1974) have given for saturated cohesive soils, correlations between N_{cor} value and consistency. This correlation is quite useful but has to be used according to the soil conditions met in the field. Table 9.4 gives the correlations.

The N_{cor} value to be used in Table 9.4 is the blow count corrected for standard energy ratio R_{es}. The present practice is to relate q_u with N_{cor} as follows,

$$q_u = \bar{k}N_{cor} \text{ kPa} \tag{9.9}$$

Table 9.3 N_{cor} and ϕ Related to Relative Density

N_{cor}	Compactness	Relative density, D_r (%)	ϕ°
0-4	Very loose	0-15	<28
4-10	Loose	15-35	28-30
10-30	Medium	35-65	30-36
30-50	Dense	65-85	36-41
>50	Very Dense	>85	> 41

Table 9.4 Relation Between N_{cor} and q_u

Consistency	N_{cor}	q_u, kPa
Very soft	0-2	<25
Soft	2-4	25-50
Medium	4-8	50-100
Stiff	8-15	100-200
Very Stiff	15-30	200-400
Hard	>30	>400

where q_u is the unconfined compressive strength.

or $$\overline{k} = \frac{q_u}{N_{cor}} \text{ kPa} \tag{9.10}$$

where, $\overline{k}$ is the proportionality factor. A value of $\overline{k} = 12$ has been recommended by Bowles (1996).

Example 9.4

For the corrected N values in Ex 9.3, determine the (a) relative density, and (b) the angle of friction. Assume the percent of fines in the deposit is less than 5%.

Solution

Per Table 9.3 the relative density and ϕ are

For $N_{60} = 42$, $D_r = 77\%$, $\phi = 39°$

For $N_{70} = 36$, $D_r = 71\%$, $\phi = 37.5°$

Per Eq (9.8)

For $D_r = 77\%$, $\phi = 30 + 0.15 \times 77 = 41.5°$

For $D_r = 71\%$, $\phi = 30 + 0.15 \times 71 = 40.7°$

Example 9.5

For the corrected values of N given in Ex 9.4, determine the unconfined compressive strength q_u in a clay deposit.

Solution

(a) From Table 9.4

$$\left.\begin{array}{l} \text{For } N_{60} = 42 \\ \text{For } N_{70} = 36 \end{array}\right\} q_u > 400 \text{ kPa - The soil is of a hard consistency.}$$

(b) Per Eq (9.9)

$q_u = \overline{k} N_{cor}$, where $\overline{k} = 12$ (Bowles, 1996)

For $N_{60} = 42$, $q_u = 12 \times 42 = 504$ kPa

For $N_{70} = 36$, $q_u = 12 \times 36 = 432$ kPa

Example 9.6

Refer to Example 9.3. Determine the corrected SPT value for R_{es} = 100 percent, and the corresponding values of D_r and ϕ. Assume the percent of fine sand in the deposit is less than 5%.

Solution

From Example 9.3, $N_{60} = 42$

$$\text{Hence } N_{100} = 2 \times \frac{0.6}{1.0} \approx 25$$

From Table 9.3, $\phi = 34.5°$ and $D_r = 57.5\%$

From Eq. (9.8) for $D_r = 57.5\%$, $\phi = 30 + 0.15 \times 57.5 = 38.6°$.

9.8 STATIC CONE PENETRATION TEST (CPT)

The static cone penetration test normally called the Dutch cone penetration test (CPT). It has gained acceptance rapidly in many countries. The method was introduced nearly 50 years ago. One of the greatest values of the CPT consists of its function as a scale model pile test. Empirical correlations established over many years permit the calculation of pile bearing capacity directly from the CPT results without the use of conventional soil parameters.

The CPT has proved valuable for soil profiling as the soil type can be identified from the combined measurement of end resistance of cone and side friction on a jacket. The test lends itself to the derivation of normal soil properties such as density, friction angle and cohesion. Various theories have been developed for foundation design.

The popularity of the CPT can be attributed to the following three important factors:

1. General introduction of the electric penetrometer providing more precise measurements, and improvements in the equipment allowing deeper penetration.
2. The need for the penetrometer testing in-situ technique in offshore foundation investigations in view of the difficulties in achieving adequate sample quality in marine environment.
3. The addition of other simultaneous measurements to the standard friction penetrometer such as pore pressure and soil temperature.

The Penetrometer

There are a variety of shapes and sizes of penetrometers being used. The one that is standard in most countries is the cone with an apex angle of 60° and a base area of 10 cm^2. The sleeve (jacket) has become a standard item on the penetrometer for most applications. On the 10 cm^2 cone penetrometer the friction sleeve should have an area of 150 cm^2 as per standard practice. The ratio of side friction and bearing resistance, the *friction ratio*, enables identification of the soil type (Schmertmann 1975) and provides useful information in particular when no bore hole data are available. Even when borings are made, the friction ratio supplies a check on the accuracy of the boring logs.

Two types of penetrometers are used which are based on the method used for measuring cone resistance and friction. They are,

1. The Mechanical Type,
2. The Electrical Type.

Mechanical Type

The Begemann Friction Cone Mechanical type penetrometer is shown in Fig. 9.9. It consists of a 60° cone with a base diameter of 35.6 mm (sectional area 10 cm^2). A sounding rod is screwed to the base. Additional rods of one meter length each are used. These rods are screwed or attached together to bear against each other. The sounding rods move inside mantle tubes. The inside diameter of the mantle tube is just sufficient for the sounding rods to move freely whereas the outside diameter is equal to or less than the base diameter of the cone. All the dimensions in Fig. 9.9 are in mm.

Jacking System

The rigs used for pushing down the penetrometer consist basically of a hydraulic system. The thrust capacity for cone testing on land varies from 20 to 30 kN for hand operated rigs and 100 to 200 kN for mechanically operated rigs as shown in Fig. 9.10. Bourden gauges are provided in the driving mechanism for measuring the pressures exerted by the cone and the friction jacket either individually or collectively during the operation. The rigs may be operated either on the ground or mounted on heavy duty trucks. In either case, the rig should take the necessary upthrust. For ground based rigs screw anchors are provided to take up the reaction thrust.

Operation of Penetrometer

The sequence of operation of the penetrometer shown in Fig. 9.11 is explained as follows:

Position 1 The cone and friction jacket assembly in a collapsed position.

Position 2 The cone is pushed down by the inner sounding rods to a depth a until a collar engages the cone. The pressure gauge records the total force Q_c to the cone. Normally $a = 40$ mm.

Position 3 The sounding rod is pushed further to a depth b. This pushes the friction jacket and the cone assembly together; the force is Q_t. Normally $b = 40$ mm.

Position 4 The outside mantle tube is pushed down a distance $a + b$ which brings the cone assembly and the friction jacket to position 1. The total movement = $a + b = 80$ mm.

The process of operation illustrated above is continued until the proposed depth is reached. The cone is pushed at a standard rate of 20 mm per second. The mechanical penetrometer has its advantage as it is simple to operate and the cost of maintenance is low. The quality of the work depends on the skill of the operator. The depth of CPT is measured by recording the length of the sounding rods that have been pushed into the ground.

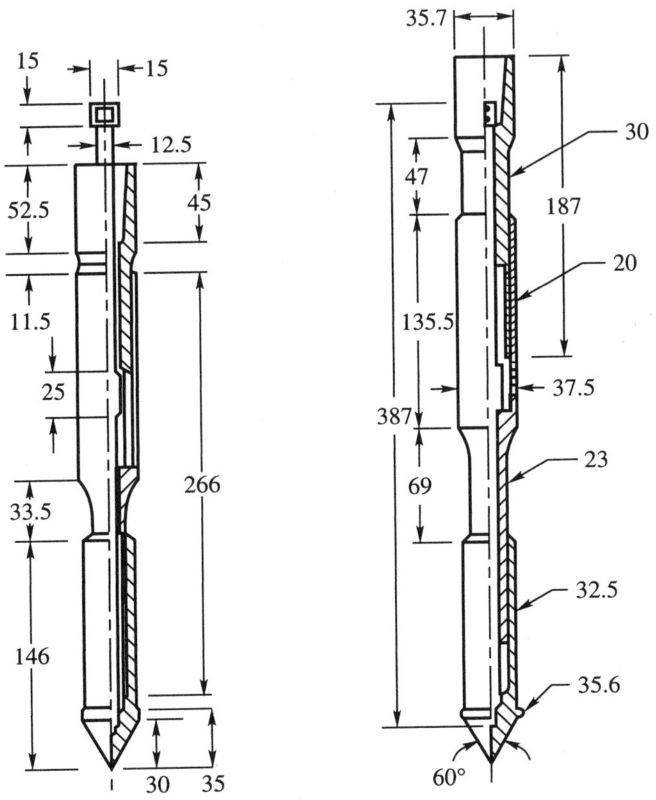

Note: All dimensions are in mm.

Figure 9.9 Begemann friction-cone mechanical type penetrometer

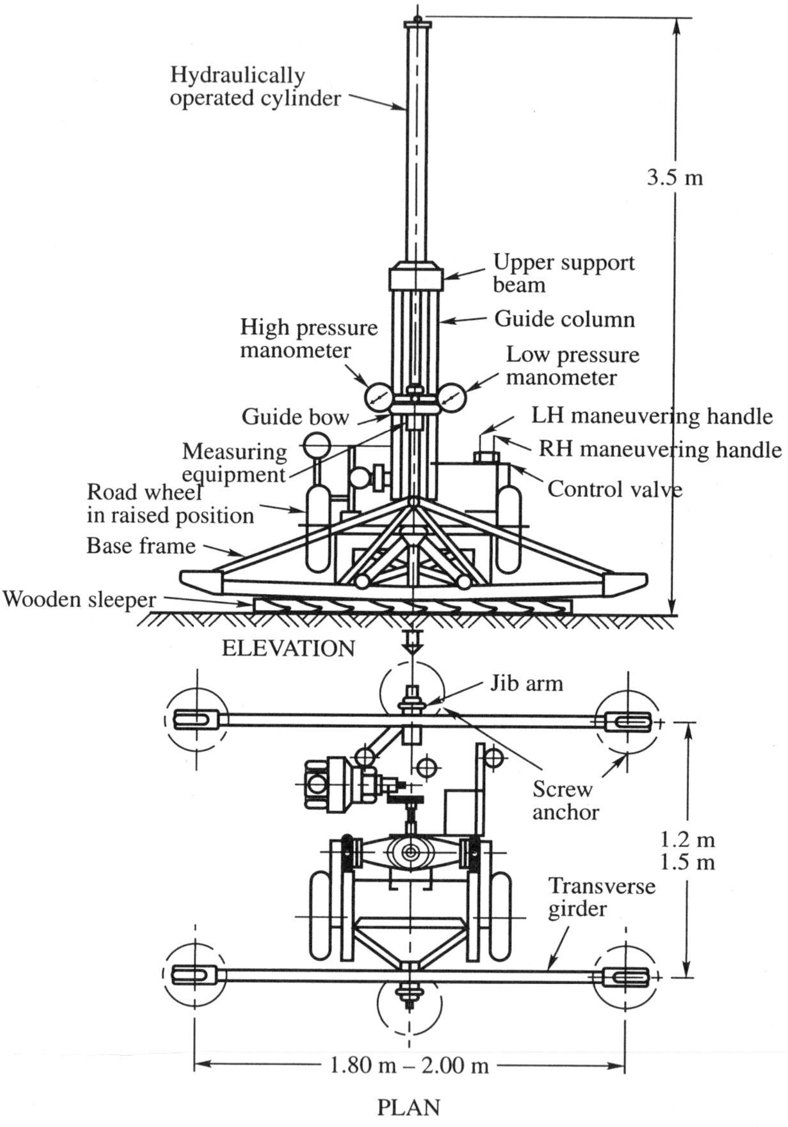

Fig. 9.10 Static cone penetration testing equipment

The Electric Penetrometer

The electric penetrometer is an improvement over the mechanical one. Mechanical penetrometers operate incrementally whereas the electric penetrometer is advanced continuously.

Figure 9.12 shows an electric-static penetrometer with the friction sleeve just above the base of the cone. The sectional area of the cone and the surface area of the friction jacket remain the same as those of a mechanical type. The penetrometer has built in load cells that record separately the cone bearing and side friction. Strain gauges are mostly used for the load cells. The load cells have a normal capacity of 50 to 100 kN for end bearing and 7.5 to 15 kN for side friction, depending on the soils to be penetrated. An electric cable inserted through the push rods (mantle tube) connect the penetrometer with the recording equipment at the surface which produces graphs of resistance versus depth.

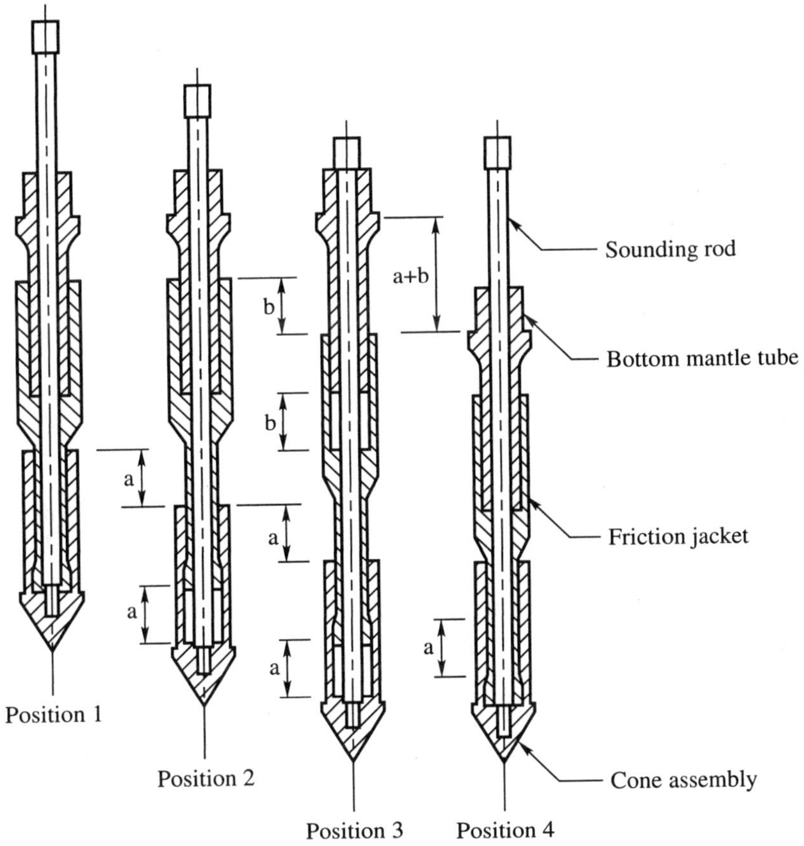

Figure 9.11 Four positions of the sounding apparatus with friction jacket

The electric penetrometer has many advantages. The repeatability of the cone test is very good. A continuous record of the penetration results reflects better the nature of the soil layers penetrated. However, electronic cone testing requires skilled operators and better maintenance. The electric penetrometer is indispensable for *offshore soil investigation.*

Operation of Penetrometer

The electric penetrometer is pushed continuously at a standard rate of 20 mm per second. A continuous record of the bearing resistance q_c and frictional resistance f_s against depth is produced in the form of a graph at the surface in the recording unit.

Piezocone

A piezometer element included in the cone penetrometer is called a *piezocone* (Fig. 9.13). There is now a growing use of the piezocone for measuring pore pressures at the tips of the cone. The porous element is mounted normally midway along the cone tip allowing pore water to enter the tip. An electric pressure transducer measures the pore pressure during the operation of the CPT. The pore pressure record provides a much more sensitive means to detect thin soil layers. This could be very important in determining consolidation rates in a clay within the sand seams.

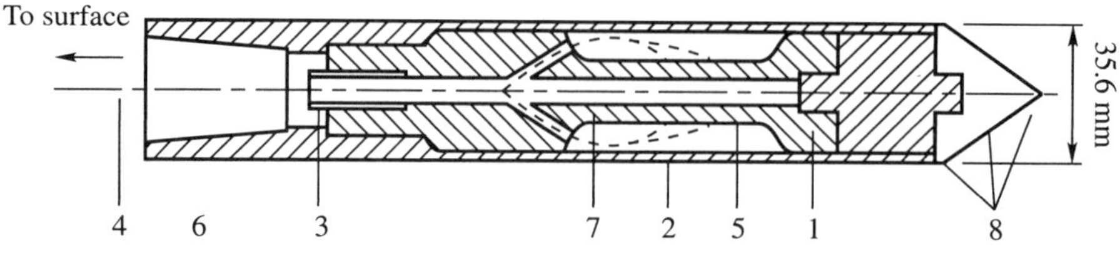

1. Load cell
2. Friction sleeve
3. Water proof bushing
4. Cable
5. Strain gases
6. Connection with rods
7. Inclinometer
8. Porous stone (piezometer)

Figure 9.12 An-electric-static cone penetrometer

Temperature Cone

The temperature of a soil is required at certain localities to provide information about environmental changes. The temperature gradient with depth may offer possibilities to calculate the heat conductivity of the soil. Measurement of the temperature during CPT is possible by incorporating a temperature sensor in the electric penetrometer. Temperature measurements have been made in permafrost, under blast furnaces, beneath undercooled tanks, along marine pipe lines, etc.

Effect of Rate of Penetration

Several studies have been made to determine the effect of the rate of penetration on cone bearing and side friction. Although the values tend to decrease for slower rates, the general conclusion is that the influence is insignificant for speeds between 10 and 30 mm per second. The standard rate of penetration has been generally accepted as 20 mm per second.

Cone Resistance q_c and Local Side Friction f_c

Cone penetration resistance q_c is obtained by dividing the total force Q_c acting on the cone by the base area A_c of the cone.

$$q_c = \frac{Q_c}{A_c}. \tag{9.11}$$

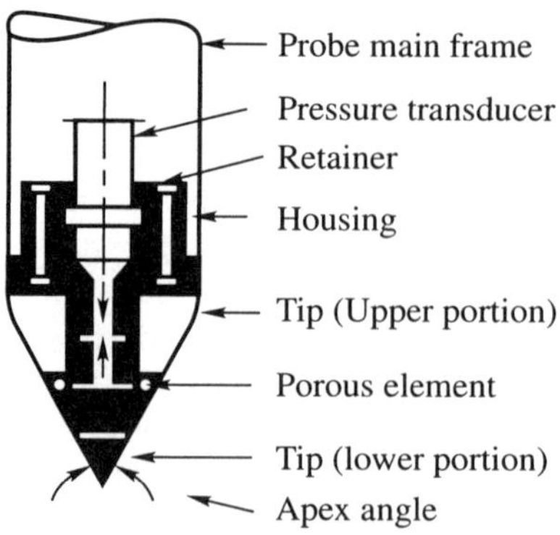

Figure 9.13 Details of 60°/10 cm² piezocone

In the same way, the local side friction f_c is

$$f_c = \frac{Q_f}{A_f}, \tag{9.12}$$

where, $Q_f = Q_t - Q_c$ = force required to push the friction jacket,
Q_t = the total force required to push the cone and friction jacket together in the case of a mechanical penetrometer,
A_f = surface area of the friction jacket.

Friction Ratio, R_f

Friction ratio, R_f is expressed as

$$R_f = \frac{f_c}{q_c}, \tag{9.13}$$

where f_c and q_c are measured at the same depth. R_f is expressed as a percentage. Friction ratio is an important parameter for classifying soil (Fig. 9.16).

Relationship Between q_o, Relative Density D_r and Friction Angle ϕ for Sand

Research carried out by many indicates that a unique relationship between cone resistance, relative density and friction angle valid for all sands does not exist. Robertson and Campanella (1983a) have provided a set of curves (Fig. 9.14) which may be used to estimate D_r based on q_c and effective

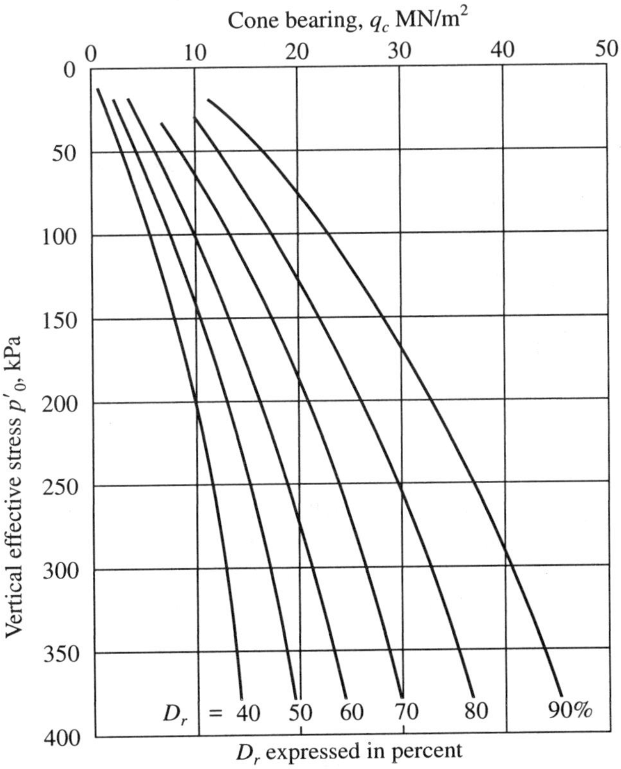

Figure 9.14 Relationship between relative density D_r and penetration resistance q_c for uncemented quartz sands (Robertson and Campanella, 1983a)

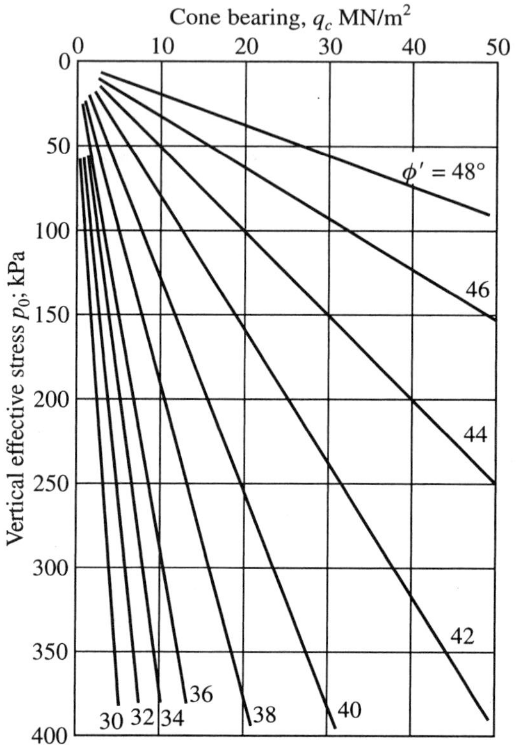

Figure 9.15 Relationship between cone point resistance q_c and angle of internal friction ϕ for uncemented quartz sands (Robertson and Campanella, 1983b)

overburden pressure. These curves are supposed to be applicable for normally consolidated clean sand. Fig. 9.15 gives the relationship between q_c and ϕ (Robertson and Campanella, 1983b).

Relationship Between q_c and Undrained Shear Strength, c_u of Clay

The cone penetration resistance q_c and c_u may be related as

$$q_c = N_k c_u + p_0 \quad \text{or} \quad c_u = \frac{q_c - p_o}{N_k} \tag{9.14}$$

where, N_k = cone factor,

$p_o = \gamma z$ = overburden pressure.

Lunne and Kelven (1981) investigated the value of the cone factor N_k for both normally consolidated and overconsolidated clays. The values of N_k as obtained are given below:

Type of clay	Cone factor
Normally consolidated	11 to 19
Overconsolidated	
At shallow depths	15 to 20
At deep depths	12 to 18

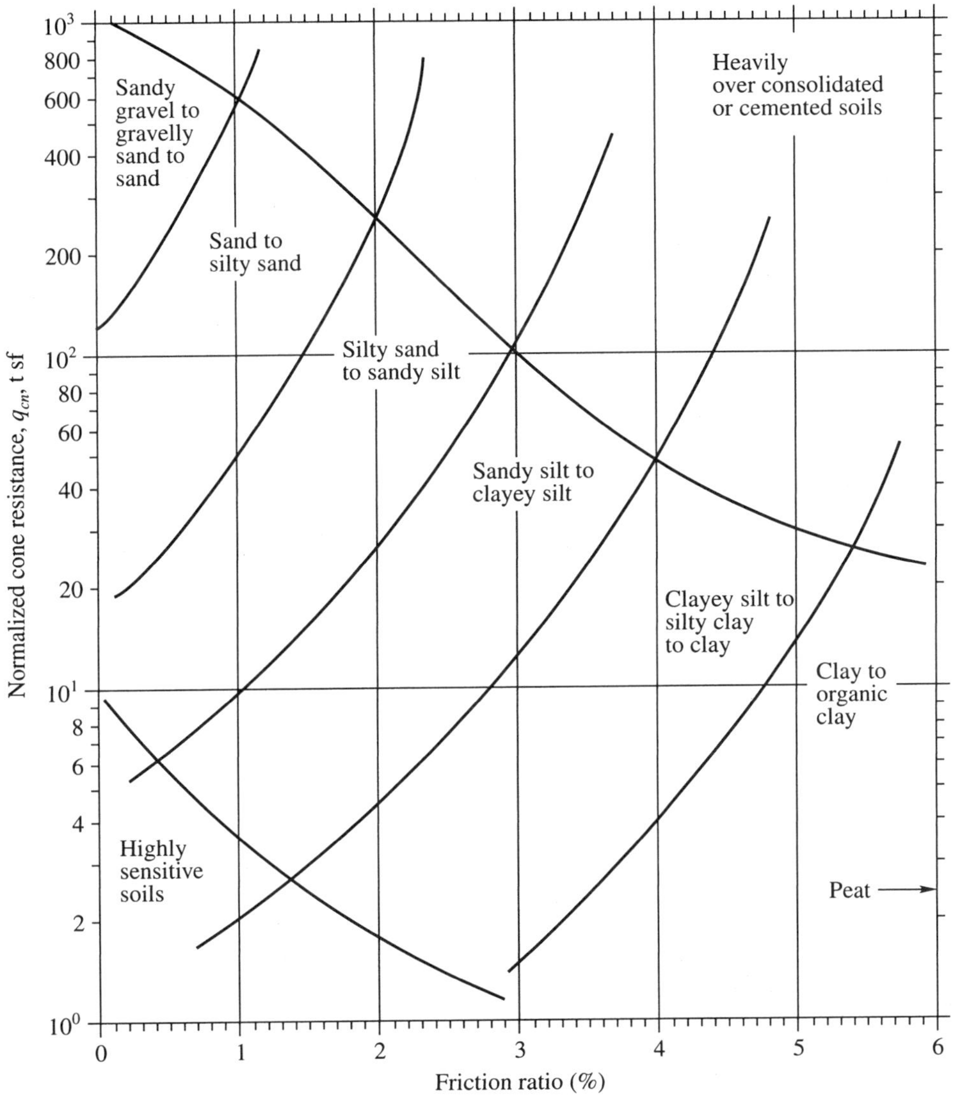

Figure 9.16 A simplified classification chart (Douglas, 1984)

Possibly a value of 20 for N_k for both types of clays may be satisfactory. Sanglerat (1972) recommends the same value for all cases where an overburden correction is of negligible value.

Soil Classification

One of the basic uses of CPT is to identify and classify soils. A CPT-Soil Behavior Type Prediction System has been developed by Douglas and Olsen (1981) using an electric-friction cone penetrometer. The classification is based on the *friction ratio* f_c/q_c. The ratio f_c/q_c varies greatly depending on whether it applies to clays or sands. Their findings have been confirmed by hundreds of tests.

For clay soils, it has been found that the friction ratio decreases with increasing liquidity index I_l. Therefore, the friction ratio is an indicator of the soil type penetrated. It permits approximate identification of soil type though no samples are recovered.

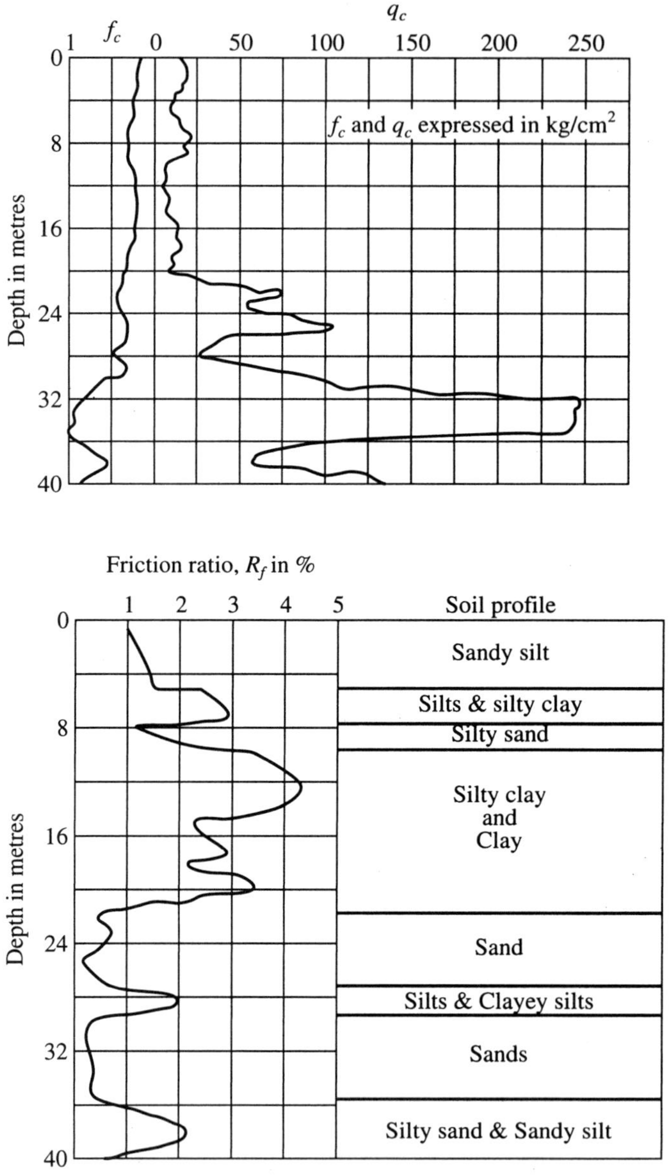

Figure 9.17 A typical sounding log

Douglas (1984) presented a simplified classification chart shown in Fig. 9.16. His chart uses cone resistance normalized (q_{cn}) for overburden pressure using the equation

$$q_{cn} = q_c(1 - 1.25 \log p'_o) \tag{9.15}$$

where, p'_o = effective overburden pressure in tsf, and q_c = cone resistance in tsf,

In conclusion, CPT data provide a repeatable index of the aggregate behavior of *in-situ* soil. The CPT classification method provides a better picture of overall subsurface conditions than is available with most other methods of exploration.

A typical sounding log is given in Fig. 9.17.

Table 9.5 Soil classification based on friction ratio R_f (Sanglerat, 1972)

R_f%	Type of soil
0–0.5	Loose gravel fill
0.5–2.0	Sands or gravels
2–5	Clay sand mixtures and silts
> 5	Clays, peats etc.

The friction ratio R_f varies greatly with the type of soil. The variation of R_f for the various types of soils is generally of the order given in Table 9.5

Correlation Between SPT and CPT

Meyerhof (1965) presented comparative data between SPT and CPT. For fine or silty medium loose to medium dense sands, he presents the correlation as

$$q_c = 0.4\,N \text{ MN/m}^2 \tag{9.16}$$

His findings are as given in Table 9.6.

Table 9.6 Approximate relationship between relative density of fine sand, the SPT, the static cone resistance and the angle of internal fraction (Meyerhof, 1965)

State of sand	D_r	N_{cor}	q_c (MPa)	$\phi°$
Very loose	<0.2	<4	<2.0	<30
Loose	0.2–0.4	4–10	2–4	30–35
Medium dense	0.4–0.6	10–30	4–12	35–40
Dense	0.6–0.8	30–50	12–20	40–45
Very dense	0.8–1.0	>50	>20	45

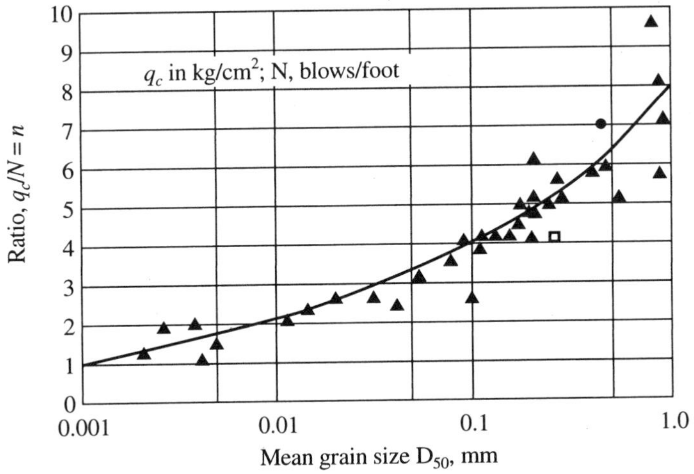

Figure 9.18 Relationship between q_c/N and mean grain size D_{50} (mm) (Robertson and Campanella, 1983a)

The lowest values of the angle of internal friction given in Table 9.6 are conservative estimates for uniform, clean sand and they should be reduced by at least 5° for clayey sand. These values, as well as the upper values of the angles of internal friction which apply to well graded sand, may be increased by 5° for gravelly sand.

Figure 9.18 shows a correlation presented by Robertson and Campanella (1983) between the ratio of q_c/N and mean grain size, D_{50}. It can be seen from the figure that the ratio varies from 1 at $D_{50} = 0.001$ mm to a maximum value of 8 at D_{50} = 1.0 mm. The soil type also varies from clay to sand.

It is clear from the above discussions that the value of $n(= q_c/N)$ is not a constant for any particular soil. Designers must use their own judgment while selecting a value for n for a particular type of soil.

Example 9.7

If a deposit at a site happens to be a saturated overconsolidated clay with a value of $q_c = 8.8$ MN/m², determine the unconfined compressive strength of clay given $p_0 = 127$ kN/m²

Solution

Per Eq (9.14)

$$c_u = \frac{q_c - p_o}{N_k} \quad \text{or} \quad q_u = \frac{2(q_c - p_o)}{N_k}$$

Assume $N_k = 20$. Substituting the known values and simplifying

$$q_u = \frac{2(8800 - 127)}{20} = 867 \text{ kN/m}^2$$

If we neglect the overburden pressure p_0

$$q_u = \frac{2 \times 8800}{20} = 880 \text{ kN/m}^2$$

It is clear that, the value of q_u is little affected by neglecting the overburden pressure in Eq. (9.14)

Example 9.8

Static cone penetration tests were carried out at a site by using an electric-friction cone penetrometer. The following data were obtained at a depth of 12.5 m.

cone resistance q_c = 19.152 MN/m² (200 tsf)

friction ratio $R_f = \dfrac{f_c}{q_c} = 1.3$

Classify the soil as per Fig. 9.16. Assume γ(effective) = 16.5 kN/m³.

Solution

The values of $q_c = 19.152 \times 10^3$ kN/m² and $R_f = 1.3$. From Eq. (9.15)

$$q_{cn} = 200 \times \left[1 - 1.25 \log\left(\frac{16.5 \times 12.5}{100}\right)\right] = 121 \text{ tsf}$$

The soil is sand to silty sand (Fig. 9.16) for q_{cn} = 121 tsf and R_f = 1.3.

Example 9.9

Static cone penetration test at a site at depth of 30 ft revealed the following

Cone resistance q_c = 125 tsf

Friction ratio R_f = 1.3%

The average effective unit weight of the soil is 115 psf. Classify the soil per Fig. 9.16.

Solution

The effective overburden pressure $p'_0 = 30 \times 115 = 3450 \text{ lb/ft}^2 = 1.725 \text{ tsf}$

From Eq (9.15)

$q_{cn} = 125\,(1–1.25 \log 1.725) = 88$ tsf

$R_f = 1.3\%$

From Fig. 9.16, the soil is classified as sand to silty sand for $q_{cn} = 88$ tsf and $R_f = 1.3\%$

Example 9.10

The static cone penetration resistance at a site at 10 m depth is 2.5 MN/m^2. The friction ratio obtained from the test is 4.25%. If the unit weight of the soil is 18.5 kN/m^3, what type of soil exists at the site.

Solution

$q_c = 2.5 \times 1000 \text{ kN/m}^2 = 2500 \text{ kN/m}^2 = 26.1$ tsf

$p'_0 = 10 \times 18.5 = 185 \text{ kN/m}^2 = 1.93$ tsf

$q_{cn} = 26.1\,(1–1.25 \log 1.93) = 16.8$ tsf

$R_f = 4.25\ \%$

From Fig 9.16, the soil is classified as clayey silt to silty clay to clay

9.9 PRESSUREMETER

A pressuremeter test is an *in-situ* stress-strain test performed on the walls of a bore hole using a cylindrical probe that can be inflated radially. The pressuremeter, which was first conceived, designed, constructed and used by Menard (1957) of France, has been in use since 1957. The test results are used either directly or indirectly for the design of foundations. The Menard test has been adopted as ASTM Test Designation 4719. The instrument as conceived by Menard consists of three independent chambers stacked one above the other (Fig. 9.19) with inflatable user membranes held together at top and bottom by steel discs with a rigid hollow tube at the center. The top and bottom chambers protect the middle chamber from the end effects caused by the finite length of the apparatus, and these are known as guard cells. The middle chamber with the end cells together is called the *Probe.* The pressuremeter consists of three parts, namely, the probe, the control unit and the tubing.

The Pressuremeter Test

The pressuremeter test involves the following:

1. Drilling of a hole
2. Lowering the probe into the hole and clamping it at the desired elevation
3. Conducting the test

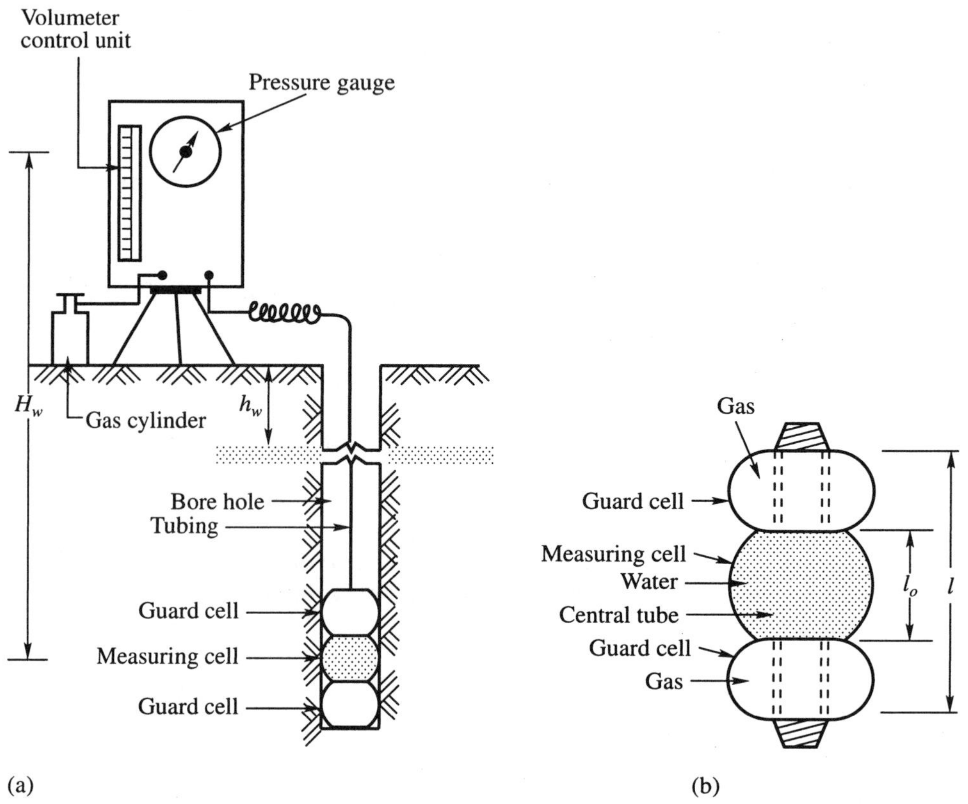

Figure 9.19 Components of Menard pressuremeter

Drilling and Positioning of Probe

A Menard pressure test is carried out in a hole drilled in advance. The drilling of the hole is completed using a suitable drilling rig which disturbs the soil the least. The diameter of the bore hole, D_h, in which the test is to be conducted shall satisfy the condition

$$1.03D_p < D_h < 1.20D_p \tag{9.17}$$

where D_p is the diameter of the probe under the deflated condition.

Typical sizes of the probe and bore hole are given in Table 9.7.

The probe is lowered down the hole soon after boring to the desired elevation and held in position by a clamping device. Pressuremeter tests are usually carried out at 1 m intervals in all the bore holes.

Conducting the Test

With the probe in position in the bore hole, the test is started by opening the valves in the control unit for admitting water and gas (or water) to the measuring cell and the guard cells respectively. The pressure in the guard cells is normally kept equal to the pressure in the measuring cell. The pressures to the soil through the measuring cell are applied by any one of the following methods:

1. Equal pressure increment method,
2. Equal volume increment method.

Table 9.7 Typical sizes of probe and bore hole for pressuremeter test

Hole dia. designation	Probe dia. (mm)	l_0 (cm)	l (cm)	Bore hole dia. Nominal (mm)	Bore hole dia. Max. (mm)
AX	44	36	66	46	52
BX	58	21	42	60	66
NX	70	25	50	72	84

Note: l_0 = length of measuring cell; l = length of probe.

If pressure is applied by the first method, each equal increment of pressure is held constant for a fixed length of time, usually one minute. Volume readings are made after one minute elapsed time. Normally ten equal increments of pressure are applied for a soil to reach the limit pressure, p_l.

If pressure is applied by the second method, the volume of the probe shall be increased in increments equal to 5 percent of the nominal volume of the probe (in the deflated condition) and held constant for 30 seconds. Pressure readings are taken after 30 seconds of elapsed time.

Steps in both the methods are continued until the maximum probe volume to be used in the test is reached. The test may continue at each position from 10 to 15 minutes. This means that the test is essentially an undrained test in clay soils and a drained test in a freely draining material.

Typical Test Result

First a typical curve based on the observed readings in the field may be plotted. The plot is made of the volume of the water read at the volumeter in the control unit, v, as abscissa for each increment of pressure, p, as ordinate. The curve is a result of the test conducted on the basis of equal increments of pressure and each pressure held constant for a period of one minute. This curve is a raw curve which requires some corrections. The pressuremeter has, therefore, to be calibrated before it is used in design. A pressuremeter has to be calibrated for

1. Pressure loss, p_c,
2. Volume loss v_c,
3. Difference in hydrostatic pressure head H_w.

Corrected Plot of Pressure-Volume Curve

A typical corrected plot of the pressure-volume curve is given in Fig. 9.20. The characteristic parts of this curve are:

1. The initial part of the curve *OA*. This curve is a result of pushing the yielded wall of the hole back to the original position. At point *A*, the at-rest condition is supposed to have been restored. The expansion of the cavity is considered only from point *A*. v_0 is the volume of water required to be injected over and above the volume V_c of the probe under the deflated condition. If V_0 is the total volume of the cavity at point *A*, we can write

   $$V_0 = V_c + v_0 \tag{9.18}$$

   where v_o is the abscissa of point *A*. The horizontal pressure at point *A* is represented as p_{om}.
2. The second part of the curve is *AB*. This is supposed to be a straight line portion of the curve and may represent the elastic range. Since *AB* gives an impression of an elastic range, it is called the *pseudo-elastic* phase of the test. Point *A* is considered to be the start of the pressuremeter test in most theories. Point *B* marks the end of the straight line portion of the curve. The coordinates of point *B* are p_f and v_f, where p_f is known as the *creep pressure*.

3. Curve *BC* marks the final phase. The plastic phase is supposed to start from point *B*, and the curve becomes asymptotic at point *C* at a large deformation of the cavity. The *limit pressure*, p_l, is usually defined as the pressure that is required to double the initial volume of the cavity. It occurs at a volume such that

$$v_l - v_0 = V_0 = V_c + v_0 \tag{9.19}$$

$$\text{or} \quad v_l = V_c + 2v_0 \tag{9.20}$$

v_0 is normally limited to about 300 cm^3 for probes used in *AX* and *BX* holes. The initial volume of these probes is on the order of 535 cm^3. This means that $(V_c + 2v_0)$ is on the order of 1135 cm^3. These values may vary according to the design of the pressuremeter.

The reservoir capacity in the control unit should be of the order of 1135 cm^3. In case the reservoir capacity is limited and p_l is not reached within its limit, the test, has to be stopped at that level. In such a case, the limit value, p_l, has to be extrapolated.

At-Rest Horizontal Pressure

The at-rest total horizontal pressure, p_{oh}, at any depth, *z*, under the *in-situ* condition before drilling a hole may be expressed as

$$p_{oh} = (\gamma z - u)K_0 + u, \tag{9.21}$$

where u = pore pressure at depth *z*,

γ = gross unit weight of the soil,

K_0 = coefficient of earth pressure for the at-rest condition.

The values of γ and K_0 are generally assumed taking into account the type and condition of the soil. The pore pressure under the hydrostatic condition is

$$u = \gamma_w (z - h_w), \tag{9.22}$$

where γ_w = unit weight of water,

h_w = depth of water table from the ground surface.

As per Fig. 9.20, p_{om} is the pressure which corresponds to the volume v_0 at the start of the straight line portion of the curve. Since it has been found that it is very difficult to determine accurately p_{om}, p_{oh} may not be equal to p_{om}. As such, p_{om} bears no relation to what the true earth pressure at-rest is. In Eq. (9.21) K_0 has to be assumed and its accuracy is doubtful. In such circumstances it is not possible to calculate p_{oh}. However, p_{om} can be used for calculating the pressuremeter modulus E_m. The experience of many investigators is that a self-boring pressuremeter gives reliable values for p_{oh}.

The Pressuremeter Modulus E_m

Since the curve between points *A* and *B* in Fig. 9.20 is approximately a straight line, the soil in this region may be assumed to behave as a more or less elastic material. The equation for the pressuremeter modulus may be expressed as

$$E_p = 2G_s(1+\mu) = 2(1+\mu)V_m \frac{\Delta p}{\Delta v} \tag{9.23}$$

where G_s is the shear modulus.

If V_m is the volume at mid point (Fig 9.20), we may write,

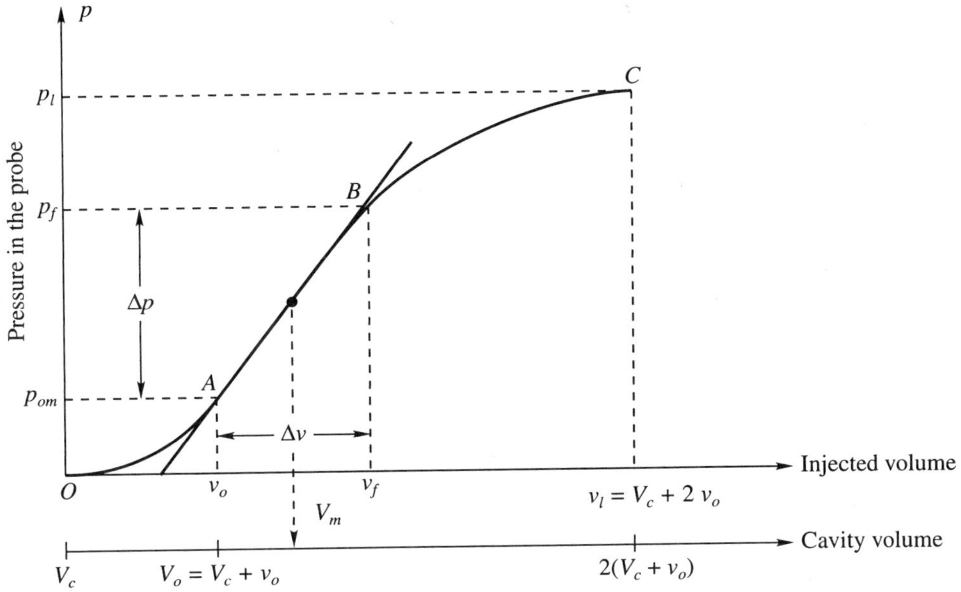

Figure 9.20 A typical corrected pressuremeter curve

$$V = V_m = V_c + \frac{v_0 + v_f}{2}, \tag{9.24}$$

where V_c is the volume of the deflated portion of the measuring cell at zero volume reading on the volumeter in the control unit.

Suitable values *for* μ may be assumed in the above equation depending on the type of soil. For saturated clay soils μ is taken as equal to 0.5 and for freely draining soils, the value is less. Since G_s (shear modulus) is not very much affected by a small variation in μ, Menard proposed a constant value of 0.33 for μ. As such the resulting deformation modulus is called Menard's Modulus E_m. The equation for E_m reduces to

$$E_m = 2.66 V_m \frac{\Delta p}{\Delta v} \tag{9.25}$$

The following empirical relationship has been established from the results obtained from pressuremeter tests. Undrained shear strength c_u as a function of the limit pressure $\overline{p}_l$ may be expressed as

$$c_u = \frac{\overline{p}_l}{9} \tag{9.26}$$

where $\overline{p}_l = p_l - p_{oh}$ and p_{oh} = total horizontal earth pressure for the at rest condition.

Amar and Jézéquel (1972) have suggested another equation of the form

$$c_u = \frac{\overline{p}_l}{10} + 25 \text{ kPa} \tag{9.27}$$

where both $\overline{p}_l$ and c_u are in kPa.

Example 9.11

A pressuremeter test was carried out at a site at a depth of 7 m below the ground surface. The water table level was at a depth of 1.5 m. The average unit weight of saturated soil is 17.3 kN/m^3. The corrected pressuremeter curve is given in Fig. Ex. 9.11 and the depleted volume of the probe is V_c = 535 cm^3. Determine the following.

(a) The coefficient of earth pressure for the at-rest condition

(b) The Menard pressuremeter modulus E_m

(c) The undrained shear strength c_u. Assume that $p_{oh} = p_{om}$ in this case

Solution

From Fig. Ex 9.11, $p_{oh} = p_{om} = 105$ kPa

The effective overburden pressure is

$$p'_0 = 17.3 \times 7 - 5.5 \times 9.81 = 67.2 \text{ kPa}$$

The effective horizontal pressure is

$$p'_{0h} = 105 - 5.5 \times 9.81 = 51.0 \text{ kPa}$$

(a) From Eq (9.21)

$$K_0 = \frac{p'_{oh}}{p'_o} = \frac{51.0}{67.2} = 0.76$$

(b) From Eq (9.25)

$$E_m = 2.66\, V_m \frac{\Delta p}{\Delta v}$$

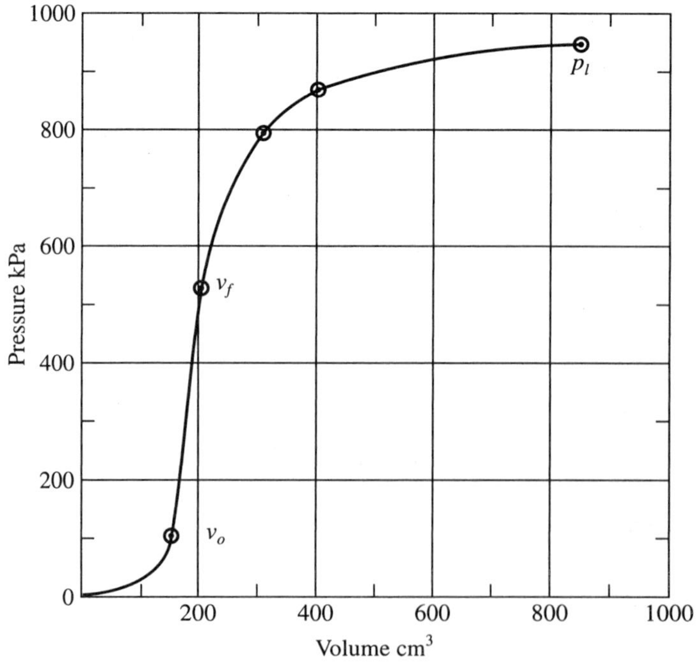

Figure Ex. 9.11

From Fig. Ex 9.11

$v_f = 200\ \text{cm}^3$ $\quad p_f = 530$ kPa

$v_o = 160\ \text{cm}^3$ $\quad p_{om} = 105$ kPa

From Eq (9.24) $V_m = 535 + \dfrac{200+160}{2} = 715\text{cm}^3$

$$\frac{\Delta p}{\Delta v} = \frac{530-105}{200-160} = 10.625$$

Now $E_m = 2.66 \times 715 \times 10.625 = 20{,}208$ kPa

(c) From Eq (9.26)

$$c_u = \frac{\overline{p_l}}{9} = \frac{p_l - p_{oh}}{9}$$

From Fig Ex 9.11

$$\overline{p_l} = 950 - 105 = 845 \text{ kPa}$$

Therefore $c_u = \dfrac{845}{9} = 94$ kPa

From Eq (9.27)

$$c_u = \frac{\overline{p_l}}{10} + 25 = \frac{845}{10} + 25 = 109.5 \text{ kPa}$$

9.10 THE FLAT DILATOMETER TEST

The *flat dilatometer* is an *in-situ* testing device developed in Italy by Marchetti (1980). It is a penetration device that includes a lateral expansion arrangement after penetration. The test, therefore, combines many of the features contained in the cone penetration test and the pressuremeter test. This test has been extensively used for reliable, economical and rapid *in-situ* determination of geotechnical parameters. The flat plate dilatometer (Fig. 9.21) consists of a stainless steel blade with a flat circular expandable membrane of 60 mm diameter on one side of the stainless steel plate, a short distance above the sharpened tip. The size of the plate is 220 mm long, 95 mm wide and 14 mm thick. When at rest the external surface of the circular membrane is flush with the surrounding flat surface of the blade.

The probe is pushed to the required depth by making use of a rig used for a static cone penetrometer (Fig. 9.10). The probe is connected to a control box at ground level through a string of drill rods, electric wires for power supply and nylon tubing for the supply of nitrogen gas. Beneath the membrane is a measuring device which turns a buzzer off in the control box. The method of conducting the DMT is as follows:

1. The probe is positioned at the required level. Nitrogen gas is pumped into the probe. When the membrane is just flush with the side of the surface, a pressure reading is taken which is called the *lift-off* pressure. Approximate zero corrections are made. This pressure is called p_1.
2. The probe pressure is increased until the membrane expands by an amount $\Delta l = 1.1$ mm. The corrected pressure is p_2.

3. The next step is to decrease the pressure until the membrane returns to the lift off position. This corrected reading is p_3. This pressure is related to excess pore water pressure (Schmertmann, 1986).

The details of the calculation lead to the following equations.

1. Material index, $I_D = \dfrac{p_z - p_1}{p_2 - u}$ (9.28)

2. The lateral stress index, $K_D = \dfrac{p_1 - u}{p'_o}$ (9.29)

3. The dilatometer modulus, $E_D = 34.7(p_2 - p_1)$ kN/m^2 (9.30)

where, p'_o = effective overburden pressure = $\gamma' z$

u = pore water pressure equal to static water level pressure

γ' = effective unit weight of soil

z = depth of probe level from ground surface

The lateral stress index K_D is related to K_0 (the coefficient of earth pressure for the at-rest condition) and to OCR (overconsolidation ratio).

Marchetti (1980) has correlated several soil properties as follows

$$E_s = (1 - \mu^2)E_D \tag{9.31}$$

$$K_0 = \left(\frac{K_D}{1.5}\right)^{0.47} - 0.6 \tag{9.32}$$

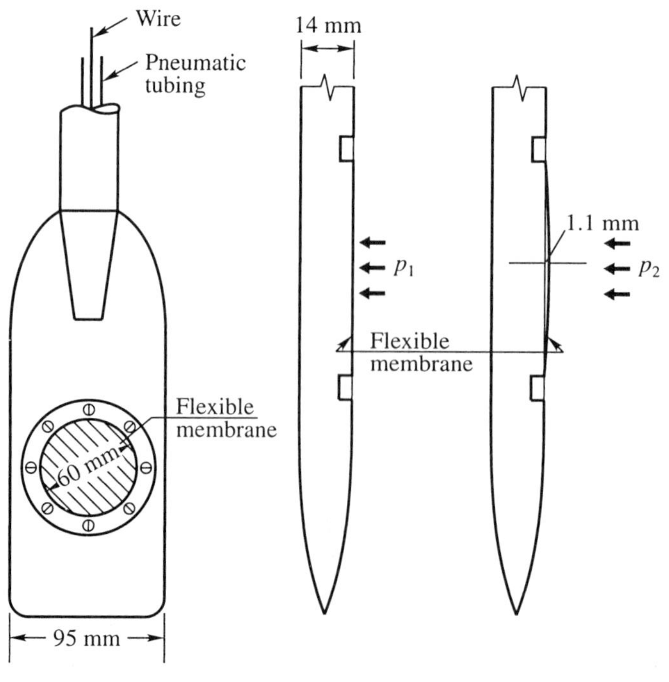

Figure 9.21 Illustration of a flat plate dilatometer (after Marchetti 1980)

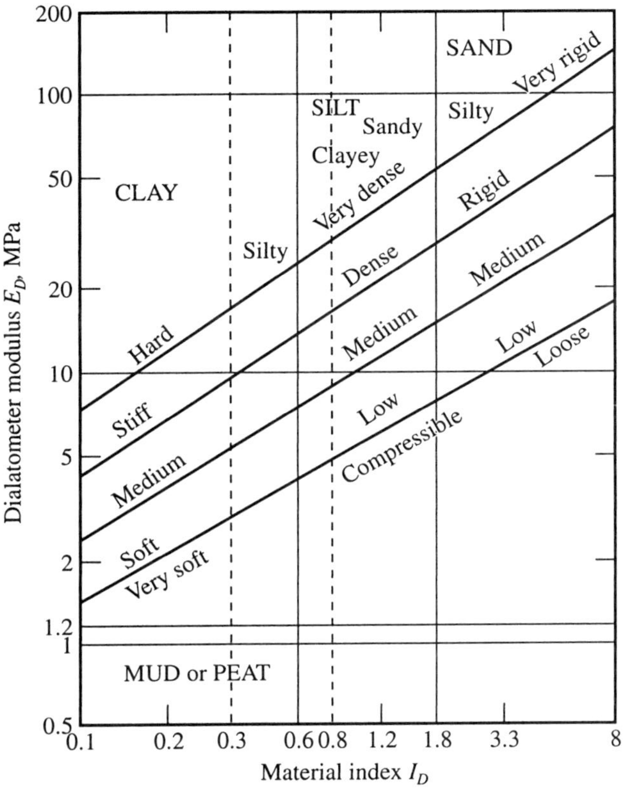

Figure 9.22 Soil profile based on dilatometer test (after Schmertmann, 1986)

$$OCR = (0.5K_D)^{1.6} \tag{9.33}$$

$$\left(\frac{c_u}{p'_o}\right)_{oc} = \left(\frac{c_u}{p'_o}\right)_{nc} \times (0.5K_D)^{1.25} \tag{9.34}$$

where E_s is the modulus of elasticity

The soil classification as developed by Schmertmann (1986) is given in Fig. 9.22. I_D is related with E_D in the development of the profile.

9.11 FIELD VANE SHEAR TEST (VST)

The vane shear test is one of the in-situ tests used for obtaining the undrained shear strength of soft sensitive clays. It is in deep beds of such material that the vane test is most valuable for the simple reason that there is at present no other method known by which the shear strength of these clays can be measured. The details of the VST have already been explained in Chapter 8.

9.12 FIELD PLATE LOAD TEST (PLT)

The field plate test is the oldest of the methods for determining either the bearing capacity or settlement of footings. The details of PLT are discussed under Shallow Foundations in Chapter 13.

9.13 GEOPHYSICAL EXPLORATION

The stratification of soils and rocks can be determined by geophysical methods of exploration which measure changes in certain physical characteristics of these materials, for example the magnetism, density, electrical resistivity, elasticity or a combination of these properties. However, the utility of these methods in the field of foundation engineering is very limited since the methods do not quantify the characteristics of the various substrata. Vital information on ground water conditions is usually lacking. Geophysical methods at best provide some missing information between widely spaced bore holes but they can not replace bore holes. Two methods of exploration which are some times useful are discussed briefly in this section. They are

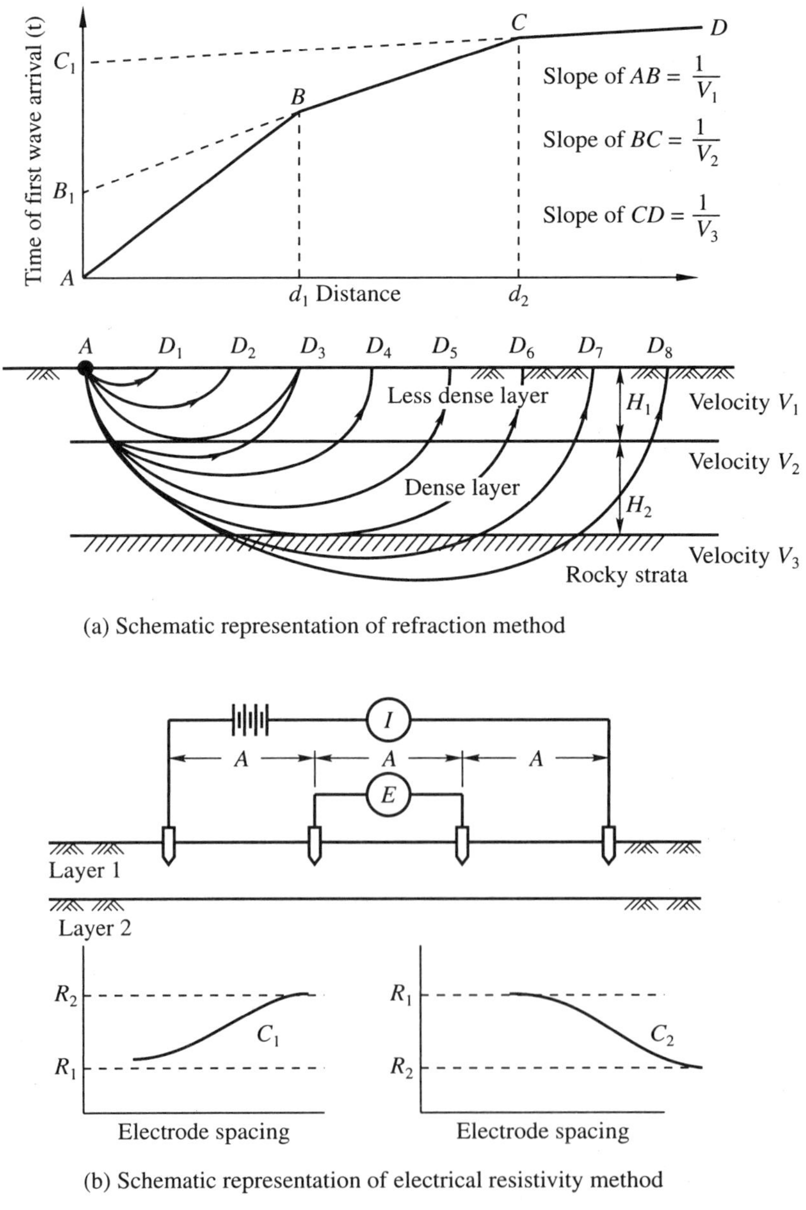

Figure 9.23 Geophysical methods of exploration

1. Seismic Refraction Method,
2. Electrical Resistivity Method.

Seismic Refraction Method

The seismic refraction method is based on the fact that seismic waves have different velocities in different types of soils (or rock). The waves refract when they cross boundaries between different types of soils. If artificial impulses are produced either by detonation of explosives or mechanical blows with a heavy hammer at the ground surface or at shallow depth within a hole, these shocks generate three types of waves. In general, only compression waves (longitudinal waves) are observed. These waves are classified as either direct, reflected or refracted. Direct waves travel in approximately straight lines from the source of the impulse to the surface. Reflected or refracted waves undergo a change in direction when they encounter a boundary separating media of different seismic velocities. The *seismic refraction method* is more suited to shallow exploration for civil engineering purposes.

The method starts by inducing impact or shock waves into the soil at a particular location. The shock waves are picked up by geophones. In Fig. 9.23(a), point A is the source of seismic impulse. The points D_1 through D_8 represent the locations of the geophones or detectors which are installed in a straight line. The spacings of the geophones are dependent on the amount of detail required and the depth of the strata being investigated. In general, the spacing must be such that the distance from D_1 to D_8 is three to four times the depth to be investigated. The geophones are connected by cable to a central recording device. A series of detonations or impacts are produced and the arrival time of the first wave at each geophone position is recorded in turn. When the distance between source and geophone is short, the arrival time will be that of a direct wave. When the distance exceeds a certain value (depending on the thickness of the stratum), the refracted wave will be the first to be detected by the geophone. This is because the refracted wave, although longer than that of the direct wave, passes through a stratum of higher seismic velocity.

A typical plot of test results for a three layer system is given in Fig. 9.23(a) with the arrival time plotted against the distance source and geophone. As in the figure, if the source-geophone spacing is more than the distance d_1, which is the distance from the source to point B, the direct wave reaches the geophone in advance of the refracted wave and the time-distance relationship is represented by a straight line AB through the origin represented by A. If on the other hand, the source geophone distance is greater than d_1, the refracted waves arrive in advance of the direct waves and the time-distance relationship is represented by another straight line BC which will have a slope different from that of AB. The slopes of the lines AB and BC are represented by $1/V_1$, and $1/V_2$ respectively, where V_1 and V_2 are the velocities of the upper and lower strata respectively. Similarly, the slope of the third line CD is represented by $1/V_3$ in the third strata.

The general types of soil or rocks can be determined from a knowledge of these velocities. The depth H_1 of the top strata (provided the thickness of the stratum is constant) can be estimated from the formula

$$H_1 = \frac{d_1}{2}\sqrt{\frac{V_2 - V_1}{V_2 + V_1}} \tag{9.35a}$$

The thickness of the second layer (H_2) is obtained from

$$H_2 = 0.85H_1 + \frac{d_2}{2}\sqrt{\frac{V_3 - V_2}{V_3 + V_2}} \tag{9.35b}$$

The procedure is continued if there are more than three layers.

If the thickness of any stratum is not constant, average thickness is taken.

Table 9.8 Range of seismic velocities in soils near the surface or at shallow depths (after Peck et al., 1974)

Material	Velocity	
	ft/sec	m/sec
1. Dry silt, sand, loose gravel, loam, loose rock talus, and moist fine-grained top soil	600-2500	180-760
2. Compact till, indurated clays, compact clayey gravel, cemented sand and sand clay	2500-7500	760-2300
3. Rock, weathered, fractured or partly decomposed	2000-10,000	600-3000
4. Shale, sound	2500-11,000	760-3350
5. Sandstone, sound	5000-14,000	1500-4300
6. Limestone, chalk, sound	6000-20,000	1800-6000
7. Igneous rock, sound	12,000-20,000	3650-6000
8. Metamorphic rock, sound	10,000-16,000	3000-4900

The following equations may be used for determining the depths H_1 and H_2 in a three layer strata:

$$H_1 = \frac{t_1 V_1}{2\cos\alpha} \tag{9.36}$$

$$H_2 = \frac{t_2 V_2}{2\cos\beta} \tag{9.37}$$

where t_1 = AB_1, (Fig. 9.23a); the point B_1 is obtained on the vertical passing through A by extending the straight line CB,

t_2 = $(AC_1 - AB_1)$; AC_1 is the intercept on the vertical through A obtained by extending the straight line DC,

α = $\sin^{-1}(V_1/V_2)$,

β = $\sin^{-1}(V_2/V_3)$. (9.38)

α and β are the angles of refraction of the first and second stratum interfaces respectively.

The formulae used to estimate the depths from seismic refraction survey data are based on the following assumptions:

1. Each stratum is homogeneous and isotropic.
2. The boundaries between strata are either horizontal or inclined planes.
3. Each stratum is of sufficient thickness to reflect a change in velocity on a time-distance plot.
4. The velocity of wave propagation for each succeeding stratum increases with depth.

Table 9.8 gives typical seismic velocities in various materials. Detailed investigation procedures for refraction studies are presented by Jakosky (1950).

Electrical Resistivity Method

The method depends on differences in the electrical resistance of different soil (and rock) types. The flow of current through a soil is mainly due to electrolytic action and therefore depends on the concentration of dissolved salts in the pores. The mineral particles of soil are poor conductors of current. The resistivity of soil, therefore, decreases as both water content and concentration of salts increase. A dense clean sand above the water table, for example, would exhibit a high resistivity due to its low degree of saturation and virtual absence of dissolved salts. A saturated clay of high void ratio, on the other hand, would exhibit a low resistivity due to the relative abundance of pore water and the free ions in that water.

There are several methods by which the field resistivity measurements are made. The most popular of the methods is the Wenner Method.

Wenner Method

The Wenner arrangement consists of four equally spaced electrodes driven approximately 20 cm into the ground as shown in Fig. 9.23(b). In this method a *dc* current of known magnitude is passed between the two outer (current) electrodes, thereby producing within the soil an electric field, whose pattern is determined by the resistivities of the soils present within the field and the boundary conditions. The potential drop E for the surface current flow lines is measured by means of the inner electrodes. The apparent resistivity, R, is given by the equation

$$R = \frac{2\pi AE}{I} \tag{9.39}$$

It is customary to express A in centimeters, E in volts, I in amperes, and R ohm-cm. The apparent resistivity represents a weighted average of true resistivity to a depth A in a large volume of soil, the soil close to the surface being more heavily weighted than the soil at greater depths. The presence of a stratum of low resistivity forces the current to flow closer to the surface resulting in a higher voltage drop and hence a higher value of apparent resistivity. The opposite is true if a stratum of low resistivity lies below a stratum of high resistivity.

The method known as *sounding* is used when the variation of resistivity with depth is required. This enables rough estimates to be made of the types and depths of strata. A series of readings are taken, the (equal) spacing of the electrodes being increased for each successive reading. However, the center of the four electrodes remains at a fixed point. As the spacing is increased, the apparent resistivity is influenced by a greater depth of soil. If the resistivity increases with the increasing electrode spacings, it can be concluded that an underlying stratum of higher resistivity is beginning to influence the readings. If increased separation produces decreasing resistivity, on the other hand, a lower resistivity is beginning to influence the readings.

Apparent resistivity is plotted against spacing, preferably, on log paper. Characteristic curves for a two layer structure are shown in Fig. 9.23(b). For curve C_1, the resistivity of layer 1 is lower than that of 2; for curve C_2, layer 1 has a higher resistivity than that of layer 2. The curves become asymptotic to lines representing the true resistance R_1, and R_2 of the respective layers. Approximate layer thickness can be obtained by comparing the observed curves of resistivity versus electrode spacing with a set of standard curves.

The procedure known as *profiling* is used in the investigation of lateral variation of soil types. A series of readings is taken, the four electrodes being moved laterally as a unit for each successive reading; the electrode spacing remains constant for each reading of the series. Apparent resistivity is plotted against the center position of the four electrodes, to natural scale; such a plot can be used to locate the position of a soil of high or low resistivity. Contours of resistivity can be plotted over a given area.

The electrical method of exploration has been found to be not as reliable as the seismic method as the apparent resistivity of a particular soil or rock can vary over a wide range of values.

Representative values of resistivity are given in Table 9.9.

Table 9.9 Representative values of resistivity. The values are expressed in units of 10^3 ohm-cm (after Peck et al., 1974)

Material	Resistivity ohm-cm
Clay and saturated silt	0-10
Sandy clay and wet silty sand	10-25
Clayey sand and saturated sand	25-50
Sand	50-150
Gravel	150-500
Weathered rock	100-200
Sound rock	150–4,000

Example 9.12

A seismic survey was carried out for a large project to determine the nature of the substrata. The results of the survey are given in Fig. Ex 9.12 in the form of a graph. Determine the depths of the strata.

Solution

Two methods may be used

1. Use of Eq (9.35)
2. Use of Eqs (9.36) and (9.37)

First we have to determine the velocities in each stratum (Fig. Ex. 9.12).

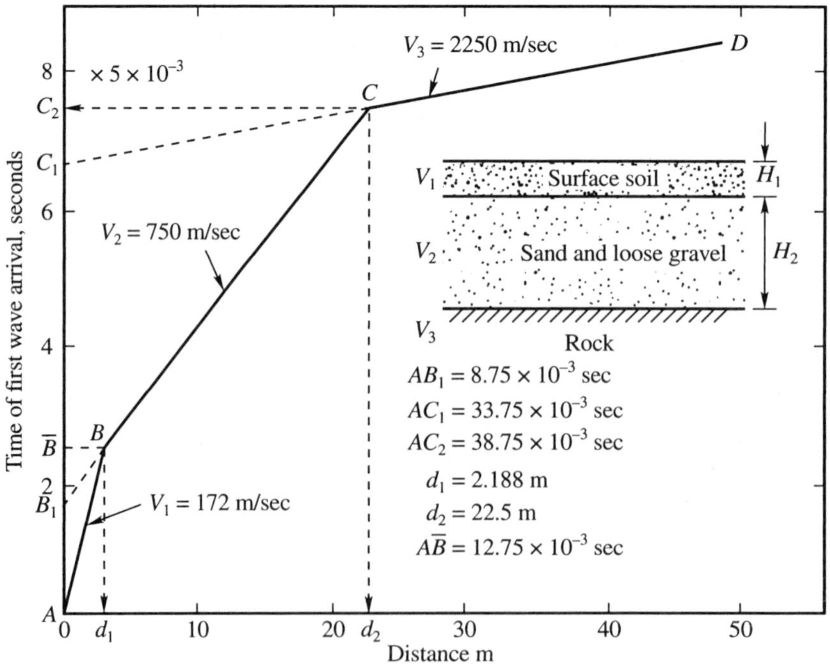

Figure Ex. 9.12

$$V_1 = \frac{\text{distance}}{A\bar{B}} = \frac{2.188}{12.75 \times 10^{-3}} = 172 \text{ m/sec}$$

$$V_2 = \frac{d_2}{AC_2 - AB_1} = \frac{22.5}{(7.75 - 1.75)5 \times 10^{-3}} = 750 \text{ m/sec}$$

In the same way, the velocity in the third stata can be determined. The velocity obtained is $V_3 = 2250$ m/sec

Method 1
From Eq (9.35 a), the thickness H_1 of the top layer is

$$H_1 = \frac{d_1}{2}\sqrt{\frac{V_2 - V_1}{V_2 + V_1}} = \frac{2.188}{2}\sqrt{\frac{750 - 172}{1000}} = 0.83 \text{ m}$$

From Eq (9.35b) the thickness H_2 is

$$H_2 = 0.85H_1 + \frac{d_2}{2}\sqrt{\frac{V_s - V_2}{V_s + V_2}}$$

$$H_2 = 0.85 \times 0.83 + \frac{22.5}{2}\sqrt{\frac{2250 - 750}{3000}}$$

$$= 0.71 + 7.955 = 8.67 \text{ m}$$

Method 2
From Eq (9.36)

$$H_1 = \frac{t_1 V_1}{2\cos\alpha}$$

$$t_1 = AB_1 = 1.75 \times 5 \times 10^{-3} \text{ sec(Fig. Ex.9.12)}$$

$$\alpha = \sin^{-1}\frac{V_1}{V_2} = \sin^{-1}\frac{172}{750} = 13.26°$$

$$\cos\alpha = 0.9733$$

$$H_1 = \frac{12.75 \times 10^{-3} \times 172}{2 \times 0.9737} = 1.13 \text{ m}$$

From Eq (9.37)

$$H_2 = \frac{t_2 V_2}{2\cos\beta}$$

$$t_2 = 5 \times 5 \times 10^{-3} \text{ sec}$$

$$\beta = \sin^{-1}\frac{750}{2250} = 19.47°; \cos\beta = 0.9428$$

$$H_2 = \frac{5 \times 5 \times 10^{-3} \times 750}{2 \times 0.9428} = 9.94 \text{m}$$

9.14 PLANNING OF SOIL EXPLORATION

The planner has to consider the following points before making a program:

1. Type, size and importance of the project.
2. Whether the site investigation is preliminary or detailed.

In the case of large projects, a preliminary investigation is normally required for the purpose of

1. Selecting a site and making a feasibility study of the project,
2. Making tentative designs and estimates of the cost of the project.

Preliminary site investigation needs only a few bore holes distributed suitably over the area for taking samples. The data obtained from the field and laboratory tests must be adequate to provide a fairly good idea of the strength characteristics of the subsoil for making preliminary drawings and design. In case a particular site is found unsuitable on the basis of the study, an alternate site may have to be chosen.

Once a site is chosen, a detailed soil investigation is undertaken. The planning of a soil investigation includes the following steps:

1. A detailed study of the geographical condition of the area which include
   (a) Collection of all the available information about the site, including the collection of existing topographical and geological maps,
   (b) General topographical features of the site,
   (c) Collection of the available hydraulic conditions, such as water table fluctuations, flooding of the site etc,
   (d) Access to the site.
2. Preparation of a layout plan of the project.
3. Preparation of a borehole layout plan which includes the depths and the number of bore holes suitably distributed over the area.
4. Marking on the layout plan any additional types of soil investigation.
5. Preparation of specifications and guidelines for the field execution of the various elements of soil investigation.
6. Preparation of specifications and guide lines for laboratory testing of the samples collected, presentation of field and laboratory test results, writing of report, etc.

The planner can make an intelligent, practical and pragmatic plan if he is conversant with the various elements of soil investigation.

Depths and Number of Bore Holes

Depths of Bore Holes

The depth up to which bore holes should be driven is governed by the depth of soil affected by the foundation bearing pressures. The standard practice is to take the borings to a depth (called the significant depth) at which the excess vertical stress caused by a fully loaded foundation is of the order of 20 per cent or less of the net imposed vertical stress at the foundation base level. The depth the borehole as per this practice works out to about 1.5 times the least width of the foundation from the base level of the foundation as shown in Fig. 9.24(a). Where strip or pad footings are closely spaced which results in the overlapping of the stressed zones, the whole loaded area becomes in effect a raft foundation with correspondingly deep borings as shown in Fig. 9.24(b) and (c). In the case of pile or pier foundations the subsoil should be explored to the depths required to cover the soil lying even below the tips of piles (or pile groups) and piers which are affected by the loads transmitted to the deeper layers, Fig. 9.24(d). In case rock is encountered at shallow depths, foundations may have to rest on rocky strata. The boring should also explore the strength characteristics of rocky strata in such cases.

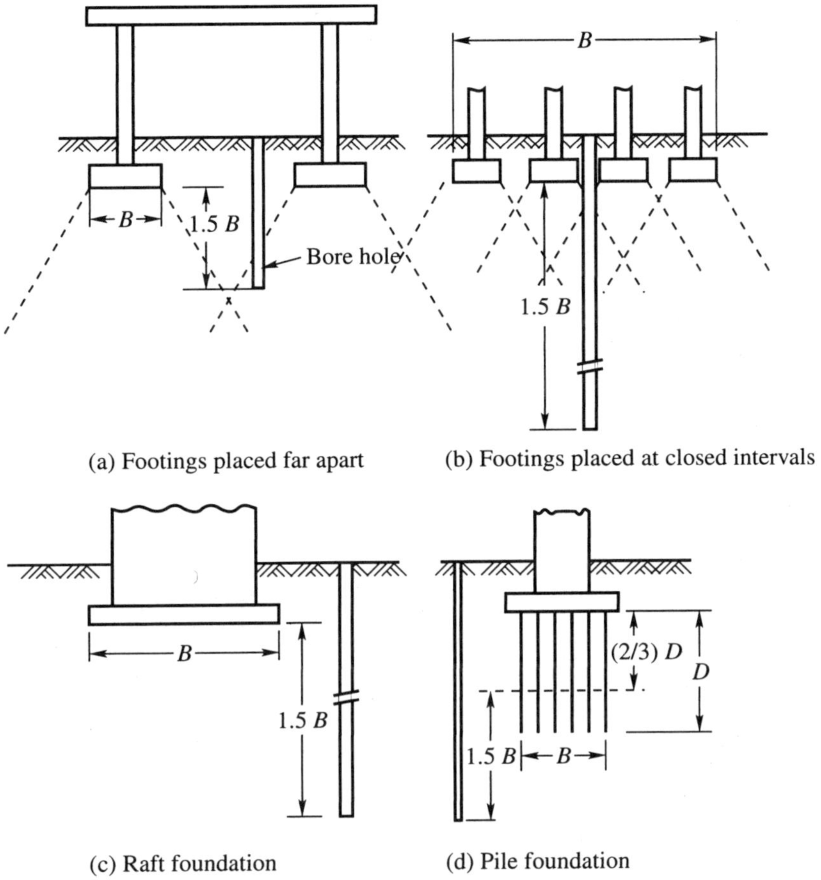

Figure 9.24 Depth of bore holes

Number of Bore Holes

An adequate number of bore holes is needed to

1. Provide a reasonably accurate determination of the contours of the proposed bearing stratum,
2. Locate any soft pockets in the supporting soil which would adversely affect the safety and performance of the proposed design.

The number of bore holes which need to be driven on any particular site is a difficult problem which is closely linked with the relative cost of the investigation and the project for which it is undertaken. When the soil is homogeneous over the whole area, the number of bore holes could be limited, but if the soil condition is erratic, limiting the number would be counter productive.

9.15 EXECUTION OF SOIL EXPLORATION PROGRAM

The three limbs of a soil exploration are

1. Planning,
2. Execution,
3. Report writing.

All three limbs are equally important for a satisfactory solution of the problem. However, the execution of the soil exploration program acts as a bridge between planning and report writing, and as such occupies an important place. No amount of planning would help report writing, if the field and laboratory works are not executed with diligence and care. It is essential that the execution part should always be entrusted to well qualified, reliable and resourceful geotechnical consultants who will also be responsible for report writing.

Deployment of Personnel and Equipment

The geotechnical consultant should have well qualified and experienced engineers and supervisors who complete the work per the requirements. The firm should have the capacity to deploy an adequate number of rigs and personnel for satisfactory completion of the job on time.

BOREHOLE LOG

Job No.	Date: 6-4-84
Project: Farakka STPP	BH No.: 1
	GL: 64.3 m
Location: WB	WTL: 63.0 m
Boring Method: Shell & Auger	Supervisor: X
Dia. of BH 15 cm	

Soil Type		Level m	Depth m	SPT 15 cm	15 cm	15 cm	N	Sample type	Remarks
Yellowish stiff clay			1.0	4	6	8	14	D U	
		62.3							
Greyish sandy silt med. dense			3.3	7	10	16	26	D W	
		59.8							
Greyish silty sand dense			5.0	14	16	21	37	D	
			7.5	15	18	23	41	D U	
		56.3							
Blackish very stiff clay			9.0	9	10	14	24	D	
		53.3	11.0						

D = disturbed sample; U = undisturbed sample;
W = water sample; N = SPT value

Figure 9.25 A typical bore-hole log

Boring Logs

A detailed record of boring operations and other tests carried out in the field is an essential part of the field work. The bore hole log is made during the boring operation. The soil is classified based on the visual examination of the disturbed samples collected. A typical example of a bore hole log is given in Fig. 9.25. The log should include the difficulties faced during boring operations including the occurrence of sand boils, and the presence of artesian water conditions if any, etc.

In-situ Tests

The field work may also involve one or more of the in-situ tests discussed earlier. The record should give the details of the tests conducted with exceptional clarity.

Laboratory Testing

A preliminary examination of the nature and type of soil brought to the laboratory is very essential before deciding upon the type and number of laboratory tests. Normally the SPT samples are used for this purpose. First the SPT samples should be arranged bore wise and depth wise. Each of the samples should be examined visually. A chart should be made giving the bore hole numbers and the types of tests to be conducted on each sample depth wise. An experienced geotechnical engineer can do this job with diligence and care.

Once the types of tests are decided, the laboratory assistant should carry out the tests with all the care required for each of the tests. The test results should next be tabulated on a suitable format bore wise and the soil is classified according to standard practice. The geotechnical consultant should examine each of the tests before being tabulated. Unreliable test results should be discarded.

Graphs and Charts

All the necessary graphs and charts are to be made based on the field and laboratory test results. The charts and graphs should present a clear insight into the subsoil conditions over the whole area. The charts made should help the geotechnical consultant to make a decision about the type of foundation, the strength and compressibility characteristics of the subsoil etc.

9.16 REPORT

A report is the final document of the whole exercise of soil exploration. A report should be comprehensive, clear and to the point. Many can write reports, but only a very few can produce a good report. A report writer should be knowledgable, practical and pragmatic. No theory, books or codes of practice provide all the materials required to produce a good report. It is the experience of a number of years of dedicated service in the field which helps a geotechnical consultant make report writing an art. A good report should normally comprise the following:

1. A general description of the nature of the project and its importance.
2. A general description of the topographical features and hydraulic conditions of the site.
3. A brief description of the various field and laboratory tests carried out.
4. Analysis and discussion of the test results
5. Recommendations
6. Calculations for determining safe bearing pressures, pile loads, etc.
7. Tables containing borelogs, and other field and laboratory test results
8. Drawings which include an index plan, a site-plan, test results plotted in the form of charts and graphs, soil profiles, etc.

9.17 PROBLEMS

9.1 Compute the area ratio of a sampling tube given the outside diameter = 100 mm and inside diameter = 94 mm. In what types of soil can this tube be used for sampling?

9.2 A standard penetration test was carried out at a site. The soil profile is given in Fig. Prob. 9.2 with the penetration values. The average soil data are given for each layer. Compute the corrected values of N and plot showing the

(a) variation of observed values with depth

(b) variation of corrected values with depth for standard energy 60%

Assume: $E_h = 0.7$, $C_d = 0.9$, $C_s = 0.85$ and $C_b = 1.05$

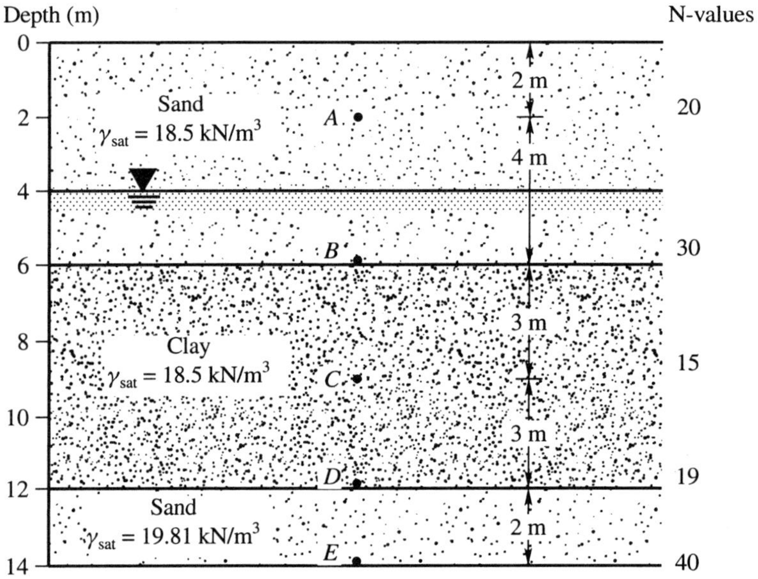

Figure Prob. 9.2

9.3 For the soil profile given in Fig. Prob 9.2, compute the corrected values of N for standard energy 70%.

9.4 For the soil profile given in Fig. Prob 9.2, estimate the average angle of friction for the sand layers based on the following:

(a) Table 9.3

(b) Eq (9.8) by assuming the profile contains less than 5% fines (D_r may be taken from Table 9.3)

Estimate the values of ϕ and D_r for 60 percent standard energy.

Assume: $N_{cor} = N_{60}$.

9.5 For the corrected values of N_{60} given in Prob 9.2, determine the unconfined compressive strengths of clay at points C and D in Fig Prob 9.2 by making use of Table 9.4 and Eq. (9.9). What is the consistency of the clay?

9.6 A static cone penetration test was carried out at a site using an electric-friction cone penetrometer. Fig. Prob 9.6 gives the soil profile and values of q_c obtained at various depths.

(a) Plot the variation of q_c with depth

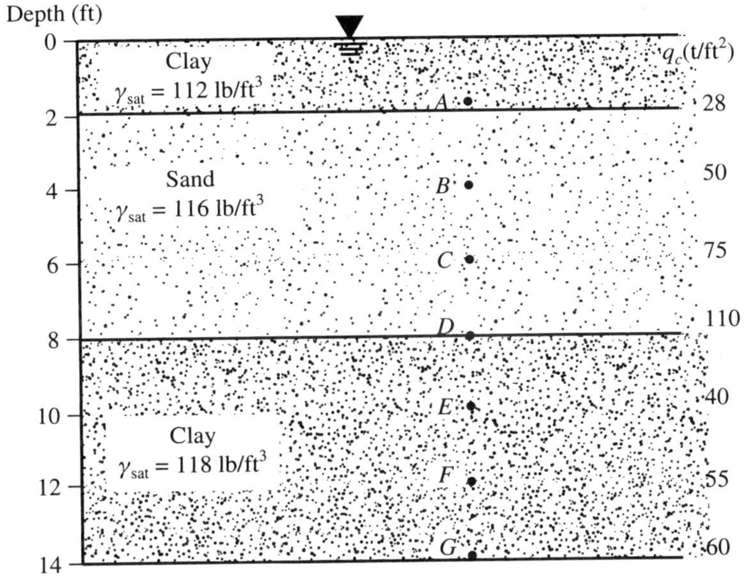

Figure Prob. 9.6

(b) Determine the relative density of the sand at the points marked in the figure by using Fig. 9.14.

(c) Determine the angle of internal friction of the sand at the points marked by using Fig. 9.15.

9.7 For the soil profile given in Fig. Prob 9.6, determine the unconfined compressive strength of the clay at the points marked in the figure using Eq (9.14).

9.8 A static cone penetration test carried out at a site at a depth of 50 ft gave the following results:

(a) cone resistance q_c = 250 t /ft^2

(b) average effective unit weight of the soil = 115 lb/ft^3

Classify the soil for friction ratios of 0.9 and 2.5 percent.

9.9 A static cone penetration test was carried out at a site using an electric-friction cone penetrometer. Classify the soil for the following data obtained from the site

q_c (MN/m^2)	Friction ratio R_f %
25	5
6.5	0.50
12.0	0.25
1.0	5.25

Assume in all the above cases that the effective overburden pressure is 50 kN/m^2.

9.10 Determine the relative density and the friction angle if the corrected SPT value N_{60} at a site is 30 from Eq (9.16) and Table 9.6. What are the values of D_r and ϕ for N_{70}?

9.11 Fig Prob 9.11 gives a corrected pressuremeter curve. The values of p_{om}, p_f and p_l and the corresponding volumes are marked on the curve. The test was conducted at a depth of 5 m below the ground surface. The average unit weight of the soil is 18.5 kN/m^3. Determine the following:

$p_{om} = 200$ kPa, $v_o = 180$ cm^3; $p_f = 660$ kPa; $v_f = 220$ cm^3;
$p_l = 1100$ kPa; $v_l = 700$ cm^3

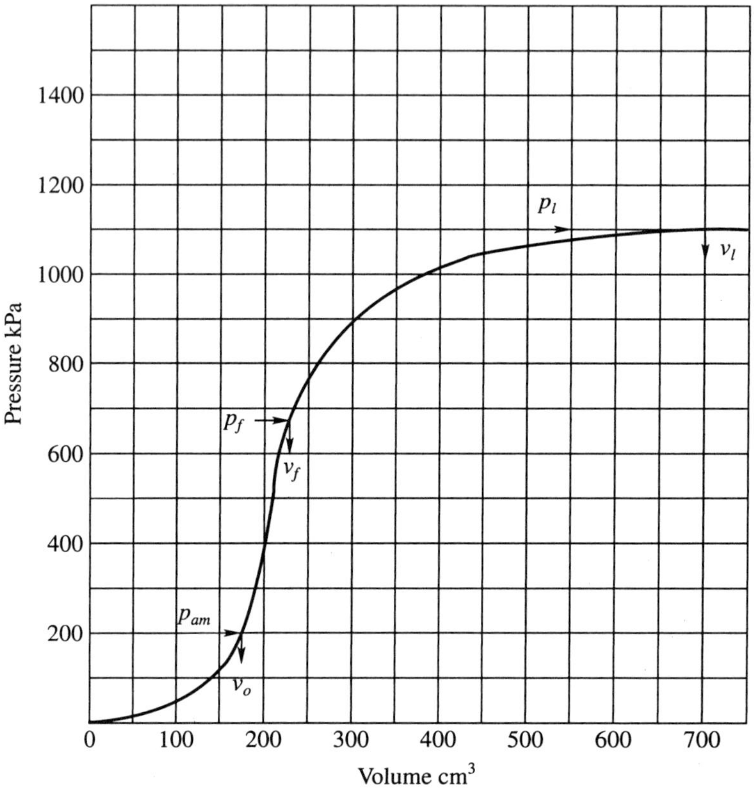

Figure Prob. 9.11

(a) The coefficient of earth pressure for the at-rest condition
(b) The Menard pressuremeter modulus
(c) The undrained shear strength c_u

9.12 A seismic refraction survey of an area gave the following data:

(i) Distance from impact point to geophone in m	15	30	60	80	100
(ii) Time of first wave arrival in sec	0.025	0.05	0.10	0.11	0.12

(a) Plot the time travel versus distance and determine velocities of the top and underlying layer of soil
(b) Determine the thickness of the top layer
(c) Using the seismic velocities evaluate the probable earth materials in the two layers

CHAPTER 10

STABILITY OF SLOPES

10.1 INTRODUCTION

Slopes of earth are of two types

1. Natural slopes
2. Man made slopes

Natural slopes are those that exist in nature and are formed by natural causes. Such slopes exist in hilly areas. The sides of cuttings, the slopes of embankments constructed for roads, railway lines, canals etc. and the slopes of earth dams constructed for storing water are examples of man made slopes. The slopes whether natural or artificial may be

1. Infinite slopes
2. Finite slopes

The term infinite slope is used to designate a constant slope of infinite extent. The long slope of the face of a mountain is an example of this type, whereas finite slopes are limited in extent. The slopes of embankments and earth dams are examples of finite slopes. The slope length depends on the height of the dam or embankment.

Slope Stability: Slope stability is an extremely important consideration in the design and construction of earth dams. The stability of a natural slope is also important. The results of a slope failure can often be catastrophic, involving the loss of considerable property and many lives.

Causes of Failure of Slopes: The important factors that cause instability in a slope and lead to failure are

1. Gravitational force
2. Force due to seepage water
3. Erosion of the surface of slopes due to flowing water

4. The sudden lowering of water adjacent to a slope
5. Forces due to earthquakes

The effect of all the forces listed above is to cause movement of soil from high points to low points. The most important of such forces is the component of gravity that acts in the direction of probable motion. The various effects of flowing or seeping water are generally recognized as very important in stability problems, but often these effects have not been properly identified. It is a fact that the seepage occurring within a soil mass causes seepage forces, which have much greater effect than is commonly realized.

Erosion on the surface of a slope may be the cause of the removal of a certain weight of soil, and may thus lead to an increased stability as far as mass movement is concerned. On the other hand, erosion in the form of undercutting at the toe may increase the height of the slope, or decrease the length of the incipient failure surface, thus decreasing the stability.

When there is a lowering of the ground water or of a freewater surface adjacent to the slope, for example in a sudden drawdown of the water surface in a reservoir there is a decrease in the buoyancy of the soil which is in effect an increase in the weight. This increase in weight causes increase in the shearing stresses that may or may not be in part counteracted by the increase in

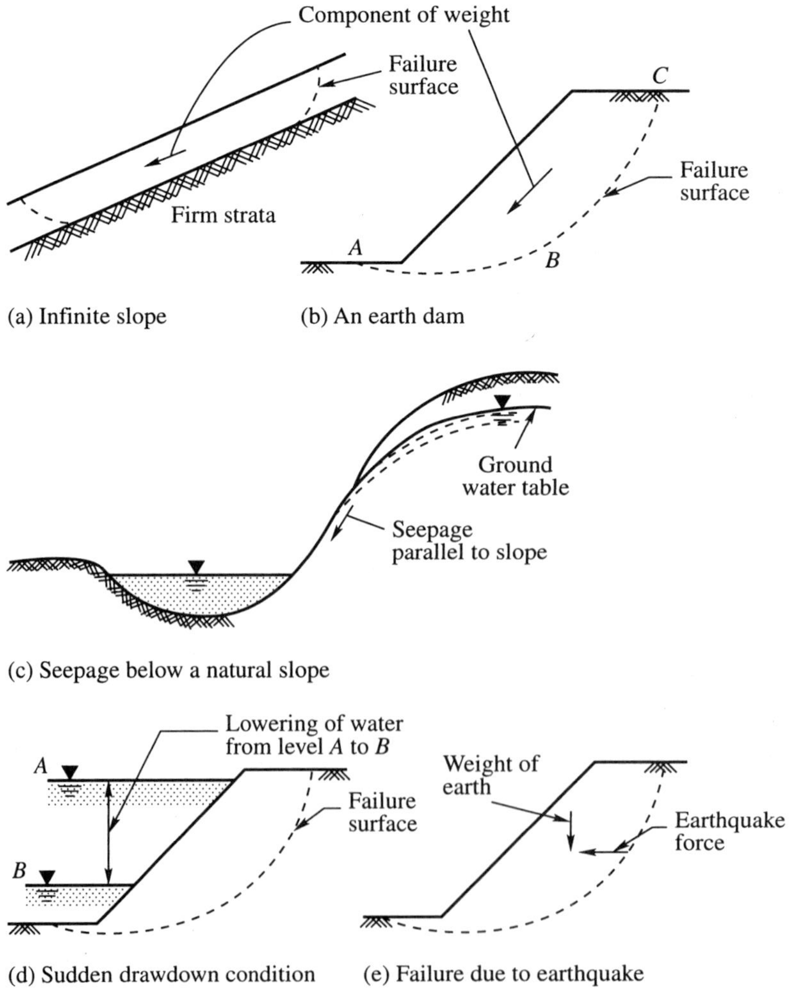

Figure 10.1 Forces that act on earth slopes

shearing strength, depending upon whether or not the soil is able to undergo compression which the load increase tends to cause. If a large mass of soil is saturated and is of low permeability, practically no volume changes will be able to occur except at a slow rate, and in spite of the increase of load the strength increase may be inappreciable.

Shear at constant volume may be accompanied by a decrease in the intergranular pressure and an increase in the neutral pressure. A failure may be caused by such a condition in which the entire soil mass passes into a state of liquefaction and flows like a liquid. A condition of this type may be developed if the mass of soil is subject to vibration, for example, due to earthquake forces.

The various forces that act on slopes are illustrated in Fig. 10.1.

10.2 GENERAL CONSIDERATIONS AND ASSUMPTIONS IN THE ANALYSIS

There are three distinct parts to an analysis of the stability of a slope. They are:

1. Testing of samples to determine the cohesion and angle of internal friction

If the analysis is for a natural slope, it is essential that the sample be undisturbed. In such important respects as rate of shear application and state of initial consolidation, the condition of testing must represent as closely as possible the most unfavorable conditions ever likely to occur in the actual slope.

2. The study of items which are known to enter but which cannot be accounted for in the computations

The most important of such items is progressive cracking which will start at the top of the slope where the soil is in tension, and aided by water pressure, may progress to considerable depth. In addition, there are the effects of the non-homogeneous nature of the typical soil and other variations from the ideal conditions which must be assumed.

3. Computation

If a slope is to fail along a surface, all the shearing strength must be overcome along that surface which then becomes a surface of rupture. Any one such as *ABC* in Fig. 10.1(b) represents one of an infinite number of possible traces on which failure might occur.

It is assumed that the problem is two dimensional, which theoretically requires a long length of slope normal to the section. However, if the cross section investigated holds for a running length of roughly two or more times the trace of the rupture, it is probable that the two dimensional case holds within the required accuracy.

The shear strength of soil is assumed to follow Coulomb's law

$$s = c' + \sigma' \tan \phi'$$

where,

c' = effective unit cohesion

σ' = effective normal stress on the surface of rupture = $(\sigma - u)$

σ = total normal stress on the surface of rupture

u = pore water pressure on the surface of rupture

ϕ' = effective angle of internal friction.

The item of great importance is the loss of shearing strength which many clays show when subjected to a large shearing strain. The stress-strain curves for such clays show the stress rising with increasing strain to a maximum value, after which it decreases and approaches an ultimate value which may be much less than the maximum. Since a rupture surface tends to develop progressively rather than with all the points at the same state of strain, it is generally the ultimate value that should be used for the shearing strength rather than the maximum value.

10.3 FACTOR OF SAFETY

In stability analysis, two types of factors of safety are normally used. They are

1. Factor of safety with respect to shearing strength.
2. Factor of safety with respect to cohesion. This is termed the factor of safety with respect to height.

Let,

F_s = factor of safety with respect to strength

F_c = factor of safety with respect to cohesion

F_H = factor of safety with respect to height

F_ϕ = factor of safety with respect to friction

c'_m = mobilized cohesion

ϕ'_m = mobilized angle of friction

τ = average value of mobilized shearing strength

s = maximum shearing strength.

The factor of safety with respect to shearing strength, F_s, may be written as

$$F_s = \frac{s}{\tau} = \frac{c' + \sigma' \tan\phi'}{\tau}$$

The shearing strength mobilized at each point on a failure surface may be written as

$$\tau = \frac{c'}{F_s} + \sigma' \frac{\tan\phi'}{F_s}$$

or $\tau = c'_m + \sigma' \tan\phi'_m$ (10.2)

where $c'_m = \frac{c'}{F_s}$, $\tan\phi'_m = \frac{\tan\phi'}{F_s}$

Actually the shearing resistance (mobilized value of shearing strength) does not develop to a like degree at all points on an incipient failure surface. The shearing strains vary considerably and the shearing stress may be far from constant. However the above expression is correct on the basis of average conditions.

If the factors of safety with respect to cohesion and friction are different, we may write the equation of the mobilized shearing resistance as

$$\tau = \frac{c'}{F_c} + \sigma' \frac{\tan\phi'}{F_\phi} \tag{10.3}$$

It will be shown later on that F_c depends on the height of the slope. From this it may be concluded that the factor of safety with respect to cohesion may be designated as the *factor of safety with respect to height*. This factor is denoted by F_H and it is the ratio between the critical height and the actual height, the critical height being the maximum height at which it is possible for a slope to be stable. We may write from Eq. (10.3)

$$\tau = \frac{c'}{F_H} + \sigma' \tan\phi' \tag{10.4}$$

where F_ϕ is arbitrarily taken equal to unity.

Example 10.1

The shearing strength parameters of a soil are

$c' = 26.7$ kN/m^2

$\phi' = 15°$

$c'_m = 17.8$ kN/m^2

$\phi'_m = 12°$

Calculate the factor of safety (a) with respect to strength, (b) with respect to cohesion and (c) with respect to friction. The average intergranular pressure σ' on the failure surface is 102.5 kN/m^2.

Solution

On the basis of the given data, the average shearing strength on the failure surface is

$$s = 26.7 + 102.5 \tan 15°$$
$$= 26.7 + 102.5 \times 0.268 = 54.2 \text{ kN/m}^2$$

and the average value of mobilized shearing resistance is

$$\tau = 17.8 + 102.5 \tan 12°$$
$$= 17.8 + 102.5 \text{ x } 0.212 = 39.6 \text{ kN/m}^2$$

$$F_s = \frac{54.2}{39.6} = 1.37; \quad F_c = \frac{26.70}{17.80} = 1.50; \quad F_\phi = \frac{\tan\phi'}{\tan\phi_m} = \frac{0.268}{0.212} = 1.26$$

The above example shows the factor of safety with respect to shear strength, F_s is 1.37, whereas the factors of safety with respect to cohesion and friction are different. Consider two extreme cases:

1. When the factor of safety with respect to cohesion is unity.
2. When the factor of safety with respect to friction is unity.

Case 1

$$\tau = 39.60 = 26.70 + \frac{102.50}{F_\phi}\tan 15° = 26.70 + \frac{102.50 \times 0.268}{F_\phi} = 26.70 + \frac{27.50}{F_\phi}$$

$$F_\phi = \frac{27.50}{12.90} = 2.13$$

Case 2

$$\tau = 39.60 = \frac{26.70}{F_c} + 102.50 \tan 15°$$

$$= \frac{26.70}{F_c} + 27.50$$

$$F_c = \frac{26.70}{12.10} = 2.20$$

We can have any combination of F_c and F_ϕ between these two extremes cited above to give the same mobilized shearing resistance of 39.6 kN/m^2. Some of the combinations of F_c and F_ϕ are given below.

Combination of F_c and F_ϕ

F_c	1.00	1.26	1.37	1.50	2.20
F_ϕ	2.12	1.50	1.37	1.26	1.00

Under Case 2, the value of $F_c = 2.20$ when $F_\phi = 1.0$. The factor of safety $F_c = 2.20$ is defined as the *factor of safety with respect to cohesion.*

Example 10.2

What will be the factors of safety with respect to average shearing strength, cohesion and internal friction of a soil, for which the shear strength parameters obtained from the laboratory tests are $c' = 32$ kN/m^2 and $\phi' = 18°$; the expected parameters of mobilized shearing resistance are $c'_m = 21$ kN/m^2 and $\phi'_m = 13°$ and the average effective pressure on the failure plane is 110 kN/m^2. For the same value of mobilized shearing resistance determine the following:

1. Factor of safety with respect to height;
2. Factor of safety with respect to friction when that with respect to cohesion is unity; and
3. Factor of safety with respect to strength.

Solution

The available shear strength of the soil is

$$s = 32 + 110 \tan 18° = 32 + 35.8 = 67.8 \text{ kN/m}^2$$

The mobilized shearing resistance of the soil is

$$\tau = 21 + 110 \tan 13° = 21 + 25.4 = 46.4 \text{ kN/m}^2$$

Factor of safety with respect to average strength, $F_s = \frac{67.8}{46.4} = 1.46$

Factor of safety with respect to cohesion, $F_c = \frac{32}{21} = 1.52$

Factor of safety with respect to friction, $F_\phi = \frac{\tan\phi'}{\tan\phi'_m} = \frac{\tan 18°}{\tan 13} = \frac{0.3249}{0.2309} = 1.41$

Factor of safety with respect to height, $F_H (= F_c)$ will be at $F_\phi = 1.0$

$$\tau = 46.4 = \frac{32}{F_c} + \frac{110 \tan 18°}{1.0}, \quad \text{therefore,} \quad F_c = \frac{32}{46.4 - 35.8} = 3.0$$

Factor of safety with respect to friction at $F_c = 1.0$ is

$$\tau = 46.4 = \frac{32}{1.0} + \frac{110 \tan 18°}{F_\phi}, \text{ therefore, } F_\phi = \frac{35.8}{46.4 - 32} = 2.49$$

Factor of safety with respect to strength F_s is obtained when $F_c = F_\phi$. We may write

$$\tau = 46.4 = \frac{32}{F_s} + \frac{110 \tan 18°}{F_s} \quad \text{or } F_s = 1.46$$

10.4 STABILITY ANALYSIS OF INFINITE SLOPES IN SAND

As an introduction to slope analysis, the problem of a slope of infinite extent is of interest. Imagine an infinite slope, as shown in Fig. 10.2, making an angle β with the horizontal. The soil is cohesionless and completely homogeneous throughout. Then the stresses acting on any vertical plane in the soil are the same as those on any other vertical plane. The stress at any point on a plane *EF* parallel to the surface at depth z will be the same as at every point on this plane.

Now consider a vertical slice of material *ABCD* having a unit dimension normal to the page. The forces acting on this slice are its weight W, a vertical reaction R on the base of the slice, and two lateral forces P_1 acting on the sides. Since the slice is in equilibrium, the weight and reaction are equal in magnitude and opposite in direction. They have a common line of action which passes through the center of the base *AB*. The lateral forces must be equal and opposite and their line of action must be parallel to the sloped surface.

The normal and shear stresses on plane *AB* are

$$\sigma'_n = \gamma z \cos^2 \beta$$

$$\tau = \gamma z \cos \beta \sin \beta$$

where σ'_n = effective normal stress,

γ = effective unit weight of the sand.

If full resistance is mobilized on plane *AB*, the shear strength, s, of the soil per Coulomb's law is

$$s = \sigma'_n \tan \phi'$$

when $\tau = s$, substituting for s and σ'_n, we have

$$\gamma z \cos \beta \sin \beta = \gamma z \cos^2 \beta \tan \phi'$$

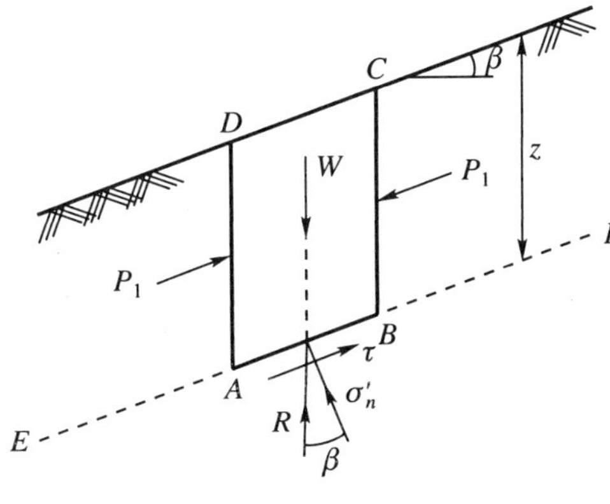

Figure 10.2 Stability analysis of infinite slope in sand

or $\tan\beta = \tan\phi'$ (10.5a)

Equation (10.5a) indicates that the maximum value of β is limited to ϕ' if the slope is to be stable. This condition holds true for cohesionless soils whether the slope is completely dry or completely submerged under water.

The factor of safety of infinite slopes in sand may be written as

$$F_s = \frac{\tan\phi'}{\tan\beta} \tag{10.5b}$$

10.5 STABILITY ANALYSIS OF INFINITE SLOPES IN CLAY

The vertical stress σ_v acting on plane AB (Fig. 10.3) where

$$\sigma_v = \gamma z \cos\beta$$

is represented by OC in Fig. 10.3 in the stress diagram. The normal stress on this plane is OE and the shearing stress is EC. The line OC makes an angle β with the σ-axis.

The Mohr strength envelope is represented by line FA whose equation is

$$s = c' + \sigma' \tan\phi'$$

According to the envelope, the shearing strength is ED where the normal stress is OE.

When β is greater than ϕ' the lines OC and FD meet. In this case the two lines meet at A. As long as the shearing stress on a plane is less than the shearing strength on the plane, there is no danger of failure. Figure 10.3 indicates that at all depths at which the direct stress is less than OB, there is no possibility of failure. However at a particular depth at which the direct stress is OB, the

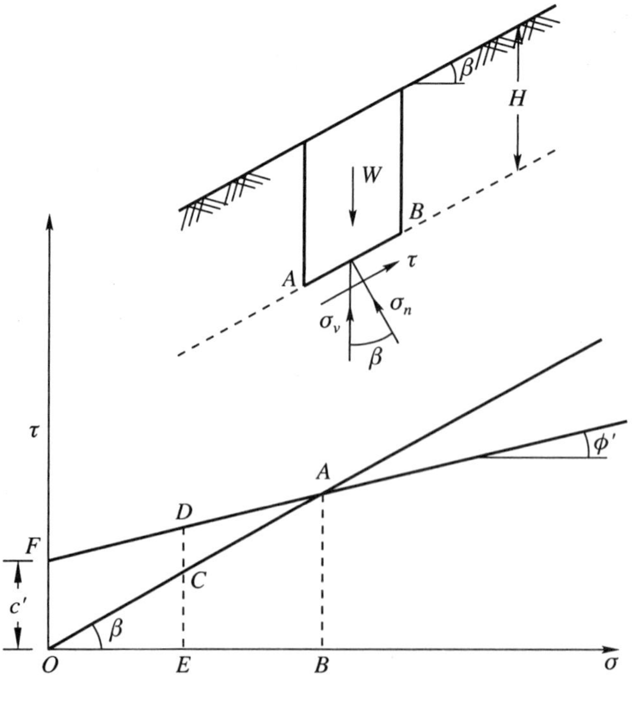

Figure 10.3 Stability analysis of infinite slopes in clay soils

shearing strength and shearing stress values are equal as represented by AB, failure is imminent. This depth at which the shearing stress and shearing strength are equal is called the *critical depth*. At depths greater than this critical value, Fig. 10.3 indicates that the shearing stress is greater than the shearing strength but this is not possible. Therefore it may be concluded that the slope may be steeper than ϕ' as long as the depth of the slope is less than the critical depth.

Expression for the Stability of an Infinite Slope of Clay of Depth *H*

Equation (10.2) gives the developed shearing stress as

$$\tau = c'_m + \sigma' \tan \phi'_m \tag{10.6}$$

Under conditions of no seepage and no pore pressure, the stress components on a plane at depth H and parallel to the surface of the slope are

$$\tau = \gamma H \sin \beta \cos \beta$$

$$\sigma' = \gamma H \cos^2 \beta$$

Substituting these stress expressions in the equation above and simplifying, we have

$$c'_m = \gamma H \cos^2 \beta (\tan \beta - \tan \phi'_m)$$

or $$N_s = \frac{c'_m}{\gamma H} = \cos^2 \beta (\tan \beta - \tan \phi'_m) \tag{10.7}$$

where H is the allowable height and the term $c'_m/\gamma H$ is a dimensionless expression called the *stability number* and is designated as N_s. This dimensionless number is proportional to the required cohesion and is inversely proportional to the allowable height. The solution is for the case when no seepage is occurring. If in Eq. (10.7) the factor of safety with respect to friction is unity, the stability number with respect to cohesion may be written as

$$N_s = \frac{c'}{F_c \gamma H} = \cos^2 \beta (\tan \beta - \tan \phi') \tag{10.8}$$

where $c'_m = \dfrac{c'}{F_c}$

The stability number in Eq. (10.8) may be written as

$$N_s = \frac{c'}{F_c \gamma H} = \frac{c'}{\gamma H_c} \tag{10.9}$$

where H_c = critical height. From Eq. (10.9), we have

$$F_c = \frac{H_c}{H} = F_H \tag{10.10}$$

Eq. (10.10) indicates that the factor of safety with respect to cohesion, F_c, is the same as the factor of safety with respect to height F_H.

If there is seepage parallel to the ground surface throughout the entire mass, with the free water surface coinciding with the ground surface, the components of effective stresses on planes parallel to the surface of slopes at depth H are given as [Fig. 10.4(a)].

Normal stress

$$\sigma'_n = (\gamma_{sat} - \gamma_w) H \cos^2 \beta = \gamma_b H \cos^2 \beta \tag{10.11a}$$

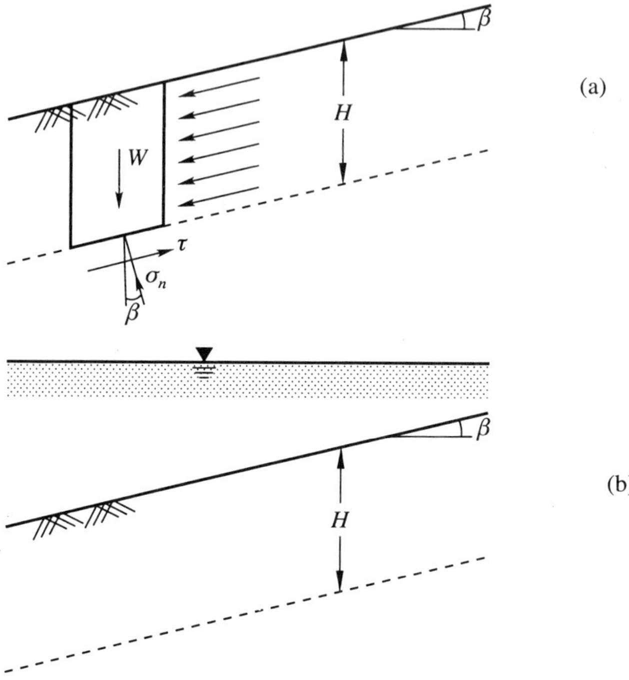

Figure 10.4 Analysis of infinite slope (a) with seepage flow through the entire mass, and (b) with completely submerged slope.

the shearing stress

$$\tau = \gamma_{sat} H \sin\beta \cos\beta \tag{10.11b}$$

Now substituting Eqs (10.11a) and (10.11b) into equation

$$\tau = c'_m + \sigma'_n \tan\phi'_m$$

and simplifying, the stability expression obtained is

$$\frac{c'_m}{\gamma_{sat} H} = \cos^2\beta \ \tan\beta - \frac{\gamma_b}{\gamma_{sat}} \tan\phi'_m \tag{10.12}$$

As before, if the factor of safety with respect to friction is unity, the stability number which represents the cohesion may be written as

$$N_s = \frac{c'}{F_c \gamma_{sat} H} = \frac{c'}{\gamma_{sat} H_c} = \cos^2\beta \ \tan\beta - \frac{\gamma_b}{\gamma_{sat}} \tan\phi' \tag{10.13}$$

If the slope is completely submerged, and if there is no seepage as in Fig. 10.4(b), then Eq. (10.13) becomes

$$N_s = \frac{c'}{F_c \gamma_b H} = \frac{c'}{\gamma_b H_c} = \cos^2\beta(\tan\beta - \tan\phi') \tag{10.14}$$

where γ_b = submerged unit weight of the soil.

Example 10.3

Find the factor of safety of a slope of infinite extent having a slope angle = 25°. The slope is made of cohesionless soil with $\phi = 30°$.

Solution

Factor of safety

$$F_s = \frac{\tan\phi'}{\tan\beta} = \frac{\tan 30°}{\tan 25°} = \frac{0.5774}{0.4663} = 1.238$$

Example 10.4

Analyze the slope of Example 10.3 if it is made of clay having $c' = 30$ kN/m², $\phi' = 20°$, $e = 0.65$ and $G_s = 2.7$ and under the following conditions: (i) when the soil is dry, (ii) when water seeps parallel to the surface of the slope, and (iii) when the slope is submerged.

Solution

For $e = 0.65$ and $G_s = 2.7$

$$\gamma_d = \frac{2.7 \times 9.81}{1+0.65} = 16.05 \text{ kN/m}^3, \quad \gamma_{sat} = \frac{(2.7+0.65)\times 9.81}{1+0.65} = 19.9 \text{ kN/m}^3,$$

$$\gamma_b = 10.09 \text{ kN/m}^3$$

(i) For dry soil the stability number N_s is

$$N_s = \frac{c'}{\gamma_d H_c} = \cos^2\beta(\tan\beta - \tan\phi') \quad \text{when } F_\phi = 1$$

$$= (\cos 25°)^2(\tan 25° - \tan 20°) = 0.084.$$

Therefore, the critical height $H_c = \dfrac{c'}{\gamma_d \times N_s} = \dfrac{30}{16.05 \times 0.084} = 22.25$ m

(ii) For seepage parallel to the surface of the slope [Eq. (10.13)]

$$N_s = \frac{c'}{\gamma_t H_c} = \cos^2 25° \ \tan 25° - \frac{10.09}{19.9}\tan 20° \ = 0.2315$$

$$H_c = \frac{c'}{\gamma_t N_s} = \frac{30}{19.9 \times 0.2315} = 6.51 \text{ m}$$

(iii) For the submerged slope [Eq. (10.14)]

$$N_s = \cos^2 25° \ (\tan 25° - \tan 20°) = 0.084$$

$$H_c = \frac{c'}{\gamma_b N_s} = \frac{30}{10.09 \times 0.084} = 35.4 \text{ m}$$

10.6 METHODS OF STABILITY ANALYSIS OF SLOPES OF FINITE HEIGHT

The stability of slopes of infinite extent has been discussed in previous sections. A more common problem is the one in which the failure occurs on curved surfaces. The most widely used method of analysis of homogeneous, isotropic, finite slopes is the *Swedish method* based on circular failure surfaces. Petterson (1955) first applied the circle method to the analysis of a soil failure in connection with the failure of a quarry wall in Goeteberg, Sweden. A Swedish National Commission, after studying a large number of failures, published a report in 1922 showing that the lines of failure of most such slides roughly approached the circumference of a circle. The failure circle might pass above the toe, through the toe or below it. By investigating the strength along the arc of a large number of such circles, it was possible to locate the circle which gave the lowest resistance to shear. This general method has been quite widely accepted as offering an approximately correct solution for the determination of the factor of safety of a slope of an embankment and of its foundation. Developments in the method of analysis have been made by Fellenius (1947), Terzaghi (1943), Gilboy (1934), Taylor (1937), Bishop (1955), and others, with the result that a satisfactory analysis of the stability of slopes, embankments and foundations by means of the circle method is no longer an unduly tedious procedure.

There are other methods of historic interest such as the *Culmann method* (1875) and the *logarithmic spiral method*. The Culmann method assumes that rupture will occur along a plane. It is of interest only as a classical solution, since actual failure surfaces are invariably curved. This method is approximately correct for steep slopes. The logarithmic spiral method was recommended by Rendulic (1935) with the rupture surface assuming the shape of logarithmic spiral. Though this method makes the problem statically determinate and gives more accurate results, the greater length of time required for computation overbalances this accuracy.

There are several methods of stability analysis based on the circular arc surface of failure. A few of the methods are described below

Methods of Analysis

The majority of the methods of analysis may be categorized as limit equilibrium methods. The basic assumption of the limit equilibrium approach is that Coulomb's failure criterion is satisfied along the assumed failure surface. A free body is taken from the slope and starting from known or assumed values of the forces acting upon the free body, the shear resistance of the soil necessary for equilibrium is calculated. This calculated shear resistance is then compared to the estimated or available shear strength of the soil to give an indication of the factor of safety.

Methods that consider only the whole free body are the (a) slope failure under undrained conditions, (b) friction-circle method (Taylor, 1937, 1948) and (c) Taylor's stability number (1948).

Methods that divide the free body into many vertical slices and consider the equilibrium of each slice are the Swedish circle method (Fellenius, 1927), Bishop method (1955), Bishop and Morgenstern method (1960) and Spencer method (1967). The majority of these methods are in chart form and cover a wide variety of conditions.

10.7 PLANE SURFACE OF FAILURE

Culmann (1875) assumed a plane surface of failure for the analysis of slopes which is mainly of interest because it serves as a test of the validity of the assumption of plane failure. In some cases this assumption is reasonable and in others it is questionable.

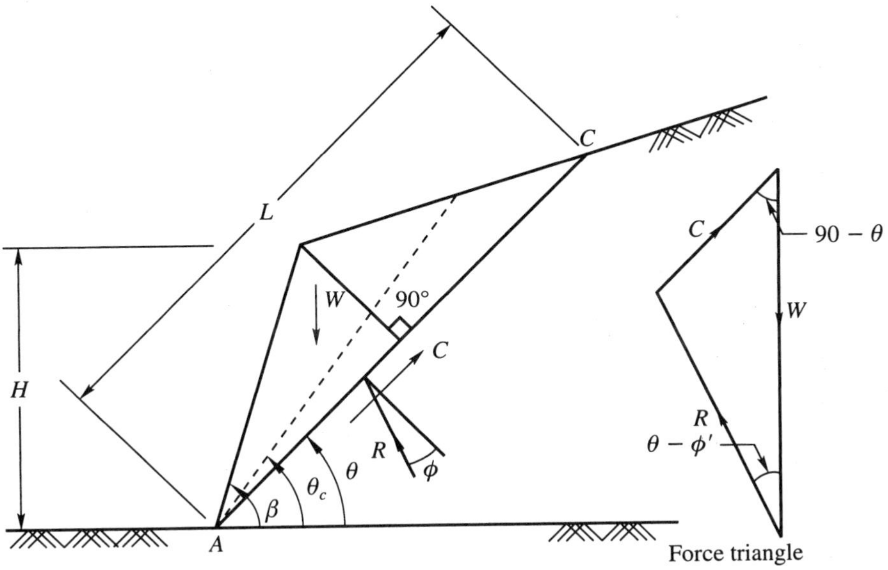

Figure 10.5 Stability of slopes by Culmann method

The method as indicated above assumes that the critical surface of failure is a plane surface passing through the toe of the dam as shown in Fig. 10.5.

The forces that act on the mass above trial failure plane *AC* inclined at angle θ with the horizontal are shown in the figure. The expression for the weight, *W*, and the total cohesion *C* are respectively,

$$W = \frac{1}{2}\gamma LH \operatorname{cosec}\beta \sin(\beta - \theta)$$

and $C = c'L$

The use of the law of sines in the force triangle, shown in the figure, gives

$$\frac{C}{W} = \frac{\sin(\theta - \phi')}{\cos\phi'}$$

Substituting herein for *C* and *W*, and rearranging we have

$$\frac{c'}{\gamma H}_{\theta} = \frac{1}{2}\operatorname{cosec}\beta \sin(\beta - \theta)\sin(\theta - \phi')\sec\phi'$$

in which the subscript θ indicates that the stability number is for the trial plane at inclination θ.

The most dangerous plane is obtained by setting the first derivative of the above equation with respect to θ equal to zero. This operation gives

$$\theta'_c = \frac{1}{2}(\beta + \phi')$$

where θ'_c is the critical angle for limiting equilibrium and the stability number for limiting equilibrium may be written as

$$\frac{c'}{\gamma H_c} = \frac{1 - \cos(\beta - \phi')}{4\sin\beta\cos\phi'} \tag{10.15}$$

where H_c is the critical height of the slope.

If we write

$$F_c = \frac{c'}{c'_m}, \quad F_\phi = \frac{\tan\phi'}{\tan\phi'_m}$$

where F_c and F_ϕ are safety factors with respect to cohesion and friction respectively, Eq. (10.15) may be modified for chosen values of c'_m and ϕ'_m as

$$\frac{c'_m}{\gamma H} = \frac{1-\cos(\beta-\phi'_m)}{4\sin\beta\cos\phi'_m} \tag{10.16}$$

The critical angle for any assumed values of c'_m and ϕ'_m is

$$\theta_c = \frac{1}{2}(\beta+\phi'_m) \tag{10.17}$$

From Eq. (10.16), the allowable height of a slope is

$$H = \frac{4c'_m \sin\beta\cos\phi'_m}{\gamma[1-\cos(\beta-\phi'_m)]} \tag{10.18}$$

Example 10.5

Determine by Culmann's method the critical height of an embankment having a slope angle of 40° and the constructed soil having $c' = 630$ psf, $\phi' = 20°$ and effective unit weight = 114 lb/ft^3. Find the allowable height of the embankment if $F_c = F_\phi = 1.25$.

Solution

$$H_c = \frac{4c'\sin\beta\cos\phi'}{\gamma[1-\cos(\beta-\phi')]} = \frac{4\times 630\times\sin 40^\circ\cos 20^\circ}{114(1-\cos 20^\circ)} = 221 \text{ ft}$$

For $F_c = F_\phi = 1.25, \quad c'_m = \frac{c'}{F_c} = \frac{630}{1.25} = 504 \text{ lb/ft}^2$

and $\tan\phi'_m = \frac{\tan\phi'}{F_\phi} = \frac{\tan 20^\circ}{1.25} = 0.291, \quad \phi'_m = 16.23^\circ$

Allowable height, $H = \frac{4\times 504\sin 40^\circ\cos 16.23^\circ}{114[1-\cos(40-16.23^\circ)]} = 128.7$ ft.

10.8 CIRCULAR SURFACES OF FAILURE

The investigations carried out in Sweden at the beginning of this century have clearly confirmed that the surfaces of failure of earth slopes resemble the shape of a circular arc. When soil slips along a circular surface, such a slide may be termed as a rotational slide. It involves downward and outward movement of a slice of earth as shown in Fig. 10.6(a) and sliding occurs along the entire surface of contact between the slice and its base. The types of failure that normally occur may be classified as

1. Slope failure

2. Toe failure
3. Base failure

In slope failure, the arc of the rupture surface meets the slope above the toe. This can happen when the slope angle β is quite high and the soil close to the toe possesses high strength. Toe failure occurs when the soil mass of the dam above the base and below the base is homogeneous. The base failure occurs particularly when the base angle β is low and the soil below the base is softer and more plastic than the soil above the base. The various modes of failure are shown in Fig. 10.6.

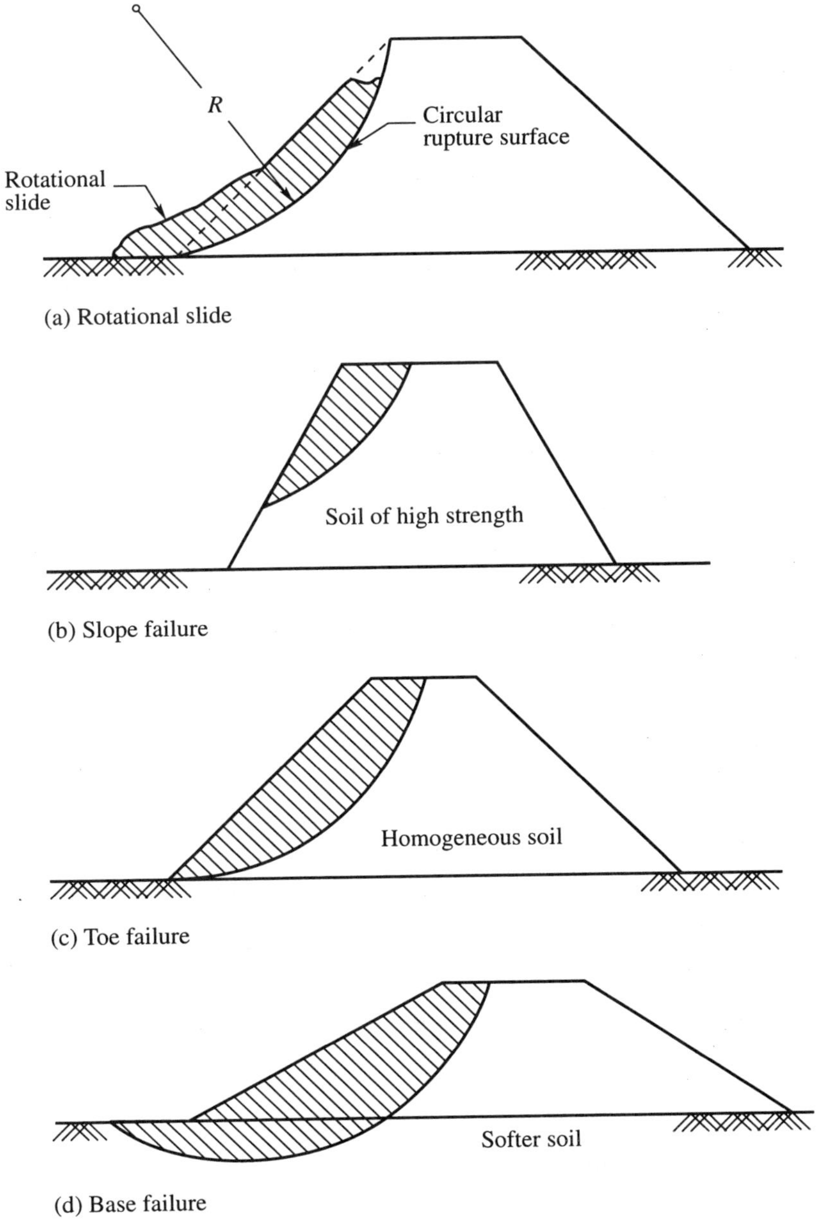

Figure 10.6 Types of failure of earth dams

10.9 FAILURE UNDER UNDRAINED CONDITIONS ($\phi_u = 0$)

A fully saturated clay slope may fail under undrained conditions ($\phi_u = 0$) immediately after construction. The stability analysis is based on the assumption that the soil is homogeneous and the potential failure surface is a circular arc. Two types of failures considered are

1. Slope failure
2. Base failure

The undrained shear strength c_u of soil is assumed to be constant with depth. A trial failure circular surface *AB* with center at *O* and radius *R* is shown in Fig. 10.7(a) for a toe failure. The slope *AC* and the chord *AB* make angles β and α with the horizontal respectively. *W* is the weight per unit

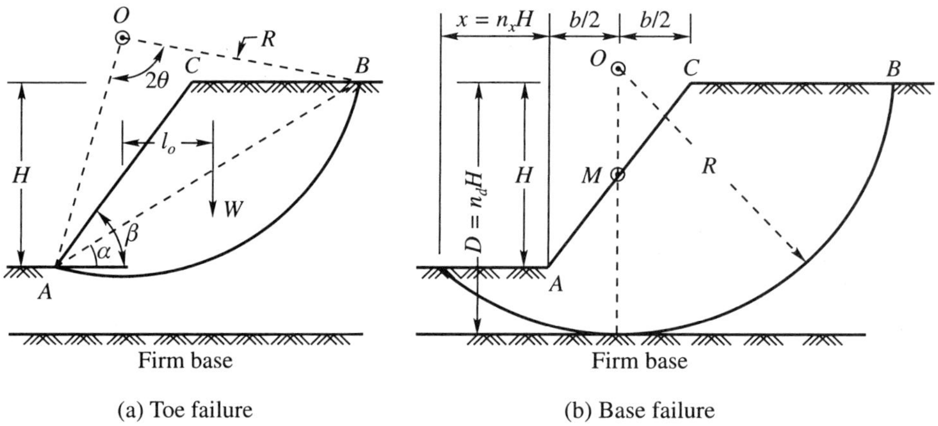

Figure 10.7 Critical circle positions for (a) slope failure (after Fellenius, 1927), (b) base failure

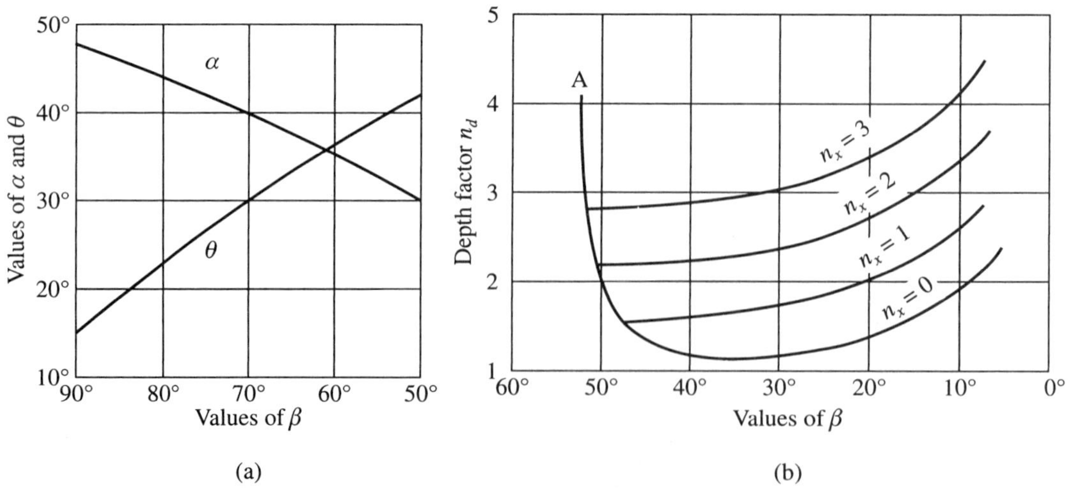

Figure 10.8 (a) Relation between slope angle β and parameters α and θ for location of critical toe circle when β is greater than 53°; (b) relation between slope angle β and depth factor n_d for various values of parameter n_x (after Fellenius, 1927)

length of the soil lying above the trial surface acting through the center of gravity of the mass. l_o is the lever arm, L_a is the length of the arc, L_c the length of the chord AB and c_m the mobilized cohesion for any assumed surface of failure.

We may express the factor of safety F_s as

$$F_s = \frac{c_u}{c_m} \tag{10.19}$$

For equilibrium of the soil mass lying above the assumed failure surface, we may write

resisting moment M_r = actuating moment M_a

The resisting moment $M_r = L_a c_m R$

Actuating moment, $M_a = Wl_o$

Equation for the mobilized c_m is

$$c_m = \frac{Wl_o}{L_a R} \tag{10.20}$$

Now the factor of safety F for the assumed trial arc of failure may be determined from Eq. (10.19). This is for one trial arc. The procedure has to be repeated for several trial arcs and the one that gives the least value is the critical circle.

If failure occurs along a toe circle, the center of the critical circle can be located by laying off the angles α and 2θ as shown in Fig. 10.7(a). Values of α and θ for different slope angles β can be obtained from Fig. 10.8(a).

If there is a base failure as shown in Fig. 10.7(b), the trial circle will be tangential to the firm base and as such the center of the critical circle lies on the vertical line passing through midpoint M on slope AC. The following equations may be written with reference to Fig. 10.7(b).

$$\text{Depth factor, } n_d = \frac{D}{H}, \quad \text{Distance factor, } n_x = \frac{x}{H} \tag{10.21}$$

Values of n_x can be estimated for different values of n_d and β by means of the chart Fig. 10.8(b).

Example 10.6

Calculate the factor of safety against shear failure along the slip circle shown in Fig. Ex. 10.6 Assume cohesion = 40 kN/m^2, angle of internal friction = zero and the total unit weight of the soil = 20.0 kN/m^3.

Solution

Draw the given slope $ABCD$ as shown in Fig. Ex. 10.6. To locate the center of rotation, extend the bisector of line BC to cut the vertical line drawn from C at point O. With O as center and OC as radius, draw the desired slip circle.

Radius $OC = R = 36.5$ m, Area $BECFB = \frac{2}{3} \times EF \times BC$

$$= \frac{2}{3} \times 4 \times 32.5 = 86.7 \text{ m}^2$$

Therefore $W = 86.7 \times 1 \times 20 = 1734$ kN

W acts through point G which may be taken as the middle of FE.

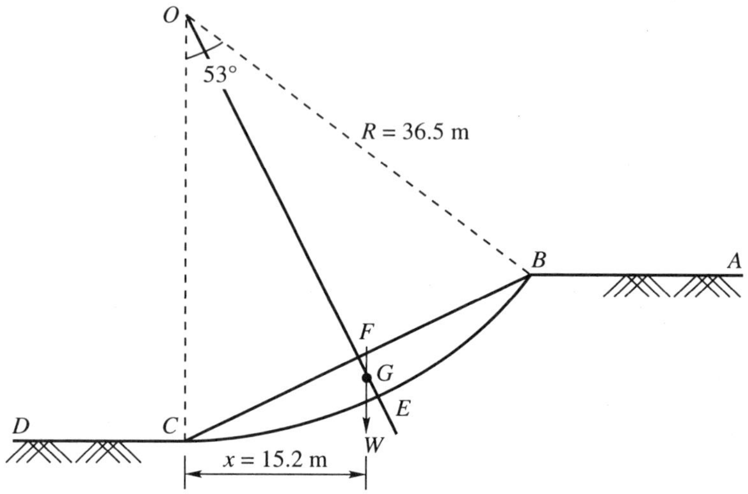

Figure. Ex. 10.6

From the figure we have, $x = 15.2$ m, and $\theta = 53°$

$$\text{Length of arc } BEC = R\theta = 36.5 \times 53° \times \frac{3.14}{180} = 33.8 \text{ m}$$

$$F_s = \frac{\text{length of arc} \times \text{cohesion} \times \text{radius}}{Wx} = \frac{33.8 \times 40 \times 36.5}{1734 \times 15.2} = 1.87$$

10.10 FRICTION-CIRCLE METHOD

Physical Concept of the Method

The principle of the method is explained with reference to the section through a dam shown in Fig. 10.9. A trial circle with center of rotation O is shown in the figure. With center O and radius

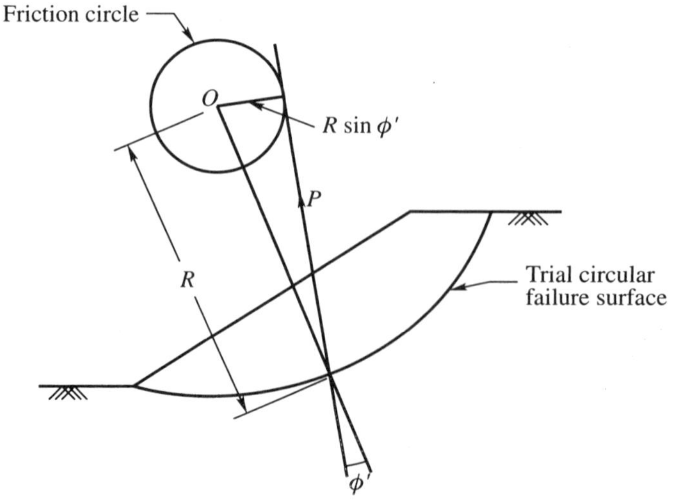

Figure 10.9 Principle of friction circle method

sin ϕ', where R is the radius of the trial circle, a circle is drawn. Any line tangent to the inner circle must intersect the trial circle at an angle ϕ' with R. Therefore, any vector representing an intergranular pressure at obliquity ϕ' to an element of the rupture arc must be tangent to the inner circle. This inner circle is called the *friction circle* or *ϕ-circle*. The friction circle method of slope analysis is a convenient approach for both graphical and mathematical solutions. It is given this name because the characteristic assumption of the method refers to the ϕ-circle.

The forces considered in the analysis are

1. The total weight W of the mass above the trial circle acting through the center of mass. The center of mass may be determined by any one of the known methods.
2. The resultant boundary neutral force U. The vector U may be determined by a graphical method from flownet construction.
3. The resultant intergranular force, P, acting on the boundary.
4. The resultant cohesive force C.

Actuating Forces

The actuating forces may be considered to be the total weight W and the resultant boundary force U as shown in Fig. 10.10.

The boundary neutral force always passes through the center of rotation O. The resultant of W and U, designated as Q, is shown in the figure.

Resultant Cohesive Force

Let the length of arc AB be designated as L_a, the length of chord AB by L_c. Let the arc length L_a be divided into a number of small elements and let the mobilized cohesive force on these elements be designated as C_1, C_2, C_3, etc. as shown in Fig. 10.11. The resultant of all these forces is shown by the force polygon in the figure. The resultant is $A'B'$ which is parallel and equal to the chord length AB. The resultant of all the mobilized cohesional forces along the arc is therefore

$$C = c'_m L_c$$

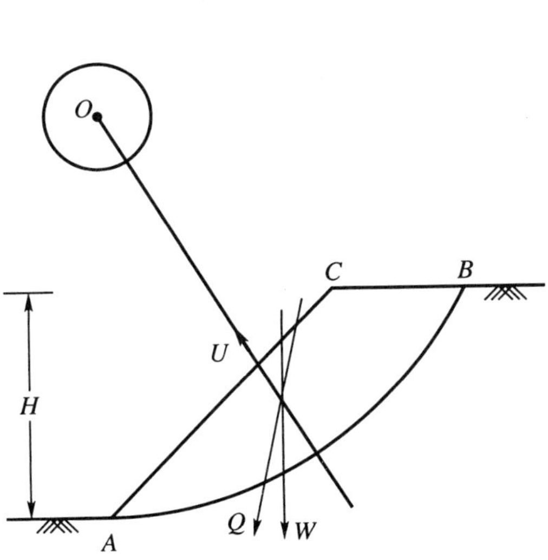

Figure 10.10 Actuating forces

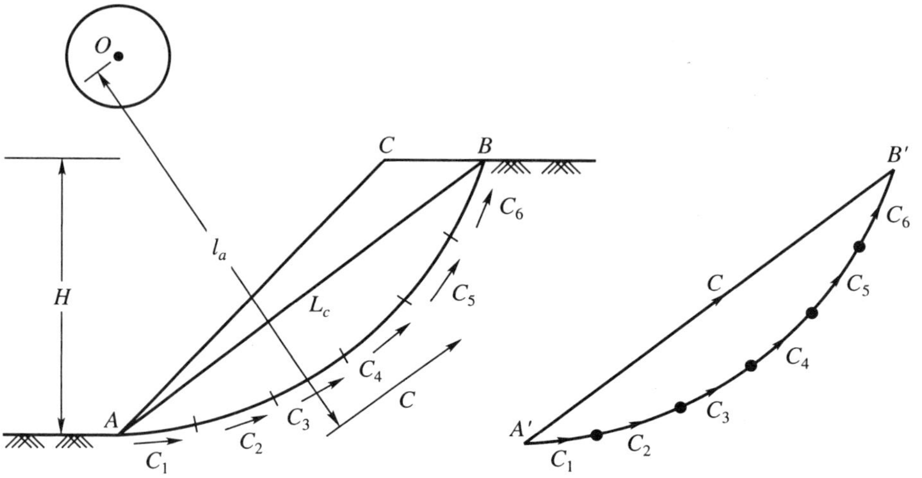

(a) Cohesive forces on a trial arc (b) Polygon of forces

Figure 10.11 Resistant cohesive forces

We may write $c'_m = \dfrac{c'}{F_c}$

wherein c' = unit cohesion, F_c = factor of safety with respect to cohesion.

The line of action of C may be determined by moment consideration. The moment of the total cohesion is expressed as

$$c'_m L_a R = c'_m L_c l_a$$

where l_a = moment arm. Therefore,

$$l_a = R\frac{L_a}{L_c} \tag{10.22}$$

It is seen that the line of action of vector C is independent of the magnitude of c'_m.

Resultant of Boundary Intergranular Forces

The trial arc of the circle is divided into a number of small elements. Let P_1, P_2, P_3, etc. be the intergranular forces acting on these elements as shown in Fig. 10.12. The friction circle is drawn with a radius of $R \sin \phi'_m$

where $\tan \phi'_m = \dfrac{\tan \phi'}{F_\phi}$

The lines of action of the intergranular forces P_1, P_2, P_3, etc. are tangential to the friction circle and make an angle of ϕ'_m at the boundary. However, the vector sum of any two small forces has a line of action through point D, missing tangency to the ϕ'_m-circle by a small amount. The resultant of all granular forces must therefore miss tangency to the ϕ'_m-circle by an amount which is not considerable. Let the distance of the resultant of the granular force P from the center of the circle be designated as $KR \sin \phi'_m$ (as shown in Fig. 10.12). The

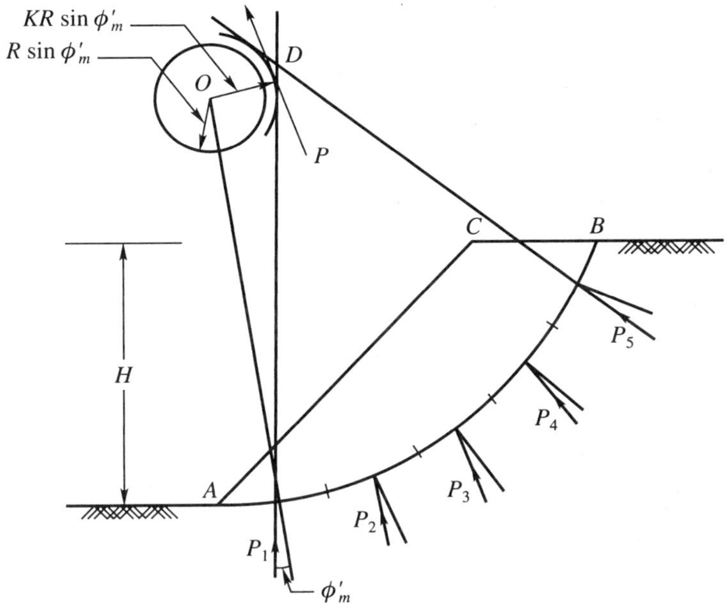

Figure 10.12 Resultant of intergranular forces

magnitude of K depends upon the type of intergranular pressure distribution along the arc. The most probable form of distribution is the sinusoidal distribution.

The variation of K with respect to the central angle α' is shown in Fig. 10.13. The figure also gives relationships between α' and K for a uniform stress distribution of effective normal stress along the arc of failure.

The graphical solution based on the concepts explained above is simple in principle. For the three forces Q, C and P of Fig. 10.14 to be in equilibrium, P must pass through the intersection of

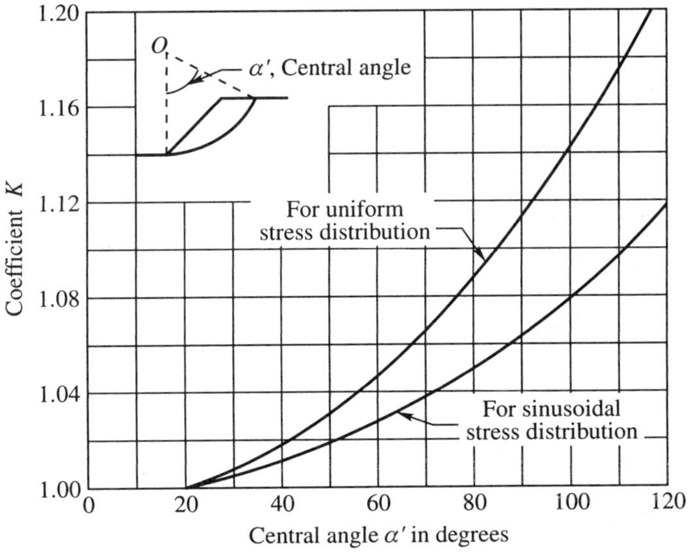

Figure 10.13 Relationship between K and central angle α'

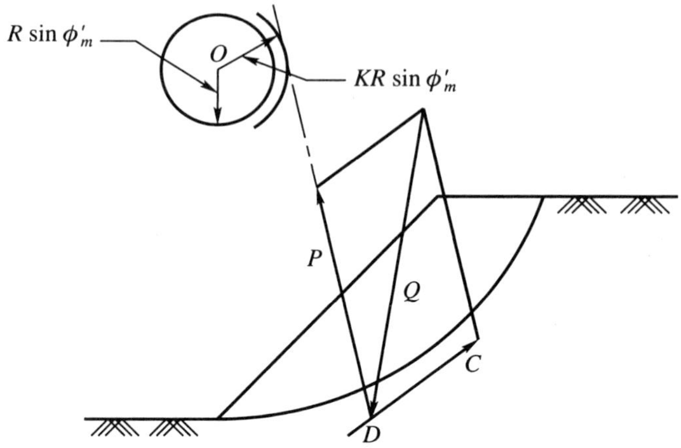

Figure 10.14 Force triangle for the friction-circle method

the known lines of action of vectors Q and C. The line of action of vector P must also be tangent to the circle of radius $KR \sin \phi'_m$. The value of K may be estimated by the use of curves given in Fig. 10.13, and the line of action of force P may be drawn as shown in Fig. 10.14. Since the lines of action of all three forces and the magnitude of force Q are known, the magnitude of P and C may be obtained by the force parallelogram construction that is indicated in the figure. The circle of radius of $KR \sin \phi'_m$ is called the *modified friction circle.*

Determination of Factor of Safety With Respect to Strength

Figure 10.15(a) is a section of a dam. AB is the trial failure arc. The force Q, the resultant of W and U is drawn as explained earlier. The line of action of C is also drawn. Let the forces Q and C

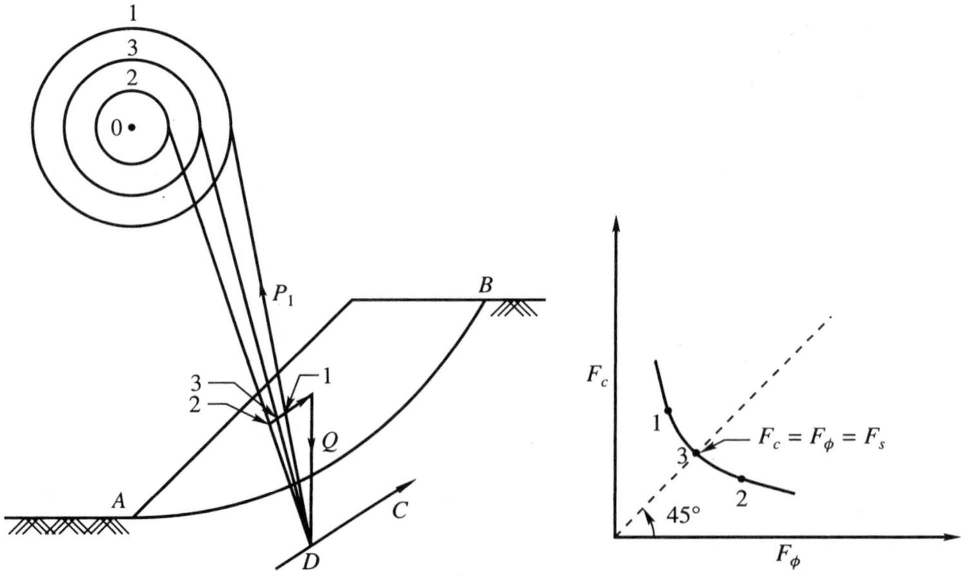

(a) Friction circle (b) Factor of safety

Figure 10.15 Graphical method of determining factor of safety with respect to strength meet at point D. An arbitrary first trial using any reasonable ϕ'_m value, which will be designated by ϕ'_{m1} is given by the use of circle 1 or radius $KR \sin \phi'_{m1}$. Subscript 1 is used for all other quantities of the first trial. The force P_1 is then drawn through D tangent to circle 1. C_1 is parallel to chord and point 1 is the intersection of forces C_1 and P_1. The mobilized cohesion is equal $c'_{m1}L_c$. From this the mobilized cohesion c'_{m1} is evaluated. The factors of safety with respect to cohesion and friction are determined from the expressions

$$F'_c = \frac{c'}{c'_{m1}}, \text{ and } F_{\phi 1} = \frac{\tan \phi'}{\tan \phi'_{m1}}$$

These factors are the values used to plot point 1 in the graph in Fig. 10.15(b). Similarly other friction circles with radii $KR \sin \phi'_{m2}$, $KR \sin \phi'_{m3}$, etc. may be drawn and the procedure repeated. Points 2, 3 etc. are obtained as shown in Fig. 10.15(b). The 45° line, representing $F_c = F_\phi$, intersects the curve to give the factor of safety F_s for this trial circle.

Several trial circles must be investigated in order to locate the critical circle, which is the one having the minimum value of F_s.

Example 10.7

An embankment has a slope of 2 (horizontal) to 1 (vertical) with a height of 10 m. It is made of a soil having a cohesion of 30 kN/m^2, an angle of internal friction of 5° and a unit weight of 20 kN/m^3. Consider any slip circle passing through the toe. Use the friction circle method to find the factor of safety with respect to cohesion.

Solution

Refer to Fig. Ex. 10.7. Let EFB be the slope and AKB be the slip circle drawn with center O and radius R = 20 m.

Length of chord $AB = L_c = 27$ m

Take J as the midpoint of AB, then

Area $AKBFEA$ = area $AKBJA$ + area $ABEA$

$$= \frac{2}{3} AB \times JK + \frac{1}{2} AB \times EL$$

$$= \frac{2}{3} \times 27 \times 5.3 + \frac{1}{2} \times 27 \times 2.0 = 122.4 \text{ m}^2$$

Therefore the weight of the soil mass = 122.4 × 1 × 20 = 2448 kN

It will act through point G, the centroid of the mass which can be taken as the mid point of FK.

Now, $\theta = 85°$,

$$\text{Length of arc } AKB = L = R\theta = 20 \times 85 \times \frac{3.14}{180} = 29.7 \text{ m}$$

$$\text{Moment arm of cohesion, } l_a = R\frac{L}{L_c} = 20 \times \frac{29.7}{27} = 22 \text{ m}$$

From center O, at a distance l_a, draw the cohesive force vector C, which is parallel to the chord AB. Now from the point of intersection of C and W, draw a line tangent to the friction circle

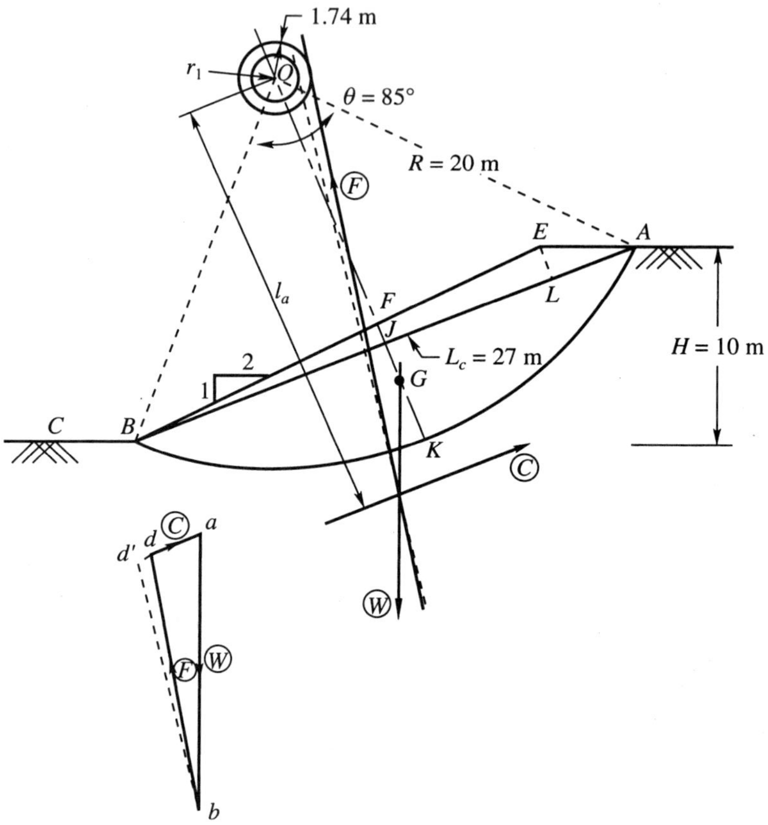

Figure Ex. 10.7

drawn at O with a radius of $R \sin \phi' = 20 \sin 5° = 1.74$ m. This line is the line of action of the third force F.

Draw a triangle of forces in which the magnitude and the direction for W is known and only the directions of the other two forces C and F are known.

Length ad gives the cohesive force $C = 520$ kN

Mobilized cohesion,

$$c'_m = \frac{C}{L} = \frac{520}{29.7} = 17.51 \text{ kN/m}^2$$

Therefore the factor of safety with respect to cohesion, F_c, is

$$F_c = \frac{c'}{c'_m} = \frac{30}{17.51} = 1.713$$

F_c will be 1.713 if the factor of safety with respect to friction, $F_\phi = 1.0$

$$\text{If, } F_s = 1.5, \text{ then } \phi'_m = \frac{\tan 5°}{F_s} = 0.058 \text{ rad; or } \phi'_m = 3.34°$$

The new radius of the friction circle is

$r_1 = R \sin \phi'_m = 20 \times \sin 3.3° = 1.16$ m.

The direction of F changes and the modified triangle of force abd' gives,

cohesive force = C = length ad' = 600 kN

Mobilised cohesino, $c'_m = \frac{C}{L} = \frac{600}{29.7} = 20.2$ kN/m^2

Therefore, $F_c = \frac{c'}{c'_m} = \frac{30}{20.2} \approx 1.5$

10.11 TAYLOR'S STABILITY NUMBER

If the slope angle β, height of embankment H, the effective unit weight of material γ, angle of internal friction ϕ', and unit cohesion c' are known, the factor of safety may be determined. In order to make unnecessary the more or less tedious stability determinations, Taylor (1937) conceived the idea of analyzing the stability of a large number of slopes through a wide range of slope angles and angles of internal friction, and then representing the results by an abstract number which he called the "*stability number*". This number is designated as N_s. The expression used is

$$N_s = \frac{c'}{F_c \gamma H} \tag{10.23}$$

From this the factor of safety with respect to cohesion may be expressed as

$$F_c = \frac{c'}{N_s \gamma H} \tag{10.24}$$

Taylor published his results in the form of curves which give the relationship between N_s and the slope angles β for various values of ϕ' as shown in Fig. 10.16. These curves are for circles passing through the toe, although for values of β less than 53°, it has been found that the most dangerous circle passes below the toe. However, these curves may be used without serious error for slopes down to β = 14°. The stability numbers are obtained for factors of safety with respect to cohesion by keeping the factor of safety with respect to friction (F_ϕ) equal to unity.

In slopes encountered in practical problems, the depth to which the rupture circle may extend is usually limited by ledge or other underlying strong material as shown in Fig. 10.17. The stability number N_s for the case when $\phi' = 0$ is greatly dependent on the position of the ledge. The depth at which the ledge or strong material occurs may be expressed in terms of a depth factor n_d which is defined as

$$n_d = \frac{D}{H} \tag{10.25}$$

where D = depth of ledge below the top of the embankment, H = height of slope above the toe.

For various values of n_d and for the $\phi = 0$ case the chart in Fig. 10.17 gives the stability number N_s for various values of slope angle β. In this case the rupture circle may pass through the toe or below the toe. The distance x of the rupture circle from the toe at the toe level may be expressed by a distance factor n_x which is defined as

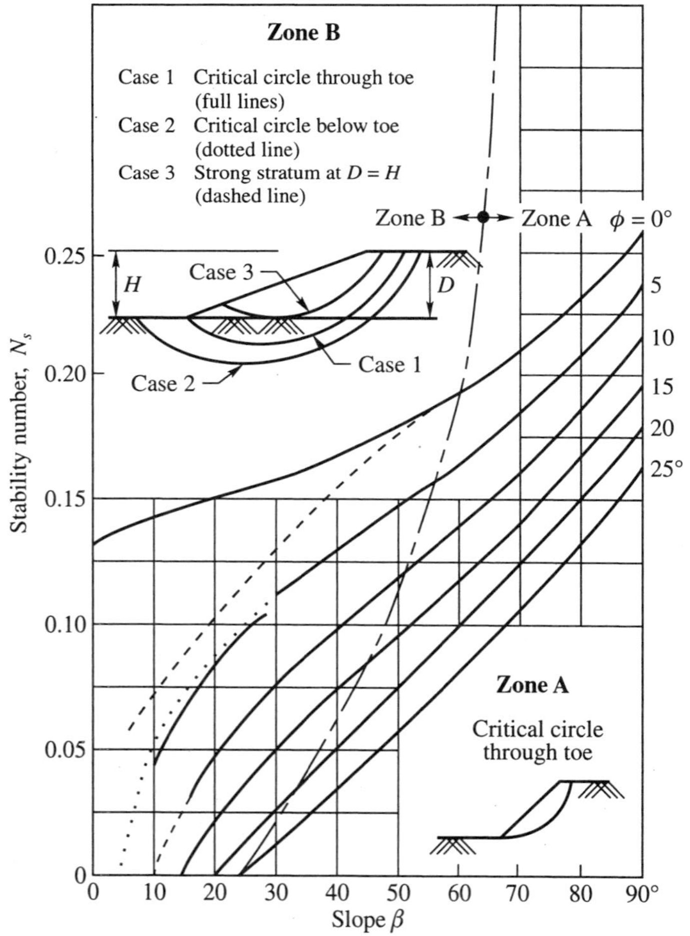

Figure 10.16 Taylor's stability numbers for circles passing through the toe and below or above the toe (after Taylor, 1937)

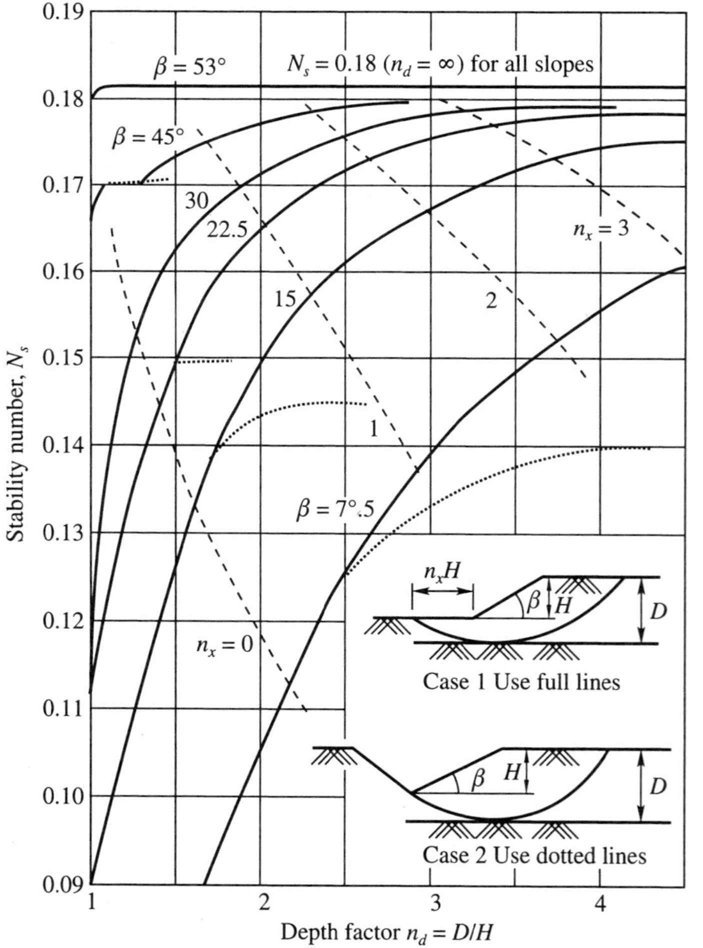

Figure 10.17 Taylor's stability numbers for $\phi' = 0$ case (after Taylor, 1937)

$$n_x = \frac{x}{H} \tag{10.26}$$

The chart in Fig. 10.17 shows the relationship between n_d and n_x. If there is a ledge or other stronger material at the elevation of the toe, the depth factor n_d for this case is unity.

Factor of Safety with Respect to Strength

The development of the stability number is based on the assumption that the factor of safety with respect to friction F_ϕ, is unity. The curves give directly the factor of safety F_c with respect to cohesion only. If a true factor of safety F_s with respect to strength is required, this factor should apply equally to both cohesion and friction. The mobilized shear strength may therefore be expressed as

$$s_m = \frac{s}{F_s} = \frac{c'}{F_s} + \frac{\sigma' \tan\phi'}{F_s}$$

In the above expression, we may write

$$\frac{c'}{F_s} = c'_m, \quad \tan\phi'_m = \frac{\tan\phi'}{F_s}, \quad \text{or } \phi'_m = \frac{\phi'}{F_s} \text{ (approx.)} \tag{10.27}$$

c'_m and ϕ'_m may be described as average values of mobilized cohesion and friction respectively.

Example 10.8

The following particulars are given for an earth dam of height 39 ft. The slope is submerged and the slope angle $\beta = 45°$.

$\gamma_b = 69$ lb/ft^3

$c' = 550$ lb/ft^2

$\phi' = 20°$

Determine the factor of safety F_s.

Solution

Assume as a first trial $F_s = 2.0$

$$\phi'_m = \frac{20}{2} = 10° \text{ (approx.)}$$

For $\phi'_m = 10°$, and $\beta = 45°$ the value of N_s from Fig. 10.16 is 0.11, we may write

From Eq. (10.23) $N_s = \dfrac{c'}{F_c \gamma H}$, substituting

$$0.11 = \frac{550}{2 \times 69 \times H}$$

$$\text{or } H = \frac{550}{2 \times 69 \times 0.11} = 36.23 \text{ ft}$$

$$\text{If } F_s = 1.9, \phi'_m = \frac{20}{1.9} = 10.53° \text{ and } N_s = 0.105$$

$$H = \frac{550}{1.9 \times 69 \times 0.105} = 40 \text{ ft}$$

The computed height 40 ft is almost equal to the given height 39 ft. The computed factor of safety is therefore 1.9.

Example 10.9

An excavation is to be made in a soil deposit with a slope of 25° to the horizontal and to a depth of 25 meters. The soil has the following properties:

$c' = 35 \text{kN/m}^2$, $\phi' = 15°$ and $\gamma = 20 \text{ kN/m}^3$

1. Determine the factor of safety of the slope assuming full friction is mobilized.
2. If the factor of safety with respect to cohesion is 1.5, what would be the factor of safety with respect to friction?

Solution

1. For $\phi' = 15°$ and $\beta = 25°$, Taylor's stability number chart gives stability number $N_s = 0.03$.

$$F_c = \frac{c'}{N_s \gamma H} = \frac{35}{0.03 \times 20 \times 25} = 2.33$$

2. For $F_c = 1.5$, $N_s = \dfrac{c'}{F_c \times \gamma \times H} = \dfrac{35}{1.5 \times 20 \times 25} = 0.047$

For $N_s = 0.047$ and $\beta = 25°$, we have from Fig. 10.16, $\phi'_m = 13°$

$$\text{Therefore, } F_\phi = \frac{\tan \phi'}{\tan \phi'_m} = \frac{\tan 15°}{\tan 13°} = \frac{0.268}{0.231} = 1.16$$

Example 10.10

An embankment is to be made from a soil having $c' = 420 \text{ lb/ft}^2$, $\phi' = 18°$ and $\gamma = 121 \text{ lb/ft}^3$. The desired factor of safety with respect to cohesion as well as that with respect to friction is 1.5. Determine

1. The safe height if the desired slope is 2 horizontal to 1 vertical.
2. The safe slope angle if the desired height is 50 ft.

Solution

$$\tan \phi' = \tan 18° = 0.325, \ \phi'_m = \tan^{-1} \frac{0.325}{1.5} = 12.23°$$

1. For $\phi' = 12.23°$ and $\beta = 26.6°$ (i.e., 2 horizontal and 1 vertical) the chart gives $N_s = 0.055$

$$\text{Therefore, } 0.055 = \frac{c'}{F_c \gamma H} = \frac{420}{1.5 \times 121 \times H}$$

$$\text{Therefore, } H_{safe} = \frac{420}{1.5 \times 121 \times 0.055} = 42 \text{ ft}$$

2. $$\text{Now, } N_s = \frac{c'}{F_c \gamma H} = \frac{420}{1.5 \times 121 \times 50} = 0.046$$

For $N_s = 0.046$ and $\phi'_m = 12.23°$, slope angle $\beta = 23.5°$

10.12 TENSION CRACKS

If a dam is built of cohesive soil, tension cracks are usually present at the crest. The depth of such cracks may be computed from the equation

$$z_0 = \frac{2c'}{\gamma} \tag{10.28}$$

where z_0 = depth of crack, c' = unit cohesion, γ = unit weight of soil.

The effective length of any trial arc of failure is the difference between the total length of arc minus the depth of crack as shown in Fig. 10.18.

10.13 STABILITY ANALYSIS BY METHOD OF SLICES FOR STEADY SEEPAGE

The stability analysis with steady seepage involves the development of the pore pressure head diagram along the chosen trial circle of failure. The simplest of the methods for knowing the pore pressure head at any point on the trial circle is by the use of flownets which is described below.

Determination of Pore Pressure with Seepage

Figure 10.19 shows the section of a homogeneous dam with an arbitrarily chosen trial arc. There is steady seepage flow through the dam as represented by flow and equipotential lines. From the equipotential lines the pore pressure may be obtained at any point on the section. For example at point *a* in Fig. 10.19 the pressure head is *h*. Point *c* is determined by setting the radial distance *ac*

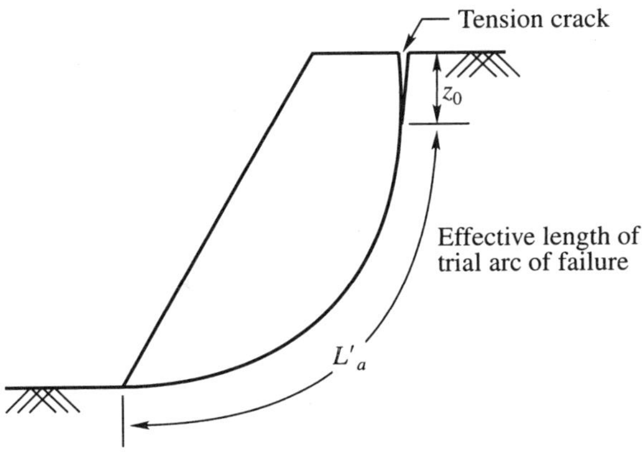

Figure 10.18 Tension crack in dams built of cohesive soils

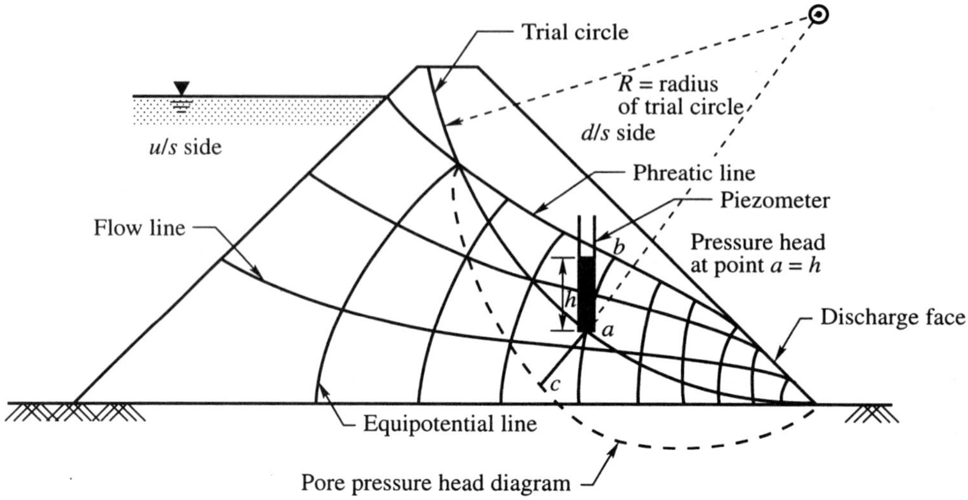

Figure 10.19 Determination of pore pressure with steady seepage

equal to *h*. A number of points obtained in the same manner as *c* give the curved line through *c* which is a pore pressure head diagram.

Method of Analysis (graphical method)

Figure 10.20(a) shows the section of a dam with an arbitrarily chosen trial arc. The center of rotation of the arc is 0. The pore pressure acting on the base of the arc as obtained from flow nets is shown in Fig. 10.20(b).

When the soil forming the slope has to be analyzed under a condition where full or partial drainage takes place the analysis must take into account both cohesive and frictional soil properties based on *effective stresses*. Since the effective stress acting across each elemental length of the assumed circular arc failure surface must be computed in this case, the method of slices is one of the convenient methods for this purpose. The method of analysis is as follows.

The soil mass above the assumed slip circle is divided into a number of vertical slices of equal width. The number of slices may be limited to a maximum of eight to ten to facilitate computation. The forces used in the analysis acting on the slices are shown in Figs. 10.20(a) and (c). The forces are:

1. The weight W of the slice.
2. The normal and tangential components of the weight W acting on the base of the slice. They are designated respectively as N and T.
3. The pore water pressure U acting on the base of the slice.
4. The effective frictional and cohesive resistances acting on the base of the slice which is designated as S.

The forces acting on the sides of the slices are statically indeterminate as they depend on the stress deformation properties of the material, and we can make only gross assumptions about their relative magnitudes.

In the conventional slice method of analysis the lateral forces are assumed equal on both sides of the slice. This assumption is not strictly correct. The error due to this assumption on the mass as a whole is about 15 percent (Bishop, 1955).

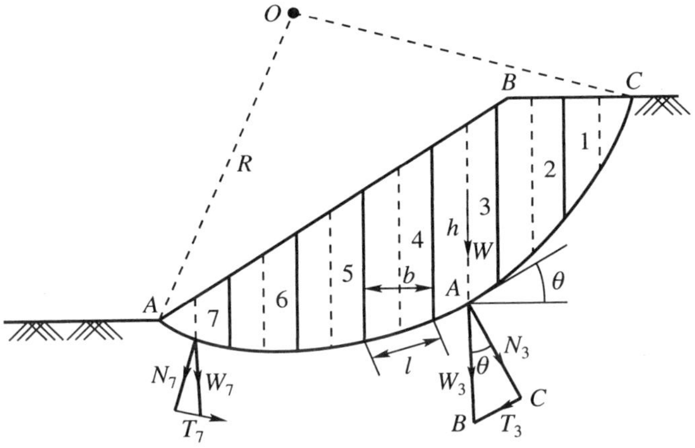

(a) Total normal and tangential components

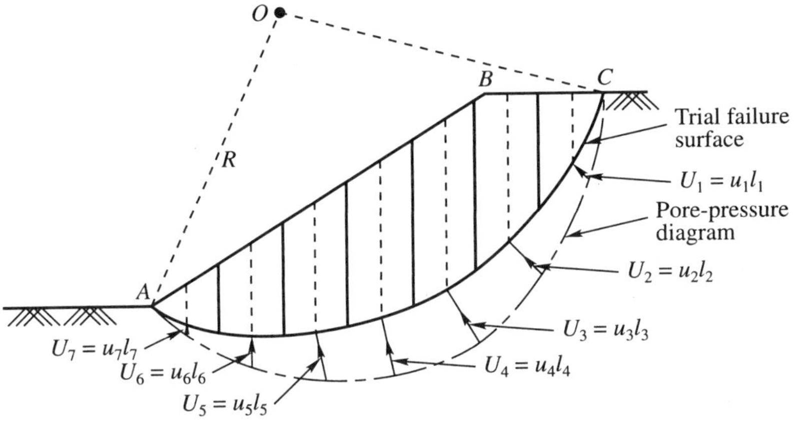

(b) Pore-pressure diagram

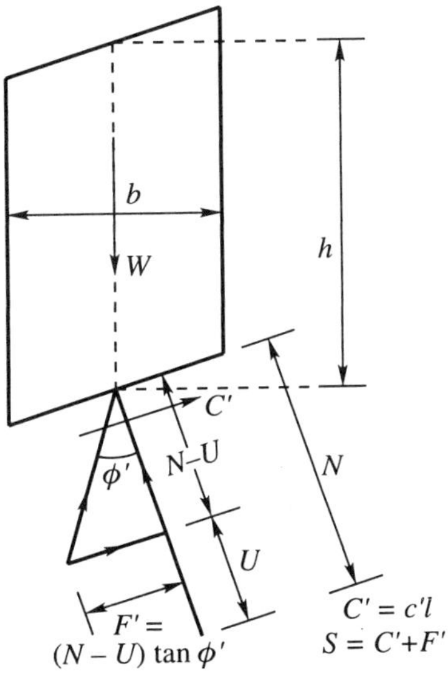

(c) Resisting forces on the base of slice

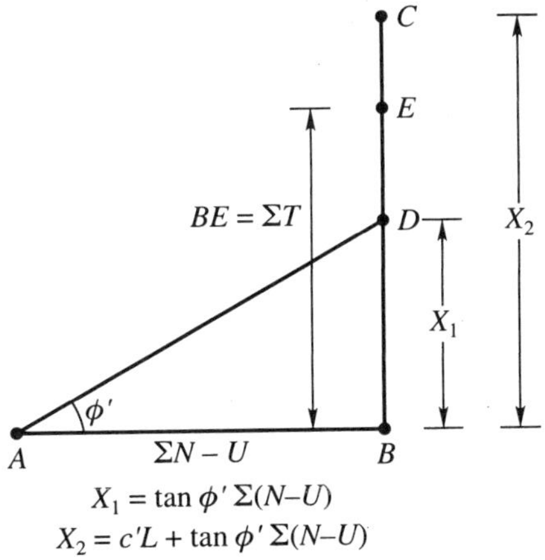

(d) Graphical representation of all the forces

Figure 10.20 Stability analysis of slope by the method of slices

The forces that are actually considered in the analysis are shown in Fig. 10.20(c). The various components may be determined as follows:

1. The weight, W, of a slice per unit length of dam may be computed from

   $W = \gamma hb$

   where, γ = total unit weight of soil, h = average height of slice, b = width of slice.

   If the widths of all slices are equal, and if the whole mass is homogeneous, the weight W can be plotted as a vector AB passing through the center of a slice as in Fig. 10.20(a). AB may be made equal to the height of the slice.
2. By constructing triangle ABC, the weight can be resolved into a normal component N and a tangential component T. Similar triangles can be constructed for all slices. The tangential components of the weights cause the mass to slide downward. The sum of all the weights cause the mass to slide downward. The sum of all the tangential components may be expressed as $\overline{T} = \Sigma T$. If the trial surface is curved upward near its lower end, the tangential component of the weight of the slice will act in the opposite direction along the curve. The algebraic sum of T should be considered.
3. The average pore pressure u acting on the base of any slice of length l may be found from the pore pressure diagram shown in Fig. 10.20(b). The total pore pressure, U, on the base of any slice is

   $U = ul$
4. The effective normal pressure N' acting on the base of any slice is

   $N' = N - U$ [Fig. 10.20(c)]
5. The frictional force F' acting on the base of any slice resisting the tendency of the slice to move downward is

   $F = (N - U) \tan \phi'$

   where ϕ' is the effective angle of friction. Similarly the cohesive force C' opposing the movement of the slice and acting at the base of the slice is

   $C' = c'l$

   where c' is the effective unit cohesion. The total resisting force S acting on the base of the slice is

   $S = C' + F' = c'l + (N - U) \tan \phi'$

   Figure 10.20(c) shows the resisting forces acting on the base of a slice.

   The sum of all the resisting forces acting on the base of each slice may be expressed as

   $S_s = c' \Sigma l + \tan \phi' \Sigma(N - U) = c'L + \tan \phi' \Sigma(N - U)$

   where $\Sigma l = L$ = length of the curved surface.

   The moments of the actuating and resisting forces about the point of rotation may be written as follows:

   Actuating moment = $R \Sigma T$

   Resisting moment = $R[c'L + \tan \phi' \Sigma(N - U)]$

   The factor of safety F_s may now be written as

$$F_s = \frac{[c'L + \tan \phi' \Sigma(N - U)]}{\Sigma T} \qquad (10.29)$$

The various components shown in Eq. (10.29) can easily be represented graphically as shown in Fig. 10.20(d). The line AB represents to a suitable scale $\Sigma(N - U)$. BC is drawn normal to AB at B and equal to $c'L + \tan\phi' \Sigma(N - U)$. The line AD drawn at an angle ϕ' to AB gives the intercept BD on BC equal to $\tan\phi' \Sigma(N - U)$. The length BE on BC is equal to ΣT. Now

$$F_s = \frac{BC}{BE} \tag{10.30}$$

Centers for Trial Circles Through Toe

The factor of safety F_s as computed and represented by Eq. (10.29) applies to one trial circle. This procedure is followed for a number of trial circles until one finds the one for which the factor of safety is the lowest. This circle that gives the least F_s is the one most likely to fail. The procedure is quite laborious. The number of trial circles may be minimized if one follows the following method.

For any given slope angle β (Fig. 10.21), the center of the first trial circle center O may be determined as proposed by Fellenius (1927). The direction angles α_A and α_B may be taken from Table 10.1. For the centers of additional trial circles, the procedure is as follows:

Mark point C whose position is as shown in Fig. 10.21. Join CO. The centers of additional circles lie on the line CO extended. This method is applicable for a homogeneous $(c - \phi)$ soil. When the soil is purely cohesive and homogeneous the direction angles given in Table 10.1 directly give the center for the critical circle.

Centers for Trial Circles Below Toe

Theoretically if the materials of the dam and foundation are entirely homogeneous, any practicable earth dam slope may have its critical failure surface below the toe of the slope. Fellenius found that the angle intersected at 0 in Fig. 10.22 for this case is about 133.5°. To find the center for the critical circle below the toe, the following procedure is suggested.

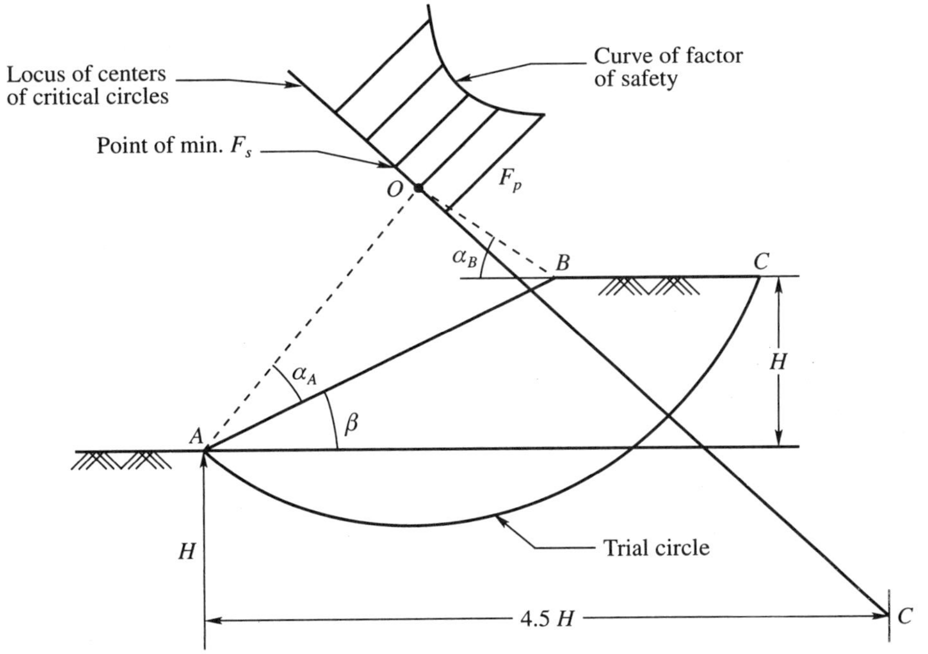

Figure 10.21 Location of centers of critical circle passing through toe of dam

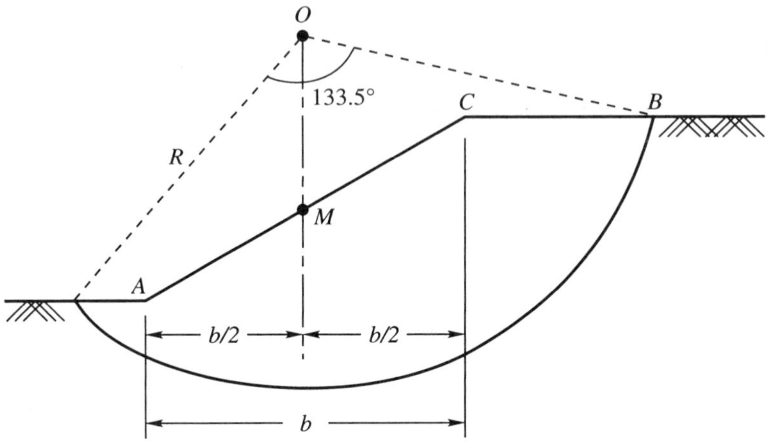

Figure 10.22 Centers of trial circles for base failure

Table 10.1 Direction angles α^o_A and α^o_B for centers of critical circles

Slope	Slope angle	Direction angles	
	β	α^o_A	α^o_B
0.6 : 1	60	29	40
1 : 1	45	28	37
1.5 : 1	33.8	26	35
2 : 1	26.6	25	35
3 : 1	18.3	25	35
5 : 1	11.3	25	37

Erect a vertical at the midpoint M of the slope. On this vertical will be the center O of the first trial circle. In locating the trial circle use an angle (133.5°) between the two radii at which the circle intersects the surface of the embankment and the foundation. After the first trial circle has been analyzed the center is some what moved to the left, the radius shortened and a new trial circle drawn and analyzed. Additional centers for the circles are spotted and analyzed.

Example 10.11

An embankment is to be made of a sandy clay having a cohesion of 30 kN/m^2, angle of internal friction of 20° and a unit weight of 18 kN/m^3. The slope and height of the embankment are 1.6 : 1 and 10 m respectively. Determine the factor of safety by using the trial circle given in Fig. Ex. 10.11 by the method of slices.

Solution

Consider the embankment as shown in Fig. Ex.10.11. The center of the trial circle O is selected by taking $\alpha_A = 26°$ and $\alpha_B = 35°$ from Table 10.1. The soil mass above the slip circle is divided into 13 slices of 2 m width each. The weight of each slice per unit length of embankment is given by $W = h_a b \gamma_t$, where h_a = average height of the slice, b = width of the slice, γ_t = unit weight of the soil.

The weight of each slice may be represented by a vector of height h_a if b and γ_t remain the same for the whole embankment. The vectors values were obtained graphically. The height vectors

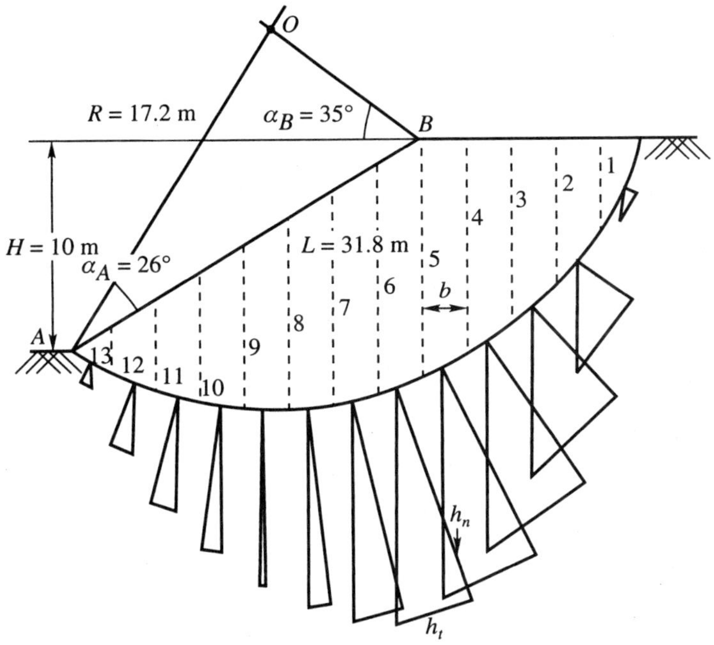

Figure Ex. 10.11

may be resolved into normal components h_n and tangential components h_t. The values of h_a, h_n and h_t for the various slices are given below in a tabular form.

Values of h_a, h_n and h_t

Slice No.	$h_a(m)$	$h_n(m)$	$h_t(m)$	Slice No.	$h_a(m)$	$h_n(m)$	$h_t(m)$
1	1.8	0.80	1.72	8	9.3	9.25	1.00
2	5.5	3.21	4.50	9	8.2	8.20	–0.20
3	7.8	5.75	5.30	10	6.8	6.82	–0.80
4	9.5	7.82	5.50	11	5.2	5.26	–1.30
5	10.6	9.62	4.82	12	3.3	3.21	–1.20
6	11.0	10.43	3.72	13	1.1	1.0	–0.50
7	10.2	10.20	2.31				

The sum of these components h_n and h_t may be converted into forces ΣN and ΣT respectively by multiplying them as given below

$$\Sigma h_n = 81.57 \text{ m}, \quad \Sigma h_t = 24.87 \text{ m}$$

Therefore, $\Sigma N = 81.57 \times 2 \times 18 = 2937$ kN

$$\Sigma T = 24.87 \times 2 \times 18 = 895 \text{ kN}$$

Length of arc = L = 31.8 m

$$\text{Factor of safety} = \frac{c'L + \tan\phi' \Sigma N}{\Sigma T} = \frac{30 \times 31.8 + 0.364 \times 2937}{895} = 2.26$$

10.14 BISHOP'S SIMPLIFIED METHOD OF SLICES

Bishop's method of slices (1955) is useful if a slope consists of several types of soil with different values of c and ϕ and if the pore pressures u in the slope are known or can be estimated. The method of analysis is as follows:

Figure 10.23 gives a section of an earth dam having a sloping surface AB. ADC is an assumed trial circular failure surface with its center at O. The soil mass above the failure surface is divided into a number of slices. The forces acting on each slice are evaluated from limit equilibrium of the slices. The equilibrium of the entire mass is determined by summation of the forces on each of the slices.

Consider for analysis a single slice $abcd$ (Fig. 10.23a) which is drawn to a larger scale in Fig. 10.23(b). The forces acting on this slice are

W = weight of the slice

N = total normal force on the failure surface cd

U = pore water pressure = ul on the failure surface cd

F_R = shear resistance acting on the base of the slice

E_1, E_2 = normal forces on the vertical faces bc and ad

T_1, T_2 = shear forces on the vertical faces bc and ad

θ = the inclination of the failure surface cd to the horizontal

The system is statically indeterminate. An approximate solution may be obtained by assuming that the resultant of E_1 and T_1 is equal to that of E_2 and T_2, and their lines of action coincide. For equilibrium of the system, the following equations hold true.

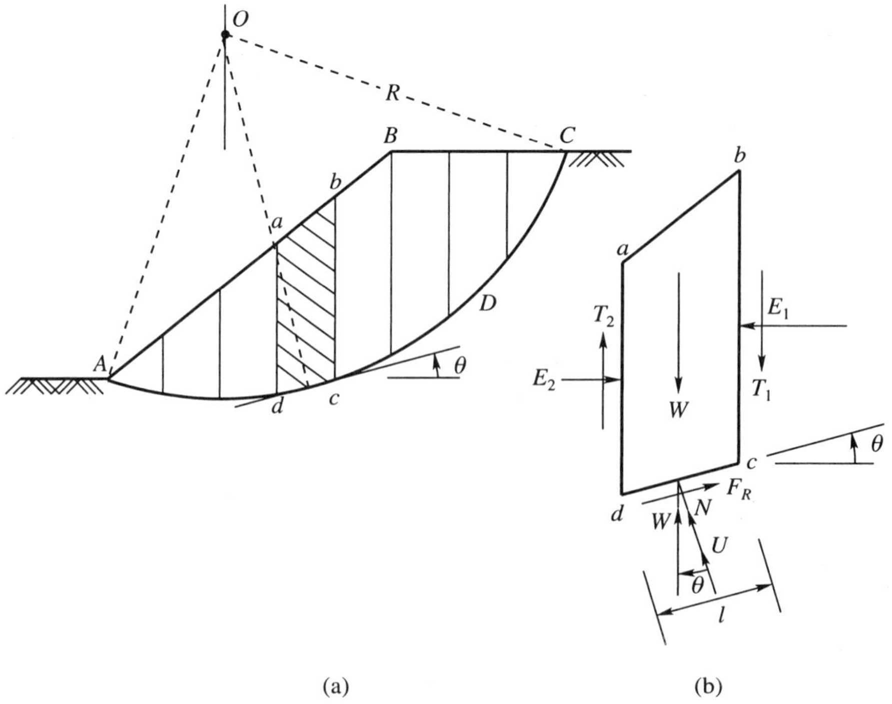

Figure 10.23 Bishop's simplified method of analysis

$$N = W\cos\theta$$
$$F_t = W\sin\theta \qquad (10.31)$$

where F_t = tangential component of W

The unit stresses on the failure surface of length, l, may be expressed as

$$\text{normal stress, } \sigma_n = \frac{W\cos\theta}{l}$$
$$\text{shear stress, } \tau_n = \frac{W\sin\theta}{l} \qquad (10.32)$$

The equation for shear strength, s, is

$$s = c' + \sigma' \tan\phi' = c' + (\sigma - u)\tan\phi'$$

where σ' = effective normal stress
c' = effective cohesion
ϕ' = effective angle of friction
u = unit pore pressure

The shearing resistance to sliding on the base of the slice is

$$sl = c'l + (W\cos\theta - ul)\tan\phi'$$

where ul = U, the total pore pressure on the base of the slice (Fig 10.23b)
$sl = F_R$

The total resisting force and the actuating force on the failure surface *ADC* may be expressed as

Total resisting force F_R is

$$F_R = \quad [c'l + (W\cos\theta - ul)\tan\phi'] \qquad (10.33)$$

Total actuating force F_t is

$$F_t = \quad W\sin\theta \qquad (10.34)$$

The factor of safety F_s is then given as

$$F_s = \frac{F_R}{F_t} = \frac{[c'l + (W\cos\theta - ul)\tan\phi']}{W\sin\theta} \qquad (10.35)$$

Eq. (10.35) is the same as Eq. (10.29) obtained by the conventional method of analysis.

Bishop (1955) suggests that the accuracy of the analysis can be improved by taking into account the forces E and T on the vertical faces of each slice. For the element in Fig. 10.23(b), we may write an expression for all the forces acting in the vertical direction for the equilibrium condition as

$$N'\cos\theta = W + (T_1 - T_2) - ul\cos\theta - F_R\sin\theta \qquad (10.36)$$

If the slope is not on the verge of failure ($F_s > 1$), the tangential force F_t is equal to the shearing resistance F_R on cd divided by F_s.

$$F_R = \frac{c'l}{F_s} + N'\frac{\tan\phi'}{F_s} \tag{10.37}$$

where, $N' = N - U$, and $U = ul$.

Substituting Eq. (10.37) into Eq. (10.36) and solving for N', we obtain

$$N' = \frac{\left(W + \Delta T - U\cos\theta - \dfrac{c'l}{F_s}\sin\theta\right)}{\cos\theta + \dfrac{\tan\phi'\sin\theta}{F_s}} \tag{10.38}$$

where, $\Delta T = T_1 - T_2$.

For equilibrium of the mass above the failure surface, we have by taking moments about O

$$\Sigma W\sin\theta R = \Sigma F_R R \tag{10.39}$$

By substituting Eqs. (10.37) and (10.38) into Eq. (10.39) and solving we obtain an expression for F_s as

$$F_s = \frac{\Sigma\left\{c'l\cos\theta + [(W - U\cos\theta) + \Delta T]\tan\phi'\right\}\frac{1}{m_\theta}}{\Sigma W\sin\theta} \tag{10.40}$$

where, $$m_\theta = \cos\theta + \frac{\tan\phi'\sin\theta}{F_s} \tag{10.41}$$

The factor of safety F_s is present in Eq. (10.40) on both sides. The quantity $\Delta T = T_1 - T_2$ has to be evaluated by means of successive approximation . Trial values of E_1 and T_1 that satisfy the equilibrium of each slice, and the conditions

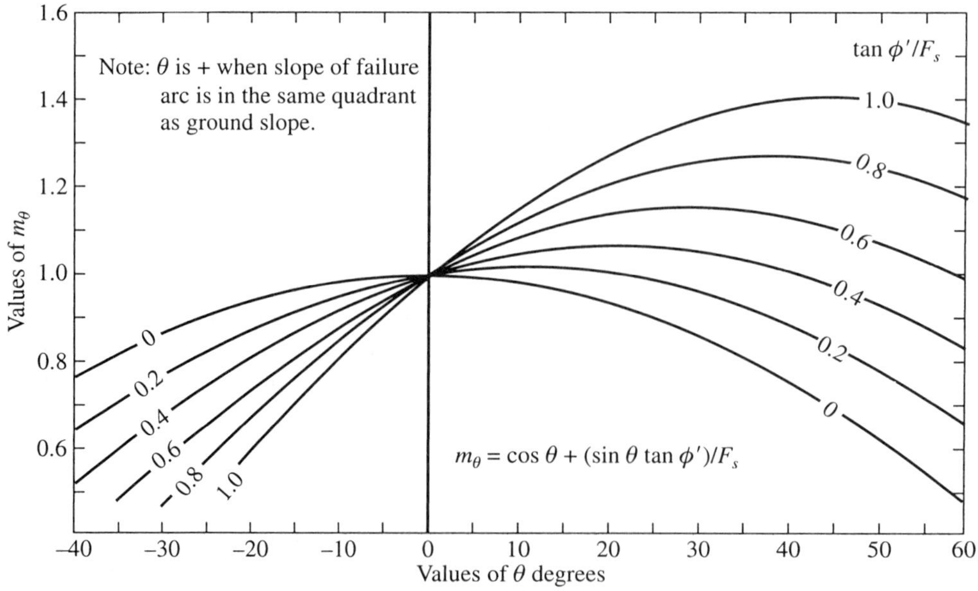

Figure 10.24 Values of m_θ (after Janbu et al., 1956)

$$(E_1 - E_2) = 0 \quad \text{and} \quad (T_1 - T_2) = 0$$

are used. The value of F_s may then be computed by first assuming an arbitrary value for F_s. The value of F_s may then be calculated by making use of Eq. (10.40). If the calculated value of F_s differs appreciably from the assumed value, a second trial is made and the computation is repeated. Figure 10.24 developed by Janbu et al. (1956) helps to simplify the computation procedure.

It is reported that an error of about 1 percent will occur if we assume $\Sigma(T_1 - T_2)\tan\phi' = 0$. But if we use the conventional method of analysis using Eq. (10.35) the error introduced is about 15 percent (Bishop, 1955).

10.15 BISHOP AND MORGENSTERN METHOD FOR SLOPE ANALYSIS

Equation (10.40) developed based on Bishop's analysis of slopes, contains the term pore pressure u. The Bishop and Morgenstern method (1960) proposes the following equation for the evaluation of u

$$r_u = \frac{u}{\gamma h} \tag{10.42}$$

where, u = pore water pressure at any point on the assumed failure surface
γ = unit weight of the soil
h = the depth of the point in the soil mass below the ground surface

The pore pressure ratio r_u is assumed to be constant throughout the cross-section, which is called a *homogeneous pore pressure distribution*. Figure 10.25 shows the various parameters used in the analysis.

The factor of safety F_s is defined as

$$F_s = m - nr_u \tag{10.43}$$

where, m, n = stability coefficients.

The m and n values may be obtained either from charts in Figs. B.1 to B.6 or Tables B1 to B6 in Appendix B. The depth factor given in the charts or tables is as per Eq. (10.25), that is $n_d = D/H$, where H = height of slope, and D = depth of firm stratum from the top of the slope. Bishop and Morgenstern (1960) limited their charts (or tables) to values of $c'/\gamma H$ equal to 0.000, 0.025, and 0.050.

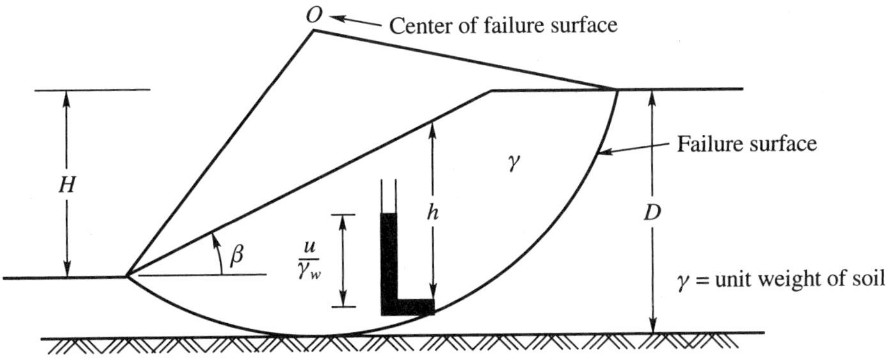

Figure 10.25 Specifications of parameters for Bishop-Morgenstern method of analysis

Extension of the Bishop and Morgenstern Slope Stability Charts

As stated earlier, Bishop and Morgenstern (1960) charts or tables cover values of $c'/\gamma H$ equal to 0.000, 0.025, and 0.050 only. These charts do not cover the values that are normally encountered in natural slopes. O' Connor and Mitchell (1977) extended the work of Bishop and Morgenstern to cover values of $c'/\gamma H$ equal to 0.075 and 0.100 for various values of depth factors n_d. The method employed is essentially the same as that adopted by the earlier authors. The extended values are given in the form of charts and tables from Figs. B.7 to B.14 and Tables B7 to B14 respectively in Appendix B.

Method of Determining F_s

1. Obtain the values of r_u and $c/\gamma H$
2. From the tables in Appendix B, obtain the values of m and n for the known values of $c/\gamma H$, ϕ and β, and for n_d = 0, 1, 1.25 and 1.5.
3. Using Eq. (10.43), determine F_s for each value of n_d.
4. The required value of F_s is the lowest of the values obtained in step 3.

Example 10.12

Figure Ex. 10.12 gives a typical section of a homogeneous earth dam. The soil parameters are: $\phi' = 30°$, $c' = 590$ lb/ft², and $\gamma = 120$ lb/ft³. The dam has a slope 4:1 and a pore pressure ratio $r_u = 0.5$. Estimate the factor of safety F_s by Bishop and Morgenstern method for a height of dam $H = 140$ ft.

Solution

Height of dam $H = 140$ ft

$$\frac{c'}{\gamma H} = \frac{590}{120 \times 140} = 0.035$$

Given: $\phi' = 30°$, slope 4:1 and $r_u = 0.5$.

Since $c'/\gamma H = 0.035$, and $n_d = 1.43$ for $H = 140$ ft, the F_s for the dam lies between $c'/\gamma H$ = 0.025 and 0.05 and n_d between 1.0 and 1.5. The equation for F_s is

$$F_s = m - nr_u$$

Using the Tables in Appendix B, the following table can be prepared for the given values of $c'/\gamma H$, ϕ, and β.

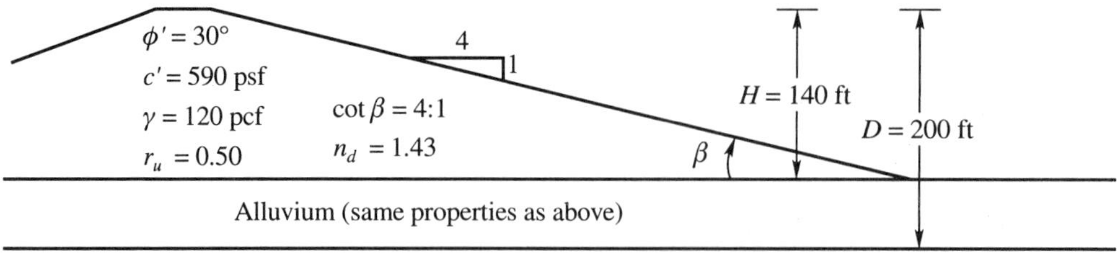

Figure Ex. 10.12

From Tables B2 and B3 for $c'/\gamma H = 0.025$

n_d	m	n	F_s	
1.0	2.873	2.622	1.562	
1.25	**2.953**	**2.806**	**1.55**	Lowest

From Table B4, B5 and B6 for $c'/\gamma H = 0.05$

n_d	m	n	F_s	
1.0	3.261	2.693	1.915	
1.25	**3.221**	**2.819**	**1.812**	Lowest
1.50	3.443	3.120	1.883	

Hence $n_d = 1.25$ is the more critical depth factor. The value of F_s for $c'/\gamma H = 0.035$ lies between 1.55 (for $c'/\gamma H = 0.025$) and 1.812 (for $c'/\gamma H = 0.05$). By proportion $F_s = 1.655$.

10.16 MORGENSTERN METHOD OF ANALYSIS FOR RAPID DRAWDOWN CONDITION

Rapid drawdown of reservoir water level is one of the critical states in the design of earth dams. Morgenstern (1963) developed the method of analysis for rapid drawdown conditions based on the Bishop and Morgenstern method of slices. The purpose of this method is to compute the factor of safety during rapid drawdown, which is reduced under no dissipation of pore water pressure. The assumptions made in the analysis are

1. Simple slope of homogeneous material
2. The dam rests on an impermeable base
3. The slope is completely submerged initially
4. The pore pressure does not dissipate during drawdown

Morgenstern used the pore pressure parameter $\overline{B}$ as developed by Skempton (1954) which states

$$\overline{B} = \frac{u}{\sigma_1} \tag{10.45}$$

where $\sigma_1 = \gamma h$

γ = total unit weight of soil or equal to twice the unit weight of water

h = height of soil above the lower level of water after drawdown

The charts developed take into account the drawdown ratio which is defined as

$$R_d = \frac{\overline{H}}{H} \tag{10.46}$$

where R_d = drawdown ratio

$\overline{H}$ = height of drawdown

H = height of dam (Fig. 10.26)

All the potential sliding circles must be tangent to the base of the section.

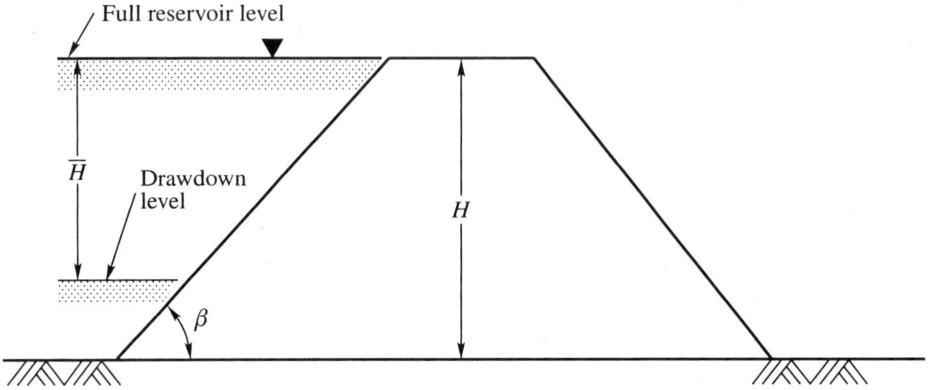

Figure 10.26 Dam section for drawdown conditions

The stability charts are given in Figs. 10.27 to 10.29 covering a range of stability numbers $c'/\gamma H$ from 0.0125 to 0.050. The curves developed are for the values of ϕ' of 20°, 30°, and 40° for different values of β.

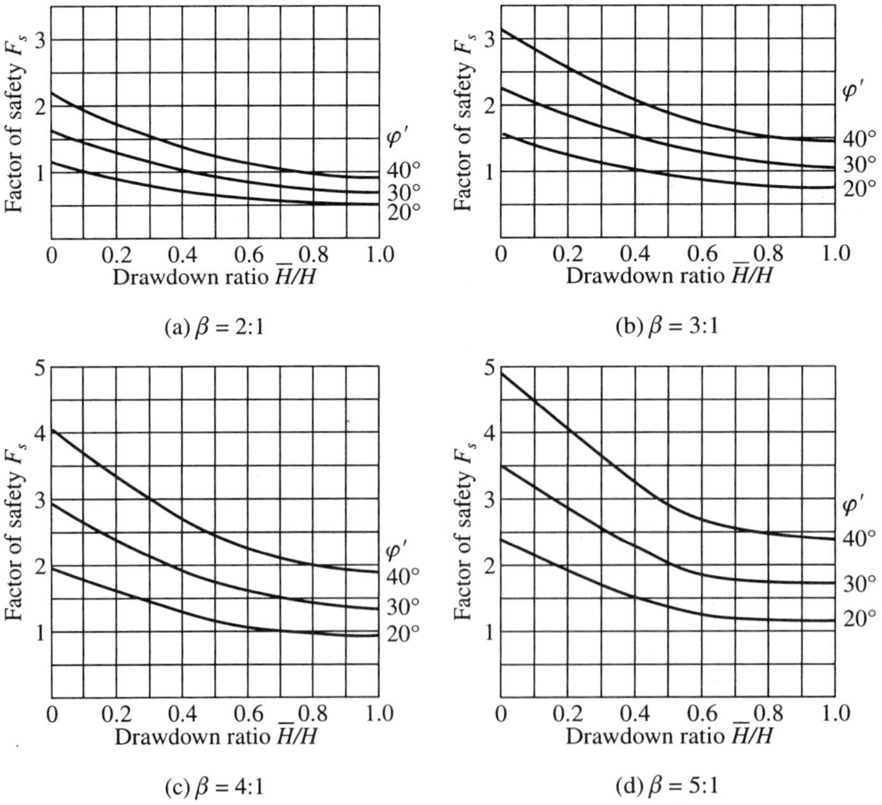

Figure 10.27 Drawdown stability chart for $c'/\gamma H = 0.0125$ (after Morgenstern, 1963)

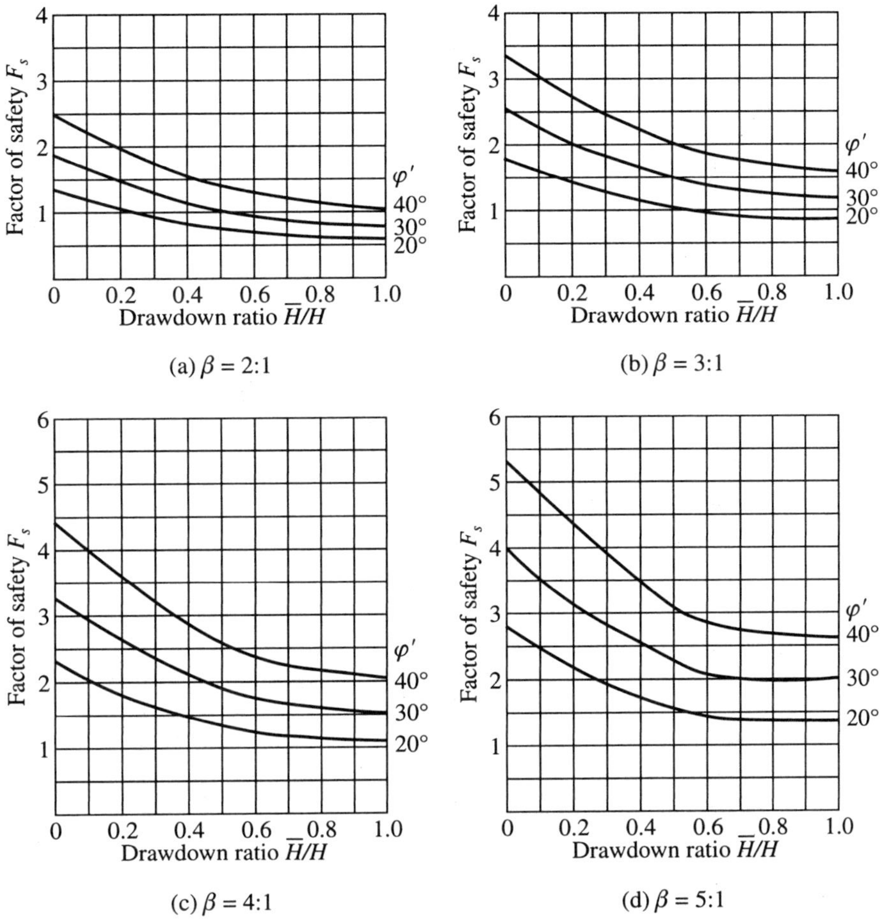

Figure 10.28 Drawdown stability chart for $c'/\gamma H = 0.025$ (after Morgenstern, 1963)

Example 10.13

It is required to estimate the minimum factor of safety for the complete drawdown of the section shown in Fig. Ex. 10.13 (Morgenstern, 1963)

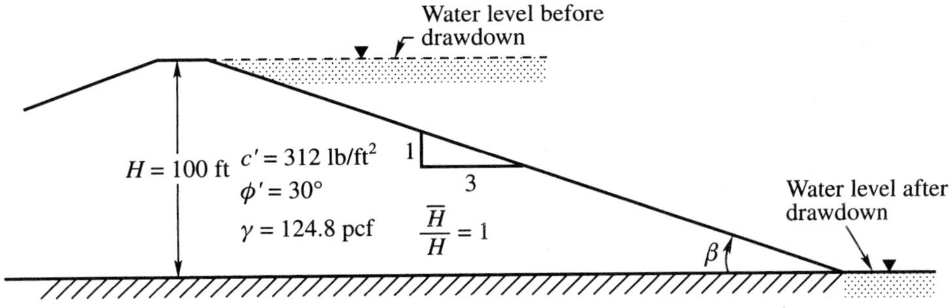

Figure Ex. 10.13

Solution

From the data given in the Fig. Ex. 10.13

$$N_s = \frac{c'}{\gamma H} = \frac{312}{124.8 \times 100} = 0.025$$

From Fig. 10.28, for $N_s = 0.025$, $\beta = 3{:}1$, $\phi' = 30°$, and $\overline{H}/H = 1$,

$F_s = 1.20$

It is evident than the critical circle is tangent to the base of the dam and no other level need be investigated since this would only raise the effective value of N_s resulting in a higher factor of safety.

10.17 SPENCER METHOD OF ANALYSIS

Spencer (1967) developed his analysis based on the method of slices of Fellenius (1927) and Bishop (1955). The analysis is in terms of effective stress and satisfies two equations of

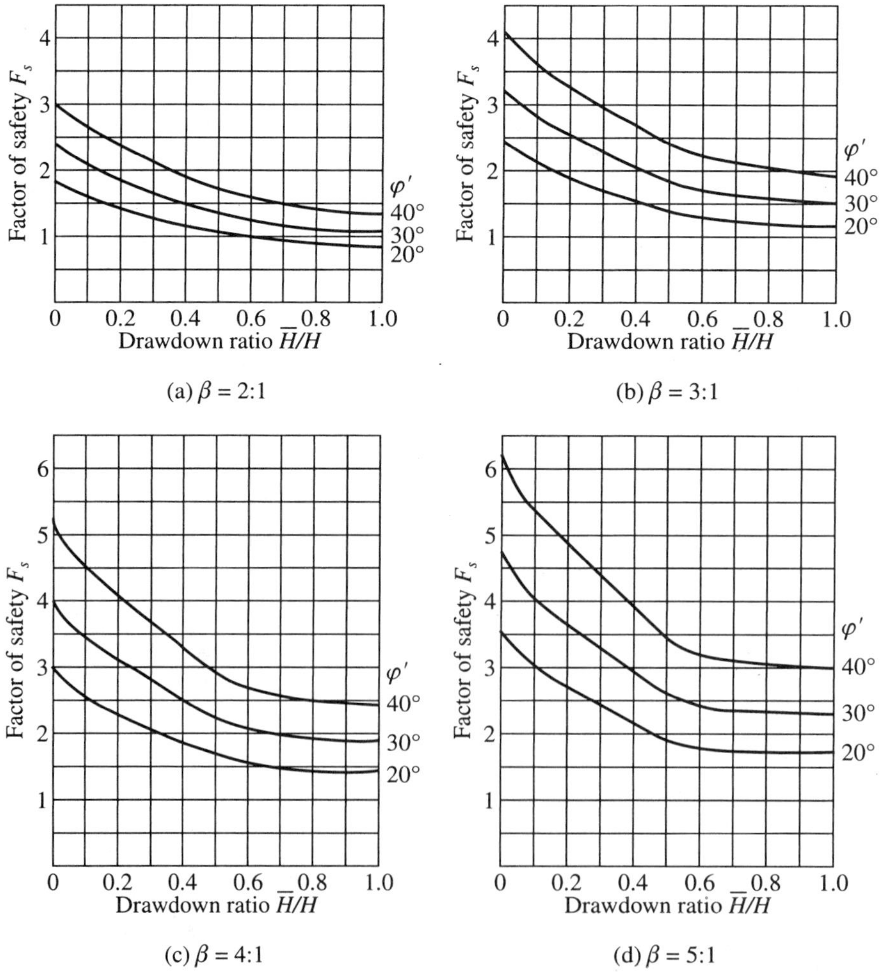

Figure 10.29 Drawdown stability chart for $c'/\gamma H = 0.05$ (after Morgenstern, 1963)

equilibrium, the first with respect to forces and the second with respect to moments. The interslice forces are assumed to be parallel as in Fig. 10.23. The factor of safety F_s is expressed as

$$F_s = \frac{\text{Shear strength available}}{\text{Shear strength mobilized}} \tag{10.47}$$

The mobilized angle of shear resistance and other factors are expressed as

$$\tan\phi'_m = \frac{\tan\phi'}{F_s} \tag{10.48}$$

$$\text{pore pressure ratio,} \quad r_u = \frac{u}{\gamma h} \tag{10.49}$$

$$\text{Stability factor,} \quad N_s = \frac{c'}{F_s \gamma H} \tag{10.50}$$

The charts developed by Spencer for different values of N_s, ϕ'_m and r_u are given in Fig. 10.30. The use of these charts will be explained with worked out examples.

Example 10.14

Find the slope corresponding to a factor of safety of 1.5 for an embankment 100 ft high in a soil whose properties are as follows:

$c' = 870$ lb/sq ft, $\gamma = 120$ lb/ft3, $\phi' = 26°$, $r_u = 0.5$

Solution (by Spencer's Method)

$$N_s = \frac{c'}{F_s \gamma H} = \frac{870}{1.5 \times 120 \times 100} = 0.048$$

$$\tan\phi'_m = \frac{\tan\phi'}{F_s} = \frac{0.488}{1.5} = 0.325$$

$$\phi'_m = 18°$$

Referring to Fig. 10.30c, for which $r_u = 0.5$, the slope corresponding to a stability number of 0.048 is 3:1.

Example 10.15

What would be the change in strength on sudden drawdown for a soil element at point P which is shown in Fig. Ex. 10.15? The equipotential line passing through this element represents loss of water head of 1.2 m. The saturated unit weight of the fill is 21 kN/m^3.

Solution

The data given are shown in Fig. Ex. 10.15. Before drawdown,

The stresses at point P are:

$$\sigma_0 = \gamma_w h_w + \gamma_{sat} h_c = 9.81 \times 3 + 21 \times 4 = 113 \text{ kN/m}^2$$

$$u_0 = \gamma_w (h_w + h_c - h') = 9.81(3 + 4 - 1.2) = 57 \text{ kN/m}^2$$

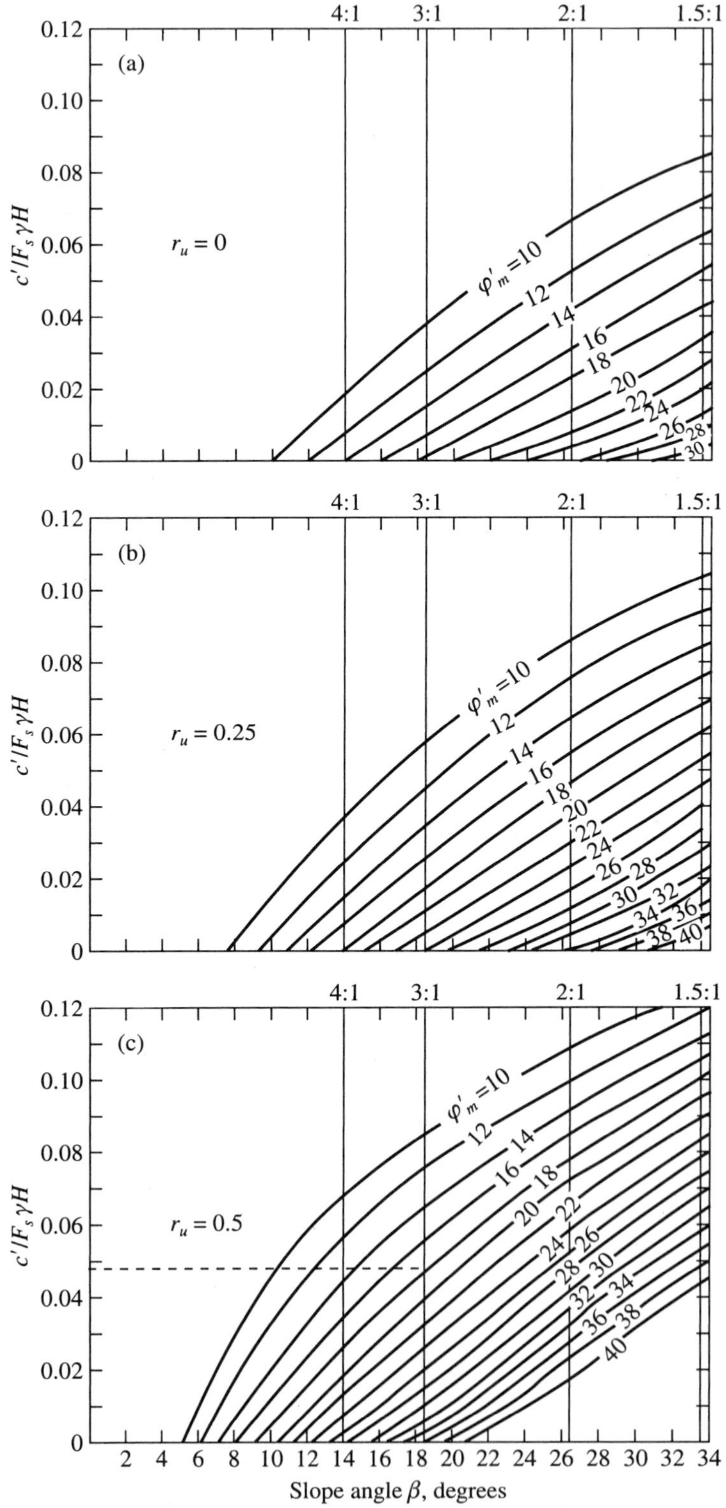

Figure 10.30 Stability charts (after Spencer, 1967)

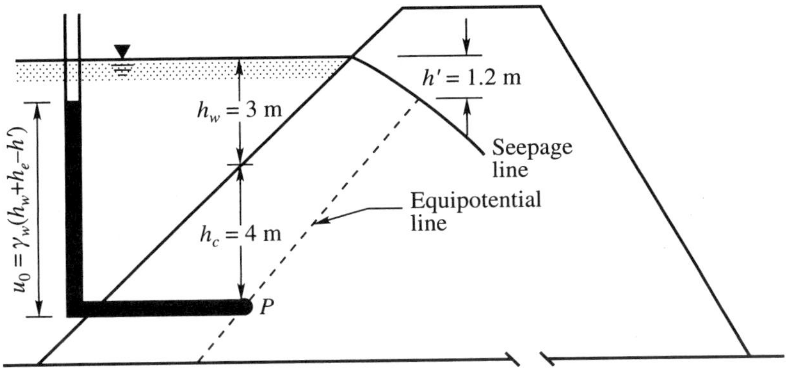

Figure Ex. 10.15

Therefore $\sigma'_0 = \sigma_0 - u_0 = 113 - 57 = 56 \text{ kN/m}^2$

After drawdown,

$\sigma = \gamma_{sat} h_c = 21 \times 4 = 84 \text{ kN/m}^2$

$u = \gamma_w (h_c - h') = 9.81(4 - 1.2) = 27.5 \text{ kN/m}^2$

$\sigma' = \sigma - u = 84 - 27.5 = 56.5 \text{ kN/m}^2$

The change in strength is zero since the effective vertical stress does not change.

Note: There is no change in strength due to sudden drawdown but the direction of forces of the seepage water changes from an inward direction before drawdown to an outward direction after drawdown and this is the main cause for the reduction in stability.

10.18 PROBLEMS

10.1 Find the critical height of an infinite slope having a slope angle of 30°. The slope is made of stiff clay having a cohesion 20 kN/m^2, angle of internal friction 20°, void ratio 0.7 and specific gravity 2.7. Consider the following cases for the analysis.

(a) the soil is dry.

(b) the water seeps parallel to the surface of the slope.

(c) the slope is submerged.

10.2 An infinite slope has an inclination of 26° with the horizontal. It is underlain by a firm cohesive soil having $G_s = 2.72$ and $e = 0.52$. There is a thin weak layer 20 ft below and parallel to the slope ($c' = 525$ lb/ft^2, $\phi' = 16°$). Compute the factors of safety when (a) the slope is dry, and (b) ground water flows parallel to the slope at the slope level.

10.3 An infinite slope is underlain with an overconsolidated clay having $c' = 210$ lb/ft^2, $\phi' = 8°$ and $\gamma_{sat} = 120$ lb/ft^3. The slope is inclined at an angle of 10° to the horizontal. Seepage is parallel to the surface and the ground water coincides with the surface. If the slope fails parallel to the surface along a plane at a depth of 12 ft below the slope, determine the factor of safety.

10.4 A deep cut of 10 m depth is made in sandy clay for a road. The sides of the cut make an angle of 60° with the horizontal. The shear strength parameters of the soil are $c' = 20$ kN/m^2, $\phi' = 25°$, and $\gamma = 18.5$ kN/m^3. If AC is the failure plane (Fig Prob. 10.4), estimate the factor of safety of the slope.

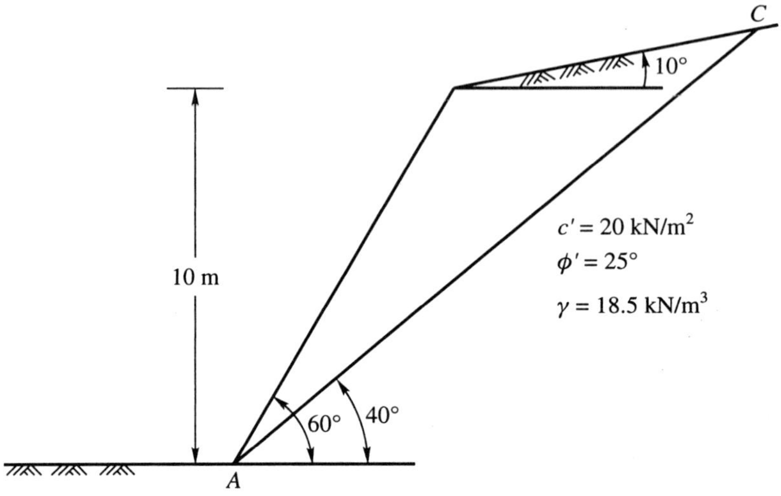

Figure Prob. 10.4

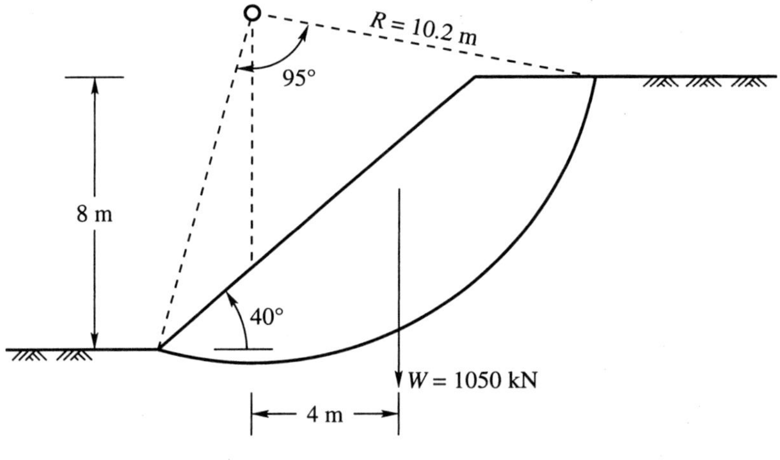

Figure Prob. 10.5

10.5 A 40° slope is excavated to a depth of 8 m in a deep layer of saturated clay having strength parameters $c = 60$ kN/m², $\phi = 0$, and $\gamma = 19$ kN/m³. Determine the factor of safety for the trial failure surface shown in Fig. Prob. 10.5.

10.6 An excavation to a depth of 8 m with a slope of 1:1 was made in a deep layer of saturated clay having $c_u = 65$ kN/m² and $\phi_u = 0$. Determine the factor of safety for a trial slip circle passing through the toe of the cut and having a center as shown in Fig. Prob. 10.6. The unit weight of the saturated clay is 19 kN/m³. No tension crack correction is required.

10.7 A 45° cut was made in a clayey silt with $c = 15$ kN/m², $\phi = 0$ and $\gamma = 19.5$ kN/m³. Site exploration revealed the presence of a soft clay stratum of 2 m thick having $c = 25$ kN/m² and $\phi = 0$ as shown in Fig. Prob. 10.7. Estimate the factor of safety of the slope for the assumed failure surface.

10.8 A cut was made in a homogeneous clay soil to a depth of 8 m as shown in Fig. Prob. 10.8. The total unit weight of the soil is 18 kN/m³, and its cohesive strength is 25 kN/m².

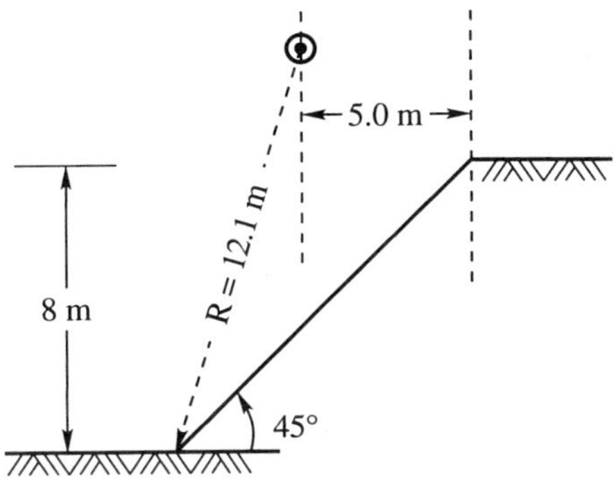

Figure Prob. 10.6

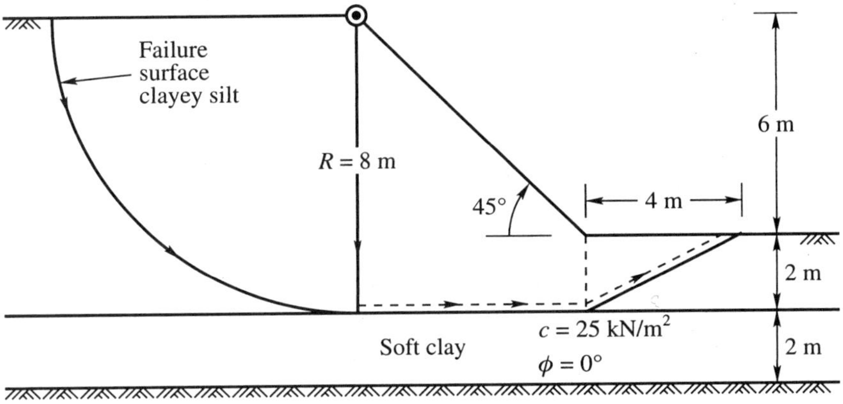

Figure Prob. 10.7

Assuming a $\phi = 0$ condition, determine the factor of safety with respect to a slip circle passing through the toe. Consider a tension crack at the end of the slip circle on the top of the cut.

10.9 A deep cut of 10 m depth is made in natural soil for the construction of a road. The soil parameters are: $c' = 35$ kN/m^2, $\phi' = 15°$ and $\gamma = 20$ kN/m^3.

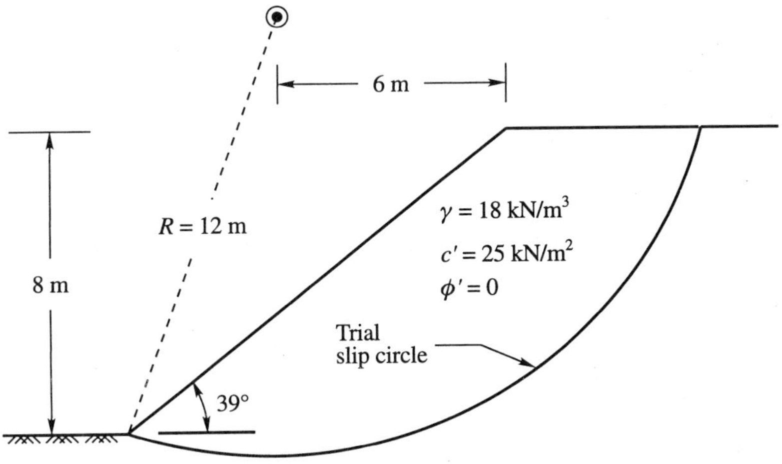

Figure Prob. 10.8

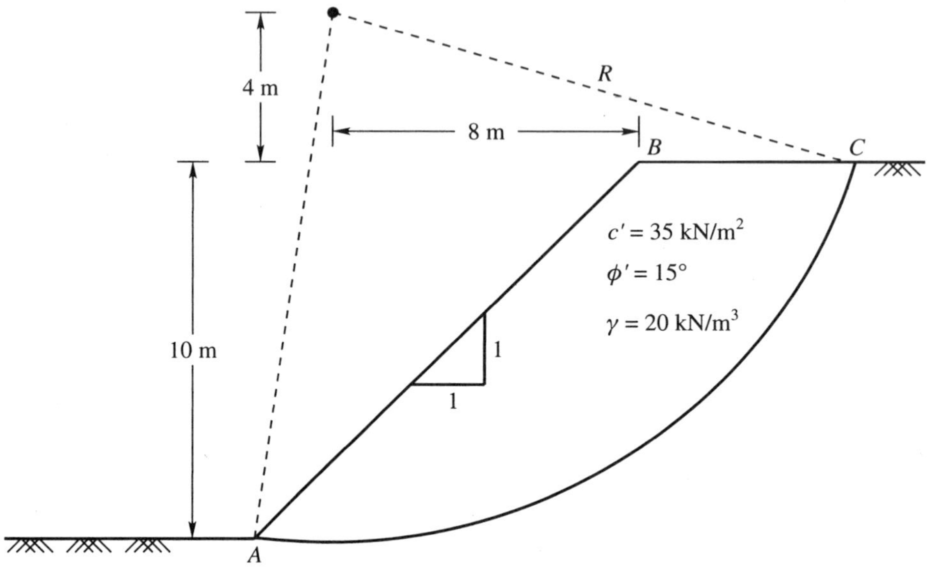

Figure Prob. 10.9

The sides of the cut make angles of 45° with the horizontal. Compute the factor of safety using friction circle method for the failure surface AC shown in Fig. Prob. 10.9.

10.10 An embankment is to be built to a height of 50 ft at an angle of 20° with the horizontal. The soil parameters are: $c' = 630$ lb/ft^2, $\phi' = 18°$ and $\gamma = 115$ lb/ft^3.

Estimate the following;

1. Factor of safety of the slope assuming full friction is mobilized.
2. Factor of safety with respect to friction if the factor of safety with respect to cohesion is 1.5.

Use Taylor's stability chart.

10.11 A cut was made in natural soil for the construction of a railway line. The soil parameters are: $c' = 700$ lb/ft^2, $\phi' = 20°$ and $\gamma = 110$ lb/ft^3.

Determine the critical height of the cut for a slope of 30° with the horizontal by making use of Taylor's stability chart.

10.12 An embankment is to be constructed by making use of sandy clay having the following properties: $c' = 35$ kN/m^2, $\phi' = 25°$ and $\gamma = 19.5$ kN/m^3.

The height of the embankment is 20 m with a slope of 30° with the horizontal as shown in Fig. Prob. 10.12. Estimate the factor of safety by the method of slices for the trial circle shown in the figure.

10.13 If an embankment of 10 m height is to be made from a soil having $c' = 25$ kN/m^2, $\phi' = 15°$, and $\gamma = 18$ kN/m^3, what will be the safe angle of slope for a factor of safety of 1.5?

10.14 An embarkment is constructed for an earth dam of 80 ft high at a slope of 3:1. The properties of the soil used for the construction are: $c' = 770$ lb/ft^2, $\phi' = 30°$, and $\gamma = 110$ lb/ft^3. The estimated pore pressuer ratio $r_u = 0.5$. Determine the factor of safety by Bishop and Morgenstern method.

10.15 For the Prob. 10.14, estimate the factor of safety for $\phi' = 20°$. All the other data remain the same.

10.16 For the Prob. 10.14, estimate the factor of safety for a slope of 2:1 with all the oother data remain the same.

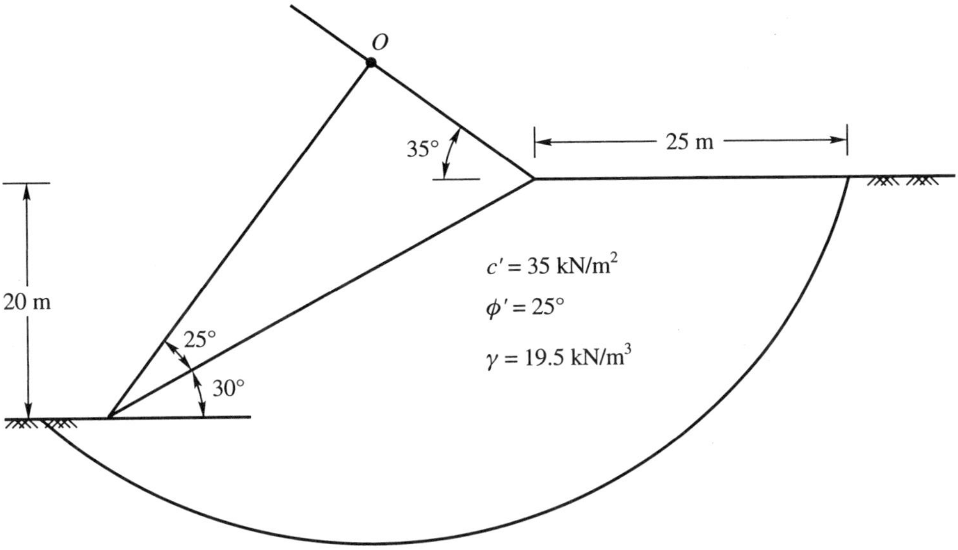

Figure Prob. 10.12

10.17 A cut of 25 m depth is made in a compacted fill having shear strength parameters of $c' = 25$ kN/m^2, and $\phi' = 20°$. The total unit weight of the material is 19 kN/m^3. The pore pressuer ratio has an average value of 0.3. The slope of the sides is 3:1. Estimate the factor of safety using the Bishop and Morgenstern method.

10.18 For the Prob. 10.17, estimate the factor of safety for $\phi' = 30°$, with all the other data remain the same.

10.19 For the Prob. 10.17, esatimate the factor of safety for a slope of 2:1 with all the other data remaining the same.

10.20 Estimate the minimum factor of safety for a complete drawdown condition for the section of dam in Fig. Prob. 10.20. The full reservoir level of 15 m depth is reduced to zero after drawdown.

10.21 What is the safety factor if the reservoir level is brought down from 15 m to 5 m depth in the Prob. 10.20?

10.22 An earth dam to be constructed at a site has the following soil parameters: $c' = 600$ lb/ft^2, $\gamma = 110$ lb/ft^3, and $\phi' = 20°$. The height of of dam $H = 50$ ft.

The pore pressure ratio $r_u = 0.5$. Determine the slope of the dam for a factor of safety of 1.5 using Spencer's method (1967).

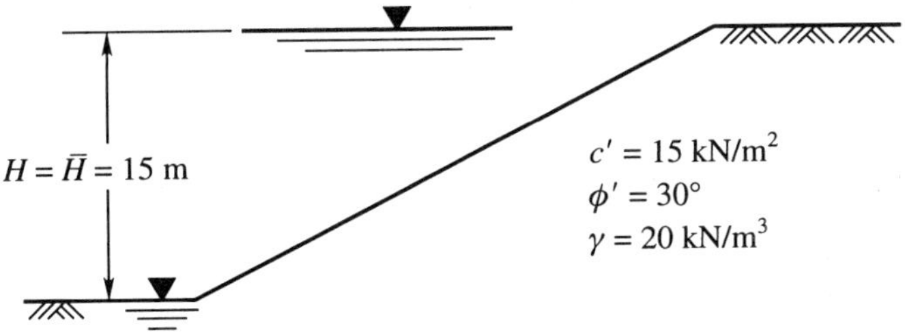

Figure Prob. 10.20

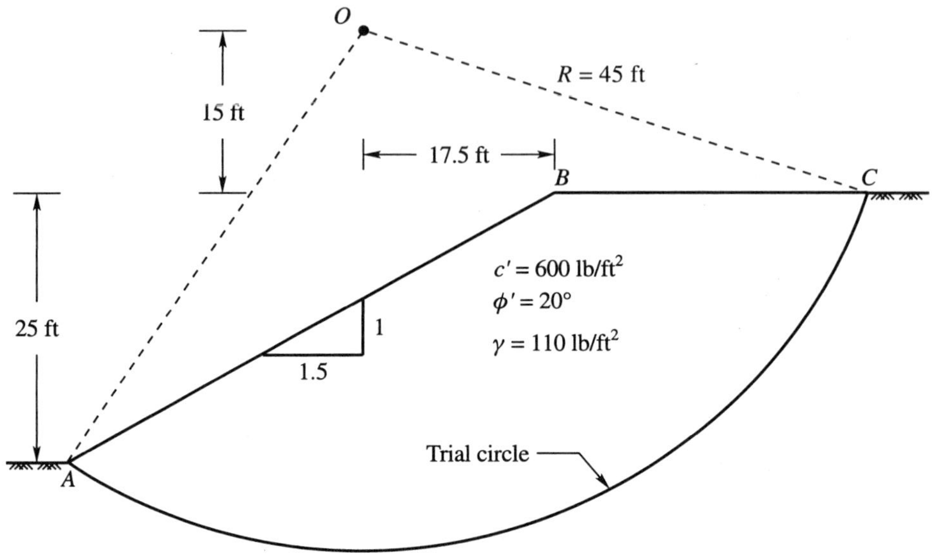

Figure Prob. 10.24

10.23 If the given pore pressure ratio is 0.25 in Prob. 10.22, what will be the slope of the dam?

10.24 An embankment has a slope of 1.5 horizontal to 1 vertical with a height of 25 feet. The soil parameters are:

$c' = 600$ lb/ft^2, $\phi' = 20°$, and $\gamma = 110$ lb/ft^3.

Determine the factor of safety using friction circle method for the failure surface *AC* shown in Fig. Prob. 10.24.

10.25 It is required to construct an embankment for a reservoir to a height of 20 m at a slope of 2 horizontal to 1 vertical. The soil parameters are:

$c' = 40$ kN/m^2, $\phi' = 18°$, and $\gamma = 17.5$ kN/m^3.

Estimate the following:

1. Factor of safety of the slope assuming full friction is mobilized.

2. Factor of safety with respect to friction if the factor of safety with respect to cohesion is 1.5.

Use Taylor's stability chart.

10.26 A cutting of 40 ft depth is to be made for a road as shown in Fig. Prob. 10.26. The soil properties are:

$c' = 500$ lb/ft^2, $\phi' = 15°$, and $\gamma = 115$ lb/ft^3.

Estimate the factor of safety by the method of slices for the trial circle shown in the figure.

10.27 An earth dam is to be constructed for a reservior. The height of the dam is 60 ft. The properties of the soil used in the construction are:

$c' = 400$ lb/ft^2, $\phi° = 20°$, and $\gamma = 115$ lb/ft^3, and $\beta = 2{:}1$.

Estimate the minimum factor of safety for the complete drawn from the full reservior level as shown in Fig. Prob. 10.27 by Morgenstern method.

10.28 What is the factor of safety if the water level is brought down from 60 ft to 20 ft above the bed level of reservoir in Prob. 10.27?

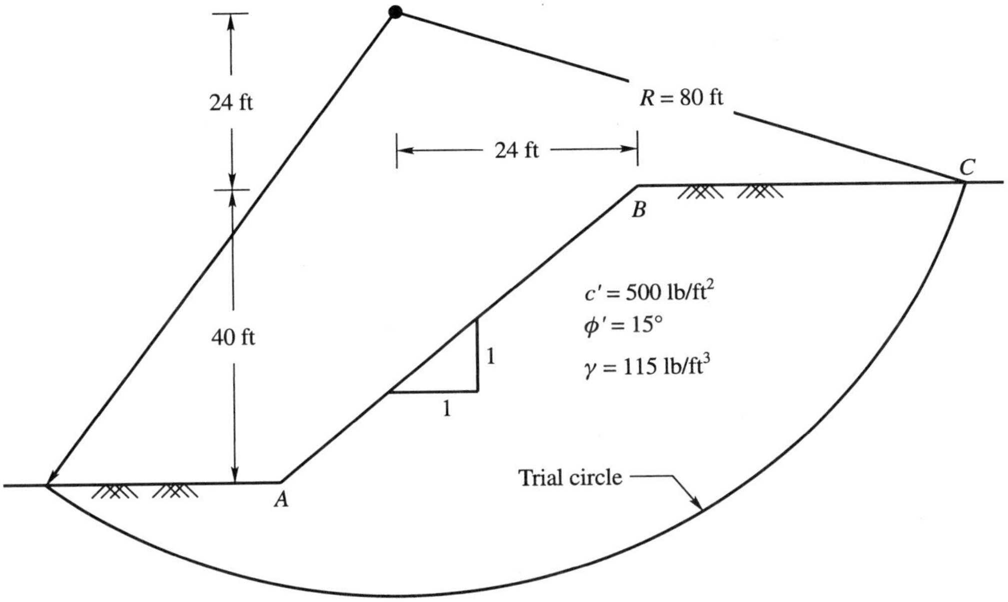

Figure Prob. 10.26

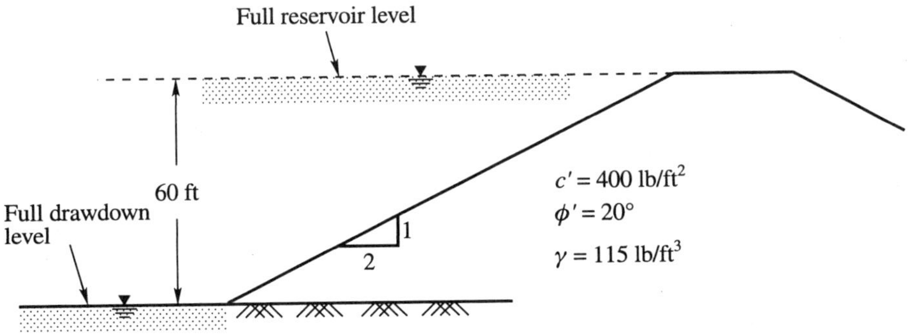

Figure Prob. 10.27

10.29 For the dam given in Prob. 10.27, determine the factor of safety for $r_u = 0.5$ by Spencer's method.

CHAPTER 11

LATERAL EARTH PRESSURE

11.1 INTRODUCTION

Structures that are built to retain vertical or nearly vertical earth banks or any other material are called *retaining walls*. Retaining walls may be constructed of masonry or sheet piles. Some of the purposes for which retaining walls are used are shown in Fig. 11.1.

Retaining walls may retain water also. The earth retained may be natural soil or fill. The principal types of retaining walls are given in Figs. 11.1 and 11.2.

Whatever may be the type of wall, all the walls listed above have to withstand lateral pressures either from earth or any other material on their faces. The pressures acting on the walls try to move the walls from their position. The walls should be so designed as to keep them stable in their position. Gravity walls resist movement because of their heavy sections. They are built of mass concrete or stone or brick masonry. No reinforcement is required in these walls. Semi-gravity walls are not as heavy as gravity walls. A small amount of reinforcement is used for reducing the mass of concrete. The stems of cantilever walls are thinner in section. The base slab is the cantilever portion. These walls are made of reinforced concrete. Counterfort walls are similar to cantilever walls except that the stem of the walls span horizontally between vertical brackets known as counterforts. The counterforts are provided on the backfill side. Buttressed walls are similar to counterfort walls except the brackets or buttress walls are provided on the opposite side of the backfill.

In all these cases, the backfill tries to move the wall from its position. The movement of the wall is partly resisted by the wall itself and partly by soil in front of the wall.

Sheet pile walls are more flexible than the other types. The earth pressure on these walls is dealt with in Chapter 20. There is another type of wall that is gaining popularity. This is mechanically stabilized reinforced earth retaining walls (MSE) which will be dealt with later on. This chapter deals with lateral earth pressures only.

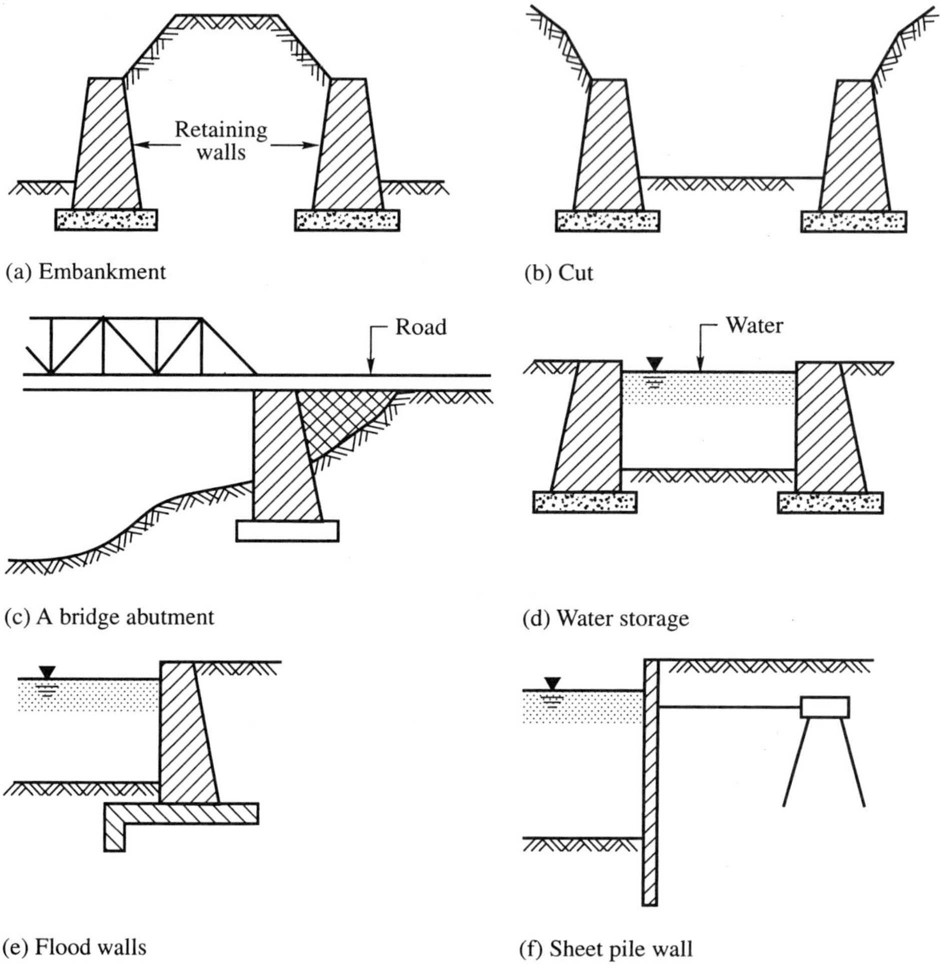

Figure 11.1 Use of retaining walls

11.2 LATERAL EARTH PRESSURE THEORY

There are two classical earth pressure theories. They are

1. Coulomb's earth pressure theory.
2. Rankine's earth pressure theory.

The first rigorous analysis of the problem of lateral earth pressure was published by Coulomb in (1776). Rankine (1857) proposed a different approach to the problem. These theories propose to estimate the magnitudes of two pressures called *active earth pressure* and *passive earth pressure* as explained below.

Consider a rigid retaining wall with a plane vertical face, as shown in Fig. 11.3(a), is backfilled with cohesionless soil. If the wall does not move even after back filling, the pressure exerted on the wall is termed as pressure for the *at rest condition* of the wall. If suppose the wall gradually rotates about point *A* and moves away from the backfill, the unit pressure on the wall is gradually reduced and after a particular displacement of the wall at the top, the pressure reaches a constant value. The pressure is the minimum possible. This pressure is termed the *active pressure* since the weight of the backfill is responsible for the movement of the wall. If the wall is smooth,

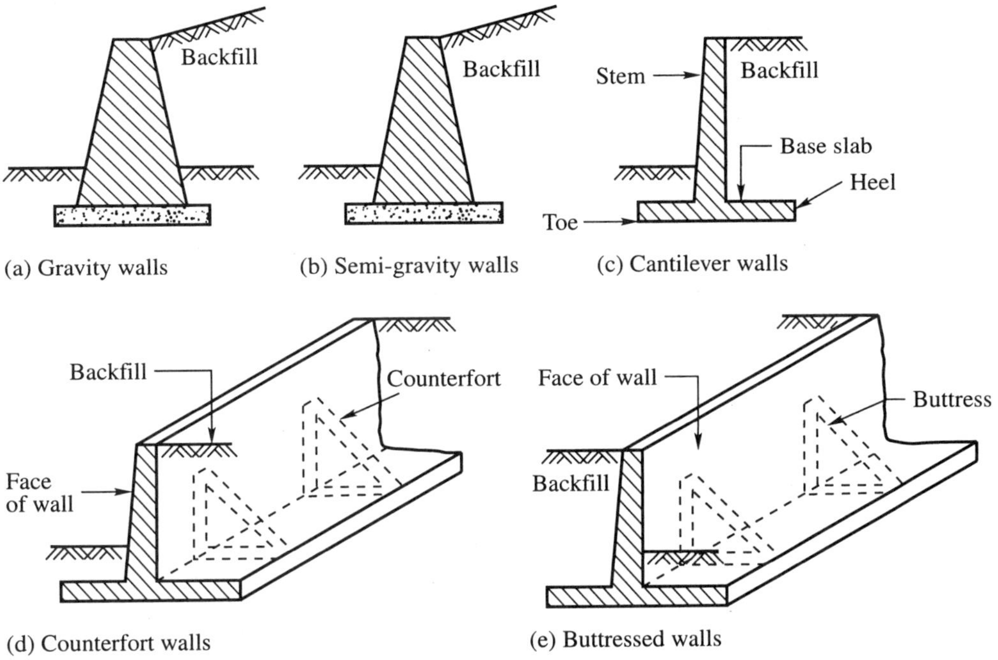

Figure 11.2 Principal types of rigid retaining walls

the resultant pressure acts normal to the face of the wall. If the wall is rough, it makes an angle δ with the normal on the wall. The angle δ is called the *angle of wall friction.* As the wall moves away from the backfill, the soil tends to move forward. When the wall movement is sufficient, a soil mass of weight W ruptures along surface ADC shown in Fig. 11.3(a). This surface is slightly curved. If the surface is assumed to be a plane surface AC, analysis would indicate that this surface would make an angle of $45° + \phi/2$ with the horizontal.

If the wall is now rotated about A *towards the backfill, the actual failure plane* ADC is also a curved surface [Fig. 11.3(b)]. However, if the failure surface is approximated as a plane AC, this makes an angle $45° - \phi/2$ with the horizontal and the pressure on the wall increases from the value of the at rest condition to the maximum value possible. The maximum pressure P_p that is developed is termed the *passive earth pressure.* The pressure is called passive because the weight of the backfill opposes the movement of the wall. It makes an angle δ with the normal if the wall is rough.

The gradual decrease or increase of pressure on the wall with the movement of the wall from the at rest condition may be depicted as shown in Fig. 11.4.

The movement Δ_p required to develop the passive state is considerably larger than Δ_a required for the active state.

11.3 LATERAL EARTH PRESSURE FOR AT REST CONDITION

If the wall is rigid and does not move with the pressure exerted on the wall, the soil behind the wall will be in a state of *elastic equilibrium.* Consider a prismatic element E in the backfill at depth z shown in Fig. 11.5.

Element E is subjected to the following pressures.

Vertical pressure $= \sigma_v = \gamma z$;   lateral pressure $= \sigma_h$

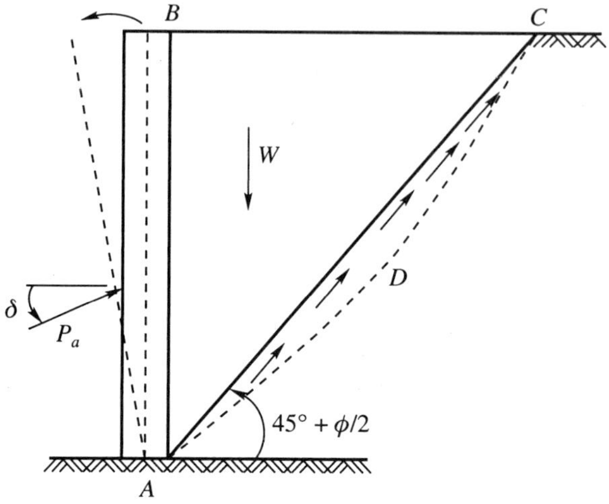

(a) Active earth pressure

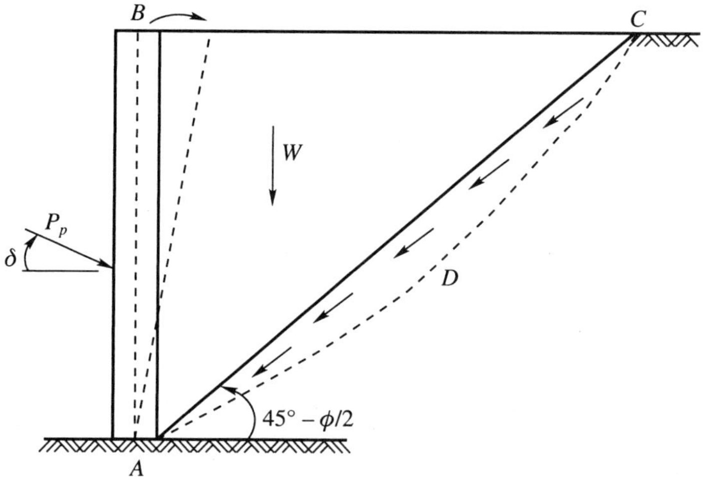

(b) Passive earth pressure

Figure 11.3 Wall movement for the development of active and passive earth pressures

where γ is the effective unit weight of the soil. If we consider the backfill is homogeneous then both σ_v and σ_h increase linearly with depth z. In such a case, the ratio of σ_h to σ_v remains constant with respect to depth, that is

$$\frac{\sigma_h}{\sigma_v} = \frac{\sigma_h}{\gamma z} = \text{constant} = K_0 \tag{11.1}$$

where K_0 is called the *coefficient of earth pressure for the at rest condition or at rest earth pressure coefficient.*

The lateral earth pressure σ_h acting on the wall at any depth z may be expressed as

$$\sigma_h = K_0 \gamma z \tag{11.1a}$$

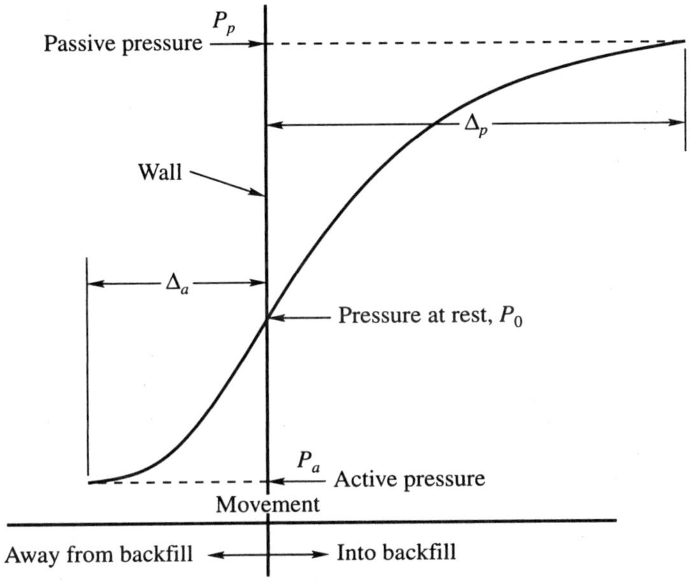

Figure 11.4 Development of active and passive earth pressures

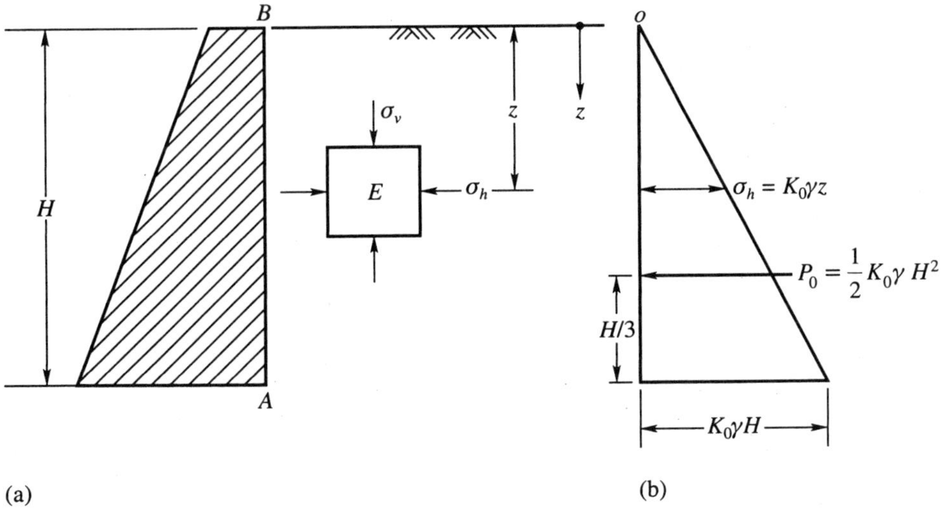

Figure 11.5 Lateral earth pressure for at rest condition

The expression for σ_h at depth H, the height of the wall, is

$$\sigma_h = K_0 \gamma H \tag{11.1b}$$

The distribution of σ_h on the wall is given in Fig. 11.5(b).

The total pressure P_0 for the soil for the at rest condition is

$$P_0 = \frac{1}{2} K_0 \gamma H^2 \tag{11.1c}$$

Table 11.1 Coefficients of earth pressure for at rest condition

Type of soil	I_p	K_0
Loose sand, saturated	–	0.46
Dense sand, saturated	–	0.36
Dense sand, dry (e = 0.6)	–	0.49
Loose sand, dry (e = 0.8)	–	0.64
Compacted clay	9	0.42
Compacted clay	31	0.60
Organic silty clay, undisturbed (w_l = 74%)	45	0.57

The value of K_0 depends upon the relative density of the sand and the process by which the deposit was formed. If this process does not involve artificial tamping the value of K_0 ranges from about 0.40 for loose sand to 0.6 for dense sand. Tamping the layers may increase it to 0.8.

The value of K_0 may also be obtained on the basis of elastic theory. If a cylindrical sample of soil is acted upon by vertical stress σ_v and horizontal stress σ_h, the lateral strain ε_1 may be expressed as

$$\varepsilon_1 = \frac{1}{E}\left[\sigma_h - \mu(\sigma_h + \sigma_v)\right] \tag{11.2}$$

where E = Young's modulus, μ = Poisson's ratio.

The lateral strain $\varepsilon_1 = 0$ when the earth is in the at rest condition. For this condition, we may write

$$\frac{1}{E}\left[\sigma_h - \mu(\sigma_h + \sigma_v)\right] = 0 \quad \text{or} \quad \frac{\sigma_h}{\sigma_v} = \frac{\mu}{1-\mu} \tag{11.3}$$

$$\text{or} \quad \sigma_h = \left(\frac{\mu}{1-\mu}\right)\sigma_v = K_0 \sigma_v = K_0 \gamma z$$

$$\text{where} \quad \frac{\mu}{1-\mu} = K_0, \; \sigma_v = \gamma z \tag{11.4}$$

According to Jaky (1944), a good approximation for K_0 is given by Eq. (11.5).

$$K_0 = 1 - \sin\phi \tag{11.5}$$

which fits most of the experimental data.

Numerical values of K_0 for some soils are given in Table 11.1.

Example 11.1

If a retaining wall 5 m high is restrained from yielding, what will be the at-rest earth pressure per meter length of the wall? Given: the backfill is cohesionless soil having $\phi = 30°$ and $\gamma = 18$ kN/m^3. Also determine the resultant force for the at-rest condition.

Solution

From Eq. (11.5)

$$K_0 = 1 - \sin\phi = 1 - \sin 30° = 0.5$$

From Eq. (11.1b), $\sigma_h = K_o \gamma H = 0.5 \times 18 \times 5 = 45$ kN/m^2

From Eq. (11.1c)

$$P_0 = \frac{1}{2}K_0\gamma H^2 = \frac{1}{2}\times 0.5\times 18\times 5^2 = 112.5 \text{ kN/m length of wall}$$

11.4 RANKINE'S STATES OF PLASTIC EQUILIBRIUM FOR COHESIONLESS SOILS

Let *XY* in Fig. 11.6(a) represent the horizontal surface of a semi-infinite mass of cohesionless soil with a unit weight γ. The soil is in an initial state of elastic equilibrium. Consider a prismatic block *ABCD*. The depth of the block is z and the cross-sectional area of the block is unity. Since the element is symmetrical with respect to a vertical plane, the normal stress on the base *AD* is

$$\sigma_v = \gamma z \tag{11.6}$$

σ_v is a principal stress. The normal stress σ_h on the vertical planes *AB* or *DC* at depth z may be expressed as a function of vertical stress.

$$\sigma_h = f(\sigma_v) = K_0\gamma z \tag{11.7}$$

where K_0 is the coefficient of earth pressure for the at rest condition which is assumed as a constant for a particular soil. The horizontal stress σ_h varies from zero at the ground surface to $K_0\gamma z$ at depth z.

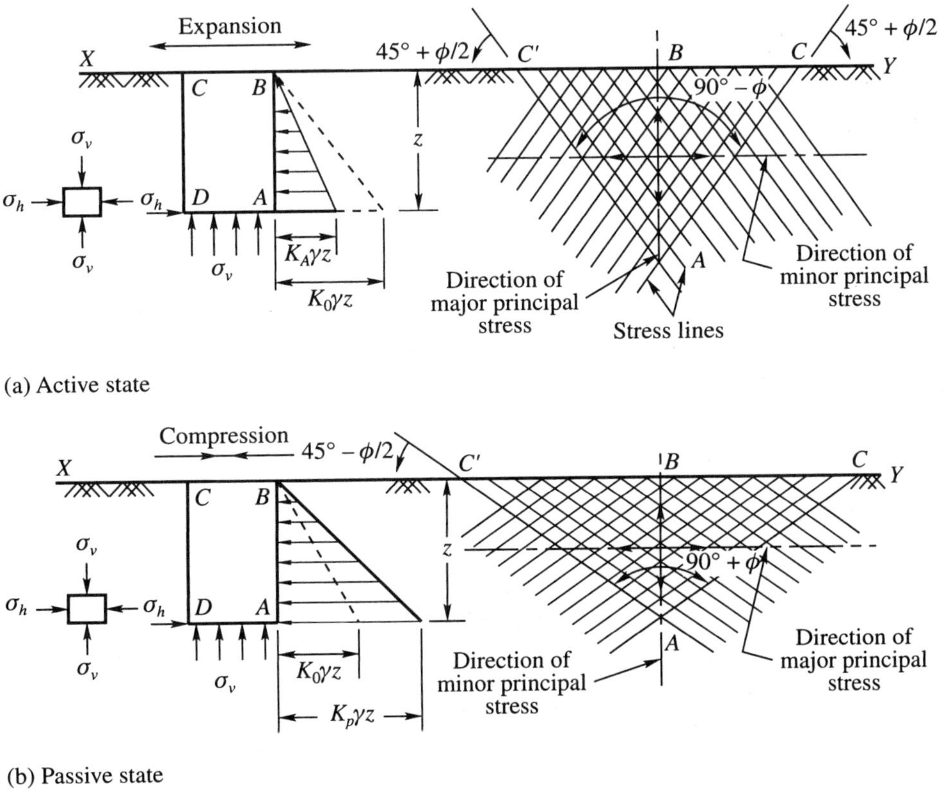

Figure 11.6(a, b) Rankine's condition for active and passive failures in a semi-infinite mass of cohesionless soil

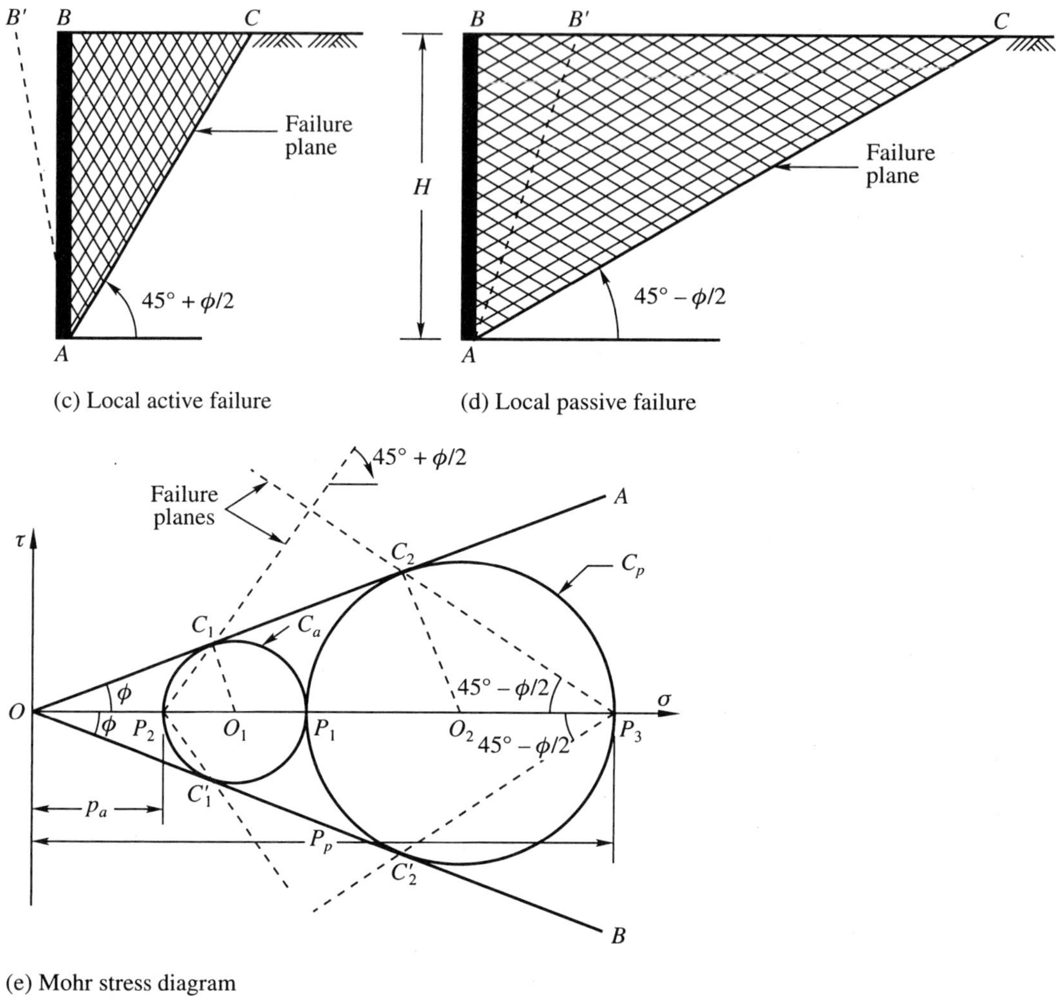

Figure 11.6(c, d, e) Rankine's condition for active and passive failures in a semi-infinite mass of cohesionless soil

If we imagine that the entire mass is subjected to horizontal deformation, such deformation is a plane deformation. Every vertical section through the mass represents a plane of symmetry for the entire mass. Therefore, the shear stresses on vertical and horizontal sides of the prism are equal to zero.

Due to the stretching, the pressure on vertical sides *AB* and *CD* of the prism decreases until the conditions of *plastic equilibrium* are satisfied, while the pressure on the base *AD* remains unchanged. Any further stretching merely causes a plastic flow without changing the state of stress. The transition from the state of *plastic equilibrium* to the state of *plastic flow* represents the failure of the mass. Since the weight of the mass assists in producing an expansion in a horizontal direction, the subsequent failure is called *active failure.*

If, on the other hand, the mass of soil is compressed, as shown in Fig. 11.6(b), in a horizontal direction, the pressure on vertical sides *AB* and *CD* of the prism increases while the pressure on its base remains unchanged at γz. Since the lateral compression of the soil is resisted by the weight of the soil, the subsequent failure by plastic flow is called a *passive failure.*

The problem now consists of determining the stresses associated with the states of plastic equilibrium in the semi-infinite mass and the orientation of the surface of sliding. The problem was solved by Rankine (1857).

The plastic states which are produced by stretching or by compressing a semi-infinite mass of soil parallel to its surface are called *active* and *passive Rankine states* respectively. The orientation of the planes may be found by Mohr's diagram.

Horizontal stretching or compressing of a semi-infinite mass to develop a state of plastic equilibrium is only a concept. However, local states of plastic equilibrium in a soil mass can be created by rotating a retaining wall about its base either away from the backfill for an active state or into the backfill for a passive state in the way shown in Figs. 11.3(c) and (d) respectively. In both cases, the soil within wedge *ABC* will be in a state of plastic equilibrium and line *AC* represents the rupture plane.

Mohr Circle for Active and Passive States of Equilibrium in Granular Soils

Point P_1 on the σ-axis in Fig. 11.6(e) represents the state of stress on base *AD* of prismatic element *ABCD* in Fig. 11.6(a). Since the shear stress on *AD* is zero, the vertical stress on the base

$$\sigma_v = \gamma z \tag{11.8}$$

is a principal stress. *OA* and *OB* are the two Mohr envelopes which satisfy the Coulomb equation of shear strength

$$s = \sigma \tan \phi \tag{11.9}$$

Two circles C_a and C_p can be drawn passing through P_1 and at the same time tangential to the Mohr envelopes *OA* and *OB*. When the semi-infinite mass is stretched horizontally, the horizontal stress on vertical faces *AB* and *CD* (Fig. 11.6 a) at depth z is reduced to the minimum possible and this stress is less than vertical stress σ_v. Mohr circle C_a gives the state of stress on the prismatic element at depth z when the mass is in active failure. The intercepts OP_1 and OP_2 are the major and minor principal stresses respectively.

When the semi-infinite mass is compressed (Fig. 11.6 b), the horizontal stress on the vertical face of the prismatic element reaches the maximum value OP_3 and circle C_p is the Mohr circle which gives that state of stress.

Active State of Stress

From Mohr circle C_a

Major principal stress = $OP_1 = \sigma_1 = \gamma z$
Minor principal stress = $OP_2 = \sigma_3$

$$OO_1 = \frac{\sigma_1 + \sigma_3}{2}, \quad O_1C_1 = \frac{\sigma_1 - \sigma_3}{2}$$

From triangle OO_1C_1, $\dfrac{\sigma_1 - \sigma_3}{2} = \dfrac{\sigma_1 + \sigma_3}{2} \sin\phi$

or $$\sigma_1 = \sigma_3 \left(\frac{1 + \sin\phi}{1 - \sin\phi}\right) = \sigma_3 \tan^2(45° + \phi/2) = \sigma_3 N_\phi \tag{11.10}$$

Therefore, $$p_a = \sigma_3 = \frac{\sigma_1}{N_\phi} = \gamma z K_A \tag{11.11}$$

where $\sigma_1 = \gamma z$, K_A = *coefficient of earth pressure for the active state* = $\tan^2(45° - \phi/2)$.

From point P_1, draw a line parallel to the base AD on which σ_1 acts. Since this line coincides with the σ-axis, point P_2 is the origin of planes. Lines P_2C_1 and $P_2C'_1$ give the orientations of the failure planes. They make an angle of $45° + \phi/2$ with the σ-axis. The lines drawn parallel to the lines P_2C_1 and $P_2C'_1$ in Fig. 11.6(a) give the shear lines along which the soil slips in the plastic state. The angle between a pair of conjugate shear lines is $(90° - \phi)$.

Passive State of Stress

C_p is the Mohr circle in Fig. (11.6e) for the passive state and P_3 is the origin of planes.

Major principal stress = $\sigma_1 = p_p = OP_3$

Minor principal stress = $\sigma_3 = OP_1 = \gamma z$

From triangle OO_2C_2, $\sigma_1 = \gamma z N_\phi$

Since $\sigma_1 = p_p$ and $\sigma_3 = \gamma z$, we have

$$p_p = \gamma z N_\phi = \gamma z K_p \tag{11.12}$$

where K_p = *coefficient of earth pressure for the passive state* = $\tan^2(45° + \phi/2)$.

The shear failure lines are P_3C_2 and $P_3C'_2$ and they make an angle of $45° - \phi/2$ with the horizontal. The shear failure lines are drawn parallel to P_3C_2 and $P_3C'_2$ in Fig. 11.6(b). The angle between any pair of conjugate shear lines is $(90° + \phi)$.

11.5 RANKINE'S EARTH PRESSURE AGAINST SMOOTH VERTICAL WALL WITH COHESIONLESS BACKFILL

Backfill Horizontal-Active Earth Pressure

Section AB in Fig. 11.6(a) in a semi-infinite mass is replaced by a smooth wall AB in Fig. 11.7(a).

The lateral pressure acting against smooth wall AB is due to the mass of soil ABC above failure line AC which makes an angle of $45° + \phi/2$ with the horizontal. The lateral pressure distribution on wall AB of height H increases in simple proportion to depth. The pressure acts normal to the wall AB [Fig. 11.7(b)].

The lateral active pressure at A is

$$p_a = \gamma H K_A \tag{11.13}$$

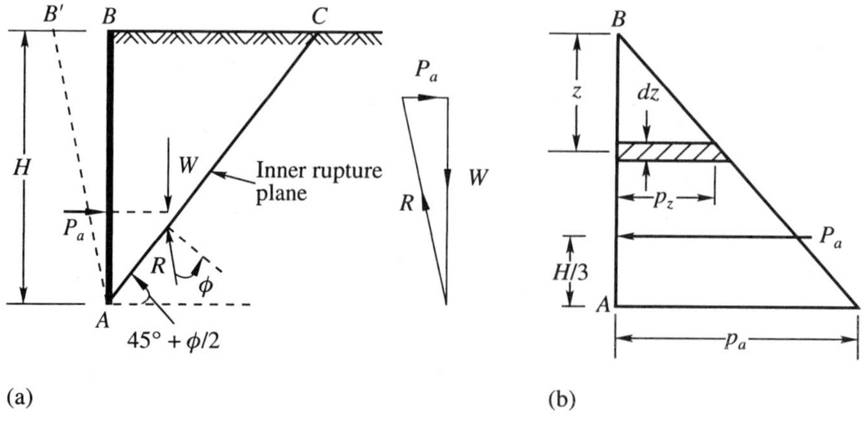

Figure 11.7 Rankine's active earth pressure in cohesionless soil

The total pressure on AB is therefore

$$P_a = \int_0^H p_z\, dz = K_A \int_0^H \gamma z\, dz = \frac{1}{2} K_A \gamma H^2 \tag{11.14}$$

$$\text{where,}\quad K_A = \tan^2(45° - \phi/2) = \frac{1 - \sin\phi}{1 + \sin\phi} \tag{11.14a}$$

P_a acts at a height $H/3$ above the base of the wall.

Backfill Horizontal-Passive Earth Pressure

If wall AB is pushed into the mass to such an extent as to impart uniform compression throughout the mass, soil wedge ABC in Fig. 11.8(a) will be in Rankine's passive state of plastic equilibrium. The inner rupture plane AC makes an angle $45° + \phi/2$ with the vertical AB. The pressure distribution on wall AB is linear as shown in Fig. 11.8(b).

The passive pressure p_p at A is

$$p_p = \gamma H K_P$$

the total pressure against the wall is

$$P_p = \int_0^H p_z\, dz = K_P \int_0^H \gamma z\, dz = \frac{1}{2} K_P \gamma H^2 \tag{11.15}$$

$$\text{where,}\quad K_P = \tan^2(45° + \phi/2) = \frac{1 + \sin\phi}{1 - \sin\phi} \tag{11.15a}$$

Relationship between K_P and K_A

The ratio of K_P and K_A may be written as

$$\frac{K_P}{K_A} = \frac{\tan^2(45° + \phi/2)}{\tan^2(45° - \phi/2)} = \tan^4(45° + \phi/2) \quad \text{or} \quad K_p = \frac{1}{K_A} \tag{11.16}$$

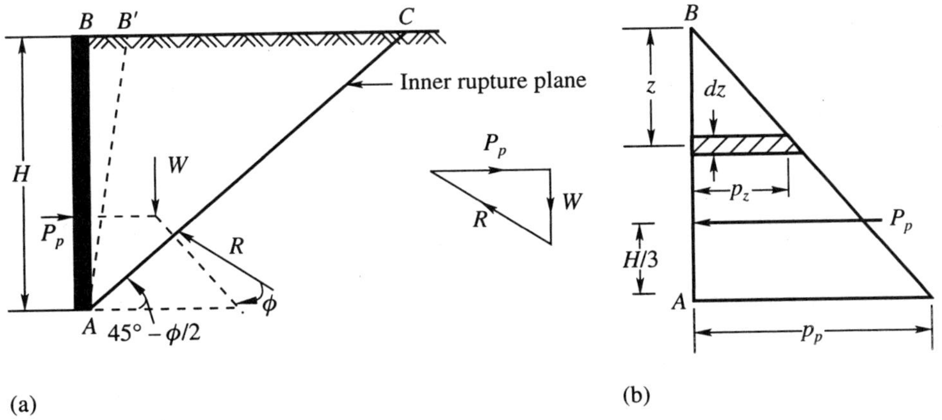

Figure 11.8 Rankine's passive earth pressure in cohesionless soil

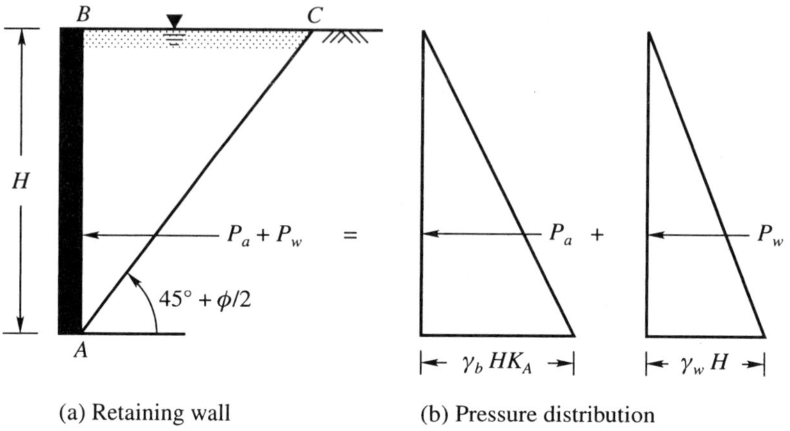

Figure 11.9 Rankine's active pressure under submerged condition in cohesionless soil

For example, if $\phi = 30°$, we have,

$$\frac{K_P}{K_A} = \tan^4 60° = 9, \quad \text{or} \quad K_P = 9K_A$$

This simple demonstration indicates that the value of K_P is quite large compared to K_A.

Active Earth Pressure-Backfill Soil Submerged with the Surface Horizontal

When the backfill is fully submerged, two types of pressures act on wall *AB*. (Fig. 11.9) They are

1. The active earth pressure due to the submerged weight of soil
2. The lateral pressure due to water

At any depth z the total unit pressure on the wall is

$$\overline{p_a} = p_a + p_w = \gamma_b z K_A + \gamma_w z$$

At depth $z = H$, we have

$$\overline{p_a} = \gamma_b H K_A + \gamma_w H$$

where γ_b is the submerged unit weight of soil and γ_w the unit weight of water. The total pressure acting on the wall at a height $H/3$ above the base is

$$\overline{P_a} = P_a + P_w = \frac{1}{2}\gamma_b H^2 K_A + \frac{1}{2}\gamma_w H^2 \tag{11.17}$$

Active Earth Pressure-Backfill Partly Submerged with a Uniform Surcharge Load

The ground water table is at a depth of H_1 below the surface and the soil above this level has an effective moist unit weight of γ. The soil below the water table is submerged with a submerged unit weight γ_b. In this case, the total unit pressure may be expressed as given below.

At depth H_1 at the level of the water table

$$\overline{p_a} = qK_A + \gamma H_1 K_A$$

At depth H we have

$$\overline{p_a} = qK_A + \gamma H_1 K_A + \gamma_b H_2 K_A + \gamma_w H_2$$

or $$\overline{p_a} = qK_A + (\gamma H_1 + \gamma_b H_2)K_A + \gamma_w H_2 \tag{11.18}$$

The pressure distribution is given in Fig. 11.10(b). It is assumed that the value of ϕ remains the same throughout the depth H.

From Fig. 11.10(b), we may say that the total pressure $\overline{P_a}$ acting per unit length of the wall may be written as equal to

$$\overline{P_a} = qHK_A + \frac{1}{2}\gamma H_1^2 K_A + \gamma H_1 H_2 K_A + \frac{1}{2}H_2^2(\gamma_b K_A + \gamma_w) \tag{11.19}$$

The point of application of $\overline{P_a}$ above the base of the wall can be found by taking moments of all the forces acting on the wall about A.

Sloping Surface-Active Earth Pressure

Figure 11.11(a) shows a smooth vertical wall with a sloping backfill of cohesionless soil. As in the case of a horizontal backfill, the active state of plastic equilibrium can be developed in the backfill by rotating the wall about A away from the backfill. Let AC be the rupture line and the soil within the wedge ABC be in an active state of plastic equilibrium.

Consider a rhombic element E within the plastic zone ABC which is shown to a larger scale outside. The base of the element is parallel to the backfill surface which is inclined at an angle β to the horizontal. The horizontal width of the element is taken as unity.

Let σ_v = the vertical stress acting on an elemental length $ab = \gamma z \cos\beta$

σ_l = the lateral pressure acting on vertical surface bc of the element

The vertical stress σ_v can be resolved into components σ_n the normal stress and τ the shear stress on surface ab of element E. We may now write

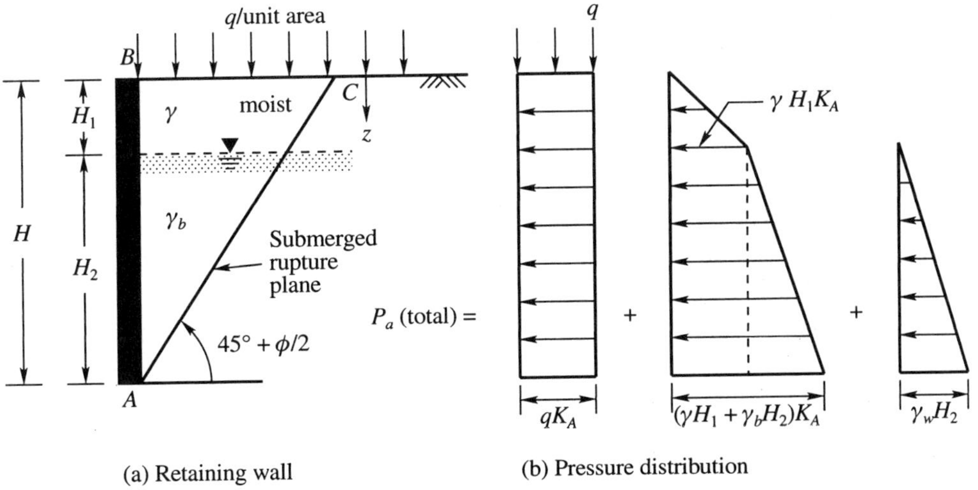

Figure 11.10 Rankine's active pressure in cohesionless backfill under partly submerged condition with surcharge load

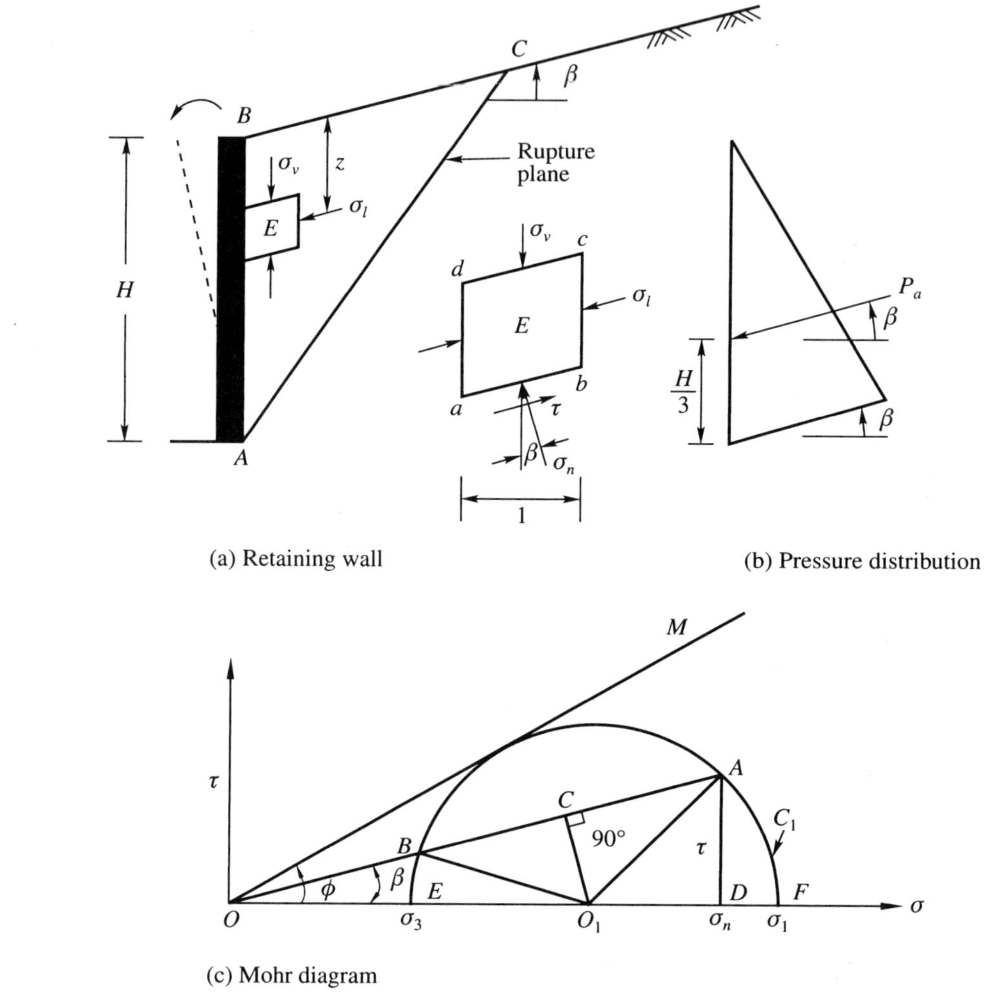

Figure 11.11 Rankine's active pressure for a sloping cohesionless backfill

$$\sigma_n = \sigma_v \cos\beta = \gamma z \cos\beta \cos\beta = \gamma z \cos^2\beta \tag{11.20}$$

$$\tau = \sigma_v \sin\beta = \gamma z \cos\beta \sin\beta \tag{11.21}$$

A Mohr diagram can be drawn as shown in Fig. 11.11(c). Here, length $OA = \gamma z \cos\beta$ makes an angle β with the σ-axis. $OD = \sigma_n = \gamma z \cos^2\beta$ and $AD = \tau = \gamma z \cos\beta \sin\beta$. OM is the Mohr envelope making an angle ϕ with the σ-axis. Now Mohr circle C_1 can be drawn passing through point A and at the same time tangential to envelope OM. This circle cuts line OA at point B and the σ-axis at E and F.

Now OB = the lateral pressure $\sigma_l = p_a$ in the active state.

The principal stresses are

$OF = \sigma_1$ and $OE = \sigma_3$

The following relationships can be expressed with reference to the Mohr diagram.

$$BC = CA = \frac{\sigma_1 + \sigma_3}{2}\sqrt{\sin^2\phi - \sin^2\beta}$$

$$\sigma_v = OA = OC + CA = \frac{\sigma_1 + \sigma_3}{2}\cos\beta + \frac{\sigma_1 + \sigma_3}{2}\sqrt{\sin^2\phi - \sin^2\beta}$$

$$\sigma_l = p_a = OC - BC = \frac{\sigma_1 + \sigma_3}{2}\cos\beta - \frac{\sigma_1 + \sigma_3}{2}\sqrt{\sin^2\phi - \sin^2\beta} \qquad (11.22)$$

Now we have (after simplification)

$$\frac{\sigma_l}{\sigma_v} = \frac{p_a}{\gamma z \cos\beta} = \frac{\cos\beta - \sqrt{\cos^2\beta - \cos^2\phi}}{\cos\beta + \sqrt{\cos^2\beta - \cos^2\phi}}$$

$$\text{or} \quad p_a = \gamma z \cos\beta \times \frac{\cos\beta - \sqrt{\cos^2\beta - \cos^2\phi}}{\cos\beta + \sqrt{\cos^2\beta - \cos^2\phi}} = \gamma z K_A \qquad (11.23)$$

$$\text{where,} \quad K_A = \cos\beta \times \frac{\cos\beta - \sqrt{\cos^2\beta - \cos^2\phi}}{\cos\beta + \sqrt{\cos^2\beta - \cos^2\phi}} \qquad (11.24)$$

is called as *the coefficient of earth pressure* for the active state or the active earth pressure coefficient.

The pressure distribution on the wall is shown in Fig. 11.11(b). The active pressure at depth H is

$$p_a = \gamma H K_A$$

which acts parallel to the surface. The total pressure P_a per unit length of the wall is

$$P_a = \frac{1}{2}\gamma H^2 K_A \qquad (11.25)$$

which acts at a height $H/3$ from the base of the wall and parallel to the sloping surface of the backfill.

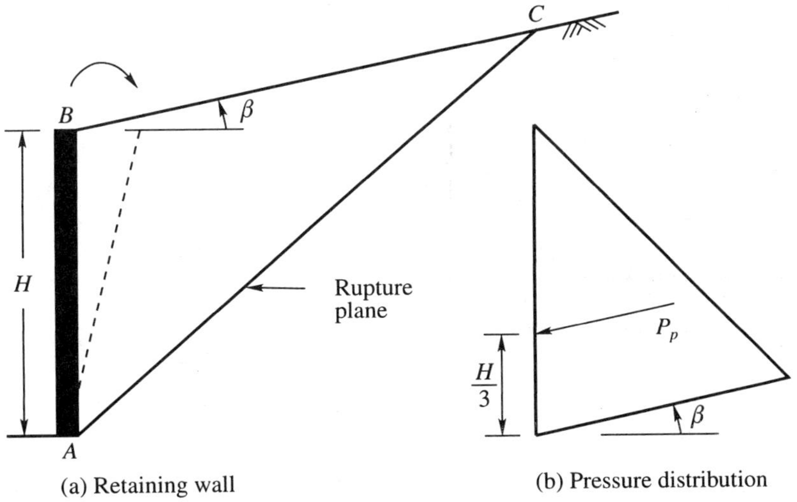

Figure 11.12 Rankine's passive pressure in sloping cohesionless backfill

Sloping Surface-Passive Earth Pressure (Fig. 11.12)

An equation for P_p for a sloping backfill surface can be developed in the same way as for an active case. The equation for P_p may be expressed as

$$P_p = \frac{1}{2}\gamma H^2 K_p \tag{11.26}$$

where, $$K_p = \cos\beta \times \frac{\cos\beta + \sqrt{\cos^2\beta - \cos^2\phi}}{\cos\beta - \sqrt{\cos^2\beta - \cos^2\phi}} \tag{11.27}$$

P_p acts at a height $H/3$ above point A and parallel to the sloping surface.

Example 11.2

A cantilever retaining wall of 7 meter height (Fig. Ex. 11.2) retains sand. The properties of the sand are: $e = 0.5$, $\phi = 30°$ and $G_s = 2.7$. Using Rankine's theory determine the active earth pressure at the base when the backfill is (i) dry, (ii) saturated and (iii) submerged, and also the resultant active force in each case. In addition determine the total water pressure under the submerged condition.

Solution

$$e = 0.5 \text{ and } G_s = 2.7, \quad \gamma_d = \frac{G_s \gamma_w}{1+e} = \frac{2.7}{1+0.5} \times 9.81 = 17.66 \text{ kN/m}^3$$

Saturated unit weight

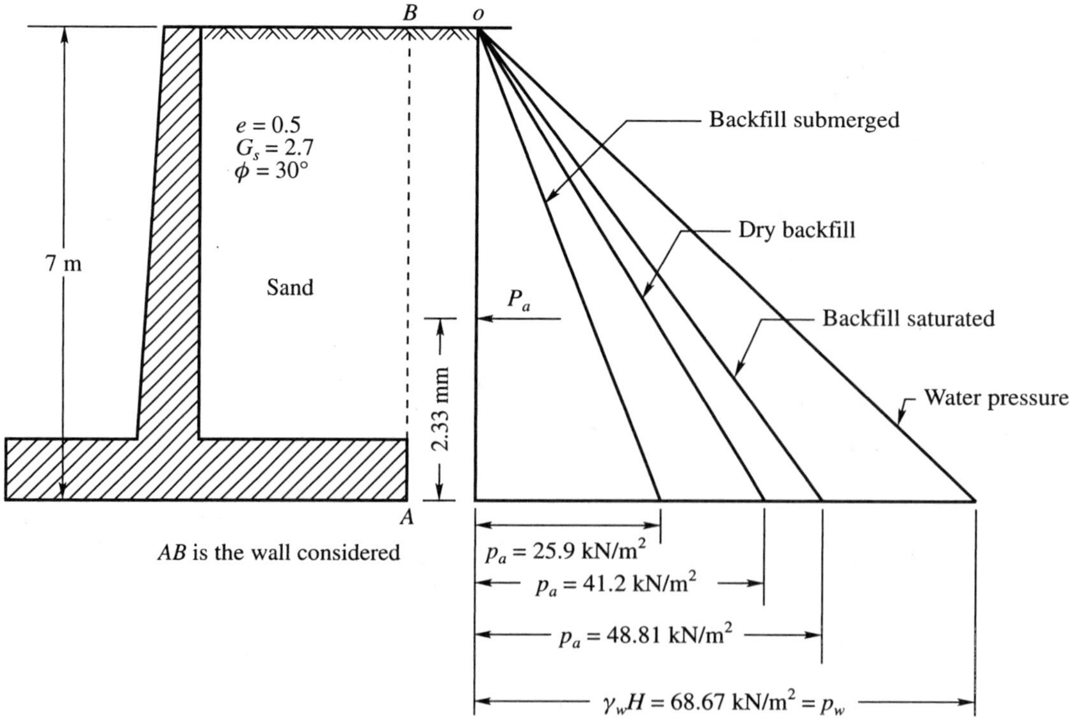

Figure Ex. 11.2

$$\gamma_{sat} = \frac{(G_s + e)\gamma_w}{1+e} = \frac{2.7+0.5}{1+0.5} \times 9.81 = 20.92 \text{ kN/m}^3$$

Submerged unit weight

$$\gamma_b = \gamma_{sat} - \gamma_w = 20.92 - 9.81 = 11.1 \text{ kN/m}^3$$

For $\phi = 30$, $K_A = \frac{1-\sin\phi}{1+\sin\phi} = \frac{1-\sin 30°}{1+\sin 30°} = \frac{1}{3}$

Active earth pressure at the base is

(i) for dry backfill

$$p_a = K_A \gamma_d H = \frac{1}{3} \times 17.66 \times 7 = 41.2 \text{ kN/m}^2$$

$$P_a = \frac{1}{2} K_A \gamma_d H^2 = \frac{1}{2} \times 41.2 \times 7 = 144.2 \text{ kN / m of wall}$$

(ii) for saturated backfill

$$p_a = K_A \gamma_{sat} H = \frac{1}{3} \times 20.92 \times 7 = 48.81 \text{ kN/m}^2$$

$$P_a = \frac{1}{2} \times 48.81 \times 7 = 170.85 \text{ kN / m of wall}$$

(iii) for submerged backfill

Submerged soil pressure

$$p_a = K_A \gamma_b H = \frac{1}{3} \times 11.1 \times 7 = 25.9 \text{ kN/m}^2$$

$$P_a = \frac{1}{2} \times 25.9 \times 7 = 90.65 \text{ kN / m of wall}$$

Water pressure

$$p_w = \gamma_w H = 9.81 \times 7 = 68.67 \text{ kN/m}^2$$

$$P_w = \frac{1}{2} \gamma_w H^2 = \frac{1}{2} \times 9.81 \times 7^2 = 240.35 \text{ kN/m of wall}$$

Example 11.3

For the earth retaining structure shown in Fig. Ex. 11.3, construct the earth pressure diagram for the active state and determine the total thrust per unit length of the wall.

Solution

For $\phi = 30°$, $K_A = \frac{1-\sin 30°}{1+\sin 30°} = \frac{1}{3}$

Dry unit weight $\gamma_d = \frac{G_s \gamma_w}{1+e} = \frac{2.65}{1+0.65} \times 62.4 = 100.22 \text{ lb/ ft}^3$

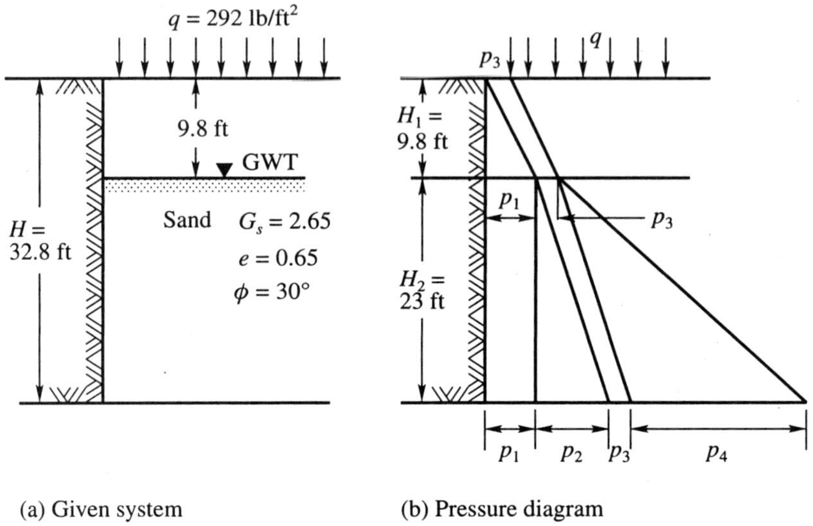

Figure Ex. 11.3

$$\gamma_b = \frac{(G_s - 1)\gamma_w}{1+e} = \frac{2.65-1}{1+0.65} \times 62.4 = 62.4 \text{ lb/ft}^3$$

Assuming the soil above the water table is dry, [Refer to Fig. Ex. 11.3(b)].

$$p_1 = K_A \gamma_d H_1 = \frac{1}{3} \times 100.22 \times 9.8 = 327.39 \text{ lb/ft}^2$$

$$p_2 = K_A \gamma_b H_2 = \frac{1}{3} \times 62.4 \times 23 = 478.4 \text{ lb/ft}^2$$

$$p_3 = K_A \times q = \frac{1}{3} \times 292 = 97.33 \text{ lb/ft}^2$$

$$p_4 = (K_A)_w \gamma_w H_2 = 1 \times 62.4 \times 23 = 1435.2 \text{ lb/ft}^2$$

Total thrust = summation of the areas of the different parts of the pressure diagram

$$= \frac{1}{2} p_1 H_1 + p_1 H_2 + \frac{1}{2} p_2 H_2 + p_3 (H_1 + H_2) + \frac{1}{2} p_4 H_2$$

$$= \frac{1}{2} \times 327.39 \times 9.8 + 327.39 \times 23 + \frac{1}{2} \times 478.4 \times 23 + 97.33(32.8) + \frac{1}{2} \times 1435.2 \times 23$$

= 34,333 lb/ft = 34.3 kips/ft of wall

Example 11.4

A retaining wall with a vertical back of height 7.32 m supports a cohesionless soil of unit weight 17.3 kN/m^3 and an angle of shearing resistance ϕ = 30°. The surface of the soil is horizontal. Determine the magnitude and direction of the active thrust per meter of wall using Rankine theory.

Solution

For the condition given here, Rankine's theory disregards the friction between the soil and the back of the wall.

The coefficient of active earth pressure K_A is

$$K_A = \frac{1-\sin\phi}{1+\sin\phi} = \frac{1-\sin 30^\circ}{1+\sin 30^\circ} = \frac{1}{3}$$

The lateral active thrust P_a is

$$P_a = \frac{1}{2}K_A\gamma H^2 = \frac{1}{2}\times\frac{1}{3}\times 17.3(7.32)^2 = 154.5 \text{ kN/m}$$

Example 11.5

A rigid retaining wall 5 m high supports a backfill of cohesionless soil with $\phi = 30°$. The water table is below the base of the wall. The backfill is dry and has a unit weight of 18 kN/m^3. Determine Rankine's passive earth pressure per meter length of the wall (Fig. Ex. 11.5).

Solution

From Eq. (11.15a)

$$K_P = \frac{1+\sin\phi}{1-\sin\phi} = \frac{1+\sin 30^\circ}{1-\sin 30^\circ} = \frac{1+0.5}{1-0.5} = 3$$

At the base level, the passive earth pressure is

$$p_p = K_P\gamma H = 3\times 18\times 5 = 270 \text{ kN/m}^2$$

From Eq. (11.15)

$$P_p = \frac{1}{2}K_P\gamma H^2 = \frac{1}{2}\times 3\times 18\times 5^2 = 675 \text{ kN/m length of wall}$$

The pressure distribution is given in Fig. Ex. 11.5.

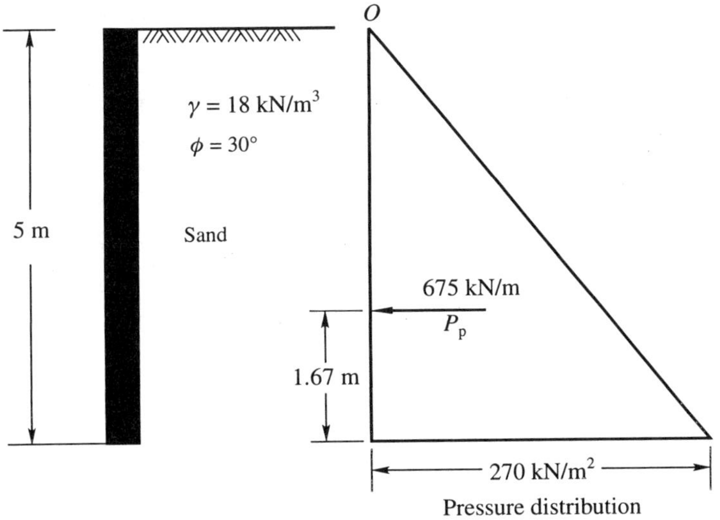

Figure Ex. 11.5

Example 11.6

A counterfort wall of 10 m height retains a non-cohesive backfill. The void ratio and angle of internal friction of the backfill respectively are 0.70 and 30° in the loose state and they are 0.40 and 40° in the dense state. Calculate and compare active and passive earth pressures for both the cases. Take the specific gravity of solids as 2.7.

Solution

(i) In the loose state, $e = 0.70$ which gives

$$\gamma_d = \frac{G_s \gamma_w}{1+e} = \frac{2.7}{1+0.7} \times 9.81 = 15.6 \text{ kN/m}^3$$

$$\text{For } \phi = 30°, \quad K_A = \frac{1-\sin\phi}{1+\sin\phi} = \frac{1-\sin 30°}{1+\sin 30°} = \frac{1}{3}, \text{and } K_P = \frac{1}{K_A} = 3$$

$$\text{Max. } p_a = K_A \gamma_d H = \frac{1}{3} \times 15.6 \times 10 = 52 \text{ kN/m}^2$$

$$\text{Max. } p_p = K_P \gamma_d H = 3 \times 15.6 \times 10 = 468 \text{ kN/m}^2$$

(ii) In the dense state, $e = 0.40$, which gives,

$$\gamma_d = \frac{2.7}{1+0.4} \times 9.81 = 18.92 \text{ kN/m}^3$$

$$\text{For } \phi = 40°, \quad K_A = \frac{1-\sin 40°}{1+\sin 40°} = 0.217, \; K_P = \frac{1}{K_A} = 4.6$$

$$\text{Max. } p_a = K_A \gamma_d H = 0.217 \times 18.92 \times 10 = 41.1 \text{ kN/m}^2$$

$$\text{and Max. } p_p = 4.6 \times 18.92 \times 10 = 870.3 \text{ kN/m}^2$$

Comment: The comparison of the results indicates that densification of soil decreases the active earth pressure and increases the passive earth pressure. This is advantageous in the sense that active earth pressure is a disturbing force and passive earth pressure is a resisting force.

Example 11.7

A wall of 8 m height retains sand having a density of 1.936 Mg/m^3 and an angle of internal friction of 34°. If the surface of the backfill slopes upwards at 15° to the horizontal, find the active thrust per unit length of the wall. Use Rankine's conditions.

Solution

There can be two solutions: analytical and graphical. The analytical solution can be obtained from Eqs. (11.25) and (11.24) viz.,

$$P_a = \frac{1}{2} K_A \gamma H^2$$

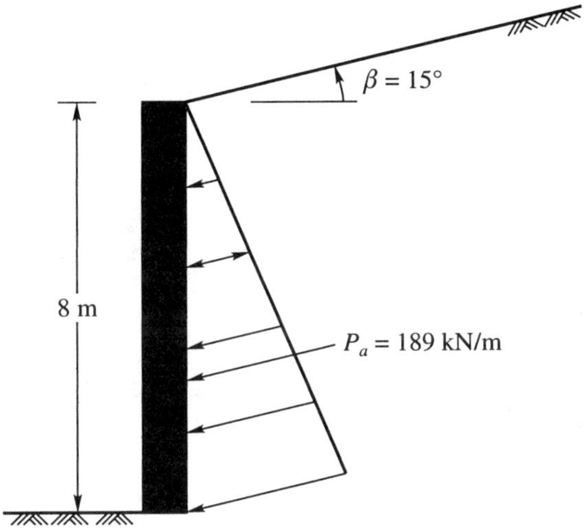

Figure Ex. 11.7a

where $K_A = \cos\beta \times \dfrac{\cos\beta - \sqrt{\cos^2\beta - \cos^2\phi}}{\cos\beta + \sqrt{\cos^2\beta - \cos^2\phi}}$

where $\beta = 15°$, $\cos\beta = 0.9659$ and $\cos^2\beta = 0.933$

and $\phi = 34°$ gives $\cos^2\phi = 0.688$

Hence $K_A = 0.966 \times \dfrac{0.966 - \sqrt{0.933 - 0.688}}{0.966 + \sqrt{0.933 - 0.688}} = 0.311$

$$\gamma = 1.936 \times 9.81 = 19.0 \text{ kN/m}^3$$

Hence $P_a = \dfrac{1}{2} \times 0.311 \times 19(8)^2 = 189$ kN/m wall

Graphical Solution

Vertical stress at a depth $z = 8$ m is

$$\gamma H \cos\beta = 19 \times 8 \times \cos 15° = 147 \text{ kN/m}^2$$

Now draw the Mohr envelope at an angle of 34° and the ground line at an angle of 15° with the horizontal axis as shown in Fig. Ex. 11.7b.

Using a suitable scale plot OP_1 = 147 kN/m².

(i) the center of circle C lies on the horizontal axis,

(ii) the circle passes through point P_1, and

(iii) the circle is tangent to the Mohr envelope

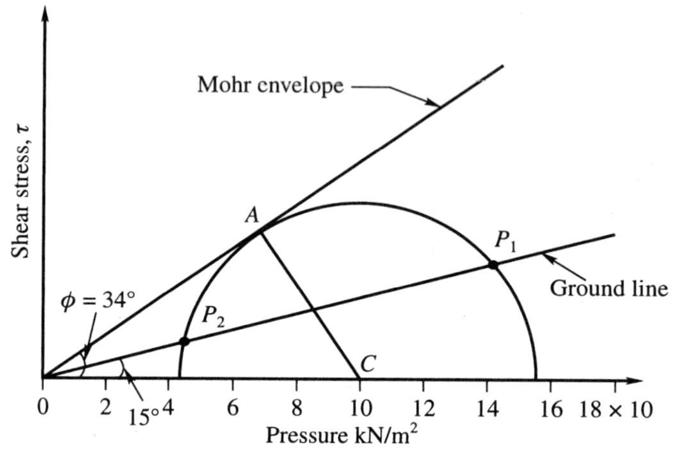

Figure Ex. 11.7b

The point P_2 at which the circle cuts the ground line represents the lateral earth pressure. The length OP_2 measures 47.5 kN/m^2.

Hence the active thrust per unit length, $P_a = \frac{1}{2} \times 47.5 \times 8 = 190$ kN/m

11.6 RANKINE'S ACTIVE EARTH PRESSURE WITH COHESIVE BACKFILL

In Fig. 11.13(a) is shown a prismatic element in a semi-infinite mass with a horizontal surface. The vertical pressure on the base *AD* of the element at depth *z* is

$$\sigma_v = \gamma z$$

The horizontal pressure on the element when the mass is in a state of plastic equilibrium may be determined by making use of Mohr's stress diagram [Fig. 11.13(b)].

Mohr envelopes *O'A* and *O'B* for cohesive soils are expressed by Coulomb's equation

$$s = c + \tan\phi \tag{11.28}$$

Point P_1 on the σ-axis represents the state of stress on the base of the prismatic element. When the mass is in the active state σ_v is the major principal stress σ_1. The horizontal stress σ_h is the minor principal stress σ_3. The Mohr circle of stress C_a passing through P_1 and tangential to the Mohr envelopes *O'A* and *O'B* represents the stress conditions in the active state. The relation between the two principal stresses may be expressed by the expression

$$\sigma_1 = \sigma_3 N_\phi + 2c\sqrt{N_\phi} \tag{11.29}$$

Substituting $\sigma_1 = \gamma z$, $\sigma_3 = p_a$ and transposing we have

$$p_a = \frac{\gamma z}{N_\phi} - \frac{2c}{\sqrt{N_\phi}} = \gamma z K_A - 2c\sqrt{K_A} \tag{11.30}$$

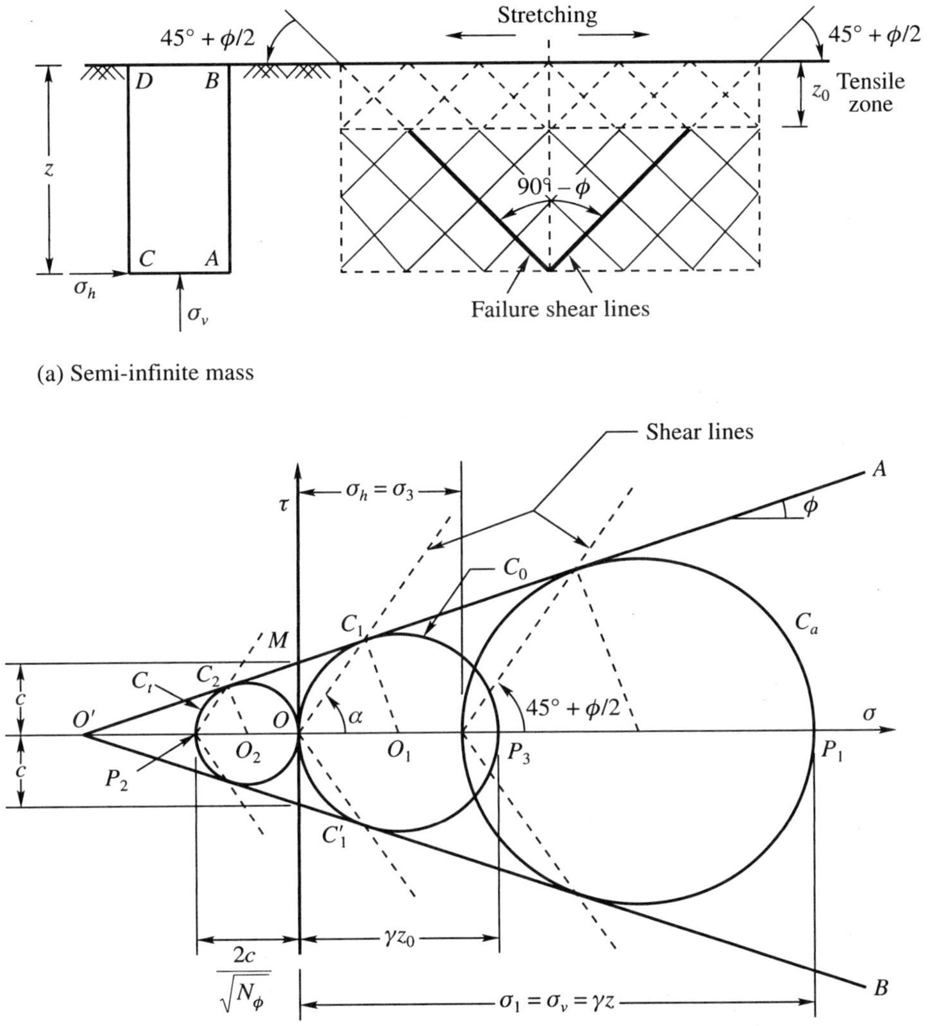

(a) Semi-infinite mass

(b) Mohr diagram

Figure 11.13 Active earth pressure of cohesive soil with horizontal backfill on a vertical wall

The active pressure $p_a = 0$ when

$$\frac{\gamma z}{N_\phi} - \frac{2c}{\sqrt{N_\phi}} = 0 \tag{11.31}$$

that is, p_a is zero at depth z, such that

$$z = z_0 = \frac{2c}{\gamma}\sqrt{N_\phi} \tag{11.32}$$

At depth $z = 0$, the pressure p_a is

$$p_a = -\frac{2c}{\sqrt{N_\phi}} \tag{11.33}$$

Equations (11.32) and (11.33) indicate that the active pressure p_a is tensile between depth 0 and z_0. The Eqs. (11.32) and (11.33) can also be obtained from Mohr circles C_0 and C_t respectively.

Shear Lines Pattern

The shear lines are shown in Fig. 11.13(a). Up to depth z_0 they are shown dotted to indicate that this zone is in tension.

Total Active Earth Pressure on a Vertical Section

If *AB* is the vertical section [11.14(a)], the active pressure distribution against this section of height *H* is shown in Fig. 11.14(b) as per Eq. (11.30). The total pressure against the section is

$$p_a = \int_0^H pzdz = \int_0^H \frac{\gamma z}{N_\phi} dz - \int_0^H \frac{2c}{\sqrt{N_\phi}} dz$$

$$= \frac{1}{2}\gamma H^2 \frac{1}{N_\phi} - 2c\frac{H}{\sqrt{N_\phi}} \tag{11.34}$$

The shaded area in Fig. 11.14(b) gives the total pressure P_a. If the wall has a height

$$H = H_c = \frac{4c}{\gamma}\sqrt{N_\phi} = 2z_0 \tag{11.35}$$

the total earth pressure is equal to zero. This indicates that a vertical bank of height smaller than H_c can stand without lateral support. H_c is called the *critical depth.* However, the pressure against the wall increases from $-2c/\sqrt{N_\phi}$ at the top to $+2c/\sqrt{N_\phi}$ at depth H_c, whereas on the vertical face of an unsupported bank the normal stress is zero at every point. Because of this difference, the greatest depth of which a cut can be excavated without lateral support for its vertical sides is slightly smaller than H_c.

For soft clay, $\phi = 0$, and $N_\phi = 1$

therefore, $$P_a = \frac{1}{2}\gamma H^2 - 2cH \tag{11.36}$$

and $$H_c = \frac{4c}{\gamma} \tag{11.37}$$

Soil does not resist any tension and as such it is quite unlikely that the soil would adhere to the wall within the tension zone of depth z_0 producing cracks in the soil. It is commonly assumed that the active earth pressure is represented by the shaded area in Fig. 11.14(c).

The total pressure on wall *AB* is equal to the area of the triangle in Fig. 11.14(c) which is equal to

$$P_a = \frac{1}{2}\left(\frac{\gamma H}{N_\phi} - \frac{2c}{\sqrt{N_\phi}}\right)(H - z_0) \tag{11.38a}$$

or $$P_a = \frac{1}{2}\left(\frac{\gamma H}{N_\phi} - \frac{2c}{\sqrt{N_\phi}}\right)\left(H - \frac{2c}{\gamma}\sqrt{N_\phi}\right) \tag{11.38b}$$

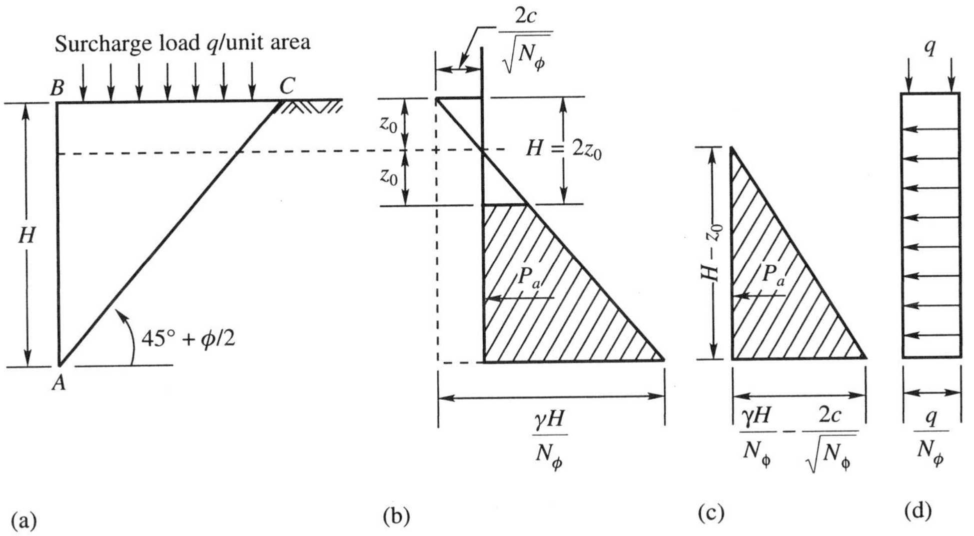

Figure 11.14 Active earth pressure on vertical sections in cohesive soils

Simplifying, we have

$$P_a = \frac{1}{2}\gamma H^2 \frac{1}{N_\phi} - 2cH\frac{1}{\sqrt{N_\phi}} + \frac{2c^2}{\gamma} \tag{11.38c}$$

For soft clay, $\phi = 0$

$$P_a = \frac{1}{2}\gamma H^2 - 2cH + \frac{2c^2}{\gamma} \tag{11.39}$$

It may be noted that $K_A = 1/N_\phi$

Effect of Surcharge and Water Table

Effect of Surcharge

When a surcharge load q per unit area acts on the surface, the lateral pressure on the wall due to surcharge remains constant with depth as shown in Fig. 11.14(d) for the active condition. The lateral pressure due to a surcharge under the active state may be written as

$$p_{aq} = \frac{q}{N_\phi}$$

The total active pressure due to a surcharge load is,

$$P_{aq} = \frac{qH}{N_\phi} \tag{11.40}$$

Effect of Water Table

If the soil is partly submerged, the submerged unit weight below the water table will have to be taken into account in both the active and passive states.

Figure 11.15(a) shows the case of a wall in the active state with cohesive material as backfill. The water table is at a depth of H_1 below the top of the wall. The depth of water is H_2.

The lateral pressure on the wall due to partial submergence is due to soil and water as shown in Fig. 11.15(b). The pressure due to soil = area of the figure *ocebo*.

The total pressure due to soil

$$P_a = oab + acdb + bde$$

$$\text{or} \quad P_a = \frac{1}{2}(H_1 - z_0)\left(\frac{\gamma_t H_1}{N_\phi} - \frac{2c}{\sqrt{N_\phi}}\right) + \left(\frac{\gamma_t H_1}{N_\phi} - \frac{2c}{\sqrt{N_\phi}}\right) H_2 + \frac{1}{2}\frac{\gamma_b H_2^2}{N_\phi} \tag{11.41}$$

After substituting for $z_0 = \dfrac{2c}{\gamma_t}\sqrt{N_\phi}$

and simplifying we have

$$P_a = \frac{1}{2N_\phi}(\gamma_t H_1^2 + \gamma_b H_2^2) - \frac{2c}{\sqrt{N_\phi}}(H_1 + H_2) + \frac{\gamma_t H_1 H_2}{N_\phi} + \frac{2c^2}{\gamma_t} \tag{11.42}$$

The total pressure on the wall due to water is

$$P_w = \frac{1}{2}\gamma_w H_2^2 \tag{11.43}$$

The point of application of P_a can be determined without any difficulty. The point of application P_w is at a height of $H_2/3$ from the base of the wall.

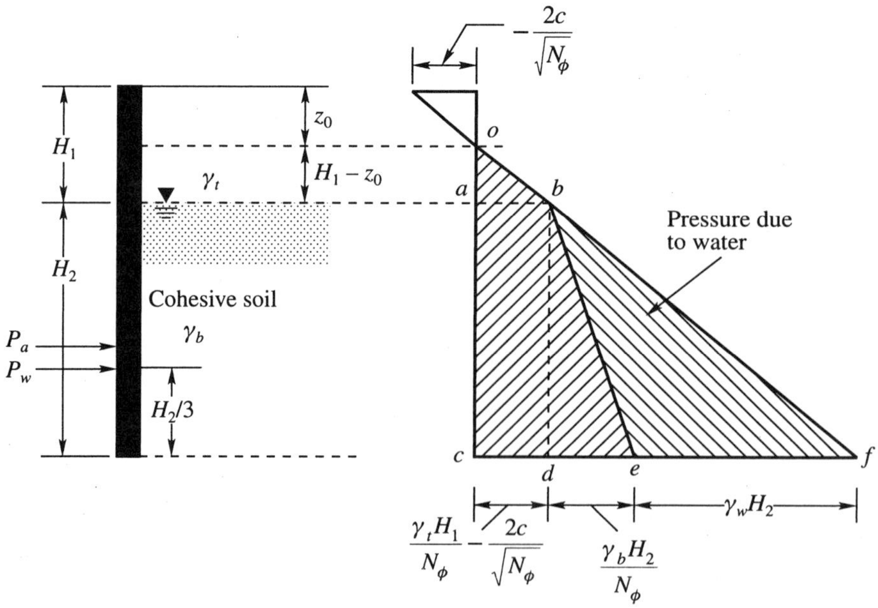

(a) Retaining wall (b) Pressure distribution

Figure 11.15 Effect of water table on lateral earth pressure

If the backfill material is cohesionless, the terms containing cohesion c in Eq. (11.42) reduce to zero.

Example 11.8

A retaining wall has a vertical back and is 7.32 m high. The soil is sandy loam of unit weight 17.3 kN/m^3. It has a cohesion of 12 kN/m^2 and $\phi = 20°$. Neglecting wall friction, determine the active thrust on the wall. The upper surface of the fill is horizontal.

Solution

(Refer to Fig. 11.14)

When the material exhibits cohesion, the pressure on the wall at a depth z is given by (Eq. 11.30)

$$p_a = \gamma z K_A - 2c\sqrt{K_A}$$

$$\text{where} \quad K_A = \frac{1-\sin\phi}{1+\sin\phi} = \frac{1-\sin 20°}{1+\sin 20°} = 0.49, \quad \sqrt{K_A} = 0.7$$

When the depth is small the expression for z is negative because of the effect of cohesion up to a theoretical depth z_0. The soil is in tension and the soil draws away from the wall.

$$z_0 = \frac{2c}{\gamma}\sqrt{N_\phi} = \frac{2c}{\gamma}\sqrt{K_P}$$

$$\text{where} \quad K_P = \frac{1+\sin\phi}{1-\sin\phi} = 2.04, \text{ and } \sqrt{K_P} = 1.43$$

$$\text{Therefore} \quad z_0 = \frac{2\times 12}{17.3}\times 1.43 = 1.98 \text{ m}$$

The lateral pressure at the surface ($z = 0$) is

$$p_a = -2c\sqrt{K_A} = -2\times 12\times 0.7 = -16.8 \text{ kN/m}^2$$

The negative sign indicates tension.

The lateral pressure at the base of the wall ($z = 7.32$ m) is

$$p_a = 17.3\times 7.32\times 0.49 - 16.8 = 45.25 \text{ kN/m}^2$$

Theoretically the area of the upper triangle in Fig. 11.14(b) to the left of the pressure axis represents a tensile force which should be subtracted from the compressive force on the lower part of the wall below the depth z_0. Since tension cannot be applied physically between the soil and the wall, this tensile force is neglected. It is therefore commonly assumed that the active earth pressure is represented by the shaded area in Fig. 11.14(c). The total pressure on the wall is equal to the area of the triangle in Fig. 11.14(c).

$$P_a = \frac{1}{2}(\gamma H K_A - 2c\sqrt{K_A})(H - z_0)$$

$$= \frac{1}{2}(17.3\times 7.32\times 0.49 - 2\times 12\times 0.7)(7.32 - 1.98) = 120.8 \text{ kN/m}$$

Example 11.9

Find the resultant thrust on the wall in Ex. 11.8 if the drains are blocked and water builds up behind the wall until the water table reaches a height of 2.75 m above the bottom of the wall.

Solution

For details refer to Fig. 11.15.

Per this figure,

$$H_1 = 7.32 - 2.75 = 4.57 \text{ m}, \ H_2 = 2.75 \text{ m}, \ H_1 - z_0 = 4.57 - 1.98 = 2.59 \text{ m}$$

The base pressure is detailed in Fig. 11.15(b)

(1) $\gamma_{sat} H_1 K_A - 2c\sqrt{K_A} = 17.3 \times 4.57 \times 0.49 - 2 \times 12 \times 0.7 = 21.94 \text{ kN/m}^2$

(2) $\gamma_b H_2 K_A = (17.3 - 9.81) \times 2.75 \times 0.49 = 10.1 \text{ kN/m}^2$

(3) $\gamma_w H_2 = 9.81 \times 2.75 = 27 \text{ kN/m}^2$

The total pressure = P_a = pressure due to soil + water

From Eqs. (11.41), (11.43), and Fig. 11.15(b)

$$P_a = oab + acdb + bde + bef$$

$$= \frac{1}{2} \times 2.59 \times 21.94 + 2.75 \times 21.94 + \frac{1}{2} \times 2.75 \times 10.1 + \frac{1}{2} \times 2.75 \times 27$$

= 28.41 + 60.34 + 13.89 +37.13 = 139.7 kN/m or say 140 kN/m

The point of application of P_a may be found by taking moments of each area and P_a about the base. Let h be the height of P_a above the base. Now

$$140 \times h = 28.41 \ \frac{1}{3} \times 2.59 + 2.75 \ + 60.34 \times \frac{2.75}{2} + 13.89 \times \frac{2.75}{3} + \frac{37.13 \times 2.75}{3}$$

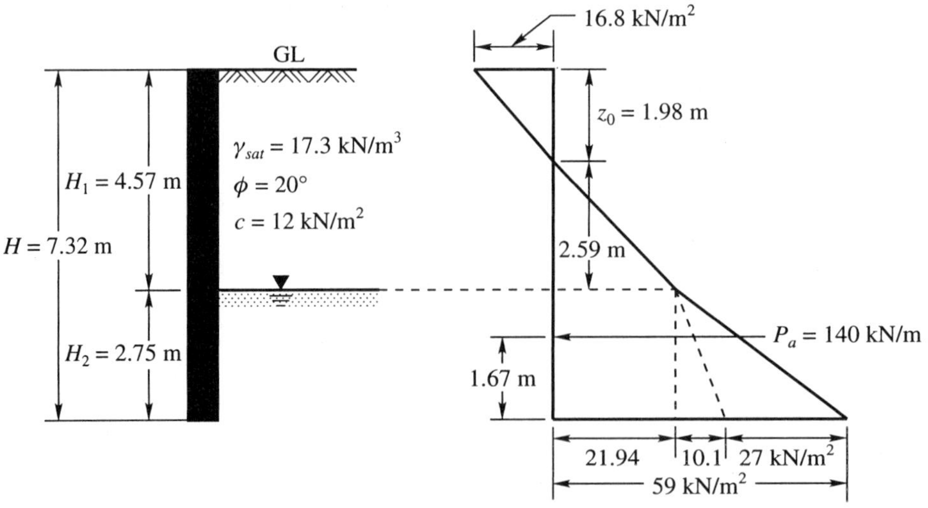

Figure Ex. 11.9

$$= 102.65 + 83.0 + 12.7 + 34.0 = 232.4$$

or $$h = \frac{232.4}{140} = 1.66 \text{ m}$$

Example 11.10

A rigid retaining wall 19.69 ft high has a saturated backfill of soft clay soil. The properties of the clay soil are $\gamma_{sat} = 111.76$ lb/ft^3, and unit cohesion $c_u = 376$ lb/ft^2. Determine (a) the expected depth of the tensile crack in the soil (b) the active earth pressure before the occurrence of the tensile crack, and (c) the active pressure after the occurrence of the tensile crack. Neglect the effect of water that may collect in the crack.

Solution

At $z = 0$, $p_a = -2c = -2 \times 376 = -752$ lb/ft^2 since $\phi = 0$

At $z = H$, $p_a = \gamma H - 2c = 111.76 \times 19.69 - 2 \times 376 = 1449$ lb/ft^2

(a) From Eq. (11.32), the depth of the tensile crack z_0 is (for $\phi = 0$)

$$z_0 = \frac{2c}{\gamma} = \frac{2 \times 376}{111.76} = 6.73 \text{ ft}$$

(b) The active earth pressure before the crack occurs.

Use Eq. (11.36) for computing P_a

$$P_a = \frac{1}{2}\gamma H^2 - 2cH$$

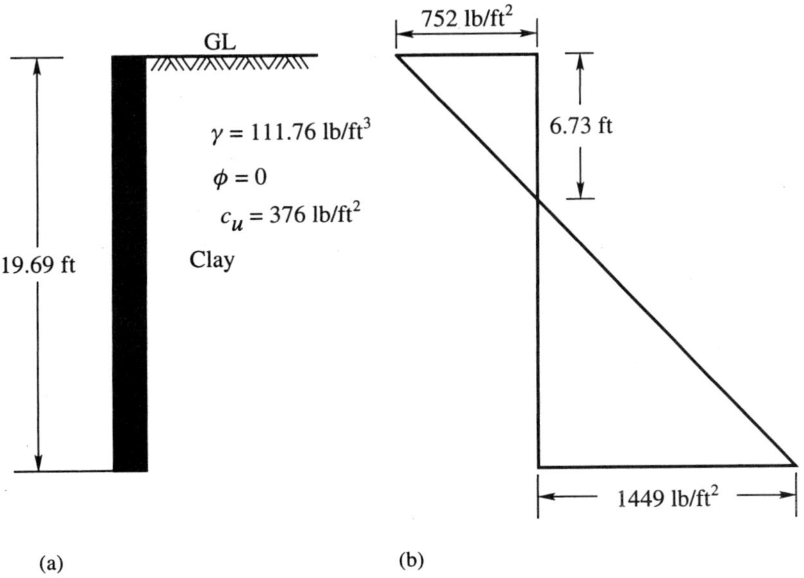

Figure Ex. 11.10

since $K_A = 1$ for $\phi = 0$. Substituting, we have

$$P_a = \frac{1}{2} \times 111.76 \times (19.69)^2 - 2 \times 376 \times 19.69 = 21{,}664 - 14{,}807 = 6857 \text{ lb/ ft}$$

(c) P_a after the occurrence of a tensile crack.

Use Eq. (11.38a),

$$P_a = \frac{1}{2}(\gamma H - 2c)(H - z_0)$$

Substituting

$$P_a = \frac{1}{2}(111.76 \times 19.69 - 2 \times 376)(19.69 - 6.73) = 9387 \text{ lb/ft}$$

Example 11.11

A rigid retaining wall of 6 m height (Fig. Ex. 11.11) has two layers of backfill. The top layer to a depth of 1.5 m is sandy clay having $\phi = 20°$, $c = 12.15$ kN/m^2 and $\gamma = 16.4$ kN/m^3. The bottom layer is sand having $\phi = 30°$, $c = 0$, and $\gamma = 17.25$ kN/m^3.

Determine the total active earth pressure acting on the wall and draw the pressure distribution diagram.

Solution

For the top layer,

$$K_A = \tan^2 45° - \frac{20}{2} = 0.49, \quad K_P = \frac{1}{0.49} = 2.04$$

The depth of the tensile zone, z_0 is

$$z_0 = \frac{2c}{\gamma}\sqrt{K_P} = \frac{2 \times 12.15\sqrt{2.04}}{16.4} = 2.12 \text{ m}$$

Since the depth of the sandy clay layer is 1.5 m, which is less than z_0, the tensile crack develops only to a depth of 1.5 m.

K_A for the sandy layer is

$$K_A = \tan^2\left(45° - \frac{\phi}{2}\right) = \tan^2\left(45° - \frac{30}{2}\right) = \frac{1}{3}$$

At a depth $z = 1.5$, the vertical pressure σ_v is

$$\sigma_v = \gamma z = 16.4 \times 1.5 = 24.6 \text{ kN/m}^2$$

The active pressure is

$$p_a = K_A \gamma z = \frac{1}{3} \times 24.6 = 8.2 \text{ kN/m}^2$$

At a depth of 6 m, the effective vertical pressure is

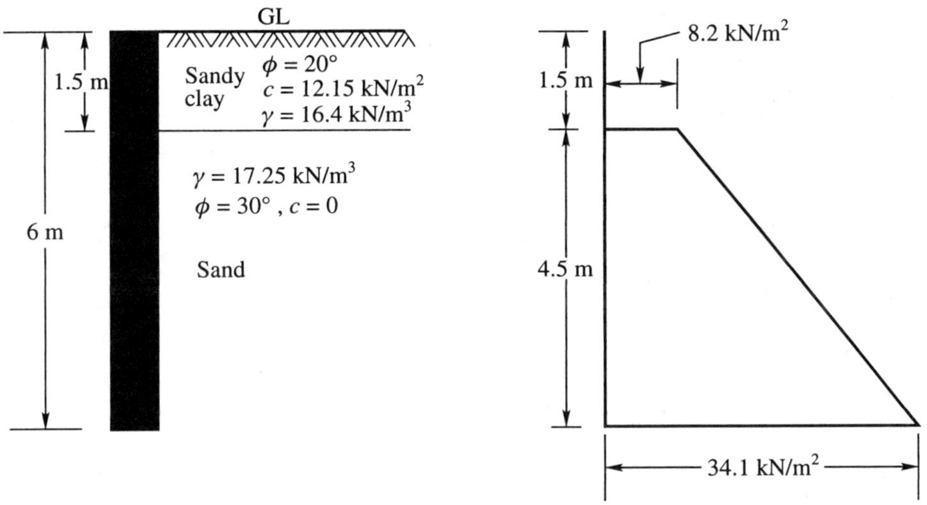

Figure Ex. 11.11

$$\sigma_v = 1.5 \times 16.4 + 4.5 \times 17.25 = 24.6 + 77.63 = 102.23 \text{ kN/m}^2$$

The active pressure p_a is

$$p_a = K_A \sigma_v = \frac{1}{3} \times 102.23 = 34.1 \text{ kN/m}^2$$

The pressure distribution diagram is given in Fig. Ex. 11.11.

11.7 RANKINE'S PASSIVE EARTH PRESSURE WITH COHESIVE BACKFILL

If the wall AB in Fig. 11.16(a) is pushed towards the backfill, the horizontal pressure p_h on the wall increases and becomes greater than the vertical pressure σ_v. When the wall is pushed sufficiently inside, the backfill attains Rankine's state of plastic equilibrium. The pressure distribution on the wall may be expressed by the equation

$$\sigma_1 = \sigma_3 N_\phi + 2c\sqrt{N_\phi}$$

In the passive state, the horizontal stress σ_h is the major principal stress σ_1 and the vertical stress σ_v is the minor principal stress σ_3. Since $\sigma_3 = \gamma z$, the passive pressure at any depth z may be written as

$$\sigma_1 = \sigma_h = p_p = \gamma z N_\phi + 2c\sqrt{N_\phi} = \gamma z K_P + 2c\sqrt{K_P} \tag{11.44a}$$

At depth $z = O,\quad p_p = 2c\sqrt{N_\phi} = 2c\sqrt{K_P}$

At depth $z = H,\quad p_p = \gamma H N_\phi + 2c\sqrt{N_\phi} = \gamma H K_P + 2c\sqrt{K_P}$ (11.44b)

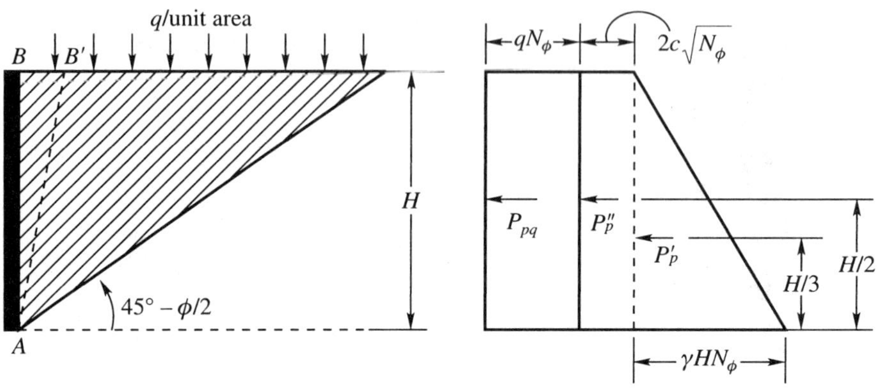

(a) Wall (b) Pressure distribution

Figure 11.16 Passive earth pressure on vertical sections in cohesive soils

The distribution of pressure with respect to depth is shown in Fig. 11.16(b). The pressure increases hydrostatically. The total pressure on the wall may be written as a sum of two pressures P_p' and P_p''

$$P_p' = \int_0^H \gamma z N_\phi \, dz = \frac{1}{2}\gamma H^2 N_\phi = \frac{1}{2}\gamma H^2 K_P \tag{11.45a}$$

This acts at a height $H/3$ from the base.

$$P_p'' = \int_0^H 2c\sqrt{N_\phi}\, dz = 2cH\sqrt{N_\phi} = 2cH\sqrt{K_P} \tag{11.45b}$$

This acts at a height of $H/2$ from the base.

$$P_p = P_p' + P_p'' = \frac{1}{2}\gamma H^2 K_p + 2cH\sqrt{K_p} \tag{11.45c}$$

The passive pressure due to a surcharge load of q per unit area is

$$p_{pq} = qN_\phi = qK_P$$

The total passive pressure due to a surcharge load is

$$P_{pq} = qHN_\phi = qHK_P \tag{11.46}$$

which acts at mid-height of the wall.

It may be noted here that $N_\phi = K_P$.

Example 11.12

A smooth rigid retaining wall 19.69 ft high carries a uniform surcharge load of 251 lb/ft^2. The backfill is clayey sand with the following properties:

γ = 102 lb/ft^3, ϕ = 25°, and c = 136 lb/ft^2.

Determine the passive earth pressure and draw the pressure diagram.

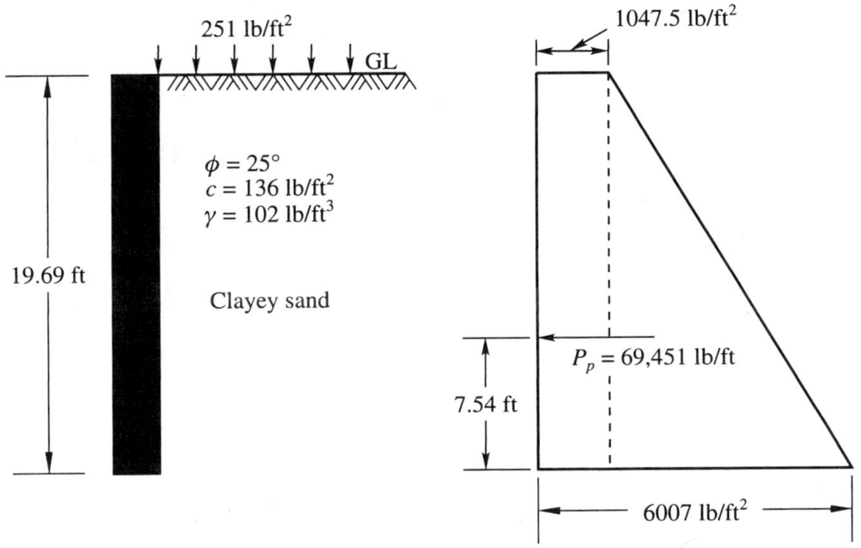

Figure Ex. 11.12

Solution

For $\phi = 25°$, the value of K_P is

$$K_P = \frac{1+\sin\phi}{1-\sin\phi} = \frac{1+0.423}{1-0.423} = \frac{1.423}{0.577} = 2.47$$

From Eq. (11.44a), p_p at any depth z is

$$p_p = \gamma z K_P + 2c\sqrt{K_P} = \sigma_v K_P + 2c\sqrt{K_P}$$

At depth $z = 0$, $\sigma_v = 251$ lb/ft²

$$p_p = 251 \times 2.47 + 2 \times 136\sqrt{2.47} = 1047.5 \text{ lb/ ft}^2$$

At $\quad z = 19.69$ ft, $\sigma_v = 251 + 19.69 \times 102 = 2259$ lb/ ft²

$$p_p = 2259 \times 2.47 + 2 \times 136\sqrt{2.47} = 6007 \text{ lb/ ft}^2$$

The pressure distribution is shown in Fig. Ex. 11.12.

The total passive pressure P_P acting on the wall is

$$P_P = 1047.5 \times 19.69 + \frac{1}{2} \times 19.69(6007 - 1047.5) = 69{,}451 \text{ lb/ ft of wall} \approx 69.5 \text{ kips/ft of wall.}$$

Location of resultant

Taking moments about the base

$$P_p \times h = \frac{1}{2} \times (19.69)^2 \times 1047.5 + \frac{1}{6} \times (19.69)^2 \times 4959.5$$

$$= 523{,}518 \text{ lb.ft.}$$

$$\text{or} \quad h = \frac{523{,}518}{P_p} = \frac{523{,}518}{69{,}451} = 7.54 \text{ ft}$$

11.8 COULOMB'S EARTH PRESSURE THEORY FOR SAND FOR ACTIVE STATE

Coulomb made the following assumptions in the development of his theory:

1. The soil is isotropic and homogeneous
2. The rupture surface is a plane surface
3. The failure wedge is a rigid body
4. The pressure surface is a plane surface
5. There is wall friction on the pressure surface
6. Failure is two-dimensional and
7. The soil is cohesionless

Consider Fig. 11.17.

1. AB is the pressure face
2. The backfill surface BE is a plane inclined at an angle β with the horizontal
3. α is the angle made by the pressure face AB with the horizontal
4. H is the height of the wall
5. AC is the assumed rupture plane surface, and
6. θ is the angle made by the surface AC with the horizontal

If AC in Fig. 17(a) is the probable rupture plane, the weight of the wedge W per unit length of the wall may be written as

$W = \gamma A$, where A = area of wedge ABC

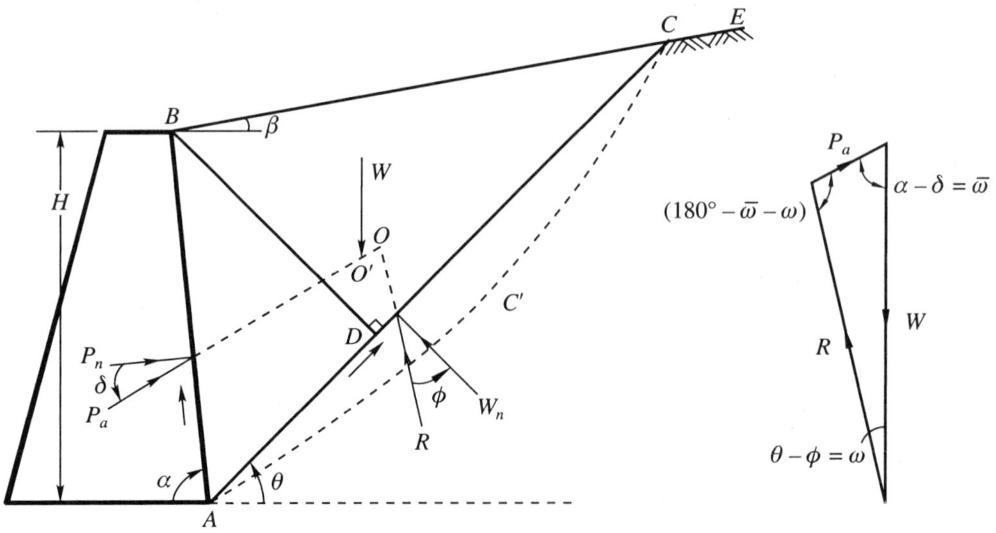

(a) Retaining wall (b) Polygon of forces

Figure 11.17 Conditions for failure under active conditions

Area of wedge $ABC = A = 1/2\, AC \times BD$

where BD is drawn perpendicular to AC.

From the law of sines, we have

$$AC = AB\frac{\sin(\alpha+\beta)}{\sin(\theta-\beta)}, \quad BD = AB\sin(\alpha+\theta), \quad AB = \frac{H}{\sin\alpha}$$

Making the substitution and simplifying we have,

$$W = \gamma A = \frac{\gamma H^2}{2\sin^2\alpha}\sin(\alpha+\theta)\frac{\sin(\alpha+\beta)}{\sin(\theta-\beta)} \tag{11.47}$$

The various forces that are acting on the wedge are shown in Fig. 11.17(a). As the pressure face AB moves away from the backfill, there will be sliding of the soil mass along the wall from B towards A. The sliding of the soil mass is resisted by the friction of the surface. The direction of the shear stress is in the direction from A towards B. If P_n is the total normal reaction of the soil pressure acting on face AB, the resultant of P_n and the shearing stress is the active pressure P_a making an angle δ with the normal. Since the shearing stress acts upwards, the resulting P_a dips below the normal. The angle δ for this condition is considered *positive.*

As the wedge ABC ruptures along plane AC, it slides along this plane. This is resisted by the frictional force acting between the soil at rest below AC, and the sliding wedge. The resisting shearing stress is acting in the direction from A towards C. If W_n is the normal component of the weight of wedge W on plane AC, the resultant of the normal W_n and the shearing stress is the reaction R. This makes an angle ϕ with the normal since the rupture takes place within the soil itself. Statical equilibrium requires that the three forces P_a, W, and R meet at a point. Since AC is not the actual rupture plane, the three forces do not meet at a point. But if the actual surface of failure $AC'C$ is considered, all three forces meet at a point. However, the error due to the nonconcurrence of the forces is very insignificant and as such may be neglected.

The polygon of forces is shown in Fig. 11.17(b). From the polygon of forces, we may write

$$\frac{P_a}{\sin(\theta-\phi)} = \frac{W}{\sin(180^\circ-\alpha-\theta+\phi+\delta)}$$

$$\text{or} \quad P_a = \frac{W\sin(\theta-\phi)}{\sin(180^\circ-\alpha-\theta+\phi+\delta)} \tag{11.48}$$

In Eq. (11.48), the only variable is θ and all the other terms for a given case are constants. Substituting for W, we have

$$P_a = \frac{\gamma H^2}{2\sin^2\alpha}\frac{\sin(\theta-\phi)}{\sin(180^\circ-\alpha-\theta+\phi+\delta)}\left(\sin(\alpha+\phi)\frac{\sin(\alpha+\beta)}{\sin(\theta-\beta)}\right) \tag{11.49}$$

The maximum value for P_a is obtained by differentiating Eq. (11.49) with respect to θ and equating the derivative to zero, i.e.

$$\frac{dP_a}{d\theta} = 0$$

The maximum value of P_a so obtained may be written as

$$P_a = \frac{1}{2}\gamma H^2 K_A \tag{11.50}$$

Table 11.2a Active earth pressure coefficients K_A for $\beta = 0$ and $\alpha = 90°$

ϕ^0	15	20	25	30	35	40
$\delta = 0$	0.59	0.49	0.41	0.33	0.27	0.22
$\delta = +\phi/2$	0.55	0.45	0.38	0.32	0.26	0.22
$\delta = +/2/3\phi$	0.54	0.44	0.37	0.31	0.26	0.22
$\delta = +\phi$	0.53	0.44	0.37	0.31	0.26	0.22

Table 11.2b Active earth pressure coefficients K_A for $\delta = 0$, β varies from −30° to +30° and α from 70° to 110°

$\beta =$		−30°	−12°	0°	+12°	+ 30°
$\phi = 20°$	α = 70°	–	0.54	0.61	0.76	–
	80°	–	0.49	0.54	0.67	–
	90°	–	0.44	0.49	0.60	–
	100	–	0.37	0.41	0.49	–
	110	–	0.30	0.33	0.38	–
$\phi = 30°$	70°	0.32	0.40	0.47	0.55	1.10
	80°	0.30	0.35	0.40	0.47	0.91
	90°	0.26	0.30	0.33	0.38	0.75
	100	0.22	0.25	0.27	0.31	0.60
	110	0.17	0.19	0.20	0.23	0.47
$\phi = 40°$	70	0.25	0.31	0.36	0.40	0.55
	80	0.22	0.26	0.28	0.32	0.42
	90	0.18	0.20	0.22	0.24	0.32
	100	0.13	0.15	0.16	0.17	0.24
	110	0.10	0.10	0.11	0.12	0.15

where K_A is the active earth pressure coefficient.

$$K_A = \frac{\sin^2(\alpha + \phi)}{\sin^2 \alpha \sin(\alpha - \delta)\left[1 + \sqrt{\dfrac{\sin(\phi + \delta)\sin(\phi - \beta)}{\sin(\alpha - \delta)\sin(\alpha + \beta)}}\right]^2} \tag{11.51}$$

The total normal component P_n of the earth pressure on the back of the wall is

$$P_n = P_a \cos\delta = \frac{1}{2}\gamma H^2 K_A \cos\delta \tag{11.52}$$

If the wall is vertical and smooth, and if the backfill is horizontal, we have

$$\beta = \delta = 0 \quad \text{and} \quad \alpha = 90°$$

Substituting these values in Eq. (11.51), we have

$$K_A = \frac{1 - \sin\phi}{1 + \sin\phi} = \tan^2\left(45° - \frac{\phi}{2}\right) = \frac{1}{N_\phi} \tag{11.53}$$

where $N_\phi = \tan^2\left(45° + \frac{\phi}{2}\right)$ (11.54)

The coefficient K_A in Eq. (11.53) is the same as Rankine's. The effect of wall friction is frequently neglected where active pressures are concerned. Table 11.2 makes this clear. It is clear from this table that K_A decreases with an increase of δ and the maximum decrease is not more than 10 percent.

11.9 COULOMB'S EARTH PRESSURE THEORY FOR SAND FOR PASSIVE STATE

In Fig. 11.18, the notations used are the same as in Fig. 11.17. As the wall moves into the backfill, the soil tries to move up on the pressure surface *AB* which is resisted by friction of the surface. Shearing stress on this surface therefore acts downward. The passive earth pressure P_p is the resultant of the normal pressure P_{pn} and the shearing stress. The shearing force is rotated upward with an angle δ which is again the angle of wall friction. In this case δ is *positive.*

As the rupture takes place along assumed plane surface *AC*, the soil tries to move up the plane which is resisted by the frictional force acting on that line. The shearing stress therefore, acts downward. The reaction *R* makes an angle ϕ with the normal and is rotated upwards as shown in the figure.

The polygon of forces is shown in (b) of the Fig. 11.18. Proceeding in the same way as for active earth pressure, we may write the following equations:

$$W = \frac{\gamma H^2}{2\sin^2\alpha}\sin(\alpha+\theta)\frac{\sin(\alpha+\beta)}{\sin(\theta-\beta)} \tag{11.55}$$

$$P_p = \frac{W\sin(\theta+\phi)}{\sin(180° - \theta - \phi - \delta - \alpha)} \tag{11.56}$$

Differentiating Eq. (11.56) with respect to θ and setting the derivative to zero, gives the minimum value of P_p as

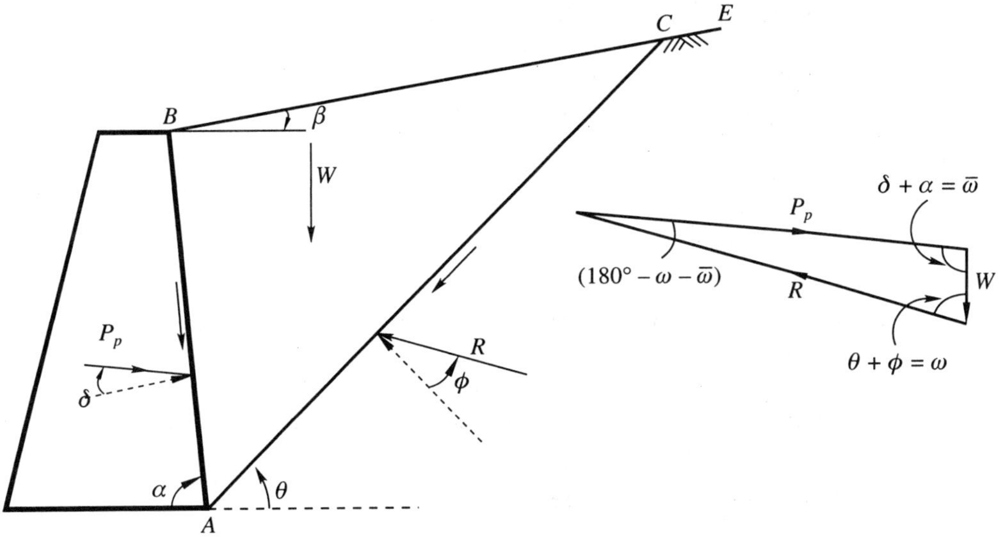

(a) Forces on the sliding wedge (b) Polygon of forces

Figure 11.18 Conditions for failure under passive state

$$P_p = \frac{1}{2}\gamma H^2 K_p \tag{11.57}$$

where K_p is called the *passive earth pressure coefficient.*

$$K_P = \frac{\sin^2(\alpha - \phi)}{\sin^2\alpha\sin(\alpha+\delta)\left[1-\sqrt{\dfrac{\sin(\phi+\delta)\sin(\phi+\beta)}{\sin(\alpha+\delta)\sin(\alpha+\beta)}}\right]^2} \tag{11.58}$$

Eq. (11.58) is valid for both positive and negative values of β and δ.

The total normal component of the passive earth pressure P_p on the back of the wall is

$$P_{pn} = \frac{1}{2}\gamma H^2 K_P \cos\delta \tag{11.59}$$

For a smooth vertical wall with a horizontal backfill, we have

$$K_P = \frac{1+\sin\phi}{1-\sin\phi} = \tan^2\left(45° + \frac{\phi}{2}\right) = N_\phi \tag{11.60}$$

Eq. (11.60) is Rankine's passive earth pressure coefficient. We can see from Eqs. (11.53) and (11.60) that

$$K_P = \frac{1}{K_A} \tag{11.61}$$

Coulomb sliding wedge theory of plane surfaces of failure is valid with respect to passive pressure, i.e., to the resistance of non-cohesive soils only. If wall friction is zero for a vertical wall and horizontal backfill, the value of K_P may be calculated using Eq. (11.59). If wall friction is considered in conjunction with plane surfaces of failure, much too high, and therefore unsafe values of earth resistance will be obtained, especially in the case of high friction angles ϕ. For example for $\phi = \delta = 40°$, and for plane surfaces of failure, $K_P = 92.3$, whereas for curved surfaces of failure $K_P = 17.5$. However, if δ is smaller than $\phi/2$, the difference between the real surface of sliding and Coulomb's plane surface is very small and we can compute the corresponding passive earth pressure coefficient by means of Eq. (11.57). If δ is greater than $\phi/2$, the values of K_P should be obtained by analyzing curved surfaces of failure.

11.10 ACTIVE PRESSURE BY CULMANN'S METHOD FOR COHESIONLESS SOILS

Without Surcharge Line Load

Culmann's (1875) method is the same as the trial wedge method. In Culmann's method, the force polygons are constructed directly on the ϕ-line *AE* taking *AE* as the load line. The procedure is as follows:

In Fig. 11.19(a) *AB* is the retaining wall drawn to a suitable scale. The various steps in the construction of the pressure locus are:

1. Draw ϕ -line *AE* at an angle ϕ to the horizontal.
2. Lay off on *AE* distances, *AV*, *A*1, A2, A3, etc. to a suitable scale to represent the weights of wedges *ABV*, A*B*1, A*B*2, A*B*3, etc. respectively.

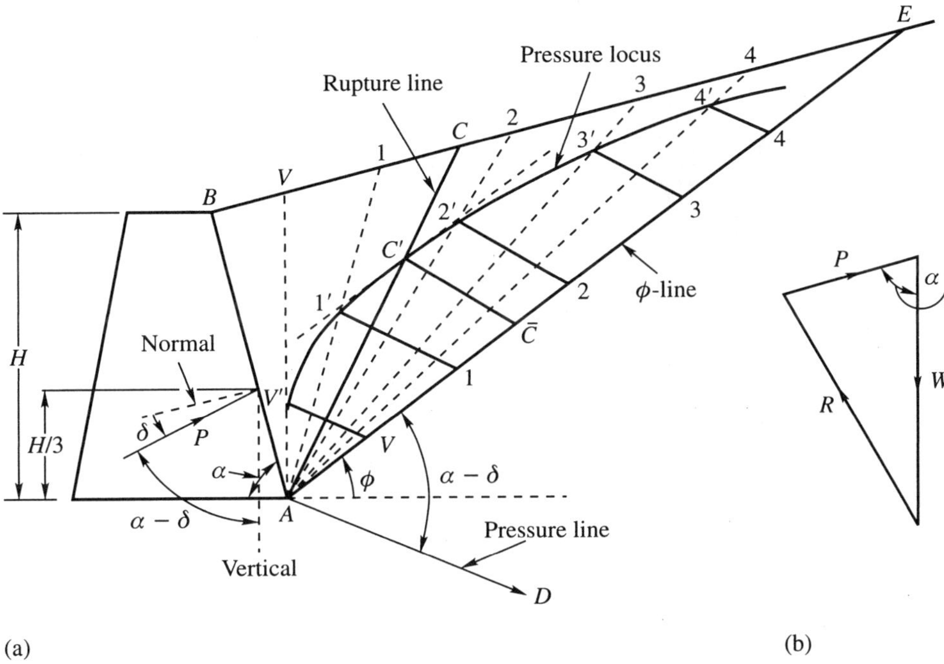

Figure 11.19 Active pressure by Culmann's method for cohesionless soils

3. Draw lines parallel to *AD* from points *V*, 1, 2, 3 to intersect assumed rupture lines *AV*, A1, A2, A3 at points *V*′, 1′, 2′, 3′, etc. respectively.
4. Join points *V*′, 1′, 2′, 3′, etc. by a smooth curve which is the pressure locus.
5. Select point *C*′ on the pressure locus such that the tangent to the curve at this point is parallel to the ϕ-line *AE*.
6. Draw $C'\overline{C}$ parallel to the pressure line *AD*. The magnitude of $C'\overline{C}$ in its natural units gives the active pressure P_a.
7. Join *AC*′ and produce to meet the surface of the backfill at *C*. *AC* is the rupture line.

For the plane backfill surface, the point of application of P_a is at a height of *H*/3 from the base of the wall.

Example 11.13

For a retaining wall system, the following data were available: (i) Height of wall = 7 m, (ii) Properties of backfill: $\gamma_d = 16$ kN/m^3, $\phi = 35°$, (iii) angle of wall friction, $\delta = 20°$, (iv) back of wall is inclined at 20° to the vertical (positive batter), and (v) backfill surface is sloping at 1 : 10.

Determine the magnitude of the active earth pressure by Culmann's method.

Solution

(a) Fig. Ex. 11.13 shows the ϕ line and pressure lines drawn to a suitable scale.
(b) The trial rupture lines Bc_1, Bc_2, Bc_3, etc. are drawn by making $Ac_1 = c_1c_2 = c_2c_3$, etc.
(c) The length of a vertical line from *B* to the backfill surface is measured.
(d) The areas of wedges BAc_1, BAc_2, BAc_3, etc. are respectively equal to 1/2(base lengths Ac_1, Ac_2, Ac_3, etc.) × perpendicular length.

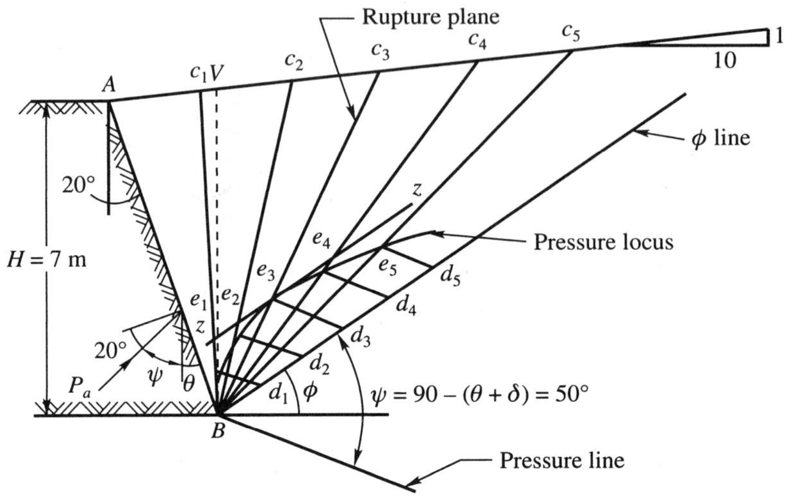

Figure Ex. 11.13

(e) The weights of the wedges in (*d*) above per meter length of wall may be determined by multiplying the areas by the unit weight of the soil. The results are tabulated below:

Wedge	Weight, kN	Wedge	Weight, kN
BAc_1	115	BAc_4	460
BAc_2	230	BAc_5	575
BAc_3	345		

(f) The weights of the wedges BAc_1, BAc_2, etc. are respectively plotted are Bd_1, Bd_2, etc. on the ϕ-line.

(g) Lines are drawn parallel to the pressure line from points d_1, d_2, d_3 etc. to meet respectively the trial rupture lines Bc_1, Bc_2, Bc_3 etc. at points e_1, e_2, e_3, etc.

(h) The pressure locus is drawn passing through points e_1, e_2, e_3, etc.

(i) Line *zz* is drawn tangential to the pressure locus at a point at which *zz* is parallel to the ϕ line. This point coincides with the point e_3.

(j) e_3d_3 gives the active earth pressure when converted to force units.
P_a = 180 kN per meter length of wall.

(k) Bc_3 is the critical rupture plane.

11.11 LATERAL PRESSURES BY THEORY OF ELASTICITY FOR SURCHARGE LOADS ON THE SURFACE OF BACKFILL

The surcharges on the surface of a backfill parallel to a retaining wall may be any one of the following

1. A concentrated load
2. A line load
3. A strip load

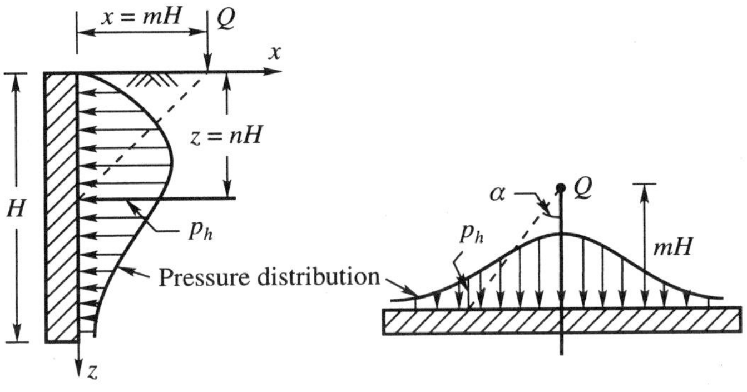

(a) Vertical section (b) Horizontal section

Figure 11.20 Lateral pressure against a rigid wall due to a point load

Lateral Pressure at a Point in a Semi-Infinite Mass due to a Concentrated Load on the Surface

Tests by Spangler (1938), and others indicate that lateral pressures on the surface of rigid walls can be computed for various types of surcharges by using modified forms of the theory of elasticity equations. Lateral pressure on an element in a semi-infinite mass at depth z from the surface may be calculated by Boussinesq theory for a concentrated load Q acting at a point on the surface. The equation may be expressed as (refer to Section 6.2 for notation)

$$p_h = \frac{Q}{2\pi z^2}\left[3\sin^2\beta\cos^2\beta - \frac{(1-2\mu)\cos^2\beta}{1+\cos\beta}\right] \tag{11.62}$$

If we write r = x in Fig. 6.1 and redefine the terms as
$x = mH$ and, $z = nH$

where H = height of the rigid wall and take Poisson's ratio $\mu = 0.5$, we may write Eq. (11.62) as

$$p_h = \frac{3Q}{2\pi H^2}\frac{m^2 n}{(m^2+n^2)^{5/2}} \tag{11.63}$$

Eq. (11.63) is strictly applicable for computing lateral pressures at a point in a semi-infinite mass. However, this equation has to be modified if a rigid wall intervenes and breaks the continuity of the soil mass. The modified forms are given below for various types of surcharge loads.

Lateral Pressure on a Rigid Wall Due to a Concentrated Load on the Surface

Let Q be a point load acting on the surface as shown in Fig. 11.20. The various equations are

(a) For $m > 0.4$

$$p_h = \frac{1.77Q}{H^2}\frac{n^2}{(m^2+n^2)^3} \tag{11.64}$$

(b) For $m \le 0.4$

$$p_h = \frac{0.28Q}{H^2} \frac{n^2}{(0.16+n^2)^3} \tag{11.65}$$

(c) Lateral pressure at points along the wall on each side of a perpendicular from the concentrated load Q to the wall (Fig. 11.20b)

$$p_h' = p_h \cos^2(1.1\alpha) \tag{11.66}$$

Lateral Pressure on a Rigid Wall due to Line Load

A concrete block wall conduit laid on the surface, or wide strip loads may be considered as a series of parallel line loads as shown in Fig. 11.21. The modified equations for computing p_h are as follows:

(a) For $m > 0.4$

$$p_h = \frac{4}{\pi}\frac{q}{H}\left[\frac{m^2 n}{(m^2+n^2)^2}\right] \tag{11.67}$$

(a) For $m \leq 0.4$

$$p_h = \frac{q}{H}\left[\frac{0.203n}{(0.16+n^2)^2}\right] \tag{11.68}$$

Lateral Pressure on a Rigid Wall due to Strip Load

A strip load is a load intensity with a finite width, such as a highway, railway line or earth embankment which is parallel to the retaining structure. The application of load is as given in Fig. 11.22.

The equation for computing p_h is

$$p_h = \frac{2q}{\pi}(\beta - \sin\beta\cos 2\alpha) \tag{11.69a}$$

The total lateral pressure per unit length of wall due to strip loading may be expressed as (Jarquio, 1981)

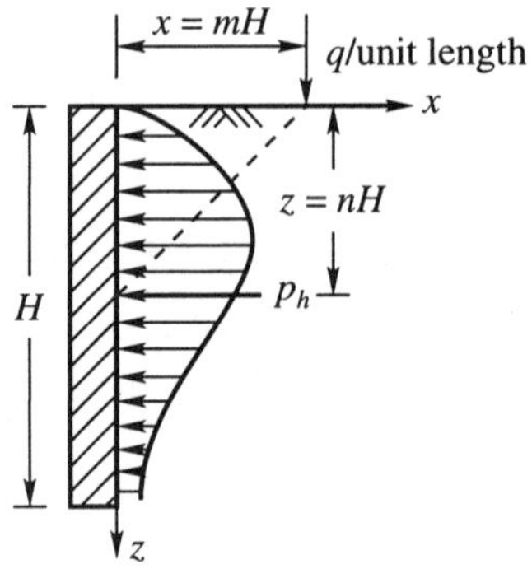

Figure 11.21 Lateral pressure against a rigid wall due to a line load

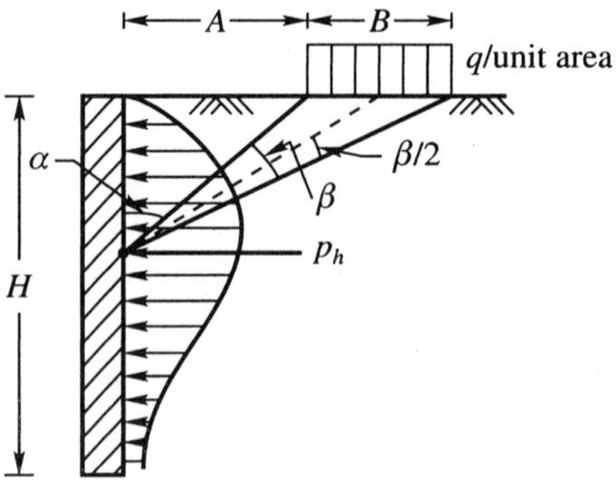

Figure 11.22 Lateral pressure against a rigid wall due to a strip load

$$P_h = \frac{q}{90}\left[H(\alpha_2 - \alpha_1)\right] \qquad (11.69b)$$

where $\alpha_1 = \tan^{-1}\left(\frac{A}{H}\right)$ and $\alpha_2 = \tan^{-1}\left(\frac{A+B}{H}\right)$

Example 11.14

A railway line is laid parallel to a rigid retaining wall as shown in Fig. Ex. 11.14. The width of the railway track and its distance from the wall is shown in the figure. The height of the wall is 10 m. Determine

(a) The unit pressure at a depth of 4m from the top of the wall due to the surcharge load
(b) The total pressure acting on the wall due to the surcharge load

Solution

(a) From Eq (11.69a)

The lateral earth pressure p_h at depth 4 m is

$$p_h = \frac{2q}{\pi}(\beta - \sin\beta\cos 2\alpha)$$

$$= \frac{2\times 60}{3.14}\left[\frac{18.44}{180}\times 3.14 - \sin 18.44^\circ \cos 2\times 36.9\right] = 8.92 \text{ kN/m}^2$$

(b) From Eq. (11.69b)

$$P_h = \frac{q}{90}\left[H(\alpha_2 - \alpha_1)\right]$$

where, $q = 60$ kN/m^2, H = 10 m

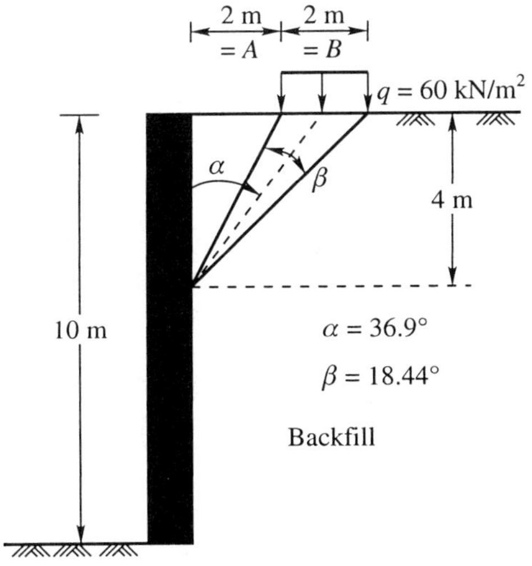

Figure Ex. 11.14

$$\alpha_1 = \tan^{-1}\frac{A}{H} = \tan^{-1}\frac{2}{10} = 11.31°$$

$$\alpha_1 = \tan^{-1}\frac{A+B}{H} = \tan^{-1}\frac{2+2}{10} = 21.80°$$

$$P_h = \frac{60}{90}[10(21.80 - 11.31)] \approx 70 \text{ kN/m}$$

11.12 CURVED SURFACES OF FAILURE FOR COMPUTING PASSIVE EARTH PRESSURE

It is customary practice to use curved surfaces of failure for determining the passive earth pressure P_p on a retaining wall with granular backfill if δ is greater than $\phi/3$. If tables or graphs are available for determining K_p for curved surfaces of failure the passive earth pressure P_p can be calculated. If tables or graphs are not available for this purpose, P_p can be calculated graphically by any one of the following methods.

1. Logarithmic spiral method
2. Friction circle method

In both these methods, the failure surface close to the wall is assumed as the part of a logarithmic spiral or a part of a circular arc with the top portion of the failure surface assumed as planar. This statement is valid for both cohesive and cohesionless materials. The methods are applicable for both horizontal and inclined backfill surfaces. However, in the following investigations it will be assumed that the surface of the backfill is horizontal.

Logarithmic Spiral Method of Determining Passive Earth Pressure of Ideal Sand

Property of a Logarithmic Spiral

The equation of a logarithmic spiral may be expressed as

$$r = r_0 e^{\theta \tan\phi} \tag{11.70}$$

where

r_0 = arbitrarily selected radius vector for reference

r = radius vector of any chosen point on the spiral making an angle θ with r_0.

ϕ = angle of internal friction of the material.

In Fig. 11.23a O is the origin of the spiral. The property of the spiral is that every radius vector such as Oa makes an angle of $90°-\phi$ to the tangent of the spiral at a or in other words, the vector Oa makes an angle ϕ with the normal to the tangent of the spiral at a.

Analysis of Forces for the Determination of Passive Pressure P_p

Fig. 11.23b gives a section through the plane contact face AB of a rigid retaining wall which rotates about point A into the backfill of cohesionless soil with a horizontal surface. BD is drawn at an angle $45° - \phi/2$ to the surface. Let O_1 be an arbitrary point selected on the line BD as the center of a logarithmic spiral, and let O_1A be the reference vector r_0. Assume a trial sliding surface Ae_1c_1 which consists of two parts. The first part is the curved part Ae_1 which is the part of the logarithmic

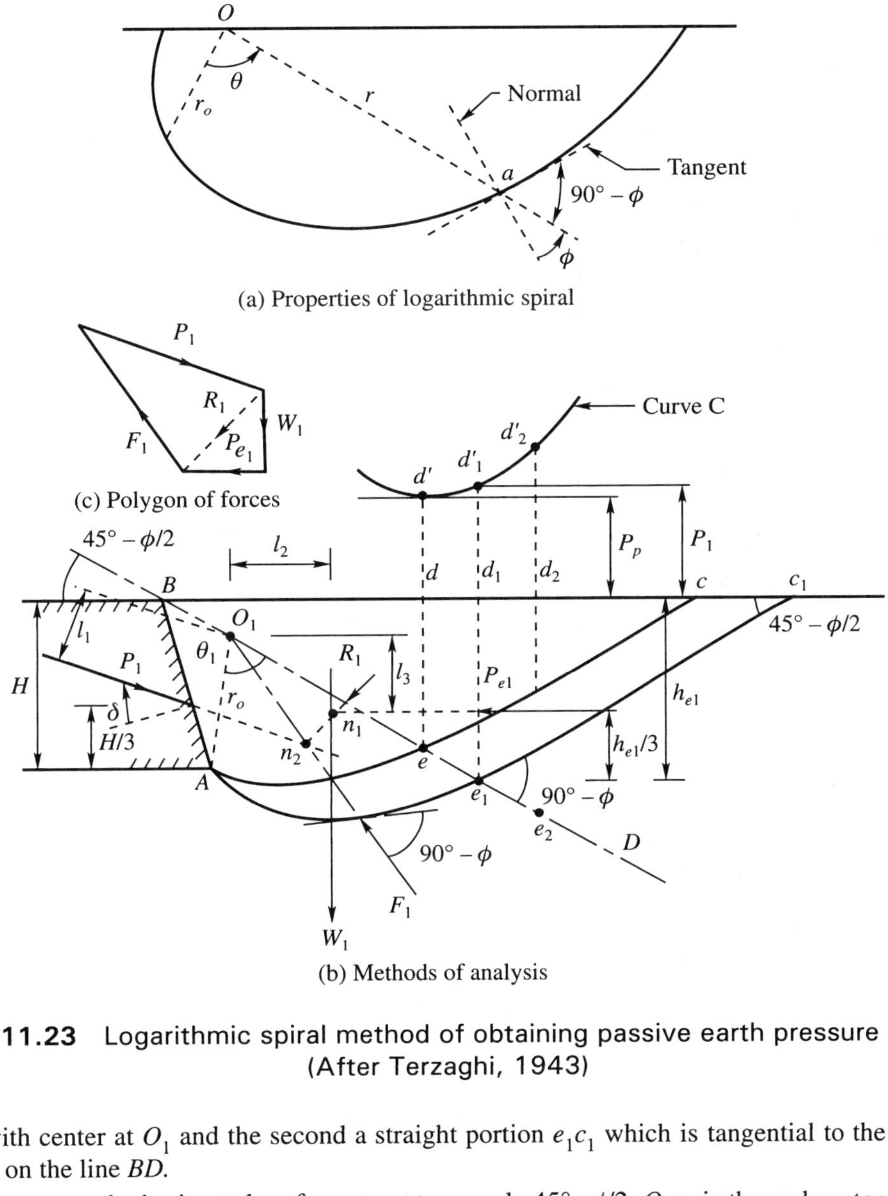

Figure 11.23 Logarithmic spiral method of obtaining passive earth pressure of sand (After Terzaghi, 1943)

spiral with center at O_1 and the second a straight portion e_1c_1 which is tangential to the spiral at point e_1 on the line BD.

e_1c_1 meets the horizontal surface at c_1 at an angle 45°– ϕ/2. O_1e_1 is the end vector r_1 of the spiral which makes an angle θ_1 with the reference vector r_0. Line BD makes an angle 90°– ϕ with line e_1c_1 which satisfies the property of the spiral.

It is now necessary to analyze the forces acting on the soil mass lying above the assumed sliding surface Ae_1c_1.

Within the mass of soil represented by triangle Be_1c_1 the state of stress is the same as that in a semi-infinite mass in a passive Rankine state. The shearing stresses along vertical sections are zero in this triangular zone. Therefore, we can replace the soil mass lying in the zone $e_1d_1c_1$ by a passive earth pressure P_{el} acting on vertical section e_1d_1 at a height $h_{el}/3$ where h_{el} is the height of the vertical section e_1d_1. This pressure is equal to

$$P_{el} = \frac{1}{2}\gamma h_{e1}^2 N_\phi \tag{11.71}$$

where $N_\phi = \tan^2(45° + \phi/2)$

The body of soil mass BAe_1d_1 (Fig. 11.23b) is acted on by the following forces:

1. The weight W_1 of the soil mass acting through the center of gravity of the mass having a lever arm l_2 with respect to O_1, the center of the spiral.
2. The passive earth pressure P_{el} acting on the vertical section e_1d_1 having a lever arm l_3.
3. The passive earth pressure P_1 acting on the surface AB at an angle δ to the normal and at a height $H/3$ above A having a lever arm l_1.
4. The resultant reaction force F_1 on the curved surface Ae_1 and passing through the center O_1.

Determination of the Force P_1 Graphically

The directions of all the forces mentioned above except that of F_1 are known. In order to determine the direction of F_1 combine the weight W_1 and the force P_{el} which gives the resultant R_1 (Fig. 11.23c). This resultant passes through the point of intersection n_1 of W_1 and P_{el} in Fig. 11.23b and intersects force P_1 at point n_2. Equilibrium requires that force F_1 pass through the same point. According to the property of the spiral, it must pass through the same point. According to the property of the spiral, it must pass through the center O_1 of the spiral also. Hence, the direction of F_1 is known and the polygon of forces shown in Fig. 11.23c can be completed. Thus we obtain the intensity of the force P_1 required to produce a slip along surface Ae_1c_1.

Determination of P_1 by Moments

Force P_1 can be calculated by taking moments of all the forces about the center O_1 of the spiral. Equilibrium of the system requires that the sum of the moments of all the forces must be equal to zero. Since the direction of F_1 is now known and since it passes through O_1, it has no moment. The sum of the moments of all the other forces may be written as

$$P_1l_1 + W_1l_2 + P_{e1}l_3 = 0 \tag{11.72}$$

Therefore,
$$P_1 = -\frac{1}{l_1}\left(W_1l_2 + P_{e1}l_3\right) \tag{11.73}$$

P_1 is thus obtained for an assumed failure surface Ae_1c_1. The next step consists in repeating the investigation for more trial surfaces passing through A which intersect line BD at points e_2, e_3 etc. The values of P_1, P_2, P_3 etc so obtained may be plotted as ordinates $d_1\,d_1'$, $d_2\,d_2'$ etc., as shown in Fig. 11.23b and a smooth curve C is obtained by joining points d_1', d_2' etc. Slip occurs along the surface corresponding to the minimum value P_p which is represented by the ordinate dd'. The corresponding failure surface is shown as Aec in Fig. 11.23b.

11.13 COEFFICIENTS OF PASSIVE EARTH PRESSURE TABLES AND GRAPHS

Concept of Coulomb's Formula

Coulomb (1776) computed the passive earth pressure of ideal sand on the simplifying assumption that the entire surface of sliding consists of a plane through the lower edge A of contact face AB as shown in Fig. 11.24a. Line AC represents an arbitrary plane section through this lower edge. The forces acting on this wedge and the polygon of forces are shown in the figure. The basic equation for computing the passive earth pressure coefficient may be developed as follows:

Consider a point on pressure surface AB at a depth z from point B (Fig 11.24a). The normal component of the earth pressure per unit area of surface AB may be expressed by the equation,

$$P_{pn} = \gamma z K_P \tag{11.74}$$

where K_P is the coefficient of passive earth pressure. The total passive earth pressure normal to surface AB, P_{pn}, is obtained from Eq. (11.74) as follows,

$$P_{pn} = \int_0^H \frac{P_{pn}}{\sin\alpha}\, dz = \frac{\gamma K_p}{\sin\alpha}\int_0^H z dz$$

$$P_{pn} = \frac{1}{2}\gamma H^2 \frac{K_P}{\sin\alpha} \tag{11.75}$$

where α is the angle made by pressure surface AB with the horizontal.

Since the resultant passive earth pressure P_p acts at an angle δ to the normal,

$$P_p = \frac{P_{pn}}{\cos\delta} = \frac{1}{2}\gamma H^2 \frac{K_P}{\sin\alpha\cos\delta} \tag{11.76}$$

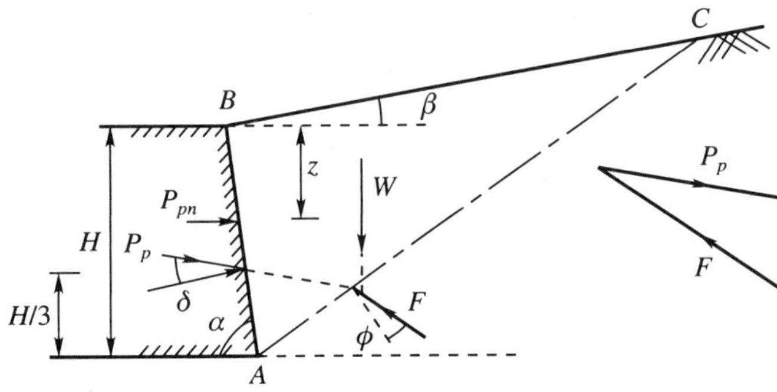

(a) Principles of Coulomb's Theory of passive earth pressure of sand

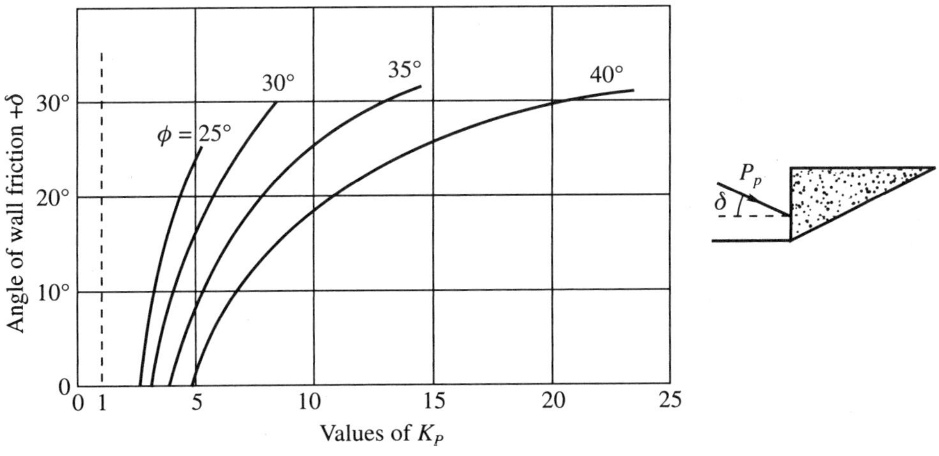

(b) Coefficient of passive earth pressure K_P

Figure 11.24 Diagram illustrating passive earth pressure theory of sand and relation between ϕ, δ and K_P (After Terzaghi, 1943)

Table 11.3 Passive earth pressure coefficient K'_p for curved surfaces of failure (After Caquot and Kerisel 1948).

ϕ =	10°	15°	20°	25°	30°	35°	40°
$\delta = 0$	1.42	1.70	2.04	2.56	3.0	3.70	4.6
$\delta = \phi/2$	1.56	1.98	2.59	3.46	4.78	6.88	10.38
$\delta = \phi$	1.65	2.19	3.01	4.29	6.42	10.20	17.50
$\delta = -\phi/2$	0.73	0.64	0.58	0.55	0.53	0.53	0.53

Eq. (11.76) may also be expressed as

$$P_p = \frac{1}{2}\gamma H^2 K'_p \tag{11.77}$$

where $$K'_p = \frac{K_P}{\sin\alpha\cos\delta} \tag{11.78}$$

Passive Earth Pressure Coefficient

Coulomb developed an analytical solution for determining K_P based on a plane surface of failure and this is given in Eq. (11.57). Figure 11.24(b) gives curves for obtaining Coulomb's values of K_P for various values of δ and ϕ for plane surfaces of failure with a horizontal backfill. They indicate that for a given value of ϕ the value of K_P increases rapidly with increasing values of δ. The limitations of plane surfaces of failure are given in Section 11.9. Curved surfaces of failure are normally used for computing P_p or K_P when the angle of wall friction δ exceeds $\phi/3$. Experience indicates that the curved surface of failure may be taken either as a part of a logarithmic spiral or a circular arc. Caquot and Kerisel (1948) computed K'_P by making use of curved surfaces of failure for various values of ϕ, δ, θ and β. Caquot and Kerisel's calculations for determining K'_P for curved surfaces of failure are available in the form of graphs.

Table 11.3 gives the values of K'_P for various values of ϕ and δ for a vertical wall with a horizontal backfill (after Caquot and Kerisel, 1948).

In the vast majority of practical cases the angle of wall friction has a positive sign, that is, the wall transmits to a soil a downward shearing force. The negative angle of wall friction might develop in the case of positive batter piles subjected to lateral loads, and also in the case of pier foundations for bridges subjected to lateral loads.

Example 11.15

A gravity retaining wall is 10 ft high with sand backfill. The backface of the wall is vertical. Given $\delta = 20°$, and $\phi = 40°$, determine the total passive thrust using Eq. (11.76) and Fig. 11.24 for a plane failure. What is the passive thrust for a curved surface of failure? Assume $\gamma = 18.5$ kN/m^3.

Solution

From Eq. (11.76)

$$P_p = \frac{1}{2}\gamma H^2 \frac{K_p}{\sin\alpha\cos\delta} \quad \text{where } \alpha = 90°$$

From Fig. 11.24 (b) for $\delta = 20°$, and $\phi = 40°$, we have $K_P = 11$

$$P_p = \frac{1}{2} \times 18.5 \times 10^2 \frac{11}{\sin 90 \cos 20^\circ} = 10{,}828 \text{ kN/m}$$

From Table 11.3 K'_p for a curved surface of failure (Caquot and Kerisel. 1948) for $\phi = 40°$ and $\delta = 20°$ is 10.38.

From Eq. (11.77)

$$P_p = \frac{1}{2}\gamma H^2 K'_p = \frac{1}{2} \times 18.5 \times 10^2 \times 10.38$$

$$= 9602 \text{kN/m}$$

Comments

For $\delta = \phi/2$, the reduction in the passive earth pressure due to a curved surface of failure is

$$\text{Reduction} = \frac{10{,}828 - 9602}{10{,}828} \times 100 = 11.32\%$$

Example 11.16

For the data given in Example 11.15, determine the reduction in passive earth pressure for a curved surface of failure if $\delta = 30°$.

Solution

For a plane surface of failure P_p from Eq. (11.76) is

$$P_p = \frac{1}{2} \times 18.5 \times 10^2 \times \frac{21}{\sin 90^\circ \cos 30^\circ} = 22{,}431 \text{ kN/m}$$

where, $K_p = 21$ from Fig. 11.24 for $\delta = 30°$ and $\phi = 40°$
From Table 11.3 for $\delta = 30°$ and $\phi = 40°$

$$K'_p = \frac{10.38 + 17.50}{2} = 13.94$$

From Eq (11.77)

$$P_p = \frac{1}{2} \times 18.5 \times 10^2 \times 13.94 = 12{,}895 \text{ kN/m}$$

$$\text{Reduction in passive pressure} = \frac{22{,}431 - 12{,}895}{22{,}431} = 42.5\%$$

It is clear from the above calculations, that the soil resistance under a passive state gives highly erroneous values for plane surfaces of failure with an increase in the value of δ. This error could lead to an unsafe condition because the computed values of P_p would become higher than the actual soil resistance.

11.14 LATERAL EARTH PRESSURE ON RETAINING WALLS DURING EARTHQUAKES

Ground motions during an earthquake tend to increase the earth pressure above the static earth pressure. Retaining walls with horizontal backfills designed with a factor of safety of 1.5 for static loading are expected to withstand horizontal accelerations up to 0.2g. For larger accelerations, and for walls with sloping backfill, additional allowances should be made for the earthquake forces. Murphy (1960) shows that when subjected to a horizontal acceleration at the base, failure occurs in the soil mass along a plane inclined at 35° from the horizontal . The analysis of Mononobe (1929) considers a soil wedge subjected to vertical and horizontal accelerations to behave as a rigid body sliding over a plane slip surface.

The current practice for earthquake design of retaining walls is generally based on design rules suggested by Seed and Whitman (1970). Richards et al. (1979) discuss the design and behavior of gravity retaining walls with unsaturated cohesionless backfill. Most of the papers make use of the popular Mononobe-Okabe equations as a starting point for their own analysis. They follow generally the pseudoplastic approach for solving the problem. Solutions are available for both the active and passive cases with as granular backfill materials. Though solutions for (c-ϕ) soils have been presented by some investigators (Prakash and Saran, 1966, Saran and Prakash, 1968), their findings have not yet been confirmed, and as such the solutions for (c-ϕ) soils have not been taken up in this chapter.

Earthquake Effect on Active Pressure with Granular Backfill

The Mononobe-Okabe method (1929, 1926) for dynamic lateral pressure on retaining walls is a straight forward extension of the Coulomb sliding wedge theory. The forces that act on a wedge under the active state are shown in Fig. 11.25

In Fig. 11.25 *AC* in the sliding surface of failure of wedge *ABC* having a weight *W* with inertial components $k_v W$ and $k_h W$. The equation for the total active thrust P_{ae} acting on the wall *AB* under dynamic force conditions as per the analysis of Mononobe-Okabe is

$$P_{ae} = \frac{1}{2}\gamma H^2 (1 - k_v) K_{Ae} \tag{11.79}$$

in which

$$K_{Ae} = \frac{\cos^2(\phi - \eta - \theta)}{\cos\eta \cos^2\theta \cos(\delta + \theta + \eta)\left[1 + \sqrt{\dfrac{\sin(\phi + \delta)\sin(\phi - \eta - \beta)}{\cos(\delta + \theta + \eta)\cos(\beta - \theta)}}\right]^2} \tag{11.80}$$

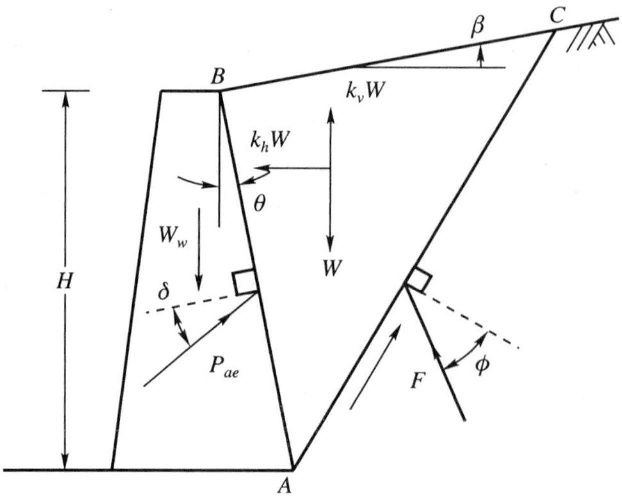

Figure 11.25 Active force on a retaining wall with earthquake forces

where $\overline{P}_{ae}$ =dynamic component of the total earth pressure P_{ae} or $P_{ae} = P_a + \overline{P}_{ae}$

K_{Ae} = the dynamic earth pressure coefficient

$$\eta = \tan^{-1}\left[\frac{k_h}{1-k_v}\right] \tag{11.81}$$

P_a = active earth pressure [Eq. (11.50)]

k_h = (horizontal acceleration)/g

k_v =(vertical acceleration)/g

g = acceleration due to gravity

γ = unit weight of soil

ϕ = angle of friction of soil

δ = angle of wall friction

β = slope of backfill

θ = slope of pressure surface of retaining wall with respect to vertical at point B (Fig. 11.25)

H = height of wall

The total resultant active earth pressure P_{ae} due to an earthquake is expressed as

$$P_{ae} = P_a + \overline{P}_{ae} \tag{11.82}$$

The dynamic component $\overline{P}_{ae}$ is expected to act at a height $0.6H$ above the base whereas the static earth pressure acts at a height $H/3$. For all practical purposes it would be sufficient to assume that the resultant force P_{ae} acts at a height $H/2$ above the base with a uniformly distributed pressure.

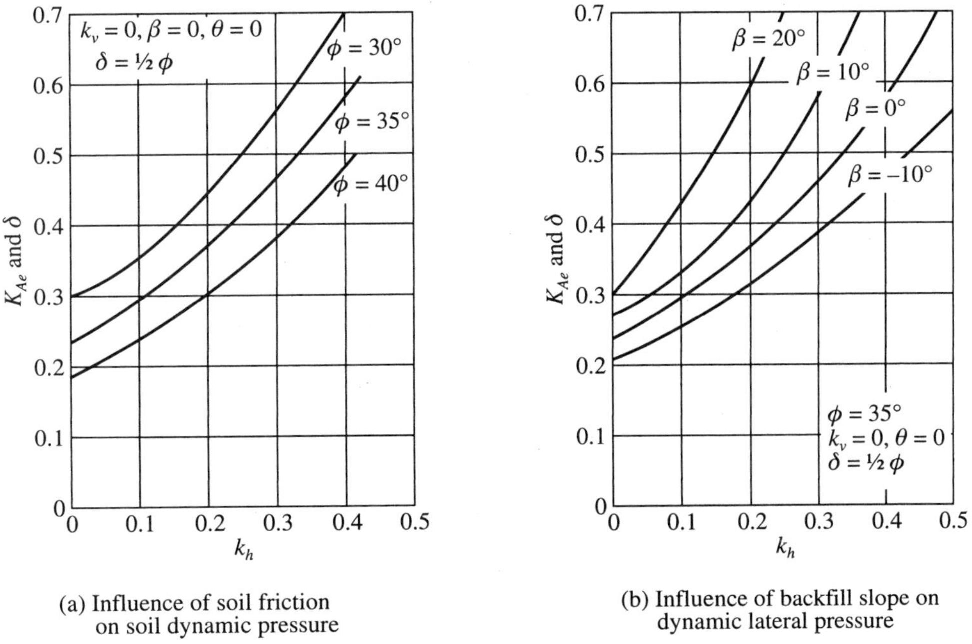

(a) Influence of soil friction on soil dynamic pressure

(b) Influence of backfill slope on dynamic lateral pressure

Figure 11.26 Dynamic lateral active pressure (after Richards et al., 1979)

It has been shown that the active pressure is highly sensitive to both the backfill slope β, and the friction angle ϕ of the soil (Fig. 11.26).

It is necessary to recognize the significance of the expression

$$\sin(\phi - \eta - \beta) \tag{11.83}$$

given under the root sign in Eq. (11.80).

a. When Eq. (11.83) is negative no real solution is possible. Hence for stability, the limiting slope of the backfill must fulfill the condition

$$\beta \le (\varphi - \eta) \tag{11.84a}$$

b. For no earthquake condition, $\eta = 0$. Therefore for stability we have

$$\beta \le \varphi \tag{11.85}$$

c. When the backfill is horizontal $\beta = 0$. For stability we have

$$\eta \le \varphi \tag{11.86}$$

d. By combining Eqs. (11.81) and (11.86), we have

$$k_h \le (1 - k_v)\tan\phi \tag{11.87a}$$

From Eq. (11.87a), we can define a critical value for horizontal acceleration k_h^* as

$$k_h^* = (1 - k_v)\tan\phi \tag{11.87b}$$

Values of critical accelerations are given in Fig 11.27 which demonstrates the sensitivity of the various quantities involved.

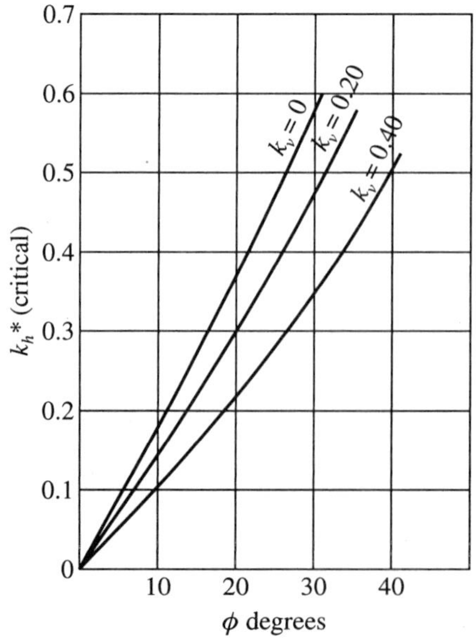

Figure 11.27 Critical values of horizontal accelerations

Effect of Wall Lateral Displacement on the Design of Retaining Wall

It is the usual practice of some designers to ignore the inertia forces of the mass of the gravity retaining wall in seismic design. Richards and Elms (1979) have shown that this approach is unconservative since it is the weight of the wall which provides most of the resistance to lateral movement. Taking into account all the seismic forces acting on the wall and at the base they have developed an expression for the weight of the wall W_w under the equilibrium condition as (for failing by sliding)

$$W_w = \frac{1}{2}\gamma H^2(1-k_v)K_{Ae}C_{IE} \tag{11.88}$$

in which,

$$C_{IE} = \frac{\cos(\delta+\theta)-\sin(\delta+\theta)\tan\delta}{(1-k_v)(\tan\delta-\tan\eta)} \tag{11.89}$$

where W_w = weight of retaining wall (Fig. 11.25)
δ = angle of friction between the wall and soil

Eq. (11.89) is considerably affected by δ. If the wall inertia factor is neglected, a designer will have to go to an exorbitant expense to design gravity walls.

It is clear that tolerable displacement of gravity walls has to be considered in the design. The weight of the retaining wall is therefore required to be determined to limit the displacement to the tolerable limit. The procedure is as follows

1. Set the tolerable displacement Δd
2. Determine the design value of k_h *by* making use of the following equation (Richards et al., 1979)

$$k_h = A_a\left[\frac{0.2\,A_v^2}{A_a(\Delta d)}\right]^{1/4} \tag{11.90}$$

where A_a, A_v = acceleration coefficients used in the Applied Technology Council (ATC) Building Code (1978) for various regions of the United States. Δd is in inches.

3. Using the values of k_h calculated above, and assuming $k_v = 0$, calculate K_{Ae} from Eq (11.80)
4. Using the value of K_{Ae}, calculate the weight, W_w, of the retaining wall by making use of Eqs. (11.88) and (11.89)
5. Apply a suitable factor of safety, say, 1.5 to W_w.

Passive Pressure During Earthquakes

Eq. (11.79) gives an expression for computing seismic active thrust which is based on the well known Mononobe-Okabe analysis for a plane surface of failure. The corresponding expression for passive resistance is

$$P_{pe} = \frac{1}{2}\gamma H^2(1-k_v)K_{Pe} \tag{11.91}$$

$$K_{Pe} = \frac{\cos^2(\phi-\eta+\theta)}{\cos\eta\cos^2\theta\cos(\delta-\theta+\eta)\left[1-\sqrt{\dfrac{\sin(\phi+\delta)\sin(\phi-\eta+\beta)}{\cos(\delta-\theta+\eta)\cos(\beta-\theta)}}\right]^2}$$

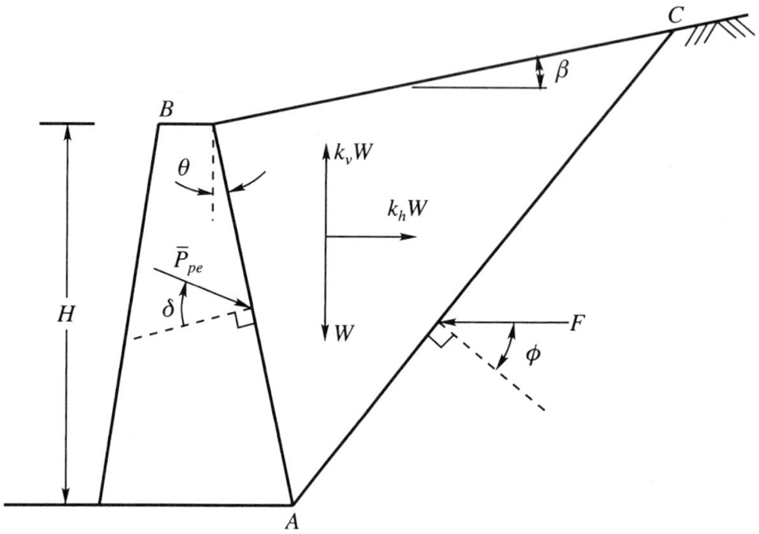

Figure 11.28 Passive pressure on a retaining wall during earthquake

Fig. 11.28 gives the various forces acting on the wall under seismic conditions. All the other notations in Fig. 11.28 are the same as those in Fig. 11.25. The effect of increasing the slope angle β is to increase the passive resistance (Fig. 11.29). The influence of the friction angle of the soil (ϕ) on the passive resistance is illustrated the Fig. 11.30.

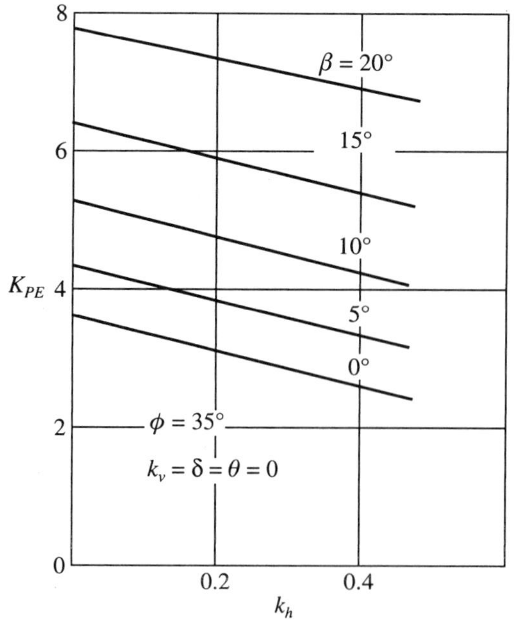

Figure 11.29 Influence of backfill slope angle on passive pressure

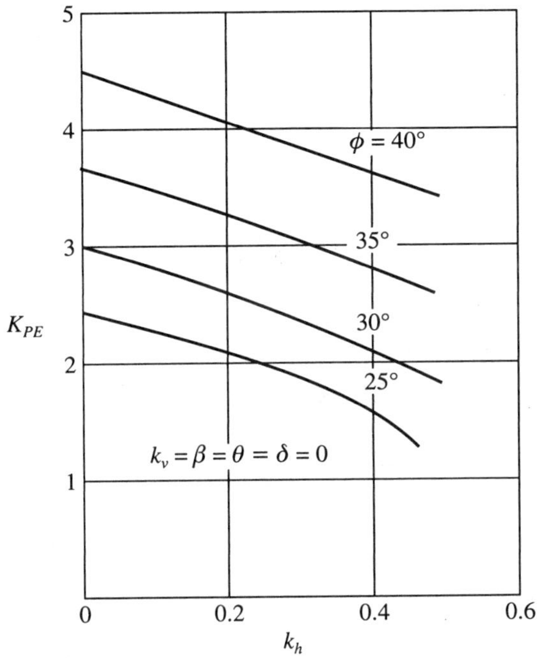

Figure 11.30 Influence of soil friction angle on passive pressure

It has been explained in earlier sections of this chapter that the passive earth pressures calculated on the basis of a plane surface of failure give unsafe results if the magnitude of δ exceeds $\phi/2$. The error occurs because the actual failure plane is curved, with the degree of curvature increasing with an increase in the wall friction angle. The dynamic Mononobe-Okabe solution assumes a linear failure surface, as does the static Coulomb formulation.

In order to set right this anomaly Morrison and Ebelling (1995) assumed the failure surface as an arc of a logarithmic spiral (Fig. 11.31) and calculated the magnitude of the passive pressure under seismic conditions.

It is assumed here that the pressure surface is vertical (θ = 0) and the backfill surface horizontal (β = 0). The following charts have been presented by Morrison and Ebelling on the basis of their analysis.

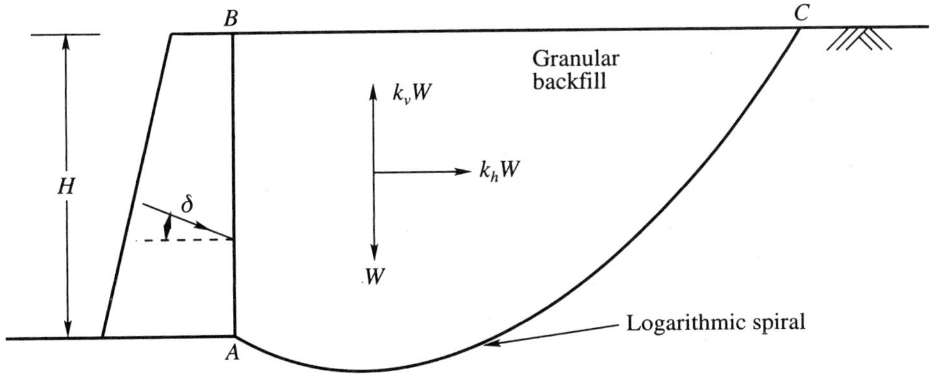

Figure 11.31 Passive pressure from log spiral failure surface during earthquakes

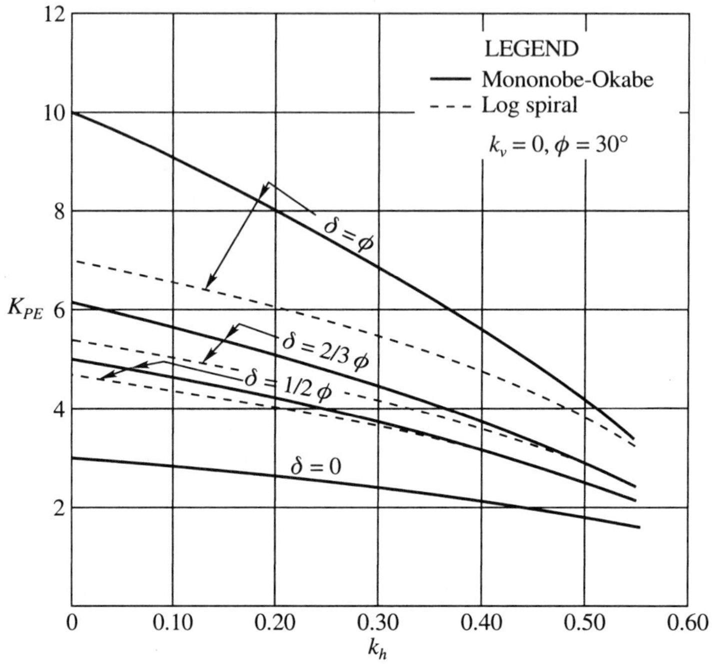

Figure 11.32 K_{Pe} versus k_h, effect of δ

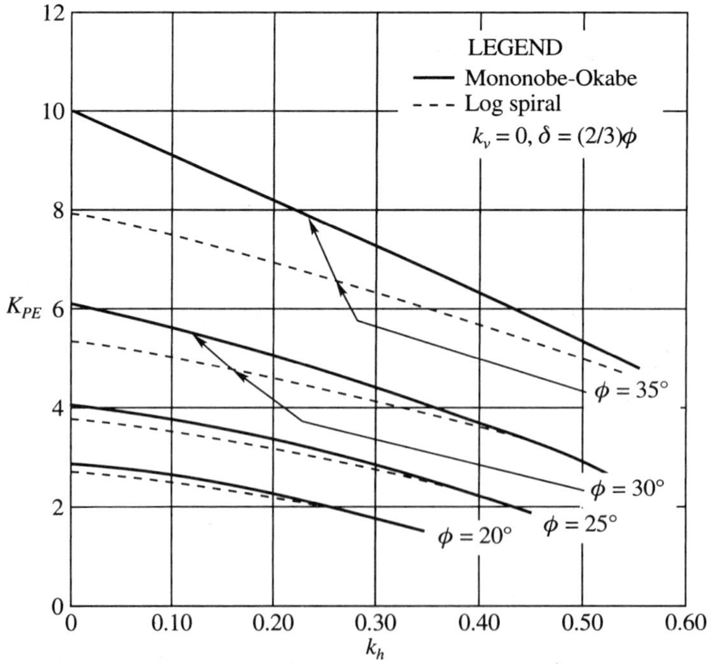

Figure 11.33 K_{Pe} versus k_h, effect of ϕ

1. Fig. 11.32 gives the effect of δ on the plot K_{Pe} versus k_h with $k_v = 0$, for $\phi = 30°$. The values of δ assumed are 0, 1/2 (ϕ) and(2/3ϕ). The plot shows clearly the difference between the Mononobe-Okabe and log spiral values. The difference between the two approaches is greatest at $k_h = 0$

2. Fig. 11.33 shows the effect of ϕ on K_{Pe}. The figure shows the difference between Mononobe-Okabe and log spiral values of K_{pe} versus k_h with $\delta = (2/3\phi)$ and $k_v = 0$. It is also clear from the figure the difference between the two approaches is greatest for $k_h = 0$ and decreases with an increase in the value of k_h.

Example 11.17

A gravity retaining wall is required to be designed for seismic conditions for the active state. The following data are given:

Height of wall = 8 m $\theta = 0°$, $\beta = 0$, $\phi = 30°$, $\delta = 15°$, $k_v = 0$, $k_h = 0.25$ and $\gamma = 19$ kN/m^3. Determine P_{ae} and the approximate point of application. What is the additional active pressure caused by the earthquake?

Solution

From Eq. (11.79)

$$P_{ae} = \frac{1}{2}\gamma H^2(1-k_v)K_{Ae} = \frac{1}{2}\gamma H^2 K_{Ae}, \text{ since } k_v = 0$$

For $\phi = 30°$, $\delta = 15°$ and $k_h = 0.25$, we have from Fig.11.26 a $K_{Ae} = 0.5$. Therefore

$$P_{ae} = \frac{1}{2}19 \times 8^2 \times 0.5 = 304 \text{ kN/m}$$

From Eq. (11.14) $P_a = \frac{1}{2}\gamma H^2 K_A$

where $K_A = \tan^2(45° - \phi/2) = \tan^2 30° = 0.33$

Therefore $P_a = \frac{1}{2} \times 19 \times 8^2 \times 0.33 = 202.7$ kN/m

ΔP_{ae} = the additional pressure due to the earthquake = 304 – 202.7 = 101.3 kN/m

For all practical purposes, the point of application of P_{ae} may be taken as equal to H/2 above the base of the wall or 4 m above the base in this case.

Example 11.18

For the wall given in Example 11.17, determine the total passive pressure P_{pe} under seismic conditions. What is the additional pressure due to the earthquake?

Solution

From Eq. (11.91),

$$P_{ae} = \frac{1}{2}\gamma H^2(1-k_v)K_{pe} = \frac{1}{2}\gamma H^2 K_{pe}, \text{ since } k_v = 0$$

From Fig 11.32, (from M-O curves), $K_{Pe} = 4.25$ for $\phi = 30°$, and $\delta = 15°$

$$\text{Now } P_{pe} = \frac{1}{2}\gamma H^2 K_{pe} = \frac{1}{2} \times 19 \times 8^2 \times 4.25 = 2584 \text{ kN/m}$$

From Eq. (11.15)

$$P_p = \frac{1}{2}\gamma H^2 K_p = \frac{1}{2} \times 19 \times 8^2 \times 3 = 1824 \text{ kN/m}$$

$$\text{where } K_p = \tan^2\left(45^\circ + \frac{30}{2}\right) = \tan^2 60^\circ = 3$$

$$\Delta P_{pe} = \left(P_{pe} - P_{pe}\right) = 2584 - 1824 = 760 \text{ kN/m}$$

11.15 PROBLEMS

11.1 Fig. Prob. 11.1 shows a rigid retaining wall prevented from lateral movements. Determine for this wall the lateral thrust for the at-rest condition and the point of application of the resultant force.

11.2 For Prob 11.1, determine the active earth pressure distribution for the following cases:

(a) when the water table is below the base and $\gamma = 17$ kN/m^3.

(b) when the water table is at 3m below ground level

(c) when the water table is at ground level

11.3 Fig. Prob. 11.3 gives a cantilever retaining wall with a sand backfill. The properties of the sand are:

$e = 0.56$, $\phi = 38°$, and $G_s = 2.65$.

Using Rankine theory, determine the pressure distribution with respect to depth, the magnitude and the point of application of the resultant active pressure with the surcharge load being considered.

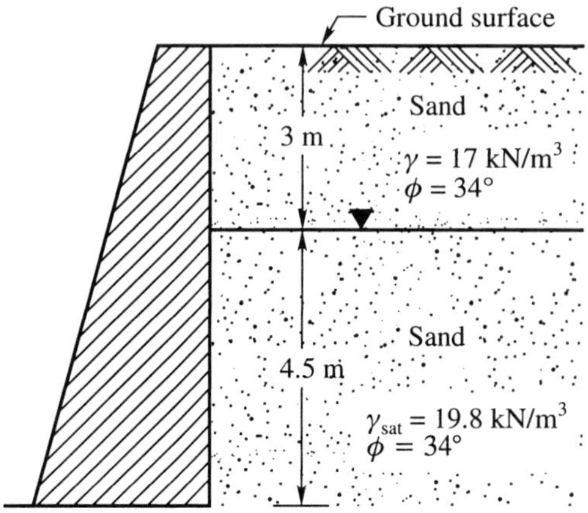

Figure Prob. 11.1

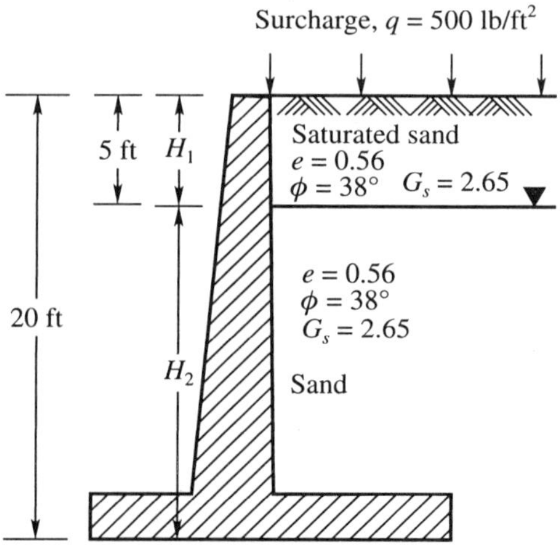

Figure Prob. 11.3

11.4 A smooth vertical wall 3.5 m high retains a mass of dry loose sand. The dry unit weight of the sand is 15.6 kN/m^3 and an angle of internal friction ϕ is 32°. Estimate the total thrust per meter acting against the wall (a) if the wall is prevented from yielding, and (b) if the wall is allowed to yield.

11.5 A wall of 6 m height retains a non-cohesive backfill of dry unit weight 18 kN/m^3 and an angle of internal friction of 30°. Use Rankine's theory and find the total active thrust per meter length of the wall. Estimate the change in the total pressure in the following circumstances:

(i) The top of the backfill carrying a uniformly distributed load of 6 kN/m^2

(ii) The backfill under a submerged condition with the water table at an elevation of 2 m below the top of the wall. Assume $G_s = 2.65$, and the soil above the water table being saturated.

11.6 For the cantilever retaining wall given in Fig. Prob 11.3 with a sand backfill, determine pressure distribution with respect to depth and the resultant thrust. Given:

$H_1 = 3$m, $H_2 = 6$m, $\gamma_{sat} = 19.5$ kN/m^3
$q = 25$ kN/m^2, and $\phi = 36°$
Assume the soil above the GWT is saturated

11.7 A retaining wall of 6 m height having a smooth back retains a backfill made up of two strata shown in Fig. Prob. 11.7. Construct the active earth pressure diagram and find the magnitude and point of application of the resultant thrust. Assume the backfill above WT remains dry.

11.8 (a) Calculate the total active thrust on a vertical wall 5 m high retaining sand of unit weight 17 kN/m^3 for which $\phi = 35°$. The surface is horizontal and the water table is below the bottom of the wall. (b) Determine the thrust on the wall if the water table rises to a level 2 m below the surface of the sand. The saturated unit weight of the sand is 20 kN/m^3.

11.9 Figure Problem 11.9 shows a retaining wall with a sloping backfill. Determine the active earth pressure distribution, the magnitude and the point of application of the resultant by the analytical method.

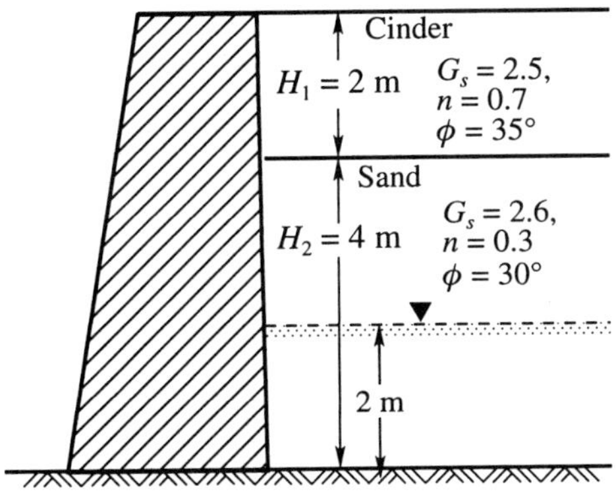

Figure Prob. 11.7

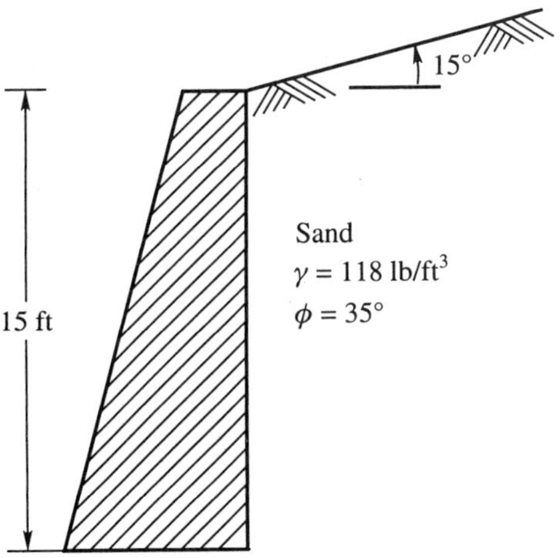

Figure Prob. 11.9

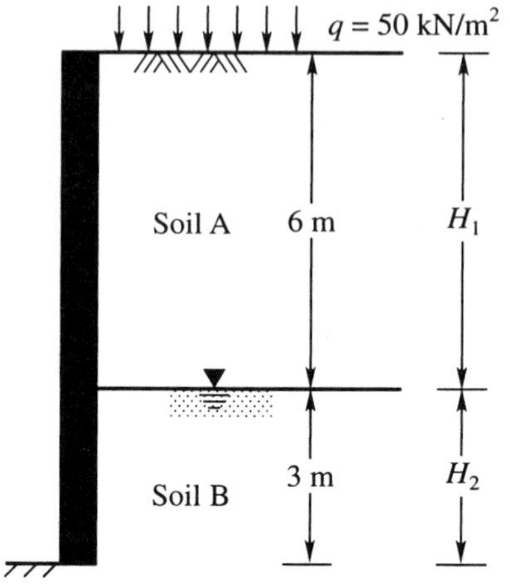

Figure Prob. 11.10

11.10 The soil conditions adjacent to a rigid retaining wall are shown in Fig. Prob. 11.10, A surcharge pressure of 50 kN/m^2 is carried on the surface behind the wall. For soil (A) above the water table, $c' = 0$, $\phi' = 38°$, $\gamma' = 18$ kN/m^3. For soil (B) below the WT, $c' = 10$ kN/m^2, $\phi' = 28°$, and $\gamma_{sat} = 20$ kN/m^3. Calculate the maximum unit active pressure behind the wall, and the resultant thrust per unit length of the wall.

11.11 For the retaining wall given in Fig. Prob. 11.10, assume the following data:

(a) surcharge load = 1000 lb/ft^2, and (b) $H_1 = 10$ ft, $H_2 = 20$ ft,

(c) Soil A: $c' = 500$ lb/ft^2, $\phi' = 30°$, $\gamma = 110$ lb/ft^3

(d) Soil B: $c' = 0$, $\phi' = 35°$, $\gamma_{sat} = 120$ lb/ft^3

Required:

(a) The maximum active pressure at the base of the wall.

(b) The resultant thrust per unit length of wall.

11.12 The depths of soil behind and in front of a rigid retaining wall are 25 ft and 10 ft respectively, both the soil surfaces being horizontal (Fig. Prob 11.12). The appropriate

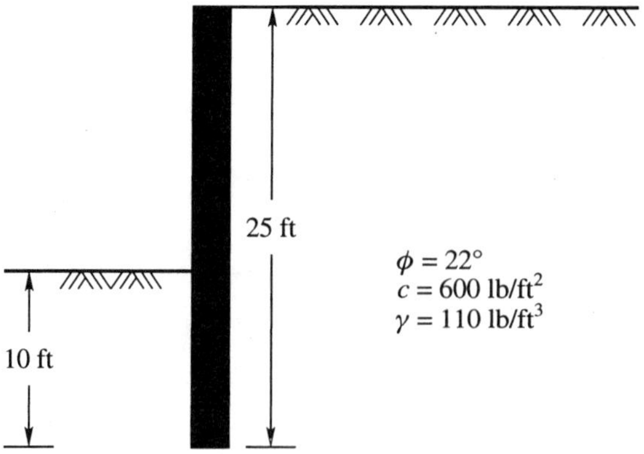

Figure Prob. 11.12

shear strength parameters for the soil are $c = 600\ \text{lb/ft}^2$, and $\phi = 22°$, and the unit weight is $110\ \text{lb/ft}^3$. Using Rankine theory, determine the total active thrust behind the wall and the total passive resistance in front of the wall. Assume the water table is at a great depth.

11.13 For the retaining wall given in Fig. Prob. 11.12, assume the water table is at a depth of 10 ft below the backfill surface. The saturated unit weight of the soil is $120\ \text{lb/ft}^3$. The soil above the GWT is also saturated. Compute the resultant active and passive thrusts per unit length of the wall.

11.14 A retaining wall has a vertical back face and is 8 m high. The backfill has the following properties:

cohesion $c = 15\ \text{kN/m}^2$, $\phi = 25°$, $\gamma = 18.5\ \text{kN/m}^3$

The water table is at great depth. The backfill surface is horizontal. Draw the pressure distribution diagram and determine the magnitude and the point of application of the resultant active thrust.

11.15 For the retaining wall given in Prob. 11.14, the water table is at a depth of 3 m below the backfill surface. Determine the magnitude of the resultant active thrust.

11.16 For the retaining wall given in Prob. 11.15, compute the magnitude of the resultant active thrust, if the backfill surface carries a surcharge load of $30\ \text{kN/m}^2$.

11.17 A smooth retaining wall is 4 m high and supports a cohesive backfill with a unit weight of $17\ \text{kN/m}^3$. The shear strength parameters of the soil are cohesion = 10 kPa and $\phi = 10°$. Calculate the total active thrust acting against the wall and the depth to the point of zero lateral pressure.

11.18 A rigid retaining wall is subjected to passive earth pressure. Determine the passive earth pressure distribution and the magnitude and point of application of the resultant thrust by Rankine theory.

Given: Height of wall = 10 m; depth of water table from ground surface = 3 m; $c = 20\ \text{kN/m}^2$, $\phi = 20°$ and $\gamma_{sat} = 19.5\ \text{kN/m}^3$. The backfill carries a uniform surcharge of $20\ \text{kN/m}^2$.

Assume the soil above the water table is saturated.

11.19 Fig. Prob. 11.19 gives a retaining wall with a vertical back face and a sloping backfill. All the other data are given in the figure. Determine the magnitude and point of application of resultant active thrust by the Culmann method.

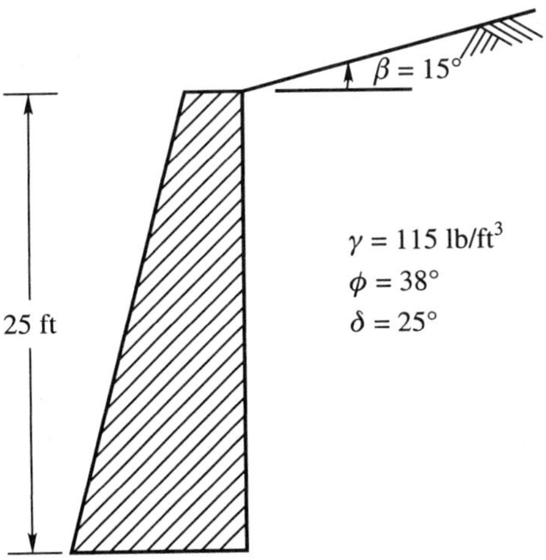

Figure Prob. 11.19

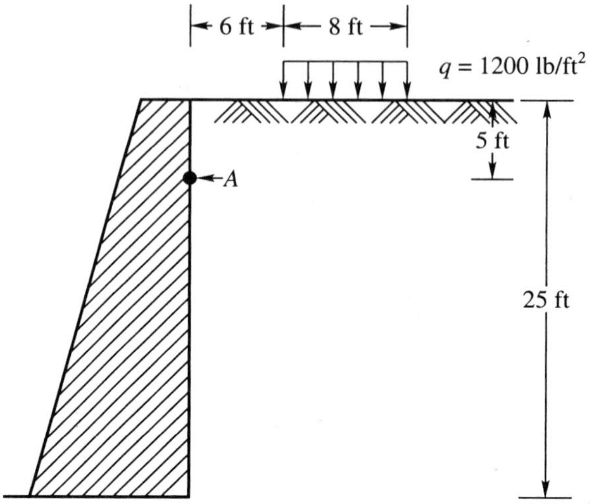

Figure Prob. 11.20

11.20 Fig. Prob. 11.20 gives a rigid retaining wall with a horizontal backfill. The backfill carries a strip load of 1200 lb/ft^2 as shown in the figure. Determine the following:

(a) The unit pressure on the wall at point *A* at a depth of 5 ft below the surface due to the surcharge load.

(b) The total thrust on the wall due to surcharge load.

11.21 A gravity retaining wall with a vertical back face is 10 m high. The following data are given:

$\phi = 25°$, $\delta = 15°$, and $\gamma = 19$ kN/m^3

Determine the total passive thrust using Eq (11.76). What is the total passive thrust for a curved surface of failure?

11.22 A gravity retaining wall is required to be designed for seismic conditions for the active state. The back face is vertical. The following data are given:

Height of wall = 30 ft, backfill surface is horizontal; $\phi = 40°$, $\delta = 20°$, $k_v = 0$, $k_h = 0.3$, $\gamma =$ 120 lb/ft^3.

Determine the total active thrust on the wall. What is the additional lateral pressure due to the earthquake?

11.23 For the wall given in Prob 11.22, determine the total passive thrust during the earthquake What is the change in passive thrust due to the earthquake? Assume $\phi = 30°$ and $\delta = 15°$.

CHAPTER 12

SHALLOW FOUNDATION I: ULTIMATE BEARING CAPACITY

12.1 INTRODUCTION

It is the customary practice to regard a foundation as shallow if the depth of the foundation is less than or equal to the width of the foundation. The different types of footings that we normally come across are given in Fig. 12.1. A foundation is an integral part of a structure. The stability of a structure depends upon the stability of the supporting soil. Two important factors that are to be considered are

1. The foundation must be stable against shear failure of the supporting soil.
2. The foundation must not settle beyond a tolerable limit to avoid damage to the structure.

The other factors that require consideration are the location and depth of the foundation. In deciding the location and depth, one has to consider the erosions due to flowing water, underground defects such as root holes, cavities, unconsolidated fills, ground water level, presence of expansive soils etc.

In selecting a type of foundation, one has to consider the functions of the structure and the load it has to carry, the subsurface condition of the soil, and the cost of the superstructure.

Design loads also play an important part in the selection of the type of foundation. The various loads that are likely to be considered are (i) dead loads, (ii) live loads, (iii) wind and earthquake forces, (iv) lateral pressures exerted by the foundation earth on the embedded structural elements, and (v) the effects of dynamic loads.

In addition to the above loads, the loads that are due to the subsoil conditions are also required to be considered. They are (i) lateral or uplift forces on the foundation elements due to high water table, (ii) swelling pressures on the foundations in expansive soils, (iii) heave pressures

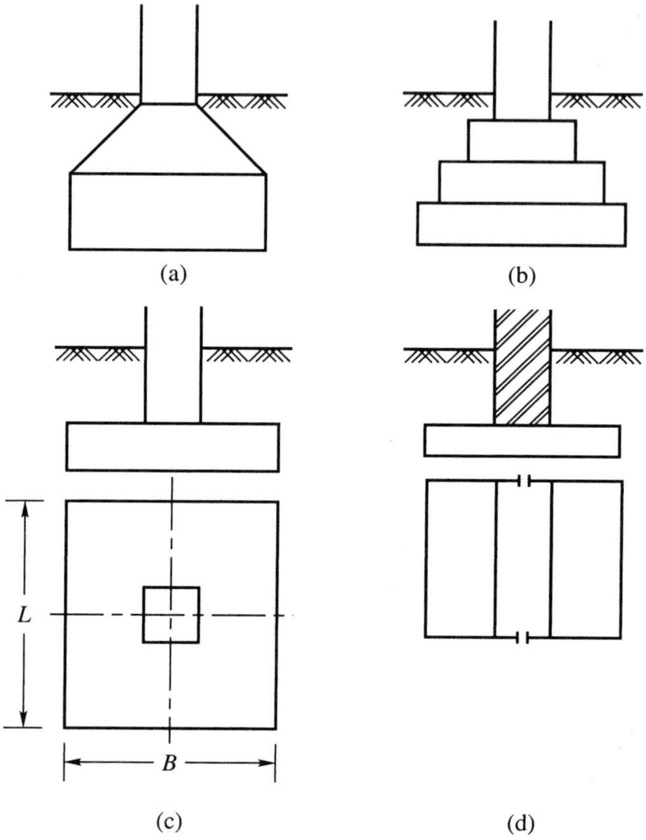

Figure 12.1 Types of shallow foundations: (a) plain concrete foundation, (b) stepped reinforced concrete foundation, (c) reinforced concrete rectangular foundation, and (d) reinforced concrete wall foundation.

on foundations in areas subjected to frost heave and (iv) negative frictional drag on piles where pile foundations are used in highly compressible soils.

Steps for the Selection of the Type of Foundation

In choosing the type of foundation, the design engineer must perform five successive steps.

1. Obtain the required information concerning the nature of the superstructure and the loads to be transmitted to the foundation.
2. Obtain the subsurface soil conditions.
3. Explore the possibility of constructing any one of the types of foundation under the existing conditions by taking into account (i) the bearing capacity of the soil to carry the required load, and (ii) the adverse effects on the structure due to differential settlements. Eliminate in this way, the unsuitable types.
4. Once one or two types of foundation are selected on the basis of preliminary studies, make more detailed studies. These studies may require more accurate determination of loads, subsurface conditions and footing sizes. It may also be necessary to make more refined estimates of settlement in order to predict the behavior of the structure.

5. Estimate the cost of each of the promising types of foundation, and choose the type that represents the most acceptable compromise between performance and cost.

12.2 THE ULTIMATE BEARING CAPACITY OF SOIL

Consider the simplest case of a shallow foundation subjected to a central vertical load. The footing is founded at a depth D_f below the ground surface [Fig. 12.2(a)]. If the settlement, S, of the footing is recorded against the applied load, Q, load-settlement curves, similar in shape to a stress-strain curve, may be obtained as shown in Fig. 12.2(b).

The shape of the curve depends generally on the size and shape of the footing, the composition of the supporting soil, and the character, rate, and frequency of loading. Normally a curve will indicate the ultimate load Q_u that the foundation can support. If the foundation soil is a dense sand or a very stiff clay, the curve passes fairly abruptly to a peak value and then drops down as shown by curve C_1 in Fig. 10.2(b). The peak load Q_u is quite pronounced in this case. On the other hand, if the soil is loose sand or soft clay, the settlement curve continues to descend on a slope as shown by curve C_2 which shows that the compression of soil is continuously taking place without giving a definite value for Q_u. On such a curve, Q_u may be taken at a point beyond which there is a constant rate of penetration.

12.3 SOME OF THE TERMS DEFINED

It will be useful to define, at this stage, some of the terms relating to bearing capacity of foundations (refer to Fig. 12.3).

(a) Total Overburden Pressure q_o

q_o is the intensity of total overburden pressure due to the weight of both soil and water at the base level of the foundation.

$$q_o = \gamma D_{w1} + \gamma_{sat} \overline{D}_w \tag{12.1}$$

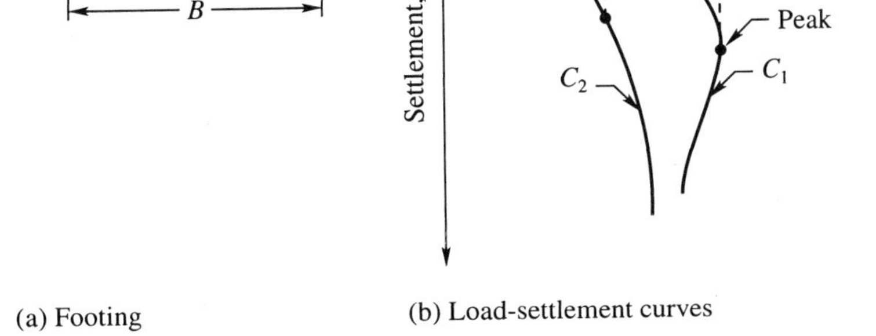

Figure 12.2 Typical load-settlement curves

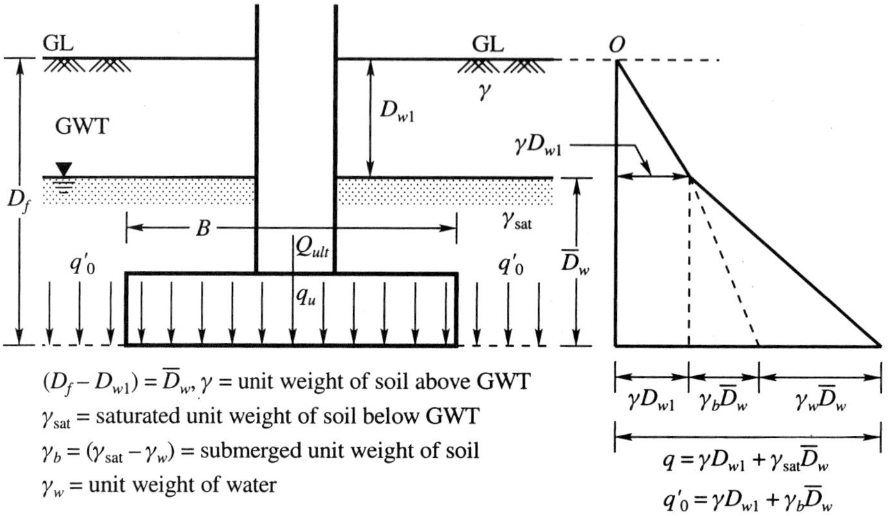

Figure 12.3 Total and effective overburden pressures

(b) Effective Overburden Pressure q'_0

q'_0 is the effective overburden pressure at the base level of the foundation.

$$q'_0 = \gamma D_{w1} + \gamma_b \overline{D}_w \tag{12.2}$$

when $\overline{D}_w = 0$, $q'_0 = \gamma D_{w1} = \gamma D_f$.

(c) The Ultimate Bearing Capacity of Soil, q_u

q_u is the maximum bearing capacity of soil at which the soil fails by shear.

(d) The Net Ultimate Bearing Capacity, q_{nu}

q_{nu} is the bearing capacity in excess of the effective overburden pressure q'_0, expressed as

$$q_{nu} = q_u - q'_0 \tag{12.3}$$

(e) Gross Allowable Bearing Pressure, q_a

q_a is expressed as

$$q_a = \frac{q_u}{F_s} \tag{12.4}$$

where F_s = factor of safety.

(f) Net Allowable Bearing Pressure, q_{na}

q_{na} is expressed as

$$q_{na} = \frac{q_u - \gamma D_f}{F_s} = \frac{q_{nu}}{F_s} \tag{12.5}$$

(g) Safe Bearing Pressure, q_s

q_s is defined as the net safe bearing pressure which produces a settlement of the foundation which does not exceed a permissible limit.

Note: In the design of foundations, one has to use the least of the two values of q_{na} and q_s.

12.4 TYPES OF FAILURE IN SOIL

Experimental investigations have indicated that foundations on dense sand with relative density greater than 70 percent fail suddenly with pronounced peak resistance when the settlement reaches about 7 percent of the foundation width. The failure is accompanied by the appearance of failure surfaces and by considerable bulging of a sheared mass of sand as shown in Fig. 12.4(a). This type of failure is designated as general shear failure by Terzaghi (1943). Foundations on sand of relative density lying between 35 and 70 percent do not show a sudden failure. As the settlement exceeds about 8 percent of the foundation width, bulging of sand starts at the surface. At settlements of about 15 percent of foundation width, a visible boundary of sheared zones at the surface appears. However, the peak of base resistance may never be reached. This type of failure is termed local shear failure, Fig. 12.4(b), by Terzaghi (1943).

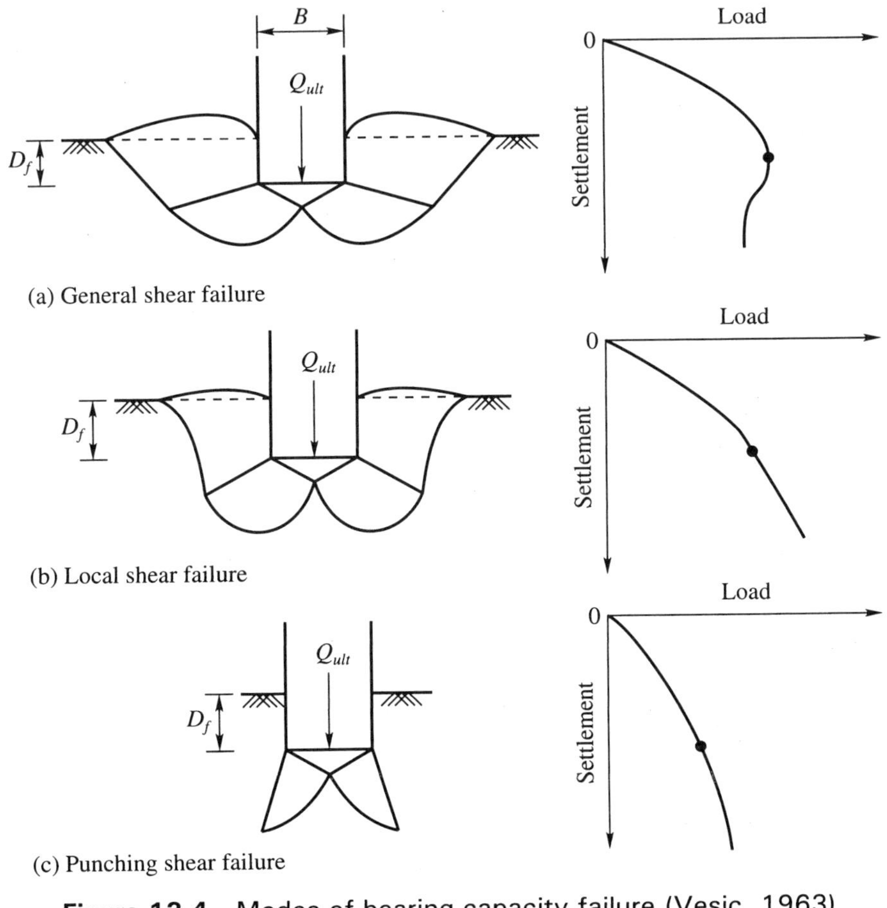

Figure 12.4 Modes of bearing capacity failure (Vesic, 1963)

Foundations on relatively loose sand with relative density less than 35 percent penetrate into the soil without any bulging of the sand surface. The base resistance gradually increases as settlement progresses. The rate of settlement, however, increases and reaches a maximum at a settlement of about 15 to 20 percent of the foundation width. Sudden jerks or shears can be observed as soon as the settlement reaches about 6 to 8 percent of the foundation width. The failure surface, which is vertical or slightly inclined and follows the perimeter of the base, never reaches the sand surface. This type of failure is designated as punching shear failure by Vesic (1963) as shown in Fig. 12.4(c).

The three types of failure described above were observed by Vesic (1963) during tests on model footings. It may be noted here that as the relative depth/width ratio increases, the limiting relative densities at which failure types change increase. The approximate limits of types of failure to be affected as relative depth D_f/B, and relative density of sand, D_r, vary are shown in Fig. 12.5 (Vesic, 1963). The same figure shows that there is a critical relative depth below which only punching shear failure occurs. For circular foundations, this critical relative depth, D_f/B, is around 4 and for long rectangular foundations around 8.

The surfaces of failures as observed by Vesic are for concentric vertical loads. Any small amount of eccentricity in the load application changes the modes of failure and the foundation tilts in the direction of eccentricity. Tilting nearly always occurs in cases of foundation failures because of the inevitable variation in the shear strength and compressibility of the soil from one point to another and causes greater yielding on one side or another of the foundation. This throws the center of gravity of the load towards the side where yielding has occurred, thus increasing the intensity of pressure on this side followed by further tilting.

A footing founded on precompressed clays or saturated normally consolidated clays will fail in general shear if it is loaded so that no volume change can take place and fails by punching shear if the footing is founded on soft clays.

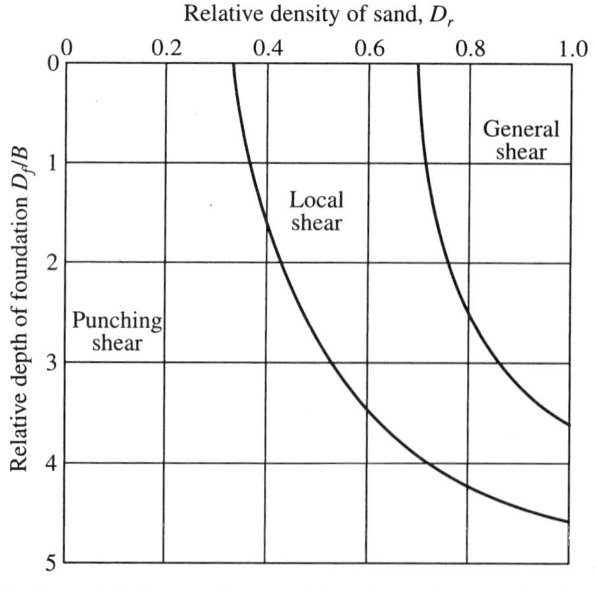

Figure 12.5 Modes of failure of model footings in sand (after Vesic, 1963)

12.5 AN OVERVIEW OF BEARING CAPACITY THEORIES

The determination of bearing capacity of soil based on the classical earth pressure theory of Rankine (1857) began with Pauker, a Russian military engineer (1889), and was modified by Bell (1915). Pauker's theory was applicable only for sandy soils but the theory of Bell took into account cohesion also. Neither theory took into account the width of the foundation. Subsequent developments led to the modification of Bell's theory to include width of footing also.

The methods of calculating the ultimate bearing capacity of shallow strip footings by plastic theory developed considerably over the years since Terzaghi (1943) first proposed a method by taking into account the weight of soil by the principle of superposition. Terzaghi extended the theory of Prandtl (1921). Prandtl developed an equation based on his study of the penetration of a long hard metal punch into softer materials for computing the ultimate bearing capacity. He assumed the material was weightless possessing only cohesion and friction. Taylor (1948) extended the equation of Prandtl by taking into account the surcharge effect of the overburden soil at the foundation level.

No exact analytical solution for computing bearing capacity of footings is available at present because the basic system of equations describing the yield problems is nonlinear. On account of these reasons, Terzaghi (1943) first proposed a semi-empirical equation for computing the ultimate bearing capacity of strip footings by taking into account cohesion, friction and weight of soil, and replacing the overburden pressure with an equivalent surcharge load at the base level of the foundation. This method was for the general shear failure condition and the principle of superposition was adopted. His work was an extension of the work of Prandtl (1921). The final form of the equation proposed by Terzaghi is the same as the one given by Prandtl.

Subsequent to the work by Terzaghi, many investigators became interested in this problem and presented their own solutions. However the form of the equation presented by all these investigators remained the same as that of Terzaghi, but their methods of determining the bearing capacity factors were different.

Of importance in determining the bearing capacity of strip footings is the assumption of plane strain inherent in the solutions of strip footings. The angle of internal friction as determined under an axially symmetric triaxial compression stress state, ϕ_t, is known to be several degrees less than that determined under plane strain conditions under low confining pressures. Thus the bearing capacity of a strip footing calculated by the generally accepted formulas, using ϕ_t, is usually less than the actual bearing capacity as determined by the plane strain footing tests which leads to a conclusion that the bearing capacity formulas are conservative.

The ultimate bearing capacity, or the allowable soil pressure, can be calculated either from bearing capacity theories or from some of the *in situ* tests. Each theory has its own good and bad points. Some of the theories are of academic interest only. However, it is the purpose of the author to present here only such theories which are of basic interest to students in particular and professional engineers in general. The application of field tests for determining bearing capacity are also presented which are of particular importance to professional engineers since present practice is to rely more on field tests for determining the bearing capacity or allowable bearing pressure of soil.

Some of the methods that are discussed in this chapter are

1. Terzaghi's bearing capacity theory
2. The general bearing capacity equation
3. Field tests

12.6 TERZAGHI'S BEARING CAPACITY THEORY

Terzaghi (1943) used the same form of equation as proposed by Prandtl (1921) and extended his theory to take into account the weight of soil and the effect of soil above the base of the foundation on the bearing capacity of soil. Terzaghi made the following assumptions for developing an equation for determining q_u for a c-ϕ soil.

(1) The soil is semi-infinite, homogeneous and isotropic, (2) the problem is two-dimensional, (3) the base of the footing is rough, (4) the failure is by general shear, (5) the load is vertical and symmetrical, (6) the ground surface is horizontal, (7) the overburden pressure at foundation level is equivalent to a surcharge load $q'_0 = \gamma D_f$, where γ is the effective unit weight of soil, and D_f, the depth of foundation less than the width B of the foundation, (8) the principle of superposition is valid, and (9) Coulomb's law is strictly valid, that is, $\sigma = c + \sigma \tan\phi$.

Mechanism of Failure

The shapes of the failure surfaces under ultimate loading conditions are given in Fig. 12.6. The zones of plastic equilibrium represented in this figure by the area *gedcf* may be subdivided into

1. Zone I of elastic equilibrium
2. Zones II of radial shear state
3. Zones III of Rankine passive state

When load q_u per unit area acting on the base of the footing of width B with a rough base is transmitted into the soil, the tendency of the soil located within zone I is to spread but this is counteracted by friction and adhesion between the soil and the base of the footing. Due to the existence of this resistance against lateral spreading, the soil located immediately beneath the base remains permanently in a state of elastic equilibrium, and the soil located within this central Zone I behaves as if it were a part of the footing and sinks with the footing under the superimposed load. The depth of this wedge shaped body of soil *abc* remains practically unchanged, yet the footing sinks. This process is only conceivable if the soil located just below point c moves vertically downwards. This type of movement requires that the surface of sliding *cd* (Fig. 12.6) through point c should start from a vertical tangent. The boundary *bc* of the zone of radial shear *bcd* (Zone II) is also the surface of sliding. As per the theory of plasticity, the potential surfaces of sliding in an ideal plastic material intersect each other in every point of the zone of plastic equilibrium at an angle $(90° - \phi)$. Therefore the boundary *bc* must rise at an angle ϕ to the horizontal provided the friction and adhesion between the soil and the base of the footing suffice to prevent a sliding motion at the base.

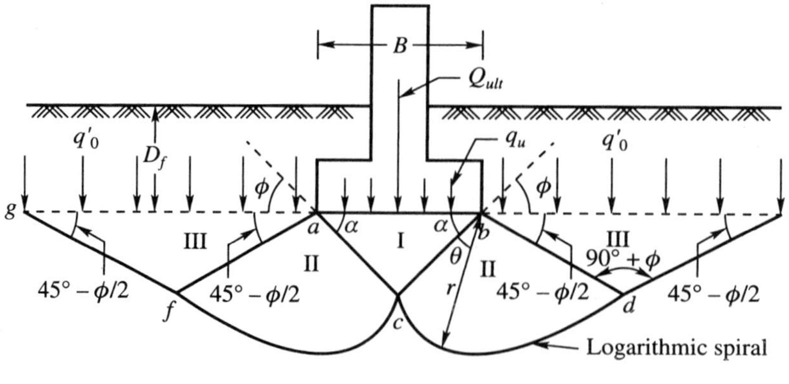

Figure 12.6 General shear failure surface as assumed by Terzaghi for a strip footing

The sinking of Zone I creates two zones of plastic equilibrium, II and III, on either side of the footing. Zone II is the radial shear zone whose remote boundaries *bd* and *af* meet the horizontal surface at angles (45° – ϕ/2), whereas Zone III is a passive Rankine zone. The boundaries *de* and *fg* of these zones are straight lines and they meet the surface at angles of (45° – ϕ/2). The curved parts *cd* and *cf* in Zone II are parts of logarithmic spirals whose centers are located at *b* and *a* respectively.

Ultimate Bearing Capacity of Soil

Strip Footings

Terzaghi developed his bearing capacity equation for strip footings by analyzing the forces acting on the wedge *abc* in Fig. 12.6. The equation for the ultimate bearing capacity q_u is

$$q_u = \frac{Q_{ult}}{B} = cN_c + \gamma D_f N_q + \frac{1}{2}\gamma B N_\gamma \tag{12.6}$$

where Q_{ult} = ultimate load per unit length of footing, c = unit cohesion, γ the effective unit weight of soil, B = width of footing, D_f = depth of foundation, N_c, N_q and N_γ are the bearing capacity factors. They are functions of the angle of friction, ϕ.

The bearing capacity factors are expressed by the following equations

$$\left.\begin{aligned} N_c &= (N_q - 1)\cot\phi \\ N_q &= \frac{a_\theta^2}{2\cos^2(45^\circ + \phi/2)} \\ \text{where}\quad a_\theta &= e^{\eta\tan\phi}, \eta = (0.75\pi - \phi/2) \\ N_\gamma &= \frac{1}{2}\tan\phi\left(\frac{K_{P\gamma}}{\cos^2\phi} - 1\right) \end{aligned}\right\} \tag{12.7}$$

where $K_{P\gamma}$ = passive earth pressure coefficient.

Table 12.1 gives the values of N_c, N_q and N_γ for various values of ϕ and Fig. 12.7 gives the same in a graphical form.

Table 12.1 Bearing capacity factors of Terzaghi

$\phi°$	N_c	N_q	$N\gamma$
0	5.7	1.0	0.0
5	7.3	1.6	0.14
10	9.6	2.7	1.2
15	12.9	4.4	1.8
20	17.7	7.4	5.0
25	25.1	12.7	9.7
30	37.2	22.5	19.7
35	57.8	41.4	42.4
40	95.7	81.3	100.4
45	172.3	173.3	360.0
50	347.5	415.1	1072.8

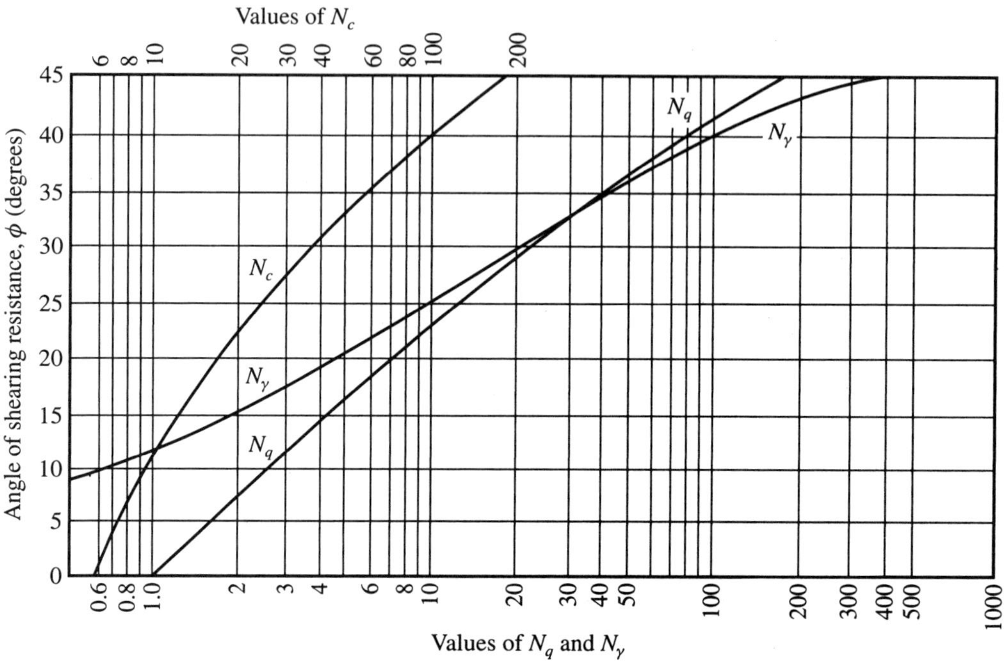

Figure 12.7 Terzaghi's bearing capacity factors for general shear failure

Equations for Square, Circular, and Rectangular Foundations

Terzaghi's bearing capacity Eq. (12.6) has been modified for other types of foundations by introducing the shape factors. The equations are

Square Foundations

$$q_u = 1.3cN_c + \gamma D_f N_q + 0.4\gamma B N_\gamma \tag{12.8}$$

Circular Foundations

$$q_u = 1.3cN_c + \gamma D_f N_q + 0.3\gamma B N_\gamma \tag{12.9}$$

Rectangular Foundations

$$q_u = cN_c\left(1 + 0.3 \times \frac{B}{L}\right) + \gamma D_f N_q + \frac{1}{2}\gamma B N_\gamma \left(1 - 0.2 \times \frac{B}{L}\right) \tag{12.10}$$

where B = width or diameter, L = length of footing.

Ultimate Bearing Capacity for Local Shear Failure

The reasons as to why a soil fails under local shear have been explained under Section 12.4. When a soil fails by local shear, the actual shear parameters c and ϕ are to be reduced as per Terzaghi (1943). The lower limiting values of c and ϕ are

$\bar{c} = 0.67c$

and $\tan \bar{\phi} = 0.67 \tan \phi$ or $\bar{\phi} = \tan^{-1}(0.67 \tan \phi)$ (12.11)

The equations for the lower bound values for the various types of footings are as given below.

Strip Foundation

$$q_u = 0.67c\bar{N}_c + \gamma D_f \bar{N}_q + \frac{1}{2}\gamma B\bar{N}_\gamma \quad (12.12)$$

Square Foundation

$$q_u = 0.867c\bar{N}_c + \gamma D_f \bar{N}_q + 0.4\gamma B\bar{N}_\gamma \quad (12.13)$$

Circular Foundation

$$q_u = 0.867c\bar{N}_c + \gamma D_f \bar{N}_q + 0.3\gamma B\bar{N}_\gamma \quad (12.14)$$

Rectangular Foundation

$$q_u = 0.67c\left(1 + 0.3 \times \frac{B}{L}\right)\bar{N}_c + \gamma D_f \bar{N}_q + \frac{1}{2}\gamma B\bar{N}_\gamma\left(1 - 0.2 \times \frac{B}{L}\right) \quad (12.15)$$

where $\bar{N}_c$, $\bar{N}_q$ and $\bar{N}_\gamma$ are the reduced bearing capacity factors for local shear failure. These factors may be obtained either from Table 12.1 or Fig. 12.7 by making use of the friction angle $\bar{\phi}$.

Ultimate Bearing Capacity q_u in Purely Cohesionless and Cohesive Soils Under General Shear Failure

Equations for the various types of footings for $(c - \phi)$ soil under general shear failure have been given earlier. The same equations can be modified to give equations for cohesionless soil (for $c = 0$) and cohesive soils (for $\phi = 0$) as follows.

It may be noted here that for $c = 0$, the value of $N_c = 0$, and for $\phi = 0$, the value of $N_c = 5.7$ for a strip footing and $N_q = 1$.

a) Strip Footing

For $c = 0$, $q_u = \gamma D_f N_q + \frac{1}{2}\gamma B N_\gamma$ (12.16)

For $\phi = 0$, $q_u = 5.7c + \gamma D_f$

b) Square Footing

For $c = 0$, $q_u = \gamma D_f N_q + 0.4\gamma B N_\gamma$ (12.17)

For $\phi = 0$, $q_u = 7.4c + \gamma D_f$

c) Circular Footing

For $c = 0$, $q_u = \gamma D_f N_q + 0.3\gamma B N_\gamma$ (12.18)

For $\phi = 0, \quad q_u = 7.4c + \gamma D_f$

d) Rectangular Footing

$$\text{For } c = 0, q_u = \gamma D_f N_q + \frac{1}{2}\gamma B N_\gamma \left(1 - 0.2 \times \frac{B}{L}\right) \tag{12.19}$$

$$\text{For } \phi = 0, \ q_u = 5.7c\left(1 + 0.3 \times \frac{B}{L}\right) + \gamma D_f$$

Similar types of equations as presented for general shear failure can be developed for local shear failure also.

Transition from Local to General Shear Failure in Sand

As already explained, local shear failure normally occurs in loose and general shear failure occurs in dense sand. There is a transition from local to general shear failure as the state of sand changes

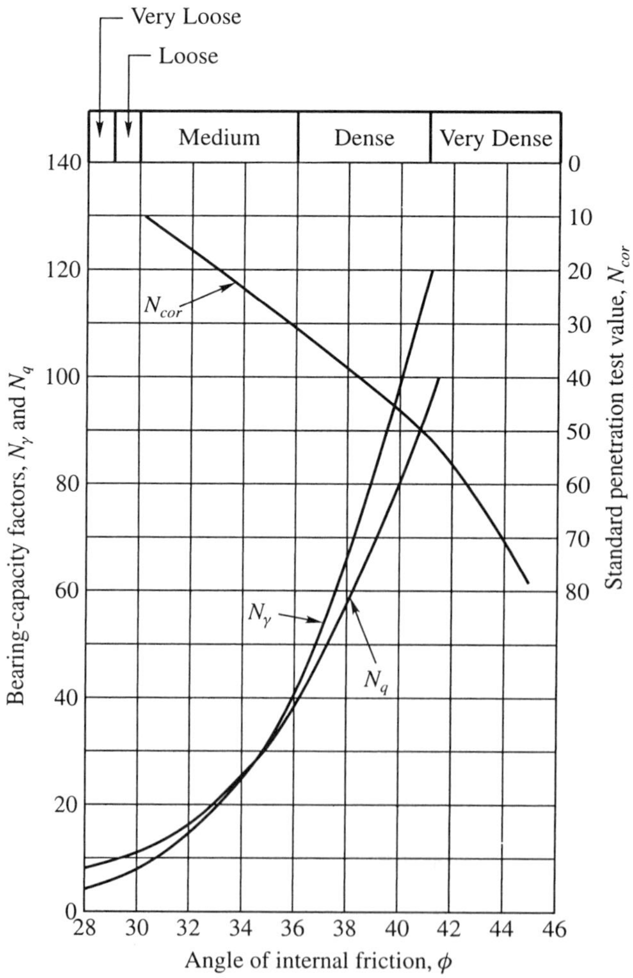

Figure 12.8 Terzaghi's bearing capacity factors which take care of mixed state of local and general shear failures in sand (Peck et al., 1974)

from loose to dense condition. There is no bearing capacity equation to account for this transition from loose to dense state. Peck et al., (1974) have given curves for N_γ and N_q which automatically incorporate allowance for the mixed state of local and general shear failures as shown in Fig. 12.8.

The curves for N_q and N_γ are developed on the following assumptions.

1. Purely local shear failure occurs when $\phi \le 28°$.
2. Purely general shear failure occurs when $\phi \ge 38°$.
3. Smooth transition curves for values of ϕ between 28° and 38° represent the mixed state of local and general shear failures.

N_q and N_γ for values of $\phi \ge 38°$ are as given in Table 12.1. Values of $\overline{N}_q$ and $\overline{N}_\gamma$ for $\phi \le 28°$ may be obtained from Table 12.1 by making use of the relationship $\overline{\phi} = \tan^{-1}(2/3)\tan\phi$.

In the case of purely cohesive soil local shear failure may be assumed to occur in soft to medium stiff clay with an unconfined compressive strength $q_u \le 100$ kPa.

Figure 12.8 also gives the relationship between SPT value N_{cor} and the angle of internal friction ϕ by means of a curve. This curve is useful to obtain the value of ϕ when the SPT value is known.

Net Ultimate Bearing Capacity and Safety Factor

The net ultimate bearing capacity q_{nu} is defined as the pressure at the base level of the foundation in excess of the effective overburden pressure $q'_0 = \gamma D_f$ as defined in Eq. (12.3). The net q_{nu} for a strip footing is

$$q_{nu} = (q_u - \gamma D_f) = cN_c + \gamma D_f(N_q - 1) + \frac{1}{2}\gamma B N_\gamma \tag{12.20}$$

Similar expressions can be written for square, circular, and rectangular foundations and also for local shear failure conditions.

Allowable Bearing Pressure

Per Eq. (12.4), the gross allowable bearing pressure is

$$q_a = \frac{q_u}{F_s} \tag{12.21a}$$

In the same way the net allowable bearing pressure q_{na} is

$$q_{na} = \frac{q_u - \gamma D_f}{F_s} = \frac{q_{nu}}{F_s} \tag{12.21b}$$

where F_s = factor of safety which is normally assumed as equal to 3.

12.7 SKEMPTON'S BEARING CAPACITY FACTOR N_c

For saturated clay soils, Skempton (1951) proposed the following equation for a strip foundation

$$q_u = cN_c + \gamma D_f \tag{12.22a}$$

$$\text{or} \quad q_{nu} = q_u - \gamma D_f = cN_c \tag{12.22b}$$

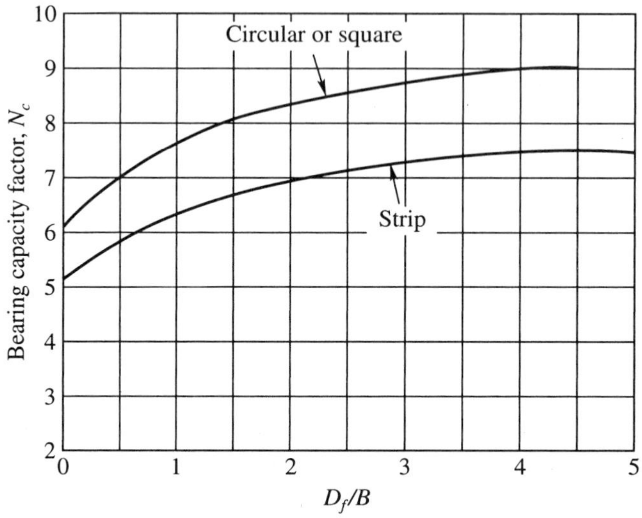

Figure 12.9 Skempton's bearing capacity factor N_c for clay soils

$$q_{na} = \frac{q_{nu}}{F_s} = \frac{cN_c}{F_s} \tag{12.22c}$$

The N_c values for strip and square (or circular) foundations as a function of the D_f/B ratio are given in Fig. 12.9. The equation for rectangular foundation may be written as follows

$$(N_c)_R = \left(0.84 + 0.16 \times \frac{B}{L}\right)(N_c)_S \tag{12.22d}$$

where $(N_c)_R = N_c$ for rectangular foundation, $(N_c)_S = N_c$ for square foundation.

The lower and upper limiting values of N_c for strip and square foundations may be written as follows:

Type of foundation	Ratio D_f/B	Value of N_c
Strip	0	5.14
	≥4	7.5
Square	0	6.2
	≥4	9.0

12.8 EFFECT OF WATER TABLE ON BEARING CAPACITY

The theoretical equations developed for computing the ultimate bearing capacity q_u of soil are based on the assumption that the water table lies at a depth below the base of the foundation equal to or greater than the width B of the foundation or otherwise the depth of the water table from ground surface is equal to or greater than $(D_f + B)$. In case the water table lies at any intermediate depth less than the depth $(D_f + B)$, the bearing capacity equations are affected due to the presence of the water table.

Two cases may be considered here.

Case 1. When the water table lies above the base of the foundation.

Case 2. When the water table lies within depth B below the base of the foundation.

We will consider the two methods for determining the effect of the water table on bearing capacity as given below.

Method 1

For any position of the water table within the depth $(D_f + B)$, we may write Eq. (12.6) as

$$q_u = cN_c + \gamma D_f N_q R_{w1} + \frac{1}{2}\gamma B N_\gamma R_{w2} \tag{12.23}$$

where R_{w1} = reduction factor for water table above the base level of the foundation,

R_{w2} = reduction factor for water table below the base level of the foundation,

γ = γ_{sat} for all practical purposes in both the second and third terms of Eq. (12.23).

Case 1: When the water table lies above the base level of the foundation or when $D_{w1}/D_f \leq 1$ (Fig. 12.10a) the equation for R_{w1} may be written as

$$R_{w1} = \frac{1}{2}\left(1 + \frac{D_{w1}}{D_f}\right) \tag{12.24a}$$

For $D_{w1}/D_f = 0$, we have $R_{w1} = 0.5$, and for $D_{w1}/D_f = 1.0$, we have $R_{w1} = 1.0$.

Case 2: When the water table lies below the base level or when $D_{w2}/B \leq 1$ (12.10b) the equation for R_{w2} is

$$R_{w2} = \frac{1}{2}\left(1 + \frac{D_{w2}}{B}\right) \tag{12.24b}$$

For $D_{w2}/B = 0$, we have $R_{w2} = 0.5$, and for $D_{w2}/B = 1.0$, we have $R_{w2} = 1.0$.

Figure 12.10 shows in a graphical form the relations D_{w1}/D_f vs. R_{w1} and D_{w2}/B vs R_{w2}.

Equations (12.23a) and (12.23b) are based on the assumption that the submerged unit weight of soil is equal to half of the saturated unit weight and the soil above the water table remains saturated.

Method 2: Equivalent effective unit weight method

Eq. (12.6) for the strip footing may be expressed as

$$q_u = cN_c + \gamma_{e1} D_f N_q + \frac{1}{2}\gamma_{e2} B N_\gamma \tag{12.25}$$

where γ_{e1} = weighted effective unit weight of soil lying above the base level of the foundation

γ_{e2} = weighted effective unit weight of soil lying within the depth B below the base level of the foundation

γ_m = moist or saturated unit weight of soil lying above WT (case 1 or case 2)

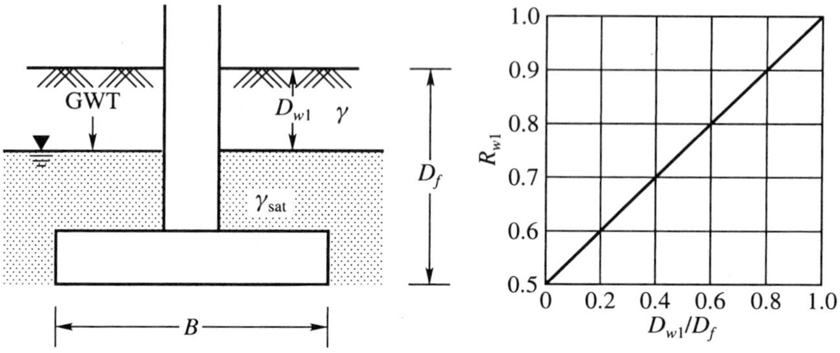

(a) Water table above base level of foundation

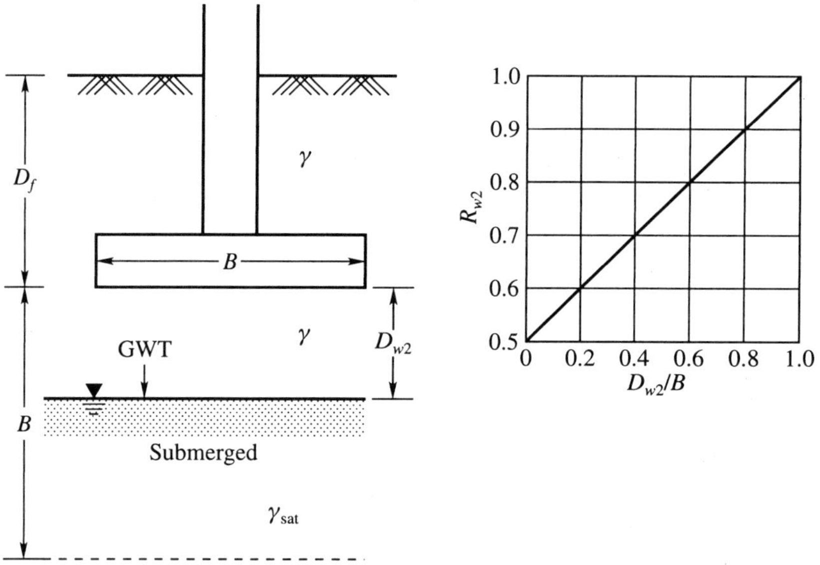

(b) Water table below base level of foundation

Figure 12.10 Effect of WT on bearing capacity

γ_{sat} = saturated unit weight of soil below the WT (case 1 or case 2)
γ_b = submerged unit weight of soil = $\gamma_{sat} - \gamma_w$

Case 1

An equation for γ_{e1} may be written as

$$\gamma_{e1} = \gamma_b + \frac{D_{w1}}{D_f}\left(\gamma_m - \gamma_b\right) \tag{12.26a}$$

$$\gamma_{e2} = \gamma_b$$

Case 2

$$\gamma_{e1} = \gamma_m$$

$$\gamma_{e2} = \gamma_b + \frac{D_{w2}}{B}\left(\gamma_m - \gamma_b\right) \tag{12.26b}$$

Example 12.1

A strip footing of width 3 m is founded at a depth of 2 m below the ground surface in a $(c - \phi)$ soil having a cohesion c = 30 kN/m^2 and angle of shearing resistance ϕ = 35°. The water table is at a depth of 5 m below ground level. The moist weight of soil above the water table is 17.25 kN/m^3. Determine (a) the ultimate bearing capacity of the soil, (b) the net bearing capacity, and (c) the net allowable bearing pressure and the load/m for a factor of safety of 3. Use the general shear failure theory of Terzaghi.

Solution

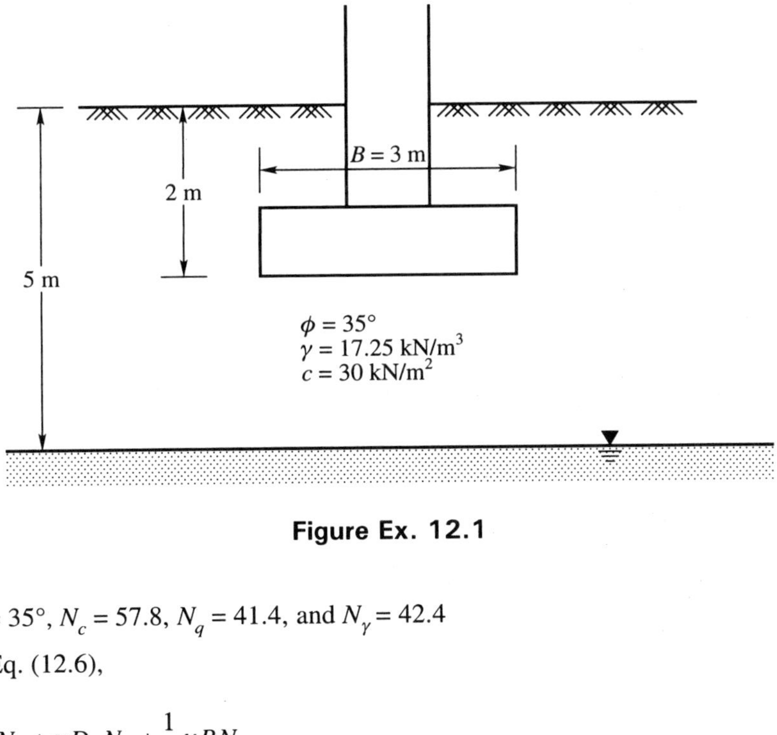

Figure Ex. 12.1

For $\phi = 35°$, $N_c = 57.8$, $N_q = 41.4$, and $N_\gamma = 42.4$

From Eq. (12.6),

$$q_u = cN_c + \gamma D_f N_q + \frac{1}{2}\gamma B N_\gamma$$

$$= 30 \times 57.8 + 17.25 \times 2 \times 41.4 + \frac{1}{2} \times 17.25 \times 3 \times 42.4 = 4259 \text{ kN/m}^2$$

$$q_{nu} = q_u - \gamma D_f = 4259 - 17.25 \times 2 \approx 4225 \text{ kN/m}^2$$

$$q_{na} = \frac{q_{nu}}{F_s} = \frac{4225}{3} \approx 1408 \text{ kN/m}^2$$

$$Q_a = q_{na} B = 1408 \times 3 = 4225 \text{ kN/m}$$

Example 12.2

If the soil in Ex. 12.1 fails by local shear failure, determine the net safe bearing pressure. All the other data given in Ex. 12.1 remain the same.

Solution

For local shear failure:

$$\overline{\phi} = \tan^{-1} 0.67 \tan 35^\circ = 25^\circ$$

$$\overline{c} = 0.67c = 0.67 \times 30 = 20 \text{ kN/m}^2$$

From Table 12.1, for $\overline{\phi} = 25^\circ$, $\overline{N}_c = 25.1$, $\overline{N}_q = 12.7$, $\overline{N}_\gamma = 9.7$

Now from Eq. (12.12)

$$q_u = 20 \times 25.1 + 17.25 \times 2 \times 12.7 + \frac{1}{2} \times 17.25 \times 3 \times 9.7 = 1191 \text{ kN/m}^2$$

$$q_{nu} = 1191 - 17.25 \times 2 = 1156.5 \text{ kN/m}^2$$

$$q_{na} = \frac{1156.50}{3} = 385.5 \text{ kN/m}^2$$

$$Q_a = 385.5 \times 3 = 1156.5 \text{ kN/m}$$

Example 12.3

If the water table in Ex. 12.1 rises to the ground level, determine the net safe bearing pressure of the footing. All the other data given in Ex. 12.1 remain the same. Assume the saturated unit weight of the soil $\gamma_{sat} = 18.5$ kN/m^3.

Solution

When the WT is at ground level we have to use the submerged unit weight of the soil.

Therefore $\gamma_b = \gamma_{sat} - \gamma_w = 18.5 - 9.81 = 8.69 \text{ kN/m}^3$

The net ultimate bearing capacity is

$$q_{nu} = 30 \times 57.8 + 8.69 \times 2(41.4 - 1) + \frac{1}{2} \times 48.69 \times 3 \times 42.4 \approx 2992 \text{ kN/m}^2$$

$$q_{na} = \frac{2992}{3} = 997.33 \text{ kN/m}^2$$

$$Q_a = 997.33 \times 3 = 2992 \text{ kN/m}$$

Example 12.4

If the water table in Ex. 12.1 occupies any of the positions (a) 1.25 m below ground level or (b) 1.25 m below the base level of the foundation, what will be the net safe bearing pressure?

Assume $\gamma_{sat} = 18.5$ kN/m^3, γ(above WT) = 17.5 kN/m^3. All the other data remain the same as given in Ex. 12.1.

Solution

Method 1—By making use of reduction factors R_{w1} and R_{w2} and using Eqs. (12.20) and (12.23), we may write

$$q_{nu} = cN_c + \gamma D_f (N_q - 1) R_{w1} + \frac{1}{2}\gamma B N_\gamma R_{w2}$$

Given: $N_q = 41.4$, $N_\gamma = 42.4$ and $N_c = 57.8$

Case 1—When the WT is 1.25 m below the GL

From Eq. (12.24), we get $R_{w1} = 0.813$ for $D_{w1}/D_f = 0.625$, $R_{w2} = 0.5$ for $D_{w2}/B = 0$.

By substituting the known values in the equation for q_{nu}, we have

$$q_{nu} = 30 \times 57.8 + 18.5 \times 2 \times 40.4 \times 0.813 + \frac{1}{2} \times 18.5 \times 3 \times 42.4 \times 0.5 = 3538 \text{ kN/m}^2$$

$$q_{na} = \frac{3538}{3} = 1179 \text{ kN/m}^2$$

Case 2—When the WT is 1.25 m below the base of the foundation

$R_{w1} = 1.0$ for $D_{w1}/D_f = 1$, $R_{w2} = 0.71$ for $D_{w2}/B = 0.42$.

Now the net bearing capacity is

$$q_{nu} = 30 \times 57.8 + 18.5 \times 2 \times 40.4 \times 1 + \frac{1}{2} \times 18.5 \times 3 \times 42.4 \times 0.71 = 4064 \text{ kN/m}^2$$

$$q_{na} = \frac{4064}{3} = 1355 \text{ kN/m}^2$$

Method 2—Using the equivalent effective unit weight method.

Submerged unit weight $\gamma_b = 18.5 - 9.81 = 8.69$ kN/m^3.

Per Eq. (12.25)

The net ultimate bearing capacity is

$$q_{nu} = cN_c + \gamma_{e1} D_f (N_q - 1) + \frac{1}{2}\gamma_{e2} B N_\gamma$$

Case 1—When D_{w1} = 1.25 m (Fig. Ex. 12.4)

From Eq. (12.26a)

$$\gamma_{e1} = \gamma_b + \frac{D_{w1}}{D_f}(\gamma_m - \gamma_b)$$

where $\gamma_m = \gamma_{sat} = 18.5 \text{ kN/m}^3$

$$\gamma_{e1} = 8.69 + \frac{1.25}{2}(18.5 - 8.69) = 14.82 \text{ kN/m}^3$$

$$\gamma_{e2} = \gamma_b = 8.69 \text{ kN/m}^3$$

$$q_{nu} = 30 \times 57.8 + 14.82 \times 2 \times 40.4 + \frac{1}{2} \times 8.69 \times 3 \times 42.4 = 3484 \text{ kN/m}^2$$

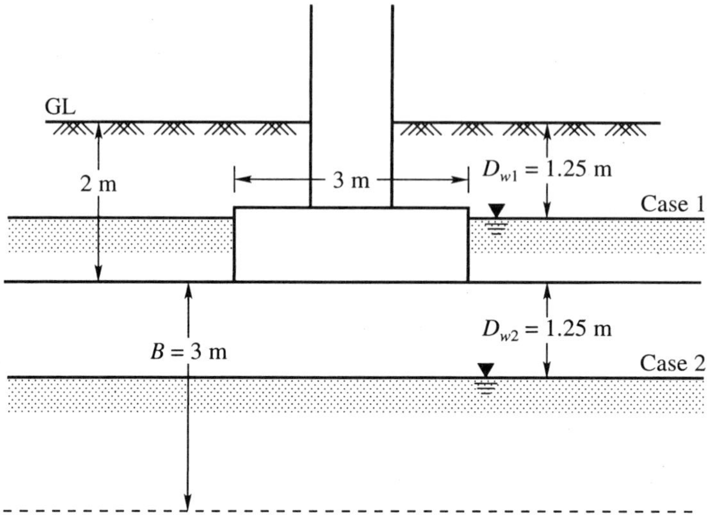

Figure Ex. 12.4 Effect of WT on bearing capacity

$$q_{na} = \frac{3484}{3} = 1161 \text{ kN/m}^2$$

Case 2—When D_{w2} = 1.25 m (Fig. Ex. 12.4)

From Eq. (12.26b)

$$\gamma_{e1} = \gamma_m = 18.5 \text{ kN/m}^3$$

$$\gamma_{e2} = 8.69 + \frac{1.25}{3}(18.5 - 8.69) = 12.78 \text{ kN/m}^3$$

$$q_{nu} = 30 \times 57.8 + 18.5 \times 2 \times 40.4 + \frac{1}{2} \times 12.78 \times 3 \times 42.4 = 4042 \text{ kN/m}^2$$

$$q_{na} = \frac{4042}{3} = 1347 \text{ kN/m}^2$$

Example 12.5

A square footing fails by general shear in a cohesionless soil under an ultimate load of Q_{ult} = 1687.5 kips. The footing is placed at a depth of 6.5 ft below ground level. Given $\phi = 35°$, and γ = 110 lb/ft^3, determine the size of the footing if the water table is at a great depth (Fig. Ex. 12.5).

Solution

For a square footing [Eq. (12.17)] for c = 0, we have

$$q_u = \gamma D_f N_q + 0.4 \gamma B N_\gamma$$

For $\phi = 35°$, N_q = 41.4, and N_γ = 42.4 from Table 12.1.

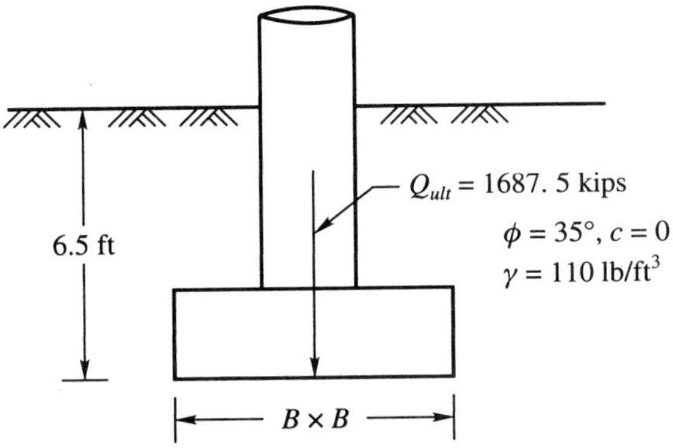

Figure Ex. 12.5

$$q_u = \frac{Q_u}{B^2} = \frac{1687.5 \times 10^3}{B^2}$$

By substituting known values, we have

$$\frac{1687.5 \times 10^3}{B^2} = 110 \times 6.5 \times 41.4 + 0.4 \times 110 \times 42.4B$$

$$= (29.601 + 1.866B)10^3$$

Simplifying and transposing, we have

$$B^3 + 15.863B^2 - 904.34 = 0$$

Solving this equation yields, $B = 6.4$ ft.

Example 12.6

A rectangular footing of size 10 × 20 ft is founded at a depth of 6 ft below the ground surface in a homogeneous cohesionless soil having an angle of shearing resistance $\phi = 35°$. The water table is at a great depth. The unit weight of soil $\gamma = 114$ lb/ft^3. Determine: (1) the net ultimate bearing capacity,

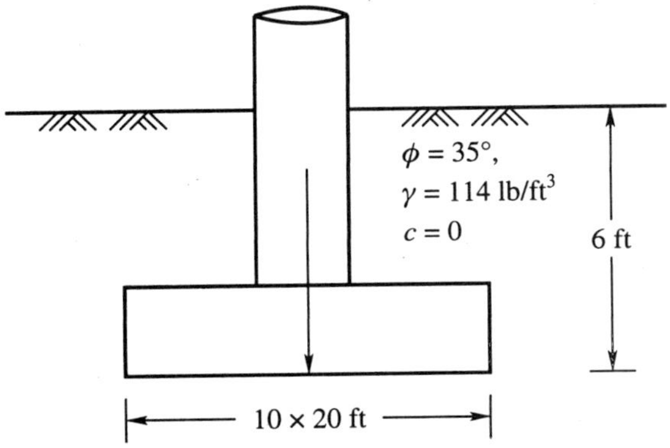

Figure Ex. 12.6

(2) the net allowable bearing pressure for $F_s = 3$, and (3) the allowable load Q_a the footing can carry. Use Terzaghi's theory. (Refer to Fig. Ex. 12.6)

Solution

Using Eq. (12.19) and Eq. (12.20) for $c = 0$, the net ultimate bearing capacity for a rectangular footing is expressed as

$$q_{nu} = \gamma D_f (N_q - 1) + \frac{1}{2}\gamma B N_\gamma \left(1 - 0.2\frac{B}{L}\right)$$

From Table 12.1, $N_q = 41.4$, $N_\gamma = 42.4$ for $\phi = 35°$

By substituting the known values,

$$q_{nu} = 114 \times 6(41.4 - 1) + \frac{1}{2} \times 114 \times 10 \times 42.4\left(1 - 0.2 \times \frac{10}{20}\right) = 49{,}385 \text{ lb/ft}^2$$

$$q_{na} = \frac{49{,}385}{3} = 16{,}462 \text{ lb/ft}^2$$

$$Q_a = (B \times L)\, q_{na} = 10 \times 20 \times 16{,}462 \approx 3{,}292 \times 10^3 \text{ lb} = 3292 \text{ kips}$$

Example 12.7

A rectangular footing of size 10 × 20 ft is founded at a depth of 6 ft below the ground level in a cohesive soil ($\phi = 0$) which fails by general shear. Given: $\gamma_{sat} = 114$ lb/ft^3, $c = 945$ lb/ft^2. The water table is close to the ground surface. Determine q_u, q_{nu} and q_{na} by (a) Terzaghi's method, and (b) Skempton's method. Use $F_s = 3$.

Solution

(a) Terzaghi's method

Use Eq. (12.19)

For $\phi = 0°$, $N_c = 5.7$, $N_q = 1$

$$q_u = cN_c\left(1 + 0.3 \times \frac{B}{L}\right) + \gamma_b D_f$$

Substituting the known values,

$$q_u = 945 \times 5.7\left(1 + 0.3 \times \frac{10}{20}\right) + (114 - 62.4) \times 6 = 6{,}504 \text{ lb/ft}^2$$

$$q_{nu} = (q_u - \gamma_b D_f) = 6504 - (114 - 62.4) \times 6 = 6195 \text{ lb/ft}^2$$

$$q_{na} = \frac{q_{nu}}{F_s} = \frac{6195}{3} = 2065 \text{ lb/ft}^2$$

(b) Skempton's method

From Eqs. (12.22a) and (12.22d) we may write

$$q_u = cN_{cr} + \gamma D_f$$

where N_{cr} = bearing capacity factor for rectangular foundation.

$$N_{cr} = \left(0.84 + 0.16 \times \frac{B}{L}\right) \times N_{cs}$$

where N_{cs} = bearing capacity factor for a square foundation.

From Fig. 12.9, $N_{cs} = 7.2$ for $D_f/B = 0.60$.

$$\text{Therefore} \quad N_{cr} = \left(0.84 + 0.16 \times \frac{10}{20}\right) \times 7.2 = 6.62$$

$$\text{Now} \quad q_u = 945 \times 6.62 + 114 \times 6 = 6940 \text{ lb/ft}^2$$

$$q_{nu} = (q_u - \gamma D_f) = 6940 - 114 \times 6 = 6{,}256 \text{ lb/ft}^2$$

$$q_{na} = \frac{q_{nu}}{F_s} = \frac{6256}{3} = 2{,}085 \text{ lb/ft}^2$$

Note: Terzaghi's and Skempton's values are in close agreement for cohesive soils.

Example 12.8

If the soil in Ex. 12.6 is cohesionless ($c = 0$), and fails in local shear, determine (i) the ultimate bearing capacity, (ii) the net bearing capacity, and (iii) the net allowable bearing pressure. All the other data remain the same.

Solution

From Eq. (12.15) and Eq. (12.20), the net bearing capacity for local shear failure for $c = 0$ is

$$q_{nu} = (q_u - \gamma D_f) = \gamma D_f(\overline{N}_q - 1) + \frac{1}{2}\gamma B \overline{N}_\gamma \left(1 - 0.2 \times \frac{B}{L}\right)$$

where $\overline{\phi} = \tan^{-1} 0.67 \tan 35° \approx 25°$, $\overline{N}_q = 12.7$, and $\overline{N}_\gamma = 9.7$ for $\overline{\phi} = 25°$ from Table 12.1.

By substituting known values, we have

$$q_{nu} = 114 \times 6(12.7 - 1) + \frac{1}{2} \times 114 \times 10 \times 9.7\left(1 - 0.2 \times \frac{10}{20}\right) = 12{,}979 \text{ lb/ft}^2$$

$$q_{na} = \frac{12{,}979}{3} = 4{,}326 \text{ lb/ft}^2$$

12.9 THE GENERAL BEARING CAPACITY EQUATION

The bearing capacity Eq. (12.6) developed by Terzaghi is for a strip footing under general shear failure. Eq. (12.6) has been modified for other types of foundations such as square, circular and rectangular by introducing shape factors. Meyerhof (1963) presented a general bearing capacity equation which takes into account the shape and the inclination of load. The general form of equation suggested by Meyerhof for bearing capacity is

$$q_u = cN_c s_c d_c i_c + q'_o N_q s_q d_q i_q + \frac{1}{2}\gamma B N_\gamma s_\gamma d_\gamma i_\gamma \tag{12.27}$$

where c = unit cohesion

q'_o = effective overburden pressure at the base level of the foundation = $\bar{\gamma} D_f$

$\bar{\gamma}$ = effective unit weight above the base level of foundation

γ = effective unit weight of soil below the foundation base

D_f = depth of foundation

s_c, s_q, s_γ = shape factors

d_c, d_q, d_γ = depth factor

i_c, i_q, i_γ = load inclination factors

B = width of foundation

N_c, N_q, N_γ = bearing capacity factors

Hansen (1970) extended the work of Meyerhof by including in Eq. (12.27) two additional factors to take care of base tilt and foundations on slopes. Vesic (1973, 1974) used the same form of equation suggested by Hansen. All three investigators use the equations proposed by Prandtl (1921) for computing the values of N_c and N_q wherein the foundation base is assumed as smooth with the angle $\alpha = 45° + \phi/2$ (Fig. 12.6). However, the equations used by them for computing the values of N_γ are different. The equations for N_c, N_q and N_γ are

$$N_q = e^{\pi \tan\phi} N_\phi$$

$$N_c = (N_q - 1)\cot\phi$$

Table 12.2 The values of N_c, N_q, and Meyerhof (M), Hansen (H) and Vesic (V) N_γ Factors

ϕ	N_c	N_q	N_γ (H)	N_γ (M)	N_γ (V)
0	5.14	1.0	0.0	0.0	0.0
5	6.49	1.6	0.1	0.1	0.4
10	8.34	2.5	0.4	0.4	1.2
15	10.97	3.9	1.2	1.1	2.6
20	14.83	6.4	2.9	2.9	5.4
25	20.71	10.7	6.8	6.8	10.9
26	22.25	11.8	7.9	8.0	12.5
28	25.79	14.7	10.9	11.2	16.7
30	30.13	18.4	15.1	15.7	22.4
32	35.47	23.2	20.8	22.0	30.2
34	42.14	29.4	28.7	31.1	41.0
36	50.55	37.7	40.0	44.4	56.2
38	61.31	48.9	56.1	64.0	77.9
40	72.25	64.1	79.4	93.6	109.4
45	133.73	134.7	200.5	262.3	271.3
50	266.50	318.50	567.4	871.7	762.84

Note: N_c and N_q are the same for all the three methods. Subscripts identify the author for N_γ.

Table 12.3 Shape, depth and load inclination factors of Meyerhof, Hansen and Vesic

Factors	Meyerhof	Hansen	Vesic
s_c	$1+0.2N_\phi\left[\frac{B}{L}\right]$	$1+\frac{N_q}{N_c}\left[\frac{B}{L}\right]$	
s_q	$1+0.1N_\phi\left[\frac{B}{L}\right]$ for $\phi>10°$	$1+\frac{B}{L}\tan\phi$	
s_γ	$s_\gamma=s_q$ for $\phi>10°$ $s_\gamma=s_q=1$ for $\phi=0$	$1-0.4\frac{B}{L}$	The shape and depth factors of Vesic are the same as those of Hansen.
d_c	$1+0.2\sqrt{N_\phi}\left[\frac{D_f}{B}\right]$	$1+0.4\left[\frac{D_f}{B}\right]$	
d_q	$1+0.1\sqrt{N_\phi}\left[\frac{D_f}{B}\right]$ for $\phi>10°$	$1+2\tan\phi(1-\sin\phi)^2\left[\frac{D_f}{B}\right]$	
d_γ	$d_\gamma=d_q$ for $\phi>10°$ $d_\gamma=d_q=1$ for $\phi=0$	1 for all ϕ Note; Vesic's s and d factors = Hansen's s and d factors	
i_c	$\left[1-\frac{\alpha°}{90}\right]^2$ for any ϕ	$i_q-\frac{1-i_q}{N_q-1}$ for $\phi>0$ $0.5\left[1-\frac{Q_h}{A_f c_a}\right]^{\frac{1}{2}}$ for $\phi=0$	Same as Hansen for $\phi>0$ $1-\frac{mQ_h}{A_f c_a N_c}$
i_q	$i_q=i_c$ for any ϕ	$\left[1-\frac{0.5Q_h}{Q_u+A_f c_a\cot\phi}\right]^5$	$\left[1-\frac{Q_h}{Q_u+A_f c_a\cot\phi}\right]^m$
i_γ	$\left[1-\frac{\alpha°}{\phi°}\right]^2$ for $\phi>0$ $i_\gamma=0$ for $\phi=0$	$\left[1-\frac{0.7Q_h}{Q_u+A_f c_a\cot\phi}\right]^5$	$\left[1-\frac{Q_h}{Q_u+A_f c_a\cot\phi}\right]^{m+1}$

$$N_\gamma=(N_q-1)\tan(1.4\phi) \quad \text{(Meyerhof)}$$

$$N_\gamma=1.5(N_q-1)\tan\phi \quad \text{(Hansen)}$$

$$N_\gamma=2(N_q+1)\tan\phi \quad \text{(Vesic)}$$

Table12.2 gives the values of the bearing capacity factors. Equations for shape, depth and inclination factors are given in Table 12.3. The tilt of the base and the foundations on slopes are not considered here.

In Table 12.3 The following terms are defined with regard to the inclination factors

Q_h = horizontal component of the inclined load

Q_u = vertical component of the inclined load

c_a = unit adhesion on the base of the footing

A_f = effective contact area of the footing

$$m = m_B = \frac{2 + B/L}{1 + B/L} \text{ with } Q_h \text{ parallel to } B$$

$$m = m_L = \frac{2 + B/L}{1 + B/L} \text{ with } Q_h \text{ parallel to } L$$

The general bearing capacity Eq. (12.27) has not taken into account the effect of the water table position on the bearing capacity. Hence, Eq. (12.27) has to be modified according to the position of water level in the same way as explained in Section 12.7.

Validity of the Bearing Capacity Equations

There is currently no method of obtaining the ultimate bearing capacity of a foundation other than as an estimate (Bowles, 1996). There has been little experimental verification of any of the methods except by using model footings. Up to a depth of $D_f \approx B$ the Meyerhof q_u is not greatly different from the Terzaghi value (Bowles, 1996). The Terzaghi equations, being the first proposed, have been quite popular with designers. Both the Meyerhof and Hansen methods are widely used. The Vesic method has not been much used. It is a good practice to use at least two methods and compare the computed values of q_u. If the two values do not compare well, use a third method.

Example 12.9

Refer to Example 12.1. Compute using the Meyerhof equation (a) the ultimate bearing capacity of the soil, (b) the net bearing capacity, and (c) the net allowable bearing pressure. All the other data remain the same.

Solution

Use Eq. (12.27). For $i = 1$ the equation for net bearing capacity is

$$q_{nu} = q_u - \gamma D_f = cN_c s_c d_c + \gamma D_f (N_q - 1) s_q d_q + \frac{1}{2} \gamma B N_\gamma s_\gamma d_\gamma$$

From Table 12.3

$$s_c = 1 + 0.2 N_\phi \left[\frac{B}{L}\right] = 1 \text{ for strip footing}$$

$$s_q = 1 + 0.1 N_q \left[\frac{B}{L}\right] = 1 \text{ for strip footing}$$

$$s_\gamma = s_q = 1$$

$$d_c = 1 + 0.2\sqrt{N_\phi}\left[\frac{D_f}{B}\right] = 1 + 0.2 \tan\left(45° + \frac{35}{2}\right)\left[\frac{2}{3}\right] = 1.257$$

$$d_q = 1 + 0.1\sqrt{N_\phi}\left[\frac{D_f}{B}\right] = 1 + 0.1 \tan\left(45° + \frac{35}{2}\right)\left[\frac{2}{3}\right] = 1.129$$

$$d_\gamma = d_q = 1.129$$

From Ex. 12.1, $c = 30$ kN/m^2, $\gamma = 17.25$ kN/m^3, $D_f = 2$ m , B = 3 m.

From Table 12.2 for $\phi = 35°$ we have $N_c = 46.35$, $N_q = 33.55$, $N_\gamma = 37.75$. Now substituting the known values, we have

$$q_{nu} = 30 \times 46.35 \times 1 \times 1.257 + 17.25 \times 2 \times (33.55 - 1) \times 1 \times 1.129$$

$$+ \frac{1}{2} \times 17.25 \times 3 \times 37.75 \times 1 \times 1.129$$

$$= 1{,}748 + 1{,}268 + 1{,}103 = 4{,}119 \text{ kN/m}^2$$

$$q_{na} = \frac{4{,}119}{3} = 1{,}373 \text{ kN/m}^2$$

There is very close agreement between Terzaghi's and Meyerhof's methods.

Example 12.10

Refer to Example 12.6. Compute by Meyerhof's method the net ultimate bearing capacity and the net allowable bearing pressure for $F_s = 3$. All the other data remain the same.

Solution

From Ex. 12.6 we have $B = 10$ ft, $L = 20$ ft, $D_f = 6$ ft, and $\gamma = 114$ lb/ft^3. From Eq. (12.27) for $c = 0$ and $i = 1$, we have

$$q_{nu} = q_u - \gamma D_f = \gamma D_f (N_q - 1) s_q d_q + \frac{1}{2} \gamma B N_\gamma s_\gamma d_\gamma$$

From Table 12.3

$$s_q = 1 + 0.1 N_\phi \left[\frac{B}{L}\right] = 1 + 0.1 \tan^2\left(45° + \frac{35}{2}\right)\left[\frac{10}{20}\right] = 1.185$$

$$s_\gamma = s_q = 1.185$$

$$d_q = 1 + 0.1\sqrt{N_\phi}\left[\frac{D_f}{B}\right] = 1 + 0.1 \tan\left(45° + \frac{35}{2}\right)\left[\frac{6}{10}\right] = 1.115$$

$$d_\gamma = d_q = 1.115$$

From Table 12.2 for $\phi = 35°$, we have $N_q = 33.55$, $N_\gamma = 37.75$. By substituting the known values, we have

$$q_{nu} = 114 \times 6(33.55 - 1) \times 1.185 \times 1.115 + \frac{1}{2} \times 114 \times 10 \times 37.75 \times 1.185 \times 1.115$$

$$= 29{,}417 + 28{,}431 = 57{,}848 \text{ lb/ft}^2$$

$$q_{na} = \frac{57{,}848}{3} = 19{,}283 \text{ lb/ft}^2$$

By Terzaghi's method $q_{na} = 16{,}462$ lb/ft^2.

Meyerhof's method gives a higher value for q_{na} by about 17%.

Example 12.11

Refer to Ex. 12.1. Compute by Hansen's method (a) net ultimate bearing capacity, and (b) the net safe bearing pressure. All the other data remain the same.

Given for a strip footing

$B = 3$ m, $D_f = 2$ m, $c = 30$ kN/m^2 and $\gamma = 17.25$ kN/m^3, $F_s = 3$.

From Eq. (12.27) for $i = 1$, we have

$$q_{nu} = q_u - \gamma D_f = cN_c s_c d_c + \gamma D_f (N_q - 1) s_q d_q + \frac{1}{2}\gamma B N_\gamma s_\gamma d_\gamma$$

From Table 12.2 for Hansen's method, we have for $\phi = 35°$

$N_c = 46.35$, $N_q = 33.55$, and $N_\gamma = 34.35$.

From Table 12.3 we have

$$s_c = 1 + \frac{N_q}{N_c}\left[\frac{B}{L}\right] = 1 \quad \text{for a strip footing}$$

$$s_q = 1 + \frac{B}{L}\tan\phi = 1 \quad \text{for a strip footing}$$

$$s_\gamma = 1 - 0.4\frac{B}{L} = 1 \quad \text{for a strip footing}$$

$$d_c = 1 + 0.4\left[\frac{D_f}{B}\right] = 1 + 0.4 \times \frac{2}{3} = 1.267$$

$$d_q = 1 + 2\tan\phi(1 - \sin\phi)^2\left[\frac{D_f}{B}\right]$$

$$= 1 + 2\tan 35°(1 - \sin 35°)^2 \times \frac{2}{3} = 1 + 2 \times 0.7(1 - 0.574)^2 \times \frac{2}{3} = 1.17$$

$$d_\gamma = 1$$

Substituting the known values, we have

$$q_{nu} = 30 \times 46.35 \times 1 \times 1.267 + 17.25 \times 2 \times (33.55 - 1) \times 1 \times 1.17$$

$$+ \frac{1}{2} \times 17.25 \times 3 \times 34.35 \times 1 \times 1$$

$$= 1{,}762 + 1{,}314 + 889 = 3{,}965 \text{ kN/m}^2$$

$$q_{na} = \frac{3{,}965}{3} = 1{,}322 \text{ kN/m}^2$$

The values of q_{na} by Terzaghi, Meyerhof and Hansen methods are

Example	Author	q_{na} kN/m^2
12.1	Terzaghi	1,408
12.9	Meyerhof	1,373
12.11	Hansen	1,322

Terzaghi's method is higher than Meyerhof's by 2.5% and Meyerhof's higher than Hansen's by 3.9%. The difference between the methods is not significant. Any of the three methods can be used.

Example 12.12

Refer to Example 12.6. Compute the net safe bearing pressure by Hansen's method. All the other data remain the same.

Solution

Given: Size 10 × 20 ft, $D_f = 6$ ft, $c = 0$, $\phi = 35°$, $\gamma = 114$ lb/ft^3, $F_s = 3$.

For $\phi = 35°$ we have from Table 12.2, $N_q = 33.55$ and $N_\gamma = 34.35$

From Table 12.3 we have

$$s_q = 1 + \frac{B}{L}\tan\phi = 1 + \frac{10}{20} \times \tan 35° = 1.35$$

$$s_\gamma = 1 - 0.4\frac{B}{L} = 1 - 0.4 \times \frac{10}{20} = 0.80$$

$$d_q = 1 + 2\tan 35°(1 - \sin 35°)^2 \times \frac{6}{10} = 1.153$$

$$d_\gamma = 1$$

Substituting the known values, we have

$$q_{nu} = q_u - \gamma D_f = \gamma D_f (N_q - 1) s_q d_q + \frac{1}{2}\gamma B N_\gamma s_\gamma d_\gamma$$

$$= 114 \times 6(33.55 - 1) \times 1.35 \times 1.153 + \frac{1}{2} \times 114 \times 10 \times 34.35 \times 0.8 \times 1$$

$$= 34{,}655 + 15{,}664 = 50{,}319 \text{ lb/ft}^2$$

$$q_{na} = \frac{50{,}319}{3} = 16{,}773 \text{ lb/ft}^2$$

The values of q_{na} by other methods are

Example	Author	q_{na} kN/m^2
12.6	Terzaghi	16,462
12.10	Meyerhof	19,283
12.12	Hansen	16,773

It can be seen from the above, the values of Terzaghi and Hansen are very close to each other, whereas the Meyerhof value is higher than that of Terzaghi by 17 percent.

12.10 EFFECT OF SOIL COMPRESSIBILITY ON BEARING CAPACITY OF SOIL

Terzaghi (1943) developed Eq. (12.6) based on the assumption that the soil is incompressible. In order to take into account the compressibility of soil, he proposed reduced strength characteristics $\bar{c}$ and $\bar{\phi}$ defined by Eq. (12.11). As per Vesic (1973) a flat reduction of ϕ in the case of local and punching shear failures is too conservative and ignores the existence of scale effects. It has been conclusively established that the ultimate bearing capacity q_u of soil does not increase in proportion to the increase in the size of the footing as shown in Fig. 12.11 or otherwise the bearing capacity factor N_γ decreases with the increase in the size of the footing as shown in Fig. 12.12.

In order to take into account the influence of soil compressibility and the related scale effects, Vesic (1973) proposed a modification of Eq. (12.27) by introducing compressibility factors as follows.

$$q_u = cN_c d_c s_c C_c + q_o' N_q d_q s_q C_q + \frac{1}{2}\gamma B N_\gamma d_\gamma s_\gamma C_\gamma \tag{12.28}$$

where, C_c, C_q and C_γ are the soil compressibility factors. The other symbols remain the same as before.

For the evaluation of the relative compressibility of a soil mass under loaded conditions, Vesic introduced a term called *rigidity index* I_r, which is defined as

$$I_r = \frac{G}{c + \overline{q}\tan\phi} \tag{12.29}$$

where, G = shear modulus of soil = $\dfrac{E_s}{2(1+\mu)}$

E_s = modulus of elasticity

$\overline{q}$ = effective overburden pressure at a depth equal to $(D_f + B/2)$

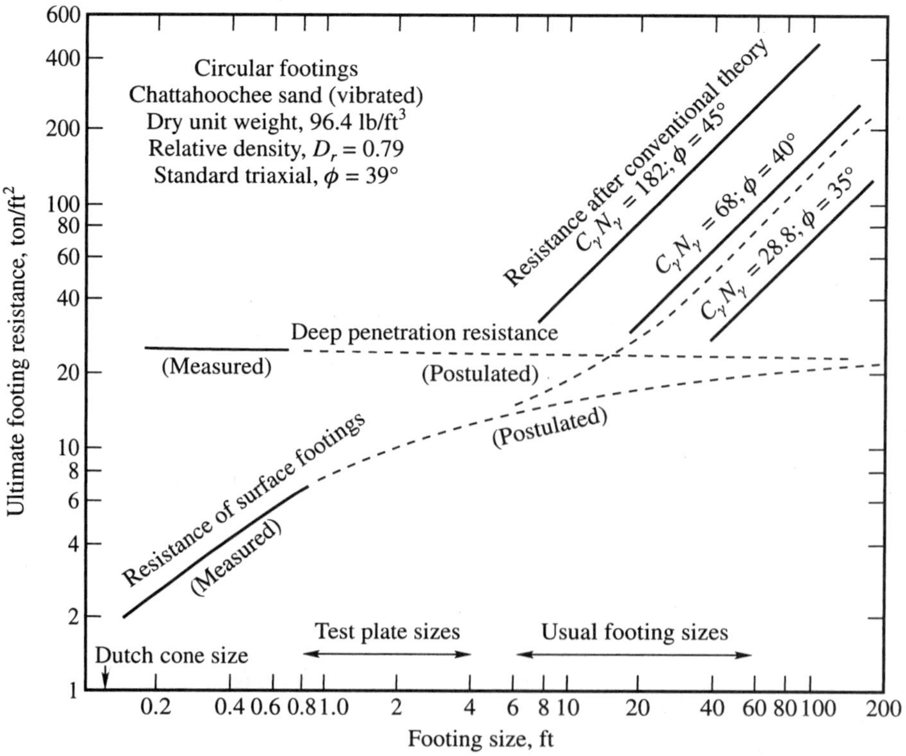

Figure 12.11 Variation of ultimate resistance of footings with size (after Vesic, 1969)

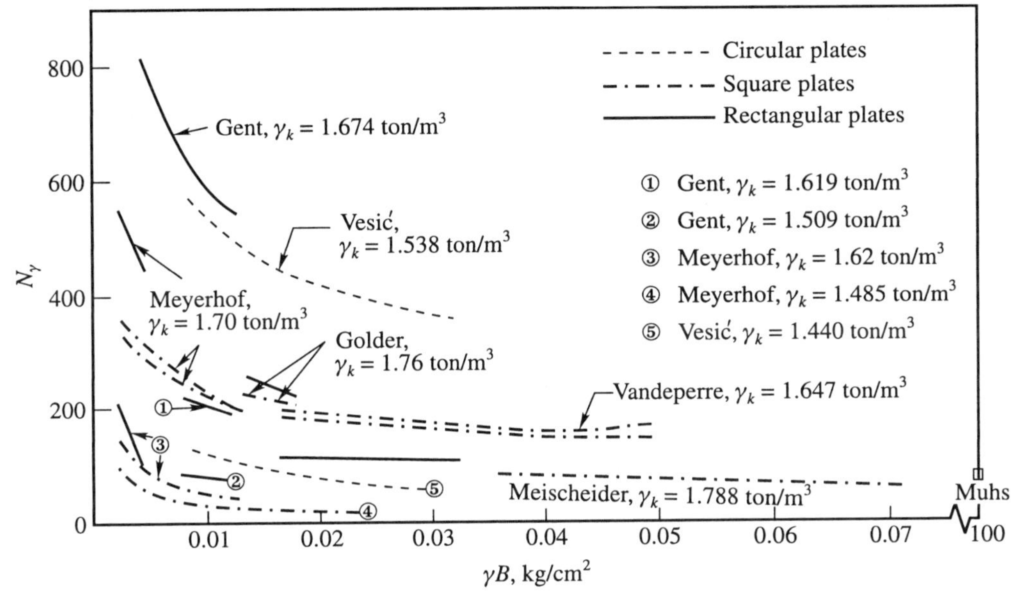

Figure 12.12 Effect of size on bearing capacity of surface footings in sand (After De Beer, 1965)

μ = Poisson's ratio
c, ϕ = shear strength parameters

Eq. (12.29) was developed on the basis of the theory of expansion of cavities in an infinite solid with the assumed ideal elastic properties behavior of soil. In order to take care of the volumetric strain Δ in the plastic zone, Vesic (1965) suggested that the value of I_r, given by Eq. (12.29), be reduced by the following equation.

$$I_{rr} = F_r I_r \tag{12.30}$$

where F_r = reduction factor $= \dfrac{1}{1+I_r\Delta}$

It is known that I_r varies with the stress level and the character of loading. A high value of I_{rr}, for example over 250, implies a relatively incompressible soil mass, whereas a low value of say 10 implies a relatively compressible soil mass.

Based on the theory of expansion of cavities, Vesic developed the following equation for the compressibility factors.

$$C_q = \exp\left\{\left[-4.4+0.6\frac{B}{L}\right]\tan\phi + \left[\frac{3.07\sin\phi\log 2I_r}{1+\sin\phi}\right]\right\} \tag{12.31}$$

For $\phi > 0$, one can determine from the theorem of correspondence

$$C_c = C_q - \frac{1-C_q}{N_q\tan\phi} \tag{12.32}$$

For $\phi = 0$, we have

$$C_c = 0.32+0.12\frac{B}{L}+0.6\log I_r \tag{12.33}$$

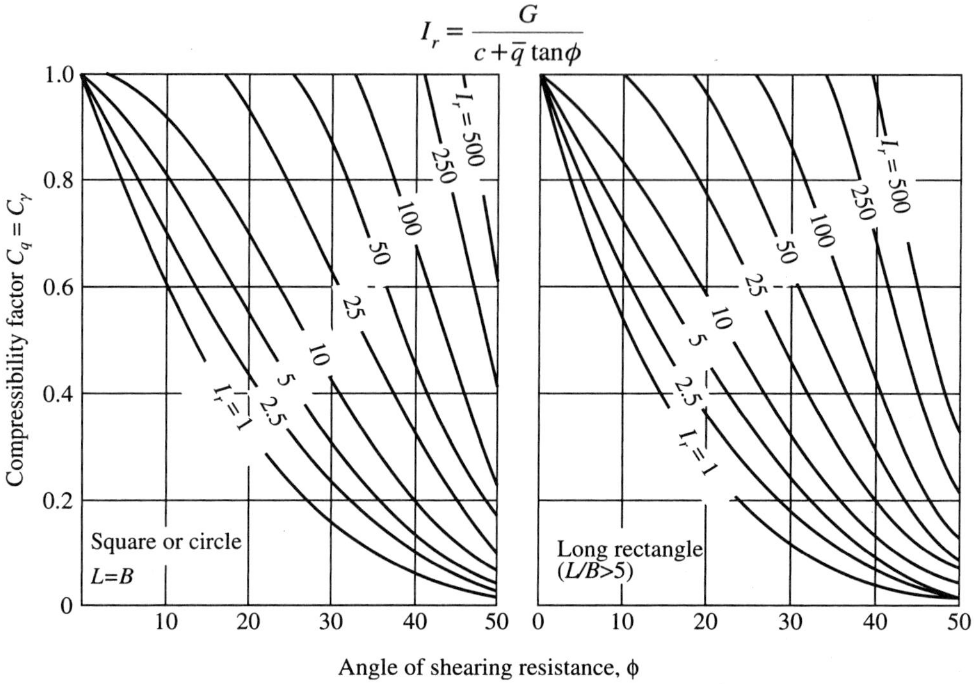

Figure 12.13 Theoretical compressibility factors (after Vesic, 1970)

For all practical purposes, Vesic suggests

$$C_q = C_\gamma \tag{12.34}$$

Equations (12.30) through Eq. (12.34) are valid as long as the values of the compressibility factors are less than unity. Fig. 12.13 shows graphically the relationship between C_q ($= C_\gamma$) and ϕ for two extreme cases of $L/B > 5$ (strip footing) and $B/L = 1$ (square) for different values of I_r (Vesic 1970). Vesic gives another expression called the *critical rigidity index* $(I_r)_{cr}$ expressed as

Table 12.4 Values of critical rigidity index

Angle of shearing resistance	Critical rigidity index for	
	Strip foundation	Square foundation
ϕ	$B/L = 0$	$B/L = 1$
0	13	8
5	18	11
10	25	15
15	37	20
20	55	30
25	89	44
30	152	70
35	283	120
40	592	225
45	1442	486
50	4330	1258

$$(I_r)_{cr} = \frac{1}{2}\exp\left[\left(3.3 - 0.45\frac{B}{L}\right)\cot(45 - \phi/2)\right] \tag{12.35}$$

The magnitude of $(I_r)_{cr}$ for any angle of ϕ and any foundation shape reduces the bearing capacity because of compressibility effects. Numerical values of $(I_r)_{cr}$ for two extreme cases of $B/L = 0$ and $B/L = 1$ are given in Table 12.4 for various values of ϕ.

Application of I_r (or I_{rr}) and $(I_r)_{crit}$

1. If I_r (or I_{rr}) $\geq (I_r)_{crit}$, assume the soil is incompressible and $C_c = C_q = C_\gamma = 1$ in Eq. (12.28).
2. If I_r (or I_{rr}) $< (I_r)_{crit}$, assume the soil is compressible. In such a case the compressibility factors C_c, C_q and C_γ are to be determined and used in Eq. (12.28).

The concept and analysis developed above by Vesic (1973) are based on a limited number of small scale model tests and need verification in field conditions.

Example 12.13

A square footing of size 12 × 12 ft is placed at a depth of 6 ft in a deep stratum of medium dense sand. The following soil parameters are available:

$\gamma = 100$ lb/ft^3, $c = 0$, $\phi = 35°$, $E_s = 100$ t/ft^2, Poissons' ratio $\mu = 0.25$.

Estimate the ultimate bearing capacity by taking into account the compressibility of the soil (Fig. Ex. 12.13).

Solution

Rigidity $I_r = \dfrac{E_s}{2(1+\mu)\overline{q}\tan\phi}$ for $c = 0$ from Eq. (12.29)

$$\overline{q} = \gamma(D_f + B/2) = 100\left(6 + \frac{12}{2}\right) = 1{,}200 \text{ lb/ft}^2 = 0.6 \text{ ton/ft}^2$$

Neglecting the volume change in the plastic zone

$$I_r = \frac{100}{2(1+0.25)0.6\tan 35°} = 95$$

From Table 12.4, $(I_r)_{crit} = 120$ for $\phi = 35°$

Since $I_r < (I_r)_{crit}$, the soil is compressible.

From Fig. 12.13, C_q $(= C_\gamma) = 0.90$ (approx) for square footing for $\phi = 35°$ and $I_r = 95$.

From Table 12.2, $N_q = 33.55$ and $N_\gamma = 48.6$ (Vesic's value)

Eq. (12.28) may now be written as

$$q_u = q_0' N_q d_q s_q C_q + \frac{1}{2}\gamma B N_\gamma d_\gamma s_\gamma C_\gamma$$

From Table 12.3

$$s_q = 1 + \frac{B}{L}\tan\phi = 1 + \tan 35° = 1.7 \quad \text{for } B = L$$

$$s_\gamma = 1 - 0.4 = 0.6 \quad \text{for } B = L$$

$$d_q = 1 + 2\tan 35°(1 - \sin 35°)^2 \times \frac{6}{12} = 1.127$$

$$d_\gamma = 1$$

$$q_0' = 100 \times 6 = 600 \text{ lb/ft}^2$$

Substituting

$$q_u = 600 \times 33.55 \times 1.127 \times 1.7 \times 0.90 + \frac{1}{2} \times 100 \times 12 \times 48.6 \times 1.0 \times 0.6 \times 0.90$$

$$= 34{,}710 + 15{,}746 = 50{,}456 \text{ lb/ft}^2$$

If the compressibility factors are not taken into account (That is, $C_q = C_\gamma = 1$) the ultimate bearing capacity q_u is

$$q_u = 38{,}567 + 17{,}496 = 56{,}063 \text{ lb/ft}^2$$

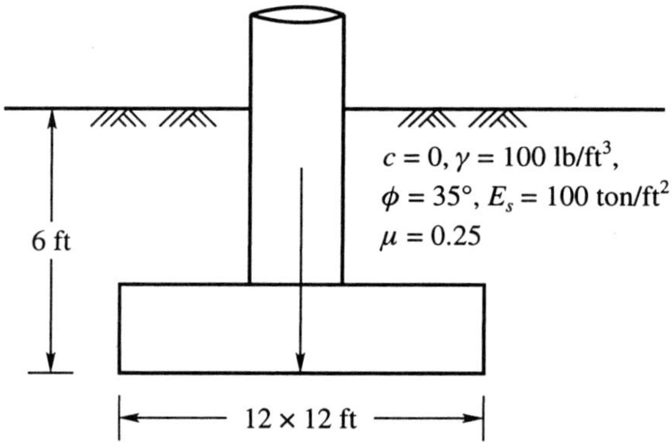

Figure Ex. 12.13

Example 12.14

Estimate the ultimate bearing capacity of a square footing of size 12 × 12 ft founded at a depth of 6 ft in a deep stratum of saturated clay of soft to medium consistency. The undrained shear strength of the clay is 400 lb/ft^2 (= 0.2 t/ft^2) The modulus of elasticity E_s = 15 ton/ft^2 under undrained conditions. Assume $\mu = 0.5$ and $\gamma = 100$ lb/ft^3.

Solution

$$\text{Rigidity } I_r = \frac{E_s}{2(1+\mu)c_u} = \frac{15}{2(1+0.5)0.2} = 25$$

From Table 12.4, $(I_r)_{crit} = 8$ for $\phi = 0$

Since $I_r > (I_r)_{crit}$, the soil is supposed to be incompressible. Use Eq. (12.28) for computing q_u by putting $C_c = C_q = 1$ for $\phi = 0$

$$q_u = cN_c s_c d_c + q_0' N_q s_q d_q$$

From Table 12.2 for $\phi = 0$, $N_c = 5.14$, and $N_q = 1$

From Table 12.3 $s_c = 1 + \frac{N_q}{N_c}\frac{B}{L} = 1 + \frac{1}{5.14} \approx 1.2$

$$d_c = 1 + 0.4\frac{D_f}{B} = 1 + 0.4\frac{6}{12} = 1.2$$

$$s_q = d_q = 1$$

Substituting and simplifying, we have

$$q_u = 400 \times 5.14 \times 1.2 \times 1.2 + 100 \times 6 \times (1)(1)(1)$$

$$= 2{,}960 + 600 = 3{,}560 \text{ lb/ft}^2 = 1.78 \text{ ton / ft}^2$$

12.11 BEARING CAPACITY OF FOUNDATIONS SUBJECTED TO ECCENTRIC LOADS

Foundations Subjected to Eccentric Vertical Loads

If a foundation is subjected to lateral loads and moments in addition to vertical loads, eccentricity in loading results. The point of application of the resultant of all the loads would lie outside the geometric center of the foundation, resulting in eccentricity in loading. The eccentricity e is measured from the center of the foundation to the point of application normal to the axis of the foundation. The maximum eccentricity normally allowed is $B/6$ where B is the width of the foundation. The basic problem is to determine the effect of the eccentricity on the ultimate bearing capacity of the foundation. When a foundation is subjected to an eccentric vertical load, as shown in Fig. 12.14(a), it tilts towards the side of the eccentricity and the contact pressure increases on the side of tilt and decreases on the opposite side. When the vertical load Q_{ult} reaches the ultimate load, there will be a failure of the supporting soil on the side of eccentricity. As a consequence, settlement of the footing will be associated with tilting of the base towards the side of eccentricity. If the eccentricity is very small, the load required to produce this type of failure is almost equal to the load required for producing a symmetrical general shear failure. Failure occurs due to intense radial shear on one side of the plane of symmetry, while the deformations in the zone of radial shear on the other side are still insignificant. For this reason the failure is always associated with a heave on that side towards which the footing tilts.

Research and observations of Meyerhof (1953, 1963) indicate that effective footing dimensions obtained (Fig. 12.14) as

$$L' = L - 2e_x, \quad B' = B - 2e_y \tag{12.36a}$$

should be used in bearing capacity analysis to obtain an effective footing area defined as

$$A' = B'L' \tag{12.36b}$$

The ultimate load bearing capacity of a footing subjected to eccentric loads may be expressed as

$$Q'_{ult} = q_u A' \tag{12.36c}$$

where q_u = ultimate bearing capacity of the footing with the load acting at the center of the footing.

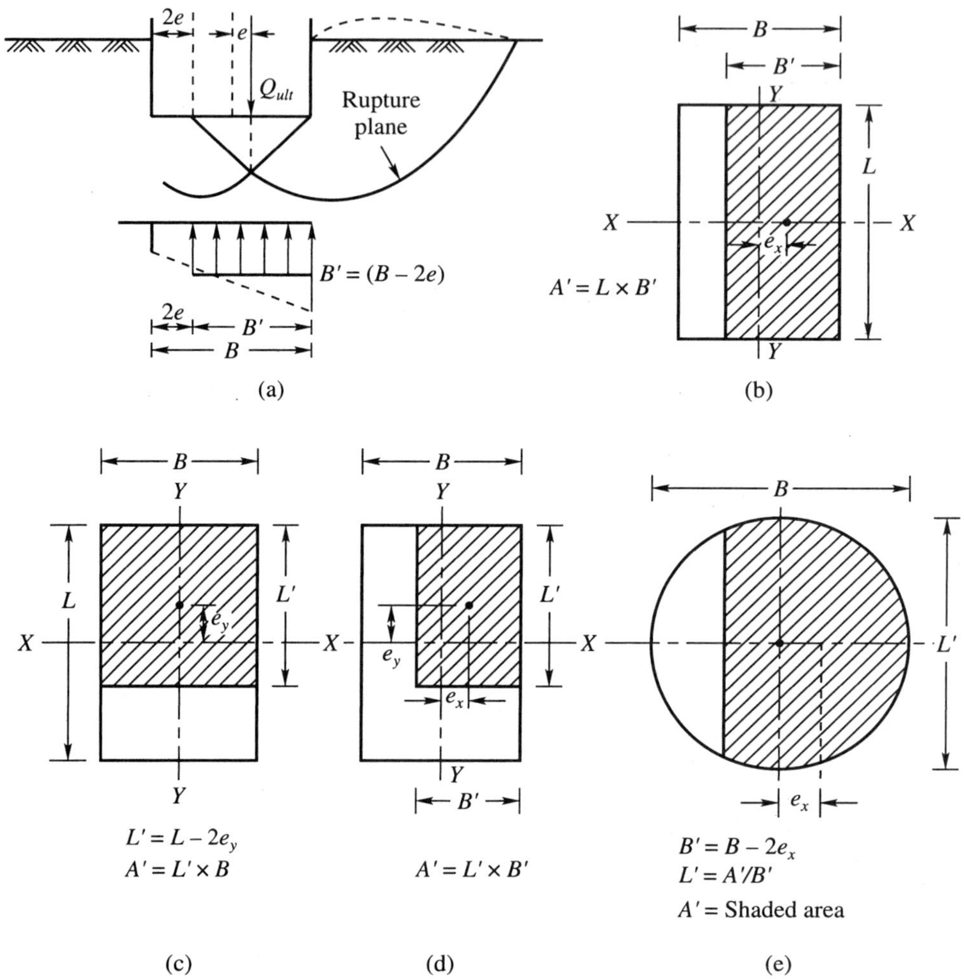

Figure 12.14 Eccentrically loaded footing (Meyerhof, 1953)

Determination of Maximum and Minimum Base Pressures Under Eccentric Loadings

The methods of determining the effective area of a footing subjected to eccentric loadings have been discussed earlier. It is now necessary to know the maximum and minimum base pressures under the same loadings. Consider the plan of a rectangular footing given in Fig. 12.15 subjected to eccentric loadings.

Let the coordinate axes *XX* and *YY* pass through the center *O* of the footing. If a vertical load passes through *O*, the footing is symmetrically loaded. If the vertical load passes through O_x on the *X*-axis, the footing is eccentrically loaded with one way eccentricity. The distance of O_x from *O*, designated as e_x, is called the eccentricity in the *X*-direction. If the load passes through O_y on the *Y*-axis, the eccentricity is e_y in the *Y*-direction. If on the other hand the load passes through O_{xy}, the eccentricity is called *two-way eccentricity* or *double eccentricity.*

When a footing is eccentrically loaded, the soil experiences a maximum or a minimum pressure at one of the corners or edges of the footing. For the load passing through O_{xy} (Fig. 12.15), the points *C* and *D* at the corners of the footing experience the maximum and minimum pressures respectively.

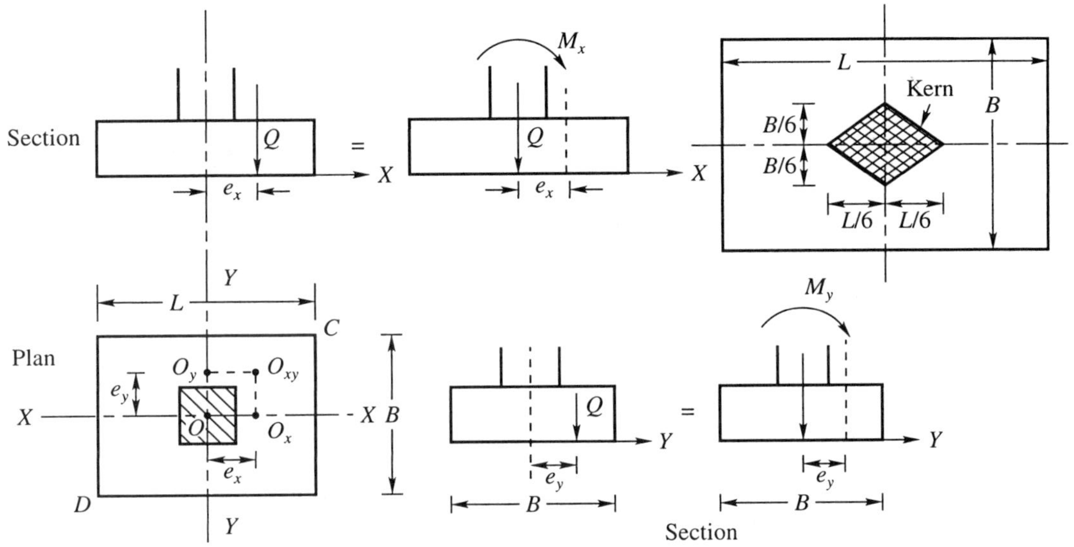

Figure 12.15 Footing subjected to eccentric loadings

The general equation for pressure may be written as

$$q = \frac{Q}{A} \pm \frac{Qe_x}{I_y} x \pm \frac{Qe_y}{I_x} y \tag{12.37a}$$

or $$q = \frac{Q}{A} \pm \frac{M_x}{I_y} x \pm \frac{M_y}{I_x} y \tag{12.37b}$$

where q = contact pressure at a given point (x, y)

Q = total vertical load

A = area of footing

Qe_x = M_x = moment about axis YY

Qe_y = M_y = moment about axis XX

I_x, I_y = moment of inertia of the footing about XX and YY axes respectively

q_{max} and q_{min} at points C and D respectively may be obtained by substituting in Eq. (12.37) for

$$I_x = \frac{LB^3}{12}, \quad I_y = \frac{BL^3}{12}, \quad x = \frac{L}{2}, \quad y = \frac{B}{2}$$

we have

$$q_{max} = \frac{Q}{A}\left[1 + 6\frac{e_x}{L} + 6\frac{e_y}{B}\right] \tag{12.39a}$$

$$q_{min} = \frac{Q}{A}\left[1 - 6\frac{e_x}{L} - 6\frac{e_y}{B}\right] \tag{12.39b}$$

Equation (12.39) may also be used for one way eccentricity by putting either $e_x = 0$, or $e_y = 0$.

When e_x or e_y exceed a certain limit, Eq. (12.39) gives a negative value of q which indicates tension between the soil and the bottom of the footing. Eqs (12.39) are applicable only when the load is applied within a limited area which is known as the *Kern* as is shown shaded in Fig 12.15 so that the load may fall within the shaded area to avoid tension. The procedure for the determination of soil pressure when the load is applied outside the kern is laborious and as such not dealt with here. However, charts are available for ready calculations in references such as Teng (1969) and Highter and Anders (1985).

12.12 ULTIMATE BEARING CAPACITY OF FOOTINGS BASED ON SPT VALUES (*N*)

Standard Energy Ratio R_{es} Applicable to *N* Value

The effects of field procedures and equipment on the field values of N were discussed in Chapter 9. The empirical correlations established in the USA between N and soil properties indicate the value of N conforms to certain standard energy ratios. Some suggest 70% (Bowles, 1996) and others 60% (Terzaghi et al., 1996). In order to avoid this confusion, the author uses N_{cor} in this book as the corrected value for standard energy.

Cohesionless Soils

Relationship Between N_{cor} and ϕ

The relation between N_{cor} and ϕ established by Peck et al., (1974) is given in a graphical form in Fig. 12.8. The value of N_{cor} to be used for getting ϕ is the corrected value for standard energy. The angle ϕ obtained by this method can be used for obtaining the bearing capacity factors, and hence the ultimate bearing capacity of soil.

Cohesive Soils

Relationship Between N_{cor} and q_u (Unconfined Compressive Strength)

Relationships have been developed between N_{cor} and q_u (the undrained compressive strength) for the $\phi = 0$ condition. This relationship gives the value of c_u for any known value of N_{cor}. The relationship may be expressed as [Eq. (9.12)]

$$q_u = 2c_u \bar{k} N_{cor} \text{(kPa)} \tag{12.40}$$

where the value of the coefficient $\bar{k}$ may vary from a minimum of 12 to a maximum of 25. A low value of 13 yields q_u given in Table 9.4.

Once q_u is determined, the net ultimate bearing capacity and the net allowable bearing pressure can be found following Skempton's approach.

12.13 THE CPT METHOD OF DETERMINING ULTIMATE BEARING CAPACITY

Cohesionless Soils

Relationship Between q_c, D_r and ϕ

Relationships between the static cone penetration resistance q_c and ϕ have been developed by Robertson and Campanella (1983b), Fig. 9.15. The value of ϕ can therefore be determined with the known value of q_c. With the known value of ϕ, bearing capacity factors can be determined and hence the ultimate bearing capacity. Experience indicates that the use of q_c for obtaining ϕ is more reliable than the use of N.

Bearing Capacity of Soil

As per Schmertmann (1978), the bearing capacity factors N_q and N_γ for use in the Terzaghi bearing capacity equation can be determined by the use of the equation

$$N_q = N_\gamma = 1.25 q_c \tag{12.41}$$

where q_c = cone point resistance in kg/cm^2 (or tsf) averaged over a depth equal to the width below the foundation.

Undrained Shear Strength

The undrained shear strength c_u under $\phi = 0$ condition may be related to the static cone point resistance q_c as [Eq. (9.16)]

$$q_c = N_k c_u + p_o$$

$$\text{or} \quad c_u = \frac{q_c - p_o}{N_k} = \frac{\bar{q}_c}{N_k} \tag{12.42}$$

where N_k = cone factor, may be taken as equal to 20 (Sanglerat, 1972) both for normally consolidated and preconsolidated clays.

p_o = total overburden pressure

When once c_u is known, the values of q_{nu} and q_{na} can be evaluated as per the methods explained in earlier sections.

Example 12.15

A water tank foundation has a footing of size 6 × 6 m founded at a depth of 3 m below ground level in a medium dense sand stratum of great depth. The corrected average SPT value obtained from the site investigation is 20. The foundation is subjected to a vertical load at an eccentricity of $B/10$ along one of the axes. Figure Ex. 12.15 gives the soil profile with the remaining data. Estimate the ultimate load, Q_{ult}, by Meyerhof's method.

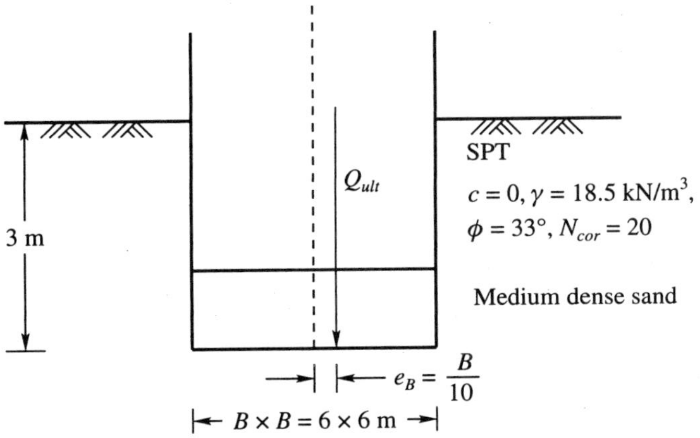

Figure Ex. 12.15

Solution

From Fig. 12.8, $\phi = 33°$ for $N_{cor} = 20$.

$B' = B - 2e = 6 - 2(0.6) = 4.8$ m

$L' = L = B = 6$ m

For $c = 0$ and $i = 1$, Eq. (12.28) reduces to

$$q'_u = \gamma D_f N_q s_q d_q + \frac{1}{2}\gamma B' N_\gamma s_\gamma d_\gamma$$

From Table 12.2 for $\phi = 33°$ we have

$N_q = 26.3$, $N_\gamma = 26.55$ (Meyerhof)

From Table 12.3 (Meyerhof)

$$s_q = 1 + 0.1 N_\phi \left[\frac{B}{L}\right] = 1 + 0.1 \tan^2\left(45° + \frac{33}{2}\right)(1) = 1.34$$

$$s_\gamma = s_q = 1.34 \qquad \text{for } \phi > 10°$$

$$d_q = 1 + 0.1\sqrt{N_\phi}\left[\frac{D_f}{B'}\right] = 1 + 0.1 \times 1.84\left[\frac{3}{4.8}\right] = 1.115$$

Substituting $d_\gamma = d_q = 1.115 \qquad$ for $\phi > 10°$

$$q'_u = 18.5 \times 3 \times 26.3 \times 1.34 \times 1.115 + \frac{1}{2} \times 18.5 \times 4.8 \times 26.55 \times 1.34 \times 1.115$$

$$= 2{,}181 + 1{,}761 = 3{,}942 \text{ kN/m}^2$$

$$Q'_{ult} = B \times B' \times q'_u = 6 \times 4.8 \times 3942 = 113{,}530 \text{ kN} \approx 114 \text{ MN}$$

Example 12.16

Figure Ex. 12.16 gives the plan of a footing subjected to eccentric load with two way eccentricity. The footing is founded at a depth 3 m below the ground surface. Given $e_x = 0.60$ m and $e_y = 0.75$ m, determine Q_{ult}. The soil properties are: $c = 0$, $N_{cor} = 20$, $\gamma = 18.5$ kN/m^3. The soil is medium dense sand. Use N_γ (Meyerhof) from Table 12.2 and Hansen's shape and depth factors from Table 12.3.

Solution

Figure Ex. 12.16 shows the two-way eccentricity. The effective lengths and breadths of the foundation from Eq. (12.36a) is

$B' = B - 2e_y = 6 - 2 \times 0.75 = 4.5$ m.

$L' = L - 2e_x = 6 - 2 \times 0.6 = 4.8$ m.

Effective area, $A' = L' \times B' = 4.5 \times 4.8 = 21.6$ m^2

As in Example 12.15

$$q'_u = \gamma D_f N_q s_q d_q + \frac{1}{2}\gamma B' N_\gamma s_\gamma d_\gamma$$

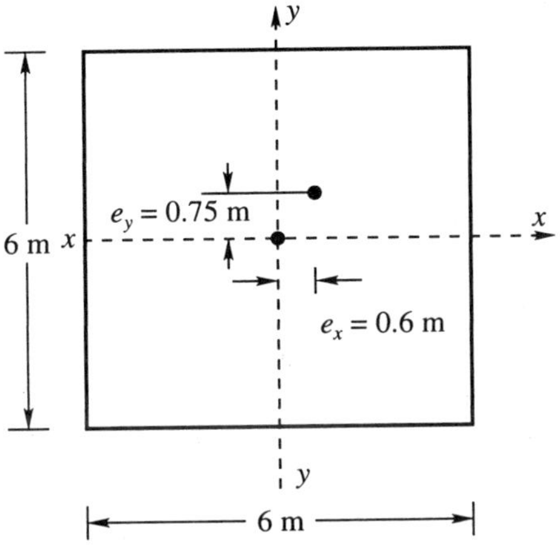

Figure Ex. 12.16

For $\phi = 33°$, $N_q = 26.3$ and $N_\gamma = 26.55$ (Meyerhof)

From Table 12.3 (Hansen)

$$s_q = 1 + \frac{B'}{L'} \tan 33° = 1 + \frac{4.5}{4.8} \times 0.65 = 1.61$$

$$s_\gamma = 1 - 0.4\frac{B'}{L'} = 1 - 0.4 \times \frac{4.5}{4.8} = 0.63$$

$$d_q = 1 + 2\tan 33° (1 - \sin 33°)^2 \times \frac{3}{4.5}$$

$$= 1 + 1.3 \times 0.21 \times 0.67 = 1.183$$

$$d_\gamma = 1$$

Substituting

$$q'_u = 18.5 \times 3 \times 26.3 \times 1.61 \times 1.183 + \frac{1}{2} \times 18.5 \times 4.5 \times 26.55 \times 0.63 \times (1)$$

$$= 2{,}780 + 696 = 3{,}476 \text{ kN/m}^2$$

$$Q_{ult} = A' q'_u = 21.6 \times 3{,}476 = 75{,}082 \text{ kN}$$

12.14 ULTIMATE BEARING CAPACITY OF FOOTINGS RESTING ON STRATIFIED DEPOSITS OF SOIL

All the theoretical analysis dealt with so far is based on the assumption that the subsoil is isotropic and homogeneous to a considerable depth. In nature, soil is generally non-homogeneous with mixtures of sand, silt and clay in different proportions. In the analysis, an average profile of such soils is normally considered. However, if soils are found in distinct layers of different compositions and strength characteristics, the assumption of homogeneity to such soils is not strictly valid if the failure surface cuts across boundaries of such layers.

The present analysis is limited to a system of two distinct soil layers. For a footing located in the upper layer at a depth D_f below the ground level, the failure surfaces at ultimate load may either lie completely in the upper layer or may cross the boundary of the two layers. Further, we may come across the upper layer strong and the lower layer weak or vice versa . In either case, a general analysis for $(c - \phi)$ will be presented and will show the same analysis holds true if the soil layers are any one of the categories belonging to sand or clay.

The bearing capacity of a layered system was first analyzed by Button (1953) who considered only saturated clay ($\phi = 0$). Later on Brown and Meyerhof (1969) showed that the analysis of Button leads to unsafe results. Vesic (1975) analyzed the test results of Brown and Meyerhof and others and gave his own solution to the problem.

Vesic considered both the types of soil in each layer, that is clay and $(c - \phi)$ soils. However, confirmations of the validity of the analysis of Vesic and others are not available. Meyerhof (1974) analyzed the two layer system consisting of dense sand on soft clay and loose sand on stiff clay and supported his analysis with some model tests. Again Meyerhof and Hanna (1978) advanced the earlier analysis of Meyerhof (1974) to encompass $(c - \phi)$ soil and supported their analysis with model tests. The present section deals briefly with the analyses of Meyerhof (1974) and Meyerhof and Hanna (1978).

Case 1: A Stronger Layer Overlying a Weaker Deposit

Figure 12.16(a) shows a strip footing of width B resting at a depth D_f below ground surface in a strong soil layer (Layer 1). The depth to the boundary of the weak layer (Layer 2) below the base of the footing is H. If this depth H is insufficient to form a full failure plastic zone in Layer 1 under the ultimate load conditions, a part of this ultimate load will be transferred to the boundary level mn. This load will induce a failure condition in the weaker layer (Layer 2). However, if the depth H is relatively large then the failure surface will be completely located in Layer 1 as shown in Fig. 12.16b.

The ultimate bearing capacities of strip footings on the surfaces of homogeneous thick beds of Layer 1 and Layer 2 may be expressed as

Layer 1

$$q_1 = c_1 N_{c1} + \frac{1}{2}\gamma_1 B N_{\gamma 1} \tag{12.43}$$

Layer 2

$$q_2 = c_2 N_{c2} + \frac{1}{2}\gamma_2 B N_{\gamma 2} \tag{12.44}$$

where N_{c1}, $N_{\gamma 1}$ = bearing capacity factors for soil in Layer 1 for friction angle ϕ_1
N_{c2}, $N_{\gamma 2}$ = bearing capacity factors for soil in Layer 2 for friction angle ϕ_2

For the footing founded at a depth D_f, if the complete failure surface lies within the upper stronger Layer 1 (Fig. 12.16(b)) an expression for ultimate bearing capacity of the upper layer may be written as

$$q_u = q_t = c_1 N_{c1} + q'_o N_{q1} + \frac{1}{2}\gamma_1 B N_{\gamma 1} \tag{12.45}$$

If q_1 is much greater that q_2 and if the depth H is insufficient to form a full failure plastic condition in Layer 1, then the failure of the footing may be considered due to pushing of soil within the boundary ad and bc through the top layer into the weak layer. The resisting force for punching may be assumed to develop on the faces ad and bc passing through the edges of the footing. The forces that act on these surfaces are (per unit length of footing),

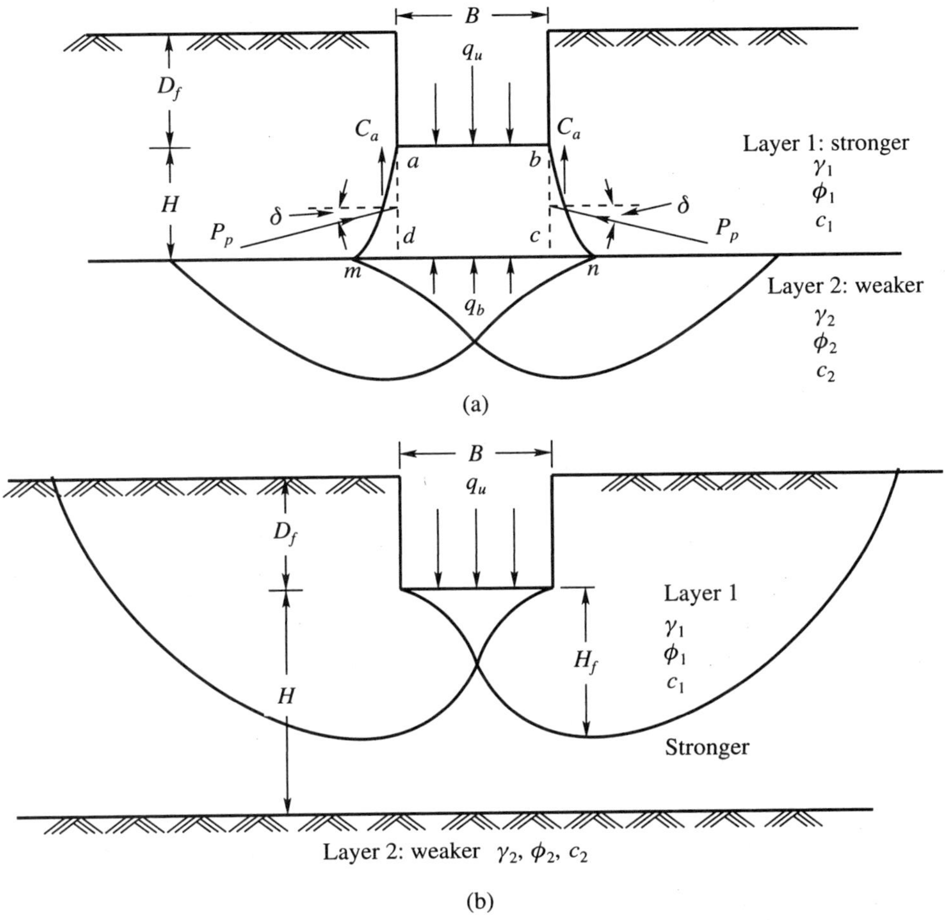

Figure 12.16 Failure of soil below strip footing under vertical load on strong layer overlying weak deposit (after Meyerhof and Hanna, 1978)

$$\text{Adhesive force,} \quad C_a = c_a H$$
$$\text{Frictional force,} \quad F_f = P_p \sin \delta \tag{12.46}$$

where c_a = unit cohesion, P_p = passive earth pressure per unit length of footing, and δ = inclination of P_p with the normal (Fig 12.16(a)).

The equation for the ultimate bearing capacity q_u for the two layer soil system may now be expressed as

$$q_u = q_b + \frac{2(C_a + P_p \sin \delta)}{B} - \gamma_1 H \tag{12.47}$$

where, q_b = ultimate bearing capacity of Layer 2

The equation for P_p may be written as

$$P_p = \frac{\gamma_1 H^2}{2 \cos \delta}\left(1 + \frac{2D_f}{H}\right) K_p \tag{12.48}$$

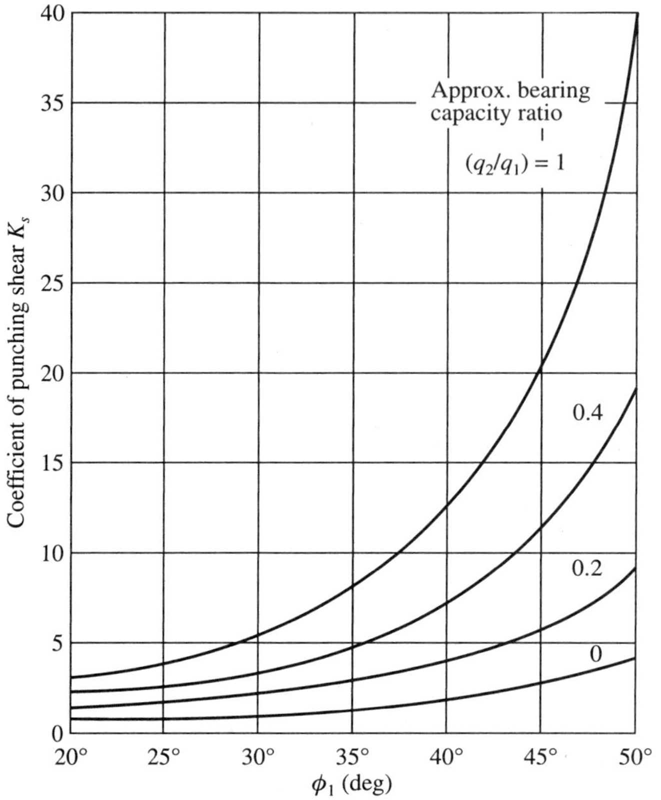

Figure 12.17 Coefficients of punching shear resistance under vertical load (after Meyerhof and Hanna, 1978)

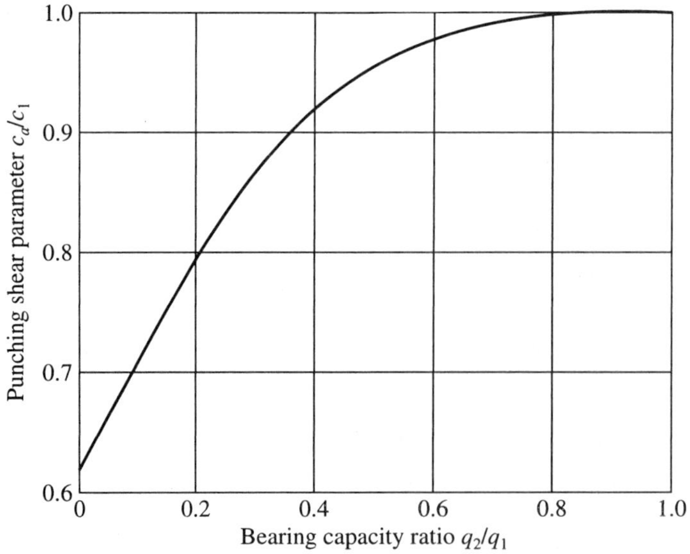

Figure 12.18 Plot of c_a/c_1 versus q_2/q_1 (after Meyerhof and Hanna, 1978)

Substituting for P_p and C_a, the equation for q_u may be written as

$$q_u = q_b + \frac{2c_a H}{B} + \frac{\gamma_1 H^2}{B}\left(1+\frac{2D_f}{H}\right)K_p \tan\delta - \gamma_1 H \tag{12.49}$$

In practice, it is convenient to use a coefficient K_s of punching shearing resistance on the vertical plane through the footing edges so that

$$K_s \tan\phi_1 = K_p \tan\delta \tag{12.50}$$

Substituting, the equation for q_u may be written as

$$q_u = q_b + \frac{2c_a H}{B} + \frac{\gamma_1 H^2}{B}\left(1+\frac{2D_f}{H}\right)K_s \tan\phi_1 - \gamma_1 H \le q_t \tag{12.51}$$

Figure 12.17 gives the value of K_s for various values of ϕ_1 as a function of q_2/q_1. The variation of c_a/c_i with q_2/q_1 is shown in Fig. 12.18.

Equation (12.45) for q_t and q_b in Eq. (12.51) are for strip footings. These equations with shape factors may be written as

$$q_t = c_1 N_{c1} s_{c1} + \gamma_1 D_f N_{q1} s_{q1} + \frac{1}{2}\gamma_1 B N_{\gamma 1} s_{\gamma 1} \tag{12.52}$$

$$q_b = c_2 N_{c2} s_{c2} + \gamma_1 (D_f + H) N_{q2} s_{q2} + \frac{1}{2}\gamma_2 B N_{\gamma 2} s_{\gamma 2} \tag{12.53}$$

where s_c, s_q and s_γ are the shape factors for the corresponding layers with subscripts 1 and 2 representing layers 1 and 2 respectively.

Eq. (12.51) can be extended to rectangular foundations by including the corresponding shape factors.

The equation for a rectangular footing may be written as

$$q_u = q_b + \left(\frac{2c_a H}{B}\right)\left(1+\frac{B}{L}\right) + \frac{\gamma_1 H^2}{B}\left(1+\frac{2D_f}{H}\right)\left(1+\frac{B}{L}\right)K_s \tan\phi_1 - \gamma_1 H \le q_t \tag{12.54}$$

Case 2: Top Layer Dense Sand and Bottom Layer Saturated Soft Clay ($\phi_2 = 0$)

The value of q_b for the bottom layer from Eq. (12.53) may be expressed as

$$q_b = c_2 N_{c2} s_{c2} + \gamma_1 (D_f + H) \tag{12.55}$$

From Table (12.3), $s_{c2} = (1+0.2\, B/L)$ (Meyerhof, 1963) and $N_c = 5.14$ for $\phi = 0$. Therefore

$$q_b = \left(1 + 0.2\frac{B}{L}\right)5.14 c_2 + \gamma_1 (D_f + H) \tag{12.56}$$

For $c_1 = 0$, q_t from Eq. (12.52) is

$$q_t = \gamma_1 D_f N_{q1} s_{q1} + \frac{1}{2}\gamma_1 B N_{\gamma 1} s_{\gamma 1} \tag{12.57}$$

We may now write an expression for q_u from Eq. (12.54) as

$$q_u = \left(1+0.2\frac{B}{L}\right)5.14c_2 + \frac{\gamma_1 H^2}{B}\left(1+\frac{2D_f}{H}\right)\left(1+\frac{B}{L}\right)K_s \tan\phi_1$$

$$+\gamma_1 D_f \le \gamma_1 D_f N_{q1} s_{q1} + \frac{1}{2}\gamma_1 B N_{\gamma 1} s_{\gamma 1} \tag{12.58}$$

The ratio of q_2/q_1 may be expressed by

$$\frac{q_2}{q_1} = \frac{c_2 N_{c2}}{0.5\gamma_1 B N_{\gamma 1}} = \frac{5.14 c_2}{0.5\gamma_1 B N_{\gamma 1}} \tag{12.59}$$

The value of K_s may be found from Fig. (12.17).

Case 3: When Layer 1 is Dense Sand and Layer 2 is Loose Sand ($c_1 = c_2 = 0$)

Proceeding in the same way as explained earlier the expression for q_u for a rectangular footing may be expressed as

$$q_u = \left[\gamma_1 (D_f + H) N_{q2} s_{q2} + \frac{1}{2}\gamma_2 B N_{\gamma 2} s_{\gamma 2}\right]$$

$$+\frac{\gamma_1 H^2}{B}\left(1+\frac{B}{L}\right)\left(1+\frac{2D_f}{H}\right)K_s \tan\phi_1 - \gamma_1 H \le q_t \tag{12.60}$$

where $q_t = \gamma_1 D_f N_{q1} s_{c1} + \frac{1}{2}\gamma_1 B N_{\gamma 1} s_{\gamma 1}$ (12.61)

$$\frac{q_2}{q_1} = \frac{\gamma_2 N_{\gamma 2}}{\gamma_1 N_{\gamma 1}} \tag{12.62}$$

Case 4: Layer 1 is Stiff Saturated Clay ($\phi_1 = 0$) and Layer 2 is Saturated Soft Clay ($\phi_2 = 0$)

The ultimate bearing capacity of the layered system can be given as

$$q_u = \left(1+0.2\frac{B}{L}\right)5.14c_2 + \left(1+\frac{B}{L}\right)\frac{2c_a H}{B} + \gamma_1 D_f \le q_t \tag{12.63}$$

$$q_t = \left(1+0.2\frac{B}{L}\right)5.14c_1 + \gamma_1 D_f \tag{12.64}$$

$$\frac{q_2}{q_1} = \frac{5.14 c_2}{5.14 c_1} = \frac{c_2}{c_1} \tag{12.65}$$

Example 12.17

A rectangular footing of size 3 × 2 m is founded at a depth of 1.5 m in a clay stratum of very stiff consistency. A clay layer of medium consistency is located at a depth of 1.5 m (= H) below the bottom of the footing (Fig. Ex. 12.17). The soil parameters of the two clay layers are as follows:

Top clay layer: $c_1 = 175$ kN/m^2

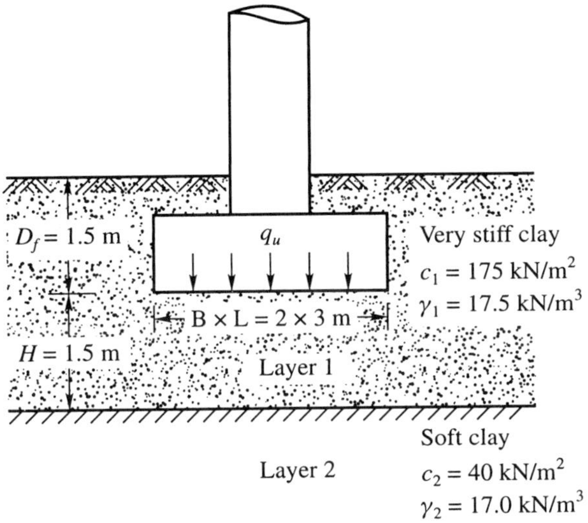

Figure Ex. 12.17

$\gamma_1 = 17.5$ kN/m^3

Bottom layer: $c_2 = 40$ kN/m^2
$\gamma_2 = 17.0$ kN/m^3

Estimate the ultimate bearing capacity and the allowable bearing pressure on the footing with a factor of safety of 3.

Solution

The solution to this problem comes under Case 4 in Section 12.14. We have to consider here Eqs (12.63), (12.64) and (12.65).

The data given are:
$B = 2$ m, $L = 3$ m, $H = 1.5$ m (Fig. 12.16a), $D_f = 1.5$ m, $\gamma_1 = 17.5$ kN/m^3.

From Fig. 12.18, for $q_2/q_1 = c_2/c_1 = 40/175 = 0.23$,
the value of $c_a/c_1 = 0.83$ or $c_a = 0.83c_1 = 0.83 \times 175 = 145.25$ kN/m^2.

From Eq. (12.63)

$$q_u = \left(1 + 0.2\frac{B}{L}\right)5.14c_2 + \left(1 + \frac{B}{L}\right)\frac{2c_a H}{B} + \gamma_1 D_f \le q_t$$

Substituting the known values

$$q_u = \left(1 + 0.2 \times \frac{2}{3}\right)5.14 \times 40 + \left(1 + \frac{2}{3}\right)\left(\frac{2 \times 145.25 \times 1.5}{2}\right) + 17.5 \times 1.5$$

$$= 233 + 364 + 26 = 623 \text{ kN/m}^2$$

From Eq. (12.64)

$$q_t = \left(1+0.2\frac{B}{L}\right)5.14c_1 + \gamma_1 D_f$$

$$= \left(1+0.2\times\frac{2}{3}\right)5.14\times 175 + 17.5\times 1.5$$

$$= 1020 + 26 = 1046 \ \text{kN/m}^2$$

It is clear from the above that $q_u < q_t$ and as such q_u is the ultimate bearing capacity to be considered. Therefore

$$q_a = \frac{q_u}{F_s} = \frac{623}{3} = 208 \ \text{kN/m}^2$$

Example 12.18

Determine the ultimate bearing capacity of the footing given in Example 12.17 in dense sand with the bottom layer being a clay of medium consistency. The parameters of the top layer are:

$\gamma_1 = 18.5$ kN/m^3, $\phi_1 = 39°$

All the other data given in Ex. 12.17 remain the same. Use Meyerhof's bearing capacity and shape factors.

Solution

Case 2 of Section 12.14 is required to be considered here.

Given: Top layer: $\gamma_1 = 18.5$ kN/m^3, $\phi_1 = 39°$, Bottom layer: $\gamma_2 = 17.0$ kN/m^3, $c_2 = 40$ kN/m^2.

From Table 12.2 $N_{\gamma 1}$ $(M) = 78.8$ for $\phi_1 = 39°$.

From Eq. (12.59)

$$\frac{q_2}{q_1} = \frac{5.14c_2}{0.5\gamma_1 BN_{\gamma 1}} = \frac{5.14\times 40}{0.5\times 18.5\times 2\times 78.8} = 0.141$$

From Fig. 12.17 $K_s = 2.9$ for $\phi = 39°$

Now from Eq. (12.58) we have

$$q_u = \left(1+0.2\frac{B}{L}\right)5.14c_2 + \frac{\gamma_1 H^2}{B}\left[1+\frac{2D_f}{H}\right]\left[1+\frac{B}{L}\right]K_s \tan\phi_1 + \gamma_1 D_f$$

$$= \left(1+0.2\times\frac{2}{3}\right)5.14\times 40 + \frac{18.5\times(1.5)^2}{2}\left[1+\frac{2\times 1.5}{1.5}\right]\left[1+\frac{2}{3}\right]2.9\tan 39° + 18.5\times 1.5$$

$$= 233 + 245 + 28 = 506 \ \text{kN/m}^2$$

$$q_u = 506 \ \text{kN/m}^2$$

From Eq. (12.58) the limiting value q_t is

$$q_t = \gamma_1 D_f N_{q1} s_{q1} + \frac{1}{2}\gamma_1 BN_{\gamma 1} s_{\gamma 1}$$

where γ_1 = 18.5 kN/m³, D_f = 1.5m, B = 2 m.

From Table 12.2 $N_{\gamma 1} = 78.8$ and $N_{q1} = 56.5$

From Table 12.3 $s_{q1} = 1 + 0.1N_\phi\left[\frac{B}{L}\right] = 1 + 0.1 \times \tan^2\left(45° + \frac{39}{2}\right) \times \frac{2}{3} = 1.29 = s_{\gamma 1}$

Now $q_t = 18.5 \times 1.5 \times 56.5 \times 1.29 + \frac{1}{2} \times 18.5 \times 2 \times 78.8 \times 1.29 = 3903 \text{ kN/m}^2 > q_u$

Hence $q_u = 506 \text{ kN/m}^2$

12.15 BEARING CAPACITY OF FOUNDATIONS ON TOP OF A SLOPE

There are occasions where structures are required to be built on slopes or near the edges of slopes. Since full formations of shear zones under ultimate loading conditions are not possible on the sides close to the slopes or edges, the supporting capacity of soil on that side get considerably reduced. Meyerhof (1957) extended his theories to include the effect of slopes on the stability of foundations.

Figure 12.19 shows a section of a foundation with the failure surfaces under ultimate loading condition. The stability of the foundation depends on the distance $\bar{b}$ of the top edge of the slope from the face of the foundation.

The form of ultimate bearing capacity equation for a strip footing may be expressed as (Meyerhof, 1957)

$$q_u = cN_{cq} + \frac{1}{2}\gamma BN_{\gamma q} \tag{12.66}$$

The upper limit of the bearing capacity of a foundation in a purely cohesive soil may be estimated from

$$q_u = cN_{cq} + \gamma D_f \tag{12.67}$$

The resultant bearing capacity factors N_{cq} and $N_{\gamma q}$ depend on the distance $\bar{b}$, β, ϕ and the D_f/B ratio. These bearing capacity factors are given in Figs 12.20 and 12.21 for strip foundation in purely cohesive and cohesionless soils respectively. It can be seen from the figures that the bearing capacity factors increase with an increase of the distance $\bar{b}$. Beyond a distance of about 2 to 6 times the foundation width B, the bearing capacity is independent of the inclination of the slope, and becomes the same as that of a foundation on an extensive horizontal surface.

For a surcharge over the entire horizontal top surface of a slope, a solution of the slope stability has been obtained on the basis of dimensionless parameters called the stability number N_s, expressed as

$$N_s = \frac{c}{\gamma H} \tag{12.68}$$

The bearing capacity of a foundation on purely cohesive soil of great depth can be represented by Eq. (12.67) where the N_{cq} factor depends on $\bar{b}$ as well as β, and the stability number N_s. This bearing capacity factor, which is given in the lower parts of Fig. 12.20, decrease considerably with greater height and to a smaller extent with the inclination of the slope. For a given height and slope angle, the bearing capacity factor increases with an increase in $\bar{b}$, and

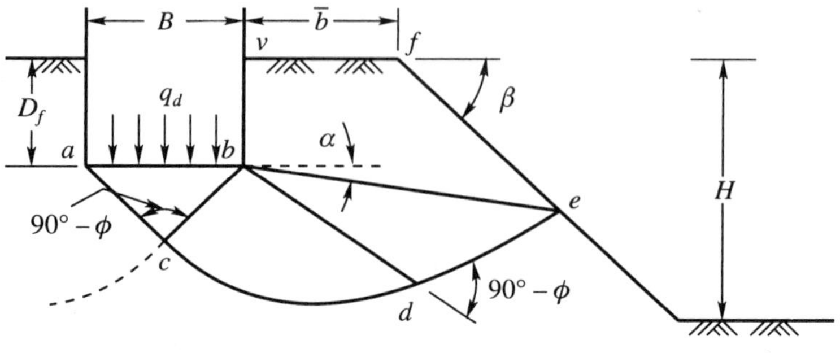

Figure 12.19 Bearing capacity of a strip footing on top of a slope (Meyerhof, 1957)

beyond a distance of about 2 to 4 times the height of the slope, the bearing capacity is independent of the slope angle. Figure 12.20 shows that the bearing capacity of foundations on top of a slope is governed by foundation failure for small slope height (N_s approaching infinity) and by overall slope failure for greater heights.

The influence of ground water and tension cracks (in purely cohesive soils) should also be taken into account in the study of the overall stability of the foundation.

Meyerhof (1957) has not supported his theory with any practical examples of failure as any published data were not available for this purpose.

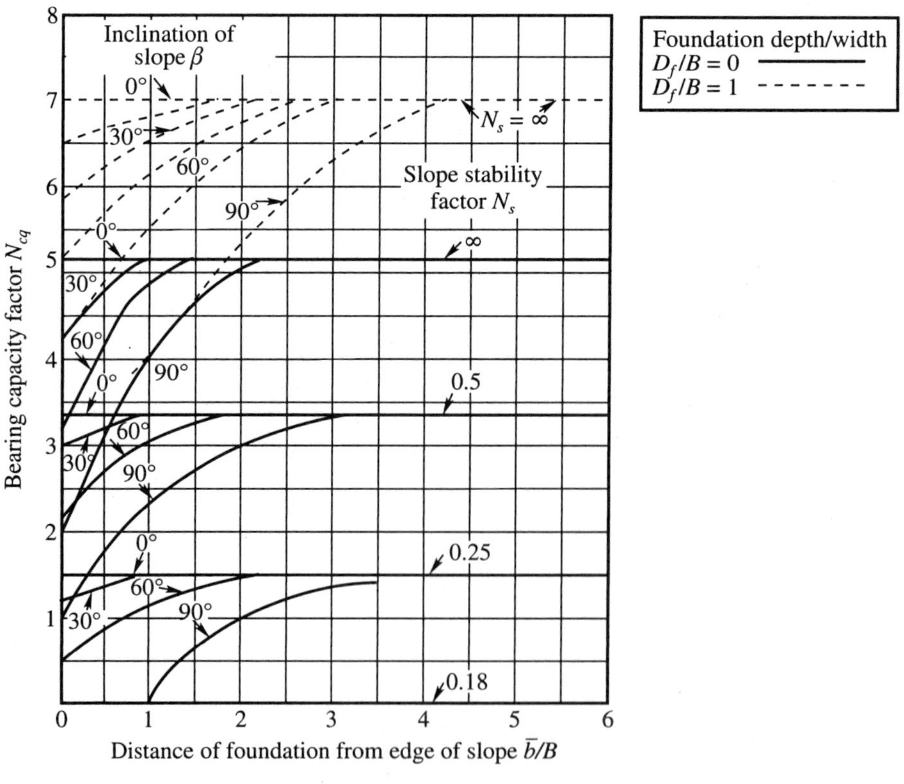

Figure 12.20 Bearing capacity factors for strip foundation on top of slope of purely cohesive material (Meyerhof, 1957)

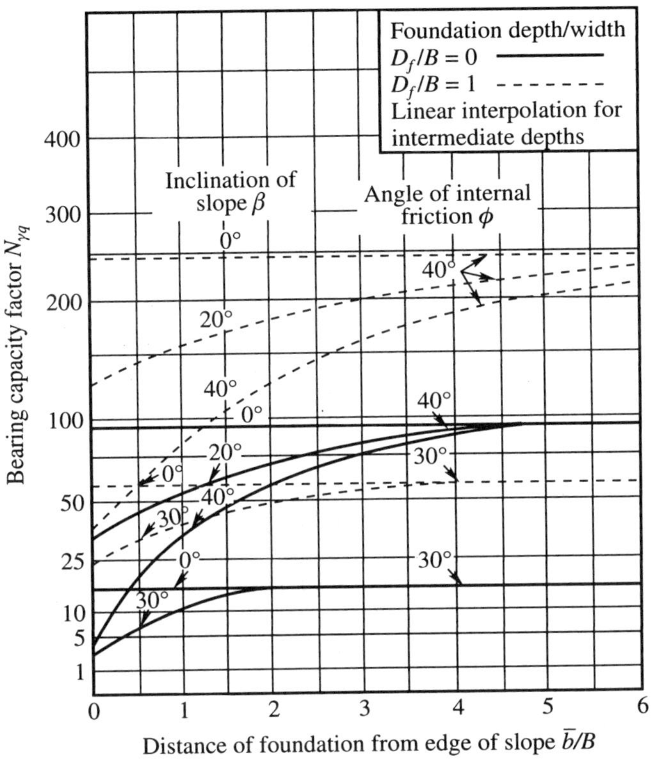

Figure 12.21 Bearing capacity factors for strip foundation on top of slope of cohesionless material (Meyerhof, 1957)

Example 12.19

A strip footing is to be constructed on the top of a slope as per Fig. 12.19. The following data are available:

$B = 3$ m, $D_f = 1.5$ m, $\bar{b} = 2$ m, $H = 8$ m, $\beta = 30°$, $\gamma = 18.5$ kN/m^3, $\phi = 0$ and $c = 75$ kN/m^2,

Determine the ultimate bearing capacity of the footing.

Solution

Per Eq. (12.67) q_u for $\phi = 0$ is

$$q_u = cN_{cq} + \gamma D_f$$

From Eq. (12.68)

$$N_s = \frac{c}{\gamma H} = \frac{75}{18.5 \times 8} = 0.51$$

and $\dfrac{\bar{b}}{B} = \dfrac{2}{3} = 0.67$

From Fig. 12.20, $N_{cq} = 3.4$ for $N_s = 0.51$, $\bar{b}/B = 0.67$, and $\beta = 30°$

Therefore

$q_u = 75 \times 3.4 + 18.5 \times 1.5 = 255+28 = 283$ kN/m^2.

12.16 FOUNDATIONS ON ROCK

Rocks encountered in nature might be igneous, sedimentary or metamorphic. Granite and basalt belong to the first group. Granite is primarily composed of feldspar, quartz and mica possesses a massive structure. Basalt is a dark-colored fine grained rock. Both basalt and granite under unweathered conditions serve as a very good foundation base. The most important rocks that belong to the second group are sandstones, limestones and shales. These rocks are easily weathered under hostile environmental conditions and as such, the assessment of bearing capacity of these types requires a little care. In the last group come gneiss, schist, slate and marble. Of these rocks gneiss serves as a good bearing material whereas schist and slate possess poor bearing quality.

All rocks weather under hostile environments. The ones that are close to the ground surface become weathered more than the deeper ones. If a rocky stratum is suspected close to the ground surface, the soundness or otherwise of these rocks must be investigated. The quality of the rocks is normally designated by RQD as explained in Chapter 9.

Joints are common in all rock masses. This word joint is used by geologists for any plane of structural weakness apart from faults within the mass. Within the sedimentary rock mass the joints are lateral in extent and form what are called bedding planes, and they are uniform throughout any one bed within igneous rock mass. Cooling joints are more closely spaced nearer the ground surface with wider spacings at deeper depths. Tension joints and tectonic joints might be expected to continue depth wise. Within metamorphic masses, open cleavage, open schistose and open gneissose planes can be of considerably further lateral extent than the bedding planes of the sedimentary masses.

Faults and fissures happen in rock masses due to internal disturbances. The joints with fissures and faults reduces the bearing strength of rocky strata.

Since most unweathered intact rocks are stronger and less compressible than concrete, the determination of bearing pressures on such materials may not be necessary. A confined rock possesses greater bearing strength than the rocks exposed at ground level.

Bearing Capacity of Rocks

Bearing capacities of rocks are often determined by crushing a core sample in a testing machine. Samples used for testing must be free from cracks and defects.

In the rock formation where bedding planes, joints and other planes of weakness exist, the practice that is normally followed is to classify the rock according to RQD (Rock Quality Designation). Table 9.2 gives the classification of the bearing capacity of rock according to RQD. Peck et al, (1974) have related the RQD to the allowable bearing pressure q_a as given in Table 12.5

The RQD for use in Table 12.5 should be the average within a depth below foundation level equal to the width of the foundation, provided the RQD is fairly uniform within that depth. If the upper part of the rock, within a depth of about $B/4$, is of lower quality, the value of this part should be used.

Table 12.5 Allowable Bearing Pressure q_a on Jointed Rock

RQD	q_a Ton/ft^2	q_a MPa
100	300	29
90	200	19
75	120	12
50	65	6.25
25	30	3
0	10	0.96

Table 12.6 Presumptive allowable bearing pressures on rocks (MPa) as recommended by various building codes in USA (after Peck et al., 1974)

Rock type	Building Codes			
	BOCA (1968)	National (1967)	Uniform (1964)	LOS Angeles (1959)
1. Massive crystalline bedrock, including granite diorite, gneiss, basalt, hard limestone and dolomite	10	10	$0.2\ q_u$*	1.0
2. Foliated rocks such as schist or slate in sound condition	4	4	$0.2\ q_u$	0.4
3. Bedded limestone in sound condition sedimentary rocks including hard shales and sandstones	2.5	1.5	$0.2\ q_u$	0.3
4. Soft or broken bedrock (excluding shale) and soft limestone	1.0	–	$0.2\ q_u$	–
5. Soft shale	0.4	–	$0.2\ q_u$	–

* q_u = unconfined compressive strength.

Another practice that is normally followed is to base the allowable pressure on the unconfined compressive strength, q_u, of the rock obtained in a laboratory on a rock sample. A factor of safety of 5 to 8 is used on q_u to obtain q_a. Past experience indicates that this method is satisfactory so long as the rocks *in situ* do not possess extensive cracks and joints. In such cases a higher factor of safety may have to be adopted.

If rocks close to a foundation base are highly fissured/fractured, they can be treated by grouting which increases the bearing capacity of the material.

The bearing capacity of a homogeneous, and discontinuous rock mass cannot be less than the unconfined compressive strength of the rock mass around the footing and this can be taken as a lower bound for a rock mass with constant angle of internal friction ϕ and unconfined compressive strength q_{ur}. Goodman (1980) suggests the following equation for determining the ultimate bearing capacity q_u.

$$q_u = q_{ur}(N_\phi + 1) \tag{12.69}$$

where $N_\phi = \tan^2(45° + \phi/2)$, q_{ur} = unconfined compressive strength of rock.

Recommendations by Building Codes

Where bedrock can be reached by excavation, the presumptive allowable bearing pressure is specified by Building Codes. Table 12.7 gives the recommendations of some buildings codes in the U.S.

12.17 CASE HISTORY OF FAILURE OF THE TRANSCONA GRAIN ELEVATOR

One of the best known foundation failures occurred in October 1913 at North Transcona, Manitoba, Canada. It was ascertained later on that the failure occurred when the foundation

Figure 12.22 The tilted Transcona grain elevator (Courtesy: UMA Engineering Ltd., Manitoba, Canada)

Figure 12.23 The straightened Transcona grain elevator (Courtesy: UMA Engineering Ltd., Manitoba, Canada)

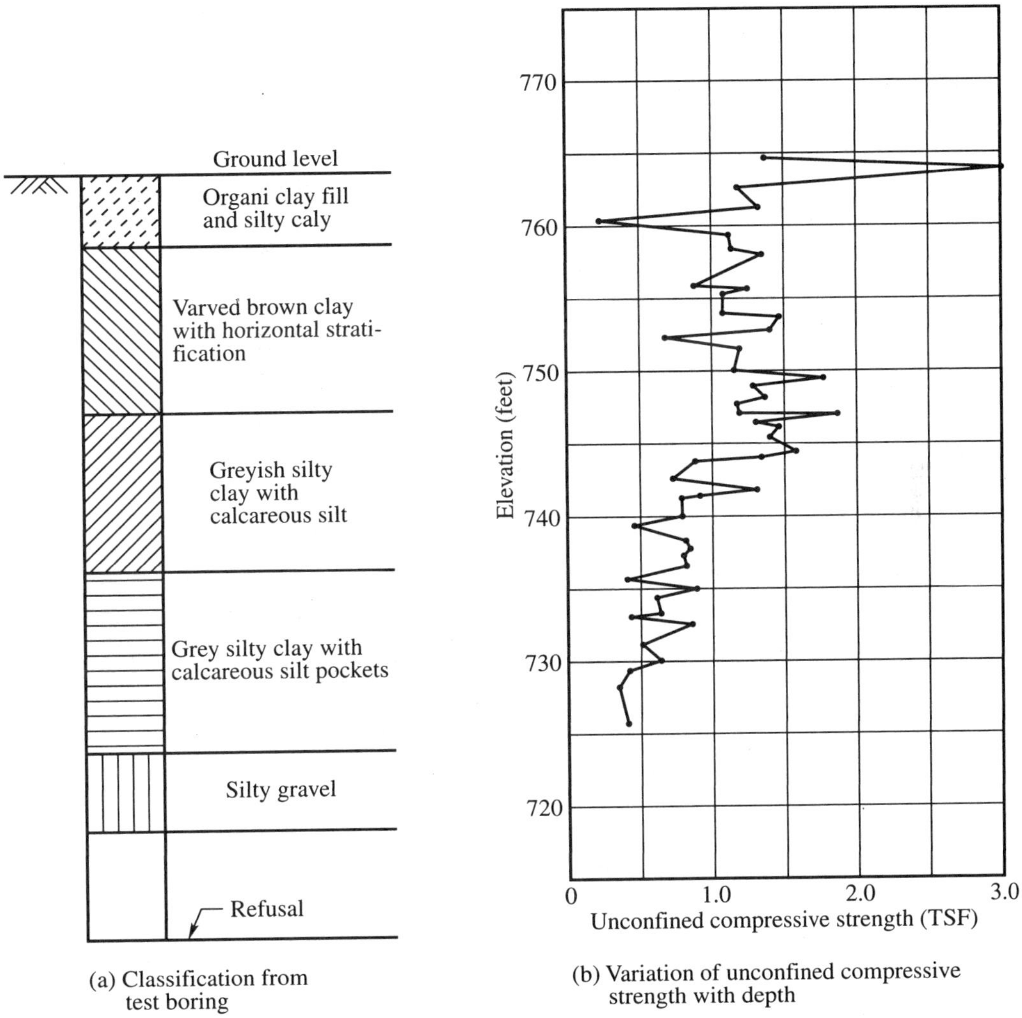

Figure 12.24 Results of test boring at site of Transcona grain elevator (Peck and Byrant, 1953)

pressure at the base was about equal to the calculated ultimate bearing capacity of an underlaying layer of plastic clay (Peck and Byrant,1953), and was essentially a shearing failure.

The construction of the silo started in 1911 and was completed in the autumn of 1913. The silo is 77 ft by 195 ft in plan and has a capacity of 1,000,000 bushels. It comprises 65 circular bins and 48 inter-bins. The foundation was a reinforced concrete raft 2 ft thick and founded at a depth of 12 ft below the ground surface. The weight of the silo was 20,000 tons, which was 42.5 percent of the total weight, when it was filled. Filling the silo with grain started in September 1913, and in October when the silo contained 875,000 bushels, and the pressure on the ground was 94 percent of the design pressure, a vertical settlement of 1 ft was noticed. The structure began to tilt to the west and within twenty four hours was at an angle of 26.9° from the vertical, the west side being 24 ft below and the east side 5 ft above the original level (Szechy, 1961). The structure tilted as a monolith and there was no damage to the structure except for a few superficial cracks. Figure 12.22 shows a view of the tilted structure. The excellent quality of the reinforced concrete structure is shown by the fact that later it was underpinned and jacked up on new piers founded on rock. The level of the new foundation is 34 ft below the ground surface. Figure 12.23 shows the view of the silo after it was straightened in 1916.

During the period when the silo was designed and constructed, soil mechanics as a science had hardly begun. The behavior of the foundation under imposed loads was not clearly understood. It was only during the year 1952 that soil investigation was carried out close to the silo and the soil properties were analyzed (Peck and Byrant, 1953). Figure 12.24 gives the soil classification and unconfined compressive strength of the soil with respect to depth. From the examination of undisturbed samples of the clay, it was determined that the average water content of successive layers of varved clay increased with their depth from 40 percent to about 60 percent. The average unconfined compressive strength of the upper stratum beneath the foundation was 1.13 tsf, that of the lower stratum was 0.65 tsf, and the weighted average was 0.93 tsf. The average liquid limit was found to be 105 percent; therefore the plasticity index was 70 percent, which indicates that the clay was highly colloidal and plastic. The average unit weight of the soil was 120 lb/ft^3.

The contact pressure due to the load from the silo at the time of failure was estimated as equal to 3.06 tsf. The theoretical values of the ultimate bearing capacity by various methods are as follows.

Methods	q_u tsf
Terzaghi [Eq. (12.19)]	3.68
Meyerhof [Eq. (12.27)]	3.30
Skempton [Eq. (12.22)]	3.32

The above values compare reasonably well with the actual failure load 3.06 tsf. Perloff and Baron (1976) give details of failure of the Transcona grain elevator.

12.18 PROBLEMS

12.1 What will be the gross and net allowable bearing pressures of a sand having $\phi = 35°$ and an effective unit weight of 18 kN/m^3 under the following cases: (a) size of footing 1 × 1 m square, (b) circular footing of 1 m dia., and (c) 1 m wide strip footing.

The footing is placed at a depth of 1 m below the ground surface and the water table is at great depth. Use $F_s = 3$. Compute by Terzaghi's general shear failure theory.

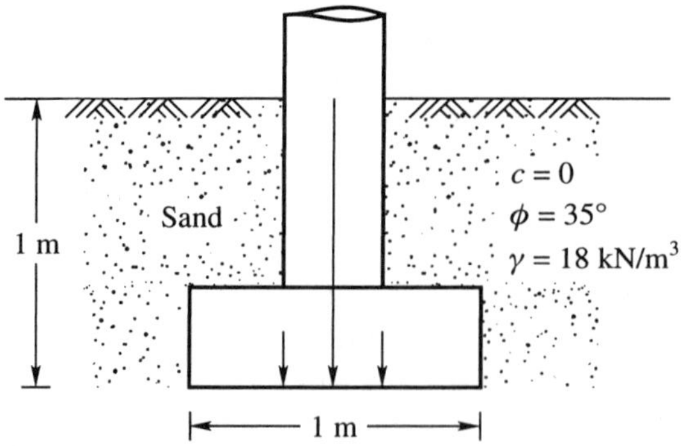

Figure Prob. 12.1

12.2 A strip footing is founded at a depth of 1.5 m below the ground surface (Fig. Prob. 12.2). The water table is close to ground level and the soil is cohesionless. The footing is supposed to carry a net safe load of 400 kN/m^2 with $F_s = 3$. Given $\gamma_{sat} = 20.85$ kN/m^3 and

$\phi = 35°$, find the required width of the footing, using Terzaghi's general shear failure criterion.

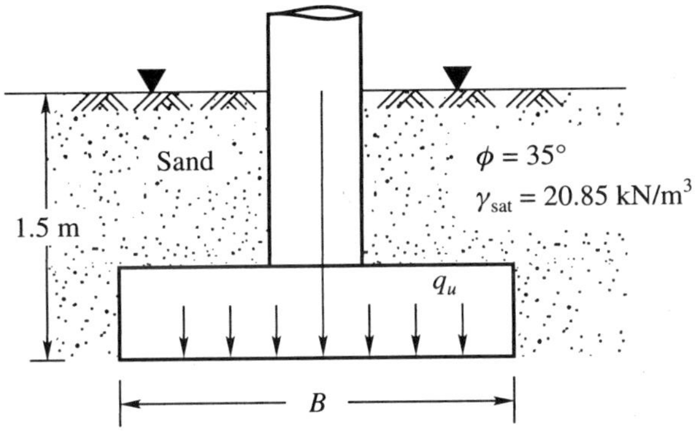

Figure Prob. 12.2

12.3 At what depth should a footing of size 2 × 3 m be founded to provide a factor of safety of 3 if the soil is stiff clay having an unconfined compressive strength of 120 kN/m^2? The unit weight of the soil is 18 kN/m^3. The ultimate bearing capacity of the footing is 425 kN/m^2. Use Terzaghi's theory. The water table is close to the ground surface (Fig. Prob. 12.3).

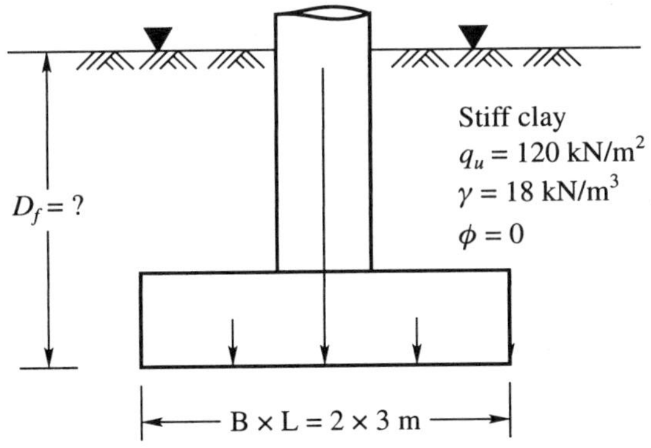

Figure Prob. 12.3

12.4 A rectangular footing is founded at a depth of 2 m below the ground surface in a $(c - \phi)$ soil having the following properties: porosity $n = 40\%$, $G_s = 2.67$, $c = 15$ kN/m^2, and $\phi = 30°$. The water table is close to the ground surface. If the width of the footing is 3 m, what is the length required to carry a gross allowable bearing pressure $q_a = 455$ kN/m^2 with a factor of safety = 3? Use Terzaghi's theory of general shear failure (Figure Prob. 12.4).

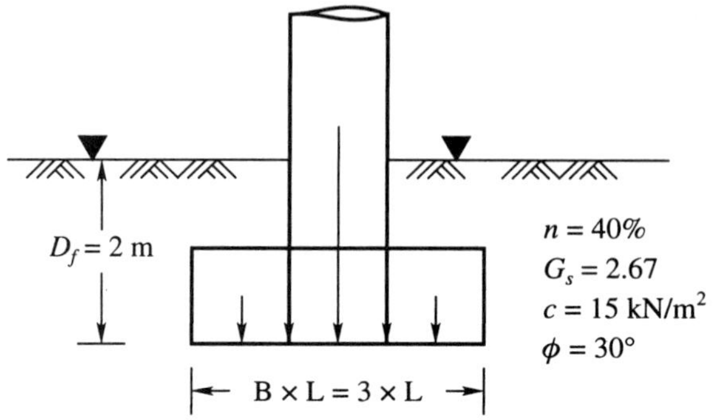

Figure Prob. 12.4

12.5 A square footing located at a depth of 5 ft below the ground surface in a cohesionless soil carries a column load of 130 tons. The soil is submerged having an effective unit weight of 73 lb/ft^3 and an angle of shearing resistance of 30°. Determine the size of the footing for $F_s = 3$ by Terzaghi's theory of general shear failure (Fig. Prob. 12.5).

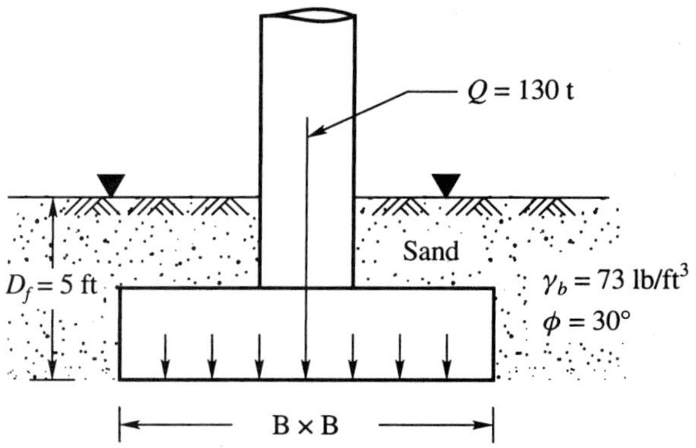

Figure Prob. 12.5

12.6 A footing of 5 ft diameter carries a safe load (including its self weight) of 80 tons in cohesionless soil (Fig. Prob. 12.6). The soil has an angle of shearing resistance $\phi = 36°$ and an effective unit weight of 80 lb/ft^3. Determine the depth of the foundation for $F_s = 2.5$ by Terzaghi's general shear failure theory.

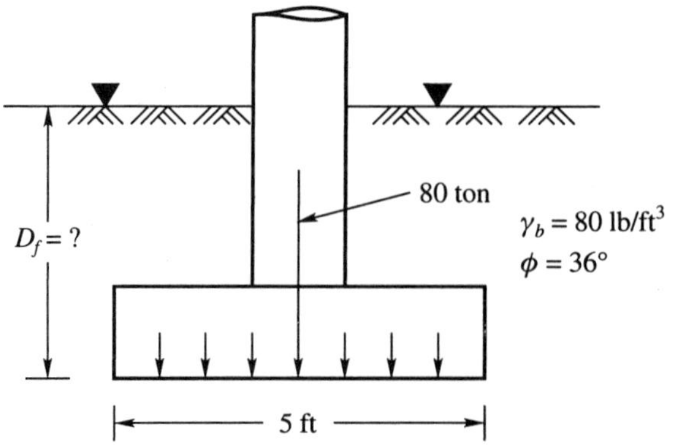

Figure Prob. 12.6

12.7 If the ultimate bearing capacity of a 4 ft wide strip footing resting on the surface of a sand is 5,250 lb/ft^2, what will be the net allowable pressure that a 10 × 10 ft square footing resting on the surface can carry with $F_s = 3$? Assume that the soil is cohesionless. Use Terzaghi's theory of general shear failure.

12.8 A circular plate of diameter 1.05 m was placed on a sand surface of unit weight 16.5 kN/m^3 and loaded to failure. The failure load was found to give a pressure of 1,500 kN/m^2. Determine the value of the bearing capacity factor N_γ. The angle of shearing resistance of the sand measured in a triaxial test was found to be 39°. Compare this value with the theoretical value of N_γ. Use Terzaghi's theory of general shear failure.

12.9 Find the net allowable bearing load per foot length of a long wall footing 6 ft wide founded on a stiff saturated clay at a depth of 4 ft. The unit weight of the clay is 110 lb/ft^3, and the shear strength is 2500 lb/ft^2. Assume the load is applied rapidly such that undrained conditions ($\phi = 0$) prevail. Use $F_s = 3$ and Skempton's method (Fig. Prob. 12.9).

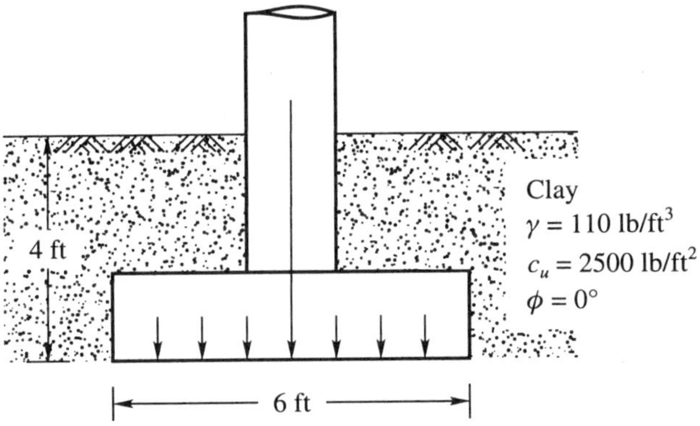

Figure Prob. 12.9

12.10 The total column load of a footing near ground level is 5000 kN. The subsoil is cohesionless soil with $\phi = 38°$ and $\gamma = 19.5$ kN/m^3. The footing is to be located at a depth of 1.50 m below ground level. For a footing of size 3 × 3 m, determine the factor of safety by Terzaghi's general shear failure theory if the water table is at a depth of 0.5 m below the base level of the foundation.

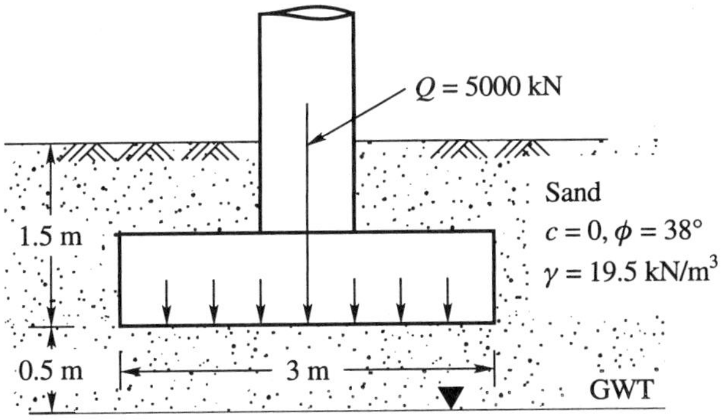

Figure Prob. 12.10

12.11 What will be the factors of safety if the water table is met (i) at the base level of the foundation, and (ii) at the ground level in the case of the footing in Prob. 12.10, keeping all the other conditions the same? Assume that the saturated unit weight of soil γ_{sat} = 19.5 kN/m^3, and the soil above the base of the foundation remains saturated even under (i) above.

12.12 If the factors of safety in Prob. 12.11 are other than 3, what should be the size of the footing for a minimum factor of safety of 3 under the worst condition?

12.13 A footing of size 10 × 10 ft is founded at a depth of 5 ft in a medium stiff clay soil having an unconfined compressive strength of 2000 lb/ft^2. Determine the net safe bearing capacity of the footing with the water table at ground level by Skempton's method. Assume F_s = 3.

12.14 If the average static cone penetration resistance, q_c, in Prob. 12.13 is 10 t/ft^2, determine q_{na} per Skempton's method. The other conditions remain the same as in Prob. 12.13. Ignore the effect of overburden pressure.

12.15 Refer to Prob. 12.10. Compute by Meyerhof theory (a) the ultimate bearing capacity, (b) the net ultimate bearing capacity, and (c) the factor of safety for the load coming on the column. All the other data given in Prob. 12.10 remain the same.

12.16 Refer to Prob. 12.10. Compute by Hansen's method (a) the ultimate bearing capacity, (b) the net ultimate bearing capacity, and (c) the factor of safety for the column load. All the other data remain the same. Comment on the results using the methods of Terzaghi, Meyerhof and Hansen.

12.17 A rectangular footing of size (Fig. 12.17) 12 × 24 ft is founded at a depth of 8 ft below the ground surface in a ($c - \phi$) soil. The following data are available: (a) water table at a depth of 4 ft below ground level, (b) c = 600 lb/ft^2, ϕ = 30°, and γ = 118 lb/ft^3. Determine the ultimate bearing capacity by Terzaghi and Meyerhof's methods.

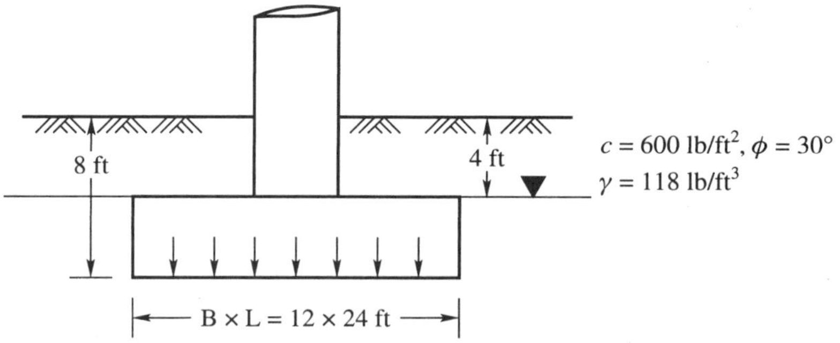

Figure Prob. 12.17

12.18 Refer to Prob. 12.17 and determine the ultimate bearing capacity by Hansen's method. All the other data remain the same.

12.19 A rectangular footing of size (Fig. Prob. 12.19) 16 × 24 ft is founded at a depth of 8 ft in a deep stratum of ($c - \phi$) soil with the following parameters:

c = 300 lb/ft^2, ϕ = 30°, E_s = 75 t/ft^2, γ = 105 lb/ft^3, μ = 0.3.

Estimate the ultimate bearing capacity by (a) Terzaghi's method, and (b) Vesic's method by taking into account the compressibility factors.

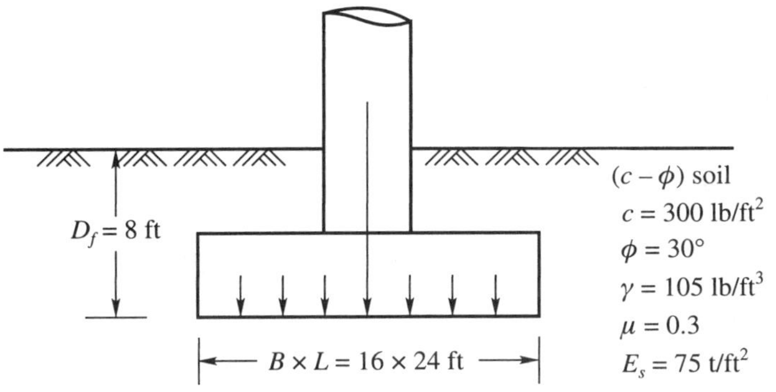

Figure Prob. 12.19

12.20 A footing of size 10 × 15 ft (Fig. Prob. 12.20) is placed at a depth of 10 ft below the ground surface in a deep stratum of saturated clay of soft to medium consistency. The unconfined compressive strength of clay under undrained conditions is 600 lb/ft^2 and $\mu = 0.5$. Assume $\gamma = 95$ lb/ft^3 and E_s = 12 t/ft^2. Estimate the ultimate bearing capacity of the soil by the Terzaghi and Vesic methods by taking into account the compressibility factors.

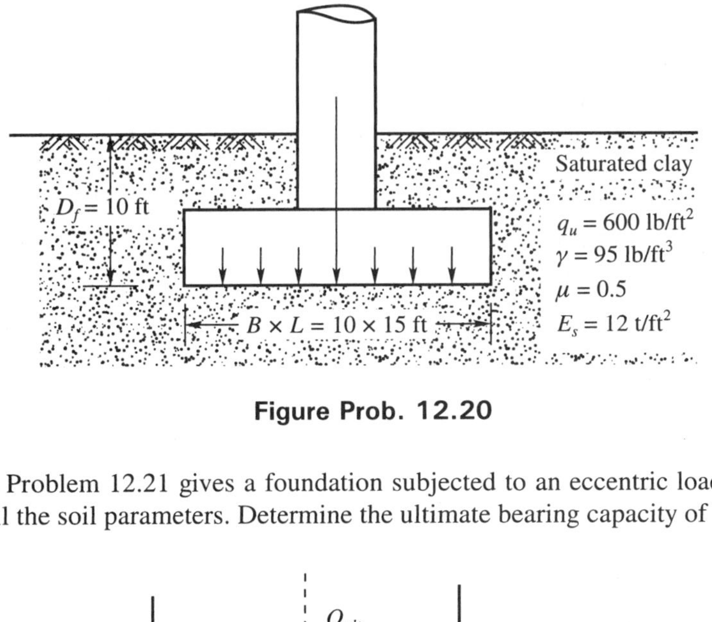

Figure Prob. 12.20

12.21 Figure Problem 12.21 gives a foundation subjected to an eccentric load in one direction with all the soil parameters. Determine the ultimate bearing capacity of the footing.

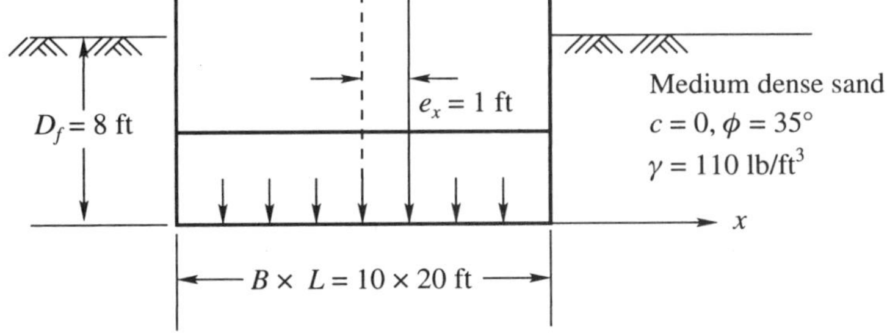

Figure Prob. 12.21

12.22 Refer to Fig. Prob. 12.22. Determine the ultimate bearing capacity of the footing if $e_x = 3$ ft and $e_y = 4$ ft. What is the allowable load for $F_s = 3$?

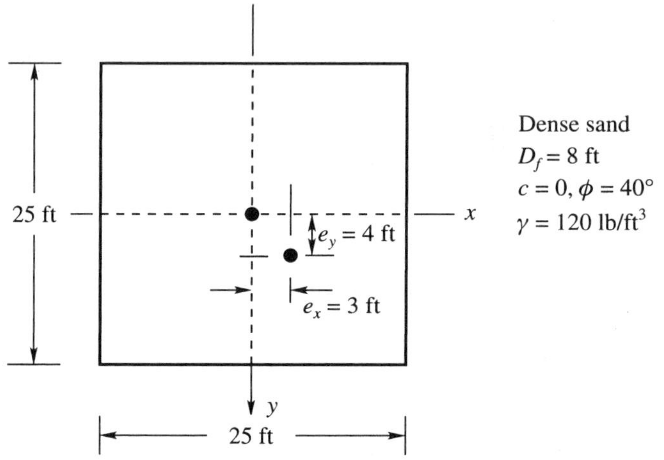

Figure Prob. 12.22

12.23 Refer to Fig. Prob. 12.22. Compute the maximum and minimum contact pressures for a column load of $Q = 800$ tons.

12.24 A rectangular footing (Fig. Prob. 12.24) of size 6 × 8 m is founded at a depth of 3 m in a clay stratum of very stiff consistency overlying a softer clay stratum at a depth of 5 m from the ground surface. The soil parameters of the two layers of soil are:
Top layer: $c_1 = 200$ kN/m^2, $\gamma_1 = 18.5$ kN/m^3
Bottom layer: $c_2 = 35$ kN/m^2, $\gamma_2 = 16.5$ kN/m^3
Estimate the ultimate bearing capacity of the footing.

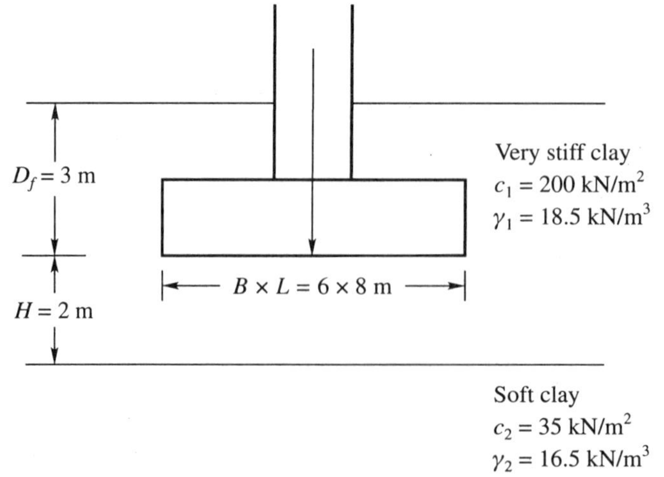

Figure Prob. 12.24

12.25 If the top layer in Prob. 12.24 is dense sand, what is the ultimate bearing capacity of the footing? The soil parameters of the top layer are:

$\gamma_1 = 19.5$ kN/m^3, $\phi_1 = 38°$

All the other data given in Prob. 12,24 remain the same.

12.26 A rectangular footing of size 3 × 8 m is founded on the top of a slope of cohesive soil as given in Fig. Prob. 12.26. Determine the ultimate bearing capacity of the footing.

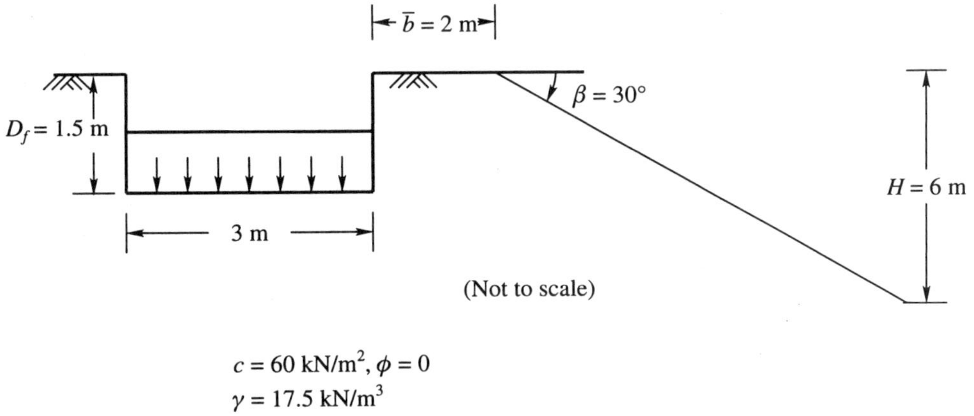

Figure Prob. 12.26

CHAPTER 13

SHALLOW FOUNDATION II: SAFE BEARING PRESSURE AND SETTLEMENT CALCULATION

13.1 INTRODUCTION

Allowable and Safe Bearing Pressures

The methods of calculating the ultimate bearing capacity of soil have been discussed at length in Chapter 12. The theories used in that chapter are based on shear failure criteria. They do not indicate the settlement that a footing may undergo under the ultimate loading conditions. From the known ultimate bearing capacity obtained from any one of the theories, the allowable bearing pressure can be obtained by applying a suitable factor of safety to the ultimate value.

When we design a foundation, we must see that the structure is safe on two counts. They are,

1. The supporting soil should be safe from shear failure due to the loads imposed on it by the superstructure,
2. The settlement of the foundation should be within permissible limits.

Hence, we have to deal with two types of bearing pressures. They are,

1. A pressure that is safe from shear failure criteria,
2. A pressure that is safe from settlement criteria.

For the sake of convenience, let us call the first the *allowable bearing pressure* and the second the *safe bearing pressure*.

In all our design, we use only the net bearing pressure and as such we call q_{na} the *net allowable bearing pressure* and q_s the net safe bearing pressure. In designing a foundation, we use the least of the two bearing pressures. In Chapter 12 we learnt that q_{na} is obtained by applying a suitable factor of safety (normally 3) to the net ultimate bearing capacity of soil. In this chapter we will learn how to obtain q_s. Even without knowing the values of q_{na} and q_s, it is possible to say from experience which of the two values should *be used in design* based upon the composition and density of soil and the size of the footing. The composition and density of the soil and the size of the footing decide the relative values of q_{na} and q_s.

The ultimate bearing capacity of footings on sand increases with an increase in the width, and in the same way the settlement of the footing increases with increases in the width. In other words for a given settlement S_1, the corresponding unit soil pressure decreases with an increase in the width of the footing. It is therefore, essential to consider that settlement will be the criterion for the design of footings in sand beyond a particular size. Experimental evidence indicates that for footings smaller than about 1.20 m, the allowable bearing pressure q_{na} is the criterion for the design of footings, whereas settlement is the criterion for footings greater than 1.2 m width.

The bearing capacity of footings on clay is independent of the size of the footings and as such the unit bearing pressure remains theoretically constant in a particular environment. However, the settlement of the footing increases with an increase in the size. It is essential to take into consideration both the shear failure and the settlement criteria together to decide the safe bearing pressure.

However, footings on stiff clay, hard clay, and other firm soils generally require no settlement analysis if the design provides a minimum factor of safety of 3 on the net ultimate bearing capacity of the soil. Soft clay, compressible silt, and other weak soils will settle even under moderate pressure and therefore settlement analysis is necessary.

Effect of Settlement on the Structure

If the structure as a whole settles uniformly into the ground there will not be any detrimental effect on the structure as such. The only effect it can have is on the service lines, such as water and sanitary pipe connections, telephone and electric cables etc. which can break if the settlement is considerable. Such uniform settlement is possible only if the subsoil is homogeneous and the load distribution is uniform. Buildings in Mexico City have undergone settlements as large as 2 m. However, the differential settlement if it exceeds the permissible limits will have a devastating effect on the structure.

According to experience, the differential settlement between parts of a structure may not exceed 75 percent of the normal absolute settlement. The various ways by which differential settlements may occur in a structure are shown in Fig. 13.1. Table 13.1 gives the absolute and permissible differential settlements for various types of structures.

Foundation settlements must be estimated with great care for buildings, bridges, towers, power plants and similar high cost structures. The settlements for structures such as fills, earthdams, levees, etc. can be estimated with a greater margin of error.

Approaches for Determining the Net Safe Bearing Pressure

Three approaches may be considered for determining the net safe bearing pressure of soil. They are,

1. Field plate load tests,
2. Charts,
3. Empirical equations.

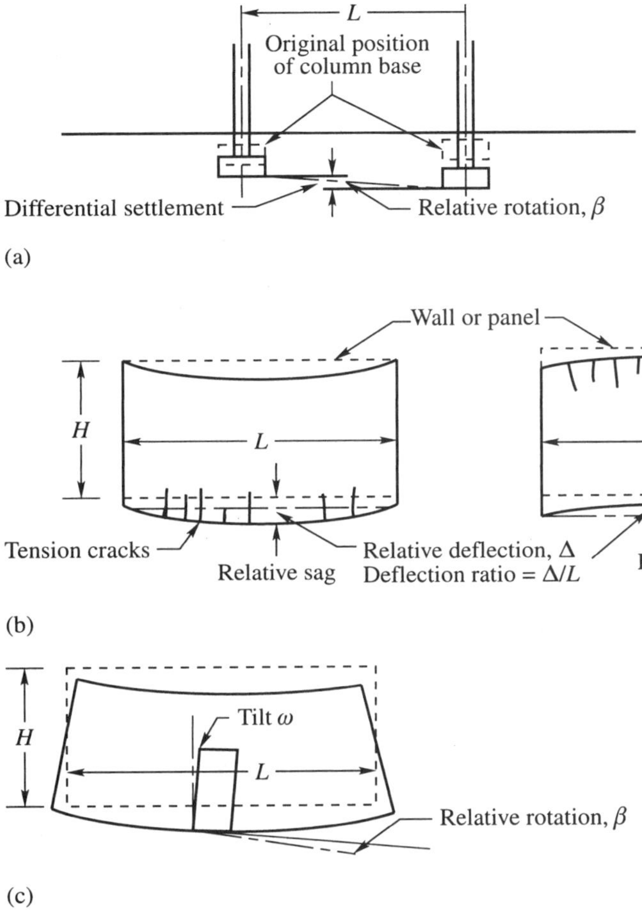

Figure 13.1 Definitions of differential settlement for framed and load-bearing wall structures (after Burland and Wroth, 1974)

Table 13.1a Maximum settlements and differential settlements of buildings in cm. (After McDonald and Skempton, 1955)

Sl. no.	Criterion	Isolated foundations	Raft
1.	Angular distortion	1/300	1/300
2.	Greatest differential settlements		
	Clays	4-5	4.5
	Sands	3-25	3.25
3.	Maximum Settlements		
	Clays	7.5	10.0
	Sands	5.0	6.25

Table 13.1b Permissible settlements (1955, U.S.S.R. Building Code)

Sl.no.	Type of building	Average settlement (cm)
1.	Building with plain brickwalls on continuous and separate foundations with wall length L to wall height H	
	$L/H \geq 2.5$	7.5
	$L/H \leq 1.5$	10.0
2.	Framed building	10.0
3.	Solid reinforced concrete foundation of blast furnaces, water towers etc.	30

Table 13.1c Permissible differential settlement (U.S.S.R Building Code, 1955)

		Type of soil	
Sl.no.	Type of structure	Sand and hard clay	Plastic clay
1.	Steel and reinforced concrete structures	$0.002L$	$0.002L$
2.	Plain brick walls in multistory buildings		
	for $L/H \leq 3$	$0.0003L$	$0.0004L$
	$L/H \geq 5$	$0.0005L$	$0.0007L$
3.	Water towers, silos etc.	$0.004L$	$0.004L$
4.	Slope of crane way as well as track		
	for bridge crane track	$0.003L$	$0.003L$

where, L = distance between two columns or parts of structure that settle different amounts, H = Height of wall.

13.2 FIELD PLATE LOAD TESTS

The plate load test is a semi-direct method to estimate the allowable bearing pressure of soil to induce a given amount of settlement. Plates, round or square, varying in size, from 30 to 60 cm and thickness of about 2.5 cm are employed for the test.

The load on the plate is applied by making use of a hydraulic jack. The reaction of the jack load is taken by a cross beam or a steel truss anchored suitably at both the ends. The settlement of the plate is measured by a set of three dial gauges of sensitivity 0.02 mm placed 120° apart. The dial gauges are fixed to independent supports which remain undisturbed during the test.

Figure 13.2a shows the arrangement for a plate load test. The method of performing the test is essentially as follows:

1. Excavate a pit of size not less than 4 to 5 times the size of the plate. The bottom of the pit should coincide with the level of the foundation.
2. If the water table is above the level of the foundation, pump out the water carefully and keep it at the level of the foundation.
3. A suitable size of plate is selected for the test. Normally a plate of size 30 cm is used in sandy soils and a larger size in clay soils. The ground should be levelled and the plate should be seated over the ground.

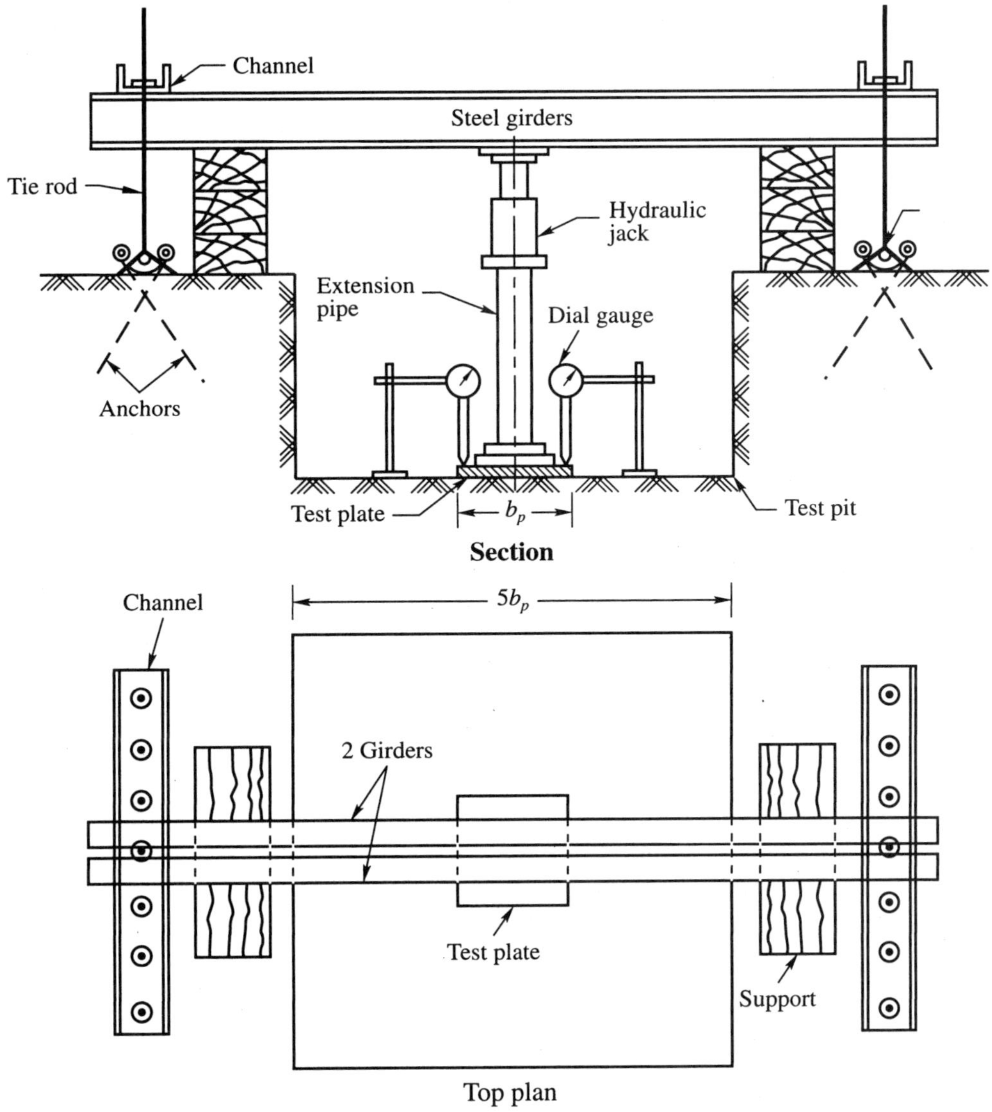

Figure 13.2a Plate load test arrangement

4. A seating load of about 70 gm/cm^2 is first applied and released after some time. A higher load is next placed on the plate and settlements are recorded by means of the dial gauges. Observations on every load increment shall be taken until the rate of settlement is less than 0.25 mm per hour. Load increments shall be approximately one-fifth of the estimated safe bearing capacity of the soil. The average of the settlements recorded by 2 or 3 dial gauges shall be taken as the settlement of the plate for each of the load increments.
5. The test should continue until a total settlement of 2.5 cm or the settlement at which the soil fails, whichever is earlier, is obtained. After the load is released, the elastic rebound of the soil should be recorded.

From the test results, a load-settlement curve should be plotted as shown in Fig. 13.2b. The allowable pressure on a prototype foundation for an assumed settlement may be found by making use of the following equations suggested by Terzaghi and Peck (1948) for square footings in granular soils.

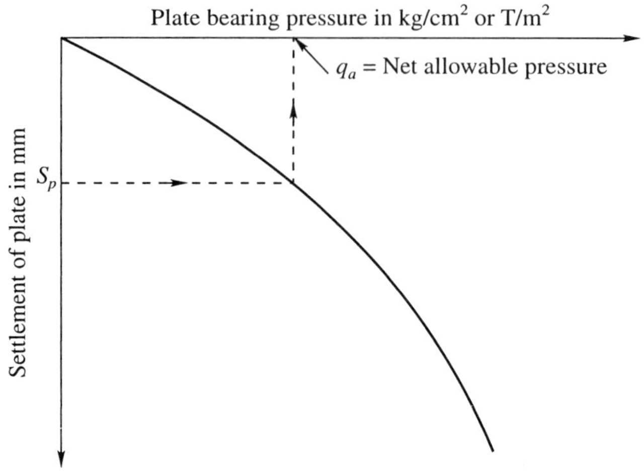

Figure 13.2b Load-settlement curve of a plate-load test

$$S_f = S_p \left(\frac{B(b_p + 0.3)}{b_p (B + 0.3)} \right)^2 \tag{13.1a}$$

in clay soils,

$$S_f = S_p \times \frac{B}{b_p} \tag{13.1b}$$

where S_f = permissible settlement of foundation in mm,
S_p = settlement of plate in mm,
B = size of foundation in meters,
b_p = size of plate in meters.

For a plate 1 ft square, Eq. (13.1a) may be expressed as

$$S_f = S_p \left[\frac{2B}{B+1} \right]^2 \tag{13.2}$$

in which S_f and S_p are expressed in inches and B in feet.

The permissible settlement S_f for a prototype foundation should be known. Normally a settlement of 2.5 cm is recommended. In Eqs (13.1a) or (13.2) the values of S_f and b_p are known. The unknowns are S_p and B. The value of S_p for any assumed size B may be found from the equation. Using the plate load settlement curve Fig. 13.3 the value of the bearing pressure corresponding to the computed value of S_p is found. This bearing pressure is the safe bearing pressure for a given permissible settlement S_f. The principal shortcoming of this approach is the unreliability of the extrapolation of Eqs (13.1a) or (13.2).

Since a load test is of short duration, consolidation settlements cannot be predicted. The test gives the value of immediate settlement only. If the underlying soil is sandy in nature immediate settlement may be taken as the total settlement. If the soil is a clayey type, the immediate settlement is only a fraction of the total settlement. Load tests, therefore, do not have much significance in clayey soils to determine allowable pressure on the basis of a settlement criterion.

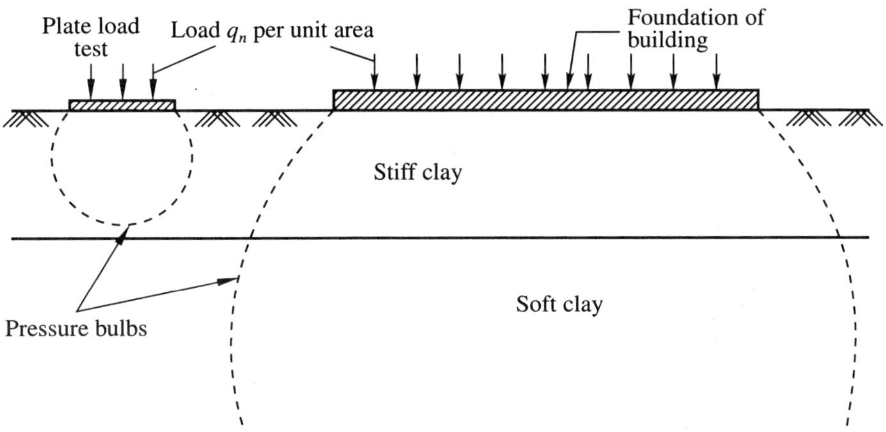

Figure 13.2c Plate load test on non-homogeneous soil

Plate load tests should be used with caution and the present practice is not to rely too much on this test. If the soil is not homogeneous to a great depth, plate load tests give very misleading results.

Assume, as shown in Fig. 13.2c, two layers of soil. The top layer is stiff clay whereas the bottom layer is soft clay. The load test conducted near the surface of the ground measures the characteristics of the stiff clay but does not indicate the nature of the soft clay soil which is below. The actual foundation of a building however has a bulb of pressure which extends to a great depth into the poor soil which is highly compressible. Here the soil tested by the plate load test gives results which are highly on the unsafe side.

A plate load test is not recommended in soils which are not homogeneous at least to a depth equal to 1½ to 2 times the width of the prototype foundation.

Plate load tests should not be relied on to determine the ultimate bearing capacity of sandy soils as the scale effect gives very misleading results. However, when the tests are carried on clay soils, the ultimate bearing capacity as determined by the test may be taken as equal to that of the foundation since the bearing capacity of clay is essentially independent of the footing size.

Housel's (1929) Method of Determining Safe Bearing Pressure from Settlement Consideration

The method suggested by Housel for determining the safe bearing pressure on settlement consideration is based on the following formula

$$Q = A_p m + P_p n \tag{13.3}$$

where Q = load applied on a given plate, A_p = contact area of plate, P_p = perimeter of plate, m = a constant corresponding to the bearing pressure, n = another constant corresponding to perimeter shear.

Objective

To determine the load Q_f and the size of a foundation for a permissible settlement S_f.

Housel suggests two plate load tests with plates of different sizes, say $B_1 \times B_1$ and $B_2 \times B_2$ for this purpose.

Procedure

1. Two plate load tests are to be conducted at the foundation level of the prototype as per the procedure explained earlier.
2. Draw the load-settlement curves for each of the plate load tests.
3. Select the permissible settlement S_f for the foundation.
4. Determine the loads Q_1 and Q_2 from each of the curves for the given permissible settlement S_f.

Now we may write the following equations

$$Q_1 = mA_{p1} + nP_{p1} \tag{13.4a}$$

for plate load test 1.

$$Q_2 = mA_{p2} + nP_{p2} \tag{13.4b}$$

for plate load test 2.

The unknown values of m and n can be found by solving the above Eqs. (13.4a) and (13.5b). The equation for a prototype foundation may be written as

$$Q_f = mA_f + nP_f \tag{13.5}$$

where A_f = area of the foundation, P_f = perimeter of the foundation.

When A_f and P_f are known, the size of the foundation can be determined.

Example 13.1

A plate load test using a plate of size 30 × 30 cm was carried out at the level of a prototype foundation. The soil at the site was cohesionless with the water table at great depth. The plate settled by 10 mm at a load intensity of 160 kN/m². Determine the settlement of a square footing of size 2 × 2 m under the same load intensity.

Solution

The settlement of the foundation S_f may be determined from Eq. (13.1a).

$$S_f = S_p\left(\frac{B(b_p + 0.3)}{b_p(B + 0.3)}\right)^2 = 10\left(\frac{2(0.3 + 0.3)}{0.3(2 + 0.3)}\right)^2 = 30.24 \text{ mm}$$

Example 13.2

For Ex. 13.1 estimate the load intensity if the permissible settlement of the prototype foundation is limited to 40 mm.

Solution

In Ex. 13.1, a load intensity of 160 kN/m² induces a settlement of 30.24 mm. If we assume that the load-settlement is linear within a small range, we may write

$$\frac{q_2}{q_1} = \frac{S_{f2}}{S_{f1}} \quad \text{or} \quad q_2 = q_1 \times \frac{S_{f2}}{S_{f1}}$$

where, q_1 = 160 kN/m^2, S_{f1} = 30.24 mm, S_{f2} = 40 mm. Substituting the known values

$$q_2 = 160 \times \frac{40}{30.24} = 211.64 \text{ kN/m}^2$$

Example 13.3

Two plate load tests were conducted at the level of a prototype foundation in cohesionless soil close to each other. The following data are given:

Size of plate	Load applied	Settlement recorded
0.3 × 0.3 m	30 kN	25 mm
0.6 × 0.6 m	90 kN	25 mm

If a footing is to carry a load of 1000 kN, determine the required size of the footing for the same settlement of 25 mm.

Solution

Use Eq. (13.3). For the two plate load tests we may write:

$$Q_1 = A_{p1}m + P_{p1}n$$

$$Q_2 = A_{p2}m + P_{p2}n$$

$$PLT1: \quad A_{p1} = 0.3 \times 0.3 = 0.09m^2; \quad P_{p1} = 0.3 \times 4 = 1.2m; \quad Q_1 = 30 \text{ kN}$$

$$PLT2: \quad A_{p2} = 0.6 \times 0.6 = 0.36m^2; \quad P_{p2} = 0.6 \times 4 = 2.4m; \quad Q_2 = 90 \text{ kN}$$

Now we have

$$30 = 0.09m + 1.2n$$

$$90 = 0.36m + 2.4n$$

On solving the equations we have

m = 166.67, and n = 12.5

For prototype foundation, we may write

$$Q_f = 166.67A_f + 12.5P_f$$

Assume the size of the footing as $B \times B$, we have

$$A_f = B^2, \quad P_f = 4B, \quad \text{and} \quad Q_f = 1000 \text{ kN}$$

Substituting we have

$$1000 = 166.67B^2 + 50B$$

$$\text{or} \quad B^2 + 0.3B - 6 = 0$$

The solution gives B = 2.3 m.

The size of the footing = 2.3 × 2.3 m.

13.3 EFFECT OF SIZE OF FOOTINGS ON SETTLEMENT

Figure 13.3a gives typical load-settlement relationships for footings of different widths on the surface of a homogeneous sand deposit. It can be seen that the ultimate bearing capacities of the footings per unit area increase with the increase in the widths of the footings. However, for a given settlement *S*, such as 25 mm, the soil pressure is greater for a footing of intermediate width B_b than for a large footing with B_c. The pressures corresponding to the three widths intermediate, large and narrow, are indicated by points *b, c* and *a* respectively.

The same data is used to plot Fig. 13.3b which shows the pressure per unit area corresponding to a given settlement S_1, as a function of the width of the footing. The soil pressure for settlement S_1 increases for increasing width of the footing, if the footings are relatively small, reaches a maximum at an intermediate width, and then decreases gradually with increasing width.

Although the relation shown in Fig. 13.3b is generally valid for the behavior of footings on sand, it is influenced by several factors including the relative density of sand, the depth at which the foundation is established, and the position of the water table. Furthermore, the shape of the curve suggests that for narrow footings small variations in the actual pressure, Fig. 13.3a, may lead to large variation in settlement and in some instances to settlements so large that the movement would be considered a bearing capacity failure. On the other hand, a small change in pressure on a wide footing has little influence on settlements as small as S_1, and besides, the value of q_1 corresponding to S_1 is far below that which produces a bearing capacity failure of the wide footing.

For all practical purposes, the actual curve given in Fig. 13.3b can be replaced by an equivalent curve *omn* where *om* is the inclined part and *mn* the horizontal part. The horizontal portion of the curve indicates that the soil pressure corresponding to a settlement S_1 is independent of the size of the footing. The inclined portion *om* indicates the pressure increasing with width for the same given settlement S_1 up to the point *m* on the curve which is the limit for a bearing capacity

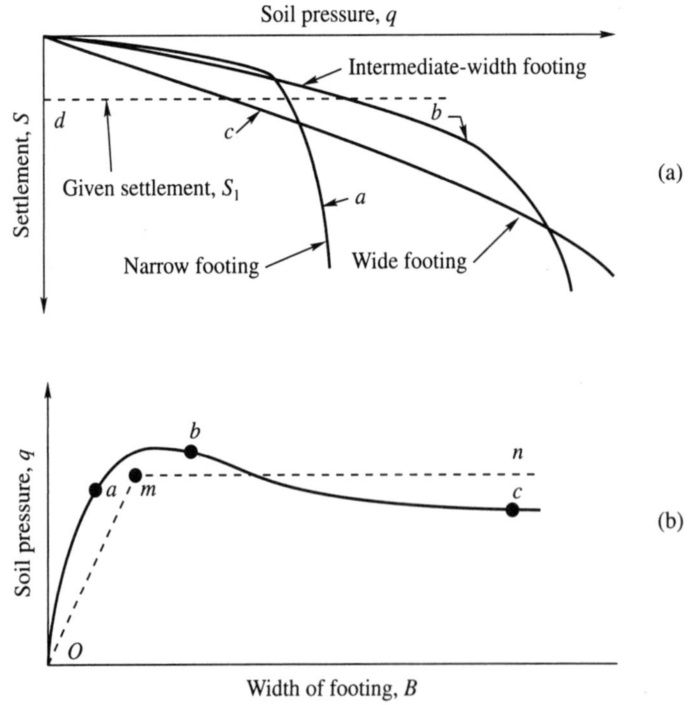

Figure 13.3 Load-settlement curves for footings of different sizes (Peck et al., 1974)

failure. This means that the footings up to size B_1 in Fig. 13.3b should be checked for bearing capacity failure also while selecting a safe bearing pressure by settlement consideration.

The position of the broken lines *omn* differs for different sand densities or in other words for different SPT N values. The soil pressure that produces a given settlement S_1 on loose sand is obviously smaller than the soil pressure that produces the same settlement on a dense sand. Since N-value increases with density of sand, q_s therefore increases with an increase in the value of N.

13.4 DESIGN CHARTS FROM SPT VALUES FOR FOOTINGS ON SAND

The methods suggested by Terzaghi et al., (1996) for estimating settlements and bearing pressures of footings founded on sand from SPT values are based on the findings of Burland and Burbidge (1985). The SPT values used are corrected to a standard energy ratio. The usual symbol N_{cor} is used in all the cases as the corrected value.

Formulas for Settlement Calculations

The following formulas were developed for computing settlements for square footings.

For normally consolidated soils and gravels

$$S_c = B^{0.75}\frac{1.7\,q_s}{N_{cor}^{1.4}} \tag{13.6}$$

For preconsolidated sand and gravels

$$\text{for } q_s > p_c \qquad S_c = B^{0.75}\frac{1.7}{N_{cor}^{1.4}}(q_s - 0.67p_c) \tag{13.7a}$$

$$\text{for } q_s < p_c \qquad S_c = \frac{1}{3}B^{0.75}\frac{1.7\,q_s}{N_{cor}^{1.4}} \tag{13.7b}$$

If the footing is established at a depth below the ground surface, the removal of the soil above the base level makes the sand below the base preconsolidated by excavation. Recompression is assumed for bearing pressures up to preconstruction effective vertical stress q'_o at the base of the foundation. Thus, for sands normally consolidated with respect to the original ground surface and for values of q_s greater than q'_o, we have,

$$\text{for } q_s > q'_o \qquad S_c = B^{0.75}\frac{1.7}{N_{cor}^{1.4}}(q_s - 0.67q'_o) \tag{13.8a}$$

$$\text{for } q_s < q'_o \qquad S_c = \tfrac{1}{3}B^{0.75}\frac{1.7}{N_{cor}^{1.4}}q_s \tag{13.8b}$$

where

S_c = settlement of footing, in mm, at the end of construction and application of permanent live load

B = width of footing in m

q_s = gross bearing pressure of footing = Q/A, in kN/m^2 based on settlement consideration

Q = total load on the foundation in kN

A = area of foundation in m^2

p_c = preconsolidation pressure in kN/m^2

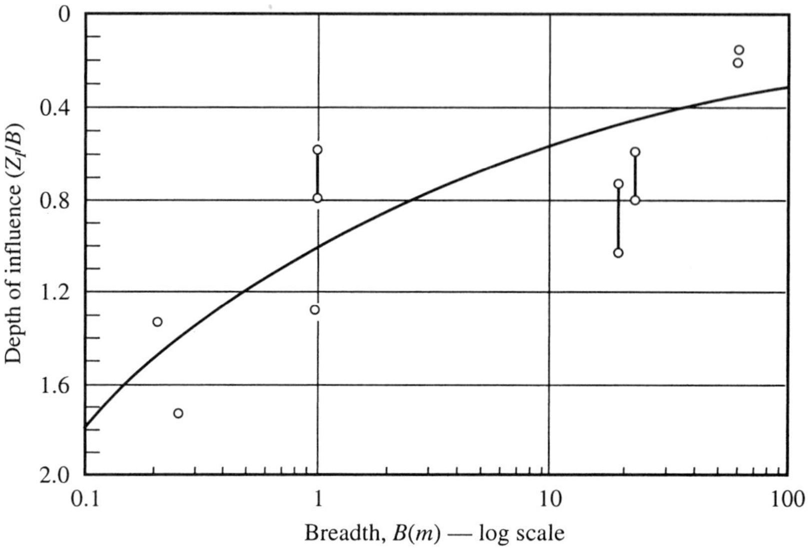

Figure 13.4 Thickness of granular soil beneath foundation contributing to settlement, interpreted from settlement profiles (after Burland and Burbidge 1985)

q'_o = effective vertical pressure at base level

N_{cor}= average corrected N value within the depth of influence Z_I below the base the of footing

The depth of influence Z_I is obtained from

$$Z_I = B^{0.75} \tag{13.9}$$

Figure 13.4 gives the variation of the depth of influence with depth based on Eq. (13.9) (after Burland and Burbidge, 1985).

The settlement of a rectangular footing of size $B \times L$ may be obtained from

$$S_c(L/B > 1) = S_c\left(\frac{L}{B} = 1\right)\left[\frac{1.25(L/B)}{L/B + 0.25}\right]^2 \tag{13.10}$$

when the ratio L/B is very high for a strip footing, we may write

$$\frac{S_c(\text{strip})}{S_c(\text{square})} = 1.56 \tag{13.11}$$

It may be noted here that the ground water table at the site may lie above or within the depth of influence Z_I. Burland and Burbidge (1985) recommend no correction for the settlement calculation even if the water table lies within the depth of influence Z_I. On the other hand, if for any reason, the water table were to rise into or above the zone of influence Z_I after the penetration tests were conducted, the actual settlement could be as much as twice the value predicted without taking the water table into account.

Chart for Estimating Allowable Soil Pressure

Fig. 13.5 gives a chart for estimating allowable bearing pressure q_s (on settlement consideration) corresponding to a settlement of 16 mm for different values of N (corrected). From Eq. (13.6) , an expression for q_s may be written as (for normally consolidated sand)

$$q_s = S_c \frac{N_{cor}^{1.4}}{1.7B^{0.75}} = 16\frac{N_{cor}^{1.4}}{1.7B^{0.75}} = 16\overline{Q} \qquad (13.12a)$$

where $$\overline{Q} = \frac{N_{cor}^{1.4}}{1.7B^{0.75}} \qquad (13,12b)$$

For sand having a preconsolidation pressure p_c, Eq. (13.7) may be written as

for $q_s > p_c$ $\quad q_s = 16\overline{Q} + 0.67p_c$ (13.13a)

for $q_s < p_c$ $\quad q_s = 3 \times 16\overline{Q}$ (13.13b)

If the sand beneath the base of the footing is preconsolidated because excavation has removed a vertical effective stress q'_o, Eq. (13.8) may be written as

for $q_s > q'_o$, $\quad q_s = 16\overline{Q} + 0.67q'_o$ (13.14a)

for $q_s < q'_o$, $\quad q_s = 3 \times 16\overline{Q}$ (13.14b)

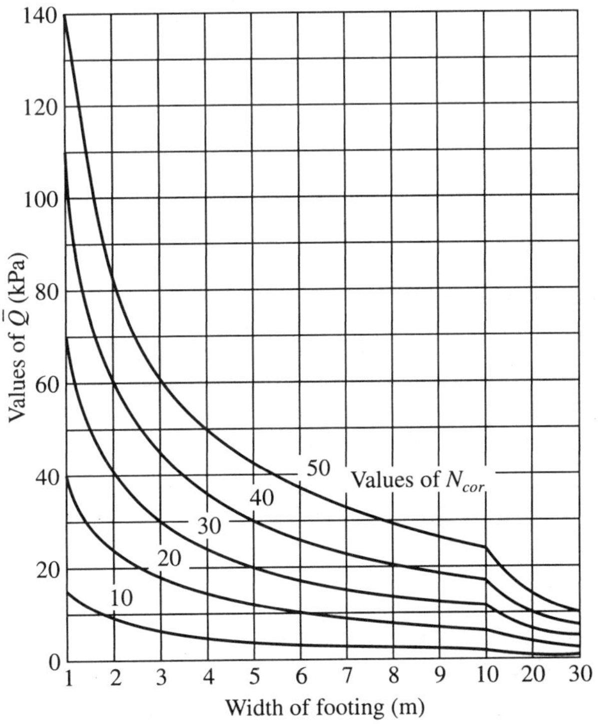

Figure 13.5 Chart for estimating allowable soil pressure for footing on sand on the basis of results of standard penetration test. (Terzaghi, et al., 1996)

The chart in Fig. 13.5 gives the relationships between B and $\overline{Q}$. The value of q_s may be obtained from $\overline{Q}$ for any given width B. The $\overline{Q}$ to be used must conform to Eqs (13.12), (13.13) and (13.14).

The chart is constructed for square footings of width B. For rectangular footings, the value of q_s should be reduced in accordance with Eq. (13.10). The bearing pressures determined by this procedure correspond to a maximum settlement of 25 mm at the end of construction.

It may be noted here that the design chart (Fig. 13.5b) has been developed by taking the SPT values corrected for 60 percent of standard energy ratio.

Example 13.4

A square footing of size 4 × 4 m is founded at a depth of 2 m below the ground surface in loose to medium dense sand. The corrected standard penetration test value N_{cor} = 11. Compute the safe bearing pressure q_s by using the chart in Fig. 13.5. The water table is at the base level of the foundation.

Solution

From Fig. 13.5 $\overline{Q}$ = 5 for B = 4 m and N_{cor} = 11.

From Eq. (13.12a)

$$q_s = 16\overline{Q} = 16 \times 5 = 80 \text{ kN/m}^2$$

Example 13.5

Refer to Example 13.4. If the soil at the site is dense sand with N_{cor} = 30, estimate q_s for B = 4 m.

Solution

From Fig. 13.5 $\overline{Q}$ = 24 for B = 4 m and N_{cor} = 30.

From Eq. (13.12a)

$$q_s = 16\overline{Q} = 16 \times 24 = 384 \text{ kN/m}^2$$

13.5 EMPIRICAL EQUATIONS BASED ON SPT VALUES FOR FOOTINGS ON COHESIONLESS SOILS

Footings on granular soils are sometimes proportioned using empirical relationships. Teng (1969) proposed an equation for a settlement of 25 mm based on the curves developed by Terzaghi and Peck (1948). The modified form of the equation is

$$q_s = 35(N_{cor} - 3)\left(\frac{B + 0.3}{2B}\right)^2 R_{w2} F_d \text{ kN/m}^2 \tag{13.15a}$$

where q_s = net allowable bearing pressure for a settlement of 25 mm in kN/m^2,

N_{cor} = corrected standard penetration value

R_{w2} = water table correction factor (Refer Section 12.7)

F_d = depth factor $= (1 + D_f / B) \leq 2.0$

B = width of footing in meters,

D_f = depth of foundation in meters.

Meyerhof (1956) proposed the following equations which are slightly different from that of Teng

$$q_s = 12 N_{cor} R_{w2} F_d \quad \text{for} \quad B \leq 1.2 \text{ m} \tag{13.15b}$$

$$q_s = 8 N_{cor} \left(\frac{B+0.3}{B} \right)^2 R_{w2} F_d \text{ for } B > 1.2 \text{ m} \tag{13.15c}$$

where $F_d = (1 + 0.33\, D_f/B) \leq 1.33$.

Experimental results indicate that the equations presented by Teng and Meyerhof are too conservative. Bowles (1996) proposes an approximate increase of 50 percent over that of Meyerhof which can also be applied to Teng's equations. Modified equations of Teng and Meyerhof are,

Teng's equation (modified),

$$q_s = 53(N_{cor} - 3) \left(\frac{B+0.3}{2B} \right)^2 R_{w2} F_d \tag{13.16a}$$

Meyerhof's equation (modified)

$$q_s = 20\, N_{cor}\, R_{w2}\, F_d \text{ for } B \leq 1.2 \text{ m} \tag{13.16b}$$

$$q_s = 12.5 N_{cor} \left(\frac{B+0.3}{B} \right)^2 R_{w2} F_d \text{ for } B > 1.2 \text{ m} \tag{13.16c}$$

If the tolerable settlement is greater than 25 mm, the safe bearing pressure computed by the above equations can be increased linearly as,

$$q_s' = \frac{S'}{25} q_s \tag{13.16d}$$

where q'_s = net safe bearing pressure for a settlement S' mm, q_s = net safe bearing pressure for a settlement of 25 mm.

13.6 SAFE BEARING PRESSURE FROM EMPIRICAL EQUATIONS BASED ON CPT VALUES FOR FOOTINGS ON COHESIONLESS SOIL

The static cone penetration test in which a standard cone of 10 cm^2 sectional area is pushed into the soil without the necessity of boring provides a much more accurate and detailed variation in the soil as discussed in Chapter 9. Meyerhof (1956) suggested a set of empirical equations based on the Terzaghi and Peck curves (1948). As these equations were also found to be conservative, modified forms with an increase of 50 percent over the original values are given below.

$$q_s = 3.6 q_c R_{w2} \text{ kPa} \quad \text{for} \quad B \leq 1.2 \text{ m} \tag{13.17a}$$

$$q_s = 2.1 q_c \left(1 + \frac{1}{B} \right)^2 R_{w2} \text{ kPa} \quad \text{for} \quad B > 1.2 \text{ m} \tag{13.17b}$$

An approximate formula for all widths

$$q_s = 2.7 q_c R_{w2} \text{ kPa} \tag{13.17c}$$

where q_c is the cone point resistance in kg/cm^2 and q_s in kPa.

The above equations have been developed for a settlement of 25 mm.

Meyerhof (1956) developed his equations based on the relationship $q_c = 4N_{cor}$ kg/cm^2 for penetration resistance in sand where N_{cor} is the corrected SPT value.

Example 13.6

Refer to Example 13.4 and compute q_s by modified (a) Teng's method, and (b) Meyerhof's method.

Solution

(a) Teng's equation (modified)—Eq. (13.16a)

$$q_s = 53(N_{cor} - 3)\left(\frac{B+0.3}{2B}\right)^2 R_{w2}F_d \text{ kN/m}^2$$

$$\text{where } R_{w2} = \frac{1}{2}\left(1+\frac{D_{w2}}{B}\right) = 0.5 \text{ since } D_{w2} = 0$$

$$F_d = \left(1+\frac{D_f}{B}\right) = \left(1+\frac{2}{4}\right) = 1.5 < 2$$

$N_{cor} = 11$, $B = 4$ m.

By substituting

$$q_s = 53(11-3)\left(\frac{4.3}{8}\right)^2 \times 0.5 \times 1.5 = 92 \text{ kN/m}^2$$

(b) Meyerhof's equation (modified)—Eq. (13.16c)

$$q_s = 12.5N\left(\frac{B+0.3}{B}\right)^2 R_{w2}F_d$$

$$\text{where } R_{w2} = 0.5, F_d = \left(1+0.33\times\frac{D_f}{B}\right) = \left(1+0.33\times\frac{2}{4}\right) = 1.165 < 1.33$$

By substituting

$$q_s = 12.5\times 11\left(\frac{4.3}{4}\right)^2 \times 0.5\times 1.165 = 93 \text{ kN/m}^2$$

Note: Both the methods give the same result.

Example 13.7

A footing of size 3 × 3 m is to be constructed at a site at a depth of 1.5 m below the ground surface. The water table is at the base of the foundation. The average static cone penetration resistance obtained at the site is 20 kg/cm^2. The soil is cohesive. Determine the safe bearing pressure for a settlement of 40 mm.

Solution

Use Eq. (13.17b)

$$q_s = 2.1q_c\left(1+\frac{1}{B}\right)^2 R_{w2}\ \text{kN/m}^2$$

where $q_c = 20$ kg/cm^2, $B = 3$ m, $R_{w2} = 0.5$.

This equation is for 25 mm settlement. By substituting, we have

$$q_s = 2.1\times 20\left(1+\frac{1}{3}\right)^2 \times 0.5 = 37.3\ \text{kN/m}^2$$

For 40 mm settlement, the value of q_s is

$$q_s = 37.3\left(\frac{40}{25}\right) \approx 60\ \text{kN/m}^2$$

13.7 FOUNDATION SETTLEMENT

Components of Total Settlement

The total settlement of a foundation comprises three parts as follows

$$S = S_e + S_c + S_s \tag{13.18}$$

where S = total settlement

S_e = elastic or immediate settlement

S_c = consolidation settlement

S_s = secondary settlement

Immediate settlement, S_e, is that part of the total settlement, S, which is supposed to take place during the application of loading. The consolidation settlement is that part which is due to the expulsion of pore water from the voids and is time-dependent settlement. Secondary settlement normally starts with the completion of the consolidation. It means, during the stage of this settlement, the pore water pressure is zero and the settlement is only due to the distortion of the soil skeleton.

Footings founded in cohesionless soils reach almost the final settlement, S, during the construction stage itself due to the high permeability of soil. The water in the voids is expelled simultaneously with the application of load and as such the immediate and consolidation settlements in such soils are rolled into one.

In cohesive soils under saturated conditions, there is no change in the water content during the stage of immediate settlement. The soil mass is deformed without any change in volume soon after the application of the load. This is due to the low permeability of the soil. With the advancement of time there will be gradual expulsion of water under the imposed excess load. The time required for the complete expulsion of water and to reach zero water pressure may be several years depending upon the permeability of the soil. Consolidation settlement may take many years to reach its final stage. Secondary settlement is supposed to take place after the completion of the consolidation settlement, though in some of the organic soils there will be overlapping of the two settlements to a certain extent.

Immediate settlements of cohesive soils and the total settlement of cohesionless soils may be estimated from elastic theory. The stresses and displacements depend on the stress-strain characteristics of the underlying soil. A realistic analysis is difficult because these characteristics are nonlinear. Results from the theory of elasticity are generally used in practice, it being assumed that the soil is homogeneous and isotropic and there is a linear relationship between stress and

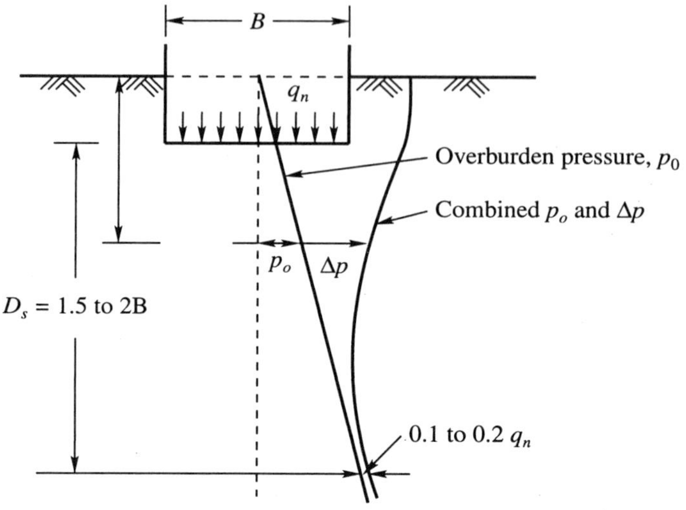

Figure 13.6 Overburden pressure and vertical stress distribution

strain. A linear stress-strain relationship is approximately true when the stress levels are low relative to the failure values. The use of elastic theory clearly involves considerable simplification of the real soil.

Some of the results from elastic theory require knowledge of Young's modulus (E_s), here called the compression or deformation modulus, E_d, and Poisson's ratio, μ, for the soil.

Seat of Settlement

Footings founded at a depth D_f below the surface settle under the imposed loads due to the compressibility characteristics of the subsoil. The depth through which the soil is compressed depends upon the distribution of effective vertical pressure p'_0 of the overburden and the vertical induced stress Δp resulting from the net foundation pressure q_n as shown in Fig. 13.6.

In the case of deep compressible soils, the lowest level considered in the settlement analysis is the point where the vertical induced stress Δp is of the order of 0.1 to $0.2q_n$, where q_n is the net pressure on the foundation from the superstructure. This depth works out to about 1.5 to 2 times the width of the footing. The soil lying within this depth gets compressed due to the imposed foundation pressure and causes more than 80 percent of the settlement of the structure. This depth D_s is called as the *zone of significant stress*. If the thickness of this zone is more than 3 m, the steps to be followed in the settlement analysis are

1. Divide the zone of significant stress into layers of thickness not exceeding 3 m,
2. Determine the effective overburden pressure p'_o at the center of each layer,
3. Determine the increase in vertical stress Δp due to foundation pressure q_n at the center of each layer along the center line of the footing by the theory of elasticity,
4. Determine the average modulus of elasticity and other soil parameters for each of the layers.

13.8 EVALUATION OF MODULUS OF ELASTICITY

The most difficult part of a settlement analysis is the evaluation of the modulus of elasticity E_s, that would conform to the soil condition in the field. There are two methods by which E_s can be evaluated. They are

1. Laboratory method,
2. Field method.

Laboratory Method

For settlement analysis, the values of E_s at different depths below the foundation base are required. One way of determining E_s is to conduct triaxial tests on representative undisturbed samples extracted from the depths required. For cohesive soils, undrained triaxial tests and for cohesionless soils drained triaxial tests are required. Since it is practically impossible to obtain undisturbed sample of cohesionless soils, the laboratory method of obtaining E_s can be ruled out. Even with regards to cohesive soils, there will be disturbance to the sample at different stages of handling it, and as such the values of E_s obtained from undrained triaxial tests do not represent the actual conditions and normally give very low values. A suggestion is to determine E_s over the range of stress relevant to the particular problem. Poulos et al., (1980) suggest that the undisturbed triaxial specimen be given a preliminary preconsolidation under K_0 conditions with an axial stress equal to the effective overburden pressure at the sampling depth. This procedure attempts to return the specimen to its original state of effective stress in the ground, assuming that the horizontal effective stress in the ground was the same as that produced by the laboratory K_0 condition. Simons and Som (1970) have shown that triaxial tests on London clay in which specimens were brought back to their original *in situ* stresses gave elastic moduli which were much higher than those obtained from conventional undrained triaxial tests. This has been confirmed by Marsland (1971) who carried out 865 mm diameter plate loading tests in 900 mm diameter bored holes in London clay. Marsland found that the average moduli determined from the loading tests were between 1.8 to 4.8 times those obtained from undrained triaxial tests. A suggestion to obtain the more realistic value for E_s is,

1. Undisturbed samples obtained from the field must be reconsolidated under a stress system equal to that in the field (K_0-condition),
2. Samples must be reconsolidated isotropically to a stress equal to 1/2 to 2/3 of the *in situ* vertical stress.

It may be noted here that reconsolidation of disturbed sensitive clays would lead to significant change in the water content and hence a stiffer structure which would lead to a very high E_s.

Because of the many difficulties faced in selecting a modulus value from the results of laboratory tests, it has been suggested that a correlation between the modulus of elasticity of soil and the undrained shear strength may provide a basis for settlement calculation. The modulus E_s may be expressed as

$$E_s = Ac_u \tag{13.19}$$

where the value of A for inorganic stiff clay varies from about 500 to 1500 (Bjerrum, 1972) and c_u is the undrained cohesion. It may generally be assumed that highly plastic clays give lower values for A, and low plasticity give higher values for A. For organic or soft clays the value of A may vary from 100 to 500. The undrained cohesion c_u can be obtained from any one of the field tests mentioned below and also discussed in Chapter 9.

Field methods

Field methods are increasingly used to determine the soil strength parameters. They have been found to be more reliable than the ones obtained from laboratory tests. The field tests that are normally used for this purpose are

1. Plate load tests (PLT)

Table 13.2 Equations for computing E_s by making use of SPT and CPT values (in kPa)

Soil	SPT	CPT
Sand (normally consolidated)	$500 (N_{cor} + 15)$	2 to 4 q_c
	$(35000 \text{ to } 50000) \log N_{cor}$	$(1+ D_r^2) q_c$
	(U.S.S.R Practice)	
Sand (saturated)	$250 (N_{cor} + 15)$	
Sand (overconsolidated)	–	6 to 30 q_c
Gravelly sand and gravel	$1200 (N_{cor} + 6)$	
Clayey sand	$320 (N_{cor} + 15)$	3 to 6 q_c
Silty sand	$300 (N_{cor} + 6)$	1 to 2 q_c
Soft clay	–	3 to 8 q_c

2. Standard penetration test (SPT)
3. Static cone penetration test (CPT)
4. Pressuremeter test (PMT)
5. Flat dilatometer test (DMT)

Plate load tests, if conducted at levels at which E_s is required, give quite reliable values as compared to laboratory tests. Since these tests are too expensive to carry out, they are rarely used except in major projects.

Many investigators have obtained correlations between E_s and field tests such as SPT, CPT and PMT. The correlations between E_s and SPT or CPT are applicable mostly to cohesionless soils and in some cases cohesive soils under undrained conditions. PMT can be used for cohesive soils to determine both the immediate and consolidation settlements together.

Some of the correlations of E_s with N and q_c are given in Table 13.2. These correlations have been collected from various sources.

13.9 METHODS OF COMPUTING SETTLEMENTS

Many methods are available for computing elastic (immediate) and consolidation settlements. Only those methods that are of practical interest are discussed here. The various methods discussed in this chapter are the following:

Computation of Elastic Settlements

1. Elastic settlement based on the theory of elasticity
2. Janbu et al., (1956) method of determining settlement under an undrained condition.
3. Schmertmann's method of calculating settlement in granular soils by using CPT values.

Computation of Consolidation Settlement

1. e-log p method by making use of oedometer test data.
2. Skempton-Bjerrum method.

13.10 ELASTIC SETTLEMENT BENEATH THE CORNER OF A UNIFORMLY LOADED FLEXIBLE AREA BASED ON THE THEORY OF ELASTICITY

The net elastic settlement equation for a flexible surface footing may be written as,

$$S_e = q_n B \frac{(1-\mu^2)}{E_s} I_f \tag{13.20a}$$

where S_e = elastic settlement
B = width of foundation,
E_s = modulus of elasticity of soil,
μ = Poisson's ratio,
q_n = net foundation pressure,
I_f = influence factor.

In Eq. (13.20a), for saturated clays, μ = 0.5, and E_s is to be obtained under undrained conditions as discussed earlier. For soils other than clays, the value of μ has to be chosen suitably and the corresponding value of E_s has to be determined. Table 13.3 gives typical values for μ as suggested by Bowles (1996).

I_f is a function of the L/B ratio of the foundation, and the thickness H of the compressible layer. Terzaghi has a given a method of calculating I_f from curves derived by Steinbrenner (1934),

for Poisson's ratio of 0.5, $I_f = F_1$,

for Poisson's ratio of zero, $I_f = F_1 + F_2$.

where F_1 and F_2 are factors which depend upon the ratios of H/B and L/B.

For intermediate values of μ, the value of I_f can be computed by means of interpolation or by the equation

$$I_f = \left[F_1 + \frac{(1-\mu-2\mu^2)F_2}{1-\mu^2}\right] \tag{13.20b}$$

The values of F_1 and F_2 are given in Fig. 13.7a. The elastic settlement at any point N (Fig. 13.7b) is given by

$$S_e \text{ at point } N = \frac{q_n(1-\mu^2)}{E_s}[I_{f1}B_1 + I_{f2}B_2 + I_{f3}B_3 + I_{f4}B_4] \tag{13.20c}$$

Table 13.3 Typical range of values for Poisson's ratio (Bowles, 1996)

Type of soil	μ
Clay, saturated	0.4–0.5
Clay, unsaturated	0.1–0.3
Sandy clay	0.2–0.3
Silt	0.3–0.35
Sand (dense)	0.2–0.4
Coarse (void ratio 0.4 to 0.7)	0.15
Fine grained (void ratio = 0.4 to 0.7)	0.25
Rock	0.1–0.4

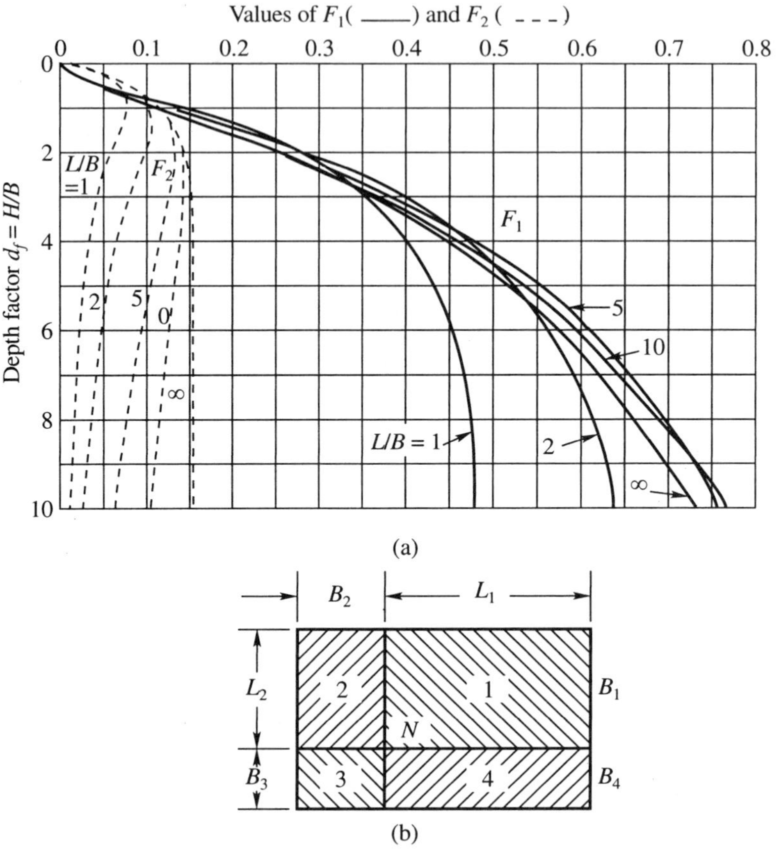

Figure 13.7 Settlement due to load on surface of elastic layer (a) F_1 and F_2 versus H/B (b) Method of estimating settlement (After Steinbrenner, 1934)

To obtain the settlement at the center of the loaded area, the principle of superposition is followed. In such a case N in Fig. 13.7b will be at the center of the area when $B_1 = B_4 = L_2 = B_3$ and $B_2 = L_1$. Then the settlement at the center is equal to four times the settlement at any one corner. The curves in Fig. 13.7a are based on the assumption that the modulus of deformation is constant with depth.

In the case of a rigid foundation, the immediate settlement at the center is approximately 0.8 times that obtained for a flexible foundation at the center. A correction factor is applied to the immediate settlement to allow for the depth of foundation by means of the depth factor d_f. Fig. 13.8 gives Fox's (1948) correction curve for depth factor. The final elastic settlement is

$$S_{ef} = C_r d_f S_e \tag{13.21}$$

where, S_{ef} = final elastic settlement
C_r = rigidity factor taken as equal to 0.8 for a highly rigid foundation
d_f = depth factor from Fig. 13.8
S_e = settlement for a surface flexible footing

Bowles (1996) has given the influence factor for various shapes of rigid and flexible footings as shown in Table 13.4.

Table 13.4 Influence factor I_f (Bowles, 1988)

Shape	I_f (average values) Flexible footing	Rigid footing
Circle	0.85	0.88
Square	0.95	0.82
Rectangle	1.20	1.06
L/B = 1.5	1.20	1.06
2.0	1.31	1.20
5.0	1.83	1.70
10.0	2.25	2.10
100.0	2.96	3.40

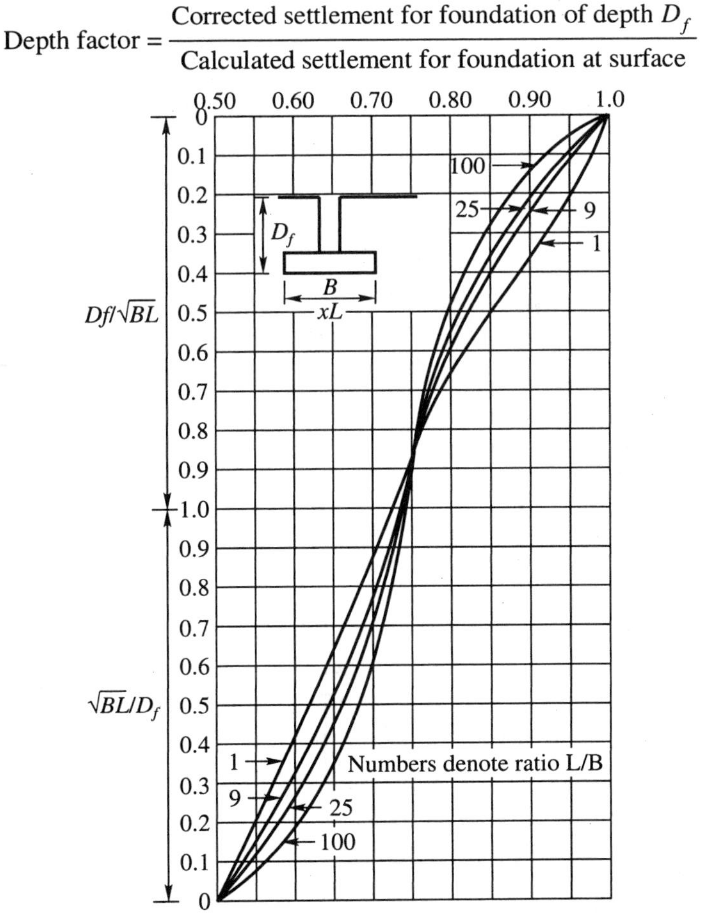

Figure 13.8 Correction curves for elastic settlement of flexible rectangular foundations at depth (Fox, 1948)

13.11 JANBU, BJERRUM AND KJAERNSLI'S METHOD OF DETERMINING ELASTIC SETTLEMENT UNDER UNDRAINED CONDITIONS

Probably the most useful chart is that given by Janbu et al., (1956) as modified by Christian and Carrier (1978) for the case of a constant E_s with respect to depth. The chart (Fig. 13.9) provides estimates of the average immediate settlement of uniformly loaded, flexible strip, rectangular, square or circular footings on homogeneous isotropic saturated clay. The equation for computing the settlement may be expressed as

$$S_e = \frac{\mu_0 \mu_1 \, q_n B}{E_s} \tag{13.22}$$

In Eq. (13.20), Poisson's ratio is assumed equal to 0.5. The factors μ_0 and μ_1 are related to the D_f/B and H/B ratios of the foundation as shown in Fig. 13.9. Values of μ_1 are given for various L/B ratios. Rigidity and depth factors are required to be applied to Eq. (13.22) as per Eq. (13.21). In Fig. 13.9 the thickness of compressible strata is taken as equal to H below the base of the foundation where a hard stratum is met with.

Generally, real soil profiles which are deposited naturally consist of layers of soils of different properties underlain ultimately by a hard stratum. Within these layers, strength and moduli generally increase with depth. The chart given in Fig. 13.9 may be used for the case of E_s increasing with depth by replacing the multilayered system with one hypothetical layer on a rigid

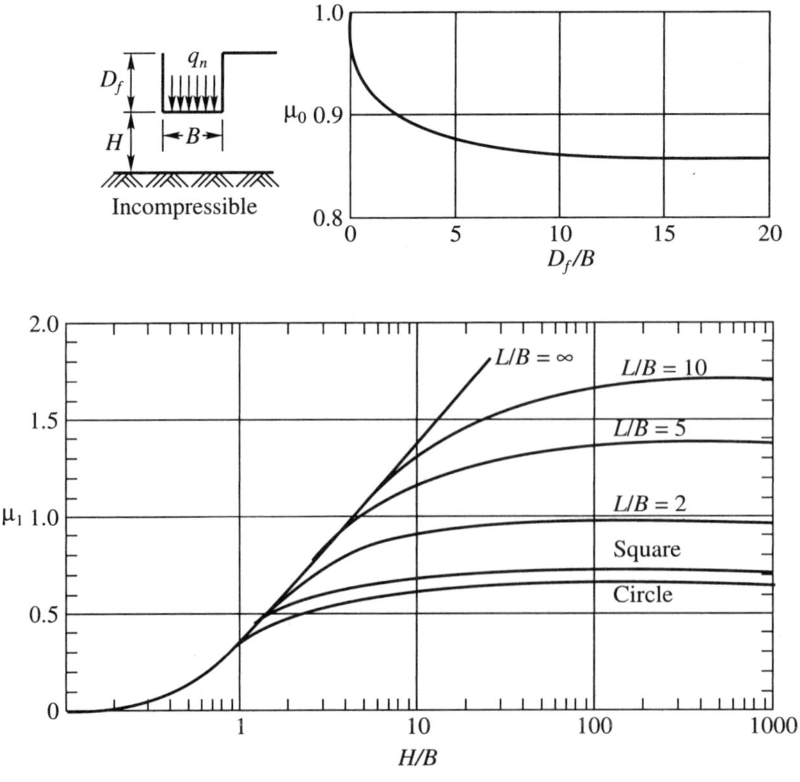

Figure 13.9 Factors for calculating the average immediate settlement of a loaded area (after Christian and Carrier, 1978)

base. The depth of this hypothetical layer is successively extended to incorporate each real layer, the corresponding values of E_s being ascribed in each case and settlements calculated. By subtracting the effect of the hypothetical layer above each real layer the separate compression of each layer may be found and summed to give the overall total settlement.

13.12 SCHMERTMANN'S METHOD OF CALCULATING SETTLEMENT IN GRANULAR SOILS BY USING CPT VALUES

It is normally taken for granted that the distribution of vertical strain under the center of a footing over uniform sand is qualitatively similar to the distribution of the increase in vertical stress. If true, the greatest strain would occur immediately under the footing, which is the position of the greatest stress increase. The detailed investigations of Schmertmann (1970), Eggestad, (1963) and others, indicate that the greatest strain would occur at a depth equal to half the width for a square or circular footing. The strain is assumed to increase from a minimum at the base to a maximum at $B/2$, then decrease and reaches zero at a depth equal to $2B$. For strip footings of L/B > 10, the maximum strain is found to occur at a depth equal to the width and reaches zero at a depth equal to $4B$. The modified triangular vertical strain influence factor distribution diagram as proposed by Schmertmann (1978) is shown in Fig. 13.10. The area of this diagram is related to the settlement. The equation (for square as well as circular footings) is

$$S = C_1 C_2 q_n \sum_0^{2B} \frac{I_z}{E_s} \Delta z \tag{13.23}$$

where,
S = total settlement,
q_n = net foundation base pressure = $(q - q'_0)$,
q = total foundation pressure,
q'_0 = effective overburden pressure at foundation level,
Δz = thickness of elemental layer,
I_z = vertical strain influence factor,
C_1 = depth correction factor,
C_2 = creep factor.

The equations for C_1 and C_2 are

$$C_1 = 1 - 0.5\left(\frac{q'_0}{q_n}\right) \tag{13.24}$$

$$C_2 = 1 + 0.2\log_{10}\left(\frac{t}{0.1}\right) \tag{13.25}$$

where t is time in years for which period settlement is required.

Equation (13.25) is also applicable for $L/B \geq 10$ except that the summation is from 0 to $4B$.

The modulus of elasticity to be used in Eq. (13.25) depends upon the type of foundation as follows:

For a square footing,

$$E_s = 2.5 q_c \tag{13.26}$$

For a strip footing, $L/B \geq 10$,

$$E_s = 3.5 q_c \tag{13.27}$$

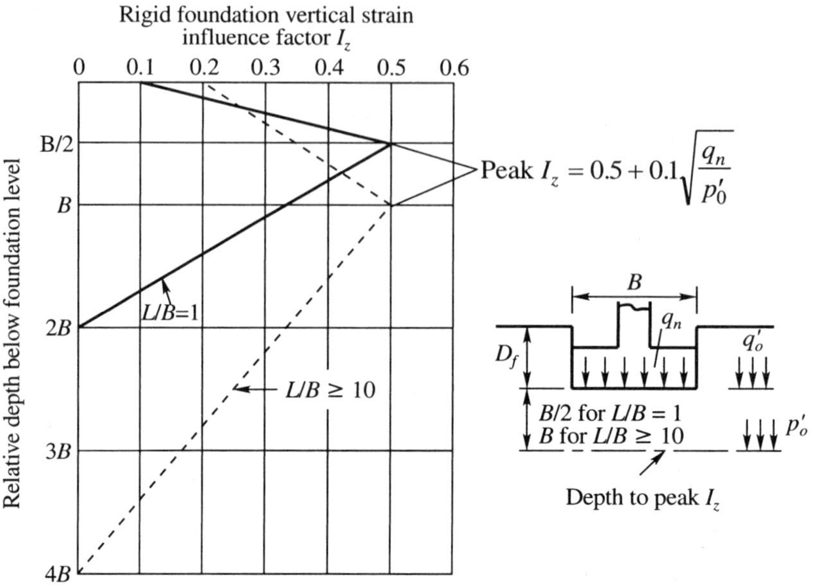

Figure 13.10 Vertical strain Influence factor diagrams (after Schmertmann et al., 1978)

Fig. 13.10 gives the vertical strain influence factor I_z distribution for both square and strip foundations if the ratio $L/B \geq 10$. Values for rectangular foundations for $L/B < 10$ can be obtained by interpolation. The depths at which the maximum I_z occurs may be calculated as follows (Fig 13.10),

$$I_z = 0.5 + 0.1\sqrt{\frac{q_n}{p'_0}} \tag{13.28}$$

where p'_0 = effective overburden pressure at depths $B/2$ and B for square and strip foundations respectively.

Further, I_z is equal to 0.1 at the base and zero at depth $2B$ below the base for square footing; whereas for a strip foundation it is 0.2 at the base and zero at depth $4B$.

Values of E_s given in Eqs. (13.26) and (13.27) are suggested by Schmertmann (1978). Lunne and Christoffersen (1985) proposed the use of the tangent modulus on the basis of a comprehensive review of field and laboratory tests as follows:

For normally consolidated sands,

$$E_s = 4\,q_c \text{ for } q_c < 10 \tag{13.29}$$

$$E_s = (2q_c + 20) \text{ for } 10 < q_c < 50 \tag{13.30}$$

$$E_s = 120 \text{ for } q_c > 50 \tag{13.31}$$

For overconsolidated sands with an overconsolidation ratio greater than 2,

$$E_s = 5\,q_c \text{ for } q_c < 50 \tag{13.32a}$$

$$E_s = 250 \text{ for } q_c > 50 \tag{13.32b}$$

where E_s and q_c are expressed in MPa.

The cone resistance diagram is divided into layers of approximately constant values of q_c and the strain influence factor diagram is placed alongside this diagram beneath the foundation which is drawn to the same scale. The settlements of each layer resulting from the net contact pressure q_n are then calculated using the values of E_s and I_z appropriate to each layer. The sum of the settlements in each layer is then corrected for the depth and creep factors using Eqs. (13.24) and (13.25) respectively.

Example 13.8

Estimate the immediate settlement of a concrete footing 1.5 × 1.5 m in size founded at a depth of 1 m in silty soil whose modulus of elasticity is 90 kg/cm^2. The footing is expected to transmit a unit pressure of 200 kN/m^2.

Solution

Use Eq. (13.20a)

Immediate settlement,

$$S_e = qB\frac{(1-\mu^2)}{E_s}I_f$$

Assume $\mu = 0.35$, $I_f = 0.82$ for a rigid footing.

Given: $q = 200$ kN/m^2, $B = 1.5$ m, $E_s = 90$ kg/cm$^2 \approx 9000$ kN/m^2.

By substituting the known values, we have

$$S_e = 200 \times 1.5 \times \frac{1-0.35^2}{9000} \times 0.82 = 0.024 \text{ m} = 2.4 \text{ cm}$$

Example 13.9

A square footing of size 8 × 8 m is founded at a depth of 2 m below the ground surface in loose to medium dense sand with $q_n = 120$ kN/m^2. Standard penetration tests conducted at the site gave the following corrected N_{60} values.

Depth below G.L. (m)	N_{cor}	Depth below G.L.	N_{cor}
2	8	10	11
4	8	12	16
6	12	14	18
8	12	16	17
		18	20

The water table is at the base of the foundation. Above the water table $\gamma = 16.5$ kN/m^3, and submerged $\gamma_b = 8.5$ kN/m^3.

Compute the elastic settlement by Eq. (13.20a). Use the equation $E_s = 250\ (N_{cor} + 15)$ for computing the modulus of elasticity of the sand. Assume $\mu = 0.3$ and the depth of the compressible layer = $2B$ = 16 m (= H).

Solution

For computing the elastic settlement, it is essential to determine the weighted average value of N_{cor}. The depth of the compressible layer below the base of the foundation is taken as equal to 16 m (= H). This depth may be divided into three layers in such a way that N_{cor} is approximately constant in each layer as given below.

Layer No.	Depth (m) From	Depth (m) To	Thickness (m)	N_{cor}
1	2	5	3	9
2	5	11	6	12
3	11	18	7	17

The weighted average

$$N_{cor}(av) = \frac{9\times3+12\times6+17\times7}{16} = 13.6 \text{ or say } 14$$

From equation $E_s = 250\ (N_{cor} + 15)$ we have

$$E_s = 250(14+15) = 7250 \text{ kN/m}^2$$

The total settlement of the center of the footing of size 8 × 8 m is equal to four times the settlement of a corner of a footing of size 4 × 4 m.

In the Eq. (13.20a), $B = 4$ m, $q_n = 120$ kN/m^2, $\mu = 0.3$.

Now from Fig. 13.7, for $H/B = 16/4 = 4$, $L/B = 1$

$F_1 = I_f = 0.4$ for $\mu = 0.5$

$F_2 = 0.03$ for $\mu = 0.5$

Now from Eq. (13.20 b) I_f for $\mu = 0.3$ is

$$I_f = F_1 + \frac{(1-\mu-2\mu^2)F_2}{1-\mu^2} = 0.40 + \frac{(1-0.3-2\times0.3^2)}{1-0.3^2}\times0.03 = 0.42$$

From Eq. (13.20a) we have settlement of a corner of a footing of size 4 × 4 m as

$$S_e = q_n B\frac{(1-\mu^2)}{E_s}I_f = \frac{120\times4(1-0.3^2)}{7250}\times0.42\times100 = 2.53 \text{ cm}$$

With the correction factor, the final elastic settlement from Eq. (13.21) is

$$S_{ef} = C_r d_f S_e$$

where C_r = rigidity factor = 1 for flexible footing d_f = depth factor

From Fig. 13.8 for

$$\frac{D_f}{\sqrt{BL}} = \frac{2}{\sqrt{4\times4}} = 0.5, \quad \frac{L}{B} = \frac{4}{4} = 1 \text{ we have } d_f = 0.85$$

Now $S_{ef} = 1 \times 0.85 \times 2.53 = 2.15$ cm

The total elastic settlement of the center of the footing is

$$S_e = 4\times2.15 = 8.6 \text{ cm} = 86 \text{ mm}$$

Per Table 13.1a, the maximum permissible settlement for a raft foundation in sand is 62.5 mm. Since the calculated value is higher, the contact pressure q_n has to be reduced.

Example 13.10

It is proposed to construct an overhead tank at a site on a raft foundation of size 8 × 12 m with the footing at a depth of 2 m below ground level. The soil investigation conducted at the site indicates that the soil to a depth of 20 m is normally consolidated insensitive inorganic clay with the water table 2 m below ground level. Static cone penetration tests were conducted at the site using a mechanical cone penetrometer. The average value of cone penetration resistance $\bar{q}_c$ was found to be 1540 kN/m² and the average saturated unit weight of the soil = 18 kN/m³. Determine the immediate settlement of the foundation using Eq. (13.22). The contact pressure q_n = 100 kN/m² (= 0.1 MPa). Assume that the stratum below 20 m is incompressible.

Solution

Computation of the modulus of elasticity

Use Eq. (13.19) with $A = 500$

$$E_s = 500\, c_u$$

where c_u = the undrained shear strength of the soil

From Eq. (9.14)

$$c_u = \frac{\bar{q}_c - p_o}{N_k}$$

where $\bar{q}_c$ = average static cone penetration resistance = 1540 kN/m²

p_o = average total overburden pressure = 10 × 18 = 180 kN/m²

N_k = 20 (assumed)

$$\text{Therefore} \quad c_u = \frac{1540 - 180}{20} = 68 \text{ kN/m}^2$$

$$E_s = 500 \times 68 = 34{,}000 \text{ kN/m}^2 = 34 \text{ MPa}$$

Eq. (13.22) for S_e is

$$S_e = \frac{\mu_0 \mu_1 q_n B}{E_s}$$

From Fig. 13.9 for $D_f/B = 2/8 = 0.25$, $\mu_0 = 0.95$, for $H/B = 16/8 = 2$ and $L/B = 12/8 = 1.5$, $\mu_1 = 0.6$.

Substituting

$$S_e\,(\text{average}) = \frac{0.95 \times 0.6 \times 0.1 \times 8}{34} = 0.0134 \text{ m} = 13.4 \text{ mm}$$

From Fig. 13.8 for $D_f/\sqrt{BL} = 2/\sqrt{8 \times 12} = 0.2$, $L/B = 1.5$ the depth factor $d_f = 0.94$

The corrected settlement S_{ef} is

$$S_{ef} = 0.94 \times 13.4 = 12.6 \text{ mm}$$

Example 13.11

Refer to Example 13.9. Estimate the elastic settlement by Schmertmann's method by making use of the relationship $q_c = 4\, N_{cor}$ kg/cm² where q_c = static cone penetration value in kg/cm². Assume settlement is required at the end of a period of 3 years.

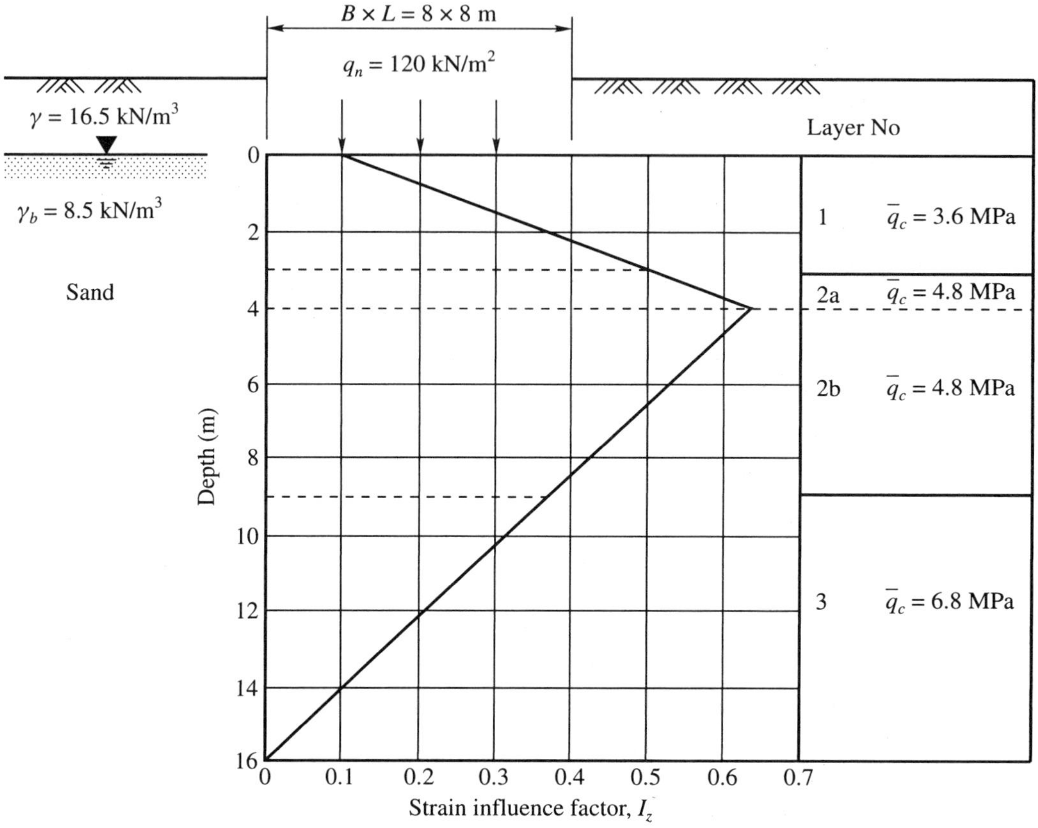

Figure Ex. 13.11

Solution

The average value of for N_{cor} each layer given in Ex. 13.9 is given below

Layer No	Average N_{cor}	Average q_c kg/cm²	MPa
1	9	36	3.6
2	12	48	4.8
3	17	68	6.8

The vertical strain influence factor I_z with respect to depth is calculated by making use of Fig. 13.10.

At the base of the foundation $I_z = 0.1$

At depth B/2, $I_z = 0.5 + 0.1\sqrt{\frac{q_n}{p'_o}}$

where q_n = 120 kpa

p'_o = effective average overburden pressure at depth = (2 + B/2) = 6 m below ground level.

= 2 × 16.5 + 4 × 8.5 = 67 kN/m².

$$I_z(\max) = 0.5 + 0.1\sqrt{\frac{120}{67}} = 0.63$$

$I_z = 0$ at $z = H = 16$ m below base level of the foundation. The distribution of I_z is given in Fig. Ex. 13.11. The equation for settlement is

$$S = C_1 C_2 q_n \sum_0^{2B} \frac{I_z}{E_s} \Delta z$$

where $C_1 = 1 - 0.5\left[\frac{q_0'}{q_n}\right] = 1 - 0.5\left[\frac{2 \times 16.5}{120}\right] = 0.86$

$$C_2 = 1 + 0.2\log\left[\frac{t}{0.1}\right] = 1 + 0.2\log\left[\frac{3}{0.1}\right] = 1.3$$

where $t = 3$ years.

The elastic modulus E_s for normally consolidated sands may be calculated by Eq. (13.29).

$E_s = 4\bar{q}_c$ for $q_c < 10$ MPa

where $\bar{q}_c$ is the average for each layer.

Layer 2 is divided into sublayers 2a and 2b for computing I_z. The average of the influence factors for each of the layers given in Fig. Ex. 13.11 are tabulated along with the other calculations

Layer No.	Δz (cm)	$\bar{q}_c$ (MPa)	E_s (MPa)	I_z (*av*)	$\frac{I_z \Delta z}{E_s}$
1	300	3.6	14.4	0.3	6.25
2a	100	4.8	19.2	0.56	2.92
2b	500	4.8	19.2	0.50	13.02
3	700	6.8	27.2	0.18	4.63
				Total	26.82

Substituting in the equation for settlement S, we have

$$S = 0.86 \times 1.3 \times 0.12 \times 26.82 = 3.6 \text{ cm} = 36 \text{ mm}$$

13.13 ESTIMATION OF CONSOLIDATION SETTLEMENT BY USING OEDOMETER TEST DATA

Equations for Computing Settlement

Settlement calculation from e-log p curves

A general equation for computing oedometer consolidation settlement may be written as follows.

Normally consolidated clays

$$S_c = H\frac{C_c}{1+e_0}\log\frac{p_0 + \Delta p}{p_0} \tag{13.33}$$

Overconsolidated clays

for $p_0 + \Delta p < p_c$

$$S_c = H\frac{C_s}{1+e_0}\log\frac{p_0+\Delta p}{p_0} \qquad (13.34)$$

for $p_0 < p_c < p_0 + \Delta p$

$$S_c = \frac{H}{1+e_0}\left[C_s \log\frac{p_c}{p_0} + C_c \log\frac{p_0+\Delta p}{p_c}\right] \qquad (13.35)$$

where C_s = swell index, and C_c = compression index

If the thickness of the clay stratum is more than 3 m the stratum has to be divided into layers of thickness less than 3 m. Further, e_0 is the initial void ratio and p_0, the effective overburden pressure corresponding to the particular layer; Δp is the increase in the effective stress at the middle of the layer due to foundation loading which is calculated by elastic theory. The compression index, and the swell index may be the same for the entire depth or may vary from layer to layer.

Settlement calculation from e-p curve

Eq. (13.35) can be expressed in a different form as follows:

$$S_c = \Sigma H m_v \Delta p \qquad (13.36)$$

where m_v = coefficient of volume compressibility

13.14 SKEMPTON-BJERRUM METHOD OF CALCULATING CONSOLIDATION SETTLEMENT (1957)

Calculation of consolidation settlement is based on one dimensional test results obtained from oedometer tests on representative samples of clay. These tests do not allow any lateral yield during the test and as such the ratio of the minor to major principal stresses, K_0, remains constant. In practice, the condition of zero lateral strain is satisfied only in cases where the thickness of the clay layer is small in comparison with the loaded area. In many practical solutions, however, significant lateral strain will occur and the initial pore water pressure will depend on the *in situ* stress condition and the value of the pore pressure coefficient A, which will not be equal to unity as in the case of a one-dimensional consolidation test. In view of the lateral yield, the ratios of the minor and major principal stresses due to a given loading condition at a given point in a clay layer do not maintain a constant K_0.

The initial excess pore water pressure at a point P (Fig. 13.11) in the clay layer is given by the expression

$$\Delta u = \Delta\sigma_3 + A(\Delta\sigma_1 - \Delta\sigma_3)$$

$$= \Delta\sigma_1\left[A + \frac{\Delta\sigma_3}{\Delta\sigma_1}(1-A)\right] \qquad (13.37)$$

where $\Delta\sigma_1$ and $\Delta\sigma_3$ are the total principal stress increments due to surface loading. It can be seen from Eq. (13.37)

$\Delta u > \Delta\sigma_3$ if A is positive

and $\Delta u = \Delta\sigma_1$ if $A = 1$

The value of A depends on the type of clay, the stress levels and the stress system.

Fig. 13.11a presents the loading condition at a point in a clay layer below the central line of circular footing. Figs. 13.11 (b), (c) and (d) show the condition before loading, immediately after loading and after consolidation respectively.

By the one-dimensional method, consolidation settlement S_{oc} is expressed as

$$S_{oc} = \int_0^H m_v \Delta\sigma_1 dz \tag{13.38}$$

By the Skempton-Bejerrum method, consolidation settlement is expressed as

$$S_{oc} = \int_0^H m_v \Delta u dz \tag{13.39}$$

or

$$S_{oc} = \int_0^H m_v \Delta\sigma_1 \left[A + \frac{\Delta\sigma_3}{\Delta\sigma_1}(1-A) \right] \tag{13.40}$$

A settlement coefficient β is used, such that $S_c = \beta S_{oc}$

The expression for β is

$$\beta = \frac{S_c}{S_{oc}} = \frac{\int_0^H m_v \Delta\sigma_1 \left[A + \frac{\Delta\sigma_3}{\Delta\sigma_1}(1-A) \right] dz}{\int_0^H m_v \Delta\sigma_1 \, dz}$$

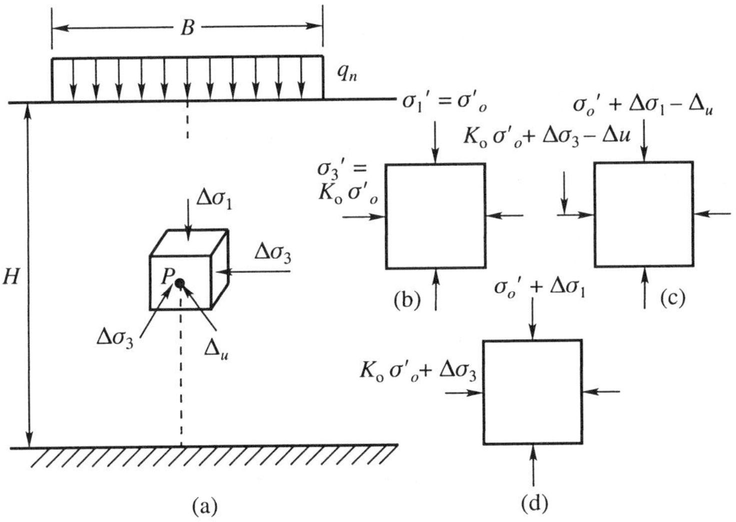

(a) Physical plane (b) Initial conditions
(c) Immediately after loading (d) After consolidation

Figure 13.11 In situ effective stresses

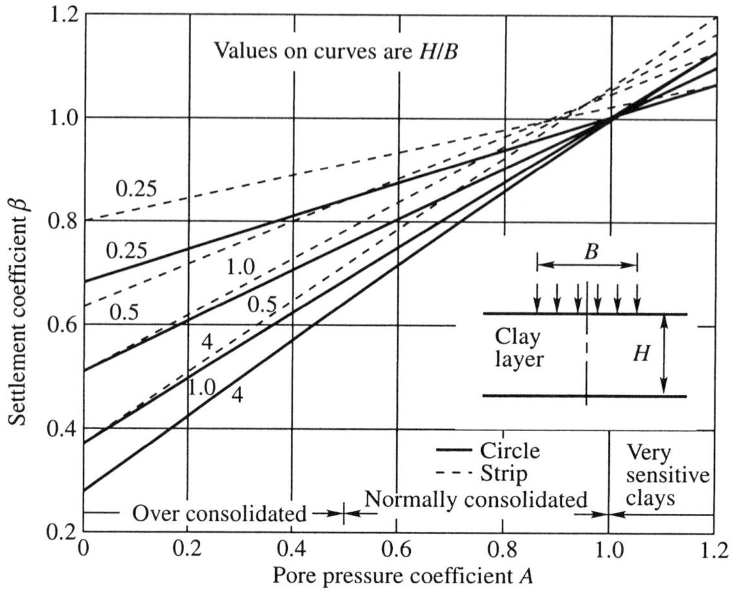

Figure 13.12 Settlement coefficient versus pore-pressure coefficient for circular and strip footings (After Skempton and Bjerrum, 1957)

Table 13.5 Values of settlement coefficient β

Type of clay	β
Very sensitive clays (soft alluvial and marine clays)	1.0 to 1.2
Normally consolidated clays	0.7 to 1.0
Overconsolidated clays	0.5 to 0.7
Heavily overconsolidated clays	0.2 to 0.5

or $$S_c = \beta S_{oc} \tag{13.41}$$

where β is called the *settlement coefficient*.

If it can be assumed that m_v and A are constant with depth (sub-layers can be used in the analysis), then β can be expressed as

$$\beta = A + (1 - A)\alpha \tag{13.42}$$

where $$\alpha = \frac{\int_0^H \Delta\sigma_3 dz}{\int_0^H \Delta\sigma_1 dz} \tag{13.43}$$

Taking Poisson's ratio μ as 0.5 for a saturated clay during loading under undrained conditions, the value of β depends only on the shape of the loaded area and the thickness of the clay layer in relation to the dimensions of the loaded area and thus β can be estimated from elastic theory.

The value of initial excess pore water pressure (Δu) should, in general, correspond to the *in situ* stress conditions. The use of a value of pore pressure coefficient A obtained from the results of a triaxial test on a cylindrical clay specimen is strictly applicable only for the condition of axial symmetry, i e., for the case of settlement under the center of a circular footing. However, the value of A so obtained will serve as a good approximation for the case of settlement under the center of a square footing (using the circular footing of the same area).

Under a strip footing plane strain conditions prevail. Scott (1963) has shown that the value of Δu appropriate in the case of a strip footing can be obtained by using a pore pressure coefficient A_s as

$$A_s = 0.866A + 0.211 \tag{13.44}$$

The coefficient A_s replaces A (the coefficient for the condition of axial symmetry) in Eq. (13.42) for the case of a strip footing, the expression for α being unchanged.

Values of the settlement coefficient β for circular and strip footings, in terms of A and ratios H/B, are given in Fig 13.12.

Typical values of β are given in Table 13.5 for various types of clay soils.

Example 13.12

For the problem given in Ex. 13.10 compute the consolidation settlement by the Skempton-Bjerrum method. The compressible layer of depth 16 m below the base of the foundation is divided into four layers and the soil properties of each layer are given in Fig. Ex. 13.12. The net contact pressure $q_n = 100$ kN/m^2.

Solution

From Eq. (13.33), the oedometer settlement for the entire clay layer system may be expressed as

$$S_{oe} = \sum H_i \frac{C_c}{1+e_o} \log \frac{p_o + \Delta p}{p_o}$$

From Eq. (13.41), the consolidation settlement as per Skempton-Bjerrum may be expressed as

$$S_c = \beta S_{oe}$$

where β = settlement coefficient which can be obtained from Fig. 13.12 for various values of A and H/B.

p_o = effective overburden pressure at the middle of each layer (Fig. Ex. 13.12)

C_c = compression index of each layer

H_i = thickness of i th layer

e_o = initial void ratio of each layer

Δp = the excess pressure at the middle of each layer obtained from elastic theory (Chapter 6)

The average pore pressure coefficient is

$$A = \frac{0.9 + 0.75 + 0.70 + 0.45}{4} = 0.7$$

The details of the calculations are tabulated below.

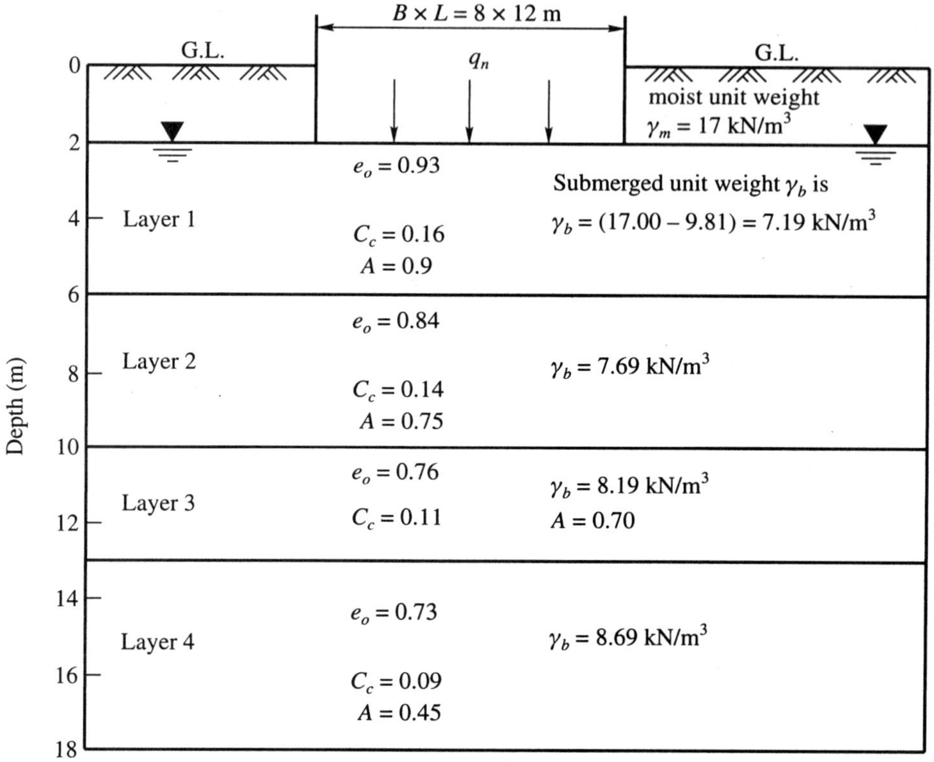

Figure Ex. 13.12

Layer No.	H_i (cm)	p_o (kN/m²)	Δp (kN/m²)	C_c	e_o	$\log\frac{p_o+\Delta p}{p_o}$	S_{oed} (cm)
1	400	48.4	75	0.16	0.93	0.407	13.50
2	400	78.1	43	0.14	0.84	0.191	5.81
3	300	105.8	22	0.11	0.76	0.082	1.54
4	500	139.8	14	0.09	0.73	0.041	1.07
						Total	21.92

For $H/B = 16/8 = 2$, $A = 0.7$, from Fig. 13.12 we have $\beta = 0.8$.

The consolidation settlement S_c is

$S_c = 0.8 \times 21.92 = 17.536$ cm $= 175.36$ mm

13.15 PROBLEMS

13.1 A plate load test was conducted in a medium dense sand at a depth of 5 ft below ground level in a test pit. The size of the plate used was 12 × 12 in. The data obtained from the test are plotted in Fig. Prob. 13.1 as a load-settlement curve. Determine from the curve the net safe bearing pressure for footings of size (a) 10 × 10 ft, and (b) 15 × 15 ft. Assume the permissible settlement for the foundation is 25 mm.

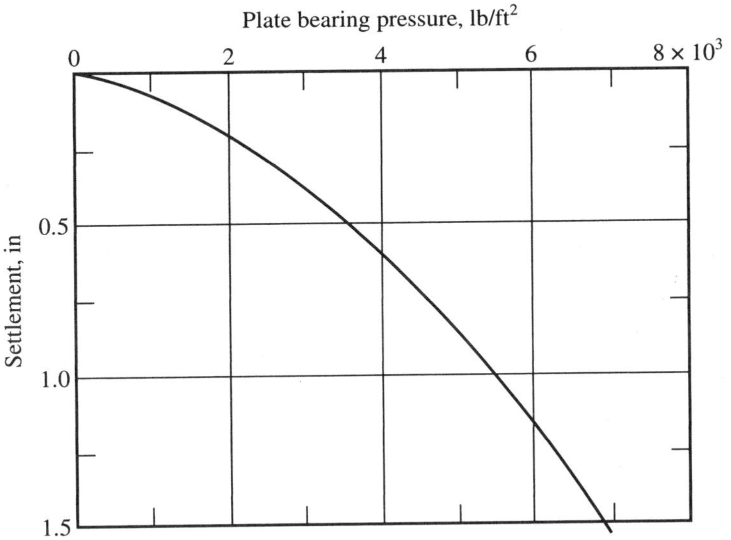

Figure Prob. 13.1

13.2 Refer to Prob. 13.1. Determine the settlements of the footings given in Prob 13.1. Assume the settlement of the plate as equal to 0.5 in. What is the net bearing pressure from Fig. Prob. 13.1 for the computed settlements of the foundations?

13.3 For Problem 13.2, determine the safe bearing pressure of the footings if the settlement is limited to 2 in.

13.4 Refer to Prob. 13.1. If the curve given in Fig. Prob. 13.1 applies to a plate test of 12 × 12 in. conducted in a clay stratum, determine the safe bearing pressures of the footings for a settlement of 2 in.

13.5 Two plate load tests were conducted in a c-ϕ soil as given below.

Size of plates (m)	Load kN	Settlement (mm)
0.3 × 0.3	40	30
0.6 × 0.6	100	30

Determine the required size of a footing to carry a load of 1250 kN for the same settlement of 30 mm.

13.6 A rectangular footing of size 4 × 8 m is founded at a depth of 2 m below the ground surface in dense sand and the water table is at the base of the foundation. N_{cor} = 30 (Fig. Prob. 13.6). Compute the safe bearing pressure q_s using the chart given in Fig. 13.5.

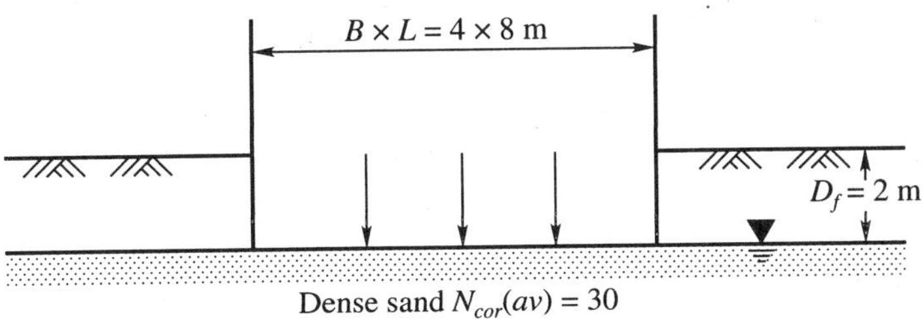

Figure Prob. 13.6

13.7 Refer to Prob. 13.6. Compute q_s by using modified (a) Teng's formula, and (b) Meyerhof's formula.

13.8 Refer to Prob. 13.6. Determine the safe bearing pressure based on the static cone penetration test value based on the relationship given in Eq. (13.7b) for $q_c = 120$ kN/m^2.

13.9 Refer to Prob. 13.6. Estimate the immediate settlement of the footing by using Eq. (13.20a). The additional data available are:

$\mu = 0.30$, $I_f = 0.82$ for rigid footing and $E_s = 11{,}000$ kN/m^2. Assume $q_n = q_s$ as obtained from Prob. 13.6.

13.10 Refer to Prob 13.6. Compute the immediate settlement for a flexible footing, given $\mu = 0.30$ and $E_s = 11{,}000$ kN/m^2. Assume $q_n = q_s$

13.11 If the footing given in Prob. 13.6 rests on normally consolidated saturated clay, compute the immediate settlement using Eq. (13.22). Use the following relationships.

$$\bar{q}_c = 120 \text{ kN/m}^2$$

$$c_u = \frac{\bar{q}_c - p_0}{N_k} \text{ and } N_k = 20$$

$$E_s = 600 c_u \text{ kN/m}^2$$

Given: $\gamma_{sat} = 18.5$ kN/m^3, $q_n = 150$ kN/m^2. Assume that the incompressible stratum lies at at depth of 10 m below the base of the foundation.

13.12 A footing of size 6 × 6 m rests in medium dense sand at a depth of 1.5 below ground level. The contact pressure $q_n = 175$ kN/m^2. The compressible stratum below the foundation base is divided into three layers. The corrected N_{cor} values for each layer is given in Fig. Prob. 13.12 with other data . Compute the immediate settlement using Eq. (13.23). Use the relationship $q_c = 400\ N_{cor}$ kN/m^2.

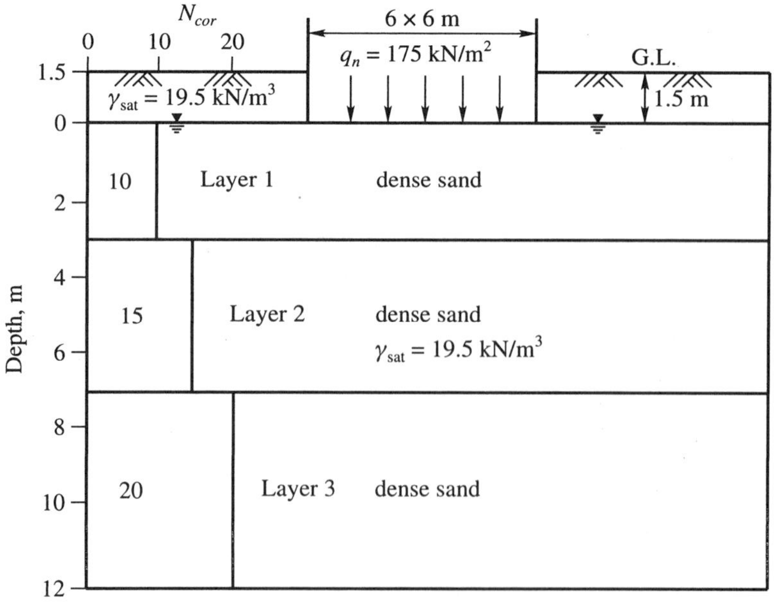

Figure Prob. 13.12

13.13 It is proposed to construct an overhead tank on a raft foundation of size 8 × 16 m with the foundation at a depth of 2 m below ground level. The subsoil at the site is a stiff homogeneous clay with the water table at the base of the foundation. The subsoil is divided into 3 layers and the properties of each layer are given in Fig. Prob. 13.13. Estimate the consolidation settlement by the Skempton-Bjerrum Method.

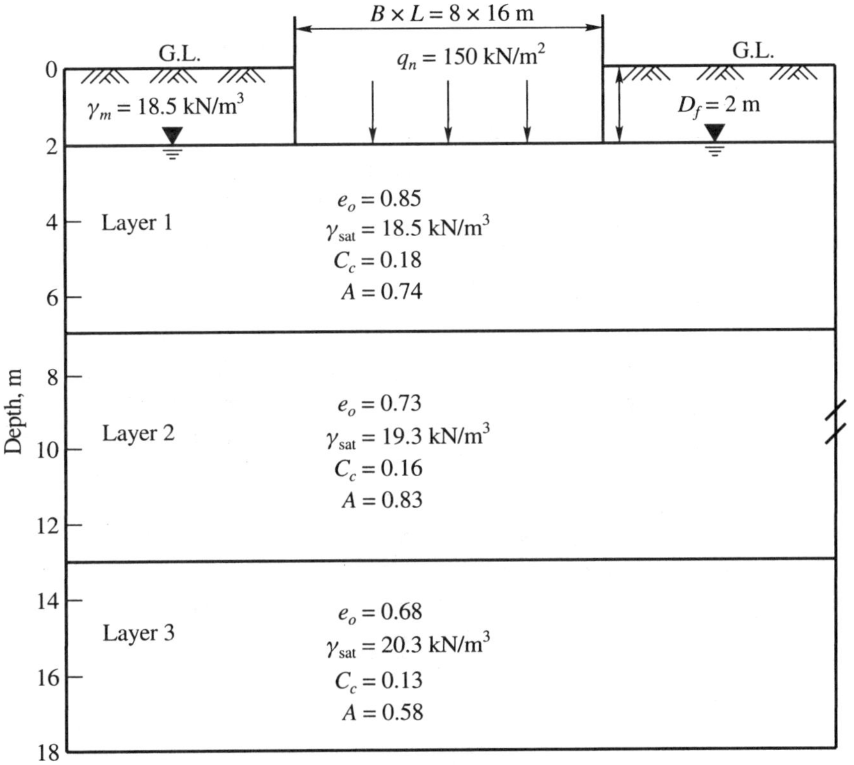

Figure Prob. 13.13

13.14 A footing of size 10 × 10 m is founded at a depth of 2.5 m below ground level on a sand deposit. The water table is at the base of the foundation. The saturated unit weight of soil from ground level to a depth of 22.5 m is 20 kN/m³. The compressible stratum of 20 m below the foundation base is divided into three layers with corrected SPT values (N) and CPT values (q_c) constant in each layer as given below.

Layer No	Depth from (m) foundation level From	To	N_{cor} (av)	q_c(av) MPa
1	0	5	20	8.0
2	5	11.0	25	10.0
3	11.0	20.0	30	12.0

Compute the settlements by Schmertmann's method.

Assume the net contact pressure at the base of the foundation is equal to 70 kPa, and t = 10 years

13.15 A square rigid footing of size 10 × 10 m is founded at a depth of 2.0 m below ground level. The type of strata met at the site is

Depth below G. L. (m)	Type of soil
0 to 5	Sand
5 to 7m	Clay
Below 7m	Sand

The water table is at the base level of the foundation. The saturated unit weight of soil above the foundation base is 20 kN/m^3. The coefficient of volume compressibility of clay, m_v, is 0.0001 m^2 /kN, and the coefficient of consolidation c_v, is 1 m^2/year. The total contact pressure q_n = 100 kN/m^2. Water table is at the base level of foundation.

Compute primary consolidation settlement.

13.16 A circular tank of diameter 3 m is founded at a depth of 1 m below ground surface on a 6 m thick normally consolidated clay. The water table is at the base of the foundation. The saturated unit weight of soil is 19.5 kN/m^3, and the *in-situ* void ratio e_0 is 1.08. Laboratory tests on representative undisturbed samples of the clay gave a value of 0.6 for the pore pressure coefficient A and a value of 0.2 for the compression index C_c. Compute the consolidation settlement of the foundation for a total contact pressure of 95 KPa. Use 2:1 method for computing Δp.

13.17 A raft foundation of size 10 × 40 m is founded at a depth of 3 m below ground surface and is uniformly loaded with a net pressure of 50 kN/m^2. The subsoil is normally consolidated saturated clay to a depth of 20 m below the base of the foundation with variable elastic moduli with respect to depth. For the purpose of analysis, the stratum is divided into three layers with constant modulus as given below:

Layer No	Depth below ground (m)		Elastic Modulus
	From	To	E_s (*MPa*)
1	3	8	20
2	8	18	25
3	18	23	30

Compute the immediate settlements by using Eqs (13.20a). Assume the footing is flexible.

CHAPTER 14

SHALLOW FOUNDATION III: COMBINED FOOTINGS AND MAT FOUNDATIONS

14.1 INTRODUCTION

Chapter 12 has considered the common methods of transmitting loads to subsoil through spread footings carrying single column loads. This chapter considers the following types of foundations:

1. Cantilever footings
2. Combined footings
3. Mat foundations

When a column is near or right next to a property limit, a square or rectangular footing concentrically loaded under the column would extend into the adjoining property. If the adjoining property is a public side walk or alley, local building codes may permit such footings to project into public property. But when the adjoining property is privately owned, the footings must be constructed within the property. In such cases, there are three alternatives which are illustrated in Fig. 14.1(a). These are

1. *Cantilever footing.* A cantilever or strap footing normally comprises two footings connected by a beam called a strap. A strap footing is a special case of a combined footing.
2. *Combined footing.* A combined footing is a long footing supporting two or more columns in one row.
3. *Mat or raft foundations.* A mat or raft foundation is a large footing, usually supporting several columns in two or more rows.

The choice between these types depends primarily upon the relative cost. In the majority of cases, mat foundations are normally used where the soil has low bearing capacity and where the total area occupied by an individual footing is not less than 50 per cent of the loaded area of the building.

When the distances between the columns and the loads carried by each column are not equal, there will be eccentric loading. The effect of eccentricity is to increase the base pressure on the side of eccentricity and decrease it on the opposite side. The effect of eccentricity on the base pressure of rigid footings is also considered here.

Mat Foundation in Sand

A foundation is generally termed as a mat if the least width is more than 6 meters. Experience indicates that the ultimate bearing capacity of a mat foundation on cohesionless soil is much higher than that of individual footings of lesser width. With the increasing width of the mat, or increasing relative density of the sand, the ultimate bearing capacity increases rapidly. Hence, the danger that a large mat may break into a sand foundation is too remote to require consideration. On account of the large size of mats the stresses in the underlying soil are likely to be relatively high to a considerable depth. Therefore, the influence of local loose pockets distributed at random throughout the sand is likely to be about the same beneath all parts of the mat and differential settlements are likely to be smaller than those of a spread foundation designed for the same soil

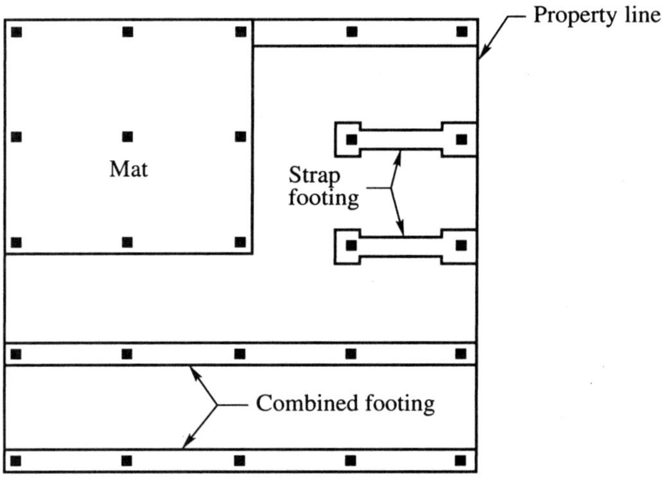

(a) Schematic plan showing mat, strap and combined footings

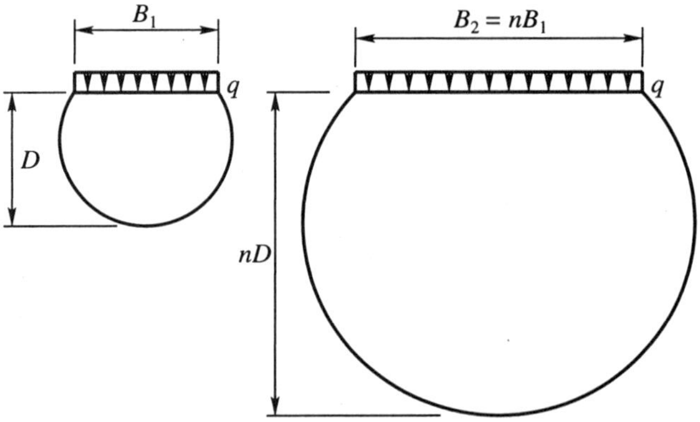

(b) Bulb of pressure for vertical stress for different beams

Figure 14.1 (a) Types of footings; (b) beams on compressible subgrade pressure. The methods of calculating the ultimate bearing capacity dealt with in Chapter 12 are also applicable to mat foundations.

Mat Foundation in Clay

The net ultimate bearing capacity that can be sustained by the soil at the base of a mat on a deep deposit of clay or plastic silt may be obtained in the same manner as for footings on clay discussed in Chapter 12. However, by using the principle of *flotation*, the pressure on the base of the mat that induces settlement can be reduced by increasing the depth of the foundation. A brief discussion on the principle of flotation is dealt with in this chapter.

Rigid and Elastic Foundation

The conventional method of design of combined footings and mat foundations is to assume the foundation as infinitely rigid and the contact pressure is assumed to have a planar distribution. In the case of an elastic foundation, the soil is assumed to be a truly elastic solid obeying Hooke's law in all directions. The design of an elastic foundation requires a knowledge of the subgrade reaction which is briefly discussed here. However, the elastic method does not readily lend itself to engineering applications because it is extremely difficult and solutions are available for only a few extremely simple cases.

14.2 SAFE BEARING PRESSURES FOR MAT FOUNDATIONS ON SAND AND CLAY

Mats on Sand

Because the differential settlements of a mat foundation are less than those of a spread foundation designed for the same soil pressure, it is reasonable to permit larger safe soil pressures on a raft foundation. Experience has shown that a pressure approximately twice as great as that allowed for individual footings may be used because it does not lead to detrimental differential settlements. The maximum settlement of a mat may be about 50 mm (2 in) instead of 25 mm as for a spread foundation.

The shape of the curve in Fig. 13.3(a) shows that the net soil pressure corresponding to a given settlement is practically independent of the width of the footing or mat when the width becomes large. The safe soil pressure for design may with sufficient accuracy be taken as twice the pressure indicated in Fig. 13.5. Peck et al., (1974) recommend the following equation for computing net safe pressure,

$$q_s = 21\ N_{cor}\ \text{kPa} \tag{14.1}$$

$$\text{for } 5 < N_{cor} < 50$$

where N_{cor} is the *SPT* value corrected for energy, overburden pressure and field procedures.

Eq. 14.1 gives q_s values above the water table. A correction factor should be used for the presence of a water table as explained in Chapter 12.

Peck et al., (1974) also recommend that the q_s values as given by Eq. 14.1 may be increased somewhat if bedrock is encountered at a depth less than about one half the width of the raft.

The value of *N* to be considered is the average of the values obtained up to a depth equal to the least width of the raft. If the average value of *N* after correction for the influence of overburden pressure and dilatancy is less than about 5, Peck et al., say that the sand is generally considered to be too loose for the successful use of a raft foundation. Either the sand should be compacted or else the foundation should be established on piles or piers.

The minimum depth of foundation recommended for a raft is about 2.5 m below the surrounding ground surface. Experience has shown that if the surcharge is less than this amount, the edges of the raft settle appreciably more than the interior because of a lack of confinement of the sand.

Safe Bearing Pressures of Mats on Clay

The quantity in Eq. 12.25(b) is the net bearing capacity q_{nu} at the elevation of the base of the raft in excess of that exerted by the surrounding surcharge. Likewise, in Eq. 12.25(c), q_{na} is the net allowable soil pressure. By increasing the depth of excavation, the pressure that can safely be exerted by the building is correspondingly increased. This aspect of the problem is considered further in Section 14.10 in *floating foundation.*

As for footings on clay, the factor of safety against failure of the soil beneath a mat on clay should not be less than 3 under normal loads, or less than 2 under the most extreme loads.

The settlement of the mat under the given loading condition should be calculated as per the procedures explained in Chapter 13. The net safe pressure should be decided on the basis of the permissible settlement.

14.3 ECCENTRIC LOADING

When the resultant of loads on a footing does not pass through the center of the footing, the footing is subjected to what is called *eccentric loading.* The loads on the footing may be vertical or inclined. If the loads are inclined it may be assumed that the horizontal component is resisted by the frictional resistance offered by the base of the footing. The vertical component in such a case is the only factor for the design of the footing. The effects of eccentricity on bearing pressure of the footings have been discussed in Chapter 12.

14.4 THE COEFFICIENT OF SUBGRADE REACTION

The coefficient of subgrade reaction is defined as the ratio between the pressure against the footing or mat and the settlement at a given point expressed as

$$k_s = \frac{q}{S}, \tag{14.2}$$

where k_s = coefficient of subgrade reaction expressed as force/length^3 (FL^{-3}),
q = pressure on the footing or mat at a given point expressed as force/length^2 (FL^{-2}),
S = settlement of the same point of the footing or mat in the corresponding unit of length.

In other words the coefficient of subgrade reaction is the unit pressure required to produce a unit settlement. In clayey soils, settlement under the load takes place over a long period of time and the coefficient should be determined on the basis of the final settlement. On purely granular soils, settlement takes place shortly after load application. Eq. (14.2) is based on two simplifying assumptions:

1. The value of k_s is independent of the magnitude of pressure.
2. The value of k_s has the same value for every point on the surface of the footing.

Both the assumptions are strictly not accurate. The value of k_s decreases with the increase of the magnitude of the pressure and it is not the same for every point of the surface of the footing as the settlement of a flexible footing varies from point to point. However the method is supposed to give realistic values for contact pressures and is suitable for beam or mat design when only a low order of settlement is required.

Factors Affecting the Value of k_s

Terzaghi (1955) discussed the various factors that affect the value of k_s. A brief description of his arguments is given below.

Consider two foundation beams of widths B_1 and B_2 such that $B_2 = nB_1$ resting on a compressible subgrade and each loaded so that the pressure against the footing is uniform and equal to q for both the beams (Fig. 14.1b). Consider the same points on each beam and, let

y_1 = settlement of beam of width B_1

y_2 = settlement of beam of width B_2

Hence $k_{s1} = \dfrac{q}{y_1}$ and $k_{s2} = \dfrac{q}{y_2}$

If the beams are resting on a subgrade whose deformation properties are more or less independent of depth (such as a stiff clay) then it can be assumed that the settlement increases in simple proportion to the depth of the pressure bulb.

Then $y_2 = ny_1$

and
$$k_{s2} = \frac{q}{ny_1} = \frac{q}{y_1}\frac{B_1}{B_2} = k_{s1}\frac{B_1}{B_2}. \tag{14.3}$$

A general expression for k_s can now be obtained if we consider B_1 as being of unit width (Terzaghi used a unit width of one foot which converted to metric units may be taken as equal to 0.30 m).

Hence by putting $B_1 = 0.30$ m, $k_s = k_{s2}$, $B = B_2$, we obtain

$$k_s = 0.3\frac{k_{s1}}{B} \tag{14.4}$$

where k_s is the coefficient of subgrade reaction of a long footing of width B meters and resting on stiff clay; k_{s1} is the coefficient of subgrade reaction of a long footing of width 0.30 m (approximately), resting on the same clay. It is to be noted here that the value of k_{s1} is derived from ultimate settlement values, that is, after consolidation settlement is completed.

If the beams are resting on clean sand, the final settlement values are obtained almost instantaneously. Since the modulus of elasticity of sand increases with depth, the deformation characteristics of the sand change and become less compressible with depth. Because of this characteristic of sand, the lower portion of the bulb of pressure for beam B_2 is less compressible than that of the sand enclosed in the bulb of pressure of beam B_1.

The settlement value y_2 lies somewhere between y_1 and ny_1. It has been shown experimentally (Terzaghi and Peck, 1948) that the settlement, y, of a beam of width B resting on sand is given by the expression

$$y = y_1\left[\frac{2B}{B+0.3}\right]^2 \tag{14.5}$$

where y_1 = settlement of a beam of width 0.30 m and subjected to the same reactive pressure as the beam of width B meters.

Hence, the coefficient of subgrade reaction k_s of a beam of width B meters can be obtained from the following equation

$$k_s = \frac{q}{y} = \frac{q}{y_1}\left[\frac{B+0.30}{2B}\right]^2 = k_{s1}\left[\frac{B+0.30}{2B}\right]^2 \tag{14.6}$$

where k_{s1} = coefficient of subgrade reaction of a beam of width 0.30 m resting on the same sand.

Measurement of k_{s1}

A value for k_{s1} for a particular subgrade can be obtained by carrying out plate load tests. The standard size of plate used for this purpose is 0.30 × 0.30 m size. Let k_1 be the subgrade reaction for a plate of size 0.30 × 0.30 m size.

From experiments it has been found that $k_{s1} \approx k_1$ for sand subgrades, but for clays k_{s1} varies with the length of the beam. Terzaghi (1955) gives the following formula for clays

$$k_{s1} = k_1\left[\frac{L+0.152}{1.5L}\right] \tag{14.7a}$$

where L = length of the beam in meters and the width of the beam = 0.30 m. For a very long beam on clay subgrade we may write

$$k_{s1} = \frac{k_1}{1.5} \tag{14.7b}$$

Procedure to Find k_s

For sand

1. Determine k_1 from plate load test or from estimation.
2. Since $ks_{s1} \approx k_1$, use Eq. (14.6) to determine k_s for sand for any given width B meter.

For clay

1. Determine k_1 from plate load test or from estimation
2. Determine k_{s1} from Eq. (14.7a) as the length of beam is known.
3. Determine k_s from Eq. (14.4) for the given width B meters.

When plate load tests are used, k_1 may be found by one of the two ways,

1. A bearing pressure equal to not more than the ultimate pressure and the corresponding settlement is used for computing k_1
2. Consider the bearing pressure corresponding to a settlement of 1.3 mm for computing k_1.

Estimation of k_1 Values

Plate load tests are both costly and time consuming. Generally a designer requires only the values of the bending moments and shear forces within the foundation. With even a relatively large error in the estimation of k_1, moments and shear forces can be calculated with little error (Terzaghi, 1955); an error of 100 per cent in the estimation of k_s may change the structural behavior of the foundation by up to 15 per cent only.

Table 14.1a k_1 values for foundations on sand (MN/m^3)

Relative density	Loose	Medium	Dense
SPT Values (Uncorrected)	<10	10–30	>30
Soil, dry or moist	15	45	175
Soil submerged	10	30	100

Table 14.1b k_1 values for foundation on clay

Consistency	Stiff	Very stiff	Hard
c_u (kN/m^2)	50-100	100-200	>200
k_1 (MN/m^3)	25	50	100

Source: Terzaghi (1955)

In the absence of plate load tests, estimated values of k_1 and hence k_s are used. The values suggested by Terzaghi for k_1 (converted into S.I. units) are given in Table 14.1.

14.5 PROPORTIONING OF CANTILEVER FOOTING

Strap or cantilever footings are designed on the basis of the following assumptions:

1. The strap is infinitely stiff. It serves to transfer the column loads to the soil with equal and uniform soil pressure under both the footings.
2. The strap is a pure flexural member and does not take soil reaction. To avoid bearing on the bottom of the strap a few centimeters of the underlying soil may be loosened prior to the placement of concrete.

A strap footing is used to connect an eccentrically loaded column footing close to the property line to an interior column as shown in Fig. 14.2.

With the above assumptions, the design of a strap footing is a simple procedure. It starts with a trial value of e, Fig. 14.2. Then the reactions R_1 and R_2 are computed by the principle of statics. The tentative footing areas are equal to the reactions R_1 and R_2 divided by the safe bearing pressure q_s. With tentative footing sizes, the value of e is computed. These steps are repeated until the trial value of e is identical with the final one. The shears and moments in the strap are determined, and the straps designed to withstand the shear and moments. The footings are assumed to be subjected to uniform soil pressure and designed as simple spread footings. Under the assumptions given above the resultants of the column loads Q_1 and Q_2 would coincide with the center of gravity of the two footing areas. Theoretically, the bearing pressure would be uniform under both the footings. However, it is possible that sometimes the full design live load acts upon one of the columns while the other may be subjected to little live load. In such a case, the full reduction of column load from Q_2 to R_2 may not be realized. It seems justified then that in designing the footing under column Q_2, only the dead load or dead load plus reduced live load should be used on column Q_1.

The equations for determining the position of the reactions (Fig. 14.2) are

$$R_1 = Q_1\left(1+\frac{e}{L_R}\right), \qquad R_2 = Q_2 - \frac{Q_1 e}{L_R} \tag{14.8}$$

where R_1 and R_2 = reactions for the column loads Q_1 and Q_2 respectively, e = distance of R_1 from Q_1, L_R = distance between R_1 and R_2.

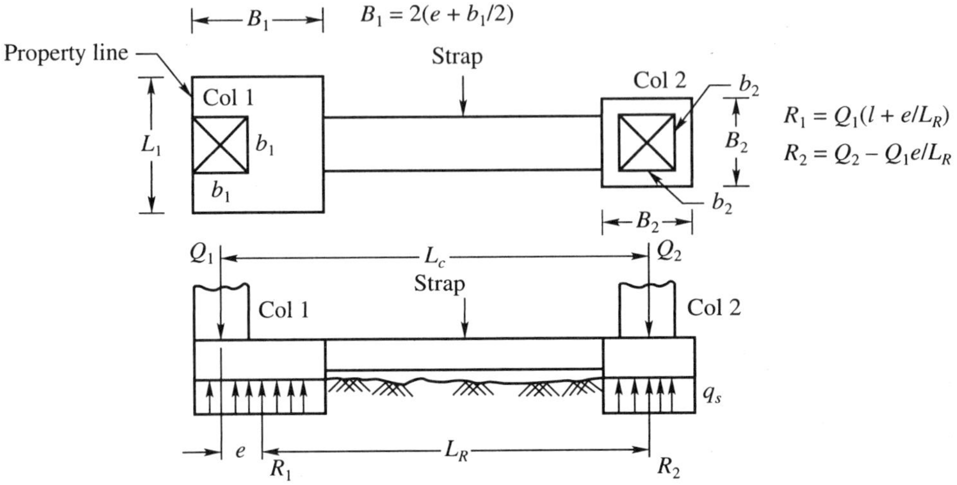

Figure 14.2 Principles of cantilever or strap footing design

14.6 DESIGN OF COMBINED FOOTINGS BY RIGID METHOD (CONVENTIONAL METHOD)

The rigid method of design of combined footings assumes that

1. The footing or mat is infinitely rigid, and therefore, the deflection of the footing or mat does not influence the pressure distribution,
2. The soil pressure is distributed in a straight line or a plane surface such that the centroid of the soil pressure coincides with the line of action of the resultant force of all the loads acting on the foundation.

Design of Combined Footings

Two or more columns in a row joined together by a stiff continuous footing form a combined footing as shown in Fig. 14.3a. The procedure of design for a combined footing is as follows:

1. Determine the total column loads $\Sigma Q = Q_1 + Q_2 + Q_3 + \ldots$ and location of the line of action of the resultant ΣQ. If any column is subjected to bending moment, the effect of the moment should be taken into account.
2. Determine the pressure distribution q per lineal length of footing.
3. Determine the width, B, of the footing.
4. Draw the shear diagram along the length of the footing. By definition, the shear at any section along the beam is equal to the summation of all vertical forces to the left or right of the section. For example, the shear at a section immediately to the left of Q_1 is equal to the area *abcd*, and immediately to the right of Q_1 is equal to $(abcd - Q_1)$ as shown in Fig. 14.3a.
5. Draw the moment diagram along the length of the footing. By definition the bending moment at any section is equal to the summation of moment due to all the forces and reaction to the left (or right) of the section. It is also equal to the area under the shear diagram to the left (or right) of the section.
6. Design the footing as a continuous beam to resist the shear and moment.
7. Design the footing for transverse bending in the same manner as for spread footings.

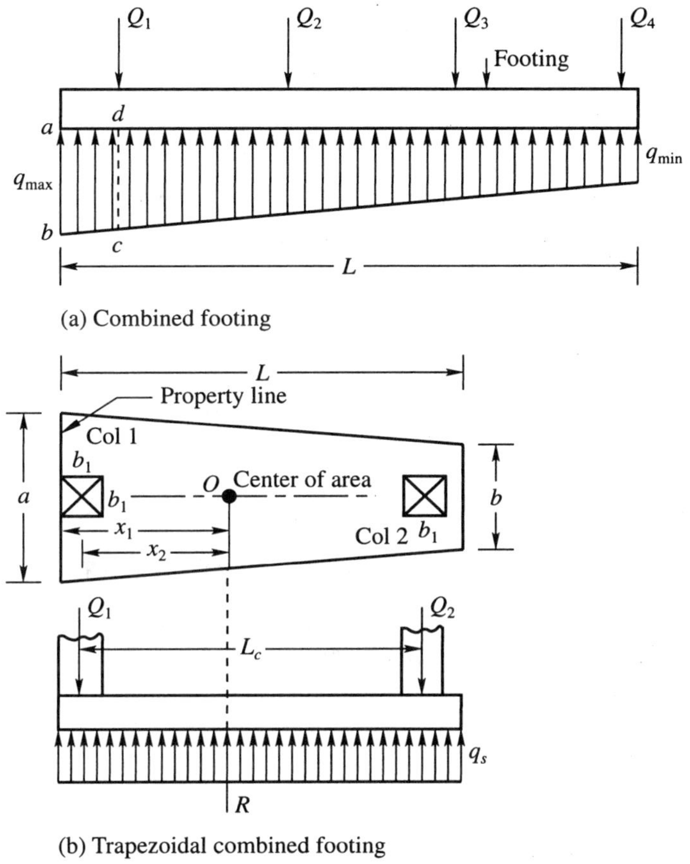

Figure 14.3 Combined or trapezoidal footing design

It should be noted here that the end column along the property line may be connected to the interior column by a rectangular or trapezoidal footing. In such a case no strap is required and both the columns together will be a combined footing as shown in Fig. 14.3b. It is necessary that the center of area of the footing must coincide with the center of loading for the pressure to remain uniform.

14.7 DESIGN OF MAT FOUNDATION BY RIGID METHOD

In the conventional rigid method the mat is assumed to be infinitely rigid and the bearing pressure against the bottom of the mat follows a planar distribution where the centroid of the bearing pressure coincides with the line of action of the resultant force of all loads acting on the mat. The procedure of design is as follows:

1. The column loads of all the columns coming from the superstructure are calculated as per standard practice. The loads include live and dead loads.
2. Determine the line of action of the resultant of all the loads. However, the weight of the mat is not included in the structural design of the mat because every point of the mat is supported by the soil under it, causing no flexural stresses.
3. Calculate the soil pressure at desired locations by the use of Eq. (12.73a)

$$q = \frac{Q_t}{A} \pm \frac{Q_t e_x}{I_y} x \pm \frac{Q_t e_y}{I_x} y$$

where $Q_t = \Sigma Q$ = total load on the mat,

A = total area of the mat,

x, y = coordinates of any given point on the mat with respect to the x and y axes passing through the centroid of the area of the mat,

e_x, e_y = eccentricities of the resultant force,

I_x, I_y = moments of inertia of the mat with respect to the x and y axes respectively.

4. The mat is treated as a whole in each of two perpendicular directions. Thus the total shear force acting on any section cutting across the entire mat is equal to the arithmetic sum of all forces and reactions (bearing pressure) to the left (or right) of the section. The total bending moment acting on such a section is equal to the sum of all the moments to the left (or right) of the section.

14.8 DESIGN OF COMBINED FOOTINGS BY ELASTIC LINE METHOD

The relationship between deflection, y, at any point on an elastic beam and the corresponding bending moment M may be expressed by the equation

$$EI\frac{d^2y}{dx^2} = M \tag{14.10}$$

The equations for shear V and reaction q at the same point may be expressed as

$$EI\frac{d^3y}{dx^3} = V \tag{14.11}$$

$$EI\frac{d^4y}{dx^4} = q \tag{14.12}$$

where x is the coordinate along the length of the beam.

From the basic assumption of an elastic foundation

$$q = -yBk_s$$

where, B = width of footing, k_s = coefficient of subgrade reaction.

Substituting for q, Eq. (14.12) may be written as

$$EI\frac{d^4y}{dx^4} = -yBk_s \tag{14.13}$$

The classical solutions of Eq. (14.13) being of closed form, are not general in their application. Hetenyi (1946) developed equations for a load at any point along a beam. The development of solutions is based on the concept that the beam lies on a bed of elastic springs which is based on Winkler's hypothesis. As per this hypothesis, the reaction at any point on the beam depends only on the deflection at that point.

Methods are also available for solving the beam-problem on an elastic foundation by the method of finite differences (Malter, 1958). The finite element method has been found to be the most efficient of the methods for solving beam-elastic foundation problem. Computer programs are available for solving the problem.

Since all the methods mentioned above are quite involved, they are not dealt with here. Interested readers may refer to Bowles (1996).

14.9 DESIGN OF MAT FOUNDATIONS BY ELASTIC PLATE METHOD

Many methods are available for the design of mat-foundations. The one that is very much in use is the finite difference method. This method is based on the assumption that the subgrade can be substituted by a bed of uniformly distributed coil springs with a spring constant k_s which is called the coefficient of subgrade reaction. The finite difference method uses the fourth order differential equation

$$\nabla^4 w = \frac{q - k_s w}{D}$$

where $$\nabla^4 w = \frac{\partial^4 w}{\partial x^4} + 2\frac{\partial^4 w}{\partial x^2 \partial y^2} + \frac{\partial^4 w}{\partial y^4} \quad (14.14)$$

q = subgrade reaction per unit area,
k_s = coefficient of subgrade reaction,
w = deflection,

D = rigidity of the mat $= \dfrac{Et^3}{12(1-\mu^2)}$

E = modulus of elasticity of the material of the footing,
t = thickness of mat,
μ = Poisson's ratio.

Eq. (14.14) may be solved by dividing the mat into suitable square grid elements, and writing difference equations for each of the grid points. By solving the simultaneous equations so obtained the deflections at all the grid points are obtained. The equations can be solved rapidly with an electronic computer. After the deflections are known, the bending moments are calculated using the relevant difference equations.

Interested readers may refer to Teng (1969) or Bowles (1996) for a detailed discussion of the method.

14.10 FLOATING FOUNDATION

General Consideration

A *floating foundation* for a building is defined as a foundation in which the weight of the building is approximately equal to the full weight including water of the soil removed from the site of the building. This principle of flotation may be explained with reference to Fig. 14.4. Fig. 14.4(a) shows a horizontal ground surface with a horizontal water table at a depth d_w below the ground surface. Fig. 14.4(b) shows an excavation made in the ground to a depth D where $D > d_w$, and Fig. 14.4(c) shows a structure built in the excavation and completely filling it.

If the weight of the building is equal to the weight of the soil and water removed from the excavation, then it is evident that the total vertical pressure in the soil below depth D in Fig. 14.4(c) is the same as in Fig. 14.4(a) before excavation.

Since the water level has not changed, the neutral pressure and the effective pressure are therefore unchanged. Since settlements are caused by an increase in effective vertical pressure, if we could move from Fig. 14.4(a) to Fig. 14.4(c) without the intermediate case of 14.4(b), the building in Fig. 14.4(c) would not settle at all.

This is the principle of a floating foundation, an exact balance of weight removed against weight imposed. The result is zero settlement of the building.

However, it may be noted, that we cannot jump from the stage shown in Fig. 14.4(a) to the stage in Fig. 14.4(c) without passing through stage 14.4(b). The excavation stage of the building is the critical stage.

Cases may arise where we cannot have a fully floating foundation. The foundations of this type are sometimes called *partly compensated foundations* (as against *fully compensated* or *fully floating* foundations).

While dealing with floating foundations, we have to consider the following two types of soils. They are:

Type 1: The foundation soils are of such a strength that shear failure of soil will not occur under the building load but the settlements and particularly differential settlements, will be too large and will constitute *failure* of the structure. A floating foundation is used to reduce settlements to an acceptable value.

Type 2: The shear strength of the foundation soil is so low that rupture of the soil would occur if the building were to be founded at ground level. In the absence of a strong layer at a reasonable depth, the building can only be built on a floating foundation which reduces the shear stresses to an acceptable value. Solving this problem solves the settlement problem.

In both the cases, a rigid raft or box type of foundation is required for the floating foundation [Fig. 14.4(d)]

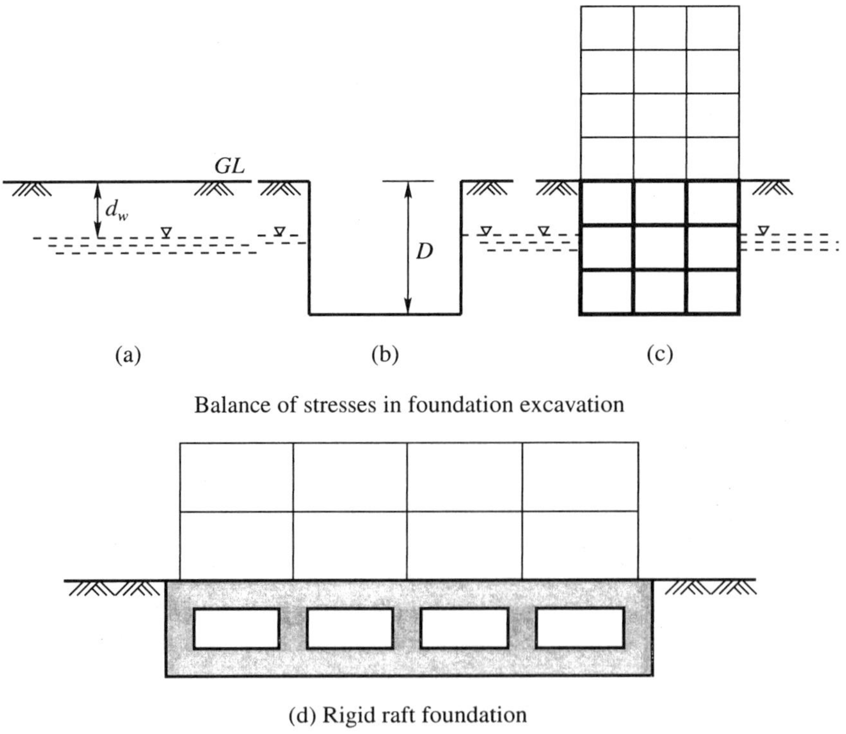

Figure 14.4 Principles of floating foundation; and a typical rigid raft foundation

Problems to be Considered in the Design of a Floating Foundation

The following problems are to be considered during the design and construction stage of a floating foundation.

1. Excavation

The excavation for the foundation has to be done with care. The sides of the excavation should suitably be supported by sheet piling, soldier piles and timber or some other standard method.

2. Dewatering

Dewatering will be necessary when excavation has to be taken below the water table level. Care has to be taken to see that the adjoining structures are not affected due to the lowering of the water table.

3. Critical depth

In Type 2 foundations the shear strength of the soil is low and there is a theoretical limit to the depth to which an excavation can be made. Terzaghi (1943) has proposed the following equation for computing the critical depth D_c,

$$D_c = \frac{5.7s}{\gamma - (s/B)\sqrt{2}} \tag{14.15}$$

for an excavation which is long compared to its width

where γ = unit weight of soil,

s = shear strength of soil = $q_u/2$,

B = width of foundation,

L = length of foundation.

Skempton (1951) proposes the following equation for D_c, which is based on actual failures in excavations

$$D_c = N_c \frac{s}{\gamma} \tag{14.16}$$

or the factor of safety F_s against bottom failure for an excavation of depth D is

$$F_s = N_c \frac{s}{\gamma D + p}$$

where N_c is the bearing capacity factor as given by Skempton, and p is the surcharge load. The values of N_c may be obtained from Fig 12.13(a). The above equations may be used to determine the maximum depth of excavation.

4. Bottom heave

Excavation for foundations reduces the pressure in the soil below the founding depth which results in the heaving of the bottom of the excavation. Any heave which occurs will be reversed and appear as settlement during the construction of the foundation and the building. Though heaving of the bottom of the excavation cannot be avoided it can be minimized to a certain extent. There are three possible causes of heave:

1. Elastic movement of the soil as the existing overburden pressure is removed.
2. A gradual swelling of soil due to the intake of water if there is some delay for placing the foundation on the excavated bottom of the foundation.

3. Plastic inward movement of the surrounding soil.

The last movement of the soil can be avoided by providing proper lateral support to the excavated sides of the trench.

Heaving can be minimized by phasing out excavation in narrow trenches and placing the foundation soon after excavation. It can be minimized by lowering the water table during the excavation process. Friction piles can also be used to minimize the heave. The piles are driven either before excavation commences or when the excavation is at half depth and the pile tops are pushed down to below foundation level. As excavation proceeds, the soil starts to expand but this movement is resisted by the upper part of the piles which go into tension. This heave is prevented or very much reduced.

It is only a *practical and pragmatic approach* that would lead to a safe and sound settlement free floating (or partly floating) foundation.

Example 14.1

A beam of length 4 m and width 0.75 m rests on stiff clay. A plate load test carried out at the site with the use of a square plate of size 0.30 m gives a coefficient of subgrade reaction k_1 equal to 25 MN/m^3. Determine the coefficient of subgrade reaction k_s for the beam.

Solution

First determine k_{s1} from Eq. (14.7a) for a beam of 0.30 m. wide and length 4 m. Next determine k_s from Eq. (14.4) for the same beam of width 0.75 m.

$$k_{s1} = k_1\left[\frac{L+0.152}{1.5L}\right] = 25\left[\frac{4+0.152}{1.5\times 4}\right] = 17.3 \text{ MN/m}^3$$

$$k_s = 0.3\frac{k_{s1}}{B} = \frac{0.3\times 17.3}{0.75} \approx 7 \text{ MN/m}^3$$

Example 14.2

A beam of length 4 m and width 0.75 m rests in dry medium dense sand. A plate load test carried out at the same site and at the same level gave a coefficient of subgrade reaction k_1 equal to 47 MN/m^3. Determine the coefficient of subgrade reaction for the beam.

Solution

For sand the coefficient of subgrade reaction, k_{s1}, for a long beam of width 0.3 m is the same as that for a square plate of size 0.3 × 0.3 m that is $k_{s1} = k_s$. k_s now can be found from Eq. (14.6) as

$$k_s = k_1\left[\frac{B+0.3}{2B}\right]^2 = 47\left[\frac{0.75+0.30}{1.5}\right]^2 = 23 \text{ MN/m}^3$$

Example 14.3

The following information is given for proportioning a cantilever footing with reference to Fig. 14.2.

Column Loads: Q_1 = 1455 kN, Q_2 = 1500 kN.

Size of column: 0.5 × 0.5 m.

L_c = 6.2 m, q_s = 384 kN/m^2

It is required to determine the size of the footings for columns 1 and 2.

Solution

Assume the width of the footing for column 1 = B_1 = 2 m.

First trial

Try e = 0.5 m. Now, L_R = 6.2 – 0.5 = 5.7 m.

Reactions

$$R_1 = Q_1\left[1+\frac{e}{L_R}\right] = 1455\left[1+\frac{0.5}{5.7}\right] = 1583 \text{ kN}$$

$$R_2 = \left[Q_2 - \frac{Q_1 e}{L_R}\right] = \left[1500 - \frac{1455\times 0.5}{5.7}\right] = 1372 \text{ kN}$$

Size of footings - First trial

Col. 1. Area of footing $A_1 = \dfrac{1583}{384} = 4.122 \text{ sq.m}$

Col. 2. Area of footing $A_2 = \dfrac{1372}{384} = 3.57 \text{ sq.m}$

Try 1.9 × 1.9 m

Second trial

New value of $e = \dfrac{B_1}{2} - \dfrac{b_1}{2} = \dfrac{2}{2} - \dfrac{0.5}{2} = 0.75 \text{ m}$

New $L_R = 6.20 - 0.75 = 5.45$ m

$$R_1 = 1455\left[1+\frac{0.75}{5.45}\right] = 1655 \text{ kN}$$

$$R_2 = \left[1500 - \frac{1455\times 0.75}{5.45}\right] = 1300 \text{ kN}$$

$$A_1 = \frac{1655}{384} = 4.31 \text{ sq.m or } 2.08\times 2.08 \text{ m}$$

$$A_2 = \frac{1300}{384} = 3.38 \text{ sq.m or } 1.84\times 1.84 \text{ m}$$

Check $e = \dfrac{B_1}{2} - \dfrac{b_1}{2} = 1.04 - 0.25 = 0.79 \approx 0.75$ m

Use 2.08 × 2.08 m for Col. 1 and 1.90 × 1.90 m for Col. 2.

Note: Rectangular footings may be used for both the columns.

Example 14.4

Figure Ex. 14.4 gives a foundation beam with the vertical loads and moment acting thereon. The width of the beam is 0.70 m and depth 0.50 m. A uniform load of 16 kN/m (including the weight of the beam) is imposed on the beam. Draw (a) the base pressure distribution, (b) the shear force diagram, and (c) the bending moment diagram. The length of the beam is 8 m.

Solution

The steps to be followed are:

1. Determine the resultant vertical force R of the applied loadings and its eccentricity with respect to the centers of the beam.
2. Determine the maximum and minimum base pressures.
3. Draw the shear and bending moment diagrams.

$R = 320 + 400 + 16 \times 8 = 848$ kN.

Taking the moment about the right hand edge of the beam, we have,

$$Rx = 848x = 320\times 7 + 400\times 1 + 16\times \frac{8^2}{2} - 160 = 2992$$

or $$x = \frac{2992}{848} = 3.528 \text{ m}$$

$e = 4.0 - 3.528 = 0.472$ m to the right of center of the beam. Now from Eqs 12.39(a) and (b), using $e_y = 0$,

$$q_{\substack{max\\min}} = \frac{\Sigma Q}{A}\left[1 \pm 6\frac{e_x}{L}\right] = \frac{848}{8\times 0.7}\left[1 \pm \frac{6\times 0.472}{8}\right] = 205.02 \text{ or } 97.83 \text{ kN/m}^2$$

Convert the base pressures per unit area to load per unit length of beam.
The maximum vertical load = 0.7 × 205.02 = 143.52 kN/m.
The minimum vertical load = 0.7 × 97.83 = 68.48 kN/m.
The reactive loading distribution is given in Fig. Ex. 14.4(b).

Shear force diagram

Calculation of shear for a typical point such as the reaction point R_1 (Fig. Ex. 14.4(a)) is explained below.

Consider forces to the left of R_1 (without 320 kN).

Shear force V = upward shear force equal to the area *abcd* – downward force due to distributed load on beam *ab*

$$= \frac{68.48 + 77.9}{2} - 16\times 1 = 57.2 \text{ kN}$$

Consider to the right of reaction point R_1 (with 320 kN).

$V = -320 + 57.2 = -262.8$ kN.

In the same away the shear at other points can be calculated. Fig. Ex. 14.4(c) gives the complete shear force diagram.

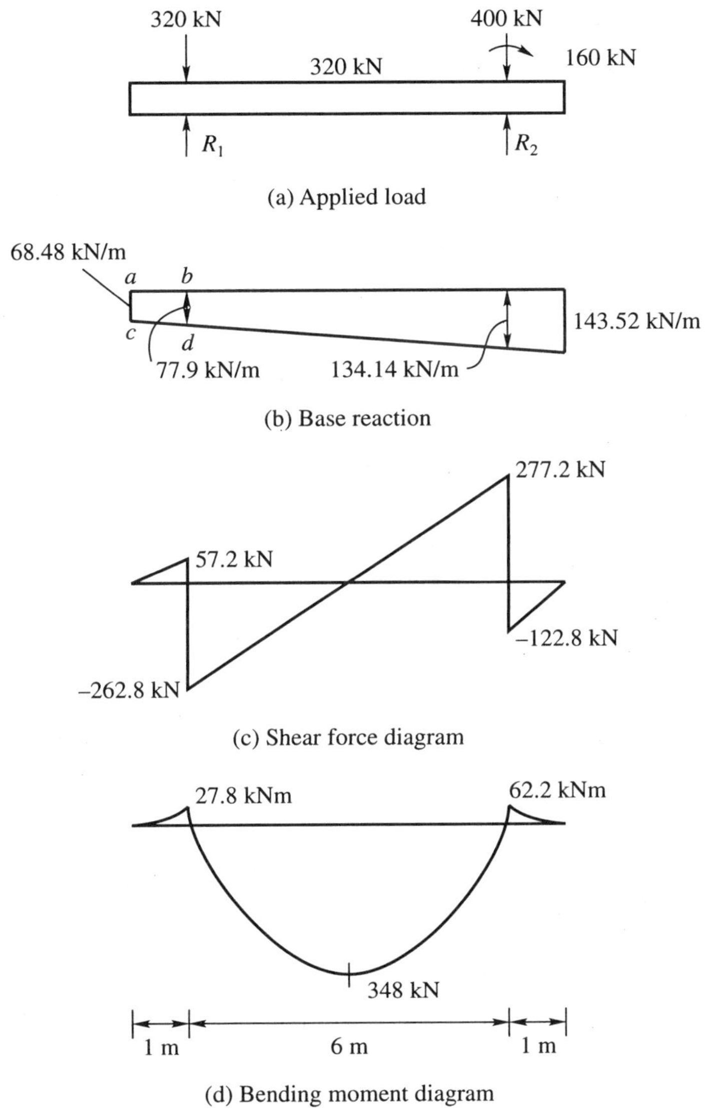

Figure Ex. 14.4

Bending Moment diagram

Bending moment at the reaction point R_1 = moment due to force equal to the area *abcd* + moment due to distributed load on beam *ab*

$$= 68.48 \times \frac{1}{2} + \frac{9.42}{2} \times \frac{1}{3} - 16 \times \frac{1}{2}$$

$= 27.8$ kN–m

The moments at other points can be calculated in the same way. The complete moment diagram is given in Fig. Ex. 14.4(d)

Example 14.5

The end column along a property line is connected to an interior column by a trapezoidal footing. The following data are given with reference to Fig. 14.3(b):

Column Loads: $Q_1 = 2016$ kN, $Q_2 = 1560$ kN.

Size of columns: 0.46 × 0.46 m.

$L_c = 5.48$ m.

Determine the dimensions a and b of the trapezoidal footing. The net allowable bearing pressure $q_{na} = 190$ kPa.

Solution

Determine the center of bearing pressure x_2 from the center of Column 1. Taking moments of all the loads about the center of Column 1, we have

$$(2016+1560)x_2 = 1560 \times 5.48$$

$$x_2 = \frac{1560 \times 5.48}{3576} = 2.39 \text{ m}$$

Now $$x_1 = 2.39 + \frac{0.46}{2} = 2.62 \text{ m}$$

Point O in Fig. 14.3(b) is the center of the area coinciding with the center of pressure. From the allowable pressure $q_a = 190$ kPa, the area of the combined footing required is

$$A = \frac{3576}{190} = 18.82 \text{ sq. m}$$

From geometry, the area of the trapezoidal footing (Fig. 14.3(b)) is

$$A = \frac{(a+b)L}{2} = \frac{(a+b)}{2}(5.94) = 18.82$$

or $$(a+b) = 6.34 \text{ m}$$

where, $$L = L_c + b_1 = 5.48 + 0.46 = 5.94 \text{ m}$$

From the geometry of the Fig. (14.3b), the distance of the center of area x_1 can be written in terms of a, b and L as

$$x_1 = \frac{L}{3}\frac{2a+b}{a+b}$$

or $$\frac{2a+b}{a+b} = \frac{3x_1}{L} = \frac{3 \times 2.62}{5.94} = 1.32 \text{ m}$$

but $a + b = 6.32$ m or $b = 6.32 - a$. Now substituting for b we have,

$$\frac{2a + 6.34 - a}{6.34} = 1.32$$

and solving, $a = 2.03$ m, from which, $b = 6.34 - 2.03 = 4.31$ m.

14.11 PROBLEMS

14.1 A beam of length 6 m and width 0.80 m is founded on dense sand under submerged conditions. A plate load test with a plate of 0.30 × 0.30 m conducted at the site gave a value for the coefficient of subgrade reaction for the plate equal to 95 MN/m^3. Determine the coefficient of subgrade reaction for the beam.

14.2 If the beam in Prob 14.1 is founded in very stiff clay with the value for k_1 equal to 45 MN/m^3, what is the coefficient of subgrade reaction for the beam?

14.3 Proportion a strap footing given the following data with reference to Fig. 14.2:
$Q_1 = 580$ kN, $Q_2 = 900$ kN
$L_c = 6.2$ m, $b_1 = 0.40$ m, $q_s = 120$ kPa.

14.4 Proportion a rectangular combined footing given the following data with reference to Fig. 14.3 (the footing is rectangular instead of trapezoidal):
$Q_1 = 535$ kN, $Q_2 = 900$ kN, $b_1 = 0.40$ m,
$L_c = 4.75$ m, $q_s = 100$ kPa.

CHAPTER 15

DEEP FOUNDATION I: PILE FOUNDATION

15.1 INTRODUCTION

Shallow foundations are normally used where the soil close to the ground surface and up to the zone of *significant stress* possesses sufficient bearing strength to carry the superstructure load without causing distress to the superstructure due to settlement. However, where the top soil is either loose or soft or of a swelling type the load from the structure has to be transferred to deeper firm strata.

The structural loads may be transferred to deeper firm strata by means of piles. Piles are long slender columns either driven, bored or *cast-in-situ*. Driven piles are made of a variety of materials such as concrete, steel, timber etc., whereas *cast-in-situ* piles are concrete piles. They may be subjected to vertical or lateral loads or a combination of vertical and lateral loads. If the diameter of a bored-*cast-in-situ* pile is greater than about 0.75 m, it is sometimes called a drilled pier, drilled caisson or drilled shaft. The distinction made between a small diameter bored *cast-in-situ* pile (less than 0.75 m) and a larger one is just for the sake of design considerations. The design of drilled piers is dealt with in Chapter 17. This chapter is concerned with driven piles and small diameter bored *cast-in-situ* piles only.

15.2 CLASSIFICATION OF PILES

Piles may be classified as long or short in accordance with the *L/d* ratio of the pile (where L = length, d = diameter of pile). A short pile behaves as a rigid body and rotates as a unit under lateral loads. The load transferred to the tip of the pile bears a significant proportion of the total vertical load on the top. In the case of a long pile, the length beyond a particular depth loses its significance under lateral loads, but when subjected to vertical load, the frictional load on the sides of the pile bears a significant part to the total load.

Piles may further be classified as vertical piles or inclined piles. Vertical piles are normally used to carry mainly vertical loads and very little lateral load. When piles are inclined at an angle to the vertical, they are called *batter piles* or *raker piles*. Batter piles are quite effective for taking lateral loads, but when used in groups, they also can take vertical loads. The behavior of vertical and batter piles subjected to lateral loads is dealt with in Chapter 16.

Types of Piles According to Their Composition

Piles may be classified according to their composition as

1. Timber Piles,
2. Concrete Piles,
3. Steel Piles.

Timber Piles: Timber piles are made of tree trunks with the branches trimmed off. Such piles shall be of sound quality and free of defects. The length of the pile may be 15 m or more. If greater lengths are required, they may be spliced. The diameter of the piles at the butt end may vary from 30 to 40 cm. The diameter at the tip end should not be less than 15 cm.

Piles entirely submerged in water last long without decay provided marine borers are not present. When a pile is subjected to alternate wetting and drying the useful life is relatively short unless treated with a wood preservative, usually creosote at 250 kg per m^3 for piles in fresh water and 350 kg/m^3 in sea water.

After being driven to final depth, all pile heads, treated or untreated, should be sawed square to sound undamaged wood to receive the pile cap. But before concrete for the pile cap is poured, the head of the treated piles should be protected by a zinc coat, lead paint or by wrapping the pile heads with fabric upon which hot pitch is applied.

Driving of timber piles usually results in the crushing of the fibers on the head (or brooming) which can be somewhat controlled by using a driving cap, or ring around the butt.

The usual maximum design load per pile does not exceed 250 kN. Timber piles are usually less expensive in places where timber is plentiful.

Concrete Piles. Concrete piles are either precast or *cast-in-situ* piles. Precast concrete piles are cast and cured in a casting yard and then transported to the site of work for driving. If the work is of a very big nature, they may be cast at the site also.

Precast piles may be made of uniform sections with pointed tips. Tapered piles may be manufactured when greater bearing resistance is required. Normally piles of square or octagonal sections are manufactured since these shapes are easy to cast in horizontal position. Necessary reinforcement is provided to take care of handling stresses. Piles may also be prestressed.

Maximum load on a prestressed concrete pile is approximately 2000 kN and on precast piles 1000 kN. The optimum load range is 400 to 600 kN.

Steel Piles. Steel piles are usually rolled *H* shapes or pipe piles. *H*-piles are proportioned to withstand large impact stresses during hard driving. Pipe piles are either welded or seamless steel pipes which may be driven either open-end or closed-end. Pipe piles are often filled with concrete after driving, although in some cases this is not necessary. The optimum load range on steel piles is 400 to 1200 kN.

15.3 TYPES OF PILES ACCORDING TO THE METHOD OF INSTALLATION

According to the method of construction, there are three types of piles. They are

1. Driven piles,
2. *Cast-in-situ* piles and
3. Driven and *cast-in-situ* piles.

Driven Piles

Piles may be of timber, steel or concrete. When the piles are of concrete, they are to be precast. They may be driven either vertically or at an angle to the vertical. Piles are driven using a pile hammer. When a pile is driven into granular soil, the soil so displaced, equal to the volume of the driven pile, compacts the soil around the sides since the displaced soil particles enter the soil spaces of the adjacent mass which leads to densification of the mass. The pile that compacts the soil adjacent to it is sometimes called a *compaction pile*. The compaction of the soil mass around a pile increases its bearing capacity.

If a pile is driven into saturated silty or cohesive soil, the soil around the pile cannot be densified because of its poor drainage qualities. The displaced soil particles cannot enter the void space unless the water in the pores is pushed out. The stresses developed in the soil mass adjacent to the pile due to the driving of the pile have to be borne by the pore water only. This results in the development of pore water pressure and a consequent decrease in the bearing capacity of the soil. The soil adjacent to the piles is remolded and loses to a certain extent its structural strength. The immediate effect of driving a pile in a soil with poor drainage qualities is, therefore, to decrease its bearing strength. However, with the passage of time, the remolded soil regains part of its lost strength due to the reorientation of the disturbed particles (which is termed *thixotrophy*) and due to consolidation of the mass. The advantages and disadvantages of driven piles are:

Advantages

1. Piles can be precast to the required specifications.
2. Piles of any size, length and shape can be made in advance and used at the site. As a result, the progress of the work will be rapid.
3. A pile driven into granular soil compacts the adjacent soil mass and as a result the bearing capacity of the pile is increased.
4. The work is neat and clean. The supervision of work at the site can be reduced to a minimum. The storage space required is very much less.
5. Driven piles may conveniently be used in places where it is advisable not to drill holes for fear of meeting ground water under pressure.
6. Drivens pile are the most favored for works over water such as piles in wharf structures or jetties.

Disadvantages

1. Precast or prestressed concrete piles must be properly reinforced to withstand handling stresses during transportation and driving.
2. Advance planning is required for handling and driving.
3. Requires heavy equipment for handling and driving.
4. Since the exact length required at the site cannot be determined in advance, the method involves cutting off extra lengths or adding more lengths. This increases the cost of the project.
5. Driven piles are not suitable in soils of poor drainage qualities. If the driving of piles is not properly phased and arranged, there is every possibility of heaving of the soil or the lifting of the driven piles during the driving of a new pile.
6. Where the foundations of adjacent structures are likely to be affected due to the vibrations generated by the driving of piles, driven piles should not be used.

Cast-in-situ Piles

Cast-in-situ piles are concrete piles. These piles are distinguished from drilled piers as small diameter piles. They are constructed by making holes in the ground to the required depth and then filling the hole with concrete. Straight bored piles or piles with one or more bulbs at intervals may be cast at the site. The latter type are called *under-reamed piles*. Reinforcement may be used as per the requirements. *Cast-in-situ* piles have advantages as well as disadvantages.

Advantages

1. Piles of any size and length may be constructed at the site.
2. Damage due to driving and handling that is common in precast piles is eliminated in this case.
3. These piles are ideally suited in places where vibrations of any type are required to be avoided to preserve the safety of the adjoining structure.
4. They are suitable in soils of poor drainage qualities since cast-in-situ piles do not significantly disturb the surrounding soil.

Disadvantages

1. Installation of *cast-in-situ* piles requires careful supervision and quality control of all the materials used in the construction.
2. The method is quite cumbersome. It needs sufficient storage space for all the materials used in the construction.
3. The advantage of increased bearing capacity due to compaction in granular soil that could be obtained by a driven pile is not produced by a *cast-in-situ* pile.
4. Construction of piles in holes where there is heavy current of ground water flow or artesian pressure is very difficult.

A straight bored pile is shown in Fig. 15.1(a).

Driven and *Cast-in-situ* Piles

This type has the advantages and disadvantages of both the driven and the *cast-in-situ* piles. The procedure of installing a *driven* and *cast-in-situ* pile is as follows:

A steel shell is driven into the ground with the aid of a mandrel inserted into the shell. The mandrel is withdrawn and concrete is placed in the shell. The shell is made of corrugated and reinforced thin sheet steel (mono-tube piles) or pipes (Armco welded pipes or common seamless pipes). The piles of this type are called a shell type. The shell-less type is formed by withdrawing the shell while the concrete is being placed. In both the types of piles the bottom of the shell is closed with a conical tip which can be separated from the shell. By driving the concrete out of the shell an enlarged bulb may be formed in both the types of piles. Franki piles are of this type. The common types of driven and cast-in-situ piles are given in Fig. 15.1. In some cases the shell will be left in place and the tube is concreted. This type of pile is very much used in piling over water.

15.4 USES OF PILES

The major uses of piles are:

1. To carry vertical compression load.
2. To resist uplift load.
3. To resist horizontal or inclined loads.

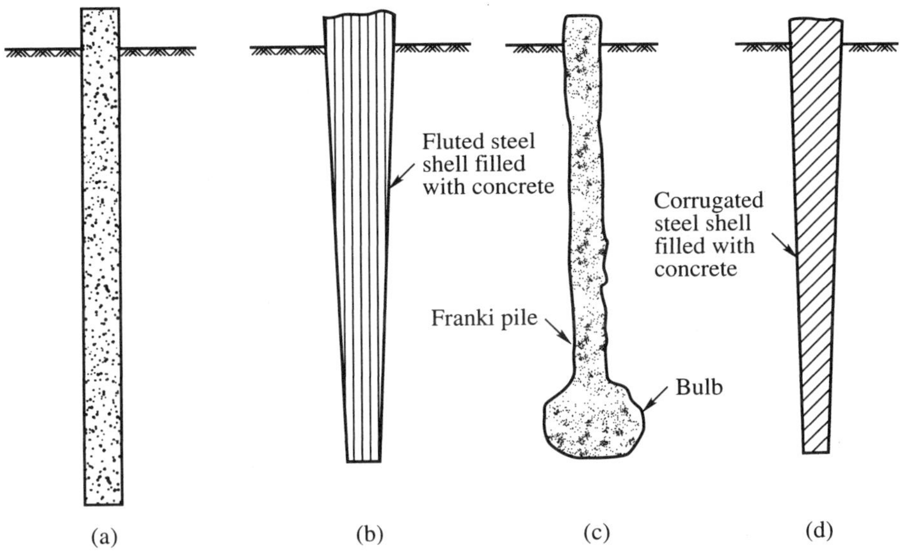

Figure 15.1 Types of *cast-in-situ* and driven *cast-in-situ* concrete piles

Normally vertical piles are used to carry vertical compression loads coming from superstructures such as buildings, bridges etc. The piles are used in groups joined together by pile caps. The loads carried by the piles are transferred to the adjacent soil. If all the loads coming on the tops of piles are transferred to the tips, such piles are called *end-bearing* or *point-bearing piles*. However, if all the load is transferred to the soil along the length of the pile such piles are called *friction piles*. If, in the course of driving a pile into granular soils, the soil around the pile gets compacted, such piles are called *compaction piles*. Fig. 15.2(a) shows piles used for the foundation of a multistoried building to carry loads from the superstructure.

Piles are also used to resist uplift loads. Piles used for this purpose are called *tension piles* or *uplift piles* or *anchor piles*. Uplift loads are developed due to hydrostatic pressure or overturning movement as shown in Fig. 15.2(a).

Piles are also used to resist horizontal or inclined forces. Batter piles are normally used to resist large horizontal loads. Fig. 15.2(b) shows the use of piles to resist lateral loads.

15.5 SELECTION OF PILE

The selection of the type, length and capacity is usually made from estimation based on the soil conditions and the magnitude of the load. In large cities, where the soil conditions are well known and where a large number of pile foundations have been constructed, the experience gained in the past is extremely useful. Generally the foundation design is made on the preliminary estimated values. Before the actual construction begins, pile load tests must be conducted to verify the design values. The foundation design must be revised according to the test results. The factors that govern the selection of piles are:

1. Length of pile in relation to the load and type of soil
2. Character of structure
3. Availability of materials
4. Type of loading
5. Factors causing deterioration

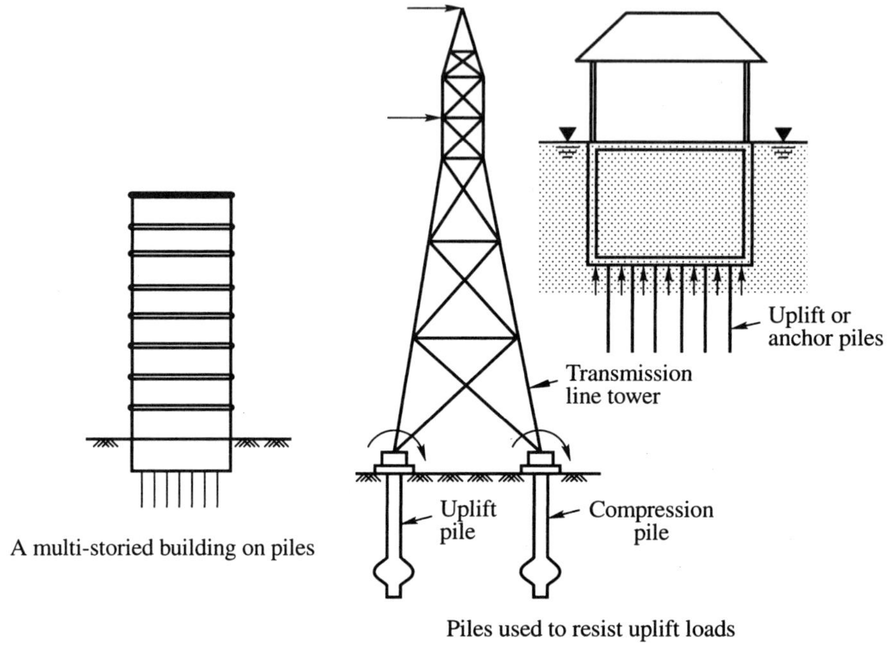

Figure 15.2(a) Principles of floating foundation; and a typical rigid raft foundation

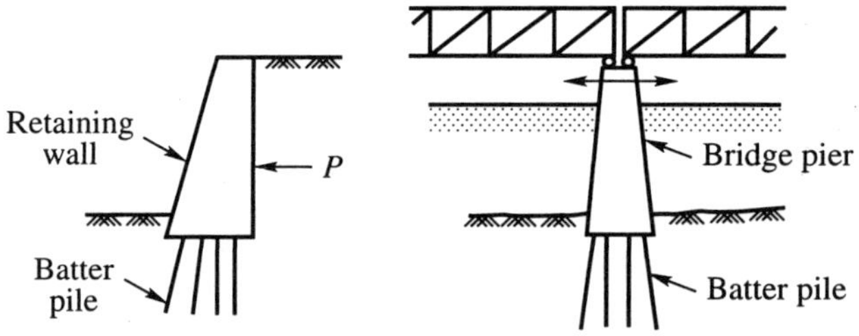

Figure 15.2(b) Piles used to resist lateral loads

6. Ease of maintenance
7. Estimated costs of types of piles, taking into account the initial cost, life expectancy and cost of maintenance
8. Availability of funds

All the above factors have to be largely analyzed before deciding up on a particular type.

15.6 INSTALLATION OF PILES

The method of installing a pile at a site depends upon the type of pile. The equipment required for this purpose varies. The following types of piles are normally considered for the purpose of installation

1. Driven piles

The piles that come under this category are,

a. Timber piles,

b. Steel piles, *H*-section and pipe piles,
c. Precast concrete or prestressed concrete piles, either solid or hollow sections.

2. Driven *cast-in-situ* piles

This involves driving of a steel tube to the required depth with the end closed by a detachable conical tip. The tube is next concreted and the shell is simultaneously withdrawn. In some cases the shell will not be withdrawn.

3. Bored *cast-in-situ* piles

Boring is done either by auguring or by percussion drilling. After boring is completed, the bore is concreted with or without reinforcement.

Pile Driving Equipment for Driven and Driven *Cast-in-situ* Piles

Pile driving equipment contains three parts. They are

1. A pile frame,
2. Piling winch,
3. Impact hammers.

Pile Frame

Pile driving equipment is required for *driven* piles or driven *cast-in-situ* piles. The driving pile frame must be such that it can be mounted on a standard tracked crane base machine for mobility on land sites or on framed bases for mounting on stagings or pontoons in offshore construction. Fig. 15.3 gives a typical pile frame for both onshore and offshore construction. Both the types must be capable of full rotation and backward or forward raking. All types of frames consist essentially of *leaders*, which are a pair of steel members extending for the full height of the frame and which guide the hammer and pile as it is driven into the ground. Where long piles have to be driven the leaders can be extended at the top by a telescopic boom.

The base frame may be mounted on swivel wheels fitted with self-contained jacking screws for leveling the frame or it may be carried on steel rollers. The rollers run on steel girders or long timbers and the frame is moved along by winching from a deadman set on the roller track, or by turning the rollers by a tommy-bar placed in holes at the ends of the rollers. Movements parallel to the rollers are achieved by winding in a wire rope terminating in hooks on the ends of rollers; the frame then skids in either direction along the rollers. It is important to ensure that the pile frame remains in its correct position throughout the driving of a pile.

Piling Winches

Piling winches are mounted on the base. Winches may be powered by steam, diesel or gasoline engines, or electric motors. Steam-powered winches are commonly used where steam is used for the piling hammer. Diesel or gasoline engines, or electric motors (rarely) are used in conjunction with drop hammers or where compressed air is used to operate the hammers.

Impact Hammers

The impact energy for driving piles may be obtained by any one of the following types of hammers. They are

1. Drop hammers,
2. Single-acting steam hammers,
3. Double-acting steam hammers,

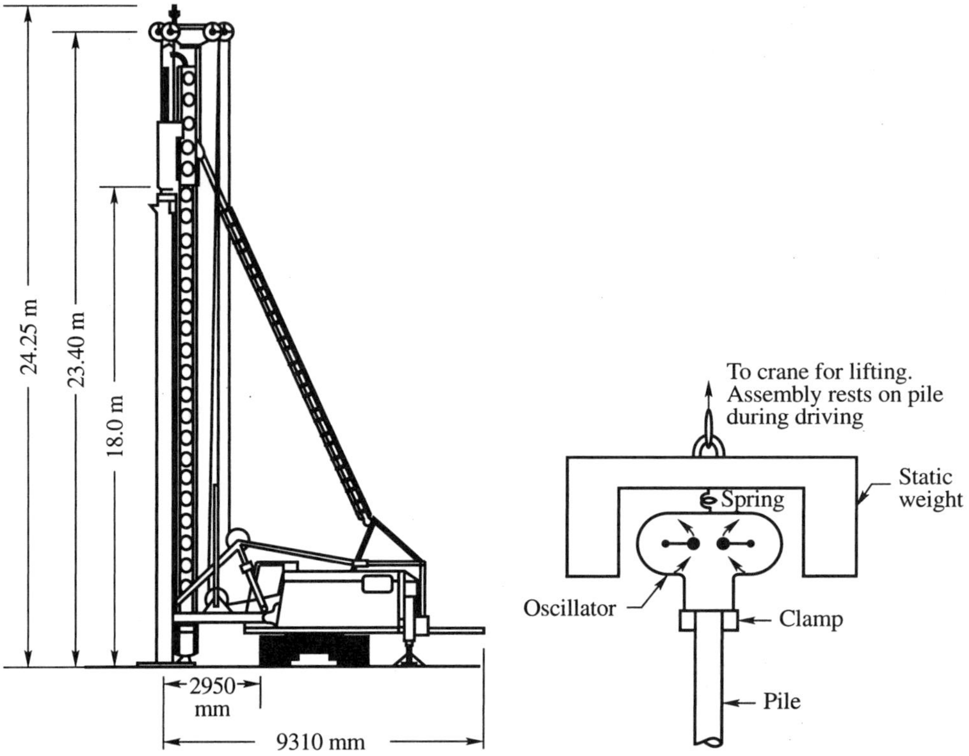

(a) The Ackermanns M14-5P pile frame (b) Diagrammatic sketch of vibratory pile driver

Figure 15.3 Pile driving equipment and vibratory pile driver

4. Diesel hammer,
5. Vibratory hammer.

D*rop hammers* are at present used for small jobs. The weight is raised and allowed to fall freely on the top of the pile. The impact drives the pile into the ground.

In the case of a *single-acting steam hammer* steam or air raises the moveable part of the hammer which then drops by gravity alone. The blows in this case are much more rapidly delivered than for a drop hammer. The weights of hammers vary from about 1500 to 10,000 kg with the length of stroke being about 90 cm. In general the ratio of ram weight to pile weight may vary from 0.5 to 1.0.

In the case of a *double-acting hammer* steam or air is used to raise the moveable part of the hammer and also to impart additional energy during the down stroke. The downward acceleration of the ram owing to gravity is increased by the acceleration due to steam pressure. The weights of hammers vary from about 350 to 2500 kg. The length of stroke varies from about 20 to 90 cm. The rate of driving ranges from 300 blows per minute for the light types, to 100 blows per minute for the heaviest types.

Diesel or internal combustion hammers utilize diesel-fuel explosions to provide the impact energy to the pile. Diesel hammers have considerable advantage over steam hammers because they are lighter, more mobile and use a smaller amount of fuel. The weight of the hammer varies from about 1000 to 2500 kg.

The advantage of the power-hammer type of driving is that the blows fall in rapid succession (50 to 150 blows per minute) keeping the pile in continuous motion. Since the pile is continuously moving, the effects of the blows tend to convert to pressure rather than impact, thus reducing damage to the pile.

The *vibration method* of driving piles is now coming into prominence. Driving is quiet and does not generate local vibrations. Vibration driving utilizes a variable speed oscillator attached to the top of the pile (Fig. 15.3(b)). It consists of two counter rotating eccentric weights which are in phase twice per cycle (180° apart) in the vertical direction. This introduces vibration through the pile which can be made to coincide with the resonant frequency of the pile. As a result, a push-pull effect is created at the pile tip which breaks up the soil structure allowing easy pile penetration into the ground with a relatively small driving effort. Pile driving by the vibration method is quite common in Russia.

Jetting Piles

Water jetting may be used to aid the penetration of a pile into dense sand or dense sandy gravel. Jetting is ineffective in firm to stiff clays or any soil containing much coarse to stiff cobbles or boulders.

Where jetting is required for pile penetration a stream of water is discharged near the pile point or along the sides of the pile through a pipe 5 to 7.5 cm in diameter. An adequate quantity of water is essential for jetting. Suitable quantities of water for jetting a 250 to 350 mm pile are

Fine sand	15-25 liters/second,
Coarse sand	25-40 liters/second,
Sandy gravels	45-600 liters/second.

A pressure of at least 5 kg/cm^2 or more is required.

PART A—VERTICAL LOAD BEARING CAPACITY OF A SINGLE VERTICAL PILE

15.7 GENERAL CONSIDERATIONS

The bearing capacity of groups of piles subjected to vertical or vertical and lateral loads depends upon the behavior of a single pile. The bearing capacity of a single pile depends upon

1. Type, size and length of pile,
2. Type of soil,
3. The method of installation.

The bearing capacity depends primarily on the method of installation and the type of soil encountered. The bearing capacity of a single pile increases with an increase in the size and length. The position of the water table also affects the bearing capacity.

In order to be able to design a safe and economical pile foundation, we have to analyze the interactions between the pile and the soil, establish the modes of failure and estimate the settlements from soil deformation under dead load, service load etc. The design should comply with the following requirements.

1. It should ensure adequate safety against failure; the factor of safety used depends on the importance of the structure and on the reliability of the soil parameters and the loading systems used in the design.

2. The settlements should be compatible with adequate behavior of the superstructure to avoid impairing its efficiency.

Load Transfer Mechanism

Statement of the Problem

Fig. 15.4(a) gives a single pile of uniform diameter d (circular or any other shape) and length L driven into a homogeneous mass of soil of known physical properties. A static vertical load is applied on the top. It is required to determine the ultimate bearing capacity Q_u of the pile.

When the ultimate load applied on the top of the pile is Q_u, a part of the load is transmitted to the soil along the length of the pile and the balance is transmitted to the pile base. The load transmitted to the soil along the length of the pile is called the *ultimate friction load or skin load* Q_f and that transmitted to the base is called the *base* or *point load* Q_b. The total ultimate load Q_u is expressed as the sum of these two, that is,

$$Q_u = Q_b + Q_f = q_b A_b + f_s A_s \tag{15.1}$$

where Q_u = ultimate load applied on the top of the pile
q_b = ultimate unit bearing capacity of the pile at the base
A_b = bearing area of the base of the pile
A_s = total surface area of pile embedded below ground surface
f_s = unit skin friction (ultimate)

Load Transfer Mechanism

Consider the pile shown in Fig. 15.4(b) is loaded to failure by gradually increasing the load on the top. If settlement of the top of the pile is measured at every stage of loading after an equilibrium condition is attained, a load settlement curve as shown in Fig. 15.4(c) can be obtained.

If the pile is instrumented, the load distribution along the pile can be determined at different stages of loading and plotted as shown in Fig. 15.4(b).

When a load Q_1 acts on the pile head, the axial load at ground level is also Q_1, but at level A_1 (Fig. 15.4(b)), the axial load is zero. The total load Q_1 is distributed as friction load within a length of pile L_1. The lower section A_1B of pile will not be affected by this load. As the load at the top is increased to Q_2, the axial load at the bottom of the pile is just zero. The total load Q_2 is distributed as friction load along the whole length of pile L. The friction load distribution curves along the pile shaft may be as shown in the figure. If the load put on the pile is greater than Q_2, a part of this load is transferred to the soil at the base as point load and the rest is transferred to the soil surrounding the pile. With the increase of load Q on the top, both the friction and point loads continue to increase. The friction load attains an ultimate value Q_f at a particular load level, say Q_m, at the top, and any further increment of load added to Q_m will not increase the value of Q_f. However, the point load, Q_p, still goes on increasing till the soil fails by punching shear failure. It has been determined by Van Wheele (1957) that the point load Q_p increases linearly with the elastic compression of the soil at the base.

The relative proportions of the loads carried by skin friction and base resistance depend on the shear strength and elasticity of the soil. Generally the vertical movement of the pile which is required to mobilize full end resistance is much greater than that required to mobilize full skin friction. Experience indicates that in bored *cast-in-situ* piles full frictional load is normally mobilized at a settlement equal to 0.5 to 1 percent of pile diameter and the full base load Q_b at 10 to 20 percent of the diameter. But, if this ultimate load criterion is applied to piles of large diameter in clay, the settlement at the working load (with a factor of safety of 2 on the ultimate load) may be excessive. A typical load-settlement relationship of friction load and base load is shown in

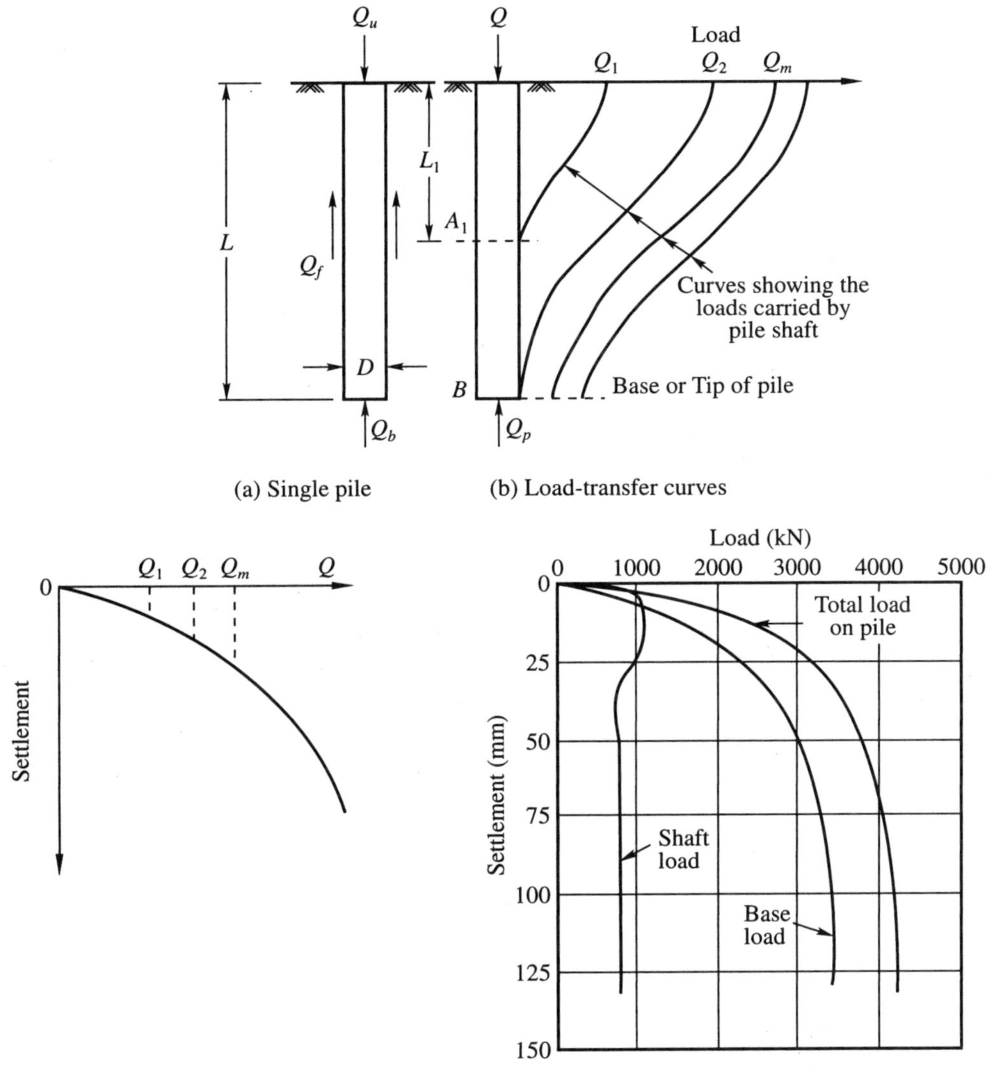

(a) Single pile (b) Load-transfer curves

(c) Load-settlement curve (d) Load-settlement relationships for large-diameter bored and cast-in-place piles (after Tomlinson, 1986)

Figure 15.4 Load transfer mechanism

Fig. 15.4(d) (Tomlinson, 1986) for a large diameter bored and *cast-in-situ* pile in clay. It may be seen from this figure that the full shaft resistance is mobilized at a settlement of only 15 mm whereas the full base resistance, and the ultimate resistance of the entire pile, is mobilized at a settlement of 120 mm. The shaft load at a settlement of 15 mm is only 1000 kN which is about 25 percent of the base resistance. If a working load of 2000 kN at a settlement of 15 mm is used for the design, at this working load, the full shaft resistance will have been mobilized whereas only about 50 percent of the base resistance has been mobilized. This means if piles are designed to carry a working load equal to 1/3 to 1/2 the total failure load, there is every likelihood of the shaft resistance being fully mobilized at the working load. This has an important bearing on the design.

The type of load-settlement curve for a pile depends on the relative strength values of the surrounding and underlying soil. Fig. 15.5 gives the types of failure (Kézdi, 1975). They are as follows:

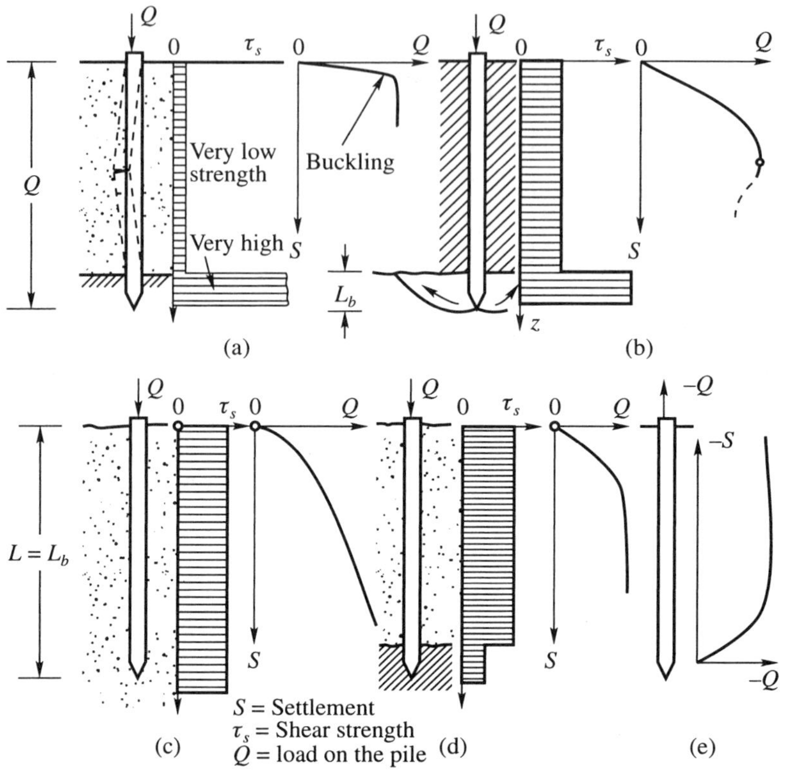

Figure 15.5 Types of failure of piles. Figures (a) to (e) indicate how strength of soil determines the type of failure: (a) buckling in very weak surrounding soil; (b) general shear failure in the strong lower soil; (c) soil of uniform strength; (d) low strength soil in the lower layer, skin friction predominant; (e) skin friction in tension (Kézdi, 1975)

Fig 15.5(a) represents a driven- pile (wooden or reinforced concrete), whose tip bears on a very hard stratum (rock). The soil around the shaft is too weak to exert any confining pressure or lateral resistance. In such cases, the pile fails like a compressed, slender column of the same material; after a more or less elastic compression buckling occurs. The curve shows a definite failure load.

Fig. 15.5(b) is the type normally met in practice. The pile penetrates through layers of soil having low shear strength down to a layer having a high strength and the layer extending sufficiently below the tip of the pile. At ultimate load Q_u, there will be a base general shear failure at the tip of the pile, since the upper layer does not prevent the formation of a failure surface. The effect of the shaft friction is rather less, since the lower dense layer prevents the occurrence of excessive settlements. Therefore, the degree of mobilization of shear stresses along the shaft will be low. The load settlement diagram is of the shape typical for a shallow footing on dense soil.

Fig. 15.5(c) shows the case where the shear strength of the surrounding soil is fairly uniform; therefore, a punching failure is likely to occur. The load-settlement diagram does not have a vertical tangent, and there is no definite failure load. The load will be carried by point resistance as well as by skin friction.

Fig 15.5(d) is a rare case where the lower layer is weaker. In such cases, the load will be carried mainly by shaft friction, and the point resistance is almost zero. The load-settlement curve shows a vertical tangent, which represents the load when the shaft friction has been fully mobilized.

Fig. 15.5(e) is a case when a pull, $-Q$, acts on the pile. Since the point resistance is again zero the same diagram, as in Fig. 15.5(d), will characterize the behavior, but heaving occurs.

Definition of Failure Load

The methods of determining failure loads based on load-settlement curves are described in subsequent sections. However, in the absence of a load settlement curve, a failure load may be defined as that which causes a settlement equal to 10 percent of the pile diameter or width (as per the suggestion of Terzaghi) which is widely accepted by engineers. However, if this criterion is applied to piles of large diameter in clay and a nominal factor of safety of 2 is used to obtain the working load, then the settlement at the working load may be excessive.

Factor of Safety

In almost all cases where piles are acting as structural foundations, the allowable load is governed solely from considerations of tolerable settlement at the working load.

The working load for all pile types in all types of soil may be taken as equal to the sum of the base resistance and shaft friction divided by a suitable factor of safety. A safety factor of 2.5 is normally used. Therefore we may write

$$Q_a = \frac{Q_b + Q_f}{2.5} \tag{15.2}$$

In case where the values of Q_b and Q_f can be obtained independently, the allowable load can be written as

$$Q_a = \frac{Q_b}{3} + \frac{Q_f}{1.5} \tag{15.3}$$

It is permissible to take a safety factor equal to 1.5 for the skin friction because the peak value of skin friction on a pile occurs at a settlement of only 3–8 mm (relatively independent of shaft diameter and embedded length but may depend on soil parameters) whereas the base resistance requires a greater settlement for full mobilization.

The least of the allowable loads given by Eqs. (15.2) and (15.3) is taken as the design working load.

15.8 METHODS OF DETERMINING ULTIMATE LOAD BEARING CAPACITY OF A SINGLE VERTICAL PILE

The ultimate bearing capacity, Q_u, of a single vertical pile may be determined by any of the following methods.

1. By the use of static bearing capacity equations.
2. By the use of SPT and CPT values.
3. By field load tests.
4. By dynamic method.

The determination of the ultimate point bearing capacity, q_b, of a deep foundation on the basis of theory is a very complex one since there are many factors which cannot be accounted for in the theory. The theory assumes that the soil is homogeneous and isotropic which is normally not the case. All the theoretical equations are obtained based on plane strain conditions. Only shape factors are applied to take care of the three-dimensional nature of the problem. Compressibility characteristics of the soil complicate the problem further. Experience and judgment are therefore very essential in applying any theory to a specific problem. The skin load Q_f depends on the nature of the surface of the pile, the method of installation of the pile and the type of soil. An exact evaluation of Q_f is a difficult job even if the soil is homogeneous over the whole length of the pile. The problem becomes all the more complicated if the pile passes through soils of variable characteristics.

15.9 GENERAL THEORY FOR ULTIMATE BEARING CAPACITY

According to Vesic (1967), only punching shear failure occurs in deep foundations irrespective of the density of the soil so long as the depth-width ratio L/d is greater than 4 where L = length of pile and d = diameter (or width of pile). The types of failure surfaces assumed by different investigators are shown in Fig. 15.6 for the general shear failure condition. The detailed experimental study of Vesic indicates that the failure surfaces do not revert back to the shaft as shown in Fig. 15.6(b).

The total failure load $\overline{Q}_u$ may be written as follows

$$\overline{Q}_u = Q_u + W_p = Q_b + Q_f + W_p \tag{15.4}$$

where Q_u = load at failure applied to the pile
Q_b = base resistance
Q_f = shaft resistance
W_p = weight of the pile.

The general equation for the base resistance may be written as

$$Q_b = \left[cN_c + q'_o N_q + \frac{1}{2}\gamma d N_\gamma\right]A_b \tag{15.5}$$

where d = width or diameter of the shaft at base level
q'_o = effective overburden pressure at the base level of the pile
A_b = base area of pile
c = cohesion of soil
γ = effective unit weight of soil
N_c, N_q, N_γ = bearing capacity factors which take into account the shape factor.

Cohesionless Soils

For cohesionless soils, $c = 0$ and the term $1/2\gamma d N_\gamma$ becomes insignificant in comparison with the term $q_o N_q$ for deep foundations. Therefore Eq. (15.5) reduces to

$$Q_b = q'_o N_q A_b = q_b A_b \tag{15.6}$$

Eq. (15.4) may now be written as

$$\overline{Q}_b = Q_u + W_p = q'_o N_q A_b + W_p + Q_f \tag{15.7}$$

The net ultimate load in excess of the overburden pressure load $q_o A_b$ is

$$Q_u + W_p - q_o A_b = q'_o N_q A_b + W_p - q'_o A_b + Q_f \tag{15.8}$$

If we assume, for all practical purposes, W_p and $q'_o A_b$ are roughly equal for straight side or moderately tapered piles, Eq. (15.8) reduces to

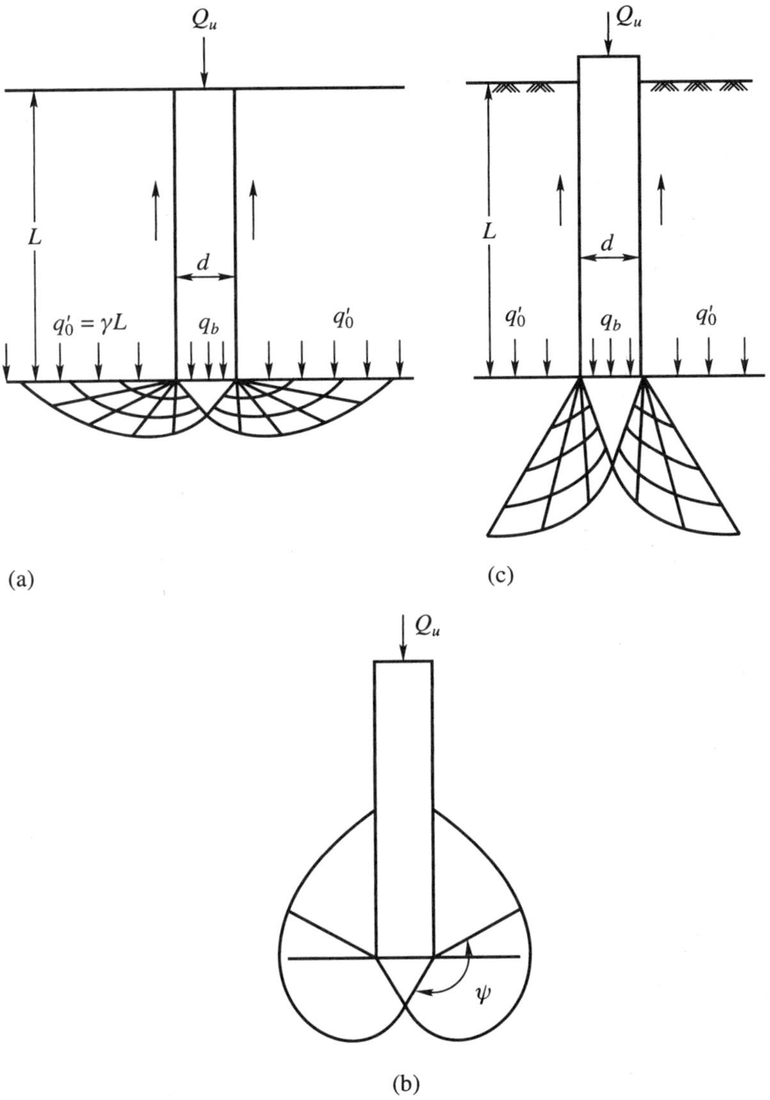

Figure 15.6 The shapes of failure surfaces at the tips of piles as assumed by (a) Terzaghi, (b) Meyerhof, and (c) Vesic

$$Q_u = q'_o N_q A_b + Q_f$$

or $$Q_u = q'_o N_q A_b + A_s \bar{q}'_o \bar{K}_s \tan\delta \qquad (15.9)$$

where A_s = surface area of the embedded length of the pile
$\bar{q}'_o$ = average effective overburden pressure over the embedded depth of the pile
$\bar{K}_s$ = average lateral earth pressure coefficient
δ = angle of wall friction.

Cohesive Soils

For cohesive soils such as saturated clays (normally consolidated), we have for $\phi = 0$, $N_q = 1$ and $N_\gamma = 0$. The ultimate base load from Eq. (15.5) is

$$\overline{Q}_b = (c_b N_c + q'_o) A_b \tag{15.10}$$

The net ultimate base load is

$$(\overline{Q}_b - q'_o A_b) = Q_b = c_b N_c A_b \tag{15.11}$$

Therefore, the net ultimate load capacity of the pile, Q_u, is

$$Q_u = c_b N_c A_b + Q_f$$

or

$$Q_u = c_b N_c A_b + A_s \alpha \bar{c}_u \tag{15.12}$$

where α = adhesion factor

$\bar{c}_u$ = average undrained shear strength of clay along the shaft

c_b = undrained shear strength of clay at the base level

N_c = bearing capacity factor

Equations (15.9) and (15.12) are used for analyzing the net ultimate load capacity of piles in cohesionless and cohesive soils respectively. In each case the following types of piles are considered.

1. Driven piles
2. Driven and cast-in-situ piles
3. Bored piles

15.10 ULTIMATE BEARING CAPACITY IN COHESIONLESS SOILS

Effect of Pile Installation on the Value of the Angle of Friction

When a pile is driven into loose sand its density is increased (Meyerhof, 1959), and the horizontal extent of the compacted zone has a width of about 6 to 8 times the pile diameter. However, in dense sand, pile driving decreases the relative density because of the dilatancy of the sand and the loosened sand along the shaft has a width of about 5 times the pile diameter (Kerisel, 1961). On the basis of field and model test results, Kishida (1967) proposed that the angle of internal friction decreases linearly from a maximum value of ϕ_2 at the pile tip to a low value of ϕ_1 at a distance of $3.5d$ from the tip where d is the diameter of the pile, ϕ_1 is the angle of friction before the installation of the pile and ϕ_2 after the installation as shown in Fig. 15.7. Based on the field data, the relationship between ϕ_1 and ϕ_2 in sands may be written as

$$\phi_2 = \frac{\phi_1 + 40}{2} \tag{15.13}$$

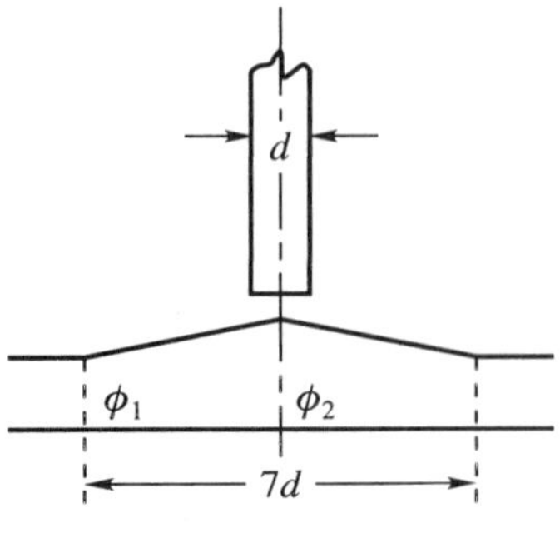

Figure 15.7 The effect of driving a pile on ϕ

An angle of $\phi_1 = \phi_2 = 40°$ in Eq. (15.13) means no change of relative density due to pile driving. Values of ϕ_1 are obtained from *insitu* penetration tests (with no correction due to overburden pressure, but corrected for field procedure) by using the relationships established between ϕ and SPT or CPT values. Kishida (1967) has suggested the following relationship between ϕ and the SPT value N_{cor} as

$$\phi° = \sqrt{20N_{cor}} + 15° \tag{15.14}$$

However, Tomlinson (1986) is of the opinion that it is unwise to use higher values for ϕ due to pile driving. His argument is that the sand may not get compacted, as for example, when piles are driven into loose sand, the resistance is so low and little compaction is given to the soil. He suggests that the value of ϕ used for the design should represent the *in situ* condition that existed before driving.

With regard to driven and *cast-in-situ* piles, there is no suggestion by any investigator as to what value of ϕ should be used for calculating the base resistance. However, it is safer to assume the *insitu* ϕ value for computing the base resistance.

With regard to bored and *cast-in-situ* piles, the soil gets loosened during boring. Tomlinson (1986) suggests that the ϕ value for calculating both the base and skin resistance should represent the loose state. However, Poulos et al., (1980) suggests that for bored piles, the value of ϕ be taken as

$$\phi = \phi_1 - 3 \tag{15.15}$$

where ϕ_1 = angle of internal friction prior to installation of the pile.

15.11 CRITICAL DEPTH

The ultimate bearing capacity Q_u in cohesionless soils as per Eq. (15.9) is

$$Q_u = q'_o N_q A_b + \bar{q}'_o \bar{K}_s \tan\delta\, A_s \tag{15.16a}$$

or

$$Q_u = q_b A_b + f_s A_s \tag{15.16b}$$

Eq. (15.16b) implies that both the point resistance q_b and the skin resistance f_s are functions of the effective overburden pressure q_o in cohesionless soils and increase linearly with the depth of embedment, *L*, of the pile. However, extensive research work carried out by Vesic (1967) has revealed that the base and frictional resistances remain almost constant beyond a certain depth of embedment which is a function of ϕ. This phenomenon was attributed to arching by Vesic. One conclusion from the investigation of Vesic is that in cohesionless soils, the bearing capacity factor, N_q, is not a constant depending on ϕ only, but also on the ratio *L/d* (where *L* = length of embedment of pile, *d* = diameter or width of pile). In a similar way, the frictional resistance, f_s, increases with the *L/d* ratio and remains constant beyond a particular depth. Let L_c be the depth, which may be called the *critical depth*, beyond which both q_b and f_s remain constant. Experiments of Vesic have indicated that L_c is a function of ϕ. The L_c/d ratio as a function of ϕ may be expressed as follows (Poulos and Davis, 1980)

For $28° < \phi < 36.5°$

$$L_c/d = 5 + 0.24\,(\phi° - 28°) \tag{15.17a}$$

For $36.5° < \phi < 42°$

$$L_c/d = 7 + 2.35(\phi° - 36.5°) \tag{15.17b}$$

The above expressions have been developed based on the curve given by Poulos and Davis, (1980) giving the relationship between L_c/d and $\phi°$.

The Eqs. (15.17) indicate

$L_c/d = 5$ at $\phi = 28°$

$L_c/d = 7$ at $\phi = 36.5°$

$L_c/d = 20$ at $\phi = 42°$

The ϕ values to be used for obtaining L_c/d are as follows (Poulos and Davis, 1980)

for driven piles $\phi = 0.75\,\phi_1 + 10°$ (15.18a)

for bored piles: $\phi = \phi_1 - 3°$ (15.18b)

where ϕ_1 = angle of internal friction prior to the installation of the pile.

15.12 TOMLINSON'S SOLUTION FOR Q_b IN SAND

Driven Piles

The theoretical N_q factor in Eq. (15.9) is a function of ϕ. There is great variation in the values of N_q derived by different investigators as shown in Fig. 15.8. Comparison of observed base resistances of piles by Nordlund (1963) and Vesic (1964) have shown (Tomlinson, 1986) that N_q values established by Berezantsev et al., (1961) which take into account the depth to width ratio of the pile,

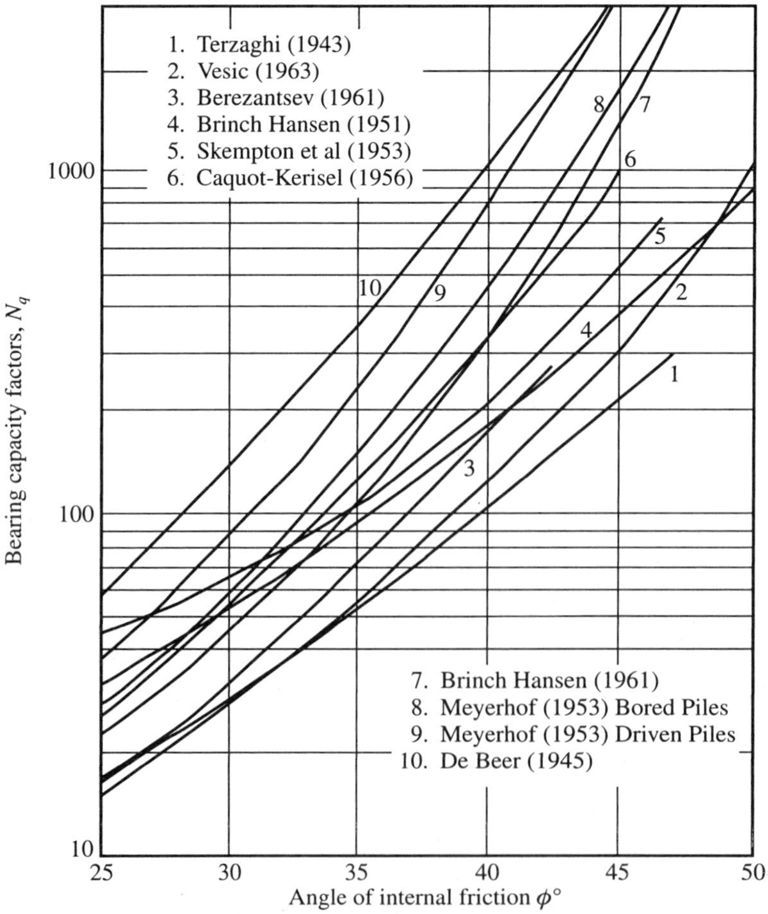

Figure 15.8 Bearing capacity factors for circular deep foundations (after Kézdi, 1975)

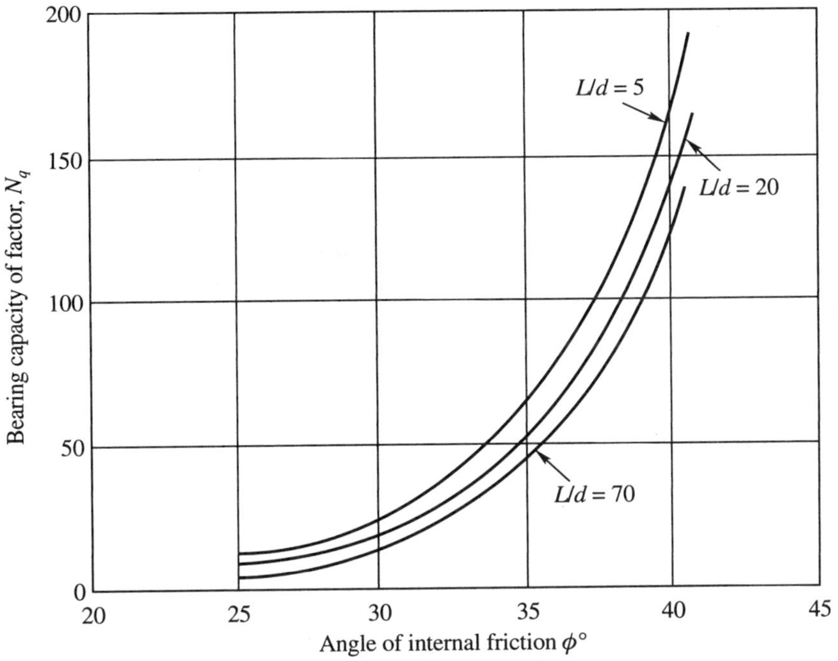

Figure 15.9 Berezantsev's bearing capacity factor, N_q (after Tomlinson, 1986)

most nearly conform to practical criteria of pile failure. Berezantsev's values of N_q as adopted by Tomlinson (1986) are given in Fig. 15.9.

It may be seen from Fig. 15.9 that there is a rapid increase in N_q for high values of ϕ, giving thereby high values of base resistance. As a general rule (Tomlinson, 1986), the allowable working load on an isolated pile driven to virtual refusal, using normal driving equipment, in a dense sand or gravel consisting predominantly of quartz particles, is given by the allowable load on the pile considered as a structural member rather than by consideration of failure of the supporting soil, or if the permissible working stress on the material of the pile is not exceeded, then the pile will not fail.

As per Tomlinson, the maximum base resistance q_b is normally limited to 11000 kN/m^2 (110 t/ft^2) whatever might be the penetration depth of the pile.

Bored and Cast-in-situ Piles in Cohesionless Soils

Bored piles are formed in cohesionless soils by drilling with rigs. The sides of the holes might be supported by the use of casing pipes. When casing is used, the concrete is placed in the drilled hole and the casing is gradually withdrawn. In all the cases the sides and bottom if the hole will be loosened as a result of the boring operations, even though it may be initially be in a dense or medium dense state. Tomlinson suggests that the values of the parameters in Eq. (15.9) must be calculated by assuming that the ϕ value will represent the loose condition.

However, when piles are installed by rotary drilling under a bentonite slurry for stabilizing the sides, it may be assumed that the ϕ value used to calculate both the skin friction and base resistance will correspond to the undisturbed soil condition (Tomlinson, 1986).

The assumption of loose conditions for calculating skin friction and base resistance means that the ultimate carrying capacity of a bored pile in a cohesionless soil will be considerably lower than that of a pile driven in the same soil type. As per De Beer (1965), the base resistance q_b of a bored and *cast-in-situ* pile is about one third of that of a driven pile.

We may write,

q_b (bored pile) = (1/3) q_b (driven pile)

So far as friction load is concerned, the frictional parameter may be calculated by assuming a value of ϕ equal to 28° which represents the loose condition of the soil.

The same Eq. (15.9) may be used to compute Q_u based on the modifications explained above.

15.13 MEYERHOF'S METHOD OF DETERMINING Q_b FOR PILES IN SAND

Meyerhof (1976) takes into account the critical depth ratio (L_c/d) for estimating the value of Q_b. Fig. 15.10 shows the variation of L_c/d for both the bearing capacity factors N_c and N_q as a function of ϕ. According to Meyerhof, the bearing capacity factors increase with L_b/d and reach a maximum value at L_b/d equal to about 0.5 (L_c/d), where L_b is the actual thickness of the bearing stratum. For example, in a homogeneous soil (15.6c) L_b is equal to L, the actual embedded length of pile; whereas in Fig. 15.6b, L_b is less than L.

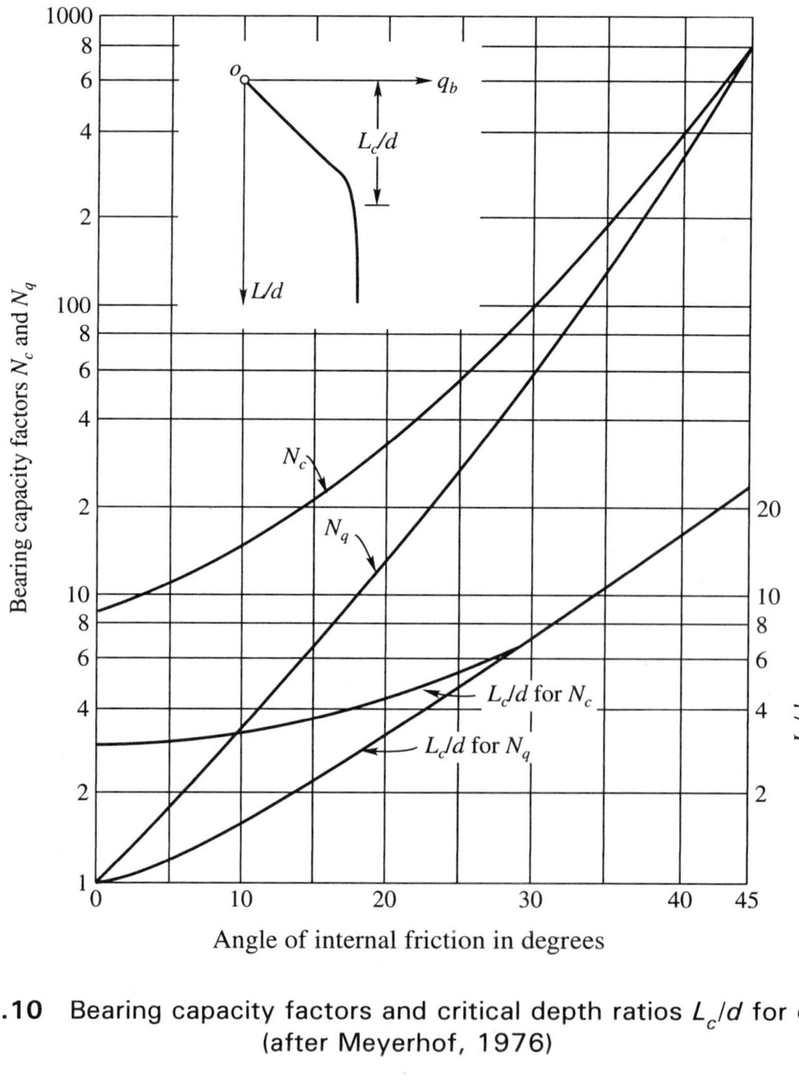

Figure 15.10 Bearing capacity factors and critical depth ratios L_c/d for driven piles (after Meyerhof, 1976)

As per Fig. 15.10, the value of L_c/d is about 25 for ϕ equal to 45° and it decreases with a decrease in the angle of friction ϕ. Normally, the magnitude of L_b/d for piles is greater than 0.5 (L_c/d) so that maximum values of N_c and N_q may apply for the calculation of q_b, the unit bearing pressure of the pile. Meyerhof prescribes a limiting value for q_b, based on his findings on static cone penetration resistance. The expression for the limiting value, q_{bl}, is

for dense sand: $$q_{bl} = 50\, N_q \tan \phi \text{ kN/m}^2 \quad (15.19a)$$

for loose sand: $$q_{bl} = 25\, N_q \tan \phi \text{ kN/m}^2 \quad (15.19b)$$

where ϕ is the angle of shearing resistance of the bearing stratum. The limiting q_{bl} values given by Eqs (15.9a and b) remain practically independent of the effective overburden pressure and groundwater conditions beyond the critical depth.

The equation for base resistance in sand may now be expressed as

$$Q_b = q'_o N_q A_b \leq q_{bl} A_b \quad (15.20)$$

where q'_o = effective overburden pressure at the tip of the pile L_c/d and N_q = bearing capacity factor (Fig. 15.10).

Eq. (15.20) is applicable only for driven piles in sand. For bored cast-in-situ piles the value of q_b is to be reduced by one third to one-half.

Clay Soil (ϕ = 0)

The base resistance Q_b for piles in saturated clay soil may be expressed as

$$Q_b = N_c c_u A_b = 9 c_u A_b \quad (15.21)$$

where $N_c = 9$, and c_u = undrained shear strength of the soil at the base level of the pile.

15.14 VESIC'S METHOD OF DETERMINING Q_b

The unit base resistance of a pile in a $(c - \phi)$ soil may be expressed as (Vesic, 1977)

$$q_b = cN_c^* + q'_o N_q^* \quad (15.22)$$

where c = unit cohesion

q'_o = effective vertical pressure at the base level of the pile

N^*_c and N^*_q = bearing capacity factors related to each other by the equation

$$N_c^* = (N_q^* - 1)\cot\phi \quad (15.23)$$

As per Vesic, the base resistance is not governed by the vertical ground pressure q'_o but by the mean effective normal ground stress σ_m expressed as

$$\sigma_m = \left(\frac{1 + 2K_o}{3}\right) q'_o \quad (15.24)$$

in which K_o = coefficient of earth pressure for the at rest condition = 1 – sin ϕ.

Now the bearing capacity in Eq. (15.22) may be expressed as

$$q_b = cN_c^* + \sigma_m N_\sigma^* \quad (15.25)$$

An equation for N^*_σ can be obtained from Eqs. (15.22), (15.24) and (15.25) as

$$N^*_\sigma = \frac{3N^*_q}{1+2K_o} \tag{15.26}$$

Vesic has developed an expression for N^*_σ based on the ultimate pressure needed to expand a spherical cavity in an infinite soil mass as

$$N^*_\sigma = \alpha_1 e^{\alpha_2} N_\phi (I_{rr})^{\alpha_3} \tag{15.27}$$

where $\alpha_1 = \dfrac{3}{3-\sin\phi}$, $\alpha_2 = \left(\dfrac{\pi}{2}-\phi\right)\tan\phi$, $\alpha_3 = \dfrac{1.33\sin\phi}{(1+\sin\phi)}$, and $N_\phi = \tan^2(45° + \phi/2)$

According to Vesic

$$I_{rr} = \frac{I_r}{1+I_r\Delta} \tag{15.28}$$

where $$I_r = \text{rigidity index} = \frac{E_s}{2(1+\mu)(c+q_o'\tan\phi)} = \frac{G}{(c+q_o'\tan\phi)} \tag{15.29}$$

where
I_{rr} = reduced rigidity index for the soil
Δ = average volumetric strain in the plastic zone below the pile point
E_s = modulus of elasticity of soil
G = shear modulus of soil
μ = Poisson's ratio of soil

Figures 15.11(a) and 15.11(b) give plots of N^*_σ versus ϕ, and N^*_c versus ϕ for various values of I_{rr} respectively.

The values of rigidity index can be computed knowing the values of shear modulus G_s and the shear strength s (= $c + q'_o \tan\phi$).

When an undrained condition exists in the saturated clay soil or the soil is cohesionless and is in a dense state we have $\Delta = 0$ and in such a case $I_r = I_{rr}$.

For $\phi = 0$ (undrained condition), we have

$$N^*_c = 1.33(\ln I_{rr} + 1) + \frac{\pi}{2} + 1 \tag{15.30}$$

The value of I_r depends upon the soil state, (a) for sand, loose or dense and (b) for clay low, medium or high plasticity. For preliminary estimates the following values of I_r may be used.

Soil type	I_r
Sand (D_r = 0.5–0.8)	75–150
Silt	50–75
Clay	150–250

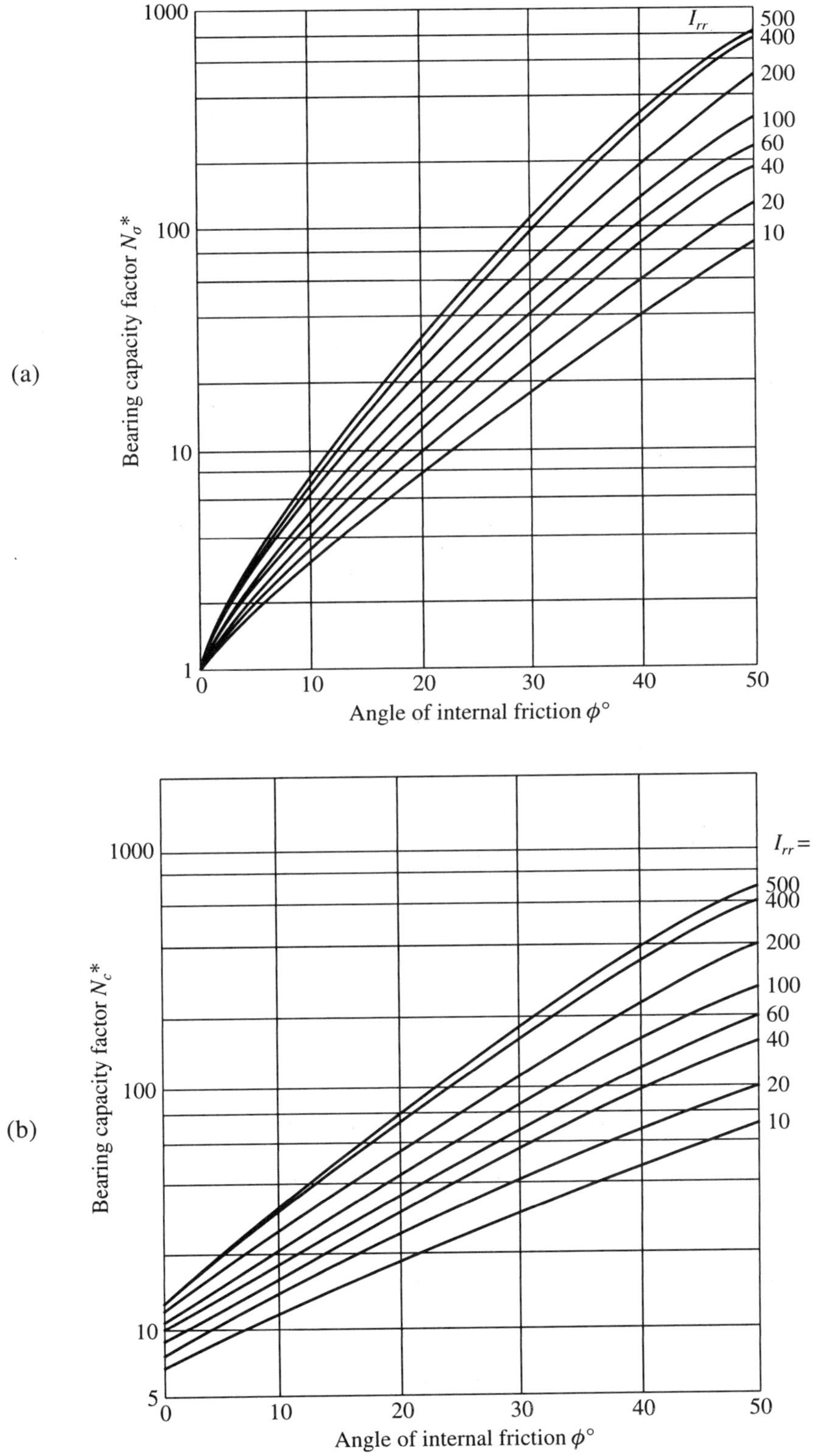

Figure 15.11 (a) Bearing capacity factor N^*_σ (b) Bearing capacity factor N^*_c (Vesic, 1977)

Table 15.1 Bearing capacity factors N^*_q and N^*_c by Janbu

ψ =	75°		90°		105°	
$\phi°$	N^*_q	N^*_c	N^*_q	N^*_c	N^*_q	N^*_c
0	1.00	5.74	1.00	5.74	1.00	5.74
5	1.50	6.25	1.57	6.49	1.64	7.33
10	2.25	7.11	2.47	8.34	2.71	9.70
20	5.29	11.78	6.40	14.83	7.74	18.53
30	13.60	21.82	18.40	30.14	24.90	41.39
35	23.08	31.53	33.30	46.12	48.04	67.18
40	41.37	48.11	64.20	75.31	99.61	117.52
45	79.90	78.90	134.87	133.87	227.68	226.68

15.15 JANBU'S METHOD OF DETERMINING Q_b

The bearing capacity equation of Janbu (1976) is the same as Eq. (15.22) and is expressed as

$$Q_b = (cN_c^* + q_o' N_q^*)A_b \tag{15.31}$$

The shape of the failure surface as assumed by Janbu is similar to that given in Fig. 15.6(b). Janbu's equation for N^*_q is

$$N_q^* = \left[\tan\phi + \sqrt{1+\tan^2\phi}\right]^2 e^{2\psi \tan\phi} \tag{15.32}$$

where ψ = angle as shown in Fig. 15.6(b). This angle varies from 60° in soft compressible soil to 105° in dense sand. The values for N^*_c used by Janbu are the same as those given by Vesic (Eq. 15.23). Table 15.1 gives the bearing capacity factors of Janbu.

Since Janbu's bearing capacity factor N^*_q depends on the angle ψ, there are two uncertainties involved in this procedure. They are

1. The difficulty in determining the values of ψ for different situations at base level.
2. The settlement required at the base level of the pile for the full development of a plastic zone.

For full base load Q_b to develop, at least a settlement of about 10 to 20 percent of the pile diameter is required which is considerable for larger diameter piles.

15.16 COYLE AND CASTELLO'S METHOD OF ESTIMATING Q_b IN SAND

Coyle and Castello (1981) made use of the results of 24 full scale pile load tests driven in sand for evaluating the bearing capacity factors. The form of equation used by them is the same as Eq. (15.6) which may be expressed as

$$Q_b = q_o' N_q^* A_b \tag{15.33}$$

where q'_o = effective overburden pressure at the base level of the pile
N^*_q = bearing capacity factor

Coyle and Castello collected data from the instrumented piles, and separated from the total load the base load and friction load. The total force at the top of the pile was applied by means of a jack. The soil at the site was generally fine sand with some percentage of silt. The lowest and the

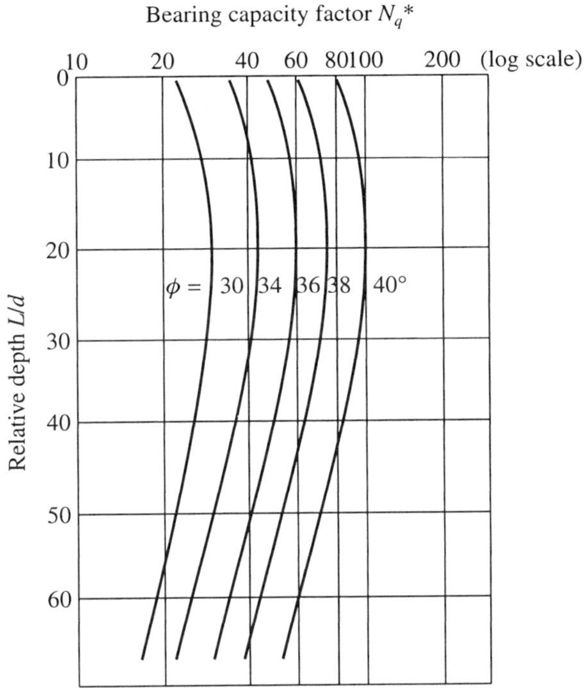

Figure 15.12 N_q^* versus L/d (after Coyle and Castello, 1981)

highest relative densities were 40 to 100 percent respectively. The pile diameter was generally around 1.5 ft and pile penetration was about 50 ft. Closed end steel pipe was used for the tests in some places and precast square piles or steel *H* piles were used at other places.

The bearing capacity factor N^*_q was evaluated with respect to depth ratio L/d in Fig. 15.12 for various values of ϕ.

15.17 THE ULTIMATE SKIN RESISTANCE OF A SINGLE PILE IN COHESIONLESS SOIL

Skin Resistance (Straight Shaft)

The ultimate skin resistance in a homogeneous soil as per Eq. (15.9) is expressed as

$$Q_f = A_s \overline{q}'_o \overline{K}_s \tan\delta \tag{15.34a}$$

In a layered system of soil $\overline{q}'_o$, $\overline{K}_s$ and δ vary with respect to depth. Equation (15.34a) may then be expressed as

$$Q_f = \int_0^L P \overline{q}'_o \overline{K}_s \tan\delta dz \tag{15.34b}$$

where $\overline{q}'_o$, $\overline{K}_s$ and δ refer to thickness dz of each layer and P is the perimeter of the pile.

As explained in Section 15.10 the effective overburden pressure does not increase linearly with depth and reaches a constant value beyond a particular depth L_c, called the critical depth which is a function of ϕ. It is therefore natural to expect the skin resistance f_s also to remain constant beyond depth L_c. The magnitude of L_c may be taken as equal to $20d$.

Table 15.2 Values of $\overline{K}_s$ and δ (Broms, 1966)

Pile material	δ	Values of $\overline{K}_s$ Low D_r	High D_r
Steel	20°	0.5	1.0
Concrete	$3/4\phi$	1.0	2.0
Wood	$2/3\phi$	1.5	4.0

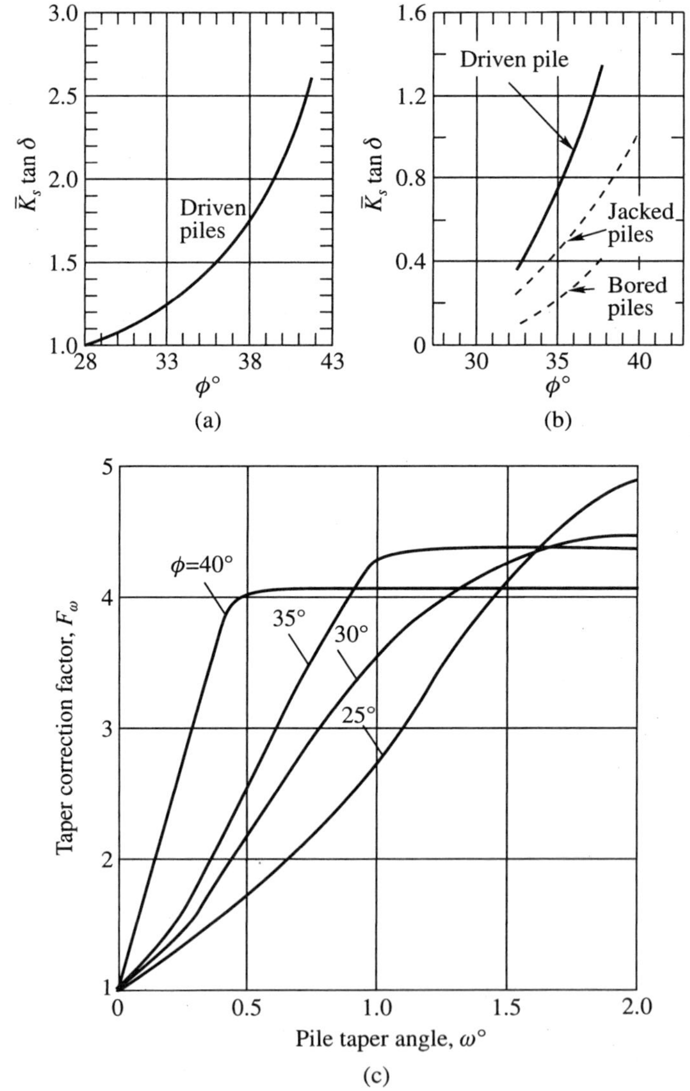

Figure 15.13 Values of $K_s \tan\delta$ in sand as per (a) Poulos and Davis 1980, (b) Meyerhof, 1976 and (c) taper factor F_ω (after Nordlund, 1963)

Eq. (15.17) can be used for determining the critical length L_c for any given set of values of ϕ and d. Q_f can be calculated from Eq. (15.34) if $\overline{K}_s$ and δ are known.

The values of $\bar{K}_s$ and δ vary not only with the relative density and pile material but also with the method of installation of the pile.

Broms (1966) has related the values of $\bar{K}_s$ and δ to the effective angle of internal friction ϕ of cohesionless soils for various pile materials and relative densities (D_r) as shown in Table 15.2. The values are applicable to driven piles. As per the present state of knowledge, the maximum skin friction is limited to 110 kN/m^2 (Tomlinson, 1986).

Eq. (15.34) may also be written as

$$Q_f = \int_0^L P\bar{q}'_o \beta dz \tag{15.35}$$

where, $\beta = \bar{K}_s \tan \delta$.

Poulos and Davis, (1980) have given a curve giving the relationship between β and $\phi°$ which is applicable for driven piles and all types of material surfaces. According to them there is not sufficient evidence to show that β would vary with the pile material. The relationship between β and ϕ is given in Fig. 15.13(a). For bored piles, Poulos et al, recommend the relationship given by Meyerhof (1976) between ϕ and β (Fig. 15.13(b)).

Skin Resistance on Tapered Piles

Nordlund (1963) has shown that even a small taper of 1° on the shaft gives a four fold increase in unit friction in medium dense sand under compression loading. Based on Nordlund's analysis, curves have been developed (Poulos and Davis, 1980) giving a relationship between taper angle $\omega°$ and a taper correction factor F_ω, which can be used in Eq. (15.35) as

$$Q_f = \int_0^L F_\omega P\bar{q}'_o \beta dz \tag{15.36}$$

Eq. (15.36) gives the ultimate skin load for tapered piles. The correction factor F_ω can be obtained from Fig. 15.13(c). The value of ϕ to be used for obtaining F_ω is as per Eq. (15.18a) for driven piles.

15.18 SKIN RESISTANCE Q_f BY COYLE AND CASTELLO METHOD (1981)

For evaluating frictional resistance, Q_f, for piles in sand, Coyle and Castello (1981) made use of the results obtained from 24 field tests on piles. The expression for Q_f is the one given in Eq. (15.34a). They developed a chart (Fig. 15.14) giving relationships between $\bar{K}_s$ and ϕ for various L/d ratios. The angle of wall friction δ is assumed equal to 0.8ϕ. The expression for Q_f is

$$Q_f = A_s \bar{q}'_o \bar{K}_s \tan \delta \tag{15.34a}$$

where $\bar{q}'_o$ = average effective overburden pressure and δ = angle of wall friction = 0.8ϕ.

The value of $\bar{K}_s$ can be obtained Fig. 15.14.

15.19 STATIC BEARING CAPACITY OF PILES IN CLAY SOIL

Equation for Ultimate Bearing Capacity

The static ultimate bearing capacity of piles in clay as per Eq. (15.12) is

$$Q_u = Q_b + Q_f = c_b N_c A_b + \alpha \bar{c}_u A_s \tag{15.37}$$

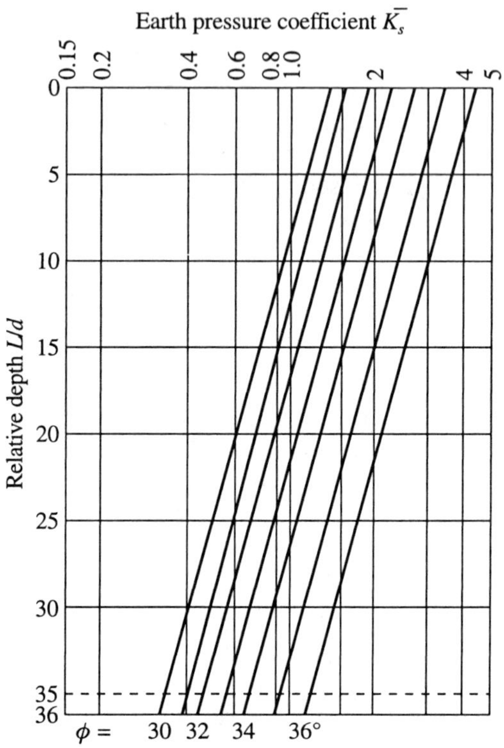

Figure 15.14 Coefficient $\overline{K}_s$ versus L/d, $\delta = 0.8\phi$ (after Coyle and Castello, 1981)

For layered clay soils where the cohesive strength varies along the shaft, Eq. (15.37) may be written as

$$Q_u = c_b N_c A_b + \sum_0^L \alpha \bar{c}_u A_s \tag{15.38}$$

Bearing Capacity Factor N_c

The value of the bearing capacity factor N_c that is generally accepted is 9 which is the value proposed by Skempton (1951) for circular foundations for a L/B ratio greater than 4. The base capacity of a pile in clay soil may now be expressed as

$$Q_b = 9c_b A_b \tag{15.39}$$

The value of c_b may be obtained either from laboratory tests on undisturbed samples or from the relationships established between c_u and field penetration tests. Eq. (15.39) is applicable for all types of pile installations.

Skin Resistance by α-Method

Tomlinson (1986) has given some empirical correlations for evaluating α in Eq. (15.37) for different types of soil conditions and L/d ratios. His procedure requires a great deal of judgment of the soil conditions in the field and may lead to different interpretations by different geotechnical engineers. A simplified approach for such problems would be needed. Dennis and Olson (1983b) made use of the information provided by Tomlinson and developed a single curve giving the relationship between α and the undrained shear strength c_u of clay as shown in Fig. 15.15.

This curve can be used to estimate the values of α for piles with penetration lengths less than 30 m. As the length of the embedment increases beyond 30 m, the value of α decreases. Piles of such great length experience elastic shortening that results in small shear strain or slip at great depth as compared to that at shallow depth. Investigation indicates that for embedment greater than about 50 m the value of α from Fig 15.15 should be multiplied by a factor 0.56. For embedments between 30 and 50 m, the reduction factor may be considered to vary linearly from 1.0 to 0.56 (Dennis and Olson, 1983a, b)

Skin Resistance by λ-Method

Vijayvergiya and Focht (1972) have suggested a different approach for computing skin load Q_f for steel-pipe piles on the basis of examination of load test results on such piles. The equation is of the form

$$Q_f = \lambda(\overline{q}'_o + 2\overline{c}_u)A_s \tag{15.40}$$

where λ = frictional capacity coefficient,

$\overline{q}'_o$ = mean effective vertical stress between the ground surface and pile tip.

The other terms are already defined. λ is plotted against pile penetration as shown in Fig. 15.16.

Eq. (15.40) has been found very useful for the design of heavily loaded pipe piles for offshore structures.

β-Method or the Effective Stress Method of Computing Skin Resistance

In this method, the unit skin friction f_s is defined as

$$f_s = \overline{K}_s \tan\delta\, \overline{q}'_o = \beta \overline{q}'_o \tag{15.41}$$

where β = the skin factor = $\overline{K}_s \tan \delta$, (15.42a)

$\overline{K}_s$ = lateral earth pressure coefficient,

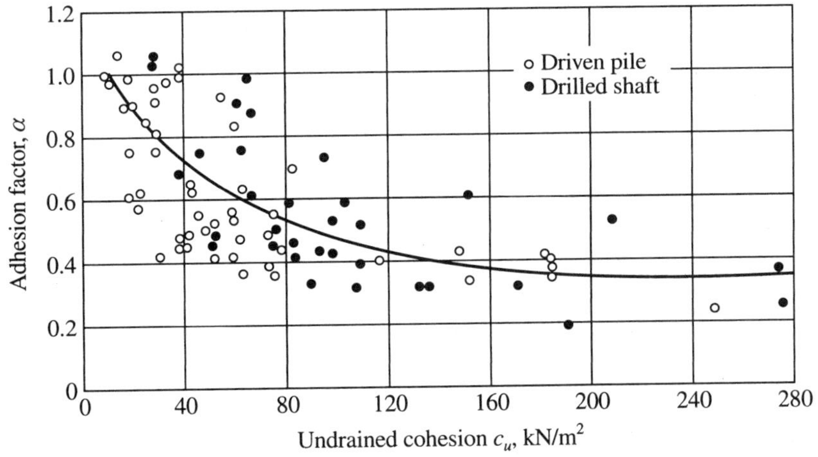

Figure 15.15 Adhesion factor α for piles with penetration lengths less than 50 m in clay. (Data from Dennis and Olson 1983 a, b; Stas and Kulhawy, 1984)

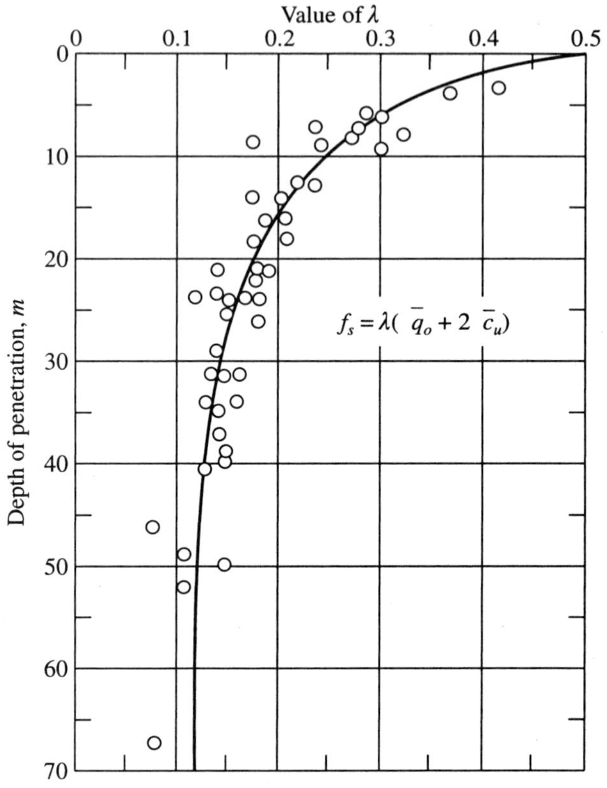

Figure 15.16 Frictional capacity coefficient λ vs pile penetration (Vijayvergiya and Focht, 1972)

δ = angle of wall friction,

$\bar{q}'_o$ = average effective overburden pressure.

Burland (1973) discusses the values to be used for β and demonstrates that a lower limit for this factor for normally consolidated clay can be written as

$$\beta_o = \bar{K}_o \tan\phi' \tag{15.42b}$$

As per Jaky (1944)

$$K_o = (1 - \sin\phi') \tag{15.42c}$$

therefore $\beta = (1 - \sin\phi')\tan\phi'$ (15.42d)

where ϕ' = effective angle of internal friction.

Since the concept of this method is based on effective stresses, the cohesion intercept on a Mohr circle is equal to zero. For driven piles in stiff overconsolidated clay, $\bar{K}_s$ is roughly 1.5 times greater than $\bar{K}_o$. For overconsolidated clays $\bar{K}_o$ may be found from the expression

$$\bar{K}_o = (1 - \sin\phi')\sqrt{R_{oc}} \tag{15.42e}$$

where R_{oc} = overconsolidation ratio of clay.

For clays, ϕ' may be taken in the range of 20 to 30 degrees. In such a case the value of β in Eq. (15.42d) varies between 0.24 and 0.29.

Meyerhof's Method (1976)

Meyerhof has suggested a semi-empirical relationship for estimating skin friction in clays.

For driven piles:

$$f_s = 1.5c_u \tan \phi' \tag{15.43}$$

For bored piles:

$$f_s = c_u \tan \phi' \tag{15.44}$$

By utilizing a value of 20° for ϕ' for the stiff to very stiff clays, the expressions reduce to

For driven piles:

$$f_s = 0.55c_u \tag{15.45}$$

For bored piles:

$$f_s = 0.36c_u \tag{15.46}$$

In practice the maximum value of unit friction for bored piles is restricted to 100 kPa.

15.20 BEARING CAPACITY OF PILES IN GRANULAR SOILS BASED ON SPT VALUE

Meyerhof (1956) suggests the following equations for single piles in granular soils based on SPT values.

For displacement piles:

$$Q_u = Q_b + Q_f = 40N_{cor}(L/d)A_b + 2\overline{N}_{cor}A_s \tag{15.47a}$$

for H-piles:

$$Q_u = 40N_{cor}(L/d)A_b + \overline{N}_{cor}A_s \tag{15.47b}$$

where $q_b = 40\ N_{cor}(L/d) \leq 400\ N_{cor}$

For bored piles:

$$Q_u = 133N_{cor}A_b + 0.67\overline{N}_{cor}A_s \tag{15.48}$$

where Q_u = ultimate total load in kN

N_{cor} = average corrected SPT value below pile tip

$\overline{N}_{cor}$ = corrected average SPT value along the pile shaft

A_b = base area of pile in m^2 (for H-piles including the soil between the flanges)

A_s = shaft surface area in m^2

In English units Q_u for a displacement pile is

$$Q_u(\text{kip}) = Q_b + Q_f = 0.80N_{cor}(L/d)A_b + 0.04\overline{N}_{cor}A_s \tag{15.49a}$$

where A_b = base area in ft^2 and A_s = surface area in ft^2

$$\text{and } 0.80N_{cor}(\frac{L}{d})A_b \leq 8N_{cor}A_b(\text{kip}) \tag{15.49b}$$

A minimum factor of safety of 4 is recommended. The allowable load Q_a is

$$Q_a = \frac{Q_u}{4} \tag{15.50}$$

Example 15.1

A concrete pile of 45 cm diameter was driven into sand of loose to medium density to a depth of 15 m. The following properties are known:

(a) Average unit weight of soil along the length of the pile, $\bar{\gamma} = 17.5$ kN/m^3, average $\phi = 30°$, (b) average $\overline{K}_s = 1.0$ and $\delta = 0.75\phi$.

Calculate (a) the ultimate bearing capacity of the pile, and (b) the allowable load with $F_s = 2.5$. Assume the water table is at great depth. Use Berezantsev's method.

Solution

From Eq. (15.9)

$$Q_u = Q_b + Q_f = q_0' A_b N_q + \bar{q}_o' A_s \overline{K}_s \tan\delta$$

where $q_0' = \bar{\gamma} L = 17.5 \times 15 = 262.5 \text{ kN/m}^2$

$$\bar{q}_o' = \frac{1}{2}\bar{\gamma} L = \frac{262.5}{2} = 131.25 \text{ kN/m}^2$$

$$A_b = \frac{3.14}{4} \times 0.45^2 = 0.159 \text{ m}^2$$

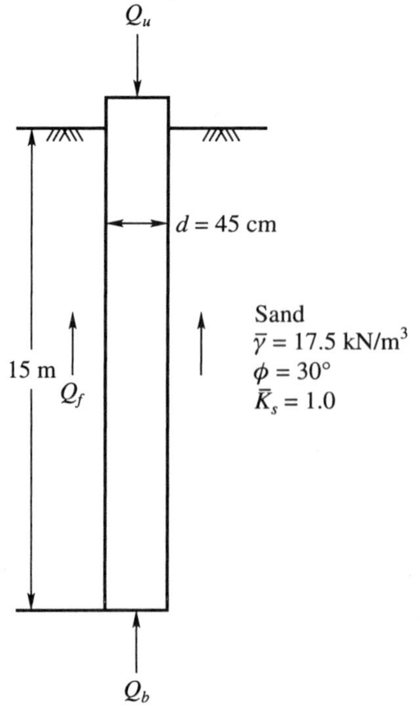

Figure Ex. 15.1

$$A_s = 3.14 \times 0.45 \times 15 = 21.195 \text{ m}^2$$

$$\delta = 0.75\phi = 0.75 \times 30 = 22.5°$$

$$\tan \delta = 0.4142$$

From Fig. 15.9, N_q for $\frac{L}{d} = \frac{15}{0.45} = 33.3$ and $\phi = 30°$ is equal to 16.5.

Substituting the known values, we have

$$Q_u = Q_b + Q_f = 262.5 \times 0.159 \times 16.5 + 131.25 \times 21.195 \times 1.0 \times 0.4142$$

$$= 689 + 1152 = 1841 \text{ kN}$$

$$Q_a = \frac{1841}{2.5} = 736 \text{ kN}$$

Example 15.2

Assume in Ex. 15.1 that the water table is at the ground surface and $\gamma_{sat} = 18.5$ kN/m^3. All the other data remain the same. Calculate Q_u and Q_a.

Solution

Water table at the ground surface $\gamma_{sat} = 18.5$ kN/m^3

$$\gamma_b = \gamma_{sat} - \gamma_w = 18.5 - 9.81 = 8.69 \text{ kN/m}^3$$

$$q_0' = 8.69 \times 15 = 130.35 \text{ kN/m}^2$$

$$\bar{q}_0' = \frac{1}{2} \times 130.35 = 65.18 \text{ kN/m}^2$$

Substituting the known values

$$Q_u = 130.35 \times 0.159 \times 16.5 + 65.18 \times 21.195 \times 1.0 \times 0.4142$$

$$= 342 + 572 = 914 \text{ kN}$$

$$Q_a = \frac{914}{2.5} = 366 \text{ kN}$$

Note: It may be noted here that the presence of a water table at the ground surface in cohesionless soil reduces the ultimate load capacity of pile by about 50 percent.

Example 15.3

A concrete pile of 45 cm diameter is driven to a depth of 16 m through a layered system of sandy soil ($c = 0$). The following data are available.

Top layer 1: Thickness = 8 m, $\gamma_d = 16.5$ kN/m^3, $e = 0.60$ and $\phi = 30°$.

Layer 2: Thickness = 6 m, $\gamma_d = 15.5$ kN/m^3, $e = 0.65$ and $\phi = 35°$.

Layer 3: Extends to a great depth, $\gamma_d = 16.00$ kN/m^3, $e = 0.65$ and $\phi = 38°$.

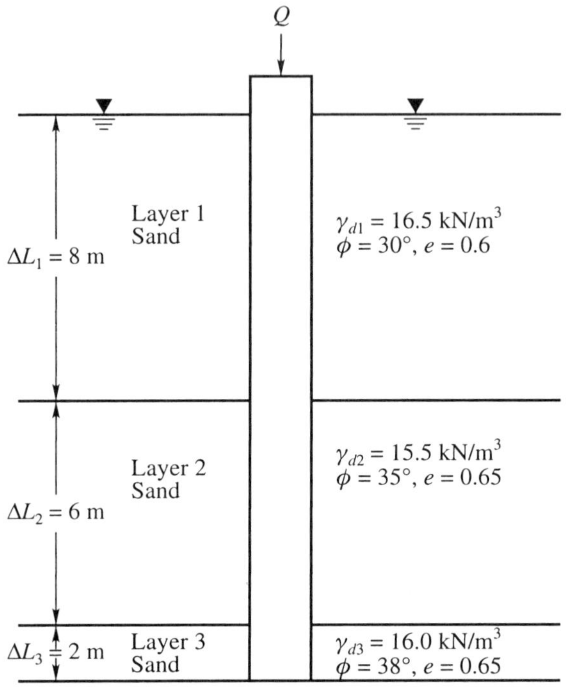

Figure Ex. 15.3

Assume that the value of δ in all the layers of sand is equal to 0.75ϕ. The value of $\overline{K}_s$ for each layer as equal to half of the passive earth pressure coefficient. The water table is at ground level.

Calculate the values of Q_u and Q_a with $F_s = 2.5$ by the conventional method for Q_f and Berezantsev's method for Q_b.

Solution

The soil is submerged throughout the soil profile. The specific gravity G_s is required for calculating γ_{sat}.

(a) Using the equation $\gamma_d = \dfrac{\gamma_w G_s}{1+e}$, calculate G_s for each layer since γ_d, γ_w and e are known.

(b) Using the equation $\gamma_{sat} = \dfrac{\gamma_w (G_s + e)}{1+e}$, calculate γ_{sat} for each layer and then $\gamma_b = \gamma_{sat} - \gamma_w$ for each layer.

(c) For a layered system of soil, the ultimate load can be determined by making use of Eq. (15.9). Now

$$Q_u = Q_b + Q_f = q_0' N_q A_b + P \sum_0^L \bar{q}_0' \overline{K}_s \tan \delta \, \Delta L$$

(d) q'_0 at the tip of the pile is

$$q_0' = \gamma_{b1} \Delta L_1 + \gamma_{b2} \Delta L_2 + \gamma_{b3} \Delta L_3$$

(e) $\overline{q}'_0$ at the middle of each layer is

$$\overline{q}'_{01} = \frac{1}{2}\Delta L_1 \gamma_{b1}$$

$$\overline{q}'_{02} = \Delta L_1 \gamma_{b1} + \frac{1}{2}\Delta L_2 \gamma_{b2}$$

$$\overline{q}'_{03} = \Delta L_1 \gamma_{b1} + \Delta L_2 \gamma_{b2} + \frac{1}{2}\Delta L_3 \gamma_{b3}$$

(f) $N_q = 95$ for $\phi = 38°$ and $\frac{L}{d} = \frac{15}{0.45} = 33.33$ from Fig. 15.9.

(g) $A_b = 0.159$ m², $P = 1.413$ m.

(h) $\overline{K}_s = \frac{1}{2}\tan^2\left(45° + \frac{\phi}{2}\right) = \frac{1}{2}K_p$. $\overline{K}_s$ for each layer can be calculated.

(i) $\delta = 0.75\phi$. The values of tan δ can be calculated for each layer. The computed values for all the layers are given below in a tabular form.

Layer no.	G_s	γ_b kN/m³	$\overline{q}'_0$ kN/m²	$\overline{K}_s$	tan δ	ΔL m
1.	2.69	10.36	41.44	1.5	0.414	8
2.	2.61	9.57	111.59	1.845	0.493	6
3.	2.69	10.05	150.35	2.10	0.543	2
	From middle of layer 3 to tip of pile = 10.05					
	At the tip of pile		$q'_0 = 160.40$ kN/m²			

$$Q_u = (160.4 \times 95 \times 0.159 + 1.413(41.44 \times 1.5 \times 0.414 \times 8$$

$$+ 111.59 \times 1.845 \times 0.493 \times 6 + 150.35 \times 2.10 \times 0.543 \times 2)$$

$$= 2423 + 1636 \approx 4059 \text{ kN}$$

$$Q_a = \frac{Q_u}{2.5} = \frac{4059}{2.5} = 1624 \text{ kN}$$

Example 15.4

If the pile in Ex. 15.2 is a bored and cast-in-situ, compute Q_u and Q_a. All the other data remain the same. Water table is close to the ground surface.

Solution

Per Tomlinson (1986), the ultimate bearing capacity of a bored and cast-in-situ-pile in cohesionless soil is reduced considerably due to disturbance of the soil. Per Section 15.12, calculate the base resistance for a driven pile and take one-third of this as the ultimate base resistance for a bored and cast-in-situ pile.

For computing δ, take $\phi = 28°$ and $\overline{K}_s = 1.0$ from Table 15.2 for a concrete pile.

Base resistance for driven pile

For $\phi = 30°$, $N_q = 16.5$ from Fig. 15.9.

$A_b = 0.159\ \text{m}^2$

$q'_0 = 130.35\ \text{kN/m}^2$ (From Ex. 15.2)

$Q_b = 130.35 \times 0.159 \times 16.5 = 342\ \text{kN}$

For bored pile

$$Q_b = \frac{1}{3} \times 342 = 114\ \text{kN}$$

Skin load

$$Q_f = A_s \bar{q}' \bar{K}_s \tan\delta$$

For $\phi = 28°$, $\delta = 0.75 \times 28 = 21°$, $\tan\delta = 0.384$

$A_s = 21.195\ \text{m}^2$ (from Ex. 15.2)

$\bar{q}'_0 = 65.18\ \text{kN/m}^2$ (Ex. 15.2)

Substituting the known values,

$$Q_f = 21.195 \times 65.18 \times 1.0 \times 0.384 = 530\ \text{kN}$$

Therefore, $Q_u = 114 + 530 = 644\ \text{kN}$

$$Q_a = \frac{644}{2.5} \approx 258\ \text{kN}$$

Example 15.5

Solve the problem given in Example 15.1 by Meyerhof's method. All the other data remain the same.

Solution

Per Table 9.3, the sand in-situ may be considered in a loose state for $\phi = 30°$. The corrected SPT value $N_{cor} = 10$.

Point bearing capacity

From Eq. (15.20)

$q_b = q'_o N_q \leq q_{bl}$

From Eq. (15.19 b)

$q_{bl} = 25\ N_q \tan\phi\ \text{kN/m}^2$

Now From Fig. 15.10 $N_q = 60$ for $\phi = 30°$

$q'_o = \gamma L = 17.5 \times 15 = 262.5\ \text{kN/m}^2$

$q_b = 262.5 \times 60 = 15{,}750\ \text{kN/m}^2$

$q_{bl} = 25 \times 60 \times \tan 30° = 866\ \text{kN/m}^2$

Hence the limiting value for q_b = 866 kN/m^2

$$\text{Now, } Q_b = A_b\, q_b = A_b q_b = \frac{3.14}{4} \times (0.45)^2 \times 866 = 138\,\text{kN}$$

Frictional resistance

Per Section 15.17, the unit skin resistance f_s is assumed to increase from 0 at ground level to a limiting value of f_{sl} at $L_c = 20\,d$ where L_c = critical depth and d = diameter. Therefore $L_c = 20 \times 0.45 = 9$ m

Now $f_{sl} = q'_0 \bar{K}_s \tan\delta = \gamma L_c \bar{K}_s \tan\delta$

Given: γ = 17.5 kN/m^3, L_c = 9 m, $\bar{K}_s$ = 1.0 and δ = 22.5° m.

Substituting and simplifying we have

$f_{sl} = 17.5 \times 9 \times 1.0 \times \tan 22.5 = 65$ kN/m^2

$$\text{The skin load } Q_f = Q_{f1} + Q_{f2} = \frac{1}{2} f_{sl} P L_c + P f_{sl}(L - L_c)$$

$$\text{Substituting } Q_f = \frac{1}{2} \times 65 \times 3.14 \times 0.45 \times 9 + 3.14 \times 0.45 \times 65(15-9)$$

$$= 413 + 551 = 964 \text{ kN}$$

The failure load Q_u is

$Q_u = Q_b + Q_f = 138 + 964 = 1{,}102$ kN

with $F_s = 2.5$,

$$Q_a = \frac{1102}{2.5} = 440 \text{ kN}$$

Example 15.6

Determine the base load of the problem in Example 15.5 by Vesic's method. Assume $I_r = I_{rr} = 50$. Determine Q_a for $F_s = 2.5$ using the value of Q_f in Ex 15.5.

Solution

From Eq. (15.25) for $c = 0$ we have

$q_b = \sigma_m N^*_\sigma$

From Eq. (15.24)

$$\sigma_m = \left[\frac{1+2K_0}{3}\right] q'_0 = \left[\frac{1+2(1-\sin\phi)}{3}\right] q'_0$$

$$q'_0 = 15 \times 17.5 = 262.5 \text{ kN/m}^2$$

$$\sigma_m = \left[\frac{1+2(1-\sin 30^0)}{3}\right] \times 262.5 = 175 \text{ kN/m}^2$$

From Fig. 15.11a, $N^*_\sigma = 36$ for $\phi = 30$ and $I_r = 50$

Substituting

$q_b = 175 \times 36 = 6300$ kN/m^2

$$Q_b = A_b q_b = \frac{3.14}{4} \times (0.45)^2 \times 6300 = 1001\,\text{kN}$$

$$Q_u = Q_b + Q_f = 1001 + 964 = 1{,}965\,\text{kN}$$

$$Q_a = \frac{1965}{2.5} = 786\ \text{kN}$$

Example 15.7

Determine the base load of the problem in Example 15.1 by Janbu's method. Use $\psi = 90°$. Determine Q_a for $F_s = 2.5$ using the Q_f estimated in Example 15.5.

Solution

From Eq (15.31), for $c = 0$ we have

$$q_b = q'_0 N_q^*$$

For $\phi = 30°$ and $\psi = 90°$, we have $N_q^* = 18.4$ from Table 15.1. $q'_0 = 262.5$ kN/m^2 as in Ex. 15.5.

Therefore $q_b = 262.5 \times 18.4 = 4830$ kN/m^2

$Q_b = A_b q_b = 0.159 \times 4830 = 768$ kN

$Q_u = Q_b + Q_f = 768 + 964 = 1732$ kN

$$\text{Q}_\text{a} = \frac{1732}{2.5} = 693\ \text{kN}$$

Example 15.8

Estimate Q_b, Q_f, Q_u and Q_a by the Coyle and Castello method using the data given in Example 15.1.

Solution

Base load Q_b from Eq. (15.33)

$$q_b = q'_o N^*_q$$

From Fig. 15.12, $N^*_q = 29$ for $\phi = 30°$ and $L/d = 33.3$

$q'_o = 262.5$ kN/m^2 as in Ex. 15.5

Therefore $q_b = 262.5 \times 29 = 7612$ kN/m^2

$Q_b = A_b q_b = 0.159 \times 7612 = 1210$ kN

From Eq. (15.34a)

$$Q_f = A_s \bar{q}'_o \bar{K}_s \tan\delta$$

where $A_s = 3.14 \times 0.45 \times 15 = 21.2\ \text{m}^2$

$$\overline{q}'_o = \frac{1}{2} \times 262.5 = 131.25\ \text{kN/m}^2$$

$$\delta = 0.8\phi = 0.8 \times 30^\circ = 24^\circ$$

From Fig. 15.14, $\overline{K}_s = 0.35$ for $\phi = 30^\circ$ and $L/d = 33.3$

Therefore $Q_f = 21.2 \times 131.25 \times 0.35 \tan 24^\circ = 434\ \text{kN}$

$$Q_u = Q_b + Q_f = 1210 + 434 = 1644\ \text{kN}$$

$$Q_a = \frac{1644}{2.5} = 658\ \text{kN}$$

Example 15.9

Determine Q_b, Q_f, Q_u and Q_a by using the SPT value for $\phi = 30^\circ$ from Fig. 12.8.

Solution

From Fig. 12.8, $N_{cor} = 10$ for $\phi = 30^\circ$. Use Eq. (15.47a) for Q_u

$$Q_u = Q_b + Q_f = 40N_{cor}\left(\frac{L}{d}\right)A_b + 2\overline{N}_{cor}A_s$$

where $Q_b \le Q_{bl} = 400\ N_{cor}A_b$

Given: $L = 15$ m, $d = 0.45$ m, $A_b = 0.159\ \text{m}^2$, $A_s = 21.2\ \text{m}^2$

$$Q_b = 40 \times 10 \times \left(\frac{15}{0.45}\right) \times 0.159 = 2120\ \text{kN}$$

$$Q_{bl} = 400 \times 10 \times 0.159 = 636\ \text{kN}$$

Since $Q_b > Q_{bl}$, use Q_{bl}

$$Q_f = 2 \times 10 \times 21.2 = 424\ \text{kN}$$

Now $Q_u = 636 + 424 = 1060\ \text{kN}$

$$Q_a = \frac{1060}{2.5} = 424\ \text{kN}$$

Example 15.10

Compare the values of Q_b, Q_f and Q_a obtained by the different methods in Examples 15.1, and 15.5 through 15.9 and make appropriate comments.

Comparison

The values obtained by different methods are tabulated below.

Method No	Example No	Investigator	Q_b kN	Q_f kN	Q_u kN	Q_a (F_s = 2.5) kN
1.	15.1	Berezantsev	689	1152	1841	736
2.	15.5	Meyerhof	138	964	1102	440
3.	15.6	Vesic	1001	964	1965	786
4.	15.7	Janbu	768	964	1732	693
5.	15.8	Coyle & Castello	1210	434	1644	658
6.	15.9	Meyerhof (Based on SPT)	636	424	1060	424

Comments

It may be seen from the table above that there are wide variations in the values of Q_b and Q_f between the different methods.

Method 1 Tomlinson (1986) recommends Berezantsev's method for computing Q_b as this method conforms to the practical criteria of pile failure. Tomlinson does not recommend the critical depth concept.

Method 2 Meyerhof's method takes into account the critical depth concept. Eq. (15.19) is based on this concept. The equation [Eq. (15.20)] $q_b = q'_o N_q$ does not consider the critical depth concept where q'_o = effective overburden pressure at the pile tip level of the pile. The value of Q_b per this equation is

$$Q_b = q_b A_b = 15{,}750 \times 0.159 = 2504 \text{ kN}$$

which is very high and this is close to the value of Q_b (= 2120 kN) by the SPT method. However Eq. (15.19 b) gives a limiting value for Q_b = 138 kN (here the sand is considered loose for $\phi = 30$.

Q_f is computed by assuming Q_f increases linearly with depth from 0 at $L = 0$ to Q_{fl} at depth $L_c = 20d$ and then remains constant to the end of the pile.

Though some investigators have accepted the critical depth concept for computing Q_b and Q_f, it is difficult to generalize this concept as applicable to all types of conditions prevailing in the field.

Method 3 Vesic's method is based on many assumptions for determining the values of I_r, I_{rr}, σ_m, N^*_σ etc. There are many assumptions in this method. Are these assumptions valid for the field conditions? Designers have to answer this question.

Method 4 The uncertainties involved in Janbu's method are given in Section 15.15 and as such difficult to assess the validity of this method.

Method 5 Coyle and Castello's method is based on full scale field tests on a number of driven piles. Their bearing capacity factors vary with depth. Of the first five methods listed above, the value of Q_b obtained by them is much higher than the other four methods whereas the value of Q_f is very much lower. But on the whole the value of Q_u is lower than the other methods.

Method 6 This method was developed by Meyerhof based on SPT values. The Q_b value (= 2120 kN) by this method is very much higher than the preceding methods, whereas the value of Q_f is the lowest of all the methods. In the table given above the limiting value of Q_{bl} = 636 kN is considered. In all the six methods $F_s = 2.5$ has been taken to evaluate Q_a whereas in method 6, $F_s = 4$ is recommended by Meyerhof.

Which Method to Use

There are wide variations in the values of Q_b, Q_f, Q_u and Q_a between the different methods. The relative proportions of loads carried by skin friction and base resistance also vary between the methods. The order of preference of the methods may be listed as follows;

Preference No.	Method No.	Name of the investigator
1	1	Berezantsev for Q_b
2	2	Meyerhof
3	5	Coyle and Castello
4	6	Meyerhof (SPT)
5	3	Vesic
6	4	Janbu

Example 15.11

A concrete pile 18 in. in diameter and 50 ft long is driven into a homogeneous mass of clay soil of medium consistency. The water table is at the ground surface. The unit cohesion of the soil under undrained condition is 1050 lb/ft^2 and the adhesion factor $\alpha = 0.75$. Compute Q_u and Q_a with $F_s = 2.5$.

Solution

Given: $L = 50$ ft, $d = 1.5$ ft, $c_u = 1050$ lb/ft^2, $\alpha = 0.75$.

From Eq. (15.37), we have

$$Q_u = Q_b + Q_f = c_b N_c A_b + A_s \alpha \bar{c}_u$$

where, $c_b = \bar{c}_u = 1050$ lb/ft^2; $N_c = 9$; $A_b = 1.766$ ft^2; $A_s = 235.5$ ft^2

Substituting the known values, we have

$$Q_u = \frac{1050 \times 9 \times 1.766}{1000} + \frac{235.5 \times 0.75 \times 1050}{1000}$$

$$= 16.69 + 185.46 = 202.15 \text{ kips}$$

$$Q_a = \frac{202.15}{2.5} = 81 \text{ kips}$$

Example 15.12

A concrete pile of 45 cm diameter is driven through a system of layered cohesive soils. The length of the pile is 16 m. The following data are available. The water table is close to the ground surface.

Top layer 1: Soft clay, thickness = 8 m, unit cohesion $\bar{c}_u$ = 30 kN/m^2 and adhesion factor $\alpha = 0.90$.

Layer 2: Medium stiff, thickness = 6 m, unit cohesion $\bar{c}_u$ = 50 kN/m^2 and $\alpha = 0.75$.

Layer 3: Stiff stratum extends to a great depth, unit cohesion $\bar{c}_u = 105$ kN/m^2 and $\alpha = 0.50$.

Compute Q_u and Q_a with $F_s = 2.5$.

Solution

Here, the pile is driven through clay soils of different consistencies.

The equations for Q_u expressed as (Eq. 15.38) yield

$$Q_u = 9c_b A_b + P\sum_0^L \alpha \bar{c}_u \Delta L$$

Here, $c_b = \bar{c}_u$ of layer 3, $P = 1.413$ m, $A_b = 0.159$ m^2.

Substituting the known values, we have

$$Q_u = 9 \times 105 \times 0.159 + 1.413\,(0.90 \times 30 \times 8 + 0.75 \times 50 \times 6 + 0.50 \times 105 \times 2)$$

$$= 150.25 + 771.5 = 921.75 \text{ kN}$$

$$Q_a = \frac{921.75}{2.5} \approx 369 \text{ kN}$$

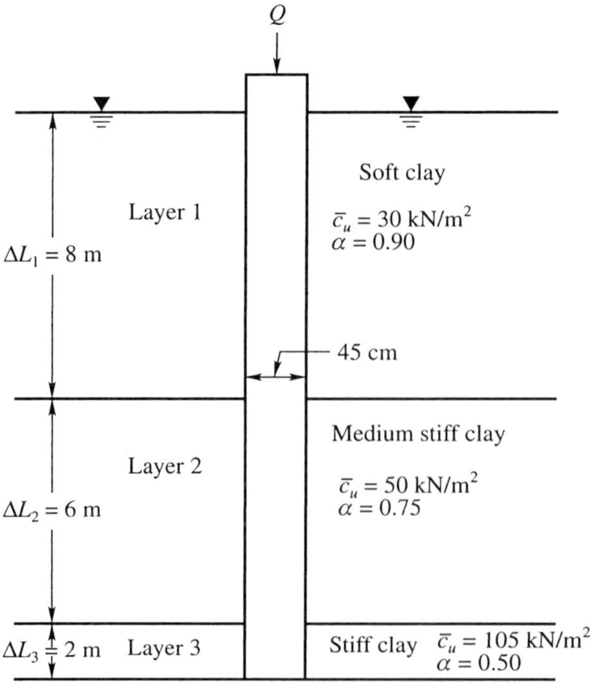

Figure Ex. 15.12

Example 15.13

A precast concrete pile of size 18 × 18 in is driven into stiff clay. The unconfined compressive strength of the clay is 4.2 kips/ft². Determine the length of pile required to carry a safe working load of 90 kips with $F_s = 2.5$.

Solution

The equation for Q_u is

$$Q_u = N_c c_u A_b + \alpha \bar{c}_u A_s$$

we have

$Q_u = 2.5 \times 90 = 225$ kips,

$N_c = 9$, $c_u = 2.1$ kips/ft²

$\alpha = 0.48$ from Fig. 15.15, $\bar{c}_u = c_u = 2.1$ kips/ft², $A_b = 2.25$ ft²

Assume the length of pile = L ft

Now, $A_s = 4 \times 1.5L = 6L$

Substituting the known values, we have

$$225 = 9 \times 2.1 \times 2.25 + 0.48 \times 2.1 \times 6L$$

or $\quad 225 = 42.525 + 6.05L$

Simplifying, we have

$$L = \frac{225 - 42.525}{6.05} = 30.2 \text{ ft}$$

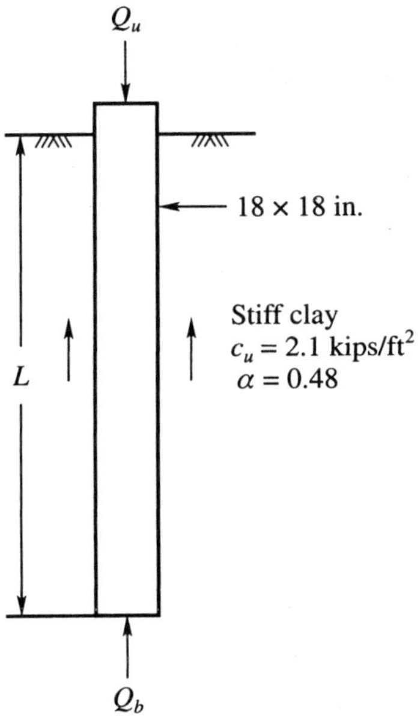

Figure Ex. 15.13

Example 15.14

For the problem given in Example 15.11 determine the skin friction load by the λ-method. All the other data remain the same. Assume the average unit weight of the soil is 110 lb/ft^3. Use Q_b given in Ex. 15.11 and determine Q_a for $F_s = 2.5$.

Solution

Per Eq. (15.40)

$$Q_f = \lambda(\overline{q}_o' + 2c_u)A_s$$

$$\overline{q}_o' = \frac{1}{2} \times 50 \times 110 = 2{,}750 \ \text{lb/ft}^2$$

Depth = 50 ft = 15.24 m

From Fig. 15.16, $\lambda = 0.2$ for depth $L = 15.24$ m

$$\text{Now } Q_f = \frac{0.2(2750 + 2 \times 1050) \times 235.5}{1000} = 228.44 \text{ kips}$$

$$\text{Now } Q_u = Q_b + Q_f = 16.69 + 228.44 = 245 \text{ kips}$$

$$Q_a = \frac{245}{2.5} = 98 \text{ kips}$$

Example 15.15

A reinforced concrete pile of size 30 x 30 cm and 10 m long is driven into coarse sand extending to a great depth. The average total unit weight of the soil is 18 kN/m^3 and the average N_{cor} value is 15. Determine the allowable load on the pile by the static formula. Use $F_s = 2.5$. The water table is close to the ground surface.

Solution

In this example only the N-value is given. The corresponding ϕ value can be found from Fig. 12.8 which is equal to 32°.

Now from Fig. 15.9, for $\phi = 32°$, and $\frac{L}{d} = \frac{10}{0.3} = 33.33$, the value of $N_q = 25$.

$$A_b = 0.3 \times 0.3 = 0.09 \text{ m}^2,$$

$$A_s = 10 \times 4 \times 0.3 = 12 \text{ m}^2$$

$$\delta = 0.75 \times 32 = 24°, \quad \tan\delta = 0.445$$

The relative density is loose to medium dense. From Table 15.2, we may take

$$\overline{K}_s = 1 + \frac{1}{3}(2-1) = 1.33$$

$$\text{Now, } Q_u = q_0' N_q A_b + \overline{q}_0' \overline{K}_s \tan\delta A_s$$

$$\gamma_b = \gamma_{sat} - \gamma_w = 18.0 - 9.81 = 8.19 \text{ kN/m}^3$$

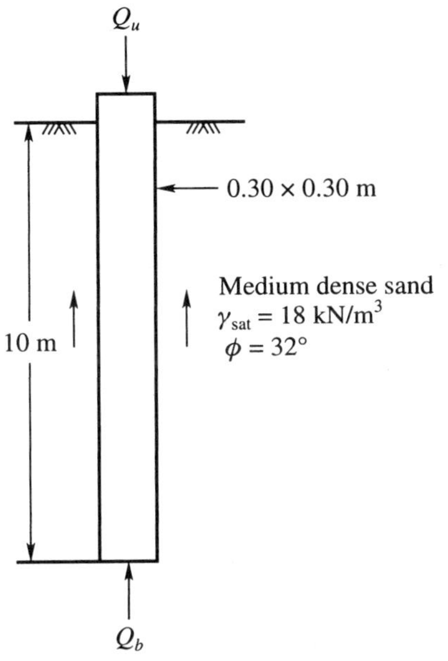

Figure Ex. 15.15

$$q_0' = \gamma_b L = 8.19 \times 10 = 81.9 \text{ kN/m}^2$$

$$\bar{q}_0' = \frac{q_0'}{2} = \frac{81.9}{2} = 40.95 \text{ kN/m}^2$$

Substituting the known values, we have

$$Q_u = 81.9 \times 25 \times 0.09 + 40.95 \times 1.33 \times 0.445 \times 12$$

$$= 184 + 291 = 475 \text{ kN}$$

$$Q_a = \frac{475}{2.5} = 190 \text{ kN}$$

Example 15.16

Determine the allowable load on the pile given in Ex. 15.15 by making use of the SPT approach by Meyerhof.

Solution

Per Ex. 15.15, $N_{cor} = 15$

Expression for Q_u is (Eq. 15.47a)

$$Q_u = Q_b + Q_f = 40 N_{cor}(L/d)A_b + 2\bar{N}_{cor}A_s$$

Here, we have to assume $N_{cor} = \bar{N}_{cor} = 15$

$$A_b = 0.3 \times 0.3 = 0.09 \text{ m}^2, \quad A_s = 4 \times 0.3 \times 10 = 12 \text{ m}^2$$

Substituting, we have

$$Q_b = 40 \times 15\left(\frac{10}{0.3}\right) \times 0.09 = 1800 \text{ kN}$$

$$Q_{bl} = 400 N_{cor} \times A_b = 400 \times 15 \times 0.09 = 540 \ \ kN \ \ < Q_b$$

Hence, Q_{bl} governs.

$$Q_f = 2\overline{N}_{cor} A_s = 2 \times 15 \times 12 = 360 \text{ kN}$$

A minimum $F_s = 4$ is recommended, thus,

$$Q_a = \frac{Q_b + Q_f}{4} = \frac{540 + 360}{4} = 225 \text{ kN}$$

Example 15.17

Precast concrete piles 16 in. in diameter are required to be driven for a building foundation. The design load on a single pile is 100 kips. Determine the length of the pile if the soil is loose to medium dense sand with an average N_{cor} value of 15 along the pile and 21 at the tip of the pile. The water table may be taken at the ground level. The average saturated unit weight of soil is equal to 120 lb/ft^3. Use the static formula and $F_s = 2.5$.

Solution

It is required to determine the length of a pile to carry an ultimate load of $Q_u = 2.5 \times 100 = 250$ kips.

The equation for Q_u is

$$Q_u = q_0' N_q A_b + \overline{q}_0' \overline{K}_s \tan\delta A_s$$

The average value of ϕ along the pile and the value at the tip may be determined from Fig. 12.8.

For $N_{cor} = 15$, $\phi = 32°$; for $N_{cor} = 21$, $\phi = 33.5°$.

Since the soil is submerged

$$\gamma_b = 120 - 62.4 = 57.6 \text{ lb/ft}^3$$

Now

$$q_0' = \gamma_b L = 57.6L \text{ lb/ft}^2$$

$$\overline{q}_0' = \frac{57.6L}{2} = 28.8L \text{ lb/ft}^2$$

$$A_b = \frac{3.14}{4} \times (1.33)^2 = 1.39 \text{ ft}^2$$

$$A_s = 3.14 \times 1.33 \times L = 4.176L \text{ ft}^2$$

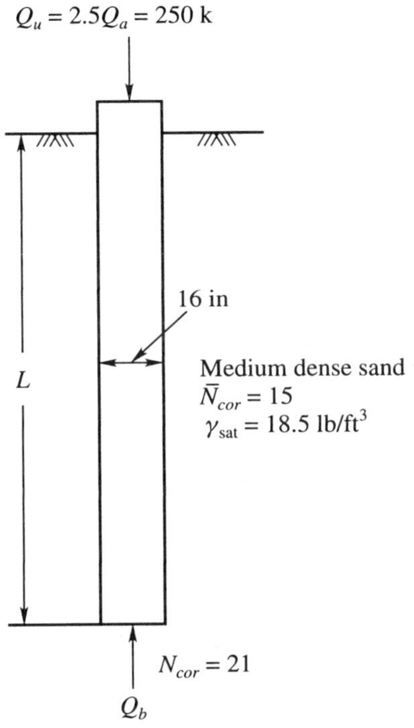

Figure Ex. 15.17

From Fig. 15.9, $N_q = 40$ for $L/d = 20$ (assumed)

From Table 15.2, $\overline{K}_s = 1.33$ for the lower side of medium dense sand

$$\delta = \frac{3}{4} \times 33.5 = 25.1°, \ \tan\delta = 0.469$$

Now by substituting the known values, we have

$$250 = \frac{(57.6L) \times 40 \times 1.39}{1000} + \frac{(28.8L) \times 0.469 \times 1.33 \times (4.176L)}{1000}$$

$$= 3.203L + 0.075L^2$$

or $\quad L^2 + 42.71L - 3333 = 0$

Solving this equation gives a value of $L = 40.2$ ft or say 41 ft.

Example 15.18

Refer to the problem in Example 15.17. Determine directly the ultimate and the allowable loads using N_{cor}. All the other data remain the same.

Solution

Use Eq. (15.49a)

$$Q_u = Q_b + Q_f = 0.8N_{cor}(L/d)A_b + 0.04\overline{N}_{cor}A_s \text{(kips)}$$

Given: $N_{cor} = 21$, $\overline{N}_{cor} = 15$, $d = 16$ in, $L = 41$ ft

$$A_b = \frac{3.14}{4}\left(\frac{16}{12}\right)^2 = 1.39\text{ft}^2;\ A_s = 3.14 \times \left(\frac{16}{12}\right) \times 41 = 171.65\ \text{ft}^2$$

Substituting

$$Q_b = 0.80 \times 21\left[\frac{41}{1.33}\right] \times 1.39 = 720 \le 8N_{cor}A_b \ \text{kips}$$

$$Q_{bl} = 8 \times 21 \times 1.39 = 234 \ \text{kips}$$

The limiting value of Q_b = 234 kips

$$Q_f = 0.04N_{cor}A_s = 0.04 \times 15 \times 171.65 = 103 \ \text{kips}$$

$$Q_u = Q_b + Q_f = 234 + 103 = 337 \ \text{kips}$$

$$Q_a = \frac{337}{4} = 84 \ \text{kips}$$

15.21 BEARING CAPACITY OF PILES BASED ON STATIC CONE PENETRATION TESTS (CPT)

Methods of Determining Pile Capacity

The cone penetration test may be considered as a small scale pile load test. As such the results of this test yield the necessary parameters for the design of piles subjected to vertical load. The types of static cone penetrometers and the methods of conducting the tests have been discussed in detail in Chapter 9. Various methods for using CPT results to predict vertical pile capacity have been proposed. The following methods will be discussed:

1. Vander Veen's method.
2. Schmertmann's method.

Vander Veen's Method for Piles in Cohesionless Soils

In the Vander Veen et al., (1957) method, the ultimate end-bearing resistance of a pile is taken, equal to the point resistance of the cone. To allow for the variation of cone resistance which normally occurs, the method considers average cone resistance over a depth equal to three times the diameter of the pile above the pile point level and one pile diameter below point level as shown in Fig. 15.17(a). Experience has shown that if a safety factor of 2.5 is applied to the ultimate end resistance as determined from cone resistance, the pile is unlikely to settle more than 15 mm under the working load (Tomlinson, 1986). The equations for ultimate bearing capacity and allowable load may be written as,

pile base resistance, $q_b = q_p$ (cone) (15.51a)

ultimate base capacity, $Q_b = A_b q_p$ (15.51b)

allowable base load, $Q_a = \frac{A_b q_p}{F_s}$ (15.51c)

where, q_p = average cone resistance over a depth $4d$ as shown in Fig. 15.17(a) and F_s = factor of safety.

The skin friction on the pile shaft in cohesionless soils is obtained from the relationships established by Meyerhof (1956) as follows.

For displacement piles, the ultimate skin friction, f_s, is given by

$$f_s = \frac{\bar{q}_c}{2} \text{(kPa)} \tag{15.52a}$$

and for H-section piles, the ultimate limiting skin friction is given by

$$f_s = \frac{\bar{q}_c}{4} \text{(kPa)} \tag{15.52b}$$

where $\bar{q}_c$ = average cone resistance in kg/cm^2 over the length of the pile shaft under consideration.

Meyerhof states that for straight sided displacement piles, the ultimate unit skin friction, f_s, has a maximum value of 107 kPa and for *H*-sections, a maximum of 54 kPa (calculated on all faces of flanges and web). The ultimate skin load is

$$Q_f = A_s f_s \tag{15.53a}$$

The ultimate load capacity of a pile is

$$Q_u = Q_b + Q_f \tag{15.53b}$$

The allowable load is

$$Q_a = \frac{Q_b + Q_f}{2.5} \tag{15.53c}$$

If the working load, Q_a, obtained for a particular position of pile in Fig. 15.17(a), is less than that required for the structural designer's loading conditions, then the pile must be taken to a greater depth to increase the skin friction f_s or the base resistance q_b.

Schmertmann's Method for Cohesionless and Cohesive Soils

Schmertmann (1978) recommends one procedure for all types of soil for computing the point bearing capacity of piles. However, for computing side friction, Schmertmann gives two different approaches, one for sand and one for clay soils.

Point Bearing Capacity Q_b in All Types of Soil

The method suggested by Schmertmann (1978) is similar to the procedures developed by De Ruiter and Beringen (1979) for sand. The principle of this method is based on the one suggested by Vander Veen (1957) and explained earlier. The procedure used in this case involves determining a representative cone point penetration value, q_p, within a depth between 0.7 to 4*d* below the tip level of the pile and 8*d* above the tip level as shown in Fig. 15.17(b) and (c). The value of q_p may be expressed as

$$q_p = \frac{(q_{c1} + q_{c2})/2 + q_{c3}}{2} \tag{15.54}$$

where q_{c1} = average cone resistance below the tip of the pile over a depth which may vary between 0.7*d* and 4*d*, where *d* = diameter of pile,

q_{c2} = minimum cone resistance recorded below the pile tip over the same depth 0.7*d* to 4*d*,

q_{c3} = average of the envelope of minimum cone resistance recorded above the pile tip to a height of $8d$.

Now, the unit point resistance of the pile, q_b, is

$$q_b \text{ (pile)} = q_p \text{ (cone)} \tag{15.55a}$$

The ultimate base resistance, Q_b, of a pile is

$$Q_b = A_b q_p \tag{15.55b}$$

The allowable base load, Q_a is

$$Q_a = \frac{A_b q_p}{F_s} \tag{15.55c}$$

Method of Computing the Average Cone Point Resistance q_p

The method of computing q_{c1}, q_{c2} and q_{c3} with respect to a typical q_c-plot shown in Fig. 15.17(b) and (c) is explained below.

Case 1: When the cone point resistance q_c below the tip of a pile is lower than that at the tip (Fig. 15.17(b)) within depth $4d$.

$$q_{c1} = \frac{d_3(q_o + q_b)/2 + d_2(q_b + q_c)/2 + d_1(q_d + q_c)/2}{4d} \tag{15.56a}$$

where q_o, q_b etc., refer to the points o, b etc, on the q_c-profile, $q_{c2} = q_c$ = minimum value below tip within a depth of $4d$ at point c on the q_c-profile.

The envelope of minimum cone resistance above the pile tip is as shown by the arrow mark along (15.17b) *aefghk*.

$$q_{c3} = \frac{d_4 q_e + d_5(q_e + q_f)/2 + d_6 q_f + d_7(q_g + q_h)/2 + d_8 q_h}{8d} \tag{15.56b}$$

where $q_a = q_e$, $q_f = q_g$, $q_h = q_k$.

Case 2: When the cone resistance q_c below the pile tip is greater than that at the tip within a depth $4d$. (Fig. 15.17(c)).

In this case q_p is found within a total depth of $0.7d$ as shown in Fig. 15.17(c).

$$q_{c1} = \frac{q_o + q_b}{2}$$

$q_{c2} = q_o$ = minimum value at the pile tip itself, q_{c3} = average of the minimum values along the envelope *ocde* as before.

In determining the average q_c above, the minimum values q_{c2} selected under Case 1 or 2 are to be disregarded.

Effect of Overconsolidation Ratio in Sand

Reduction factors have been developed that should be applied to the theoretical end bearing of a pile as determined from the CPT if the bearing layer consists of overconsolidated sand. The problem in many cases will be to make a reasonable estimate of the overconsolidation ratio in sand. In sands with a high q_c, some conservation in this respect is desirable, in particular for shallow foundations. The influence of overconsolidation on pile end bearing is one of the reasons for applying a limiting value to pile end bearing, irrespective of the cone resistances recorded in the

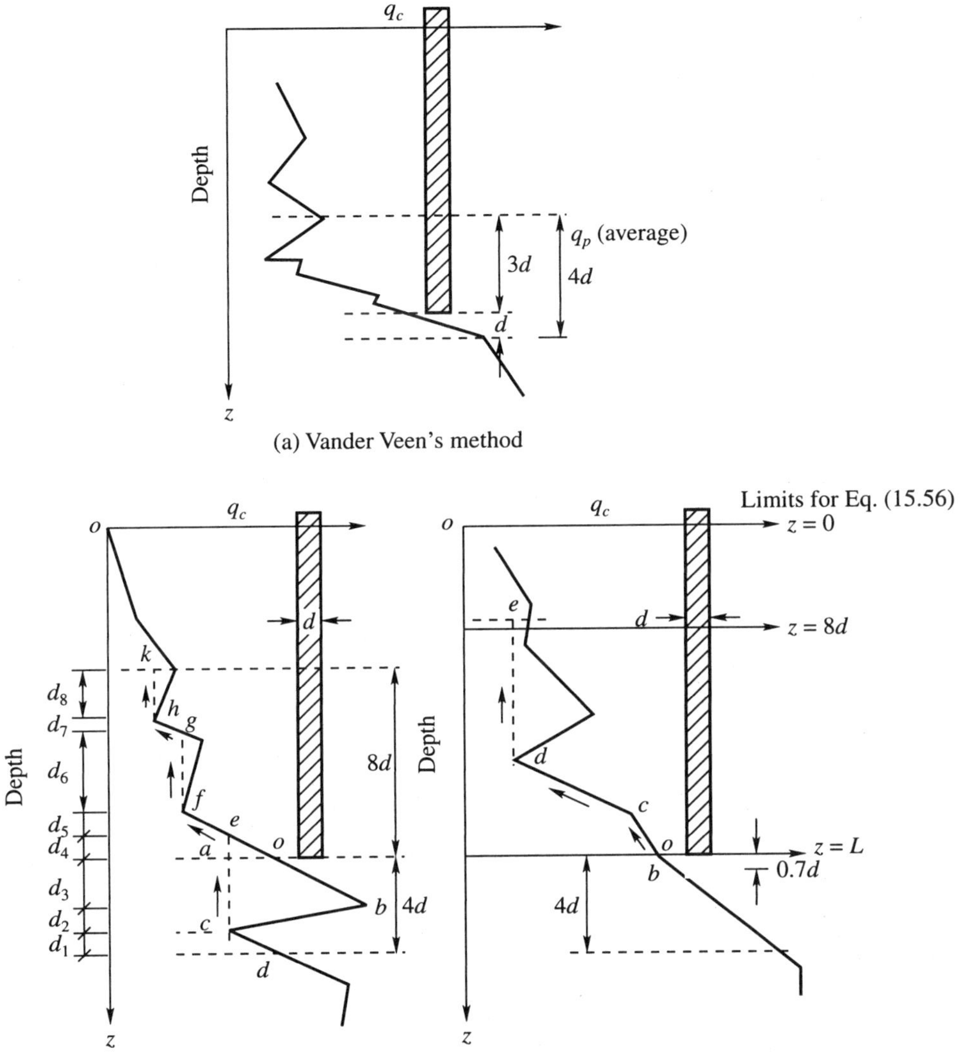

(a) Vander Veen's method

(b) Resistance below pile tip lower than that at pile tip within depth 4d

(c) Resistance below pile tip greater than that at pile tip within 0.7d depth

Figure 15.17 Pile capacity by use of CPT values (a) Vander Veen's method, and (b,c) Schmertmann's method

bearing layer. A limit pile end bearing of 15 MN/m^2 is generally accepted (De Ruiter and Beringen, 1979), although in dense sands cone resistance may be greater than 50 MN/m^2. It is unlikely that in dense normally consolidated sand ultimate end bearing values higher than 15 MN/m^2 can occur but this has not been adequately confirmed by load tests.

Design CPT Values for Sand and Clay

The application of CPT in evaluating the design values for skin friction and bearing as recommended by De Ruiter and Beringen (1979) is summarized in Table. 15.3.

Table 15.3 Application of CPT in Pile Design (After De Ruiter and Beringen, 1979)

Item	Sand	Clay	Legend
Unit Friction f_s	Minimum of $f_1 = 012$ MPa f_2 = CPT sleeve friction $f_3 = q_c/300$ (compression) or $f_3 = q_c/400$ (tension)	$f_s = \alpha' c_u$, where $\alpha' = 1$ in N.C. clay $= 0.5$ in O.C. clay	q_c = cone resistance below pile tip $c_u = q_c/N_k$
Unit end bearing q_p	minimum of q_p from Fig. 15.17(b) and c	$q_p = N_c c_u$ $N_c = 9$	q_p = ultimate resistance of pile

Ultimate Skin Load Q_f in Cohesionless Soils

For the computation of skin load, Q_f, Schmertmann (1978) presents the following equation

$$Q_f = K\left[\sum_{z=0}^{z=8d} \frac{z}{8d} f_c A_s + \sum_{z=8d}^{z=L} f_c A_s\right] \tag{15.56c}$$

where $K = f_s/f_c$ = correction factor for f_c
f_s = unit pile friction
f_c = unit sleeve friction measured by the friction jacket
z = depth to f_c value considered from ground surface
d = pile diameter or width
A_s = pile-soil contact area per f_c depth interval
L = embedded depth of pile.

When f_c does not vary significantly with depth, Eq. (15.56c) can be written in a simplified form as

$$Q_f = K\left[\frac{1}{2}(\bar{f}_c A_s)_{o-8d} + (\bar{f}_c A_s)_{8d-L}\right] \tag{15.56d}$$

where $\bar{f}_c$ is the average value within the depths specified. The correction factor K is given in Fig. 15.18(a).

Ultimate Skin Load Q_f for Piles in Clay Soil

For piles in clay Schmertmann gives the expression

$$Q_f = \alpha' \bar{f}_c A_s \tag{15.57}$$

where, α' = ratio of pile to penetrometer sleeve friction,
f_c = average sleeve friction,
A_s = pile to soil contact area.

Fig. 15.18(b) gives values of α'.

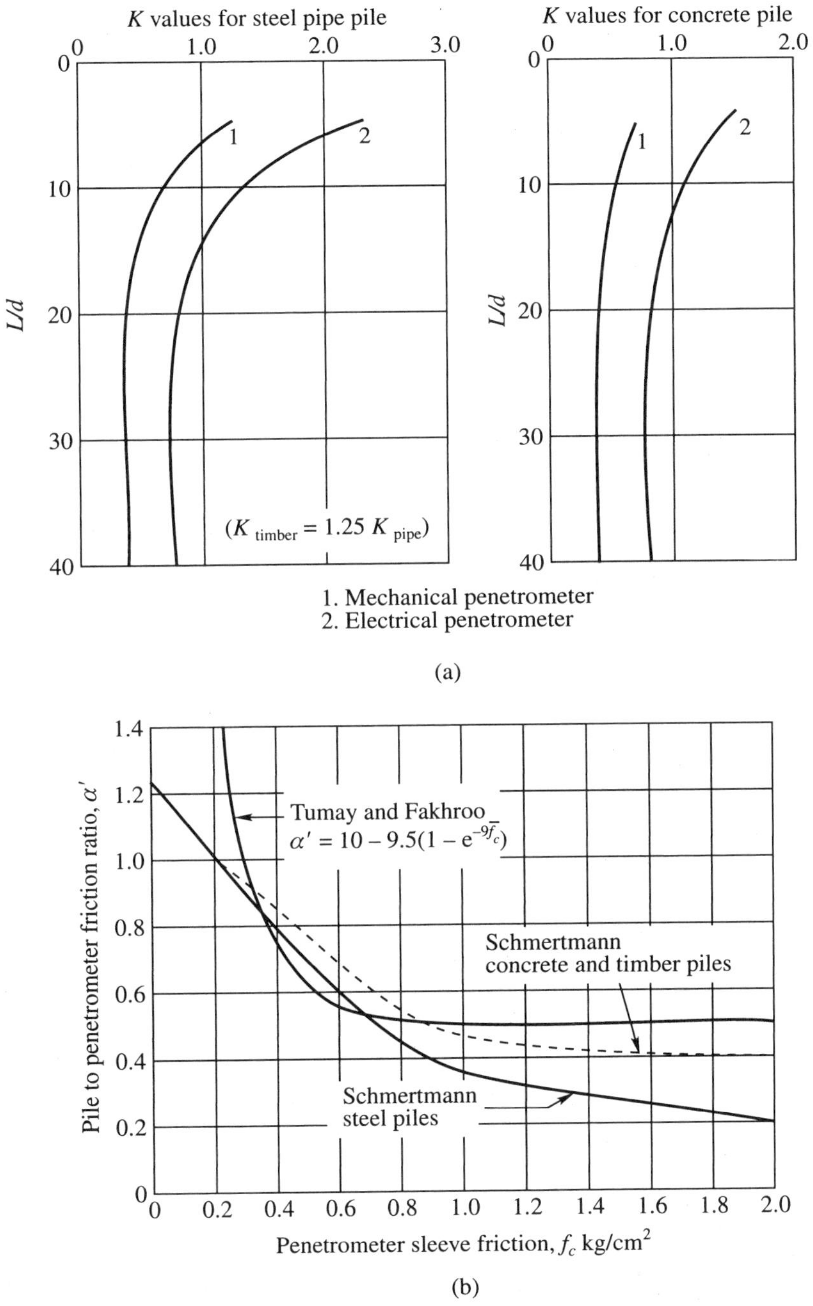

Figure 15.18 Penetrometer design curves (a) for pile side friction in sand (Schmertmann, 1978), and (b) for pile side friction in clay (Schmertmann, 1978)

Example 15.19

A concrete pile of 40 cm diameter is driven into a homogeneous mass of cohesionless soil. The pile carries a safe load of 650 kN. A static cone penetration test conducted at the site indicates an average value of $q_c = 40$ kg/cm^2 along the pile and 120 kg/cm^2 below the pile tip. Compute the length of the pile with $F_s = 2.5$. (Fig. Ex. 15.19)

Solution

From Eq. (15.51a)

q_b (pile) = q_p (cone)

Given, q_p = 120 kg/cm^2, therefore,

q_b = 120 kg/cm^2 = 120 × 100 = 12000 kN/m^2

Per Section 15.12, q_b is restricted to 11,000 kN/m^2.
Therefore,

$$Q_b = A_b q_b = \frac{3.14}{4} \times 0.4^2 \times 11000 = 1382 \text{ kN}$$

Assume the length of the pile = L m

The average, $\overline{q}_c = 40 \text{ kg/cm}^2$

Per Eq. (15.52a),

$$f_s = \frac{\overline{q}_c}{2} \text{ kN/m}^2 = \frac{40}{2} = 20 \text{ kN/m}^2$$

Now, $Q_f = f_s A_s = 20 \times 3.14 \times 0.4 \times L = 25.12L$ kN

Given Q_a = 650 kN. With F_s = 2.5, Q_u = 650 × 2.5 = 1625 kN.

Now, $1625 = Q_b + Q_f = (1382 + 25.12L)$ kN

or $L = \dfrac{1625 - 1382}{25.12} = 9.67$ m or say 10 m

The pile has to be driven to a depth of 10 m to carry a safe load of 650 kN with F_s = 2.5.

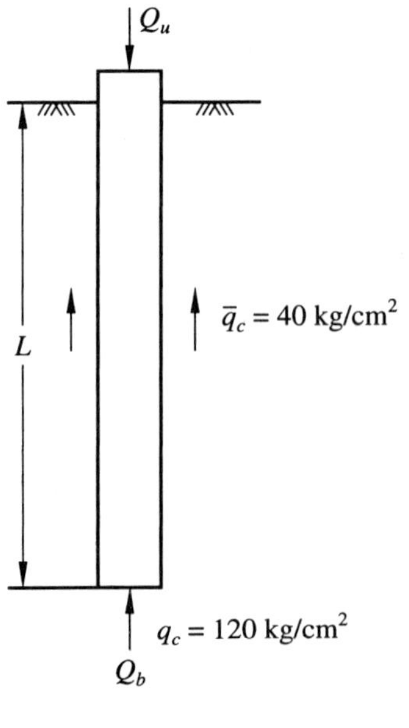

Figure Ex. 15.19

Example 15.20

A concrete pile of size 0.4 × 0.4 m is driven to a depth of 12 m into medium dense sand. The water table is close to the ground surface. Static cone penetration tests were carried out at this site by using an electric cone penetrometer. The values of q_c and f_c as obtained from the test have been plotted against depth and shown in Fig. Ex. 15.20. Determine the safe load on this pile with $F_s = 2.5$ by Schmertmann's method (Section 15.21).

Solution

First determine the representative cone penetration value q_p by using Eq. (15.54)

$$q_p = \frac{(q_{c1} + q_{c2}) + q_{c3}}{2}$$

Now from Fig. Ex. 15.20 and Eq (15.56a)

$$q_{c1} = \frac{d_3(q_o + q_b)/2 + d_2(q_b + q_d)/2 + d_1(q_d + q_c)/2}{4d}$$

$$= \frac{0.7(76+85)/2 + 0.3(85+71)/2 + 0.6(71+80)/2}{4 \times 0.4}$$

$$= 78 \text{ kg/cm}^2$$

$q_{c2} = q_d =$ the minimum value below the tip of pile within $4d$ depth = 71 kg/cm².

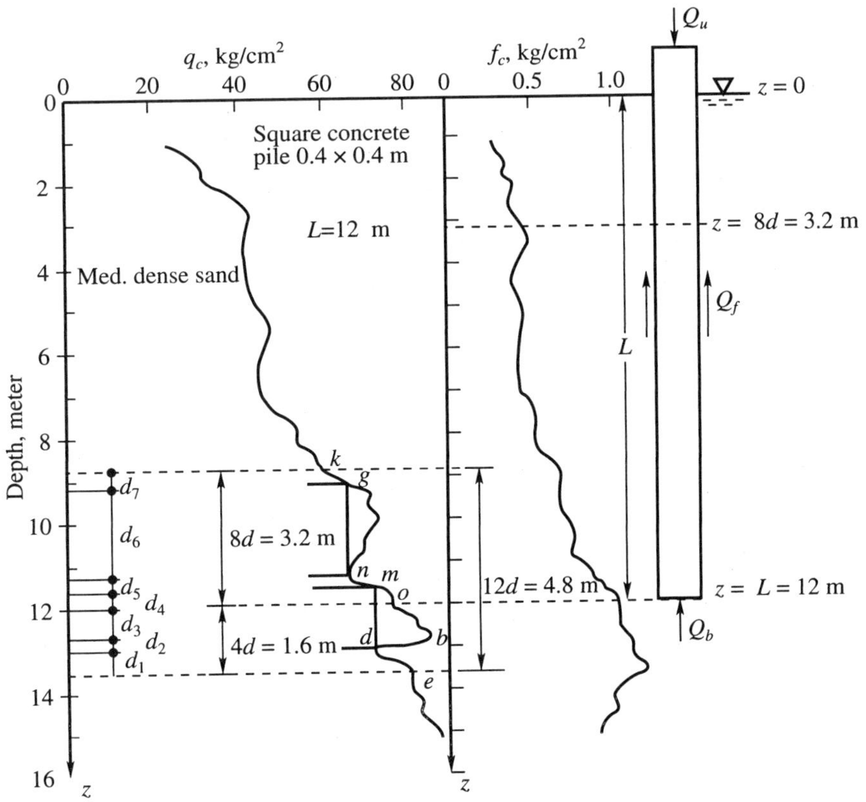

Figure Ex. 15.20

From Eq. (15.56b)

$$q_{c3} = \frac{d_4 q_m + d_5(q_m + q_n)/2 + d_6 q_n + d_7(q_g + q_k)/2}{8d}$$

$$= \frac{0.4 \times 71 + 0.3(71 + 65)/2 + 2.1 \times 65 + 0.4(65 + 60)/2}{8 \times 0.4}$$

$= 66$ kg/cm^2 = 660 t/m^2 (metric)

From Eq. (15.54)

$$q_p = \frac{(78 + 71)/2 + 66}{2} = 70 \text{ kg/cm}^2 \approx 700 \text{ t/m}^2 \text{ (metric)}$$

Ultimate Base Load

$$Q_b = q_b A_b = q_p A_b = 700 \times 0.4^2 = 112 \text{ t (metric)}$$

Frictional Load Q_f

From Eq. (15.56d)

$$Q_f = K\left[\frac{1}{2}(\bar{f}_c A_s)_{0-8d} + (\bar{f}_c A_s)_{8d-L}\right]$$

where K = correction factor from Fig. (15.18a) for electrical penetrometer.

For $\frac{L}{d} = \frac{12}{0.4} = 30$, $K = 0.75$ for concrete pile. It is now necessary to determine the average sleeve friction $\bar{f}_c$ between depths $z = 0$ and $z = 8d$, and $z = 8d$ and $z = L$ from the top of pile from f_c profile given in Fig. Ex. 15.20.

$$Q_f = 0.75\left[\tfrac{1}{2} \times 0.34 \times 10 \times 4 \times 0.4 \times 3.2 + 0.71 \times 10 \times 4 \times 0.4 \times 8.8\right]$$

$= 0.75$ [8.7 + 99.97] = 81.5 t (metric)

$Q_u = Q_b + Q_f = 112 + 81.5 = 193.5$ t

$$Q_a = \frac{Q_b + Q_f}{2.5} = \frac{193.5}{2.5} = 77.4 \text{ t (metric)} = 759 \text{ kN}$$

Example 15.21

A concrete pile of section 0.4 × 0.4 m is driven into normally consolidated clay to a depth of 10 m. The water table is at ground level. A static cone penetration test (CPT) was conducted at the site with an electric cone penetrometer. Fig. Ex. 15.21 gives a profile of q_c and f_c values with respect to depth. Determine safe loads on the pile by the following methods:

(a) α-method (b) λ-method, given: $\gamma_b = 8.5$ kN/m^3 and (c) Schmertmann's method. Use a factor of safety of 2.5.

Solution

(a) α-method

The α-method requires the undrained shear strength of the soil. Since this is not given, it has to be determined by using the relation between q_c and c_u given in Eq. (9.14).

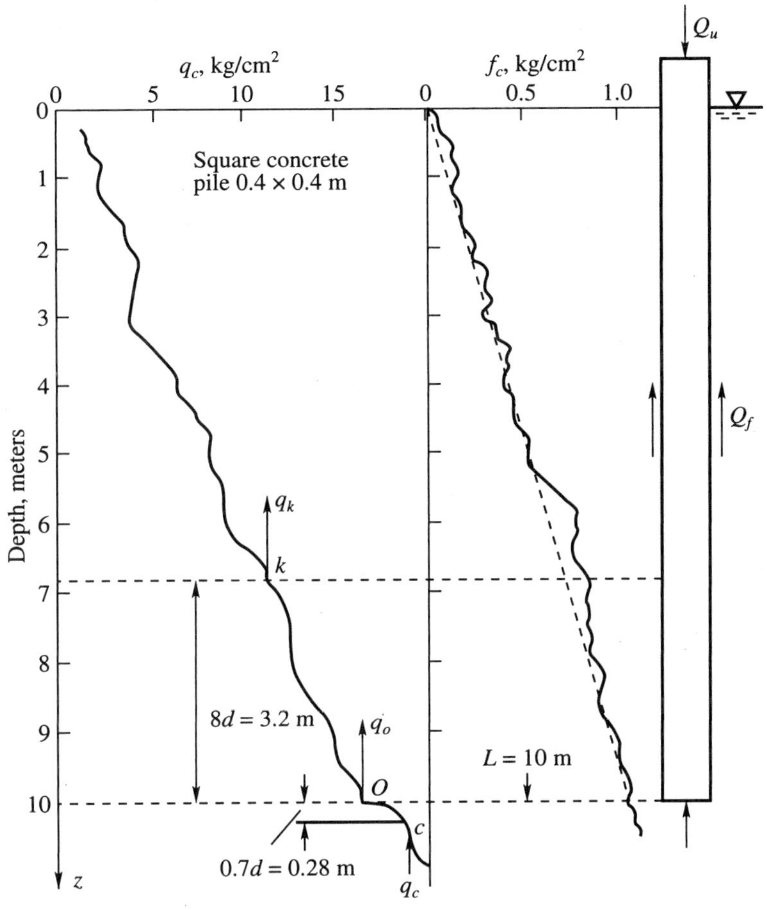

Figure Ex. 15.21

$$c_u = \frac{q_c}{N_k} \text{ by neglecting the overburden effect,}$$

where N_k = cone factor = 20.

It is necessary to determine the average $\bar{c}_u$ along the pile shaft and $\bar{c}_b$ at the base level of the pile. For this purpose find the corresponding f_c (sleeve friction) values from Fig. Ex. 15.21.

$$\text{Average } \bar{q}_c \text{ along the shaft, } \bar{q}_c = \frac{1+16}{2} = 8.5 \text{ kg/cm}^2.$$

Average of q_c within a depth $3d$ above the base and d below the base of the pile (Refer to Fig. 15.17a)

$$q_p = \frac{15+18.5}{2} = 17 \text{ kg/cm}^2$$

$$\bar{c}_u = \frac{8.5}{20} = 0.43 \text{ kg/cm}^2 \approx 43 \text{ kN/m}^2$$

$$c_b = \frac{17}{20} = 0.85 \text{ kg/cm}^2 \approx 85 \text{ kN/m}^2.$$

Ultimate Base Load, Q_b

From Eq. (15.39)

$$Q_b = 9c_bA_b = 9 \times 85 \times 0.40^2 = 122 \text{ kN}.$$

Ultimate Friction Load, Q_f

From Eq. (15.37)

$$Q_f = \alpha \bar{c}_u A_s$$

From Fig. 15.15 for $\bar{c}_u = 43$ kN/m^2, $\alpha = 70$

$Q_f = 0.70 \times 43 \times 10 \times 4 \times 0.4 = 481.6$ kN or say 482 kN

$Q_u = 122 + 482 = 604$ kN

$$Q_a = \frac{604}{2.5} = 241.6 \text{ kN or say 242 kN}$$

(b) λ-method

Base Load Q_b

In this method the base load is the same as in (a) above. That is

$Q_b = 122$ kN

Friction Load

From Fig. 15.17 $\lambda = 0.25$ for $L = 10$ m (= 32.4 ft). From Eq. 15.40

$$f_s = \lambda(\bar{q}_o + 2\bar{c}_u)$$

$$\bar{q}_o = \frac{1}{2} \times 10 \times 8.5 = 42.5 \text{ kN/m}^2$$

$$f_s = 0.25(42.5 + 2 \times 43) = 32 \text{ kN/m}^2$$

$$Q_f = f_s A_s = 32 \times 10 \times 4 \times 0.4 = 512 \text{ kN}$$

$$Q_u = 122 + 512 = 634 \text{ kN}$$

$$Q_a = \frac{634}{2.5} = 254 \text{ kN}.$$

(c) Schmertmann's Method

Base load Q_b

Use Eq. (15.54) for determining the representative value for q_p. Here, the minimum value for q_c is at point O on the q_c-profile in Fig. Ex. 15.21 which is the base level of the pile. Now q_{c1} is the average q_c at the base and 0.7d below the base of the pile, that is,

$$q_{c1} = \frac{q_o + q_e}{2} = \frac{16 + 18.5}{2} = 17.25 \text{ kg/cm}^2$$

$$q_{c2} = q_o = 16 \text{ kg/cm}^2$$

q_{c3} = The average of q_c within a depth 8d above the base level

$$= \frac{q_o + q_k}{2} = \frac{16+11}{2} = 13.5 \text{ kg/cm}^2$$

From Eq. (15.45), $q_p = \dfrac{(17.25+16)/2+13.5}{2} = 15 \text{ kg/cm}^2$.

From Eq. (15.55a)

$q_b(\text{pile}) = q_p\,(\text{cone}) = 15 \text{ kg/cm}^2 \approx 1500 \text{ kN/m}^2$

$Q_b = q_b A_b = 1500 \times (0.4)^2 = 240 \text{ kN}$

Friction Load Q_f

Use Eq. (15.57)

$Q_f = \alpha' \bar{f}_c A_s$

where α' = ratio of pile to penetrometer sleeve friction.

From Fig. Ex. 15.21 $\bar{f}_c = \dfrac{0+1.15}{2} = 0.58 \text{ kg/cm}^2 \approx 58 \text{ kN/m}^2$.

From Fig. 16.18b for $f_c = 58 \text{ kN/m}^2$, $\alpha' = 0.70$

$Q_f = 0.70 \times 58 \times 10 \times 4 \times 0.4 = 650 \text{ kN}$

$Q_u = 240 + 650 = 890 \text{ kN}$

$Q_a = \dfrac{890}{2.5} = 356 \text{ kN}$

Note: The values given in the examples are only illustrative and not factual.

15.22 BEARING CAPACITY OF A SINGLE PILE BY LOAD TEST

A pile load test is the most acceptable method to determine the load carrying capacity of a pile. The load test may be carried out either on a driven pile or a *cast-in-situ* pile. Load tests may be made either on a single pile or a group of piles. Load tests on a pile group are very costly and may be undertaken only in very important projects.

Pile load tests on a single pile or a group of piles are conducted for the determination of

1. Vertical load bearing capacity,
2. Uplift load capacity,
3. Lateral load capacity.

Generally load tests are made to determine the bearing capacity and to establish the load-settlement relationship under a compressive load. The other two types of tests may be carried out only when piles are required to resist large uplift or lateral forces.

Usually pile foundations are designed with an estimated capacity which is determined from a thorough study of the site conditions. At the beginning of construction, load tests are made for the purpose of verifying the adequacy of the design capacity. If the test results show an inadequate factor of safety or excessive settlement, the design must be revised before construction is under way.

Load tests may be carried out either on

1. A working pile or
2. A test pile.

A *working pile* is a pile driven or *cast-in-situ* along with the other piles to carry the loads from the superstructure. The maximum test load on such piles should not exceed one and a half times the design load.

A *test pile* is a pile which does not carry the loads coming from the structure. The maximum load that can be put on such piles may be about 2½ times the design load or the load imposed must be such as to give a total settlement not less than one-tenth the pile diameter.

Method of Carrying Out Vertical Pile Load Test

A vertical pile load test assembly is shown in Fig. 15.19(a). It consists of

1. An arrangement to take the reaction of the load applied on the pile head,
2. A hydraulic jack of sufficient capacity to apply load on the pile head, and
3. A set of three dial gauges to measure settlement of the pile head.

Load Application

A load test may be of two types:

1. Continuous load test.
2. Cyclic load test.

In the case of a continuous load test, continuous increments of load are applied to the pile head. Settlement of the pile head is recorded at each load level.

In the case of the cyclic load test, the load is raised to a particular level, then reduced to zero, again raised to a higher level and reduced to zero. Settlements are recorded at each increment or decrement of load. Cyclic load tests help to separate frictional load from point load.

The total elastic recovery or settlement S_e, is due to

1. The total plastic recovery of the pile material,
2. Elastic recovery of the soil at the tip of the pile, $\overline{S}_e$

The total settlement S due to any load can be separated into elastic and plastic settlements by carrying out cyclic load tests as shown in Fig. 15.19(b).

A pile loaded to Q_1 gives a total settlement S_1. When this load is reduced to zero, there is an elastic recovery which is equal to S_{e1}. This elastic recovery is due to the elastic compression of the pile material and the soil. The net settlement or plastic compression is S_{p1}. The pile is loaded again from zero to the next higher load Q_2 and reduced to zero thereafter. The corresponding settlements may be found as before. The method of loading and unloading may be repeated as before.

Allowable Load from Single Pile Load Test Data

There are many methods by which allowable loads on a single pile may be determined by making use of load test data. If the ultimate load can be determined from load-settlement curves, allowable loads are found by dividing the ultimate load by a suitable factor of safety which varies from 2 to 3. A factor of safety of 2.5 is normally recommended. A few of the methods that are useful for the determination of ultimate or allowable loads on a single pile are given below:

1. The ultimate load, Q_u, can be determined as the abscissa of the point where the curved part of the load-settlement curve changes to a falling straight line, Fig. 15.20(a).
2. Q_u is the abscissa of the point of intersection of the initial and final tangents of the load-settlement curve, Fig. 15.20(b).
3. The allowable load Q_a is 50 percent of the ultimate load at which the total settlement amounts to one-tenth of the diameter of the pile for uniform diameter piles.

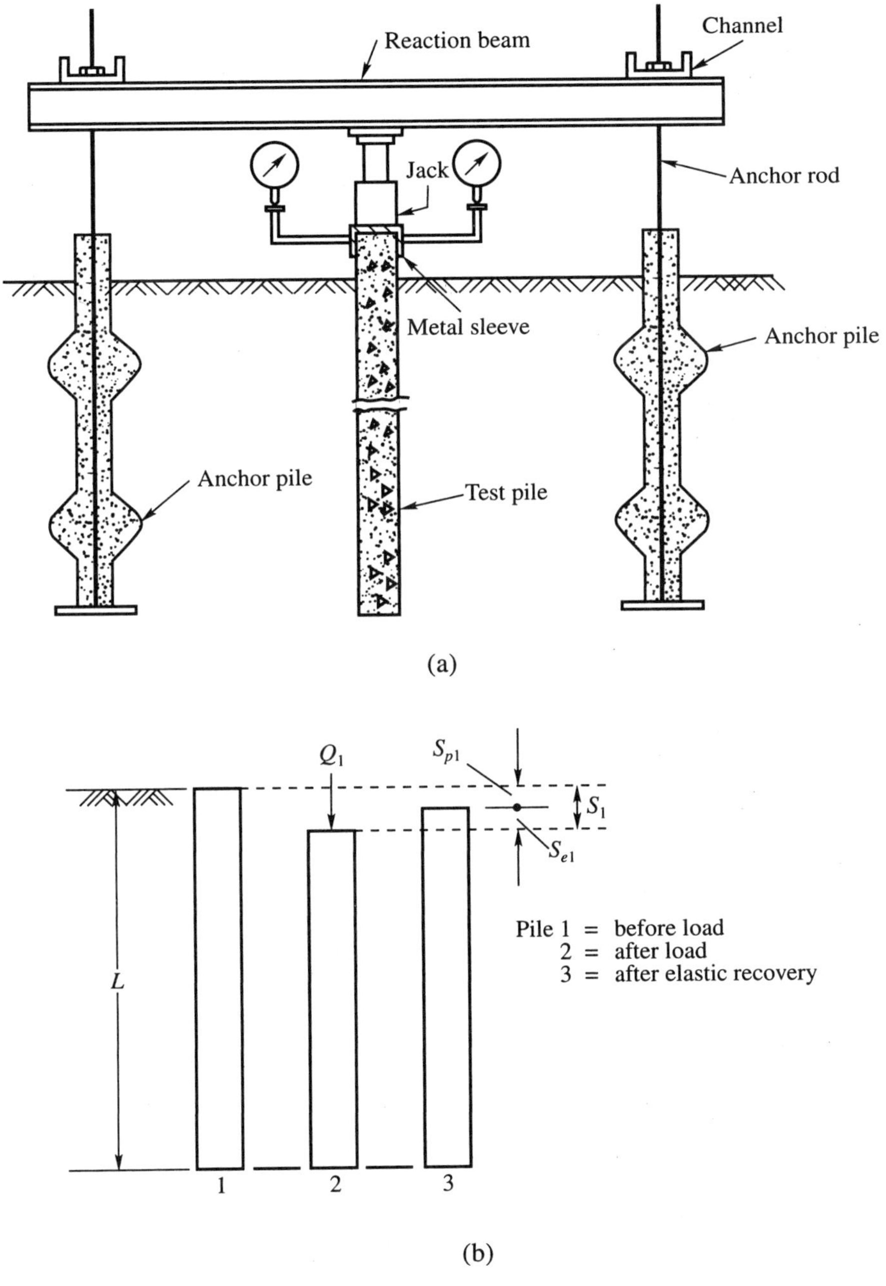

Figure 15.19 (a) Vertical pile load test assembly, and (b) elastic compression at the base of the pile

4. The allowable load Q_a is sometimes taken as equal to two-thirds of the load which causes a total settlement of 12 mm.
5. The allowable load Q_a is sometimes taken as equal to two-thirds of the load which causes a net (plastic) settlement of 6 mm.

If pile groups are loaded to failure, the ultimate load of the group, Q_{gu}, may be found by any one of the first two methods mentioned above for single piles. However, if the groups are subjected to only

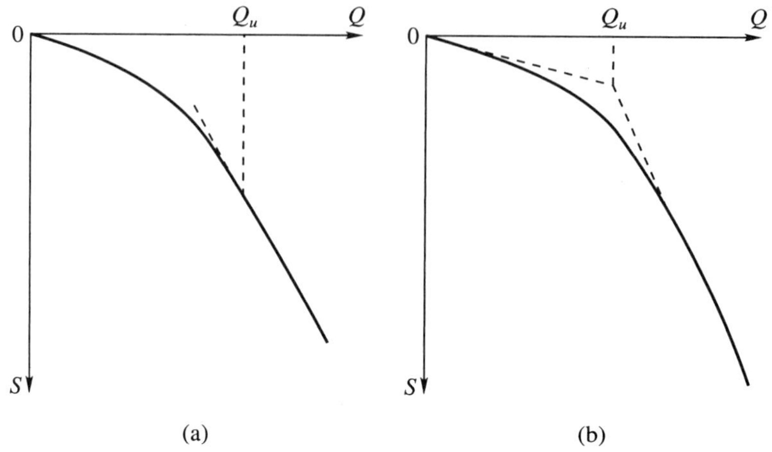

Figure 15.20 Determination of ultimate load from load-settlement curves

one and a half-times the design load of the group, the allowable load on the group cannot be found on the basis of 12 or 6 mm settlement criteria applicable to single piles. In the case of a group with piles spaced at less than 6 to 8 times the pile diameter, the stress interaction of the adjacent piles affects the settlement considerably. The settlement criteria applicable to pile groups should be the same as that applicable to shallow foundations at design loads.

15.23 PILE BEARING CAPACITY FROM DYNAMIC PILE DRIVING FORMULAS

The resistance offered by a soil to penetration of a pile during driving gives an indication of its bearing capacity. Qualitatively speaking, a pile which meets greater resistance during driving is capable of carrying a greater load. A number of dynamic formulae have been developed which equate pile capacity in terms of driving energy.

The basis of all these formulae is the simple energy relationship which may be stated by the following equation. (Fig. 15.21).

$$Wh = Q_u s$$

or

$$Q_u = \frac{Wh}{s} \tag{15.58}$$

where W = weight of the driving hammer

h = height of fall of hammer

Wh = energy of hammer blow

Q_u = ultimate resistance to penetration

s = pile penetration under one hammer blow

$Q_u s$ = resisting energy of the pile

Hiley Formula

Equation (15.58) holds only if the system is 100 percent efficient. Since the driving of a pile involves many losses, the energy of the system may be written as

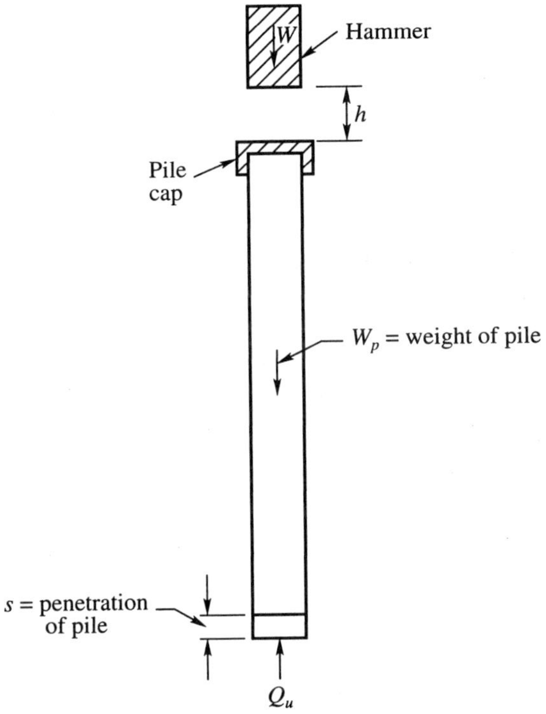

Figure 15.21 Basic energy relationship

Energy input = Energy used + Energy losses

or Energy used = Energy input – Energy losses.

The expressions for the various energy terms used are

1. Energy used = $Q_u s$,
2. Energy input = $\eta_h W h$, where η_h is the efficiency of the hammer.
3. The energy losses are due to the following:
   (i) The energy loss E_1 due to the elastic compressions of the pile cap, pile material and the soil surrounding the pile. The expression for E_1 may be written as

$$E_1 = \frac{1}{2} Q_u (c_1 + c_2 + c_3) = Q_u C$$

where c_1 = elastic compression of the pile cap
c_2 = elastic compression of the pile
c_3 = elastic compression of the soil.

   (ii) The energy loss E_2 due to the interaction of the pile hammer system (impact of two bodies). The expression for E_2 may be written as

$$E_2 = WhW_p \frac{1 - C_r^2}{W + W_p}$$

where W_p = weight of pile
C_r = coefficient of restitution.

Substituting the various expressions in the energy equation and simplifying, we have

$$Q_u = \frac{\eta_h Wh}{s+C} \times \frac{1+C_r^2 R}{1+R} \tag{15.59}$$

where $R = \dfrac{W_p}{W}$

Equation (15.59) is called the Hiley formula. The allowable load Q_u may be obtained by dividing Q_u by a suitable factor of safety.

If the pile tip rests on rock or relatively impenetrable material, Eq. (15.59) is not valid. Chellis (1961) suggests for this condition that the use of $W_p/2$ instead of W_p may be more correct. The various coefficients used in the Eq. (15.59) are as given below:

1. *Elastic compression c_1 of cap and pile head*

Pile Material	Range of Driving Stress kg/cm^2	Range of c_1
Precast concrete pile with packing inside cap	30-150	0.12-0-50
Timber pile without cap	30-150	0.05-0.20
Steel *H*-pile	30-150	0.04-0.16

2. *Elastic compression c_2 of pile.*

This may be computed using the equation

$$c_2 = \frac{Q_u L}{AE}$$

where L = embedded length of the pile,
A = average cross-sectional area of the pile,
E = Young's modulus.

3. *Elastic compression c_3 of soil.*

The average value of c_3 may be taken as 0.1 (the value ranges from 0.0 for hard soil to 0.2 for resilient soils).

4. *Pile-hammer efficiency*

Hammer Type	η_h
Drop	1.00
Single acting	0.75–0.85
Double acting	0.85
Diesel	1.00

5. *Coefficient of restitution C_r*

Material	C_r
Wood pile	0.25
Compact wood cushion on steel pile	0.32
Cast iron hammer on concrete pile without cap	0.40
Cast iron hammer on steel pipe without cushion	0.55

Engineering News Record (ENR) Formula

The general form of the Engineering News Record Formula for the allowable load Q_a may be obtained from Eq. (15.59) by putting

$\eta_h = 1$ and $C_r = 1$ and a factor of safety equal to 6. The formula proposed by A.M. Wellington, editor of the Engineering News, in 1886, is

$$Q_a = \frac{Wh}{6(s+C)} \tag{15.60}$$

where Q_a = allowable load in kg,
W = weight of hammer in kg,
h = height of fall of hammer in cm,
s = final penetration in cm per blow (which is termed as set). The set is taken as the average penetration per blow for the last 5 blows of a drop hammer or 20 blows of a steam hammer,
C = empirical constant,
= 2.5 cm for a drop hammer,
= 0.25 cm for single and double acting hammers.

The equations for the various types of hammers may be written as:

1. Drop hammer

$$Q_a = \frac{Wh}{6(s+2.5)} \tag{15.61}$$

2. Single-acting hammer

$$Q_a = \frac{Wh}{6(s+0.25)} \tag{15.62}$$

3. Double-acting hammer

$$Q_a = \frac{(W+ap)}{6(s+0.25)} \tag{15.63}$$

a = effective area of the piston in sq. cm,
p = mean effective steam pressure in kg/cm^2.

Comments on the Use of Dynamic Formulae

1. Detailed investigations carried out by Vesic (1967) on deep foundations in granular soils indicate that the Engineering News Record Formula applicable to drop hammers, Eq. (15.61), gives pile loads as low as 44 % of the actual loads. In order to obtain better agreement between the one computed and observed loads, Vesic suggests the following values for the coefficient C in Eq. (15.60).
   For steel pipe piles, C = 1 cm.
   For precast concrete piles C = 1.5 cm.
2. The tests carried out by Vesic in granular soils indicate that Hiley's formula does not give consistent results. The values computed from Eq. (15.59) are sometimes higher and sometimes lower than the observed values.

3. Dynamic formulae in general have limited value in pile foundation work mainly because the dynamic resistance of soil does not represent the static resistance, and because often the results obtained from the use of dynamic equations are of questionable dependability. However, engineers prefer to use the Engineering News Record Formula because of its simplicity.
4. Dynamic formulae could be used with more confidence in freely draining materials such as coarse sand. If the pile is driven to saturated loose fine sand and silt, there is every possibility of development of liquefaction which reduces the bearing capacity of the pile.
5. Dynamic formulae are not recommended for computing allowable loads of piles driven into cohesive soils. In cohesive soils, the resistance to driving increases through the sudden increase in stress in pore water and decreases because of the decreased value of the internal friction between soil and pile because of pore water. These two oppositely directed forces do not lend themselves to analytical treatment and as such the dynamic penetration resistance to pile driving has no relationship to static bearing capacity.

There is another effect of pile driving in cohesive soils. During driving the soil becomes remolded and the shear strength of the soil is reduced considerably. Though there will be a regaining of shear strength after a lapse of some days after the driving operation, this will not be reflected in the resistance value obtained from the dynamic formulae.

Example 15.22

A 40 × 40 cm reinforced concrete pile 20 m long is driven through loose sand and then into dense gravel to a final set of 3 mm/blow, using a 30 kN single-acting hammer with a stroke of 1.5 m. Determine the ultimate driving resistance of the pile if it is fitted with a helmet, plastic dolly and 50 mm packing on the top of the pile. The weight of the helmet and dolly is 4 kN. The other details are: weight of pile = 74 kN; weight of hammer = 30 kN; pile hammer efficiency η_h = 0.80 and coefficient of restitution C_r = 0.40.

Use the Hiley formula. The sum of the elastic compression $\overline{C}$ is

$\overline{C} = c_1 + c_2 + c_3 = 19.6$ mm.

Solution

Hiley Formula

Use Eq. (15.59)

$$Q_u = \frac{\eta_h Wh}{s+C} \times \frac{1+C_r^2 R}{1+R}$$

where $\eta_h = 0.80$, $W = 30$ kN, $h = 1.5$ m, $R = \frac{W_p}{W} = \frac{(74+4)}{30} = 2.6$, $C_r = 0.40$, $s = 0.30$ cm.

Substituting we have,

$$Q_u = \frac{0.8 \times 30 \times 150}{0.3 + 1.96/2} \times \frac{1+0.4^2 \times 2.6}{1+2.6} = 2813 \times 0.393 = 1105 \text{ kN}$$

15.24 BEARING CAPACITY OF PILES FOUNDED ON A ROCKY BED

Piles are at times required to be driven through weak layers of soil until the tips meet a hard strata for bearing. If the bearing strata happens to be rock, the piles are to be driven to refusal in order to obtain the maximum carrying capacity from the piles. If the rock is strong at its surface, the pile will refuse further driving at a negligible penetration. In such cases the carrying capacity of the piles is governed by the strength of the pile shaft regarded as a column as shown in Fig. 15.6(a). If the soil mass through which the piles are driven happens to be stiff clays or sands, the piles can be regarded as being supported on all sides from buckling as a strut. In such cases, the carrying capacity of a pile is calculated from the safe load on the material of the pile at the point of minimum cross-section. In practice, it is necessary to limit the safe load on piles regarded as short columns because of the likely deviations from the vertical and the possibility of damage to the pile during driving.

If piles are driven to weak rocks, working loads as determined by the available stress on the material of the pile shaft may not be possible. In such cases the frictional resistance developed over the penetration into the rock and the end bearing resistance are required to be calculated. Tomlinson (1986) suggests an equation for computing the end bearing resistance of piles resting on rocky strata as

$$q_u = 2N_\phi q_{ur} \tag{15.64}$$

where $N_\phi = \tan^2 (45 + \phi/2)$,
q_{ur} = unconfined compressive strength of the rock.

Boring of a hole in rocky strata for constructing bored piles may weaken the bearing strata of some types of rock. In such cases low values of skin friction should be used and normally may not be more than 20 kN/m^2 (Tomlinson, 1986) when the boring is through friable chalk or mud stone. In the case of moderately weak to strong rocks where it is possible to obtain core samples for unconfined compression tests, the end bearing resistance can be calculated by making use of Eq. (15.64).

15.25 UPLIFT RESISTANCE OF PILES

Piles are also used to resist uplift loads. Piles used for this purpose are called tension piles, uplift piles or anchor piles. Uplift forces are developed due to hydrostatic pressure or overturning moments as shown in Fig. 15.22.

Figure 15.22 shows a straight edged pile subjected to uplift force. The equation for the uplift force P_{ul} may be written as

$$P_{ul} = W_p + A_s f_r \tag{15.65}$$

where, P_{ul} = uplift capacity of pile,
W_p = weight of pile,
f_r = unit resisting force
A_s = effective area of the embedded length of pile.

Uplift Resistance of Pile in Clay

For piles embedded in clay, Eq. (15.65) may written as

$$P_{ul} = W_p + A_s \alpha \bar{c}_u \tag{15.66}$$

where, $\bar{c}_u$ = average undrained shear strength of clay along the pile shaft,
α = adhesion factor (= c_a/c_u),
c_a = average adhesion.

Figure 15.23 gives the relationship between α and c_u based on pull out test results as collected by Sowa (1970). As per Sowa, the values of c_a agree reasonably well with the values for piles subjected to compression loadings.

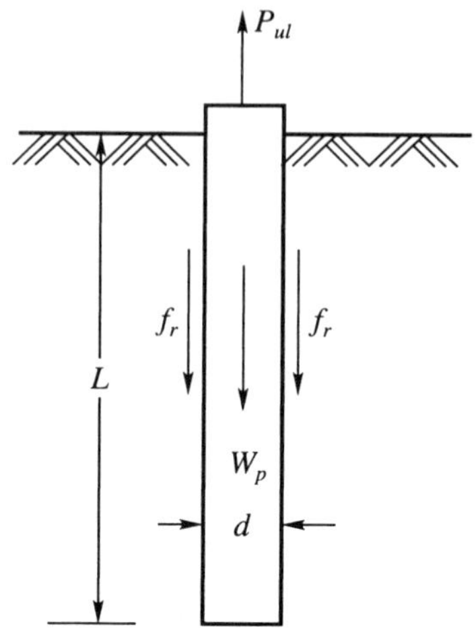

Figure 15.22 Single pile subjected to uplift

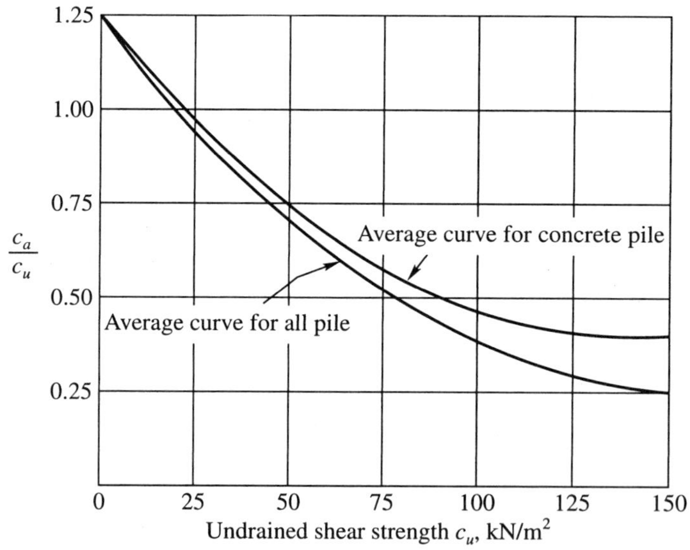

Figure 15.23 Relationship between adhesion factor α and undrained shear strength c_u (Source: Poulos and Davis, 1980)

Uplift Resistance of Pile in Sand

Adequate confirmatory data are not available for evaluating the uplift resistance of piles embedded in cohesionless soils. Ireland (1957) reports that the average skin friction for piles under compression loading and uplift loading are equal, but data collected by Sowa (1970) and Downs and Chieurzzi (1966) indicate lower values for upward loading as compared to downward loading especially for *cast-in-situ* piles. Poulos and Davis (1980) suggest that the skin friction of upward loading may be taken as two-thirds of the calculated shaft resistance for downward loading.

A safety factor of 3 is normally assumed for calculating the safe uplift load for both piles in clay and sand.

Example 15.23

A reinforced concrete pile 30 ft long and 15 in. in diameter is embedded in a saturated clay of very stiff consistency. Laboratory tests on samples of undisturbed soil gave an average undrained cohesive strength $c_u = 2500$ lb/ft^2. Determine the net pullout capacity and the allowable pullout load with $F_s = 3$.

Solution

Given: $L = 30$ ft, $d = 15$ in. diameter, $c_u = 2500$ lb/ft^2, $F_s = 3$.

From Fig. 15.23 $c_a/c_u = 0.41$ for $c_u = 2500 \times 0.0479 \approx 120$ kN/m^2 for concrete pile.

From Eq. (15.66)

$$P_{ul}(\text{net}) = \alpha \bar{c}_u A_s$$

where $\alpha = c_a / c_u = 0.41, c_u = 2500$ lb/ft^2

$$A_s = 3.14 \times \frac{15}{12} \times 30 = 117.75 \text{ ft}^2$$

Substituting

$$P_{ul}(\text{net}) = \frac{0.41 \times 2500 \times 117.75}{1000} = 120.69 \text{ kips}$$

$$P_{ul}(\text{allowed}) = \frac{120.69}{3} \approx 40 \text{ kips}$$

Example 15.24

Refer to Ex. 15.23. If the pile is embedded in medium dense sand, determine the net pullout capacity and the net allowable pullout load with $F_s = 3$.

Given: $L = 30$ ft, $\phi = 38°$, $\bar{K}_s = 1.5$, and $\delta = 25°$, γ (average) = 110 lb/ft^3.

The water table is at great depth. Refer to Section 15.25.

Solution

Downward skin resistance Q_f

$$Q_f = \bar{q}'_o \bar{K}_s \tan \delta A_s$$

where $\bar{q}'_o = \frac{1}{2} \times 30 \times 110 = 1650$ lb/ft^2

$$A_s = 3.14 \times 1.25 \times 30 = 117.75 \text{ ft}^2$$

$$Q_f(\text{down}) = \frac{1650 \times 1.5 \tan 25° \times 117.75}{1000} = 136 \text{ kips}$$

Based on the recommendations of Poulos and Davis (1980)

$$Q_f(\text{up}) = \frac{2}{3} Q_f(\text{down}) = \frac{2}{3} \times 136 = 91 \text{ kips}$$

$$Q_{fa}(\text{up}) = \frac{91}{3} \approx 30 \text{ kips}$$

PART B—PILE GROUP

15.26 NUMBER AND SPACING OF PILES IN A GROUP

Very rarely are structures founded on single piles. Normally, there will be a minimum of three piles under a column or a foundation element because of alignment problems and inadvertent eccentricities. The spacing of piles in a group depends upon many factors such as

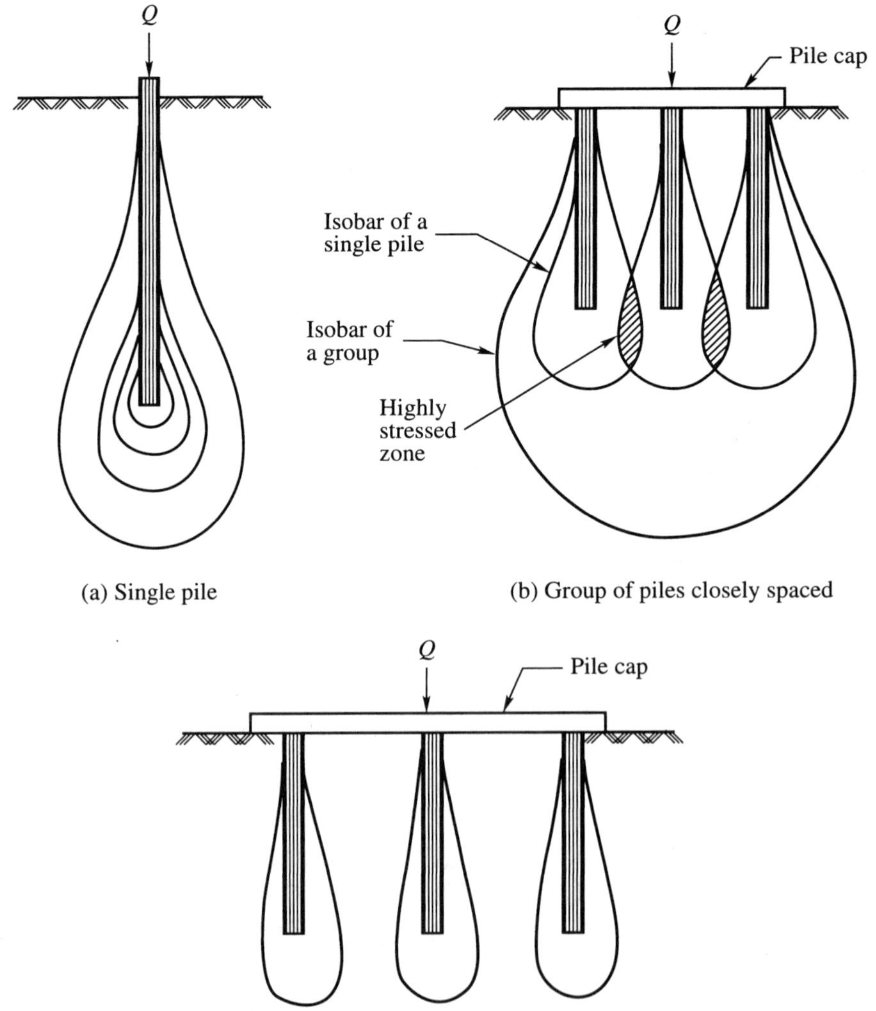

Figure 15.24 Pressure isobars of (a) single pile, (b) group of piles, closely spaced, and (c) group of piles with piles far apart.

1. overlapping of stresses of adjacent piles,
2. cost of foundation,
3. efficiency of the pile group.

The pressure isobars of a single pile with load Q acting on the top are shown in Fig. 15.24(a). When piles are placed in a group, there is a possibility the pressure isobars of adjacent piles will overlap each other as shown in Fig. 15.24(b). The soil is highly stressed in the zones of overlapping of pressures. With sufficient overlap, either the soil will fail or the pile group will settle excessively since the combined pressure bulb extends to a considerable depth below the base of the piles. It is possible to avoid overlap by installing the piles further apart as shown in Fig. 15.24(c). Large spacings are not recommended sometimes, since this would result in a larger pile cap which would increase the cost of the foundation.

The spacing of piles depends upon the method of installing the piles and the type of soil. The piles can be driven piles or *cast-in-situ* piles. When the piles are driven there will be greater overlapping of stresses due to the displacement of soil. If the displacement of soil compacts the soil in between the piles as in the case of loose sandy soils, the piles may be placed at closer intervals.

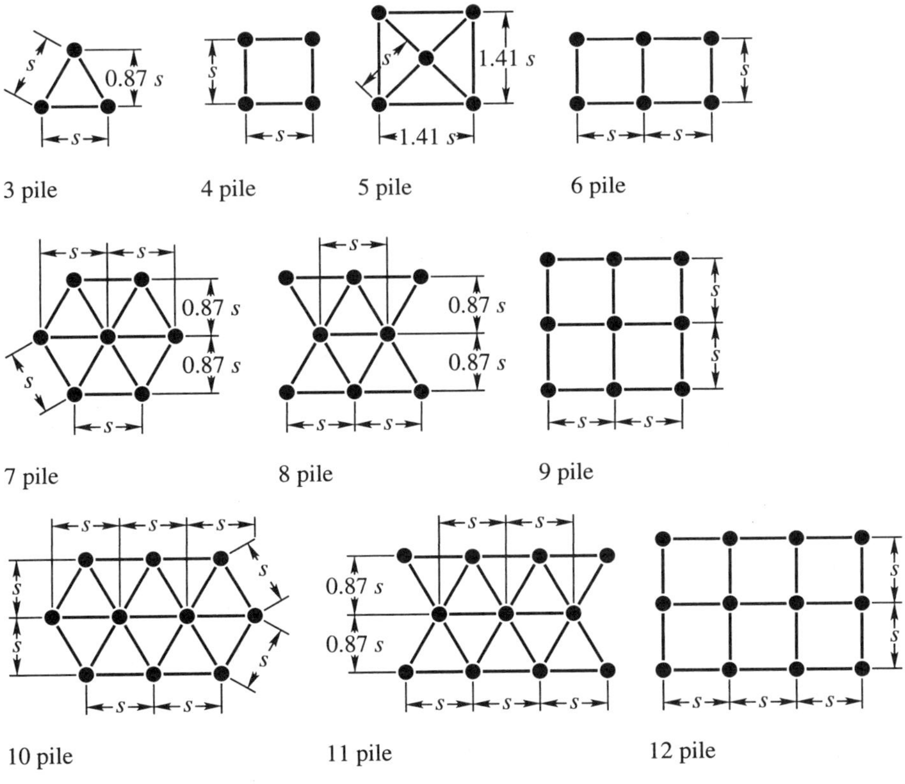

Figure 15.25 Typical arrangements of piles in groups

But if the piles are driven into saturated clay or silty soils, the displaced soil will not compact the soil between the piles. As a result the soil between the piles may move upwards and in this process lift the pile cap. Greater spacing between piles is required in soils of this type to avoid lifting of piles. When piles are *cast-in-situ*, the soils adjacent to the piles are not stressed to that extent and as such smaller spacings are permitted.

Generally, the spacing for point bearing piles, such as piles founded on rock, can be much less than for friction piles since the high-point-bearing stresses and the superposition effect of overlap of the point stresses will most likely not overstress the underlying material nor cause excessive settlements.

The minimum allowable spacing of piles is usually stipulated in building codes. The spacings for straight uniform diameter piles may vary from 2 to 6 times the diameter of the shaft. For friction piles, the minimum spacing recommended is $3d$ where d is the diameter of the pile. For end bearing piles passing through relatively compressible strata, the spacing of piles shall not be less than $2.5d$. For end bearing piles passing through compressible strata and resting in stiff clay, the spacing may be increased to $3.5d$. For compaction piles, the spacing may be $2d$. Typical arrangements of piles in groups are shown in Fig. 15.25.

15.27 PILE GROUP EFFICIENCY

The spacing of piles is usually predetermined by practical and economical considerations. The design of a pile foundation subjected to vertical loads consists of

1. The determination of the ultimate load bearing capacity of the group Q_{gu}.
2. Determination of the settlement of the group, S_g, under an allowable load Q_{ga}.

The ultimate load of the group is generally different from the sum of the ultimate loads of individual piles Q_u.

The factor

$$E_g = \frac{Q_{gu}}{\Sigma Q_u} \tag{15.67}$$

is called group efficiency which depends on parameters such as type of soil in which the piles are embedded, method of installation of piles i.e. either driven or *cast-in-situ* piles, and spacing of piles.

There is no acceptable "*efficiency formula*" for group bearing capacity. There are a few formulae such as the Converse-Labarre formula that are sometimes used by engineers. These formulae are empirical and give efficiency factors less than unity. But when piles are installed in sand, efficiency factors greater than unity can be obtained as shown by Vesic (1967) by his experimental investigation on groups of piles in sand. There is not sufficient experimental evidence to determine group efficiency for piles embedded in clay soils.

Efficiency of Pile Groups in Sand

Vesic (1967) carried out tests on 4 and 9 pile groups driven into sand under controlled conditions. Piles with spacings 2, 3, 4, and 6 times the diameter were used in the tests. The tests were conducted in homogeneous, medium dense sand. His findings are given in Fig. 15.26. The figure gives the following:

1. The efficiencies of 4 and 9 pile groups when the pile caps do not rest on the surface.
2. The efficiencies of 4 and 9 pile groups when the pile caps rest on the surface.

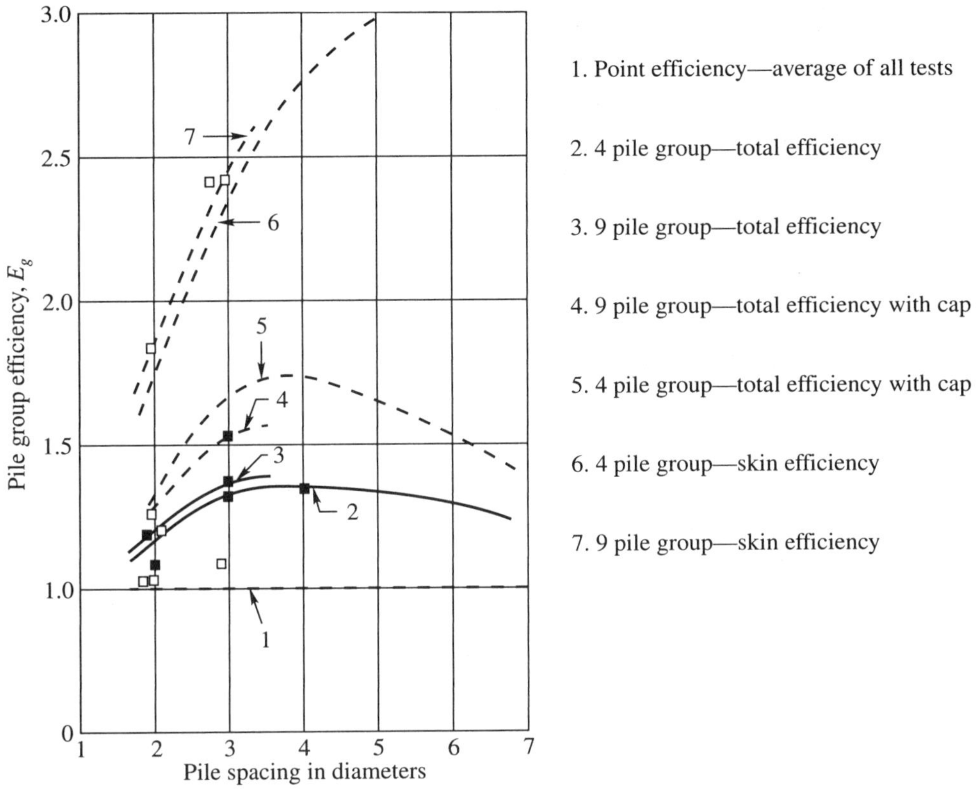

Figure 15.26 Efficiency of pile groups in sand (Vesic, 1967)

3. The skin efficiency of 4 and 9 pile groups.
4. The average point efficiency of all the pile groups.

It may be mentioned here that a pile group with the pile cap resting on the surface takes more load than one with free standing piles above the surface. In the former case, a part of the load is taken by the soil directly under the cap and the rest is taken by the piles. The pile cap behaves the same way as a shallow foundation of the same size. Though the percentage of load taken by the group is quite considerable, building codes have not so far considered the contribution made by the cap.

It may be seen from the Fig. 15.26 that the overall efficiency of a four pile group with a cap resting on the surface increases to a maximum of about 1.7 at pile spacings of 3 to 4 pile diameters, becoming somewhat lower with a further increase in spacing. A sizable part of the increased bearing capacity comes from the caps. If the loads transmitted by the caps are reduced, the group efficiency drops to a maximum of about 1.3.

Very similar results are indicated from tests with 9 pile groups. Since the tests in this case were carried out only up to a spacing of 3 pile diameters, the full picture of the curve is not available. However, it may be seen that the contribution of the cap for the bearing capacity is relatively smaller.

Vesic measured the skin loads of all the piles. The skin efficiencies for both the 4 and 9-pile groups indicate an increasing trend. For the 4-pile group the efficiency increases from about 1.8 at 2 pile diameters to a maximum of about 3 at 5 pile diameters and beyond. In contrast to this, the average point load efficiency for the groups is about 1.01. Vesic showed for the first time that the increasing bearing capacity of a pile group for piles driven in sand comes primarily from an increase in skin loads. The point loads seem to be virtually unaffected by group action.

Pile Group Efficiency Equation

There are many pile group equations. These equations are to be used very cautiously, and may in many cases be no better than a good guess. The Converse-Labarre Formula is one of the most widely used group-efficiency equations which is expressed as

$$E_g = 1 - \frac{\theta(n-1)m + (m-1)n}{90\ mn} \tag{15.68}$$

where m = number of columns of piles in a group,
n = number of rows,
$\theta = \tan^{-1}(d/s)$ in degrees,
d = diameter of pile,
s = spacing of piles center to center.

15.28 VERTICAL BEARING CAPACITY OF PILE GROUPS EMBEDDED IN SANDS AND GRAVELS

Driven piles. If piles are driven into loose sands and gravel, the soil around the piles to a radius of at least three times the pile diameter is compacted. When piles are driven in a group at close spacing, the soil around and between them becomes highly compacted. When the group is loaded, the piles and the soil between them move together as a unit. Thus, the pile group acts as a pier foundation having a base area equal to the gross plan area contained by the piles. The efficiency of the pile group will be greater than unity as explained earlier. It is normally assumed that the efficiency falls to unity when the spacing is increased to five or six diameters. Since present knowledge is not sufficient to evaluate the efficiency for different spacing of piles, it is conservative to assume an efficiency factor of unity for all practical purposes. We may, therefore, write

$$Q_{gu} = nQ_u \tag{15.69}$$

where n = the number of piles in the group.

The procedure explained above is not applicable if the pile tips rest on compressible soil such as silts or clays. When the pile tips rest on compressible soils, the stresses transferred to the compressible soils from the pile group might result in over-stressing or extensive consolidation. The carrying capacity of pile groups under these conditions is governed by the shear strength and compressibility of the soil, rather than by the '*efficiency*' of the group within the sand or gravel stratum.

Bored Pile Groups In Sand And Gravel

Bored piles are *cast-in-situ* concrete piles. The method of installation involves

1. Boring a hole of the required diameter and depth,
2. Pouring in concrete.

There will always be a general loosening of the soil during boring and then too when the boring has to be done below the water table. Though bentonite slurry (sometimes called as *drilling mud*) is used for stabilizing the sides and bottom of the bores, loosening of the soil cannot be avoided. Cleaning of the bottom of the bore hole prior to concreting is always a problem which will never be achieved quite satisfactorily. Since bored piles do not compact the soil between the piles,

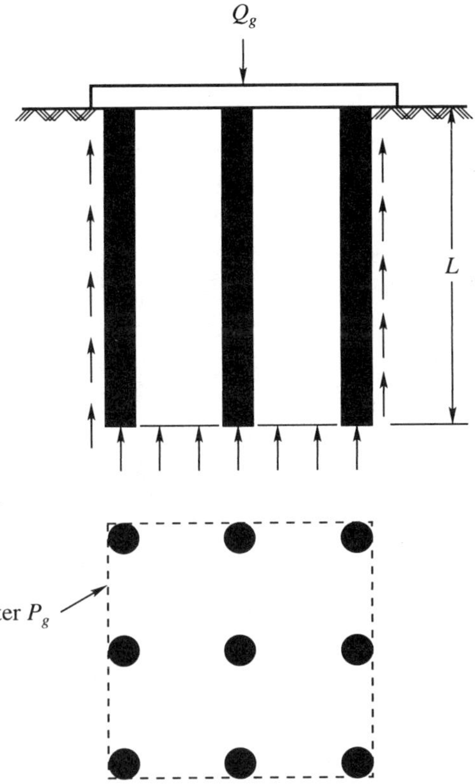

Figure 15.27 Block failure of a pile group in clay soil

the efficiency factor will never be greater than unity. However, for all practical purposes, the efficiency may be taken as unity.

Pile Groups In Cohesive Soils

The effect of driving piles into cohesive soils (clays and silts) is very different from that of cohesionless soils. It has already been explained that when piles are driven into clay soils, particularly when the soil is soft and sensitive, there will be considerable remolding of the soil. Besides there will be heaving of the soil between the piles since compaction during driving cannot be achieved in soils of such low permeability. There is every possibility of lifting of the pile during this process of heaving of the soil. Bored piles are, therefore, preferred to driven piles in cohesive soils. In case driven piles are to be used, the following steps should be favored:

1. Piles should be spaced at greater distances apart.
2. Piles should be driven from the center of the group towards the edges, and
3. The rate of driving of each pile should be adjusted as to minimize the development of pore water pressure.

Experimental results have indicated that when a pile group installed in cohesive soils is loaded, it may fail by any one of the following ways:

1. May fail as a block (called block failure).
2. Individual piles in the group may fail.

When piles are spaced at closer intervals, the soil contained between the piles move downward with the piles and at failure, piles and soil move together to give the typical '*block failure*'. Normally this type of failure occurs when piles are placed within 2 to 3 pile diameters. For wider spacings, the piles fail individually. The efficiency ratio is less than unity at closer spacings and may reach unity at a spacing of about 8 diameters.

The equation for block failure may be written as (Fig. 15.27).

$$Q_{gu} = cN_c A_g + P_g L\bar{c} \tag{15.70}$$

where c = cohesive strength of clay beneath the pile group,
$\bar{c}$ = average cohesive strength of clay around the group,
L = length of pile,
P_g = perimeter of pile group,
A_g = sectional area of group,
N_c = bearing capacity factor which may be assumed as 9 for deep foundations.

The bearing capacity of a pile group on the basis of individual pile failure may be written as

$$Q_{gu} = nQ_u \tag{15.71}$$

where n = number of piles in the group,
Q_u = bearing capacity of an individual pile.

The bearing capacity of a pile group is normally taken as the smaller of the two given by Eqs. (15.70) and (15.71).

Example 15.25

A group of 9 piles with 3 piles in a row was driven into a soft clay extending from ground level to a great depth. The diameter and the length of the piles were 30 cm and 10 m respectively. The unconfined compressive strength of the clay is 70 kPa. If the piles were placed 90 cm center to center, compute the allowable load on the pile group on the basis of a shear failure criterion for a factor of safety of 2.5.

Solution

The allowable load on the group is to be calculated for two conditions: (a) block failure and (b) individual pile failure. The least of the two gives the allowable load on the group.

(a) Block failure (Fig. 15.27). Use Eq. (15.70),

$$Q_{gu} = cN_c A_g + P_g L\bar{c} \quad \text{where} \quad N_c = 9,\ c = \bar{c} = 70/2 = 35 \text{ kN/m}^2$$

$$A_g = 2.1 \times 2.1 = 4.4 \text{ m}^2, \quad P_g = 4 \times 2.1 = 8.4 \text{ m}, \quad L = 10 \text{ m}$$

$$Q_{gu} = 35 \times 9 \times 4.4 + 8.4 \times 10 \times 35 = 4326 \text{ kN}, \quad Q_a = \frac{4326}{2.5} = 1730 \text{ kN}$$

(b) Individual pile failure

$Q_u = Q_b + Q_f = q_b A_b + \alpha \bar{c} A_s$. Assume $\alpha = 1$.

Now, $q_b = cN_c = 35 \times 9 = 315 \text{ kN/m}^2$, $A_b = 0.07 \text{ m}^2$,

$A_s = 3.14 \times 0.3 \times 10 = 9.42 \text{ m}^2$

Substituting, $Q_u = 315 \times 0.07 + 1 \times 35 \times 9.42 = 352 \text{ kN}$

$Q_{gu} = nQ_u = 9 \times 352 = 3168 \text{ kN}, \quad Q_a = \frac{3168}{2.5} = 1267 \text{ kN}$

The allowable load is 1267 kN.

15.29 SETTLEMENT OF PILES AND PILE GROUPS IN SANDS AND GRAVELS

Normally it is not necessary to compute the settlement of a single pile as this settlement under a working load will be within the tolerable limits. However, settlement analysis of a pile group is very much essential. The total settlement analysis of a pile group does not bear any relationship with that of a single pile since in a group the settlement is very much affected due to the interaction stresses between piles and the stressed zone below the tips of piles.

Settlement analysis of single piles by Poulos and Davis (1980) indicates that immediate settlement contributes the major part of the final settlement (which includes the consolidation settlement for saturated clay soils) even for piles in clay. As far as piles in sand is concerned, the immediate settlement is almost equal to the final settlement.

However, it may be noted here that consolidation settlement becomes more important for pile groups in saturated clay soils.

Immediate settlement of a single pile may be computed by making use of semi-empirical methods. The method as suggested by Vesic (1977) has been discussed here.

In recent years, with the advent of computers, more sophisticated methods of analysis have been developed to predict the settlement and load distribution in a single pile. The following three methods are often used.

1. 'Load transfer' method which is also called as the 't-z' method.
2. Elastic method based on Mindlin's (1936) equations for the effects of subsurface loadings within a semi-infinite mass.
3. The finite element method.

This chapter discusses only the 't-z' method. The analysis of settlement by the elastic method is quite complicated and is beyond the scope of this book. Poulos and Davis, (1980) have discussed this procedure in detail. The finite element method of analysis of a single pile axially loaded has been discussed by many investigators such as Ellison et al., (1971), Desai (1974), Balaam et al., (1975), etc. The finite element approach is a generalization of the elastic approach. The power of this method lies in its capability to model complicated conditions and to represent non-linear stress/strain behavior of the soil over the whole zone of the soil modelled. Use of computer programs is essential and the method is more suited to research or investigation of particularly complex problems than to general design.

Present knowledge is not sufficient to evaluate the settlements of piles and pile groups. For most engineering structures, the loads to be applied to a pile group will be governed by consideration of consolidation settlement rather than by bearing capacity of the group divided by an arbitrary factor of safety of 2 or 3. It has been found from field observation that the settlement of a pile group is many times the settlement of a single pile at the corresponding working load. The settlement of a group is affected by the shape and size of the group, length of piles, method of installation of piles and possibly many other factors.

Semi-Empirical Formulas and Curves

Vesic (1977) proposed an equation to determine the settlement of a single pile. The equation has been developed on the basis of experimental results he obtained from tests on piles. Tests on piles of diameters ranging from 2 to 18 inches were carried out in sands of different relative densities. Tests were carried out on driven piles, jacked piles, and bored piles (jacked piles are those that are pushed into the ground by using a jack). The equation for total settlement of a single pile may be expressed as

$$S = S_p + S_f \tag{15.72}$$

where S = total settlement,
S_p = settlement of the pile tip,
S_f = settlement due to the deformation of the pile shaft.

The equation for S_p is

$$S_p = \frac{C_w Q_p}{\left(1 + D_r^2\right) q_{pu}} \tag{15.73}$$

The equation for S_f is

$$S_f = \left(Q_p + \alpha Q_f\right)\frac{L}{AE} \tag{15.74}$$

where Q_p = point load,
d = diameter of the pile at the base,
q_{pu} = ultimate point resistance per unit area,
D_r = relative density of the sand,
C_w = settlement coefficient, = 0.04 for driven piles
= 0.05 for jacked piles
= 0.18 for bored piles,
Q_f = friction load,
L = pile length,
A = cross-sectional area of the pile,
E = modulus of deformation of the pile shaft,
α = coefficient which depends on the distribution of skin friction along the shaft and can be taken equal to 0.6.

Settlement of piles cannot be predicted accurately by making use of equations such as the ones given here. One should use such equations with caution. It is better to rely on load tests for piles in sands.

Settlement of Pile Groups in Sand

The relation between the settlement of a group and a single pile at corresponding working loads may be expressed as

$$F_g = \frac{S_g}{S} \tag{15.75}$$

where F_g = group settlement factor,
S_g = settlement of group,
S = settlement of a single pile.

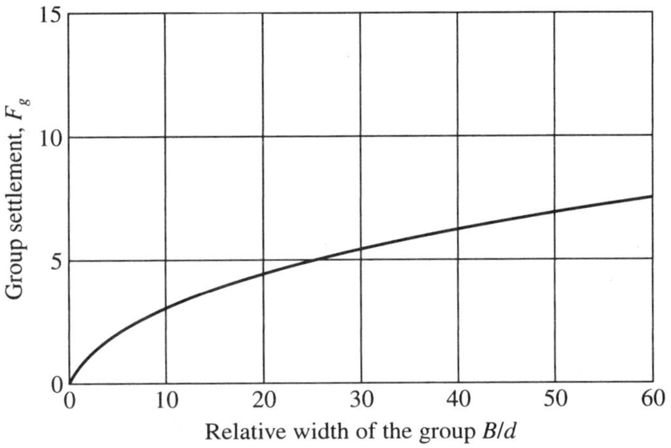

Figure 15.28 Curve showing the relationship between group settlement ratio and relative widths of pile groups in sand (Vesic, 1967)

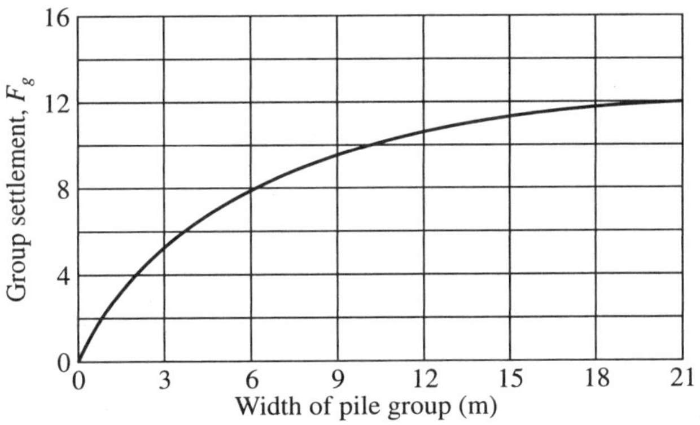

Figure 15.29 Curve showing relationship between F_g and pile group width (Skempton, et al., 1953)

Vesic (1967) obtained the curve given in Fig. 15.28 by plotting F_g against B/d where d is the diameter of the pile and B, the distance between the center to center of the outer piles in the group (only square pile groups are considered). It should be remembered here that the curve is based on the results obtained from tests on groups of piles embedded in medium dense sand. It is possible that groups in much looser or much denser deposits might give somewhat different behavior. The group settlement ratio is very likely be affected by the ratio of the pile point settlement S_p to total pile settlement.

Skempton et al., (1953) published curves relating F_g with the width of pile groups as shown in Fig. 15.29. These curves can be taken as applying to driven or bored piles. Since the abscissa for the curve in Fig. 15.29 is not expressed as a ratio, this curve cannot directly be compared with Vesic's curve given in Fig. 15.28. According to Fig. 15.29 a pile group 3 m wide would settle 5 times that of a single test pile.

t-z Method

Consider a floating vertical pile of length L and diameter d subjected to a vertical load Q (Fig. 15.30). This load will be transferred to the surrounding and underlying soil layers as described in

Section 15.7. The pile load will be carried partly by skin friction (which will be mobilized through the increasing settlement on the mantle surface and the compression of the pile shaft) and partly through the pile tip, in the form of point resistance as can be seen from Fig. 15.30. Thus the load is taken as the sum of these components. The distribution of the point resistance is usually considered as uniform; however, the distribution of the mantle friction depends on many factors. Load transfer in pile-soil system is a very complex phenomenon involving a number of parameters which are difficult to evaluate in numerical terms. Yet, some numerical assessment of load transfer characteristics of a pile soil system is essential for the rational design of pile foundation.

The objective of a load transfer analysis is to obtain a load-settlement curve. The basic problem of load transfer is shown in Fig. 15.30. The following are to be determined:

1. The vertical movements s, of the pile cross-sections at any depth z under loads acting on the top,
2. The corresponding pile load Q_z at depth z acting on the pile section,
3. The vertical movement of the base of the pile and the corresponding point stress.

The mobilization of skin shear stress τ at any depth z, from the ground surface depends on the vertical movement of the pile cross section at that level. The relationship between the two may be linear or non-linear. The shear stress τ, reaches the maximum value, τ_f, at that section when the vertical movement of the pile section is adequate. It is, therefore, essential to construct $(\tau - s)$ curves at various depths z as required.

There will be settlement of the tip of pile after the full mobilization of skin friction. The movement of the tip, s_b, may be assumed to be linear with the point pressure q_p. When the movement of the tip is adequate, the point pressure reaches the maximum pressure q_b (ultimate base pressure). In order to solve the load–transfer problem, it is esential to construct a $(q_p - s)$ curve.

$(\tau - s)$ and $(q_p - s)$ curves

Coyle and Reese (1966) proposed a set of average curves of load transfer based on laboratory test piles and instrumented field piles. These curves are limited to the case of steel-pipe friction piles in clay with an embedded depth not exceeding 100 ft. Coyle and Sulaiman (1967) have also given load transfer curves for piles in sand. These curves are meant for specific cases and therefore meant to solve specific problems and as such this approach cannot be considered as a general case.

Verbrugge (1981) proposed an elastic-plastic model for the $(\tau - s)$ and $(q_p - s)$ curves based on CPT results. The slopes of the elastic portion of the curves given are

$$\frac{\tau}{s} = \frac{0.22E_s}{2R} \tag{15.76}$$

$$\frac{q_p}{s} = \frac{3.125E_s}{2R} \tag{15.77}$$

The value of elastic modulus E_s of cohesionless soils may be obtained by the following expressions (Verbrugge);

$$\text{for bored piles } E_s = (36 + 2.2\, q_c) \text{ kg/cm}^2 \tag{15.78}$$

$$\text{for driven piles } E_s = 3(36 + 2.2\, q_c) \text{ kg/cm}^2; \tag{15.79}$$

where q_c = point resistance of static cone penetrometer in kg/cm^2.

The relationship recommended for E_s is for $q_c > 4$ kg/cm^2. The maximum values of τ, and τ_f, for the plastic portion of $(\tau - s)$ curves are given in Table 15.4 and 15.5 for cohesionless and cohesive soils respectively. In Eqs (15.76) and (15.77) the value of E_s can be obtained by any one of

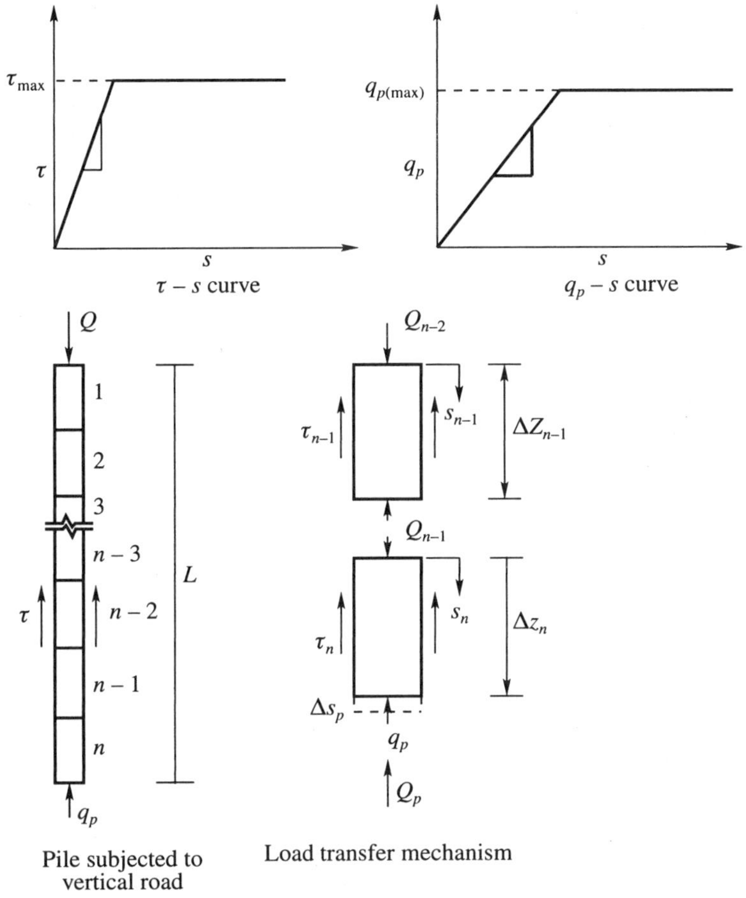

Figure 15.30 *t–z* method of analysis of pile load-settlement relationship

the known methods. The maximum value of q_p (q_b) may be obtained by any one of the known methods such as

1. From the relationship
   $q_b = q'_o N_q$ for cohesionless soils
   $q_b = 9c_u$ for cohesive soils.

Table 15.4 Recommended maximum skin shear stress τ_{max} for piles in cohesionless soils (After Verbrugge, 1981)

τ_{max} kN/m^2	Pile type
0.011 q_c	Driven concrete piles
0.009 q_c	Driven steel piles
0.005 q_c	Bored concrete piles
Limiting values	
τ_{max} = 80 kN/m^2 for bored piles	
τ_{max} = 120 kN/m^2 for driven piles	

Table 15.5 Recommended maximum skin shear stress τ_{max} for piles in cohesive soils (After Verbrugge, 1981)
[Values recommended are for Dutch cone penetrometer]

Type of pile	Material	Range of q_c, kN/m^2	τ_{max}
Driven	Concrete	$q_c \le 375$	$0.053\ q_c$
		$375 \le q_c < 4500$	$18 + 0.006\ q_c$
		$4500 < q_c$	$0.01\ q_c$
	Steel	$q_c \le 450$	$0.033\ q_c$
		$450 \le q_c \le 1500$	15
		$1500 < q_c$	$0.01\ q_c$
Bored	Concrete	$q_c \le 600$	$0.037\ q_c$
		$600 \le q_c < 4500$	$18 + 0.006\ q_c$
		$4500 < q_c$	$0.01\ q_c$
	Steel	$q_c \le 500$	$0.03\ q_c$
		$500 \le q_c < 1500$	15
		$1500 < q_c$	$0.01\ q_c$

2. From static cone penetration test results
3. From pressuremeter test results

Method of obtaining load-settlement curve (Fig. 15.30)

The approximate load-settlement curve is obtained point by point in the following manner:

1. Divide the pile into any convenient segments (possibly10 for computer programming and less for hand calculations).
2. Assume a point pressure q_p less than the maximum q_b.
3. Read the corresponding displacement s_p from the $(q_p - s)$ curve.
4. Assume that the load in the pile segment closest to the point (segment n) is equal to the point load.
5. Compute the compression of the segment n under that load by

$$\Delta s_n = \frac{Q_p L}{A E_p}$$

where, $Q_p = q_p A_b$,

A = cross-sectional area of segment,

E_p = modulus of elasticity of the pile material.

6. Calculate the settlement of the top of segment n by

$$s_n = s_p + \Delta s_n$$

7. Use the $(\tau - s)$ curves to read the friction τ_n on segment n, at displacement s_n.
8. Calculate the load in pile segment $(n - 1)$ by:

$$Q_{n-2} = \tau_n \Delta z_n \pi d_n + Q_p$$

where Δz_n = length of segment n,

d_n = average diameter of pile in segment n (applies to tapered piles).

9. Do 4 through 8 up to the top segment. The load and displacement at the top of the pile provide one point on the load-settlement curve.
10. Repeat 1 through 9 for the other assumed values of the point pressure, q_p.

E_p may also be obtained from the relationship established between E_s and the field tests such as SPT, CPT and PMT. It may be noted here that the accuracy of the results obtained depends upon the accuracy with which the values of E_s simulate the field conditions.

Example 15.26

A concrete pile of section 30 × 30 cm is driven into medium dense sand with the water table at ground level. The depth of embedment of the pile is 18 m. Static cone penetration test conducted at the site gives an average value q_c = 50 kg/cm². Determine the load transfer curves and then calculate the settlement. The modulus of elasticity of the pile material E_p is 21 × 10⁴ kg/cm² (Fig. Ex. 15.26).

Solution

It is first necessary to draw the ($q_p - s$) and ($\tau - s$) curves (see section 15.29). The curves can be constructed by determining the ratios of q_p/s and t/s from Eqs (15.77) and (15.76) respectively.

$$\frac{q_p}{s} = \frac{3.125E}{2R}$$

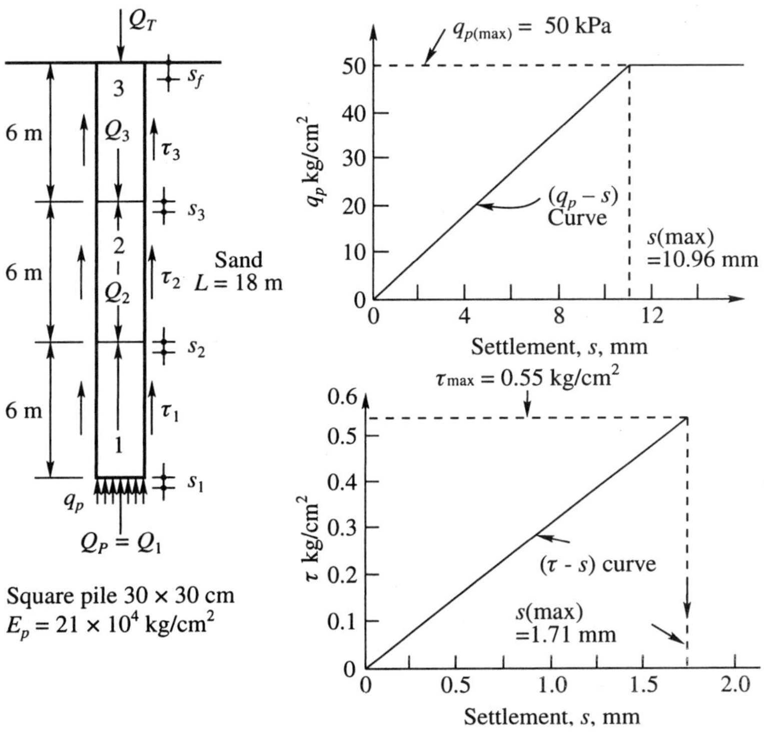

Figure Ex. 15.26

$$\frac{\tau}{s} = \frac{0.22E}{2R}$$

where R = radius or width of pile

The value of E_s for a driven pile may be determined from Eq. (15.78).

$$E_s = 3(36 + 2.2\, q_c) \text{ kg/cm}^2$$

$$= 3(36 + 2.2 \times 50) = 438 \text{ kg/cm}^2$$

Now $\dfrac{q_p}{s} = \dfrac{3.125 \times 438}{2 \times 15} = 45.62 \text{ kg/cm}^3$

$$\frac{\tau}{s} = \frac{0.22 \times 438}{2 \times 0.15} = 3.21 \text{ kg/cm}^3$$

To construct the $(q_p - s)$ and $(\tau - s)$ curves, we have to know the maximum values of q_p/s and τ. Given $q_b = q_p = 50$ kg/cm^2 – the maximum value.

From Table 15.4 $\tau_{max} = 0.011\, q_c = 0.011 \times 50 = 0.55$ kg/cm^2

Now the theoretical maximum settlement s for q_p(max) = 50 kg/cm^2 is

$$s_{(max)} = \frac{50}{45.62} = 1.096 \text{ cm} = 10.96 \text{ mm}$$

The curve $(q_p - s)$ may be drawn as shown in Fig. Ex. 15.26. Similarly for $\tau_{(max)} = 0.55$ kg/cm^2

$$s_{(max)} = \frac{0.55}{3.21} = 0.171 \text{ cm} = 1.71 \text{ mm.}$$

Now the $(\tau - s)$ curve can be constructed as shown in Fig. Ex. 15.26.

Calculation of pile settlement

The various steps in the calculations are

1. Divide the pile length 18 m into three equal parts of 6 m each.
2. To start with assume a base pressure $q_p = 5$ kg/cm^2.
3. From the $(q_p - s)$ curve $s_1 = 0.12$ cm for $q_p = 5$ kg/cm^2.
4. Assume that a load $Q_1 = Q_p = 5 \times 900 = 4500$ kg acts axially on segment 1.
5. Now the compression Δs_1 of segment 1 is

$$\Delta s_1 = \frac{Q_1 \Delta L}{AE_p} = \frac{4500 \times 600}{30 \times 30 \times 21 \times 10^4} = 0.014 \text{ cm}$$

6. Settlement of the top of segment 1 is
   $s_2 = s_1 + \Delta s_1 = 0.12 + 0.014 = 0.134$ cm.
7. Now from $(\tau - s)$ curve Fig. Ex. 15.26, $\tau = 0.43$ kg/cm^2.
8. The pile load in segment 2 is
   $Q_2 = 4 \times 30 \times 600 \times 0.43 + 4500 = 30{,}960 + 4500 = 35{,}460$ kg
9. Now the compression of segment 2 is

$$\Delta s_2 = \frac{Q_2 \Delta L}{AE_p} = \frac{35{,}460 \times 600}{900 \times 21 \times 10^4} = 0.113 \text{ cm}$$

10. Settlement of the top of segment 2 is

$s_3 = s_2 + \Delta s_2 = 0.134 + 0.113 = 0.247$ cm.

11. Now from $(\tau - s)$ curve, Fig. Ex. 15.26 $\tau = 0.55$ kg/cm^2 for $s_3 = 0.247$ cm. This is the maximum shear stress.
12. Now the pile load in segment 3 is

    $Q_3 = 4 \times 30 \times 600 \times 0.55 + 35{,}460 = 39{,}600 + 35{,}460 = 75{,}060$ kg

13. The compression of segment 3 is

$$\Delta s_3 = \frac{Q_3 \Delta L}{AE_p} = \frac{75{,}060 \times 600}{900 \times 21 \times 10^4} = 0.238 \text{ cm}$$

14. Settlement of top of segment 3 is

    $s_4 = s_t = s_3 + \Delta s_3 = 0.247 + 0.238 = 0.485$ cm.

15. Now from $(\tau - s)$ curve, $\tau_{max} = 0.55$ kg/cm^2 for $s \geq 0.17$ cm.
16. The pile load at the top of segment 3 is

    $Q_T = 4 \times 30 \times 600 \times 0.55 + 75{,}060 = 39{,}600 + 75{,}060 = 114{,}660$ kg

    ≈ 115 tones (metric)

    The total settlement $s_t = 0.485$ cm ≈ 5 mm.

    Total pile load $Q_T = 115$ tones.

    This yields one point on the load settlement curve for the pile. Other points can be obtained in the same way by assuming different values for the base pressure q_p in Step 2 above. For accurate results, the pile should be divided into smaller segments.

15.30 SETTLEMENT OF PILE GROUPS IN COHESIVE SOILS

The total settlements of pile groups may be calculated by making use of consolidation settlement equations. The problem involves evaluating the increase in stress Δp beneath a pile group when the group is subjected to a vertical load Q_g. The computation of stresses depends on the type of soil through which the pile passes. The methods of computing the stresses are explained below:

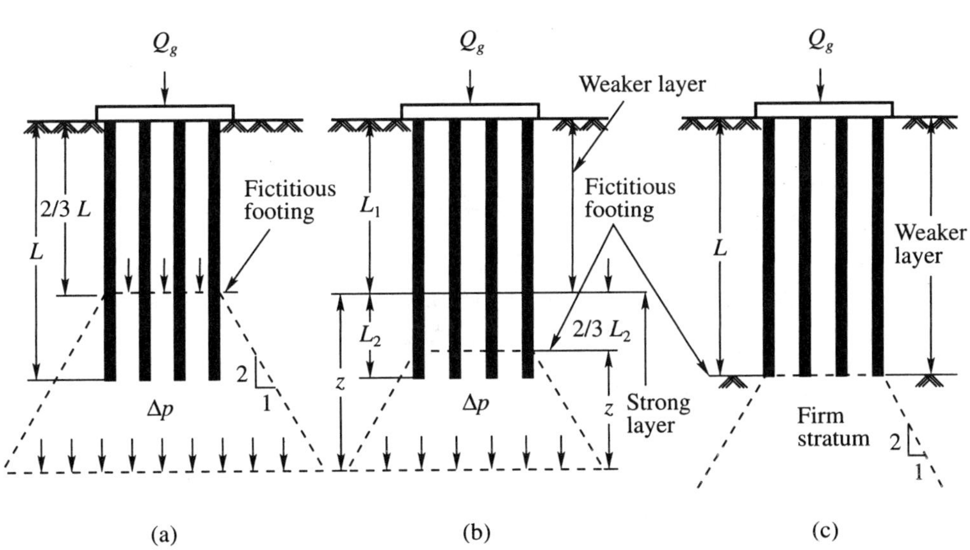

Figure 15.31 Settlement of pile groups in clay soils

1. The soil in the first group given in Fig. 15.31 (a) is homogeneous clay. The load Q_g is assumed to act on a fictitious footing at a depth 2/3L from the surface and distributed over the sectional area of the group. The load on the pile group acting at this level is assumed to spread out at a 2 Vert : 1 Horiz slope. The stress Δp at any depth z below the fictitious footing may be found as explained in Chapter 6.
2. In the second group given in (b) of the figure, the pile passes through a very weak layer of depth L_1 and the lower portion of length L_2 is embedded in a strong layer. In this case, the load Q_g is assumed to act at a depth equal to 2/3 L_2 below the surface of the strong layer and spreads at a 2 : 1 slope as before.
3. In the third case shown in (c) of the figure, the piles are point bearing piles. The load in this case is assumed to act at the level of the firm stratum and spreads out at a 2 : 1 slope.

15.31 ALLOWABLE LOADS ON GROUPS OF PILES

The basic criterion governing the design of a pile foundation should be the same as that of a shallow foundation, that is, the settlement of the foundation must not exceed some permissible value. The permissible values of settlements assumed for shallow foundations in Chapter 13 are also applicable to pile foundations. The allowable load on a group of piles should be the least of the values computed on the basis of the following two criteria.

1. Shear failure,
2. Settlement.

Procedures have been given in earlier chapters as to how to compute the allowable loads on the basis of a shear failure criterion. The settlement of pile groups should not exceed the permissible limits under these loads.

Example 15.27

It is required to construct a pile foundation comprised of 20 piles arranged in 5 columns at distances of 90 cm center to center. The diameter and lengths of the piles are 30 cm and 9 m respectively. The bottom of the pile cap is located at a depth of 2.0 m from the ground surface. The details of the soil properties etc. are as given below with reference to ground level as the datum. The water table was found at a depth of 4 m from ground level.

Depth (m)		Soil properties
From	To	
0	2	Silt, saturated, $\gamma = 16$ kN/m^3
2	4	Clay, saturated, $\gamma = 19.2$ kN/m^3
4	12	Clay, saturated, $\gamma = 19.2$ kN/m^3, $q_u = 120$ kN/m^2, $e_0 = 0.80$, $C_c = 0.23$
12	14	Clay, $\gamma = 18.24$ kN/m^3, $q_u = 90$ kN/m^2, $e_0 = 1.08$, $C_c = 0.34$.
14	17	Clay, $\gamma = 20$ kN/m^3, $q_u = 180$ kN/m^2, $e_0 = 0.70$, $C_c = 0.2$
17	–	Rocky stratum

Compute the consolidation settlement of the pile foundation if the total load imposed on the foundation is 2500 kN.

Solution

Assume that the total load 2500 kN acts at a depth $(2/3)L = (2/3) \times 9 = 6$ m from the bottom of the pile cap on a fictitious footing as shown in Fig. 15.31(a). This fictitious footing is now at a depth of 8 m below ground level. The size of the footing is 3.9 × 3.0 m. Now three layers are assumed to contribute to the settlement of the foundation. They are: *Layer 1*—from 8 m to 12 m (= 4 m thick) below ground level; *Layer 2*—from 12 m to 14 m = 2 m thick; *Layer 3*—from 14 m to 17 m = 3 m thick. The increase in pressure due to the load on the fictitious footing at the centers of each layer is computed on the assumption that the load is spread at an angle of 2 vertical to 1 horizontal [Fig. 15.31(a)] starting from the edges of the fictitious footing. The settlement is computed by making use of the equation

$$S_t = \Sigma H_i \frac{C_c}{1+e_o} \log \frac{p_o + \Delta p}{p_o}$$

where p_o = the effective overburden pressure at the middle of each layer,
Δp = the increase in pressure at the middle of each layer

Computation of p_o

For Layer 1, $p_o = 2 \times 16 + 2 \times 19.2 + (10-4)(19.2-9.81) = 126.74 \text{ kN/m}^2$

For Layer 2, $p_o = 126.74 + 2(19.2-9.81) + 1 \times (18.24-9.81) = 153.95 \text{ kN/m}^2$

For Layer 3, $p_o = 153.95 + 1(18.24-9.81) + 1.5 \times (20.0-9.81) = 177.67 \text{ kN/m}^2$

Computation of Δp

For Layer 1

Area at 2 m depth below fictitious footing = (3.9 + 2) × (3 + 2) = 29.5 m^2

$$\Delta p = \frac{2500}{29.5} = 84.75 \text{ kN/m}^2$$

For Layer 2

Area at 5 m depth below fictitious footing = (3.9 + 5) × (3 + 5) = 71.2 m^2

$$\Delta p = \frac{2500}{71.2} = 35.1 \text{ kN/m}^2$$

For Layer 3

Area at 7.5 m below fictitious footing = (3.9 + 7.5) × (3 + 7.5) = 119.7 m^2

$$\Delta p = \frac{2500}{119.7} = 20.9 \text{ kN/m}^2$$

Settlement computation

Layer 1 $$S_1 = \frac{4 \times 0.23}{1+0.80} \log \frac{126.74 + 84.75}{126.74} = 0.113 \text{ m}$$

Layer 2 $$S_2 = \frac{2 \times 0.34}{1+1.08} \log \frac{153.95 + 35.1}{153.95} = 0.029 \text{ m}$$

$$\text{Layer 3} \quad S_3 = \frac{3 \times 0.2}{1+0.7} \log \frac{177.67 + 20.9}{177.67} = 0.017 \text{ m}$$

Total = 0.159 m ≈ 16 cm.

15.32 NEGATIVE FRICTION

Figure 15.32(a) shows a single pile and (b) a group of piles passing through a recently constructed cohesive soil fill. The soil below the fill had completely consolidated under its overburden pressure.

When the fill starts consolidating under its own overburden pressure, it develops a drag on the surface of the pile. This drag on the surface of the pile is called '*negative friction*'. Negative friction

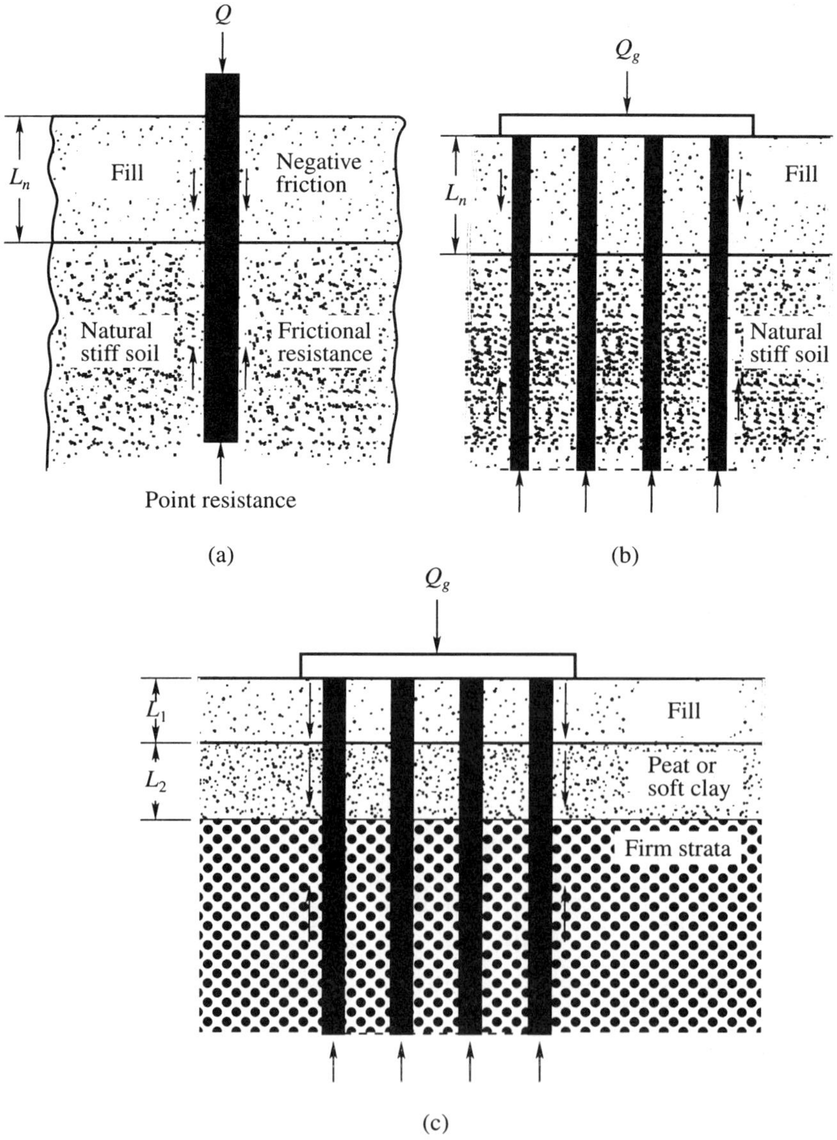

Figure 15.32 Negative friction on piles

may develop if the fill material is loose cohesionless soil. Negative friction can also occur when fill is placed over peat or a soft clay stratum as shown in Fig. 15.32c. The superimposed loading on such compressible stratum causes heavy settlement of the fill with consequent drag on piles.

Negative friction may develop by lowering the ground water which increases the effective stress causing consolidation of the soil with resultant settlement and friction forces being developed on the pile.

Negative friction must be allowed when considering the factor of safety on the ultimate carrying capacity of a pile. The factor of safety, F_s, where negative friction is likely to occur may be written as

$$F_s = \frac{\text{Ultimate carrying capacity of a single pile or group of piles}}{\text{Working load + Negative skin friction load}}$$

Computation of Negative Friction on a Single Pile

The magnitude of negative friction F_n for a single pile in a fill may be taken as (Fig. 15.32(a)).

(a) For cohesive soils

$$F_n = PL_n s. \tag{15.80}$$

(b) For cohesionless soils

$$F_n = \frac{1}{2} PL_n^2 \gamma K \tan \delta \tag{15.81}$$

where L_n = length of piles in the compressible material,
s = shear strength of cohesive soils in the fill,
P = perimeter of pile,
K = earth pressure coefficient normally lies between the active and the passive earth pressure coefficients,
δ = angle of wall friction which may vary from $\phi/2$ to ϕ.

Negative Friction on Pile Groups

When a group of piles passes through a compressible fill, the negative friction, F_{ng}, on the group may be found by any of the following methods [Fig. 15.32b].

(a) $F_{ng} = nF_n$ (15.82)

(b) $F_{ng} = sL_n P_g + \gamma L_n A_g$ (15.83)

where n = number of piles in the group,
γ = unit weight of soil within the pile group to a depth L_n,
P_g = perimeter of pile group,
A_g = sectional area of pile group within the perimeter P_g,
s = shear strength of soil along the perimeter of the group.

Equation (15.82) gives the negative friction forces of the group as equal to the sum of the friction forces of all the single piles.

Eq. (15.83) assumes the possibility of block shear failure along the perimeter of the group which includes the volume of the soil $\gamma L_n A_g$ enclosed in the group. The maximum value obtained from Eqs (15.82) or (15.83) should be used in the design.

When the fill is underlain by a compressible stratum as shown in Fig. 15.32(c), the total negative friction may be found as follows:

$$F_{ng} = n(F_{n1} + F_{n2}) \tag{15.84}$$

$$F_{ng} = s_1 L_1 P_g + s_2 L_2 P_g + \gamma_1 L_1 A_g + \gamma_2 L_2 A_g$$
$$= P_g(s_1 L_1 + s_2 L_2) + A_g(\gamma_1 L_1 + \gamma_2 L_2) \tag{15.85}$$

where L_1 = depth of fill,
L_2 = depth of compressible natural soil,
s_1, s_2 = shear strengths of the fill and compressible soils respectively,
γ_1, γ_2 = unit weights of fill and compressible soils respectively,
F_{n1} = negative friction of a single pile in the fill,
F_{n2} = negative friction of a single pile in the compressible soil.

The maximum value of the negative friction obtained from Eqs. (15.84) or (15.85) should be used for the design of pile groups.

Example 15.28

A square pile group similar to the one shown in Fig. 15.27 passes through a recently constructed fill. The depth of fill L_n = 3 m. The diameter of the pile is 30 cm and the piles are spaced 90 cm center to center. If the soil is cohesive with q_u = 60 kN/m^2, and γ = 15 kN/m^3, compute the negative frictional load on the pile group.

Solution

The negative frictional load on the group is the maximum of [(Eqs (15.82) and (15.83)]

(a) $F_{ng} = nF_n$, and (b) $F_{ng} = sL_nP_g + \gamma L_nA_g$,

where $P_g = 4 \times 3 = 12$ m, $A_g = 3 \times 3 = 9$ m^2, $c_u = 60/2 = 30$ kN/rn^2

(a) $F_{ng} = 9 \times 3.14 \times 0.3 \times 3 \times 30 = 763$ kN

(b) $F_{ng} = 30 \times 3 \times 12 + 15 \times 3 \times 9 = 1485$ kN

The negative frictional load on the group = 1485 kN.

15.33 UPLIFT CAPACITY OF A PILE GROUP

The uplift capacity of a pile group, when the vertical piles are arranged in a closely spaced groups may not be equal to the sum of the uplift resistances of the individual piles. This is because, at ultimate load conditions, the block of soil enclosed by the pile group gets lifted. The manner in which the load is transferred from the pile to the soil is quite complex. A simplified way of calculating the uplift capacity of a pile group embedded in cohesionless soil is shown in Fig. 15.33(a). A spread of load of 1 Horiz : 4 Vert from the pile group base to the ground surface may be taken as the volume of the soil to be lifted by the pile group (Tomlinson, 1977). For simplicity in calculation, the weight of the pile embedded in the ground is assumed to be equal to that of the volume of soil it displaces. If the pile group is partly or fully submerged, the submerged weight of soil below the water table has to be taken.

In the case of cohesive soil, the uplift resistance of the block of soil in undrained shear enclosed by the pile group given in Fig. 15.33(b) has to be considered. The equation for the total uplift capacity P_{gu} of the group may be expressed by

$$P_{gu} = 2L(\bar{L} + \bar{B})\bar{c}_u + W \tag{15.86}$$

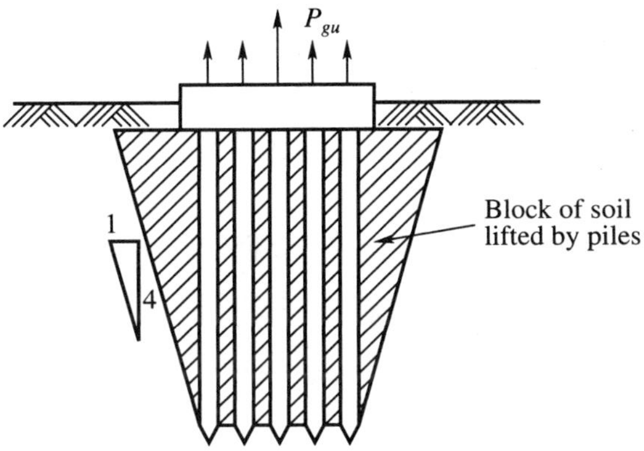

(a) Uplift of a group of closely-spaced piles in cohesionless soils

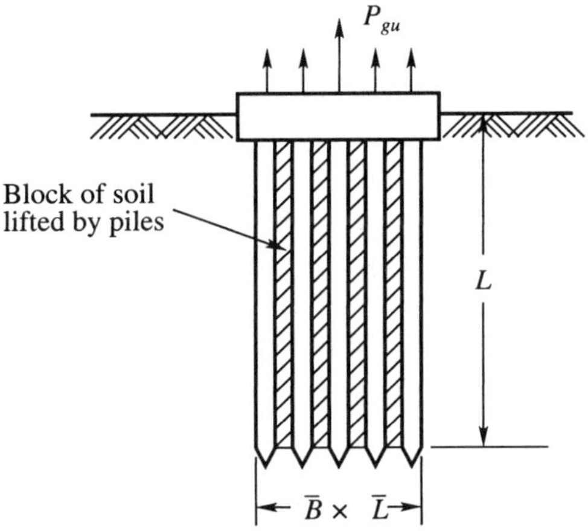

(b) Uplift of a group of piles in cohesive soils

Figure 15.33 Uplift capacity of a pile group

where L = depth of the pile block
$\overline{L}$ and $\overline{B}$ = overall length and width of the pile group
$\overline{c}_u$ = average undrained shear strength of soil around the sides of the group
W = combined weight of the block of soil enclosed by the pile group plus the weight of the piles and the pile cap.

A factor of safety of 2 may be used in both cases of piles in sand and clay.

The uplift efficiency E_{gu} of a group of piles may be expressed as

$$E_{gu} = \frac{P_{gu}}{nP_{us}} \tag{15.89}$$

where P_{us} = uplift capacity of a single pile
n = number of piles in the group

The efficiency E_{gu} varies with the method of installation of the piles, length and spacing and the type of soil. The available data indicate that E_{gu} increases with the spacing of piles. Meyerhof and Adams (1968) presented some data on uplift efficiency of groups of two and four model circular footings in clay. The results indicate that the uplift efficiency increases with the spacing of the footings or bases and as the depth of embedment decreases, but decreases as the number of footings or bases in the group increases. How far the footings would represent the piles is a debatable point. For uplift loading on pile groups in sand, there appears to be little data from full scale field tests.

15.34 PROBLEMS

15.1 A 45 cm diameter pipe pile of length 12 m with closed end is driven into a cohesionless soil having $\phi = 35°$. The void ratio of the soil is 0.48 and $G_s = 2.65$. The water table is at the ground surface. Estimate (a) the ultimate base load Q_b, (b) the frictional load Q_f, and (c) the allowable load Q_a with $F_s = 2.5$.

Use the Berezantsev method for estimating Q_b. For estimating Q_f, use $\overline{K}_s = 0.75$ and $\delta = 20°$.

15.2 Refer to Problem 15.1. Compute Q_b by Meyerhof's method. Determine Q_f using the critical depth concept, and Q_a with $F_s = 2.5$. All the other data given in Prob. 15.1 remain the same.

15.3 Estimate Q_b by Vesic's method for the pile given in Prob. 15.1. Assume $I_r = I_{rr} = 60$. Determine Q_a for $F_s = 2.5$ and use Q_f obtained in Prob. 15.1.

15.4 For Problem 15.1, estimate the ultimate base resistance Q_b by Janbu's method. Determine Q_a with $F_s = 2.5$. Use Q_f obtained in Prob. 15.1. Use $\psi = 90°$.

15.5 For Problem 15.1, estimate Q_b, Q_f, and Q_a by Coyle and Castello method. All the data given remain the same.

15.6 For problem 15.1, determine Q_b, Q_f, and Q_a by Meyerhof's method using the relationship between N_{cor} and ϕ given in Fig 12.8.

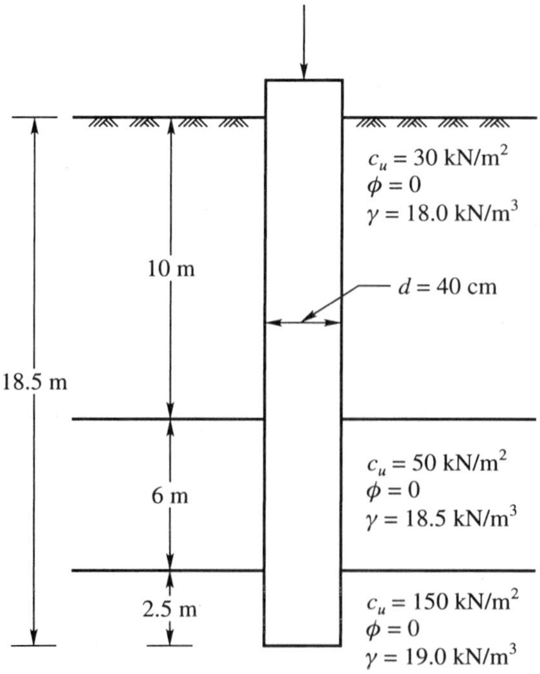

Figure Prob. 15.11

15.7 A concrete pile 40 cm in diameter is driven into homogeneous sand extending to a great depth. Estimate the ultimate load bearing capacity and the allowable load with $F_s = 3.0$ by Coyle and Castello's method. Given: $L = 15$ m, $\phi = 36°$, $\gamma = 18.5$ kN/m^3.

15.8 Refer to Prob. 15.7. Estimate the allowable load by Meyerhof's method using the relationship between ϕ and N_{cor} given in Fig. 12.8.

15.9 A concrete pile of 15 in. diameter, 40 ft long is driven into a homogeneous's stratum of clay with the water table at ground level. The clay is of medium stiff consistency with the undrained shear strength $c_u = 600$ lb/ft^2. Compute Q_b by Skempton's method and Q_f by the α-method. Determine Q_a for $F_s = 2.5$.

15.10 Refer to Prob 15.9. Compute Q_f by the λ-method. Determine Q_a by using Q_b computed in Prob. 15.9. Assume $\gamma_{sat} = 120$ lb/ft^2.

15.11 A pile of 40 cm diameter and 18.5 m long passes through two layers of clay and is embedded in a third layer. Fig. Prob. 15.11 gives the details of the soil system. Compute Q_f by the α - method and Q_b by Skempton's method. Determine Q_a for $F_s = 2.5$.

15.12 A concrete pile of size 16 × 16 in. is driven into a homogeneous clay soil of medium consistency. The water table is at ground level. The undrained shear strength of the soil is 500 lb/ft^2. Determine the length of pile required to carry a safe load of 50 kips with $F_s = 3$. Use the α - method.

15.13 Refer to Prob. 15.12. Compute the required length of pile by the λ-method. All the other data remain the same. Assume $\gamma_{sat} = 120$ lb/ft^3.

15.14 A concrete pile 50 cm in diameter is driven into a homogeneous mass of cohesionless soil. The pile is required to carry a safe load of 700 kN. A static cone penetration test conducted at the site gave an average value of $q_c = 35$ kg/cm^2 along the pile and 60 kg/cm^2 below the base of the pile. Compute the length of the pile with $F_s = 3$.

15.15 Refer to Problem 15.14. If the length of the pile driven is restricted to 12 m, estimate the ultimate load Q_u and safe load Q_a with $F_s = 3$. All the other data remain the same.

15.16 A reinforced concrete pile 20 in. in diameter penetrates 40 ft into a stratum of clay and rests on a medium dense sand stratum. Estimate the ultimate load.

Given: for sand- $\phi = 35°$, $\gamma_{sat} = 120$ lb/ft^3

for clay $\gamma_{sat} = 119$ lb/ft^3, $c_u = 800$ lb/ft^2.

Use (a) the α-method for computing the frictional load, (b) Meyerhof's method for estimating Q_b. The water table is at ground level.

15.17 A ten-story building is to be constructed at a site where the water table is close to the ground surface. The foundation of the building will be supported on 30 cm diameter pipe piles. The bottom of the pile cap will be at a depth of 1.0 m below ground level. The soil investigation at the site and laboratory tests have provided the saturated unit weights, the shear strength values under undrained conditions (average), the corrected SPT values, and the soil profile of the soil to a depth of about 40 m. The soil profile and the other details are given below.

Depth (m) From	To	Soil	γ_{sat} kN/m^3	N_{cor}	$\phi°$	c (average) kN/m^2
0	6	Sand	19	18	33°	-
6.0	22	Med. stiff clay	18	-	-	60
22	30	sand	19.6	25	35°	-
30	40	stiff clay	18.5	-	-	75

Determine the ultimate bearing capacity of a single pile for lengths of (a) 15 m, and (b) 25 m below the bottom of the cap.

Use $\alpha = 0.50$ and $\bar{K}_s = 1.2$. Assume $\delta = 0.8\ \phi$.

15.18 For a pile designed for an allowable load of 400 kN driven by a steam hammer (single acting) with a rated energy of 2070 kN-cm, what is the approximate terminal set of the pile using the ENR formula?

15.19 A reinforced concrete pile of 40 cm diameter and 25 m long is driven through medium dense sand to a final set of 2.5 mm, using a 40 kN single - acting hammer with a stroke of 150 cm. Determine the ultimate driving resistance of the pile if it is fitted with a helmet, plastic dolly and 50 mm packing on the top of the pile. The weight of the helmet, with dolly is 4.5 kN. The other particulars are: weight of pile = 85 kN, weight of hammer 35 kN; pile hammer efficiency $\eta_h = 0.85$; the coefficient restitution $C_r = 0.45$. Use Hiley's formula. The sum of elastic compression $\bar{C} = c_1 + c_2 + c_3 = 20.1$ mm.

15.20 A reinforced concrete pile 45 ft long and 20 in. in diameter is driven into a stratum of homogeneous saturated clay having $c_u = 800$ lb/ft^2. Determine (a) the ultimate load capacity and the allowable load with $F_s = 3$; (b) the pullout capacity and the allowable pullout load with $F_s = 3$. Use the α-method for estimating the compression load.

15.21 Refer to Prob. 15.20. If the pile is driven to medium dense sand, estimate (a) the ultimate compression load and the allowable load with $F_s = 3$, and (b) the pullout capacity and the allowable pullout load with $F_s = 3$. Use the Coyle and Castello method for computing Q_b and Q_f. The other data available are: $\phi = 36°$, and $\gamma = 115$ lb/ft^3. Assume the water table is at a great depth.

15.22 A group of nine friction piles arranged in a square pattern is to be proportioned in a deposit of medium stiff clay. Assuming that the piles are 30 cm diameter and 10 m long, find the optimum spacing for the piles. Assume $\alpha = 0.8$ and $c_u = 50$ kN/m^2.

15.23 A group of 9 piles with 3 in a row was driven into sand at a site. The diameter and length of the piles are 30 cm and 12 m respectively. The properties of the soil are: $\phi = 30$, $e = 0.7$, and $G_s = 2.64$.

If the spacing of the piles is 90 cm, compute the allowable load on the pile group on the basis of shear failure for $F_s = 2.0$ with respect to skin resistance, and $F_s = 2.5$ with respect to base resistance. For $\phi = 30°$, assume $N_q = 22.5$ and $N_\gamma = 19.7$. The water table is at ground level.

15.24 Nine RCC piles of diameter 30 cm each are driven in a square pattern at 90 cm center to center to a depth of 12 m into a stratum of loose to medium dense sand. The bottom of the pile cap embedding all the piles rests at a depth of 1.5 m below the ground surface. At a depth of 15 m lies a clay stratum of thickness 3 m and below which lies sandy strata. The liquid limit of the clay is 45%. The saturated unit weights of sand and clay are 18.5 kN/m^3 and 19.5 kN/m^3 respectively. The initial void ratio of the clay is 0.65. Calculate the consolidation settlement of the pile group under the allowable load. The allowable load $Q_a = 120$ kN.

15.25 A square pile group consisting of 16 piles of 40 cm diameter passes through two layers of compressible soils as shown in Fig. 15.32(c). The thicknesses of the layers are : $L_1 = 2.5$ m and $L_2 = 3$ m. The piles are spaced at 100 cm center to center. The properties of the fill material are: top fill $c_u = 25$ kN/m^2; the bottom fill (peat), $c_u = 30$ kN/m^2. Assume $\gamma = 14$ kN/m^3 for both the fill materials. Compute the negative frictional load on the pile group.

CHAPTER 16

DEEP FOUNDATION II: BEHAVIOR OF LATERALLY LOADED VERTICAL AND BATTER PILES

16.1 INTRODUCTION

When a soil of low bearing capacity extends to a considerable depth, piles are generally used to transmit vertical and lateral loads to the surrounding soil media. Piles that are used under tall chimneys, television towers, high rise buildings, high retaining walls, offshore structures, etc. are normally subjected to high lateral loads. These piles or pile groups should resist not only vertical movements but also lateral movements. The requirements for a satisfactory foundation are,

1. The vertical settlement or the horizontal movement should not exceed an acceptable maximum value,
2. There must not be failure by yield of the surrounding soil or the pile material.

Vertical piles are used in foundations to take normally vertical loads and small lateral loads. When the horizontal load per pile exceeds the value suitable for vertical piles, *batter piles* are used in combination with vertical piles. Batter piles are also called *inclined piles* or *raker piles*. The degree of batter, is the angle made by the pile with the vertical, may up to 30°. If the lateral load acts on the pile in the direction of batter, it is called an *in-batter* or *negative batter* pile. If the lateral load acts in the direction opposite to that of the batter, it is called an *out-batter* or *positive batter* pile. Fig. 16.1a shows the two types of batter piles.

Extensive theoretical and experimental investigation has been conducted on single vertical piles subjected to lateral loads by many investigators. Generalized solutions for laterally loaded vertical piles are given by Matlock and Reese (1960). The effect of vertical loads in addition to lateral loads has been evaluated by Davisson (1960) in terms of non-dimensional parameters. Broms (1964a, 1964b) and Poulos and Davis (1980) have given different approaches for solving laterally loaded pile problems. Brom's method is ingenious and is based primarily on the use of limiting values of soil resistance. The method of Poulos and Davis is based on the theory of elasticity.

The finite difference method of solving the differential equation for a laterally loaded pile is very much in use where computer facilities are available. Reese et al., (1974) and Matlock (1970) have developed the concept of (p-y) curves for solving laterally loaded pile problems. This method is quite popular in the USA and in some other countries.

However, the work on batter piles is limited as compared to vertical piles. Three series of tests on single 'in' and 'out' batter piles subjected to lateral loads have been reported by Matsuo (1939). They were run at three scales. The small and medium scale tests were conducted using timber piles embedded in sand in the laboratory under controlled density conditions. Loos and Breth (1949) reported a few model tests in dry sand on vertical and batter piles. Model tests to determine the effect of batter on pile load capacity have been reported by Tschebotarioff (1953), Yoshimi (1964), and Awad and Petrasovits (1968). The effect of batter on deflections has been investigated by Kubo (1965) and Awad and Petrasovits (1968) for model piles in sand.

Full-scale field tests on single vertical and batter piles, and also groups of piles, have been made from time to time by many investigators in the past. The field test values have been used mostly to check the theories formulated for the behavior of vertical piles only. Murthy and Subba Rao (1995) made use of field and laboratory data and developed a new approach for solving the laterally loaded pile problem.

Reliable experimental data on batter piles are rather scarce compared to that of vertical piles. Though Kubo (1965) used instrumented model piles to study the deflection behavior of batter piles, his investigation in this field was quite limited. The work of Awad and Petrasovits (1968) was based on non-instrumented piles and as such does not throw much light on the behavior of batter piles.

The author (Murthy, 1965) conducted a comprehensive series of model tests on instrumented piles embedded in dry sand. The batter used by the author varied from –45° to +45°. A part of the author's study on the behavior of batter piles, based on his own research work, has been included in this chapter.

16.2 WINKLER'S HYPOTHESIS

Most of the theoretical solutions for laterally loaded piles involve the concept of *modulus of subgrade reaction* or otherwise termed as *soil modulus* which is based on Winkler's assumption that a soil medium may be approximated by a series of closely spaced independent elastic springs. Fig. 16.1(b) shows a loaded beam resting on a elastic foundation. The reaction at any point on the base of the beam is actually a function of every point along the beam since soil material exhibits varying degrees of continuity. The beam shown in Fig. 16.1(b) can be replaced by a beam in Fig. 16.1(c). In this figure the beam rests on a bed of elastic springs wherein each spring is independent of the other. According to Winkler's hypothesis, the reaction at any point on the base of the beam in Fig. 16.1(c) depends only on the deflection at that point. Vesic (1961) has shown that the error inherent in Winkler's hypothesis is not significant.

The problem of a laterally loaded pile embedded in soil is closely related to the beam on an elastic foundation. A beam can be loaded at one or more points along its length, whereas in the case of piles the external loads and moments are applied at or above the ground surface only.

The nature of a laterally loaded pile-soil system is illustrated in Fig. 16.1(d) for a vertical pile. The same principle applies to batter piles. A series of nonlinear springs represents the force-deformation characteristics of the soil. The springs attached to the blocks of different sizes indicate reaction increasing with deflection and then reaching a yield point, or a limiting value that depends on depth; the taper on the springs indicates a nonlinear variation of load with deflection. The gap between the pile and the springs indicates the molding away of the soil by repeated loadings and the

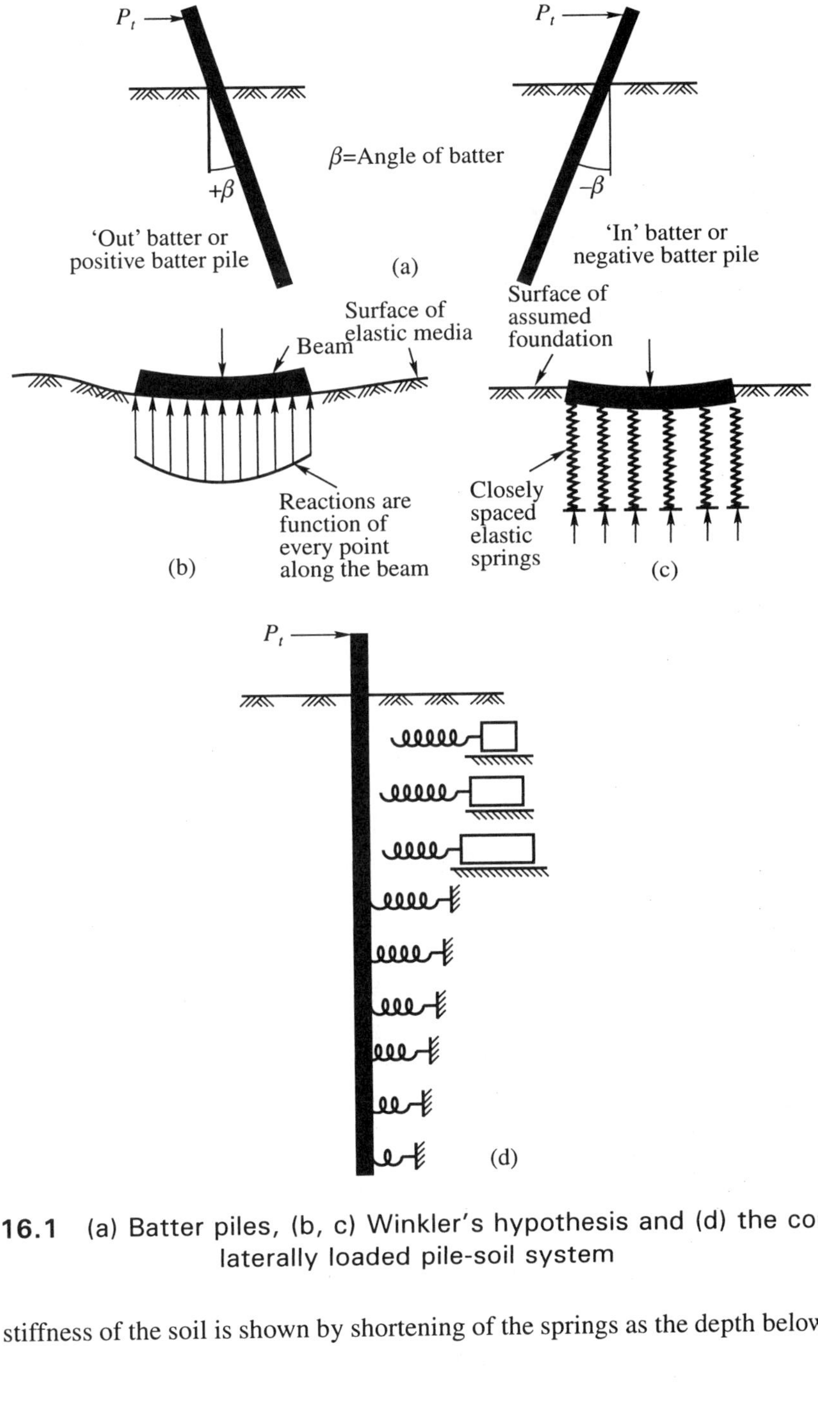

Figure 16.1 (a) Batter piles, (b, c) Winkler's hypothesis and (d) the concept of laterally loaded pile-soil system

increasing stiffness of the soil is shown by shortening of the springs as the depth below the surface increases.

16.3 THE DIFFERENTIAL EQUATION

Compatibility

As stated earlier, the problem of the laterally loaded pile is similar to the beam-on-elastic foundation problem. The interaction between the soil and the pile or the beam must be treated quantitatively in the problem solution. The two conditions that must be satisfied for a rational analysis of the problem are,

1. Each element of the structure must be in equilibrium and
2. Compatibility must be maintained between the superstructure, the foundation and the supporting soil.

If the assumption is made that the structure can be maintained by selecting appropriate boundary conditions at the top of the pile, the remaining problem is to obtain a solution that insures equilibrium and compatibility of each element of the pile, taking into account the soil response along the pile. Such a solution can be made by solving the differential equation that describes the pile behavior.

The Differential Equation of the Elastic Curve

The standard differential equations for slope, moment, shear and soil reaction for a beam on an elastic foundation are equally applicable to laterally loaded piles.

The deflection of a point on the elastic curve of a pile is given by y. The x-axis is along the pile axis and deflection is measured normal to the pile-axis.

The relationships between y, slope, moment, shear and soil reaction at any point on the deflected pile may be written as follows.

deflection of the pile $= y$

slope of the deflected pile $S = \frac{dy}{dx}$ (16.1)

moment of pile $M = EI\frac{d^2y}{dx^2}$ (16.2)

shear $V = EI\frac{d^3y}{dx^3}$ (16.3)

soil reaction, $p = EI\frac{d^4y}{dx^4}$ (16.4)

where EI is the flexural rigidity of the pile material.

The soil reaction p at any point at a distance x along the axis of the pile may be expressed as

$$p = -E_s y \tag{16.5}$$

where y is the deflection at point x, and E_s is the soil *modulus*. Eqs (16.4) and (16.5) when combined gives

$$EI\frac{d^4y}{dx^4} + E_s y = 0 \tag{16.6}$$

which is called the *differential equation* for the *elastic curve* with zero axial load.

The key to the solution of laterally loaded pile problems lies in the determination of the value of the modulus of subgrade reaction (soil modulus) with respect to depth along the pile. Fig. 16.2(a) shows a vertical pile subjected to a lateral load at ground level. The deflected position of the pile and the corresponding soil reaction curve are also shown in the same figure. The soil modulus E_s at any point x below the surface along the pile as per Eq. (16.5) is

$$E_s = -\frac{p}{y} \tag{16.7}$$

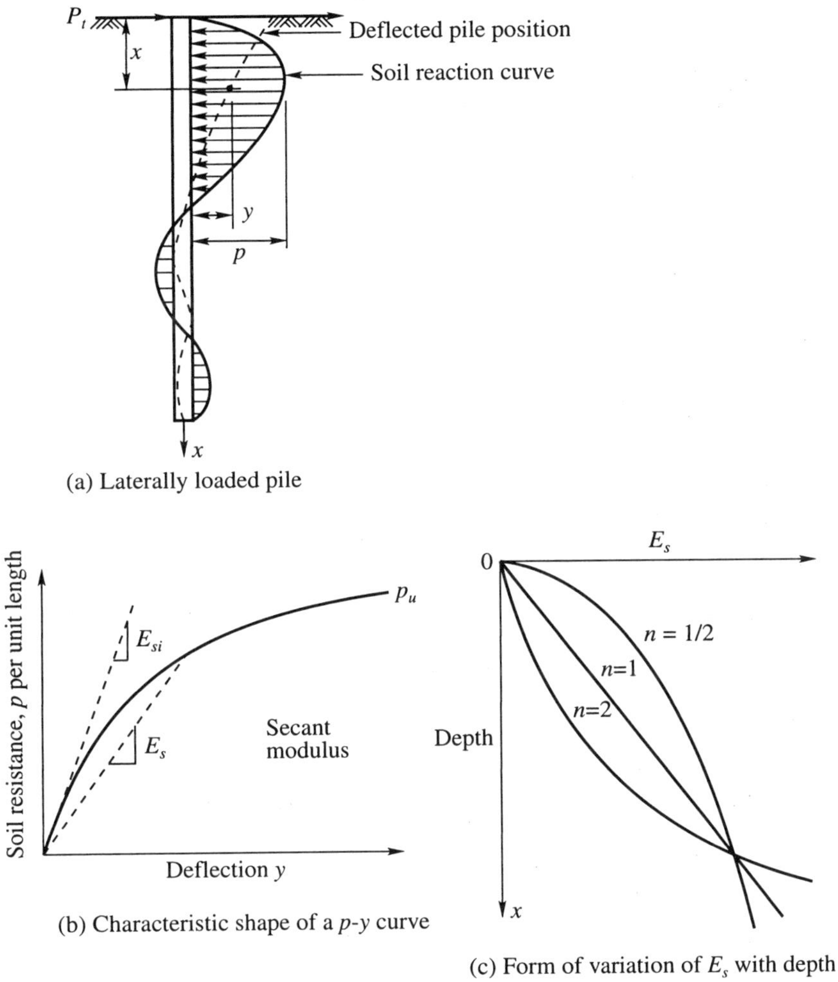

Figure 16.2 The concept of (p-y) curves: (a) a laterally loaded pile, (b) characteristic shape of a p–y curve, and (c) the form of variation of E_s with depth

As the load P_t at the top of the pile increases the deflection y and the corresponding soil reaction p increase. A relationship between p and y at any depth x may be established as shown in Fig. 16.2(b). It can be seen that the curve is strongly non-linear, changing from an initial tangent modulus E_{si} to an ultimate resistance p_u. E_s is not a constant and changes with deflection.

There are many factors that influence the value of E_s such as the pile width d, the flexural stiffness EI, the magnitude of loading P_t and the soil properties.

The variation of E_s with depth for any particular load level may be expressed as

$$E_s = n_h x^n \tag{16.8a}$$

in which n_h is termed the *coefficient of soil modulus variation.* The value of the power n depends upon the type of soil and the batter of the pile. Typical curves for the form of variation of E_s with depth for values of n equal to 1/2, 1, and 2 are given 16.2(c). The most useful form of variation of E_s is the linear relationship expressed as

$$E_s = n_h x \tag{16.8b}$$

which is normally used by investigators for vertical piles.

Table 16.1 Typical values of n_h for cohesive soils (Taken from Poulos and Davis, 1980)

Soil type	n_h lb/in^3	Reference
Soft NC clay	0.6 to 12.7	Reese and Matlock, 1956
	1.0 to 2.0	Davisson and Prakash, 1963
NC organic clay	0.4 to 1.0	Peck and Davisson, 1962
	0.4 to 3.0	Davisson, 1970
Peat	0.2	Davisson, 1970
	0.1 to 0.4	Wilson and Hills, 1967
Loess	29 to 40	Bowles, 1968

Table 16.1 gives some typical values for cohesive soils for n_h and Fig. 16.3 gives the relationship between n_h and the relative density of sand (Reese, 1975).

16.4 NON-DIMENSIONAL SOLUTIONS FOR VERTICAL PILES SUBJECTED TO LATERAL LOADS

Matlock and Reese (1960) have given equations for the determination of y, S, M, V, and p at any point x along the pile based on dimensional analysis. The equations are

deflection, $$y = \left[\frac{P_t T^3}{EI}\right] A_y + \left[\frac{M_t T^2}{EI}\right] B_y \tag{16.9}$$

slope, $$S = \left[\frac{P_t T^2}{EI}\right] A_s + \left[\frac{M_t T}{EI}\right] B_s \tag{16.10}$$

moment, $$M = \left[P_t T\right] A_m + \left[M_t\right] B_m \tag{16.11}$$

shear, $$V = \left[P_t\right] A_v + \left[\frac{M_t}{T}\right] B_v \tag{16.12}$$

soil reaction, $$p = \frac{P_t}{T} A_p + \frac{M_t}{T^2} B_p \tag{16.13}$$

where T is the relative stiffness factor expressed as

$$T = \left[\frac{EI}{n_h}\right]^{\frac{1}{5}} \tag{16.14a}$$

for $E_s = n_h x$

For a general case $$T = \left[\frac{EI}{n_h}\right]^{\frac{1}{n+4}} \tag{16.14b}$$

In Eqs (16.9) through (16.13), A and B are the sets of non-dimensional coefficients whose values are given in Table 16.2. The principle of superposition for the deflection of a laterally loaded

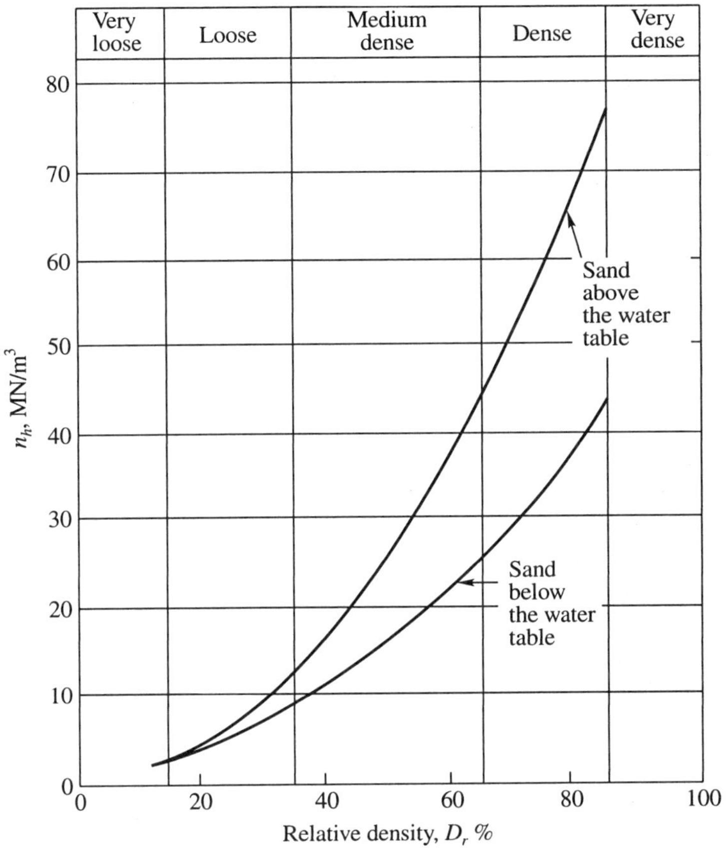

Figure 16.3 Variation of n_h with relative density (Reese, 1975)

pile is shown in Fig. 16.4. The *A* and *B* coefficients are given as a function of the depth coefficient, *Z*, expressed as

$$Z = \frac{x}{T} \tag{16.14c}$$

The *A* and *B* coefficients tend to zero when the depth coefficient *Z* is equal to or greater than 5 or otherwise the length of the pile is more than 5*T*. Such piles are called long or flexible piles. The length of a pile loses its significance beyond 5*T*.

Normally we need deflection and slope at ground level. The corresponding equations for these may be expressed as

$$y_g = 2.43\frac{P_t T^3}{EI} + 1.62\frac{M_t T^2}{EI} \tag{16.15a}$$

$$S_g = 1.62\frac{P_t T^2}{EI} + 1.75\frac{M_t T}{EI} \tag{16.15b}$$

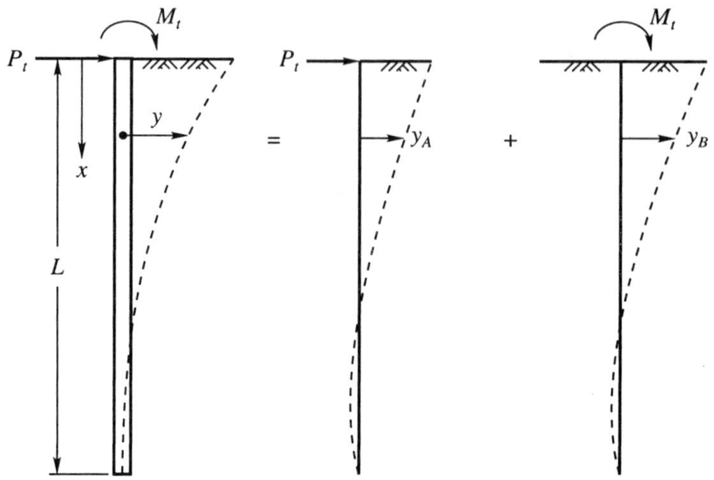

Figure 16.4 Principle of superposition for the deflection of laterally loaded piles

y_g for fixed head is

$$y_g = 0.93 \frac{P_t T^3}{EI} \tag{16.16a}$$

Moment at ground level for fixed head is

$$M_t = -0.93[P_t T] \tag{16.16b}$$

16.5 *p-y* CURVES FOR THE SOLUTION OF LATERALLY LOADED PILES

Section 16.4 explains the methods of computing deflection, slope, moment, shear and soil reaction by making use of equations developed by non-dimensional methods. The prediction of the various curves depends primarily on the single parameter n_h. If it is possible to obtain the value of n_h independently for each stage of loading P_t, the *p-y* curves at different depths along the pile can be constructed as follows:

1. Determine the value of n_h for a particular stage of loading P_t.
2. Compute T from Eq. (16.14a) for the linear variation of E_s with depth.
3. Compute y at specific depths $x = x_1$, $x = x_2$, etc. along the pile by making use of Eq. (16.9), where A and B parameters can be obtained from Table 16.2 for various depth coefficients Z.
4. Compute p by making use of Eq. (16.13), since T is known, for each of the depths $x = x_1$, $x = x_2$, etc.
5. Since the values of p and y are known at each of the depths x_1, x_2 etc., one point on the *p-y* curve at each of these depths is also known.
6. Repeat steps 1 through 5 for different stages of loading and obtain the values of p and y for each stage of loading and plot to determine *p-y* curves at each depth.

The individual *p-y* curves obtained by the above procedure at depths x_1, x_2, etc. can be plotted on a common pair of axes to give a family of curves for the selected depths below the surface. The *p-y* curve shown in Fig. 16.2b is strongly non-linear and this curve can be predicted only if the

Table 16.2 The A and B coefficients as obtained by Reese and Matlock (1956) for long vertical piles on the assumption $E_s = n_h x$

Z	A_y	A_s	A_m	A_v	A_p
0.0	2.435	–1.623	0.000	1.000	0.000
0.1	2.273	–1.618	0.100	0.989	–0.227
0.2	2.112	–1.603	0.198	0.966	–0.422
0.3	1.952	–1.578	0.291	0.906	–0.586
0.4	1.796	–1.545	0.379	0.840	–0.718
0.5	1.644	–1.503	0.459	0.764	–0.822
0.6	1.496	–1.454	0.532	0.677	–0.897
0.7	1.353	–1.397	0.595	0.585	–0.947
0.8	1.216	–1.335	0.649	0.489	–0.973
0.9	1.086	–1.268	0.693	0.392	–0.977
1.0	0.962	–1.197	0.727	0.295	–0.962
1.2	0.738	–1.047	0.767	0.109	–0.885
1.4	0.544	–0.893	0.772	–0.056	–0.761
1.6	0.381	–0.741	0.746	–0.193	–0.609
1.8	0.247	–0.596	0.696	–0.298	–0.445
2.0	0.142	–0.464	0.628	–0.371	–0.283
3.0	–0.075	–0.040	0.225	–0.349	0.226
4.0	–0.050	0.052	0.000	–0.016	0.201
5.0	–0.009	0.025	–0.033	0.013	0.046
Z	B_y	B_s	B_m	B_v	B_p
0.0	1.623	–1.750	1.000	0.000	0.000
0.1	1.453	–1.650	1.000	–0.007	–0.145
0.2	1.293	–1.550	0.999	–0.028	–0.259
0.3	1.143	–1.450	0.994	–0.058	–0.343
0.4	1.003	–1.351	0.987	–0.095	–0.401
0.5	0.873	–1.253	0.976	–0.137	–0.436
0.6	0.752	–1.156	0.960	–0.181	–0.451
0.7	0.642	–1.061	0.939	–0.226	–0.449
0.8	0.540	–0.968	0.914	–0.270	–0.432
0.9	0.448	–0.878	0.885	–0.312	–0.403
1.0	0.364	–0.792	0.852	–0.350	–0.364
1.2	0.223	–0.629	0.775	–0.414	–0.268
1.4	0.112	–0.482	0.668	–0.456	–0.157
1.6	0.029	–0.354	0.594	–0.477	–0.047
1.8	–0.030	–0.245	0.498	–0.476	0.054
2.0	–0.070	–0.155	0.404	–0.456	0.140
3.0	–0.089	0.057	0.059	–0.0213	0.268
4.0	–0.028	0.049	0.042	0.017	0.112
5.0	0.000	0.011	0.026	0.029	–0.002

values of n_h are known for each stage of loading. Further, the curve can be extended until the soil reaction, p, reaches an ultimate value, p_u, at any specific depth x below the ground surface.

If n_h values are not known to start with at different stages of loading, the above method cannot be followed. Supposing p-y curves can be constructed by some other independent method, then p-y curves are the starting points to obtain the curves of deflection, slope, moment and shear. This means we are proceeding in the reverse direction in the above method. The methods of constructing p-y curves and predicting the non-linear behavior of laterally loaded piles are beyond the scope of this book. This method has been dealt with in detail by Reese (1985).

Example 16.1

A steel pipe pile of 61 cm outside diameter with a wall thickness of 2.5 cm is driven into loose sand (D_r = 30%) under submerged conditions to a depth of 20 m. The submerged unit weight of the soil is 8.75 kN/m^3 and the angle of internal friction is 33°. The EI value of the pile is 4.35×10^{11} kg-cm^2 (4.35×10^2 MN-m^2). Compute the ground line deflection of the pile under a lateral load of 268 kN at ground level under a free head condition using the non-dimensional parameters of Matlock and Reese. The n_h value from Fig. 16.3 for D_r = 30% is 6 MN/m^3 for a submerged condition.

Solution

From Eq. (16.15a)

$$y_g = 2.43\frac{P_t T^3}{EI} \text{ for } M_t = 0$$

From Eq. (16.14a),

$$T = \left[\frac{EI}{n_h}\right]^{\frac{1}{5}}$$

where, P_t = 0.268 MN

$EI = 4.35 \times 10^2$ MN-m^2

n_h = 6 MN/m^3

$$T = \left[\frac{4.35 \times 10^2}{6}\right]^{\frac{1}{5}} = 2.35 \text{ m}$$

$$\text{Now } y_g = \frac{2.43 \times 0.268 \times (2.35)^3}{4.35 \times 10^2} = 0.0194 \text{ m} = 1.94 \text{ cm}$$

Example 16.2

If the pile in Ex. 16.1 is subjected to a lateral load at a height 2 m above ground level, what will be the ground line deflection?

Solution

From Eq. (16.15a)

$$y_g = 2.43\frac{P_t T^3}{EI} + 1.62\frac{M_t T^2}{EI}$$

As in Ex. 16.1 $T = 2.35$ m, $M_t = 0.268 \times 2 = 0.536$ MN-m

$$\text{Substituting, } y_g = \frac{2.43 \times 0.268 \times (2.35)^3}{4.35 \times 10^2} + \frac{1.62 \times 0.536 \times (2.35)^2}{4.35 \times 10^2}$$

$$= 0.0194 + 0.0110 = 0.0304 \text{ m} = 3.04 \text{ cm.}$$

Example 16.3

If the pile in Ex. 16.1 is fixed against rotation, calculate the deflection at the ground line.

Solution

Use Eq. (16.16a)

$$y_g = \frac{0.93 P_t T^3}{EI}$$

The values of P_t, T and El are as given in Ex. 16.1. Substituting these values

$$y_g = \frac{0.93 \times 0.268 \times (2.35)^3}{4.35 \times 10^2} = 0.0075 \text{ m} = 0.75 \text{ cm}$$

16.6 BROMS' SOLUTIONS FOR LATERALLY LOADED PILES

Broms' (1964a, 1964b) solutions for laterally loaded piles deal with the following:

1. Lateral deflections of piles at ground level at working loads
2. Ultimate lateral resistance of piles under lateral loads

Broms' provided solutions for both short and long piles installed in cohesive and cohesionless soils respectively. He considered piles fixed or free to rotate at the head. Lateral deflections at working loads have been calculated using the concept of subgrade reaction. It is assumed that the deflection increases linearly with the applied loads when the loads applied are less than one-half to one-third of the ultimate lateral resistance of the pile.

Lateral Deflections at Working Loads

Lateral deflections at working loads can be obtained from Fig. 16.5 for cohesive soil and Fig. 16.6 for cohesionless soils respectively. For piles in saturated cohesive soils, the plot in Fig. 16.5 gives the relationships between the dimensionless quantity βL and $(y_o kdL)/P_t$ for free-head and restrained piles, where

$$\beta = \left(\frac{kd}{4EI}\right)^{1/4} \tag{16.17}$$

EI = stiffness of pile section
k = coefficient of horizontal subgrade reaction
d = width or diameter of pile
L = length of pile

A pile is considered long or short on the following conditions

Free-head Pile

Long pile when $\beta L > 2.50$
Short pile when $\beta L < 2.50$

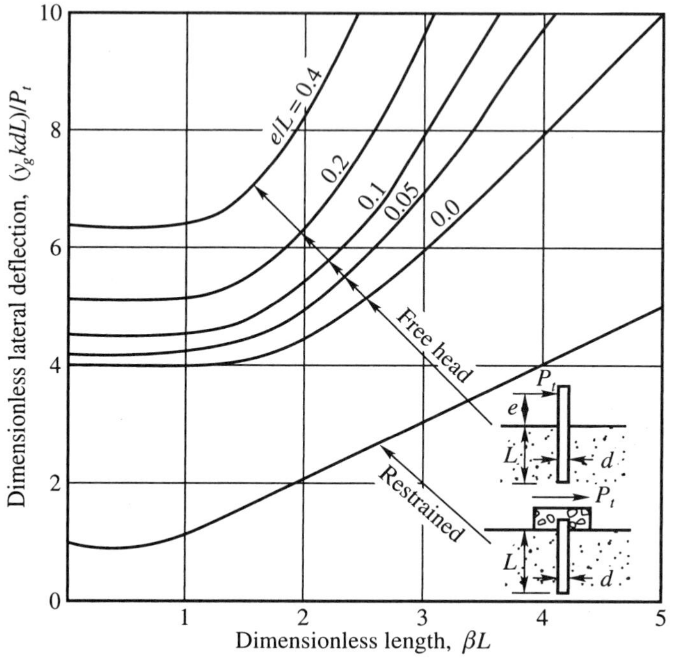

Figure 16.5 Charts for calculating lateral deflection at the ground surface of horizontally loaded pile in cohesive soil (after Broms 1964a)

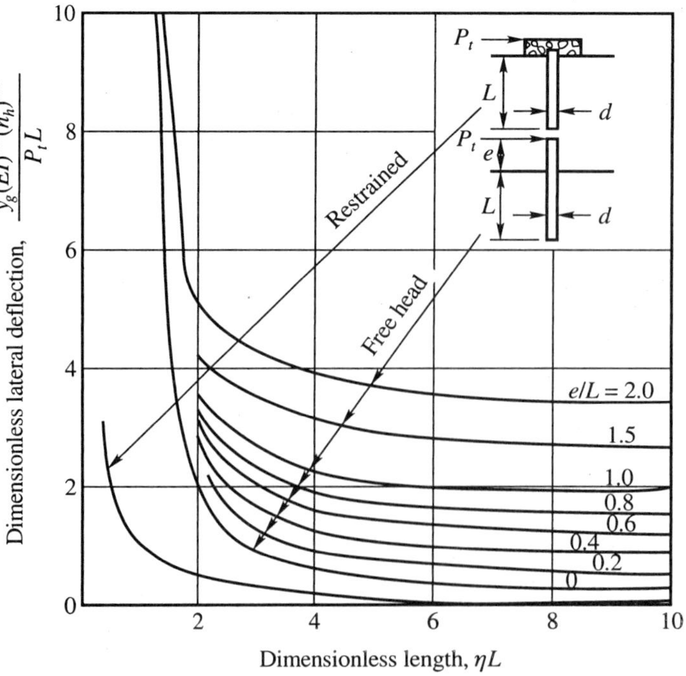

Figure 16.6 Charts for calculating lateral deflection at the ground surface of horizontally loaded piles in cohesionless soil (after Broms 1964b)

Fixed-head Pile

Long pile when $\beta L > 1.5$
Short pile when $\beta L < 1.5$

Tomlinson (1977) suggests that it is sufficiently accurate to take the value of k in Eq. (16.17) as equal to k_1 given in Table 14.1(b).

Lateral deflections at working loads of piles embedded in cohesionless soils may be obtained from Fig. 16.6 Non-dimensionless factor $[y_g(EI)^{3/5}(n_h)^{2/5}]/P_tL$ is plotted as a function of ηL for various values of e/L

where y_g = deflection at ground level

$$\eta = \left(\frac{n_h}{EI}\right)^{1/5} \tag{16.18}$$

n_h = coefficient of soil modulus variation
P_t = lateral load applied at or above ground level
L = length of pile
e = eccentricity of load.

Ultimate Lateral Resistance of Piles in Saturated Cohesive Soils

The ultimate soil resistance of piles in cohesive soils increases with depth from $2c_u$ (c_u = undrained shear strength) to 8 to 12 c_u at a depth of three pile diameters (3d) below the surface. Broms (1964a) suggests a constant value of $9c_u$ below a depth of $1.5d$ as the ultimate soil resistance. Figure 16.7 gives solutions for short piles and Fig. 16.8 for long piles. The solution for long piles

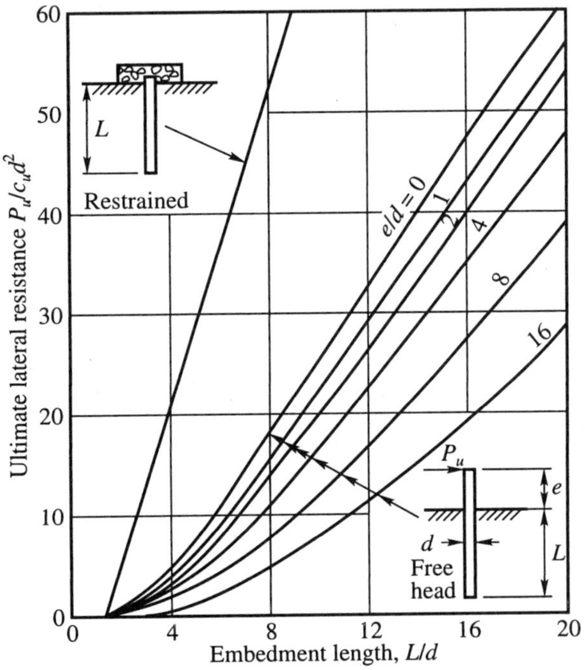

Figure 16.7 Ultimate lateral resistance of a short pile in cohesive soil related to embedded length (after Broms (1964a))

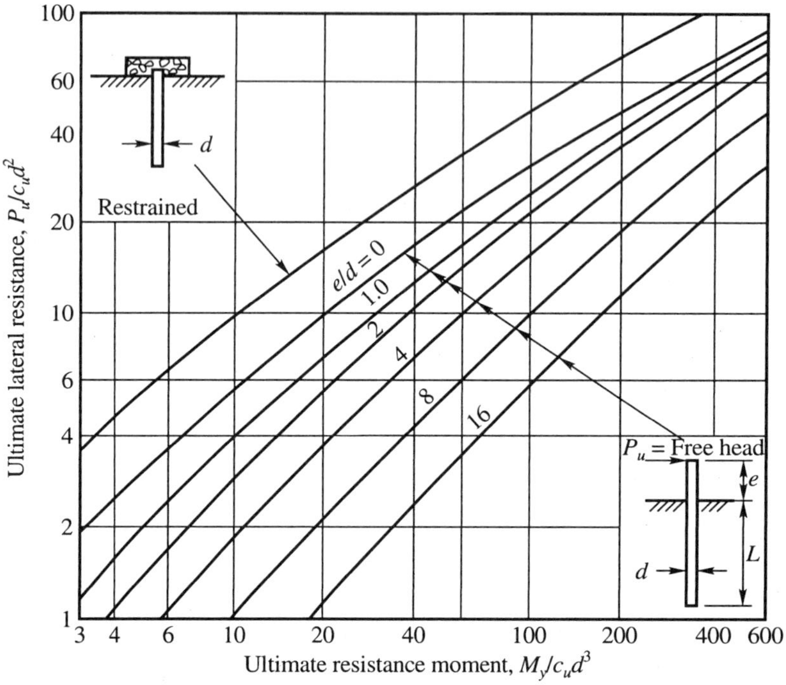

Figure 16.8 Ultimate lateral resistance of a long pile in cohesive soil related to embedded length (after Broms (1964a))

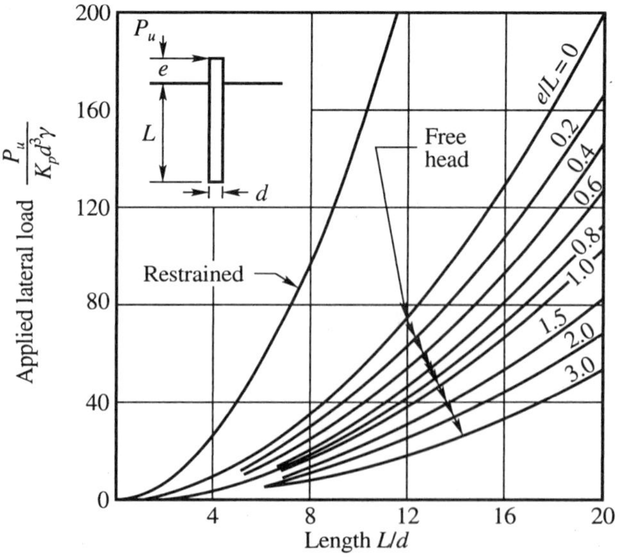

Figure 16.9 Ultimate lateral resistance of a short pile in cohesionless soil related to embedded length (after Broms (1964b))

involves the yield moment M_y for the pile section. The equations suggested by Broms for computing M_y are as follows:

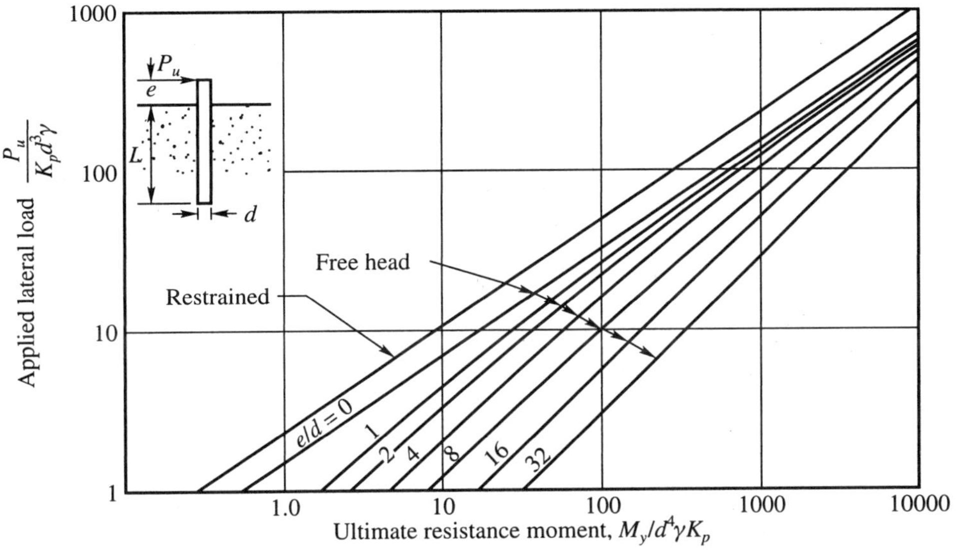

Figure 16.10 Ultimate lateral resistance of a long pile in cohesionless soil related to embedded length (after Broms (1964b))

For a cylindrical steel pipe section

$$M_y = 1.3 f_y Z \tag{16.19a}$$

For an H-section

$$M_y = 1.1 f_y Z_{max} \tag{16.19b}$$

where f_y = yield strength of the pile material

Z = section modulus of the pile section

The ultimate strength of a reinforced concrete pile section can be calculated in a similar manner.

Ultimate Lateral Resistance of Piles in Cohesionless Soils

The ultimate lateral resistance of a short piles embedded in cohesionless soil can be estimated making use of Fig. 16.9 and that of long piles from Fig. 16.10. In Fig. 16.9 the dimensionless quantity $P_u/\gamma d^3 K_P$ is plotted against the L/d ratio for short piles and in Fig. 16.10 $P_u/\gamma d^3 K_P$ is plotted against $M_y/\gamma d^4 K_P$. In both cases the terms used are

γ = effective unit weight of soil

K_P = Rankine's passive earth pressure coefficient = $\tan^2(45°+\phi/2)$

Example 16.4

A steel pipe pile of 61 cm outside diameter with 2.5 cm wall thickness is driven into saturated cohesive soil to a depth of 20 m. The undrained cohesive strength of the soil is 85 kPa. Calculate the ultimate lateral resistance of the pile by Broms' method with the load applied at ground level.

Solution

The pile is considered as a long pile. Use Fig. 16.8 to obtain the ultimate lateral resistance P_u of the pile.

The non-dimensional yield moment $= \dfrac{M_y}{c_u d^3}$

where M_y = yield resistance of the pile section

= $1.3 f_y Z$

f_y = yield strength of the pile material

= 2800 kg/cm^2 (assumed)

Z = section modulus = $\dfrac{\pi}{64R}[d_0^4 - d_i^4]$

I = moment of inertia,

d_o = outside diameter = 61 cm,

d_i = inside diameter = 56 cm,

R = outside radius = 30.5 cm

$$Z = \frac{3.14}{64 \times 30.5}[61^4 - 56^4] = 6{,}452.6 \text{ cm}^3$$

M_y = 1.3 × 2,800 × 6,452.6 =23.487 × 10^6 kg-cm.

$$\frac{M_y}{c_u d_0^3} = \frac{23.487 \times 10^6}{0.85 \times 61^3} = 122$$

From Fig. 16.8 for $e/d = 0$, $\dfrac{M_y}{c_u d^3} = 122$, $\dfrac{P_u}{c_u d_0^2} \approx 35$

$P_u = 35\, c_u d_0^2 = 35 \times 85 \times 0.61^2 = 1{,}107$ kN

Example 16.5

If the pile given in Ex. 16.4 is restrained against rotation, calculate the ultimate lateral resistance P_u.

Solution

Per Ex. 16.4 $\dfrac{M_y}{c_u d_0^3} = 122$

From Fig. 16.8, for $\dfrac{M_y}{c_u d_0^3} = 122$, for restrained pile $\dfrac{P_u}{c_u d_0^2} \approx 50$

Therefore $P_u = \dfrac{50}{35} \times 1{,}107 = 1{,}581$ kN

Example 16.6

A steel pipe pile of outside diameter 61 cm and inside diameter 56 cm is driven into a medium dense sand under submerged conditions. The sand has a relative density of 60% and an angle of internal friction of 38°. Compute the ultimate lateral resistance of the pile by Brom`s method. Assume that the yield resistance of the pile section is the same as that given in Ex 16.4. The submerged unit weight of the soil γ_b =8.75 kN/m^3.

Solution

From Fig. 16.10

$$\text{Non-dimensional yield moment} = \frac{M_y}{\gamma d^4 K_P}$$

where, $K_p = \tan^2(45 + \phi/2) = \tan^2 64 = 4.20$,

$M_y = 23.487 \times 10^6$ kg-cm,

$\gamma = 8.75 \text{ kN/m}^3 \approx 8.75 \times 10^{-4} \text{ kg/cm}^3$,

$d = 61$ cm.

Substituting,

$$\frac{M_y}{\gamma d^4 K_P} = \frac{23.487 \times 10^6 \times 10^4}{8.75 \times 61^4 \times 4.2} = 462$$

From Fig. 16.10, for $\frac{M_y}{\gamma d^4 K_P} = 462$, for $e/d = 0$ we have $\frac{P_u}{\gamma d^3 K_P} \approx 80$

Therefore $P_u = 80\ \gamma d^3 K_p = 80 \times 8.75 \times 0.61^3 \times 4.2 = 667$ kN

Example 16.7

If the pile in Ex. 16.6 is restrained, what is the ultimate lateral resistance of the pile?

Solution

From Fig. 16.10, for $\frac{M_y}{\gamma d^4 K_P} = 462$, the value $\frac{P_u}{\gamma d^3 K_P} = 135$

$P_u = 135\ \gamma d^3 K_p = 135 \times 8.75 \times 0.61^3 \times 4.2 = 1{,}126$ kN.

Example 16.8

Compute the deflection at ground level by Broms' method for the pile given in Ex. 16.1.

Solution

From Eq. (16.18)

$$\eta = \left(\frac{n_h}{EI}\right)^{1/5} = \left[\frac{6}{4.35 \times 10^2}\right]^{1/5} = 0.424$$

$\eta L = 0.424 \times 20 = 8.5$.

From Fig. 16.6, for $\eta L = 8.5$, $e/L = 0$, we have

$$\frac{y_g (EI)^{3/5} (n_h)^{2/5}}{P_t L} = 0.2$$

$$y_g = \frac{0.2 P_t L}{(EI)^{3/5} (n_h)^{2/5}} = \frac{0.2 \times 0.268 \times 20}{(4.35 \times 10^2)^{3/5} (6)^{2/5}} = 0.014 \text{ m} = 1.4 \text{ cm}$$

Example 16.9

If the pile given in Ex. 16.1 is only 4 m long, compute the ultimate lateral resistance of the pile by Broms' method.

Solution

From Eq. (16.18)

$$\eta = \left(\frac{n_h}{EI}\right)^{1/5} = \left[\frac{6}{4.35 \times 10^2}\right]^{1/5} = 0.424$$

$\eta L = 0.424 \times 4 = 1.696.$

The pile behaves as an infinitely stiff member since $\eta L < 2.0$, $L/d = 4/0.61 = 6.6$.
From Fig. 16.9, for $L / d = 6.6$, $e / L = 0$, , we have

$P_u / \gamma d^3 K_p = 25.$

$\phi = 33°$, $\gamma = 8.75$ kN/m^3, $d = 61$ cm, $K_p = \tan^2 (45° + \phi/2) = 3.4$.

Now $P_u = 25 \gamma d^3 K_p = 25 \times 8.75 \times (0.61)^3 \times 3.4 = 169$ kN

If the sand is medium dense, as given in Ex. 16.6, then $K_p = 4.20$, and the ultimate lateral resistance P_u is

$$P_u = \frac{4.2}{3.4} \times 169 = 209 \text{ kN}$$

As per Ex. 16.6, P_u for a long pile = 667 kN, which indicates that the ultimate lateral resistance increases with the length of the pile and remains constant for a long pile.

16.7 A DIRECT METHOD FOR SOLVING THE NON-LINEAR BEHAVIOR OF LATERALLY LOADED FLEXIBLE PILE PROBLEMS

Key to the Solution

The key to the solution of a laterally loaded vertical pile problem is the development of an equation for n_h. The present state of the art does not indicate any definite relationship between n_h, the properties of the soil, the pile material, and the lateral loads. However it has been recognized that n_h depends on the relative density of soil for piles in sand and undrained shear strength c for piles in clay. It is well known that the value of n_h decreases with an increase in the deflection of the pile. It was Palmer et al (1948) who first showed that a change of width d of a pile will have an effect on deflection, moment and soil reaction even while EI is kept constant for all the widths. The selection of an initial value for n_h for a particular problem is still difficult and many times quite arbitrary. The available recommendations in this regard (Terzaghi 1955, and Reese 1975) are widely varying.

The author has been working on this problem since a long time (Murthy, 1965). An explicit relationship between n_h and the other variable soil and pile properties has been developed on the principles of dimensional analysis (Murthy and Subba Rao, 1995).

Development of Expressions for n_h

The term n_h may be expressed as a function of the following parameters for piles in sand and clay.

(a) Piles in sand

$$n_h = f_s\,(EI, \text{d}, P_e, \gamma, \phi) \tag{16.20}$$

(b) Piles in clay

$$n_h = f_c\,(EI, \text{d}, P_e, \gamma, c) \tag{16.21}$$

The symbols used in the above expressions have been defined earlier.

In Eqs (16.20) and (16.21), an equivalent lateral load P_e at ground level is used in place of P_t acting at a height e above ground level. An expression for P_e may be written from Eq. (16.15) as follows.

$$P_e = P_t(1 + 0.67\frac{e}{T}) \tag{16.22}$$

Now the equation for computing groundline deflection y_g is

$$y_g = \frac{2.43 P_e T^3}{EI} \tag{16.23}$$

Based on dimensional analysis the following non-dimensional groups have been established for piles in sand and clay.

Piles in Sand

$$F_n = \frac{n_h P_e^{1/3}}{C_\phi d \gamma^{4/3}} \text{ and } F_p = \frac{(EI)\gamma^{1/3}}{d P_e^{4/3}} \tag{16.24}$$

where C_ϕ = correction factor for the angle of friction ϕ. The expression for C_ϕ has been found separately based on a critical study of the available data. The expression for C_ϕ is

$$C_\phi = 3 \times 10^{-5}\,(1.316)^{\phi^\circ} \tag{16.25}$$

Fig. 16.11 gives a plot of C_ϕ versus ϕ.

Piles in Clay

The nondimensional groups developed for piles in clay are

$$F_n = \frac{n_h \sqrt{P_e}\,(1 + e/d)^{1.5}}{c^{1.5}}; \qquad F_p = \frac{\sqrt{EI\gamma d}}{P_e} \tag{16.26}$$

In any lateral load test in the field or laboratory, the values of EI, γ, ϕ (for sand) and c (for clay) are known in advance. From the lateral load tests, the ground line deflection curve P_t versus y_g is known, that is, for any applied load P_t, the corresponding measured y_g is known. The values of T, n_h and P_e can be obtained from Eqs (16.14a), (16.15) and (16.22) respectively. C_ϕ is obtained from Eq. (16.25) for piles in sand or from Fig. 16.11. Thus the right hand side of functions F_n and F_p are known at each load level.

A large number of pile test data were analyzed and plots of $\sqrt{F_n}$ versus F_p were made on log-log scale for piles in sand, Fig. (16.12) and F_n versus F_p for piles in clay, Fig. (16.13). The method of least squares was used to determine the linear trend. The equations obtained are as given below.

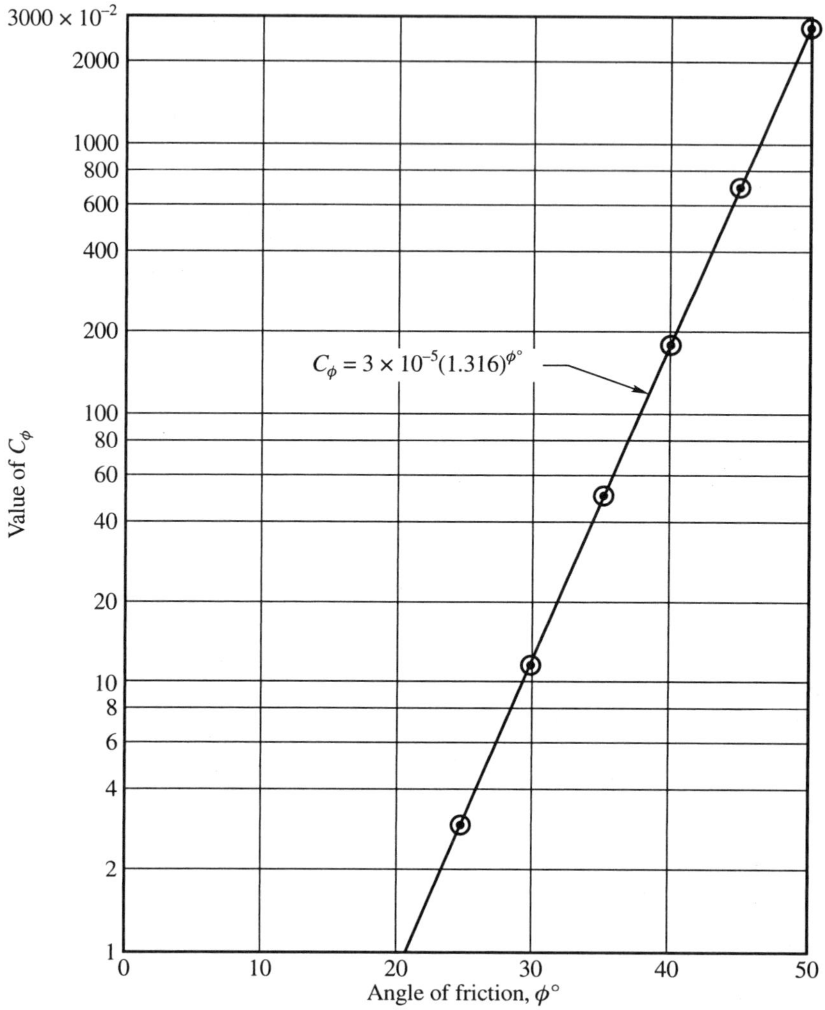

Figure 16.11 C_ϕ versus $\phi°$

Piles in Sand

$$F_n = 150\sqrt{F_p} \tag{16.27}$$

Piles in Clay

$$F_n = 125\, F_p \tag{16.28}$$

By substituting for F_n and F_p, and simplifying, the expressions for n_h for piles in sand and clay are obtained as

$$\text{for piles in sand,} \quad n_h = \frac{150 C_\phi \gamma^{1.5}\sqrt{EId}}{P_e} \tag{16.29}$$

$$\text{for piles in clay,} \quad n_h = \frac{125 c^{1.5}\sqrt{EI\gamma d}\Big/\left(1+e/d\right)^{1.5}}{P_e^{1.5}} \tag{16.30}$$

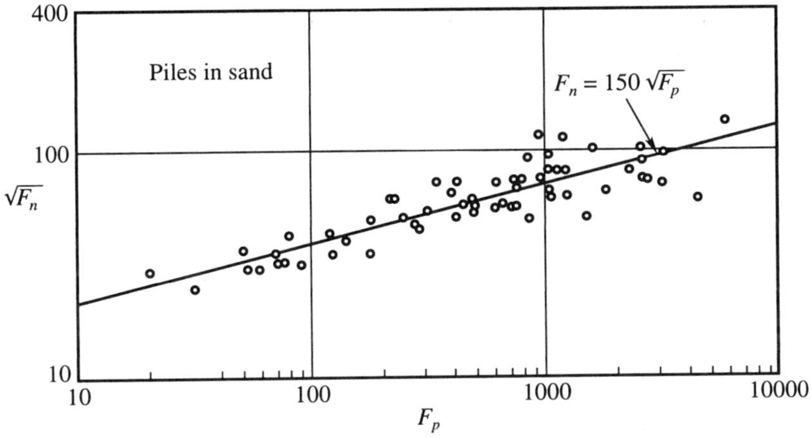

Figure 16.12 Nondimensional plot for piles in sand

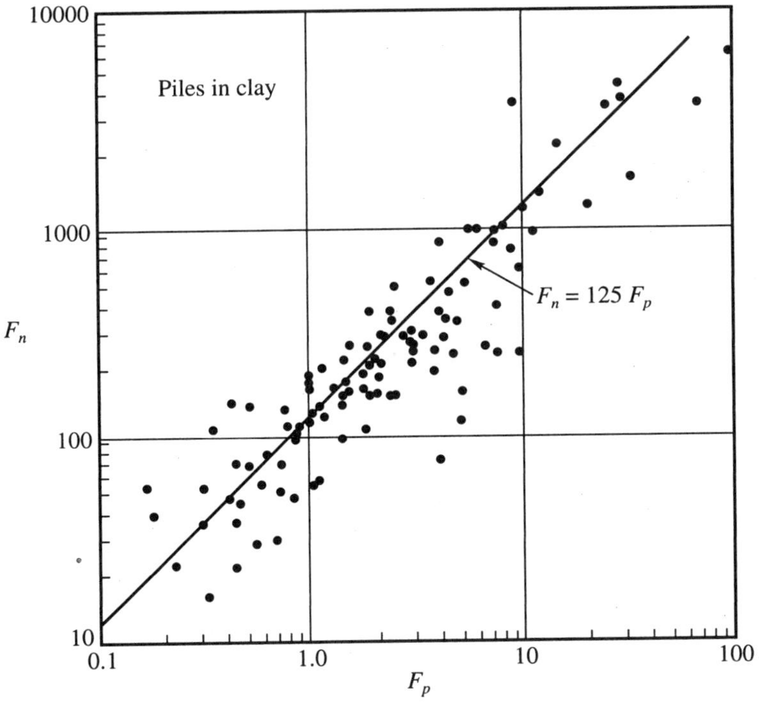

Figure 16.13 Nondimensional plot for piles in clay

It can be seen in the above equations that the numerators in both cases are constants for any given set of pile and soil properties.

The above two equations can be used to predict the non-linear behavior of piles subjected to lateral loads very accurately.

Example 16.10

Solve the problem in Example 16.1 by the direct method (Murthy and Subba Rao, 1995). The soil is loose sand in a submerged condition.

Given; $EI = 4.35 \times 10^{11}$ kg- cm^2 = 4.35×10^5 kN-m^2

d = 61 cm, L = 20 m, $\gamma_b = 8.75$ kN/m^3

ϕ = 33°, P_t = 268 kN (since e = 0)

Required y_g at ground level

Solution

For a pile in sand for the case of $e = 0$, use Eq. (16.29)

$$n_h = \frac{150 C_\phi \gamma^{1.5} \sqrt{EId}}{P_e}$$

For $\phi = 33°$, $C_\phi = 3 \times 10^{-5} (1.316)^{33} = 0.26$ from Eq. (16.25)

$$n_h = \frac{150 \times 0.26 \times (8.75)^{1.5} \sqrt{4.65 \times 10^5 \times 0.61}}{P_e} = \frac{54 \times 10^4}{P_e} = \frac{54 \times 10^4}{268} = 2{,}015 \text{ kN / m}^3$$

$$T = \left[\frac{EI}{n_h}\right]^{1/5} = \left[\frac{43.5 \times 10^4}{2015}\right]^{1/5} = 2.93 \text{ m}$$

Now, using Eq. (16.23)

$$y_g = \frac{2.43 \times 268 \times (2.93)^3}{4.35 \times 10^5} = 0.0377 \text{ m} = 3.77 \text{ cm}$$

It may be noted that the direct method gives a greater ground line deflection (= 3.77 cm) as compared to the 1.96 cm in Ex. 16.1.

Example 16.11

Solve the problem in Example 16.2 by the direct method. In this case P_t is applied at a height 2 m above ground level All the other data remain the same.

Solution

From Example 16.10

$$n_h = \frac{54 \times 10^4}{P_e}$$

For $P_e = P_t = 268$ kN, we have $n_h = 2{,}015$ kN/m^3, and $T = 2.93$ m

From Eq. (16.22)

$$P_e = P_t\left(1 + 0.67\frac{e}{T}\right) = 268\left(1 + 0.67 \times \frac{2}{2.93}\right) = 391 \text{ kN}$$

$$\text{For } P_e = 391 \text{ kN}, n_h = \frac{54 \times 10^4}{391} = 1{,}381 \text{ kN/m}^3$$

Now $T = \left[\dfrac{43.5\times 10^4}{1{,}381}\right]^{1/5} = 3.16 \text{ m}$

As before $P_e = 268\left(1+0.67\times\dfrac{2}{3.16}\right) = 382 \text{ kN}$

For P_e = 382 kN, n_h = 1,414 kN/m^3, T = 3.14 m

Convergence will be reached after a few trials. The final values are

P_e = 387 kN, n_h = 1718 kN/m^3, T = 3.025 m

Now from Eq. (16.23)

$$y_g = \frac{2.43 P_e T^3}{EI} = \frac{2.43\times 382\times(3.14)^3}{4.35\times 10^5} = 0.066 \text{ m} = 6.6 \text{ cm}$$

The n_h value from the direct method is 1,414 kN/m^3 whereas from Fig. 16.3 it is 6,000 kN/m^3. The n_h from Fig. 16.3 gives y_g which is 50 percent of the probable value and is on the unsafe side.

Example 16.12

Compute the ultimate lateral resistance for the pile given in Example 16.4 by the direct method. All the other data given in the example remain the same.

Given: EI = 4.35×10^5 kN-m^2, d = 61 cm, L = 20 m

c_u = 85 kN/m^3, γ_b = 10 kN/m^3 (assumed for clay)

M_y = 2,349 kN-m; e = 0

Required: The ultimate lateral resistance P_u.

Solution

Use Eqs (16.30) and (16.14)

$$n_h = \frac{125 c^{1.5}\sqrt{EI\gamma d}}{P_t^{1.5}} \quad \text{for } e = 0$$

$$T = \left[\frac{EI}{n_h}\right]^{0.2} \tag{a}$$

Substituting the known values and simplifying

$$n_h = \frac{1{,}600\times 10^5}{P_t^{1.5}} \tag{b}$$

Step 1

Let P_t = 1,000 kN, $n_h = \dfrac{1{,}600\times 10^5}{(1000)^{1.5}} = 5{,}060 \text{ kN/m}^3$

$$T = \left[\frac{4.35\times 10^5}{5060}\right]^{0.2} = 2.437 \text{ m}$$

For $e = 0$, from Table 16.2 and Eq. (16.11) we may write

$$M_{max} = 0.77[P_t T]$$

where $A_m = 0.77$ (max) correct to two decimal places.

For $P_t = 1000$ kN, and $T = 2.437$ m

$$M_{max} = 0.77 \times 1000 \times 2.437 = 1876 \text{ kN-m} < M_y.$$

Step 2

Let P_t	=	1500 kN
n_h	=	2754 kN/m^3 from Eq. (b)
and T	=	2.75 m from Eq. (a)
Now M_{max}	=	$0.77 \times 1500 \times 2.75 = 3179$ kN-m $> M_y$.
P_u for M_y	=	2349 kN-m can be determined as

$$P_u = 1{,}000 + (1{,}500 - 1{,}000) \times \frac{(2{,}349 - 1{,}876)}{(3{,}179 - 1{,}876)} = 1{,}182 \text{ kN}$$

$P_u = 1{,}100$ kN by Brom's method which agrees with the direct method.

16.8 CASE STUDIES FOR LATERALLY LOADED VERTICAL PILES IN SAND

Case 1: Mustang Island Pile LoadTest (Reese et al., 1974)

Data:

Pile diameter, d	=	24 in, steel pipe (driven pile)
EI	=	4.854×10^{10} lb-in^2
L	=	69 ft
e	=	12 in.
ϕ	=	39°
γ	=	66 lb/ft^3 (= 0.0382 lb/in^3)
M_y	=	7×10^6 in-lbs

The soil was fine silty sand with WT at ground level

Required:

(a) Load-deflection curve (P_t vs. y_g) and n_h vs. y_g curve
(b) Load-max moment curve (P_t Vs M_{max})
(c) Ultimate load P_u

Solutions:

For pile in sand, $$n_h = \frac{150\, C_\phi \gamma^{1.5} \sqrt{EId}}{P_e}$$

For $\phi = 39°$, $$C_\phi = 3 \times 10^{-5} (1.316)^{39°} = 1.34$$

After substitution and simplifying

$$n_h = \frac{1631 \times 10^3}{P_e} \tag{a}$$

From Eqs (16.22) and (16.14a), $$P_e = P_t\left(1 + 0.67\frac{e}{T}\right) \quad \text{(b)}$$

$$T = \left[\frac{EI}{n_h}\right]^{\frac{1}{5}} \quad \text{(c)}$$

(a) Calculation of Groundline Deflection, y_g

Step 1

Since T is not known to start with, assume $e = 0$, and $P_e = P_t = 10{,}000$ lbs

Now, from Eq. (a), $$n_h = \frac{1631 \times 10^3}{10 \times 10^3} = 163 \text{ lb/in}^3$$

from Eq. (c), $$T = \left[\frac{4.854 \times 10^{10}}{163}\right]^{\frac{1}{5}} = 49.5 \text{ in}$$

from Eq. (b), $$P_e = 10 \times 10^3\left(1 + 0.67 \times \frac{12}{49.5}\right) = 11.624 \times 10^3 \text{ lbs}$$

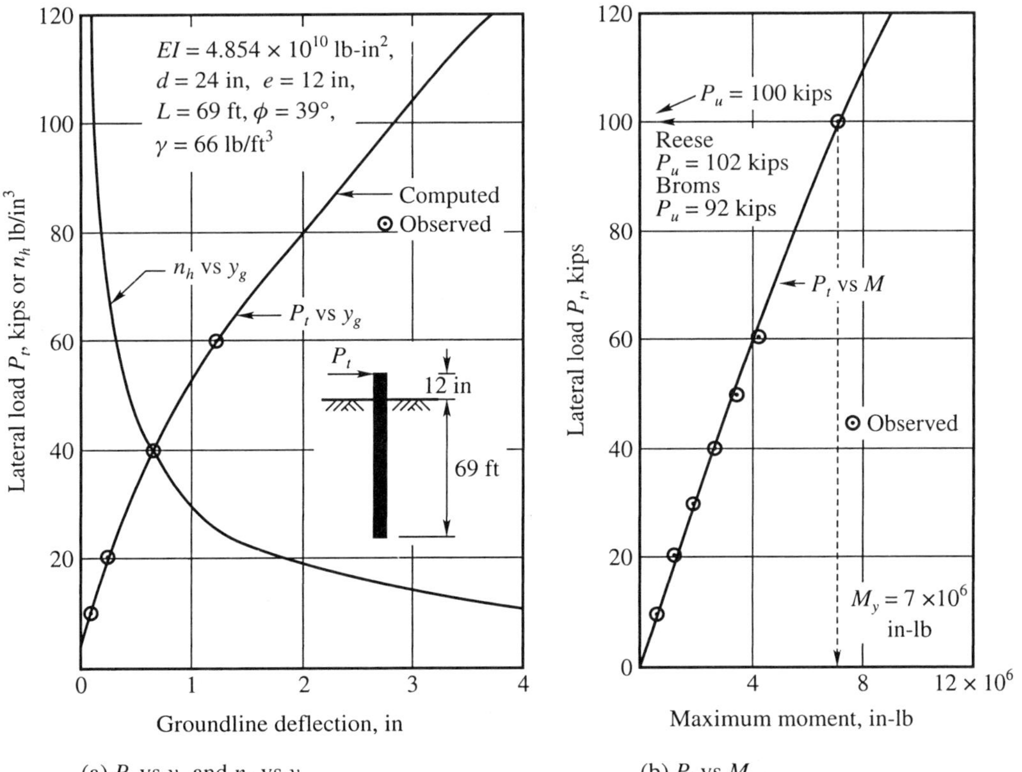

(a) P_t vs y_g and n_h vs y_g (b) P_t vs M_{max}

Figure 16.14 Mustang Island lateral load test

Step 2

For $P_e = 11.62 \times 10^3$ lb, $n_h = \dfrac{1631 \times 10^3}{11.624 \times 10^3} = 140 \text{ lb/in}^3$

As in Step 1 $T = 51$ ins, $P_e = 12.32 \times 10^3$ lbs

Step 3

Continue Step 1 and Step 2 until convergence is reached in the values of T and P_e. The final values obtained for $P_t = 10 \times 10^3$ lb are $T = 51.6$ in, and $P_e = 12.32 \times 10^3$ lbs

Step 4

The ground line deflection may be obtained from Eq (16.23).

$$y_g = \frac{2.43 P_e T^3}{EI} = \frac{2.43 \times 12.32 \times 10^3 \times (51.6)^3}{4.854 \times 10^{10}} = 0.0845 \text{ in}$$

This deflection is for $P_t = 10 \times 10^3$ lbs. In the same way the values of y_g can be obtained for different stages of loadings. Fig. 16.14(a) gives a plot P_t vs. y_g. Since n_h is known at each stage of loading, a curve of n_h vs. y_g can be plotted as shown in the same figure.

(b) Maximum Moment

The calculations under (a) above give the values of T for various loads P_t. By making use of Eq. (16.11) and Table 16.2, moment distribution along the pile for various loads P_t can be calculated. From these curves the maximum moments may be obtained and a curve of P_t vs. M_{max} may be plotted as shown in Fig. 16.14b.

(c) Ultimate Load P_u

Figure 16.14(b) is a plot of M_{max} vs. P_t. From this figure, the value of P_u is equal to 100 kips for the ultimate pile moment resistance of 7×10^6 in-lb. The value obtained by Broms' method and by computer (Reese, 1986) are 92 and 102 kips respectively

Comments:

Figure 16.14a gives the computed P_t vs. y curve by the direct approach method (Murthy and Subba Rao 1995) and the observed values. There is an excellent agreement between the two. In the same way the observed and the calculated moments and ultimate loads agree well.

Case 2: Florida Pile Load Test (Davis, 1977)

Data

Pile diameter, d = 56 in steel tube filled with concrete

$EI = 132.5 \times 10^{10}$ lb-in^2

L = 26 ft

e = 51 ft

$\phi = 38°$,

$\gamma = 60$ lb/ft^3

M_y = 4630 ft-kips.

The soil at the site was medium dense and with water table close to the ground surface.

Required

(a) P_t vs. y_g curve and n_h vs. y_g curve

(b) Ultimate lateral load P_u

Solution

The same procedure as given for the Mustang Island load test has been followed for calculating the P_t vs. y_g and n_h vs. y_g curves. For getting the ultimate load P_u the P_t vs. M_{max} curve

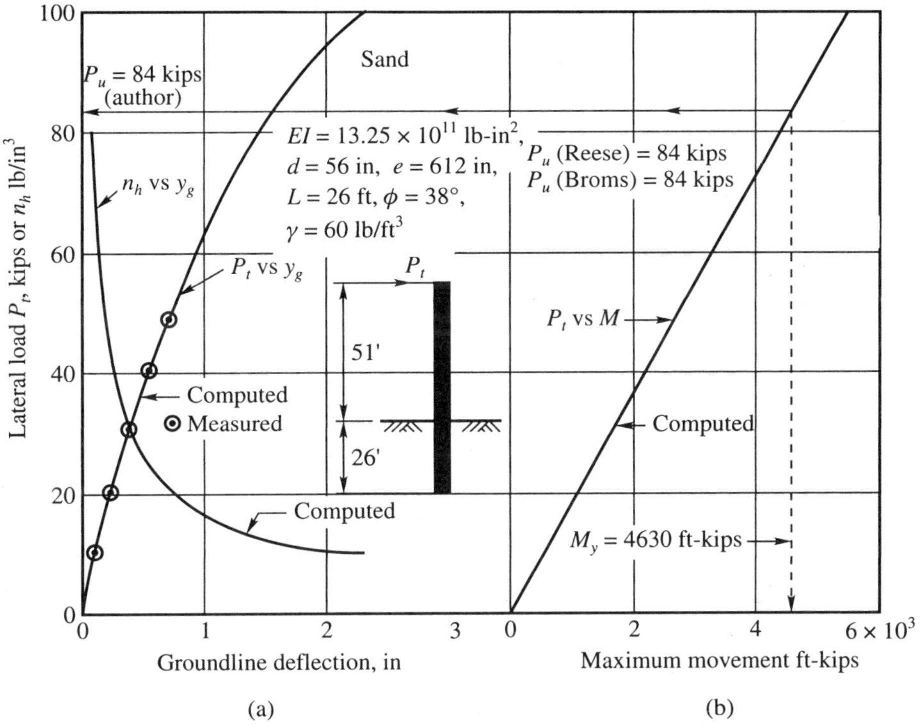

Figure 16.15 Florida pile test (Davis, 1977)

is obtained. The value of P_u obtained is equal to 84 kips which is the same as the ones obtained by Broms (1964) and Reese (1985) methods. There is a very close agreement between the computed and the observed test results as shown in Fig. 16.15.

Case 3: Model Pile Tests in Sand (Murthy, 1965)

Data

Model pile tests were carried out to determine the behavior of vertical piles subjected to lateral loads. Aluminum alloy tubings, 0.75 in diameter and 0.035 in wall thickness, were used for the test. The test piles were instrumented. Dry clean sand was used for the test at a relative density of 67%. The other details are given in Fig. 16.16.

Solution

Fig. 16.16 gives the predicted and observed

(a) load-ground line deflection curve

(b) deflection distribution curves along the pile

(c) moment and soil reaction curves along the pile

There is an excellent agreement between the predicted and the observed values. The direct approach method has been used.

16.9 CASE STUDIES FOR LATERALLY LOADED VERTICAL PILES IN CLAY

Case 1: Pile load test at St. Gabriel (Capazzoli, 1968)

Data

Pile diameter, d = 10 in, steel pipe filled with concrete

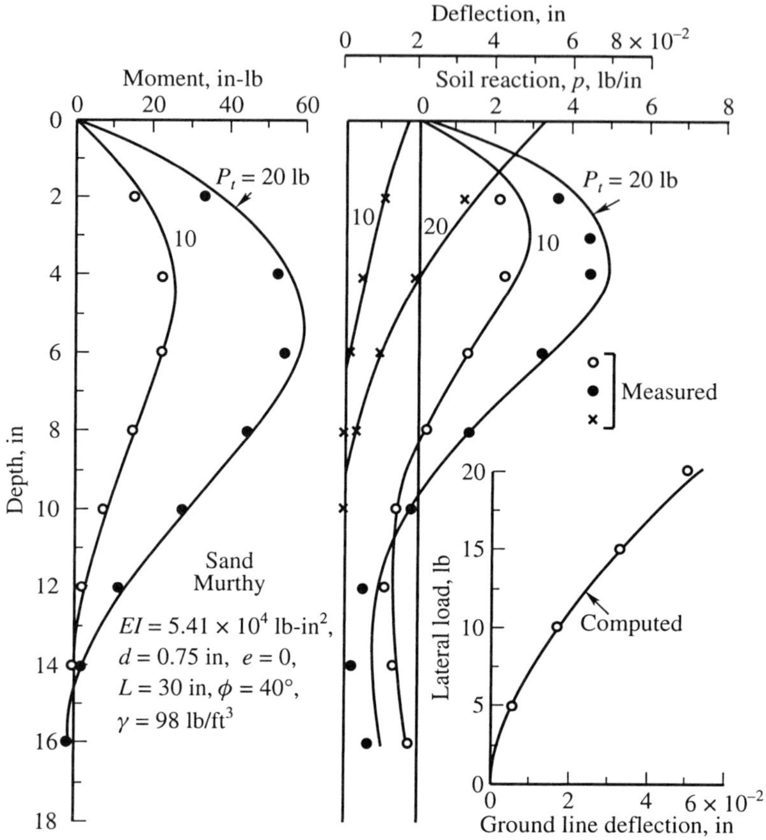

Figure 16.16 Curves of bending moment, deflection and soil reaction for a model pile in sand (Murthy, 1965)

$$
\begin{aligned}
\text{EI} &= 38 \times 10^8 \text{ lb-in}^2\\
L &= 115 \text{ ft}\\
e &= 12 \text{ in.}\\
c &= 600 \text{ lb/ft}^2\\
\gamma &= 110 \text{ lb/ft}^3\\
M_y &= 116 \text{ ft-kips}
\end{aligned}
$$

Water table was close to the ground surface.

Required

(a) P_t vs. y_g curve

(b) the ultimate lateral load, P_u

Solution

We have,

$$\text{(a)}\ n_h = \frac{125c^{1.5}\sqrt{EI\gamma d}}{(1+e/d)^{1.5}P_e^{1.5}} \quad \text{(b)}\ P_e = P_t\left(1+0.67\frac{e}{T}\right) \quad \text{(c)}\ T = \left[\frac{EI}{n_h}\right]^{\frac{1}{5}}$$

After substituting the known values in Eq. (a) and simplifying, we have

$$n_h = \frac{16045 \times 10^3}{P_e^{1.5}}$$

(a) Calculation of groundline deflection

1. Let $P_e = P_t = 500$ lbs
   From Eqs (a) and (c), $n_h = 45$ lb/in^3, $T = 38.51$ in
   From Eq. (d), $P_e = 6044$ lb.
2. For $P_e = 6044$ lb, $n_h = 34$ lb/in^3 and $T = 41$ in
3. For $T = 41$ in, $P_e = 5980$ lb, and $n_h = 35$ lb/in^3
4. For $n_h = 35$ lb/in^3, $T = 40.5$ in, $P_e = 5988$ lb
5. $y_g = \dfrac{2.43 P_e T^3}{EI} = \dfrac{2.43 \times 5988 \times (40.5)^3}{38 \times 10^8} = 0.25$ in
6. Continue steps 1 through 5 for computing y_g for different loads P_t. Fig. 16.17 gives a plot of P_t vs. y_g which agrees very well with the measured values.

(b) Ultimate load P_u

A curve of M_{max} vs. P_t is given in Fig. 16.17 following the procedure given for the Mustang Island Test. From this curve $P_u = 23$ k for $M_y = 116$ ft kips. This agrees well with the values obtained by the methods of Reese (1985) and Broms (1964a).

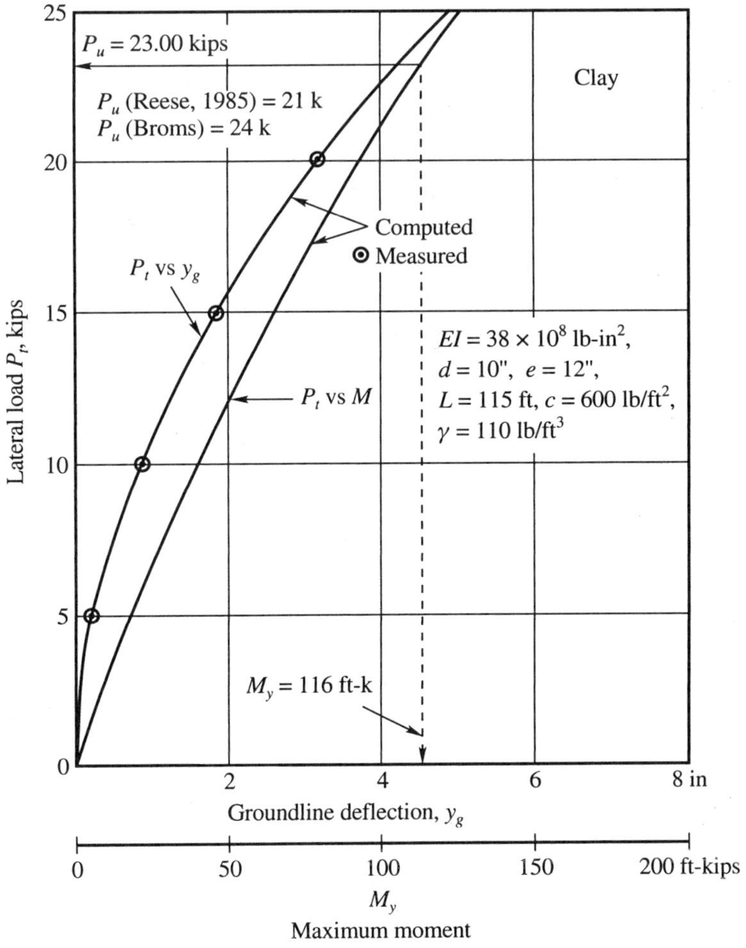

Figure 16.17 St. Gabriel pile load test in clay

Case 2: Pile Load Test at Ontario (Ismael and Klym, 1977)

Data

Pile diameter, d = 60 in, concrete pile (Test pile 38)
$EI = 93 \times 10^{10}$ lb-in^2
L = 38 ft
e = 12 in.
c = 2000 lb/ft^2
γ = 60 lb/ft^3

The soil at the site was heavily overconsolidated

Required:

(a) P_t vs. y_g curve
(b) n_h vs. y_g curve

Solution

By substituting the known quantities in Eq. (16.30) and simplifying,

$$n_h = \frac{68495 \times 10^5}{P_e^{1.5}}, \quad T = \left[\frac{EI}{n_h}\right]^{\frac{1}{5}}, \text{ and } P_e = P_t\left(1 + 0.67\frac{e}{T}\right)$$

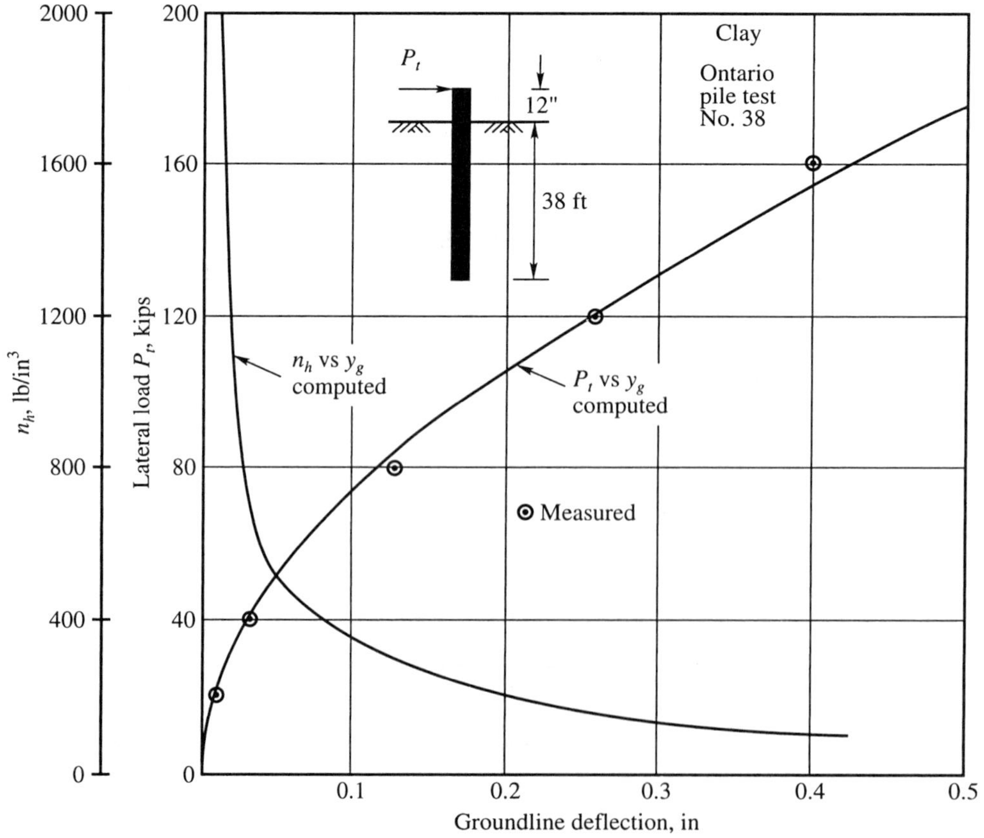

Figure 16.18 Ontario pile load test (38)

Follow the same procedure as given for Case 1 to obtain values of y_g for the various loads P_t. The load deflection curve can be obtained from the calculated values as shown in Fig. 16.18. The measured values are also plotted. It is clear from the curve that there is a very close agreement between the two. The figure also gives the relationship between n_h and y_g.

Case 3: Restrained Pile at the Head for Offshore Structure (Matlock and Reese, 1961)

Data

The data for the problem are taken from Matlock and Reese (1961). The pile is restrained at the head by the structure on the top of the pile. The pile considered is below the sea bed. The undrained shear strength c and submerged unit weights are obtained by working back from the known values of n_h and T. The other details are

$$\begin{aligned} \text{Pile diameter, } d &= 33 \text{ in, pipe pile} \\ EI &= 42.35 \times 10^{10} \text{ lb-in}^2 \\ c &= 500 \text{ lb/ft}^2 \\ \gamma &= 40 \text{ lb/ft}^3 \\ P_t &= 150{,}000 \text{ lbs} \end{aligned}$$

$$\text{(a) } \frac{M_t}{P_t T} = \frac{-T}{12.25 + 1.078T}, \quad \text{(b) } T = \left[\frac{EI}{n_h}\right]^{\frac{1}{5}}, \quad \text{(c) } P_e = P_t\left(1 - 0.67\frac{e}{T}\right)$$

Required

(a) deflection at the pile head

(b) moment distribution diagram

Solution

Substituting the known values in Eq (16.30) and simplifying,

$$n_h = \frac{458 \times 10^6}{P_e^{1.5}(1 + e/d)^{1.5}} \qquad \text{(d)}$$

Calculations

1. Assume $e = 0$, $P_e = P_t = 150{,}000$ lb

From Eqs (d) and (b) $n_h = 7.9 \text{ lb/in}^2, \quad T = 140 \text{ in}$

From Eq. (a) $$\frac{M_t}{P_t T} = \frac{-140}{12.25 + 1.078 \times 140} = -0.858$$

or $M_t = -0.858 \; P_t T = P_t e$

Therefore $e = 0.858 \times 140 = 120$ in

2. $$P_e = P_t\left(1 - 0.67\frac{e}{T}\right) = 1.5 \times 10^5\left(1 - 0.67 \times \frac{120}{140}\right) = 63{,}857 \text{ lb}$$

$$\left(1 + \frac{e}{d}\right)^{1.5} = \left(1 + \frac{120}{33}\right)^{1.5} = 10$$

Now from Eq. (d), $n_h = 2.84$ lb/in^3, from Eq. (b) $T = 171.64$ in
After substitution in Eq. (a)

$$\frac{M_t}{P_t T} = -0.875, \text{ and } e = 0.875 \times 171.64 = 150.2 \text{ in}$$

$$P_e = \left(1 - 0.67 \times \frac{150.2}{171.64}\right) \times 1.5 \times 10^5 = 62{,}205 \text{ lbs}$$

3. Continuing this process for a few more steps there will be convergence of values of n_h, T and P_e. The final values obtained are

$$n_h = 2.1 \text{ lb/in}^3, \; T = 182.4 \text{ in, and } P_e = 62{,}246 \text{ lb}$$

$$M_t = -P_t e = -150{,}000 \times 150.2 = -22.53 \times 10^6 \text{ lb-in}^2$$

$$y_g = \frac{2.43\, P_e T^3}{EI} = \frac{2.43 \times 62{,}246 \times (182.4)^3}{42.35 \times 10^{10}} = 2.17 \text{ in}$$

Moment distribution along the pile may now be calculated by making use of Eq. (16.11) and Table 16.2. Please note that M_t has a negative sign. The moment distribution curve is given in Fig. 16.19. There is a very close agreement between the computed values by direct method and the Reese and Matlock method. The deflection and the negative bending moment as obtained by Reese and Matlock are

$y_m = 2.307$ in and $M_t = -24.75 \times 10^6$ lb-in^2

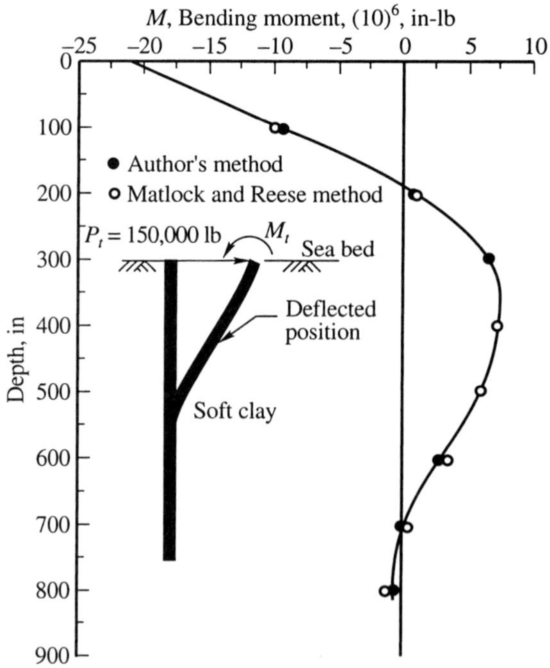

Figure 16.19 Bending moment distribution for an offshore pile supported structure (Matlock and Reese, 1961)

16.10 BEHAVIOR OF LATERALLY LOADED BATTER PILES IN SAND

General Considerations

The earlier sections dealt with the behavior of long vertical piles. The author has so far not come across any rational approach for predicting the behavior of batter piles subjected to lateral loads. He has been working on this problem for a long time (Murthy, 1965). Based on the work done by the author and others, a method for predicting the behavior of long batter piles subjected to lateral load has now been developed.

Model Tests on Piles in Sand (Murthy, 1965)

A series of seven instrumented model piles were tested in sand with batters varying from 0 to ±45°. Aluminum alloy tubings of 0.75 in outside diameter and 30 in long were used for the tests. Electrical resistance gauges were used to measure the flexural strains at intervals along the piles at different load levels. The maximum load applied was 20 lbs. The pile had a flexural rigidity $EI = 5.14 \times 10^4$ lb-in^2. The tests were conducted in dry sand, having a unit weight of 98 lb/ft^3 and angle of friction ϕ equal to 40°. Two series of tests were conducted-one series with loads horizontal and the other with loads normal to the axis of the pile. The batters used were 0°, ± 15°, ±30° and ±45°. Pile movements at ground level were measured with sensitive dial gauges. Flexural strains were converted to moments. Successive integration gave slopes and deflections and successive differentiations gave shears and soil reactions respectively. A very high degree of accuracy was maintained throughout the tests. Based on the test results a relationship was established between the n^b_h values of batter piles and n^o_h

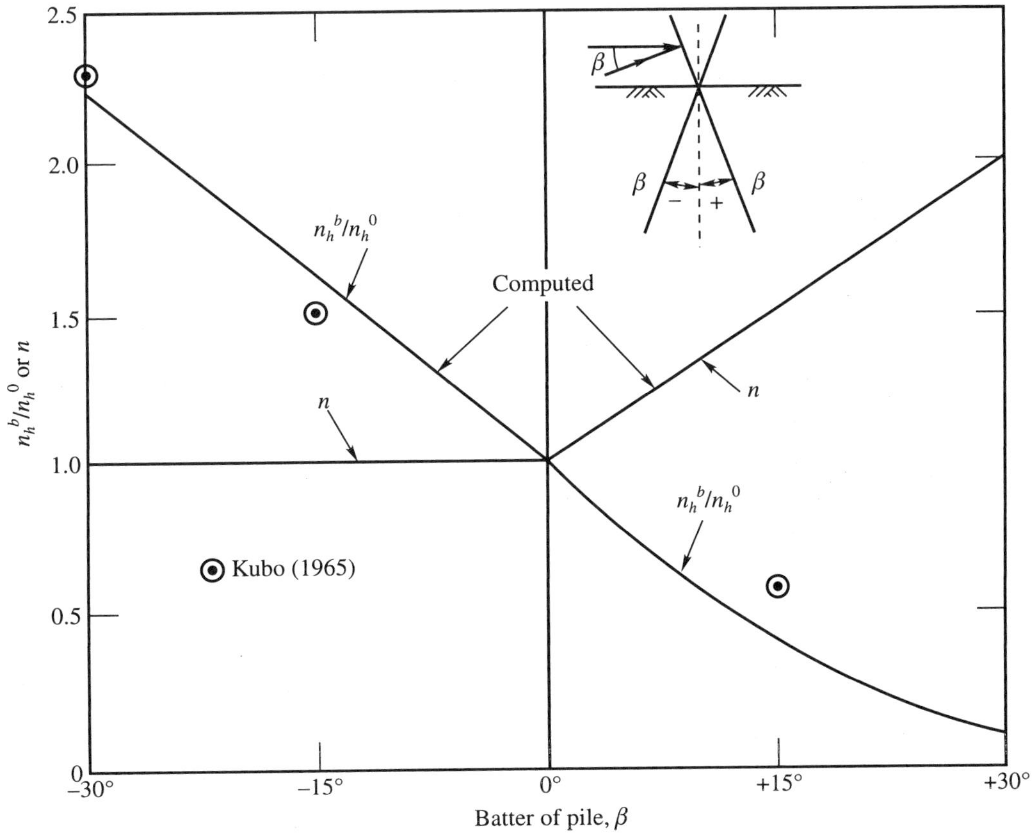

Figure 16.20 Effect of batter on n^b_h/n^o_h and n (after Murthy, 1965)

values of vertical piles. Fig. 16.20 gives this relationship between $n^b{}_h/n^o{}_h$ and the angle of batter β. It is clear from this figure that the ratio increases from a minimum of 0.1 for a positive 30° batter pile to a maximum of 2.2 for a negative 30° batter pile. The values obtained by Kubo (1965) are also shown in this figure. There is close agreement between the two.

The other important factor in the prediction is the value of n in Eq. (16.8a). The values obtained from the experimental test results are also given in Fig. 16.20. The values of n are equal to unity for vertical and negative batter piles and increase linearly for positive batter piles up to a maximum of 2.0 at + 30° batter.

In the case of batter piles the loads and deflections are considered normal to the pile axis for the purpose of analysis. The corresponding loads and deflections in the horizontal direction may be written as

$$P_t(\text{Hor}) = \frac{P_t(Nor)}{\cos\beta} \tag{16.31}$$

$$y_g(\text{Hor}) = \frac{y_g(\text{Nor})}{\cos\beta} \tag{16.32}$$

where P_t and y_g are normal to the pile axis; P_t(Hor) and y_g(Hor) are the corresponding horizontal components.

Application of the Use of $n^b{}_h/n^o{}_h$ and n

It is possible now to predict the non-linear behavior of laterally loaded batter piles in the same way as for vertical piles by making use of the ratio $n^b{}_h/n^o{}_h$ and the value of n. The validity of this method is explained by considering a few case studies.

Case Studies

Case 1: Model Pile Test (Murthy, 1965).

Piles of +15° and +30° batters have been used here to predict the P_t vs. y_g and P_t vs. M_{max} relationships. The properties of the pile and soil are given below.

$EI = 5.14 \times 10^4$ lb in^2, $d = 0.75$ in, $L = 30$ in; $e = 0$

$\gamma = 98$ lb/ft^3 and $\phi = 40°$

For $\phi = 40°$, $C_\phi = 1.767$ [$= 3 \times 10^{-5} (1.316)^\phi$]

From Eq. (16.29), $n_h^o = \dfrac{150 C_\phi \gamma^{1.5}\sqrt{EId}}{P_t}$

After substituting the known values and simplifying we have

$$n_h^o = \frac{700}{P_t}$$

Solution: +15° batter pile

From Fig. 16.20 $n^b{}_h/n^o{}_h = 0.4$, $n = 1.5$

From Eq. (16.14b), $T_b = \left[\dfrac{EI}{n_h^b}\right]^{\frac{1}{1.5+4}} = 5.33\left[\dfrac{5.14}{n_h^b}\right]^{0.1818}$

Calculations of Deflection y_g

For P_t = 5 lbs, n^o_h= 141 lbs/in^3, n^b_h= 141 × 0.4 = 56 lb/in^3 and T_b = 3.5 in

$$y_g = \frac{2.43 P_t^3 T_b^3}{5.14 \times 10^4} = 0.97 \times 10^{-2} \text{in}$$

Similarly, y_g can be calculated for P_t = 10, 15 and 20 lbs.

The results are plotted in Fig. 16.21 along with the measured values of y_g. There is a close agreement between the two.

Calculation of Maximum Moment, $M_{\max}$

For P_t = 5 lb, T_b = 3.5 in, The equation for M is [Eq. (16.11)]

$$M = A_m P_t T_b = 0.77 P_t T_b$$

where A_m = 0.77 (max) from Table 16.2

By substituting and calculating, we have

$$M_{(\max)} = 13.5 \text{ in-lb}$$

Similarly $M_{(\max)}$ can be calculated for other loads. The results are plotted in Fig. 16.21 along with the measured values of $M_{(\max)}$. There is very close agreement between the two.

+30° Batter Pile

From Fig. 16.20, n^b_h/n^o_h = 0.1, and n = 2; $T_b = \left[\frac{EI}{n_h^b}\right]^{\frac{1}{2+4}} = 4.64\left[\frac{5.14}{n_h^b}\right]^{0.1667}, \quad n_h^o = \frac{700}{P_t}$

For P_t = 5 lbs, n^o_h = 141 lbs/in^3, n^b_h= 0.1 × 141 = 14.1 lb/in^3, T_b = 3.93 in.

For P_t = 5 lbs, T_b = 3.93 in, we have, $y_g = 1.43 \times 10^{-2}$ in

As before, $M_{(\max)}$ = 0.77 × 5 × 3.93 = 15 in-lb.

The values of y_g and $M_{(\max)}$ for other loads can be calculated in the same way. Fig. 16.21 gives P_t vs. y_g and P_t vs. $M_{(\max)}$ along with measured values. There is close agreement up to about

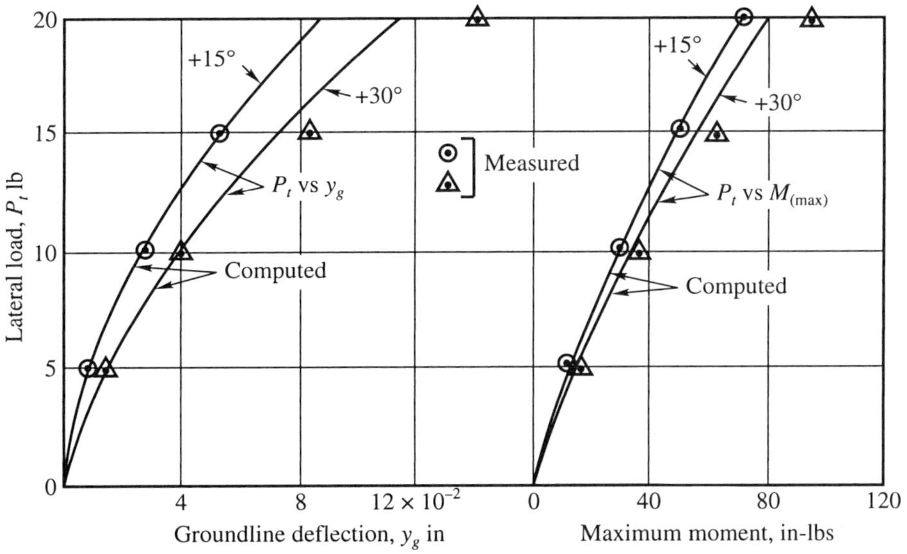

Figure 16.21 Model piles of batter +15° and +30° (Murthy, 1965)

$P_t = 10$ lb, and beyond this load, the measured values are greater than the predicted by about 25 percent which is expected since the soil yields at a load higher than 10 lb at this batter and there is a plastic flow beyond this load.

Case 2: Arkansas River Project (Pile 12) (Alizadeh and Davisson, 1970).

Given:

$EI = 278.5 \times 10^8$ lb $-$in^2, $d = 14$ in, $e = 0.$

$\phi = 41°$, $\gamma = 63$ lb/ft^3, $\beta = 18.4°$ (–ve)

From Fig. 16.11, $C_\phi = 2.33$, from Fig. 16.20 $n^b{}_h/n^o{}_h = 1.7$, $n = 1.0$

From Eq. (16.29), after substituting the known values and simplifying, we have,

$$\text{(a)}\ n_h^o = \frac{1528 \times 10^3}{P_t}, \text{ and } \text{(b)}\ T_b = 39.8\left[\frac{278.5}{n_h^b}\right]^{0.2}$$

Calculation for $P_t = 12.6^k$

From Eq. (a), $n^o{}_h = 121$ lb/in^3; now $n^b{}_h = 1.7 \times 121 = 206$ lb/in^3

From Eq. (b), $T_b = 42.27$ in

$$y_g = \frac{2.43 \times 12{,}600(42.27)^3}{278.5 \times 10^8} = 0.083 \text{ in}$$

$M_{(max)} = 0.77\, P_t T = 0.77 \times 12.6 \times 3.52 = 34$ ft-kips.

The values of y_g and $M_{(max)}$ for $P_t = 24.1^k, 35.5^k, 42.0^k, 53.5^k, 60^k$ can be calculated in the same way the results are plotted the Fig. 16.22 along with the measured values. There is a very close agreement between the computed and measured values of y but the computed values of M_{max}

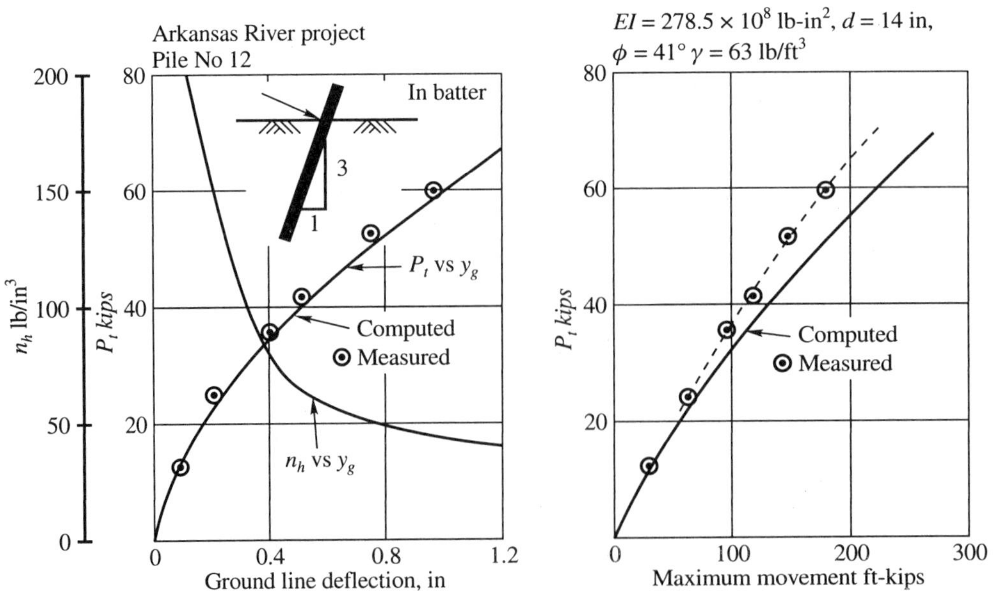

Figure 16.22 Lateral load test-batter pile 12-Arkansas River Project (Alizadeh and Davisson, 1970)

are higher than the measured values at higher loads. At a load of 60 kips, $M_{(max)}$ is higher than the measured by about 23% which is quite reasonable.

Case 3: Arkansas River Project (Pile 13) (Alizadeh and Davisson, 1970).

Given:

$EI = 288 \times 10^8$ lb-ins, $d = 14''$, $e = 6$ in.

$\gamma = 63$ lbs/ft^3, $\phi = 41°(C_\phi = 2.33)$

$\beta = 18.4°$ (+ve), $n = 1.6$, $n^b_h/n^o_h = 0.3$

$$T_b = \left[\frac{EI}{n_h^b}\right]^{\frac{1}{1.6+4}} = 27\left[\frac{288}{n_h^b}\right]^{0.1786} \quad \text{(a)}$$

After substituting the known values in the equation for n^o_h [Eq. (16.29)] and simplifying, we have

$$n_h^0 = \frac{1597 \times 10^3}{P_e} \quad \text{(b)}$$

Calculations for y_g for $P_t = 141.4$k

1. From Eq (b), $n^o_h = 39$ lb/in^3, hence $n^b_h = 0.3 \times 39 = 11.7$ lb/in^3
   From Eq. (a), $T_b \approx 48$ in.

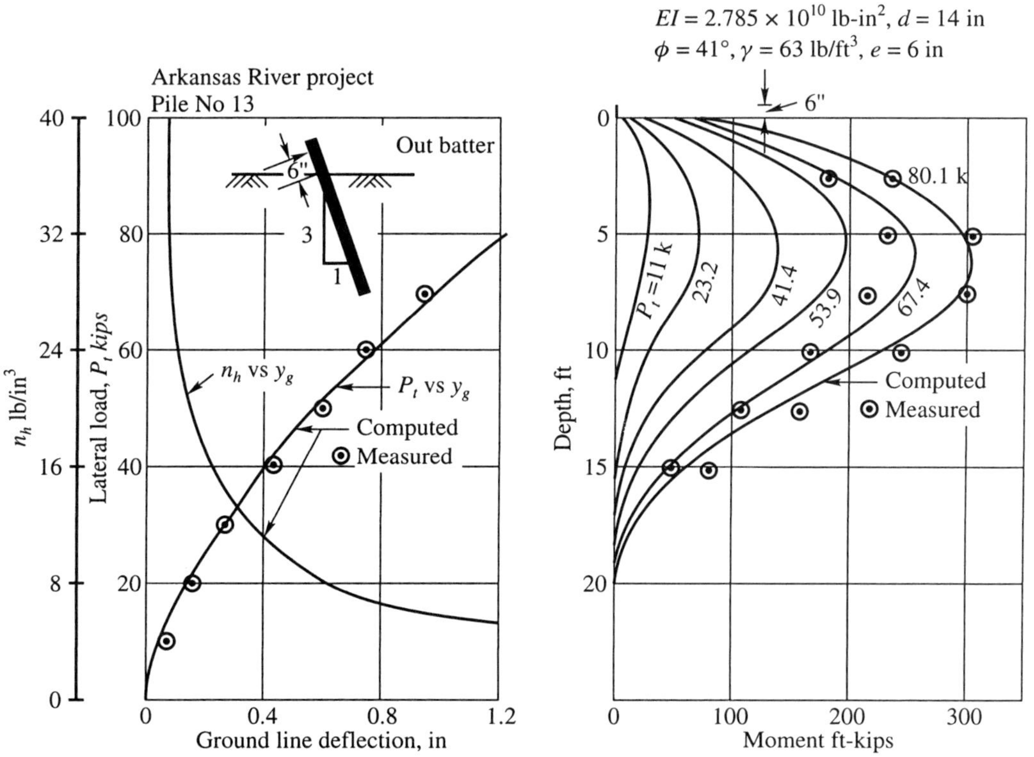

Figure 16.23 Lateral load test-batter pile 13-Arkansas River Project (Alizadeh and Davisson, 1970)

2. $P_e = P_t\left(1+0.67\frac{e}{T}\right) = 41.4\left(1+0.67\times\frac{6}{48}\right) = 44.86$ kips

3. For $P_e = 44.86$ kips, $n_h^o = 36\ \text{lb/in}^3$, and $n_h^b = 11\ \text{lb/in}^3$, $T_b \approx 48$ in

4. Final values: $P_e = 44.86$ kips, $n_h^b = 11\ \text{lb/in}^3$, and $T_b = 48$ in

5. $y_g = \frac{2.43 P_e T_b^3}{EI} = \frac{2.43\times 44{,}860\times(48)^3}{288\times 10^8} = 0.42$ in.

6. Follow Steps 1 through 5 for other loads. Computed and measured values of y are plotted in Fig. 16.23 and there is a very close agreement between the two. The n_h values against y_g are also plotted in the same figure.

Calculation of Moment Distribution

The moment at any distance x along the pile may be calculated by the equation

$$M = \left[P_t T\right]A_m + \left[M_t\right]B_m$$

As per the calculations shown above, the value of T will be known for any lateral load level P_t. This means $[P_t T]$ will be known. The values of A_m and B_m are functions of the depth coefficient Z which can be taken from Table 16.2 for the distance $x(Z = x/T)$. The moment at distance x will be known from the above equation. In the same way moments may be calculated for other distances. The same procedure is followed for other load levels. Fig. 16.23 gives the computed moment distribution along the pile axis. The measured values of M are shown for two load levels $P_t = 67.4$ and 80.1 kips. The agreement between the measured and the computed values is very good.

Example 16.13

A steel pipe pile of 61 cm diameter is driven vertically into a medium dense sand with the water table close to the ground surface. The following data are available:

Pile: $EI = 43.5 \times 10^4$ kN-m², $L = 20$ m, the yield moment M_y of the pile material = 2349 kN-m.

Soil: Submerged unit weight $\gamma_b = 8.75$ kN/m³, $\phi = 38°$.

Lateral load is applied at ground level ($e = 0$)

Determine:

(a) The ultimate lateral resistance P_u of the pile

(b) The groundline deflection y_g at the ultimate lateral load level.

Solution

From Eq. (16.29) the expression for n_h is

$$n_h = \frac{150 C_\phi \gamma^{1.5}\sqrt{EId}}{P_t} \quad \text{since } P_e = P_t \text{ for } e = 0 \qquad \text{(a)}$$

From Eq. (16.25) $C_\phi = 3\times 10^{-5}(1.326)^{38°} = 1.02$

Substituting the known values for n_h we have

$$n_h = \frac{150\times 1.02\times(8.75)^{1.5}\sqrt{43.5\times 10^4\times 0.61}}{P_t} = \frac{204\times 10^4}{P_t}\ \text{kN/m}^3$$

(a) Ultimate lateral load P_u

Step 1:

Assume $P_u = P_t = 1000$ kN (a)

Now from Eq. (a) $n_h = \dfrac{204\times 10^4}{1000} = 2040 \text{ kN/m}^3$

From Eq. (16.14a) $T = \left[\dfrac{EI}{n_h}\right]^{\frac{1}{n+4}} = \left[\dfrac{EI}{n_h}\right]^{\frac{1}{1+4}} = \left[\dfrac{EI}{n_h}\right]^{\frac{1}{5}}$

Substituting and simplyfiing $T = \left[\dfrac{43.5\times 10^4}{2040}\right]^{\frac{1}{5}} = 2.92$ m

The moment equation for $e = 0$ may be written as (Eq. 16.11)

$M = A_m[P_tT]$

Substituting and simplifying we have (where $A_m(\max) = 0.77$)

$M_{\max} = 0.77(1000\times 2.92) = 2248$ kN - m

which is less than $M_y = 2349$ kN-m.

Step 2:

Try $P_t = 1050$ kN.

Following the procedure given in Step 1

$T = 2.95$ m for $P_t = 1050$ kN

Now $M_{\max} = 0.77(1050\times 2.95) = 2385$ kN - m

which is greater than $M_y = 2349$ kN-m.

The actual value P_u lies between 1000 and 1050 kN which can be obtained by proportion as

$$P_u = 1000 + (1050 - 1000)\times\frac{(2349-2248)}{(2385-2248)} = 1037 \text{ kN}$$

(b) Groundline deflection for P_u = 1037 kN

For this the value T is required at $P_u = 1037$ kN. Following the same procedure as in Step 1, we get $T = 2.29$ m.

Now from Eq. (16.15a) for $e = 0$

$$y_g = 2.43\frac{P_tT^3}{EI} = \frac{2.43\times 1037\times (2.944)^3}{43.5\times 10^4} = 0.1478 \text{ m} = 14.78 \text{ cm}$$

Example 16.14

Refer to Ex. 16.13. If the pipe pile is driven at an angle of 30° to the vertical, determine the ultimate lateral resistance and the corresponding groundline deflection for the load applied (a) against batter, and (b) in the direction of batter.

In both the cases the load is applied normal to the pile axis.

All the other data given in Ex. 16.13 remain the same.

Solution

From Ex. 16.13, the expression for n_h for vertical pile is

$$n_h = n_h^o = \frac{204 \times 10^4}{P_t} \text{ kN/m}^3 \tag{a}$$

+ 30° Batter pile

$$\text{From Fig. 16.20 } \frac{n_h^{+b}}{n_h^o} = 0.1 \text{ and } n = 2 \tag{b}$$

$$\text{From Eq. (16.14b) } T = T_b = \left[\frac{EI}{n_h^b}\right]^{\frac{1}{n+4}} = \left[\frac{EI}{n_h^b}\right]^{\frac{1}{2+4}} = \left[\frac{EI}{n_h^b}\right]^{\frac{1}{6}} \tag{c}$$

Determination of P_u

Step 1

Assume $P_e = P_t = 500$ kN.

Following the Step 1 in Ex. 16.13, and using Eq. (a) above

$n^o_h = 4{,}083 \text{ kN/m}^3$, hence $n_h^{+b} = 4083 \times 0.1 \approx 408 \text{ kN/m}^3$

$$\text{Form Eq (c), } T_b = \left[\frac{43.5 \times 10^4}{408}\right]^{\frac{1}{6}} = 3.2 \text{ m}$$

As before, $M_{max} = 0.77 P_t T_b = 0.77 \times 500 \times 3.2 = 1232$ kN-m $< M_y$

Step 2

Try $P_t = 1{,}000$ kN

Proceeding in the same way as given in Step 1 we have $T_b = 3.59$ m, $M_{max} = 2764$ kN-m which is more than M_y. The actual P_u is

$$P_u = 500 + (1000 - 500) \times \frac{(2349 - 1232)}{(2764 - 1232)} = 865 \text{ kN}$$

Step 3

As before the corresponding T_b for $P_u = 865$ kN is 3.5 m.

Step 4

The groundline deflection is

$$y_g^{+b} = \frac{2.43 \times 865 \times (3.5)^3}{43.5 \times 10^4} = 0.2072 \text{ m} = 20.72 \text{ cm}$$

– 30° Batter pile

$$\text{From Fig. 16.20, } \frac{n_h^{-b}}{n_h^o} = 2.2 \text{ and } n = 1.0 \tag{d}$$

$$\text{and } T_b = \left[\frac{EI}{n_h^{-b}}\right]^{\frac{1}{n+4}} = \left[\frac{EI}{n_h^{-b}}\right]^{\frac{1}{1+4}} = \left[\frac{EI}{n_h^{-b}}\right]^{\frac{1}{5}} \quad \text{(e)}$$

Determination of P_u

Step 1
Try $P_t = 1000$

From Eq. (a) $n_h^o = 2040$ kN/m^3 and from Eq. (d) $n_h^{-b} = 2.2 \times 2040 = 4488$ kN/m^3

Now from Eq. (e), $T_b = 2.5$ m

As before $M_{max} = 0.77 \times 1000 \times 2.5 = 1925$ kN-m

which is less than $M_y = 2349$ kN-m

Step 2

Try $P_t = 1{,}500$ kN

Proceeding as in Step 1, $T_b = 2.71$ m, and $M_{max} = 0.77 \times 1500 \times 2.71 = 3130$ kN-m which is greater than M_y.

Step 3
The actual value of P_u is therefore

$$P_u = 1000 + (1500 - 1000) \times \frac{(2349 - 1925)}{(2764 - 1925)} = 1253 \text{ kN}$$

Step 4
Groundline deflection

$T_b = 2.58$ m for $P_u = 1253$ kN

$$\text{Now } y_g^{-b} = \frac{2.43 \times 1176 \times (2.58)^3}{43.5 \times 10^4} = 0.1202 \text{ m} = 12.0 \text{ cm}$$

The above calculations indicate that the negative batter piles are more resistant to lateral loads than vertical or positive batter piles. Besides, the groundline deflections of the negative batter piles are less than the vertical and corresponding positive batter piles.

16.11 PROBLEMS

16.1 A reinforced concrete pile 50 cm square in section is driven into a medium dense sand to a depth of 20 m . The sand is in a submerged state. A lateral load of 50 kN is applied on the pile at a height of 5 m above the ground level. Compute the lateral deflection of the pile at ground level. Given: $n_h = 15$ MN/m^3, $EI = 115 \times 10^9$ kg-cm^2. The submerged unit weight of the soil is 8.75 kN/m^3.

16.2 If the pile given in Prob. 16.1 is fully restrained at the top, what is the deflection at ground level?

16.3 If the pile given in Prob. 16.1 is 3 m long, what will be the deflection at ground level (a) when the top of the pile is free, and (b) when the top of the pile is restrained? Use Broms' method.

16.4 Refer to Prob. 16.1. Determine the ultimate lateral resistance of the pile by Broms' method. Use $\phi = 38°$. Assume $M_y = 250$ kN-m.

16.5 If the pile given in Prob. 16.1 is driven into saturated normally consolidated clay having an unconfined compressive strength of 70 kPa, what would be the ultimate lateral resistance of the soil under (a) a free-head condition, and (b) a fixed-condition? Make necessary assumptions for the yield strength of the material.

16.6 Refer to Prob. 16.1. Determine the lateral deflection of the pile at ground level by the direct method. Assume $m_y = 250$ kN-m and $e = 0$

16.7 Refer to Prob. 16.1. Determine the ultimate lateral resistance of the pile by the direct method.

16.8 A precast reinforced concrete pile of 30 cm diameter is driven to a depth of 10 m in a vertical direction into a medium dense sand which is in a semi-dry state. The value of the coefficient of soil modulus variation (n_h) may be assumed as equal to 0.8 kg/cm^3. A lateral load of 40 kN is applied at a height of 3 m above ground level. Compute (a) the deflection at ground level, and (b) the maximum bending moment on the pile (assume $E = 2.1 \times 10^5$ kg/cm^2).

16.9 Refer to Prob. 16.8. Solve the problem by the direct method. All the other data remain the same. Assume $\phi = 38°$ and $\gamma = 16.5$ kN/m^3.

16.10 If the pile in Prob. 16.9 is driven at a batter of 22.5° to the vertical, and lateral load is applied at ground level, compute the normal deflection at ground level, for the cases of the load acting in the direction of batter and against the batter.

CHAPTER 17

DEEP FOUNDATION III: DRILLED PIER FOUNDATIONS

17.1 INTRODUCTION

Chapter 15 dealt with piles subjected to vertical loads and Chapter 16 with piles subjected to lateral loads. Drilled pier foundations, the subject matter of this chapter, belong to the same category as pile foundations. Because piers and piles serve the same purpose, no sharp deviations can be made between the two. The distinctions are based on the method of installation. A pile is installed by driving, a pier by excavating. Thus, a foundation unit installed in a drill-hole may also be called a bored cast-in-situ concrete pile. Here, distinction is made between a small diameter pile and a large diameter pile. A pile, cast-in-situ, with a diameter less than 0.75 m (or 2.5 ft) is sometimes called a small diameter pile. A pile greater than this size is called a large diameter bored-cast-in-situ pile. The latter definition is used in most non-American countries whereas in the USA, such large-diameter bored piles are called drilled piers, drilled shafts, and sometimes drilled *caissons*. Chapter 15 deals with small diameter bored-cast-in situ piles in addition to driven piles.

17.2 TYPES OF DRILLED PIERS

Drilled piers may be described under four types. All four types are similar in construction technique, but differ in their design assumptions and in the mechanism of load transfer to the surrounding earth mass. These types are illustrated in Figure 17.1.

Straight-shaft end-bearing piers develop their support from end-bearing on strong soil, "hardpan" or rock. The overlying soil is assumed to contribute nothing to the support of the load imposed on the pier (Fig. 17.1(a)).

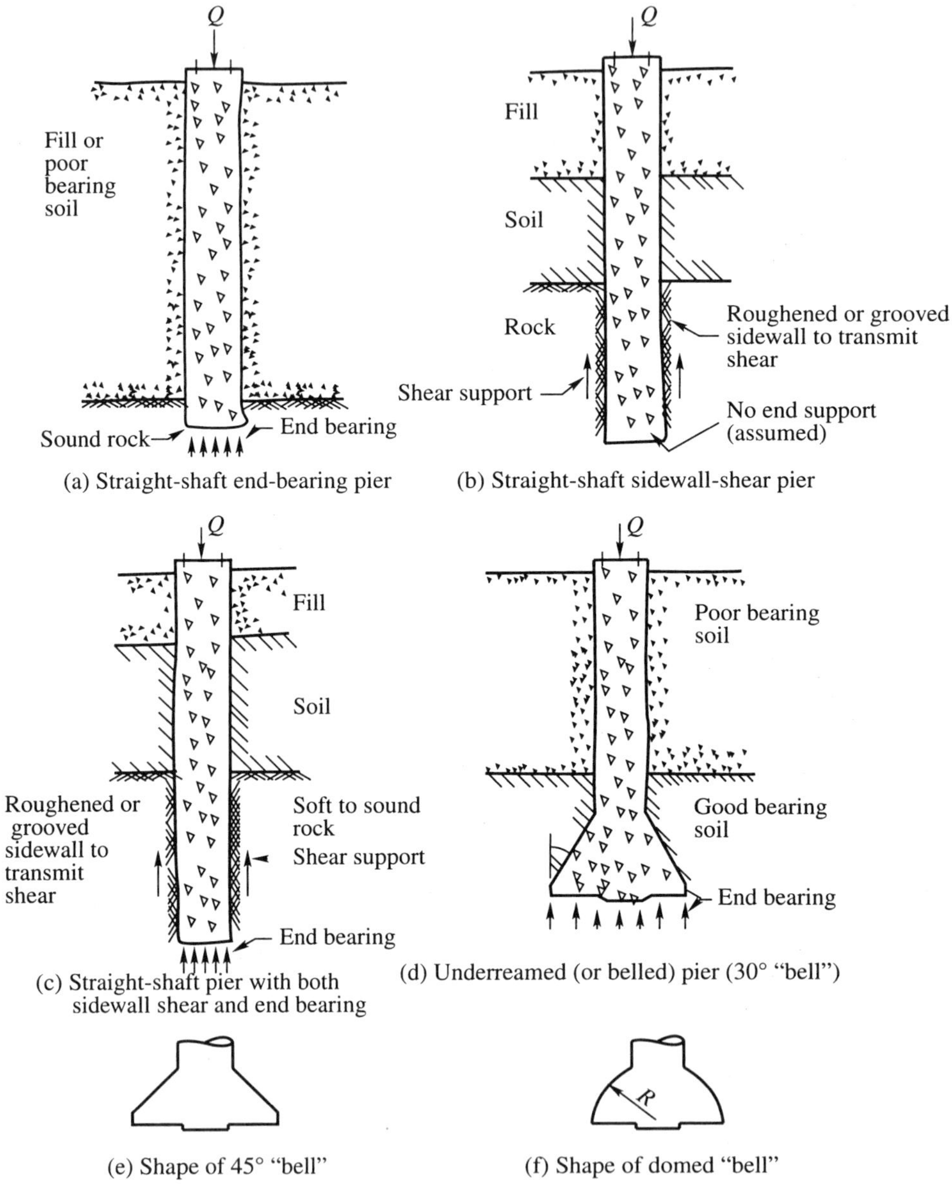

Figure 17.1 Types of drilled piers and underream shapes (Woodward et al., 1972)

Straight-shaft side wall friction piers pass through overburden soils that are assumed to carry none of the load, and penetrate far enough into an assigned bearing stratum to develop design load capacity by side wall friction between the pier and bearing stratum (Fig. 17.1(b)).

Combination of straight shaft side wall friction and end bearing piers are of the same construction as the two mentioned above, but with both side wall friction and end bearing assigned a role in carrying the design load. When carried into rock, this pier may be referred to as a socketed pier or a "drilled pier with rock socket" (Fig. 17.1(c)).

Belled or underreamed piers are piers with a bottom bell or underream (Fig. 17.1(d)). A greater percentage of the imposed load on the pier top is assumed to be carried by the base.

17.3 ADVANTAGES AND DISADVANTAGES OF DRILLED PIER FOUNDATIONS

Advantages

1. Pier of any length and size can be constructed at the site
2. Construction equipment is normally mobile and construction can proceed rapidly
3. Inspection of drilled holes is possible because of the larger diameter of the shafts
4. Very large loads can be carried by a single drilled pier foundation thus eliminating the necessity of a pile cap
5. The drilled pier is applicable to a wide variety of soil conditions
6. Changes can be made in the design criteria during the progress of a job
7. Ground vibration that is normally associated with driven piles is absent in drilled pier construction
8. Bearing capacity can be increased by underreaming the bottom (in non-caving materials)

Disadvantages

1. Installation of drilled piers needs a careful supervision and quality control of all the materials used in the construction
2. The method is cumbersome. It needs sufficient storage space for all the materials used in the construction
3. The advantage of increased bearing capacity due to compaction in granular soil that could be obtained in driven piles is not there in drilled pier construction
4. Construction of drilled piers at places where there is a heavy current of ground water flow due to artesian pressure is very difficult

17.4 METHODS OF CONSTRUCTION

Earlier Methods

The use of drilled piers for foundations started in the United States during the early part of the twentieth century. The two most common procedures were the Chicago and Gow methods shown in Fig. 17.2. In the Chicago method a circular pit was excavated to a convenient depth and a cylindrical shell of vertical boards or staves was placed by making use of an inside compression ring. Excavation then continued to the next board length and a second tier of staves was set and the procedure continued. The tiers could be set at a constant diameter or stepped in about 50 mm. The Gow method, which used a series of telescopic metal shells, is about the same as the current method of using casing except for the telescoping sections reducing the diameter on successive tiers.

Modern Methods of Construction

Equipment

There has been a phenomenal growth in the manufacture and use of heavy duty drilling equipment in the United States since the end of World War II. The greatest impetus to this development occurred in two states, Texas and California (Woodward et al., 1972). Improvements in the machines were made responding to the needs of contractors. Commercially produced drilling rigs of sufficient size and capacity to drill pier holes come in a wide variety of mountings and driving arrangements. Mountings are usually truck crane, tractor or skid. Fig. 17.3 shows a tractor mounted rig. Drilling machine ratings as presented in manufacturer's catalogs and technical data sheets are

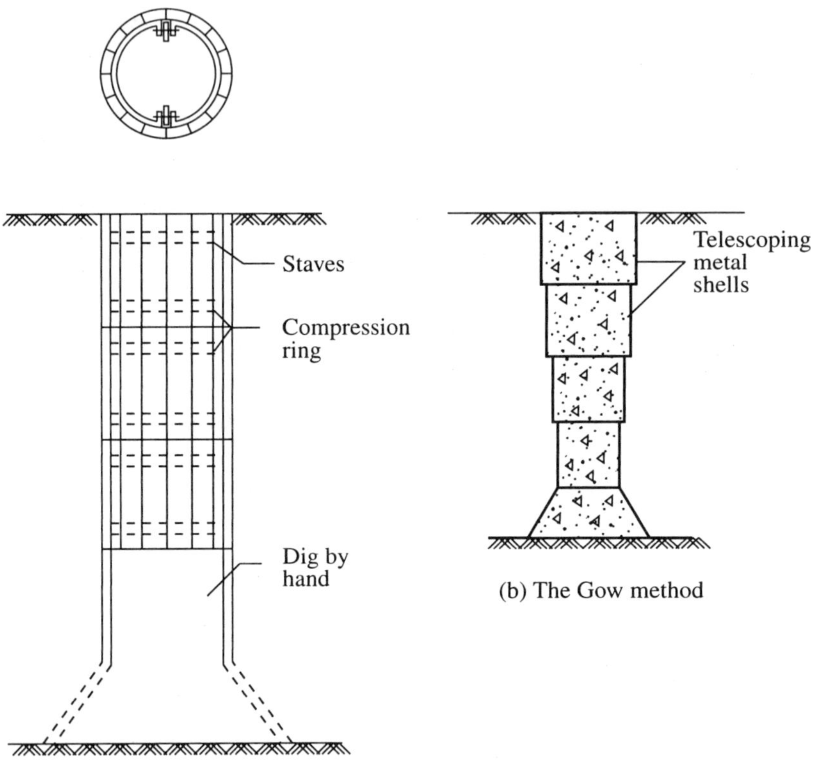

(a) The Chicago method

(b) The Gow method

Figure 17.2 Early methods of caisson construction

usually expressed as maximum hole diameter, maximum depth, and maximum torque at some particular rpm.

Many drilled pier shafts through soil or soft rock are drilled with the open-helix auger. The tool may be equipped with a knife blade cutting edge for use in most homogeneous soil or with hard-surfaced teeth for cutting stiff or hard soils, stony soils, or soft to moderately hard rock. These augers are available in diameters up to 3 m or more. Fig. 17.4 shows commercially available models.

Underreaming tools (or buckets) are available in a variety of designs. Figure 17.5 shows a typical 30° underreamer with blade cutter for soils that can be cut readily. Most such underreaming tools are limited in size to a diameter three times the diameter of the shaft.

When rock becomes too hard to be removed with auger-type tools, it is often necessary to resort to the use of a core barrel. This tool is a simple cylindrical barrel, set with tungsten carbide teeth at the bottom edge. For hard rock which cannot be cut readily with the core barrel set with hard metal teeth, a calyx or shot barrel can be used to cut a core of rock.

General Construction Methods of Drilled Pier Foundations

The rotary drilling method is the most common method of pier construction in the United States. The methods of drilled pier construction can be classified in three categories as

1. The dry method
2. The casing method
3. The slurry method

Figure 17.3 Tractor mounted hydraulic drilling rig (Courtesy: Kelly Tractor Co, USA)

Dry Method of Construction

The dry method is applicable to soil and rock that are above the water table and that will not cave or slump when the hole is drilled to its full depth. The soil that meets this requirement is a homogeneous, stiff clay. The first step in making the hole is to position the equipment at the desired location and to select the appropriate drilling tools. Fig. 17.6(a) gives the initial location. The drilling is next carried out to its fill depth with the spoil from the hole removed simultaneously.

After drilling is complete, the bottom of the hole is underreamed if required. Fig. 17.6(b) and (c) show the next steps of concreting and placing the rebar cage. Fig 17.6(d) shows the hole completely filled with concrete.

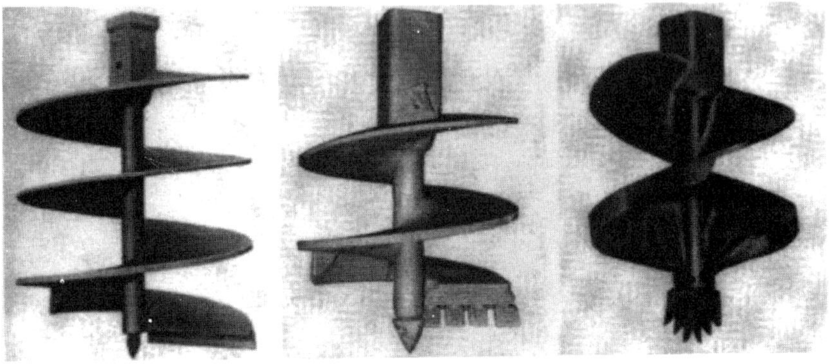

(a) (b) (c)

Figure 17.4 (a) Single-flight auger bit with cutting blade for soils, (b) single-flight auger bit with hard-metal cutting teeth for hard soils, hardpan, and rock, and (c) cast steel heavy-duty auger bit for hardpan and rock (Source: Woodward et al., 1972)

Figure 17.5 A 30° underreamer with blade cutters for soils that can be cut readily (Source: Woodward et al., 1972)

Casing Method of Construction

The casing method is applicable to sites where the soil conditions are such that caving or excessive soil or rock deformation can occur when a hole is drilled. This can happen when the boring is made in dry soils or rocks which are stable when they are cut but will slough soon afterwards. In such a

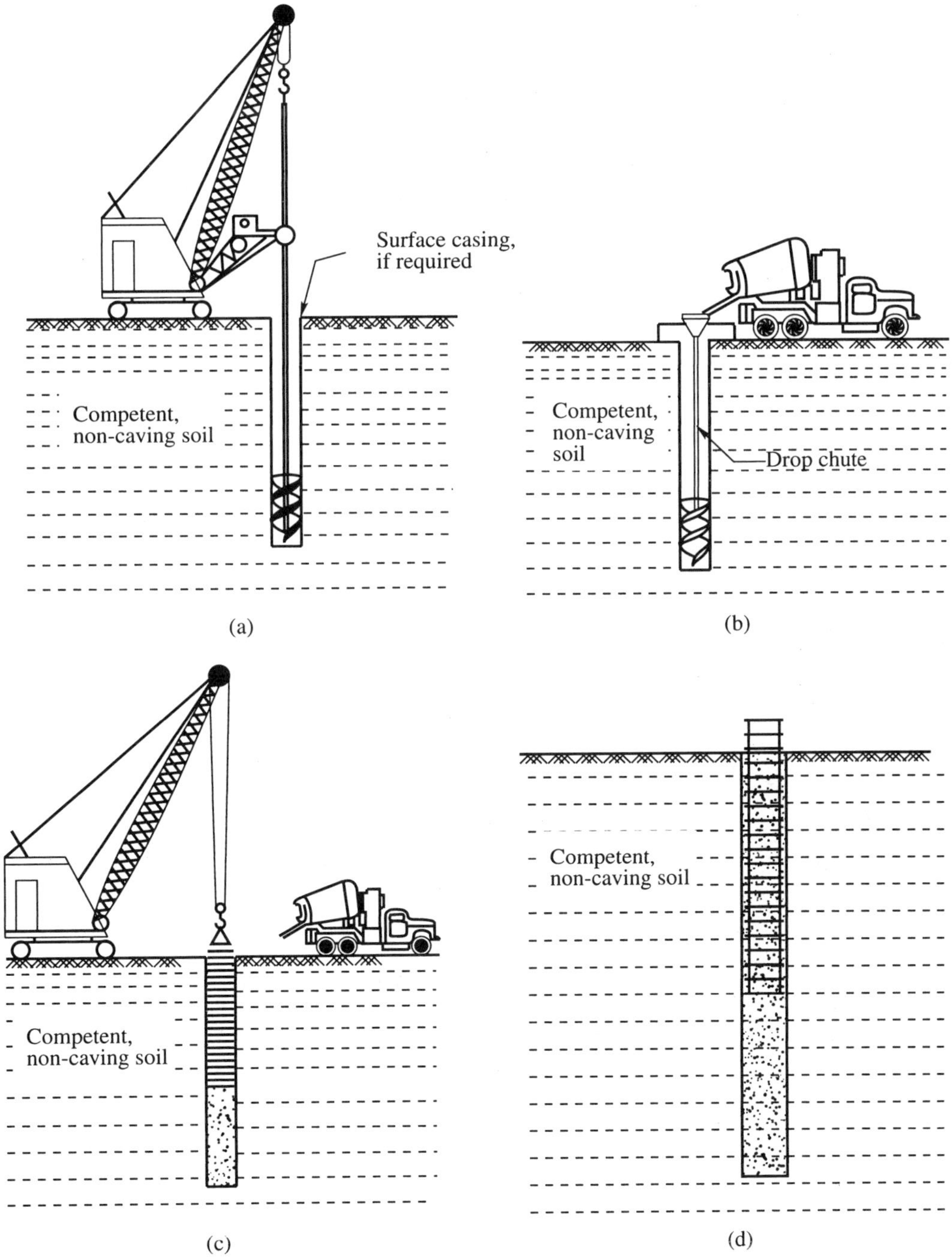

Figure 17.6 Dry method of construction: (a) initiating drilling, (b) starting concrete pour, (c) placing rebar cage, and (d) completed shaft (O'Neill and Reese, 1999)

case, the bore hole is drilled, and a steel pipe casing is quickly set to prevent sloughing. Casing is also required if drilling is required in clean sand below the water table underlain by a layer of impermeable stones into which the drilled shaft will penetrate. The casing is removed soon after the concrete is deposited. In some cases, the casing may have to be left in place permanently. It may be noted here that until the casing is inserted, a slurry is used to maintain the stability of the hole. After the casing is seated, the slurry is bailed out and the shaft extended to the required depth. Figures 17.7(a) to (h) give the sequence of operations. Withdrawl of the casing, if not done carefully, may lead to voids or soil inclusions in the concrete, as illustrated in Fig. 17.8.

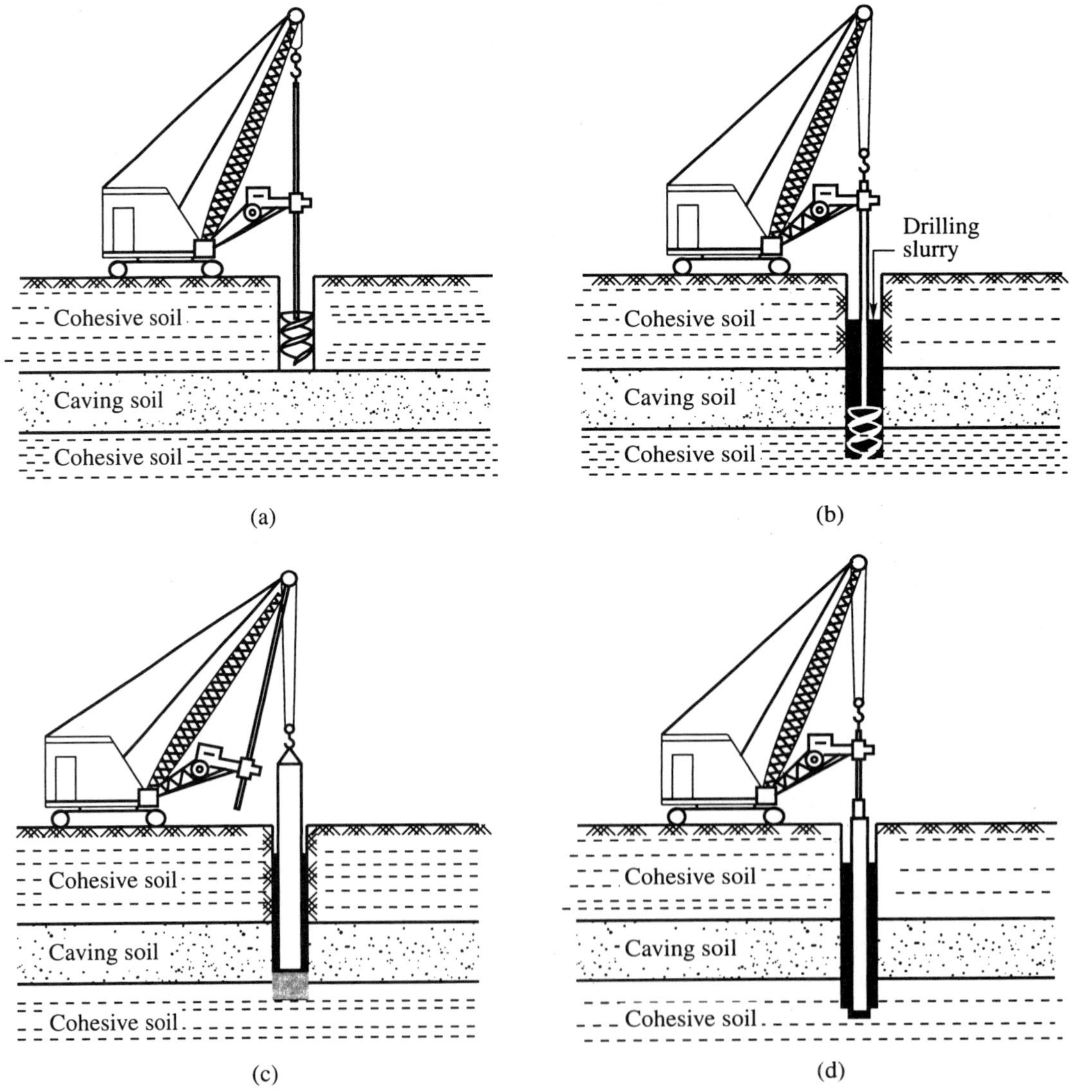

Figure 17.7 Casing method of construction: (a) initiating drilling, (b) drilling with slurry; (c) introducing casing, (d) casing is sealed and slurry is being removed from interior of casing (continued)

Slurry Method of Construction

The slurry method of construction involves the use of a prepared slurry to keep the bore hole stable for the entire depth of excavation. The soil conditions for which the slurry displacement method is applicable could be any of the conditions described for the casing method. The slurry method is a

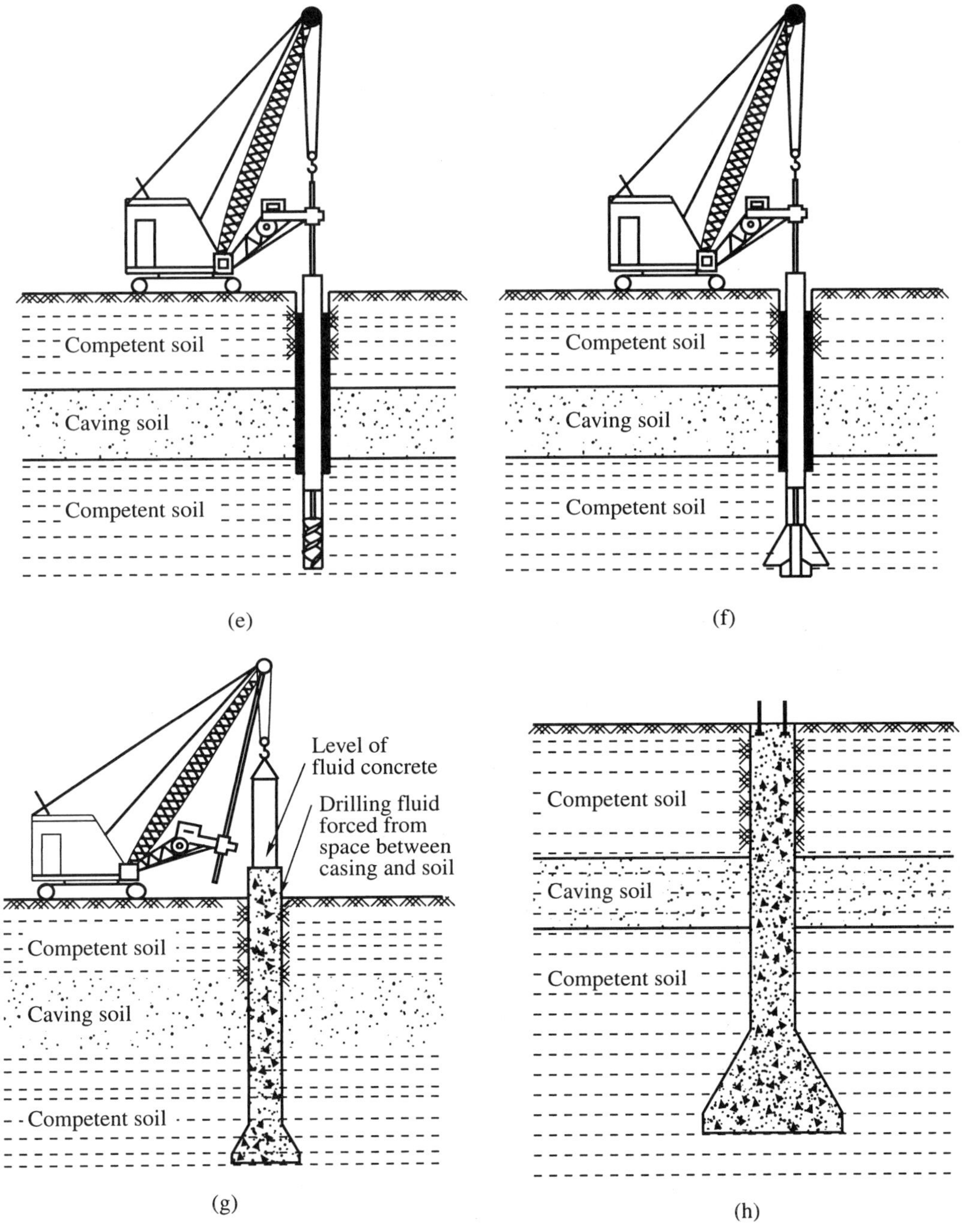

Figure 17.7 (continued) casing method of construction: (e) drilling below casing, (f) underreaming, (g) removing casing, and (h) completed shaft (O'Neill and Reese, 1999)

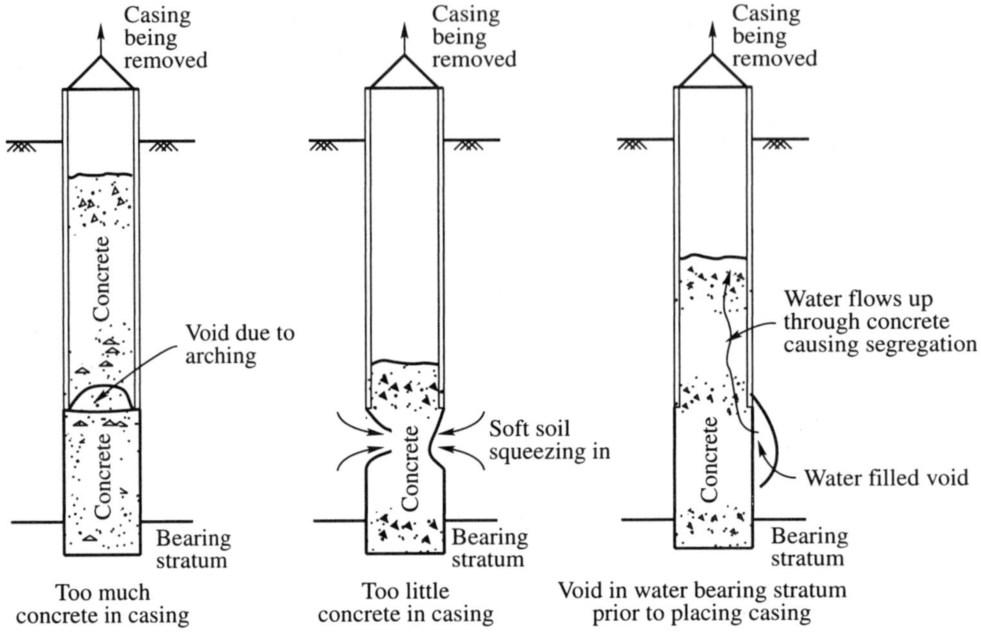

Figure 17.8 Potential problems leading to inadequate shaft concrete due to removal of temporary casing without care (D'Appolonia, et al., 1975)

viable option at any site where there is a caving soil, and it could be the only feasible option in a permeable, water bearing soil if it is impossible to set a casing into a stratum of soil or rock with low permeability. The various steps in the construction process are shown in Fig. 17.9. It is essential in this method that a sufficient slurry head be available so that the inside pressure is greater than that from the GWT or from the tendency of the soil to cave.

Bentonite is most commonly used with water to produce the slurry. Polymer slurry is also employed. Some experimentation may be required to obtain an optimum percentage for a site, but amounts in the range of 4 to 6 percent by weight of admixture are usually adequate.

The bentonite should be well mixed with water so that the mixture is not lumpy. The slurry should be capable of forming a filter cake on the side of the bore hole. The bore hole is generally not underreamed for a bell since this procedure leaves unconsolidated cuttings on the base and creates a possibility of trapping slurry between the concrete base and the bell roof.

If reinforcing steel is to be used, the rebar cage is placed in the slurry as shown in Fig 17.9(b). After the rebar cage has been placed, concrete is placed with a tremie either by gravity feed or by pumping. If a gravity feed is used, the bottom end of the tremie pipe should be closed with a closure plate until the base of the tremie reaches the bottom of the bore hole, in order to prevent contamination of the concrete by the slurry. Filling of the tremie with concrete, followed by subsequent slight lifting of the tremie, will then open the plate, and concreting proceeds. Care must be taken that the bottom of the tremie is buried in concrete at least for a depth of 1.5 m (5 ft). The sequence of operations is shown in Fig 17.9(a) to (d).

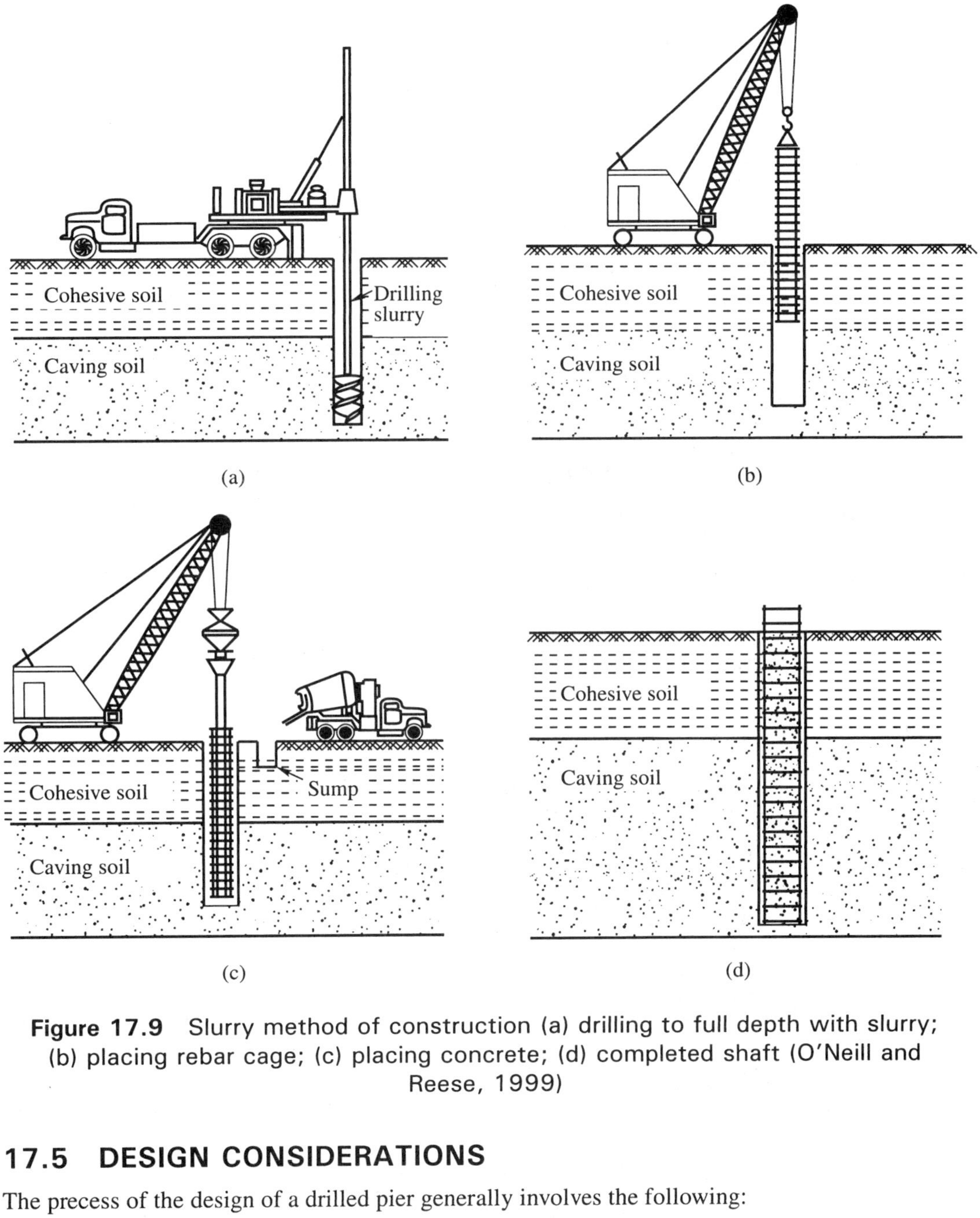

Figure 17.9 Slurry method of construction (a) drilling to full depth with slurry; (b) placing rebar cage; (c) placing concrete; (d) completed shaft (O'Neill and Reese, 1999)

17.5 DESIGN CONSIDERATIONS

The precess of the design of a drilled pier generally involves the following:

1. The objectives of selecting drilled pier foundations for the project.
2. Analysis of loads coming on each pier foundation element.
3. A detailed soil investigation and determining the soil parameters for the design.
4. Preparation of plans and specifications which include the methods of design, tolerable settlement, methods of construction of piers, etc.
5. The method of execution of the project.

In general the design of a drilled pier may be studied under the following headings.

1. Allowable loads on the piers based on ultimate bearing capacity theories.
2. Allowable loads based on vertical movement of the piers.
3. Allowable loads based on lateral bearing capacity of the piers.

In addition to the above, the uplift capacity of piers with or without underreams has to be evaluated.

The following types of strata are considered.

1. Piers embedded in homogeneous soils, sand or clay.
2. Piers in a layered system of soil.
3. Piers socketed in rocks.

It is better that the designer select shaft diameters that are multiples of 150 mm (6 in) since these are the commonly available drilling tool diameters.

17.6 LOAD TRANSFER MECHANISM

Figure 17.10(a) shows a single drilled pier of diameter *d*, and length *L* constructed in a homogeneous mass of soil of known physical properties. If this pier is loaded to failure under an ultimate load Q_u, a part of this load is transmitted to the soil along the length of the pier and the balance is transmitted to the pier base. The load transmitted to the soil along the pier is called the ultimate *friction load or skin load,* Q_f and that transmitted to the base is the ultimate base or point load Q_b. The total ultimate load, Q_u, is expressed as (neglecting the weight of the pier)

$$Q_u = Q_b + Q_f = q_b A_b + \sum_{i=1}^{N} f_{si} P_i \Delta z_i \tag{17.1a}$$

where
q_b = net ultimate bearing pressure
A_b = base area
f_{si} = unit skin resistance (ultimate) of layer *i*
P_i = perimeter of pier in layer *i*
Δz_i = thickness of layer *i*
N = number of layers

If the pier is instrumented, the load distribution along the pier can be determined at different stages of loading. Typical load distribution curves plotted along a pier are shown in Fig 17.10(b) (O'Neill and Reese, 1999). These load distribution curves are similar to the one shown in Fig. 15.5(b). Since the load transfer mechanism for a pier is the same as that for a pile, no further discussion on this is necessary here. However, it is necessary to study in this context the effect of settlement on the mobilization of side shear and base resistance of a pier. As may be seen from Fig. 17.11, the maximum values of base and side resistance are not mobilized at the same value of displacement. In some soils, and especially in some brittle rocks, the side shear may develop fully at a small value of displacement and then decrease with further displacement while the base resistance is still being mobilized (O'Neill and Reese, 1999). If the value of the side resistance at point *A* is added to the value of the base resistance at point *B*, the total resistance shown at level *D* is overpredicted. On the other hand, if the designer wants to take advantage primarily of the base resistance, the side resistance at point *C* should be added to the base resistance at point *B* to evaluate Q_u. Otherwise, the designer may wish to design for the side resistance at point *A* and disregard the base resistance entirely.

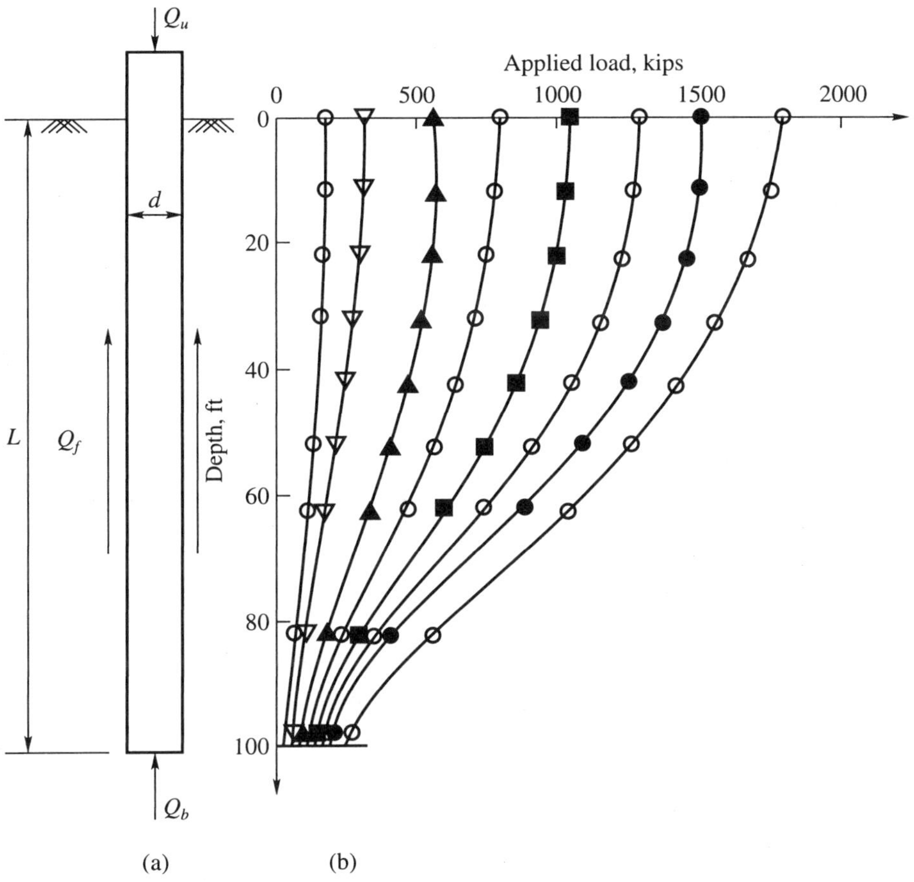

Figure 17.10 Typical set of load distribution curves (O'Neill and Reese, 1999)

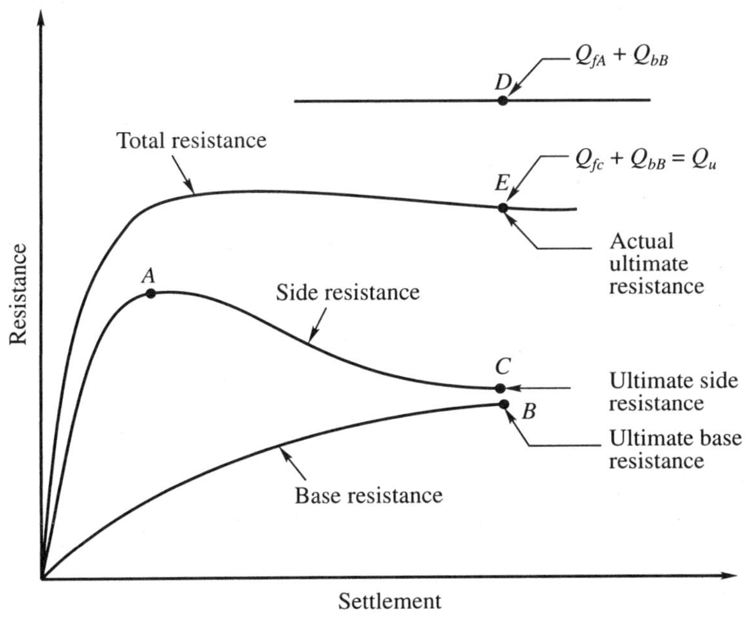

Figure 17.11 Condition in which $(Q_b + Q_f)$ is not equal to actual ultimate resistance

17.7 VERTICAL BEARING CAPACITY OF DRILLED PIERS

For the purpose of estimating the ultimate bearing capacity, the subsoil is divided into layers (Fig. 17.12) based on judgment and experience (O'Neill and Reese, 1999). Each layer is assigned one of four classifications.

1. Cohesive soil [clays and plastic silts with undrained shear strength $c_u \leq 250$ kN/m^2 (2.5 t/ft^2)] .
2. Granular soil [cohesionless geomaterial, such as sand, gravel or nonplastic silt with uncorrected SPT(N) values of 50 blows per 0.3/m or less].
3. Intermediate geometerial [cohesive geometerial with undrained shear strength c_u between 250 and 2500 kN/m^2 (2.5 and 25 tsf), or cohesionless geomaterial with SPT(N) values > 50 blows per 0.3 m].
4. Rock [highly cemented geomaterial with unconfined compressive strength greater than 5000 kN/m^2 (50 tsf)].

The unit side resistance f_s ($=f_{max}$) is computed in each layer through which the drilled shaft passes, and the unit base resistance q_b ($=q_{max}$) is computed for the layer on or in which the base of the drilled shaft is founded.

The soil along the whole length of the shaft is divided into four layers as shown in Fig. 17.12.

Effective Length for Computing Side Resistance in Cohesive Soil

O'Neill and Reese (1999) suggest that the following effective length of pier is to be considered for computing side resistance in cohesive soil.

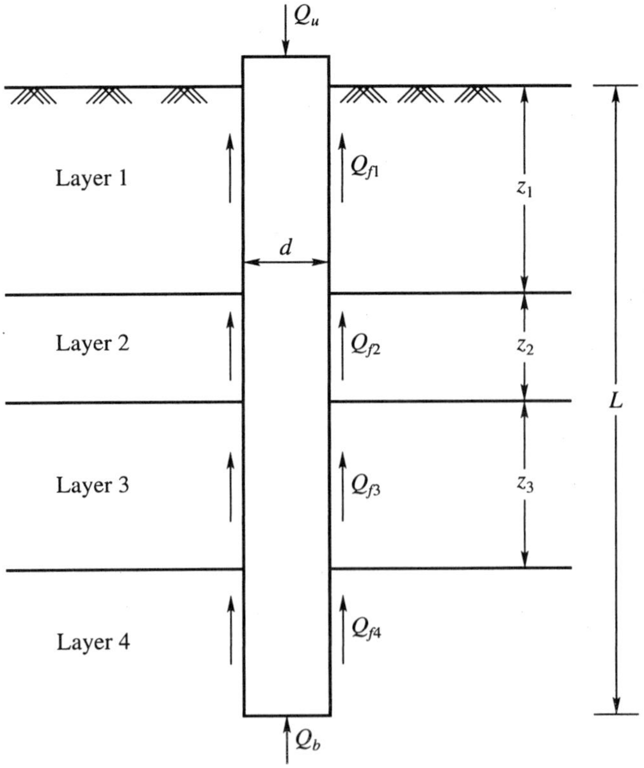

Figure 17.12 Idealized geomaterial layering for computation of compression load and resistance (O'Neill and Reese, 1999)

Straight shaft: One diameter from the bottom and 1.5 m (5 feet) from the top are to be excluded from the embedded length of pile for computing side resistance as shown in Fig. 17.13(a) .

Belled shaft: The height of the bell plus the diameter of the shaft from the bottom and 1.5 m (5 ft) from the top are to be excluded as shown in Fig 17.13(b).

17.8 THE GENERAL BEARING CAPACITY EQUATION FOR THE BASE RESISTANCE q_b (= q_{max})

The equation for the ultimate base resistance may be expressed as

$$q_b = s_c d_c N_c c + s_q d_q (N_q - 1) q'_o + \frac{1}{2} \gamma d\ s_\gamma d_\gamma N_\gamma \tag{17.2}$$

where N_c, N_q and N_γ = bearing capacity of factors for long footings

s_c, s_q and s_γ = shape factors

d_c, d_q and d_γ = depth factors

q'_o = effective vertical pressure at the base level of the drilled pier

γ = effective unit weight of the soil below the bottom of the drilled shaft to a depth = 1.5 d where d = width or diameter of pier at base level

c = average cohesive strength of soil just below the base.

For deep foundations the last term in Eq. (17.2) becomes insignificant and may be ignored. Now Eq. (17.2) may be written as

$$q_b = s_c d_c N_c c + s_q d_q (N_q - 1) q'_o \tag{17.3}$$

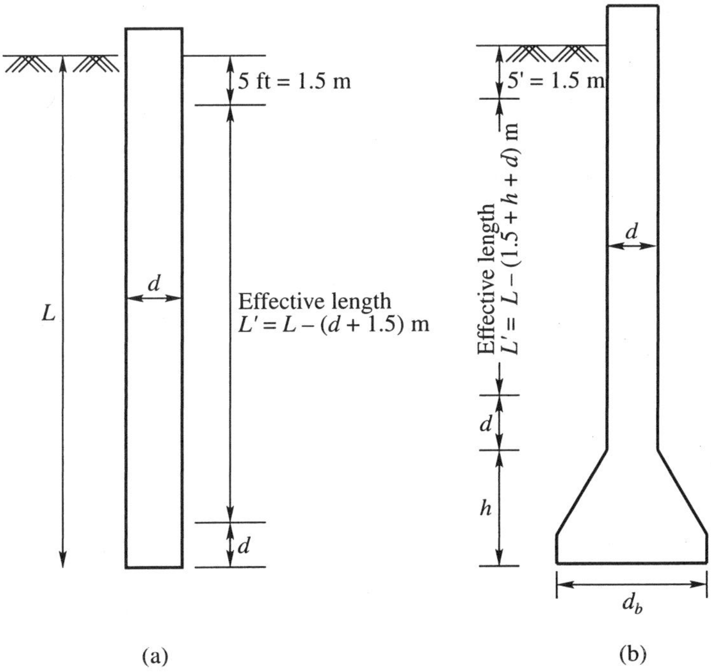

Figure 17.13 Exclusion zones for estimating side resistance for drilled shafts in cohesive soils

17.9 BEARING CAPACITY EQUATIONS FOR THE BASE IN COHESIVE SOIL

When the Undrained Shear Strength, $c_u \leq 250$ kN/m^2 (2.5 t/ft^2)

For $\phi = 0$, $N_q = 1$ and $(N_q - 1) = 0$, here Eq. (17.3) can be written as (Vesic, 1972)

$$q_b = N_c^* c_u \tag{17.4}$$

in which

$$N_c^* = \frac{4}{3}(\ln I_r + 1) \tag{17.5}$$

I_r = rigidity index of the soil

Eq. (17.4) is applicable for $c_u \leq 96$ kPa and $L \geq 3d$ (base width)

For $\phi = 0$, I_r may be expressed as (O'Neill and Reese, 1999)

$$I_r = \frac{E_s}{3c_u} \tag{17.6}$$

where E_s = Young's modulus of the soil in undrained loading. Refer to Section 13.8 for the methods of evaluating the value of E_s.

Table 17.1 gives the values of I_r and N_c^* as a function of c_u.
If the depth of base $(L) < 3d$ (base)

$$q_b(= q_{\max}) = \frac{2}{3}\left[1 + \frac{L}{6d}\right] N_c^* c_u \tag{17.7}$$

When $c_u \geq 96$ kPa (2000 lb/ft^2), the equation for q_b may be written as

$$q_b = 9c_u \tag{17.8}$$

for depth of base $(= L) \geq$ 3d (base width).

17.10 BEARING CAPACITY EQUATION FOR THE BASE IN GRANULAR SOIL

Values N_c and N_q in Eq. (17.3) are for strip footings on the surface of rigid soils and are plotted as a function of ϕ in Fig. 17.14. Vesic (1977) explained that during bearing failure, a plastic failure zone develops beneath a circular loaded area that is accompanied by elastic deformation in the surrounding elastic soil mass. The confinement of the elastic soil surrounding the plastic soil has an effect on q_b $(= q_{\max})$. The values of N_c and N_q are therefore dependent not only on ϕ, but also on I_r. They must be corrected for soil rigidity as given below.

Table 17.1 Values of $I_r = E_s/3c_u$ and N_c^*

c_u	I_r	N_c^*
24 kPa (500 lb/ft^2)	50	6.5
48 kPa (1000 lb/ft^2)	150	8.0
$\geq$ 96 kPa (2000 lb/ft^2)	250–300	9.0

$$N_c\,(\text{corrected}) = N_c C_c$$
$$N_q\,(\text{corrected}) = N_q C_q \qquad (17.9)$$

where C_c and C_q are the correction factors. As per Chen and Kulhawy (1994)

Eq (17.3) may now be expressed as

$$q_b = c\,N_c s_c d_c C_c + (N_q - 1)q'_o s_q d_q C_q \qquad (17.10)$$

$$C_c = C_q - \frac{1 - C_q}{N_c \tan\phi} \qquad (17.11a)$$

$$C_q = \exp\{[-3.8\tan\phi] + [(3.07\sin\phi)\log_{10} 2I_{rr}]/(1+\sin\phi)\} \qquad (17.11b)$$

where ϕ is an effective angle of internal friction. I_{rr} is the reduced rigidity index expressed as [Eq. (15.28)]

$$I_{rr} = \frac{I_r}{1 + \Delta I_r} \qquad (17.12)$$

and
$$I_r = \frac{E_d}{2(1+\mu_d)q'_o \tan\phi} \qquad (17.13)$$

by ignoring cohesion, where,

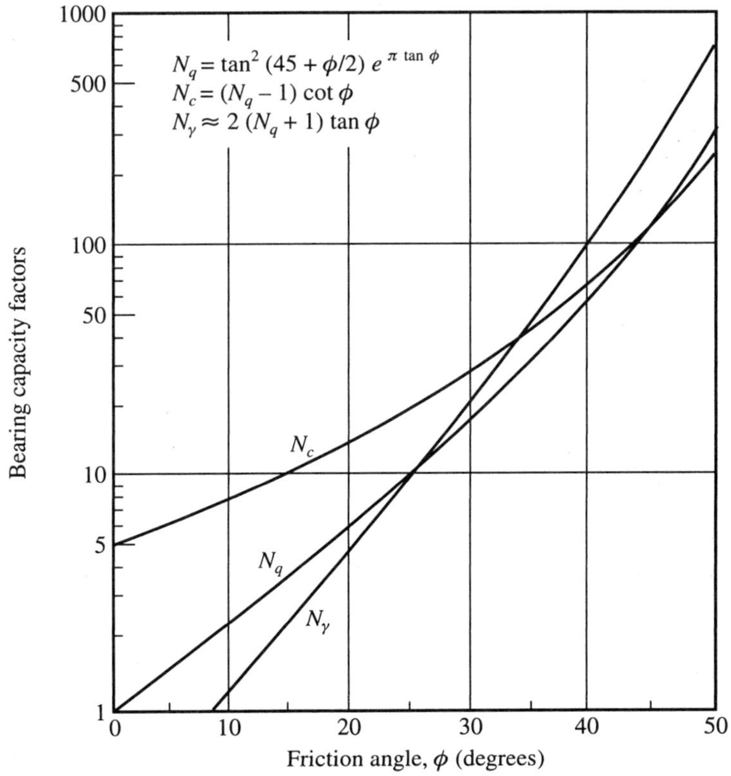

Figure 17.14 Bearing capacity factors (Chen and Kulhawy, 1994)

E_d = drained Young's modulus of the soil
μ_d = drained Poisson's ratio
Δ = volumetric strain within the plastic zone during the loading process

The expressions for μ_d and Δ may be written as (Chen and Kulhawy, 1994)

$$\mu_d = 0.1 + 0.3\phi_{rel} \tag{17.14}$$

$$\Delta = \frac{0.005(1-\phi_{rel})q'_o}{p_a} \tag{17.15}$$

where $\phi_{rel} = \dfrac{(\phi^\circ - 25^\circ)}{45^\circ - 25^\circ}$ for $25^\circ \le \phi^\circ \le 45^\circ$ (17.16)

= relative friction angle factor, p_a = atmospheric pressure = 101 kPa.

Chen and Kulhawy (1994) suggest that, for granular soils, the following values may be considered.

loose soil, $E_d = 100 \text{ to } 200 p_a$ (17.17)

medium dense soil, $E_d = 200 \text{ to } 500 p_a$

dense soil, $E_d = 500 \text{ to } 1000 p_a$

The correction factors C_c and C_q indicated in Eq. (17.9) need be applied only if I_{rr} is less than the critical rigidity index $(I_r)_{crit}$ expressed as follows

$$(I_r)_{cr} = \frac{1}{2}\exp\left[2.85\cot\left(45^\circ - \frac{\phi^\circ}{2}\right)\right] \tag{17.18}$$

The values of critical rigidity index may be obtained from Table 12.4 for piers circular or square in section.

If $I_{rr} > (I_r)_{crit}$, the factors C_c and C_q may be taken as equal to unity.

The shape and depth factors in Eq. (17.3) can be evaluated by making use of the relationships given in Table 17.2.

Table 17.2 Shape and depth factors (Eq. 17.3) (Chen and Kulhawy, 1994)

Factors	Value
s_c	$1 + \dfrac{N_q}{N_c}$
s_d	$d_q - \dfrac{1-d_q}{N_c \tan\phi}$
s_q	$1 + \tan\phi$
d_q	$1 + 2\tan\phi(1-\sin\phi)^2\left[\dfrac{\pi}{180}\tan^{-1}\dfrac{L}{d}\right]$

Base in Cohesionless Soil

The theoretical approach as outlined above is quite complicated and difficult to apply in practice for drilled piers in granular soils. Direct and simple empirical correlations have been suggested by O'Neill and Reese (1999) between SPT N value and the base bearing capacity as given below for cohesionless soils.

$$q_b (= q_{max}) = 57.5N \text{ kPa } \leq 2900 \text{ kN/m}^2 \tag{17.19a}$$

$$q_b (= q_{max}) = 0.60N \text{ tsf } \leq 30 \text{ tsf} \tag{17.19b}$$

where N = SPT value $\leq$ 50 blows / 0.3 m.

Base in Cohesionless IGM

Cohesionless IGM's are characterized by SPT blow counts if more than 50 per 0.3 m. In such cases, the expression for q_b is

$$q_b (= q_{max}) = 0.60\left[N_{60} \frac{p_a}{q'_o}\right]^{0.8} q'_o \tag{17.20}$$

where N_{60} = average SPT corrected for 60 percent standard energy within a depth of $2d$ (base) below the base. The value of N_{60} is limited to 100. No correction for overburden pressure

p_a = atmospheric pressure in the units used for q'_o (= 101 kPa in the SI System)

q'_o = vertical effective stress at the elevation of the base of the drilled shaft.

17.11 BEARING CAPACITY EQUATIONS FOR THE BASE IN COHESIVE IGM OR ROCK (O'NEILL AND REESE, 1999)

Massive rock and cohesive intermediate materials possess common properties. They possess low drainage qualities under normal loadings but drain more rapidly under large loads than cohesive soils. It is for these reasons undrained shear strengths are used for rocks and IGMs.

If the base of the pier lies in cohesive IGM or rock (RQD = 100 percent) and the depth of socket, D_s, in the IGM or rock is equal to or greater than $1.5d$, the bearing capacity may be expressed as

$$q_b (= q_{max}) = 2.5 q_u \tag{17.21}$$

where q_u = unconfined compressive strength of IGM or rock below the base

For RQD between 70 and 100 percent,

$$q_b (= q_{max}) = 4.83 (q_u)^{0.51} \text{ MPa} \tag{17.22}$$

For jointed rock or cohesive IGM

$$q_b (= q_{max}) = [s^{0.5} + (ms^{0.5} + s)^{0.5}] q_u \tag{17.23}$$

where q_u is measured on intact cores from within $2d$ (base) below the base of the drilled pier. In all the above cases q_b and q_u are expressed in the same units and s and m indicate the properties of the rock or IGM mass that can be estimated from Tables 17.3 and 17.4.

Table 17.3 Descriptions of rock types

Rock type	Description
A	Carbonate rocks with well-developed crystal cleavage (eg., dolostone, limestone, marble)
B	Lithified argillaeous rocks (mudstone, siltstone, shale, slate)
C	Arenaceous rocks (sandstone, quartzite)
D	Fine-grained igneous rocks (andesite, dolerite, diabase, rhyolite)
E	Coarse-grained igneous and metamorphic rocks (amphibole, gabbro, gneiss, granite, norite, quartz-diorite)

Table 17.4 Values of s and m (dimensionless) based on rock classification (Carter and Kulhawy, 1988)

Quality of rock mass	Joint description and spacing	s	Value of m as function of rock type (A–E) from				
			A	B	C	D	E
Excellent	Intact (closed); spacing > 3 m (10 ft)	1	7	10	15	17	25
Very good	Interlocking; spacing of 1 to 3 m (3 to 10 ft)	0.1	3.5	5	7.5	8.5	12.5
Good	Slightly weathered; spacing of 1 to 3 m (3 to 10 ft)	4×10^{-2}	0.7	1	1.5	1.7	2.5
Fair	Moderately weathered; spacing of 0.3 to 1m (1 to 3 ft)	10^{-4}	0.14	0.2	0.3	0.34	0.5
Poor	Weathered with gouge (soft material); spacing of 30 to 300 mm (1 in. to 1 ft)	10^{-5}	0.04	0.05	0.08	0.09	0.13
Very poor	Heavily weathered; spacing of less than 50 mm (2 in.)	0	0.007	0.01	0.015	0.017	0.025

17.12 THE ULTIMATE SKIN RESISTANCE OF COHESIVE AND INTERMEDIATE MATERIALS

Cohesive Soil

The process of drilling a borehole for a pier in cohesive soil disturbs the natural condition of the soil all along the side to a certain extent. There is a reduction in the soil strength not only due to boring but also due to stress relief and the time spent between boring and concreting. It is very difficult to quantify the extent of the reduction in strength analytically. In order to take care of the disturbance, the unit frictional resistance on the surface of the pier may be expressed as

$$f_s = \alpha c_u \tag{17.24}$$

where α = adhesion factor

c_u = undrained shear strength

Relationships have been developed between c_u and α by many investigators based on field load tests. Fig 17.15 gives one such relationship in the form of a curve developed by Chen and Kulhawy (1994). The curve has been developed on the following assumptions (Fig. 17.15).

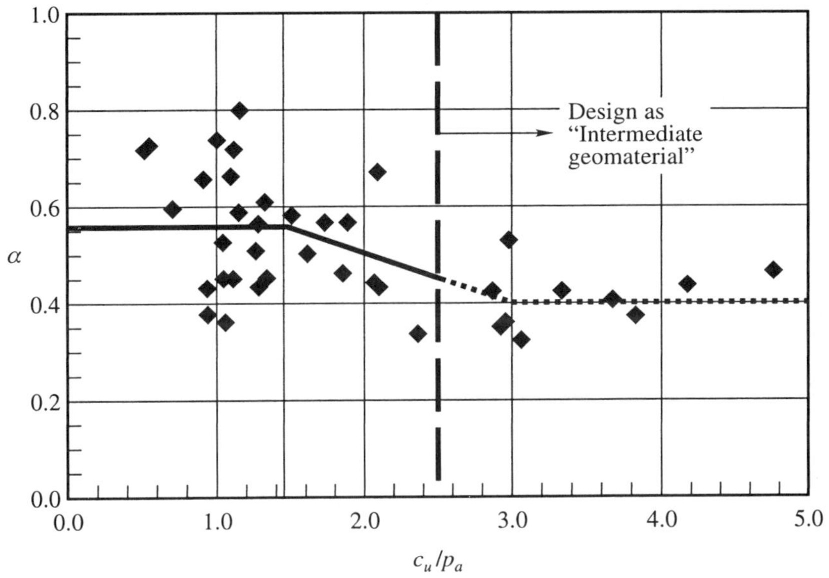

Figure 17.15 Correlation between α and c_u/p_a

$f_s = 0$ up to 1.5 m (= 5 ft) from the ground level.

$f_s = 0$ up to a height equal to $(h + d)$ as per Fig 17.13

O'Neill and Reese (1999) recommend the chart's trend line given in Fig. 17.15 for designing drilled piers. The suggested relationships are:

$$\alpha = 0.55 \quad \text{for } c_u / p_a \leq 1.5 \tag{17.25a}$$

$$\text{and } \alpha = 0.55 - 0.1\left[\frac{c_u}{p_a} - 1.5\right] \quad \text{for } 1.5 \leq c_u / p_a \leq 2.5 \tag{17.25b}$$

Cohesive Intermediate Geomaterials

Cohesive IGM's are very hard clay-like materials which can also be considered as very soft rock (O'Neill and Reese, 1999). IGM's are ductile and failure may be sudden at peak load. The value of f_a (please note that the term f_a is used instead of f_s for ultimate unit resistance at infinite displacement) depends upon the side condition of the bore hole, that is, whether it is rough or smooth. For design purposes the side is assumed as smooth. The expression for f_a may be written as

$$f_a = \alpha q_u \tag{17.26}$$

where, q_u = unconfined compressive strength

f_a = the value of ultimate unit side resistance which occurs at infinite displacement.

Figure 17.16 gives a chart for evaluating α. The chart is prepared for an effective angle of friction between the concrete and the IGM (assuming that the intersurface is drained) and S_t denotes the settlement of piers at the top of the socket. Further, the chart involves the use of σ_n/p_a where σ_n is the normal effective pressure against the side of the borehole by the drilled pier and p_a is the atmospheric pressure (101 kPa).

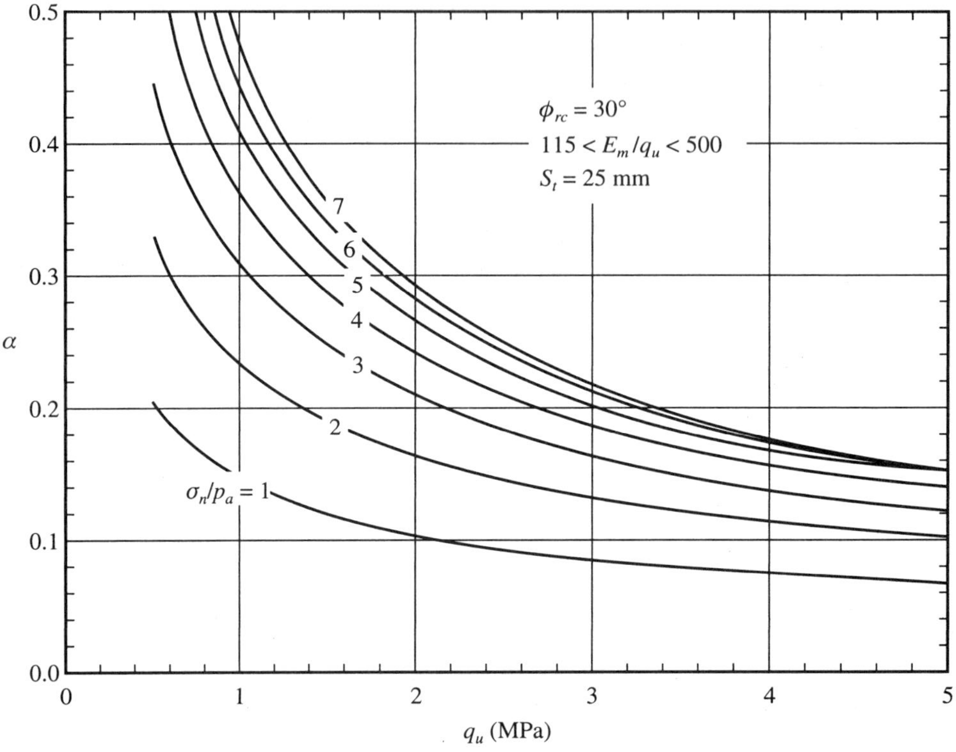

Figure 17.16 Factor α for cohesive IGM's (O'Neill and Reese, 1999)

Table 17.5 Estimation of E_m/E_i based on RQD (Modified after Carter and Kulhawy, 1988)

	E_m/E_i	
RQD (percent)	Closed joints	Open joints
100	1.00	0.60
70	0.70	0.10
50	0.15	0.10
20	0.05	0.05

Note: Values intermediate between tabulated values may be obtained by linear interpolation.

Table 17.6 f_{aa}/f_a based on E_m/E_i (O'Neill et al., 1996)

E_m/E_i	f_{aa}/f_a
1.0	1.0
0.5	0.8
0.3	0.7
0.1	0.55
0.05	0.45

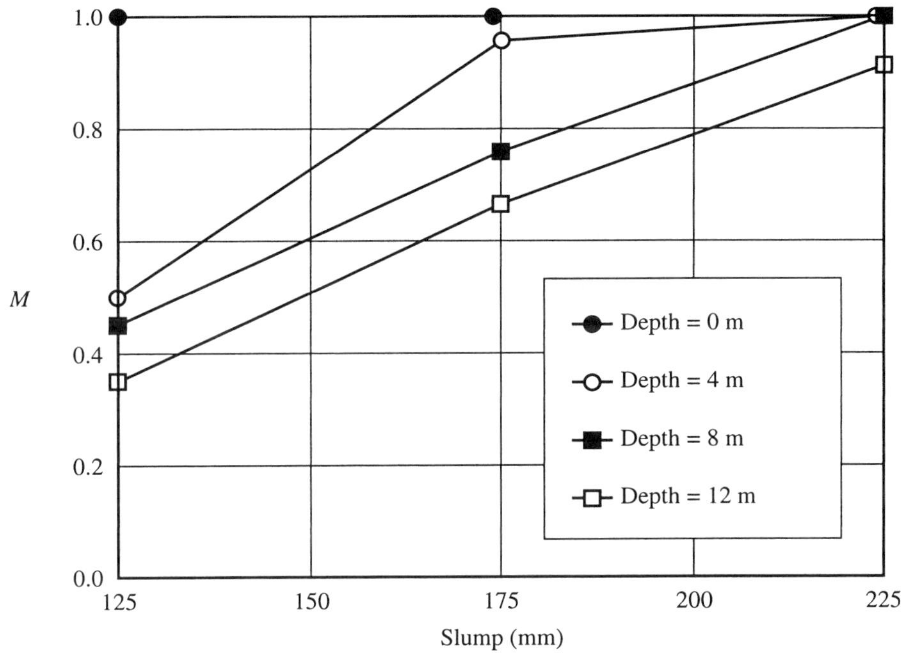

Figure 17.17 Factor *M* versus concrete slump (O'Neill et al., 1996)

O'Neill and Reese (1999) give the following equation for computing σ_n

$$\sigma_n = M\gamma_c z_c \tag{17.27}$$

where γ_c = the unit weight of the fluid concrete used for the construction
z_c = the depth of the point at which σ_n is required
M = an empirical factor which depends on the fluidity of the concrete as indexed by the concrete slump

Figure 17.17 gives the values of *M* for various slumps.

The mass modulus of elasticity of the IGM (E_m) should be determined before proceeding, in order to verify that the IGM is within the limits of Fig 17.16. This requires the average Young's modulus of intact IGM core (E_i) which can be determined in the laboratory. Table 17.5 gives the ratios of E_m/E_i for various values of RQD. Values of E_m/E_i less than unity indicate that soft seams and/or joints exist in the IGM. These discontinuities reduce the value of f_a. The reduced value of f_a may be expressed as

$$f_{aa} = f_a R_a \tag{17.28}$$

where the ratio $R_a = f_{aa}/f_a$ can be determined from Table 17.6.

If the socket is classified as smooth, it is sufficiently accurate to set $f_s = f_{\max} = f_{aa}$

17.13 ULTIMATE SKIN RESISTANCE IN COHESIONLESS SOIL AND GRAVELLY SANDS (O'NEILL AND REESE, 1999)

In Sands

A general expression for total skin resistance in cohesionless soil may be written as [Eq. (17.1)]

$$Q_{fi} = \sum_{i=1}^{N} P_i f_{si} \Delta z_i = \sum_{i=1}^{N} P_i q'_{oi} K_{si} \tan\delta_i \Delta z_i \tag{17.29}$$

$$\text{or} \quad Q_{fi} = \sum_{i=1}^{N} P_i \beta_i q'_{oi} \Delta z_i \tag{17.30}$$

where $f_{si} = \beta_i q'_{oi}$ (17.31)

$\beta_i = K_{si} \tan \delta_i$

δ_i = angle of skin friction of the i th layer

The following equations are provided by O'Neill and Reese (1999) for computing β_i.

For SPT N_{60} (uncorrected) $\geq$ 15 blows / 0.3 m

$$\beta_i = 1.5 - 0.245[z_i]^{0.5} \tag{17.32}$$

For SPT N_{60} (uncorrected) < 15 blows / 0.3 m

$$\beta_i = \frac{N_{60}}{15}\left[1.5 - 0.245(z_i)^{0.5}\right] \tag{17.33}$$

In Gravelly Sands or Gravels

For SPT $N_{60} \geq$ 15 blows / 0.3 m

$$\beta_i = 2.0 - 0.15[z_i]^{0.75} \tag{17.34}$$

In gravelly sands or gravels, use the method for sands if N_{60} < 15 blows / 0.3 m.

The definitions of various symbols used above are

β_i = dimensionless correlation factor applicable to layer i. Limited to 1.2 in sands and 1.8 in gravelly sands and gravel. Minimum value is 0.25 in both types of soil; f_{si} is limited 200 kN/m^2 (2.1 tsf)

q'_{oi} = vertical effective stress at the middle of each layer

N_{60} = design value for SPT blow count, uncorrected for depth, saturation or fines corresponding to layer i

z_i = vertical distance from the ground surface, in meters, to the middle of layer i. The layer thickness Δz_i is limited to 9 m.

17.14 ULTIMATE SIDE AND TOTAL RESISTANCE IN ROCK (O'NEILL AND REESE, 1999)

Ultimate Skin Resistance (for Smooth Socket)

Rock is defined as a cohesive geomaterial with q_u > 5 MPa (725 psi). The following equations may be used for computing f_s ($= f_{max}$) when the pier is socketed in rock. Two methods are proposed.

Method 1

$$f_s(= f_{max}) = 0.65 p_a \left[\frac{q_u}{p_a}\right]^{0.5} \leq 0.65 p_a \left[\frac{f_c}{p_a}\right]^{0.5} \tag{17.35}$$

where, p_a = atmospheric pressure (= 101 kPa)

q_u = unconfined compressive strength of rock mass

f_c = 28 day compressive cylinder strength of concrete used in the drilled pier

Method 2

$$f_s(= f_{\max}) = 1.42 p_a \left[\frac{q_u}{p_a}\right]^{0.5} \tag{17.36}$$

Carter and Kulhawy (1988) suggested equation (17.36) based on the analysis of 25 drilled shaft socket tests in a very wide variety of soft rock formations, including sandstone, limestone, mudstone, shale and chalk.

Ultimate Total Resistance Q_u

If the base of the drilled pier rests on sound rock, the side resistance can be ignored. In cases where significant penetration of the socket can be made, it is a matter of engineering judgment to decide whether Q_f should be added directly to Q_b to obtain the ultimate value Q_u, When the rock is brittle in shear, much side resistance will be lost as the settlement increases to the value required to develop the full value of q_b (= $q_{\max}$). If the rock is ductile in shear, there is no question that the two values can be added directly (O'Neill and Reese, 1999).

17.15 ESTIMATION OF SETTLEMENTS OF DRILLED PIERS AT WORKING LOADS

O'Neill and Reese (1999) suggest the following methods for computing axial settlements for isolated drilled piers:

1. Simple formulas
2. Normalized load-transfer methods

The total settlement S_t at the pier head at working loads may be expressed as (Vesic, 1977)

$$S_t = S_e + S_{bb} + S_{bs} \tag{17.37}$$

where, S_e = elastic compression
S_{bb} = settlement of the base due to the load transferred to the base
S_{bs} = settlement of the base due to the load transferred into the soil along the sides.

The equations for the settlements are

$$S_e = \frac{L(Q_a - 0.5Q_{fm})}{A_b E}$$

$$S_{bb} = C_p \left(\frac{Q_{mb}}{dq_b}\right) \tag{17.38}$$

$$S_{bs} = \left[0.93 + 0.16\sqrt{\frac{L}{d}}\right] C_p \left(\frac{Q_{fm}}{Lq_b}\right) \tag{17.39}$$

where L = length of the drilled pier
A_b = base cross-sectional area
E = Young's modulus of the drilled pier
Q_a = load applied to the head
Q_{fm} = mobilized side resistance when Q_a is applied

Table 17.7 Values of C_p for various soils (Vesic, 1977)

Soil	C_p
Sand (dense)	0.09
Sand (loose)	0.18
Clay (stiff)	0.03
Clay (soft)	0.06
Silt (dense)	0.09
Silt (loose)	0.12

Q_{bm} = mobilized base resistance
d = pier width or diameter
C_p = soil factor obtained from Table 17.7

Normalized Load-Transfer Methods

Reese and O'Neill (1988) analyzed a series of compression loading test data obtained from full-sized drilled piers in soil. They developed normalized relations for piers in cohesive and cohesionless soils. Figures 17.18 and 17.19 can be used to predict settlements of piers in cohesive soils and Figs. 17.20 and 17.21 in cohesionless soils including soil mixed with gravel.

The boundary limits indicated for gravel in Fig. 17.20 have been found to be approximately appropriate for cemented fine-grained desert IGM's (Walsh et al., 1995). The range of validity of the normalized curves are as follows:

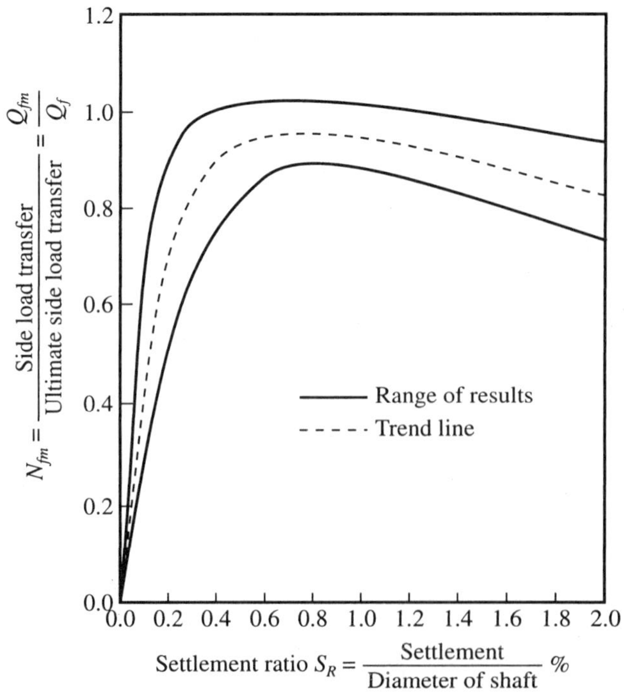

Figure 17.18 Normalized side load transfer for drilled shaft in cohesive soil (O'Neill and Reese, 1999)

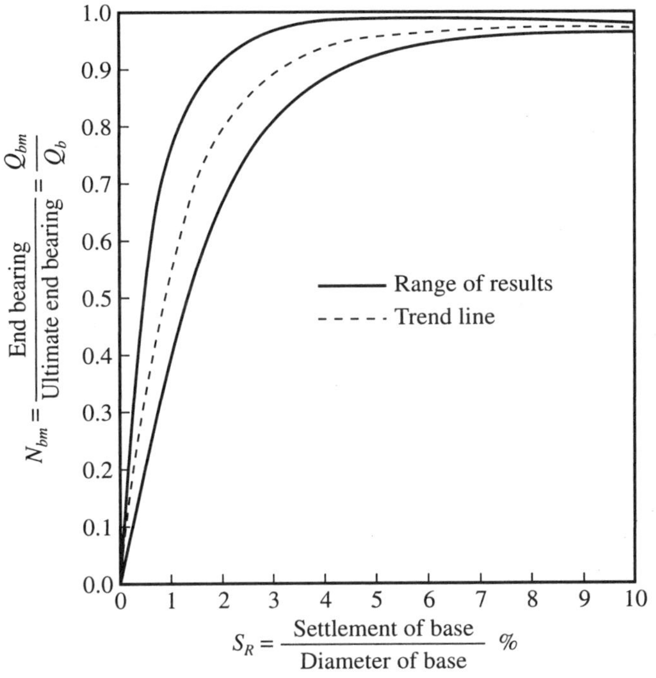

Figure 17.19 Normalized base load transfer for drilled shaft in cohesive soil (O'Neill and Reese, 1999)

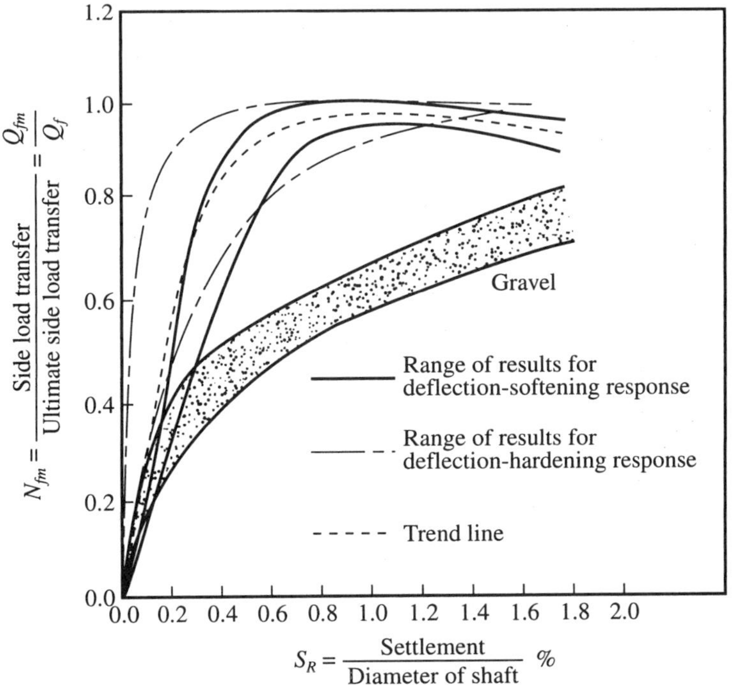

Figure 17.20 Normalized side load transfer for drilled shaft in cohesionless soil (O'Neill and Reese, 1999)

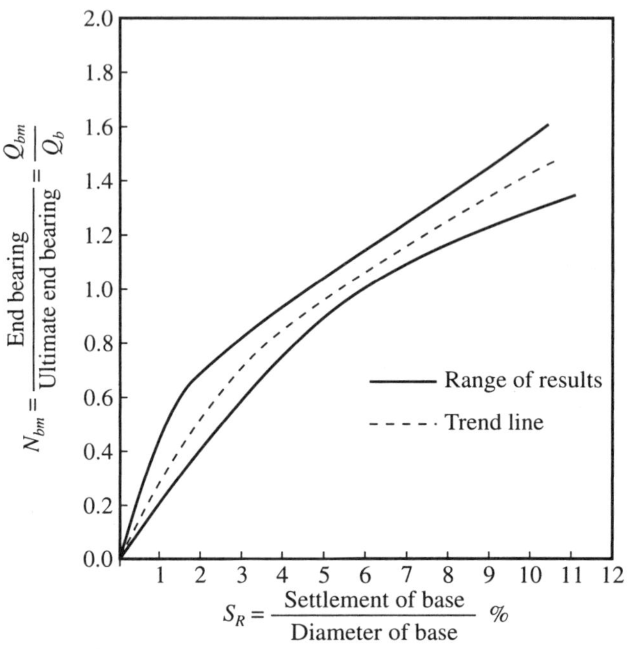

Figure 17.21 Normalized base load transfer for drilled shaft in cohesionless soil (O'Neill and Reese, 1999)

Figures 17.18 and 17.19

Normalizing factor	=	shaft diameter d
Range of d	=	0.46 m to 1.53 m

Figures 17.20 and 17.21

Normalizing factor	=	base diameter
Range of d	=	0.46 m to 1.53 m

The following notations are used in the figures:

S_R	=	Settlement ratio = S_a/d
S_a	=	Allowable settlement
N_{fm}	=	Normalized side load transfer ratio = Q_{fm}/Q_f
N_{bm}	=	Normalize base load transfer ratio = Q_{bm}/Q_b

Example 17.1

A multistory building is to be constructed in a stiff to very stiff clay. The soil is homogeneous to a great depth. The average value of undrained shear strength c_u is 150 kN/m^2. It is proposed to use a drilled pier of length 25 m and diameter 1.5 m. Determine (a) the ultimate load capacity of the pier, and (b) the allowable load on the pier with $F_s = 2.5$. (Fig. Ex. 17.1)

Solution

Base load

When $c_u \geq 96$ kPa (2000 lb/ft^2), use Eq. (17.8) for computing q_b. In this case $c_u > 96$ kPa.

$$q_b = 9c_u = 9 \times 150 = 1350 \text{ kN/m}^2$$

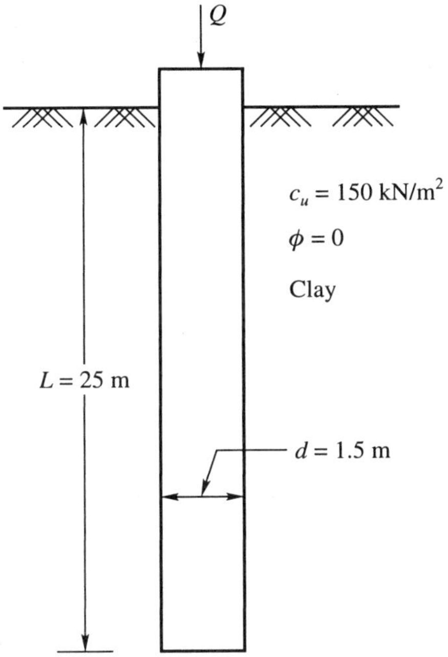

Figure Ex. 17.1

Base load $Q_b = A_b\, q_b = \dfrac{3.14 \times 1.5^2}{4} \times 1350 = 1.766 \times 1350 = 2384$ kN

Frictional load

The unit ultimate frictional resistance f_s is determined using Eq. (17.24)

$f_s = \alpha\, c_u$

From Fig. (17.15) $\alpha = 0.55$ for $c_u/p_a = 150/101 = 1.5$

where p_a is the atmospheric pressure = 101 kPa

Therefore $f_s = 0.55 \times 150 = 82.5$ kN/m²

The effective length of the shaft for computing the frictional load (Fig. 17.13 a) is

$L' = [L - (d + 1.5)]$ m $= 25 - (1.5 + 1.5) = 22$ m

The effective surface area $A_s = \pi\, d\, L' = 3.14 \times 1.5 \times 22 = 103.62$ m²

Therefore $Q_f = f_s A_s = 82.5 \times 103.62 = 8{,}549$ kN

The total ultimate load is

$Q_u = Q_f + Q_b = 8{,}549 + 2{,}384 = 10{,}933$ kN

The allowable load may be determined by applying an overall factor of safety to Q_u. Normally $F_s = 2.5$ is sufficient.

$$Q_a = \frac{10{,}933}{2.5} = 4{,}373 \text{ kN}$$

Example 17.2

For the problem given in Ex. 17.1, determine the allowable load for a settlement of 10 mm (= S_a). All the other data remain the same.

Solution

Allowable skin load

$$\text{Settlement ratio } S_R = \frac{S_a}{d} = \frac{10}{1.5 \times 10^3} \times 100 = 0.67\%$$

From Fig. 17.18 for $S_R = 0.67\%$, $N_{fm} = \dfrac{Q_{fm}}{Q_f} = 0.95$ by using the trend line.

$Q_{fm} = 0.95\, Q_f = 0.95 \times 8{,}549 = 8{,}122$ kN.

Allowable base load for S_a = 10 mm

From Fig. 17.19 for $S_R = 0.67$ %, $N_{bm} = \dfrac{Q_{bm}}{Q_b} = 0.4$

$Q_{bm} = 0.4\, Q_b = 0.4 \times 2{,}384 = 954$ kN.

Now the allowable load Q_{as} based on settlement consideration is

$Q_{as} = Q_{fm} + Q_{bm} = 8{,}122 + 954 = 9{,}076$ kN

Q_{as} based on settlement consideration is very much higher than Q_a (Ex. 17.1) and as such Q_a governs the criteria for design.

Example 17.3

Figure Ex. 17.3 depicts a drilled pier with a belled bottom. The details of the pile and the soil properties are given in the figure. Estimate (a) the ultimate load, and (b) the allowable load with $F_s = 2.5$.

Solution

Based load

Use Eq. (17.8) for computing q_b

$q_b = 9c_u = 9 \times 200 = 1{,}800$ kN/m²

$$\text{Base load } Q_b = \frac{\pi d_b^2}{4} \times q_b = \frac{3.14 \times 3^2}{4} \times 1{,}800 = 12{,}717 \text{ kN}$$

Frictional load

The effective length of shaft $L' = 25 - (2.75 + 1.5) = 20.75$ m

From Eq. (17.24) $f_s = \alpha c_u$

$$\text{For } \frac{c_u}{p_a} = \frac{100}{101} \approx 1.0, \quad \alpha = 0.55 \quad \text{from Fig. 17.15}$$

Hence $f_s = 0.55 \times 100 = 55$ kN/m²

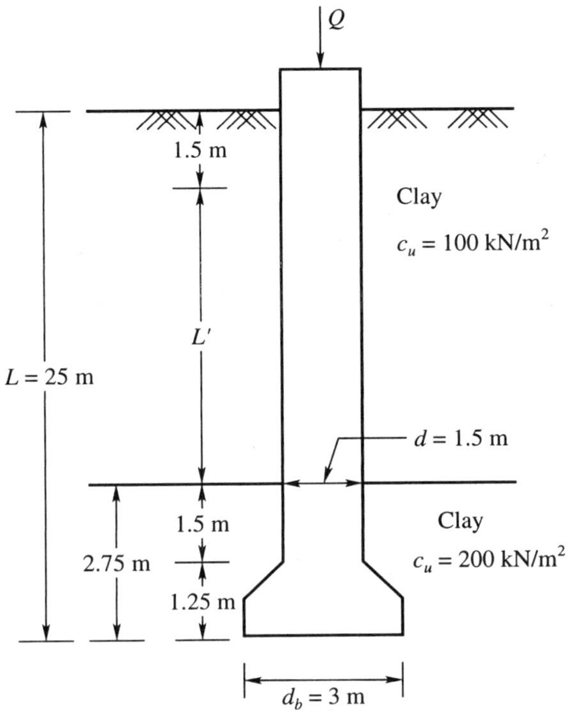

Figure Ex. 17.3

$$Q_f = PL'f_s = 3.14 \times 1.5 \times 20.75 \times 55 = 5{,}375 \text{ kN}$$

$$Q_u = Q_b + Q_f = 12{,}717 + 5{,}375 = 18{,}092 \text{ kN}$$

$$Q_a = \frac{18{,}092}{2.5} = 7{,}237 \text{ kN}$$

Example 17.4

For the problem given in Ex. 17.3, determine the allowable load Q_{as} for a settlement $S_a = 10$ mm.

Solution

Skin load Q_{fm} (mobilized)

Settlement ratio $S_R = \dfrac{10}{1.5 \times 10^3} \times 100 = 0.67\%$

From Fig. 17.18 for $S_R = 0.67$, $N_{fm} = 0.95$ from the trend line.

Therefore $Q_{fm} = 0.95 \times 5{,}375 = 5{,}106$ kN

Base load Q_{bm} (mobilized)

$$S_R = \frac{10}{3 \times 10^3} \times 100 = 0.33\%$$

From Fig. 17.19 for $S_R = 0.33\%$, $N_{bm} = 0.3$ from the trend line.

Hence $Q_{bm} = 0.3 \times 12{,}717 = 3815$ kN

$Q_{as} = Q_{fm} + Q_{bm} = 5{,}106 + 3{,}815 = 8{,}921$ kN

The factor of safety with respect to Q_u is (from Ex. 17.3)

$$F_s = \frac{18{,}092}{8{,}921} = 2.03$$

This is low as compared to the normally accepted value of $F_s = 2.5$. Hence Q_a rules the design.

Example 17.5

Figure Ex. 17.5 shows a straight shaft drilled pier constructed in homogeneous loose to medium dense sand. The pile and soil properties are:

$$L = 25 \text{ m},\ d = 1.5 \text{ m},\ c = 0,\ \phi = 36° \text{ and } \gamma = 17.5 \text{ kN/m}^3$$

Estimate (a) the ultimate load capacity, and (b) the allowable load with $F_s = 2.5$. The average SPT value $N_{cor} = 30$ for $\phi = 36°$.

Use (i) Vesic's method, and (ii) the O'Neill and Reese method.

Solution

(i) Vesic's method

From Eq. (17.10) for $c = 0$

$$q_b = (N_q - 1)\,q'_o s_q d_q C_q \qquad \text{(a)}$$

$$q'_o = 25 \times 17.5 = 437.5 \text{ kN/m}^2$$

From Eq. (17.16) $$\phi_{rel} = \frac{\phi° - 25°}{45° - 25°} = \frac{36 - 25}{20} = 0.55$$

From Eq. (17.15) $$\Delta = \frac{0.005(1 - 0.55) \times 437.5}{101} \approx 0.01$$

From Eq. (17.14) $$\mu_d = 0.1 + 0.3\phi_{rel} = 0.1 + 0.3 \times 0.55 = 0.265$$

From Eq. (17.17) $$E_d = 200 p_a = 200 \times 101 = 20{,}200 \text{ kN/m}^2$$

From Eq. (17.13) $$I_r = \frac{E_d}{2(1 + \mu_d) q'_o \tan\phi} = \frac{20{,}200}{2(1 + 0.265) \times 437.5 \tan 36°} = 25$$

From Eq. (17.12) $$I_{rr} = \frac{I_r}{1 + \Delta I_r} = \frac{25}{1 + 0.01 \times 25} = 20$$

From Eq. (17.11b)

$$C_q = \exp\left\{[-3.8 \tan 36°] + \left[\frac{3.07 \sin 36° \log_{10} 2 \times 20}{1 + \sin 36°}\right]\right\} = \exp-(0.9399) = 0.391$$

From Fig. 17.14, $N_q = 30$ for $\phi = 36°$

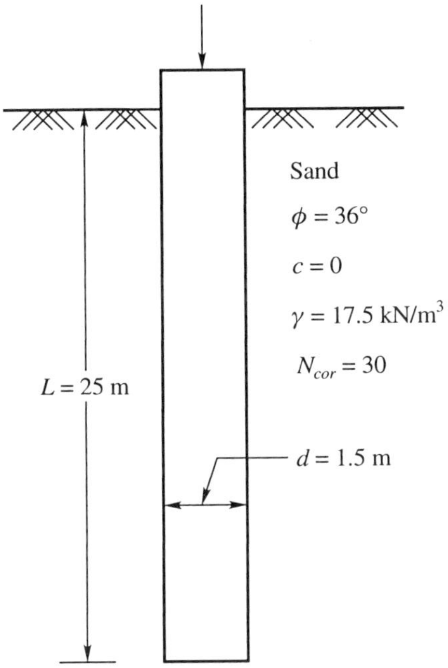

Figure Ex. 17.5

From Table (17.2) $s_q = 1 + \tan 36° = 1.73$

$$d_q = 1 + 2\tan 36° (1 - \sin 36°)^2 \left[\frac{3.14}{180} \times \tan^{-1}\frac{25}{1.5}\right] = 1.373$$

Substituting in Eq. (a)

$q_b = (30 - 1) \times 437.5 \times 1.73 \times 1.373 \times 0.391 = 11{,}783$ kN/m^2 > 11,000 kN/m^2

As per Tomlinson (1986) the computed q_b should be less than 11,000 kN/m^2.

Hence $Q_b = \dfrac{3.14}{4} \times (1.5)^2 \times 11{,}000 = 19{,}429$ kN

Skin load Q_f

From Eqs (17.31) and (17.32)

$f_s = \beta q'_o$, $\beta = 1.5 - 0.245 z^{0.5}$, where $z = \dfrac{L}{2} = \dfrac{25}{2} = 12.5$ m

Substituting

$\beta = 1.5 - 0.245 \times (12.5)^{0.5} = 0.63$

Hence $f_s = 0.63 \times 437.5 = 275.62$ kN/m^2

Per Tomlinson (1986) f_s should be limited to 110 kN/m^2. Hence f_s = 110 kN/m^2

Therefore $Q_f = \pi d L f_s = 3.14 \times 1.5 \times 25 \times 110 = 12{,}953$ kN

Ultimate load $Q_u = 19{,}429 + 12{,}953 = 32{,}382$ kN

$$Q_a = \frac{32{,}382}{2.5} = 12{,}953 \text{ kN}$$

O'Neill and Reese method

This method relates q_b to the SPT N value as per Eq. (17.19a)

$$q_b = 57.5N \ \ \text{kN/m}^2 = 57.5 \times 30 = 1{,}725 \ \ \text{kN/m}^2$$

$$Q_b = A_b q_b = 1.766 \times 1{,}725 = 3{,}046 \ \ \text{kN}$$

The method for computing Q_f remains the same as above.

Now $Q_u = 3{,}406 + 12{,}953 = 15{,}999$

$$Q_a = \frac{15{,}999}{2.5} = 6{,}400 \ \ \text{kN}$$

Example 17.6

Compute Q_u and Q_a for the pier given in Ex. 17.5 by the following methods.

1. Use the SPT value [Eq. (15.48)] for bored piles
2. Use the Tomlinson method of estimating Q_b and Table 15.2 for estimating Q_f. Compare the results of the various methods.

Solution

Use of the SPT value[Meyerhof Eq. (15.48)]

$$q_b = 133N_{cor} = 133 \times 30 = 3{,}990 \ \ \text{kN/m}^2$$

$$Q_b = \frac{3.14 \times (1.5)^2}{4} \times 3{,}990 = 7{,}047 \ \ \text{kN}$$

$$f_s = 0.67N_{cor} = 0.67 \times 30 = 20 \ \ \text{kN/m}^2$$

$$Q_f = 3.14 \times 1.5 \times 25 \times 20 = 2{,}355 \ \ \text{kN}$$

$$Q_u = 7{,}047 + 2{,}355 = 9{,}402 \ \ \text{kN}$$

$$Q_a = \frac{9{,}402}{2.5} = 3{,}760 \ \ \text{kN}$$

Tomlinson Method for Q_b

For a driven pile

From Fig. 15.9 $N_q = 65$ for $\phi = 36°$ and $\frac{L}{d} = \frac{25}{1.5} \approx 17$

Hence $q_b = q_o' N_q = 437.5 \times 65 = 28{,}438$ kN/m^2

For bored pile

$$q_b = \frac{1}{3} q_b \text{(driven pile)} = \frac{1}{3} \times 28{,}438 = 9{,}479 \ \text{kN/m}^2$$

$$Q_b = A_b q_b = 1.766 \times 9{,}479 = 16{,}740 \ \text{kN}$$

Q_f from Table 15.2

For $\phi = 36°$, $\delta = 0.75 \times 36 = 27$, and $\overline{K}_s = 1.5$ (for medium dense sand).

$$f_s = \overline{q}'_o \overline{K}_s \tan\delta = \frac{437.5}{2} \times 1.5 \tan 27° = 167 \ \text{kN/m}^2$$

As per Tomlinson (1986) f_s is limited to 110 kN/m^2. Use f_s = 110 kN/m^2.

Therefore $Q_f = 3.14 \times 1.5 \times 25 \times 110 = 12{,}953 \ \text{kN}$

$$Q_u = Q_b + Q_f = 16{,}740 + 12{,}953 = 29{,}693 \ \text{kN}$$

$$Q_a = \frac{29{,}693}{2.5} = 11{,}877 \ \text{kN}$$

Comparison of estimated results (F_s = 2.5)

Example No	Name of method	Q_b (kN)	Q_f (kN)	Q_u (kN)	Q_a (kN)
17.5	Vesic	19,429	12,953	32,382	12,953
17.5	O'Neill and Reese, for Q_b and Vesic for Q_f	3,046	12,953	15,999	6,400
17.6	Meyerhof Eq. (15.49)	7,047	2,355	9,402	3,760
17.6	Tomlinson for Q_b (Fig. 15.9) Table 15.2 for Q_f	16,740	12,953	29,693	11,877

Which method to use

The variation in the values of Q_b and Q_f are very large between the methods. Since the soils encountered in the field are generally heterogeneous in character no theory holds well for all the soil conditions. Designers have to be practical and pragmatic in the selection of any one or combination of the theoretical approaches discussed earlier.

Example 17.7

For the problem given in Example 17.5 determine the allowable load for a settlement of 10 mm. All the other data remain the same. Use (a) the values of Q_f and Q_b obtained by Vesic's method, and (b) Q_b from the O'Neill and Reese method.

Solution

(a) Vesic's values Q_f and Q_b

Settlement ratio for S_a = 10 mm is

$$S_R = \frac{S_a}{d} = \frac{10 \times 10^2}{1.5 \times 10^3} = 0.67\%$$

From Fig. 17.20 for $S_R = 0.67\%$ $N_{fm} = 0.96$ (approx.) using the trend line.

$$Q_{fm} = 0.96 \times Q_f = 0.96 \times 12{,}953 = 12{,}435 \text{ kN}$$

From Fig. 17.21 for $S_R = 0.67\%$

$$N_{bm} = 0.20, \text{ or } Q_{bm} = 0.20 \times 19{,}429 = 3{,}886 \text{ kN}$$

$$Q_{as} = 12{,}435 + 3{,}886 = 16{,}321 \text{ kN}$$

Shear failure theory give Q_a = 12,953 kN which is much lower than Q_{as}. As such Q_a determines the criteria for design.

(b) O'Neill and Reese Q_b = 3,046 kN

As above, Q_{bm} = 0.20 × 3,046 = 609 kN

Using Q_{fm} in (a) above,

Q_{as} = 609 + 12,435 = 13,044 kN

The value of Q_{as} is closer to Q_a (Vesic) but much higher than Q_a calculated by all the other methods.

Example 17.8

Figure Ex. 17.8 shows a drilled pier penetrating an IGM: clay-shale to a depth of 8 m. Joints exists within the IGM stratum. The following data are available: L_s = 8 m (= z_c), d = 1.5 m, q_u (rock) = 3 × 10^3 kN/m^2, E_i (rock) = 600 × 10^3 kN/m^2, concrete slump = 175 mm, unit weight of concrete γ_c = 24 kN/m^3, E_c (concrete) = 435 × 10^6 kN/m^2, and RQD = 70 percent, q_u (concrete) = 435 × 10^6 kN/m^2. Determine the ultimate frictional load Q_f (max).

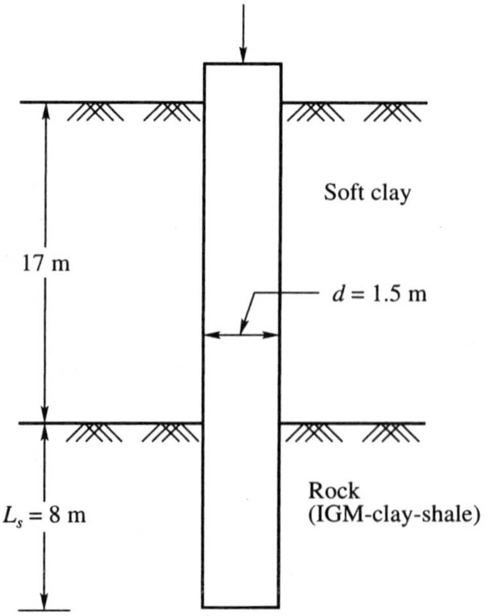

Figure Ex. 17.8

Solution

(a) Determine α in Eq. (17.26)

$f_a = \alpha q_u$ where $q_u = 3$ MPa for rock

For the depth of socket $L_s = 8$ m, and slump $= 175$ mm

$M = 0.76$ from Fig. 17.17.

From Eq. (17.27)

$$\sigma_n = M\gamma_c z_c = 0.76 \times 24 \times 8 = 146 \text{ kN/m}^2$$

$$p_a = 101 \text{ kN/m}^2, \; \sigma_n / p_a = 146/101 = 1.45$$

From Fig. 17.16 for $q_u = 3$ MPa and $\sigma_n / p_a = 1.45$, we have $\alpha = 0.11$

(b) Determination of f_a

$f_a = 0.11 \times 3 = 0.33$ MPa

(c) Determination f_{aa} in Eq. (17.28)

For RQD = 70%, $E_m/E_i = 0.1$ from Table 17.5 for open joints, and f_{aa}/f_a (= R_a) = 0.55 from Table 17.6

$$f_{\max} = f_{aa} = 0.55 \times 0.33 = 0.182 \text{ MPa} = 182 \text{ kN/m}^2$$

(d) Ultimate friction load Q_f

$$Q_f = PLf_{aa} = 3.14 \times 1.5 \times 8 \times 182 = 6{,}858 \text{ kN}$$

Example 17.9

For the pier given in Ex. 17.8, determine the ultimate bearing capacity of the base. Neglect the frictional resistance. All the other data remain the same.

Solution

For RQD between 70 and 100 percent

from Eq. (17.22)

$$q_b (= q_{\max}) = 4.83 (q_u)^{0.5} \text{MPa} = 4.83 \times (3)^{0.5} = 8.37 \text{ MPa}$$

$$Q_b(\max) = \frac{3.14}{4} \times 1.5^2 \times 8.37 = 14.78 \text{ MN} = 14{,}780 \text{ kN}$$

17.16 UPLIFT CAPACITY OF DRILLED PIERS

Structures subjected to large overturning moments can produce uplift loads on drilled piers if they are used for the foundation. The design equation for uplift is similar to that of compression. Figure 17.22 shows the forces acting on the pier under uplift-load Q_{ul}. The equation for Q_{ul} may be expressed as

$$Q_{ul} = Q_{fr} + W_p = A_s f_r + W_p \tag{17.40}$$

where, Q_{fr} = total side resistance for uplift
W_p = effective weight of the drilled pier
A_s = surface area of the pier
f_r = frictional resistance to uplift

Uplift Capacity of Single Pier (straight edge)

For a drilled pier in cohesive soil, the frictional resistance may expressed as (Chen and Kulhawy, 1994)

$$f_r = \alpha c_u \tag{17.41a}$$

$$\alpha = 0.31 + 0.17 \frac{c_u}{p_a} \tag{17.41b}$$

where, α = adhesion factor
c_u = undrained shear strength of cohesive soil
p_a = atmospheric pressure (101 kPa)

Poulos and Davis, (1980) suggest relationships between c_u and α as given in Fig. 17.23. The curves in the figure are based on pull out test data collected by Sowa (1970).

Uplift Resistance of Piers in Sand

There are no confirmatory methods available for evaluating uplift capacity of piers embedded in cohesionless soils. Poulos and Davis, (1980) suggest that the skin frictional resistance for pull out may be taken as equal to two-thirds of the shaft resistance for downward loading.

Uplift Resistance of Piers in Rock

According to Carter and Kulhawy (1988), the frictional resistance offered by the surface of the pier under uplift loading is almost equal to that for downward loading if the drilled pier is rigid relative

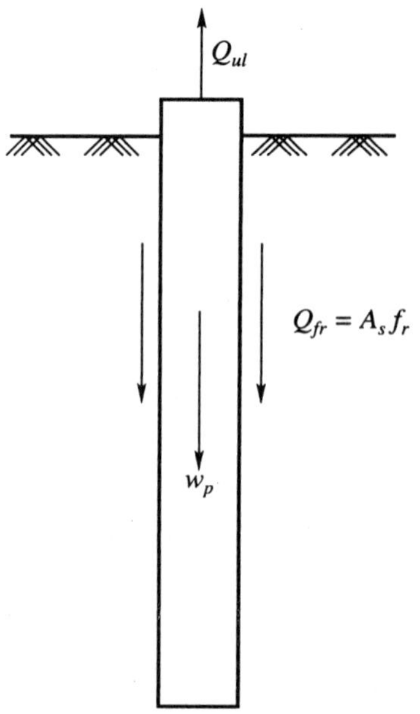

Figure 17.22 Uplift forces for a straight edged pier

to the rock. The effective rigidity is defined as $(E_c/E_m)(d/D_s)^2$, in which E_c and E_m are the Young's modulus of the drilled pier and rock mass respectively, d is the socket diameter and D_s is the depth of the socket. A socket is rigid when $(E_c/E_m)(d/D_s)^2 \geq 4$. When the effective rigidity is less than 4, the frictional resistance f_r for upward loading may be taken as equal to 0.7 times the value for downward loading.

Example 17.10

Determine the uplift capacity of the drilled pier given in Fig. Ex. 17.10. Neglect the weight of the pier.

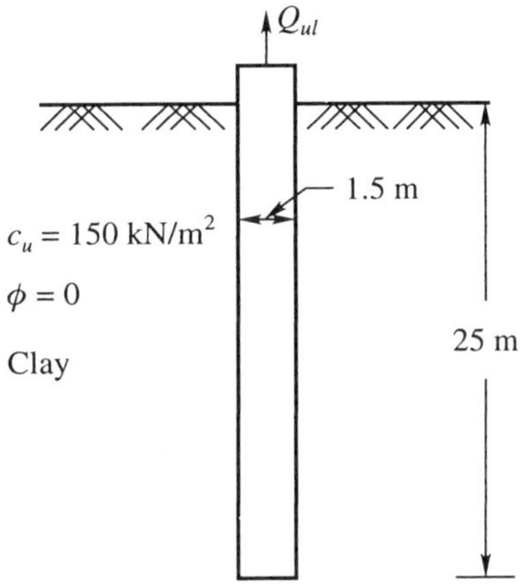

Figure Ex. 17.10

Solution

From Eq. (17.40)

$$Q_{ul} = A_s f_r$$

From Eq. (17.41a) $f_r = \alpha c_u$

From Eq. (17.41b) $\alpha = 0.31 + 0.17\dfrac{c_u}{p_a}$

Given: $L = 25$ m, $d = 1.5$ m, $c_u = 150$ kN/m^2

Hence $\alpha = 0.31 + 0.17 \times \dfrac{150}{101} = 0.56$

$$f_r = 0.56 \times 150 = 84 \text{ kN/m}^2$$

$$Q_{ul} = 3.14 \times 1.5 \times 25 \times 84$$

$$= 9{,}891 \text{ kN}$$

It may be noted here that $f_s = 82.5$ kN/m^2 for downward loading and $f_r = 84$ kN/m^2 for uplift. The two values are very close to each other.

17.17 LATERAL BEARING CAPACITY OF DRILLED PIERS

It is quite common that drilled piers constructed for bridge foundations and other similar structures are also subjected to lateral loads and overturning moments. The methods applicable to piles are applicable to piers also. Chapter 16 deals with such problems. This chapter deals with one more method as recommended by O'Neill and Reese (1999). This method is called **Characteristic load method** and is described below.

Characteristic Load Method (Duncan et al., 1994)

The characteristic load method proceeds by defining a characteristic or normalizing shear load (P_c) and a characteristic or normalizing bending moment (M_c) as given below.

For clay

$$P_c = 7.34 d^2 (ER_I)\left[\frac{c_u}{ER_I}\right]^{0.68} \tag{17.42}$$

$$M_c = 3.86 d^3 (ER_I)\left[\frac{c_u}{ER_I}\right]^{0.46} \tag{17.43}$$

For sand

$$P_c = 1.57d^2(ER_I)\left[\frac{\gamma' d\phi' K_p}{ER_I}\right]^{0.57} \tag{17.44}$$

$$M_c = 1.33d^3(ER_I)\left[\frac{\gamma' d\phi' K_p}{ER_I}\right]^{0.40} \tag{17.45}$$

where

d = shaft diameter
E = Young's modulus of the shaft material
R_I = ratio of moment of inertia of drilled shaft to moment of inertia of solid section (= 1 for a normal uncracked drilled shaft without central voids)
c_u = average value of undrained shear strength of the clay in the top 8d below the ground surface
γ' = average effective unit weight of the sand (total unit weight above the water table, buoyant unit weight below the water table) in the top 8d below the ground surface
ϕ' = average effective stress friction angle for the sand in the top 8d below ground surface
K_p = Rankine's passive earth pressure coefficient = $\tan^2(45° + \phi'/2)$

In the design method, the moments and shears are resolved into groundline values, P_t and M_t, and then divided by the appropriate characteristic load values [Equations (17.42) through (17.45)]. The lateral deflections at the shaft head, y_t are determined from Figures 17.23 and 17.24, considering the conditions of pile-head fixity. The value of the maximum moment in a free- or fixed-headed drilled shaft can be determined through the use of figure 17.25 if the only load that is applied is ground line shear. If both a moment and a shear are applied, one must compute y_t (combined), and then solve Eq. (17.46) for the "characteristic length" T (relative stiffness factor).

$$y_t(\text{combined}) = 2.43\frac{P_t T^3}{EI} + 1.62\frac{M_t T^2}{EI} \tag{17.46}$$

where I is the moment of inertia of the cross-section of the drilled shaft.

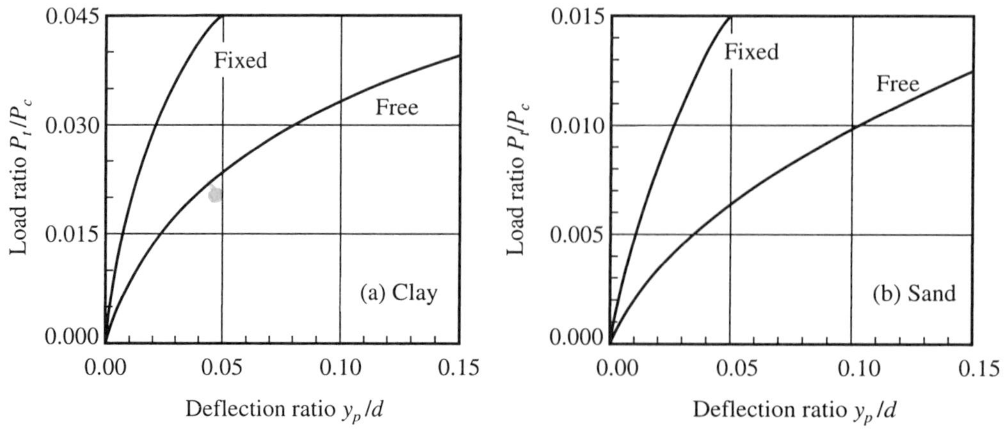

Figure 17.23 Groundline shear-deflection curves for (a) clay and (b) sand (Duncan et al., 1994)

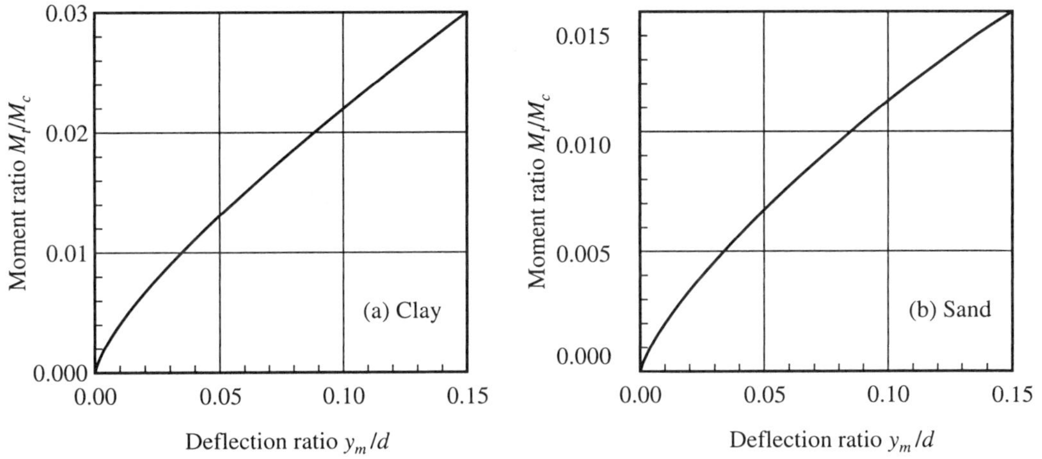

Figure 17.24 Groundline moment-deflection curves for (a) clay and (b) sand (Duncan et al., 1994)

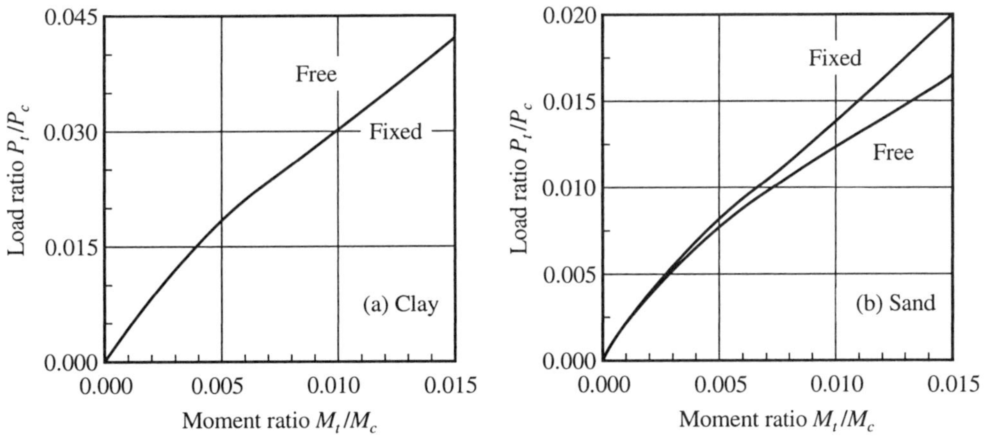

Figure 17.25 Groundline shear-maximum moment curves for (a) clay and (b) sand (Duncan et al., 1994)

The principle of superposition is made use of for computing ground line deflections of piers (or piles) subjected to groundline shears and moments at the pier head. The explanation given here applies to a free-head pier. The same principle applies for a fixed head pile also.

Consider a pier shown in Fig. 17.26(a) subjected to a shear load P_t and moment M_t at the pile head at ground level. The total deflection y_t caused by the combined shear and moment may be written as

$$y_t = y_p + y_m \tag{17.47}$$

where y_p = deflection due to shear load P_t alone with $M_t = 0$
y_m = deflection due to moment M_t alone with $P_t = 0$

Again consider Fig. 17.26(b). The shear load P_t acting alone at the pile head causes a deflection y_p (as above) which is equal to deflection y_{pm} caused by an equivalent moment M_p acting alone.

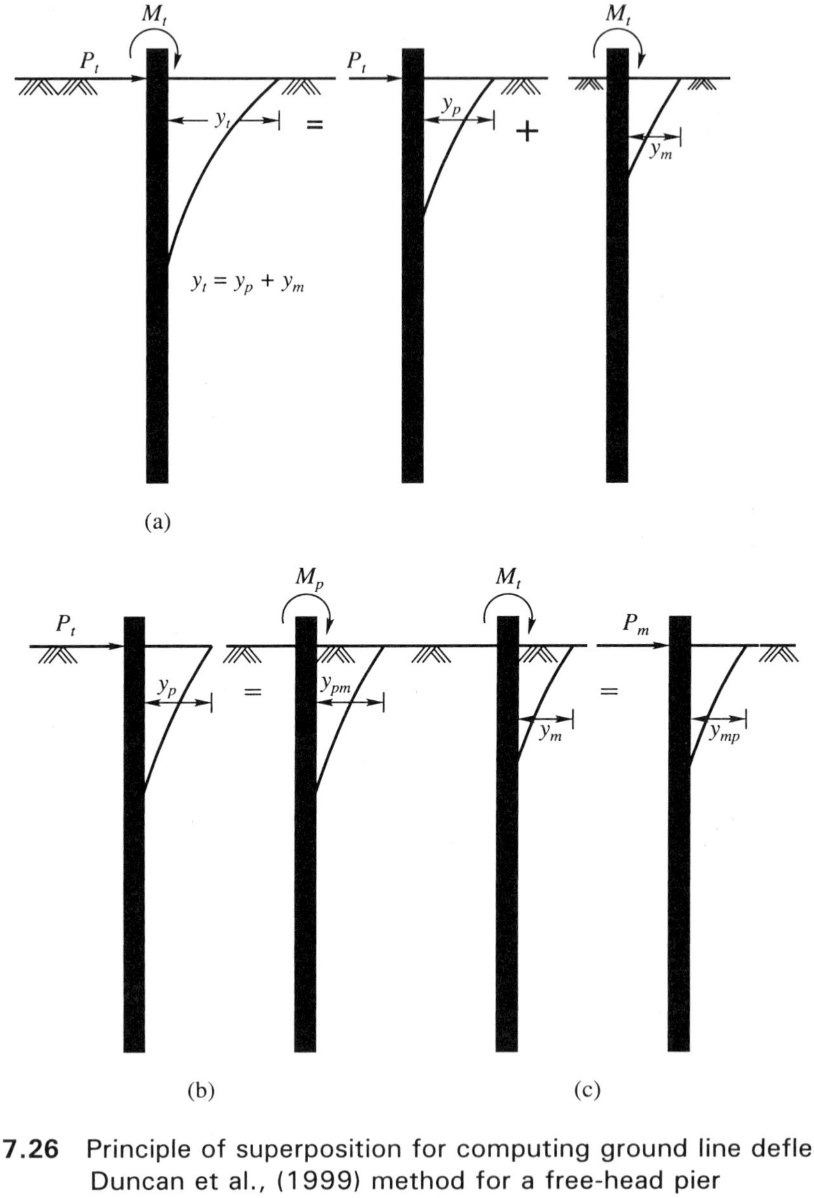

Figure 17.26 Principle of superposition for computing ground line deflection by Duncan et al., (1999) method for a free-head pier

In the same way Fig. 17.26(c) shows a deflection y_m caused by moment M_t at the pile head. An equivalent shear load P_m causes the same deflection y_m which is designated here as y_{mp}. Based on the principles explained above, groundline deflection at the pile head due to a combined shear load and moment may be explained as follows.

1. Use Figs 17.23 and 17.24 to compute groundline deflections y_p and y_m due to shear load and moment respectively.
2. Determine the groundline moment M_p that will produce the same deflection as by a shear load P_t (Fig. 17.26(b)). In the same way, determine a groundline shear load P_m, that will produce the same deflection as that by the groundline moment M_t (Fig. 17.26(c)).
3. Now the deflections caused by the shear loads $P_t + P_m$ and that caused by the moments $M_t + M_p$ may be written as follows:

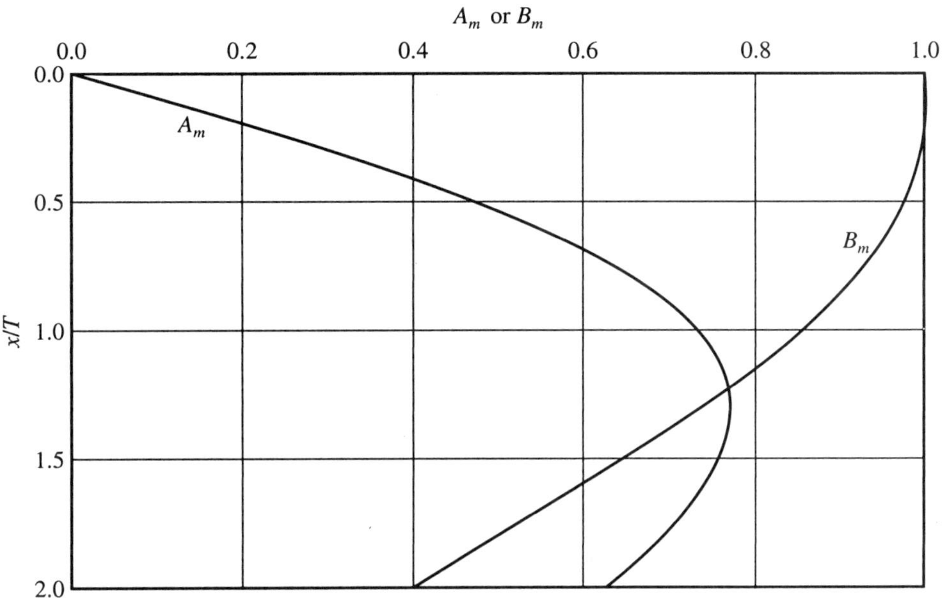

Figure 17.27 Parameters A_m and B_m (Matlock and Reese, 1961)

$$y_{tp} = y_p + y_{mp} \tag{17.48}$$

$$y_{tm} = y_m + y_{pm} \tag{17.49}$$

Theoretically $y_{tp} = y_{tm}$

4. Lastly the total deflection y_t is obtained as

$$y_t = \frac{y_{tp} + y_{tm}}{2} = \frac{(y_p + y_{mp}) + (y_m + y_{pm})}{2} \tag{17.50}$$

The distribution of moment along a pier may be determined using Eq. (16.11) and Table 16.2 or Fig. 17.27.

Direct Method by Making Use of n_h

The direct method developed by Murthy and Subba Rao (1995) for long laterally loaded piles has been explained in Chapter 16. The application of this method for long drilled piers will be explained with a case study.

Example 17.11 (O'Neill and Reese, 1999)

Refer to Fig. Ex. 17.11. Determine for a free-head pier (a) the groundline deflection, and (b) the maximum bending moment. Use the Duncan et al., (1994) method. Assume $R_I = 1$ in the Eqs (17.44) and (17.45).

Solution

Substituting in Eqs (17.42) and (17.43)

$$P_c = 7.34 \times (0.80)^2 [25 \times 10^3 \times (1)] \left[\frac{0.06}{25 \times 10^3} \right]^{0.68} = 17.72 \text{ MN}$$

$$M_c = 3.86(0.80)^3[25\times10^3 \times(1)]\left[\frac{0.06}{25\times10^3}\right]^{0.46} = 128.5 \text{ MN - m}$$

$$\text{Now } \frac{P_t}{P_c} = \frac{0.080}{17.72} = 0.0045, \quad \frac{M_t}{M_c} = \frac{0.4}{128.5} = 0.0031$$

Step 1

From Fig. 17.23a for $\frac{P_t}{P_c} = 0.0045$

$$\frac{y_p}{d} = 0.003 \text{ or } y_p = 0.003\times0.8\times10^3 = 2.4 \text{ mm}$$

From Fig. 17.24a, $M_t/M_c = 0.0031$

$$\frac{y_m}{d} = 0.006 \text{ or } y_m = 0.006d = 0.006\times0.8\times10^3 = 4.8 \text{ mm}$$

Step 2

From Fig. 17.23(a) for $y_m/d = 0.006$, $P_m/P_c = 0.0055$

From Fig. 17.24(a), for $y_p/d = 0.003$, $M_p/M_c = 0.0015$.

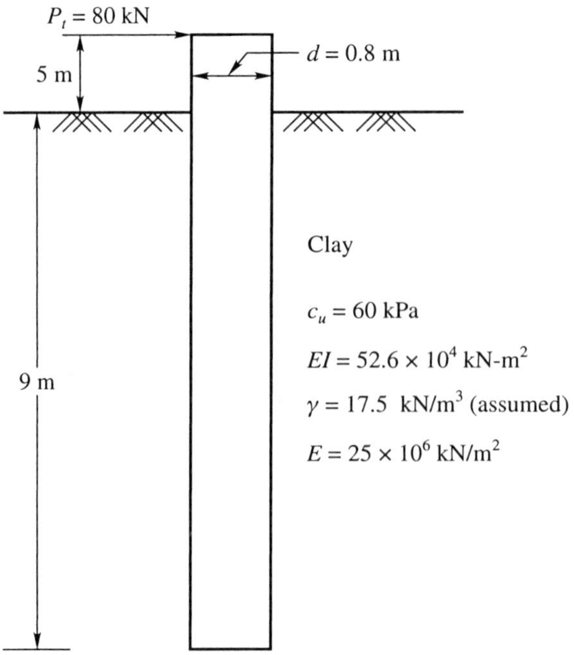

Figure Ex. 17.11

Step 3

The shear loads P_t and P_m applied at ground level, may be expressed as

$$\frac{P_t}{P_c}+\frac{P_m}{P_c}=0.0045+0.0055=0.01$$

From Fig. 17.23,

$$\frac{y_{tp}}{d}=0.013 \text{ for } \frac{P_t+P_m}{P_c}=0.01$$

$$\text{or } y_{tp}=0.013\times(0.80)\times10^3=10.4 \text{ mm}$$

Step 4

In the same way as in Step 3

$$\frac{M_t}{M_c}+\frac{M_p}{M_c}=0.0031+0.0015=0.0046$$

From Fig. 17.24a $\dfrac{y_{tm}}{d}=\dfrac{y_m+y_{pm}}{d}=0.011$

Hence $y_{tm}=0.011\times0.8\times10^3=8.8$ mm

Step 5

From Eq. (17.50)

$$y_t=\frac{y_{tp}+y_{tm}}{2}=\frac{10.4+8.8}{2}=9.6 \text{ mm}$$

Step 6

The maximum moment for the combined shear load and moment at the pier head may be calculated in the same way as explained in Chapter 16. $M_{(max)}$ as obtained is

$M_{max}=470.5$ kN-m

This occurs at a depth of 1.3 m below ground level.

Example 17.12

Solve the problem in Ex. 17.11 by the direct method.

Given: $EI=52.6\times10^4$ kN-m^2, $d=80$ cm, $\gamma=17.5$ kN/m^3, $e=5$ m, $L=9$ m, $c=60$ kN/m^2 and $P_t=80$ kN.

Solution

Groundline deflection

From Eq. (16.30) for piers in clay

$$n_h=\frac{125c^{1.5}\sqrt{EI\gamma d}}{\left(1+\dfrac{e}{d}\right)^{1.5}\times P_e^{1.5}}$$

Substituting and simplifying

$$n_h = \frac{125\times 60^{1.5}\sqrt{52.6\times 10^4\times 17.5\times 0.8}}{\left(1+\frac{5}{0.8}\right)^{1.5}\times P_e^{1.5}} = \frac{808\times 10^4}{P_e^{1.5}}\ \text{kN/m}^3 \qquad \text{(a)}$$

Step 1

Assume $P_e = P_t = 80$ kN,

From Eq. (a), $n_h = 11{,}285$ kN/m^3 and

$$T = \left[\frac{EI}{n_h}\right]^{0.2} = \left[\frac{52.6\times 10^4}{11{,}285}\right]^{0.2} = 2.16\ \text{m}$$

Step 2

From Eq. (16.22) $P_e = P_t\times\left(1+0.67\frac{e}{T}\right) = 80\left[1+0.67\times\frac{5}{2.16}\right] = 204$ kN

From Eq. (a), $n_h = 2772$ kN/m^3, hence $T = 2.86$ m

Step 3

Continue the above process till convergence is reached. The final values are

$P_e = 177$ kN, $n_h = 3410$ kN/m^3 and $T = 2.74$ m

For $P_e = 190$ kN, we have $n_h = 8{,}309$ kN/m^3 and $T = 2.29$ m

Step 4

From Eq. (17.46)

$$y_t = \frac{2.43\times 177\times (2.74)^3}{52.6\times 10^4}\times 1000 = 16.8\ \text{mm}$$

By Duncan et al, method $y_t = 9.6$ mm

Maximum moment from Eq. (16.12)

$$M = [P_tT]A_m + [M_t]B_m = [80\times 2.74]A_m + [400]B_m = 219.2A_m + 400B_m$$

Depth $x/T = Z$	A_m	B_m	M_1	M_2	M (kN-m)
0	0	1	0	400	400
0.4	0.379	0.99	83	396	479
0.5	0.46	0.98	101	392	493
0.6	0.53	0.96	116	384	500
0.7	0.60	0.94	132	376	**508** (max)
0.8	0.65	0.91	142	364	506

The maximum bending moment occurs at $x/T = 0.7$ or $x = 0.7\times 2.74 = 1.91$ m (6.26 ft). As per Duncan et al., method M(max) = 470.5 kN-m. This occurs at a depth of 1.3 m.

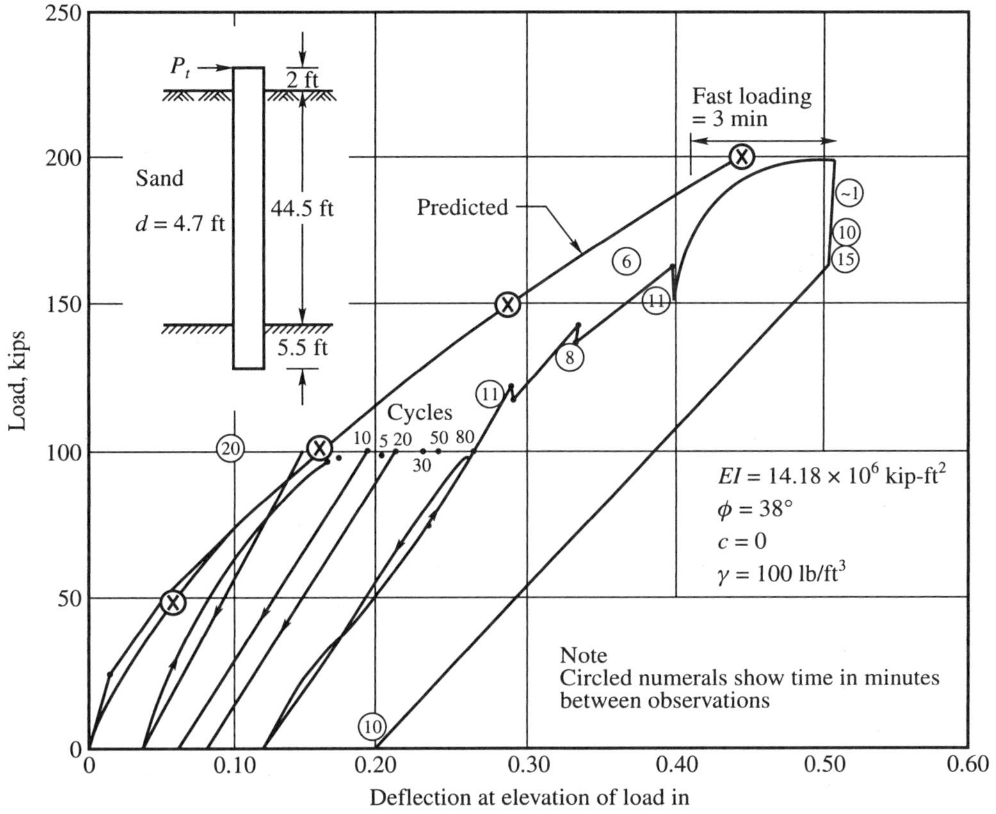

Figure 17.28 Load deflection relationship, Pier 2*S* (Davisson et al., 1969)

17.18 CASE STUDY OF A DRILLED PIER SUBJECTED TO LATERAL LOADS

Lateral load test was performed on a circular drilled pier by Davisson and Salley (1969). Steel casing pipe was provided for the concrete pier. The details of the pier and the soil properties are given in Fig. 17.28. The pier was instrumented and subjected to cyclic lateral loads. The load deflection curve as obtained by Davison et al., is shown in the same figure.

Direct method (Murthy and Subba Rao, 1995) has been used here to predict the load displacement relationship for a continuous load increase by making use of Eq. (16.29). The predicted curve is also shown in Fig. 17.28. There is an excellent agreement between the predicted and the observed values.

17.19 PROBLEMS

17.1 Fig. Prob. 17.1 shows a drilled pier of diameter 1.25 m constructed for the foundation of a bridge. The soil investigation at the site revealed soft to medium stiff clay extending to a great depth. The other details of the pier and the soil are given in the figure. Determine (a) the ultimate load capacity, and (b) the allowable load for $F_s = 2.5$. Use Vesic's method for base load and α method for the skin load.

17.2 Refer to Prob. 17.1. Given $d = 3$ ft, $L = 30$ ft, and $c_u = 1050$ lb/ft^2. Determine the ultimate (a) base load capacity by Vesic's method, and (b) the frictional load capacity by the α-method.

17.3 Fig. Prob. 17.3 shows a drilled pier with a belled bottom constructed for the foundation of a multistory building. The pier passes through two layers of soil. The details of the pier and the properties of the soil are given in the figure. Determine the allowable load Q_a for $F_s = 2.5$. Use (a) Vesic's method for the base load, and (b) the O'Neill and Reese method for skin load.

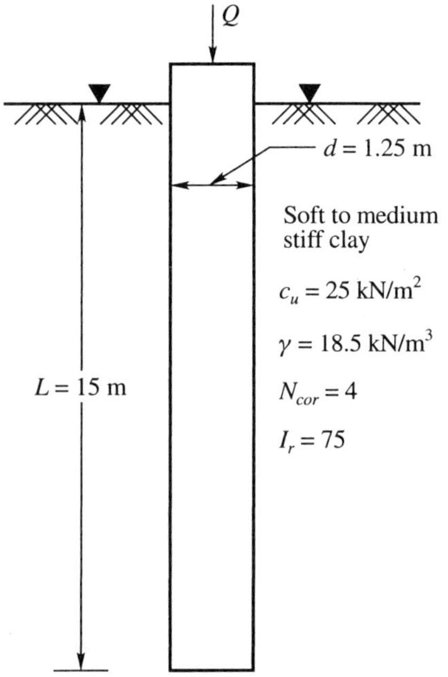

Figure Prob. 17.1

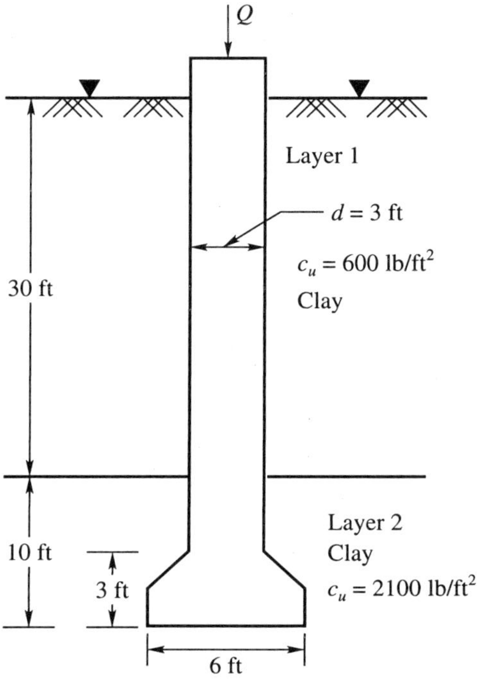

Figure Prob. 17.3

17.4 For the drilled pier given in Fig. Prob. 17.1, determine the working load for a settlement of 10 mm. All the other data remain the same. Compare the working load with the allowable load Q_a.

17.5 For the drilled pier given in Prob 17.2, compute the working load for a settlement of 0.5 in. and compare this with the allowable load Q_a.

17.6 If the drilled pier given in Fig. Prob. 17.6 is to carry a safe load of 2500 kN, determine the length of the pier for $F_s = 2.5$. All the other data are given in the figure.

17.7 Determine the settlement of the pier given in Prob. 17.6 by the O'Neill and Reese method. All the other data remain the same.

17.8 Fig. Prob. 17.8 depicts a drilled pier with a belled bottom constructed in homogeneous clay extending to a great depth. Determine the

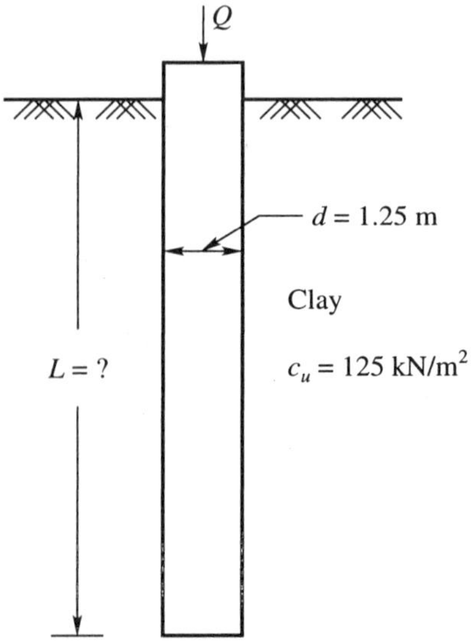

Figure Prob. 17.6

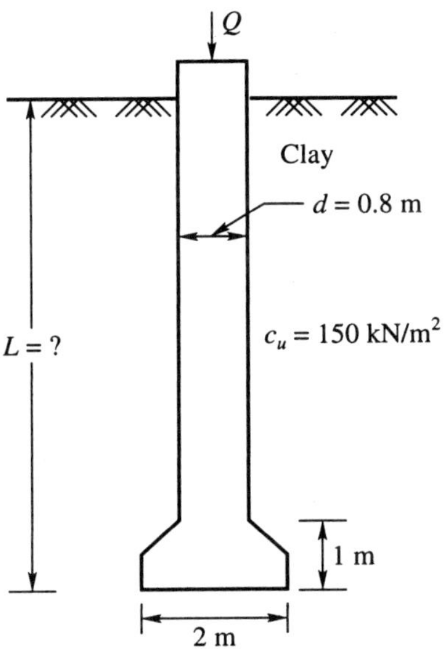

Figure Prob. 17.8

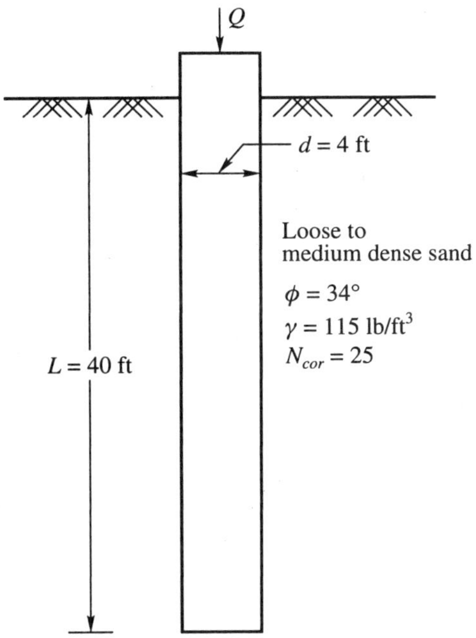

Figure Prob. 17.10

length of the pier to carry an allowable load of 3000 kN with a $F_s = 2.5$. The other details are given in the figure.

17.9 Determine the settlement of the pier in Prob 17.8 for a working load of 3000 kN. All the other data remain the same. Use the length L computed.

17.10 Fig. Prob 17.10 shows a drilled pier. The pier is constructed in homogeneous loose to medium dense sand. The pier details and the properties of the soil are given in the figure. Estimate by Vesic's method the ultimate load bearing capacity of the pier.

17.11 For Problem 17.10 determine the ultimate base capacity by the O'Neill and Reese method. Compare this value with the one computed in Prob. 17.10.

17.12 Compute the allowable load for the drilled pier given in Fig. 17.10 based on the SPT value. Use Meyerhof's method.

17.13 Compute the ultimate base load of the pier in Fig. Prob. 17.10 by Tomlinson's method.

17.14 A pier is installed in a rocky stratum. Fig. Prob. 17.14 gives the details of the pier and the properties of the rock materials. Determine the ultimate frictional load Q_f (max).

17.15 Determine the ultimate base resistance of the drilled pier in Prob. 17.14. All the other data remain the same. What is the allowable load with $F_s = 4$ by taking into account the frictional load Q_f computed in Prob. 17.14?

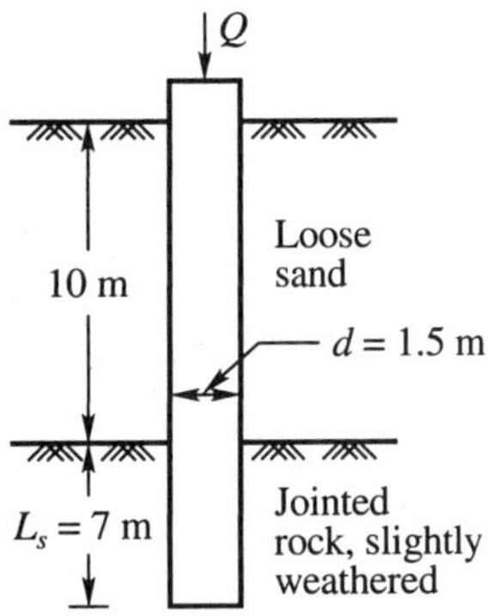

q_u (rock) = 2×10^3 kN/m²
E_i (rock) = 40×10^4 kN/m²
E_c (concrete) = 435×10^6 kN/m²
RQD = 50%
q_u (concrete) = 40×10^4 kN/m²
slump = 175 mm, γ_c = 23.5 kN/m³

Figure Prob. 17.14

17.16 Determine the ultimate point bearing capacity of the pier given in Prob. 17.14 if the base rests on sound rock with RQD = 100%.

17.17 Determine the uplift capacity of the drilled pier given in Prob. 17.1. Given: L = 15 m, d = 1.25 m, and c_u = 25 kN/m². Neglect the weight of the pile.

17.18 The drilled pier given in Fig. Prob. 17.18 is subjected to a lateral load of 120 kips. The soil is homogeneous clay. Given: d = 60 in., EI = 93 × 10^{10} lb-in.², L = 38 ft, e = 12 in., c = 2000 lb ft², and γ_b = 60 lb/ft³. Determine by the Duncan et al., method the groundline deflection.

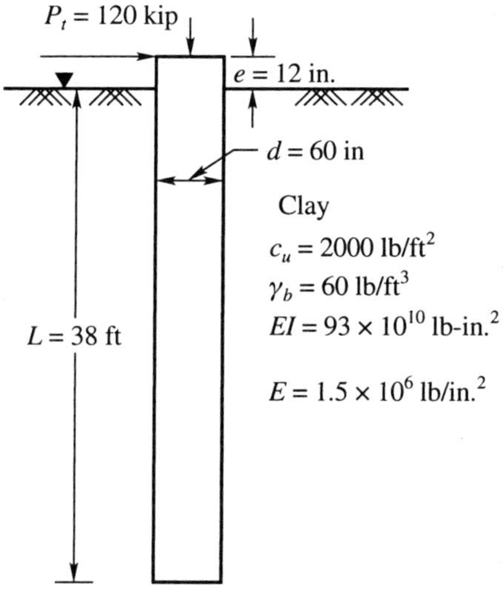

Figure Prob. 17.18

CHAPTER 18

FOUNDATIONS ON COLLAPSIBLE AND EXPANSIVE SOILS

18.1 GENERAL CONSIDERATIONS

The structure of soils that experience large loss of strength or great increase in compressibility with comparatively small changes in stress or deformations is said to be *metastable* (Peck et al., 1974). Metastable soils include (Peck et al., 1974):

1. Extra-sensitive clays such as quick clays,
2. Loose saturated sands susceptible to liquefaction,
3. Unsaturated primarily granular soils in which a loose state is maintained by apparent cohesion, cohesion due to clays at the intergranular contacts or cohesion associated with the accumulation of soluble salts as a binder, and
4. Some saprolites either above or below the water table in which a high void ratio has been developed as a result of leaching that has left a network of resistant minerals capable of transmitting stresses around zones in which weaker minerals or voids exist.

Footings on quick clays can be designed by the procedures applicable for clays as explained in Chapter 12. Very loose sands should not be used for support of footings. This chapter deals only with soils under categories 3 and 4 listed above.

There are two types of soils that exhibit volume changes under constant loads with changes in water content. The possibilities are indicated in Fig. 18.1 which represent the result of a pair of tests in a consolidation apparatus on identical undisturbed samples. Curve a represents the e-log p curve for a test started at the natural moisture content and to which no water is permitted access. Curves b and c, on the other hand, correspond to tests on samples to which water is allowed access under all loads until equilibrium is reached. If the resulting e-log p curve, such as curve b, lies entirely below curve a, the soil is said to have *collapsed*. Under field conditions, at present overburden pressure p_1

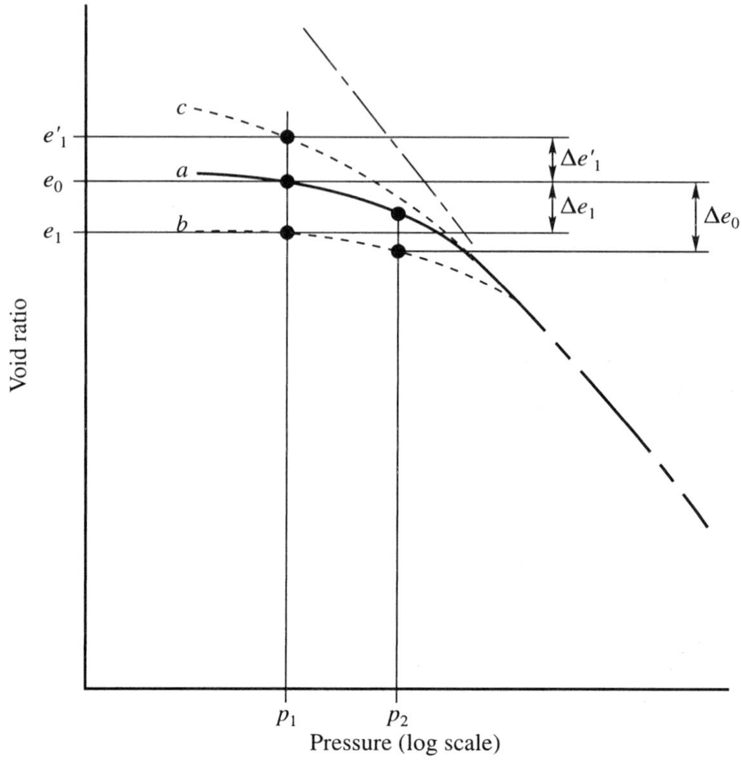

Figure 18.1 Behavior of soil in double oedometer or paired confined compression test (a) relation between void ratio and total pressure for sample to which no water is added, (b) relation for identical sample to which water is allowed access and which experiences collapse, (c) same as (b) for sample that exhibits swelling (after Peck et al., 1974)

and void ratio e_0, the addition of water at the commencement of the tests to sample 1, causes the void ratio to decrease to e_1. The collapsible settlement S_c may be expressed as

$$S_c = \frac{H\Delta e_1}{1+e_0} \tag{18.1a}$$

where H = the thickness of the stratum in the field.

Soils exhibiting this behavior include true loess, clayey loose sands in which the clay serves merely as a binder, loose sands cemented by soluble salts, and certain residual soils such as those derived from granites under conditions of tropical weathering.

On the other hand, if the addition of water to the second sample leads to curve *c*, located entirely above *a*, the soil is said to have *swelled*. At a given applied pressure p_1, the void ratio increases to e'_1, and the corresponding rise of the ground is expressed as

$$S_s = \frac{H\Delta e'_1}{1+e_0} \tag{18.1b}$$

Soils exhibiting this behavior to a marked degree are usually montmorillonitic clays with high plasticity indices.

PART A—COLLAPSIBLE SOILS

18.2 GENERAL OBSERVATIONS

According to Dudley (1970), and Barden et al., (1973), four factors are needed to produce collapse in a soil structure:

1. An open, partially unstable, unsaturated fabric
2. A high enough net total stress that will cause the structure to be metastable
3. A bonding or cementing agent that stabilizes the soil in the unsaturated condition
4. The addition of water to the soil which causes the bonding or cementing agent to be reduced, and the interaggregate or intergranular contacts to fail in shear, resulting in a reduction in total volume of the soil mass.

Collapsible behavior of compacted and cohesive soils depends on the percentage of fines, the initial water content, the initial dry density and the energy and the process used in compaction.

Current practice in geotechnical engineering recognizes an unsaturated soil as a four phase material composed of air, water, soil skeleton, and contractile skin. Under the idealization, two phases can flow, that is air and water, and two phases come to equilibrium under imposed loads, that is the soil skeleton and contractile skin. Currently, regarding the behavior of compacted collapsing soils, geotechnical engineering recognized that

1. Any type of soil compacted at *dry of optimum* conditions and at a low dry density may develop a collapsible fabric or metastable structure (Barden et al., 1973).
2. A compacted and metastable soil structure is supported by microforces of shear strength, that is bonds, that are highly dependent upon capillary action. The bonds start losing strength with the increase of the water content and at a critical degree of saturation, the soil structure collapses (Jennings and Knight 1957; Barden et al., 1973).

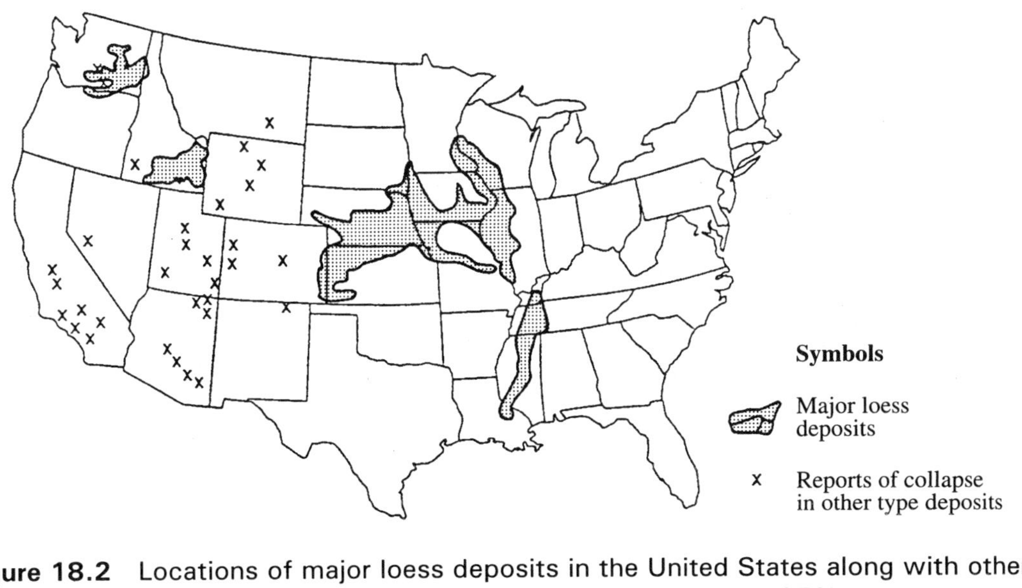

Figure 18.2 Locations of major loess deposits in the United States along with other sites of reported collapsible soils (after Dudley, 1970)

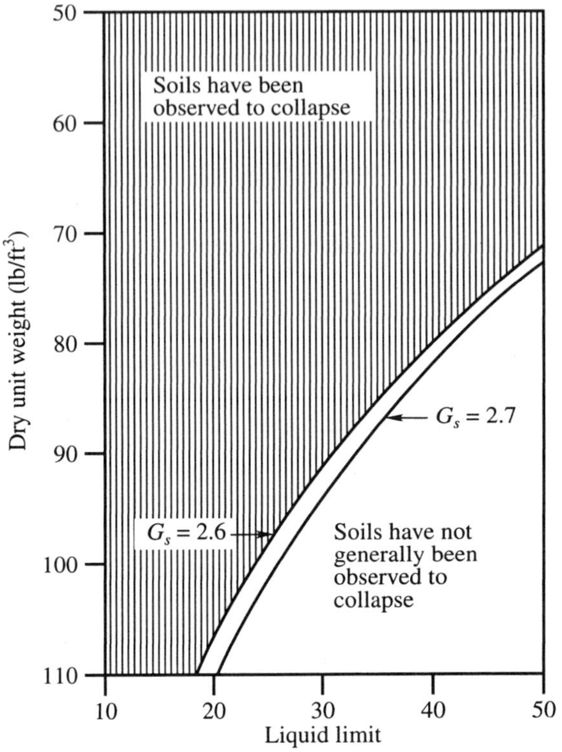

Figure 18.3 Collapsible and noncollapsible loess (after Holtz and Hilf, 1961)

3. The soil collapse progresses as the degree of saturation increases. There is, however, a critical degree of saturation for a given soil above which negligible collapse will occur regardless of the magnitude of the prewetting overburden pressure (Jennings and Burland, 1962; Houston et al., 1989).
4. The collapse of a soil is associated with localized shear failures rather than an overall shear failure of the soil mass.
5. During wetting induced collapse, under a constant vertical load and under K_o-oedometer conditions, a soil specimen undergoes an increase in horizontal stresses.
6. Under a triaxial stress state, the magnitude of volumetric strain resulting from a change in stress state or from wetting, depends on the mean normal total stress and is independent of the principal stress ratio.

The geotechnical engineer needs to be able to identify readily the soils that are likely to collapse and to determine the amount of collapse that may occur. Soils that are likely to collapse are loose fills, altered windblown sands, hillwash of loose consistency, and decomposed granites and acid igneous rocks.

Some soils at their natural water content will support a heavy load but when water is provided they undergo a considerable reduction in volume. The amount of collapse is a function of the relative proportions of each component including degree of saturation, initial void ratio, stress history of the materials, thickness of the collapsible strata and the amount of added load.

Collapsing soils of the loessial type are found in many parts of the world. Loess is found in many parts of the United States, Central Europe, China, Africa, Russia, India, Argentina and elsewhere. Figure 18.2 gives the distribution of collapsible soil in the United States.

Holtz and Hilf (1961) proposed the use of the natural dry density and liquid limit as criteria for predicting collapse. Figure 18.3 shows a plot giving the relationship between liquid limit and dry unit weight of soil, such that soils that plot above the line shown in the figure are susceptible to collapse upon wetting.

18.3 COLLAPSE POTENTIAL AND SETTLEMENT

Collapse Potential

A procedure for determining the collapse potential of a soil was suggested by Jennings and Knight (1975). The procedure is as follows:

A sample of an undisturbed soil is cut and fit into a consolidometer ring and loads are applied progressively until about 200 kPa (4 kip/ft^2) is reached. At this pressure the specimen is flooded with water for saturation and left for 24 hours. The consolidation test is carried on to its maximum loading. The resulting e-log p curve plotted from the data obtained is shown in Fig. 18.4.

The collapse potential C_p is then expressed as

$$C_p = \frac{\Delta e_c}{1 + e_o} \tag{18.2a}$$

in which Δe_c = change in void ratio upon wetting, e_o = natural void ratio.

The collapse potential is also defined as

$$C_p = \frac{\Delta H_o}{H_c} \tag{18.2b}$$

where, ΔH_c = change in the height upon wetting, H_c = initial height.

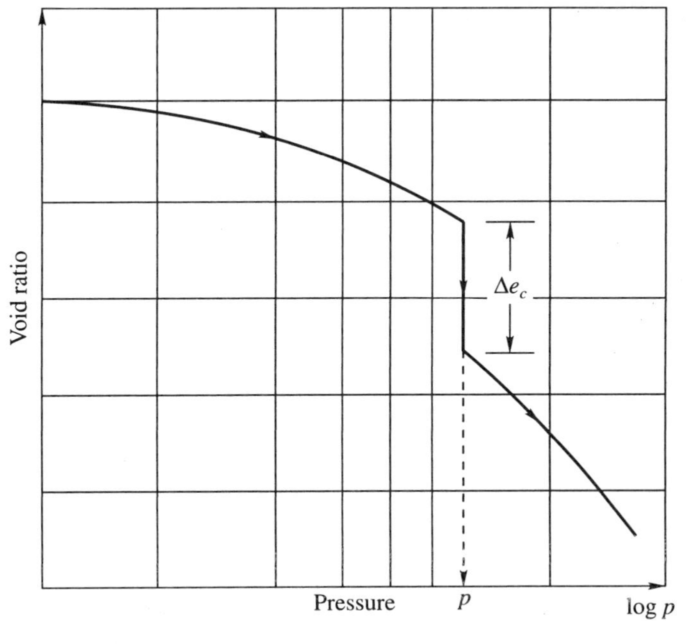

Figure 18.4 Typical collapse potential test result

Table 18.1 Collapse potential values

C_p%	Severity of problem
0–1	No problem
1–5	Moderate trouble
3–10	Trouble
10–20	Severe trouble
> 20	Very severe trouble

Jennings and Knight have suggested some values for collapse potential as shown in Table 18.1. These values are only qualitative to indicate the severity of the problem.

18.4 COMPUTATION OF COLLAPSE SETTLEMENT

The double oedometer method was suggested by Jennings and Knight (1975) for determining a quantitative measure of collapse settlement. The method consists of conducting two consolidation tests. Two identical undisturbed soil samples are used in the tests. The procedure is as follows:

1. Insert two identical undisturbed samples into the rings of two oedometers.

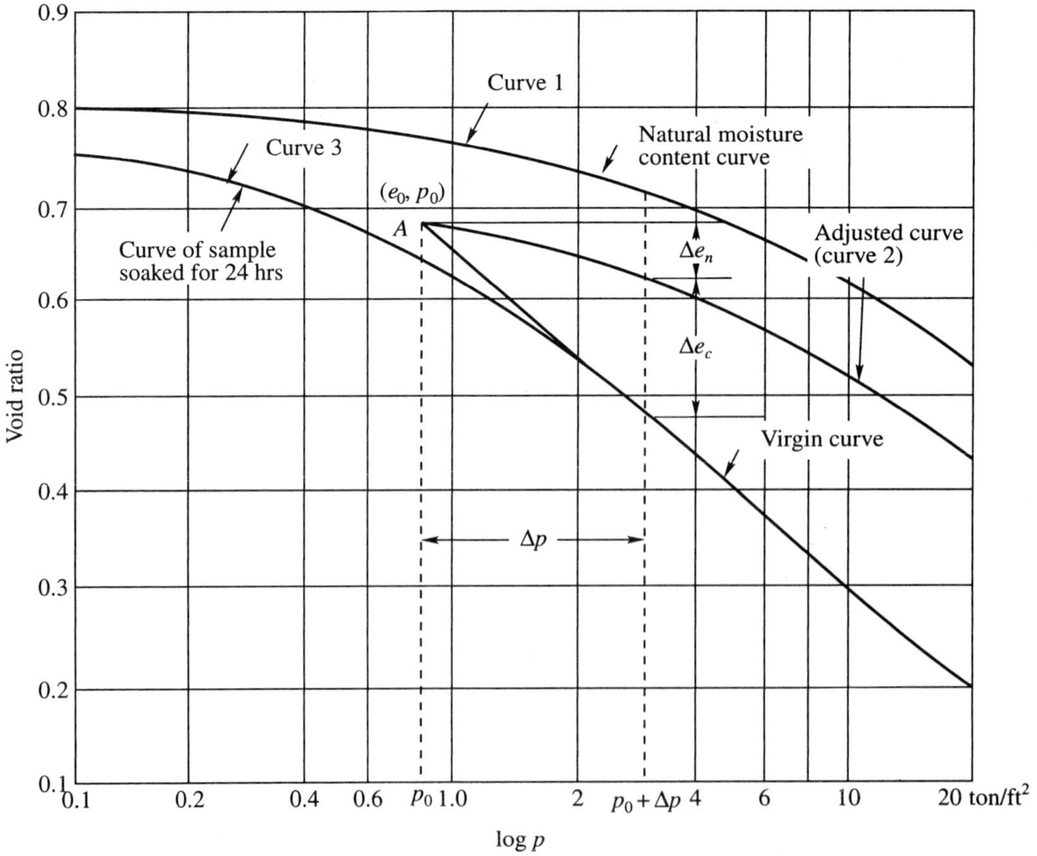

Figure 18.5 Double consolidation test and adjustments for normally consolidated soil (Clemence and Finbarr, 1981)

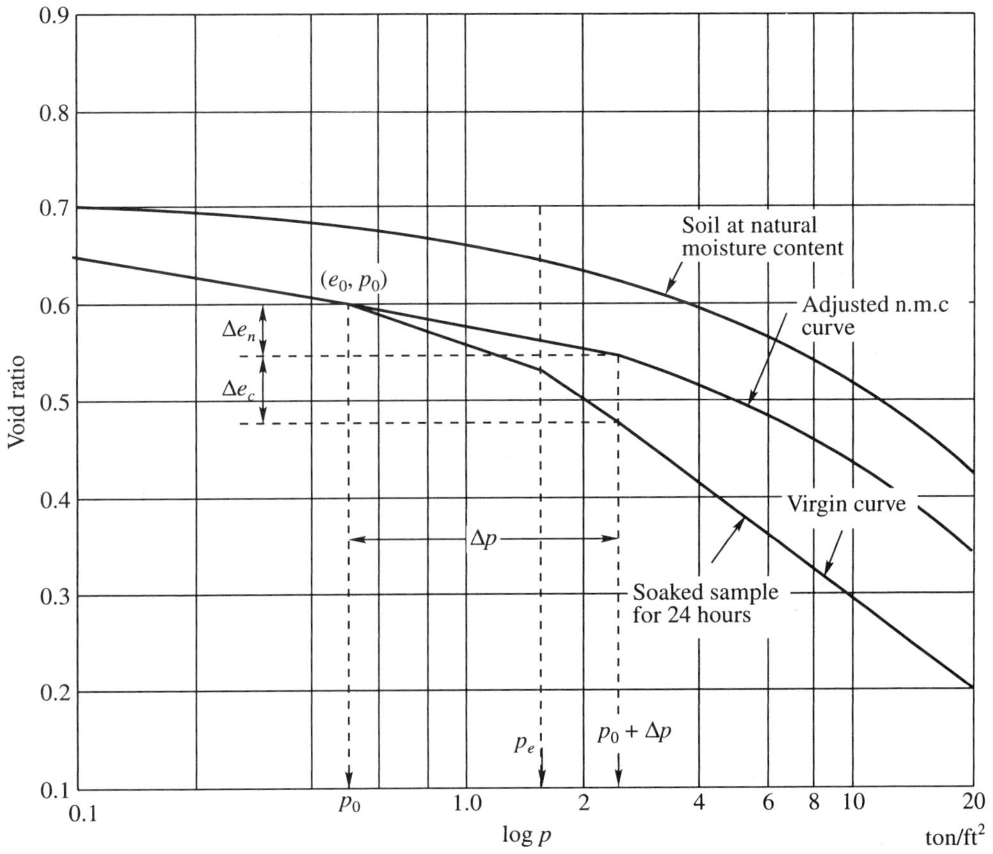

Figure 18.6 Double consolidation test and adjustments for overconsolidated soil (Clemence and Finbarr, 1981)

2. Keep both the specimens under a pressure of 1 kN/m^2 (= 0.15 lb/in^2) for a period of 24 hours.
3. After 24 hours, saturate one specimen by flooding and keep the other at its natural moisture content
4. After the completion of 24 hour flooding, continue the consolidation tests for both the samples by doubling the loads. Follow the standard procedure for the consolidation test.
5. Obtain the necessary data from the two tests, and plot *e*-log *p* curves for both the samples as shown in Fig. 18.5 for normally consolidated soil.
6. Follow the same procedure for overconsolidated soil and plot the *e*-log *p* curves as shown in Fig. 18.6.

From *e*-log *p* plots, obtain the initial void ratios of the two samples after the first 24 hour of loading. It is a fact that the two curves do not have the same initial void ratio. The total overburden pressure p_0 at the depth of the sample is obtained and plotted on the *e*-log *p* curves in Figs 18.5 and 18.6. The preconsolidation pressures p_c are found from the soaked curves of Figs 18.5 and 18.6 and plotted.

Normally Consolidated Case

For the case in which p_c/p_0 is about unity, the soil is considered normally consolidated. In such a case, compression takes place along the virgin curve. The straight line which is tangential to the soaked *e*-log *p* curve passes through the point (e_0, p_0) as shown in Fig. 18.5. Through the point (e_0, p_0) a curve is drawn parallel to the *e*-log *p* curve obtained from the sample tested at natural moisture content. The settlement for any increment in pressure Δp due to the foundation load may be expressed in two parts as

$$S_1 = \frac{\Delta e_n H_c}{1+e_0} \tag{18.3a}$$

$$S_2 = \frac{\Delta e_c H_c}{1+e_0} \tag{18.3b}$$

where Δe_n = change in void ratio due to load Δp as per the *e*-log *p* curve without change in moisture content

Δe_c = change in void ratio at the same load Δp with the increase in moisture content (settlement caused due to collapse of the soil structure)

H_c = thickness of soil stratum susceptible to collapse.

From Eqs (18.3a) and (18.3b), the total settlement due to the collapse of the soil structure is

$$S_c = S_1 + S_2 = \frac{H_c}{1+e_0}(\Delta e_n + \Delta e_c) \tag{18.4}$$

Overconsolidated Case

In the case of an overconsolidated soil the ratio p_c/p_0 is greater than unity. Draw a curve from the point (e_0, p_0) on the soaked soil curve parallel to the curve which represents no change in moisture content during the consolidation stage. For any load $(p_0 + \Delta p) > p_c$, the settlement of the foundation may be determined by making use of the same Eq. (18.4). The changes in void ratios Δe_n and Δe_c are defined in Fig. 18.6.

Example 18.1

A footing of size 10 × 10 ft is founded at a depth of 5 ft below ground level in collapsible soil of the loessial type. The thickness of the stratum susceptible to collapse is 30 ft. The soil at the site is normally consolidated. In order to determine the collapse settlement, double oedometer tests were conducted on two undisturbed soil samples as per the procedure explained in Section 18.4. The *e*-log *p* curves of the two samples are given in Fig. 18.5. The average unit weight of soil γ = 106.6 lb/ft^3 and the induced stress Δp, at the middle of the stratum due to the foundation pressure, is 4,400 lb/ft^2 (= 2.20 t/ft^2). Estimate the collapse settlement of the footing under a soaked condition.

Solution

Double consolidation test results of the soil samples are given in Fig. 18.5. Curve 1 was obtained with natural moisture content. Curve 3 was obtained from the soaked sample after 24 hours. The virgin curve is drawn in the same way as for a normally loaded clay soil (Fig. 7.9a).

The effective overburden pressure p_0 at the middle of the collapsible layer is

p_0 = 15 × 106.6 = 1,599 lb/ft^2 or 0.8 ton/ft^2

A vertical line is drawn in Fig. 18.5 at $p_0 = 0.8$ ton/ft^2. Point A is the intersection of the vertical line and the virgin curve giving the value of $e_0 = 0.68$. $p_0 + \Delta p = 0.8 + 2.2 = 3.0$ t/ft^2. At $(p_0 + \Delta p) = 3$ ton/ft^2, we have (from Fig. 18.5)

$$\Delta e_n = 0.68 - 0.62 = 0.06$$

$$\Delta e_c = 0.62 - 0.48 = 0.14$$

From Eq. (18.3)

$$S_1 = \frac{\Delta e_n H_c}{1 + e_0} = \frac{0.06 \times 30 \times 12}{1 + 0.68} = 12.86 \text{ in.}$$

$$S_2 = \frac{\Delta e_c H_c}{1 + e_0} = \frac{0.14 \times 30 \times 12}{1 + 0.68} = 30.00 \text{ in.}$$

Total settlement $S_c = 42.86$ in.

The total settlement would be reduced if the thickness of the collapsible layer is less or the foundation pressure is less.

Example 18.2

Refer to Example 18.1. Determine the expected collapse settlement under wetted conditions if the soil stratum below the footing is overconsolidated. Double oedometer test results are given in Fig. 18.6. In this case $p_0 = 0.5$ ton/ft^2, $\Delta p = 2$ ton/ft^2, and $p_c = 1.5$ ton/ft^2.

Solution

The virgin curve for the soaked sample can be determined in the same way as for an overconsolidated clay (Fig. 7.9b). Double oedometer test results are given in Fig. 18.6. From this figure:

$$e_0 = 0.6,\ \Delta e_n = 0.6 - 0.55 = 0.05,\ \Delta e_c = 0.55 - 0.48 = 0.07$$

As in Ex. 18.1

$$S_1 = \frac{\Delta e_n}{1 + e_0} H_c = \frac{0.05 \times 30 \times 12}{1 + 0.6} = 11.25 \text{ in.}$$

$$S_2 = \frac{\Delta e_c}{1 + e_0} H_c = \frac{0.07 \times 30 \times 12}{1.6} = 15.75 \text{ in.}$$

Total $S_c = 27.00$ in.

18.5 FOUNDATION DESIGN

Foundation design in collapsible soil is a very difficult task. The results from laboratory or field tests can be used to predict the likely settlement that may occur under severe conditions. In many cases, deep foundations, such as piles, piers etc, may be used to transmit foundation loads to deeper bearing strata below the collapsible soil deposit. In cases where it is feasible to support the structure on shallow foundations in or above the collapsing soils, the use of continuous strip footings may provide a more economical and safer foundation than isolated footings (Clemence and Finbarr, 1981). Differential settlements between columns can be minimized, and a more equitable distribution of stresses may be achieved with the use of strip footing design as shown in Fig. 18.7 (Clemence and Finbarr, 1981).

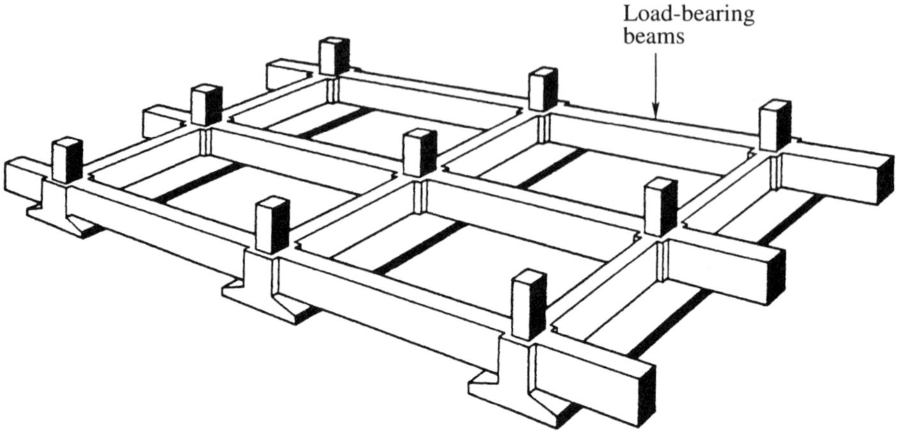

Figure 18.7 Continuous footing design with load-bearing beams for collapsible soil (after Clemence and Finbarr, 1981)

18.6 TREATMENT METHODS FOR COLLAPSIBLE SOILS

On some sites, it may be feasible to apply a pretreatment technique either to stabilize the soil or cause collapse of the soil deposit prior to construction of a specific structure. A great variety of treatment methods have been used in the past. Moistening and compaction techniques, with either conventional impact, or vibratory rollers may be used for shallow depths up to about 1.5 m. For deeper depths, vibroflotation, stone columns, and displacement piles may be tried. Heat treatment to solidify the soil in place has also been used in some countries such as Russia. Chemical stabilization with the use of sodium silicate and injection of carbon dioxide have been suggested (Semkin et al., 1986).

Field tests conducted by Rollins et al., (1990) indicate that dynamic compaction treatment provides the most effective means of reducing the settlement of collapsible soils to tolerable limits. Prewetting, in combination with dynamic compaction, offers the potential for increasing compaction efficiency and uniformity, while increasing vibration attenuation. Prewetting with a 2 percent solution of sodium silicate provides cementation that reduces the potential for settlement. Prewetting with water was found to be the easiest and least costly treatment, but it proved to be completely ineffective in reducing collapse potential for shallow foundations. Prewetting must be accompanied by preloading, surcharging or excavation in order to be effective.

PART B—EXPANSIVE SOILS

18.7 DISTRIBUTION OF EXPANSIVE SOILS

The problem of expansive soils is widespread throughout world. The countries that are facing problems with expansive soils are Australia, the United States, Canada, China, Israel, India, and Egypt. The clay mineral that is mostly responsible for expansiveness belongs to the montmorillonite group. Fig. 18.8 shows the distribution of the montmorillonite group of minerals in the United States. The major concern with expansive soils exists generally in the western part of the United States. In the northern and central United States, the expansive soil problems are primarily related to highly overconsolidated shales. This includes the Dakotas, Montana, Wyoming and Colorado (Chen, 1988). In Minneapolis, the expansive soil problem exists in the Cretaceous

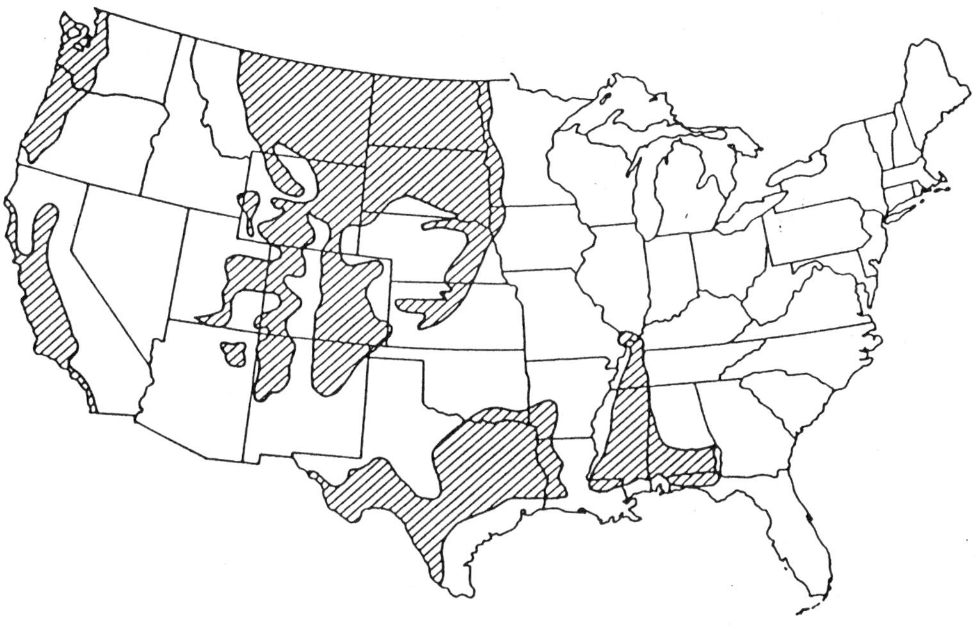

Figure 18.8 General abundance of montmorillonite in near outcrop bedrock formations in the United States (Chen, 1988)

deposits along the Mississippi River and a shrinkage/swelling problem exists in the lacustrine deposits in the Great Lakes Area. In general, expansive soils are not encountered regularly in the eastern parts of the central United States.

In eastern Oklahoma and Texas, the problems encompass both shrinking and swelling. In the Los Angeles area, the problem is primarily one of desiccated alluvial and colluvial soils. The weathered volcanic material in the Denver formation commonly swells when wetted and is a cause of major engineering problems in the Denver area.

The six major natural hazards are earthquakes, landslides, expansive soils, hurricane, tornado and flood. A study points out that expansive soils tie with hurricane wind/storm surge for second place among America's most destructive natural hazards in terms of dollar losses to buildings. According to the study, it was projected that by the year 2000, losses due to expansive soil would exceed 4.5 billion dollars annually (Chen, 1988).

18.8 GENERAL CHARACTERISTICS OF SWELLING SOILS

Swelling soils, which are clayey soils, are also called expansive soils. When these soils are partially saturated, they increase in volume with the addition of water. They shrink greatly on drying and develop cracks on the surface. These soils possess a high plasticity index. Black cotton soils found in many parts of India belong to this category. Their color varies from dark grey to black. It is easy to recognize these soils in the field during either dry or wet seasons. Shrinkage cracks are visible on the ground surface during dry seasons. The maximum width of these cracks may be up to 20 mm or more and they travel deep into the ground. A lump of dry black cotton soil requires a hammer to break. During rainy seasons, these soils become very sticky and very difficult to traverse.

Expansive soils are residual soils which are the result of weathering of the parent rock. The depths of these soils in some regions may be up to 6 m or more. Normally the water table is met at great depths in these regions. As such the soils become wet only during rainy seasons and are dry or

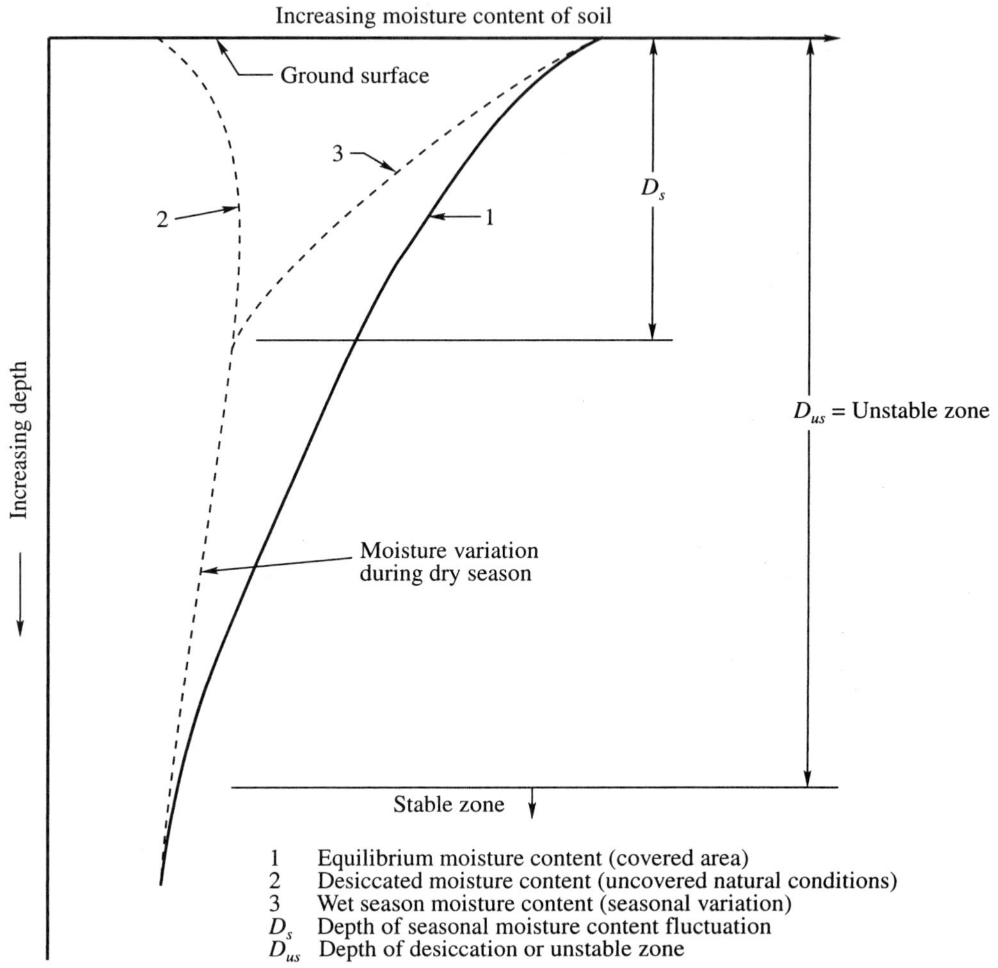

Figure 18.9 Moisture content variation with depth below ground surface (Chen, 1988)

partially saturated during the dry seasons. In regions which have well-defined, alternately wet and dry seasons, these soils swell and shrink in regular cycles. Since moisture change in the soils bring about severe movements of the mass, any structure built on such soils experiences recurring cracking and progressive damage. If one measures the water content of the expansive soils with respect to depth during dry and wet seasons, the variation is similar to the one shown in Fig. 18.9.

During dry seasons, the natural water content is practically zero on the surface and the volume of the soil reaches the shrinkage limit. The water content increases with depth and reaches a value w_n at a depth D_{us}, beyond which it remains almost constant. During the wet season the water content increases and reaches a maximum at the surface. The water content decreases with depth from a maximum of w_n at the surface to a constant value of w_n at almost the same depth D_{us}. This indicates that the intake of water by the expansive soil into its lattice structure is a maximum at the surface and nil at depth D_{us}. This means that the soil lying within this depth D_{us} is subjected to drying and wetting and hence cause considerable movements in the soil. The movements are considerable close to the ground surface and decrease with depth. The cracks that are developed in the dry seasons close due to lateral movements during the wet seasons.

The zone which lies within the depth D_{us} may be called the *unstable zone* (or active zone) and the one below this the *stable zone*. Structures built within this unstable *zone* are likely to move up and down according to seasons and hence suffer damage if differential movements are considerable.

If a structure is built during the dry season with the foundation lying within the unstable zone, the base of the foundation experiences a *swelling pressure* as the partially saturated soil starts taking in water during the wet season. This swelling pressure is due to constraints offered by the foundation for free swelling. The maximum swelling pressure may be as high as 2 MPa (20 tsf). If the imposed bearing pressure on the foundation by the structure is less than the swelling pressure, the structure is likely to be lifted up at least locally which would lead to cracks in the structure. If the imposed bearing pressure is greater than the swelling pressure, there will not be any problem for the structure. If on the other hand, the structure is built during the wet season, it will definitely experience settlement as the dry season approaches, whether the imposed bearing pressure is high or low. However, the imposed bearing pressure during the wet season should be within the allowable bearing pressure of the soil. The *better practice* is to *construct a structure during the dry season and complete it before the wet season.*

In covered areas below a building there will be very little change in the moisture content except due to lateral migration of water from uncovered areas. The moisture profile is depicted by curve 1 in Fig. 18.9.

18.9 CLAY MINERALOGY AND MECHANISM OF SWELLING

Clays can be divided into three general groups on the basis of their crystalline arrangement. They are:

1. Kaolinite group
2. Montmorillonite group (also called the smectite group)
3. Illite group.

The kaolinite group of minerals are the most stable of the groups of minerals. The kaolinite mineral is formed by the stacking of the crystalline layers of about 7 Å thick one above the other with the base of the silica sheet bonding to hydroxyls of the gibbsite sheet by hydrogen bonds. Since hydrogen bonds are comparatively strong, the kaolinite crystals consists of many sheet stackings that are difficult to dislodge. The mineral is, therefore, stable and water cannot enter between the sheets to expand the unit cells.

The structural arrangement of the montmorillonite mineral is composed of units made of two silica tetrahedral sheets with a central alumina-octahedral sheet. The silica and gibbsite sheets are combined in such a way that the tips of the tetrahedrons of each silica sheet and one of the hydroxyl layers of the octahedral sheet form a common layer. The atoms common to both the silica and gibbsite layers are oxygen instead of hydroxyls. The thickness of the silica-gibbsite-silica unit is about 10 Å. In stacking of these combined units one above the other, oxygen layers of each unit are adjacent to oxygen of the neighboring units, with a consequence that there is a weak bond and excellent cleavage between them. Water can enter between the sheets causing them to expand significantly and thus the structure can break into 10 Å thick structural units. The soils containing a considerable amount of montmorillonite minerals will exhibit high swelling and shrinkage characteristics. The illite group of minerals has the same structural arrangement as the montmorillonite group. The presence of potassium as the bonding materials between units makes the illite minerals swell less.

18.10 DEFINITION OF SOME PARAMETERS

Expansive soils can be classified on the basis of certain inherent characteristics of the soil. It is first necessary to understand certain basic parameters used in the classification.

Swelling Potential

Swelling potential is defined as the percentage of swell of a laterally confined sample in an oedometer test which is soaked under a surcharge load of 7 kPa (1 lb/in^2) after being compacted to maximum dry density at optimum moisture content according to the AASHTO compaction test.

Swelling Pressure

The swelling pressure p_s, is defined as the pressure required for preventing volume expansion in soil in contact with water. It should be noted here that the swelling pressure measured in a laboratory oedometer is different from that in the field. The actual field swelling pressure is always less than the one measured in the laboratory.

Free Swell

Free swell S_f is defined as

$$S_f = \frac{V_f - V_i}{V_i} \times 100 \tag{18.5}$$

where V_i = initial dry volume of poured soil
V_f = final volume of poured soil

According to Holtz and Gibbs (1956), 10 cm^3 (V_i) of dry soil passing thorough a No. 40 sieve is poured into a 100 cm^3 graduated cylinder filled with water. The volume of settled soil is measured after 24 hours which gives the value of V_f. Bentonite-clay is supposed to have a free swell value ranging from 1200 to 2000 percent. The free swell value increases with plasticity index. Holtz and Gibbs suggested that soils having a free-swell value as low as 100 percent can cause considerable damage to lightly loaded structures and soils heaving a free swell value below 50 percent seldom exhibit appreciable volume change even under light loadings.

18.11 EVALUATION OF THE SWELLING POTENTIAL OF EXPANSIVE SOILS BY SINGLE INDEX METHOD (CHEN, 1988)

Simple soil property tests can be used for the evaluation of the swelling potential of expansive soils (Chen, 1988). Such tests are easy to perform and should be used as routine tests in the investigation of building sites in those areas having expansive soil. These tests are

1. Atterberg limits tests
2. Linear shrinkage tests
3. Free swell tests
4. Colloid content tests

Atterberg Limits

Holtz and Gibbs (1956) demonstrated that the plasticity index, I_p, and the liquid limit, w_l, are useful indices for determining the swelling characteristics of most clays. Since the liquid limit and the

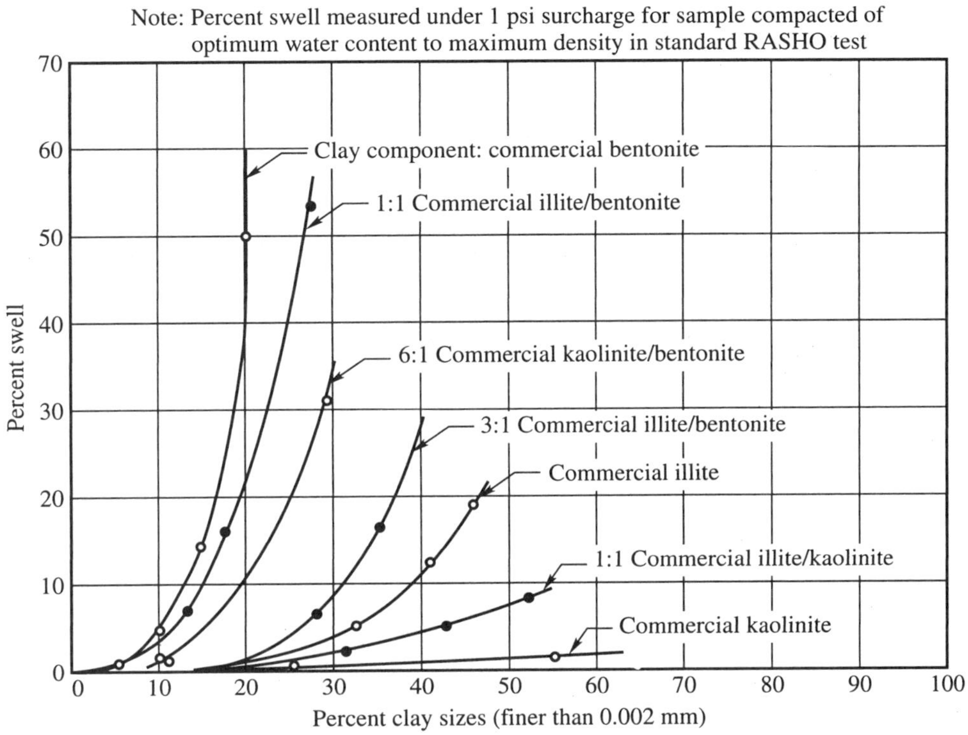

Figure 18.10 Relationship between percentage of swell and percentage of clay sizes for experimental soils (after Seed et al., 1962)

swelling of clays both depend on the amount of water a clay tries to absorb, it is natural that they are related. The relation between the swelling potential of clays and the plasticity index has been established as given in Table 18.2

Linear Shrinkage

The swell potential is presumed to be related to the opposite property of linear shrinkage measured in a very simple test. Altmeyer (1955) suggested the values given in Table 18.3 as a guide to the determination of potential expansiveness based on shrinkage limits and linear shrinkage.

Colloid Content

There is a direct relationship between colloid content and swelling potential as shown in Fig. 18.10 (Chen, 1988). For a given clay type, the amount of swell will increase with the amount of clay present in the soil.

Table 18.2 Relation between swelling potential and plasticity index, I_p

Plasticity index I_p (%)	Swelling potential
0–15	Low
10–35	Medium
20–55	High
35 and above	Very high

Table 18.3 Relation between swelling potential, shrinkage limits, and linear shrinkage

Shrinkage limit %	Linear shrinkage %	Degree of expansion
< 10	> 8	Critical
10–12	5–8	Marginal
> 12	0–5	Non-critical

18.12 CLASSIFICATION OF SWELLING SOILS BY INDIRECT MEASUREMENT

By utilizing the various parameters as explained in Section 18.11, the swelling potential can be evaluated without resorting to direct measurement (Chen, 1988).

USBR Method

Holtz and Gibbs (1956) developed this method which is based on the simultaneous consideration of several soil properties. The typical relationships of these properties with swelling potential are shown in Fig. 18.11. Table 18.4 has been prepared based on the curves presented in Fig. 18.11 by Holtz and Gibbs (1956).

The relationship between the swell potential and the plasticity index can be expressed as follows (Chen, 1988)

$$S_p = Be^{A(I_p)} \tag{18.6}$$

where, A = 0.0838
B = 0.2558
I_p = plasticity index.

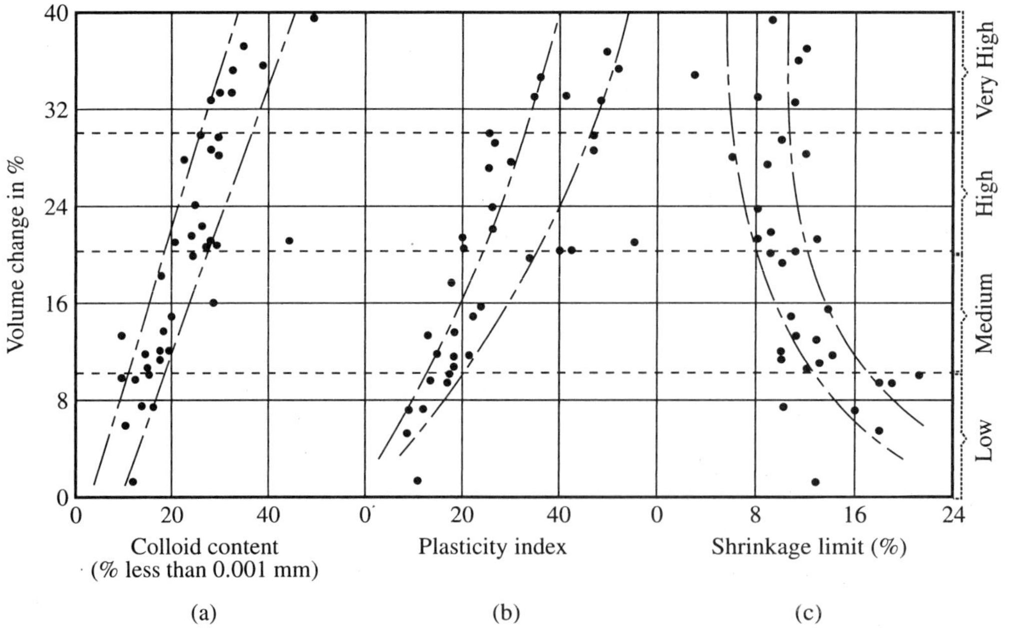

Figure 18.11 Relation of volume change to (a) colloid content, (b) plasticity index, and (c) shrinkage limit (air-dry to saturated condition under a load of 1 lb per sq in) (Holtz and Gibbs, 1956)

Table 18.4 Data for making estimates of probable volume changes for expansive soils (Source: Chen, 1988)

Data from index tests*			Probable expansion, percent total vol. change	Degree of expansion
Colloid content, percent minus 0.001 mm	Plasticity index	Shrinkage limit		
> 28	> 35	< 11	> 30	Very high
20–13	25–41	7–12	20–30	High
13–23	15–28	10–16	10–30	Medium
< 15	< 18	> 15	<10	Low

*Based on vertical loading of 1.0 psi. (after Holtz and Gibbs, 1956)

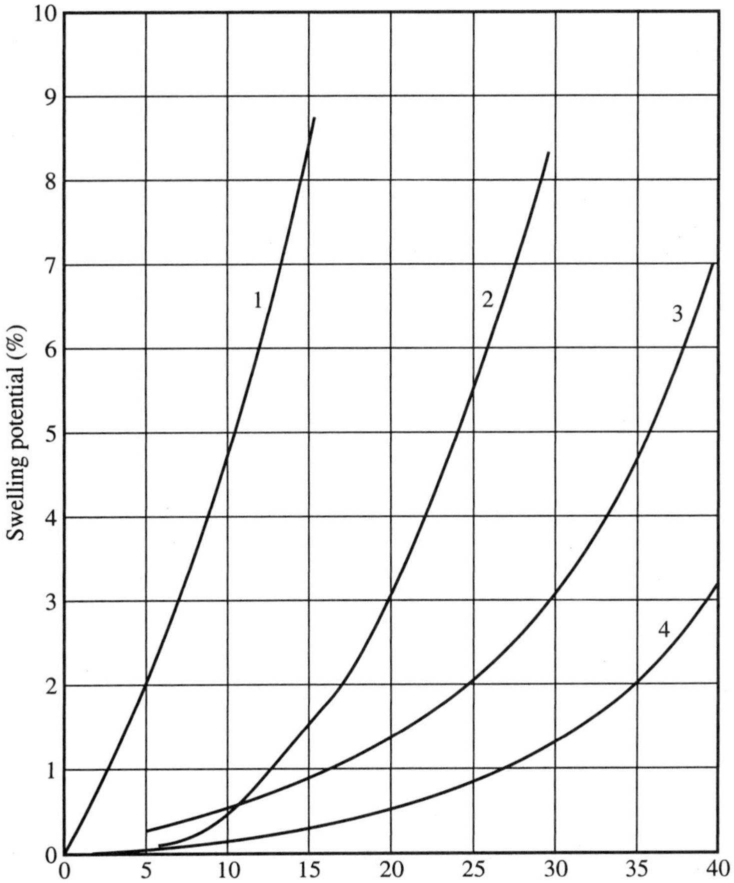

1. Holtz and Gibbs (Surcharge pressure 1 psi)
2. Seed, Woodward and Lundgren (Surcharge pressure 1 psi)
3. Chen (Surcharge pressure 1 psi)
4. Chen (Surcharge pressure 6.94 psi)

Figure 18.12 Relationships of volume change to plasticity index (Source: Chen, 1988)

Figure 18.12 shows that with an increase in plasticity index, the increase of swelling potential is much less than predicted by Holtz and Gibbs. The curves given by Chen (1988) are based on thousands of tests performed over a period of 30 years and as such are more realistic.

Activity Method

Skempton (1953) defined activity by the following expression

$$A = \frac{I_p}{C} \tag{18.7}$$

where I_p = plasticity index
C = percentage of clay size finer than 0.002 mm by weight.

The activity method as proposed by Seed, Woodward, and Lundgren, (1962) was based on remolded, artificially prepared soils comprising of mixtures of bentonite, illite, kaolinite and fine sand in different proportions. The activity for the artificially prepared sample was defined as

$$\text{activity } A = \frac{I_p}{C - n} \tag{18.9}$$

where n = 5 for natural soils and, n = 10 for artificial mixtures.

The proposed classification chart is shown in Fig. 18.13. This method appears to be an improvement over the USBR method.

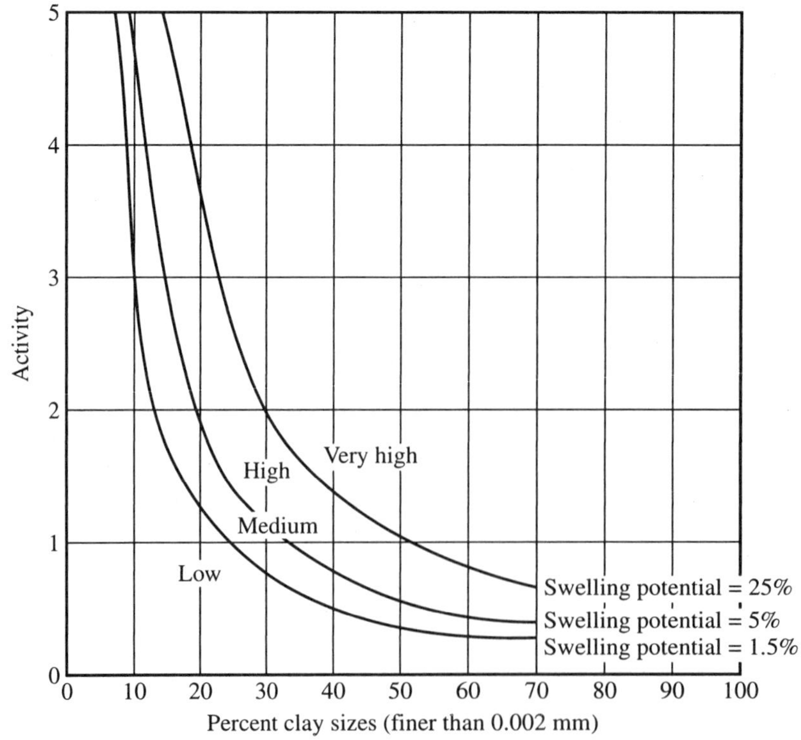

Figure 18.13 Classification chart for swelling potential (after Seed, Woodward, and Lundgren, 1962)

The Potential Volume Change Method (PVC)

A determination of soil volume change was developed by Lambe under the auspices of the Federal Housing Administration (Source: Chen, 1988). Remolded samples were specified. The procedure is as given below.

The sample is first compacted in a fixed ring consolidometer with a compaction effect of 55,000 ft-lb per cu ft. Then an initial pressure of 200 psi is applied, and water added to the sample which is partially restrained by vertical expansion by a proving ring. The proving ring reading is taken at the end of 2 hours. The reading is converted to pressure and is designated as the *swell index*. From Fig. 18.14, the swell index can be converted to potential volume change. Lambe established the categories of PVC rating as shown in Table 18.5.

The PVC method has been widely used by the Federal Housing Administration as well as the Colorado State Highway Department (Chen, 1988).

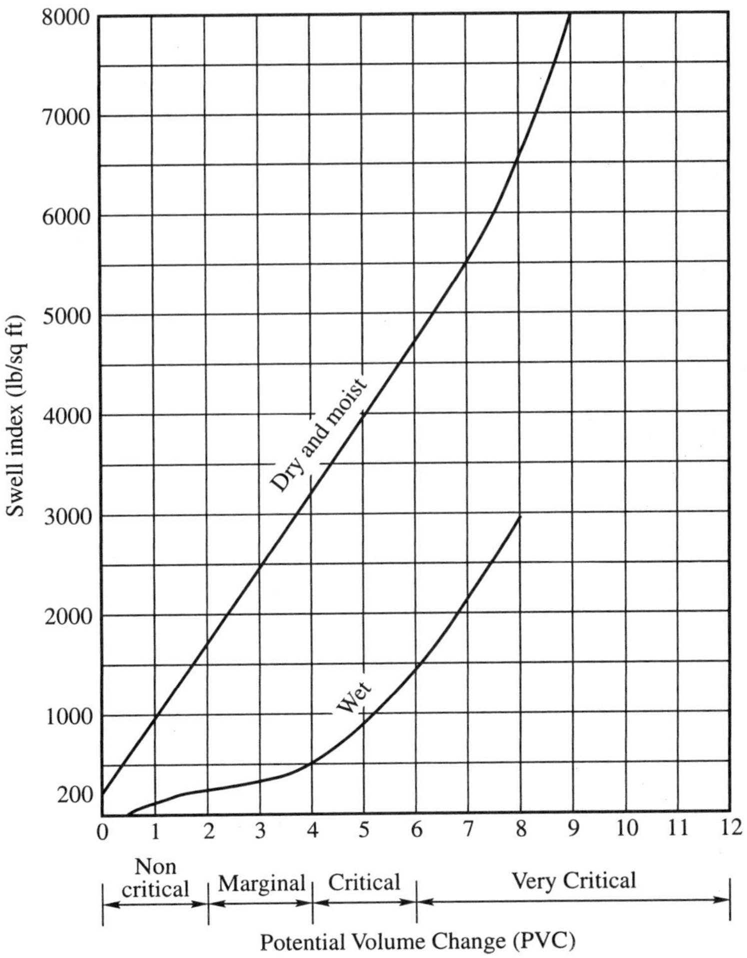

Figure 18.14 Swell index versus potential volume change (from 'FHA soil PVC meter publication,' Federal Housing Administration Publication no. 701) (Source: Chen, 1988)

Table 18.5 Potential volume change rating (PVC)

PVC rating	Category
Less than 2	Non-critical
2–4	Marginal
4–6	Critical
> 6	Very critical

(Source: Chen, 1988)

Figure 18.15(a) shows a soil volume change meter (ELE International Inc). This meter measures both shrinkage and swelling of soils, ideal for measuring swelling of clay soils, and fast and easy to operate.

Expansion Index (EI)—Chen (1988)

The ASTM Committee on Soil and Rock suggested the use of an *Expansion Index* (EI) as a unified method to measure the characteristics of swelling soils. It is claimed that the EI is a basic index property of soil such as the liquid limit, the plastic limit and the plasticity index of the soil.

The sample is sieved through a No 4 sieve. Water is added so that the degree of saturation is between 49 and 51 percent. The sample is then compacted into a 4 inch diameter mold in two layers to give a total compacted depth of approximates 2 inches. Each layer is compacted by 15 blows of 5.5 lb hammer dropping from a height of 12 inches. The prepared specimen is allowed to consolidate under 1 lb/in^2 pressure for a period of 10 minutes, then inundated with water until the rate of expansion ceases.

The expansion index is expressed as

$$EI = \frac{\Delta h}{h_i} \times 1000 \tag{18.9}$$

where Δh = change in thickness of sample, in.

h_i = initial thickness of sample, in.

The classification of a potentially expansive soil is based on Table 18.6.

This method offers a simple testing procedure for comparing expansive soil characteristics.

Figure 18.15(b) shows an ASTMD-829 expansion index test apparatus (ELE International Inc). This is a completely self-contained apparatus designed for use in determining the expansion index of soils.

Table 18.6 Classification of potentially expansive soil

Expansion Index, *EI*	Expansion potential
0–20	Very low
21–50	Low
51–90	Medium
91–130	High
> 130	Very high

(Source: Chen, 1988)

Swell Index

Vijayvergiya and Gazzaly (1973) suggested a simple way of identifying the swell potential of clays, based on the concept of the swell index. They defined the swell index, I_s, as follows

$$I_s = \frac{w_n}{w_l} \tag{18.10}$$

where w_n = natural moisture content in percent
w_l = liquid limit in percent

The relationship between I_s and swell potential for a wide range of liquid limit is shown in Fig 18.16. Swell index is widely used for the design of post-tensioned slabs on expansive soils.

Prediction of Swelling Potential

Plasticity index and shrinkage limit can be used to indicate the swelling characteristics of expansive soils. According to Seed at al., (1962), the swelling potential is given as a function of the plasticity index by the formula

$$S_p = 60k\, I_p^{2.44} \tag{18.11}$$

(a) (b)

Figure 18.15 (a) Soil volume change meter, and (b) Expansion index test apparatus (Courtesy: Soiltest)

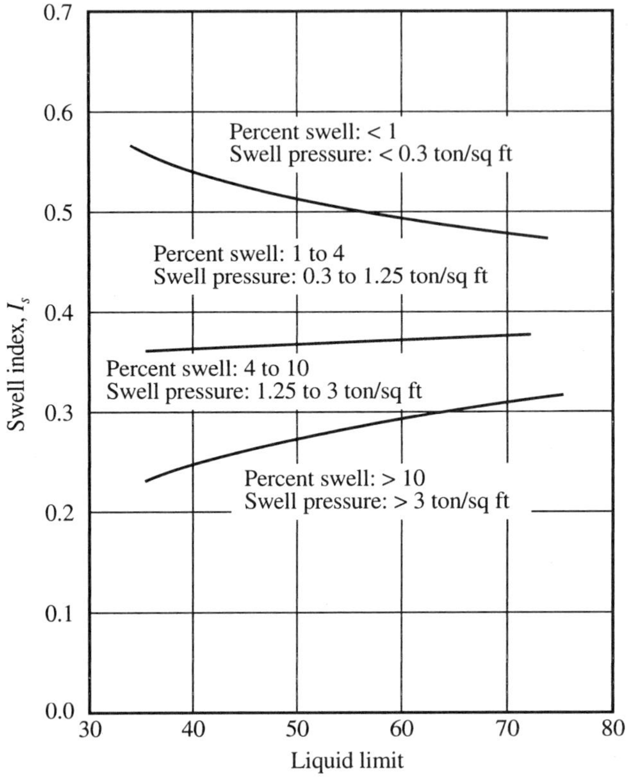

Figure 18.16 Relationship between swell index and liquid limit for expansive clays (Source: Chen, 1988)

where S_p = swelling potential in percent
I_p = plasticity index in percent
k = 3.6×10^{-5}, a factor for clay content between 8 and 65 percent.

18.13 SWELLING PRESSURE BY DIRECT MEASUREMENT

ASTM defines swelling pressure which prevents the specimen from swelling or that pressure which is required to return the specimen to its original state (void ratio, height) after swelling. Essentially, the methods of measuring swelling pressure can be either stress controlled or strain controlled (Chen, 1988).

In the stress controlled method, the conventional oedometer is used. The samples are placed in the consolidation ring trimmed to a height of 0.75 to 1 inch. The samples are subjected to a vertical pressure ranging from 500 psf to 2000 psf depending upon the expected field conditions. On the completion of consolidation, water is added to the sample. When the swelling of the sample has ceased the vertical stress is increased in increments until it has been compressed to its original height. The stress required to compress the sample to its original height is commonly termed the *zero volume change swelling pressure*. A typical consolidation curve is shown in Fig 18.17.

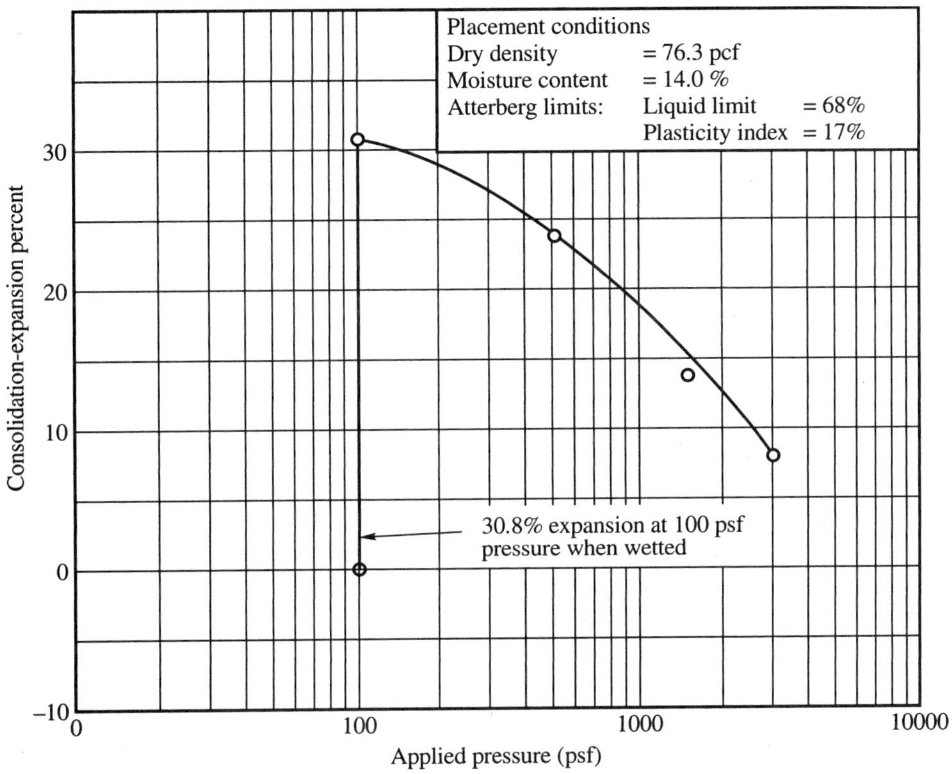

Figure 18.17 Typical stress controlled swell-consolidation curve

Prediction of Swelling Pressure

Komornik et al., (1969) have given an equation for predicting swelling pressure as

$$\log p_s = 2.132 + 0.0208 w_l + 0.00065 \gamma_d - 0.0269 w_n \tag{18.12}$$

where p_s = swelling pressure in kg/cm^2
w_l = liquid limit (%)
w_n = natural moisture content (%)
γ_d = dry density of soil in kg/cm^3

18.14 EFFECT OF INITIAL MOISTURE CONTENT AND INITIAL DRY DENSITY ON SWELLING PRESSURE

The capability of swelling decreases with an increase of the initial water content of a given soil because its capacity to absorb water decreases with the increase of its degree of saturation. It was found from swelling tests on black cotton soil samples, that the initial water content has a small effect on swelling pressure until it reaches the shrinkage limit, then its effect increases (Abouleid, 1982). This is depicted in Fig. 18.18(a).

The effect of initial dry density on the swelling percent and the swelling pressure increases with an increase of the dry density because the dense soil contains more clay particles in a unit volume and consequently greater movement will occur in a dense soil than in a loose soil upon

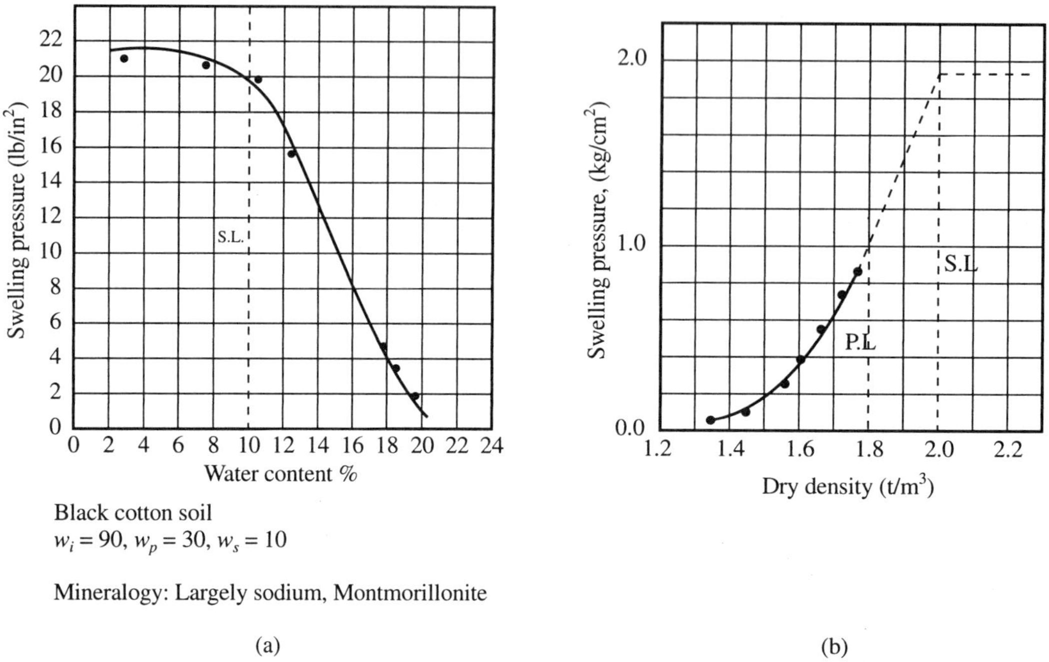

Figure 18.18 (a) effect of initial water content on swelling pressure of black cotton soil, and (b) effect of initial dry density on swelling pressure of black cotton soil (Source: Abouleid, 1982)

wetting (Abouleid, 1982). The effect of initial dry density on swelling pressure is shown in Fig. 18.18(b).

18.15 ESTIMATING THE MAGNITUDE OF SWELLING

When footings are built in expansive soil, they experience lifting due to the swelling or heaving of the soil. The amount of total heave and the rate of heave of the expansive soil on which a structure is founded are very complex. The heave estimate depends on many factors which cannot be readily determined. Some of the major factors that contribute to heaving are:

1. Climatic conditions involving precipitation, evaporation, and transpiration affect the moisture in the soil. The depth and degree of desiccation affect the amount of swell in a given soil horizon.
2. The thickness of the expansive soil stratum is another factor. The thickness of the stratum is controlled by the depth to the water table.
3. The depth to the water table is responsible for the change in moisture of the expansive soil lying above the water table. No swelling of soil takes place when it lies below the water table.
4. The predicted amount of heave depends on the nature and degree of desiccation of the soil immediately after construction of a foundation.
5. The single most important element controlling the swelling pressure as well as the swell potential is the in-situ density of the soil. On the completion of excavation, the stress condition in the soil mass undergoes changes, such as the release of stresses due to elastic

rebound of the soil. If construction proceeds without delay, the structural load compensates for the stress release.

6. The permeability of the soil determines the rate of ingress of water into the soil either by gravitational flow or diffusion, and this in turn determines the rate of heave.

Various methods have been proposed to predict the amount of total heave under a given structural load. The following methods, however, are described here.

1. The Department of Navy method (1982)
2. The South African method [also known as the Van Der Merwe method (1964)]

The Department of Navy Method

Procedure for Estimating Total Swell under Structural Load

1. Obtain representative undisturbed samples of soil below the foundation level at intervals of depth. The samples are to be obtained during the dry season when the moisture contents are at their lowest.
2. Load specimens (at natural moisture content) in a consolidometer under a pressure equal to the ultimate value of the overburden plus the weight of the structure. Add water to saturate the specimen. Measure the swell.
3. Compute the final swell in terms of percent of original sample height.
4. Plot swell versus depth.
5. Compute the total swell which is equal to the area under the percent swell versus depth curve.

Procedure for Estimating Undercut

The procedure for estimating undercut to reduce swell to an allowable value is as follows:

1. From the percent swell versus depth curve, plot the relationship of total swell versus depth at that height. Total swell at any depth equals area under the curve integrated upward from the depth of zero swell.
2. For a given allowable value of swell, read the amount of undercut necessary from the total swell versus depth curve.

Van Der Merwe Method (1964)

Probably the nearest practical approach to the problem of estimating swell is that of Van Der Merwe. This method starts by classifying the swell potential of soil into very high to low categories as shown in Fig. 18.19. Then assign potential expansion (PE) expressed in in./ft of thickness based on Table 18.7.

Table 18.7 Potential expansion

Swell potential	Potential expansion (PE) in./ft
Very high	1
High	1/2
Medium	1/4
Low	0

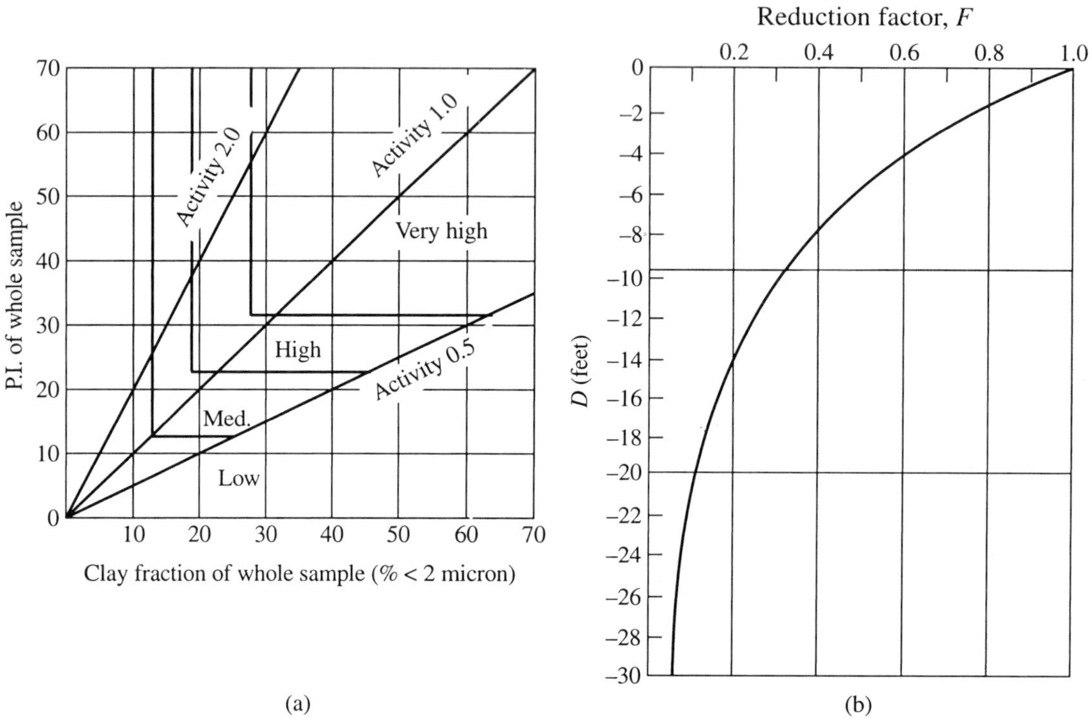

Figure 18.19 Relationships for using Van Der Merwe's prediction method: (a) potential expansiveness, and (b) reduction factor (Van der Merwe, 1964)

Procedure for Estimating Swell

1. Assume the thickness of an expansive soil layer or the lowest level of ground water.
2. Divide this thickness (z) into several soil layers with variable swell potential.
3. The total expansion is expressed as

$$\Delta H_e = \sum_{i=1}^{i=n} \Delta_i \tag{18.13}$$

where ΔH_e = total expansion (in.)

$$\Delta_i = (PE)_i (\Delta D)_i (F)_i \tag{18.14}$$

$$(F)_i = \log^{-1}\left(-\frac{D_i}{20}\right) = \text{reduction factor for layer } i.$$

z = total thickness of expansive soil layer (ft)

D_i = depth to midpoint of i th layer (ft)

$(\Delta D)_i$ = thickness of i th layer (ft)

Fig. 18.19(b) gives the reduction factor plotted against depth.

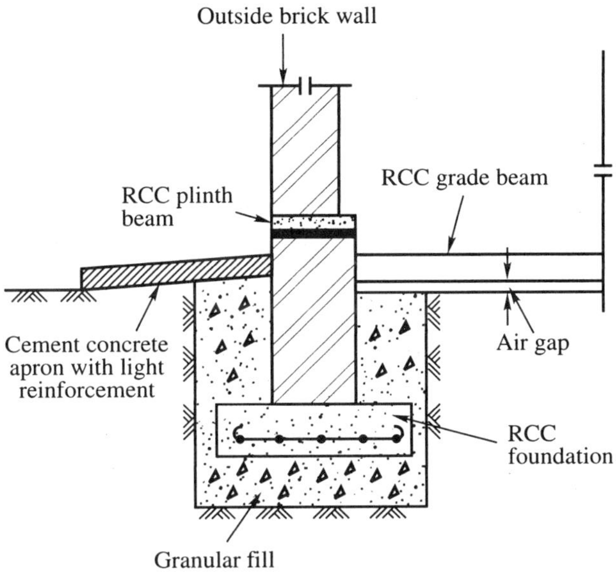

Figure 18.20 Foundation in expansive soil

18.16 DESIGN OF FOUNDATIONS IN SWELLING SOILS

It is necessary to note that all parts of a building will not equally be affected by the swelling potential of the soil. Beneath the center of a building where the soil is protected from sun and rain the moisture changes are small and the soil movements the least. Beneath outside walls, the movement are greater. Damage to buildings is greatest on the outside walls due to soil movements.

Three general types of foundations can be considered in expansive soils. They are

1. Structures that can be kept isolated from the swelling effects of the soils
2. Designing of foundations that will remain undamaged in spite of swelling
3. Elimination of swelling potential of soil.

All three methods are in use either singly or in combination, but the first is by far the most widespread. Fig. 18.20 show a typical type of foundation under an outside wall. The granular fill provided around the shallow foundation mitigates the effects of expansion of the soils.

18.17 DRILLED PIER FOUNDATIONS

Drilled piers are commonly used to resist uplift forces caused by the swelling of soils. Drilled piers, when made with an enlarged base, are called, *belled piers* and when made without an enlarged base are referred to as *straight-shaft piers*.

Woodward, et al., (1972) commented on the empirical design of piers: "Many piers, particularly where rock bearing is used, have been designed using strictly empirical considerations which are derived from regional experience". They further stated that "when surface conditions are well established and are relatively uniform, and the performance of past constructions well documented, the design by experience approach is usually found to be satisfactory."

The principle of drilled piers is to provide a relatively inexpensive way of transferring the structural loads down to stable material or to a stable zone where moisture changes are improbable.

There should be no direct contact between the soil and the structure with the exception of the soils supporting the piers.

Straight-shaft Piers in Expansive Soils

Figure 18.21(a) shows a straight-shaft drilled pier embedded in expansive soil. The following notations are used.

L_1 = length of shaft in the unstable zone (active zone) affected by wetting.
L_2 = length of shaft in the stable zone unaffected by wetting
d = diameter of shaft
Q = structural dead load = qA_b
q = unit dead load pressure and
A_b = base area of pier

When the soil in the unstable zone takes water during the wet season, the soil tries to expand which is partially or wholly prevented by the rough surface of the pile shaft of length L_1. As a result there will be an upward force developed on the surface of the shaft which tries to pull the pile out of its position. The upward force can be resisted in the following ways.

1. The downward dead load Q acting on the pier top
2. The resisting force provided by the shaft length L_2 embedded in the stable zone.

Two approaches for solving this problem may be considered. They are

1. The method suggested by Chen (1988)
2. The O'Neill (1988) method with belled pier.

Two cases may be considered. They are

1. The stability of the pier when no downward load Q is acting on the top. For this condition a factor of safety of 1.2 is normally found sufficient.
2. The stability of the pier when Q is acting on the top. For this a value of F_s = 2.0 is used.

Equations for Uplift Force Q_{up}

Chen (1988) suggested the following equation for estimating the uplift force Q_{up}

$$Q_{up} = \pi d \alpha_u p_s L_1 \tag{18.15}$$

where d = diameter of pier shaft
α_u = coefficient of uplift between concrete and soil = 0.15
p_s = swelling pressure
= 10,000 psf (480 kN/m^2) for soil with high degree of expansion
= 5,000 (240 kN/m^2) for soil with medium degree of expansion

The depth (L_1) of the unstable zone (wetting zone) varies with the environmental conditions. According to Chen (1988) the wetting zone is limited only to the upper 5 feet of the pier. It is possible for the wetting zone to extend beyond 10-15 feet in some countries and limiting the depth of unstable zone to a such a low value of 5 ft may lead to unsafe conditions for the stability of structures. However, it is for designers to decide this depth L_1 according to local conditions. With regards to swelling pressure p_s, it is unrealistic to fix any definite value of 10,000 or 5,000 psf for all types of expansive soils under all conditions of wetting. It is also not definitely known if the results obtained from laboratory tests truly represent the in situ swelling pressure. Possibly one way of overcoming this complex problem is to relate the uplift resistance to undrained cohesive strength of soil just as in the case of friction piers under compressive loading. Equation (18.15) may be written as

$$Q_{up} = \pi d \alpha_s c_u L_1 \tag{18.16}$$

where α_s = adhesion factor between concrete and soil under a swelling condition

c_u = unit cohesion under undrained conditions

It is possible that the value of α_s may be equal to 1.0 or more according to the swelling type and environmental conditions of the soil. Local experience will help to determine the value of α_s This approach is simple and pragmatic.

Resisting Force

The length of pier embedded in the stable zone should be sufficient to keep the pier being pulled out of the ground with a suitable factor of safety. If L_2 is the length of the pier in the stable zone, the resisting force Q_R is the frictional resistance offered by the surface of the pier within the stable zone. We may write

$$Q_R = \pi d L_2 \alpha c_u \tag{18.17}$$

where α = adhesion factor under compression loading

c_u = undrained unit cohesion of soil

The value of α may be obtained from Fig. 17.15.

Two cases of stability may be considered:

1. Without taking into account the dead load Q acting on the pier top, and using $F_s = 1.2$

$$Q_{up} = \frac{Q_R}{1.2} \tag{18.18a}$$

2. By taking into account the dead load Q and using $F_s = 2.0$

$$(Q_{up} - Q) = \frac{Q_R}{2.0} \tag{18.18b}$$

For a given shaft diameter d equations (18.18a) and (18.18b) help to determine the length L_2 of the pier in the stable zone. The one that gives the maximum length L_2 should be used.

Belled Piers

Piers with a belled bottom are normally used when large uplift forces have to be resisted. Fig. 18.21(b) shows a belled pier with all the forces acting.

The uplift force for a belled pier is the same as that applicable for a straight shaft. The resisting force equation for the pier in the stable zone may be written as (O'Neill, 1988)

$$Q_{R1} = \pi d \, L_2 \, \alpha \, c_u \tag{18.19a}$$

$$Q_{R2} = \frac{\pi}{4}\left[d_b^2 - d^2\right]\left[c\,N_c + \gamma\,L_2\right] \tag{18.19b}$$

where

d_b = diameter of the underream

N_c = bearing capacity factor

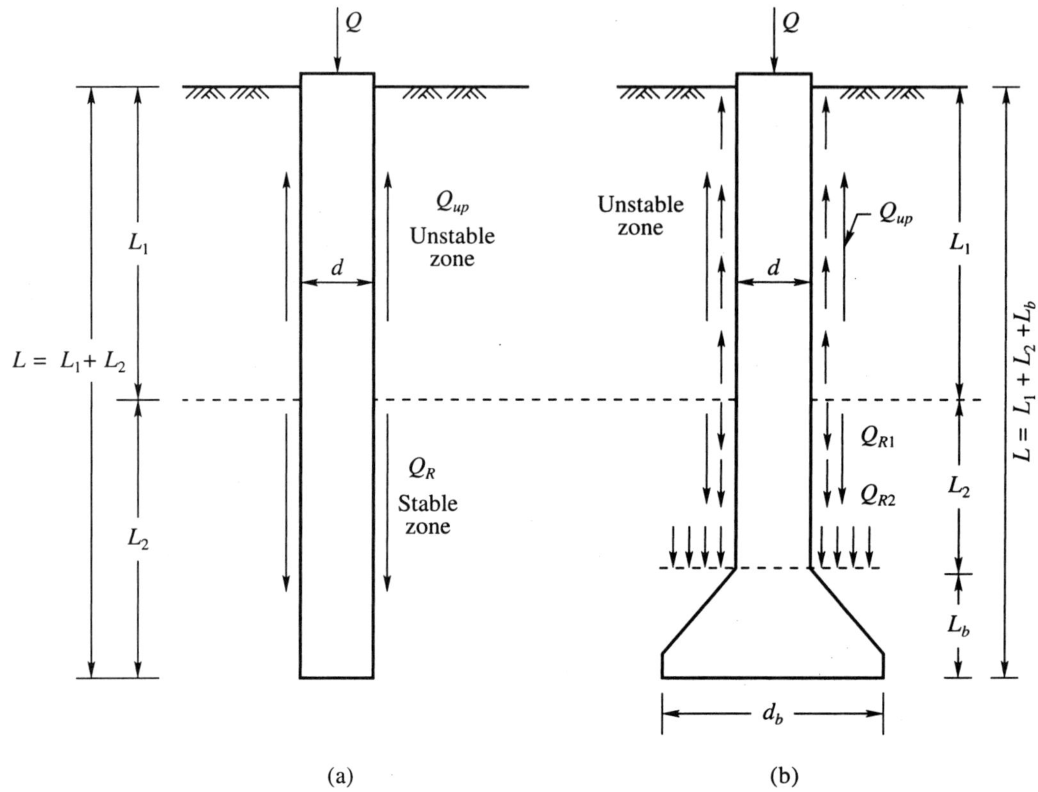

Figure 18.21 Drilled pier in expansive soil

c = unit cohesion under undrained condition

γ = unit weight of soil

The values of N_c are given in Table 18.8 (O'Neill, 1988)

Two cases of stability may be written as before.

1. Without taking the dead load Q and using $F_s = 1.2$

$$Q_{up} = \frac{1}{1.2}\left[Q_{R1} + Q_{R2}\right]$$

2. By taking into account the dead load and $F_s = 2.0$

$$(Q_{up} - Q) = \frac{1}{2.0}\left[Q_{R1} + Q_{R2}\right]$$

Table 18.8 Values of N_c

L_2/d_b	N_c
1.7	4
2.5	6
≥5.0	9

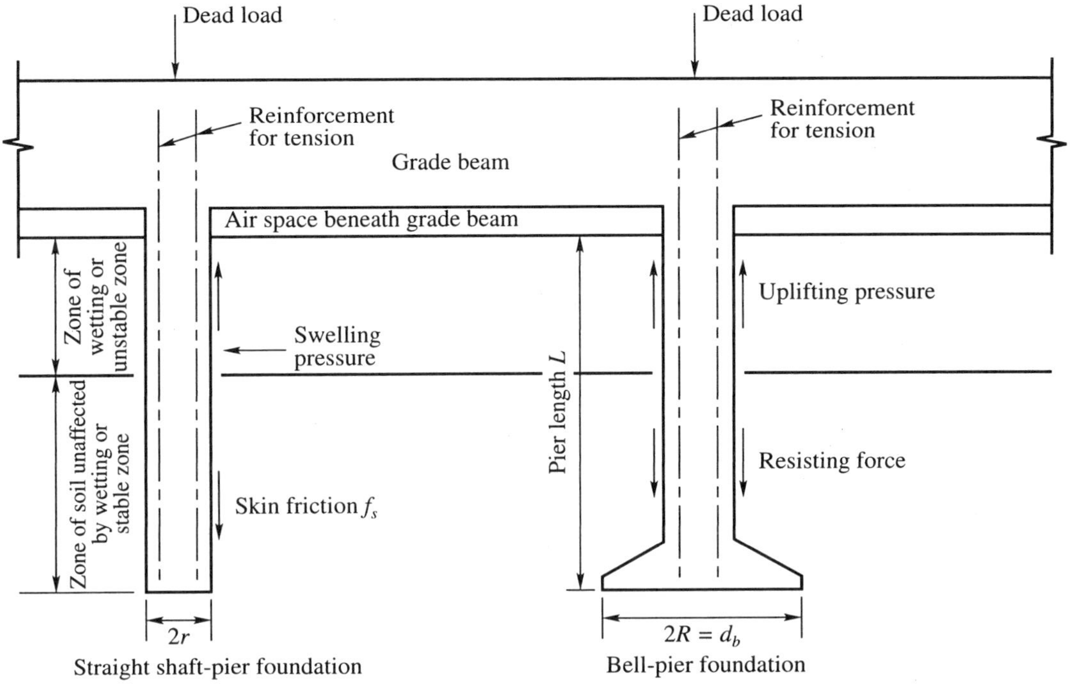

Figure 18.22 Grade beam and pier system (Chen, 1988)

For a given shaft diameter d and base diameter d_b, the above equations help to determine the value of L_2. The one that gives the maximum value for L_2 has to be used in the design.

Fig. 18.22 gives a typical foundation design with grade beams and drilled piers (Chen, 1988). The piers should be taken sufficiently below the unstable zone of wetting in order to resist the uplift forces.

Example 18.3

A footing founded at a depth of 1 ft below ground level in expansive soil was subjected to loads from the superstructure. Site investigation revealed that the expansive soil extended to a depth of 8 ft below the base of the foundation, and the moisture contents in the soil during the construction period were at their lowest. In order to determine the percent swell, three undisturbed samples at depths of 2, 4 and 6 ft were collected and swell tests were conducted per the procedure described in Section 18.16. Fig. Ex. 18.3a shows the results of the swell tests plotted against depth. A line passing through the points is drawn. The line indicates that the swell is zero at 8 ft depth and maximum at a base level equal to 3%. Determine (a) the total swell, and (b) the depth of undercut necessary for an allowable swell of 0.03 ft.

Solution

(a) The total swell is equal to the area under the percent swell versus depth curve in Fig. Ex. 18.3a.

Total swell = $1/2 \times 8 \times 3 \times 1/100 = 0.12$ ft

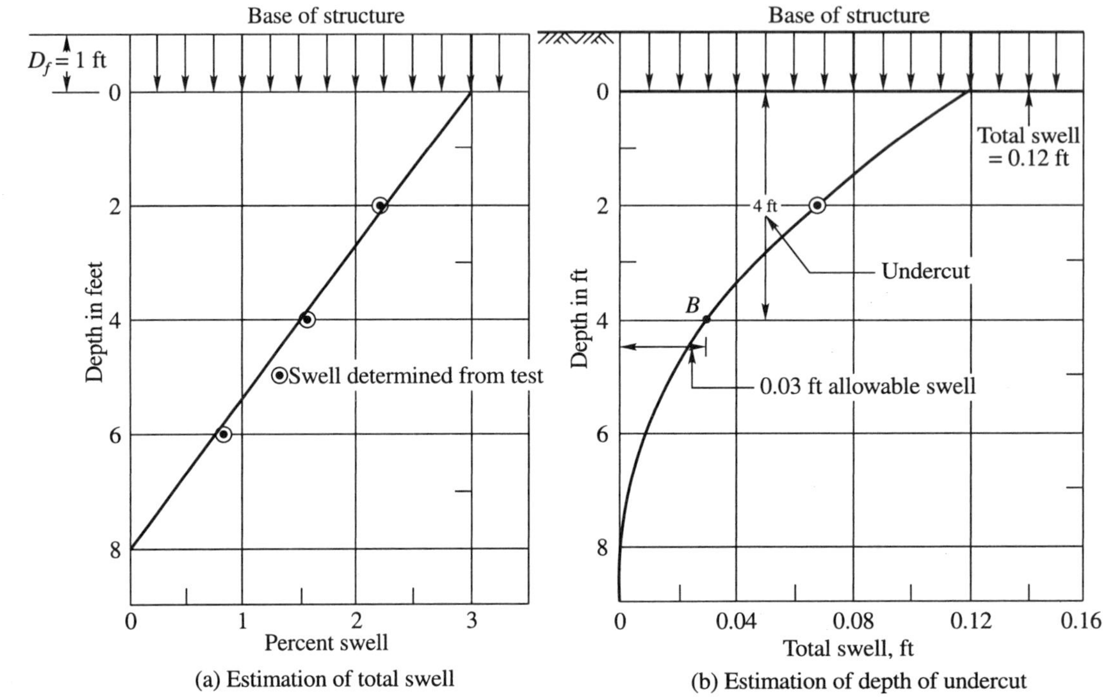

Figure Ex. 18.3

(b) Depth of undercut

From the percent swell versus depth relationship given by the curve in Fig. 18.3a, total swell at different depths are calculated and plotted against depth in Fig. 18.3b. For example the total swell at depth 2 ft below the foundation base is

Total swell = 1/2 (8 – 2) × 2.25 × 1/100 = 0.067 ft plotted against depth 2 ft. Similarly total swell at other depths can be calculated and plotted. Point *B* on curve in Fig. 18.3b gives the allowable swell of 0.03 ft at a depth of 4 ft below foundation base. That is, the undercut necessary in clay is 4 ft which may be replaced by an equivalent thickness of nonswelling compacted fill.

Example 18.4

Fig. Ex. 18.4 shows that the soil to a depth of 20 ft is an expansive type with different degrees of swelling potential. The soil mass to a 20 ft depth is divided into four layers based on the swell potential rating given in Table 18.7. Calculate the total swell per the Van Der Merwe method.

Solution

The procedure for calculating the total swell is explained in Section 18.16. The details of the calculated results are tabulated below.

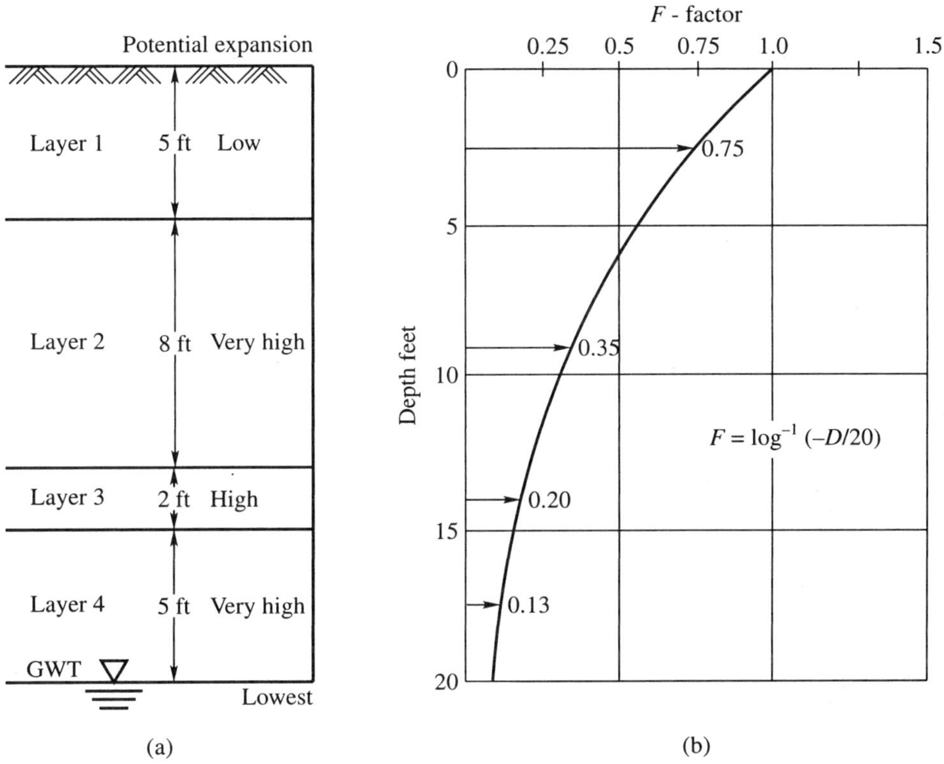

Figure Ex. 18.4

The details of calculated results

Layer No.	Thickness ΔD(ft)	PE	D ft	F	ΔH_s (in.)
1	5	0	2.5	0.75	0
2	8	1.0	9.0	0.35	2.80
3	2	0.5	14.0	0.20	0.20
4	5	1.0	17.5	0.13	0.65
				Total	**3.65**

In the table above D = depth from ground level to the mid-depth of the layer considered. F = reduction factor.

Example 18.5

A drilled pier [refer to Fig. 18.21(a)] was constructed in expansive soil. The water table was not encountered. The details of the pier and soil are:

$L = 20$ ft, $d = 12$ in., $L_1 = 5$ ft, $L_2 = 15$ ft, $p_s = 10{,}000$ lb/ft^2, $c_u = 2089$ lb/ft^2, SPT(N) = 25 blows per foot,

Required:

(a) total uplift capacity Q_{up}

(b) total resisting force due to surface friction

(c) factor of safety without taking into account the dead load Q acting on the top of the pier

(d) factor of safety with the dead load acting on the top of the pier

Assume Q = 10 kips. Calculate Q_{up} by Chen's method (Eq. 18.15).

Solution

(a) Uplift force Q_{up} from Eq. (18.15)

$$Q_{up} = \pi d \alpha_u p_s L_1 = \frac{3.14 \times (1) \times 0.15 \times 10{,}000 \times 5}{1000} = 23.55 \text{ kips}$$

(b) Resisting force Q_R

From Eq. (18.17)

$Q_R = \pi d \, (L - L_1) \, \alpha \, c_u$

where c_u = 2089 lb/ft² ≈ 100 kN/m²

$$\frac{c_u}{p_a} = \frac{100}{101} \approx 1.0 \text{ where } p_a = \text{atmospheric pressure} = 101 \text{ kPa}$$

From Fig. 17.15, $\alpha = 0.55$ for $c_u/p_a \approx 1.0$

Now substituting the known values

$Q_R = 3.14 \times 1 \times (20 - 5) \times 0.55 \times 2000 = 51{,}810$ lb ≈ 52 kips

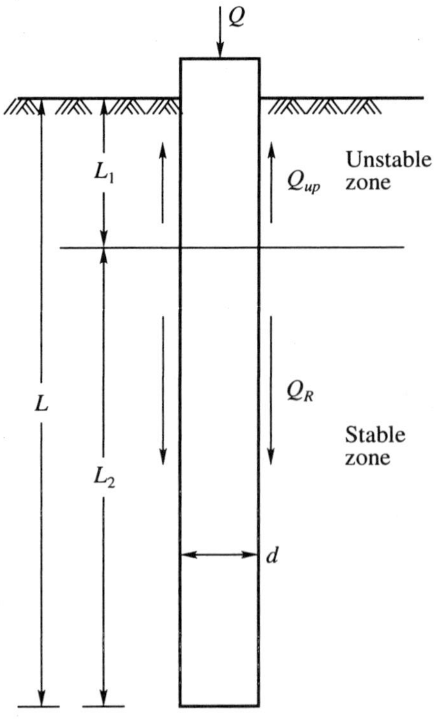

Figure Ex. 18.5

(c) Factor of safety with $Q = 0$
From Eq. (18.18a)

$$F_s = \frac{Q_R}{Q_{up}} = \frac{52}{23.5} = 2.2 > 1.2 \text{ required--OK.}$$

(d) Factor of safety with $Q = 10$ kips
From Eq. (18.18b)

$$F_s = \frac{Q_R}{(Q_{up} - Q)} = \frac{52}{(23.5 - 10)} = \frac{52}{13.5} = 3.9 > 2.0 \text{ required--OK.}$$

Example 18.6

Solve Example 18.5 with $L_1 = 10$ ft. All the other data remain the same.

Solution

(a) Uplift force Q_{up}

$Q_{up} = 23.5 \times (10/5) = 47.0$ kips

where $Q_{up} = 23.5$ kips for $L_1 = 5$ ft

(b) Resisting force Q_R

$Q_R = 52 \times (10/15) \approx 34.7$ kips

where $Q_R = 52$ kips for $L_2 = 15$ ft

(c) Factor of safety for $Q = 0$

$$F_s = \frac{34.7}{47.0} = 0.74 < 1.2 \text{ as required -- not OK.}$$

(d) Factor of safety for $Q = 10$ kips

$$F_s = \frac{34.7}{(47 - 10)} = \frac{34.7}{37} = 0.94 < 2.0 \text{ as required -- not OK.}$$

The above calculations indicate that if the wetting zone (unstable zone) is 10 ft thick the structure will not be stable for $L = 20$ ft.

Example 18.7

Determine the length of pier required in the stable zone for $F_s = 1.2$ where $Q = 0$ and $F_s = 2.0$ when $Q = 10$ kips. All the other data given in Example 18.6 remain the same.

Solution

(a) Uplifting force Q_{up} for L_1 (10 ft) = 47 kips

(b) Resisting force for length L_2 in the stable zone.

$$Q_R = \pi_d \,\alpha\, c_u\, L_2 = 3.14 \times 1 \times 0.55 \times 2000\, L_2 = 3{,}454\, L_2 \text{ lb/ft}^2$$

(c) $Q = 0$, minimum $F_s = 1.2$

$$\text{or } 1.2 = \frac{\alpha_R}{Q_{up}} = \frac{3{,}454\ L_2}{47{,}000}$$

solving we have $L_2 = 16.3$ ft.

(d) $Q = 10$ kips. Minimum $F_s = 2.0$

$$F_s = 2.0 = \frac{Q_R}{(Q_{up} - Q)} = \frac{3{,}454\ L_2}{(47{,}000 - 10{,}000)} = \frac{3{,}454\ L_2}{37{,}000}$$

Solving we have $L_2 = 21.4$ ft.

The above calculations indicate that the minimum $L_2 = 21.4$ ft or say 22 ft is required for the structure to be stable with $L_1 = 10$ ft. The total length $L = 10 + 22 = 32$ ft.

Example 18.8

Figure Ex. 18.8 shows a drilled pier with a belled bottom constructed in expansive soil. The water table is not encountered. The details of the pier and soil are given below:

$L_1 = 10$ ft, $L_2 = 10$ ft, $L_b = 2.5$ ft, $d = 12$ in., $d_b = 3$ ft, $c_u = 2000$ lb/ft^2, $p_s = 10{,}000$ lb/ft^2, $\gamma = 110$ lb/ft^3.

Required

(a) Uplift force Q_{up}
(b) Resisting force Q_R
(c) Factor of safety for $Q = 0$ at the top of the pier
(d) Factor of safety for $Q = 20$ kips at the top of the pier

Solution

(a) Uplift force Q_{up}

As in Ex. 18.6 $Q_{up} = 47$ kips

(b) Resisting force Q_R

$Q_{R1} = \pi d\, L_2\, \alpha\, c_u$

$\alpha = 0.55$ as in Ex. 18.5

Substituting known values

$Q_{R1} = 3.14 \times 1 \times 10 \times 0.55 \times 2000 = 34540$ lbs $= 34.54$ kips

$$Q_{R2} = \frac{\pi}{4}\left[d_b^2 - d^2\right]\left[cN_c + \gamma\ L_2\right]$$

where $d_b = 3$ ft, $c = 2000$ lb/ft^2, $N_c = 7.0$ from Table 18.8 for $L_2/d_b = 10/3 = 3.33$

Substituting known values

$$Q_{R2} = \frac{3.14}{4}\left[3^2 - (1)^2\right]\left[2000 \times 7.0 + 110 \times 10\right]$$

$= 6.28\ [15{,}100] = 94{,}828$ lbs $= 94.8$ kips

$Q_R = Q_{R1} + Q_{R2} = 34.54 + 94.8 = 129.\ 3$ kips

(c) Factor of safety for $Q = 0$

$$F_s = \frac{Q_R}{Q_{up}} = \frac{129.3}{47.0} = 2.75 > 1.2 \text{ --OK}$$

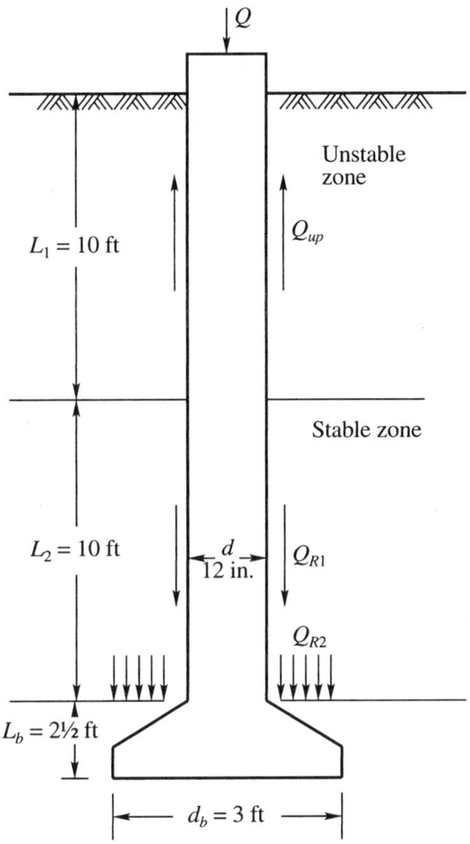

Figure Ex. 18.8

(d) Factor of safety for $Q = 20$ kips

$$F_s = \frac{Q_R}{(Q_{up} - Q)} = \frac{129.3}{(47 - 20)} = 4.79 > 2.0 \quad \text{-- as required OK.}$$

The above calculations indicate that the design is over conservative. The length L_2 can be reduced to provide an acceptable factor of safety.

18.18 ELIMINATION OF SWELLING

The elimination of foundation swelling can be achieved in two ways. They are

1. Providing a granular bed and cover below and around the foundation (Fig. 18.19)
2. Chemical stabilization of swelling soils

Figure 18.19 gives a typical example of the first type. In this case, the excavation is carried out up to a depth greater than the width of the foundation by about 20 to 30 cms. Freely draining soil, such as a mixture of sand and gravel, is placed and compacted up to the base level of the foundation. A Reinforced concrete footing is constructed at this level. A mixture of sand and gravel is filled up loosely over the fill. A reinforced concrete apron about 2 m wide is provided around the building to prevent moisture directly entering the foundation. A cushion of granular soils below the foundation absorbs the effect of swelling, and thereby its effect on the foundation will considerably be reduced. A foundation of this type should be constructed only during the dry season when the soil has shrunk to its lowest level. Arrangements should be made to drain away the water from the granular base during the rainy seasons.

Chemical stabilization of swelling soils by the addition of lime may be remarkably effective if the lime can be mixed thoroughly with the soil and compacted at about the optimum moisture content. The appropriate percentage usually ranges from about 3 to 8 percent. The lime content is estimated on the basis of pH tests and checked by compacting, curing and testing samples in the laboratory. The lime has the effect of reducing the plasticity of the soil, and hence its swelling potential.

18.19 PROBLEMS

18.1 A building was constructed in a loessial type normally consolidated collapsible soil with the foundation at a depth of 1 m below ground level. The soil to a depth of 6 m below the foundation was found to be collapsible on flooding. The average overburden pressure was 56 kN/m^2. Double consolidometer tests were conducted on two undisturbed samples taken at a depth of 4 m below ground level, one with its natural moisture content and the other under soaked conditions per the procedure explained in Section 18.4. The following data were available.

Applied pressure, kN/m^2	10	20	40	100	200	400	800
Void ratios at natural moisture content	0.80	0.79	0.78	0.75	0.725	0.68	0.61
Void ratios in the soaked condition	0.75	0.71	0.66	0.58	0.51	0.43	0.32

Plot the e-log p curves and determine the collapsible settlement for an increase in pressure Δp = 34 kN/m^2 at the middle of the collapsible stratum.

18.2 Soil investigation at a site indicated overconsolidated collapsible loessial soil extending to a great depth. It is required to construct a footing at the site founded at a depth of 1.0 m below ground level. The site is subject to flooding. The average unit weight of the soil is 19.5 kN/m^3. Two oedometer tests were conducted on two undisturbed samples taken at a depth of 5 m from ground level. One test was conducted at its natural moisture content and the other on a soaked condition per the procedure explained in Section 18.4.

The following test results are available.

Applied pressure, kN/m^2	10	20	40	100	200	400	800	2000
Void ratio under natural moisture condition	0.795	0.79	0.787	0.78	0.77	0.74	0.71	0.64
Void ratio under soaked condition	0.775	0.77	0.757	0.730	0.68	0.63	0.54	0.37

The swell index determined from the rebound curve of the soaked sample is equal to 0.08.

Required:

(a) Plots of e-log p curves for both tests.

(b) Determination of the average overburden pressure at the middle of the soil stratum.

(c) Determination of the preconsolidation pressure based on the curve obtained from the soaked sample

(d) Total collapse settlement for an increase in pressure Δp = 710 kN/m^2?

18.3 A footing for a building is founded 0.5 m below ground level in an expansive clay stratum which extends to a great depth. Swell tests were conducted on three undisturbed samples taken at different depths and the details of the tests are given below.

Depth (m) below GL	Swell %
1	2.9
2	1.75
3	0.63

Required:

(a) The total swell under structural loadings

(b) Depth of undercut for an allowable swell of 1 cm

18.4 Fig. Prob. 18.4 gives the profile of an expansive soil with varying degrees of swelling potential. Calculate the total swell per the Van Der Merwe method.

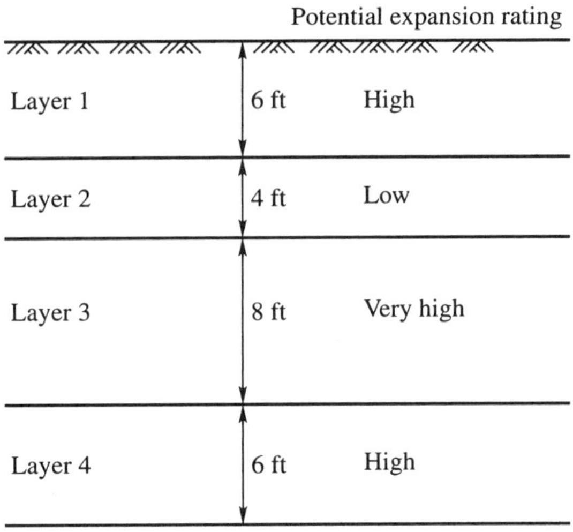

Figure Prob. 18.4

18.5 Fig. Prob. 18.5 depicts a drilled pier embedded in expansive soil. The details of the pier and soil properties are given in the figure.

Determine:

(a) The total uplift capacity.

(b) Total resisting force.

(c) Factor of safety with no load acting on the top of the pier.

(d) Factor of safety with a dead load of 100 kN on the top of the pier.

Calculate Q_{up} by Chen's method.

18.6 Solve problem 18.5 using Eq. (18.16)

18.7 Fig. Prob. 18.7 shows a drilled pier with a belled bottom. All the particulars of the pier and soil are given in the figure.

Required:

(a) The total uplift force.

(b) The total resisting force

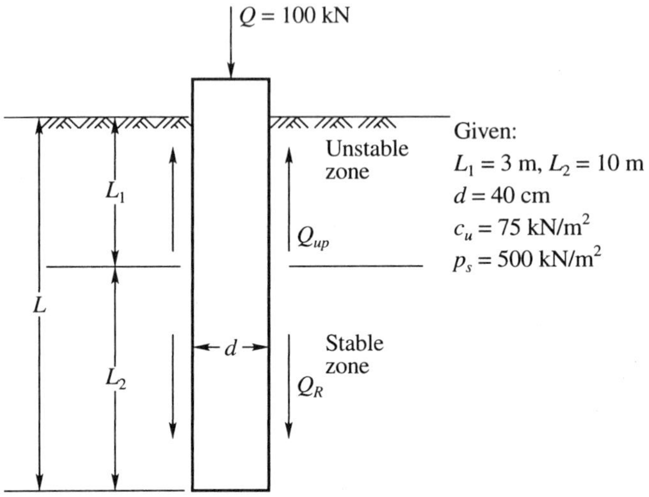

Figure Prob. 18.5

(c) Factor of safety for $Q = 0$

(d) Factor of safety for Q = 200 kN.

Use Chen's method for computing Q_{up}.

18.8 Solve Prob. 18.7 by making use of Eq. (18.16) for computing Q_{up}.

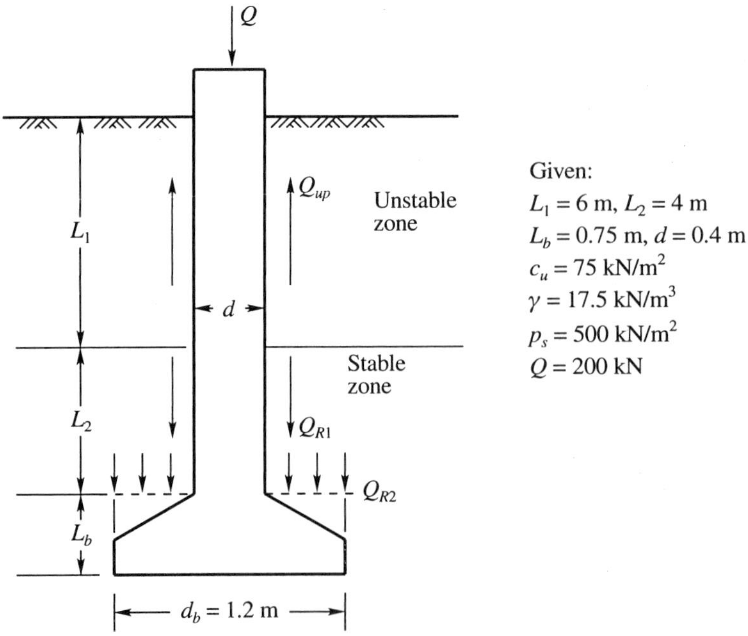

Figure Prob. 18.7

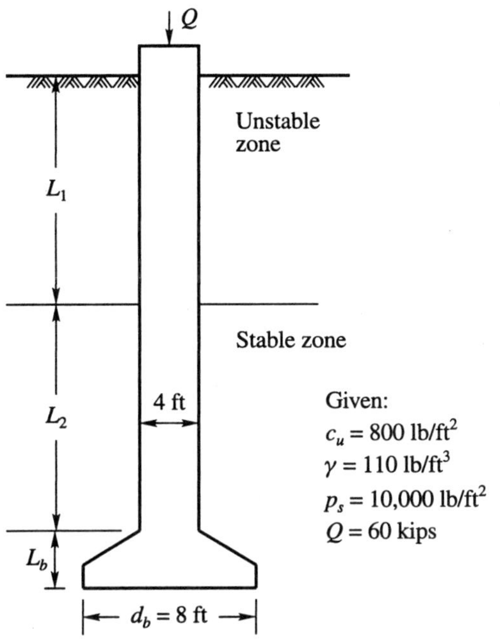

Figure Prob. 18.9

18.9 Refer to Fig. Prob. 18.9. The following data are available:
$L_1 = 15$ ft, $L_2 = 13$ ft, $d = 4$ ft, $d_b = 8$ ft and $L_b = 6$ ft.
All the other data are given in the figure.
Required:
(a) The total uplift force
(b) The total resisting force
(c) Factor of safety for $Q = 0$
(d) Factor of safety for $Q = 60$ kips.
Use Chen's method for computing Q_{up}.

18.10 Solve Prob. 18.9 using Eq. (18.16) for computing Q_{up}.

18.11 If the length L_2 is not sufficient in Prob. 18.10, determine the required length to get $F_s = 3.0$.

CHAPTER 19

CONCRETE AND MECHANICALLY STABILIZED EARTH RETAINING WALLS

PART A—CONCRETE RETAINING WALLS

19.1 INTRODUCTION

The common types of concrete retaining walls and their uses were discussed in Chapter 11. The lateral pressure theories and the methods of calculating the lateral earth pressures were described in detail in the same chapter. The two classical earth pressure theories that have been considered are those of Rankine and Coulomb. In this chapter we are interested in the following:

1. Conditions under which the theories of Rankine and Coulomb are applicable to cantilever and gravity retaining walls under the active state.
2. The common minimum dimensions used for the two types of retaining walls mentioned above.
3. Use of charts for the computation of active earth pressure.
4. Stability of retaining walls.
5. Drainage provisions for retaining walls.

19.2 CONDITIONS UNDER WHICH RANKINE AND COULOMB FORMULAS ARE APPLICABLE TO RETAINING WALLS UNDER ACTIVE STATE

Conjugate Failure Planes Under Active State

When a backfill of cohesionless soil is under an active state of plastic equilibrium due to the stretching of the soil mass at every point in the mass, two failure planes called *conjugate rupture planes* are formed. These are further designated as the *inner failure plane* and the *outer failure plane* as shown in Fig. 19.1. These failure planes make angles of α_i and α_0 with the vertical. The equations for these angles may be written as (for a sloping backfill)

$$\alpha_i = \frac{90-\phi}{2} + \frac{\varepsilon-\beta}{2} \tag{19.1a}$$

$$\alpha_0 = \frac{90-\phi}{2} - \frac{\varepsilon-\beta}{2} \tag{19.1b}$$

$$\text{where} \quad \sin\varepsilon = \frac{\sin\beta}{\sin\phi} \tag{19.2}$$

$$\text{when} \quad \beta = 0, \quad \alpha_i = \frac{90-\phi}{2} = 45° - \frac{\phi}{2}, \quad \alpha_0 = \frac{90-\phi}{2} = 45° - \frac{\phi}{2}$$

The angle between the two failure planes = $90 - \phi$.

Conditions for the Use of Rankine's Formula

1. Wall should be vertical with a smooth pressure face.
2. When walls are inclined, it should not come in the way of the formation of the outer failure plane. Figure 19.1 shows the formation of failure planes. Since the sloping face AB′ of the retaining wall makes an angle α_w greater than α_o, the wall does not interfere with the formation of the outer failure plane. The plastic state exists within wedge *ACC′*.

The method of calculating the lateral pressure on *AB′* is as follows.

1. Apply Rankine's formula for the vertical section *AB*.
2. Combine P_a with W_s, the weight of soil within the wedge *ABB′*, to give the resultant P_R.

Let the resultant P_R in this case make an angle δ_r with the normal to the face of the wall. Let the maximum angle of wall friction be δ_m. If $\delta_r > \delta_m$, the soil slides along the face *AB′* of the wall.

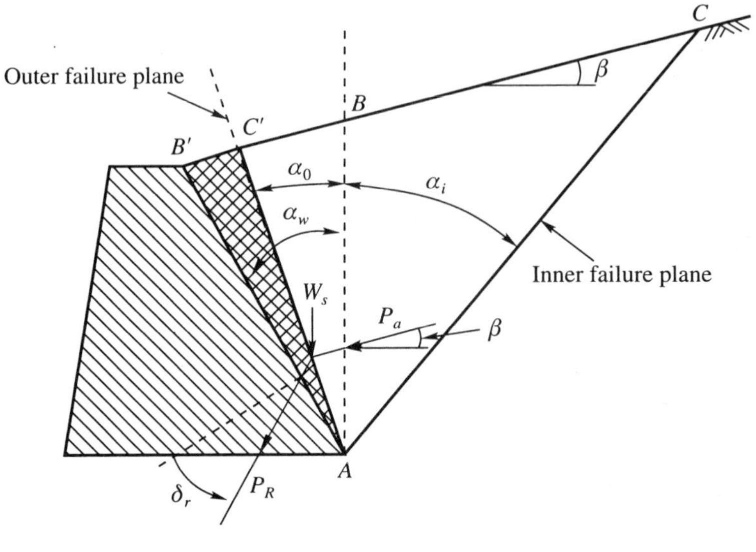

Figure 19.1 Application of Rankine's active condition to gravity walls

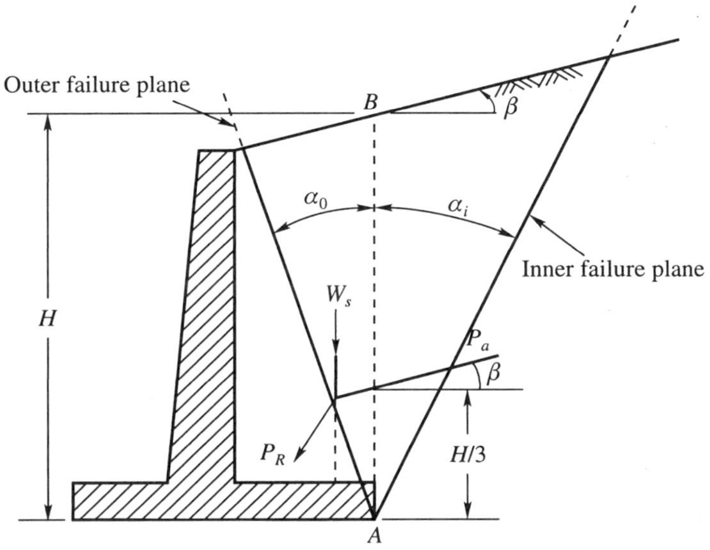

Figure 19.2 Lateral earth pressure on cantilever walls under active condition

In such an eventuality, the Rankine formula is not recommended but the Coulomb formula may be used.

Conditions for the Use of Coulomb's Formula

1. The back of the wall must be plane or nearly plane.
2. Coulomb's formula may be applied under all other conditions where the surface of the wall is not smooth and where the soil slides along the surface.

In general the following recommendations may be made for the application of the Rankine or Coulomb formula without the introduction of significant errors:

1. Use the Rankine formula for cantilever and counterfort walls.
2. Use the Coulomb formula for solid and semisolid gravity walls.

In the case of cantilever walls (Fig. 19.2), P_a is the active pressure acting on the vertical section *AB* passing through the heel of the wall. The pressure is parallel to the backfill surface and acts at a height *H*/3 from the base of the wall where *H* is the height of the section *AB*. The resultant pressure P_R is obtained by combining the lateral pressure P_a with the weight of the soil W_s between the section *AB* and the wall.

19.3 PROPORTIONING OF RETAINING WALLS

Based on practical experience, retaining walls can be proportioned initially which may be checked for stability subsequently. The common dimensions used for the various types of retaining walls are given below.

Gravity Walls

A gravity walls may be proportioned in terms of its height given in Fig. 19.3(a). The minimum top width suggested is 0.30 m. The tentative dimensions for a cantilever wall are given in Fig. 19.3(b) and those for a counterfort wall are given in Fig. 19.3(c).

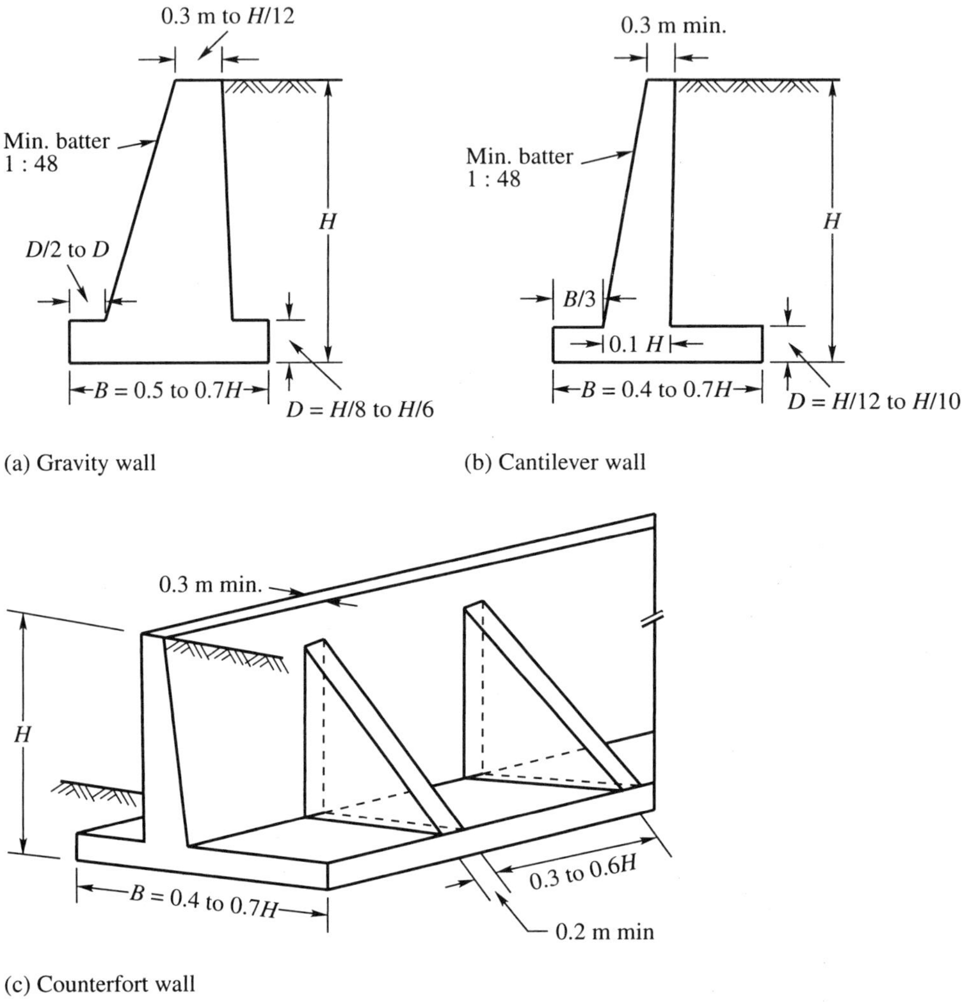

Fig. 19.3 Tentative dimensions for retaining walls

19.4 EARTH PRESSURE CHARTS FOR RETAINING WALLS

Charts have been developed for estimating lateral earth pressures on retaining walls based on certain assumed soil properties of the backfill materials. These semi empirical methods represent a body of valuable experience and summarize much useful information. The charts given in Fig. 19.4 are meant to produce a design of retaining walls of heights not greater than 6 m. The charts have been developed for five types of backfill materials given in Table 19.1. The charts are applicable to the following categories of backfill surfaces. They are

1. The surface of the backfill is plane and carries no surcharge
2. The surface of the backfill rises on a slope from the crest of the wall to a level at some elevation above the crest.

The chart is drawn to represent a concrete wall but it may also be used for a reinforced soil wall. All the dimensions of the retaining walls are given in Fig. 19.4. The total horizontal and vertical pressures on the vertical section of $A\ B$ of height H are expressed as

$$P_h = \frac{1}{2} K_h H^2 \tag{19.3}$$

Table 19.1 Types of backfill for retaining walls

Type	Backfill material
1	Coarse-grained soil without admixture of fine soil particles, very permeable (clean sand or gravel)
2	Coarse-grained soil of low permeability due to admixture of particles of silt size
3	Residual soil with stones fine silty sand, and granular materials with conspicuous clay content
4	Very soft or soft clay, organic silts, or silty clays
5	Medium or stiff clay

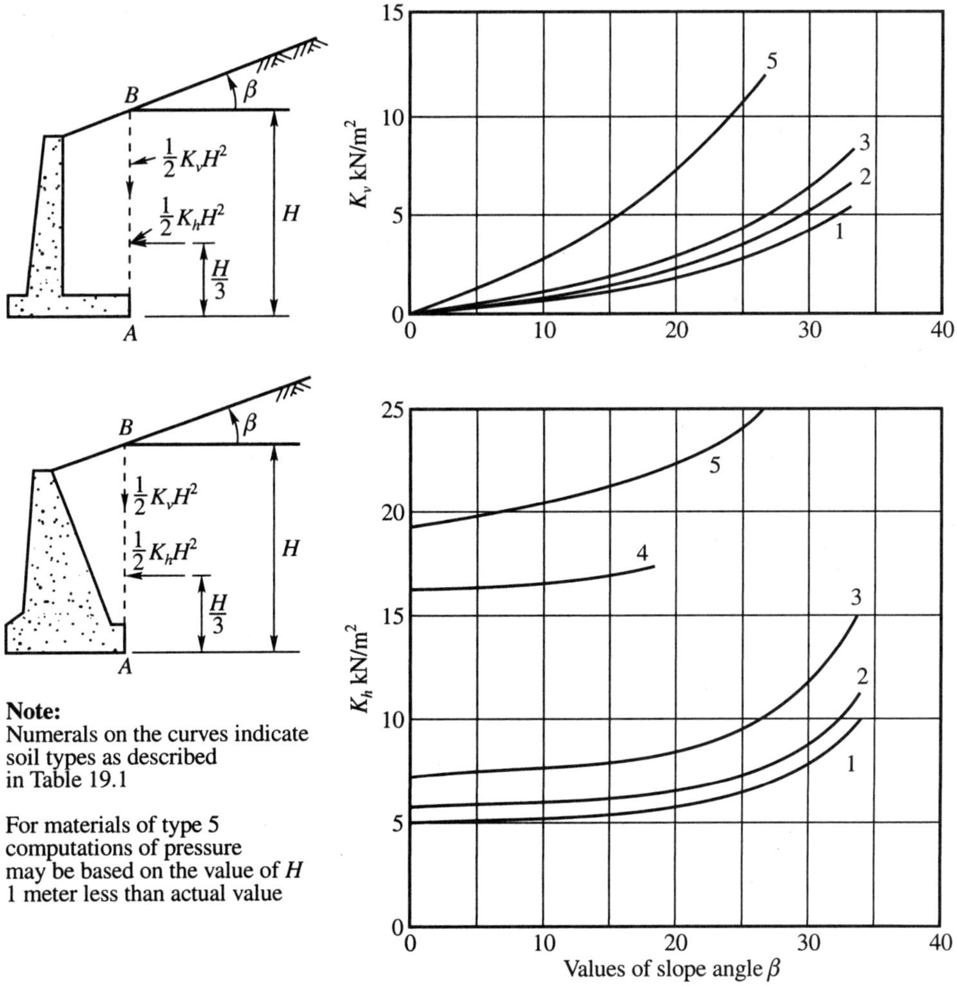

Figure 19.4 Chart for estimating pressure of backfill against retaining walls supporting backfills with a plane surface. (Terzaghi, Peck, and Mesri, 1996)

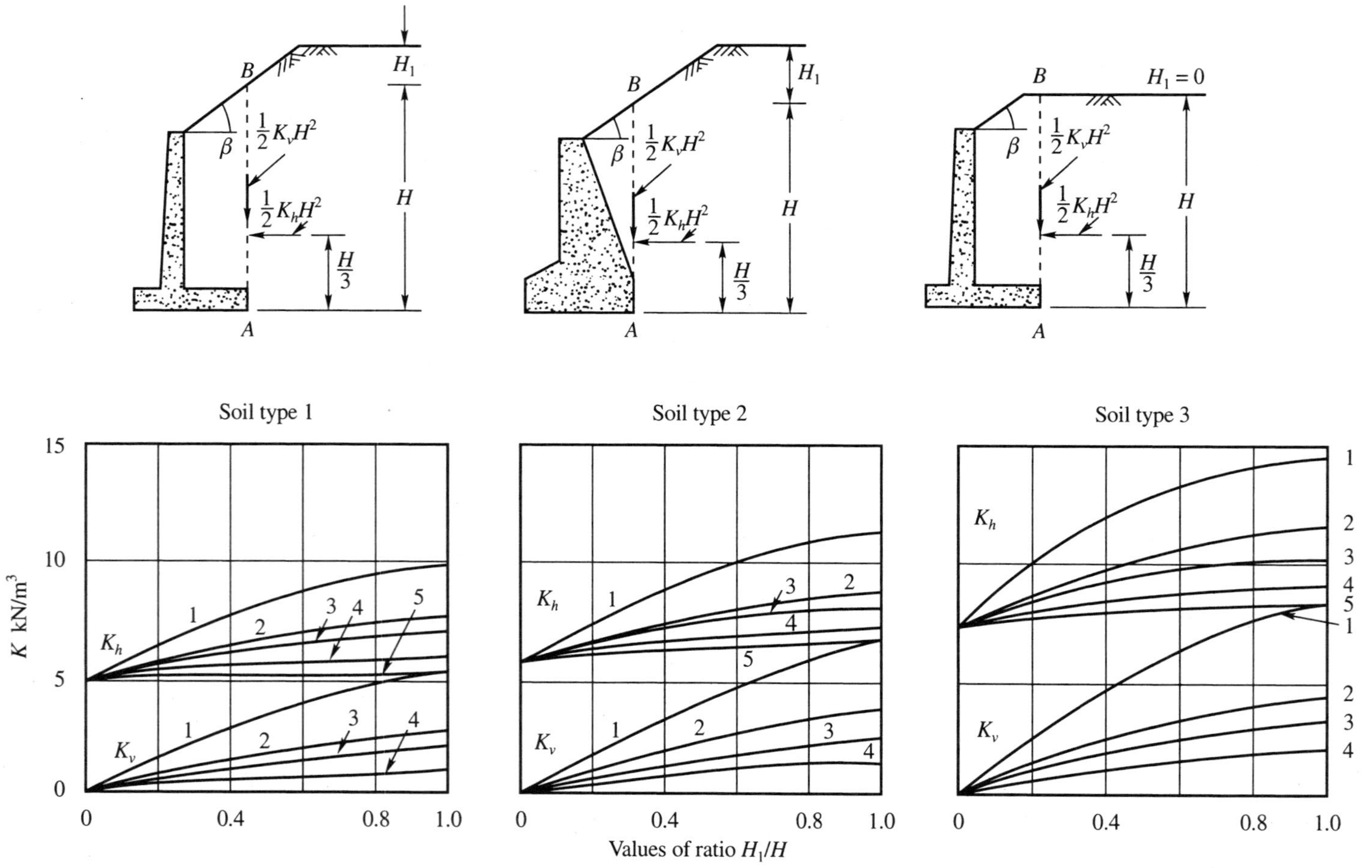

Figure 19.5 Chart for estimating pressure of backfill against retaining walls supporting backfills with a surface that slopes upward from the crest of the wall for limited distance and then becomes horizontal. (Terzaghi et al., 1996)

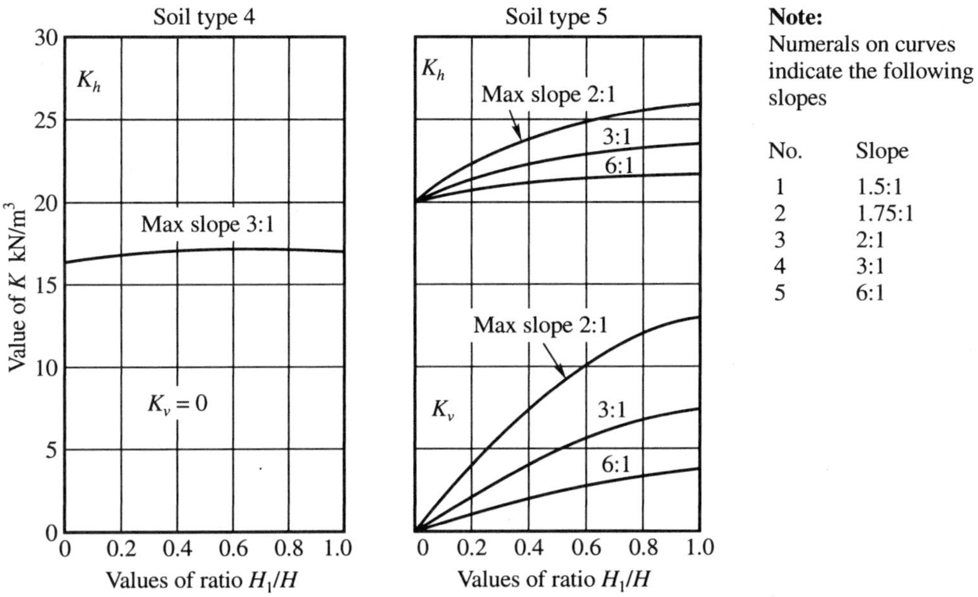

Figure 19.5 Continued

$$P_v = \frac{1}{2} K_v H^2 \tag{19.4}$$

Values of K_h and K_v are plotted against slope angle β in Fig. 19.4 and the ratio H_1/H in Fig. 19.5.

19.5 STABILITY OF RETAINING WALLS

The stability of retaining walls should be checked for the following conditions:

1. Check for sliding
2. Check for overturning
3. Check for bearing capacity failure
4. Check for base shear failure

The minimum factors of safety for the stability of the wall are:

1. Factor of safety against sliding = 1.5
2. Factor of safety against overturning = 2.0
3. Factor of safety against bearing capacity failure = 3.0

Stability Analysis

Consider a cantilever wall with a sloping backfill for the purpose of analysis. The same principle holds for the other types of walls.

Fig. 19.6 gives a cantilever wall with all the forces acting on the wall and the base, where

P_a = active earth pressure acting at a height $H/3$ over the base on section AB
P_h = $P_a \cos \beta$
P_v = $P_a \sin \beta$
β = slope angle of the backfill

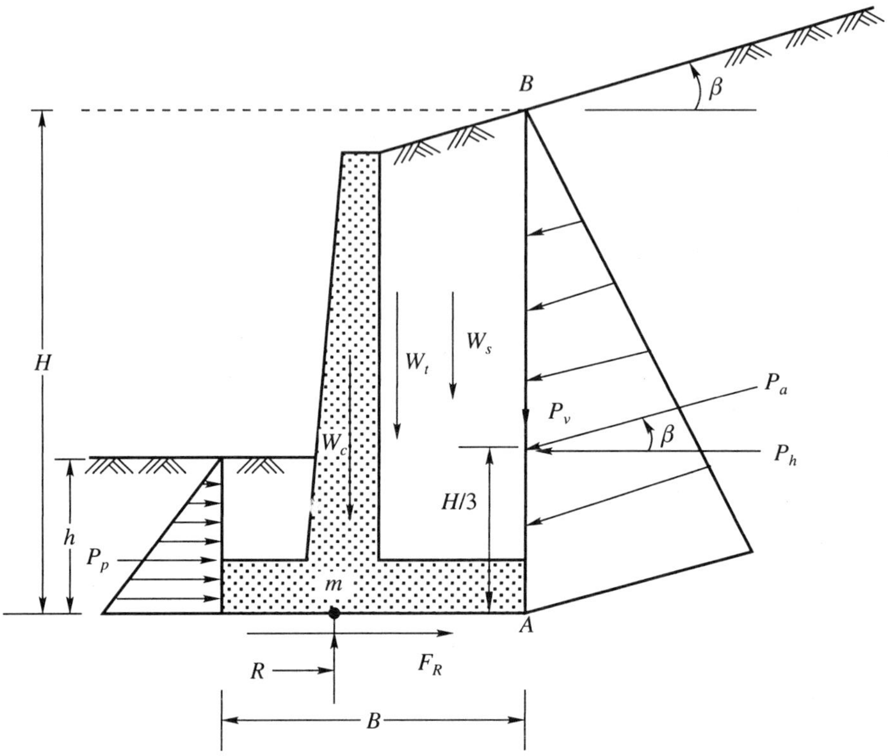

(a) Forces acting on the wall

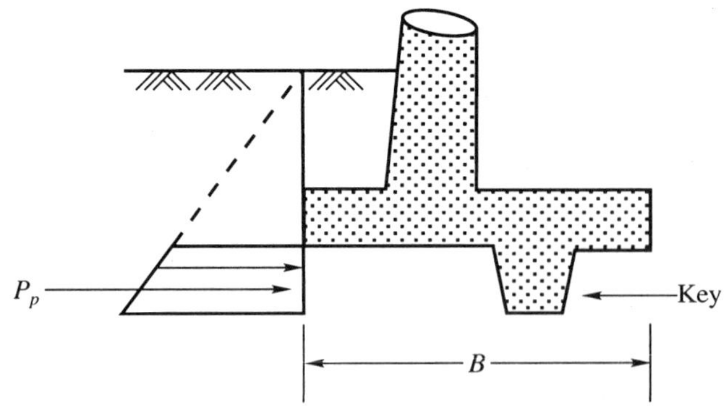

(b) Provision of key to increase sliding resistance

Figure 19.6 Check for sliding

W_s	=	weight of soil
W_c	=	weight of wall including base
W_t	=	the resultant of W_s and W_c
P_P	=	passive earth pressure at the toe side of the wall.
F_R	=	base sliding resistance

Check for Sliding (Fig. 19.6)

The force that moves the wall = horizontal force P_h

The force that resists the movement is

$$F_R = c_a B + R \tan \delta + P_p \tag{19.5}$$

R = total vertical force = $W_s + W_c + P_v$,

δ = angle of wall friction

c_a = unit adhesion

If the bottom of the base slab is rough, as in the case of concrete poured directly on soil, the coefficient of friction is equal to tan ϕ, ϕ being the angle of internal friction of the soil.

The factor of safety against sliding is

$$F_s = \frac{F_R}{P_h} \geq 1.5 \tag{19.6}$$

In case $F_s < 1.5$, additional factor of safety can be provided by constructing one or two keys at the base level shown in Fig. 19.6b. The passive pressure P_p (Fig. 19.6a) in front of the wall should not be relied upon unless it is certain that the soil will always remain firm and undisturbed.

Check for Overturning

The forces acting on the wall are shown in Fig. 19.7. The overturning and stabilizing moments may be calculated by taking moments about point O. The factor of safety against overturning is therefore

$$F_o = \frac{\text{Sum of moments that resist overturning}}{\text{Sum of overturning moments}} = \frac{M_R}{M_o} \tag{19.7a}$$

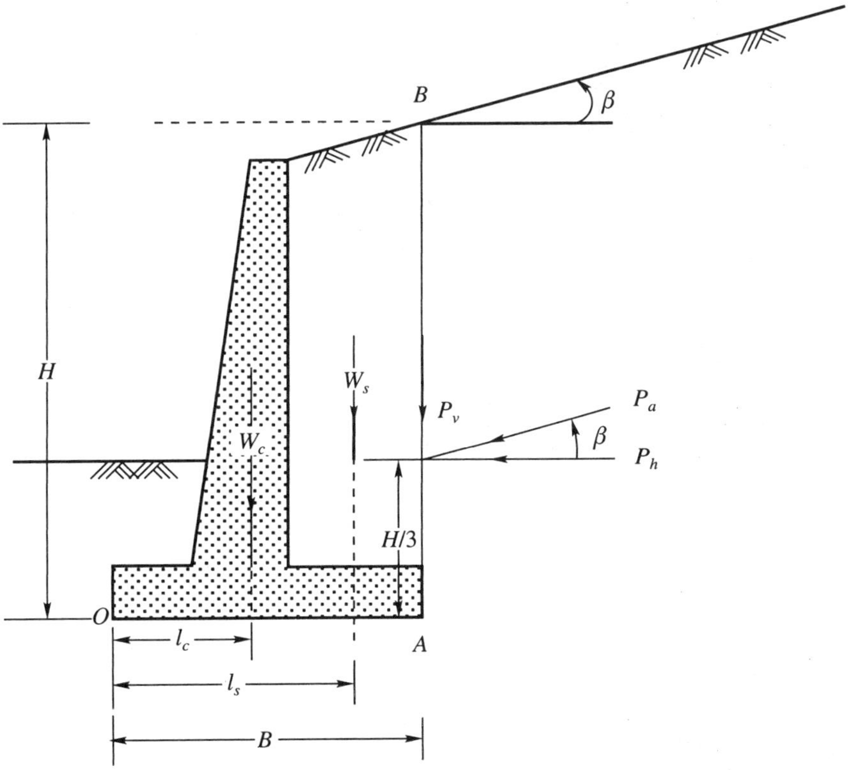

Figure 19.7 Check for overturning

we may write (Fig. 19.7)

$$F_o = \frac{W_c l_c + W_s l_s + P_v B}{P_h (H/3)} \quad (19.7b)$$

where F_o should not be less than 2.0.

Check for Bearing Capacity Failure (Fig. 19.8)

In Fig. 19.8, W_t is the resultant of W_s and W_c. P_R is the resultant of P_a and W_t and P_R meets the base at m. R is the resultant of all the vertical forces acting at m with an eccentricity e. Fig. 19.8 shows the pressure distribution at the base with a maximum q_t at the toe and a minimum q_h at the heel.

An expression for e may be written as

$$e = \frac{B}{2} - \frac{(M_R - M_o)}{\Sigma V} \quad (19.8a)$$

where $R = \Sigma V$ = sum of all vertical forces

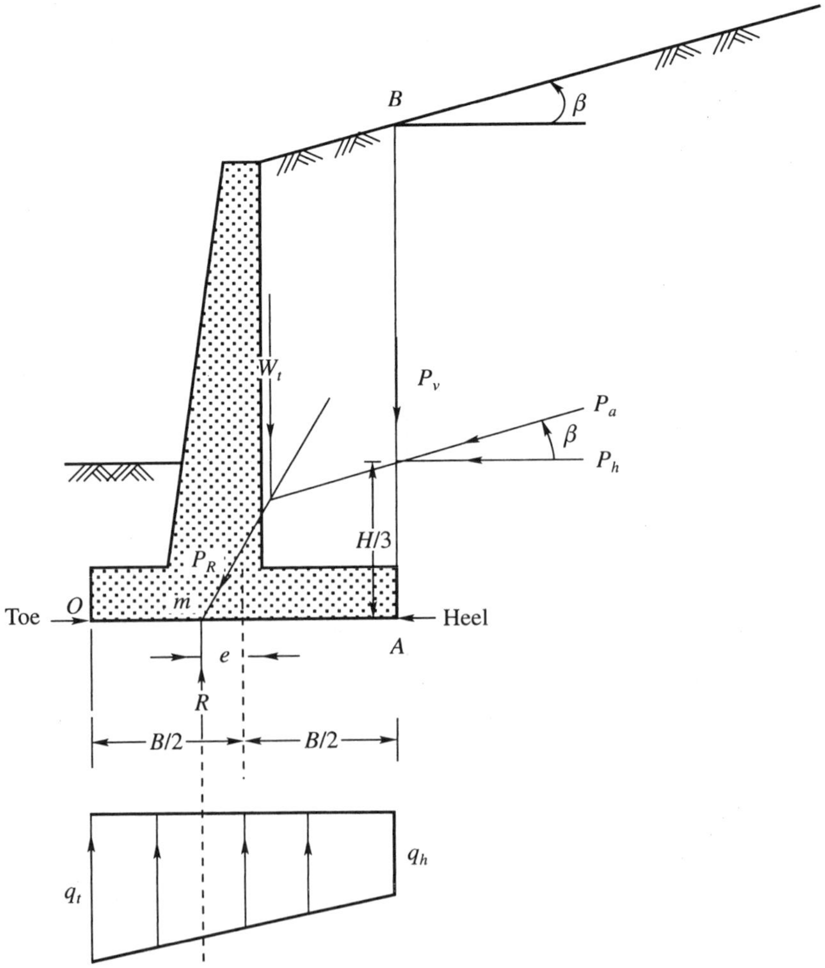

Figure 19.8 Stability against bearing capacity failure

The values of q_t and q_h may be calculated by making use of the equations

$$q_t = \frac{R}{B}\left[1+\frac{6e}{B}\right] = q_a\left[1+\frac{6e}{B}\right] \tag{19.8b}$$

$$q_h = q_a\left[1-\frac{6e}{B}\right] \tag{19.8c}$$

where, q_a = R/B = allowable bearing pressure.

Equation (19.8) is valid for $e \leq$ B/6. When e = B/6, $q_t = 2q_a$ and $q_h = 0$. The base width B should be adjusted to satisfy Eq. (19.8) . When the subsoil below the base is of a low bearing capacity, the possible alternative is to use a pile foundation.

The ultimate bearing capacity q_u may be determined using Eq. (12.27) taking into account the eccentricity. It must be ensured that

$$q_t \leq \frac{q_u}{F_s} = \frac{q_u}{3}$$

Base Failure of Foundation (Fig. 19.9)

If the base soil consists of medium to soft clay, a circular slip surface failure may develop as shown in Fig. 19.9. The most dangerous slip circle is actually the one that penetrates deepest into the soft material. The critical slip surface must be located by trial. Such stability problems may be analyzed either by the method of slices or any other method discussed in Chapter 10.

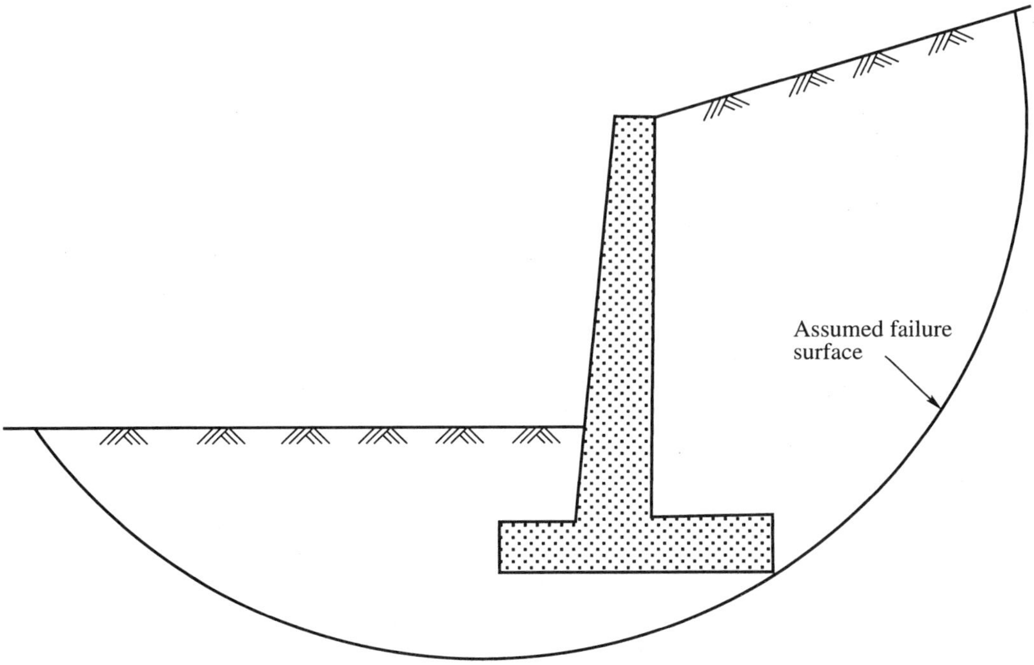

Figure 19.9 Stability against base slip surface shear failure

Drainage Provision for Retaining Walls (Fig. 19.10)

The saturation of the backfill of a retaining wall is always accompanied by a substantial hydrostatic pressure on the back of the wall. Saturation of the soil increases the earth pressure by increasing the unit weight. It is therefore essential to eliminate or reduce pore pressure by providing suitable drainage. Four types of drainage are given in Fig. 19.10. The drains collect the water that enters the backfill and this may be disposed of through outlets in the wall called weep holes. The graded filter material should be properly designed to prevent clogging by fine materials. The present practice is to use geotextiles or geogrids.

The weep holes are usually made by embedding 100 mm (4 in.) diameter pipes in the wall as shown in Fig. 19.10. The vertical spacing between horizontal rows of weep holes should not exceed 1.5 m. The horizontal spacing in a given row depends upon the provisions made to direct the seepage water towards the weep holes.

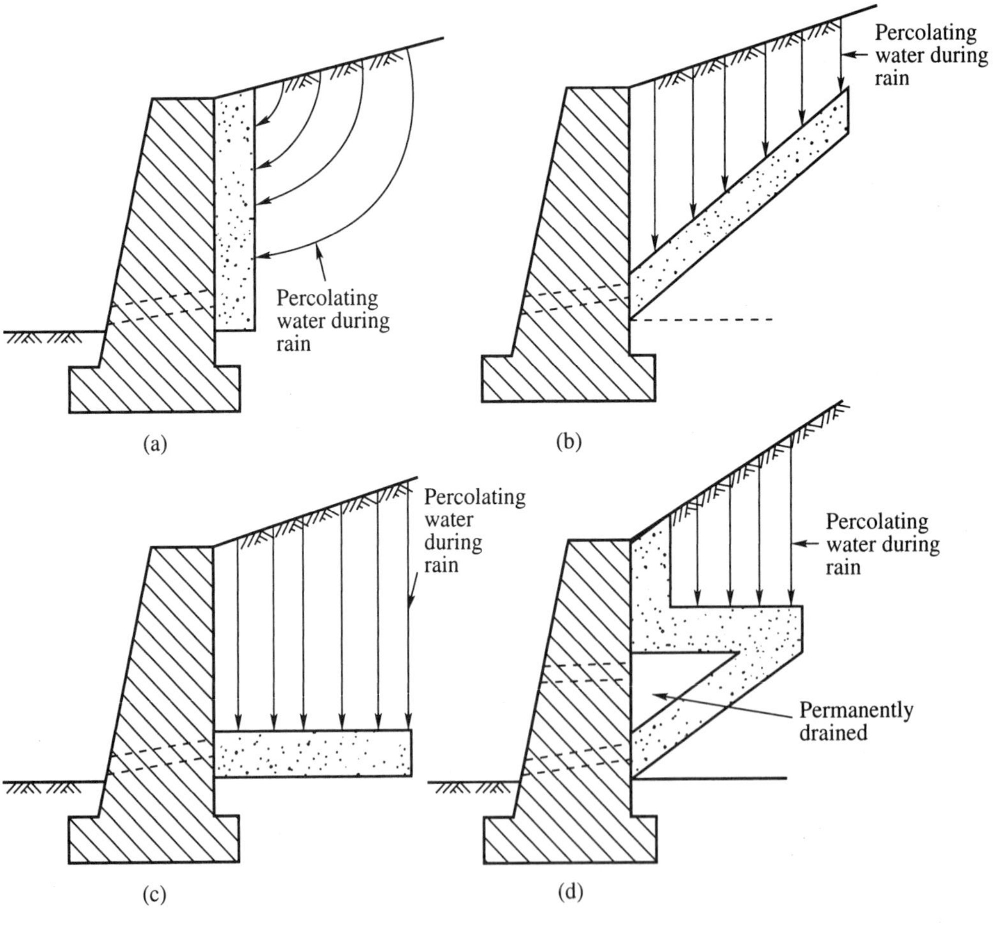

Figure 19.10 Diagram showing provisions for drainage of backfill behind retaining walls: (a) vertical drainage layer (b) inclined drainage layer for cohesionless backfill, (c) bottom drain to accelerate consolidation of cohesive back fill, (d) horizontal drain and seal combined with inclined drainage layer for cohesive backfill (Terzaghi et al., 1996)

Example 19.1

Figure Ex. 19.1(a) shows a section of a cantilever wall with dimensions and forces acting thereon. Check the stability of the wall with respect to (a) overturning, (b) sliding, and (c) bearing capacity.

Solution

Check for Rankine's condition

From Eq. (19.1b)

$$\alpha_o = \frac{90-\phi}{2} - \frac{\varepsilon-\beta}{2}$$

where $\sin\varepsilon = \frac{\sin\beta}{\sin\phi} = \frac{\sin 15°}{\sin 30°} = 0.5176$

or $\varepsilon \approx 31°$

$$\alpha_0 = \frac{90-30}{2} - \frac{31-15}{2} = 22°$$

The outer failure line AC is drawn making an angle 22° with the vertical AB. Since this line does not cut the wall Rankine's condition is valid in this case.

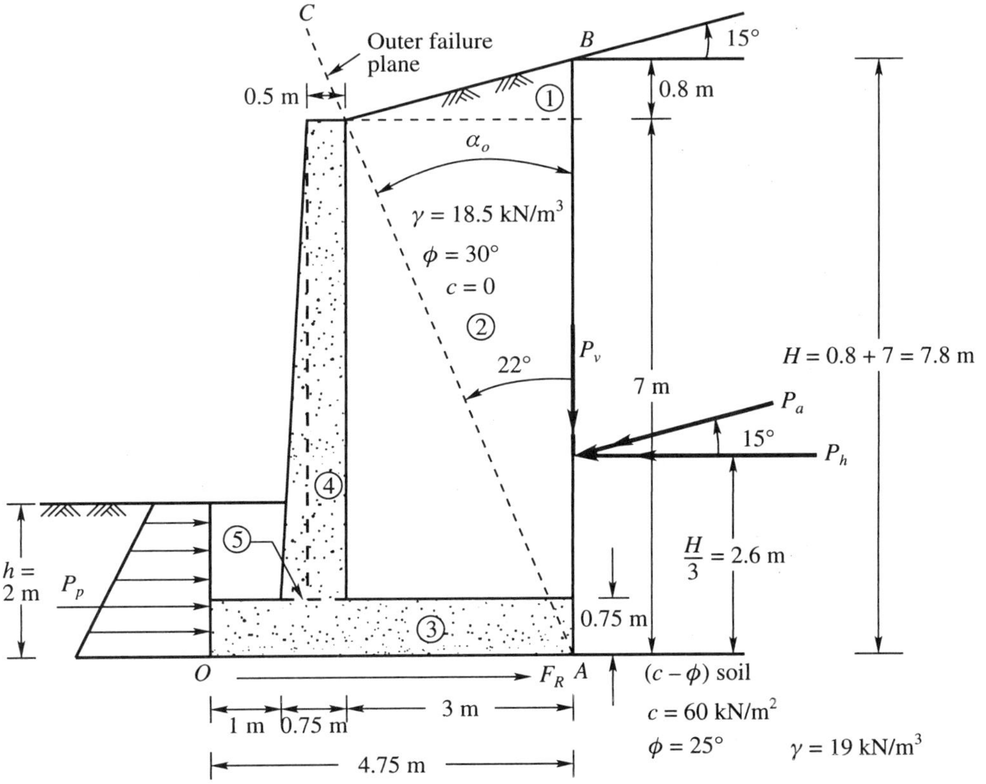

Figure Ex. 19.1(a)

Rankine active pressure

Height of wall = $AB = H = 7.8$ m (Fig. Ex. 19.1(a))

$$P_a = \frac{1}{2}\gamma H^2 K_A$$

where $K_A = \tan^2(45° - \phi/2) = \frac{1}{3}$

substituting

$$P_a = \frac{1}{2} \times 18.5 \times (7.8)^2 \times \frac{1}{3} = 187.6 \text{ kN/m of wall}$$

$$P_h = P_a \cos\beta = 187.6 \cos 15° = 181.2 \text{ kN/m}$$

$$P_v = P_a \sin\beta = 187.6 \sin 15° = 48.6 \text{ kN/m}$$

Check for overturning

The forces acting on the wall in Fig. Ex. 19.1(a) are shown. The overturning and stabilizing moments may be calculated by taking moments about point O.

The whole section is divided into 5 parts as shown in the figure. Let these forces be represented by $w_1, w_2, \ldots w_5$ and the corresponding lever arms as $l_1, l_2, \ldots l_5$. Assume the weight of concrete $\gamma_c = 24$ kN/m^3. The equation for the resisting moment is

$$M_R = w_1 l_1 + w_2 l_2 + \ldots w_5 l_5$$

The overturning moment is

$$M_o = P_h \frac{H}{3}$$

The details of calculations are tabulated below.

Section No.	Area (m^2)	Unit weight kN/m^3	Weight kN/m	Lever arm(m)	Moment kN-m
1	1.20	18.5	22.2	3.75	83.25
2	18.75	18.5	346.9	3.25	1127.40
3	3.56	24.0	85.4	2.38	203.25
4	3.13	24.0	75.1	1.50	112.65
5	0.78	24.0	18.7	1.17	21.88
			$P_v = 48.6$	4.75	230.85
			$\Sigma_v = 596.9$		$\Sigma_M = 1{,}779.3 = M_R$

$$M_O = 181.2 \times 2.6 = 471.12 \text{ kN–m}$$

$$F_s = \frac{M_R}{M_O} = \frac{1{,}779.3}{471.12} = 3.78 > 2.0 \quad \text{-- OK.}$$

Check for sliding (Fig. 19.1a)

The force that resists the movement as per Eq. (19.5) is

$$F_R = c_a B + R \tan\delta + P_p$$

where B = width = 4.75 m

$c_a = \alpha c_u$, α = adhesion factor = 0.55 from Fig. 17.15

R = total vertical force Σ_V = 596.9 kN

For the foundation soil:

δ = angle of wall friction $\approx \phi = 25°$

From Eq. (11.45c)

$$P_p = \frac{1}{2}\gamma h^2 K_p + 2ch\sqrt{K_p}$$

where h = 2 m, γ = 19 kN/m^3, c = 60 kN/m^2

$K_p = \tan^2(45° + \phi/2) = \tan^2(45° + 25/2) = 2.46$

substituting

$$P_p = \frac{1}{2}\times 19\times 2^2\times 2.46 + 2\times 60\times 2\times\sqrt{2.46} = 470 \text{ kN/m}$$

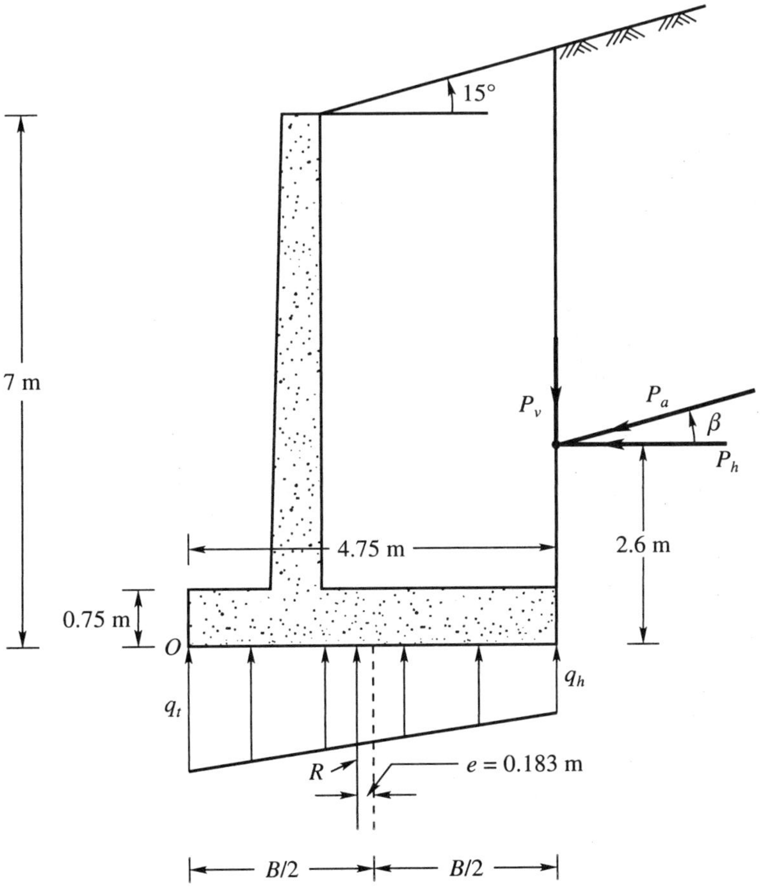

Figure Ex. 19.1(b)

$F_R = 60 \times 4.75 + 596.9 \tan 25° + 470 = 285 + 278 + 470 = 1033$ kN/m

$P_h = 181.2$ kN/m

$$F_s = \frac{F_R}{P_h} = \frac{1033}{181.2} = 5.7 > 1.5 \text{ - OK.}$$

Normally the passive earth pressure P_p is not considered in the analysis. By neglecting P_p, the factor of safety is

$$F_s = \frac{1033 - 470}{181.2} = \frac{563}{181.2} = 3.1 > 1.5 \text{ - OK}$$

Check for bearing capacity failure (Fig. 19.1b)

From Eq. (19.8b and c), the pressures at the toe and heel of the retaining wall may be written as

$$q_t = \frac{R}{B}\left[1 + \frac{6e}{B}\right]$$

$$q_h = \frac{R}{B}\left[1 - \frac{6e}{B}\right]$$

where e = eccentricity of the total load R (= ΣV) acting on the base. From Eq. (19.8a), the eccentricity e may be calculated.

$$e = \frac{B}{2} - \frac{(M_R - M_O)}{R} = \frac{4.75}{2} - \frac{(1{,}779.3 - 471.12)}{596.9} = 0.183 \text{ m}$$

Now $q_t = \frac{596.9}{4.75}\left[1 + \frac{6 \times 0.183}{4.75}\right] = 154.7$ kN/m^2

$$q_h = \frac{596.9}{4.75}\left[1 - \frac{6 \times 0.183}{4.75}\right] = 96.6 \text{ kN/m}^2$$

The ultimate bearing capacity q_u may be determined by Eq. (12.27). It has to be ensured that

$$q_t \le \frac{q_u}{F_s}$$

where $F_s = 3$

PART B—MECHANICALLY STABILIZED EARTH RETAINING WALLS

19.6 GENERAL CONSIDERATIONS

Reinforced earth is a construction material composed of soil fill strengthened by the inclusion of rods, bars, fibers or nets which interact with the soil by means of frictional resistance. The concept of strengthening soil with rods or fibers is not new. Throughout the ages attempts have been made to improve the quality of adobe brick by adding straw. The present practice is to use thin metal strips, geotextiles, and geogrids as reinforcing materials for the construction of reinforced earth retaining walls. A new era of retaining walls with reinforced earth was introduced by Vidal (1969). Metal strips were used as reinforcing material as shown in Fig. 19.11(a). Here the metal strips extend from the panel back into the soil to serve the dual role of anchoring the facing units and being restrained through the frictional stresses mobilized between the strips and the backfill soil. The backfill soil creates the lateral pressure and interacts with the strips to resist it. The walls are relatively flexible compared to massive gravity structures. These flexible walls offer many advantages including significant lower cost per square meter of exposed surface. The variations in the types of facing units, subsequent to Vidal's introduction of the reinforced earth walls, are many. A few of the types that are currently in use are (Koerner, 1999)

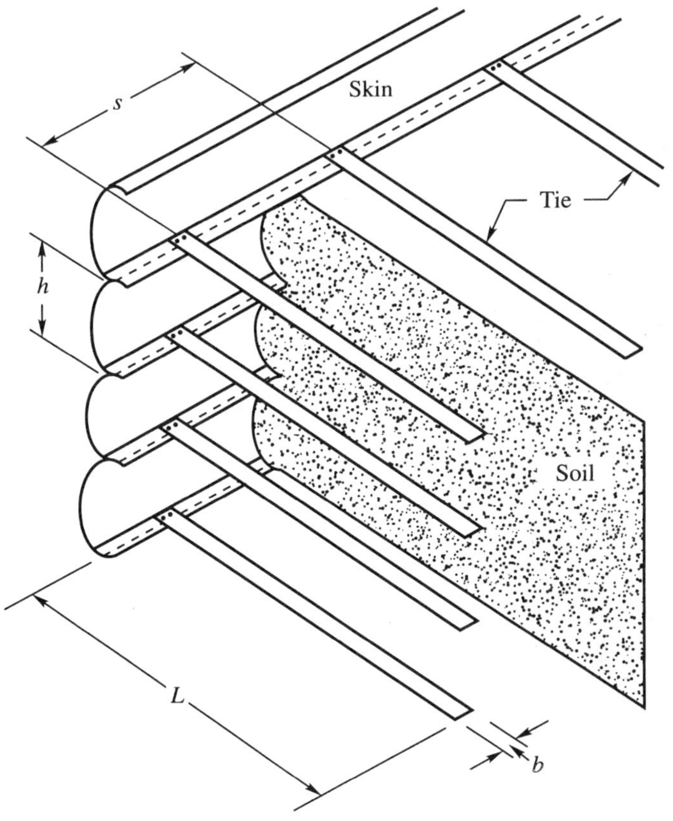

Figure 19.11(a) Component parts and key dimensions of reinforced earth wall (Vidal, 1969)

1. Facing panels with metal strip reinforcement
2. Facing panels with wire mesh reinforcement
3. Solid panels with tie back anchors
4. Anchored gabion walls
5. Anchored crib walls

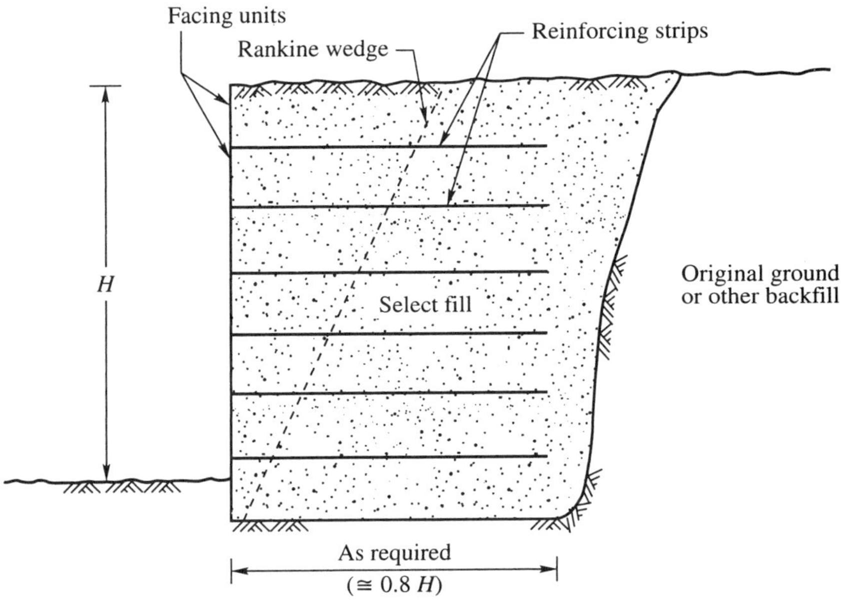

(b) Line details of a reinforced earth wall in place

(c) Front face of a reinforced earth wall under construction for a bridge approach fill using patented precast concrete wall face units

Figure 19.11(b) and (c) Reinforced earth walls (Bowles, 1996)

6. Geotextile reinforced walls
7. Geogrid reinforced walls

In all cases, the soil behind the wall facing is said to be *mechanically stabilized earth* (MSE) and the wall system is generally called an MSE wall.

The three components of a MSE wall are the facing unit, the backfill and the reinforcing material. Figure 19.11(b) shows a side view of a wall with metal strip reinforcement and Fig. 19.11(c) the front face of a wall under construction (Bowles, 1996).

Modular concrete blocks, currently called segmental retaining walls (SRWS, Fig. 19.12(a)) are most common as facing units. Some of the facing units are shown in Fig. 19.12. Most interesting in regard to SRWS are the emerging block systems with openings, pouches, or planting areas within them. These openings are soil-filled and planted with vegetation that is indigenous to the area (Fig. 19.12(b)). Further possibilities in the area of reinforced wall systems could be in the use of polymer rope, straps, or anchor ties to the facing in units or to geosynthetic layers, and extending them into the retained earth zone as shown in Fig. 19.12(c).

A recent study (Koerner 2000) has indicated that geosynthetic reinforced walls are the least expensive of any wall type and for all wall height categories (Fig. 19.13).

19.7 BACKFILL AND REINFORCING MATERIALS

Backfill

The backfill, is limited to cohesionless, free draining material (such as sand), and thus the key properties are the density and the angle of internal friction.

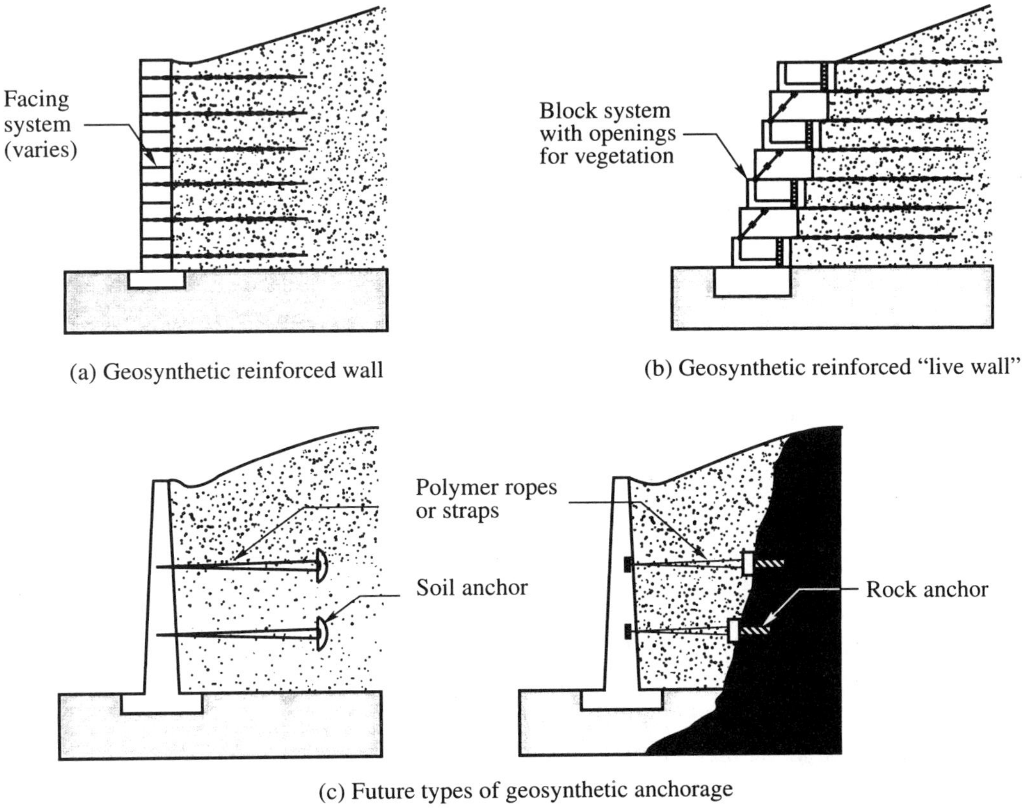

Figure 19.12 Geosynthetic use for reinforced walls and bulkheads (Koerner, 2000)

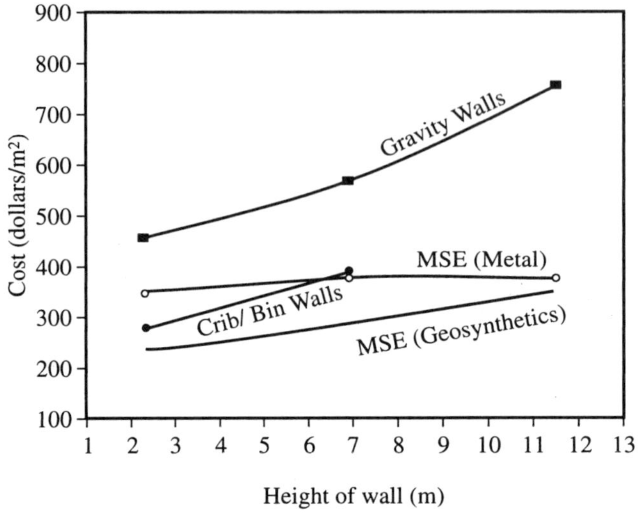

Figure 19.13 Mean values of various categories of retaining wall costs (Koerner, 2000)

Reinforcing material

The reinforcements may be strips or rods of metal or sheets of geotextile, wire grids or geogrids (grids made from plastic).

Geotextile is a permeable geosynthetic comprised solely of textiles. Geotextiles are used with foundation soil, rock, earth or any other geotechnical engineering-related material as an integral part of a human made project, structure, or system (Koerner, 1999). AASHTO (M288-96) provides (Table 19.2) geotextile strength requirements (Koerner, 1999). The tensile strength of geotextile varies with the geotextile designation as per the design requirements. For example, a woven slit-film polypropylene (weighing 240 g/m^2) has a range of 30 to 50 kN/m. The friction angle between soil and geotextiles varies with the type of geotextile and the soil. Table 19.3 gives values of geotextile friction angles (Koerner, 1999).

The test properties represent an idealized condition and therefore result in the maximum possible numerical values when used directly in design. Most laboratory test values cannot generally be used directly and must be suitably modified for in-situ conditions. For problems dealing with geotextiles the ultimate strength T_u should be reduced by applying certain reduction factors to obtain the allowable strength T_a as follows (Koerner, 1999).

$$T_a = T_u\left(\frac{1}{RF_{ID} \times RF_{CR} \times RF_{CD} \times RF_{BD}}\right) \tag{19.9}$$

where

T_a = allowable tensile strength
T_u = ultimate tensile strength
RF_{ID} = reduction factor for installation damage
RF_{CR} = reduction factor for creep
RF_{BD} = reduction factor for biological degradation and
RF_{CD} = reduction factor for chemical degradation

Typical values for reduction factors are given in Table 19.4.

Table 19.2 AASHTO M288-96 Geotextile strength property requirements

			Geotextile Classification* † ‡					
			Case 1		Case 2		Case 3	
	Test methods	Units	Elongation < 50 %	Elongation ≥ 50 %	Elongation < 50 %	Elongation ≥ 50 %	Elongation < 50 %	Elongation ≥50 %
Grab strength	ASTM D4632	N	1400	900	1100	700	800	500
Sewn seam strength ‡	ASTM D4632	N	1200	810	990	630	720	450
Tear strength	ASTM D4533	N	500	350	400	250	300	180
Puncture strength	ASTM D4833	N	500	350	400	2505	300	180
Burst strength	ASTM D3786	kPa	3500	1700	2700	1300	2100	950

† As measured in accordance with ASTM D4632. Woven geotextiles fail at elongations (strains)< 50%, while nonwovens fail at elongation (strains) > 50%.
‡ When sewnseams are required. Overlap seam requirements are application specific.
§ The required MARV tear strength for woven monofilament geotextiles is 250 N.

Table 19.3 Peak soil-to-geotextile friction angles and efficiencies in selected cohesionless soils*

Geotextile type	Concrete sand (ϕ = 30°)	Rounded sand (ϕ = 28°)	Silty sand (ϕ = 26°)
Woven, monofilament	26° (84 %)	–	–
Woven, slit-film	24° (77%)	24° (84 %)	23° (87 %)
Nonwoven, heat-bonded	26° (84 %)	–	–
Nonwoven, needle-punched	30° (100 %)	26° (92 %)	25° (96 %)

* Numbers in parentheses are the efficiencies. Values such as these should not be used in final design. Site specific geotextiles and soils must be individually tested and evaluated in accordance with the particular project conditions: saturation, type of liquid, normal stress, consolidation time, shear rate, displacement amount, and so on. (Koerner, 1999)

Table 19.4 Recommended reduction factor values for use in [Eq. (19.9)]

	Range of Reduction Factors			
Application Area	Installation Damage	Creep*	Chemical Degradation	Biological Degradation
Separation	1.1 to 2.5	1.5 to 2.5	1.0 to 1.5	1.0 to 1.2
Cushioning	1.1 to 2.0	1.2 to 1.5	1.0 to 2.0	1.0 to 1.2
Unpaved roads	1.1 to 2.0	1.5 to 2.5	1.0 to 1.5	1.0 to 1.2
Walls	1.1 to 2.0	2.0 to 4.0	1.0 to 1.5	1.0 to 1.3
Embankments	1.1 to 2.0	2.0 to 3.5	1.0 to 1.5	1.0 to 1.3
Bearing capacity	1.1 to 2.0	2.0 to 4.0	1.0 to 1.5	1.0 to 1.3
Slope stabilization	1.1 to 1.5	2.0 to 3.0	1.0 to 1.5	1.0 to 1.3
Pavement overlays	1.1 to 1.5	1.0 to 2.0	1.0 to 1.5	1.0 to 1.1
Railroads (filter/sep.)	1.5 to 3.0	1.0 to 1.5	1.5 to 2.0	1.0 to 1.2
Flexible forms	1.1 to 1.5	1.5 to 3.0	1.0 to 1.5	1.0 to 1.1
Silt fences	1.1 to 1.5	1.5 to 2.5	1.0 to 1.5	1.0 to 1.1

* The low end of the range refers to applications which have relatively short service lifetimes and / or situations where creep deformations are not critical to the overall system performance. (Koerner, 1999)

Table 19.5 Recommended reduction factor values for use in Eq. (19.10) for determining allowable tensile strength of geogrids

Application Area	RF_{ID}	RF_{CR}	RF_{CD}	RF_{BD}
Unpaved roads	1.1 to 1.6	1.5 to 2.5	1.0 to 1.5	1.0 to 1.1
Paved roads	1.2 to 1.5	1.5 to 2.5	1.1 to 1.6	1.0 to 1.1
Embankments	1.1 to 1.4	2.0 to 3.0	1.1 to 1.4	1.0 to 1.2
Slopes	1.1 to 1.4	2.0 to 3.0	1.1 to 1.4	1.0 to 1.2
Walls	1.1 to 1.4	2.0 to 3.0	1.1 to 1.4	1.0 to 1.2
Bearing capacity	1.2 to 1.5	2.0 to 3.0	1.1 to 1.6	1.0 to 1.2

Geogrid

A geogrid is defined as a geosynthetic material consisting of connected parallel sets of tensile ribs with apertures of sufficient size to allow strike-through of surrounding soil, stone, or other geotechnical material (Koerner, 1999).

Geogrids are matrix like materials with large open spaces called apertures, which are typically 10 to 100 mm between the ribs, called *longitudinal* and *transverse* respectively. The primary function of geogrids is clearly reinforcement. The mass of geogrids ranges from 200 to 1000 g/m^2 and the open area varies from 40 to 95 %. It is not practicable to give specific values for the tensile strength of geogrids because of its wide variation in density. In such cases one has to consult manufacturer's literature for the strength characteristics of their products. The allowable tensile strength, T_a, may be determined by applying certain reduction factors to the ultimate strength T_u as in the case of geotextiles. The equation is

$$T_a = T_u\left(\frac{1}{RF_{ID} \times RF_{CR} \times RF_{BD} \times RF_{CD}}\right) \tag{19.10}$$

The definition of the various terms in Eq (19.10) is the same as in Eq. (19.9). However, the reduction factors are different. These values are given in Table 19.5 (Koerner, 1999).

Metal Strips

Metal reinforcement strips are available in widths ranging from 75 to 100 mm and thickness on the order of 3 to 5 mm, with 1 mm on each face excluded for corrosion (Bowles, 1996). The yield strength of steel may be taken as equal to about 35000 lb/in^2 (240 MPa) or as per any code of practice.

19.8 CONSTRUCTION DETAILS

The method of construction of MSE walls depends upon the type of facing unit and reinforcing material used in the system. The facing unit which is also called the *skin* can be either flexible or stiff, but must be strong enough to retain the backfill and allow fastenings for the reinforcement to be attached. The facing units require only a small foundation from which they can be built, generally consisting of a trench filled with mass concrete giving a footing similar to those used in domestic housing. The segmental retaining wall sections of dry-laid masonry blocks, are shown in Fig. 19.12(a). The block system with openings for vegetation is shown in Fig. 19.12(b).

The construction procedure with the use of geotextiles is explained in Fig. 19.14(a). Here, the geotextile serve both as a reinforcement and also as a facing unit. The procedure is described below (Koerner, 1985) with reference to Fig. 19.14(a).

1. Start with an adequate working surface and staging area (Fig. 19.14a).
2. Lay a geotextile sheet of proper width on the ground surface with 4 to 7 ft at the wall face draped over a temporary wooden form (b).
3. Backfill over this sheet with soil. Granular soils or soils containing a maximum 30 percent silt and /or 5 percent clay are customary (c).
4. Construction equipment must work from the soil backfill and be kept off the unprotected geotextile. The spreading equipment should be a wide-tracked bulldozer that exerts little pressure against the ground on which it rests. Rolling equipment likewise should be of relatively light weight.

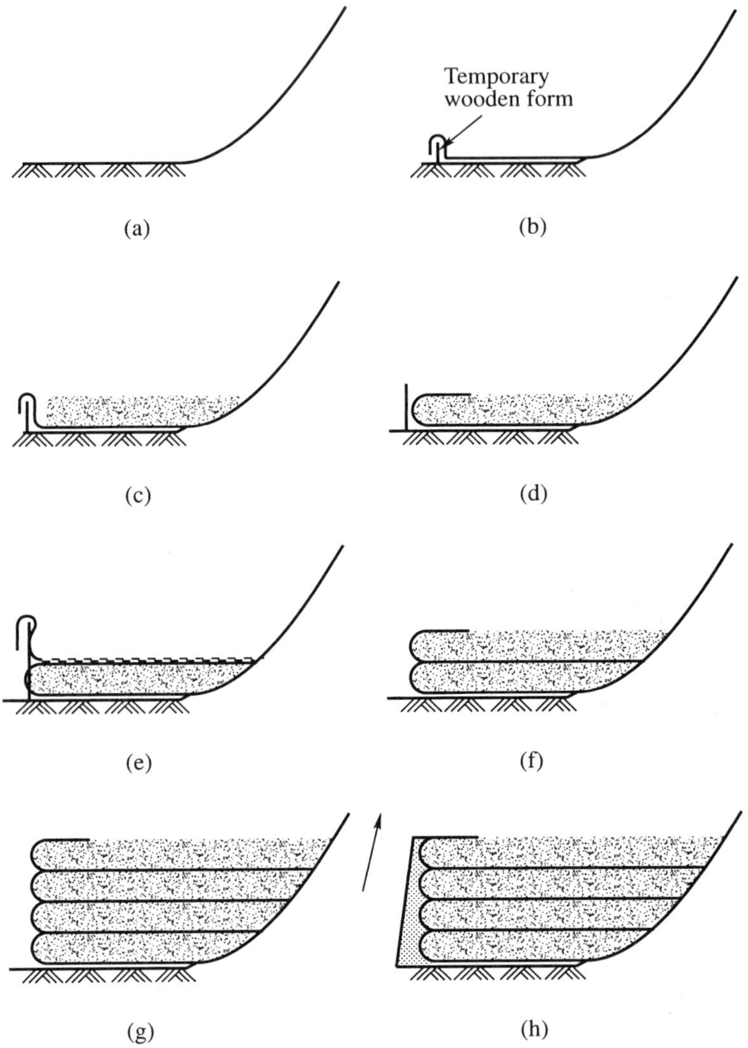

Figure 19.14(a) General construction procedures for using geotextiles in fabric wall construction (Koerner, 1985)

5. When the first layer has been folded over the process should be repeated for the second layer with the temporary facing form being extended from the original ground surface or the wall being stepped back about 6 inches so that the form can be supported from the first layer. In the latter case, the support stakes must penetrate the fabric.
6. This process is continued until the wall reaches its intended height.
7. For protection against ultraviolet light and safety against vandalism the faces of such walls must be protected. Both shotcrete and gunite have been used for this purpose.

Figure 19.14(b) shows complete geotextile walls (Koerner, 1999).

Figure 19.14(b) Geotextile walls (Koerner, 1999)

19.9 DESIGN CONSIDERATIONS FOR A MECHANICALLY STABILIZED EARTH WALL

The design of a MSE wall involves the following steps:

1. Check for internal stability, addressing reinforcement spacing and length.
2. Check for external stability of the wall against overturning, sliding, and foundation failure.

The general considerations for the design are:

1. Selection of backfill material: granular, freely draining material is normally specified. However, with the advent of geogrids, the use of cohesive soil is gaining ground.
2. Backfill should be compacted with care in order to avoid damage to the reinforcing material.
3. Rankine's theory for the active state is assumed to be valid.
4. The wall should be sufficiently flexible for the development of active conditions.
5. Tension stresses are considered for the reinforcement outside the assumed failure zone.
6. Wall failure will occur in one of three ways

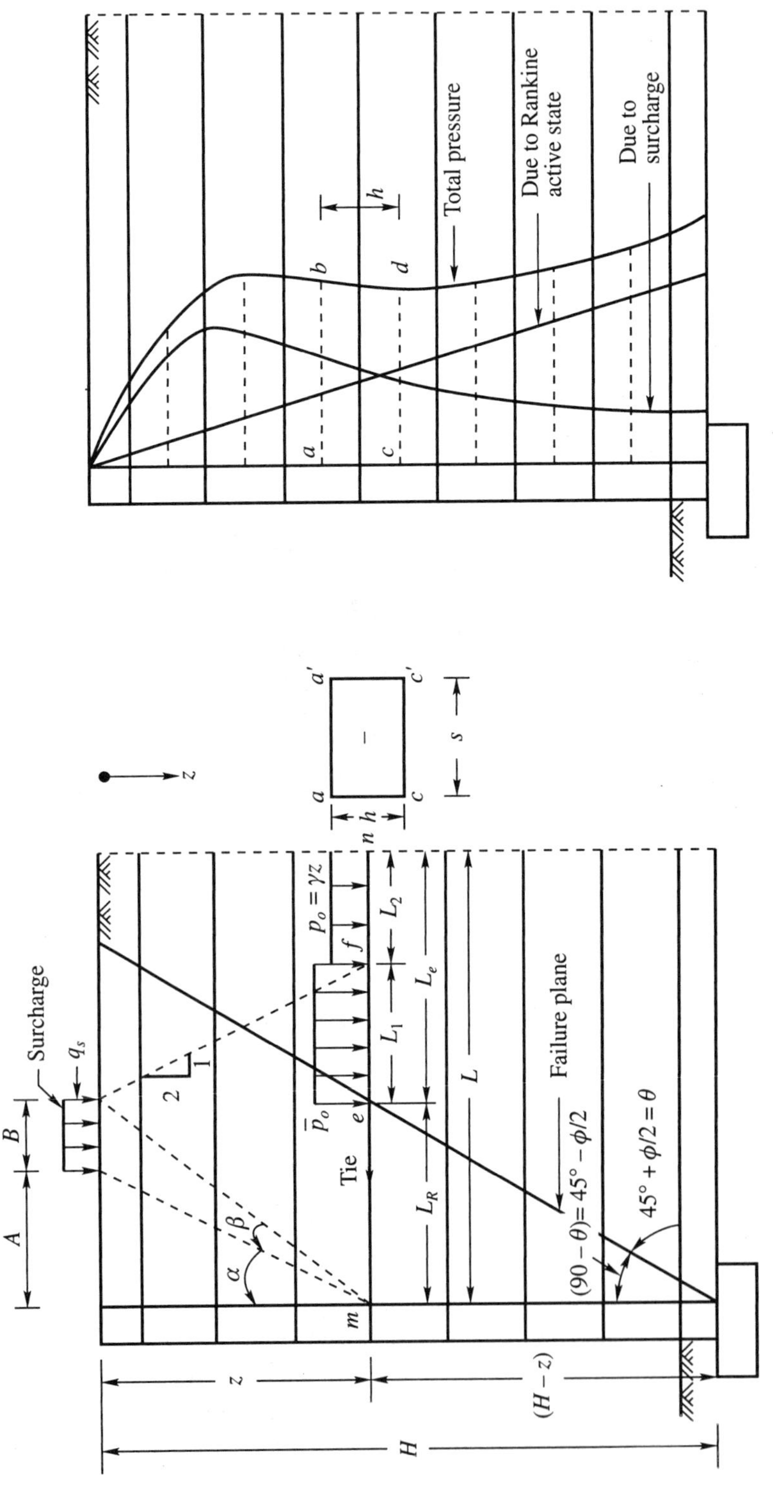

(a) Reinforced earth-wall profile with surcharge load

(b) Lateral pressure distribution diagrams

Figure 19.15 Principles of MSE wall design

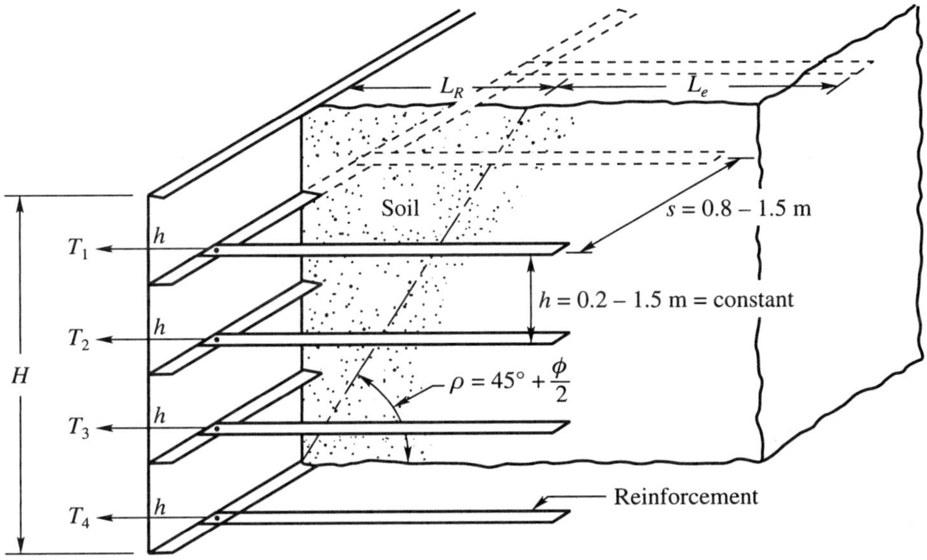

Figure 19.16 Typical range in strip reinforcement spacing for reinforced earth walls (Bowles, 1996)

a. tension in reinforcements
b. bearing capacity failure
c. sliding of the whole wall soil system.

7. Surcharges are allowed on the backfill. The surcharges may be permanent (such as a roadway) or temporary.
   a. Temporary surcharges within the reinforcement zone will increase the lateral pressure on the facing unit which in turn increases the tension in the reinforcements, but does not contribute to reinforcement stability.
   b. Permanent surcharges within the reinforcement zone will increase the lateral pressure and tension in the reinforcement and will contribute additional vertical pressure for the reinforcement friction.
   c. Temporary or permanent surcharges outside the reinforcement zone contribute lateral pressure which tends to overturn the wall.
8. The total length L of the reinforcement goes beyond the failure plane AC by a length L_e. Only length L_e (effective length) is considered for computing frictional resistance. The length L_R lying within the failure zone will not contribute for frictional resistance (Fig. 19.15a).
9. For the propose of design the total length L remains the same for the entire height of wall H. Designers, however, may use their discretion to curtail the length at lower levels. Typical ranges in reinforcement spacing are given in Fig. 19.16.

19.10 DESIGN METHOD

The following forces are considered:

1. Lateral pressure on the wall due to backfill
2. Lateral pressure due to surcharge if present on the backfill surface.

3. The vertical pressure at any depth z on the strip due to
   a) overburden pressure p_o only
   b) overburden pressure p_o and pressure due to surcharge.

Lateral Pressure

Pressure due to Overburden

Lateral earth pressure due to overburden

At depth z $\quad p_a = p_{oz} K_A = \gamma z K_A$ (19.11a)

At depth H $\quad p_a = p_{oH} K_A = \gamma H K_A$ (19.11b)

Total active earth pressure

$$P_a = \frac{1}{2} \gamma H^2 K_A \tag{19.12}$$

Pressure Due to Surcharge (a) of Limited Width, and (b) Uniformly Distributed

(a) From Eq. (11.69)

$$q_h = \frac{2q_s}{\pi} (\beta - \sin\beta \cos 2\alpha) \tag{19.13a}$$

(b) $\quad q_h = q_s K_A$ (19.13b)

Total lateral pressure due to overburden and surcharge at any depth z

$$p_h = p_a + q_h = (\gamma z K_A + q_h) \tag{19.14}$$

Vertical pressure

Vertical pressure at any depth z due to overburden only

$$p_o = \gamma z \tag{19.15a}$$

due to surcharge (limited width)

$$\Delta q = \frac{q_s B}{B + z} \tag{19.15b}$$

where the 2:1 (2 vertical : 1 horizontal) method is used for determining Δq at any depth z.

Total vertical pressure due to overburden and surcharge at any depth z.

$$\overline{p}_o = p_o + \Delta q \tag{19.15c}$$

Reinforcement and Distribution

Three types of reinforcements are normally used. They are

1. Metal strips
2. Geotextiles
3. Geogrids.

Galvanized steel strips of widths varying from 5 to 100 mm and thickness from 3 to 5 mm are generally used. Allowance for corrosion is normally made while deciding the thickness at the rate of 0.001 in. per year and the life span is taken as equal to 50 years. The vertical spacing may range from 20 to 150 cm (8 to 60 in.) and can vary with depth. The horizontal lateral spacing may be on the order of 80 to 150 cm (30 to 60 in.). The ultimate tensile strength may be taken as equal to 240 MPa (35,000 lb/in.^{2}). A factor of safety in the range of 1.5 to 1.67 is normally used to determine the allowable steel strength f_a.

Figure 19.16 depicts a typical arrangement of metal reinforcement. The properties of geotextiles and geogrids have been discussed in Section 19.7. However, with regards to spacing, only the vertical spacing is to be considered. Manufacturers provide geotextiles (or geogrids) in rolls of various lengths and widths. The tensile force per unit width must be determined.

Length of Reinforcement

From Fig. 19.15(a)

$$L = L_R + L_e = L_R + L_1 + L_2 \tag{19.16}$$

where L_R = $(H - z)\tan(45° - \phi/2)$
L_e = effective length of reinforcement outside the failure zone
L_1 = length subjected to pressure $(p_o + \Delta q) = \bar{p}_o$
L_2 = length subjected to p_o only.

Strip Tensile Force at any Depth *z*

The equation for computing T is

$$T = p_h \times h \times s/\text{strip} = (\gamma z K_A + q_h) h \times s \tag{19.17a}$$

The maximum tie force will be

$$T(\max) = (\gamma H K_A + q_{hH}) h \times s \tag{19.17b}$$

where p_h = $\gamma z K_A + q_h$
q_h = lateral pressure at depth z due to surcharge
q_{hH} = q_h at depth H
h = vertical spacing
s = horizontal spacing

$$T = P_a + P_q \tag{19.18}$$

where P_a = $1/2\gamma H^2 K_A$—Rankine's lateral force
P_q = lateral force due to surcharge

Frictional Resistance

In the case of strips of width b both sides offer frictional resistance. The frictional resistance F_R offered by a strip at any depth z must be greater than the pullout force T by a suitable factor of safety. We may write

$$F_R = 2b[(p_o + \Delta q)L_1 + p_o L_2]\tan\delta \le TF_s \tag{19.19}$$

or $$F_R = 2b[\bar{p}_o L_1 + p_o L_2]\tan\delta \le TF_s \tag{19.20}$$

where F_s may be taken as equal to 1.5.

The friction angle δ between the strip and the soil may be taken as equal to ϕ for a rough strip surface and for a smooth surface δ may lie between 10 to 25°.

Sectional Area of Metal Strips

Normally the width b of the strip is assumed in the design. The thickness t has to be determined based on T (max) and the allowable stress f_a in the steel. If f_y is the yield stress of steel, then

$$f_a = \frac{f_y}{F_s(\text{steel})} \tag{19.21}$$

Normally F_s (steel) ranges from 1.5 to 1.67. The thickness t may be obtained from

$$t = \frac{T(\max)}{bf_a} \tag{19.22}$$

The thickness of t is to be increased to take care of the corrosion effect. The rate of corrosion is normally taken as equal to 0.001 in/yr for a life span of 50 years.

Spacing of Geotextile Layers

The tensile force T per unit width of geotextile layer at any depth z may be obtained from

$$T = p_h h = (\gamma z K_A + q_h)h \tag{19.23}$$

where q_h = lateral pressure either due to a stripload or due to uniformly distributed surcharge.

The maximum value of the computed T should be limited to the allowable value T_a as per Eq. (19.9). As such we may write Eq. (19.23) as

$$T_a = TF_s = (\gamma z K_A + q_h)hF_s \tag{19.24}$$

$$\text{or } h = \frac{T_a}{(\gamma z K_A + q_h)F_s} = \frac{T_a}{p_h F_s} \tag{19.25}$$

where F_s = factor of safety (1.3 to 1.5) when using T_a.

Equation (19.25) is used for determining the vertical spacing of geotextile layers.

Frictional Resistance

The frictional resistance offered by a geotextile layer for the pullout force T_a may be expressed as

$$F_R = 2\left[(\gamma z + \Delta q)L_1 + \gamma z L_z\right]\tan\delta \geq T_a F_s \tag{19.26}$$

Equation (19.26) expresses frictional resistance per unit width and both sides of the sheets are considered.

Design with Geogrid Layers

A tremendous number of geogrid reinforced walls have been constructed in the past 10 years (Koerner, 1999). The types of permanent geogrid reinforced wall facings are as follows (Koerner, 1999):

1. *Articulated precast panels* are discrete precast concrete panels with inserts for attaching the geogrid.
2. *Full height precast panels* are concrete panels temporarily supported until backfill is complete.
3. *Cast-in-place concrete panels* are often wrap-around walls that are allowed to settle and, after 1/2 to 2 years, are covered with a cast-in-place facing panel.

4. *Masonry block facing walls* are an exploding segment of the industry with many different types currently available, all of which have the geogrid embedded between the blocks and held by pins, nubs, and/or friction.
5. *Gabion facings* are polymer or steel-wire baskets filled with stone, having a geogrid held between the baskets and fixed with rings and/or friction.

The frictional resistance offered by a geogrid against pullout may be expressed as (Koerner, 1999)

$$F_R = 2C_i C_r L_e p_o \tan\phi \geq TF_s \tag{19.27}$$

where C_i = interaction coefficient = 0.75 (may vary)
C_r = coverage ratio = 0.8 (may vary)

All the other notations are already defined. The spacing of geogrid layers may be obtained from

$$h = \frac{T_a C_r}{p_h} \tag{19.28}$$

where p_h = lateral pressure per unit length of wall.

19.11 EXTERNAL STABILITY

The MSE wall system consists of three zones. Thye are

1. The reinforced earth zone.
2. The backfill zone.

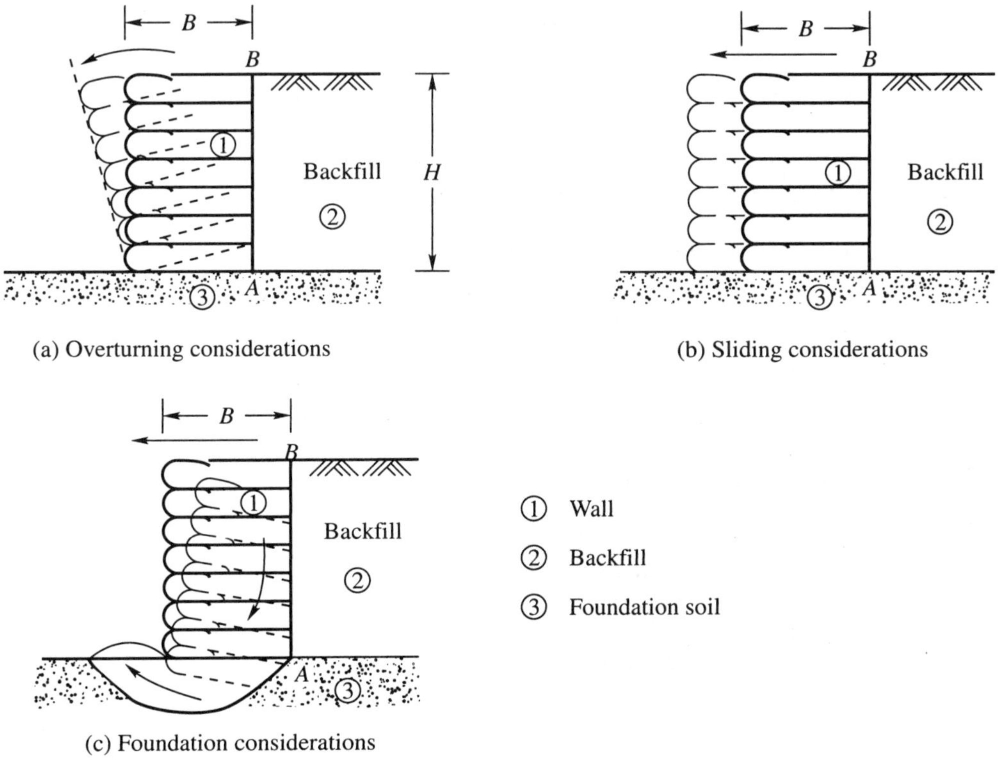

Figure 19.17 External stability considerations for reinforced earth walls

3. The foundation soil zone.

The reinforced earth zone is considered as the wall for checking the internal stability whereas all three zones are considered for checking the external stability. The soils of the first two zones are placed in layers and compacted whereas the foundation soil is a normal one. The properties of the soil in each of the zones may be the same or different. However, the soil in the first two zones is normally a free draining material such as sand.

It is necessary to check the reinforced earth wall (width = B) for external stability which includes overturning, sliding and bearing capacity failure. These are illustrated in Fig. 19.17. Active earth pressure of the backfill acting on the internal face AB of the wall is taken in the stability analysis. The resultant earth thrust P_a is assumed to act horizontally at a height $H/3$ above the base of the wall. The methods of analysis are the same as for concrete retaining walls.

Example 19.2

A typical section of a retaining wall with the backfill reinforced with metal strips is shown in Fig. Ex. 19.2. The following data are available:

Height $H = 9$ m; $b = 100$ mm; $t = 5$ mm; $f_y = 240$ MPa; F_s for steel = 1.67; F_s on soil friction = 1.5; $\phi = 36°$; $\gamma = 17.5$ kN/m^3; $\delta = 25°$; $h \times s = 1 \times 1$ m.

Required:

(a) Lengths L and L_e at varying depths.

(b) The largest tension T in the strip.

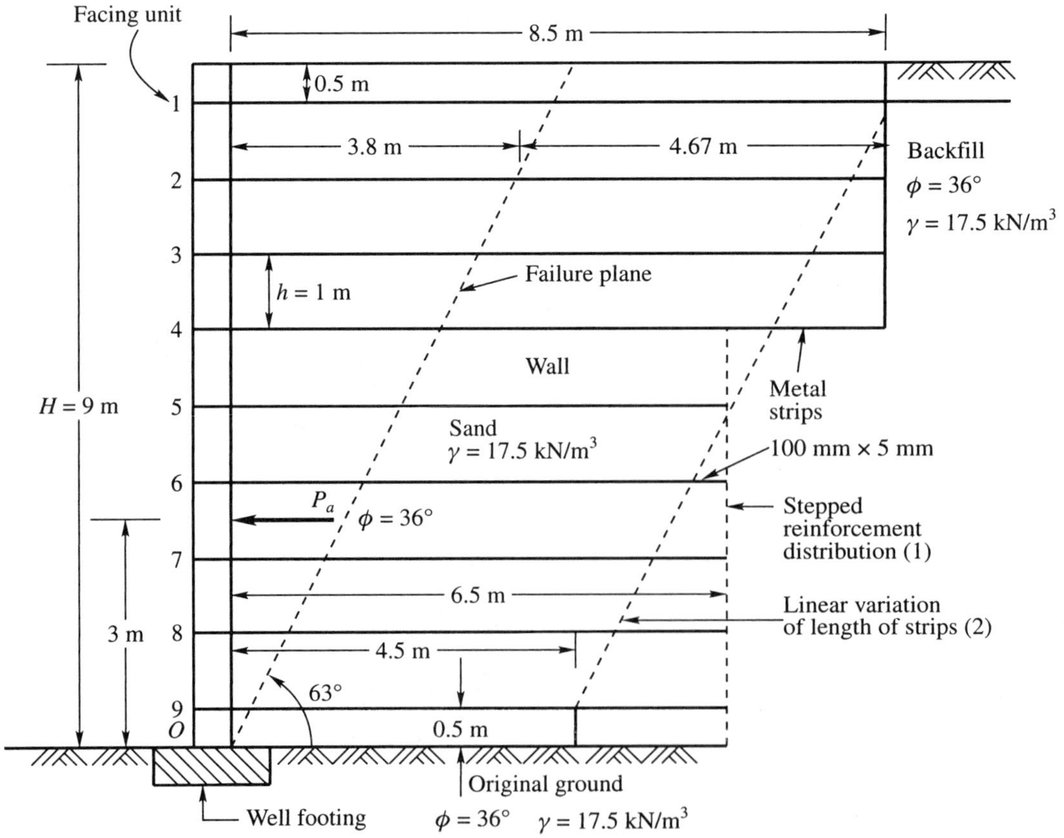

Figure Ex. 19.2

(c) The allowable tension in the strip.
(d) Check for external stability.

Solution

From Eq. (19.17a), the tension in a strip at depth z is

$T = \gamma z\, K_A sh$ for $q_h = 0$

where $\gamma = 17.5$ kN/m^3, $K_A = \tan^2(45° - 36/2) = 0.26$, $s = 1$m; $h = 1$ m.

Substituting

$T = 17.5 \times 0.26\ (1)\ [1]\ z = 4.55z$ kN/strip.

$$L_e = \frac{F_s T}{2\gamma z b \tan\delta} = \frac{1.5 \times 4.55z}{2 \times 17.5 \times 0.1 \times 0.47 \times z} = 4.14 \text{ m}$$

This shows that the length $L_e = 4.14$ m is a constant with depth. Fig. Ex. 19.2 shows the positions of L_e for strip numbers 1, 2 ... 9. The first strip is located 0.5 m below the backfill surface and the 9th at 8.5 m below with spacings at 1 m apart. Tension in each of the strips may be obtained by using the equation $T = 4.55\ z$. The total tension ΣT as computed is

$\Sigma T = 184.29$ kN/m since $s = 1$ m.

As a check the total active earth pressure is

$$P_a = \frac{1}{2}\gamma H^2 K_A = \frac{1}{2} 17.5 \times 9^2 \times 0.26 = 184.28 \text{ kN/m} = \Sigma T$$

The maximum tension is in the 9th strip, that is, at a depth of 8.5 m below the backfill surface. Hence

$T = \gamma z\, K_A sh = 17.5 \times 8.5 \times 0.26 \times 1 \times 1 = 38.68$ kN/strip

The allowable tension is

$T_a = f_a tb$

where $f_a = \dfrac{240 \times 10^3}{1.67} = 143.7 \times 10^3 \text{ kN/m}^2$

Substituting $T_a = 143.7 \times 10^3 \times 0.005 \times 0.1$ » 72 kN > T – OK.

The total length of strip L at any depth z is

$L = L_R + L_e = (H - z) \tan (45 - \phi/2) + 4.14 = 0.51\ (9 - z) + 4.14$ m

where $H = 9$ m.

The lengths as calculated have been shown in Fig. Ex. 19.2. It is sometimes convenient to use the same length L with depth or stepped in two or more blocks or use a linear variation as shown in the figure.

Check for External Stability

Check of bearing capacity

It is necessary to check the base of the wall with the backfill for the bearing capacity per unit length of the wall. The width of the wall may be taken as equal to 4.5 m (Fig. Ex. 19.2). The procedure as explained in Chapter 12 may be followed. For all practical purposes, the shape, depth, and inclination factors may be taken as equal to 1.

Check for sliding resistance

$$F_s = \frac{\text{Sliding resistance } F_R}{\text{Driving force } P_a}$$

where $F_R = W \tan\delta = \frac{4.5+8.5}{2} \times 17.5 \times 9 \tan 36°$

$= 1024 \times 0.73 = 744$ kN

where $\delta = \phi = 36°$ for the foundation soil, and W = weight of the reinforced wall

$P_a = 184.28$ kN

$$F_s = \frac{744}{184.28} = 4 > 1.5 \text{ -- OK}$$

Check for overturning

$$F_s = \frac{M_R}{M_o}$$

From Fig. Ex. 19.2 taking moments of all forces about O, we have

$$M_R = 4.5 \times 9 \times 17.5 \times \frac{4.5}{2} + \frac{1}{2} \times 9 \times (8.5-4.5)(4.5+\frac{4}{3}) \times 17.5$$

$= 1595 + 1837 = 3432$ kN–m

$$M_o = P_a \times \frac{H}{3} = 184.28 \times \frac{9}{3} = 553 \text{ kN} - \text{m}$$

$$F_s = \frac{3432}{553} = 6.2 > 2 - \text{OK}.$$

Example 19.3

A section of a retaining wall with a reinforced backfill is shown in Fig. Ex. 19.3. The backfill surface is subjected to a surcharge of 30 kN/m^2. Required:

(a) The reinforcement distribution.
(b) The maximum tension in the strip.
(c) Check for external stability.

Given: b = 100 mm, t = 5 mm, f_a = 143.7 MPa, c = 0, ϕ = 36°, δ = 25°, γ = 17.5 kN/m^3, s = 0.5 m, and h = 0.5 m.

Solution

From Eq. (19.17a)

$$T = (\gamma z K_A + q_h) h \times s = (p_o + q_h) A_c$$

where γ = 17.5 kN/m^3, K_A = 0.26, $A_c = h \times s$ = (0.5 × 0.5) m^2

From Eq. (19.13a)

$$q_h = \frac{2q_s}{\pi}[\beta - \sin\beta \cos 2\alpha]$$

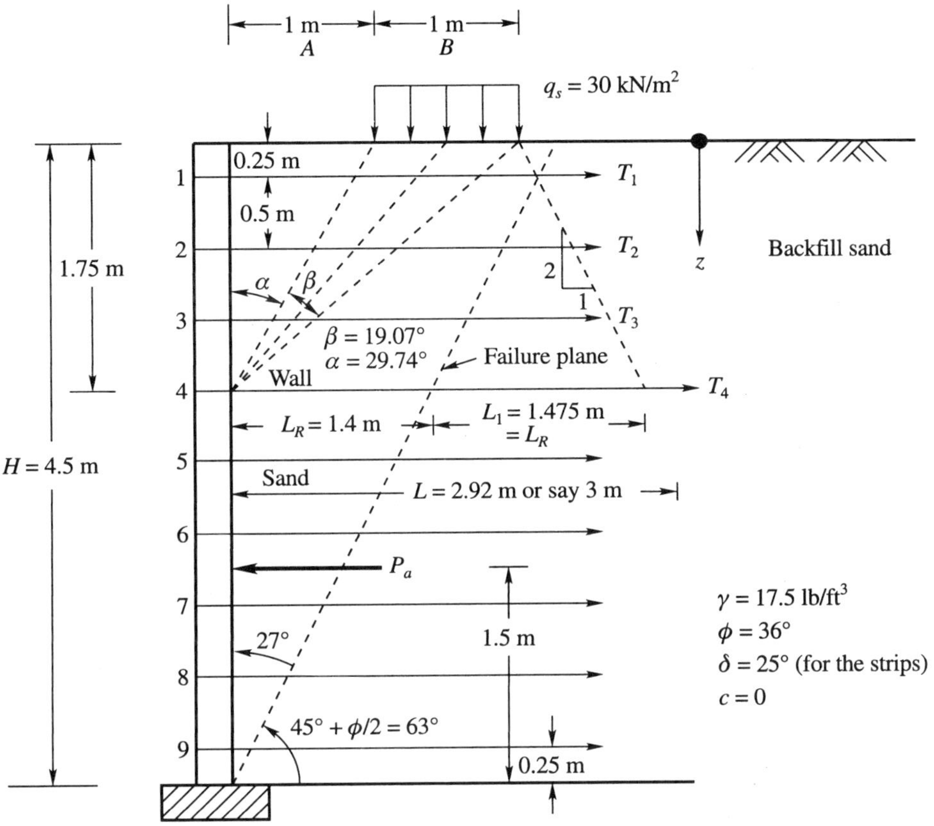

Figure Ex. 19.3

Refer to Fig. Ex. 19.3 for the definition of α and β.

$q_s = 30$ kN/m²

The procedure for calculating length L of the strip for one depth $z = 1.75$ m (strip number 4) is explained below. The same method is valid for the other strips.

Strip No. 4. Depth z = 1.75 m

$p_a = \gamma z K_A = 17.5 \times 1.75 \times 0.26 = 7.96$ kN/m²

From Fig. Ex. 19.3, $\beta = 19.07° = 0.3327$ radians

$\alpha = 29.74°$

$q_s = 30$ kN/m²

$$q_h = \frac{2 \times 30}{3.14}\left[0.3327 - \sin 19.07° \cos 59.5°\right] = 3.19 \text{ kN/m}^2$$

Figure Ex. 19.3 shows the surcharge distribution at a 2 (vertical) to 1 (horizontal) slope. Per the figure at depth $z = 1.75$ m, $L_1 = 1.475$ m from the failure line and $L_R = (H - z)\tan(45° - \phi/2) = 2.75 \tan(45° - 36°/2) = 1.4$m from the wall to the failure line. It is now necessary to determine L_2 (Refer to Fig. 19.15a).

Now $T = (7.96 + 3.19) \times 0.5 \times 0.5 = 2.79$ kN/strip.

The equation for the frictional resistance per strip is

$F_R = 2b\,(\gamma z + \Delta q)\, L_1 \tan\delta + (\gamma z\, L_2 \tan\delta)\, 2b$

From the 2:1 distribution Δq at $z = 1.75$ m is

$$\Delta q = \frac{Q}{B+z} = \frac{30\times 1}{1+1.75} = 10.9 \quad \text{kN/m}^2$$

$p_o = 17.5\times 1.75 = 30.63 \quad \text{kN/m}^2$

Hence $\bar{p}_o = 10.9 + 30.63 = 41.53$ kN/m^2

Now equating frictional resistance F_R to tension in the strip with $F_s = 1.5$, we have $F_R = 1.5\,T$. Given $b = 100$ mm. Now from Eq. (19.20)

$$F_R = 2b\tan\delta(\bar{p}_o L_1 + p_o L_2) = 1.5\ T$$

Substituting and taking $\delta = 25°$, we have

$$2\times 0.1\times 0.47\ [41.53\times 1.475 + 30.63\ L_2] = 1.5\times 2.79$$

Simplifying

$L_2 = -0.546 \text{ m} \approx 0$

Hence $L_e = L_1 + 0 = 1.475$ m

$L = L_R + L_e = 1.4 + 1.475 = 2.875$ m

L can be calculated in the same way at other depths.

Maximum tension T

The maximum tension is in strip number 9 at depth $z = 4.25$m

Allowable $T_a = f_a bt = 143.7 \times 10^3 \times 0.1 \times 0.005 = 71.85$ kN

$T = (\gamma z K_A + q_h) sh$

where $\gamma z K_A = 17.50 \times 4.25 \times 0.26 = 19.34$ kN/m^2

$q_h = 0.89$ kN/m^2 from equation for q_h at depth $z = 4.25$m.

Hence $T = (19.34 + 0.89) \times 1/2 \times 1/2 = 5.05$ kN/strip < 71.85 kN – OK

Example 19.4 (Koerner, 1999)

Figure Ex. 19.4 shows a section of a retaining wall with geotextile reinforcement. The wall is backfilled with a granular soil having $\gamma = 18$ kN/m^3 and $\phi = 34°$.

A woven slit-film geotextile with warp (machine) direction ultimate wide-width strength of 50 kN/m and having $\delta = 24°$ (Table 19.3) is intended to be used in its construction.

The orientation of the geotextile is perpendicular to the wall face and the edges are to be overlapped to handle the weft direction. A factor of safety of 1.4 is to be used along with site-specific reduction factors (Table 19.4).

Required:

(a) Spacing of the individual layers of geotextile.

(b) Determination of the length of the fabric layers.

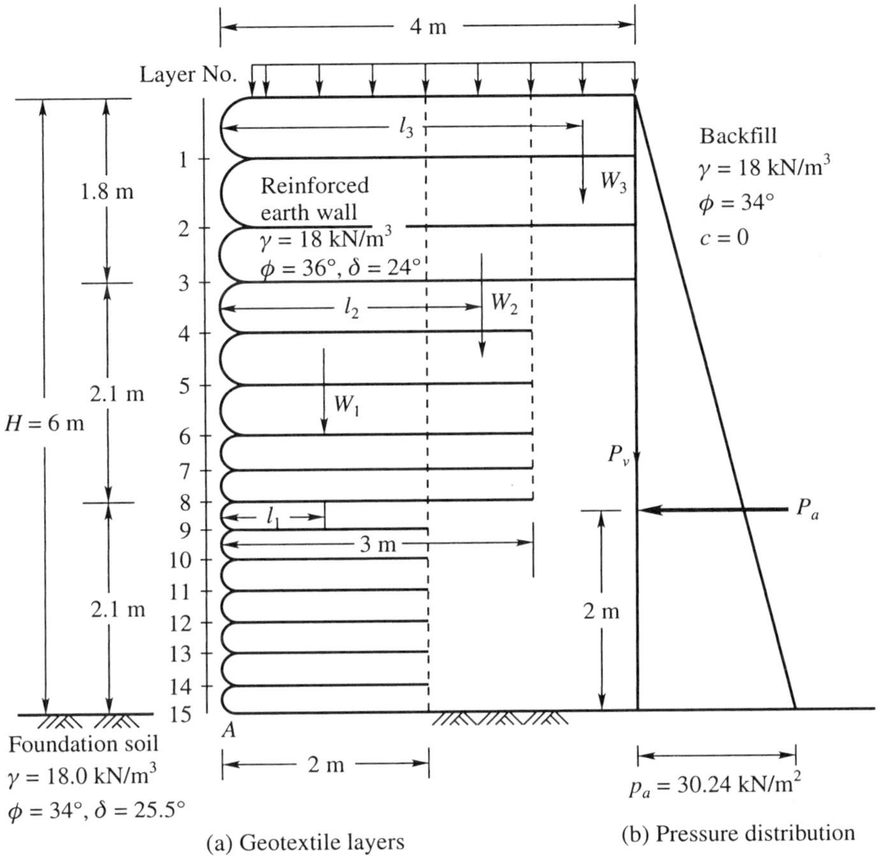

Figure Ex. 19.4

(c) Check the overlap.
(d) Check for external stability.

The backfill surface carries a uniform surcharge dead load of 10 kN/m^2

Solution

(a) The lateral pressure p_h at any depth z is expressed as

$$p_h = p_a + q_h$$

where $p_a = \gamma z K_A$, $q_h = q K_A$, $K_A = \tan^2(45° - 36/2) = 0.26$

Substituting

$$p_h = 18 \times 0.26 z + 0.26 \times 10 = 4.68 z + 2.60$$

From Eq. (19.9), the allowable geotextile strength is

$$T_a = T_u \left[\frac{1}{RF_{ID} \times RF_{CR} \times RF_{CD} \times RF_{BD}} \right]$$

$$= 50 \left[\frac{1}{1.2 \times 2.5 \times 1.15 \times 1.1} \right] = 13.2 \text{ kN/m}$$

From Eq. (19.17a), the expression for allowable stress in the geotextile at any depth z may be expressed as

$$T = T_a = p_h \, h \, F_s$$

$$h = \frac{T_a}{p_h F_s}$$

where h = vertical spacing (lift thickness)

T_a = allowable stress in the geotextile

p_h = lateral earth pressure at depth z

F_s = factor of safety = 1.4

Now substituting

$$h = \frac{13.2}{[4.68(z) + 2.60]1.4} = \frac{13.2}{6.55(z) + 3.64}$$

At z = 6m, $h = \dfrac{13.2}{6.55 \times 6 + 3.64} = 0.307$ m or say 0.30 m

At z = 3.3m, $h = \dfrac{13.2}{6.55 \times 3.3 + 3.64} = 0.52$ m or say 0.50 m

At z = 1.3m, $h = \dfrac{13.2}{6.55 \times 1.3 + 3.64} = 1.08$ m, but use 0.65 m for a suitable distribution.

The depth 3.3 m or 1.3 m are used just as a trial and error process to determine suitable spacings. Figure Ex. 19.4 shows the calculated spacings of the geotextiles.

(b) Length of the Fabric Layers

From Eq. (19.26) we may write

$$L_e = \frac{T \, F_s}{2\gamma \, z \tan\delta} = \frac{p_h \, h \, F_s}{2\gamma \, z \tan\delta} = \frac{h\,(4.68z + 2.60)1.4}{2 \times 18 \; z \tan 24^\circ} = L_e = \frac{h(6.55z + 3.64)}{16z}$$

From Fig. (19.15) the expression for L_R is

$$L_R = (H - z)\tan(45^\circ - \phi/2) = (H - z)\tan(45^\circ - 36/2) = (6.0 - z)\ (0.509)$$

The total length L is

$$L = L_R + L_e$$

The computed L and suggested L are given in a tabular form below.

Layer No	Depth z (m)	Spacing h (m)	L_e (m)	L_e (min) (m)	L_R (m)	L (cal) (m)	L (suggested) (m)
1	0.65	0.65	0.49	1.0	2.72	3.72	4.0
2	1.30	0.65	0.38	1.0	2.39	3.39	–
3	1.80	0.50	0.27	1.0	2.14	3.14	–
4	2.30	0.50	0.26	1.0	1.88	2.88	3.0
5	2.80	0.50	0.25	1.0	1.63	2.63	–
6	3.30	0.50	0.24	1.0	1.37	2.37	–
7	3.60	0.30	0.14	1.0	1.22	2.22	–
8	3.90	0.30	0.14	1.0	1.07	2.07	–
9	4.20	0.30	0.14	1.0	0.92	1.92	2.0
10	4.50	0.30	0.14	1.0	0.76	1.76	–
11	4.80	0.30	0.14	1.0	0.61	1.61	–
12	5.10	0.30	0.14	1.0	0.46	1.46	–
13	5.40	0.30	0.14	1.0	0.31	1.31	–
14	5.70	0.30	0.14	1.0	0.15	1.15	–
15	6.00	0.30	0.13	1.0	0.00	1.00	–

It may be noted here that the calculated values of L_e are very small and a minimum value of 1.0 m should be used.

(c) Check for the overlap

When the fabric layers are laid perpendicular to the wall, the adjacent fabric should overlap a length L_o. The minimum value of L_o is 1.0m. The equation for L_o may be expressed as

$$L_o = \frac{h\,p_h F_s}{2\times 2\gamma\, z \tan\delta} = \frac{h[4.68(z)+2.60]1.4}{4\times 18(z)\tan 24°}$$

The maximum value of L_o is at the upper layer at $z = 0.65$. Substituting for z we have

$$L_o = \frac{0.65[4.68(0.65)+2.60]1.4}{4\times 18(0.65)\tan 24°} = 0.25 \text{ m}$$

Since this value of L_o calculated is quite low, use $L_o = 1.0$m for all the layers.

(d) Check for external stability

The total active earth pressure P_a is

$$P_a = \frac{1}{2}\gamma H^2 K_A = \frac{1}{2}\times 18\times 6^2\times 0.28 = 90.7 \text{ kN/m}$$

$$F_s = \frac{\text{Resisting moment } M_R}{\text{Driving moment } M_o} = \frac{W_1 l_1 + W_2 l_2 + W_3 l_3 + P_v l_4}{P_a(H/3)}$$

where $W_1 = 6\times 2\times 18 = 216$ kN and $l_1 = 2/2 = 1$m

$W_2 = (6 - 2.1)\times(3 - 2)(18) = 70.2$ kN, and $l_2 = 2.5$m

$W_3 = (6 - 4.2)(4 - 3)(18) = 32.4$ kN and $l_3 = 3.5$m

$$F_s = \frac{213\times(1)+70.2\times(2.5)+32.4\times(3.5)}{90.7\times(2)} = 2.78 > 2 \quad -\text{OK}$$

Check for sliding

$$F_s = \frac{\text{Total resisting force } F_R}{\text{Total driving force } F_d}$$

$$F_R = \left(W_1 + W_2 + W_3\right)\tan\delta$$

$$= \left(216 + 70.2 + 32.4\right)\tan 25.5°$$

$$= 318.6 \times 0.477 = 152 \text{ kN}$$

$$F_d = P_a = 90.7 \text{ kN}$$

Hence $F_s = \dfrac{152}{90.7} = 1.68 > 1.5 \quad -\text{OK}$

Check for a foundation failure

Consider the wall as a surface foundation with $D_f = 0$. Since the foundation soil is cohesionless, we may write

$$q_u = \frac{1}{2}\gamma\, B\, N_\gamma$$

Use Terzaghi's theory. For $\phi = 34°$, $N_\gamma = 38$, and $B = 2$m

$$q_u = \frac{1}{2}\times 18\times 2\times 38 = 684 \text{ kN/m}^2$$

The actual load intensity on the base of the backfill

$q(\text{actual}) = 18 \times 6 + 10 = 118 \text{ kN/m}^2$

$$F_s = \frac{684}{`118} = 5.8 > 3 \text{ which is acceptable}$$

Example 19.5 (Koerner, 1999)

Design a 7m high geogrid-reinforced wall when the reinforcement vertical maximum spacing must be 1.0 m. The coverage ratio is 0.80 (Refer to Fig. Ex. 19.5). Given: T_u = 156 kN/m, C_r = 0.80, C_i = 0.75. The other details are given in the figure.

Solution

Internal Stability

From Eq. (19.14)

$$p_h = (\gamma\, z\, K_A + q_h) = \gamma\, z\, K_A + q_s\, K_A$$

$$K_A = \tan^2(45° - \phi/2) = \tan^2(45° - 32/2) = 0.31$$

$$p_h = (18\times z\times 0.31) + (15\times 0.31) = 5.58z + 4.65$$

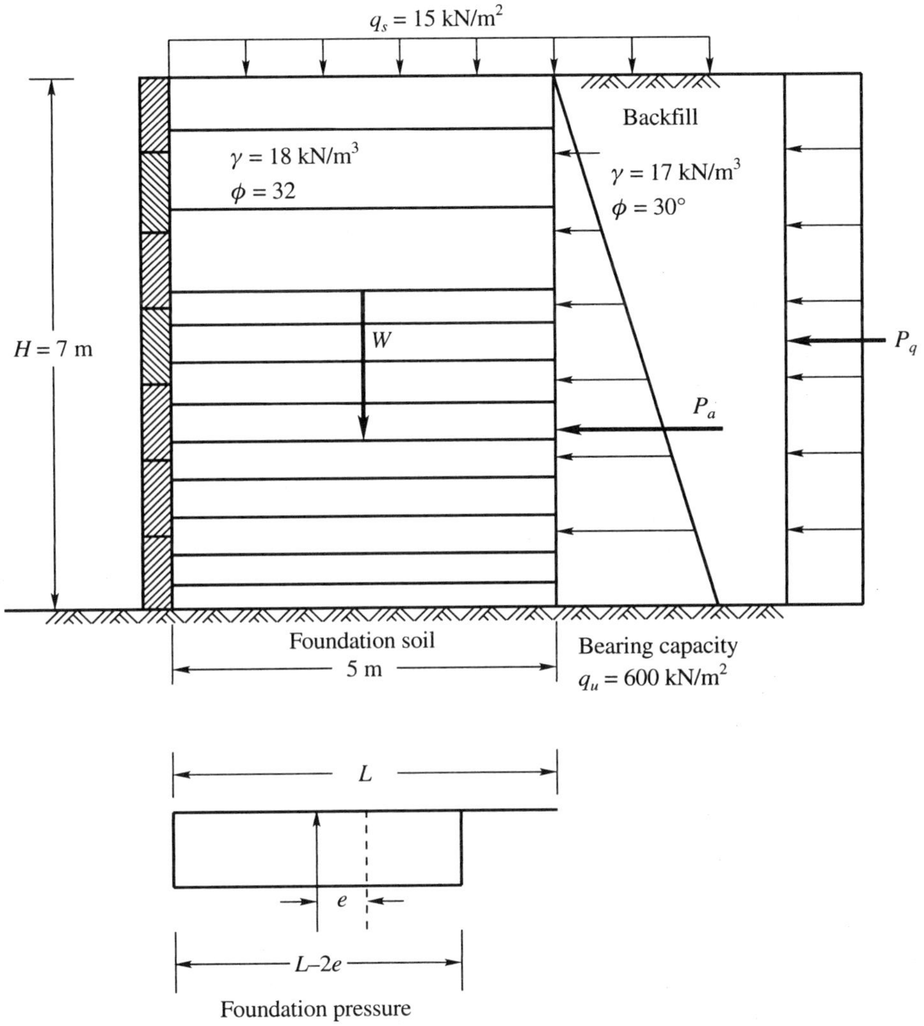

Figure Ex. 19.5

1. For geogrid vertical spacing.

   Given $T_u = 156$ kN/m

   From Eq. (19.10) and Table 19.5, we have

$$T_a = T_u \left[\frac{1}{RF_{ID} \times RF_{CR} \times RF_{BD} \times RF_{CD}} \right]$$

$$T_a = 156 \, \frac{1}{1.2 \times 2.5 \times 1.3 \times 1.0} = 40 \text{ kN/m}$$

   But use $T_{\text{design}} = 28.6$ kN/m with $F_s = 1.4$ on T_a

   From Eq. (19.28)

$$T_{\text{design}} = \frac{h\,p_h}{C_r}$$

$$28.6 = h\frac{5.58z + 4.65}{0.8}$$

$$\text{or} \quad h = \frac{22.9}{5.58z + 4.65}$$

Maximum depth for h = 1m is

$$1.0 = \frac{22.9}{5.58z + 4.65} \quad \text{or } z = 3.27\text{m}$$

Maximum depth for h = 0.5m

$$0.5 = \frac{22.9}{5.58z + 4.65} \quad \text{or } z = 7.37 \text{ m}$$

The distribution of geogrid layers is shown in Fig. Ex. 19.5.

2. Embedment length of geogrid layers.
   From Eqs (19.27) and (19.24)

   $2\ C_1\ C_r\ L_e\ p_o \tan\phi\ = T_H\ F_s = p_h \text{ h } F_s$

   Substituting known values

   $2 \times 0.75 \times 0.8 \times (L_e) \times 18 \times (z) \tan 32° = h\ (5.58z + 4.65)\ 1.5$

   Simplfying $\quad L_e = \dfrac{(0.62z + 0.516)h}{z}$

   The equation for L_R is

   $$L_R = (H - z)\tan(45° - \phi/2) = (7 - z)\tan(45° - 32/2)$$

   $$= 3.88 - 0.554(z)$$

   From the above relationships the spacing of geogrid layers and their lengths are given below.

Layer No.	Depth (m)	Spacing h (m)	L_e (m)	L_e (min) (m)	L_R (m)	L (cal) (m)	L (required) (m)
1	0.75	0.75	0.98	1.0	3.46	4.46	5.0
2	1.75	1.00	0.92	1.0	2.91	3.91	5.0
3	2.75	1.00	0.81	1.0	2.36	3.36	5.0
4	3.25	0.50	0.39	1.0	2.08	3.08	5.0
5	3.75	0.50	0.38	1.0	1.80	2.80	5.0
6	4.25	0.50	0.37	1.0	1.52	2.52	5.0
7	4.75	0.50	0.36	1.0	1.25	2.25	5.0
8	5.25	0.50	0.36	1.0	0.97	1.97	5.0
9	5.75	0.50	0.36	1.0	0.69	1.69	5.0
10	6.25	0.50	0.35	1.0	0.42	1.42	5.0
11	6.75	0.50	0.35	1.0	0.14	1.14	5.0

External Stability

(a) Pressure distribution

$$P_a = \frac{1}{2}\gamma H^2 K_A = \frac{1}{2}\times 17\times 7^2 \tan^2(45^\circ - 30/2) = 138.8 \text{ kN/m}$$

$$P_q = q_s K_A H = 15\times 0.33\times 7 = 34.7 \text{ kN/m}$$

$$\text{Total } \approx 173.5 \text{ kN/m}$$

1. Check for sliding (neglecting effect of surcharge)

$$F_R = W\tan\delta = \gamma\times H\times L\tan 25^\circ = 18\times 7\times 5.0\times 0.47 = 293.8 \text{ kN/m}$$

$$P = P_a + P_q = 173.5 \text{ kN/m}$$

$$F_s = \frac{293.8}{173.5} = 1.69 > 1.5 \quad \text{OK}$$

2. Check for overturning

$$\text{Resisting moment} \quad M_R = W\times\frac{L}{2} = 18\times 7\times 5\times\frac{5}{2} = 1575 \text{ kN-m}$$

$$\text{Overturning moment } M_O = P_a\times\frac{H}{3} + P_q\times\frac{H}{2}$$

$$\text{or } M_O = 138.8\times\frac{7}{3} + 34.7\times\frac{7}{2} = 445.3 \text{ kN-m}$$

$$F_s = \frac{1575}{445.3} = 3.54 > 2.0 \quad \text{OK}$$

3. Check for bearing capacity

$$\text{Eccentricity } e = \frac{M_O}{W + q_s L} = \frac{445.3}{18\times 7\times 5 + 15\times 5} = 0.63$$

$$e = 0.63 < \frac{L}{6} = \frac{5}{6} = 0.83 \quad \text{Ok}$$

Effective length = $L - 2e = 5 - 2\times 0.63 = 3.74$m

$$\text{Bearing pressure} = [18\times 7 + 15]\left(\frac{5}{3.74}\right) = 189 \text{ kN/m}^2$$

$$F_s = \frac{600}{189} = 3.17 > 3.0 \text{ OK}$$

19.12 EXAMPLES OF MEASURED LATERAL EARTH PRESSURES

Backfill Reinforced with Metal Strips

Laboratory tests were conducted on retaining walls with backfills reinforced with metal strips (Lee et al., 1973). The walls were built within a 30 in. × 48 in. × 2 in. wooden box. Skin elements

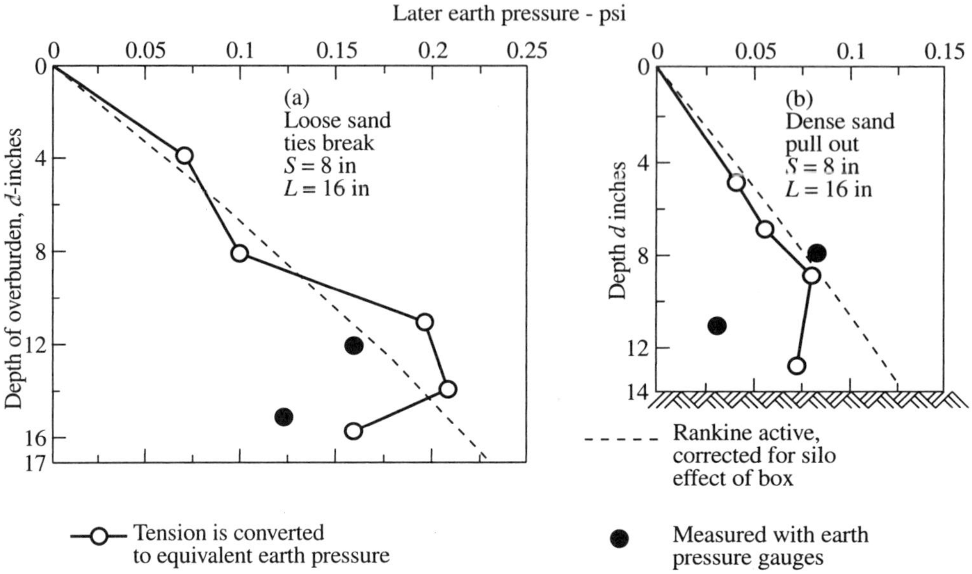

Figure 19.18 Typical examples of measured lateral earth pressures just prior to wall failure (1 in. = 25.4 mm; 1 psi = 6.89 kN/m^2) (Lee et al., 1973)

were made from 0.012 in aluminum sheet. The strips (ties) used for the tests were 0.155 in wide and 0.0005 in thick aluminum foil. The backfill consisted of dry Ottawa No. 90 sand. The small walls built of these materials in the laboratory were constructed in much the same way as the larger walls in the field. Two different sand densities were used: loose, corresponding to a relative density, $D_r = 20\%$, and medium dense, corresponding to $D_r = 63\%$, and the corresponding angles of internal friction were 31° and 44° respectively. SR-4 strain gages were used on the ties to determine tensile stresses in the ties during the tests.

Examples of the type of earth pressure data obtained from two typical tests are shown in Fig. 19.18. Data in Fig. 19.18(a) refer to a typical test in loose sand whereas data in 19.18(b) refer to test in dense sand. The ties lengths were different for the two tests. For comparison, Rankine lateral earth pressure variation with depth is also shown. It may be seen from the curve that the measured values of the earth pressures follow closely the theoretical earth pressure variation up to two thirds of the wall height but fall off comparatively to lower values in the lower portion.

Field Study of Retaining Walls with Geogrid Reinforcement

Field studies of the behavior of geotextile or geogrid reinforced permanent wall studies are fewer in number. Berg et al., (1986) reported the field behavior of two walls with geogrid reinforcement. One wall in Tucson, Arizona, 4.6 m high, used a cumulative reduction factor of 2.6 on ultimate strength for allowable strength T_a and a value of 1.5 as a global factor of safety. The second wall was in Lithonia, Georgia, and was 6 m high. It used the same factors and design method. Fig. 19.19 presents the results for both the walls shortly after construction was complete. It may be noted that the horizontal pressures at various wall heights are overpredicted for each wall, that is, the wall designs that were used appear to be quite conservative.

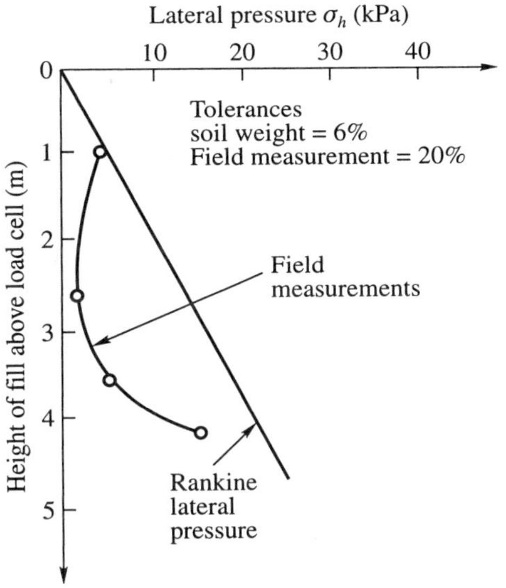

(a) Results of Tucson, Arizona, wall

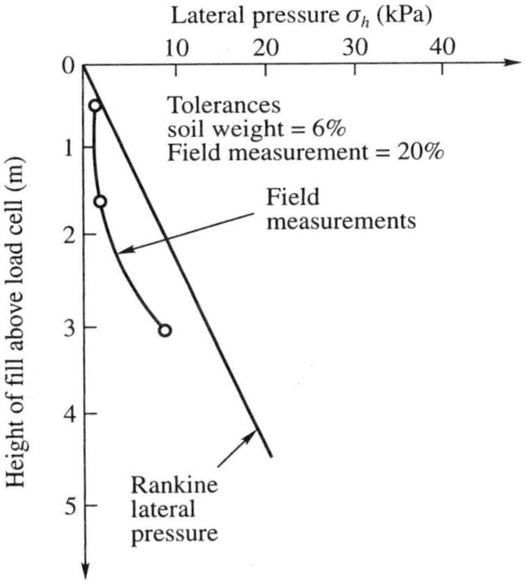

(b) Results of Lithonia, Georgia, wall

Figure 19.19 Comparison of measured stresses to design stresses for two geogrid reinforced walls (Berg et al., 1986)

19.13 PROBLEMS

19.1 Fig. Prob. 19.1 gives a section of a cantilever wall. Check the stability of the wall with respect to (a) overturning, (b) sliding, and (c) bearing capacity.

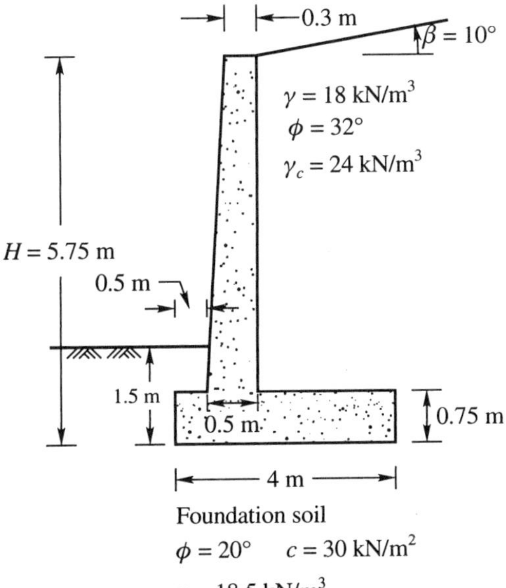

Figure Prob. 19.1

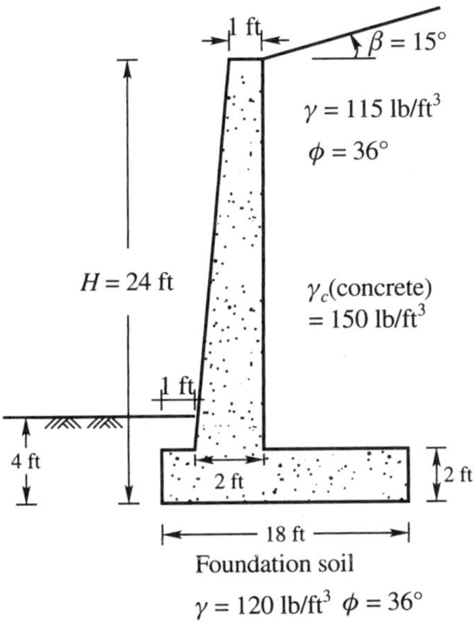

Figure Prob. 19.3

19.2 Check the stability of the wall given in Prob. 19.1 for the condition that the slope is horizontal and the foundation soil is cohesionless with $\phi = 30°$. All the other data remain the same.

19.3 Check the stability of the cantilever wall given in Fig. Prob. 19.3 for (a) overturning, (b) sliding, and (c) bearing capacity failure.

19.4 Check the stability of the wall in Prob. 19.3 assuming (a) $\beta = 0$, and (b) the foundation soil has $c = 300$ lb/ft^2, $\gamma = 115$ lb/ft^3, and $\phi = 26°$.

19.5 Fig. Prob. 19.5 depicts a gravity retaining wall. Check the stability of the wall for sliding, and overturning.

19.6 Check the stability of the wall given in Fig. Prob. 19.6. All the data are given on the figure.

19.7 Check the stability of the gravity wall given in Prob. 19.6 with the foundation soil having properties $\phi = 30°$, $\gamma = 110$ lb/ft^3 and $c = 500$ lb/ft^2. All the other data remain the same.

19.8 Check the stability of the gravity retaining wall given in Fig. Prob. 19.8.

19.9 Check the stability of the gravity wall given in Prob. 19.8 for Coulomb's condition. Assume $\delta = 2/3\phi$.

19.10 A typical section of a wall with granular backfill reinforced with metal strips is given in Fig. Prob. 19.10. The following data are available.

$H = 6$ m, $b = 75$ mm, $t = 5$ mm, $f_y = 240$ MPa, F_s for steel = 1.75, F_s on soil friction = 1.5. The other data are given in the figure. Spacing: $h = 0.6$ m, and $s = 1$ m.

Required

(a) Lengths of tie at varying depths

(b) Check for external stability

19.11 Solve the Prob. 19.10 with a uniform surcharge acting on the backfill surface. The intensity of surcharge is 20 kN/m^2.

19.12 Figure Prob. 19.12 shows a section of a MSE wall with geotextile reinforcement.

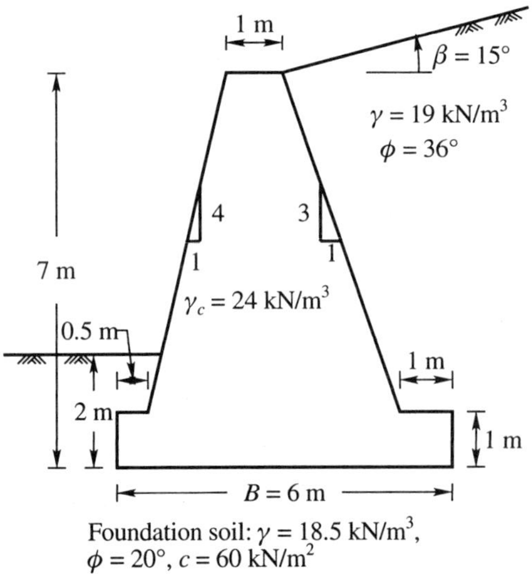

Figure Prob. 19.5

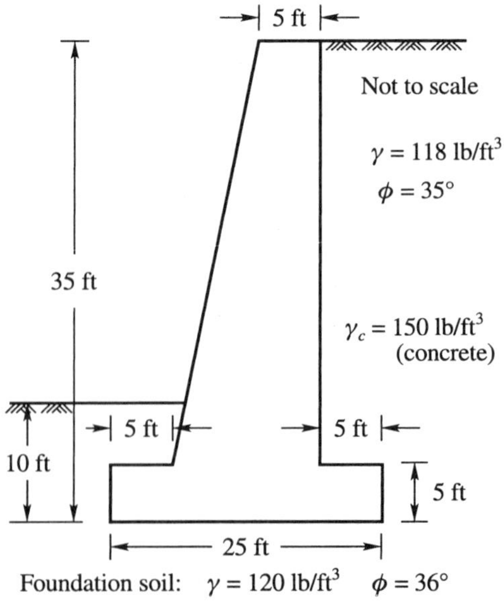

Figure Prob. 19.6

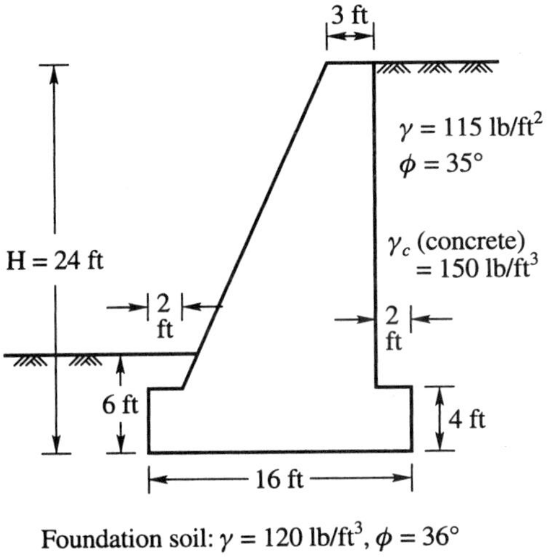

Figure Prob. 19.8

Required:

(a) Spacing of the individual layers of geotextile

(b) Length of geotextile in each layer

(c) Check for external stability

19.13 Design a 6 m high geogrid-reinforced wall (Fig. Prob. 19.13), where the reinforcement maximum spacing must be at 1.0 m. The coverage ratio C_r = 0.8 and the interaction coefficient C_i = 0.75, and T_a = 26 kN/m. (T_{design})

Given : Reinforced soil properties : γ = 18 kN/m³ ϕ = 32°

Foundation soil : γ = 17.5 kN/m³ ϕ = 34°

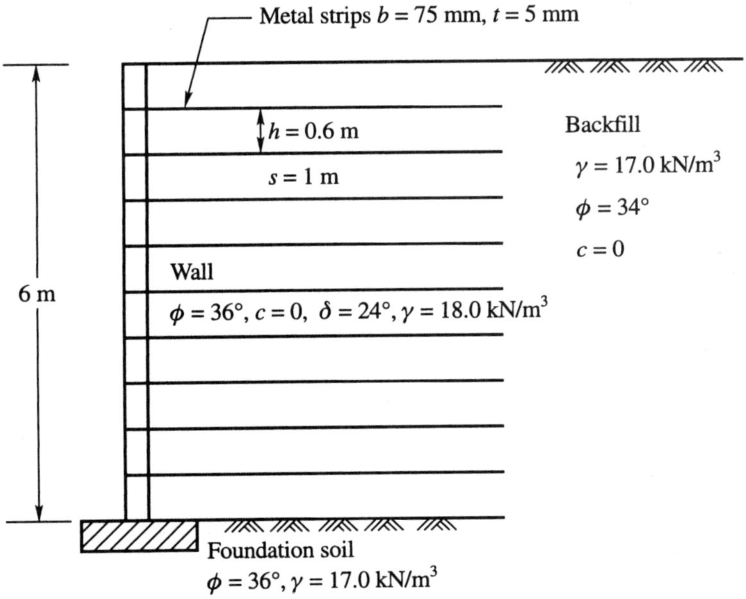

Figure Prob. 19.10

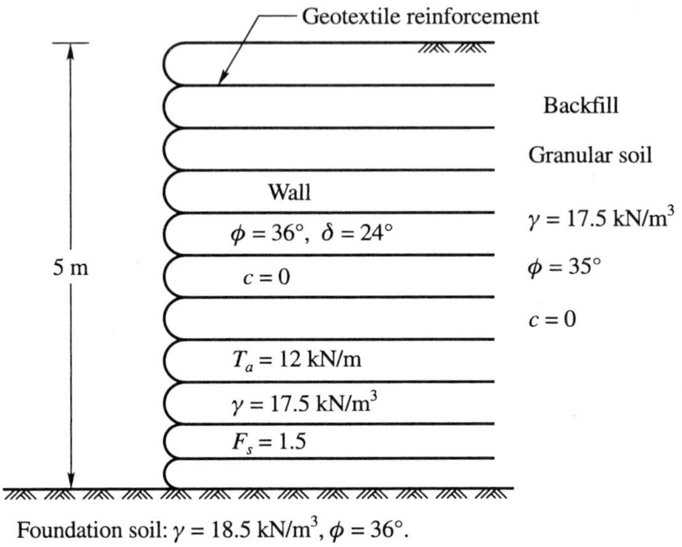

Figure Prob. 19.12

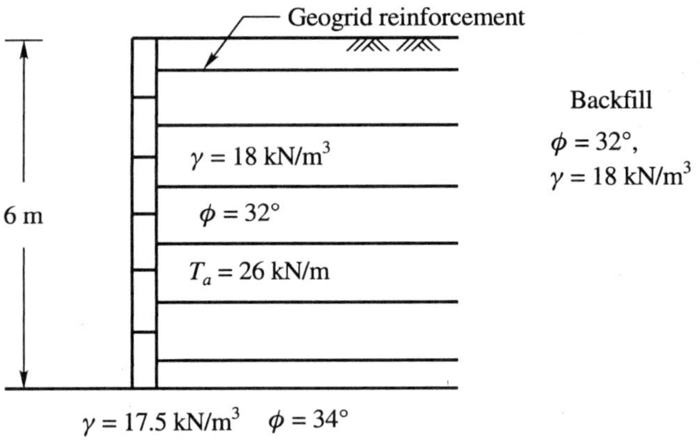

Figure Prob. 19.13

CHAPTER 20

SHEET PILE WALLS AND BRACED CUTS

20.1 INTRODUCTION

Sheet pile walls are retaining walls constructed to retain earth, water or any other fill material. These walls are thinner in section as compared to masonry walls described in Chapter 19. Sheet pile walls are generally used for the following:

1. Water front structures, for example, in building wharfs, quays, and piers
2. Building diversion dams, such as cofferdams
3. River bank protection
4. Retaining the sides of cuts made in earth

Sheet piles may be of timber, reinforced concrete or steel. Timber piling is used for short spans and to resist light lateral loads. They are mostly used for temporary structures such as braced sheeting in cuts. If used in permanent structures above the water level, they require preservative treatment and even then, their span of life is relatively short. Timber sheet piles are joined to each other by tongue-and-groove joints as indicated in Fig. 20.1. Timber piles are not suitable for driving in soils consisting of stones as the stones would dislodge the joints.

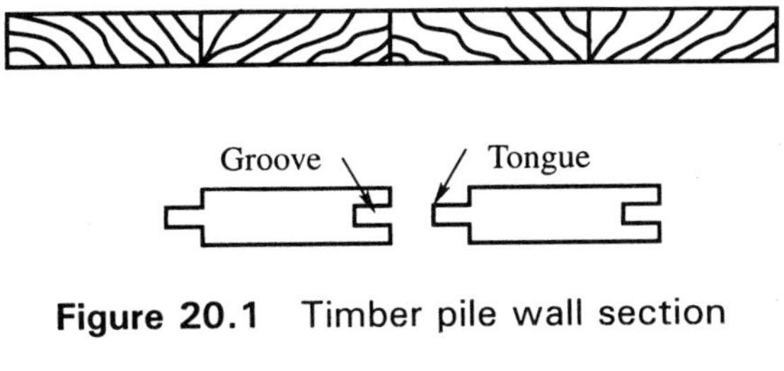

Figure 20.1 Timber pile wall section

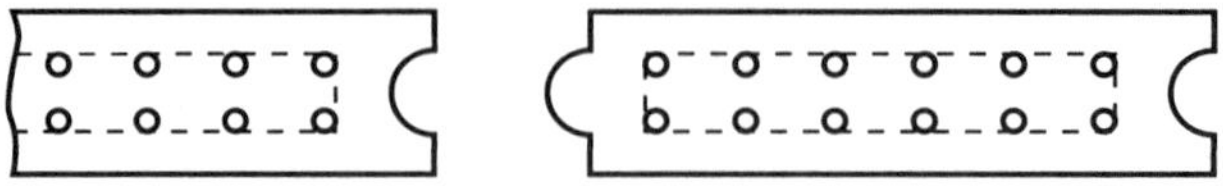

Figure 20.2 Reinforced concrete Sheet pile wall section

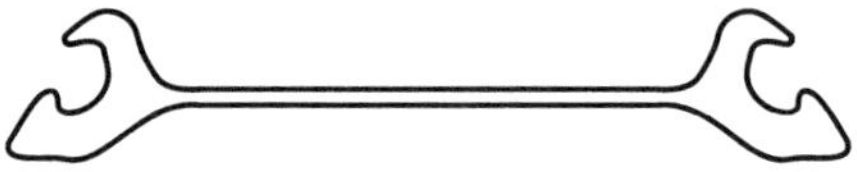

(a) Straight sheet piling

(b) Shallow arch-web piling

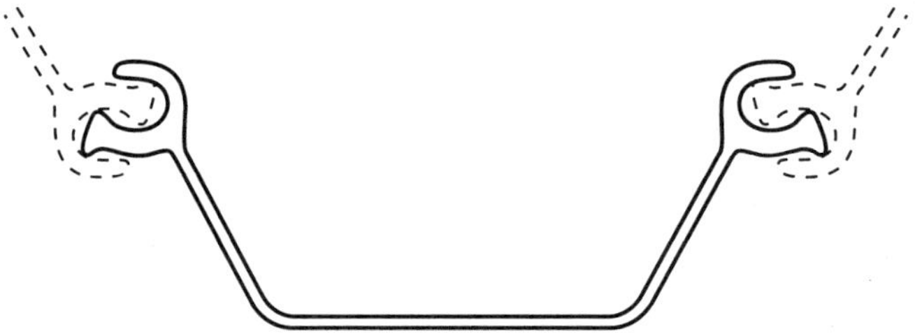

(c) Arch-web piling

(d) Z-pile

Figure 20.3 Sheet pile sections

Reinforced concrete sheet piles are precast concrete members, usually with a tongue-and-groove joint. Typical section of piles are shown in Fig. 20.2. These piles are relatively heavy and bulky. They displace large volumes of solid during driving. This large volume displacement of soil tends to increase the driving resistance. The design of piles has to take into account the large driving stresses and suitable reinforcement has to be provided for this purpose.

The most common types of piles used are steel sheet piles. Steel piles possess several advantages over the other types. Some of the important advantages are:

1. They are resistant to high driving stresses as developed in hard or rocky material
2. They are lighter in section
3. They may be used several times

4. They can be used either below or above water and possess longer life
5. Suitable joints which do not deform during driving can be provided to have a continuous wall
6. The pile length can be increased either by welding or bolting

Steel sheet piles are available in the market in several shapes. Some of the typical pile sections are shown in Fig. 20.3. The archweb and Z-piles are used to resist large bending moments, as in anchored or cantilever walls. Where the bending moments are less, shallow-arch piles with corresponding smaller section moduli can be used. Straight-web sheet piles are used where the web will be subjected to tension, as in cellular cofferdams. The ball-and-socket type of joints, Fig. 20.3 (d), offer less driving resistance than the thumb-and-finger joints, Fig. 20.3 (c).

20.2 SHEET PILE STRUCTURES

Steel sheet piles may conveniently be used in several civil engineering works. They may be used as:

1. Cantilever sheet piles
2. Anchored bulkheads
3. Braced sheeting in cuts
4. Single cell cofferdams
5. Cellular cofferdams, circular type
6. Cellular cofferdams (diaphragm)

Anchored bulkheads Fig. 20.4 (b) serve the same purpose as retaining walls. However, in contrast to retaining walls whose weight always represent an appreciable fraction of the weight of the sliding wedge, bulkheads consist of a single row of relatively light sheet piles of which the lower ends are driven into the earth and the upper ends are anchored by tie or anchor rods. The anchor rods are held in place by anchors which are buried in the backfill at a considerable distance from the bulkhead.

Anchored bulkheads are widely used for dock and harbor structures. This construction provides a vertical wall so that ships may tie up alongside, or to serve as a pier structure, which may jet out into the water. In these cases sheeting may be required to laterally support a fill on which railway lines, roads or warehouses may be constructed so that ship cargoes may be transferred to other areas. The use of an anchor rod tends to reduce the lateral deflection, the bending moment, and the depth of the penetration of the pile.

Cantilever sheet piles depend for their stability on an adequate embedment into the soil below the dredge line. Since the piles are fixed only at the bottom and are free at the top, they are called *cantilever sheet piles*. These piles are economical only for moderate wall heights, since the required section modulus increases rapidly with an increase in wall height, as the bending moment increases with the cube of the cantilevered height of the wall. The lateral deflection of this type of wall, because of the cantilever action, will be relatively large. Erosion and scour in front of the wall, i.e., lowering the dredge line, should be controlled since stability of the wall depends primarily on the developed passive pressure in front of the wall.

20.3 FREE CANTILEVER SHEET PILE WALLS

When the height of earth to be retained by sheet piling is small, the piling acts as a cantilever. The forces acting on sheet pile walls include:

1. The active earth pressure on the back of the wall which tries to push the wall away from the backfill

2. The passive pressure in front of the wall below the dredge line. The passive pressure resists the movements of the wall

The active and passive pressure distributions on the wall are assumed hydrostatic. In the design of the wall, although the Coulomb approach considering wall friction tends to be more realistic, the Rankine approach (with the angle of wall friction $\delta = 0$) is normally used.

The pressure due to water may be neglected if the water levels on both sides of the wall are the same. If the difference in level is considerable, the effect of the difference on the pressure will have to be considered. Effective unit weights of soil should be considered in computing the active and passive pressures.

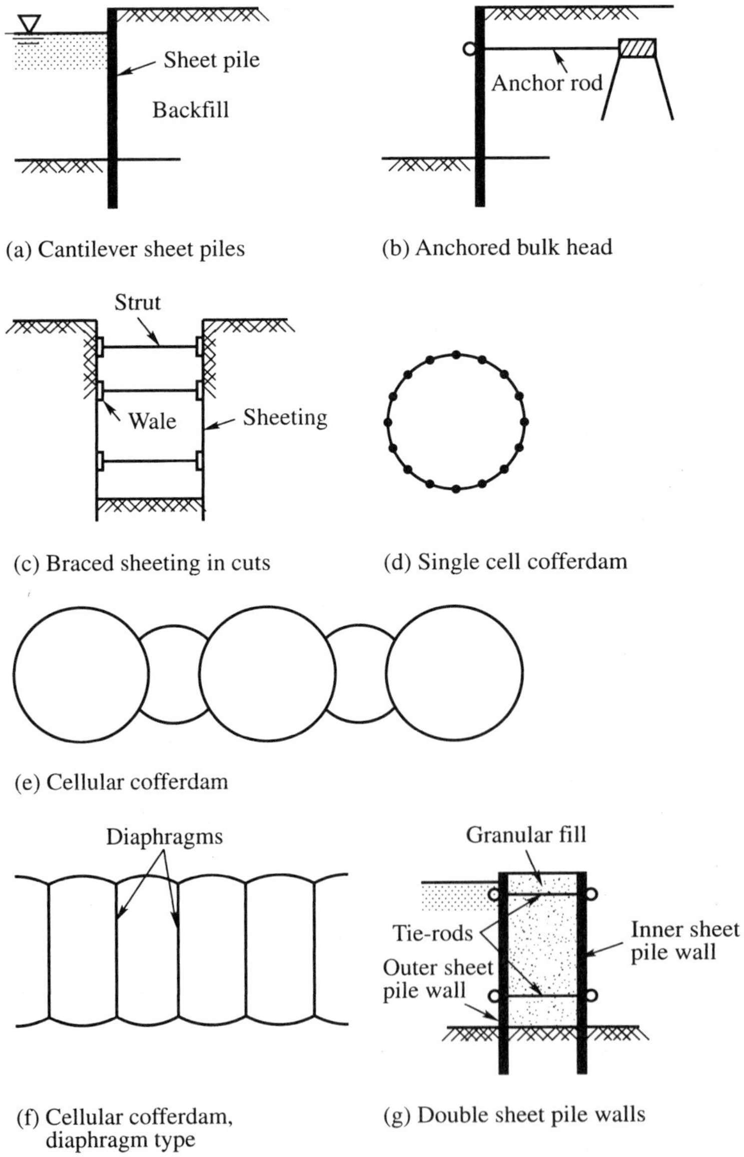

Figure 20.4 Use of sheet piles

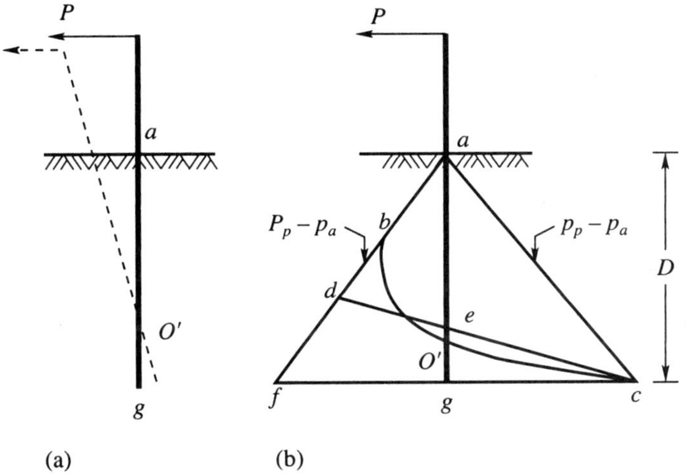

Figure 20.5 Example illustrating earth pressure on cantilever sheet piling

General Principle of Design of Free Cantilever Sheet Piling

The action of the earth pressure against cantilever sheet piling can be best illustrated by a simple case shown in Fig. 20.5 (a). In this case, the sheet piling is assumed to be perfectly rigid. When a horizontal force P is applied at the top of the piling, the upper portion of the piling tilts in the direction of P and the lower portion moves in the opposite direction as shown by a dashed line in the figure. Thus the piling rotates about a stationary point O'. The portion above O' is subjected to a passive earth pressure from the soil on the left side of the piles and an active pressure on the right side of the piling, whereas the lower portion $O'g$ is subjected to a passive earth pressure on the right side and an active pressure on the left side of the piling. At point O' the piling does not move and therefore is subjected to equal and opposite earth pressures (at-rest pressure from both sides) with a net pressure equal to zero. The net earth pressure (the difference between the passive and the active) is represented by $abO'c$ in Fig. 20.5 (b). For the purpose of design, the curve $bO'c$ is replaced by a straight line dc. Point d is located at such a location on the line af that the sheet piling is in static equilibrium under the action of force P and the earth pressures represented by the areas ade and ecg. The position of point d can be determined by a trial and error method.

This discussion leads to the conclusion that cantilever sheet piling derives its stability from passive earth pressure on both sides of the piling. However, the distribution of earth pressure is different between sheet piling in granular soils and sheet piling in cohesive soils. The pressure distribution is likely to change with time for sheet pilings in clay.

20.4 DEPTH OF EMBEDMENT OF CANTILEVER WALLS IN SANDY SOILS

Case 1: With Water Table at Great Depth

The active pressure acting on the back of the wall tries to move the wall away from the backfill. If the depth of embedment is adequate the wall rotates about a point O' situated above the bottom of the wall as shown in Fig. 20.6 (a). The types of pressure that act on the wall when rotation is likely to take place about O' are:

1. Active earth pressure at the back of wall from the surface of the backfill down to the point of rotation, O'. The pressure is designated as P_{a1}.

2. Passive earth pressure in front of the wall from the point of rotation O' to the dredge line. This pressure is designated as P_{p_1}.
3. Active earth pressure in front of the wall from the point of rotation to the bottom of the wall. This pressure is designated as P_{a_2}.
4. Passive earth pressure at the back of wall from the point of rotation O' to the bottom of the wall. This pressure is designated as P_{p2}.

The pressures acting on the wall are shown in Fig. 20.6 (a).

If the passive and active pressures are algebraically combined, the resultant pressure distribution below the dredge line will be as given in Fig. 20.6 (b). The various notations used are:

D = minimum depth of embedment with a factor of safety equal to 1

K_A = Rankine active earth pressure coefficient

K_P = Rankine passive earth pressure coefficient

K = $K_P - K_A$

$\bar{p}_a$ = effective active earth pressure acting against the sheet pile at the dredge line $= \gamma H K_A$

$\bar{p}_p$ = effective passive earth pressure at the base of the pile wall and acting towards the backfill $= \gamma D_o K$

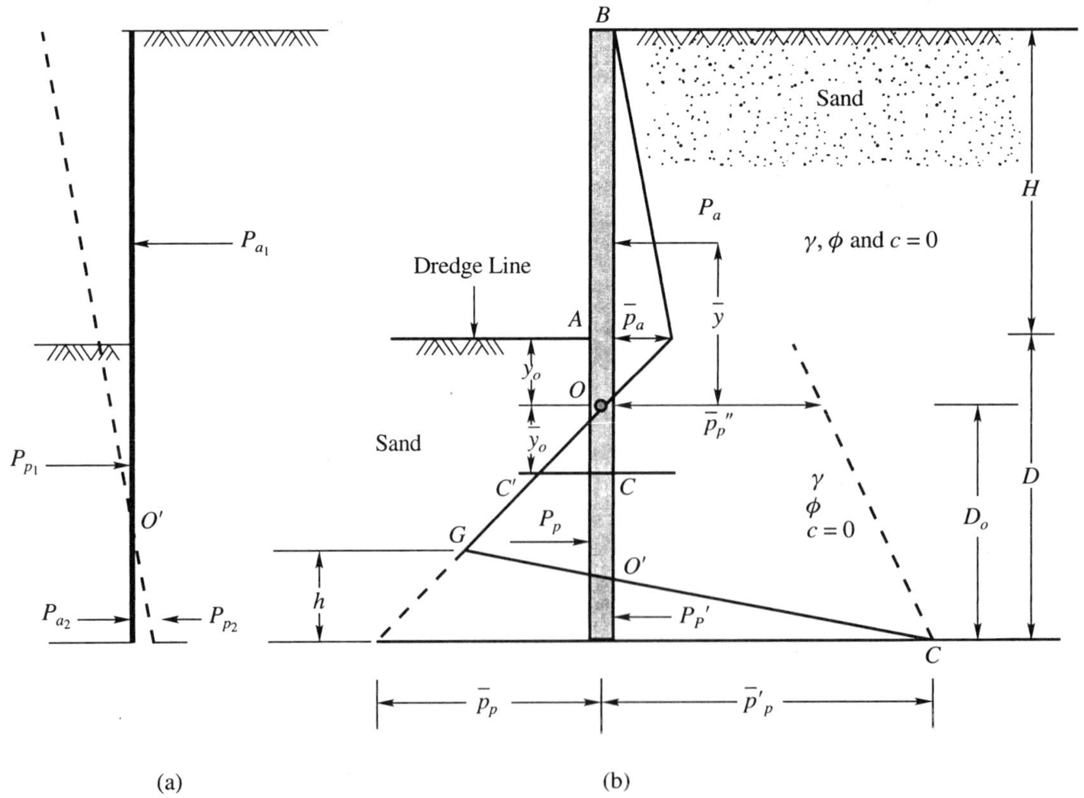

Figure 20.6 Pressure distribution on a cantilever wall.

$\overline{p}'_p$ = effective passive earth pressure at the base of the sheet pile wall acting against the backfill side of the wall = $p''_p + \gamma K D_o$

$\overline{p}''_p$ = effective passive earth pressure at level of $O = \gamma y_0 K + \gamma H K_P$

γ = effective unit weight of the soil assumed the same below and above dredge line

y_0 = depth of point O below dredge line where the active and passive pressures are equal

$\overline{y}$ = height of point of application of the total active pressure P_a above point O

h = height of point G above the base of the wall

D_0 = height of point O above the base of the wall

Expression for y_0

At point O, the passive pressure acting towards the right should equal the active pressure acting towards the left, that is

$$\gamma y_0 K_p = \gamma (H + y_0) K_A$$

Therefore, $\gamma y_0 (K_p - K_A) = \gamma H K_A$

$$y_0 = \frac{\gamma H K_A}{\gamma (K_p - K_A)} = \frac{\overline{p}_a}{\gamma K} \tag{20.1}$$

Expression for h

For static equilibrium, the sum of all the forces in the horizontal direction must equal zero. That is

$$P_a - \frac{1}{2}\overline{p}_p\ (D - y_0) + \frac{1}{2}\ (\overline{p}_p + \overline{p}'_p)h = 0$$

Solving for h,

$$h = \frac{\overline{p}_p (D - y_0) - 2P_a}{\overline{p}_p + \overline{p}'_p} \tag{20.2}$$

Taking moments of all the forces about the bottom of the pile, and equating to zero,

$$P_a (D_0 + \overline{p}) - \frac{1}{2}\ \overline{p}_p \times D_0 \times \frac{D_0}{3} + \frac{1}{2}(\overline{p}_p + \overline{p}'_p) \times h \times \frac{h}{3} = 0$$

$$\text{or}\quad 6P_a (D_0 + \overline{y}) - \overline{p}_p D_0^2 + \left(\overline{p}_p + \overline{p}'_p\right) h^2 = 0 \tag{20.3}$$

Therefore,

$$\overline{p}_p = \gamma\ K\ D_0$$

$$\overline{p}'_p = \overline{p}''_p + \gamma\ K D_0$$

Substituting in Eq. (20.3) for $\overline{p}_p$, $\overline{p}'_p$ and h and simplifying,

$$D_0^4 + C_1 D_0^3 + C_2 D_0^2 + C_3 D_0 + C_4 = 0 \tag{20.4}$$

where,

$$C_1 = \frac{\bar{p}_p''}{\gamma K}$$

$$C_2 = -\frac{8P_a}{\gamma K}$$

$$C_3 = -\left[\frac{6P_a}{(\gamma K)^2}\left(2\bar{y}\gamma K + \bar{p}_p''\right)\right]$$

$$C_4 = -\left[\frac{6P_a \bar{y}\bar{p}_p'' + 4P_a^2}{(\gamma K)^2}\right]$$

The solution of Eq. (20.4) gives the depth D_0. The method of trial and error is generally adopted to solve this equation. The minimum depth of embedment D with a factor of safety equal to 1 is therefore

$$D = D_0 + y_0 \tag{20.5}$$

A minimum factor of safety of 1.5 to 2 may be obtained by increasing the minimum depth D by 20 to 40 percent.

Maximum Bending Moment

The maximum moment on section AB in Fig. 20.6(b) occurs at the point of zero shear. This point occurs below point O in the figure. Let this point be represented by point C at a depth $\bar{y}_0$ below point O. The net pressure (passive) of the triangle OCC' must balance the net active pressure P_a acting above the dredge line. The equation for P_a is

$$P_a = \frac{1}{2}\bar{y}_0^2 \gamma (K_p - K_A) = \frac{1}{2}\bar{y}_0^2 \gamma K$$

or
$$\bar{y}_0 = \sqrt{\frac{2P_a}{\gamma K}} \tag{20.6}$$

where γ = effective unit weight of the soil. If the water table lies above point O, γ will be equal to γ_b, the submerged unit weight of the soil.

Once the point of zero shear is known, the magnitude of the maximum bending moment may be obtained as

$$M_{\max} = P_a(\bar{y} + \bar{y}_o) - \frac{1}{3}\left[\frac{1}{2}\bar{y}_o^2 \gamma K\right]\bar{y}_o = P_a(\bar{y} + \bar{y}_o) - \frac{1}{6}\bar{y}_o^3 \gamma K \tag{20.7}$$

The section modulus Z_s of the sheet pile may be obtained from the equation

$$Z_s = \frac{M_{\max}}{f_b} \tag{20.8}$$

where, f_b = allowable flexural stress of the sheet pile.

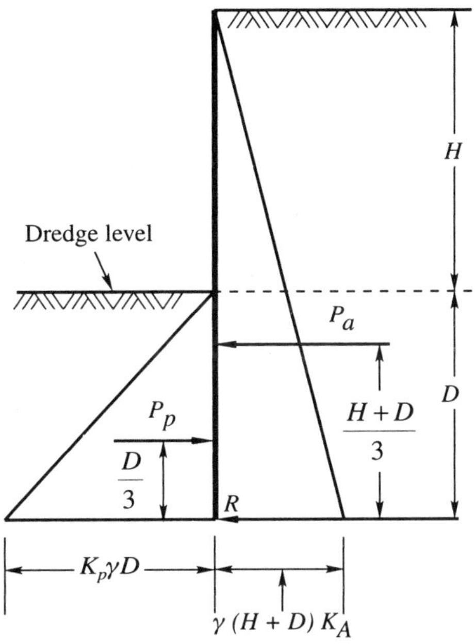

Figure 20.7 Simplified method of determining D for cantilever sheet pile

Simplified Method

The solution of the fourth degree equation is quite laborious and the problem can be simplified by assuming the passive pressure $\overline{p'_p}$ (Fig. 20.6) as a concentrated force R acting at the foot of the pile. The simplified arrangement is shown in Fig. 20.7.

For equilibrium, the moments of the active pressure on the right and passive resistance on the left about the point of reaction R must balance.

$$\frac{1}{3}P_p D - \frac{P_a}{3}(H+D) = 0$$

Now, $P_p = \frac{1}{2}K_p \gamma D^2$ and $P_a = \frac{1}{2}K_A \gamma (H+D)^2$

Therefore, $K_p D^3 - K_A (H+D)^3 = 0$

$$\text{or} \quad KD^3 - 3HD(H+D)K_A = 0. \tag{20.9}$$

The solution of Eq. (20.9) gives a value for D which is at least a guide to the required depth. The depth calculated should be increased by at least 20 percent to provide a factor of safety and to allow extra length to develop the passive pressure R. An approximate depth of embedment may be obtained from Table 20.1.

Case 2: With Water Table Within the Backfill

Figure 20.8 gives the pressure distribution against the wall with a water table at a depth h_1 below the ground level. All the notations given in Fig. 20.8 are the same as those given in Fig. 20.6. In this case the soil above the water table has an effective unit weight γ and a saturated unit weight γ_{sat} below the water table. The submerged unit weight is

Table 20.1 Approximate penetration (D) of sheet piling

Relative density	Depth, D
Very loose	2.0 H
Loose	1.5 H
Firm	1.0 H
Dense	0.75 H

Source: Teng, 1969.

$\gamma_b = (\gamma_{sat} - \gamma_w)$.

The active pressure at the water table is

$$\bar{p}_1 = \gamma h_1 K_A$$

and $\bar{p}_a$ at the dredge line is

$$\bar{p}_a = \gamma h_1 K_A + \gamma_b h_2 K_A = (\gamma h_1 + \gamma_b h_2) K_A$$

The other expressions are

$$\bar{p}_p = \gamma_b D_o K$$

$$\bar{p}'_p = \bar{p}''_p + \gamma_b D_o K$$

$$\bar{p}''_p = \gamma_b y_o K + (\gamma h_1 + \gamma_b h_2) K_p$$

$$y_o = \frac{\bar{p}_a}{\gamma_b K}$$

$$P_a = \frac{1}{2}\bar{p}_p (D - y_o) - \frac{1}{2}(\bar{p}_p + \bar{p}'_p)h$$

$$h = \frac{\bar{p}_p (D - y_o) - 2P_a}{\bar{p}_p + \bar{p}'_p}$$

The fourth degree equation in terms of D_o is

$$D_o^4 + C_1 D_o^3 + C_2 D_o^2 + C_3 D_o + C_4 = 0 \qquad (20.10)$$

where, $C_1 = \dfrac{\bar{p}''_p}{\gamma_b K}$

$$C_2 = -\frac{8P_a}{\gamma_b K}$$

$$C_3 = -\frac{6P_a}{(\gamma_b K)^2}\left(2\bar{y}\gamma_b K + \bar{p}''_p\right)$$

$$C_4 = -\frac{6P_a \bar{y}\,\bar{p}''_p + 4P_a^2}{(\gamma_b K)^2}$$

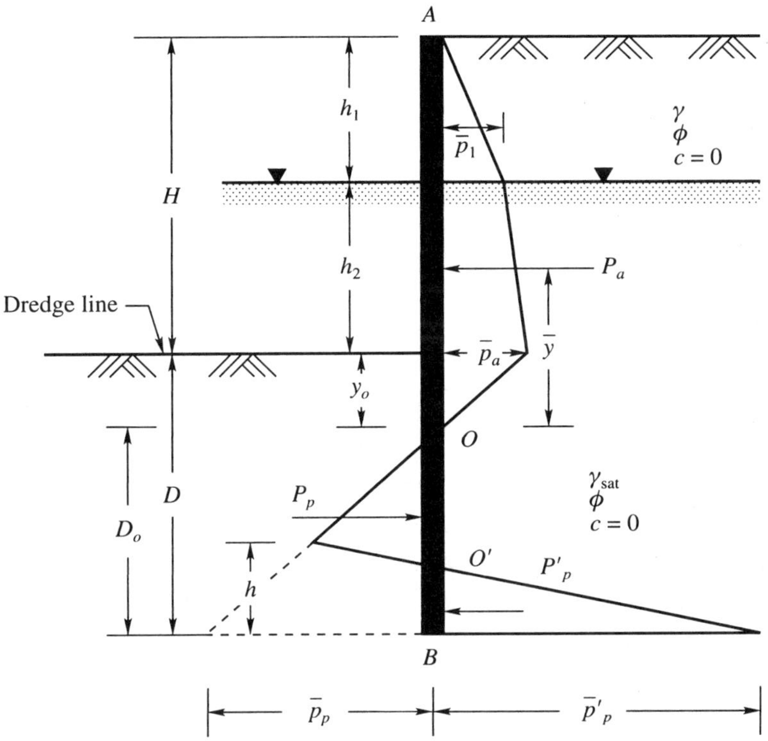

Figure 20.8 Pressure distribution on a cantilever wall with a water table in the backfill

The depth of embedment can be determined as in the previous case and also the maximum bending moment can be calculated. The depth D computed should be increased by 20 to 40 percent.

Case 3: When the Cantilever is Free Standing with No Backfill (Fig. 20.9)

The cantilever is subjected to a line load of P per unit length of wall. The expressions can be developed on the same lines explained earlier for cantilever walls with backfill. The various expressions are

$$h = \frac{\gamma KD^2 - 2P}{2\gamma DK}$$

where $K = (K_p - K_A)$

The fourth degree equation in D is

$$D^4 + C_1 D^2 + C_2 D + C_3 = 0 \tag{20.11}$$

where $C_1 = -\dfrac{8P}{\gamma K}$

$$C_2 = -\frac{12PH}{\gamma K}$$

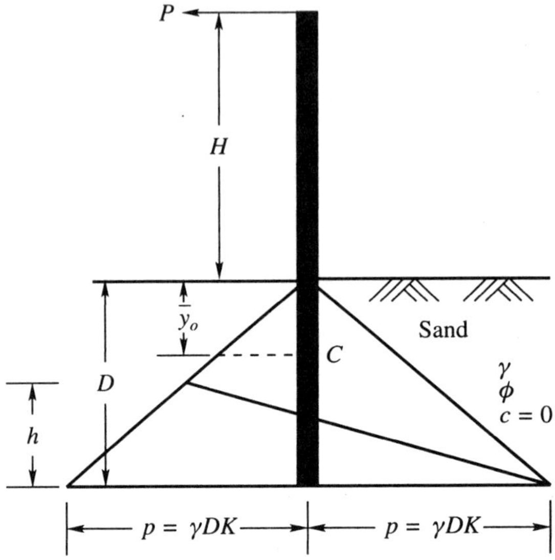

Figure 20.9 Free standing cantilever with no backfill

$$C_3 = -\frac{4P^2}{(\gamma K)^2}$$

Equation (20.11) gives the theoretical depth D which should be increased by 20 to 40 percent. Point C in Fig. 20.9 is the point of zero shear. Therefore,

$$M_{\max} = P(H + \bar{y}_o) - \frac{\gamma \bar{y}_o^3 K}{6} \tag{20.12}$$

$$\bar{y}_o = \sqrt{\frac{2P}{\gamma K}}$$

where γ = effective unit weight of the soil

Example 20.1

Determine the depth of embedment for the sheet-piling shown in Fig. Ex. 20.1a by rigorous analysis. Determine also the minimum section modulus. Assume an allowable flexural stress f_b = 175 MN/m². The soil has an effective unit weight of 17 kN/m³ and angle of internal friction of 30°.

Solution

For $\phi = 30°$, $K_A = \tan^2 (45° - \phi/2) = \tan^2 30 = \frac{1}{3}$

$$K_p = \frac{1}{K_A} = 3, \quad K = K_P - K_A = 3 - \frac{1}{3} = 2.67.$$

The pressure distribution along the sheet pile is assumed as shown in Fig. Ex. 20.1(b)

$$\bar{p}_a = \gamma H K_A = 17 \times 6 \times \frac{1}{3} = 34 \ \text{kN/m}^2$$

From Eq (20.1)

$$y_0 = \frac{\bar{p}_a}{\gamma(K_p - K_A)} = \frac{34}{17 \times 2.67} = 0.75 \text{ m.}$$

$$P_a = \frac{1}{2}\bar{p}_a H + \frac{1}{2}\bar{p}_a y_0 = \frac{1}{2} \times 34 \times 6 + \frac{1}{2} \times 34 \times 0.75$$

= 102 + 12.75 = 114.75 kN/meter length of wall or say 115 kN/m.

$$\bar{p}_p = \gamma D(K_p - K_A) - \bar{p}_a = 17 \times D \times 2.67 - 34 = 45.4D - 34$$

$$\bar{p}'_p = \gamma HK_p + \gamma D(K_p - K_A) = 17 \times 6 \times 3 + 17 \times D \times 2.67 = 306 + 45.4D$$

$$\bar{p}''_p = \gamma HK_p + \gamma y_0 (K_p - K_A) = 17 \times 6 \times 3 + 17 \times 0.75 \times 2.67 = 340 \text{ kN/m}^2$$

To find $\bar{y}$

$$P_a \bar{y} = \frac{1}{2}\bar{p}_a H\left(\frac{H}{3} + y_0\right) + \frac{1}{2}\bar{p}_a y_0\left(\frac{2}{3}y_0\right)$$

$$= \frac{1}{2} \times 34 \times 6 \times (2 + 0.75) + \frac{1}{2} \times 34 \times 0.75 \times \left(\frac{2}{3} \times 0.75\right) = 286.9$$

Therefore, $\bar{y} = \frac{286.9}{P_a} = \frac{286.9}{115} = 2.50$ m.

Now D_0 can be found from Eq. (20.4), namely

$$D_0^4 + C_1 D_0^3 + C_2 D_0^2 + C_3 D_0 + C_4 = 0$$

$$C_1 = \frac{\bar{p}''_p}{\gamma K} = \frac{340}{17 \times 2.67} = 7.49, \quad C_2 = -\frac{8P_a}{\gamma K} = -\frac{8 \times 115}{17 \times 2.67} = -20.3$$

$$C_3 = -\frac{6P_a}{(\gamma K)^2}(2\bar{y}\gamma K + \bar{p}''_p) = -\frac{6 \times 115}{(17 \times 2.67)^2}(2 \times 2.50 \times 17 \times 2.67 + 340) = -189.9$$

$$C_4 = -\frac{6 P_a \bar{y} \bar{p}''_p + 4P_a^2}{(\gamma K)^2} = -\frac{6 \times 115 \times 2.50 \times 340 + 4 \times (115)^2}{(17 \times 2.67)^2} = -310.4$$

Substituting for C_1, C_2, C_3 and C_4, and simplifying we have

$$D_0^4 + 7.49D_0^3 - 20.3D_0^2 - 189.9D_0 - 310.4 = 0$$

This equation when solved by the method of trial and error gives

$$D_0 \approx 5.3 \text{ m}$$

Depth of Embedment

$$D = D_0 + y_0 = 5.3 + 0.75 = 6.05 \text{ m}$$

Increasing D by 40%, we have

D (design) = 1.4 × 6.05 = 8.47 m or say 8.5 m.

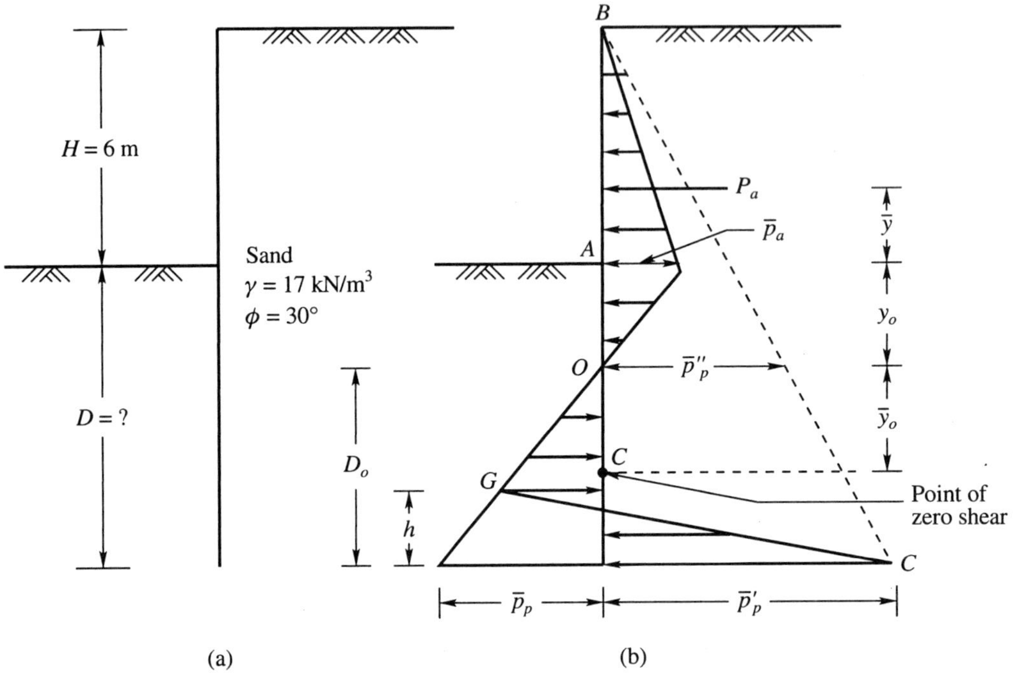

Figure Ex. 20.1

(c) Section modulus

From Eq. (20.6) (The point of zero shear)

$$\bar{y}_o = \sqrt{\frac{2P_a}{\gamma K}} = \sqrt{\frac{2 \times 115}{17 \times 2.67}} = 2.25 \text{ m}$$

$$M_{max} = P_a(\bar{y} + \bar{y}_o) - \frac{1}{6}\bar{y}_o^3 \gamma K$$

$$= 115(2.50 + 2.25) - \frac{1}{6}(2.25)^3 \times 17 \times 2.67$$

$$= 546.3 - 86.2 = 460 \text{ kN-m/m}$$

From Eq. (20.8)

Section modulus

$$Z_s = \frac{M_{max}}{f_b} = \frac{460}{175 \times 10^3} = 26.25 \times 10^{-2} \text{ m}^3\text{/m of wall}$$

Example 20.2

Fig. Ex. 20.2 shows a free standing cantilever sheet pile with no backfill driven into homogeneous sand. The following data are available:

H = 20 ft, P = 3000 lb/ft of wall, γ = 115 lb/ft³, ϕ = 36°.

Determine: (a) the depth of penetration, D, and (b) the maximum bending moment M_{max}.

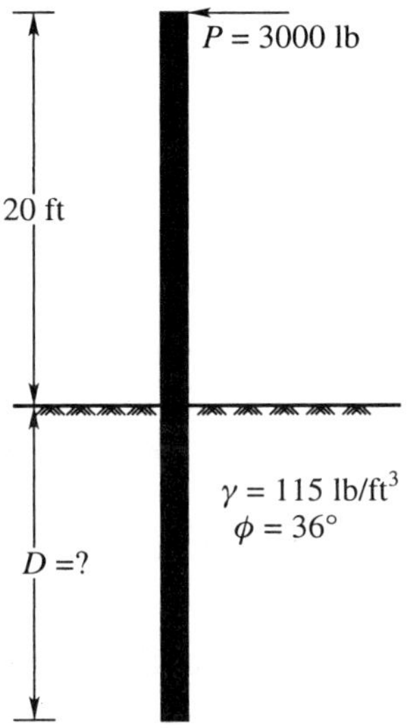

Fig. Ex. 20.2

Solution

$$K_p = \tan^2\left(45^\circ + \frac{\phi}{2}\right) = \tan^2\left(45^\circ + \frac{36}{2}\right) = 3.85$$

$$K_A = \frac{1}{K_p} = \frac{1}{3.85} = 0.26$$

$K = K_p - K_A = 3.85 - 0.26 = 3.59$

The equation for D is (Eq 20.11)

$D^4 + C_1 D^2 + C_2 D + C_3 = 0$

where $C_1 = -\dfrac{8P}{\gamma K} = -\dfrac{8 \times 3000}{115 \times 3.59} = -58.133$

$$C_2 = -\frac{12PH}{\gamma K} = -\frac{12 \times 3000 \times 20}{115 \times 3.59} = -1744$$

$$C_3 = -\frac{4P^2}{(\gamma K)^2} = -\frac{4 \times 3000^2}{(115 \times 3.59)^2} = -211.2$$

Substituting and simplifying, we have

$D^4 - 58.133\, D^2 - 1744\, D - 211.2 = 0$

From the above equation D ≈13.5 ft.
From Eq. (20.6)

$$\bar{y}_0 = \sqrt{\frac{2P_a}{\gamma K}} = \sqrt{\frac{2P}{\gamma K}} = \sqrt{\frac{2 \times 3000}{115 \times 3.59}} = 3.81 \text{ ft}$$

From Eq. (20.12)

$$M_{max} = P_a(H + \bar{y}_o) - \frac{\gamma \bar{y}_0^3 K}{6}$$

$$= 3000(20 + 3.81) - \frac{115 \times (3.81)^3 \times 3.59}{6}$$

$= 71{,}430 - 3{,}806 = 67{,}624$ lb-ft/ft of wall

$M_{max} = 67{,}624$ lb-ft/ft of wall

20.5 DEPTH OF EMBEDMENT OF CANTILEVER WALLS IN COHESIVE SOILS

Case 1: When the Backfill is Cohesive Soil

The pressure distribution on a sheet pile wall is shown in Fig. 20.10.

The active pressure p_a at any depth z may be expressed as

$$p_a = \sigma_v K_A - 2c\sqrt{K_A}$$

where

σ_v = vertical pressure, γz

z = depth from the surface of the backfill.

The passive pressure p_p at any depth y below the dredge line may be expressed as

$$p_p = \sigma_v K_P + 2c\sqrt{K_P}$$

The active pressure distribution on the wall from the backfill surface to the dredge line is shown in Fig. 20.10. The soil is supposed to be in tension up to a depth of z_0 and the pressure on the wall is zero in this zone. The net pressure distribution on the wall is shown by the shaded triangle. At the dredge line (at point A)

(a) The active pressure $\bar{p}_a$ acting towards the left is

$$\bar{p}_a = \gamma H K_A - 2c\sqrt{K_A}$$

when $\phi = 0$ $\quad \bar{p}_a = \gamma H - 2c = \gamma H - q_u$ (20.13a)

where q_u = unconfined compressive strength of the clay soil = $2c$.

(b) The passive pressure acting towards the right at the dredge line is

$$\bar{p}_p = 2c \qquad \text{since } \phi = 0$$

or $\bar{p}_p = q_u$

The resultant of the passive and active pressures at the dredge line is

$$\bar{p}_p = \bar{p}_a = q_u - (\gamma H - q_u) = 2q_u - \gamma H - \bar{p} \tag{20.13b}$$

The resultant of the passive and active pressures at any depth y below the dredge line is passive pressure, $p_p = \gamma y + q_u$

active pressure, $p_a = \gamma (H + y) - q_u$

The resultant pressure is

$$p_p - p_a = \overline{p} = (\gamma y + q_u) - [\gamma(H + y) - q_u] = 2q_u - \gamma H \qquad (20.14)$$

Equations (20.13b) and (20.14) indicate that the resultant pressure remains constant at $(2q_u - \gamma H)$ at all depths.

If passive pressure is developed on the backfill side at the bottom of the pile (point B), then

$p_p \;=\; \gamma (H + D) + q_u$ acting towards the left

$p_a \;=\; \gamma D - q_u$ acting towards the right

The resultant is

$$p_p - p_a = \gamma(H + D) + q_u - \gamma D + q_u = \gamma H + 2q_u = \overline{p}' \qquad (20.15)$$

For static equilibrium, the sum of all the horizontal forces must be equal zero, that is,

$$P_a - (2q_u - \gamma H)D + \frac{1}{2}(2q_u + 2q_u)h = 0$$

Simplifying,

$P_a + 2q_u h - 2q_u D + \gamma HD = 0$, therefore,

$$h = \frac{D(2q_u - \gamma H) - P_a}{2q_u} \qquad (20.16)$$

Also, for equilibrium, the sum of the moments at any point should be zero. Taking moments about the base,

$$P_a(\overline{y} + D) + \frac{h^2}{6}(2q_u) - \frac{(2q_u - \gamma H)D^2}{2} = 0 \qquad (20.17)$$

Substituting for h in (Eq. 20.17) and simplifying,

$$C_1 D^2 + C_2 D + C_3 = 0 \qquad (20.18)$$

where $C_1 = (2q_u - \gamma H)$

$C_2 = -2P_a$

$$C_3 = -\frac{P_a(6q_u\overline{y} + P_a)}{(q_u + \gamma H)}$$

The depth computed from Eq. (20.18) should be increased by 20 to 40 percent so that a factor of safety of 1.5 to 2.0 may be obtained. Alternatively the unconfined compressive strength q_u may be divided by a factor of safety.

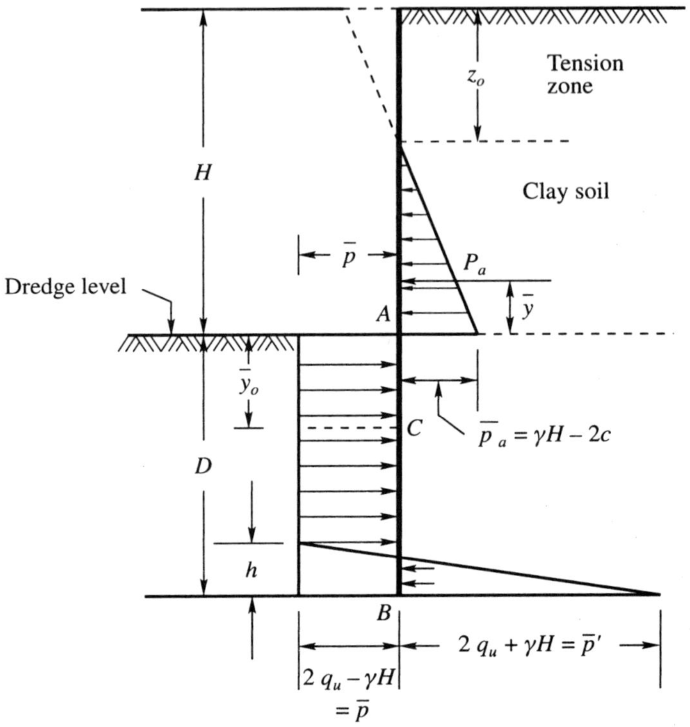

Figure 20.10 Depth of embedment of a cantilever wall in cohesive soil

Limiting Height of Wall

Equation (20.14) indicates that when $(2q_u - \gamma H) = 0$ the resultant pressure is zero. The wall will not be stable. In order that the wall may be stable, the condition that must be satisfied is

$$\frac{2q_u}{F} \geq \gamma H \tag{20.19}$$

where F = factor of safety.

Maximum Bending Moment

As per Fig. 20.10, the maximum bending moment may occur within the depth $(D–h)$ below the dredge line. Let this depth be $\bar{y}_0$ below the dredge line for zero shear. We may write,

$$P_a - \bar{p}\,\bar{y}_0 = 0$$

or
$$\bar{y}_o = \frac{P_a}{\bar{p}} \tag{20.20a}$$

The expression for maximum bending moment is,

$$M_{max} = P_a(\bar{y}_o + \bar{y}) - \frac{\bar{p}\bar{y}_o^2}{2} \tag{20.20b}$$

where $\bar{p} = 2q_u - \gamma H$

The section modulus of the sheet pile may now be calculated as before.

Case 2: When the Backfill is Sand with Water Table at Great Depth

Figure 20.11 gives a case where the backfill is sand with no water table within. The following relationships may be written as:

$$\bar{p}_a = \gamma H K_A$$
$$\bar{p} = 2q_u - \gamma H = 4c - \gamma H$$
$$\bar{p}' = 2q_u + \gamma H = 4c + \gamma H$$

$$P_a = \frac{1}{2}\gamma H^2 K_A$$

$$h = \frac{D(2q_u - \gamma H) - \frac{1}{2}\gamma H^2 K_A}{2q_u} \tag{20.21}$$

A second degree equation in D can be developed as before

$$C_1 D^2 + C_2 D + C_3 = 0 \tag{20.22}$$

where $C_1 = (2q_u - \gamma H)$

$C_2 = -2P_a$

$$C_3 = -\frac{P_a(P_a + 6q_u\bar{y})}{\gamma H + q_u}$$

where $\bar{y} = \dfrac{H}{3}$, $q_u = 2c$

An expression for computing maximum bending moment may be written as

$$M_{max} = P_a(\bar{y} + \bar{y}_o) - \frac{(2q_u - \gamma H)\bar{y}_o^2}{2} \tag{20.23}$$

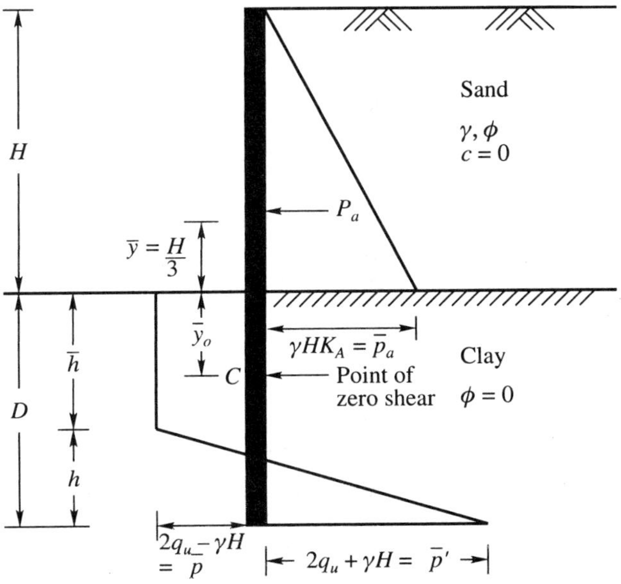

Figure 20.11 Sheet pile wall embedded in clay with sand backfill.

where $$\bar{y}_o = \frac{P_a}{(2q_u - \gamma H)} = \frac{\gamma H^2 K_A}{2(2q_u - \gamma H)} \tag{20.24}$$

Case 3: Cantilever Wall with Sand Backfill and Water Table Above Dredge Line [Fig. 20.12]

The various expressions for this case may be developed as in the earlier cases. The various relationships may be written as

$$\bar{p}_1 = \gamma h_1 K_A$$
$$\bar{p}_a = \gamma_{hl} K_A + \gamma_b h_2 K_A = (\gamma h_l + \gamma_b h_2) K_A$$
$$\bar{p} = 2q_u - \gamma H$$
$$\bar{p}' = 2q_u + (\gamma h_l + \gamma_b h_2)$$

$$P_a = \frac{1}{2}(\bar{p}_1 H + \bar{p}_a h_2) \tag{20.25}$$

$$h = \frac{[2q_u - (\gamma h_1 + \gamma_b h_2)]D - P_a}{2q_u} \tag{20.26}$$

The expression for the second degree equation in D is

$$C_1 D^2 + C_2 D + C_3 = 0 \tag{20.27}$$

where

$$C_1 = [2q_u - (\gamma h_l + \gamma_b h_2)]$$
$$C_2 = -2P_a$$

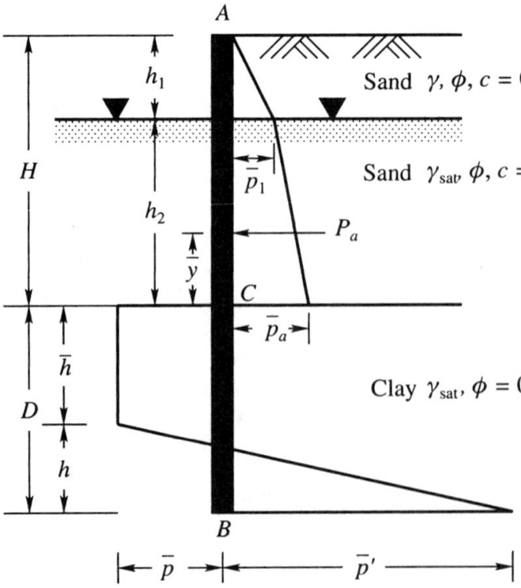

Figure 20.12 Cantilever wall with sand backfill and water table

$$C_3 = -\frac{(P_a + 6q_u\bar{y})P_a}{q_u + (\gamma h_1 + \gamma_b h_2)}$$

Eq (20.27) may be solved for D. The depth computed should be increased by 20 to 40% to obtain a factor of safety of 1.5 to 2.0.

Case 4: Free-Standing Cantilever Sheet Pile Wall Penetrating Clay

Figure 20.13 shows a freestanding cantilever wall penetrating clay. An expression for D can be developed as before. The various relationships are given below.

$$\bar{p} = 2\,q_u = \bar{p}'$$

The expression for D is

$$C_1\,D^2 + C_2\,D + C_3 = 0 \qquad (20.28)$$

where

$$C_1 = 2\,q_u$$
$$C_2 = -2P$$
$$C_3 = -\frac{(P + 6q_u H)P}{q_u}$$

The expression for h is

$$h = \frac{2q_u D - P}{2q_u} \qquad (20.29)$$

The maximum moment may be calculated per unit length of wall by using the expression

$$M_{\max} = P(H + \bar{y}_o) - \frac{2q_u\bar{y}_o^2}{2} \qquad (20.30)$$

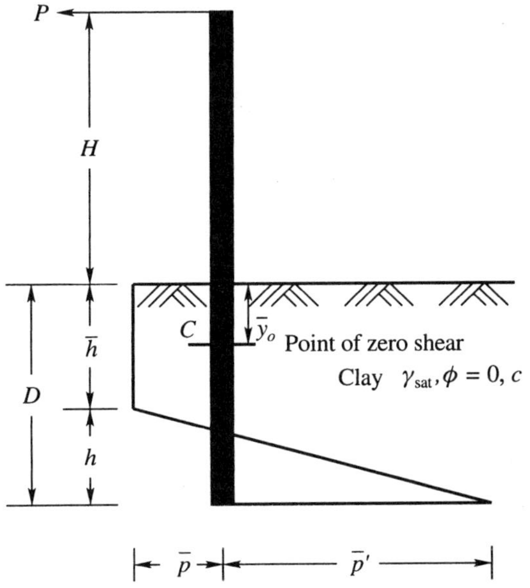

Figure 20.13 Free standing cantilever wall penetrating clay

where $\bar{y}_o = \dfrac{P}{2q_u}$ = depth to the point of zero shear. (20.31)

Example 20.3

Solve Example 20.1, if the soil is clay having an unconfined compressive strength of 70 kN/m² and a unit weight of 17 kN/m³. Determine the maximum bending moment.

Solution

The pressure distribution is assumed as shown in Fig. Ex. 20.3.

$$\text{For } \phi_u = 0,\ \bar{p}_a = \gamma H - q_u = 17 \times 6 - 70 = 32 \text{ kN/m}^2$$

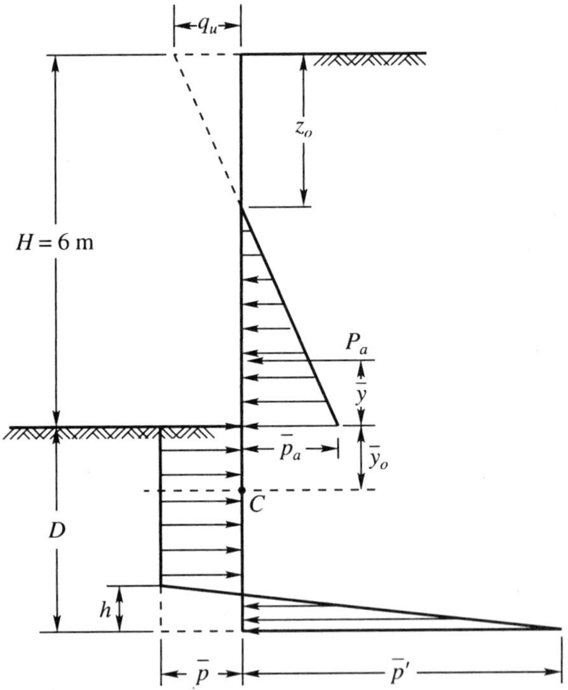

Figure Ex. 20.3

$$z_0 = \frac{q_u}{\gamma} = \frac{70}{17} \text{ m} = 4.12 \text{ m}$$

$$P_a = \frac{1}{2}\bar{p}_a(H - z_0) = \frac{1}{2} \times 32 \times (6.0 - 4.12) = 30 \text{ kN/m of wall}$$

$$\bar{p} = 2q_u - \gamma H = 2 \times 70 - 17 \times 6 = 38 \text{ kN/m}^2$$

$$\bar{p}' = 2q_u + \gamma H = 2 \times 70 + 17 \times 6 = 242 \text{ kN/m}^2$$

$$\bar{y} = \frac{1}{3}(H - z_0) = \frac{1}{3}(6 - 4.12) = 0.63 \text{ m}$$

For the determination of h, equate the summation of all horizontal forces to zero, thus

$$P_a - \bar{p} \times D + \frac{1}{2}(\bar{p} + \bar{p}')h = 0$$

or $$30 - 38 \times D + \frac{1}{2}(38 + 242)h = 0$$

Therefore $$h = \frac{3.8D - 3}{14}$$

For the determination of D, taking moments of all the forces about the base of the wall, we have

$$P_a \times (D + \bar{y}) - \bar{p} \times \frac{D^2}{2} + (\bar{p} + \bar{p}') \times \frac{h}{2} \times \frac{h}{3} = 0$$

or $$30(D + 0.63) - 38 \times \frac{D^2}{2} + (38 + 242) \times \frac{h^2}{6} = 0$$

Substituting for h we have,

$$3D + 1.89 - 1.9D^2 + 4.7\left[\frac{3.8D - 3}{14}\right]^2 = 0$$

Simplifying, we have

$$D^2 - 1.57D + 1.35 = 0$$

Solving $D = 2.2$ m; Increasing D by 40%, we have $D = 1.4(2.2) = 3.1$ m.

Maximum bending moment

From Eq. (20.20)

$$M_{max} = P_a(\bar{y}_0 + \bar{y}) - \frac{\bar{p}\bar{y}_0^2}{2}$$

$$\bar{y}_0 = \frac{P_a}{\bar{p}} = \frac{30}{38} = 0.79 \text{ m}$$

$$\bar{y} = 0.63 \text{ m}$$

$$M_{max} = 30(0.79 + 0.63) - \frac{38 \times (0.79)^2}{2}$$

$$= 42.6 - 11.9 = 30.7 \text{ kN-m/m of wall}$$

Example 20.4

Solve Example 20.1 if the soil below the dredge line is clay having a cohesion of 35 kN/m^2 and the backfill is sand having an angle of internal friction of 30°. The unit weight of both the soils may be assumed as 17 kN/m^3. Determine the maximum bending moment.

Solution

Refer to Fig. Ex. 20.4

$$\overline{p}_a = \gamma HK_A = 17 \times 6 \times \frac{1}{3} = 34 \text{ kN / m}^2$$

$$P_a = \frac{1}{2} \times 34 \times 6 = 102 \text{ kN/m of wall}$$

$$\overline{p} = 2q_u - \gamma H = 2 \times 2 \times 35 - 17 \times 6 = 38 \text{ kN / m}^2$$

From Eq. (20.22)

$$C_1 D^2 + C_2 D + C_3 = 0$$

where $C_1 = (2q_u - \gamma \text{ H}) = (2 \times 2 \times 35 - 17 \times 6) = 38 \text{ kN/m}^2 = \overline{p}$

$C_2 = -2P_a = -2 \times 102 = -204$ kN

$$C_3 = -\frac{P_a(P_a + 6q_u \overline{y})}{\gamma H + q_u} = -\frac{102(102 + 6 \times 70 \times 2)}{17 \times 6 + 70} = -558.63$$

Substituting and simplifying

$38D^2 - 204D - 558.63 = 0$

or $D^2 - 5.37D - 14.7 = 0$

Solving the equations, we have $D \approx 7.37$ m

Increasing D by 40%, we have

D (design) = 1.4 (7.37) = 10.3 m

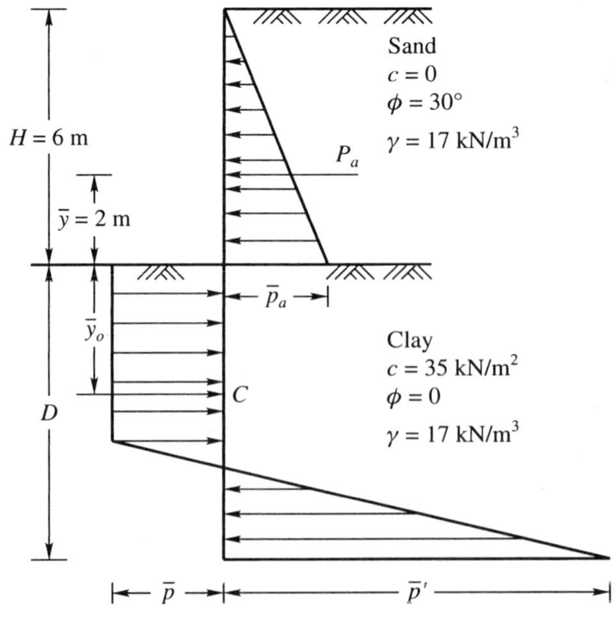

Fig. Ex. 20.4

Maximum Bending Moment

From Eq. (20.23)

$$M_{max} = P_a(\bar{y} + \bar{y}_0) - \frac{\bar{p}\,\bar{y}_0^2}{2}$$

$$\bar{y} = \frac{H}{3} = \frac{6}{3} = 2\,\text{m}, \; \bar{p} = 38\ \text{kN/m}^2$$

$$\bar{y}_0 = \frac{P_a}{\bar{p}} = \frac{\gamma H^2 K_A}{2\bar{p}} = \frac{17 \times 6^2 \times 0.33}{2 \times 38} = 2.66\,\text{m}$$

$$M_{max} = 102(2 + 2.66) - \frac{38 \times (2.66)^2}{2} = 475.32 - 134.44 = 340.9\ \text{kN-m/m of wall.}$$

$$M_{max} = 340.9\ \text{kN-m/m of wall}$$

Example 20.5

Refer to Fig. Ex. 20.5. Solve the problem in Ex. 20.4 if the water table is above the dredge line.

Given: $h_1 = 2.5$ m, $\gamma_{sat} = 17$ kN/m^3

Assume the soil above the water table remains saturated. All the other data given in Ex. 20.4 remain the same.

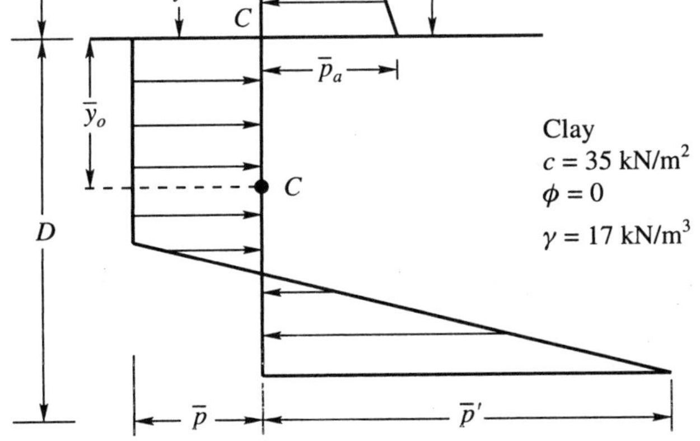

Figure Ex. 20.5

Solution

$h_1 = 2.5$ m, $h_2 = 6 - 2.5 = 3.5$ m, $\gamma_b = 17 - 9.81 = 7.19$ kN/m^3

$p_1 = \gamma h_1 K_A = 17 \times 2.5 \times 1/3 = 14.17$ kN/m^2

$\bar{p}_a = p_1 + \gamma_b h_2 K_A = 14.17 + 7.19 \times 3.5 \times 1/3 = 22.56$ kN/m^2

$$P_a = \frac{1}{2} p_1 h_1 + p_1 h_2 + \frac{1}{2} (\bar{p}_a - p_1) h_2$$

$$= \frac{1}{2} \times 14.17 \times 2.5 + 14.17 \times 3.5 + \frac{1}{2} (22.56 - 14.17) \times 3.5$$

$= 17.71 + 49.6 + 14.7 = 82$ kN/m

Determination of $\bar{y}$ (Refer to Fig. 20.12)

Taking moments of all the forces above dredge line about C we have

$$82\bar{y} = 17.71\left(3.5 + \frac{2.5}{3}\right) + 49.6 \times \frac{3.5}{2} + 14.7 \times \frac{3.5}{3}$$

$$= 76.74 + 86.80 + 17.15 = 180.69$$

$$\bar{y} = \frac{180.69}{82} = 2.20 \text{ m}$$

From Eq. (20.27), the equation for D is

$C_1 D^2 + C_2 D + C_3 = 0$

where $C_1 = [2 q_u - (\gamma h_1 + \gamma_b h_2)]$

$= [140 - (17 \times 2.5 + 7.19 \times 3.5)] = 72.3$

$C_2 = -2 P_a = -2 \times 82 = -164$

$$C_3 = -\frac{(P_a + 6 q_u \bar{y}) P_a}{q_u + (\gamma h_1 + \gamma_b h_2)} = -\frac{(82 + 6 \times 70 \times 2.2) \times 82}{70 + 17 \times 2.5 + 7.19 \times 3.5)} = -599$$

Substituting we have,

$72.3 D^2 - 164 D - 599 = 0$

or $D^2 - 2.27 D - 8.285 = 0$

solving we have $D \approx 4.23$ m

Increasing D by 40%; the design value is

D (design) $= 1.4(4.23) = 5.92$ m

Example 20.6

Fig. Example 12.6 gives a freestanding sheet pile penetrating clay. Determine the depth of penetration. Given: $H = 5$ m, $P = 40$ kN/m, and $q_u = 30$ kN/m^2.

Solution

From Eq. (20.13a)

$\bar{p} = 2 q_u - \gamma H = 2 q_u = 2 \times 30 = 60$ kN/m^2

From Eq. (20.28), The expression for D is

$C_1 D^2 + C_2 D + C_3 = 0$

where $C_1 = 2\, q_u = 60$

$C_2 = -2\, P = -2 \times 40 = -80$

$$C_3 = -\frac{(P + 6 q_u H) P}{q_u} = -\frac{(40 + 6 \times 30 \times 5) 40}{30} = -1253$$

Substituting and simplifying

$60\, D^2 - 80\, D - 1253 = 0$

or $D^2 - 1.33\, D - 21 = 0$

Solving $D \approx 5.3$ m. Increasing by 40% we have

D (design) = 1.4(5.3) = 7.42 m

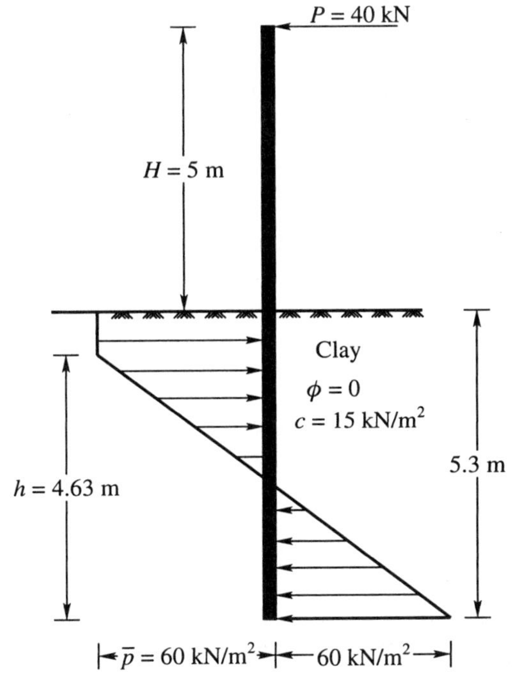

Figure Ex. 20.6

From Eq. (20.29)

$$h = \frac{2q_u D - P}{2q_u} = \frac{2 \times 30 \times 5.3\text{m} - 40}{2 \times 30} = 4.63\text{m}$$

20.6 ANCHORED BULKHEAD: FREE-EARTH SUPPORT METHOD—DEPTH OF EMBEDMENT OF ANCHORED SHEET PILES IN GRANULAR SOILS

If the sheet piles have been driven to a shallow depth, the deflection of a bulkhead is somewhat similar to that of a vertical elastic beam whose lower end B is simply supported and the other end is fixed as shown in Fig. 20.14. Bulkheads which satisfy this condition are called bulkheads with *free earth support.* There are two methods of applying the factor of safety in the design of bulkheads.

1. Compute the minimum depth of embedment and increase the value by 20 to 40 percent to give a factor of safety of 1.5 to 2.
2. The alternative method is to apply the factor of safety to K_p and determine the depth of embedment.

Method 1: Minimum Depth of Embedment

The water table is assumed to be at a depth h_1 from the surface of the backfill. The anchor rod is fixed at a height h_2 above the dredge line. The sheet pile is held in position by the anchor rod and the tension in the rod is T_a. The forces that are acting on the sheet pile are

1. Active pressure due to the soil behind the pile,
2. Passive pressure due to the soil in front of the pile, and
3. The tension in the anchor rod.

The problem is to determine the minimum depth of embedment D. The forces that are acting on the pile wall are shown in Fig. 20.15.

The resultant of the passive and active pressures acting below the dredge line is shown in Fig. 20.15. The distance y_0 to the point of zero pressure is

$$y_0 = \frac{\bar{p}_a}{\gamma_b K}$$

The system is in equilibrium when the sum of the moments of all the forces about any point is zero. For convenience if the moments are taken about the anchor rod,

$$P_p h_4 = P_a \bar{y}_a$$

But $$P_p = \frac{1}{2}\gamma_b K D_0^2$$

$$h_4 = h_3 + y_o + \frac{2}{3}D_o$$

Therefore,

$$P_a \bar{y}_a = \frac{1}{2}\gamma_b K D_0^2\left(h_3 + y_0 + \frac{2}{3}D_0\right)$$

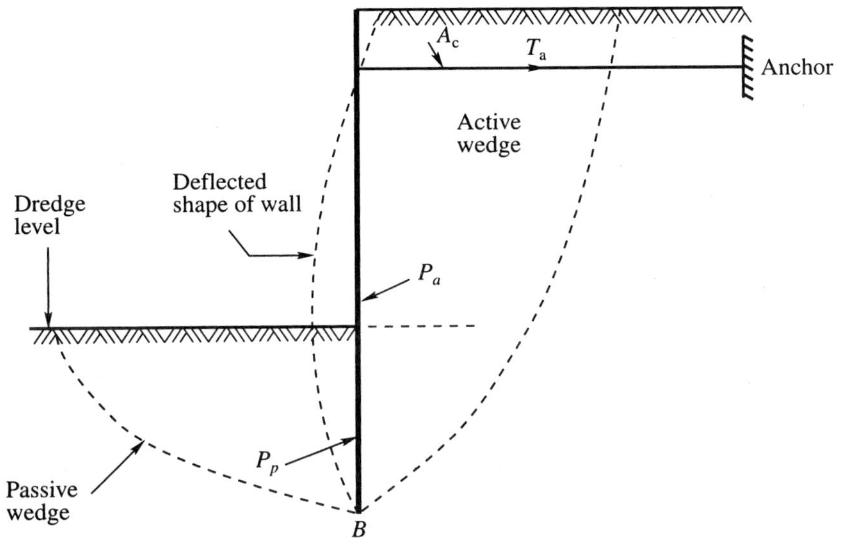

Figure 20.14 Conditions for free-earth support of an anchored bulkhead

Simplifying the equation,

$$C_1D_0^3 + C_2D_0^2 + C_3 = 0 \tag{20.32}$$

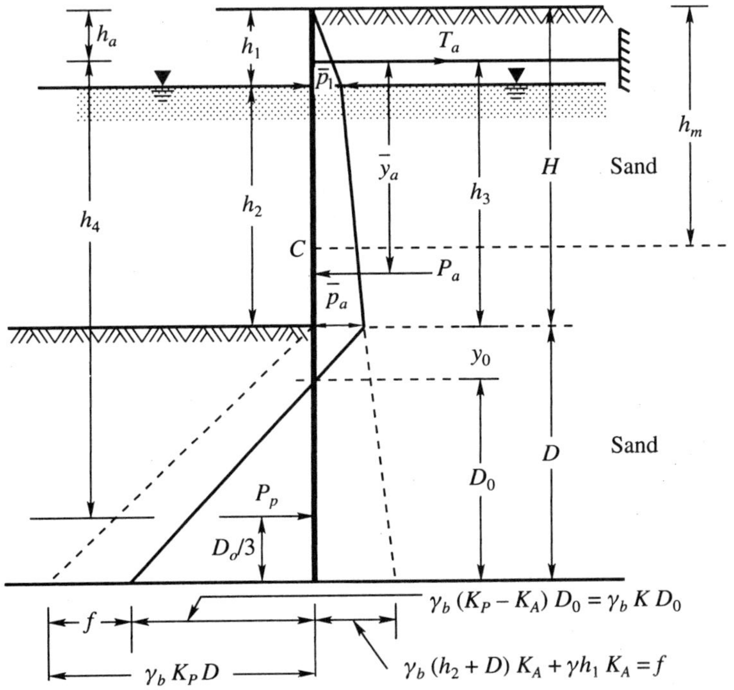

Figure 20.15 Depth of embedment of an anchored bulkhead by the free-earth support method (method 1)

where $$C_1 = \frac{\gamma_b K}{3}$$

$$C_2 = \left(\frac{\gamma_b K}{2}\right)\left(h_2 + y_0\right)$$

$$C_3 = -P_a \bar{y}_a$$

γ_b = submerged unit weight of soil

$K = K_P - K_A$

The force in the anchor rod, T_a, is found by summing the horizontal forces as

$$T_a = P_a - P_P \tag{20.33}$$

The minimum depth of embedment is

$$D = D_0 + y_0 \tag{20.34}$$

Increase the depth D by 20 to 40% to give a factor of safety of 1.5 to 2.0.

Maximum Bending Moment

The maximum theoretical moment in this case may be at a point C any depth h_m below ground level which lies between h_1 and H where the shear is zero. The depth h_m may be determined from the equation

$$\frac{1}{2}\bar{p}_1 h_1 - T_a + \bar{p}_1(h_m - h_1) + \frac{1}{2}\gamma_b (h_m - h_1)^2 K_A = 0 \tag{20.35}$$

Once h_m is known the maximum bending moment can easily be calculated.

Method 2: Depth of Embedment by Applying a Factor of Safety to K_p

(a) Granular Soil Both in the Backfill and Below the Dredge Line

The forces that are acting on the sheet pile wall are as shown in Fig. 20.16. The maximum passive pressure that can be mobilized is equal to the area of triangle ABC shown in the figure. The passive pressure that has to be used in the computation is the area of figure $ABEF$ (shaded). The triangle ABC is divided by a vertical line EF such that

$$\text{Area } ABEF = \frac{\text{Area } ABC}{\text{Factor of safety}} = P'_p$$

The width of figure $ABEF$ and the point of application of P'_p can be calculated without any difficulty.

Equilibrium of the system requires that the sum of all the horizontal forces and moments about any point, for instance, about the anchor rod, should be equal to zero.

Hence, $$P'_p + T_a - P_a = 0 \tag{20.36}$$

$$P_a \bar{y}_a - P'_p h_4 = 0 \tag{20.37}$$

where, $$P'_p = \frac{1}{2}\gamma_b K_P D^2 \times \frac{1}{F_s}$$

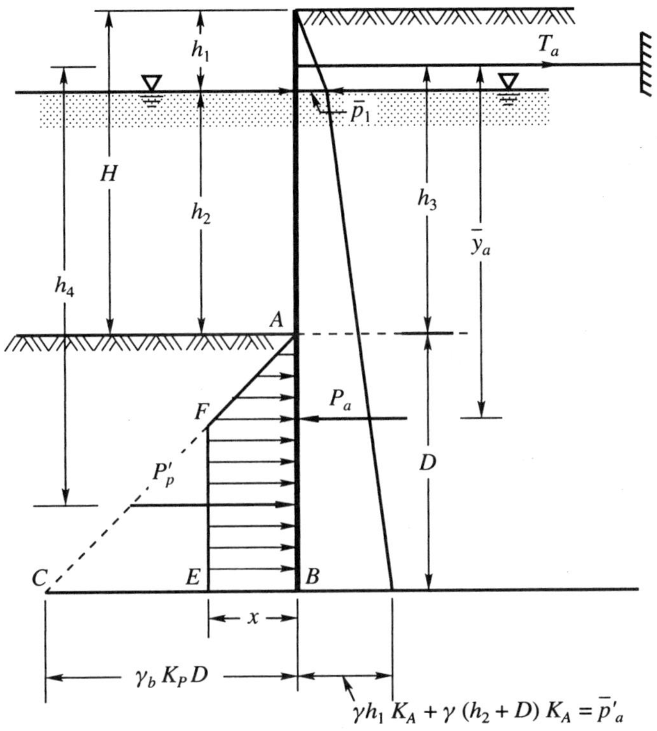

Figure 20.16 Depth of embedment by free-earth support method (method 2)

and F_s = assumed factor of safety.

The tension in the anchor rod may be found from Eq. (20.36) and from Eq. (20.37) D can be determined.

(b) Depth of Embedment when the Soil Below Dredge Line is Cohesive and the Backfill Granular

Figure 20.17 shows the pressure distribution.

The surcharge at the dredge line due to the backfill may be written as

$$q = \gamma h_1 + \gamma_b h_2 = \gamma_e H \tag{20.38}$$

where h_3 = depth of water above the dredge line, γ_e effective equivalent unit weight of the soil, and $H = h_1 + h_2$.

The active earth pressure acting towards the left at the dredge line is (when $\phi = 0$)

$$\bar{p}_a = q - q_u$$

The passive pressure acting towards the right is

$$\bar{p}_p = q_u$$

The resultant of the passive and active earth pressures is

$$\bar{p}_p - \bar{p}_a = 2q_u - q = \bar{p} \tag{20.39}$$

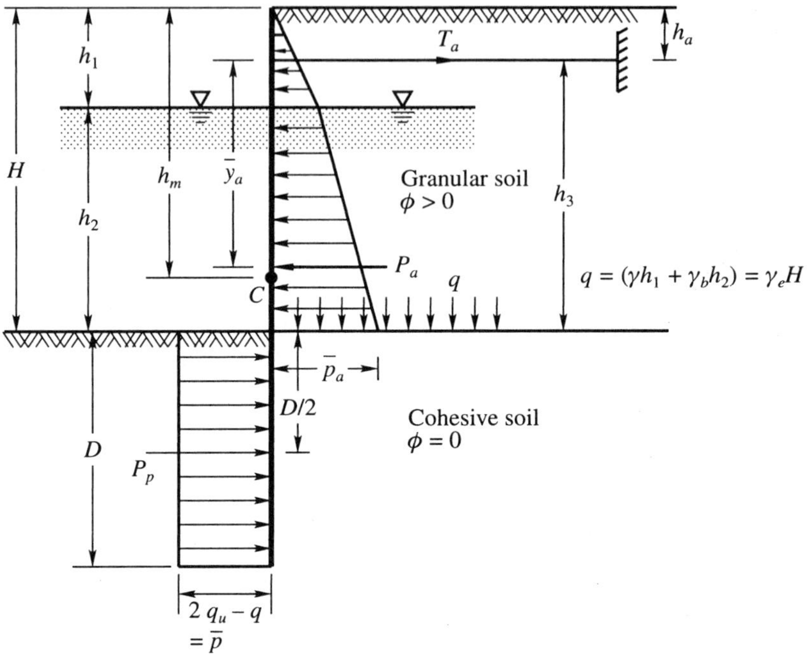

Figure 20.17 Depth of embedment when the soil below the dredge line is cohesive

The pressure remains constant with depth. Taking moments of all the forces about the anchor rod,

$$P_a \bar{y}_a - D(2q_u - q)(h_3 + D/2) = 0 \tag{20.40}$$

where $\bar{y}_a$ = the distance of the anchor rod from P_a.

Simplifying Eq. (15.40),

$$D^2 + C_1 D + C_2 = 0 \tag{20.41}$$

where $C_1 = 2h_3$

$$C_2 = \frac{2\bar{y}_a P_a}{2q_u - q}$$

The force in the anchor rod is given by Eq. (20.33).

It can be seen from Eq. (20.39) that the wall will be unstable if

$$2q_u - q = 0$$

or $4c - q = 0$

For all practical purposes $q = \gamma_e H = \gamma H$, then Eq. (20.39) may be written as

$$4c - \gamma H = 0$$

or $$N_S = \frac{c}{\gamma H} = \frac{1}{4} = 0.25 \tag{20.42}$$

Eq. (20.42) indicates that the wall is unstable if the ratio $c/\gamma H$ is equal to 0.25. N_s is termed is *Stability Number.* The stability is a function of the wall height H, but is relatively independent of the material used in developing q. If the wall adhesion c_a is taken into account the stability number N_S becomes

$$N_S = \frac{c}{\gamma H}\sqrt{1+\frac{c_a}{c}} \tag{20.43}$$

At passive failure $\sqrt{1+c_a/c}$ is approximately equal to 1.25.

The stability number for sheet pile walls embedded in cohesive soils may be written as

$$N_S = \frac{1.25c}{\gamma H} \tag{20.44}$$

When the factor of safety $F_s = 1$ and $\dfrac{c}{\gamma H} = 0.25,\ N_S = 0.30.$

The stability number N_s required in determining the depth of sheet pile walls is therefore

$$N_s = 0.30 \times F_s \tag{20.45}$$

The maximum bending moment occurs as per Eq. (20.35) at depth h_m which lies between h_1 and H.

20.7 DESIGN CHARTS FOR ANCHORED BULKHEADS IN SAND

Hagerty and Nofal (1992) provided a set of design charts for determining

1. The depth of embedment
2. The tensile force in the anchor rod and
3. The maximum moment in the sheet piling

The charts are applicable to sheet piling in sand and the analysis is based on the free-earth support method. The assumptions made for the preparation of the design charts are:

1. For active earth pressure, Coulomb's theory is valid
2. Logarithmic failure surface below the dredge line for the analysis of passive earth pressure.
3. The angle of friction remains the same above and below the dredge line
4. The angle of wall friction between the pile and the soil is $\phi/2$

The various symbols used in the charts are the same as given in Fig. 20.15

where,

h_a = the depth of the anchor rod below the backfill surface

h_1 = the depth of the water table from the backfill surface

h_2 = depth of the water above dredge line

H = height of the sheet pile wall above the dredge line

D = the minimum depth of embedment required by the free-earth support method

T_a = tensile force in the anchor rod per unit length of wall

Hagerty and Nofal developed the curves given in Fig. 20.18 on the assumption that the water table is at the ground level, that is $h_1 = 0$. Then they applied correction factors for $h_1 > 0$. These correction factors are given in Fig. 20.19. The equations for determining D, T_a and $M_{(max)}$ are

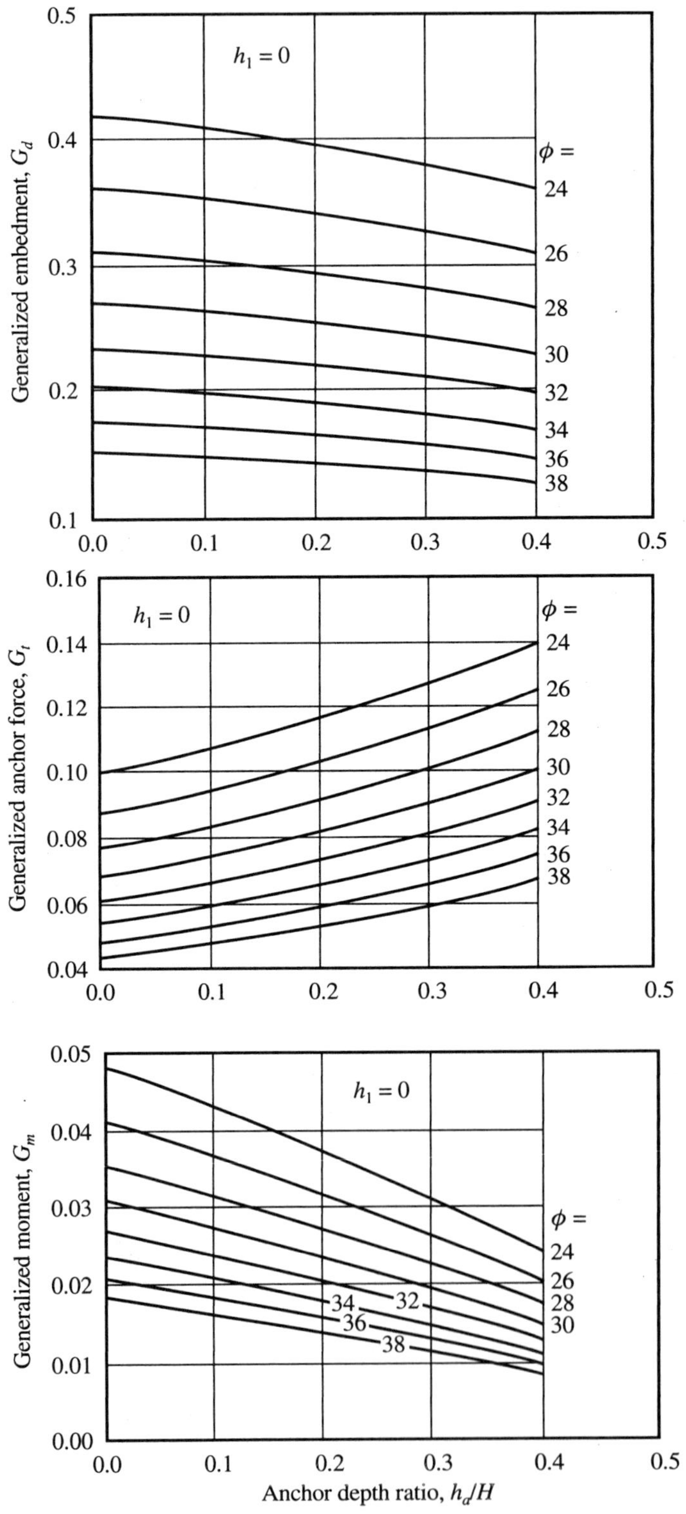

Figure 20.18 Generalized (a) depth of embedment, G_d, (b) anchor force G_t, and (c) maximum moment G_m (after Hagerty and Nofal, 1992)

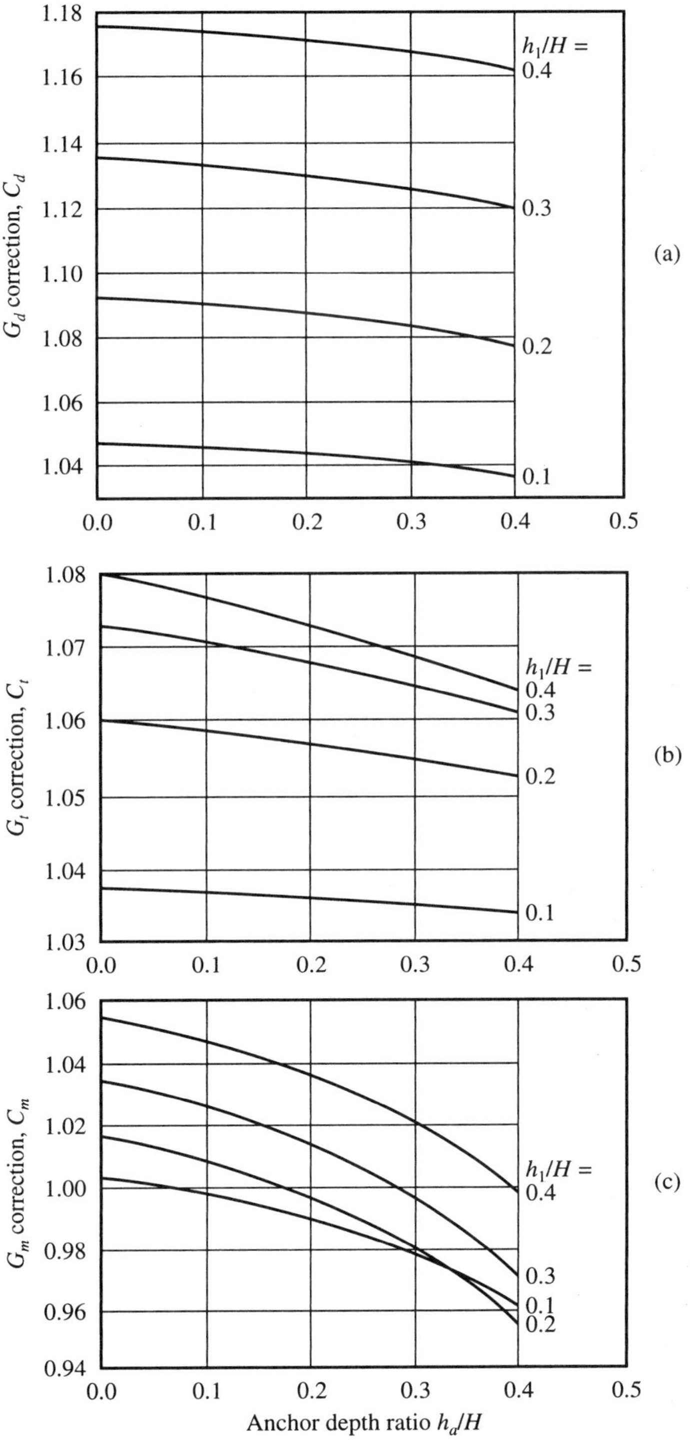

Figure 20.19 Correction factors for variation of depth of water h_1, (a) depth correction C_d, (b) anchor force correction C_t and (c) moment correction C_m (after Hagerty and Nofal, 1992)

$$D = G_d C_d H \tag{20.46 a}$$

$$T_a = G_t C_t \gamma_a H^2 \tag{20.46 b}$$

$$M_{(max)} = G_m C_m \gamma_a H^3 \tag{20.46 c}$$

where,

G_d = generalized non-dimensional embedment = D/H for $h_1 = 0$

G_t = generalized non-dimensional anchor force = $T_a / (\gamma_a H^2)$ for $h_1 = 0$

G_m = generalized non-dimensional moment = $M_{(max)} / \gamma_a (H^3)$ for $h_1 = 0$

C_d, C_t, C_m = correction factors for $h_1 > 0$

γ_a = average effective unit weight of soil

$= (\gamma_m h_1^2 + \gamma_b h_2^2 + 2\gamma_m h_1 h_2)/H^2$

γ_m = moist or dry unit weight of soil above the water table

γ_b = submerged unit weight of soil

The theoretical depth D as calculated by the use of design charts has to be increased by 20 to 40% to give a factor of safety of 1.5 to 2.0 respectively.

20.8 MOMENT REDUCTION FOR ANCHORED SHEET PILE WALLS

The design of anchored sheet piling by the free-earth method is based on the assumption that the piling is perfectly rigid and the earth pressure distribution is hydrostatic, obeying classical earth pressure theory. In reality, the sheet piling is rather flexible and the earth pressure differs considerably from the hydrostatic distribution.

As such the bending moments M(max) calculated by the lateral earth pressure theories are higher than the actual values. Rowe (1952) suggested a procedure to reduce the calculated moments obtained by the *free earth support method.*

Anchored Piling in Granular Soils

Rowe (1952) analyzed sheet piling in granular soils and stated that the following significant factors are required to be taken in the design

1. The relative density of the soil
2. The relative flexibility of the piling which is expressed as

$$\rho = 109 \times 10^{-6} \frac{\overline{H}^4}{EI} \tag{20.47a}$$

where,

ρ = flexibility number

$\overline{H}$ = the total height of the piling in m

EI = the modulus of elasticity and the moment of inertia of the piling (MN–m^2) per m of wall

Eq. (20.47a) may be expressed in English units as

$$\rho = \frac{\overline{H}^4}{EI} \tag{20.47b}$$

where, $\overline{H}$ is in *ft*, E is in lb/in^2 and I is in in^4/*ft*-of wall

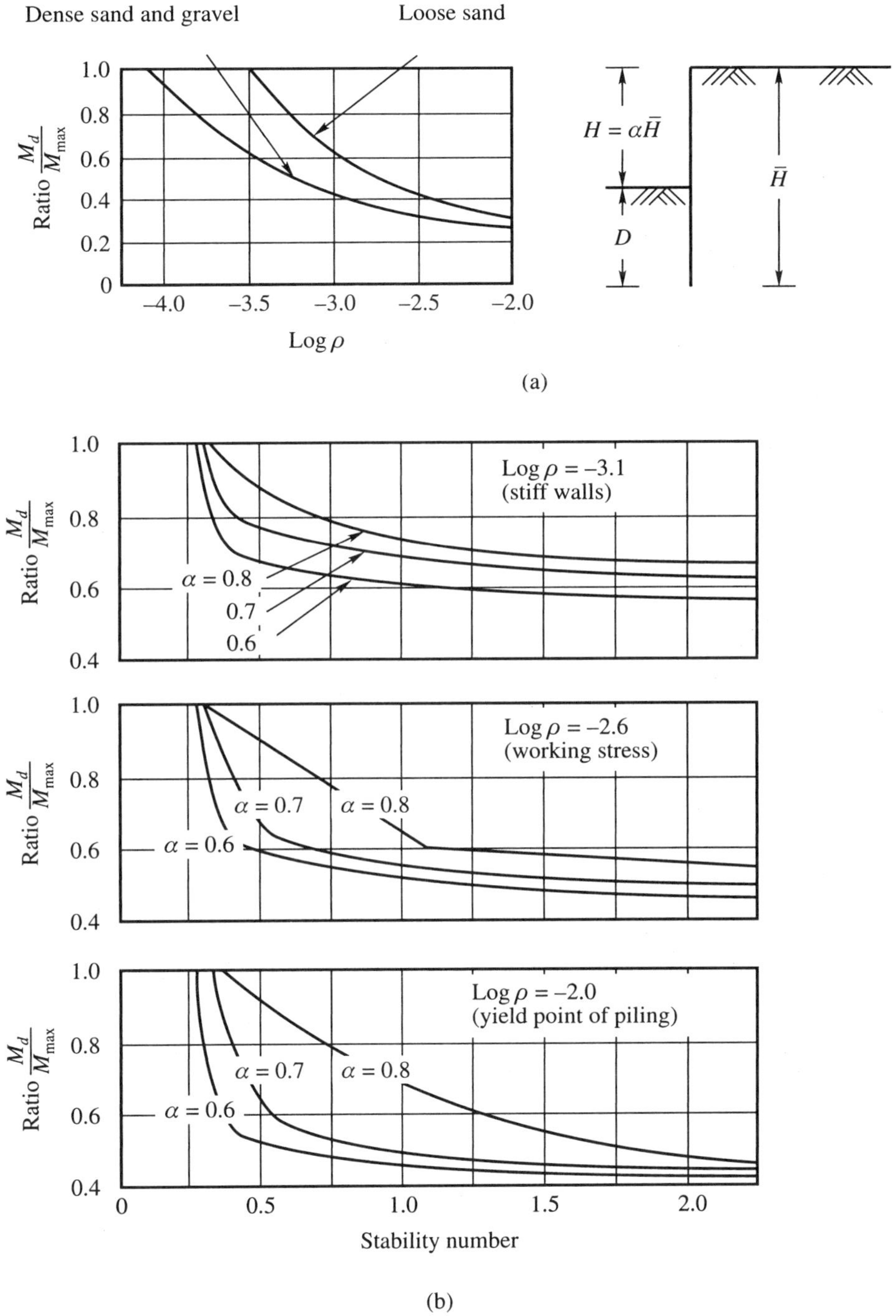

Figure 20.20 Bending moment in anchored sheet piling by free-earth support method, (a) in granular soils, and (b) in cohesive soils (Rowe, 1952)

Anchored Piling in Cohesive Soils

For anchored piles in cohesive soils, the most significant factors are (Rowe, 1957)

1. The stability number

$$N_s = \frac{c}{\gamma H}\sqrt{1+\frac{c_a}{c}} = 1.25\frac{c}{\gamma H} \tag{20.48}$$

2. The relative height of piling α

where,

H = height of piling above the dredge line in meters

γ = effective unit weight of the soil above the dredge line = moist unit weight above water level and buoyant unit weight below water level, kN/m^3

c = the cohesion of the soil below the dredge line, kN/m^2

c_a = adhesion between the soil and the sheet pile wall, kN/m^2

$\sqrt{1+\frac{c_a}{c}}$ = 1.25 for design purposes

α = ratio between H and $\overline{H}$

M_d = design moment

$M_{\max}$ = maximum theoretical moment

Fig. 20.20 gives charts for computing design moments for pile walls in granular and cohesive soils.

Example 20.7

Determine the depth of embedment and the force in the tie rod of the anchored bulkhead shown in Fig. Ex. 20.7(a). The backfill above and below the dredge line is sand, having the following properties

$G_s = 2.67$, $\gamma_{sat} = 18$ kN/m^3, $\gamma_d = 13$ kN/m^3 and $\phi = 30°$

Solve the problem by the free-earth support method. Assume the backfill above the water table remains dry.

Solution

Assume the soil above the water table is dry

For $\phi = 30°, \quad K_A = \frac{1}{3}, \quad K_P = 3.0$

and $K = K_P - K_A = 3 - \frac{1}{3} = 2.67$

$\gamma_b = \gamma_{\text{sat}} - \gamma_w = 18 - 9.81 = 8.19$ kN/m^3.

where $\gamma_w = 9.81$ kN/m^3.

The pressure distribution along the bulkhead is as shown in Fig. Ex. 20.7(b)

$$\overline{p}_1 = \gamma_d h_1 K_A = 13 \times 2 \times \frac{1}{3} = 8.67 \text{ kN/m}^2 \text{ at GW level}$$

$$\overline{p}_a = \overline{p}_1 + \gamma_b h_2 K_A = 8.67 + 8.19 \times 3 \times \frac{1}{3} = 16.86 \text{ kN/m}^2 \text{ at dredge line level}$$

$$y_0 = \frac{\overline{p}_a}{\gamma_b \times K} = \frac{16.86}{8.19 \times 2.67} = 0.77 \text{ m}$$

$$P_a = \frac{1}{2} \times \overline{p}_1 \times h_1 + \overline{p}_1 \times h_2 + \frac{1}{2}(\overline{p}_a - \overline{p}_1)h_2 + \frac{1}{2}\overline{p}_a y_0$$

$$= \frac{1}{2} \times 8.67 \times 2 + 8.67 \times 3 + \frac{1}{2}(16.86 - 8.67)3$$

$$+ \frac{1}{2} \times 16.86 \times 0.77 = 53.5 \text{ kN / m of wall}$$

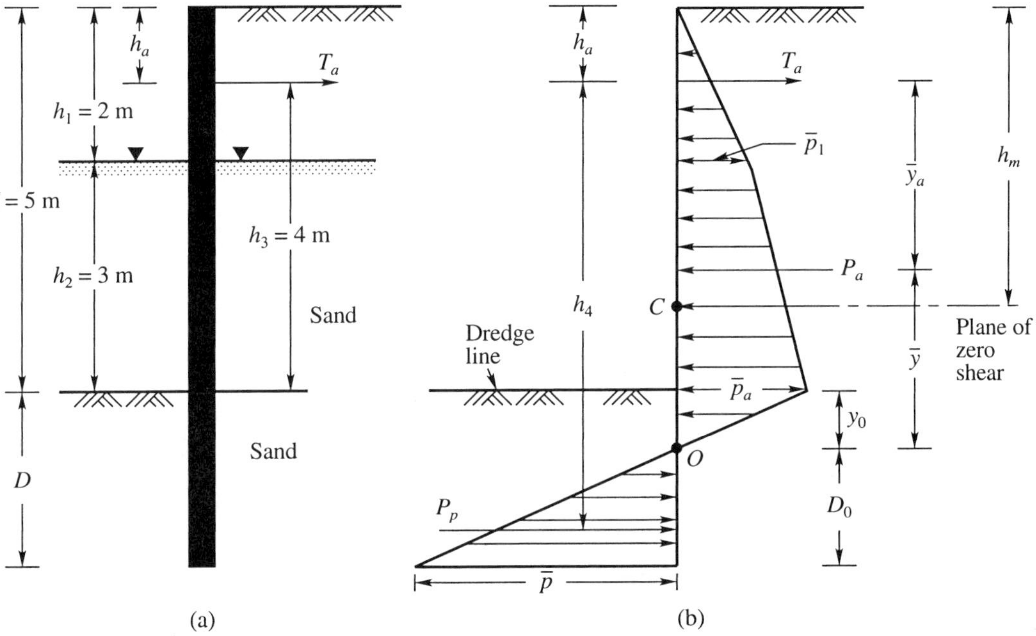

Figure Ex. 20.7

To find $\overline{y}$, taking moments of areas about 0, we have

$$53.5 \times \overline{y} = \frac{1}{2} \times 8.67 \times 2\left(\frac{2}{3} + 3 + 0.77\right) + 8.67 \times 3(3/2 + 0.77)$$

$$+ \frac{1}{2}(16.86 - 8.67) \times 3(3/3 + 0.77) + \frac{1}{2} \times 16.86 \times \frac{2}{3} \times 0.77^2 = 122.6$$

We have $\overline{y} = \dfrac{122.6}{53.5} = 2.3$ m, $\quad \overline{y}_a = 4 + 0.77 - 2.3 = 2.47$ m

Now $\quad P_p = \dfrac{1}{2} \times \gamma_b \times K \times D_0^2 = \dfrac{1}{2} \times 8.19 \times 2.67\ D_0^2 = 10.93\ D_0^2$

and its distance from the anchor rod is

$$h_4 = h_3 + y_0 + 2/3D_0 = 4 + 0.77 + 2/3D_0 = 4.77 + 0.67D_0$$

Now, taking the moments of the forces about the tie rod, we have

$$P_a \times \bar{y}_a = P_p \times h_4$$

$$53.5 \times 2.47 = 10.93 D_0^2 \times (4.77 + 0.67 D_0)$$

Simplifying, we have

$D_0 \approx 1.5$ m, $D = y_0 + D_0 = 0.77 + 1.5 = 2.27$ m

D (design) = 1.4 × 2.27 = 3.18 m

For finding the tension in the anchor rod, we have

$$P_a - P_p - T_a = 0$$

Therefore, $T_a = P_a - P_p = 53.5 - 10.93(1.5)^2 = 28.9$ kN/m of wall for the calculated depth D_o.

Example 20.8

Solve Example 20.7 by applying $F_s = 2$ to the passive earth pressure.

Solution

Refer to Fig. Ex. 20.8

The following equations may be written

$$P_p' = \frac{1}{2}\gamma_b K_P D^2 \frac{1}{F_s} = \frac{1}{2} \times 8.19 \times 3D^2 \times \frac{1}{2} = 6.14D^2$$

$$p_p = \gamma_b K_p D = 8.19 \times 3D \approx 24.6D$$

$$\frac{FG}{BC} = \frac{\alpha p_p}{p_p} = \frac{D-h}{D} \quad \text{or} \quad h = D(1-\alpha)$$

$$\text{Area } ABEF = \left(\frac{D+h}{2}\right)\alpha p_p = \left(\frac{D+h}{2}\right)\alpha \times 24.6D$$

or $6.14\, D^2 = \alpha\, (D + h) \times 12.3\, D$

Substituting for $h = D\,(1 - \alpha)$ and simplifying we have

$$2\,\alpha^2 - 4\,\alpha + 1 = 0$$

Solving the equation, we get $\alpha = 0.3$.

Now $h = D\,(1 - 0.3) = 0.7\,D$ and $AG = D - 0.7\,D = 0.3\,D$

Taking moments of the area $ABEF$ about the base of the pile, and assuming $\alpha p_p = 1$ in Fig. Ex. 20.8 we have

$$\frac{1}{2}(1) \times 0.3D\left(\frac{1}{3} \times 0.3D + 0.7D\right) + (1) \times 0.7D \times \frac{0.7D}{2}$$

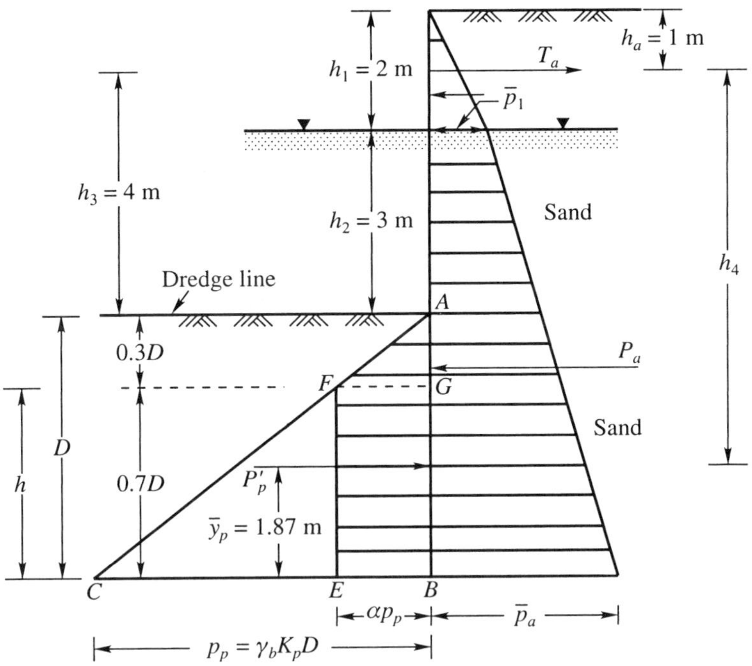

Figure Ex. 20.8

$$= \left[\frac{1}{2}(1)0.3D + (1)\times 0.7D\right]\bar{y}_p$$

simplifying we have $\bar{y}_p = 0.44D$

$\bar{y}_p = 0.44\ D$

Now $h_4 = h_3 + (D - \bar{y}_p) = 4 + (D - 0.44\ D) = 4 + 0.56D$

From the active earth pressure diagram (Fig. Ex. 20.8) we have

$$\bar{p}_1 = \gamma_d\, h_1\, K_A = 13\times 2\times\frac{1}{3} = 8.67\ \text{ kN/m}^2$$

$$\bar{p}_a = p_1 + \gamma_b\,(h_2 + D)\,K_A = 8.67 + \frac{8.19\,(3+D)}{3} = 16.86 + 2.73D$$

$$P_a = \frac{1}{2}p_1\, h_1 + \frac{(\bar{p}_1 + \bar{p}_a)}{2}(h_2 + D)$$

$$= \frac{1}{2}\times 8.67\times 2 + \frac{8.67 + 16.86 + 2.73D}{2}(3 + D) = 1.36D^2 + 16.86D + 47$$

Taking moments of active and passive forces about the tie rod, and simplifying , we have

(a) for moments due to active forces $= 0.89D^3 + 13.7D^2 + 66.7D + 104$

(b) for moments due to passive forces $= 6.14D^2(4 + 0.56D) = 24.56D^2 + 3.44D^3$

Since the sum of the moments about the anchor rod should be zero, we have

$0.89\ D^3 + 13.7\ D^2 + 66.7D + 104 = 24.56\ D^2 + 3.44\ D^3$

Simplifying we have

$D^3 + 4.26\ D^2 - 26.16\ D - 40.8 = 0$

By solving the equation we obtain $D = 4.22$ m with $F_s = 2.0$

Force in the anchor rod

$T_a = P_a - P'_p$

where $P_a = 1.36\ D^2 + 16.86\ D + 47 = 1.36 \times (4.22)^2 + 16.86 \times 4.25 + 47 = 143$ kN

$P'_p = 6.14\ D^2 = 6.14 \times (4.22)^2 = 109$ kN

Therefore $T_a = 143 - 109 = 34$ kN/m length of wall.

Example 20.9

Solve Example 20.7, if the backfill is sand with $\phi = 30°$ and the soil below the dredge line is clay having $c = 20$ kN/m². For both the soils, assume $G_s = 2.67$.

Solution

The pressure distribution along the bulkhead is as shown in Fig. Ex. 20.9.

$\bar{p}_1 = 8.67\ \text{kN/m}^2$ as in Ex. 20.7, $\bar{p}_a = 16.86\ \text{kN/m}^2$

$$P_a = \frac{1}{2}\bar{p}_1 h_1 + \bar{p}_1 h_2 + \frac{1}{2}(\bar{p}_a - \bar{p}_1)h_2$$

$$= \frac{1}{2} \times 8.67 \times 2 + 8.67 \times 3 + \frac{1}{2}(16.86 - 8.67) \times 3 = 47.0\ \text{kN/m}$$

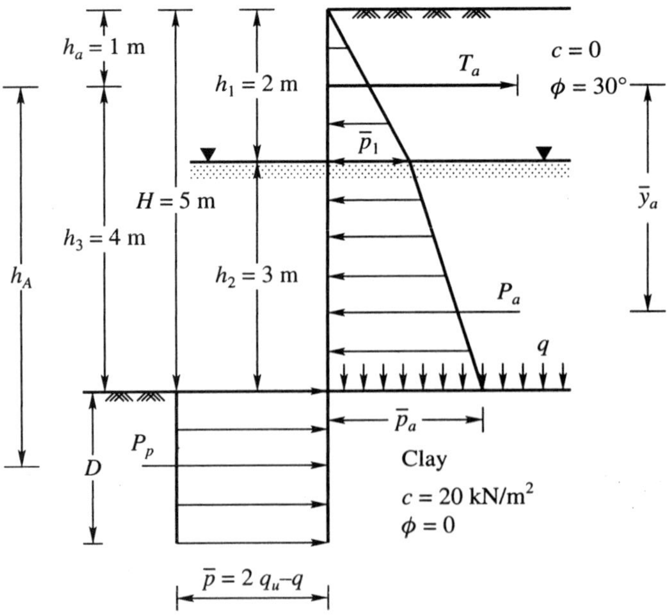

Figure Ex. 20.9

To determine $\bar{y}_a$, take moments about the tie rod.

$$P_a \times \bar{y}_a = \frac{1}{2} \times 8.67 \times 2\left(\frac{2}{3} \times 2 - 1\right) + 8.67 \times 3 \times 2.5$$

$$+\frac{1}{2}(16.86 - 8.67) \times 3 \times 3 = 104.8$$

Therefore $\bar{y}_a = \frac{104.8}{P_a} = \frac{104.8}{46.7} = 2.23$ m

Now, $q = \gamma_d h_1 + \gamma_b h_2 = 13 \times 2 + 8.19 \times 3 = 50.6$ kN/m^2

$\bar{p} = 2 \times 2 \times 20 - 50.6 = 29.4$ kN/m^2

Therefore, $P_p = \bar{p} \times D = 29.4\, D$ kN/m

and $h_4 = \left(4 + \frac{D}{2}\right)$

Taking moments of forces about the tie rod, we have

$$P_a \times \bar{y}_a - P_p \times h_4 = 0$$

or $47 \times 2.23 - 29.4D\left(4 + \frac{D}{2}\right) = 0$

or $1.47D^2 + 11.76D - 10.48 = 0$ or $D \approx 0.8$ m.

$D = 0.8$ m is obtained with a factor of safety equal to one. It should be increased by 20 to 40 percent to increase the factor of safety from 1.5 to 2.0. For a factor of safety of 2, the depth of embedment should be at least 1.12 m. However the suggested depth (design) = 2 m.

Hence P_p for D (design) = 2 m is

$P_p = 29.4 \times 2 = 58.8$ kN/m length of wall

The tension in the tie rod is

$T_a = P_a - P_p = 47 - 59 = -12$ kN/m of wall.

This indicates that the tie rod will not be in tension under a design depth of 2 m. However there is tension for the calculated depth $D = 0.8$ m.

Example 20.10

Determine the depth of embedment for the sheet pile given in Fig. Example 20.7 using the design charts given in Section 20.7.

Given: $H = 5$ m, $h_1 = 2$ m, $h_2 = 3$ m, $h_a = 1$ m, $h_3 = 4$ m, $\phi = 30°$

Solution

$$\frac{h_a}{H} = \frac{1}{5} = 0.2$$

From Fig. 20.18

For $h_a / H = 0.2$, and $\phi = 30°$ we have

$G_d = 0.26$, $G_t = 0.084$, $G_m = 0.024$

From Fig. 20.19 for $\phi = 30$, $h_a / H = 0.2$ and $h_1 / H = 0.4$ we have

$C_d = 1.173$, $C_t = 1.073$, $C_m = 1.036$

Now from Eq. (20.46 a)

$D = G_d C_d H = 0.26 \times 1.173 \times 5 = 1.52$ m

D (design) $= 1.4 \times 1.52 = 2.13$ m.

$T_a = G_t C_t \gamma_a H^2$

$$\text{where} \quad \gamma_a = \frac{(\gamma_m h_1^2 + \gamma_b h_2^2 + 2\gamma_m h_1 h_2)}{H^2}$$

$$= \frac{13 \times 2^2 + 8.19 \times 3^2 + 2 \times 13 \times 2 \times 3}{5^2} = 11.27 \text{ kN/m}^2$$

Substituting and simplifying

$T_a = 0.084 \times 1.073 \times 11.27 \times 5^2 = 25.4$ kN

$M_{max} = G_m C_m \gamma_a H^3$

$= 0.024 \times 1.036 \times 11.27 \times 5^3 = 35.03$ kN-m/m of wall

The values of D (design) and T_a from Ex. 20.7 are

D (design) $= 3.18$ m

$T_a = 28.9$ kN

The design chart gives less by 33% in the value of D (design) and 12% in the value of T_a.

Example 20.11

Refer to Example 20.7. Determine for the pile (a) the bending moment M_{max}, and (b) the reduced moment by Rowe's method.

Solution

Refer to Fig. Ex. 20.7. The following data are available

$H = 5$ m, $h_1 = 2$ m, $h_2 = 3$ m, $h_3 = 4$ m

$\gamma_d = 13$ kN/m^3, $\gamma_b = 8.19$ kN/m^3 and $\phi = 30°$

The maximum bending moment occurs at a depth h_m from ground level where the shear is zero. The equation which gives the value of h_m is

$$\frac{1}{2}\overline{p}_1 h_1 - T_a + \overline{p}_1(h_m - h_1) + \frac{1}{2}\gamma_b K_A (h_m - h_1)^2 = 0$$

where $\overline{p}_1 = 8.67$ kN/m^2

$h_1 = 2$ m, $T_a = 28.9$ kN/m

Substituting

$$\frac{1}{2}\times 8.67\times 2 - 28.9 + 8.67(h_m - 2) + \frac{1}{2}\times 8.19\times\frac{1}{3}(h_m - 2)^2 = 0$$

Simplifying, we have

$$h_m^2 + 2.35 h_m - 23.5 = 0$$

or $h_m \approx 3.81$ m

Taking moments about the point of zero shear,

$$M_{\max} = -\frac{1}{2}\overline{p}_1 h_1\left(h_m - \frac{2}{3}h_1\right) + T_a(h_m - h_a) - \overline{p}_1\frac{(h_m - h_1)^2}{2} - \frac{1}{6}\gamma_b K_A (h_m - h_1)^3$$

$$= -\frac{1}{2}\times 8.67\times 2\left(3.81 - \frac{2}{3}\times 2\right) + 28.9(3.81 - 1) - 8.67\frac{(3.81-2)^2}{2} - \frac{1}{6}8.19\times\frac{1}{3}(3.81-2)^3$$

$$= -21.41 + 81.2 - 14.2 - 2.70 \approx 42.8 \text{ kN-m/m}$$

From Ex. 20.7

D (design) = 3.18 m, $H = 5$ m

Therefore $\overline{H} = 8.18$ m

From Eq. (20.47a) $\rho = 109\times 10^{-6}\dfrac{\overline{H}^4}{EI}$

Assume $E = 20.7 \times 10^4$ MN/m^2

For a section $P_z - 27$, $I = 25.2 \times 10^{-5}$ m^4/m

Substituting $\rho = \dfrac{10.9\times 10^{-5}\times(8.18)^4}{2.07\times 10^5\times 25.2\times 10^{-5}} = 9.356\times 10^{-3}$

$$\log\rho = \log\left[\frac{9.36}{10^3}\right] = -2.0287 \text{ or say } 2.00$$

Assuming the sand backfill is loose, we have from Fig. 20.20 (a)

$$\frac{M_d}{M_{\max}} = 0.32 \text{ for } \log\rho = -2.00$$

Therefore M_d (design) = 0.32 × 42.8 = 13.7 kN-m/m

20.9 ANCHORAGE OF BULKHEADS

Sheet pile walls are many times tied to some kind of anchors through tie rods to give them greater stability as shown in Fig. 20.21. The types of anchorage that are normally used are also shown in the same figure.

Anchors such as anchor walls and anchor plates which depend for their resistance entirely on passive earth pressure must be given such dimensions that the anchor pull does not exceed a certain fraction of the pull required to produce failure. The ratio between the tension in the anchor T_a and the maximum pull which the anchor can stand is called the factor of safety of the anchor.

The types of anchorages given in Fig. 20.21 are:

1. *Deadmen, anchor plates, anchor beams etc.:* Deadmen are short concrete blocks or continuous concrete beams deriving their resistance from passive earth pressure. This type is suitable when it can be installed below the level of the original ground surface.
2. *Anchor block supported by battered piles:* Fig (20.21b) shows an anchor block supported by two battered piles. The force T_a exerted by the tie rod tends to induce compression in pile P_1 and tension in pile P_2. This type is employed where firm soil is at great depth.
3. *Sheet piles:* Short sheet piles are driven to form a continuous wall which derives its resistance from passive earth pressure in the same manner as deadmen.
4. *Existing structures:* The rods can be connected to heavy foundations such as buildings, crane foundations etc.

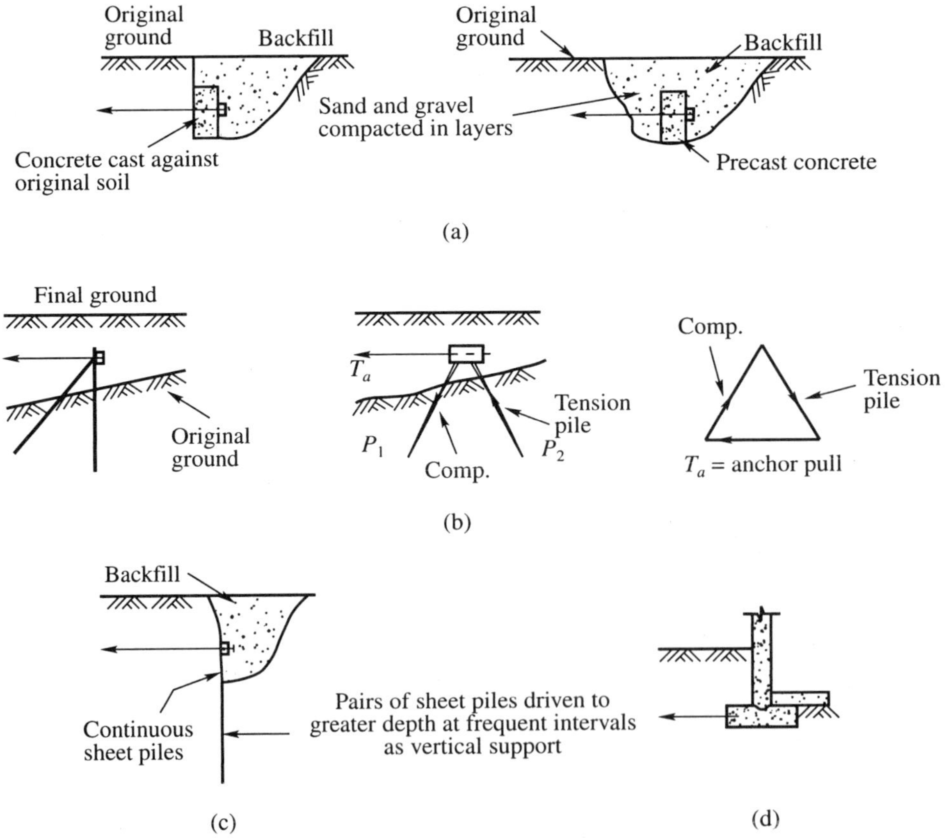

Figure 20.21 Types of anchorage: (a) deadman; (b) braced piles; (c) sheet piles; (d) large structure (after Teng, 1969)

Location of Anchorage

The minimum distance between the sheet pile wall and the anchor blocks is determined by the failure wedges of the sheet pile (under free-earth support condition) and deadmen. The anchorage does not serve any purpose if it is located within the failure wedge *ABC* shown in Fig. 20.22a.

If the failure wedges of the sheet pile and the anchor interfere with each other, the location of the anchor as shown in Fig. 20.22b reduces its capacity. Full capacity of the anchorage will be available if it is located in the shaded area shown in Fig. 20.22c. In this case

1. The active sliding wedge of the backfill does not interfere with the passive sliding wedge of the deadman.
2. The deadman is located below the slope line starting from the bottom of the sheet pile and making an angle ϕ with the horizontal, ϕ being the angle of internal friction of the soil.

Capacity of Deadman (After Teng, 1969)

A series of deadmen (anchor beams, anchor blocks or anchor plates) are normally placed at intervals parallel to the sheet pile walls. These anchor blocks may be constructed near the ground surface or at great depths, and in short lengths or in one continuous beam. The holding capacity of these anchorages is discussed below.

Continuous Anchor Beam Near Ground Surface (Teng, 1969)

If the length of the beam is considerably greater than its depth, it is called a continuous deadman. Fig. 20.23(a) shows a deadman. If the depth to the top of the deadman, $\bar{h}$, is less than about one-third to one-half of H (where H is depth to the bottom of the deadman), the capacity may be calculated by assuming that the top of the deadman extends to the ground surface. The ultimate capacity of a deadman may be obtained from (per unit length)

For granular soil ($c = 0$)

$$T_u = P_p - P_a = \frac{1}{2}\gamma H^2 K_p - \frac{1}{2}\gamma H^2 K_A$$

or $$T_u = \frac{1}{2}\gamma H^2 (K_P - K_A) \qquad (20.49)$$

For clay soil ($\phi = 0$)

$$T_u = P_p - P_a = \left(q_u H + \frac{1}{2}\gamma H^2\right) - \left(\frac{1}{2}\gamma H^2 - q_u H + \frac{2c^2}{\gamma}\right) = 2q_u H - \frac{2c^2}{\gamma} \qquad (20.50)$$

where q_u = unconfined compressive strength of soil,

γ = effective unit weight of soil, and

K_P, K_A = Rankine's active and passive earth pressure coefficients.

It may be noted here that the active earth pressure is assumed to be zero at a depth = $2c/\gamma$ which is the depth of the tension cracks. It is likely that the magnitude and distribution of earth pressure may change slowly with time. For lack of sufficient data on this, the design of deadmen in cohesive soils should be made with a conservative factor of safety.

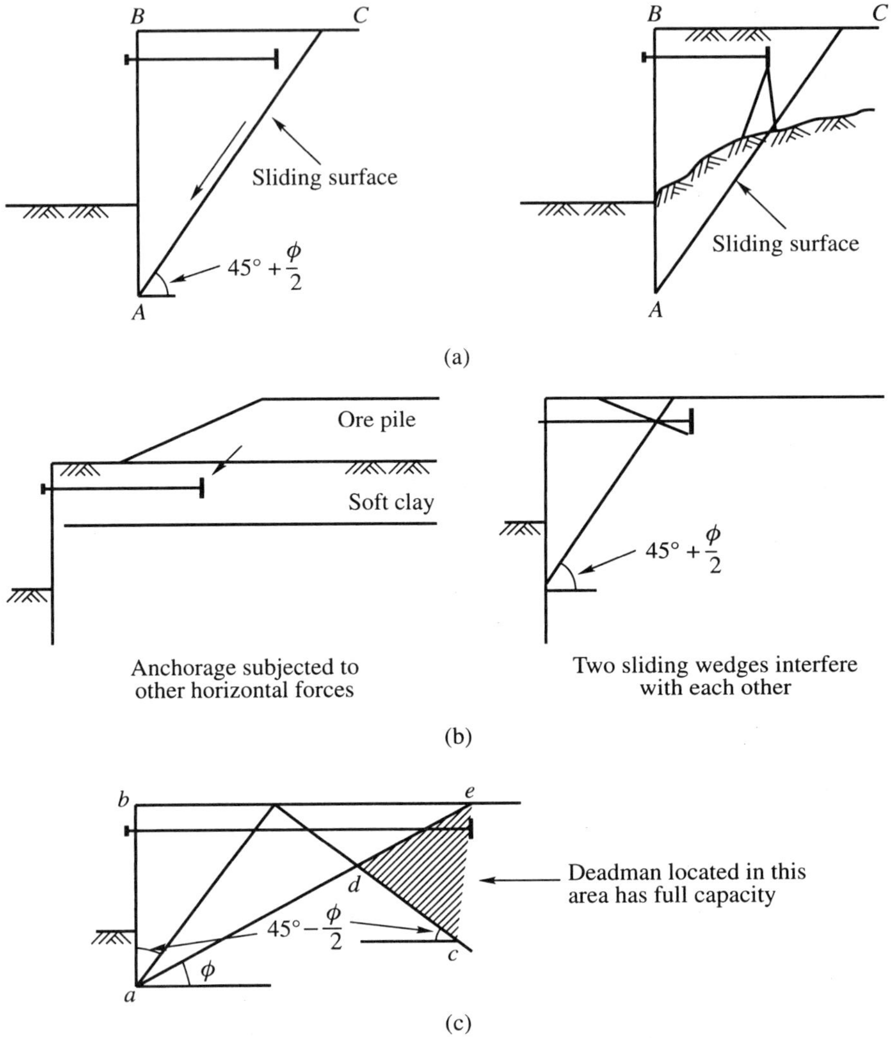

Figure 20.22 Location of deadmen: (a) offers no resistance; (b) efficiency greatly impaired; (c) full capacity. (after Teng, 1969)

Short Deadman Near Ground Surface in Granular Soil (Fig. 20.23b)

If the length of a deadman is shorter than $5h$ (h = height of deadman) there will be an end effect with regards to the holding capacity of the anchor. The equation suggested by Teng for computing the ultimate tensile capacity T_u is

$$T_u = L(P_p - P_a) + \frac{1}{3} K_o \gamma (\sqrt{K_P} + \sqrt{K_A}) H^3 \tan\phi \tag{20.51}$$

where

h	=	height of deadman
$\bar{h}$	=	depth to the top of deadman
L	=	length of deadman
H	=	depth to the bottom of the dead man from the ground surface
P_p, P_a	=	total passive and active earth pressures per unit length

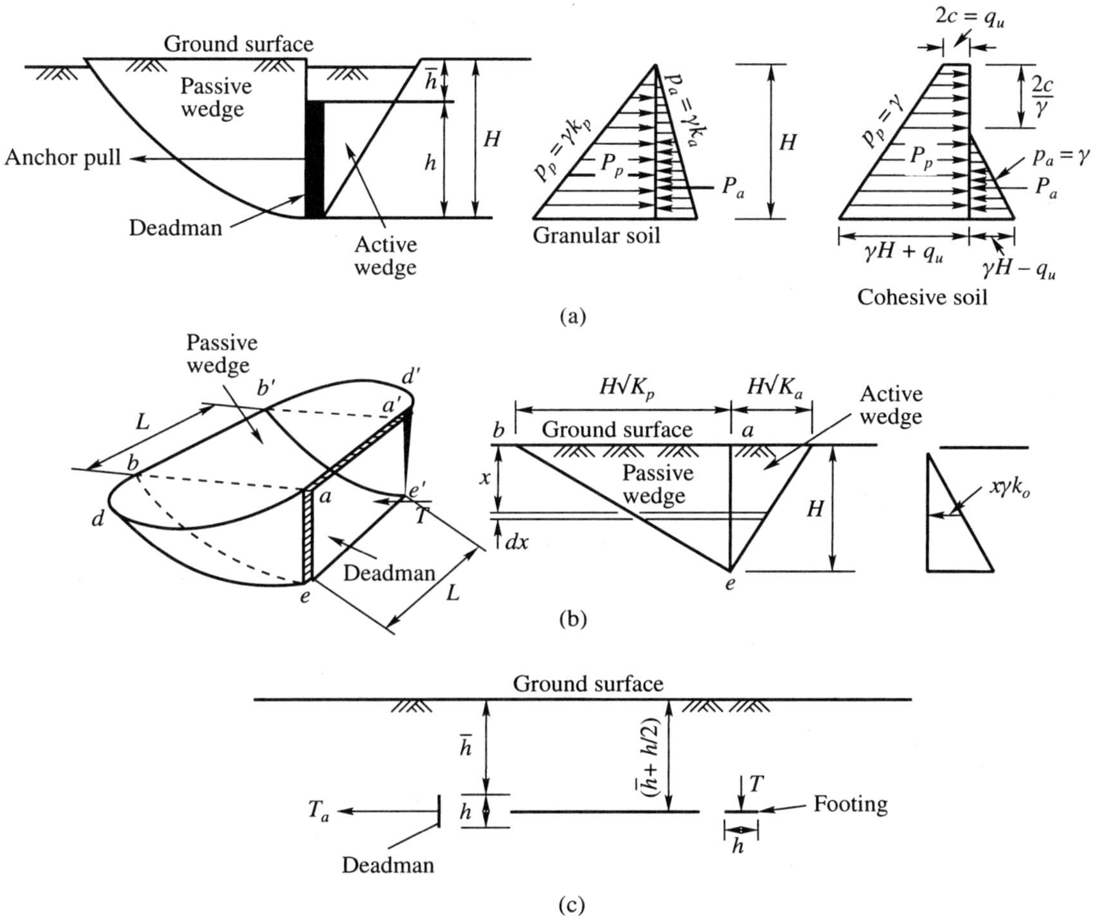

Figure 20.23 Capacity of deadmen: (a) continuous deadmen near ground surface ($\bar{h}/H < 1/3 \sim 1/2$); (b) short deadmen near ground surface; (c) deadmen at great depth below ground surface (after Teng, 1969)

K_o	=	coefficient of earth pressures at-rest, taken equal to 0.4
γ	=	effective unit weight of soil
K_P, K_A	=	Rankine's coefficients of passive and active earth pressures
ϕ	=	angle of internal friction

Anchor Capacity of Short Deadman in Cohesive Soil Near Ground Surface

In cohesive soils, the second term of Eq. (20.51) should be replaced by the cohesive resistance

$$T_u = L\,(P_p - P_a) + q_u H^2 \tag{20.52}$$

where q_u = unconfined compressive strength of soil.

Deadman at Great Depth

The ultimate capacity of a deadman at great depth below the ground surface as shown in Fig. (20.23c) is approximately equal to the bearing capacity of a footing whose base is located at a depth ($\bar{h} + h/2$), corresponding to the mid height of the deadman (Terzaghi, 1943).

Ultimate Lateral Resistance of Vertical Anchor Plates in Sand

The load-displacement behavior of horizontally loaded vertical anchor plates was analyzed by Ghaly (1997). He made use of 128 published field and laboratory test results and presented equations for computing the following:

1. Ultimate horizontal resistance T_u of single anchors
2. Horizontal displacement u at any load level T

The equations are

$$T_u = \frac{C_A \gamma AH}{\tan\phi}\left[\frac{H^2}{A}\right]^{\alpha} \tag{20.53}$$

$$\frac{T}{T_u} = 2.2\left[\frac{u}{H}\right]^{0.3} \tag{20.54}$$

where

A = area of anchor plate = h_L
H = depth from the ground surface to the bottom of the plate
h = height of plate and L = width
γ = effective unit weight of the sand
ϕ = angle of friction

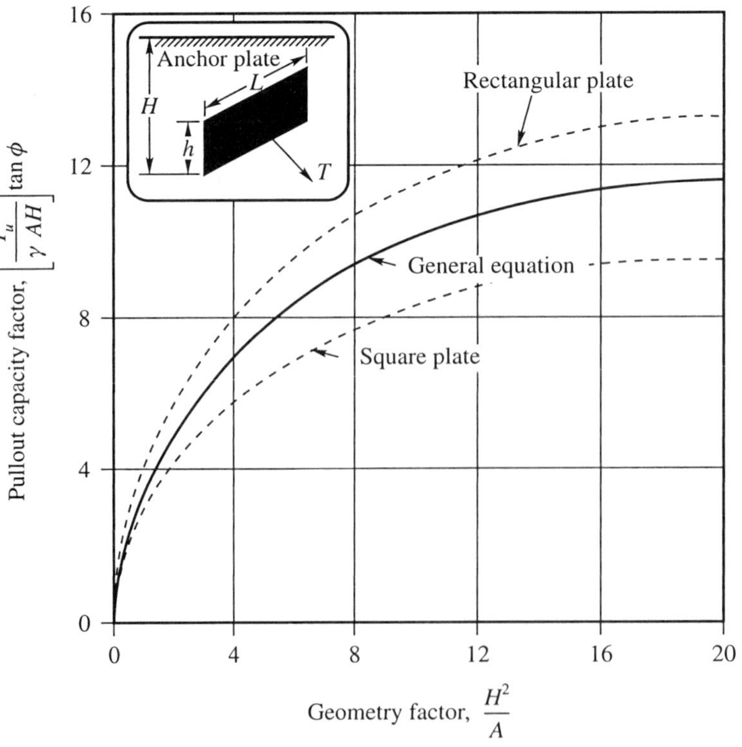

Figure 20.24 Relationship of pullout-capacity factor versus geometry factor (after Ghaly, 1997)

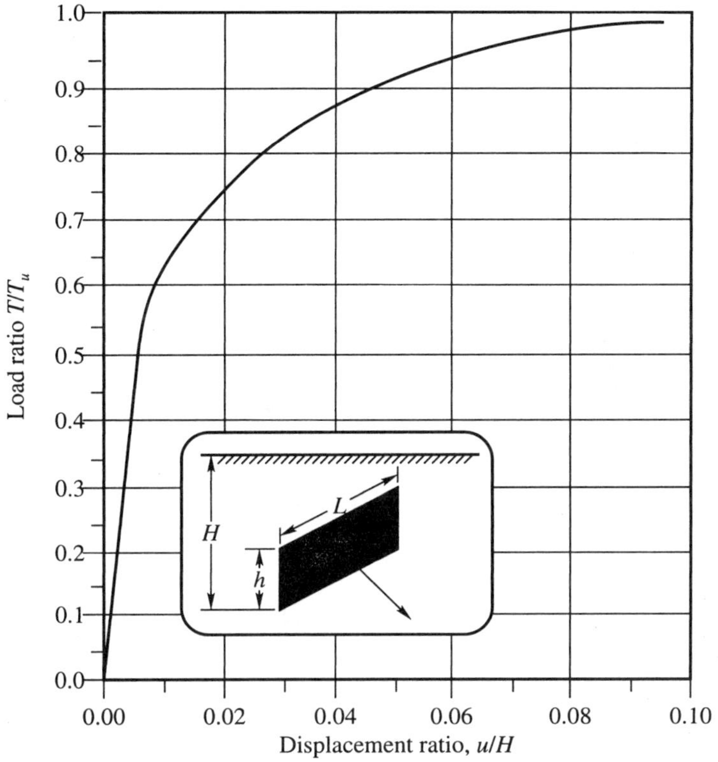

Figure 20.25 Relationship of load ratio versus displacement ratio [from data reported by Das and Seeley, 1975)] (after Ghaly, 1997)

C_A = 5.5 for a rectangular plate
= 5.4 for a general equation
= 3.3 for a square plate.
α = 0.31 for a rectangular plate
0.28 for a general equation
0.39 for a square plate

The equations developed are valid for relative depth ratios $(H/h) \leq 5$. Figs 20.24 and 20.25 give relationships in non-dimensional form for computing T_u and u respectively. Non-dimensional plots for computing T_u for square and rectangular plates are also given in Fig. 20.24.

20.10 BRACED CUTS

General Considerations

Shallow excavations can be made without supporting the surrounding material if there is adequate space to establish slopes at which the material can stand. The steepest slopes that can be used in a given locality are best determined by experience. Many building sites extend to the edges of the property lines. Under these circumstances, the sides of the excavation have to be made vertical and must usually be supported by bracings.

Common methods of bracing the sides when the depth of excavation does not exceed about 3 m are shown in Figs 20.26(a) and (b). The practice is to drive vertical timber planks known as sheeting along the sides of the excavation. The sheeting is held in place by means of horizontal beams called *wales* that in turn are commonly supported by horizontal *struts* extending from side to side of the excavation. The struts are usually of timber for widths not exceeding about 2 m. For greater widths metal pipes called *trench braces* are commonly used.

When the excavation depth exceeds about 5 to 6 m, the use of vertical timber sheeting will become uneconomical. According to one procedure, *steel sheet piles* are used around the boundary of the excavation. As the soil is removed from the enclosure, wales and struts are inserted. The wales are commonly of steel and the struts may be of steel or wood. The process continues until the

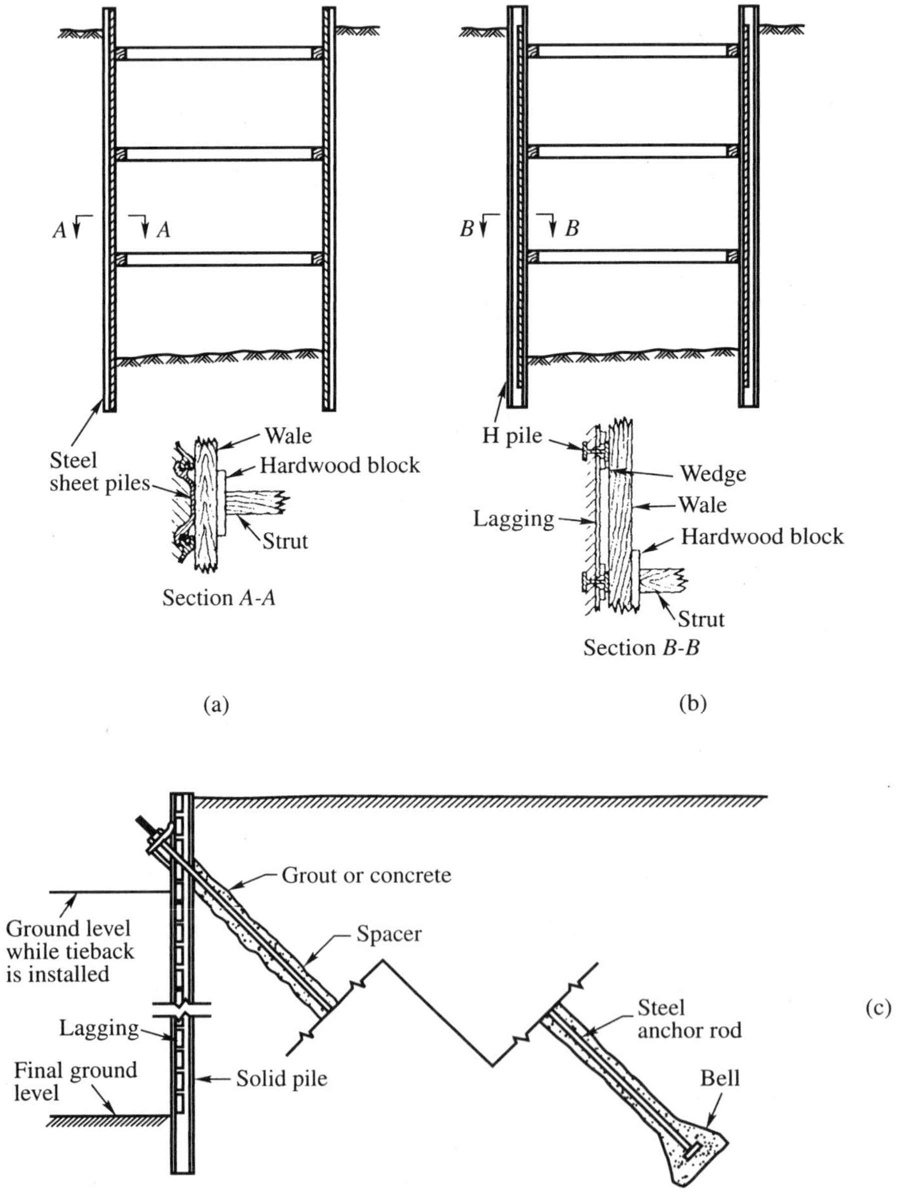

Figure 20.26 Cross sections, through typical bracing in deep excavation. (a) sides retained by steel sheet piles, (b) sides retained by *H* piles and lagging, (c) one of several tieback systems for supporting vertical sides of open cut. several sets of anchors may be used, at different elevations (Peck, 1969)

excavation is complete. In most types of soil, it may be possible to eliminate sheet piles and to replace them with a series of H piles spaced 1.5 to 2.5 m apart. The H piles, known as *soldier piles* or *soldier beams*, are driven with their flanges parallel to the sides of the excavation as shown in Fig. 20.26(b). As the soil next to the piles is removed horizontal boards known as *lagging* are introduced as shown in the figure and are wedged against the soil outside the cut. As the general depth of excavation advances from one level to another, wales and struts are inserted in the same manner as for steel sheeting.

If the width of a deep excavation is too great to permit economical use of struts across the entire excavation, *tiebacks* are often used as an alternative to cross-bracings as shown in Fig. 20.26(c). Inclined holes are drilled into the soil outside the sheeting or H piles. Tensile reinforcement is then inserted and concreted into the hole. Each tieback is usually prestressed before the depth of excavation is increased.

Example 20.12

Fig. Ex. 20.12 gives an anchor plate fixed vertically in medium dense sand with the bottom of the plate at a depth of 3 ft below the ground surface. The size of the plate is 2 × 12 ft. Determine the ultimate lateral resistance of the plate. The soil parameters are $\gamma = 115$ lb/ft^3, $\phi = 38°$.

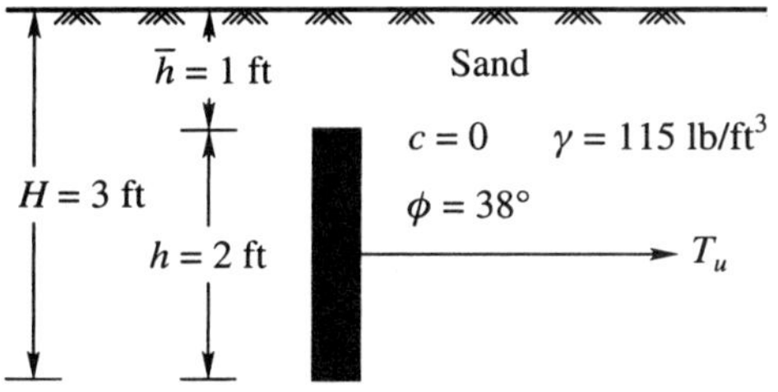

Figure Ex. 20.12

Solution

For all practical purposes if $L/\bar{h} \geq 5$, the plate may be considered as a long beam. In this case $L/h = 12/2 = 6 > 5$. If the depth $\bar{h}$ to the top of the plate is less than about 1/3 to 1/2 of H (where H is the depth to the bottom of the plate), the lateral capacity may be calculated using Eq. (20.49). In this case $\bar{h}/H = 1/3$, As such

$$T_u = \frac{1}{2}\gamma H^2 (K_p - K_A)$$

$$\text{where } K_p = \tan^2\left(45° + \frac{38}{2}\right) = 4.204$$

$$K_A = \frac{1}{4.204} = 0.238$$

$$T_u = \frac{1}{2} \times 115 \times 3^2 (4.204 - 0.238) = 2051 \text{ lb/ft}$$

Example 20.13

Solve Example 20.12 if $\phi = 0$ and c = 300 lb/ft^2. All the other data remain the same.

Solution

Use Eq. (20.50)

$$T_u = 4cH - \frac{2c^2}{\gamma} = 4\times 300\times 3 - \frac{2\times 300^2}{115} = 2035 \text{ lb/ft}$$

Example 20.14

Solve the problem in Example 20.12 for a plate length of 6 ft. All the other data remain the same.

Solution

Use Eq. (20.51) for a shorter length of plate

$$T_u = L(P_p - P_a) + \frac{1}{3}K_o\gamma\left(\sqrt{K_P} + \sqrt{K_A}\right)H^3 \tan\phi$$

where $(P_p - P_a) = 2051$ lb/ft from Ex. 20.12

$K_o = 1 - \sin\phi = 1 - \sin 38° = 0.384$

$\tan\phi = \tan 38° = 0.78$

$$\sqrt{K_P} = \sqrt{4.204} = 2.05,\quad \sqrt{K_A} = \sqrt{0.238} = 0.488$$

substituting

$$T_u = 6\times 2051 + \frac{1}{3}\times 0.384\times 115(2.05+0.488)\times 3^3\times 0.78 = 12{,}306 + 787 = 13{,}093 \text{ lb}$$

Example 20.15

Solve Example 20.13 if the plate length $L = 6$ ft. All the other data remain the same.

Solution

Use Eq. (20.52)

$$T_u = L\,(P_P - P_u) + q_u H^2$$

where $P_p - P_u = 2035$ lb/ft from Ex. 20.13

$q_u = 2 \times 300 = 600 \text{ lb/ft}^2$

Substituting

$T_u = 6 \times 2035 + 600 \times 3^2 = 12{,}210 + 5{,}400 = 17{,}610$ lb

Example 20.16

Solve the problem in Example 20.14 using Eq. (20.53). All the other data remain the same.

Solution

Use Eq. (20.53)

$$T_u = \frac{5.4\gamma AH}{\tan\phi}\left[\frac{H^2}{A}\right]^{0.28}$$

where $A = 2 \times 6 = 12$ sq ft, $H = 3$ ft, $C_A = 5.4$ and $\alpha = 0.28$ for a general equation

$\tan \phi = \tan 38° = 0.78$

Substituting

$$T_u = \frac{5.4 \times 115 \times 12 \times 3}{0.78} \left[\frac{3^2}{12} \right]^{0.28} = 26{,}443 \text{ lb/ft}$$

20.11 LATERAL EARTH PRESSURE DISTRIBUTION ON BRACED-CUTS

Since most open cuts are excavated in stages within the boundaries of sheet pile walls or walls consisting of soldier piles and lagging, and since struts are inserted progressively as the excavation precedes, the walls are likely to deform as shown in Fig. 20.27. Little inward movement can occur at the top of the cut after the first strut is inserted. The pattern of deformation differs so greatly from that required for Rankine's state that the distribution of earth pressure associated with retaining walls is not a satisfactory basis for design (Peck et al, 1974). The pressures against the upper portion of the walls are substantially greater than those indicated by the equation.

$$p_a = \frac{1 - \sin \phi}{1 + \sin \phi} p_v \tag{20.55}$$

for Rankine's condition

where

p_v = vertical pressure,
ϕ = friction angle

Apparent Pressure Diagrams

Peck (1969) presented pressure distribution diagrams on braced cuts. These diagrams are based on a wealth of information collected by actual measurements in the field. Peck called these pressure diagrams *apparent pressure envelopes* which represent fictitious pressure distributions for estimating strut loads in a system of loading. Figure 20.28 gives the apparent pressure distribution diagrams as proposed by Peck.

Deep Cuts in Sand

The apparent pressure diagram for sand given in Fig. 20.28 was developed by Peck (1969) after a great deal of study of actual pressure measurements on braced cuts used for subways.

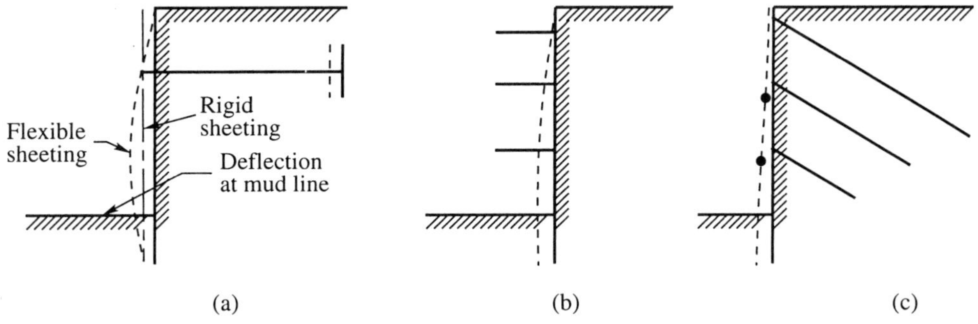

Figure 20.27 Typical pattern of deformation of vertical walls (a) anchored bulkhead, (b) braced cut, and (c) tieback cut (Peck et al., 1974)

The pressure diagram given in Fig. 20.28(b) is applicable to both loose and dense sands. The struts are to be designed based on this apparent pressure distribution. The most probable value of any individual strut load is about 25 percent lower than the maximum (Peck, 1969). It may be noted here that this apparent pressure distribution diagram is based on the assumption that the water table is below the bottom of the cut.

The pressure p_a is uniform with respect to depth. The expression for p_a is

$$p_a = 0.65\,\gamma H K_A \tag{20.56}$$

where,

K_A = $\tan^2 (45° - \phi/2)$
γ = unit weight of sand

Cuts in Saturated Clay

Peck (1969) developed two apparent pressure diagrams, one for soft to medium clay and the other for stiff fissured clay. He classified these clays on the basis of non-dimensional factors (stability number N_s) as follows.

Stiff Fissured clay

$$N_s = \frac{\gamma H}{c} \le 4 \tag{20.57a}$$

Soft to Medium clay

$$N_s = \frac{\gamma H}{c} > 4 \tag{20.57b}$$

where γ = unit weight of clay, c = undrained cohesion ($\phi = 0$)

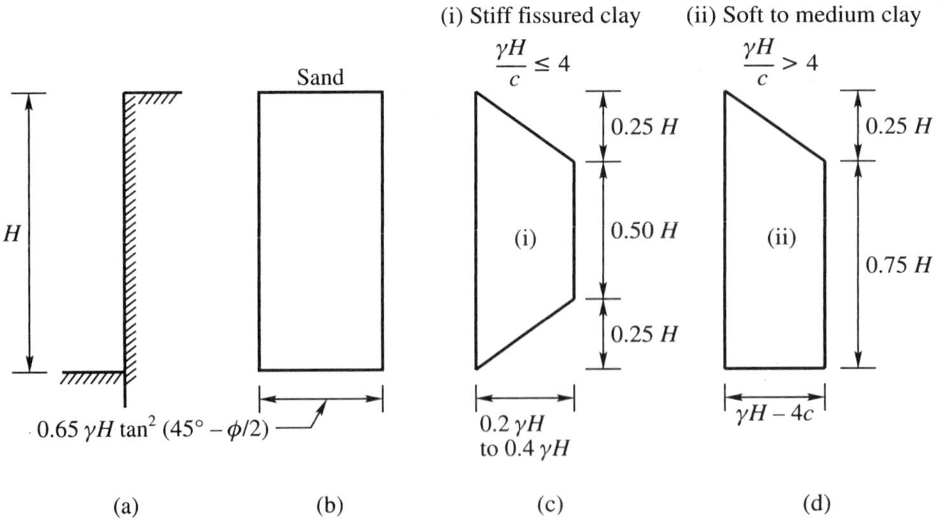

Figure 20.28 Apparent pressure diagram for calculating loads in struts of braced cuts: (a) sketch of wall of cut, (b) diagram for cuts in dry or moist sand, (c) diagram for clays if $\gamma H/c$ is less than 4 (d) diagram for clays if $\gamma H/c$ is greater than 4 where c is the average undrained shearing strength of the soil (Peck, 1969)

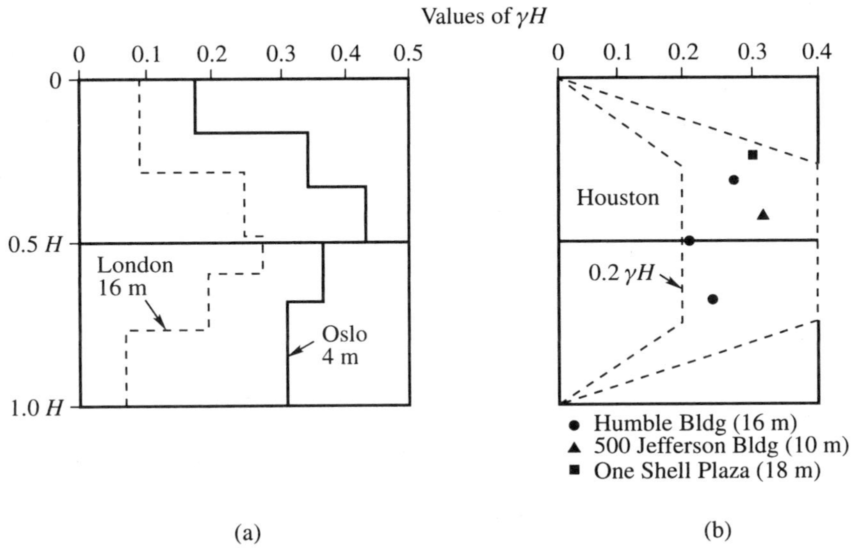

Figure 20.29 Maximum apparent pressures for cuts in stiff clays: (a) fissured clays in London and Oslo, (b) stiff slickensided clays in Houston (Peck, 1969)

The pressure diagrams for these two types of clays are given in Fig. 20.28(c) and (d) respectively. The apparent pressure diagram for soft to medium clay (Fig. 20.28(d)) has been found to be conservative for estimating loads for design supports. Fig. 20.28(c) shows the apparent pressure diagram for stiff-fissured clays. Most stiff clays are weak and contain fissures. Lower pressures should be used only when the results of observations on similar cuts in the vicinity so indicate. Otherwise a lower limit for $p_a = 0.3\,\gamma H$ should be taken. Fig. 20.29 gives a comparison of measured and computed pressures distribution for cuts in London, Oslo and Houston clays.

Cuts in Stratified Soils

It is very rare to find uniform deposits of sand or clay to a great depth. Many times layers of sand and clays overlying one another the other are found in nature. Even the simplest of these conditions does not lend itself to vigorous calculations of lateral earth pressures by any of the methods

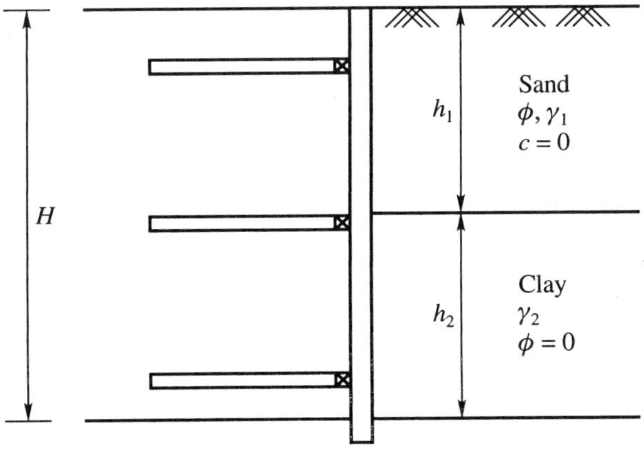

Figure 20.30 Cuts in stratified soils

available. Based on field experience, empirical or semi-empirical procedures for estimating apparent pressure diagrams may be justified. Peck (1969) proposed the following unit pressure for excavations in layered soils (sand and clay) with sand overlying as shown in Fig. 20.30.

When layers of sand and soft clay are encountered, the pressure distribution shown in Fig. 20.28(d) may be used if the unconfined compressive strength q_u is substituted by the average $\bar{q}_u$ and the unit weight of soil γ by the average value $\bar{\gamma}$ (Peck, 1969). The expressions for $\bar{q}_u$ and $\bar{\gamma}$ are

$$\bar{q}_u = \frac{1}{H}\left[\gamma_1 K_s h_1^2 \tan\phi + h_2 n q_u\right] \tag{20.58}$$

$$\bar{\gamma} = \frac{1}{H}\left[\gamma_1 h_1 + \gamma_2 h_2\right] \tag{20.59}$$

where

H = total depth of excavation
γ_1, γ_2 = unit weights of sand and clay respectively
h_1, h_2 = thickness of sand and clay layers respectively
K_s = hydrostatic pressure ratio for the sand layer, may be taken as equal to 1.0 for design purposes
ϕ = angle of friction of sand
n = coefficient of progressive failure varies from 0.5 to 1.0 which depends upon the creep characteristics of clay. For Chicago clay n varies from 0.75 to 1.0.
q_u = unconfined compression strength of clay

20.12 STABILITY OF BRACED CUTS IN SATURATED CLAY

A braced-cut may fail as a unit due to unbalanced external forces or heaving of the bottom of the excavation. If the external forces acting on opposite sides of the braced cut are unequal, the stability of the entire system has to be analyzed. If soil on one side of a braced cut is removed due to some unnatural forces the stability of the system will be impaired. However, we are concerned here about the stability of the bottom of the cut. Two cases may arise. They are

1. Heaving in clay soil
2. Heaving in cohesionless soil

Heaving in Clay Soil

The danger of heaving is greater if the bottom of the cut is soft clay. Even in a soft clay bottom, two types of failure are possible. They are

Case 1: When the clay below the cut is homogeneous at least up to a depth equal 0.7 B where B is the width of the cut.

Case 2: When a hard stratum is met within a depth equal to 0.7 B.

In the first case a full plastic failure zone will be formed and in the second case this is restricted as shown in Fig. 20.31. A factor of safety of 1.5 is recommended for determining the resistance here. Sheet piling is to be driven deeper to increase the factor of safety. The stability analysis of the bottom of the cut as developed by Terzaghi (1943) is as follows.

Case 1: Formation of Full Plastic Failure Zone Below the Bottom of Cut.

Figure 20.31(a) is a vertical section through a long cut of width B and depth H in saturated cohesive soil ($\phi = 0$). The soil below the bottom of the cut is uniform up to a considerable depth for the formation of a full plastic failure zone. The undrained cohesive strength of soil is c. The weight of the blocks of clay on either side of the cut tends to displace the underlying clay toward the excavation. If the underlying clay experiences a bearing capacity failure, the bottom of the excavation heaves and the earth pressure against the bracing increases considerably.

The anchorage load block of soil $a\ b\ c\ d$ in Fig. 20.31(a) of width $\overline{B}$ (assumed) at the level of the bottom of the cut per unit length may be expressed as

$$Q = \gamma H \overline{B} - cH = \overline{B} H \left(\gamma - \frac{c}{\overline{B}} \right) \tag{20.60}$$

The vertical pressure q per unit length of a horizontal, ba, is

$$q = \frac{Q}{\overline{B}} = H \left(\gamma - \frac{c}{\overline{B}} \right) \tag{20.61}$$

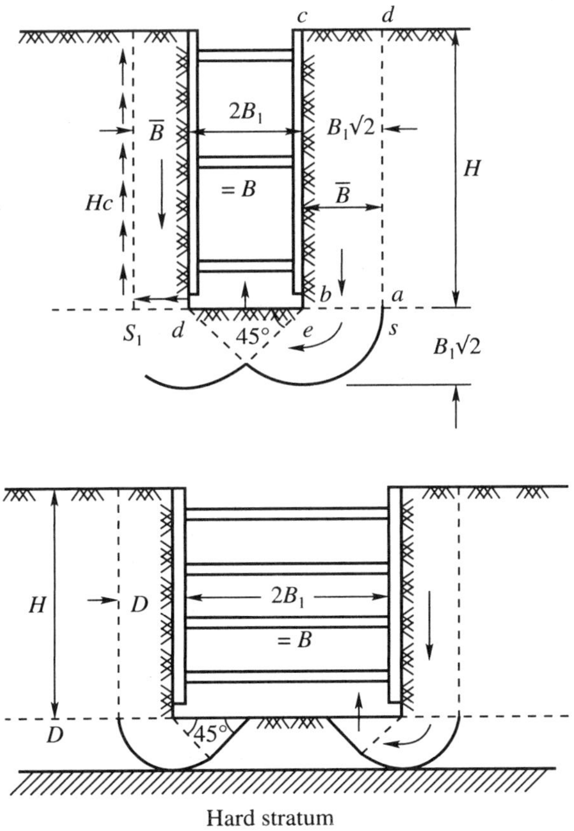

Figure 20.31 Stability of braced cut: (a) heave of bottom of timbered cut in soft clay if no hard stratum interferes with flow of clay, (b) as before, if clay rests at shallow depth below bottom of cut on hard stratum (after Terzaghi, 1943)

The bearing capacity q_u per unit area at level *ab* is

$$q_u = N_c c = 5.7c \tag{20.62}$$

where $N_c = 5.7$

The factor of safety against heaving is

$$F_s = \frac{q_u}{q} = \frac{5.7c}{H\left(\gamma - \frac{c}{\overline{B}}\right)} \tag{20.63}$$

Because of the geometrical condition, it has been found that the width $\overline{B}$ cannot exceed 0.7 *B*. Substituting this value for $\overline{B}$,

$$F_s = \frac{5.7c}{H\left(\gamma - \frac{c}{0.7B}\right)} \tag{20.64}$$

This indicates that the width of the failure slip is equal to $\overline{B}\sqrt{2} = 0.7B$.

Case 2: When the Formation of Full Plastic Zone is Restricted by the Presence of a Hard Layer

If a hard layer is located at a depth *D* below the bottom of the cut (which is less than 0.7*B*), the failure of the bottom occurs as shown in Fig. 20.31(b). The width of the strip which can sink is also equal to *D*.

Replacing 0.7*B* by *D* in Eq. (20.64), the factor of safety is represented by

$$F_s = \frac{5.7c_u}{H\left(\gamma - \frac{c}{D}\right)} \tag{20.65}$$

For a cut in soft clay with a constant value of c_u below the bottom of the cut, *D* in Eq. (20.65) becomes large, and F_s approaches the value

$$F_s = \frac{5.7c_u}{\gamma H} = \frac{5.7}{N_s} \tag{20.66}$$

where $$N_s = \frac{\gamma H}{c_u} \tag{20.67}$$

is termed the *stability number*. The stability number is a useful indicator of potential soil movements. The soil movement is smaller for smaller values of N_s.

The analysis discussed so far is for long cuts. For short cuts, square, circular or rectangular, the factor of safety against heave can be found in the same way as for footings.

20.13 BJERRUM AND EIDE (1956) METHOD OF ANALYSIS

The method of analysis discussed earlier gives reliable results provided the width of the braced cut is larger than the depth of the excavation and that the braced cut is very long. In the cases where the braced cuts are rectangular, square or circular in plan or the depth of excavation exceeds the width of the cut, the following analysis should be used.

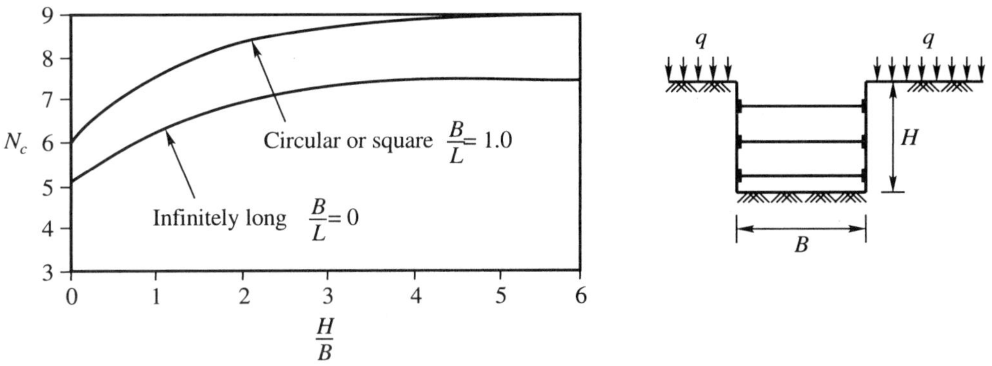

Figure 20.32 Stability of bottom excavation (after Bjerrum and Eide, 1956)

In this analysis the braced cut is visualized as a deep footing whose depth and horizontal dimensions are identical to those at the bottom of the braced cut. This deep footing would fail in an identical manner to the bottom braced cut failed by heave. The theory of Skempton for computing N_c (bearing capacity factor) for different shapes of footing is made use of. Figure 20.32 gives values of N_c as a function of H/B for long, circular or square footings. For rectangular footings, the value of N_c may be computed by the expression

$$N_c \text{ (rect)} = (0.84 + 0.16\, B/L)\, N_c \text{ (sq)} \tag{20.68}$$

where

L = length of excavation

B = width of excavation

The factor of safety for bottom heave may be expressed as

$$F_s = \frac{cN_c}{\gamma H + q} \geq 1.5$$

where

γ = effective unit weight of the soil above the bottom of the excavation

q = uniform surcharge load (Fig. 20.32)

Example 20.17

A long trench is excavated in medium dense sand for the foundation of a multistorey building. The sides of the trench are supported with sheet pile walls fixed in place by struts and wales as shown in Fig. Ex. 20.17. The soil properties are:

$\gamma = 18.5$ kN/m^3, $c = 0$ and $\phi = 38°$

Determine: (a) The pressure distribution on the walls with respect to depth.

(b) Strut loads. The struts are placed horizontally at distances $L = 4$ m center to center.

(c) The maximum bending moment for determining the pile wall section.

(d) The maximum bending moments for determining the section of the wales.

Solution

(a) For a braced cut in sand use the apparent pressure envelope given in Fig. 20.28 b. The equation for p_a is

$p_a = 0.65\, \gamma H\, K_A = 0.65 \times 18.5 \times 8 \tan^2 (45 - 38/2) = 23$ kN/m^2

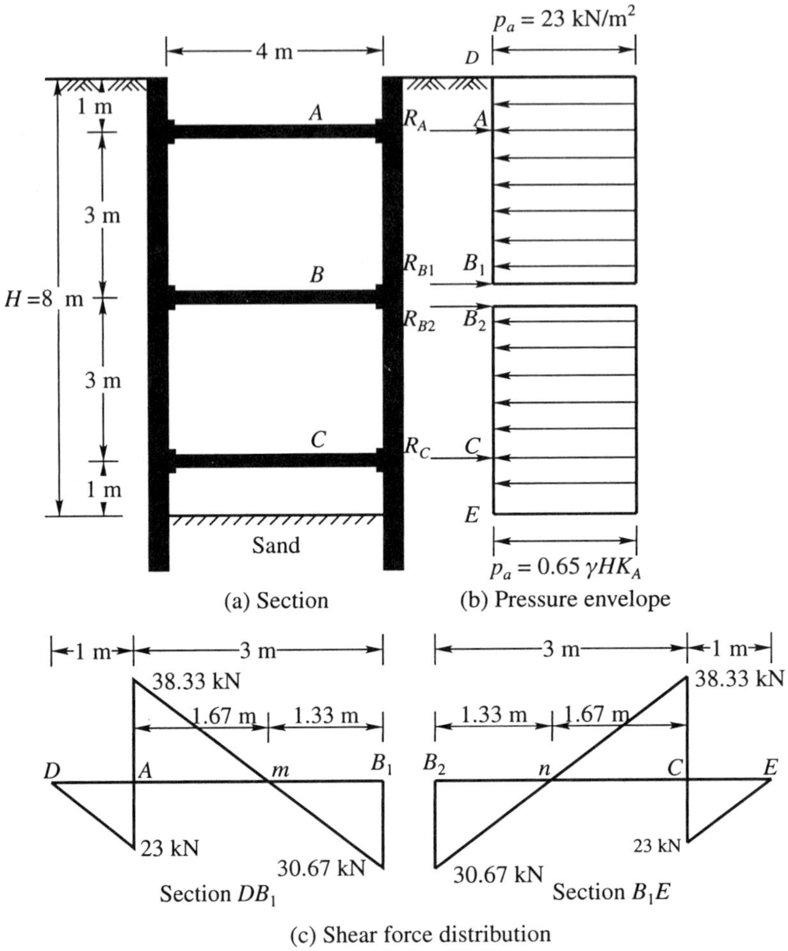

Figure Ex. 20.17

Fig. Ex. 20.17b shows the pressure envelope.

(b) Strut loads

The reactions at the ends of struts A, B and C are represented by R_A, R_B and R_c respectively.

For reaction R_A, take moments about B

$$R_A \times 3 = 4 \times 23 \times \frac{4}{2} \text{ or } R_A = \frac{184}{3} = 61.33 \text{ kN}$$

$R_{B1} = 23 \times 4 - 61.33 = 30.67$ kN

Due to the symmetry of the load distribution,

$R_{B1} = R_{B2} = 30.67$ kN, and $R_A = R_C = 61.33$ kN.

Now the strut loads are (for $L = 4$ m)

Strut A, $P_A = 61.33 \times 4 \approx 245$ kN

Strut B, $P_B = (R_{B1} + R_{B2}) \times 4 = 61.34 \times 4 \approx 245$ kN

Strut C, $P_C = 245$ kN

(c) Moment of the pile wall section

To determine moments at different points it is necessary to draw a diagram showing the shear force distribution.

Consider sections DB_1 and B_2E of the wall in Fig. Ex. 20.17(b). The distribution of the shear forces are shown in Fig. 20.17(c) along with the points of zero shear.

The moments at different points may be determined as follows

$$M_A = \frac{1}{2} \times 1 \times 23 = 11.5 \text{ kN- m}$$

$$M_C = \frac{1}{2} \times 1 \times 23 = 11.5 \text{ kN- m}$$

$$M_m = \frac{1}{2} \times 1.33 \times 30.67 = 20.4 \text{ kN- m}$$

$$M_n = \frac{1}{2} \times 1.33 \times 30.67 = 20.4 \text{ kN- m}$$

The maximum moment M_{max} = 20.4 kN-m. A suitable section of sheet pile can be determined as per standard practice.

(d) Maximum moment for wales

The bending moment equation for wales is

$$M_{max} = \frac{RL^2}{8}$$

where R = maximum strut load = 245 kN

L = spacing of struts = 4 m

$$M_{max} = \frac{245 \times 4^2}{8} = 490 \text{ kN-m}$$

A suitable section for the wales can be determined as per standard practice.

Example 20.18

Fig. Ex. 20.18a gives the section of a long braced cut. The sides are supported by steel sheet pile walls with struts and wales. The soil excavated at the site is stiff clay with the following properties

c = 800 lb/ft^2, $\phi = 0$, γ = 115 lb/ft^3

Determine: (a) The earth pressure distribution envelope.

(b) Strut loads.

(c) The maximum moment of the sheet pile section.

The struts are placed 12 ft apart center to center horizontally.

Solution

(a) The stability number N_s from Eq. (20.57a) is

$$N_s = \frac{\gamma H}{c} = \frac{115 \times 25}{800} = 3.6 < 4$$

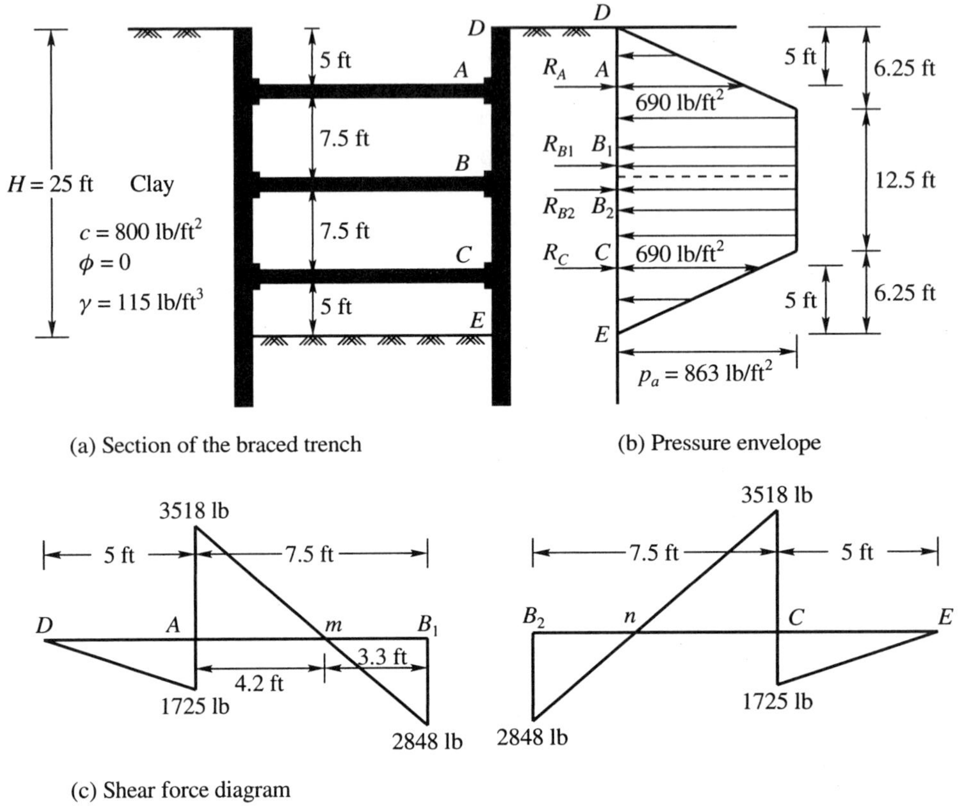

Figure Ex. 20.18

The soil is stiff fissured clay. As such the pressure envelope shown in Fig. 20.28(c) is applicable. Assume $p_a = 0.3\,\gamma H$

$p_a = 0.3 \times 115 \times 25 = 863\ \text{lb/ft}^2$

The pressure envelope is drawn as shown in Fig. Ex. 20.18(b).

(b) Strut loads

Taking moments about the strut head B_1 (B)

$$R_A \times 7.5 = \frac{1}{2} 863 \times 6.25 \left(\frac{6.25}{3} + 6.25 \right) + 863 \times \frac{(6.25)^2}{2}$$

$= 22.47 \times 10^3 + 16.85 \times 10^3 = 39.32 \times 10^3$

$R_A = 5243$ lb/ft

$$R_{B1} = \frac{1}{2} \times 863 \times 6.25 + 863 \times 6.25 - 5243 = 2848 \text{ lb/ft}$$

Due to symmetry

$R_A = R_C = 5243$ lb/ft

$R_{B2} = R_{B1} = 2848$ lb/ft

Strut loads are:

$P_A = 5243 \times 12 = 62{,}916$ lb = 62.92 kips

$P_B = 2 \times 2848 \times 12 = 68{,}352$ lb = 68.35 kips

$P_C = 62.92$ kips

(c) Moments

The shear force diagram is shown in Fig. 20.18c for sections DB_1 and B_2 E

$$\text{Moment at } A = \frac{1}{2} \times 5 \times 690 \times \frac{5}{3} = 2{,}875 \text{ lb-ft/ft of wall}$$

$$\text{Moment at } m = 2848 \times 3.3 - 863 \times 3.3 \times \frac{3.3}{2} = 4699 \text{ lb-ft/ft}$$

Because of symmetrical loading

Moment at A = Moment at C = 2875 lb-ft/ft of wall

Moment at m = Moment at n = 4699 lb-ft/ft of wall

Hence, the maximum moment = 4699 lb-ft/ft of wall.

The section modulus and the required sheet pile section can be determined in the usual way.

20.14 PIPING FAILURES IN SAND CUTS

Sheet piling is used for cuts in sand and the excavation must be dewatered by pumping from the bottom of the excavation. Sufficient penetration below the bottom of the cut must be provided to reduce the amount of seepage and to avoid the danger of piping.

Piping is a phenomenon of water rushing up through pipe-shaped channels due to large upward seepage pressure. When piping takes place, the weight of the soil is counteracted by the upward hydraulic pressure and as such there is no contact pressure between the grains at the bottom of the excavation. Therefore, it offers no lateral support to the sheet piling and as a result the sheet piling may collapse. Further the soil will become very loose and may not have any bearing power. It is therefore, essential to avoid piping. For further discussions on piping, see Chapter 4 on Soil Permeability and Seepage. Piping can be reduced by increasing the depth of penetration of sheet piles below the bottom of the cut.

20.15 PROBLEMS

20.1 Figure Prob. 20.1 shows a cantilever sheet pile wall penetrating medium dense sand with the following properties of the soil:

$\gamma = 115$ lb/ft^3, $\phi = 38°$,

All the other data are given in the figure.

Determine: (a) the depth of embedment for design, and (b) the maximum theoretical moment of the sheet pile.

20.2 Figure Prob. 20.2 shows a sheet pile penetrating medium dense sand with the following data:

$h_1 = 6$ ft, $h_2 = 18$ ft, $\gamma_{sat} = 120$ lb/ft^3, $\phi = 38°$,

Determine: (a) the depth of embedment for design, and (b) the maximum theoretical moment of the sheet pile. The sand above the water table is saturated.

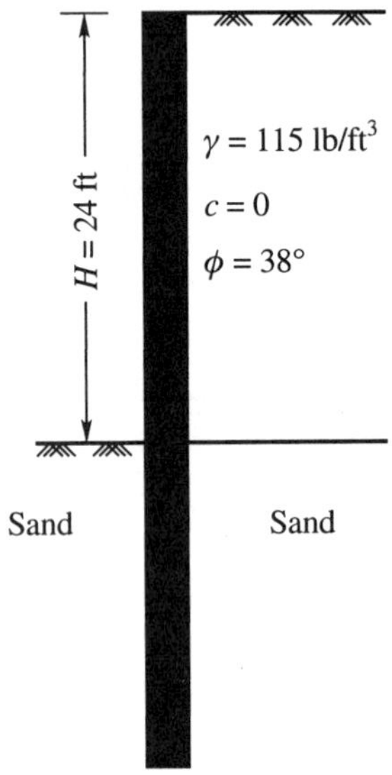

Figure Prob. 20.1

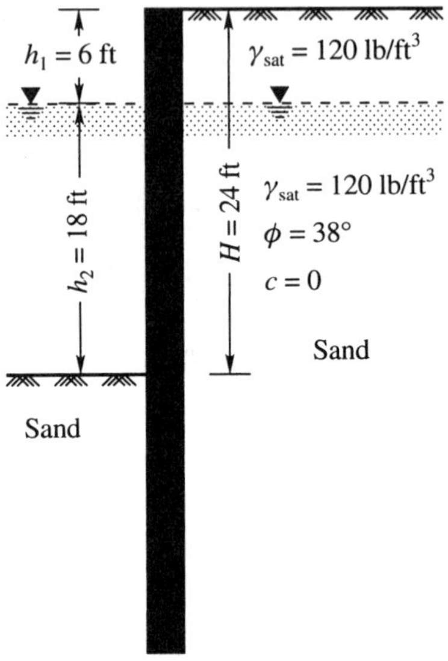

Figure Prob. 20.2

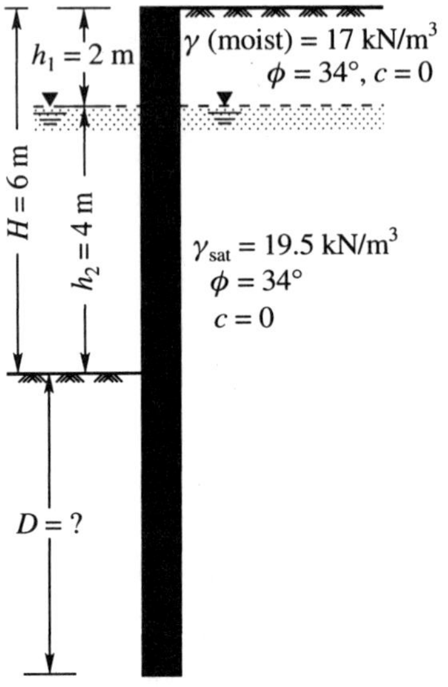

Figure Prob. 20.3

20.3 Figure Prob. 20.3 shows a sheet pile penetrating loose to medium dense sand with the following data:
$h_1 = 2$ m, $h_2 = 4$ m, γ (moist) = 17 kN/m^3, $\gamma_{sat} = 19.5$ kN/m^3, $\phi = 34°$,
Determine: (a) the depth of embedment, and (b) the maximum bending moment of the sheet pile.

20.4 Solve Problem 20.3 for the water table at great depth. Assume $\gamma = 17$ kN/m^3. All the other data remain the same.

20.5 Figure Problem 20.5 shows freestanding cantilever wall with no backfill. The sheet pile penetrates medium dense sand with the following data:
$H = 5$ m, $P = 20$ kN/m, $\gamma = 17.5$ kN/m^3 , $\phi = 36°$.
Determine: (a) the depth of embedment, and (b) the maximum moment

20.6 Solve Prob 20.5 with the following data:
$H = 20$ ft, P = 3000 lb/ft, $\gamma = 115$ lb/ft^3, $\phi = 36°$

20.7 Figure Prob. 20.7 shows a sheet pile wall penetrating clay soil and the backfill is also clay. The following data are given.
H = 5 m, c = 30 kN/m^2, $\gamma = 17.5$ kN/m^3
Determine: (a) the depth of penetration, and (b) the maximum bending moment.

20.8 Figure Prob. 20.8 has sand backfill and clay below the dredge line. The properties of the Backfill are:
$\phi = 32°$, $\gamma = 17.5$ kN/m^3;
Determine the depth of penetration of pile.

20.9 Solve Prob. 20.8 with the water table above dredge line: Given $h_1 = 3$ m, $h_2 = 3$ m, $\gamma_{sat} = 18$ kN/m^3, where h_1 = depth of water table below backfill surface and $h_2 = (H - h_1)$. The soil above the water table is also saturated. All the other data remain the same.

20.10 Figure Prob. 20.10 shows a free-standing sheet pile penetrating clay. The following data are available:
$H = 5$ m, $P = 50$ kN/m, $c = 35$ kN/m^2, $\gamma = 17.5$ kN/m^3,
Determine: (a) the depth of embedment, and (b) the maximum bending moment.

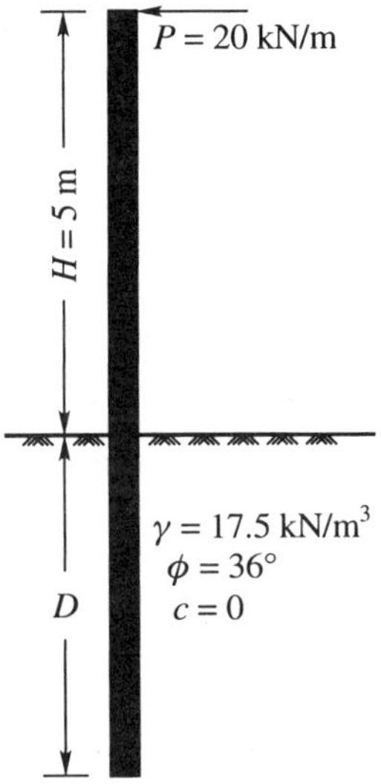

Figure Prob. 20.5

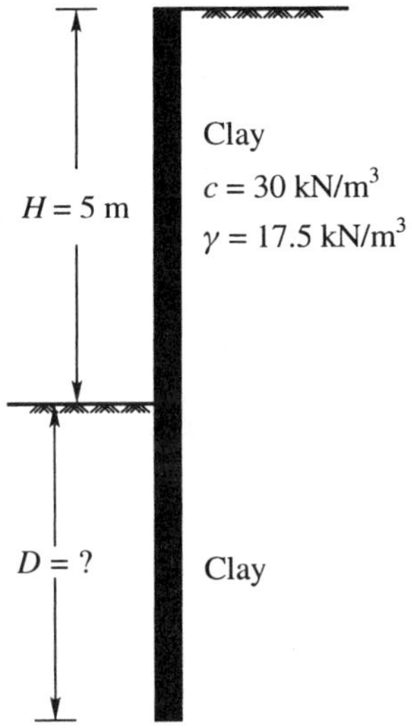

Figure Prob. 20.7

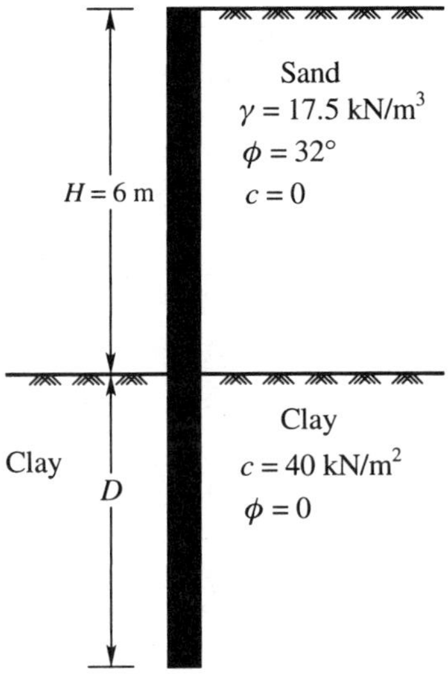

Figure Prob. 20.8

20.11 Solve Prob. 20.10 with the following data:
$H = 15$ ft, $P = 3000$ lb/ft, $c = 300$ lb/ft^2, $\gamma = 100$ lb/ft^3.

20.12 Figure Prob. 20.12 shows an anchored sheet pile wall for which the following data are given
$H = 8$ m, $h_a = 1.5$ m, $h_1 = 3$ m, $\gamma_{sat} = 19.5$ kN/m^3, $\phi = 38°$
Determine: the force in the tie rod.
Solve the problem by the free-earth support method.

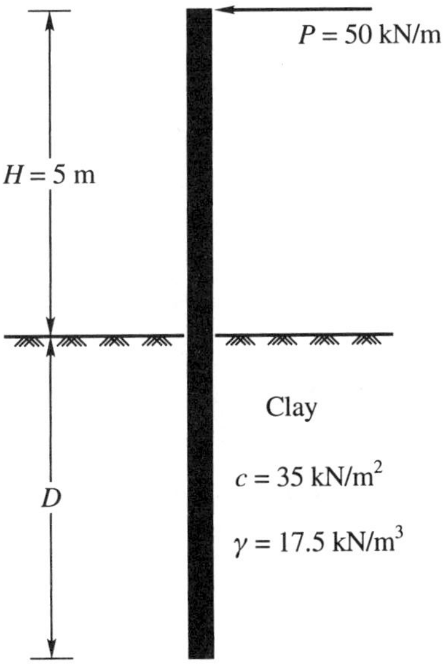

Figure Prob. 20.10

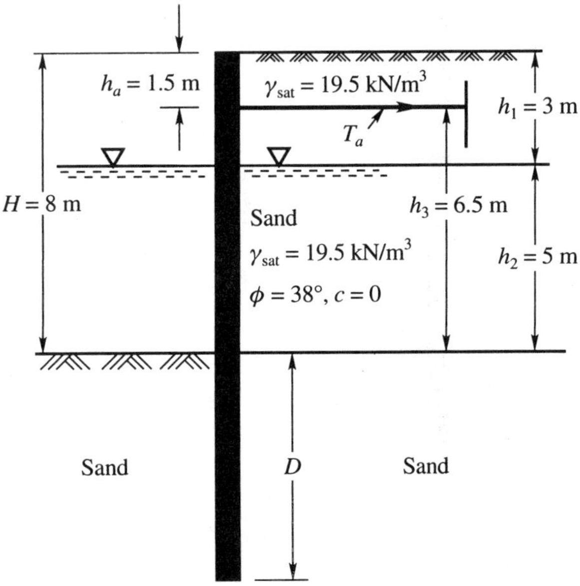

Figure Prob. 20.12

20.13 Solve the Prob. 20.12 with the following data: $H = 24$ ft, $h_1 = 9$ ft, $h_2 = 15$ ft, and $h_a = 6$ ft. Soil properties: $\phi = 32°$, $\gamma_{sat} = 120$ lb/ft^3. The soil above the water table is also saturated.

20.14 Figure Prob. 20.14 gives an anchored sheet pile wall penetrating clay. The backfill is sand. The following data are given:

$H = 24$ ft, $h_a = 6$ ft, $h_1 = 9$ ft,

For sand: $\phi = 36°$, $\gamma_{sat} = 120$ lb/ft^3,

The soil above the water table is also saturated.

For clay: $\phi = 0$, $c = 600$ lb/ft^2

Determine: (a) the depth of embedment, (b) the force in the tie rod, and (c) the maximum moment.

20.15 Solve Prob. 20.14 with the following data:

$H = 8$ m, $h_a = 1.5$ m, $h_1 = 3$ m

For sand: $\gamma_{sat} = 19.5$ kN/m^3, $\phi = 36°$

The sand above the WT is also saturated.

For clay: $c = 30$ kN/m^2

20.16 Solve Prob. 20.12 by the use of design charts given in Section 20.7.

20.17 Refer to Prob. 20.12. Determine for the pile (a) the bending moment M_{max}, and (b) the reduced moment by Rowe's method.

20.18 Figure Prob. 20.18 gives a rigid anchor plate fixed vertically in medium dense sand with the bottom of the plate at a depth of 6 ft below the ground surface. The height (h) and length (L) of the plate are 3 ft and 18 ft respectively. The soil properties are: $\gamma = 120$ lb/ft^3, and $\phi = 38°$. Determine the ultimate lateral resistance per unit length of the plate.

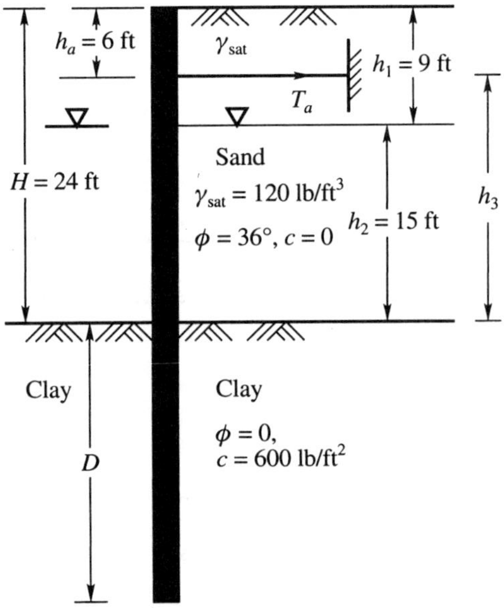

Figure Prob. 20.14

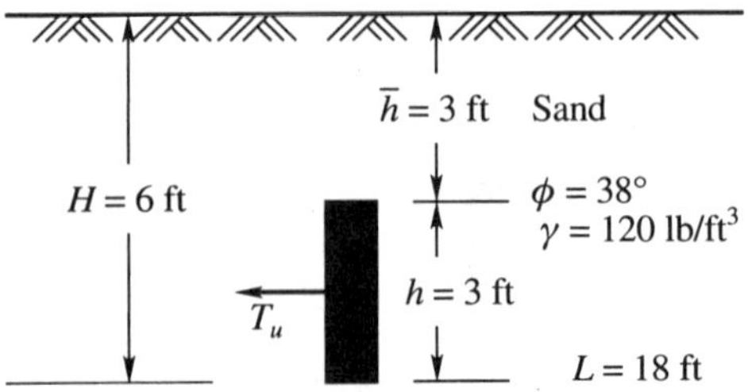

Figure Prob. 20.18

20.19 Solve Prob. 20.18 for a plate of length = 9 ft. All the other data remain the same.

20.20 Solve Prob. 20.18 for the plate in clay ($\phi = 0$) having $c = 400$ lb/ft^2. All the other data remain the same.

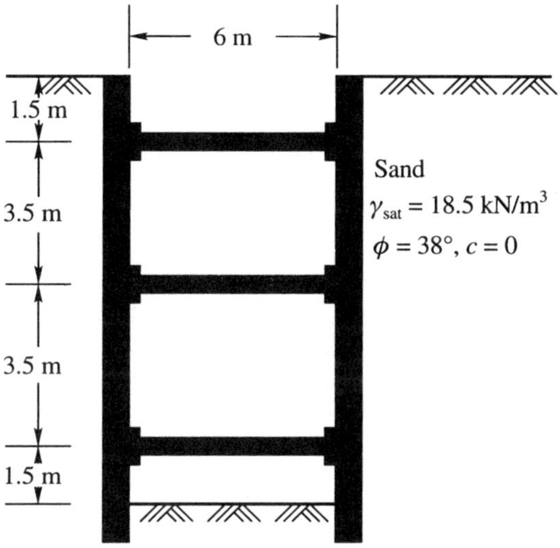

Fig. Prob. 20.23

20.21 Solve Prob. 20.20 for a plate of length 6 ft. All the other data remain the same.

20.22 Solve the prob. 20.19 by Eq (20.53). All the other determine the same.

20.23 Figure Prob. 20.23 shows a braced cut in medium dense sand. Given γ= 18.5 kN/m^3, $c = 0$ and $\phi = 38°$.

(a) Draw the pressure envelope, (b) determine the strut loads, and (c) determine the maximum moment of the sheet pile section.

The struts are placed laterally at 4 m center to center.

20.24 Figure Prob. 20.24 shows the section of a braced cut in clay. Given: c = 650 lb/ft^2, γ= 115 lb/ft^3.

(a) Draw the earth pressure envelope, (b) determine the strut loads, and (c) determine the maximum moment of the sheet pile section.

Assume that the struts are placed laterally at 12 ft center to center.

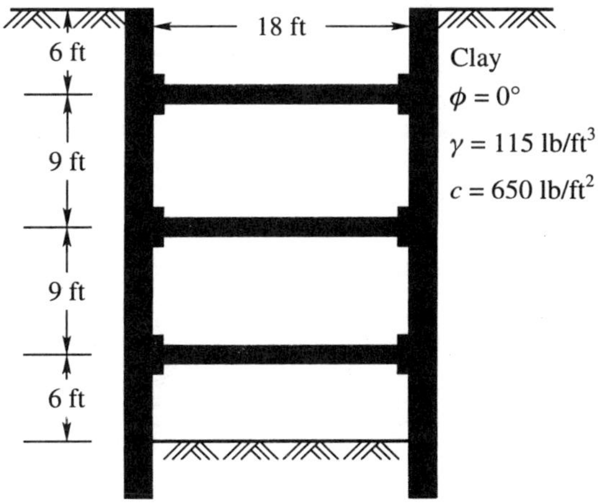

Fig. Prob. 20.24

CHAPTER 21

SOIL IMPROVEMENT

21.1 INTRODUCTION

General practice is to use shallow foundations for the foundations of buildings and other such structures, if the soil close to the ground surface possesses sufficient bearing capacity. However, where the top soil is either loose or soft, the load from the superstructure has to be transferred to deeper firm strata. In such cases, pile or pier foundations are the obvious choice.

There is also a third method which may in some cases prove more economical than deep foundations or where the alternate method may become inevitable due to certain site and other environmental conditions. This third method comes under the heading *foundation soil improvement*. In the case of earth dams, there is no other alternative than compacting the remolded soil in layers to the required density and moisture content. The soil for the dam will be excavated at the adjoining areas and transported to the site. There are many methods by which the soil at the site can be improved. Soil improvement is frequently termed *soil stabilization*, which in its broadest sense is alteration of any property of a soil to improve its engineering performance. Soil improvement

1. Increases shear strength
2. Reduces permeability, and
3. Reduces compressibility

The methods of soil improvement considered in this chapter are

1. Mechanical compaction
2. Dynamic compaction
3. Vibroflotation
4. Preloading
5. Sand and stone columns

6. Use of admixtures
7. Injection of suitable grouts
8. Use of geotextiles

21.2 MECHANICAL COMPACTION

Mechanical compaction is the least expensive of the methods and is applicable in both cohesionless and cohesive soils. The procedure is to remove first the weak soil up to the depth required, and refill or replace the same in layers with compaction. If the soil excavated is cohesionless or a sand-silt clay mixture, the same can be replaced suitably in layers and compacted. If the soil excavated is a fine sand, silt or soft clay, it is not advisable to refill the same as these materials, even under compaction, may not give sufficient bearing capacity for the foundations. Sometimes it might be necessary to transport good soil to the site from a long distance. The cost of such a project has to be studied carefully before undertaking the same.

The compaction equipment to be used on a project depends upon the size of the project and the availability of the compacting equipment. In projects where excavation and replacement are confined to a narrow site, only tampers or surface vibrators may be used. On the other hand, if the whole area of the project is to be excavated and replaced in layers with compaction, suitable roller types of heavy equipment can be used. Cohesionless soils can be compacted by using vibratory rollers and cohesive soils by sheepsfoot rollers.

The control of field compaction is very important in order to obtain the desired soil properties. Compaction of a soil is measured in terms of the dry unit weight of the soil. The dry unit weight, γ_d, may be expressed as

$$\gamma_d = \frac{\gamma_t}{1+w} \tag{21.1}$$

where,

γ_t = total unit weight
w = moisture content

Factors Affecting Compaction

The factors affecting compaction are

1. The moisture content
2. The compactive effort

The compactive effort is defined as the amount of energy imparted to the soil. With a soil of given moisture content, increasing the amount of compaction results in closer packing of soil particles and increased dry unit weight. For a particular compactive effort, there is only one moisture content which gives the maximum dry unit weight. The moisture content that gives the maximum dry unit weight is called the *optimum moisture content.* If the compactive effort is increased, the maximum dry unit weight also increases, but the optimum moisture content decreases. If all the desired qualities of the material are to be achieved in the field, suitable procedures should be adopted to compact the earthfill. The compactive effort to the soil is imparted by mechanical rollers or any other compacting device. Whether the soil in the field has attained the required maximum dry unit weight can be determined by carrying out appropriate laboratory tests on the soil. The following tests are normally carried out in a laboratory.

1. Standard Proctor test (ASTM Designation D-698), and
2. Modified Proctor test (ASTM Designation D-1557)

21.3 LABORATORY TESTS ON COMPACTION

Standard Proctor Compaction Test

Proctor (1933) developed this test in connection with the construction of earth fill dams in California. The standard size of the apparatus used for the test is given in Fig 21.1. Table 21.1 gives the standard specifications for conducting the test (ASTM designation D-698). Three alternative procedures are provided based the soil material used for the test.

Test Procedure

A soil at a selected water content is placed in layers into a mold of given dimensions (Table 21.1 and Fig. 21.1), with each layer compacted by 25 or 56 blows of a 5.5 lb (2.5 kg) hammer dropped from a height of 12 in (305 mm), subjecting the soil to a total compactive effort of about 12,375 fl-lb/ft^3 (600 kNm/m^3). The resulting dry unit weight is determined. The procedure is repeated for a sufficient number of water contents to establish a relationship between the dry unit weight and the water content of the soil. This data, when, plotted, represents a curvilinear relationship known as the compaction curve or moisture-density curve. The values of water content and standard maximum dry unit weight are determined from the compaction curve as shown in Fig. 21.2.

Table 21.1 Specification for standard Proctor compaction test

	Procedure		
Item	A	B	C
1. Diameter of mold	4 in. (101.6 mm)	4 in. (101.6 mm)	6 in. (152.4 mm)
2. Height of mold	4.584 in. (116.43 mm)	4.584 in. (116.43 mm)	4.584 in. (116.43 mm)
3. Volume of mold	0.0333 ft^3 (944 cm^3)	0.0333 ft^3 (944 cm^3)	0.075 ft^3 (2124 cm^3)
4. Weight of hammer	5.5 lb (2.5 kg)	5.5 lb (2.5 kg)	5.5 lb (2.5 kg)
5. Height of drop	12.0 in. (304.8 mm)	12.0 in. (304.8 mm)	12.0 in. (304.8 mm)
6. No. of layers	3	3	3
7. Blows per layer	25	25	56
8. Energy of compaction	12,375 ft-lb/ft^3 (600 kN-m/m^3)	12,375 ft-lb/ft^3 (600 kN-m/m^3)	12,375 ft-lb/ft^3 (600 kN-m/m^3)
9. Soil material	Passing No. 4 sieve (4.75 mm). May be used if 20% or less retained on No. 4 siere	Passing No 4 sieve (4.75 mm). Shall be used if 20% or more retained on No. 4 sieve and 20% or less retained on 3/8 in (9.5 mm) sieve	Passing No. 4 sieve (4.75 mm). Shall be used if 20% or more retained on 3/8 in. (9.5 mm) sieve and less than 30% retained on 3/4 in. (19 mm) sieve

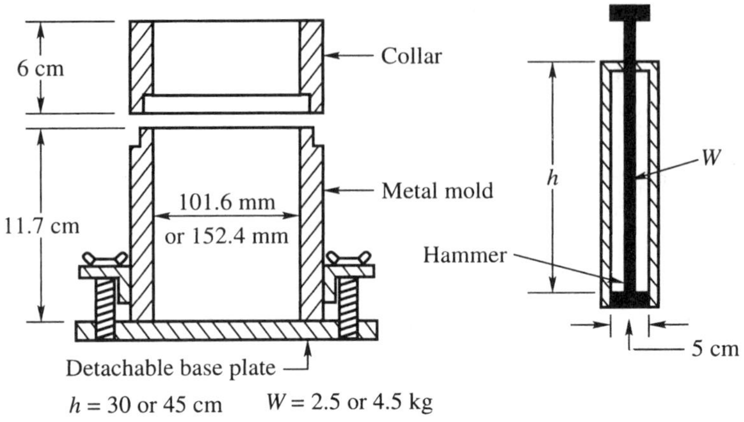

(a)

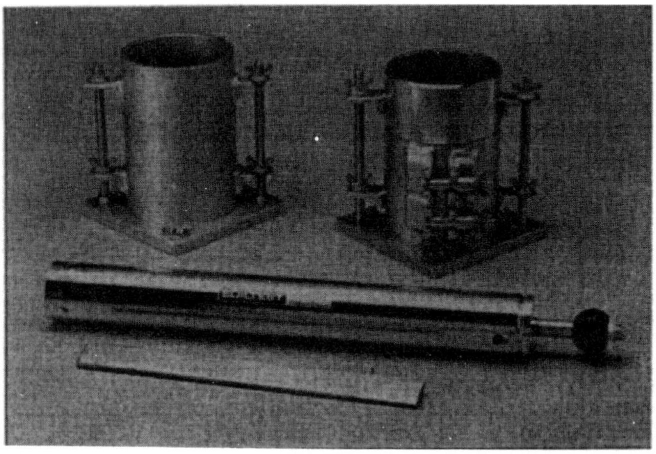

(b)

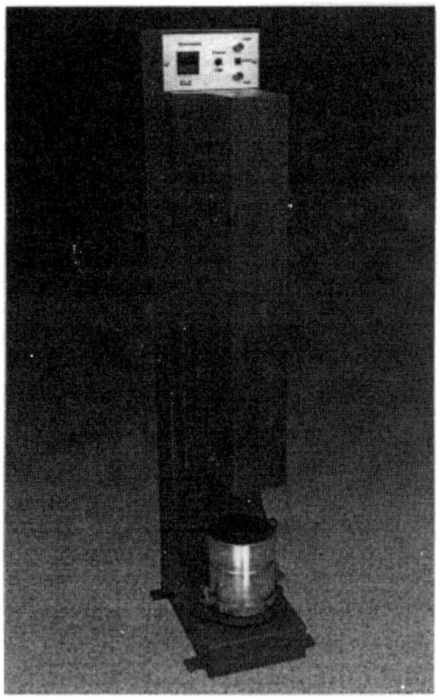

(c)

Figure 21.1 Proctor compaction apparatus: (a) diagrammatic sketch, and (b) photograph of mold, and (c) automatic soil compactor (Courtesy: Soiltest)

Modified Proctor Compaction Test (ASTM Designation: D1557)

This test method covers laboratory compaction procedures used to determine the relationship between water content and dry unit weight of soils (compaction curve) compacted in a 4 in. or 6 in. diameter mold with a 10 lb (5 kg) hammer dropped from a height of 18 in. (457 mm) producing a compactive effort of 56,250 ft-lb/ft^3 (2,700 kN-m/m^3). As in the case of the standard test, the code provides three alternative procedures based on the soil material tested. The details of the procedures are given in Table 21.2.

Table 21.2 Specification for modified Proctor compaction test

Item		Procedure A	Procedure B	Procedure C
1.	Mold diameter	4 in. (101.6mm)	4 in. (101.6 mm)	6 in. (101.6 mm)
2.	Volume of mold	0.0333 ft^3 (944 cm^3)	0.0333 ft^3 (944 cm^3)	0.075 ft^3 (2124 cm^3)
3.	Weight of hammer	10 lb (4.54 kg)	10 lb (4.54 kg)	10 lb (4.54 kg)
4.	Height of drop	18 in. (457.2mm)	18 in. (457.2mm)	18 in. (457.2mm)
5.	No. of layers	5	5	5
6.	Blows / layer	25	25	56
7.	Energy of compaction	56,250 ft lb/ft^3 (2700 kN-m/m^3)	56,250 ft lb/ft^3 (2700 kN-m/m^3)	56,250 ft lb/ft^3 (2700 kN-m/m^3)
8.	Soil material	May be used if 20% or less retained on No. 4 sieve.	Shall be used if 20% or more retained on No. 4 sieve and 20% or less retained on the 1/8 in. sieve	Shall be used if more-than 20% retained on 3/8 in. sieve and less-than 30% retained on the 3/4 in. sieve (19 mm)

Test Procedure

A soil at a selected water content is placed in five layers into a mold of given dimensions, with each layer compacted by 25 or 56 blows of a 10 lb (4.54 kg) hammer dropped from a height of 18 in. (457 mm) subjecting the soil to a total compactive effort of about 56,250 ft-lb/ft^3 (2700 kN-m/m^3). The resulting dry unit weight is determined. The procedure is repeated for a sufficient number of water contents to establish a relationship between the dry unit weight and the water content for the soil. This data, when plotted, represents a curvilinear relationship known as the compaction curve or moisture-dry unit weight curve. The value of the optimum water content and maximum dry unit weight are determined from the compaction curve as shown in Fig. 21.2.

Determination of Zero Air Voids Line

Referring to Fig. 21.3, we have

Degree of saturation, $$S = \frac{V_w}{V_v}$$

Water content, $$w = \frac{W_w}{W_s}$$

Dry weight of solids, $$W_s = V_s G_s \gamma_w = G_s \gamma_w \quad \text{since } V_s = 1$$

$$V_w = \frac{W_w}{\gamma_w} = \frac{wG_s\gamma_w}{\gamma_w} = wG_s$$

Therefore $$S = \frac{wG_s}{V_v} \tag{21.2}$$

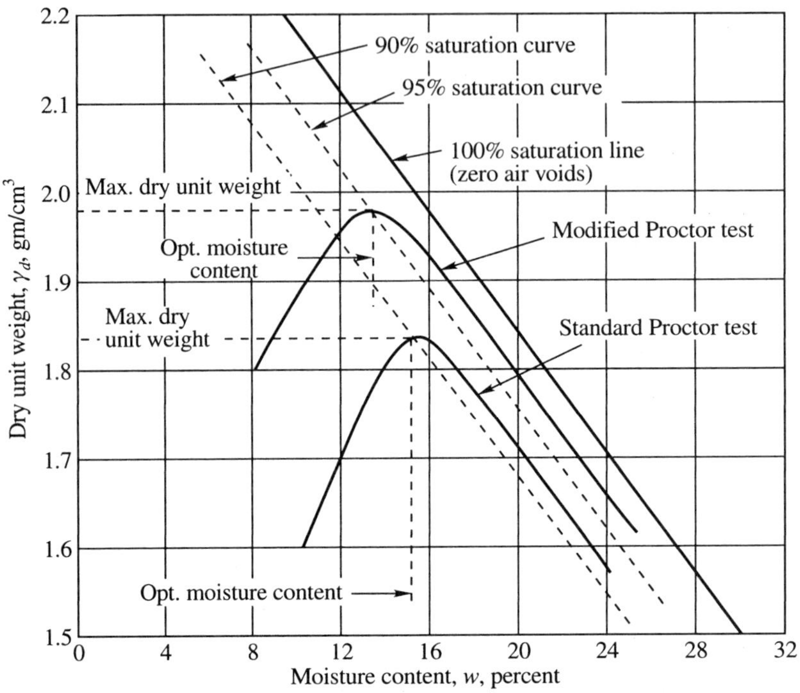

Figure 21.2 Moisture-dry unit weight relationship

or

$$V_v = \frac{wG_s}{S}$$

Dry unit weight

$$\gamma_d = \frac{W_s}{1+V_v} = \frac{G_s \gamma_w}{1+\dfrac{wG_s}{S}} \tag{21.3}$$

In Eq. (21.3), since G_s and γ_w, remain constant for a particular soil, the dry unit weight is a function of water content for any assumed degree of saturation. If $S = 1$, the soil is fully saturated (zero air voids). A curve giving the relationship between γ_d and w may be drawn by making use of Eq. (20.3) for $S = 1$. Curves may be drawn for different degrees of saturation such as 95, 90, 80 etc

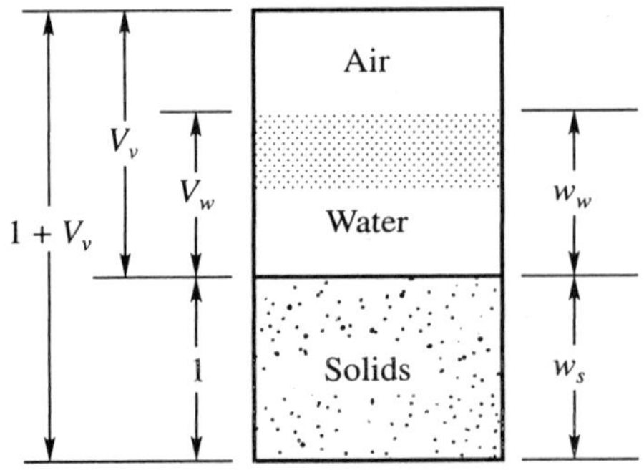

Figure 21.3 Block diagram for determining zero air voids line

percents. Fig. 21.2 gives typical curves for different degrees of saturation along with moisture-dry unit weight curves obtained by different compactive efforts.

Example 21.1

A proctor compaction test was conducted on a soil sample, and the following observations were made:

Water content, percent	7.7	11.5	14.6	17.5	19.7	21.2
Mass of wet soil, g	1739	1919	2081	2033	1986	1948

If the volume of the mold used was 950 cm^3 and the specific gravity of soils grains was 2.65, make necessary calculations and draw, (i) compaction curve and (ii) 80% and 100% saturation lines.

Solution

From the known mass of the wet soil sample and volume of the mold, wet density or wet unit weight is obtained by the equations,

$$\rho_t(\text{g}/\text{cm}^3) = \frac{M}{V} = \frac{\text{Mass of wet sample in gm}}{950\ \text{cm}^3} \quad \text{or } \gamma_t = (\text{kN}/\text{m}^3) \approx 9.81 \times p_t(\text{g}/\text{cm}^3)$$

Then from the wet density and corresponding moisture content, the dry density or dry unit weight is obtained from,

$$\rho_d = \frac{\rho_t}{1+w} \quad \text{or} \quad \gamma_d = \frac{\gamma_t}{1+w}$$

Thus for each observation, the wet density and then the dry density are calculated and tabulated as follows:

Water content, percent	7.7	11.5	14.6	17.5	19.7	21.2
Mass of wet sample, g	1739	1919	2081	2033	1986	1948
Wet density, g/cm^3	1.83	2.02	2.19	2.14	2.09	2.05
Dry density, g/cm^3	1.70	1.81	1.91	1.82	1.75	1.69
Dry unit weight kN/m^3	16.7	17.8	18.7	17.9	17.2	16.6

Hence the compaction curve, which is a plot between the dry unit weight and moisture content can be plotted as shown in the Fig. Ex. 21.1. The curve gives,

Maximum dry unit weight, MDD = 18.7 kN/m^3

Optimum moisture content, OMC = 14.7 percent

For drawing saturation lines, make use of Eq. (21.3), viz.,

$$\gamma_d = \frac{G_s \gamma_w}{1 + \dfrac{wG_s}{S}}$$

where, G_s = 2.65, given, S = degree of saturation 80% and 100%, w = water content, may be assumed as 8%, 12%, 16%, 20% and 24%.

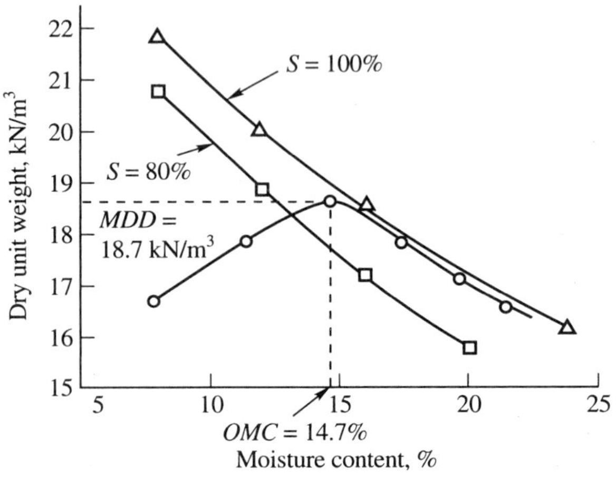

Fig. Ex. 21.1

Hence for each value of saturation and water content, find γ_d, and tabulate:

Water content, percentage	8	12	16	20	24
γ_d kN/m^3 for S = 100%	21.45	19.73	18.26	17.0	15.69
γ_d kN/m^3 for S = 80%	20.55	18.61	17.00	15.64	14.49

With these calculations, saturation lines for 100% and 80% are plotted, as shown in the Fig. Ex. 21.1.

Also the saturation, corresponding to MDD = 18.7 kN/m^3 and OMC = 14.7% can be calculated as,

$$18.7 = \gamma_d = \frac{G_s \gamma_w}{1 + \dfrac{wG_s}{S}} = \frac{2.65 \times 9.81}{1 + \dfrac{0.147 \times 2.65}{S}}$$

which gives S = 99.7%

Example 21.2

A small cylinder having volume of 600 cm^3 is pressed into a recently compacted fill of embankment filling the cylinder. The mass of the soil in the cylinder is 1100 g. The dry mass of the soil is 910 g. Determine the void ratio and the saturation of the soil. Take the specific gravity of the soil grains as 2.7.

Solution

Wet density of soil

$$\rho_t = \frac{1100}{600} = 1.83 \text{ g/cm}^3 \text{ or } \gamma_t = 17.99 \text{ kN/m}^3$$

Water content, $$w = \frac{1100 - 910}{910} = \frac{190}{910} = 0.209 = 20.9\%$$

Dry unit weight, $\gamma_d = \frac{\gamma_t}{1+w} = \frac{17.99}{1+0.209} = 14.88 \text{ kN/m}^3$

From, $\gamma_d = \frac{G_s \gamma_w}{1+e}$ we have $e = \frac{G_s \gamma_w}{\gamma_d} - 1$

Substituting and simplifying

$$e = \frac{2.7 \times 9.81}{14.88} - 1 = 0.78$$

From, $Se = wG_s$, or $S = \frac{wG_s}{e} = \frac{0.209 \times 2.7 \times 100}{0.78} = 72.35\%$

21.4 EFFECT OF COMPACTION ON ENGINEERING BEHAVIOR

Effect of Moisture Content on Dry Density

The moisture content affects the behavior of the soil. When the moisture content is low, the soil is stiff and difficult to compress. Thus, low unit weight and high air contents are obtained (Fig. 21.2). As the moisture content increases, the water acts as a lubricant, causing the soil to soften and become more workable. This results in a denser mass, higher unit weights and lower air contents under compaction. The water and air combination tend to keep the particles apart with further compaction, and prevent any appreciable decrease in the air content of the total voids, however, continue to increase with moisture content and hence the dry unit weight of the soil falls.

To the right of the peak of the dry unit weight-moisture content curve (Fig. 21.2), lies the saturation line. The theoretical curve relating dry density with moisture content with no air voids is approached but never reached since it is not possible to expel by compaction all the air entrapped in the voids of the soil.

Effect of Compactive Effort on Dry Unit Weight

For all types of soil with all methods of compaction, increasing the amounts of compaction, that is, the energy applied per unit weight of soil, results in an increase in the maximum dry unit weight and a corresponding decrease in the optimum moisture content as can be seen in Fig. 21.4.

Shear Strength of Compacted Soil

The shear strength of a soil increases with the amount of compaction applied. The more the soil is compacted, the greater is the value of cohesion and the angle of shearing resistance. Comparing the shearing strength with the moisture content for a given degree of compaction, it is found that the greatest shear strength is attained at a moisture content lower than the optimum moisture content for maximum dry unit weight. Fig. 21.5 shows the relationship between shear strength and moisture-dry unit weight curves for a sandy clay soil. It might be inferred from this that it would be an advantage to carry out compaction at the lower value of the moisture content. Experiments, however, have indicated that soils compacted in this way tend to take up moisture and become saturated with a consequent loss of strength.

Effect of Compaction on Structure

Fig. 21.6 illustrates the effects of compaction on clay structure (Lambe, 1958a). Structure (or fabric) is the term used to describe the arrangement of soil particles and the electric forces between adjacent particles.

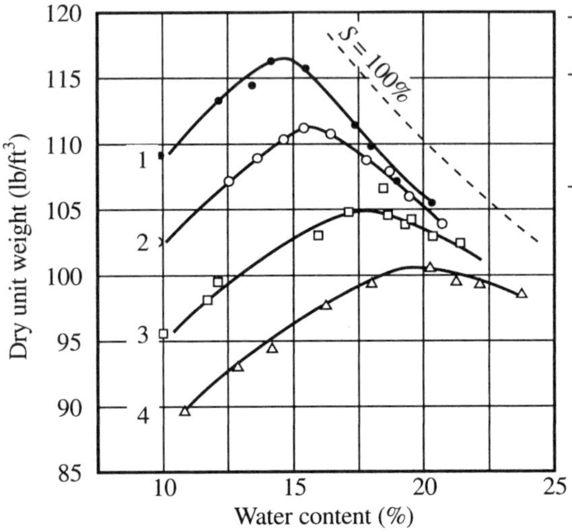

No.	Layers	Blows per Layer	Hammer weight(lb)	Hammer drop (in.)
1	5	55	10	18 (mod. AASHTO)
2	5	26	10	18
3	5	12	10	18 (std. AASHTO)
4	3	25	5½	12

Note: 6 in. diameter mold used for all tests

Figure 21.4 Dynamic compaction curves for a silty clay (from Turnbull, 1950)

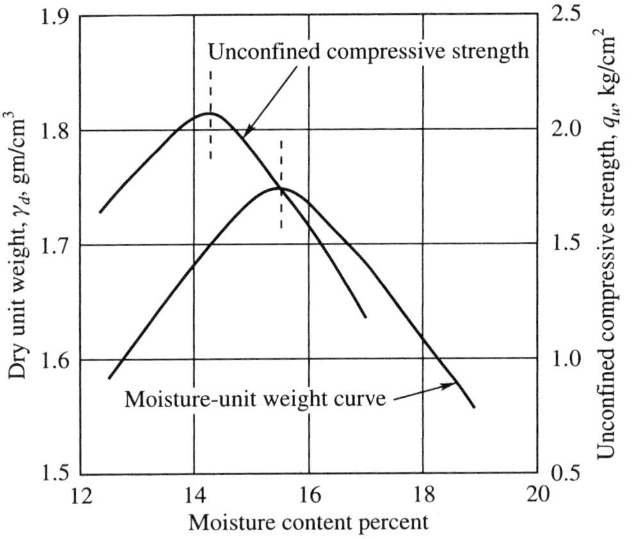

Figure 21.5 Relationship between compaction and shear strength curves

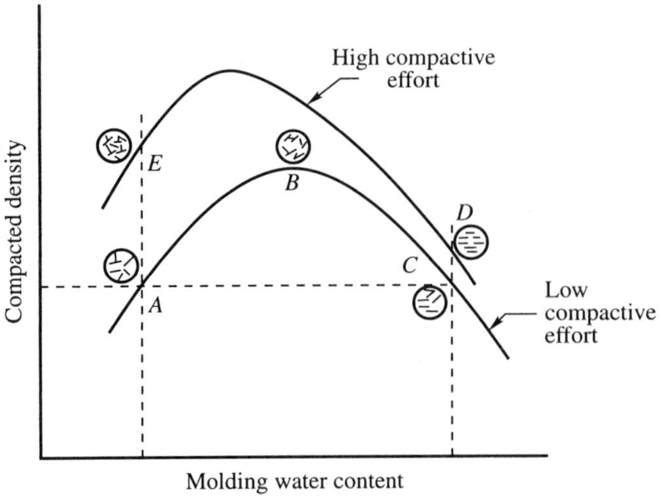

Figure 21.6 Effects of compaction on structure (from Lambe, 1958a)

The effects of compaction conditions on soil structure, and thus on the engineering behavior of the soil, vary considerably with soil type and the actual conditions under which the behavior is determined.

At low water content, w_A in Fig. 21.6, the repulsive forces between particles are smaller than the attractive forces, and as such the particles flocculate in a disorderly array. As the water content increases beyond w_A, the repulsion between particles increases, permitting the particles to disperse, making particles arrange themselves in an orderly way. Beyond w_B the degree of particle parallelism increases, but the density decreases. Increasing the compactive effort at any given water content increases the orientation of particles and therefore gives a higher density as indicated in Fig. 21.6.

Effect of Compaction on Permeability

Fig. 21.7 depicts the effect of compaction on the permeability of a soil. The figure shows the typical marked decrease in permeability that accompanies an increase in molding water content on the dry side of the optimum water content. A minimum permeability occurs at water contents slightly above optimum moisture content (Lambe, 1958a), after which a slight increase in permeability occurs. Increasing the compactive effort decreases the permeability of the soil.

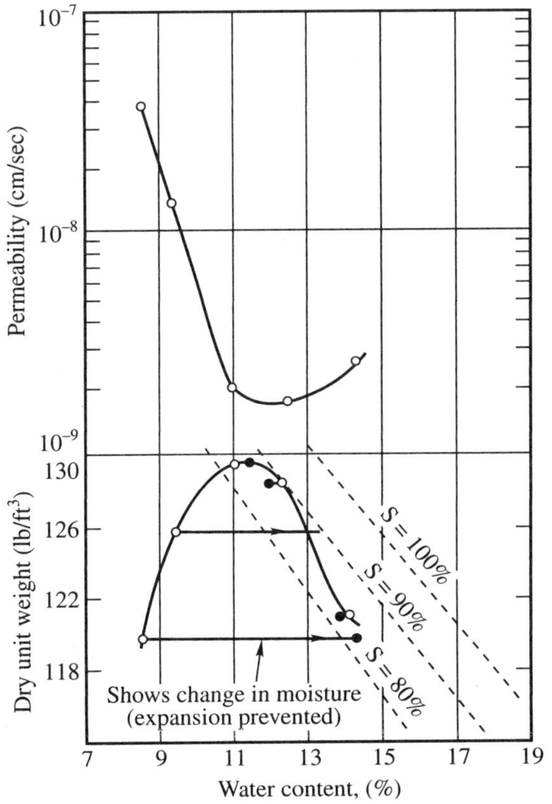

Figure 21.7 Compaction-permeability tests on Siburua clay (from Lambe, 1962)

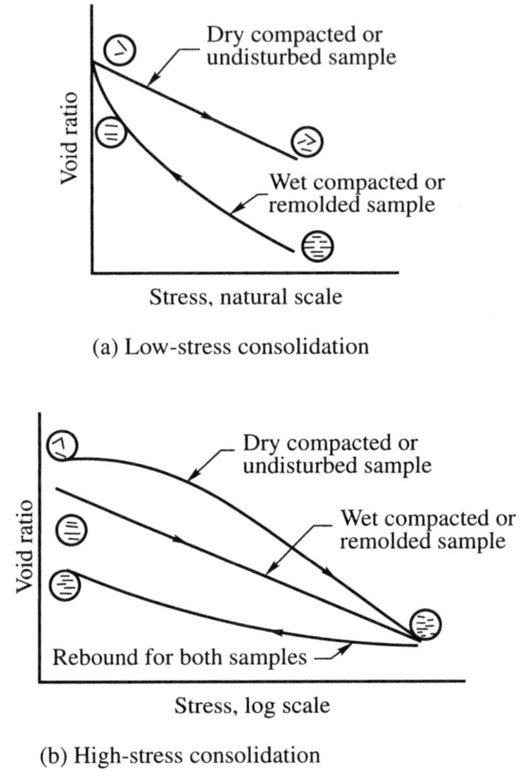

Figure 21.8 Effect of one-dimensional compression on structure (Lambe, 1958b)

Effect of Compaction on Compressibility

Figure 21.8 illustrates the difference in compaction characteristics between two saturated clay samples at the same density, one compacted on the dry side of optimum and one compacted on the wet side (Lambe, 1958b). At low stresses the sample compacted on the wet side is more compressible than the one compacted on the dry side. However, at high applied stresses the sample compacted on the dryside is more compressible than the sample compacted on the wet side.

21.5 FIELD COMPACTION AND CONTROL

The necessary compaction of subgrades of roads, earth fills, and embankments may be obtained by mechanical means. The equipment that are normally used for compaction consists of

1. Smooth wheel rollers
2. Rubber tired rollers
3. Sheepsfoot rollers
4. Vibratory rollers

Laboratory tests on the soil to be used for construction in the field indicate the maximum dry density that can be reached and the corresponding optimum moisture content under specified methods of compaction. The field compaction method should be so adjusted as to translate

Figure 21.9 Smooth wheel roller (Courtesy: Caterpillar, USA)

laboratory condition into practice as far as possible. The two important factors that are necessary to achieve the objectives in the field are

1. The adjustment of the natural moisture content in the soil to the value at which the field compaction is most effective.
2. The provision of compacting equipment suitable for the work at the site.

The equipment used for compaction are briefly described below:

Smooth Wheel Roller

There are two types of smooth wheel rollers. One type has two large wheels, one in the rear and a similar single drum in the front. This type is generally used for compacting base courses. The equipment weighs from 50 to 125 kN (Fig.21.9). The other type is the tandem roller normally used for compacting paving mixtures. This roller has large single drums in the front and rear and the weights of the rollers range from 10 to 200 kN.

Figure 21.10 Sheepsfoot roller (Courtesy: Vibromax America Inc.)

Rubber Tired Roller

The maximum weight of this roller may reach 2000 kN. The smaller rollers usually have 9 to 11 tires on two axles with the tires spaced so that a complete coverage is obtained with each pass. The tire loads of the smaller roller are in the range of 7.5 kN and the tire pressures in the order of 200 kN/m^2. The larger rollers have tire loads ranging from 100 to 500 kN per tire, and tire pressures range from 400 to 1000 kN/m^2.

Sheepsfoot Roller

Sheepsfoot rollers are available in drum widths ranging from 120 to 180 cm and in drum diameters ranging from 90 to 180 cm. Projections like a sheepsfoot are fixed on the drums. The lengths of these projections range from 17.5 cm to 23 cm. The contact area of the tamping foot ranges from 35 to 56 sq. cm. The loaded weight per drum ranges from about 30 kN for the smaller sizes to 130 kN for the larger sizes (Fig. 21.10).

Vibratory Roller

The weights of vibratory rollers range from 120 to 300 kN. In some units vibration is produced by weights placed eccentrically on a rotating shaft in such a manner that the forces produced by the rotating weights are essentially in a vertical direction. Vibratory rollers are effective for compacting granular soils (Fig. 21.11).

Figure 21.11 Vibratory drum on smooth wheel roller (Courtesy: Caterpillar, USA)

Selection of Equipment for Compaction in the Field

The choice of a roller for a given job depends on the type of soil to be compacted and percentage of compaction to be obtained. The types of rollers that are recommended for the soils normally met are:

Type of soil	Type of roller recommended
Cohesive soil	Sheepsfoot roller, or Rubber tired roller
Cohesionless soils	Rubber-tired roller or Vibratory roller.

Method of Compaction

The first approach to the problem of compaction is to select suitable equipment. If the compaction is required for an earth dam, the number of passes of the roller required to compact the given soil to the required density at the optimum moisture content has to be determined by conducting a field trial test as follows:

The soil is well mixed with water which would give the optimum water content as determined in the laboratory. It is then spread out in a layer. The thickness of the layer normally varies from 15 to 22.5 cm. The number of passes required to obtain the specified density has to be found by determining the density of the compacted material after every definite number of passes. The density may be checked for different thickness in the layer. The suitable thickness of the layer and the number of passes required to obtain the required density will have to be determined.

In cohesive soils, densities of the order of 95 percent of standard Proctor can be obtained with practically any of the rollers and tampers; however, vibrators are not effective in cohesive soils. Where high densities are required in cohesive soils in the order of 95 percent of modified Proctor, rubber tired rollers with tire loads in the order of 100 kN and tire pressure in the order of 600 kN/m^2 are effective.

In cohesionless sands and gravels, vibrating type equipment is effective in producing densities up to 100 percent of modified Proctor. Where densities are needed in excess of 100 percent of modified Proctor such as for base courses for heavy duty air fields and highways, rubber tired rollers with tire loads of 130 kN and above and tire pressure of 1000 kN/m^2 can be used to produce densities up to 103 to 104 percent of modified Proctor.

Field Control of Compaction

Methods of Control of Density

The compaction of soil in the field must be such as to obtain the desired unit weight at the optimum moisture content. The field engineer has therefore to make periodic checks to see whether the compaction is giving desired results. The procedure of checking involves:

1. Measurement of the dry unit weight, and
2. Measurement of the moisture content.

There are many methods for determining the dry unit weight and/or moisture content of the soil in-situ. The important methods are:

1. Sand cone method,
2. Rubber balloon method,
3. Nuclear method, and
4. Proctor needle method.

Sand Cone Method (ASTM Designation D-1556)

The sand for the sand cone method consists of a sand pouring jar shown in Fig. 21.12. The jar contains uniformly graded clean and dry sand. A hole about 10 cm in diameter is made in the soil to be tested up to the depth required. The weight of soil removed from the hole is determined and its water content is also determined. Sand is run into the hole from the jar by opening the valve above the cone until the hole and the cone below the valve is completely filled. The valve is closed. The jar is calibrated to give the weight of the sand that just fills the hole, that is, the difference in weight of the jar before and after filling the hole after allowing for the weight of sand contained in the cone is the weight of sand poured into the hole.

Let

W_s = weight of dry sand poured into the hole

G_s = specific gravity of sand particles

W = weight of soil taken out of the hole

w = water content of the soil

Volume of sand in the hole = volume of soil taken out of the hole

that is, $$V = \frac{W_s}{G_s \gamma_w} \tag{21.4a}$$

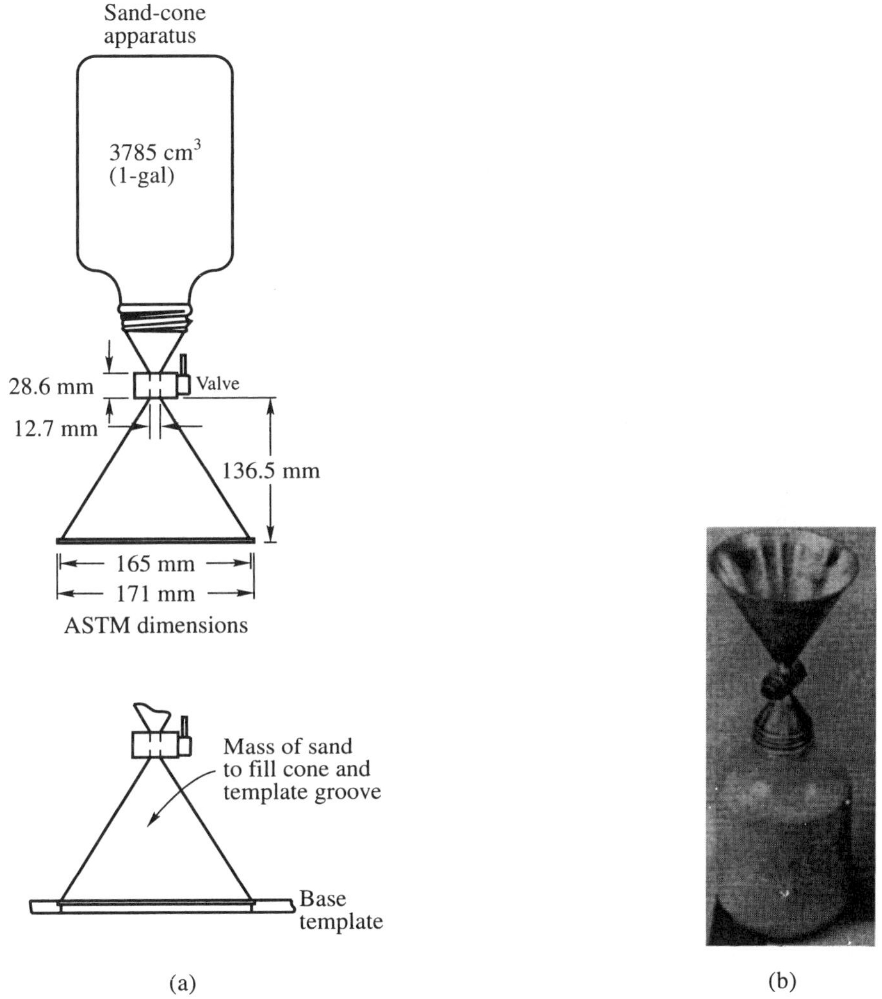

Fig. 21.12 Sand-cone apparatus: (a) Schematic diagram, and (b) Photograph

The bulk unit weight of soil, $\gamma_t = \dfrac{W}{V} = \dfrac{WG_s\gamma_w}{W_s}$ (21.4b)

The dry unit weight of soil, $\gamma_d = \dfrac{\gamma_t}{1+w}$

Rubber Balloon Method (ASTM Designation: D 2167)

The volume of an excavated hole in a given soil is determined using a liquid-filled calibrated cylinder for filling a thin rubber membrane. This membrane is displaced to fill the hole. The in-place unit weight is determined by dividing the wet mass of the soil removed by the volume of the hole. The water (moisture) content and the in-place unit weight are used to calculate the in-place dry unit weight. The volume is read directly on the graduated cylinder. Fig. 21.13 shows the equipment.

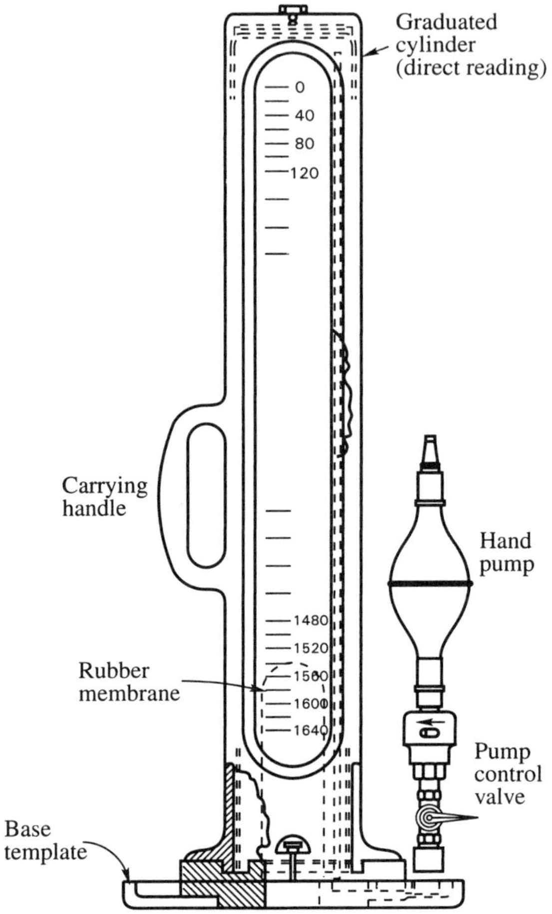

(a)

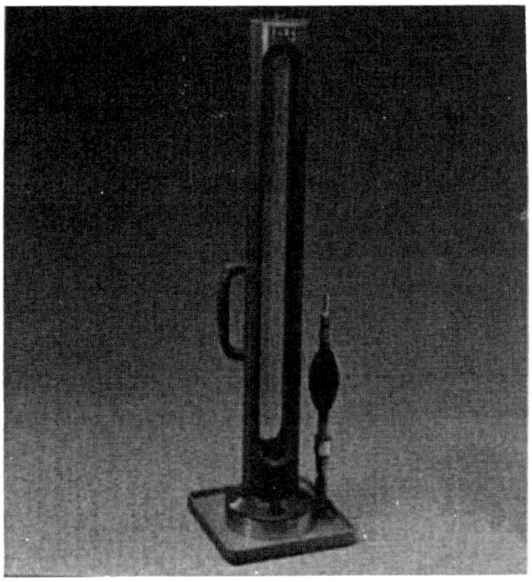

(b)

Figure 21.13 Rubber balloon density apparatus, (a) diagrammatic sketch, and (b) a photograph

Nuclear Method

The modern instrument for rapid and precise field measurement of moisture content and unit weight is the Nuclear density/Moisture meter. The measurements made by the meter are non-destructive and require no physical or chemical processing of the material being tested. The instrument may be used either in drilled holes or on the surface of the ground. The main advantage of this equipment is that a single operator can obtain an immediate and accurate determination of the *in-situ* dry density and moisture content.

Proctor Needle Method

The Proctor needle method is one of the methods developed for rapid determination of moisture contents of soils *in-situ*. It consists of a needle attached to a spring loaded plunger, the stem of which is calibrated to read the penetration resistance of the needle in lbs/in^2 or kg/cm^2. The needle is supplied with a series of bearing points so that a wide range of penetration resistances can be measured. The bearing areas that are normally provided are 0.05, 0.1, 0.25, 0.50 and 1.0 sq. in. The apparatus is shown in Fig. 21.14. A Proctor penetrometer set is shown in Fig. 21.15 (ASTMD–1558).

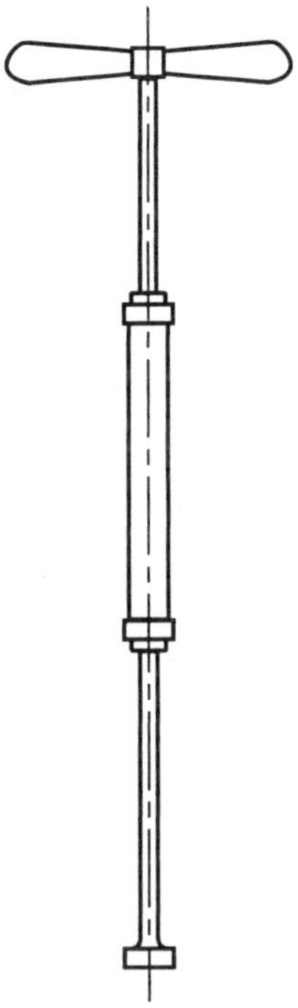

Figure 21.14 Proctor needle

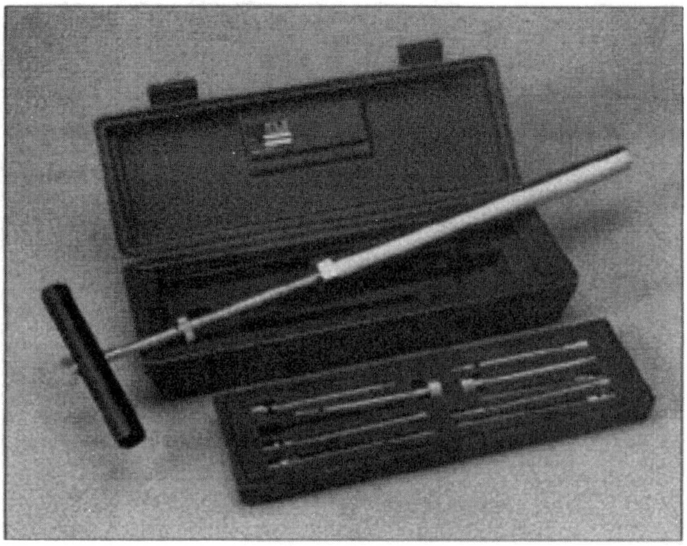

Figure 21.15 Proctor penetrometer set (Courtesy: Soiltest)

Laboratory Penetration Resistance Curve

A suitable needle point is selected for a soil to be compacted. If the soil is cohesive, a needle with a larger bearing area is selected. For cohesionless soils, a needle with a smaller bearing area will be sufficient. The soil sample is compacted in the mold.

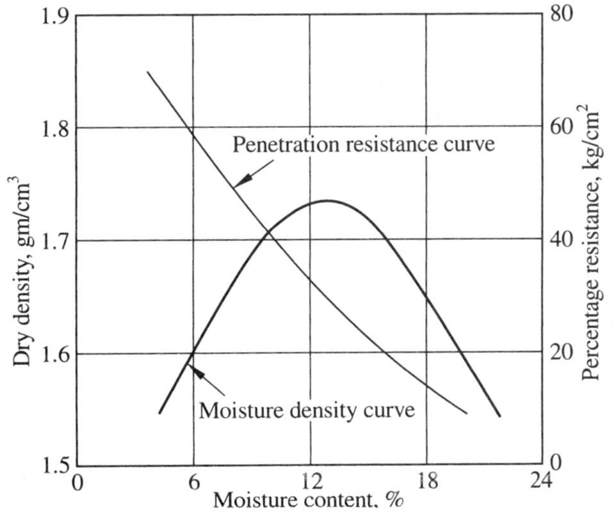

Figure 21.16 Field method of determining water content by Proctor needle method

The penetrometer with a known bearing area of the tip is forced with a gradual uniform push at a rate of about 1.25 cm per sec to a depth of 7.5 cm into the soil. The penetration resistance in kg/cm^2 is read off the calibrated shaft of the penetrometer. The water content of the soil and the corresponding dry density are also determined. The procedure is repeated for the same soil compacted at different moisture contents. Curves giving the moisture-density and penetration resistance-moisture content relationship are plotted as shown in Fig. 21.16.

To determine the moisture content in the field, a sample of the wet soil is compacted into the mold under the same conditions as used in the laboratory for obtaining the penetration resistance curve. The Proctor needle is forced into the soil and its resistance is determined. The moisture content is read from the laboratory calibration curve.

This method is quite rapid, and is sufficiently accurate for fine-grained cohesive soils. However, the presence of gravel or small stones in the soil makes the reading on the Proctor needle less reliable. It is not very accurate in cohesionless sands.

Example 21.3

The following observations were recorded when a sand cone test was conducted for finding the unit weight of a natural soil:

Total density of sand used in the test = 1.4 g/cm^3

Mass of the soil excavated from hole = 950 g.

Mass of the sand filling the hole = 700 g.

Water content of the natural soil = 15 percent.

Specific gravity of the soil grains = 2.7

Calculate: (i) the wet unit weight, (ii) the dry unit weight, (iii) the void ratio, and (iv) the degree of saturation.

Solution

Volume of the hole $$V_p = \frac{700}{1.4} = 500 \text{ cm}^3$$

Wet density of natural soil, $$\rho_t = \frac{950}{500} = 1.9 \text{ g/cm}^3 \text{ or } \gamma_t = 18.64 \text{ kN/m}^3$$

Dry density $$\rho_d = \frac{\rho_t}{1+w} = \frac{1.9}{1+0.15} = 1.65 \text{ g/cm}^3$$

$$\rho_d = \frac{G_s}{1+e}\rho_w = \frac{2.7}{1+e} \times 1 \quad \text{or} \quad 1.65 + 1.65e = 2.7$$

Therefore $$e = \frac{2.7 - 1.65}{1.65} = 0.64$$

And $$S = \frac{wG_s}{e} = \frac{0.15 \times 2.7 \times 100}{0.64} = 63\%$$

Example 21.4

Old records of a soil compacted in the past gave compaction water content of 15% and saturation 85%. What might be the dry density of the soil?

Solution

The specific gravity of the soil grains is not known, but as it varies in a small range of 2.6 to 2.7, it can suitably be assumed. An average value of 2.65 is considered here.

Hence $$e = \frac{wG_s}{S} = \frac{0.15 \times 2.65}{0.85} = 0.47$$

and dry density $$\rho_d = \frac{G_s}{1+e}\rho_w = \frac{2.65}{1+0.47} \times 1 = 1.8 \text{ g/cm}^3 \text{ or dry unit weight } = 17.66 \text{ kN/m}^3$$

Example 21.5

The following data are available in connection with the construction of an embankment:

(a) **soil from borrow pit:** Natural density = 1.75 Mg/m^3, Natural water content = 12%
(b) **soil after compaction:** density = 2 Mg/m^3, water content = 18%.

For every 100 m^3 of compacted soil of the embankment, estimate:

(i) the quantity of soil to be excavated from the borrow pit, and
(ii) the amount of water to be added

Note: 1 g/cm^3 = 1000 kg/m^3 = $10^3 \times 10^3$ g/m^3 = 1 Mg/m^3 where Mg stands for Megagram = 10^6 g.

Solution

The soil is compacted in the embankment with density of 2 Mg/m^3 and with 18% water content.

Hence, for 100 m^3 of soil

Mass of compacted wet soil = 100 × 2.0 = 200 Mg = 200 × 10^3 kg.

$$\text{Mass of compacted dry soil} = \frac{200}{1+w} = \frac{200}{1+0.18} = 169.5 \text{ Mg} = 169.5 \times 10^3 \text{ kg}$$

$$\text{Mass of wet soil to be excavated} = 169.5(1+w) = 169.5(1+0.12) = 189.84 \text{ Mg}$$

$$\text{Volume of the wet soil to be excavated} = \frac{189.84}{1.75} = 108.48 \text{ m}^3$$

Now, in the natural state, the moisture present in 169.5 × 10^3 kg of dry soil would be

169.5 × 10^3 × 0.12 = 20.34 × 10^3 kg

and the moisture which the soil will possess during compaction is

169.5 × 10^3 × 0. 18 = 30.51 × 10^3 kg

Hence mass of water to be added for every 100 m^3 of compacted soil is

(30.51 – 20.34) 10^3 = 10.17 × 10^3 kg.

Example 21.6

A sample of soil compacted according to the standard Proctor test has a density of 2.06 g/cm^3 at 100% compaction and at an optimum water content of 14%. What is the dry unit weight? What is the dry unit weight at zero air-voids? If the voids become filled with water what would be the saturated unit weight? Assume $G_s = 2.67$.

Solution

Refer to Fig. Ex. 21.6. Assume V = total volume = 1 cm^3. Since water content is 14% we may write,

$$\frac{M_w}{M_s} = 0.14 \quad \text{or} \quad M_w = 0.14\, M_s$$

and since, $M_w + M_s = 2.06$ g

$$0.14 M_s + M_s = 1.14 M_s = 2.06$$

$$\text{or} \qquad M_s = \frac{2.06}{1.14} = 1.807 \text{ g}$$

$$M_w = 0.14 \times 1.807 = 0.253 \text{ g.}$$

$$\text{By definition, } \rho_d = \frac{M_s}{V} = \frac{1.807}{1} = 1.807 \text{ g/cm}^3 \text{ or } \gamma_d = 1.807 \times 9.81 = 17.73 \text{ kN/m}^3$$

The volume of solids (Fig. Ex. 21.6) is

$$V_s = \frac{1.807}{2.67} = 0.68 \text{ g/cm}^3$$

The volume of voids = 1 – 0.68 = 0.32 cm^3

The volume of water = 0.253 cm^3

The volume of air = 0.320 – 0.253 = 0.067 cm^3

If all the air is squeezed out of the samples the dry density at zero air voids would be, by definition,

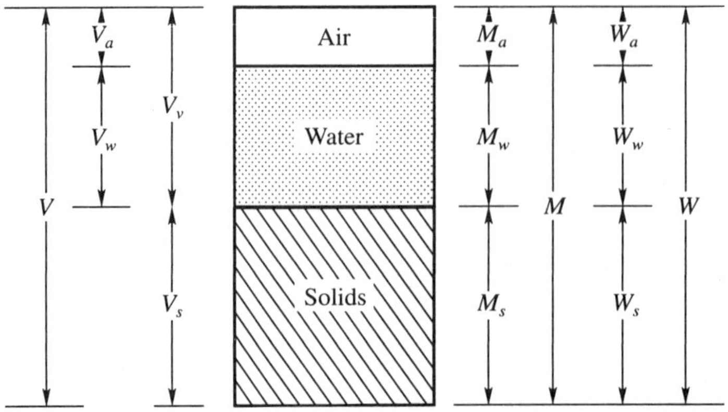

Figure Ex. 21.6

$$\rho_d = \frac{1.807}{0.68+0.253} = 1.94 \text{ g/cm}^3 \text{ or } \gamma_d = 1.94 \times 9.81 = 19.03 \text{ kN/m}^3$$

on the other hand, if the air voids also were filled with water,

The mass of water would be = 0.32 × 1 = 0.32 g

The saturated density is

$$\rho_{sat} = \frac{1.807+0.32}{1} = 2.13 \text{ g/cm}^3 \text{ or } \gamma_{sat} = 2.13 \times 9.81 = 20.90 \text{ kN/m}^3$$

21.6 COMPACTION FOR DEEPER LAYERS OF SOIL

Three types of dynamic compaction for deeper layers of soil are discussed here. They are:

1. Vibroflotation.
2. Dropping of a heavy weight.
3. Blasting.

Vibroflotation

The vibroflotation technique is used for compacting granular soil only. The vibroflot is a cylindrical tube containing water jets at top and bottom and equipped with a rotating eccentric weight, which develops a horizontal vibratory motion as shown in Fig. 21.17. The vibroflot is sunk into the soil using the lower jets and is then raised in successive small increments, during which the surrounding material is compacted by the vibration process. The enlarged hole around the vibroflot is backfilled with suitable granular material. This method is very effective for increasing the density of a sand deposit for depths up to 30 m. Probe spacings of compaction holes should be on a grid pattern of about 2 m to produce relative densities greater than 70 percent over the entire area. If the sand is coarse, the spacings may be somewhat larger.

In soft cohesive soil and organic soils the vibroflotation technique has been used with gravel as the backfill material. The resulting densified stone column effectively reinforces softer soils and acts as a bearing pile for foundations.

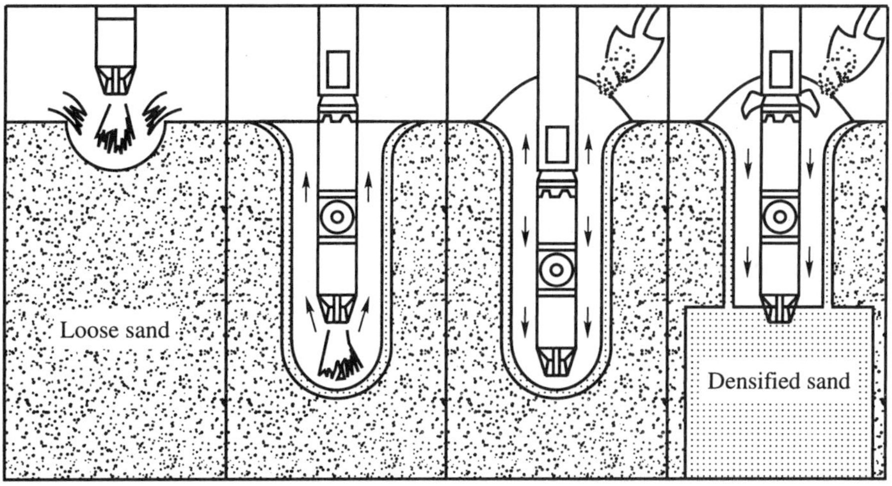

Figure 21.17 Compaction by using vibroflot (Brown, 1977)

Dropping of a Heavy Weight

The repeated dropping of a heavy weight on to the ground surface is one of the simplest of the methods of compacting loose soil.

The method, known as deep dynamic compaction or deep dynamic consolidation may be used to compact cohesionless or cohesive soils. The method uses a crane to lift a concrete or steel block, weighing up to 500 kN and up to heights of 40 to 50 m, from which height it is allowed to fall freely on to the ground surface. The weight leaves a deep pit at the surface. The process is then repeated either at the same location or sequentially over other parts of the area to be compacted. When the required number of repetitions is completed over the entire area, the compaction at depth is completed. The soils near the surface, however, are in a greatly disturbed condition. The top soil may then be levelled and compacted, using normal compactiing equipment. The principal claims of this method are:

1. Depth of recompaction can reach up to 10 to 12 m.
2. All soils can be compacted.
3. The method produces equal settlements more quickly than do static (surcharge type) loads.

The depth of recompaction, D, in meters is approximately given by Leonards, et al., (1980) as

$$D \approx \frac{1}{2}(Wh)^{1/2} \tag{21.5}$$

where W = weight of falling mass in metric tons,
h = height of drop in meters.

Blasting

Blasting, through the use of buried, time-delayed explosive charges, has been used to densify loose, granular soils. The sands and gravels must be essentially cohesionless with a maximum of 15 percent of their particles passing the No. 200 sieve size and 3 percent passing 0.005 mm size. The moisture condition of the soil is also important for surface tension forces in the partially saturated state limit the effectiveness of the technique. Thus the soil, as well as being granular, must be dry or saturated, which requires sometimes prewetting the site via construction of a dike and reservoir system.

The technique requires careful planning and is used at a remote site. Theoretically, an individual charge densifies the surrounding adjacent soil and soil beneath the blast. It should not lift the soil situated above the blast, however, since the upper soil should provide a surcharge load. The charge should not create a crater in the soil. Charge delays should be timed to explode from the bottom of the layer being densified upward in a uniform manner. The uppermost part of the stratum is always loosened, but this can be surface-compacted by vibratory rollers. Experience indicates that repeated blasts of small charges are more effective than a single large charge for achieving the desired results.

21.7 PRELOADING

Preloading is a technique that can successfully be used to densify soft to very soft cohesive soils. Large-scale construction sites composed of weak silts and clays or organic materials (particularly marine deposits), sanitary land fills, and other compressible soils may often be stabilized effectively and economically by preloading. Preloading compresses the soil. Compression takes place when the water in the pores of the soil is removed which amounts to artificial consolidation of soil in the field. In order to remove the water squeezed out of the pores and hasten the period of consolidation, horizontal and vertical drains are required to be provided in the mass. The preload is

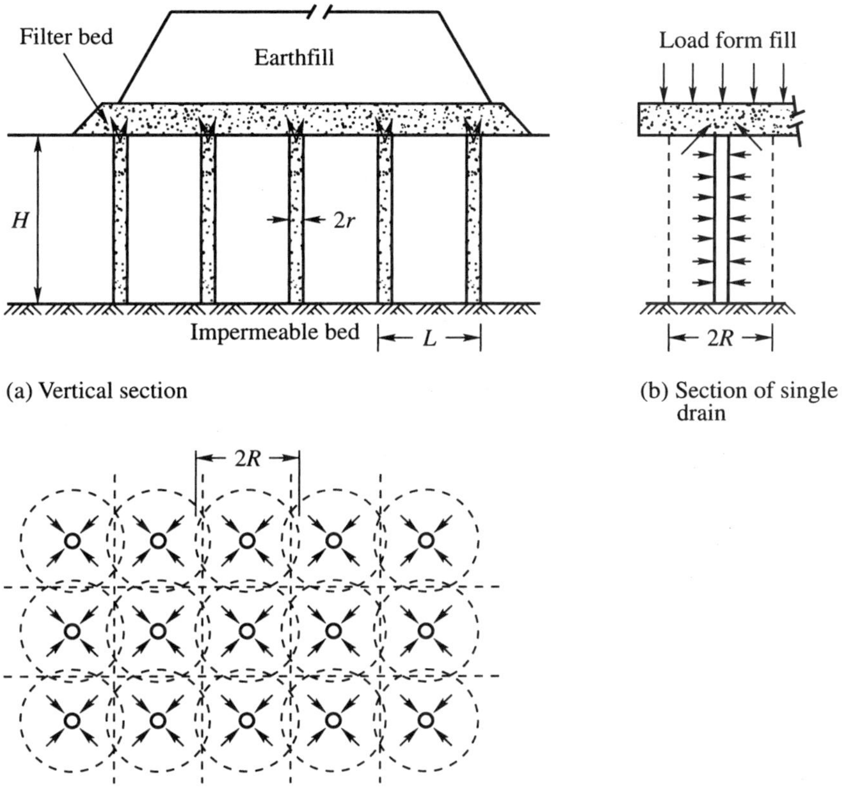

(c) plan of sand drains

Figure 21.18 Consolidation of soil using sand drains

generally in the form of an imposed earth fill which must be left in place long enough to induce consolidation. The process of consolidation can be checked by providing suitable settlement plates and piezometers. The greater the surcharge load, shorter the time for consolidation. This is a case of three-dimension consolidation.

Two types of vertical drains considered are

1. Cylindrical sand drains
2. Wick (prefabricated vertical) drains

Sand Drains

Vertical and horizontal and drains are normally used for consolidating very soft clay, silt and other compressible materials. The arrangement of sand drains shown in Fig. 21.18 is explained below:

1. It consists of a series of vertical sand drains or piles. Normally medium to coarse sand is used.
2. The diameter of the drains are generally not less than 30 cm and the drains are placed in a square grid pattern at distances of 2 to 3 meters apart. Economy requires a careful study of the effect of spacing the sand drains on the rate of consolidation.
3. Depth of the vertical drains should extend up to the thickness of the compressible stratum.

4. A horizontal blanket of free draining sand should be placed on the top of the stratum and the thickness of this may be up to a meter, and
5. Soil surcharge in the form of an embankment is constructed on top of the sand blanket in stages.

The height of surcharge should be so controlled as to keep the development of pore water pressure in the compressible strata at a low level. Rapid loading may induce high pore water pressures resulting in the failure of the stratum by rupture. The lateral displacement of the soil may shear off the sand drains and block the drainage path.

The application of surcharge squeezes out water in radial directions to the nearest sand drain and also in the vertical direction to the sand blanket. The dashed lines shown in Fig. 21.18(b) are drawn midway between the drains. The planes passing through these lines may be considered as impermeable membranes and all the water within a block has to flow to the drain at the center. The problem of computing the rate of radial drainage can be simplified without appreciable error by assuming that each block can be replaced by a cylinder of radius R such that

$$\pi R^2 = L^2$$

where L is the side length of the prismatic block.

The relation between the time t and degree of consolidation $U_z\%$ is determined by the equation

$$U_z\% = 100 f(\mathrm{T})$$

wherein,

$$T = \frac{c_v t}{H^2} \tag{21.6}$$

If the bottom of the compressible layer is impermeable, then H is the full thickness of the layer.

For radial drainage, Rendulic (1935) has shown that the relation between the time t and the degree of consolidation $U_r\%$ can be expressed as

$$U_r\% = 100 f(\mathrm{T}) \tag{21.7}$$

wherein,

$$T_r = \frac{c_v r}{4R^2} t \tag{21.8}$$

is the time factor. The relation between the degree of consolidation $U_r\%$ and the time factor T_r depends on the value of the ratio R/r. The relation between T_r and $U_r\%$ for ratios of R/r equal to 1, 10 and 100 in Fig. 21.19 are expressed by curves C_1, C_{10} and C_{100} respectively.

Installation of Vertical Sand Drains

The sand drains are installed as follows

1. A casing pipe of the required diameter with the bottom closed with a loose-fit-cone is driven up to the required depth,
2. The cone is slightly separated from the casing by driving a mandrel into the casing, and
3. The sand of the required gradation is poured into the pipe for a short depth and at the same time the pipe is pulled up in steps. As the pipe is pulled up, the sand is forced out of the pipe by applying pressure on to the surface of the sand. The procedure is repeated till the holes is completely filled with sand.

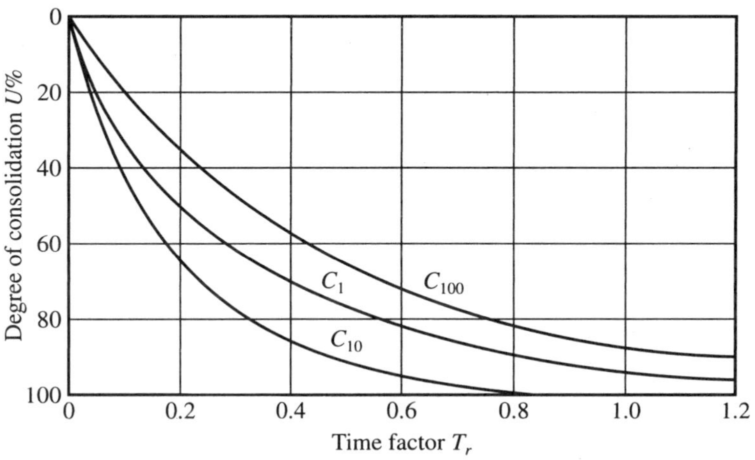

Figure 21.19 T_r versus U_r

The sand drains may also be installed by jetting a hole in the soil or by driving an open casing into the soil, washing the soil out of the casing, and filling the hole with sand afterwards.

Sand drains have been used extensively in many parts of the world for stabilizing soils for port development works and for foundations of structures in reclaimed areas on the sea coasts. It is possible that sand drains may not function satisfactorily if the soil surrounding the well gets remolded. This condition is referred to as *smear.* Though theories have been developed by considering different thickness of smear and different permeability, it is doubtful whether such theories are of any practical use since it would be very difficult to evaluate the quality of the smear in the field.

Wick (Prefabricated Vertical) Drains

Geocomposites used as drainage media have completely taken over certain geotechnical application areas. Wick drains, usually consisting of plastic fluted or nubbed cores that are surrounded by a geotextile filter, have considerable tensile strength. Wick drains do not require any sand to transmit flow. Most synthetic drains are of a strip shape. The strip drains are generally 100 mm wide and 2 to 6 mm thick. Fig. 21.20 shows typical core shapes of strip drains (Hausmann, 1990).

Wick drains are installed by using a hollow lance. The wick drain is threaded into a hollow lance, which is pushed (or driven) through the soil layer, which collapses around it. At the ground surface the ends of the wick drains (typically at 1 to 2 m spacing) are interconnected by a granular soil drainage layer or geocomposite sheet drain layer. There are a number of commercially available wick drain manufacturers and installation contractors who provide information on the current products, styles, properties, and estimated costs (Koerner, 1999).

With regards to determining wick drain spacings, the initial focal point is on the time for the consolidation of the subsoil to occur. Generally the time for 90% consolidation (t_{90}) is desired. In order to estimate the time t, it is first necessary to estimate an equivalent sand drain diameter for the wick drain used. The equations suggested by Koerner (1999) are

$$d_{sd} = \sqrt{\frac{d_v^2}{n_s}} \tag{21.9a}$$

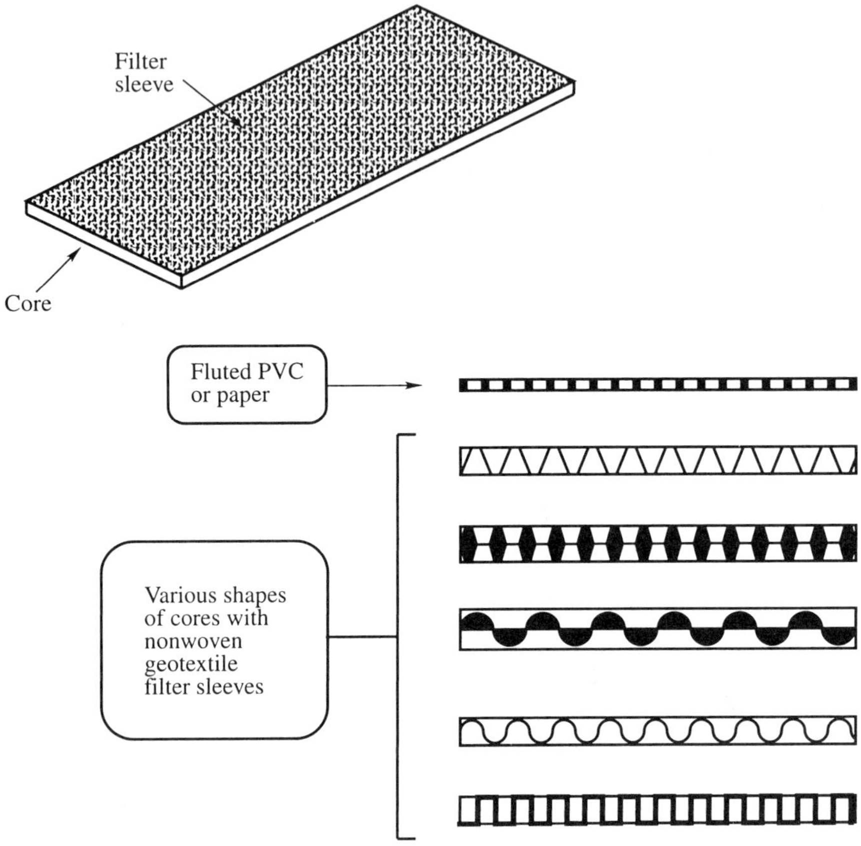

Figure 21.20 Typical core shapes of strip drains (Hausmann, 1990)

$$d_v = \sqrt{\frac{4btn_d}{\pi}} \tag{21.9b}$$

where d_{sd} = equivalent sand drain diameter
d_v = equivalent void circle diameter
b, t = width and thickness of the wick drain
n_s = porosity of sand drain

$$n_d = \frac{\text{Void area of wick drain}}{\text{total cross sectional area of strip}} = \frac{\text{Void area of wick drain}}{b \times t}$$

It may be noted here that equivalent sand drain diameters for various commercially available wick drains vary from 30 to 50 mm (Korner, 1999).

The equation for estimating the time t for consolidation is (Koerner, 1999)

$$t = \frac{D^2}{8c_h}\left[\ln\frac{D}{d} - 0.75\right]\ln\frac{1}{1-U} \tag{21.10}$$

where t = time for consolidation
c_h = coefficient of consolidation of soil for horizontal flow

d = equivalent diameter of strip drain

$$= \frac{\text{circumference}}{\pi}$$

D = sphere of influence of the strip drain;
a) for a triangular pattern, $D = 1.05 \times$ spacing D_t
b) for a square pattern, $D = 1.13 \times$ spacing D_s

D_t = distance between drains in triangular spacing and

D_s = distance for square pattern

U = average degree of consolidation

Advantages of Using Wick Drains (Koerner, 1999)

1. The analytic procedure is available and straightforward in its use.
2. Tensile strength is definitely afforded to the soft soil by the installation of the wick drains.
3. There is only nominal resistance to the flow of water if it enters the wick drain.
4. Construction equipment is generally small.
5. Installation is simple, straightforward and economic.

Example 21.7

What is the equivalent sand drain diameter of a wick drain measuring 96 mm wide and 2.9 mm thick that is 92% void in its cross section? Use an estimated sand porosity of 0.3 for typical sand in a sand drain.

Solution

Total area of wick drain = $b \times t = 96 \times 2.9 = 279 \text{ mm}^2$

Void area of wick drain = $n_d \times b \times t = 0.92 \times 279 = 257 \text{ mm}^2$

The equivalent circle diameter (Eq. 21.9b) is

$$d_v = \sqrt{\frac{4btn_d}{\pi}} = \sqrt{\frac{4 \times 257}{3.14}} = 18.1 \text{ mm}$$

The equivalent sand drain diameter (Eq. 21.19a) is

$$d_{sd} = \sqrt{\frac{d_v^2}{n_s}} = \sqrt{\frac{18.1^2}{0.3}} = 33 \text{ mm}$$

Example 21.8

Calculate the times required for 50, 70 and 90% consolidation of a saturated clayey silt soil using wick drains at various triangular spacings. The wick drains measure 100×4 mm and the soil has a $c_h = 6.5\ 10^{-6} \text{ m}^2/\text{min}$.

Solution

In the simplified formula the equivalent diameter d of a strip drain is

$$d = \frac{\text{circumference}}{\pi} = \frac{100 + 100 + 4 + 4}{3.14} = 66.2 \text{ mm}$$

Using Eq. (21.10)

$$t = \frac{D^2}{8c_h}\left[\ln\frac{D}{d} - 0.75\right]\ln\left(\frac{1}{1-U}\right)$$

substituting the known values

$$t = \frac{D^2}{8(6.5\times 10^{-6})}\left[\ln\frac{D}{0.0062} - 0.75\right]\ln\left(\frac{1}{1-U}\right)$$

The times required for the various degrees of consolidation are tabulated below for assumed theoretical spacings of wick drains.

Wick drain spacings D(m)	Time in days for various degrees of consolidation (U)		
	50%	70%	90%
2.1	110	192	367
1.8	77	133	254
1.5	49	86	164
1.2	29	50	95
0.9	14	24	46
0.6	4.8	8.4	16
0.3	0.6	1.1	2.1

For the triangular pattern, the spacing D_t is

$$D_t = \frac{D}{1.05}.$$

21.8 SAND COMPACTION PILES AND STONE COLUMNS

Sand Compaction Piles

Sand compaction piles consists of driving a hollow steel pipe with the bottom closed with a collapsible plate down to the required depth; filling it with sand, and withdrawing the pipe while air pressure is directed against the sand inside it. The bottom plate opens during withdrawal and the sand backfills the voids created earlier during the driving of the pipe. The in-situ soil is densified while the pipe is being withdrawn, and the sand backfill prevents the soil surrounding the compaction pipe from collapsing as the pipe is withdrawn. The maximum limits on the amount of fines that can be present are 15 percent passing the No. 200 sieve (0.075 mm) and 3 percent passing 0.005 mm. The distance between the piles may have to be planned according to the site conditions.

Stone Columns

The method described for installing sand compaction piles or the vibroflot described earlier can be used to construct stone columns. The size of the stones used for this purpose range from about 6 to 40 mm. Stone columns have particular application in soft inorganic, cohesive soils and are generally inserted on a volume displacement basis.

The diameter of the pipe used either for the construction of sand drains or sand compaction piles can be increased according to the requirements. Stones are placed in the pipe instead of sand, and the technique of constructing stone columns remains the same as that for sand piles.

Stone columns are placed 1 to 3 m apart over the whole area. There is no theoretical procedure for predicting the combined improvement obtained, so it is usual to assume the foundation loads are carried only by the several stone columns with no contribution from the intermediate ground (Bowles, 1996).

Bowles (1996) gives an approximate formula for the allowable bearing capacity of stone columns as

$$q_a = \frac{K_P}{F_s}(4c + \sigma_r') \tag{21.11}$$

where $K_P = \tan^2(45° + \phi'/2)$,

ϕ' = drained angle of friction of stone,

c = either drained cohesion (suggested for large areas) or the undrained shear strength c_u,

σ_r' = effective radial stress as measured by a pressuremeter (but may use $2c$ if pressuremeter data are not available),

F_s = factor of safety, 1.5 to 2.0.

The total allowable load on a stone column of average cross-section area A_c is

$$Q_a = q_a A_c \tag{21.10}$$

Stone columns should extend through soft clay to firm strata to control settlements. There is no end bearing in Eq. (21.11) because the principal load carrying mechanism is local perimeter shear.

Settlement is usually the principal concern with stone columns since bearing capacity is usually quite adequate (Bowles, 1996). There is no method currently available to compute settlement on a theoretical basis.

Stone columns are not applicable to thick deposits of peat or highly organic silts or clays (Bowles, 1996). Stone columns can be used in loose sand deposits to increase the density.

21.9 SOIL STABILIZATION BY THE USE OF ADMIXTURES

The physical properties of soils can often economically be improved by the use of admixtures. Some of the more widely used admixtures include lime, portland cement and asphalt. The process of soil stabilization first involves mixing with the soil a suitable additive which changes its property and then compacting the admixture suitably. This method is applicable only for soils in shallow foundations or the base courses of roads, airfield pavements, etc.

Soil-lime Stabilization

Lime stabilization improves the strength, stiffness and durability of fine grained materials. In addition, lime is sometimes used to improve the properties of the fine grained fraction of granular soils. Lime has been used as a stabilizer for soils in the base courses of pavement systems, under concrete foundations, on embankment slopes and canal linings.

Adding lime to soils produces a maximum density under a higher optimum moisture content than in the untreated soil. Moreover, lime produces a decrease in plasticity index.

Lime stabilization has been extensively used to decrease swelling potential and swelling pressures in clays. Ordinarily the strength of wet clay is improved when a proper amount of lime is added. The improvement in strength is partly due to the decrease in plastic properties of the clay and partly to the pozzolanic reaction of lime with soil, which produces a cemented material that increases in strength with time. Lime-treated soils, in general, have greater strength and a higher modulus of elasticity than untreated soils.

Recommended percentages of lime for soil stabilization vary from 2 to 10 percent. For coarse soils such as clayey gravels, sandy soils with less than 50 per cent silt-clay fraction, the per cent of lime varies from 2 to 5, whereas for soils with more than 50 percent silt-clay fraction, the percent of lime lies between 5 and 10. Lime is also used with fly ash. The fly ash may vary from 10 to 20 per cent, and the percent of lime may lie between 3 and 7.

Soil-Cement Stabilization

Soil-cement is the reaction product of an intimate mixture of pulverized soil and measured amounts of portland cement and water, compacted to high density. As the cement hydrates, the mixture becomes a hard, durable structural material. Hardened soil-cement has the capacity to bridge over local weak points in a subgrade. When properly made, it does not soften when exposed to wetting and drying, or freezing and thawing cycles.

Portland cement and soil mixed at the proper moisture content has been used increasingly in recent years to stabilize soils in special situations. Probably the main use has been to build stabilized bases under concrete pavements for highways and airfields. Soil cement mixtures are also used to provide wave protection on earth dams. There are three categories of soil-cement (Mitchell and Freitag, 1959). They are:

1. Normal soil-cement usually contains 5 to 14 percent cement by weight and is used generally for stabilizing low plasticity soils and sandy soils.
2. Plastic soil-cement has enough water to produce a wet consistency similar to mortar. This material is suitable for use as water proof canal linings and for erosion protection on steep slopes where road building equipment may not be used.
3. Cement-modified soil is a mix that generally contains less than 5 percent cement by volume. This forms a less rigid system than either of the other types, but improves the engineering properties of the soil and reduces the ability of the soil to expand by drawing in water.

The cement requirement depends on the gradation of the soil. A well graded soil containing gravel, coarse sand and fine sand with or without small amounts of silt or clay will require 5 percent or less cement by weight. Poorly graded sands with minimal amount of silt will require about 9 percent by weight. The remaining sandy soils will generally require 7 percent. Non-plastic or moderately plastic silty soils generally require about 10 percent, and plastic clay soils require 13 percent or more.

Bituminous Soil Stabilization

Bituminous materials such as asphalts, tars, and pitches are used in various consistencies to improve the engineering properties of soils. Mixed with cohesive soils, bituminous materials improve the bearing capacity and soil strength at low moisture content. The purpose of incorporating bitumen into such soils is to water proof them as a means to maintain a low moisture content. Bituminous materials added to sand act as a cementing agent and produces a stronger, more coherent mass. The amount of bitumen added varies from 4 to 7 percent for cohesive materials and 4 to 10 percent for sandy materials. The primary use of bituminous materials is in road construction where it may be the primary ingredient for the surface course or be used in the subsurface and base courses for stabilizing soils.

21.10 SOIL STABILIZATION BY INJECTION OF SUITABLE GROUTS

Grouting is a process whereby fluid like materials, either in suspension, or solution form, are injected into the subsurface soil or rock.

The purpose of injecting a grout may be any one or more of the following:

1. To decrease permeability.
2. To increase shear strength.
3. To decrease compressibility.

Suspension-type grouts include soil, cement, lime, asphalt emulsion, etc., while the solution type grouts include a wide variety of chemicals. Grouting proves especially effective in the following cases:

1. When the foundation has to be constructed below the ground water table. The deeper the foundation, the longer the time needed for construction, and therefore, the more benefit gained from grouting as compared with dewatering.
2. When there is difficult access to the foundation level. This is very often the case in city work, in tunnel shafts, sewers, and subway construction.
3. When the geometric dimensions of the foundation are complicated and involves many boundaries and contact zones.
4. When the adjacent structures require that the soil of the foundation strata should not be excavated (extension of existing foundations into deeper layers).

Grouting has been extensively used primarily to control ground water flow under earth and masonry dams, where rock grouting is used. Since the process fills soil voids with some type of stabilizing material grouting is also used to increase soil strength and prevent excessive settlement.

Many different materials have been injected into soils to produce changes in the engineering properties of the soil. In one method a casing is driven and injection is made under pressure to the soil at the bottom of the hole as the casing is withdrawn. In another method, a grouting hole is drilled and at each level in which injection is desired, the drill is withdrawn and a collar is placed at the top of the area to be grouted and grout is forced into the soil under pressure. Another method is to perforate the casing in the area to be grouted and leave the casing permanently in the soil.

Penetration grouting may involve portland cement or fine grained soils such as bentonite or other materials of a particulate nature. These materials penetrate only a short distance through most soils and are primarily useful in very coarse sands or gravels. Viscous fluids, such as a solution of sodium silicate, may be used to penetrate fine grained soils. Some of these solutions form gels that restrict permeability and improve compressibility and strength properties.

Displacement grouting usually consists of using a grout like portland cement and sand mixture which when forced into the soil displaces and compacts the surrounding material about a central core of grout. Injection of lime is sometimes used to produce lenses in the soil that will block the flow of water and reduce compressibility and expansion properties of the soil. The lenses are produced by hydraulic fracturing of the soil.

The injection and grouting methods are generally expensive compared with other stabilization techniques and are primarily used under special situations as mentioned earlier. For a detailed study on injections, readers may refer to Caron et al., (1975).

21.11 PROBLEMS

21.1 Differentiate: (i) Compaction and consolidation, and (ii) Standard Proctor and modified Proctor tests.

21.2 Draw an ideal 'compaction curve' and discuss the effect of moisture on the dry unit weight of soil.

21.3 Explain: (i) the unit, in which the compaction is measured, (ii) 95 percent of Proctor density, (iii) zero air-voids line, and (iv) effect of compaction on the shear strength of soil.

21.4 What are the types of rollers used for compacting different types of soils in the field? How do you decide the compactive effort required for compacting the soil to a desired density in the field?

21.5 What are the methods adopted for measuring the density of the compacted soil? Briefly describe the one which will suit all types of soils.

21.6 A soil having a specific gravity of solids $G_s = 2.75$, is subjected to Proctor compaction test in a mold of volume $V = 945$ cm^3. The observations recorded are as follows:

Observation number	1	2	3	4	5
Mass of wet sample, g	1389	1767	1824	1784	1701
Water content, percentage	7.5	12.1	17.5	21.0	25.1

What are the values of maximum dry unit weight and the optimum moisture content? Draw 100% saturation line.

21.7 A field density test was conducted by sand cone method. The observation data are given below:

(a) Mass of jar with cone and sand (before use) = 4950 g, (b) mass of jar with cone and sand (after use) = 2280 g, (c) mass of soil from the hole = 2925 g, (d) dry density of sand = 1.48 g/cm^3, (e) water content of the wet soil = 12%. Determine the dry unit weight of compacted soil.

21.8 If a clayey sample is saturated at a water content of 30%, what is its density? Assume a value for specific gravity of solids.

21.9 A soil in a borrow pit is at a dry density of 1.7 Mg/m^3 with a water content of 12%. If a soil mass of 2000 cubic meter volume is excavated from the pit and compacted in an embankment with a porosity of 0.32, calculate the volume of the embankment which can be constructed out of this material. Assume $G_s = 2.70$.

21.10 In a Proctor compaction test, for one observation, the mass of the wet sample is missing. The oven dry mass of this sample was 1800 g. The volume of the mold used was 950 cm^3. If the saturation of this sample was 80 percent, determine (i) the moisture content, and (ii) the total unit weight of the sample. Assume $G_s = 2.70$.

21.11. A field-compacted sample of a sandy loam was found to have a wet density of 2.176 Mg/m^3 at a water content of 10%. The maximum dry density of the soil obtained in a standard Proctor test was 2.0 Mg/m^3. Assume $G_s = 2.65$. Compute ρ_d, S, n and the percent of compaction of the field sample.

21.12 A proposed earth embankment is required to be compacted to 95% of standard Proctor dry density. Tests on the material to be used for the embankment give $\rho_{max} = 1.984$ Mg/m^3 at an optimum water content of 12%. The borrow pit material in its natural condition has a void ratio of 0.60. If $G_s = 2.65$, what is the minimum volume of the borrow required to make 1 cu.m of acceptable compacted fill?

21.13 The following data were obtained from a field density test on a compacted fill of sandy clay. Laboratory moisture density tests on the fill material indicated a maximum dry density of 1.92 Mg/m^3 at an optimum water content of 11%. What was the percent compaction of the fill? Was the fill water content above or below optimum.

Mass of the moist soil removed from the test hole = 1038 g
Mass of the soil after oven drying = 914 g
Volume of the test hole = 478.55 cm^3

21.14 A field density test performed by sand-cone method gave the following data.

Mass of the soil removed + pan = 1590 g
Mass of the pan = 125 g
Volume of the test hole = 750 cm^3
Water content information
Mass of the wet soil + pan = 404.9 g
Mass of the dry soil + pan = 365.9 g
Mass of the pan = 122.0 g
Compute: ρ_d, γ_d, and the water content of the soil. Assume G_s = 2.67

APPENDIX A

SI UNITS IN GEOTECHNICAL ENGINEERING

Introduction

There has always been some confusion with regards to the system of units to be used in engineering practices and other commercial transactions. FPS (Foot-pound-second) and MKS (Meter-Kilogram-second) systems are still in use in many parts of the world. Sometimes a mixture of two or more systems are in vogue making the confusion all the greater. Though the SI (Le System International d'Unites or the International System of Units) units was first conceived and adopted in the year 1960 at the Eleventh General Conference of Weights and Measures held in Paris, the adoption of this coherent and systematically constituted system is still slow because of the past association with the FPS system. The conditions are now gradually changing and possibly in the near future the SI system will be the only system of use in all academic institutions in the world over. It is therefore essential to understand the basic philosophy of the SI units.

The Basics of the SI System

The SI system is a fully coherent and rationalized system. It consists of six basic units and two supplementary units, and several derived units. (Table A.1)

Table A.1 Basic units of interest in geotechnical engineering

	Quantity	Unit	SI symbol
1.	Length	Meter	m
2.	Mass	Kilogram	kg
3.	Time	Second	S
4.	Electric current	Ampere	A
5.	Thermodynamic temperature	Kelvin	K

Supplementary Units

The supplementary units include the *radian* and *steradian*, the units of plane and solid angles, respectively.

Derived Units

The derived units used by geotechnical engineers are tabulated in Table A.2.

Prefixes are used to indicate *multiples* and *submultiples* of the basic and derived units as given below.

Factor	Prefix	Symbol
10^6	mega	M
10^3	kilo	k
10^{-3}	milli	m
10^{-6}	micro	μ

Table A.2 Derived units

Quantity	Unit	SI symbol	Formula
acceleration	meter per second squared	–	m/sec^2
area	square meter	–	m^2
density	kilogram per cubic meter	–	kg/m^3
force	newton	N	$kg\text{-}m/s^2$
pressure	pascal	Pa	N/m^2
stress	pascal	Pa	N/m^2
moment or torque	newton-meter	N-m	$kg\text{-}m^2/s^2$
unit weight	newton per cubic meter	N/m^3	kg/s^2m^2
frequency	hertz	Hz	cycle/sec
volume	cubic meter	m^3	–
volume	liter	L	$10^{-3}m^3$
work (energy)	joule	J	N-m

Mass

Mass is a measure of the amount of matter an object contains. The mass remains the same even if the object's temperature and its location change. Kilogram, kg, is the unit used to measure the quantity of mass contained in an object. Sometimes *Mg (megagram)* and gram (*g*) are also used as a measure of mass in an object.

Time

Although the second (s) is the basic SI time unit, minutes (min), hours (h), days (d) etc. may be used as and where required.

Force

As per Newton's second law of motion, force, F, is expressed as $F = Ma$, where, M = mass expressed in kg, and a is acceleration in units of m/sec^2. If the acceleration is g, the standard value of which is $9.80665\ m/sec^2 \approx 9.81\ m/s^2$, the force F will be replaced by W, the weight of the body. Now the above equation may be written as $W = Mg$.

The correct unit to express the weight W, of an object is the *newton* since the weight is the gravitational force that causes a downward acceleration of the object.

Newton, N, is defined as the force that causes a 1 kg mass to accelerate 1 m/s²

$$\text{or} \quad 1\,\text{N} = 1\frac{\text{kg - m}}{\text{s}^2}$$

Since, a *newton*, is too small a unit for engineering usage, multiples of newtons expressed as *kilonewton, kN,* and *meganewton, MN,* are used. Some of the useful relationships are

1 kilonewton, kN = 10^3 newton = 1000 N

meganewton, MN = 10^6 newton = 10^3 kN = 1000 kN

Stress and Pressure

The unit of *stress* and *pressure* in SI units is the *pascal (Pa)* which is equal to 1 newton per square meter (N/m^2). Since a *pascal* is too small a unit, multiples of pascals are used as *prefixes* to express the unit of stress and pressure. In engineering practice kilopascals or megapascals are normally used. For example,

1 kilopascal = 1 kPa = 1 kN/m^2 = 1000 N/m^2

1 megapascal = 1 MPa = 1 MN/m^2 = 1000 kN/m^2

Density

Density is defined as mass per unit volume. In the SI system of units, mass is expressed in kg/m^3. In many cases, it may be more convenient to express density in megagrams per cubic meter or in gm per cubic centimeter. The relationships may be expressed as

1 g/cm^3 = 1000 kg/m^3 = 10^6 g/m^3 = 1 Mg/m^3

It may be noted here that the density of water, ρ_w, is exactly 1.00 g/cm^3 at 4 °C, and the variation is relatively small over the range of temperatures in ordinary engineering practice. It is sufficiently accurate to write

$$\rho_w = 1.00\ \text{g/cm}^3 = 10^3\ \text{kg/m}^3 = 1\ \text{Mg/m}^3$$

Unit weight

Unit weight is still the common measurement in geotechnical engineering practice. The relationship between unit weight, γ, and density ρ, may be expressed as $\gamma = \rho g$.
For example, if the density of water, ρ_w = 1000 kg/m^3, then

$$\gamma_w = \rho_w g = 1000\frac{\text{kg}}{\text{m}^3} \times 9.81\frac{\text{m}}{\text{s}^2} = 9810\frac{\text{kg}}{\text{m}^3}\cdot\frac{\text{m}}{\text{s}^2}$$

$$\text{Since,}\ 1\,\text{N} = 1\,\frac{\text{kg m}}{\text{s}^2}, \quad \gamma_w = 9810\frac{\text{N}}{\text{m}^3} = 9.81\ \text{kN/m}^3$$

Table A.3 Conversion factors

To convert	SI to FPS			FPS to SI		
	From	To	Multiply by	From	To	Multiply by
Length	m	ft	3.281	ft	m	0.3048
	m	in	39.37	in	m	0.0254
	cm	in	0.3937	in	cm	2.54
	mm	in	0.03937	in	mm	25.4
Area	m^2	ft^2	10.764	ft^2	m^2	929.03×10^{-4}
	m^2	in^2	1550	in^2	m^2	6.452×10^{-4}
	cm^2	in^2	0.155	in^2	cm^2	6.452
	mm^2	in^2	0.155×10^{-2}	in^2	mm^2	645.16
Volume	m^3	ft^3	35.32	ft^3	m^3	28.317×10^{-3}
	m^3	in^3	61,023.4	in^3	m^3	16.387×10^{-6}
	cm^3	in^3	0.06102	in^3	cm^3	16.387
Force	N	lb	0.2248	lb	N	4.448
	kN	lb	224.8	lb	kN	4.448×10^{-3}
	kN	kip	0.2248	kip	kN	4.448
	kN	US ton	0.1124	US ton	kN	8.896
Stress	N/m^2	lb/ft^2	20.885×10^{-3}	lb/ft^2	N/m^2	47.88
	kN/m^2	lb/ft^2	20.885	lb/ft^2	kN/m^2	0.04788
	kN/m^2	US ton/ft^2	0.01044	US ton/ft^2	kN/m^2	95.76
	kN/m^2	kip/ft^2	20.885×10^{-3}	kip/ft^2	kN/m^2	47.88
	kN/m^2	lb/in^2	0.145	lb/in^2	kN/m^2	6.895
Unit weight	kN/m^3	lb/ft^3	6.361	lb/ft^3	kN/m^3	0.1572
	kN/m^3	lb/in^3	0.003682	lb/in^3	kN/m^3	271.43
Moment	N-m	lb-ft	0.7375	lb-ft	N-m	1.3558
	N-m	lb-in	8.851	lb-in	N-m	0.11298
Moment of inertia	mm^4	in^4	2.402×10^{-6}	in^4	mm^4	0.4162×10^{6}
	m^4	in^4	2.402×10^{6}	in^4	m^4	0.4162×10^{-6}
Section modulus	mm^3	in^3	6.102×10^{-5}	in^3	mm^3	0.16387×10^{5}
	m^3	in^3	6.102×10^{4}	in^3	m^3	0.16387×10^{-4}
Hydraulic conductivity	m/min	ft/min	3.281	ft/min	m/min	0.3048
	cm/min	ft/min	0.03281	ft/min	cm/min	30.48
	m/sec	ft/sec	3.281	ft/sec	m/sec	0.3048
	cm/sec	in/sec	0.3937	in/sec	cm/sec	2.54
Coefficient of consolidation	cm^2/sec	in^2/sec	0.155	in^2/sec	cm^2/sec	6.452
	$m^2/year$	in^2/sec	4.915×10^{-5}	in^2/sec	$m^2/year$	20.346×10^{3}
	cm^2/sec	ft^2/sec	1.0764×10^{-3}	ft^2/sec	cm^2/sec	929.03

Table A.4 Conversion factors—general

To convert from	To	Multiply by
Angstrom units	inches feet millimeters centimeters meters	$3.9370079\ 10^{-9}$ 3.28084×10^{-10} 1×10^{-7} 1×10^{-8} 1×10^{-10}
Microns	inches	3.9370079×10^{-5}
US gallon (gal)	cm^3 m^3 ft^3 liters	3785 3.785×10^{-3} 0.133680 3.785
Pounds	dynes grams kilograms	4.44822×10^5 453.59243 0.45359243
Tons (short or US tons)	kilograms pounds kips	907.1874 2000 2
Tons (metric)	grams kilograms pounds kips tons (short or US tons)	1×10^6 1000 2204.6223 2.2046223 1.1023112
kips/ft^2	lbs/in^2 lbs/ft^2 US tons/ft^2 kg/cm^2 metric ton/ft^2	6.94445 1000 0.5000 0.488244 4.88244
Pounds/in^3	gms/cm^3 kg/m^3 lbs/ft^3	27.6799 27679.905 1728
Poise	kN-sec/m^2	10^{-4}
millipoise	poise kN-sec/m^2 gm-sec/cm^2	10^{-3} 10^{-7} 10^{-6}
ft/min	ft/day ft/year	1440 5256×10^2
ft/year	ft/min	1.9025×10^{-6}
cm/sec	m/min ft/min ft/year	0.600 1.9685 1034643.6

APPENDIX B

SLOPE STABILITY CHARTS AND TABLES

As per Eq.(10.43), the factor of safety F_s is defined as

$$F_s = m - nr_u$$

where, m, n = stability coefficients, and r_u = pore pressure ratio. The values of m and n may be obtained from Figs. B.1 to B.14

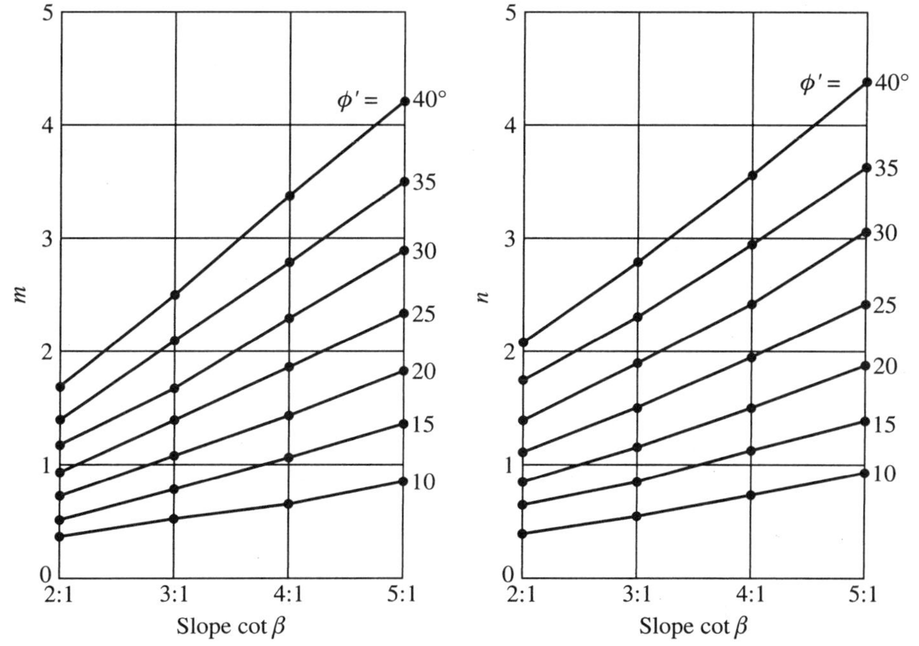

Figure B.1 Stability coefficients m and n for $c'/\gamma H = 0$ (Bishop and Morgenstern, 1960)

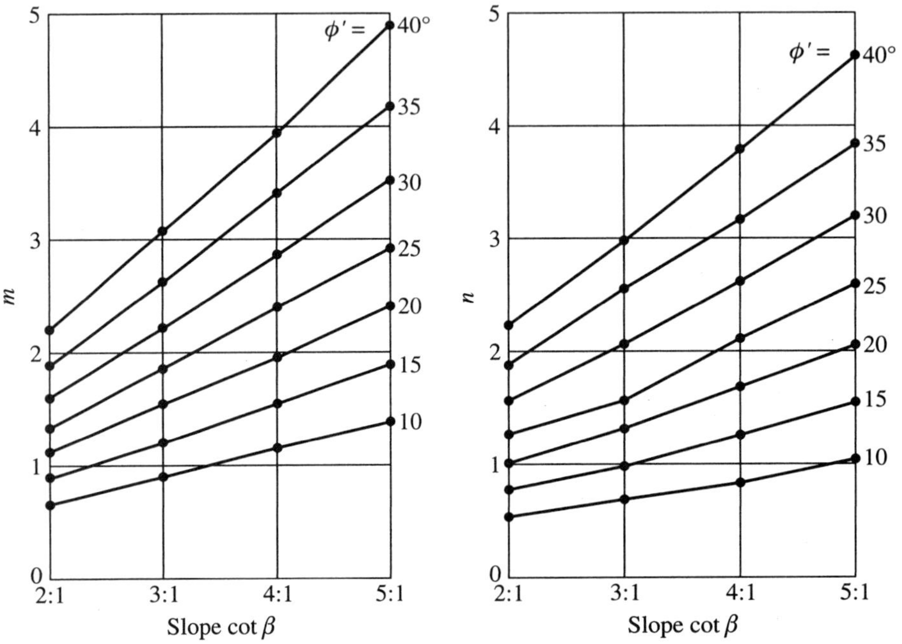

Figure B.2 Stability coefficients for $c'/\gamma H = 0.025$ and $n_d = 1.00$ (Bishop and Morgenstern, 1960)

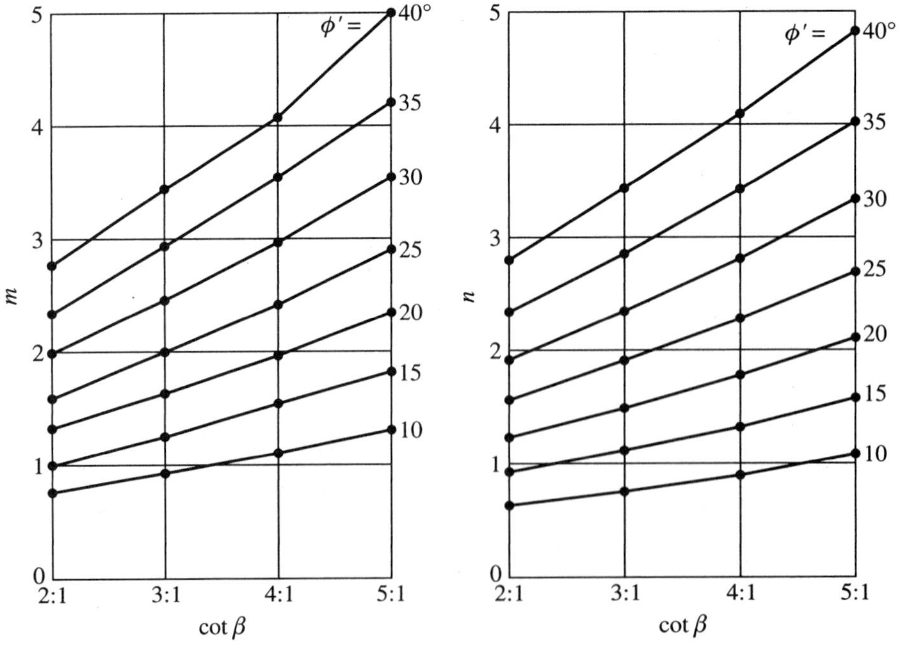

Figure B.3 Stability coefficients m and n for $c'/\gamma H = 0.025$ and $n_d = 1.25$ (Bishop and Morgenstern, 1960)

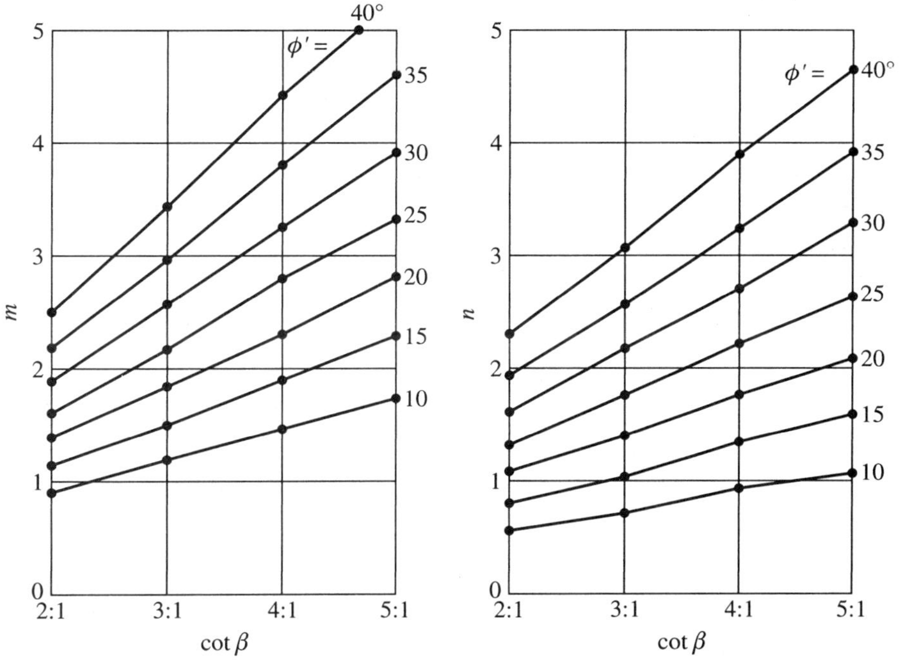

Figure B.4 Stability coefficients m and n for $c'/\gamma H = 0.05$ and $n_d = 1.00$ (Bishop and Morgenstern, 1960)

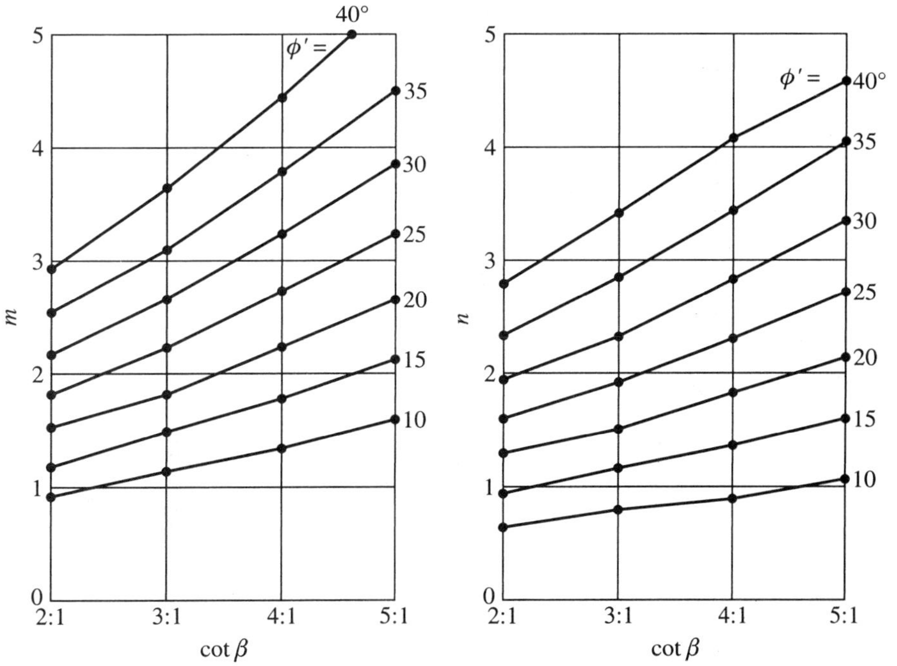

Figure B.5 Stability coefficients m and n for $c'/\gamma H = 0.05$ and $n_d = 1.25$ (Bishop and Morgenstern, 1960)

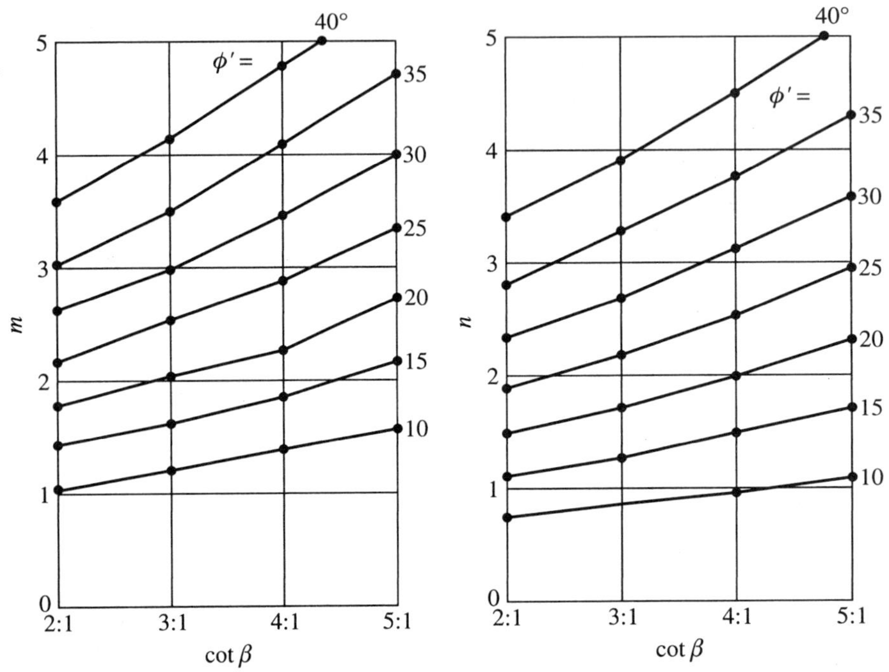

Figure B.6 Stability coefficients m and n for $c'/\gamma H = 0.05$ and $n_d = 1.50$ (Bishop and Morgenstern, 1960)

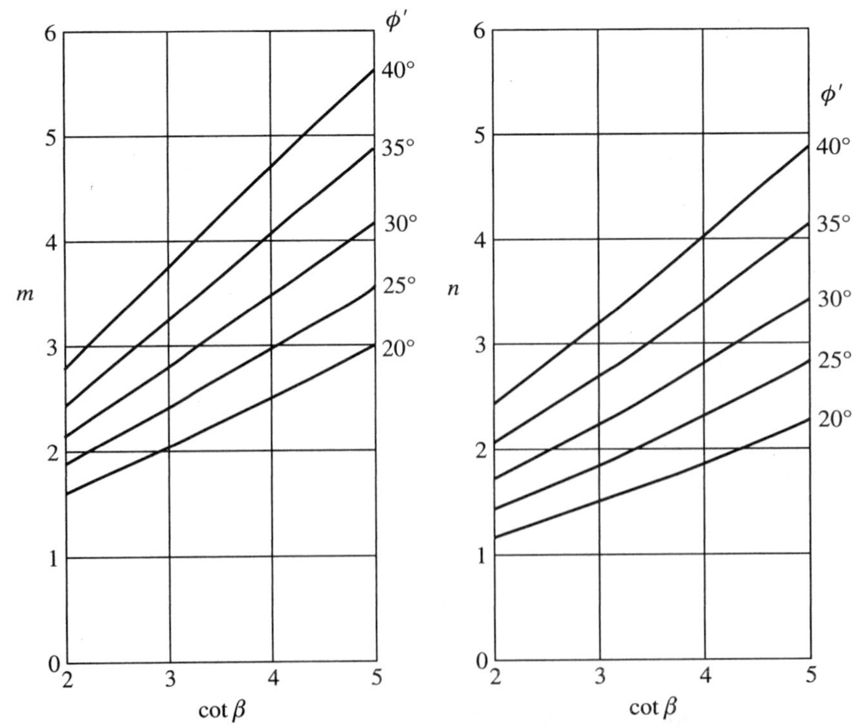

Figure B.7 Stability coefficients m and n for $c'/\gamma H = 0.075$ toe circles (O'Connor and Mitchell, 1977)

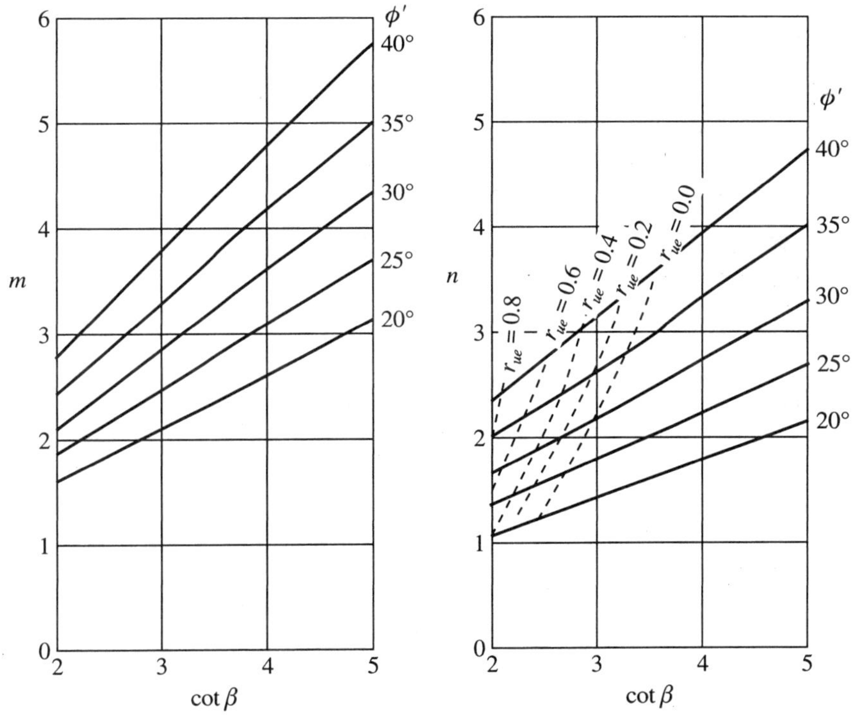

Figure B.8 Stability coefficients m and n for $c'/\gamma H = 0.075$ and $n_d = 1.00$ (O'Connor et al., 1977)

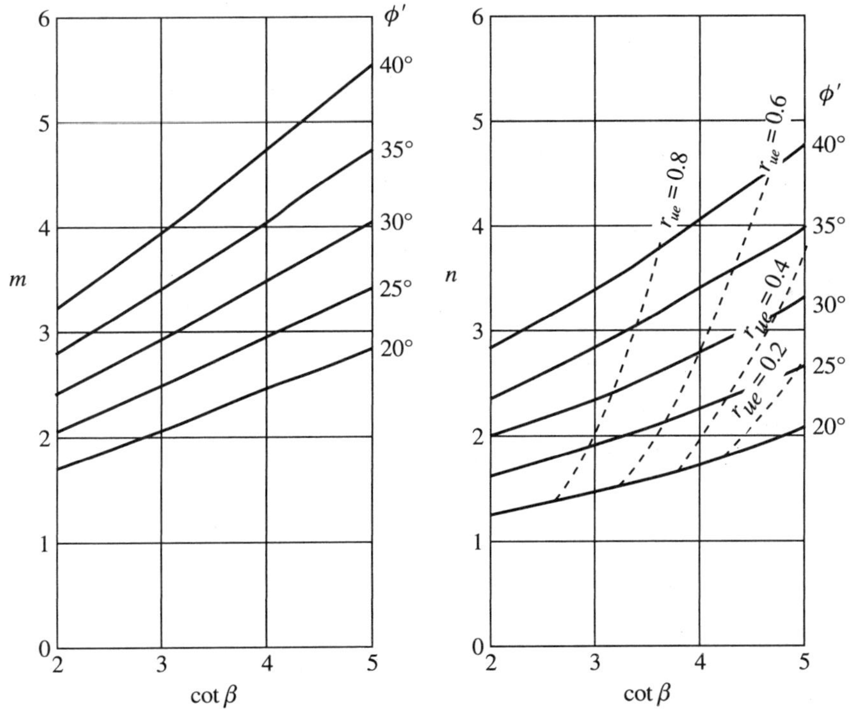

Figure B.9 Stability coefficients m and n for $c'/\gamma H = 0.075$ and $n_d = 1.25$ (O'Connor and Mitchell, 1977)

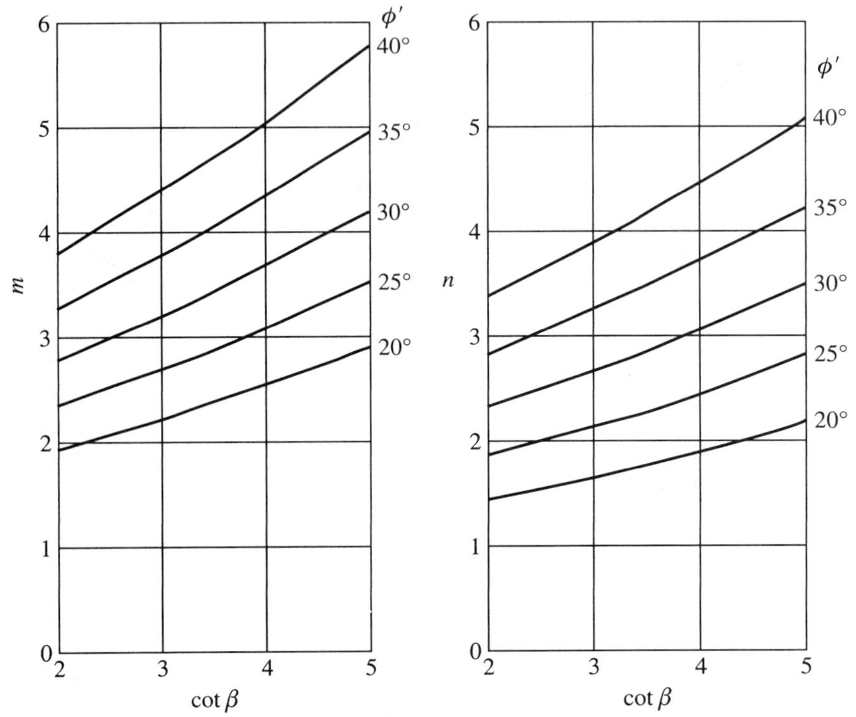

Figure B.10 Stability coefficients m and n for $c'/\gamma H = 0.075$ and $n_d = 1.50$ (O'Connor and Mitchell, 1977)

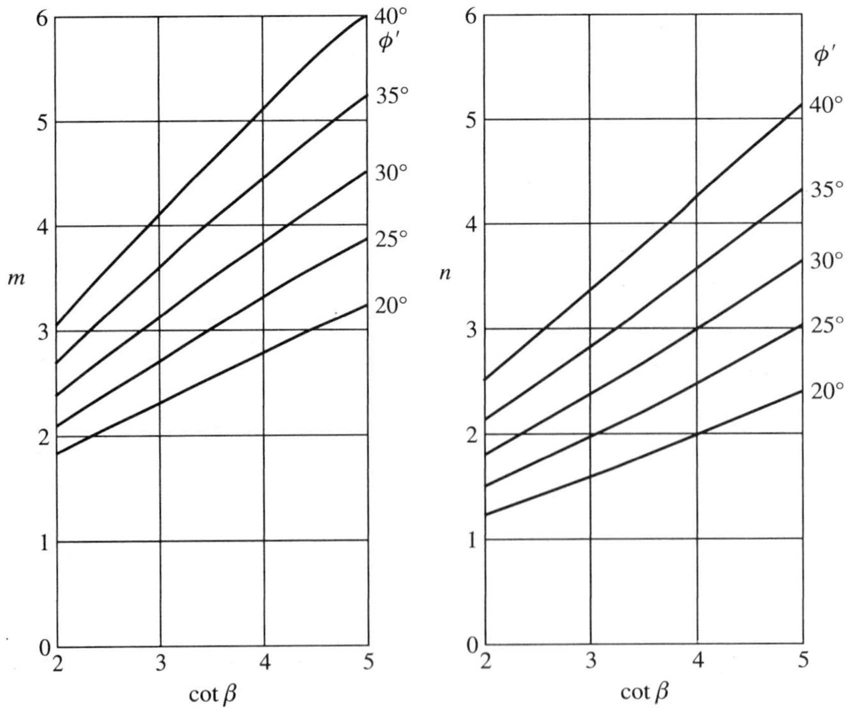

Figure B.11 Stability coefficients m and n for $c'/\gamma H = 0.100$ toe circles (O'Connor and Mitchell, 1977)

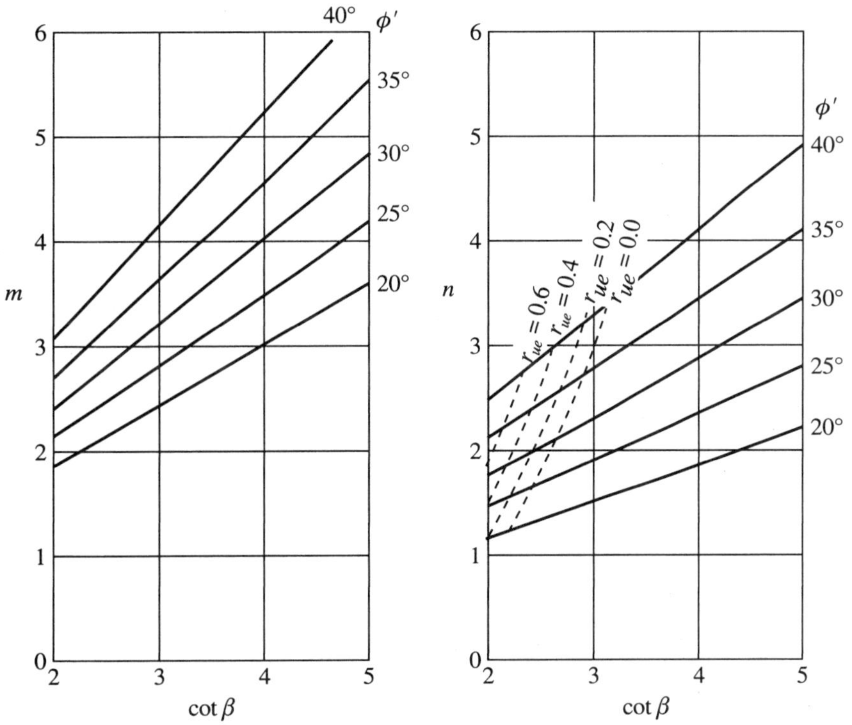

Figure B.12 Stability coefficients m and n for $c'/\gamma H = 0.100$ and $n_d = 1.00$ (O'Connor and Mitchell, 1977)

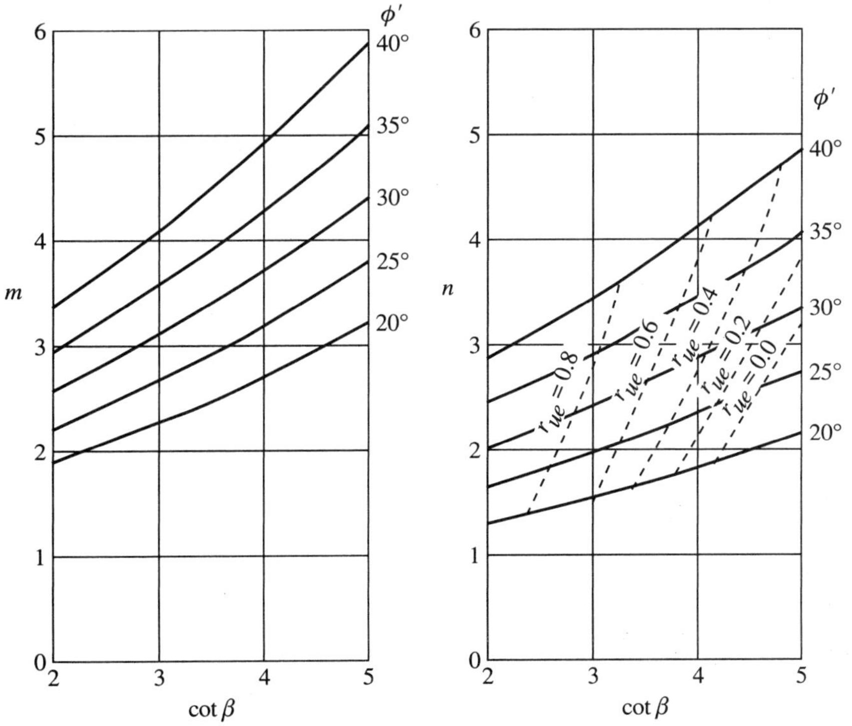

Figure B.13 Stability coefficients m and n for $c'/\gamma H = 0.100$ and $n_d = 1.25$ (O'Connor and Mitchell, 1977)

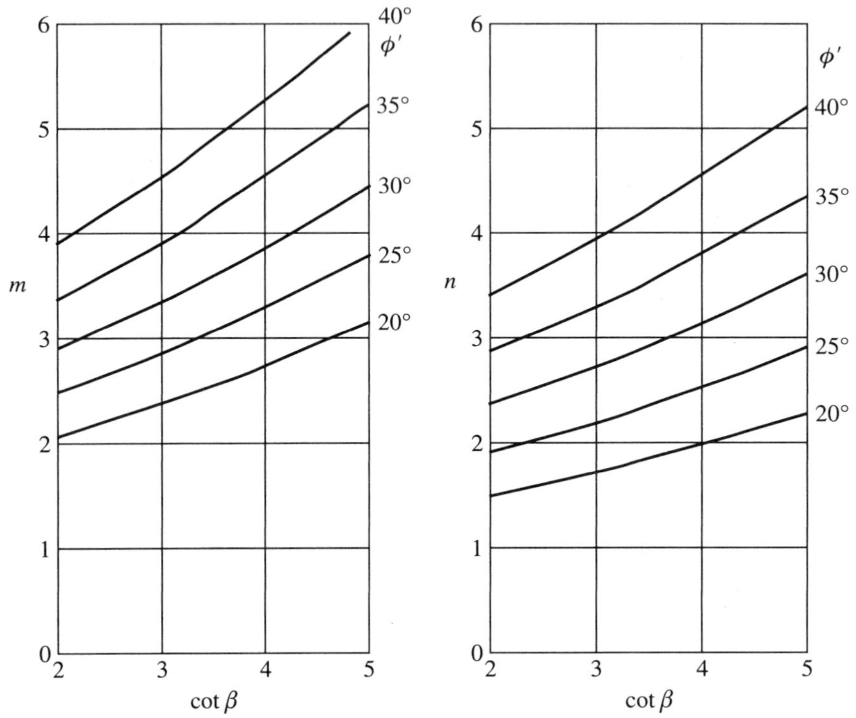

Figure B.14 Stability coefficients m and n for $c'/\gamma H = 0.100$ and $n_d = 1.50$ (O'Connor and Mitchell, 1977)

Bishop and Morgenstern (1960) Stability Coefficients are Presented in Tabular Form

$F_s = m - n.r_u$

Table B1 Stability coefficients m and n for $\frac{c'}{\gamma H} = 0$

Stability coefficients for earth slopes								
ϕ'	Slope 2:1		Slope 3:1		Slope 4:1		Slope 5:1	
	m	n	m	n	m	n	m	n
10.0	0.353	0.441	0.529	0.588	0.705	0.749	0.882	0.917
12.5	0.443	0.554	0.665	0.739	0.887	0.943	1.109	1.153
15.0	0.536	0.670	0.804	0.893	1.072	1.139	1.340	1.393
17.5	0.631	0.789	0.946	1.051	1.261	1.340	1.577	1.639
20.0	0.728	0.910	1.092	1.213	1.456	1.547	1.820	1.892
22.5	0.828	1.035	1.243	1.381	1.657	1.761	2.071	2.153
25.0	0.933	1.166	1.399	1.554	1.865	1.982	2.332	2.424
27.5	1.041	1.301	1.562	1.736	2.082	2.213	2.603	2.706
30.0	1.155	1.444	1.732	1.924	2.309	2.454	2.887	3.001
32.5	1.274	1.593	1.911	2.123	2.548	2.708	3.185	3.311
35.0	1.400	1.750	2.101	2.334	2.801	2.977	3.501	3.639
37.5	1.535	1.919	2.302	2.588	3.069	3.261	3.837	3.989
40.0	1.678	2.098	2.517	2.797	3.356	3.566	4.196	4.362

Table B2 Stability coefficients m and n for $\frac{c'}{\gamma H} = 0.025$ and $n_d = 1.00$

ϕ'	Slope 2:1		Slope 3:1		Slope 4:1		Slope 5:1	
	m	n	m	n	m	n	m	n
10.0	0.678	0.534	0.906	0.683	1.130	0.846	1.365	1.031
12.5	0.790	0.655	1.066	0.849	1.337	1.061	1.620	1.282
15.0	0.901	0.776	1.224	1.014	1.544	1.273	1.868	1.534
17.5	1.012	0.898	1.380	1.179	1.751	1.485	5.121	1.789
20.0	1.124	1.022	1.542	1.347	1.962	1.698	2.380	2.050
22.5	1.239	1.150	1.705	1.518	2.177	4.916	2.646	2.317
25.0	1.356	1.282	1.875	1.696	2.400	2.141	2.921	2.596
27.5	1.478	1.421	2.050	1.882	2.631	2.375	3.207	2.886
30.0	1.606	1.567	2.235	2.078	2.873	2.622	3.508	3.191
32.5	1.739	1.721	2.431	2.285	3.127	2.883	3.823	3.511
35.0	1.880	1.885	2.635	2.505	3.396	3.160	4.156	3.849
37.5	2.030	2.060	2.855	2.741	3.681	3.458	4.510	4.209
40.0	2.190	2.247	3.090	2.933	3.984	3.778	4.885	4.592

Table B3 Stability coefficients m and n for $\frac{c'}{\gamma H} = 0.025$ and $n_d = 1.25$

ϕ'	Slope 2:1		Slope 3:1		Slope 4:1		Slope 5:1	
	m	n	m	n	m	n	m	n
10.0	0.737	0.614	0.901	0.726	1.085	0.867	1.285	1.014
12.5	0.878	0.759	1.076	0.908	1.299	1.089	1.543	1.278
15.0	1.019	0.907	1.253	1.093	1.515	1.312.	1.803	1.545
17.5	1.162	1.059	1.433	1.282	1.736	1.541	2.065	1.814
20.0	1.309	1.216	1.618	1.478	1.961	1.775	2.344	2.090
22.5	1.461	1.379	1.808	1.680	2.194	2.017	2.610	2.373
25.0	1.619	1.547	2.007	1.891	2.437	2.269	2.897	2.669
27.5	1.783	1.728	2.213	2.111	2.689	2.531	3.196	2.976
30.0	1.956	1.915	2.431	2.342	2.953	2.806	3.511	3.299
32.5	2.139	2.112	2.659	2.588	3.231	3.095	3.841	3.638
35.0	2.331	2.321	2.901	2.841	3.524	3.400	4.191	3.998
37.5	2.536	2.541	3.158	3.112	3.835	3.723	4.563	4.379
40.0	2.753	2.775	3.431	3.399	4.164	4.064	4.988	4.784

Table B4 Stability coefficients m and n for $\frac{c'}{\gamma H} = 0.05$ and $n_d = 1.00$

ϕ'	Slope 2:1		Slope 3:1		Slope 4:1		Slope 5:1	
	m	n	m	n	m	n	m	n
10.0	0.913	0.563	1.181	0.717	1.469	0.910	1.733	1.069
12.5	1.030	0.690	1.343	0.878	1.688	1.136	1.995	1.316
15.0	1.145	0.816	1.506	1.043	1.904	1.353	2.256	1.576
17.5	1.262	0.942	1.671	1.212	2.117	1.565	2.517	1.825
20.0	1.380	1.071	1.840	1.387	2.333	1.776	2.783	2.091
22.5	1.500	1.202	2.014	1.568	2.551	1.989	3.055	2.365
25.0	1.624	1.338	2.193	1.757	2.778	2.211	3.336	2.651
27.5	1.753	1.480	2.380	1.952	3.013	2.444	3.628	2.948
30.0	1.888	1.630	2.574	2.157	3.261	2.693	3.934	3.259
32.5	2.029	1.789	2.777	2.370	3.523	2.961	4.256	3.585
35.0	2.178	1.958	2.990	2.592	3.803	3.253	4.597	3.927
37.5	2.336	2.138	3.215	2.826	4.103	3.574	4.959	4.288
40.0	2.505	2.332	3.451	3.071	4.425	3.926	5.344	4.668

Table B5 Stability coefficients m and n for $\frac{c'}{\gamma H} = 0.05$ and $n_d = 1.25$

ϕ'	Slope 2:1		Slope 3:1		Slope 4:1		Slope 5:1	
	m	n	m	n	m	n	m	n
10.0	0.919	0.633	1.119	0.766	1.344	0.886	1.594	1.042
12.5	1.065	0.792	1.294	0.941	1.563	1.112	1.850	1.300
15.0	1.211	0.950	1.471	1.119	1.782	1.338	2.109	1.562
17.5	1.359	1.108	1.650	1.303	2.004	1 .567	2.373	1.831
20.0	1.509	1.266	1.834	1.493	2.230	1.799	2.643	2.107
22.5	1.663	1.428	2.024	1.690	2.463	2.038	2.921	2.392
25.0	1.822	1.595	2.222	1.897	2.705	2.287	3.211	2.690
27.5	1.988	1.769	2.428	2.113	2.957	2.546	3.513	2.999
30.0	2.161	1.950	2.645	2.342	3.221	2.819	3.829	3.324
32.5	2.343	2.141	2.873	2.583	3.500	3.107	4.161	3.665
35.0	2.535	2.344	3.114	2.839	3.795	3.413	4.511	4.025
37.5	2.738	2.560	3.370	3.111	4.109	3.740	4.881	4.405
40.0	2.953	2.791	3.642	3.400	4.442	4.090	5.273	4.806

Table B6 Stability coefficients m and n for $\frac{c'}{\gamma H} = 0.05$ and $n_d = 1.50$

ϕ'	Slope 2:1		Slope 3:1		Slope 4:1		Slope 5:1	
	m	n	m	n	m	n	m	n
10.0	1.022	0.751	1.170	0.828	1.343	0.974	1.547	1.108
12.5	1.202	0.936	1.376	1.043	1.589	1.227	1.829	1.399
15.0	1.383	1.122	1.583	1.260	1.835	1.480	2.112	1.690
17.5	1.565	1.309	1.795	1.480	2.084	1.734	2.398	1.983
20.0	1.752	1.501	2.011	1.705	2.337	1.993	2.690	2.280
22.5	1.943	1.698	2.234	1.937	2.597	2.258	2.990	2.585
25.0	2.143	1.903	2.467	2.179	2.867	2.534	3.302	2.902
27.5	2.350	2.117	2.709	2.431	3.148	2.820	3.626	3.231
30.0	2.568	2.342	2.964	2.696	3.443	3.120	3.967	3.577
32.5	2.798	2.580	3.232	2.975	3.753	3.436	4.326	3.940
35.0	3.041	2.832	3.515	3.269	4.082	3.771	4.707	4.325
37.5	3.299	3.102	3.817	3.583	4.431	4.128	5.112	4.735
40.0	3.574	3.389	4.136	3.915	4.803	4.507	5.543	5.171

Extension of the Bishop and Morgenstern Stability Coefficients (O'Connor and Mitchell, 1977)

Table B7 Stability coefficients m and n for $\dfrac{c'}{\gamma H} = 0.075$ and toe circle

ϕ'	Slope 2:1		Slope 3:1		Slope 4:1		Slope 5:1	
	m	n	m	n	m	n	m	n
20	1.593	1.158	2.055	1.516	2.498	1.903	2.934	2.301
25	1.853	1.430	2.426	1.888	2.980	2.361	3.520	2.861
30	2.133	1.730	2.826	2.288	3.496	2.888	4.150	3.461
35	2.433	2.058	3.253	2.730	4.055	3.445	4.846	4.159
40	2.773	2.430	3.737	3.231	4.680	4.061	5.609	4.918

Table B8 Stability coefficients m and n for $\dfrac{c'}{\gamma H} = 0.075$ and $n_d = 1.00$

ϕ'	Slope 2:1		Slope 3:1		Slope 4:1		Slope 5:1	
	m	n	m	n	m	n	m	n
20	1.610	1.100	2.141	1.443	2.664	1.801	3.173	2.130
25	1.872	1.386	2.502	1.815	3.126	2.259	3.742	2.715
30	2.142	1.686	2.884	2.201	3.623	2.758	4.357	3.331
35	2.443	2.030	3.306	2.659	4.177	3.331	5.024	4.001
40	2.772	2.386	3.775	3.145	4.785	3.945	5.776	4.759

Table B9 Stability coefficients m and n for $\dfrac{c'}{\gamma H} = 0.075$ and $n_d = 1.25$

ϕ'	Slope 2:1		Slope 3:1		Slope 4:1		Slope 5:1	
	m	n	m	n	m	n	m	n
20	1.688	1.285	2.071	1.543	2.492	1.815	2.954	2.173
25	2.004	1.641	2.469	1.975	2.792	2.315	3.523	2.730
30	2.352	2.015	2.888	2.385	3.499	2.857	4.149	3.357
35	2.782	2.385	3.357	2.870	4.079	3.457	4.831	4.043
40	3.154	2.841	3.889	3.428	4.729	4.128	5.063	4.830

Table B10 Stability coefficients m and n for $\dfrac{c'}{\gamma H} = 0.075$ and $n_d = 1.50$

ϕ'	Slope 2:1		Slope 3:1		Slope 4:1		Slope 5:1	
	m	n	m	n	m	n	m	n
20	1.918	1.514	2.199	1.728	2.548	1.985	2.931	2.272
25	2.308	1.914	2.660	2.200	3.083	2.530	3.552	2.915
30	2.735	2.355	3.158	2.714	3.659	3.128	4.218	3.585
35	3.211	2.854	3.708	3.285	4.302	3.786	4.961	4.343
40	3.742	3.397	4.332	3.926	5.026	4.527	5.788	5.185

Table B11 Stability coefficients m and n for $\frac{c'}{\gamma H} = 0.100$ and toe circle

ϕ'	Slope 2:1		Slope 3:1		Slope 4:1		Slope 5:1	
	m	n	m	n	m	n	m	n
20	1.804	1.201	2.286	1.588	2.748	1.974	3.190	2.361
25	2.076	1.488	2.665	1.945	3.246	2.459	3.796	2.959
30	2.362	1.786	3.076	2.359	3.770	2.961	4.442	3.576
35	2.673	2.130	3.518	2.803	4.339	3.518	5.146	4.249
40	3.012	2.486	4.008	3.303	4.984	4.173	5.923	5.019

Table B12 Stability coefficients m and n for $\frac{c'}{\gamma H} = 0.100$ and $n_d = 1.00$

ϕ'	Slope 2:1		Slope 3:1		Slope 4:1		Slope 5:1	
	m	n	m	n	m	n	m	n
20	1.841	1.143	2.421	1.472	2.982	1.815	3.549	2.157
25	2.102	1.430	2.785	1.845	3.458	2.303	4.131	2.743
30	2.378	1.714	3.183	2.258	3.973	2.830	4.751	3.372
35	2.692	2.086	3.612	2.715	4.516	3.359	5.426	4.059
40	3.025	2.445	4.103	3.230	5.144	4.001	6.187	4.831

Table B13 Stability coefficients m and n for $\frac{c'}{\gamma H} = 0.100$ and $n_d = 1.25$

ϕ'	Slope 2:1		Slope 3:1		Slope 4:1		Slope 5:1	
	m	n	m	n	m	n	m	n
20	1.874	1.301	2.283	1.558	2.751	1.843	3.253	2.158
25	2.197	1.642	2.681	1.972	3.233	2.330	3.833	2.758
30	2.540	2.000	3.112	2.415	3.753	2.858	4.451	3:372
35	2.922	2.415	3.588	2.914	4.333	3.458	5.141	4.072
40	3.345	2.855	4.119	3.457	4.987	4.142	5.921	4.872

Table B14 Stability coefficients m and n for $\frac{c'}{\gamma H} = 0.100$ and $n_d = 1.50$

ϕ'	Slope 2:1		Slope 3:1		Slope 4:1		Slope 5:1	
	m	n	m	n	m	n	m	n
20	2.079	1.528	2.387	1.742	2.768	2.014	3.158	2.285
25	2.477	1.942	2.852	2.215	3.297	2.542	3.796	2.927
30	2.908	2.385	3.349	2.728	3.881	3.143	4.468	3.614
35	3.385	2.884	3.900	3.300	4.520	3.800	5.211	4.372
40	3.924	3.441	4.524	3.941	5.247	4.542	6.040	5.200

REFERENCES

Abduljauwad, S.N., Al-Sulaimani, G.J., Basunbul, A., and Al-Buraim, I. (1998). "Laboratory and Field Studies of Response of Structures to Heave of Expansive Clay," *Geotechnique,* 48 No. 1.

Abouleid, A.F. (1982). "Measurement of Swelling and Collapsible Soil Properties," *Foundation Engineering,* Edited by George Pilot. Presses de le cole nationale des Ponts et chaussees, Paris, France.

Alizadeh, M., and Davisson, M.T. (1970). "Lateral Load Test on Piles-Arkansas River Project," *J. of S.M.F.D.*, SM5, Vol. 96.

Alpan, I. (1967). "The Empirical Evaluation of the coefficient K_o and K_{or} Soil and Foundation," *J. Soc. Soil Mech. Found. Eng.*

Altmeyer, W.T. (1955). "Discussion of Engineering properties of expansive clays," *Proc. ASCE*, Vol. 81, No. SM2.

Amar, S., Jézéquel, J.F. (1972). "Essais en place et en laboratoire sus sols coberents Comparaisons des resultats," *Bulletin de Liaison des Laboratoires des Ponts et Chaussees*, No. 58.

American Association of State Highway and Transportation Officials (1982). AASHTO Materials, Part I, Specifications, Washington, D.C.

American Society for Testing and Materials (1994). Annual Book of ASTM Standards, Vol. 04.08, Philadelphia, Pa.

American Society of Civil Engineers, New York (1972). "Performance of Earth and Earth-supported Structures," Vol. 1, Part 2, *Proc. of the specialty Conf.* Purdue University, Lafayette, Indiana.

Anderson, P. (1956). *Substructure Analysis and Design,* Second Ed., The Rotand Press Co., New York.

Applied Technology Council, (1978). "Tentative Provisions for the Developing of Seismic Regulations for Buildings," Publication ATC-3-06.

Atterberg, A. (1911). "Uber die Physikalisehe Bodenuntersuchung Und Uber die Plastizitat der Tone," *Int. Mitt. Für Bodenkunde*, Vol. 1, Berlin.

Awad, A. and Petrasovits, G. (1968). "Consideration on the Bearing Capacity of Vertical and Batter Piles Subjected to Forces Acting in Different Directions," *Proc. 3rd Budapest Conf. SM and FE*, Budapest.

Azzouz, A.S., Kriezek, R.J., and Corotis, R.B. (1976). "Regression Analysis of soil compressibility," *Soils Found.*, Tokyo, Vol. 16, No. 2.

Baguelin, F., Jezequel, J.F., and Shields, D.H. (1978) *"The Pressuremeter and Foundation Engineering,* Trans Tech. Publications, Clausthal, Germany.

Balaam, N.P., Poulos, H.G., and Booker, J.R. (1975). "Finite Element Analysis of the Effects of Installation on Pile Load-settlement Behaviour," *Geot. Eng.*, Vol. 6, No. 1.

Barden, L., McGown, A., and Collins, K. (1973). "The Collapse Mechanism in Partly Saturated Soil," *Engineering Geology*, Vol. 7.

Baver, L.D. (1940). *Soil Physics*, John Wiley and Sons, New York.

Bazaraa, A.R., (1967). "Use of Standard Penetration Test for Estimating Settlements of Shallow Foundations on Sand," Ph.D. Thesis, Univ of Illinois, U.S.A.

Bell, A.L. (1915). "The Lateral Pressure and Resistance of Clay and Supporting Power of Clay Foundations," *Century of Soil Mechanics*, ICF, London.

Berezantsev, V.G., Khristoforov, V., and Golubkov, V. (1961). Load Bearing Capacity and Deformation of Piled Foundations, *Proc. 5th Int. Conf. SM and FE*, Vol. 2.

Berg, R.R., Bonaparte, R., Anderson, R.P., and Chouery, V.E. (1986). "Design Construction and Performance of Two Tensar Geogrid Reinforced Walls" *Proc. 3rd Intl. Conf. Geotextiles*, Vienna: Austrian Society of Engineers.

Bergado, D.T., Anderson, L.R., Miura, N., Balasubramaniam, A.S. (1996). *Soft Ground Improvement*, Published by ASCE Press, ASCE, New York.

Biot, M.A., and Clingan, F.M. (1941). "General Theory of Three-Dimensional Consolidation," *J. Applied Physics*, Vol. 12.

Bishop, A.W. (1954). "The use of Pore Pressure coefficients in Practice," *Geotechnique*, Vol. 4, No. 4, London.

Bishop, A.W. (1955). "The Use of Slip Circle in the Stability Analysis of Earth Slopes," *Geotechnique*, Vol. 5, No. I.

Bishop, A.W, (1960–61). "The Measurement of Pore Pressure in The Triaxial Tests," *Proc. Conf. Pore Pressure and Suction in Soils*, Butterworths, London.

Bishop, A.W., and Bjerrum, L. (1960). "The Relevance of the Triaxial Test to the Solution of Stability Problems," *Proc. Res. Conf. Shear Strength of Soils*, ASCE.

Bishop, A.W., and Henkel, D.J. (1962).*The Measurement of Soil Properties in the Triaxial Test*, Second Ed., Edward Arnold, London.

Bishop, A.W., and Morgenstern, N. (1960). "Stability Coefficients for Earth Slopes," *Geotechnique*, Vol. 10, No. 4, London.

Bjerrum, L. (1972). "Embankments on Soft Ground," *ASCE Conf. on Performance of Earth and Earth-supported Structures*, Purdue University.

Bjerrum, L. (1973). "Problems of soil mechanics and construction on soft days," *Proc. 8th Int. Conf. on Soil Mech. and Found. Eng.*, Moscow, 3.

Bjerrum, L., and Eide, O. (1956). "Stability of Strutted Excavations in Clay," *Geotechnique*, Vol. 6, No. 1.

Bjerrum, L., Casagrande, A., Peck, R.B., and Skempton, A.W., eds. (1960). *From Theory to Practice in Soil Mechanics*, Selections from the writings of Karl Terzaghi, John Wiley and Sons.

Bjerrum, L., and Simons, N.E. (1960). "Comparison of Shear Strength Characteristics of Normally Consolidated Clay," *Proc. Res. Conf. Shear Strength Cohesive Soils*, ASCE.

Boussinesq, J.V. (1885). *Application des Potentiels á L'Etude de L'Equilibre et due mouvement des solides Elastiques*, Gauthier-Villard, Paris.

Bowles, J.E. (1992). *Engineering Properties of Soils and their Measurement*, McGraw Hill, New York.

Bowles, J.E. (1996). *Foundation Analysis and Design*, McGraw-Hill, New York.

Briaud, J.L., Smith, T.D., and Tucker, L.M. (1985). "A Pressuremeter Method for Laterally Loaded Piles," *Int. Conf. of SM and FE*, San Francisco.

Broms, B.B. (1964a). "Lateral Resistance of Piles in Cohesive Soils," *JSMFD*, ASCE, Vol. 90, SM 2.

Broms, B.B. (1964b). "Lateral Resistance of Piles in Cohesionless Soils," *JSMFD*, ASCE, Vol. 90, SM 3.

Broms, B.B. (1966). "Methods of Calculating the Ultimate Bearing Capacity of Piles—A Summary," *Soils-Soils*, No. 18-19: 21-32.

Broms, B.B. (1987). "Soil Improvement Methods in Southeast Asia for Soft Soil," *Proc. 8ARC on SM and FE*, Kyoto, Vol. 2.

Broms, B.B. (1995). "Fabric Reinforced Soil," *Developments in Deep Foundations and Ground Improvement Schemes,* Balasubramanium, et al., (eds), Balkema Rotterdam, ISBN 905410 5933.

Brown, J.D., and Meyerhof, G.G. (1969). "Experimental study of Bearing Capacity in Layered Clays," *7th ICSFME*, Vol. 2.

Brown, R.E. (1977). "Drill Rod Influence on Standard Penetration Test," *JGED*, ASCE, Vol. 103, SM 3.

Brown, R.E. (1977). "Vibroflotation Compaction of Cohesioinless Soils," *J. of Geotech. Engg. Div.*, ASCE, Vol. 103, No. GT12.

Burland, J.B. (1973). "Shaft Friction of Piles in Clay–A Simple Fundamental Approach," *Ground Engineering*, Vol.6, No. 3.

Burland, J.B., and Burbidge, M.C. (1985). "Settlement of Foundations on Sand and Gravel," *Proc. ICE*, Vol. 78.

Burland, J.B., and Wroth, C.P. (1974). "Allowable and Differential Settlement of Structures Including Damage and Soil structure Interaction," *Proceedings*, Conf. on Settlement of Structures, Cambridge University, England.

Burmister, D.M. (1943). "The Theory of Stresses and Displacements in Layer Systems and Application to Design of Airport Runways," *Proc. Highway Res. Board*, Vol. 23.

Button, S.J. (1953). "The Bearing Capacity of Footings on a two-Layer Cohesive Sub-soil," *3rd ICEMFE*, Vol. 1.

Capazzoli, L. (1968). "Test Pile Program at St. Gabriel, Louisiana, Louis," *J. Capazzoli and Associates.*

Caquot, A., and Kerisel, J. (1948). *Tables for the Calculation of Passive Pressure, Active Pressure, and Bearing Capacity of Foundations* (Translated by M.A. Bec, London), Gauthier-Villars Paris.

Caquot, A., and Kerisel, J. (1956). *Traité de méchanique des sols,* 2nd ed., Gauthier Villars, Paris.

Caron, P.C., and Thomas, F.B. (1975). "Injection," *Foundation Engineering Hand book*, Edited by Winterkorn and Fang, Van Nastrand Reinhold Company, New York.

Carrillo, N. (1942). "Simple Two and Three-Dimensional Cases in the Theory of Consolidation of Soils," *J. Math. Phys.* Vol. 21.

Carter, J.P., and Kulhawy, F.H. (1988). "Analysis and Design of Drilled Shaft Foundations Socketed into Rock," Final Report Project 1493–4, EPRI EL–5918, Geotechnical Group, Cornell University, Ithaca.

Casagrande, A. (1931). *The Hydrometer Method for Mechanical Analysis of Soils and other Granular Materials*, Cambridge, Massachusetts.

Casagrande, A. (1932). "Research of Atterberg Limits of Soils," *Public Roads*, Vol. 13, No. 8.

Casagrande, A. (1936). "The determination of the Preconsolidation Load and its Practical Significance," *Proc. First Int. Conf. Soil Mech. Found Eng*, Vol. 3.

Casagrande, A. (1937). "Seepage through dams," *J. N. Engl.* Water Works Assoc. L1(2).

Casagrande, A. (1958–59). "Review of Past and Current Works on Electro-osmotic Stabilisation of Soils," *Harvard Soil Mech.,* Series No. 45. Harvard Univ., Cambridge, Mass.

Casagrande, A. (1967). "Classification and Identification of Soils," *Proc. ASCE*, Vol. 73, No. 6, Part 1, New York.

Casagrande, A., and Fadum, R.E. (1940). "Notes on Soil Testing for Engineering Purposes," *Soil Mechanics Series,* Graduate School of Engineering, Harvard University, Cambridge. M.A., Vol. 8.

Casagrande, L. (1932) "Naeherungsmethoden zur Bestimmurg von Art'und Menge der sickerung durch geschuettete Daemme," Thesis, Techniche Hochschule, Vienna.

Chellis, R.D. (1961). *Pile Foundations*, McGraw Hill, New York.

Chen, F.H. (1988). *Foundations on Expansive Soils*. Elsevier Science Publishing Company Inc., New York.

Chen, Y.J., and Kulhawy, F.H. (1994). "Case History Evaluation of the Behavior of Drilled Shaft Under Axial and Lateral Loading," Final Report, Project 1493-04, EPRI TR-104601, Geotechnical Group, Cornell University, Ithacka.

Christian, J.T., and Carrier, W.D. (1978). "Janbu, Bjerrum, and Kjaernsli's Chart Reinterpreted," *Canadian Geotechnical Journal*, Vol. 15.

Clemence, S.P., and Finbarr, A.O. (1981). "Design Considerations for Collapsible Soils," *J. of the Geotech. Eng. Div.*, ASCE, Vol. 107, No. GT3.

Coulomb, C.A. (1776). "Essai Sur Une Application des regles des Maximis et Minimum a Queiques *Problemes de Statique Relatifs a L'Architecture*," Mem. Acad. Roy Pres. Divers Savants, Paris, Vol. 3, Paris.

Coyle, H.M., and Reese, L.C. (1966). "Load Transfer for Axially Loaded Piles in Clay," *J. of the S.M. and F. Div.*, ASCE, Vol. 92, SM2.

Coyle, H.M., and Sulaiman, I.H. (1967). "Skin Friction for Steel Piles in Sand," *JSMFD*, ASCE, Vol. 93, SM 6.

Coyle, H.M., and Castello, R.R. (1981). "New Design Correlations for Piles in Sand," *J. of the Geotech. Eng. Div.*, ASCE, Vol. 107, No. GT17.

Culmann, C. (1875). *Die Graphische Statik*, Zurich, Meyer and Zeller.

D'Appolonia, E., Ellison, R.D., and D'Appolonia, D.J. (1975). "Drilled Piers," *Foundation Engineering Handbook*, Edited: H.F. Winterkorn and H.Y. Fang, Van Nostrand Reinhold Company, NewYork.

Darcy, H. (1856). *Les Fontaines Publiques de la Ville de Dijon*, Paris.

Das, B.M., and Seeley, G.R. (1975). "Load-Displacement Relationships for Vertical Anchor Plates," *J. of the Geotech. Engg. Div.*, ASCE, Vol. 101, GT 7.

Davis, L.H. (1977). "Tubular Steel Foundation," Test Report RD-1517, Florida Power and Light company, Miami, Florida.

Davis, M.C.R., and Schlosser, F. (1997). "Ground Improvement Geosystems," *Proc. Third Int. Conf. on Ground Improvement Geosystems*, London.

Davisson, M.T. (1960). "Behavior of Flexible Vertical Piles Subjected to Moments, Shear and Axial Load," Ph.D. Thesis, Univ. of Illinois, Urbana, U.S.A.

Davisson, M.T., and Salley, J.R. (1969). "Lateral Load Tests on Drilled Piers," Performance of Deep Foundations, ASTM STP 444, ASTM

De Beer, E.E. (1965). "Bearing Capacity and Settlement of Shallow Foundations on Sand," *Proc.* of Symp held at Duke University.

De Ruiter, J., and Beringen, F.L. (1979). "Pile Foundations for Large North Sea Structures," *Marine Geotechnology*, Vol. 3, No. 3.

Dennis, N.D., and Olson, R.E. (1983a). "Axial capacity of steel Pipe Piles in sand," *Proc. ASCE Conf. on Geotech. Practice in Offshore Eng*, Austin, Texas.

Dennis, N.D., and Olson, R.E., (1983b). "Axial Capacity of Steel Pipe Piles in Clay," *Proc. ASCE Conf. on Geotech. Practice in Offshore Eng.*, Austin, Texas.

Department of the Navy, Naval Facilities Engineering Command. (1982). *Design Manual*, NAVFAC DM-71.

Desai, C.S. (1974). "Numerical Design-Analysis for Piles in Sands," *J. Geotech. Eng. Div.*, ASCE, Vol. 100, No. GT6.

Douglas, B.J. (1984). *The Electric Cone Penetrometer Test: A User's Guide to Contracting for Services, Quality Assurance, Data Analysis*, The Earth Technology Corporation, Long Beach, California, U.S.A.

Douglas, B.J., and Olsen, R.S. (1981). "Soil Classification Using the Electric Cone Penetrometer, Cone Penetration Testing and Experience," *ASCE Fall Convention.*

Downs, D.I., and Chieurzzi, R. (1966). "Transmission Tower Foundations," *J. Power Div.*, ASCE, Vol. 92, PO2.

Dudley, J.H. (1970). "Review of Collapsible Soils," *J. of S.M. and F. Div.*, ASCE, Vol. 96, No. SM3.

Duncan, J. M., Evans, L. T., Jr., and Ooi P. S. K. (1994). "Lateral Load Analysis of Single Piles and Drilled Shafts," *J. of Geotech. Engg*. ASCE, Vol 120, No. 6.

Dunn, I.S., Anderson, I.R., and Kiefer, F.W. (1980). *Fundamentals of Geotechnical Analysis*, John Wiley and Sons.

Dupuit, J. (1863). *Etudes Theoriques et practiques sur le Mouvement des eaux dans les Canaux Decouverts et a travers les Terrains Permeables*, Dunod, Paris.

Eggestad, A. (1963). "Deformation Measurement Below a Model Footing on the Surface of Dry Sand," *Proc. European Conf. Soil Mech.*, Wiesbaden, 1.

Ellison, R.D., D'Appolonia, E., and Thiers, G.R. (1971). "Load-Deformation Mechanism for Bored Piles," *JSMFD*, ASCE, Vol. 97, SM4.

Fadum, R.E. (1948). "Influence Values for Estimating Stresses in Elastic Foundations," *Proc. Second Int. Conf. SM and FE*, Vol. 3.

Fellenius, W. (1927). *"Erdstatische Berechnungen*," (revised edition, 1939), W. Ernstu, Sons, Berlin.

Fellenius, W. (1947). *Erdstatische Berechnungen mit Reibung and Kohasion and unter Annahme keissy Undrischer Gleitf lachen*, Wilhelm Ernt and Sohn, 11 Ed., Berlin.

Forchheimer, P. (1930). *General Hydraulics*, Third Ed., Leipzig.

Foster, C.R., and Ashlvin, R.G. (1954). "Stresses and deflections induced by a uniform circular load," *Proceedings*, HRB.

Fox, E.N. (1948). "The mean Elastic Settlements of Uniformly Loaded Area at Depth Below the Ground Surface," *2nd ICSMEF*, Vol. 1.

Frohlick, O.K. (1955). "General Theory of Stability of Slopes," *Geotechnique*, Vol. 5.

Ghaly, A.M. (1997). "Load-Displacement Prediction for Horizontally Loaded Vertical Plates," *J. of Geotechnical and Environmental Engineering*, ASCE, Vol. 123, No. 1.

Gibbs, H.J., and Holtz, W.G. (1957). "Research on Determining The Density of Sands by Spoon Penetration Testing," *4th ICSMEF*, Vol. 1.

Gibson, R.E. (1953). "Experimental Determination of the True Cohesion and True angle of Internal Friction in Clays," *Proc. Third Int. Conf. Soil Mech. Found Eng.*, Zurich, Vol. 1.

Gibson, R.E., and Lo, K.Y. (1961). *A Theory of Consolidation for Soils Exhibiting Secondary Compression*, Norwegian Geotechnical Institute, Publication No. 41.

Gilboy, G. (1934). "Mechancis of Hydraulic Fill Dams," *Journal Boston Society of Civil Engineers*, Boston.

Glesser, S.M. (1953). *Lateral Load Tests on Vertical Fixed-Head and Free-Head Pile*, ASTM, STP 154.

Goodman, R.E. (1980). *Introduction to Rock Mechanics*, Wiley, New York.

Grim, R.E. (1959). "Physico-Chemical Properties of Soils," *ASCE J. Soil Mech.*, 85, SM2, 1–17.

Grim, R.E. (1962). *Applied Clay Mineralogy*, McGraw Hill.

Hagerty, D.J., and Nofal, M.M. (1992). "Design Aids: Anchored Bulkheads in Sand," *Canadian Geotech. Journal*, Vol. 29, No. 5.

Hansen, J.B. (1970). "A Revised and Extended Formula for Bearing Capacity," *Danish Geotechnical Institute Bul.* No. 28, Copenhagen.

Hanzawa, H., and Kishida, T. (1982). "Determination of *in situ* undrained strength of soft clay deposits," *Soils and Foundations*, 22, No. 2.

Harr, M.E. (1962). *Ground Water and Seepage*, McGraw Hill, New York.

Harr, M.E. (1966). *Foundations of Theoretical Soil Mechanics*, McGraw Hill, New York.

Hausmann, M.R. (1990). *Engineering Principles of Ground Modification,* McGraw Hill, New York.

Hazen, A. (1893). "Some Physical Properties of Sands and gravels with special reference to their use in filtration," 24th Annual Report, Massachusetts, State Board of Health.

Hazen, A. (1911). "Discussion of Dams and Sand Foundations," by A.C. Koenig, Trans ASCE, Vol. 73.

Hazen, A. (1930). *Hydraulics,* Third Ed., Leipzig,

Henkel, D.J. (1960). "The Shear Strength of Saturated Remoulded Clays," *Proc. Re. Conf. Shear Strength Cohesive Soils*, ASCE.

Hetenyi, M. (1946). *Beams on Elastic Foundations*, The Univ. of Michigan Press, Ann Arbor, MI.

Highter, W.H., and Anders, J.C. (1985). "Dimensioning Footings Subjected to Eccentric Loads," *J. of Geotech. Eng.*, ASCE, Vol. III, No. GT5.

Holl, D.L. (1940). "Stress Transmission on Earths," *Proc. Highway Res. Board*, Vol. 20.

Holl, D.L. (1941). "Plane Strain Distribution of Stress in Elastic Media," Iowa Engg. Expts Station, Iowa.

Holtz, R.D., and Kovacs, W.D. (1981). *An Introduction to Geotechnical Engineering*, Prentice Hall Inc., New Jersey.

Holtz, W.G. and Gibbs, H.J. (1956). "Engineering Properties of Expansive Clays." Trans ASCE. 121.

Holtz, W. G. and Hilf, J. W. (1961). "Settlement of Soil Foundations Due to Saturation," *Proc. 5th Internation on SMFE*, Paris, Vol 1.

Hough, B.K. (1957). *Basic Soil Engineering*, Ronald Press, New York.

Housel, W.S. (1929). "A Practical Method for the Selection of Foundations Based on Fundamental Research in Soil Mechanics," *Research Bulletin,* No. 13, University of Michigan, AnnArbor.

Houston, W.N., and Houston, S.L. (1989). "State-of-the-Practice Mitigation Measures for Collapsible Soil Sites," *Proceedings, Foundation Engineering*: *Current Principles and Practices*, ASCE, Vol. 1.

Hrennikoff, A. (1949). "Analysis of Pile Foundations with Batter Piles," *Proc. ASCE,* Vol. 75.

Huntington, W.C. (1957). *Earth Pressures and Retaining Walls*, John Wiley and Sons, New York.

Hvorslev, M.J. (1949). *Surface Exploration and Sampling of Soils for Civil Engineering Purposes. Water ways Experimental Station, Engineering Foundations,* New York.

Hvorslev, M.J. (1960). "Physical Components of The Shear Strength of Saturated Clays," *Proc. Res. Conf. on Shear Strength of Cohesive Soils*, ASCE, Boulder, Colorado.

Ireland, H.O. (1957). "Pulling Tests on Piles in Sand," *Proc. 4th Int. Conf. SM and FE*, Vol. 2.

Ismael, N.F., and Klym, T.W. (1977). "Behavior of Rigid Piers in Layered cohesive Soils". *J. Geotech. Eng. Div.*, ASCE. Vol. 104, No. GT8.

Jakosky, J.J. (1950). *Exploration Geophysics*, Trija Publishing Company, California. U.S.A.

Jaky, J. (1944). "The Coefficient of Earth Pressure at Rest," *J. Soc. Hungarian Arch. Eng.*

Janbu, N. (1976). "Static Bearing Capacity of Friction Piles," *Proc. 6th European Conf. on Soil Mech. and F. Engg.*Vol. 1–2.

Janbu, N., Bjerrum, L., and Kjaernsli, B. (1956). "Veiledning Ved losning av fundamentering-soppgaver," Pub. No. 16, Norwegian Geotechnical Institute.

Jarquio, R. (1981). "Total Lateral Surcharge Pressure Due to Strip Load," *J. Geotech. Eng. Div.*, ASCE, 107, 10.

Jennings, J.E., and Knight, K. (1957). "An Additional Settlement of Foundation Due to a Collapse Structure of Sandy Subsoils on Wetting," *Proc. 4th Int. Conf on S.M. and F. Eng.*, Paris, Vol. 1.

Jennings, J.E., and Knight, K. (1975). "A Guide to Construction on or with Materials Exhibiting Additional Settlements Due to 'Collapse' of Grain Structure," *Proc. Sixth Regional Conference for Africa on S.M and F. Engineering*, Johannesburg.

Jumikis, A.R. (1962). *Soil Mechanics*, D. Van Nostrand Co. Inc., New York.

Kamon, M., and Bergado, D.T. (1991). "Ground Improvement Techniques," *Proc. 9 ARC on SMFE*, Bangkok, Vol. 2.

Karlsson, R., Viberg, L. (1967). "Ratio c_u/p in relation to liquid limit and plasticity index with special reference to Swedish clays," *Proc. Geotech. Conf*, Olso, Norway, Vol. 1.

Kenney, T.C., Lau, D., and Ofoegbu, G.I. (1984). "Permeability of Compacted Granular Materials," *Canadian Geotech. J.*, 21 No. 4.

Kerisel, J. (1961). "Fondations Profondes en Milieu Sableux," *Proc. 5th Int. Conf. SM and FE,* Vol. 2.

Kezdi, A. (1957). "The Bearing Capacity of Piles and Pile Groups," *Proc. 4th Int. Conf. SM and FE,* Vol. 2.

Kezdi, A. (1965). "Foundations Profondes en Milies Sableux," *Proc. 5th Int. Conf. SM and FE,* Vol. 2.

Kezdi, A. (1965). "General Report on Deep Foundations," *Proc. 6th Int. Conf. on SM and FE,* Montreal.

Kezdi, A. (1975). "Pile Foundations," *Foundations Engineering Handbook*, Edited by H.F. Winterkorn and H.Y. Fang, Van Nostrand, New York.

Kishida, H. (1967). "Ultimate Bearing Capacity of Piles Driven into Loose Sand," *Soil and Foundations,* Vol. 7, No. 3.

Koerner, R.M. (1985). *Construction and Geotechnical Methods in Foundation Engineering*, McGraw Hill, New York.

Koerner, R.M. (1999). *Design with Geosynthetics*. Fourth Edition, Prentice Hall, New Jersey.

Koerner, R.M. (2000). "Emerging and Future Developments of Selected Geosynthetic Applications" Thirty-Second Terzaghi Lecture, Drexel University, Philadelphia.

Kogler, F., and Scheidig, A. (1962). *Soil Mechanics* by A. R. Jumikis, D. Van Nostrand.

Komornik, A., and David, D. (1969). "Prediction of Swelling Pressure of Clays," *JSMFD*, ASCE, Vol. 95, SM1.

Kovacs, W.D., and Salomone (1982). "SPT Hammer Energy Measurement," *JGED*, ASCE, GT4.

Kozeny, J.S. (1931). "Ground wasserbewegung bei freim spiegel, Fluss and Kanalversicheringi," *Wasserkraft and Wasserwitrschaft*, No. 3.

Kubo, J. (1965). "Experimental Study of the Behaviour of Laterally Loaded Piles," *Proc. 6th Int.Conf. SM and FE*, Vol. 2.

Ladd, C.C., and Foott, R. (1974). "New Design Procedure for Stability of Soft Clays," *J. Geotech Eng. Div.*, ASCE, Vol. 100, No. GT7.

Ladd, C.C., Foott, R., Ishihara, K., Schlosser, F., and Poulos, H.G. (1977). "Stress Deformation and strength Characteristics," *Proceedings 9th Int.Conf. SM and FE*, Tokyo, Vol. 2.

Lambe, T.W. (1958a). "The Structure of Compacted Clay," *J. of S.M. & F. Div.*, ASCE, Vol. 84, No. SM2.

Lambe, T.W. (1958b). "The Engineering Behavior of Compacted Clay," *J. Soil Mech, Found, Div.*, ASCE, Vol. 84, No. SM2.

Lambe, T.W. (1962). "Soil Stabilization," Chapter 4 of *Foundation Engineering*, G.A. Leonards (ed.), McGraw Hill, New York.

Lambe, T.W., and Whitman, R.V. (1969). *Soil Mechanics*, John Wiley and Sons Inc.

Lee, K.L., and Singh, A. (1971). "Relative Density and Relative Compaction," *J. Soil Mech., Found, Div.,* ASCE, Vol. 97, SM 7.

Lee, K.L., Adams, B.D., and Vagneron, J.J. (1973). "Reinforced Earth Retaining Walls," *Jou. of SMFD.*, ASCE, Vol. 99, No. SM 10.

Leonards, G.A. (1962). *Foundation Engineering*, McGraw Hill.

Leonards, G.A., Cutter, W.A., and Holtz, R.D. (1980). "Dynamic Compaction of Granular Soils," *Jou. of Geotech. Eng. Div.*, ASCE, Vol. 106, No. GT1.

Liao, S.S., and Whitman, R.V. (1986). "Overburden Correction Factors for SPT in Sand," *JGED*, Vol. 112, No. 3.

Littlejohn, G.S., and Bruce, D.A. (1975). "Rock Anchors: State of the Art, Part I – Design," *Ground Engineering*, Vol. 8, No. 3.

Loos, W., and Breth, H. (1949). *Modellversuche uber Biege Beanspruch - ungen Von Pfahlen and Spunwenden*, Der Bauingeniur, Vol. 28.

Louden, A.G. (1952). "The Computation of Permeability from Soil Tests," *Geotech.*, 3, No. 4.

Lowe, J. (1974). "New Concepts in Consolidation and Settlement Analysis," *J. Geotech. Eng. Div.*, ASCE, Vol. 100.

Lunne, T., and Christoffersen, H.P., (1985). *Interpretation of Cone Penetrometer data for Offshore Sands,* Norwegian Geotechnical Institute, Pub. No. 156.

Lunne, T., and Kelven, A. (1981). "Role of CPT in North Sea Foundation Engineering," *Symposium on Cone Penetration Testing and Experience,* Geotech. Eng. Dev. ASCE, St. Louis.

Malter, H. (1958). "Numerical solutions for Beams on Elastic Foundations," *J. Soil Mech and Found Div.*, LXXXIV, No. ST2, Part 1, ASCE.

Marchetti, S. (1980). "*In situ* Test by Flat Dilatometer," *J. of Geotech. Eng. Div.*, ASCE, Vol. 106, No. GT3.

Marsland, A. (1971). "Large *in situ* tests to measure the properties of stiff fissure clay," *Proc. 1st Australian Conf. an Geomech*, Melbourne, 1.

Matlock, H. (1970). "Correlations for Design of Laterally Loaded Piles in Soft Clay," *Proc. 2nd Offshore Tech. Conf.,* Houston, Vol. 1.

Matlock, H., and Reese, L.C. (1960). "Generalized Solutions for Laterally Loaded Piles," *JSMFD,* ASCE, Vol. 86, N. SM5, Part 1.

Matlock, H., and Reese, L.C. (1961). "Foundation Analysis of Offshore Supported Structures," *Proc. 5th Int. Conf. SM and FE*, Vol. 2.

Matsuo, H. (1939). Tests on the Lateral Resistance of Piles (in Japanese), Research Inst. of Civ. Bugg, Min, of Home Affairs, Report No. 46.

McDonald, D.H,. and Skempton, A.W. (eds) (1955). "A Survey of Comparison between calculated and observed settlements of structures on clay," *Conf. on Correlation of Calculated and Observed Stresses and Displacements*, Institution of Civil Engineers, London.

Means, R.E., Parcher, J.V. (1965). *Physical Properties of Soils.* Prentice Hall of India, New Delhi.

Menard, L. (1957). An Apparatus for Measuring the Strength of Soils in Place, M.Sc. Thesis, Univ. Illinois, Urbana, U.S.A.

Menard, L. (1976). "Interpretation and Application of pressuremeter Test Results to Foundation Design, Soils" *Soils*, No. 26.

Mesri, G. (1973). "Coefficient of Secondary Compression," *J. Soil Mech. Found. Div.*, ASCE, Vol. 99, SMI.

Mesri, G., and Godlewski, P.M. (1977). "Time and stress-compressibility interrelationship," *J. Geotech. Eng.,* ASCE, 103, No. 5.

Mesri, G., and Choi, Y.K. (1985). "Settlement analysis of embankments on soft clays," *J. Geotech. Eng.,* ASCE, III, No. 4.

Mesri, G., and Choi, Y.K. (1985b). "The Uniqueness of the End-of-Primary (EOP) Void ratio-effective Stress relationship," *Proc. 11th Int. Conf. on Soil Mech. and Found Engg.*, San Francisco, No. 2.

Mesri, G., and Feng, T.W. (1986). "Stress-strain–strain rate relation for the compressibility of sensitive natural clays," Discussion, *Geotech.*, 36, No. 2.

Mesri, G., Feng, T.W., Ali, S., and Hayat, T.M. (1994). "Permeablity Characteristics of Soft Clays," *Proc. 13th Int. Conf. On Soil Mech and Found. Eng.*, New Delhi.

Meyerhof, G.G. (1951). "The Ultimate Bearing Capacity of Foundation," *Geotechnique*, Vol. 2, No. 4.

Meyerhof, G.G. (1953). "The Bearing Capacity of Foundations Under Eccentric and Inclined Loads," *3rd ICSMFE*, Vol. 1.

Meyerhof, G.G. (1956). "Penetration Tests and Bearing Capacity of Cohesionless Soils," *JSMFD*, ASCE, Vo. 82, SM1.

Meyerhof, G.G. (1957). "Discussions on Sand Density by Spoon Penetration," *4th ICSMFE*, Vol. 3.

Meyerhof, G.G. (1957). "The Ultimate Bearing Capacity of Foundations on Slopes," *4th ICSMFC*, Vol. 1, London.

Meyerhof, G.G. (1959). "Compaction of Sands and Bearing Capacity of Piles," *JSMFD*, ASCE, Vol. 85, SM6.

Meyerhof, G.G. (1963). *Some Recent Research on Bearing Capacity of Foundation,* CGJ Ottawa, Vol. 1.

Meyerhof, G.G. (1965). "Shallow Foundations," *JSMFD*, ASCE, Vol. 91, SM2.

Meyerhof, G.G. (1974). "Ultimate Bearing Capacity of Footings on Sand Layer Overlying Clay," *Canadian Geotech. J.*, Vol. II, No. 2.

Meyerhof, G.G. (1975). "Penetration Testing Outside Europe," General Report, *Proc. of the European Symp. on Penetration Testing*, Vol. 2, Stockholm.

Meyerhof, G.G. (1976). "Bearing Capacity and Settlement of Pile Foundations," *JGED*, ASCE, Vol. 102, GT 3.

Meyerhof, G.G., and Adams J.I. (1968). "The Ultimate Uplift Capacity of Foundations," *Canadian Geotech. J.,* Vol. 5, No. 4.

Meyerhof, G.G., and Hanna, A.M. (1978). "Ultimate Bearing Capacity of Foundations on Layered Soil Under Inclined Load," *Canadian Geotech. J.*, Vol. 15, No. 4.

Mindlin, R.D. (1936). "Forces at a point in the Interior of a Semi-Infinite Solid," *Physics*, Vol. 7.

Mindlin, R.D. (1939). "Stress distribution around a Tunnel," Trans ASCE. *Am. Soc. Civil Engrs.* Vol. 104.

Mitchell, J.K., and Freitag, D.R. (1959). "A Review and Evaluation of Soil-Cement Pavements," *J. of SM and FD*, ASCE, Vol. 85, No. SM6.

Mohr, O. (1900). Die Elastizitatsgrenze Und Bruseh eines Materials (The elastic limit and the failure of a Material), Zeitschrift Veneins Deuesche Ingenieure, Vol. 44.

Mononobe, N. (1929). "On the Determination of Earth Pressures During Earthquakes" *Proceedings World Engineering Conference*, Vol. 9.

Morgenstern, N.R. (1963). "Stability Charts for Earth slopes During Rapid Drawdown," *Geotechnique*, Vol. 13, No. 2.

Morrison, C.S., and Reese L.C. (1986). "A Lateral load test of a full scale pile group in sand." Geotechnical Engineering center, Bureau of Engineering Research, University of Texas, Austin.

Morrison, E.E. Jr., and Ebeling, R.M. (1995). "Limit Equilibrium Computation of Dynamic Passive Earth Pressure," *Canadian Geotech. J.*, Vol. 32, No. 3.

Murphy, V.A. (1960). "The effect of ground water characteristics on the sismic design of structures," *Proce. 2nd World Conf. on Earthquake Engineering*, Tokyo, Japan.

Murthy, V.N.S, and Subba Rao, K.S. (1995). "Prediction of Nonlinear Behaviour of Laterally Loaded Long Piles," *Foundation Engineer*, Vol. 1, No. 2, New Delhi.

Murthy, V.N.S. (1965). "Behaviour of Batter Piles Subjected to Lateral Loads," Ph.D. Thesis, Indian Institute of Technology, Kharagpur, India.

Murthy, V.N.S. (1982). "Report on Soil Investigation for the construction of cooling water system," Part II of Farakha Super thermal Power Project, National Thermal Power Corporation, New Delhi.

Muskat, M. (1946). *The Flow of homogeneous fluids through porous media*, McGraw Hill Co.

Nagaraj, T.S., and Murthy, B.R.S. (1985). "Prediction of the Preconsolidation Pressure and Recompression Index of Soils," *Geotech. Testing Journal*, Vol. 8. No. 4.

Nelson, J.D., Miller, D.J. (1992). *Expansive Soils: Problems and Practice in Foundation and Pavement Engineering*, John Wiley & Sons, Inc., New York.

Newmark, N.M. (1942). "Influence Charts for Computation of Stresses in *Elastic Soils*," Univ. of Illinois Expt. Stn., *Bulletin* No. 338.

Nordlund, R.L. (1963). "Bearing Capacity of Piles in Cohesionless Soils," *JSMFD*, ASCE, Vol. 89, SM 3.

Norris, G.M., and Holtz, R.D. (1981). "Cone penetration Testing and Experience," *Proc. ASCE National Convention,* St. Louis, Missouri, Published by ASCE, NY.

O' Connor, M.J., and Mitchell, R.J. (1977). "An Extension of the Bishop and Morgenstern Slope Stability Charts," *Canadian Geotech. J.* Vol. 14.

O' Neill, M.W., and Reese, L.C. (1999). "Drilled Shafts: Construction Procedures and Design Methods" Report No. FHWA-IF-99-025, Federal Highway Administration Office of Infrastructure/Office of Bridge Technology, HIBT, Washington D.C.

O'Neill, M.W., Townsend, F. C., Hassan, K. H., Buttler A., and Chan, P. S. (1996). "Load Transfer for Intermediate Geomaterials," Publication No. FHWA-RD-95-171, Federal Highway Administration, Office of Enggniering and Highway Operations R & D, McLean, V. A.

O'Neill, M.W. (1988). "Special Topics in Foundations," *Proc. Geotech. Eng. Div.*, ASCE National Convention, Nashville.

Okabe, S. (1926). "General Theory of Earth Pressures," *J. of the Jap. Soc. of Civil Engineers,* Tokyo, Vol. 12, No. 1.

Osterberg, J.O. (1952). "New Piston Type Soil Sampler," *Eng. News. Rec.*, 148.

Osterberg, J.O. (1957). "Influence Values for Vertical Stresses in Semi-Infinite Mass due to Embankment Loading," *Proc. Fourth Int. Conf. Conf. SM and FE*, Vol. 1.

Ostwald, W. (1919). *A Handbook of Colloid Chemistry*, Trans. by M.H. Fisher, Philadelphia, P. Balkiston's Son and Co.

Palmer, L.A., and Thompson, J.B. (1948). "The Earth Pressure and Deflection along Embedded Lengths of Piles Subjected to Lateral Thrusts," *Proc. 2nd Int. Conf. SM and FE*, Rottterdam, Vo. 5.

Pauker, H.E. (1889). "An explanatory report on the project of a sea-battery" (in Russian), *Journal of the Ministry of Ways and Communications*, St. Petersburg,

Peck, R.B. (1969). "Deep excavations and Tunneling in Soft Ground," *Proc., 7th Int. Conf. on SMFE*, Mexico city.

Peck, R.B., and Byrant, F.G. (1953). "The Bearing Capacity Failure of the Transcona Elevator," *Geotechnique*, Vol. 3, No. 5.

Peck, R.B., Hanson, W.E., and Thornburn, T.H., (1974). *Foundation Engineering,* John Wiley and Sons Inc., New York.

Perloff, W.H., and Baron, W. (1976). *Soil Mechanics, Principles and Applications,* John Wiley & Sons, New York.

Petterson, K.E. (1955). "The Early History of Circular Sliding Surfaces," *Geotechnique.* The Institution of Engineers, London, Vol. 5.

Poncelet, J.V. (1840). "Mémoire sur la stabilité des rerétments et de leurs fondation," Note additionelle sur les relations analytiques qui lient entre elles la poussée et la butee de la terre. Memorial de l'officier du genie, Paris, Vol. 13.

Poulos, H.G. (1974). *Elastic Solutions for Soil and Rock Mechanics,* John Wiley & Sons, New York.

Poulos, H.G., and Davis, E.H. (1980). *Pile Foundation Analysis and Design*, John Wiley & Sons, New York.

Prakash, S., and Saran, S. (1966). "Static and Dynamic Earth Pressure Behind Retaining Walls," *Proc. 3rd Symp. on Earthquake Engineering,* Roorke, India, Vol. 1.

Prandtl, L. (1921). "Uber die Eindringungsflstigkeit plastischer Baustoffe und die Festigkeit von Schneiden," Zeit. angew. Math., 1, No. 1.

Proctor, R.R. (1933). "Four Articles on the Design and Construction of Rolled Earth-Dams," *Eng. News Record*, Vol. 3.

Rankine, W.J.M. (1857). "On the Stability of Loose Earth Dams," Phil. Trans. Royal Soc., Vol. 147, London.

Reese, L.C. (1975). "Laterally Loaded Piles," *Proc. of the Seminar Series*, Design, Construction and Performance of Deep Foundations: Geotech. Group and Continuing Education Committee, San Francisco Section, ASCE, Berkeley.

Reese, L.C. (1985). "Behavior of Piles and Pile Groups under Lateral Load," Technical Report No. FHWA/RD-85/106., Federal Highway Administration. Office of Engineering and Highway operations, Research and Development, Washington D.C.

Reese, L.C., and Matlock, H. (1956). "Non-dimensional solutions for Laterally Loaded Piles with Soil Modulus assumed Proportional to Depth," *Proc. 8th Texas Conf. S.M. and F.E.* Spec. Pub 29, Bureau of Eng. Res., Univ. of Texas, Austin.

Reese, L.C., Cox, W.R., and Koop, F.D. (1974). "Analysis of Laterally Loaded Piles in Sand," *Proc. 6th Offshore Tech. Conf.* Houston, Texas.

Reese, L.C., and Welch, R.C. (1975). "Lateral Loadings of Deep Foundations, in Stiff Clay," *JGED*, ASCE, Vol. 101, GT 7.

Reese, L.C., and O'Neill, M.W. (1988). "Field Load Tests of Drilled Shafts," *Proc., International Seminar on Deep Foundations and Auger Piles*, Van Impe (ed.), Balkema, Rotterdam.

Rendulic, L. (1935). "Der Hydrodynamische spannungsaugleich in zentral entwasserten Tonzylindern," Wasserwirtsch. u. Technik, Vol. 2.

Reynolds, O. (1883). "An Experimental investigation of the circumstances which determine whether the motion of water shall be Direct or Sinuous and the Law of Resistance in Parallel Channel," Trans. Royal Soc. Vol. 174, London.

Richards, R., and Elms, D.G. (1979). "Seismic Behavior of Gravity Retaining Walls," *Jou. of the Geotech. Eng. Div.*, ASCE, Vol. 105, No. GT4.

Robertson, P.K., and Campanella, R.G. (1983a). "Interpretation of Cone Penetration Tests," Part I–Sand, *CBJ*, Ottawa, Vol. 20, No. 4.

Robertson, P.K., and Companella, R.G. (1983b). "SPT-CPT Correlations," *JGED,* ASCE, Vol. 109.

Rowe, P.W. (1952) "Anchored Sheet Pile Walls," *Proc. Inst. of Civ. Engrs.*, Vol. 1, Part 1.

Sanglerat, G. (1972). *The Penetrometer and Soil Exploration*, Elsevier Publishing Co., Amsterdam.

Saran, S., and Prakash, S. (1968). "Dimensionless parameters for Static and Dynamic Earth Pressure for Retaining Walls," Indian Geotechnical Journal, Vol. 7, No. 3.

Scheidig, A. (1934). Der Loss (Loess), Dresden.

Schmertmann, J.H. (1955). "The Undisturbed Consolidation Behavior of Clay," Trans. ASCE, No. 120.

Schmertmann, J.H. (1970). "Static Cone to Compute Static Settlement Over Sand," *JSMFD*, ASCE, Vol. 96, SM 3.

Schmertmann, J.H. (1978). *Guidelines for Cone Penetration Test: Performance and Design.* U.S. Dept. of Transportation, Washington, D.C.

Schmertmann, J.H. (1986). "Suggested Method for Performing the Flat Dilatometer Test," *Geotech. Testing Journal*, ASTM, Vol. 9, No. 2.

Schmertmann, J.M. (1975). "Measurement of *in situ* shear strength," *Proc. ASCE Specially Conf. on In Situ Measurement of Soil Properties*, Raleigh, 2.

Scott, R.F. (1963). *Principles of Soil Mechanics*, Addison-Wesely Publishing Co., Inc. London, .

Seed, H.B., Woodward, R.J., and Lundgren, R. (1962). "Prediction of Swelling Potential for Compacted Clays," *Journal ASCE*, SMFD, Vol. 88, No. SM 3.

Seed, H.B., and Whitmann, R.V. (1970). "Design of Earth Retaining Structures for Dynamic Loads," ASCE, Spec. Conf. Lateral stresses in the ground and design of earth retaining structures.

Semkin, V.V., Ermoshin, V.M., and Okishev, N.D. (1986). "Chemical Stabilization of Loess Soils in Uzbekistan," Soil Mechanics and Foundation Engineering (trans. from Russian), Vol. 23, No. 5.

Sichardt, W. (1930). "Grundwasserabsenkung bei Fundierarbeiten," Berlin, Julius Springer.

Simons, N.E. (1960). "Comprehensive Investigation of the Shear Strength of an Undisturbed Drammen Clay," *Proc. Res. Conf. Shear Strength Cohesive Soils*, ASCE.

Simons, N.E., and Som, N.N. (1970). Settlement of Structures on Clay with Particular emphasis on London Clay Industry Research Institute, Assoc. Report 22.

Skempton, A.W. (1944). "Notes on the compressibilities of clays," *Q.J. Geological. Soc*, London, C (C: Parts 1 & 2).

Skempton, A.W. (1951), "The Bearing Capacity of Clays," *Proc. Building Research Congress,* Vol. 1, London.

Skempton, A.W. (1953). "The Colloidal Activity of Clays," *Proceedings, 3rd Int. Conf. SM and FE,* London, Vol. 1.

Skempton, A.W. (1954), "The Pore Pressure Coefficients A and B," *Geotechnique*, Vol. 4.

Skempton, A.W. (1957), "The planning and design of the new Hongkong airport Discussion," *Proc. Int. of Civil Engineers* (London), Vol. 7.

Skempton, A.W. (1959). "*Cast-in-situ* Bored Piles in London Clay," *Geotech*, Vol. 9.

Skempton, A.W., and Northey, R.D. (1952). "The Sensitivity of Clays," *Geotechnique*, Vol. III.

Skempton, A.W., and Northey, R.D. (1954)., "Sensitivity of Clays," *Geotechnique*, Vol. 3, London.

Skempton, A.W., and Bjerrum, L. (1957). "A Contribution to the Settlement Analysis of Foundations on Clay," *Geotechnique* 7, London.

Skempton, A.W., Yassin, A.A., and Gibson, R.E. (1953). "Theorie de la force Portaute des pieuse dans le sable," Annales de l," Institut Technique, Batiment 6.

Smith, R.E., and Wahls, H.E. (1969). "Consolidation under Constant Rate of Strain" *J. Soil Mech. Found. Div.,* ASCE, Vol. 95, SM2.

Sowa, V.A. (1970). "Pulling Capacity of Concrete *cast-in-situ* Bored Pile," *Can. Geotech. J.*, Vol. 7.

Spangler, M.G. (1938). "Horizontal Pressures on Retaining Walls due to Concentrated Surface Loads," IOWA State University Engineering Experiment Station, *Bulletin*, No. 140.

Spencer, E. (1967). "A Method of Analysis of the Stability of Embankments, Assuming Parallel Inter-slice Forces," *Geotechnique*, Vol. 17, No. 1.

Stas, C.V., and Kulhawy, F.H. (1984). "Critical Evaluation of Design Methods for Foundations under Axial Uplift and Compression Loading," EPRI Report EL-3771, Palo Alto, California.

Steinbrenner, W. (1934). "Tafeln zur Setzungsberechnung," Vol. 1., No. 4, Schriftenreihe der strasse 1, Strasse.

Stokes, G.G. (1856). "On the effect of the Internal Friction of Fludis on the Motion of Pendulium" Trans. Cambridge Philosophical Society, Vol. 9, Part 2.

Szechy, C. (1961). *Foundation Failures,* Concrete Publications Limited, London.

Tavenas, F. and Leroueil, S. (1987). "Laboratory and Stress-strain-time behaviour of soft clays," *Proc. Int. Symp on Geotech. Eng. of soft soils*, Mexico City, 2.

Taylor, D.W. (1937). "Stability of Earth Slopes," *J. of Boston Soc. of Civil Engineers*, Vol. 24.

Taylor, D.W. (1948). *Fundamentals of Soil Mechanics*, John Wiley and Sons, New York.

Teng, W.C. (1969). *Foundation Design*, Prentice Hall, Englewood Cliffs, N.J.

Terzaghi, K. (1925). "Erdbaumechanik." Franz, Deuticke, Vienna.

Terzaghi, K. (1943). *Theoretical Soil Mechanics*, Wiley & Sons, New York.

Terzaghi, K. (1951). "The Influence of Modern Soil Studies on the Design and Construction of Foundations" Bldg. Research Congr., London, 1951, Div. 1, Part III.

Terzaghi, K. (1955). "Evaluation of the Co-efficient of Sub-grade Reaction," *Geotechnique*, Institute of Engineers, Vol. 5, No. 4, London.

Terzaghi, K., and Frohlich, O.K. (1936). Theorie der setzung von Tonschichten (Theory of Settlement of Clay Layers), Leipzig, Deuticke.

Terzaghi, K., and Peck, R.B. (1948). *Soil Mechanics in Engineering Practice*, Wiley, New York.

Terzaghi, K., and Peck, R.B. (1967). *Soil Mechanics in Engineering Practice,* John Wiley & Sons, N.Y.

Terzaghi, K., Peck, R.B., and Mesri, G. (1996). *Soil Mechanics in Engineering Practice*, John Wiley & Sons, Inc., Third Edition, New York.

Timoshenko, S.P., and Goodier, J.N. (1970). *Theory of Elasticity,* Third Ed., McGraw Hill, New York.

Tomlinson, M.J. (1977). *Pile Design and Construction Practice*, Viewpoint Publications, London.

Tomlinson, M.J. (1986). *Foundation Design and Construction*, 5th Ed., New York, John Wiley and Sons.

Tschebotarioff, G.P. (1953). "The Resistance of Lateral Loading of Single Piles and of Pile Group." ASTM special Publication No. 154, Vol. 38.

Tschebotarioff, G.P. (1958). *Soil Mechanics Foundation and Earth Structures*, McGraw Hill, New York.

Tsytovich, N. (1986). *Soil Mechanics,* Mir Publishers, Moscow.

Turnbull, W.J. (1950). "Compaction and Strength Tests on Soil," Presented at Annual Meeting, ASCE (January).

U.S. Army Engineer Water Ways Experiments Station. (1957). The Unified Soil Classification system Technical memo No.3-357, Corps of Engineers, Vicksburg, U.S.A.

U.S. Corps of Engineers. (1953). Technical Memo 3-360, Filter Experiments and Design Criteria, U.S Waterways Experimental Station, Vicksburg, Miss.

U.S. Department of the Navy. (1971). "Design Manual – Soil Machanics, Foundations, and Earth Structures," NAVF AC DM-7, Washington, D.C.

Van Der Merwe, D.H. (1964). "The Prediction of Heave from the Plasticity Index and Percentage clay Fraction of Soils," *Civil Engineer in South Africa,* Vol. 6, No. 6.

Van Wheele, A.F. (1957). "A Method of separating the Bearing Capacity of a Test Pile into Skin Friction and Point Resistance," *Proc. 4th Int. Conf. SM and FE*, Vol. 2.

Vander Veen, C., and Boersma, L. (1957). "The Bearing Capacity of a Pile Predetermined by a Cone Penetration Test," *Proc. 4th Int. Conf S.M. and F.E.*, Vol 2.

Verbrugge, J.C. (1981). "Evaluation du Tassement des Pieure a Partir de l' Essai de Penetration Statique," *Revue Francaise de Geotechnique*, No. 15.

Vesic, A.S. (1956). "Contribution a L'Extude des Foundations Sur Pieux Verticaux et Inclines," Annales des Travaux Publics de Belgique, No.6.

Vesic, A.S. (1961). "Bending of Beam Resting on Isotropic Elastic Solid," *Jou. Eng. Mechs. Div.*, ASCE, Vol. 87, No. EM2.

Vesic, A.S. (1963). "Bearing Capacity of Deep Foundations in Sand," Highway Research Record, No. 39, Highway Research Board, Washington D.C.

Vesic, A.S. (1963). "Discussion-Session III," *Proc. of the 1st. Int. Conf. on Structural Design of Asphalt Pavements,* University of Michigau.

Vesic, A.S. (1964). "Investigation of Bearing Capacity of Piles in Sand," *Proc. No. Amer. Conf. on Deep Foundations*, Mexico city, Vol 1.

Vesic, A.S. (1965). "Ultimate Loads and Settlements of Deep Foundations," *Proc. of a Symp* held at Duke University.

Vesic, A.S. (1967). "A Study of Bearing Capacity of Deep Foundations," Final Report, School of Civil Engg., Georgia Inst. Tech., Atlanta, U.S.A.

Vesic, A.S. (1969). "Effect of Scale and Compressibility on Bearing Capacity of Surface Foundations," Discussions, *Proceedings, 7th Int. Conf. SM and FE.* Mexico City, Volume III.

Vesic, A.S. (1970). "Tests on Instrumented Piles - Ogeechee River Site," *JSMFD*, ASCE, Vol.96, No SM2.

Vesic, A.S. (1972). "Expansion of cavities in Infinite Soil Mass," *J.S.M.F.D.*, ASCE, Vol. 98.

Vesic, A.S. (1973). "Analysis of Ultimate Loads of Shallow Foundations," *JSMFD*, ASCE, Vol. 99, SM1.

Vesic, A.S. (1975). "Bearing Capacity of Shallow Foundations," *Foundation Engineering Hand Book*, Van Nostrand Reinhold Book Co., N.Y.

Vesic, A.S. (1977). *Design of Pile Foundations*, Synthesis of Highway Practice 42, Res. Bd., Washington D.C.

Vidal, H. (1969). "The Principle of Reinforced Earth," HRR No. 282.

Vijayvergiya, V.N., and Focht, J.A. Jr., (1972). *A New way to Predict the Capacity of Piles in Clay, 4th Annual Offshore Tech Conf.*, Houston, Vol. 2.

Vijayvergiya, V.N., and Ghazzaly, O.I. (1973). "Prediction of Swelling Potential of Natural Clays," *Third International Research and Engineering Conference on Expansive Clays.*

Walsh, K.D., Houston, S.L., and Houston, W.N. (1995). "Development of *t-z* Curves for Comented Fine-Grained Soil Deposits," *Journal of Geotechnical Engineering*, ASCE, Vol. 121.

Westergaard, H.M. (1917). "The Resistance of a group of Piles," *J. Western Society of Engineers,* Vol. 22.

Westergaard, H.M. (1926). Stresses in Concrete Pavement Computed by Theoretical Analysis Public Road, Vol. 7, No. 12, Washington.

Westergaard, H.M. (1938). "A Problem of Elasticity Suggested by Problem in Soil Mechanics, Soft Material Reinforced by Numerous Strong Horizontal Sheets," in *Contribution to the Mechanic of Solids*, Stephen Timoshenko 60th Anniversary Vol., Macmillan, New York.

Winterkorn, H.F., and Hsai-Yang Fang. (1975). *Foundation Engineering Hand Book,* Van Nostrand Reinhold Company, New York.

Woodward, Jr., R.J., Gardner, W.S., and Greer, D.M. (1972). *Drilled Pier Foundations*, McGraw Hill, New York.

Yoshimi, Y. (1964). "Piles in Cohesionless Soils Subjected to Oblique Pull," *JSMFD*, ASCE, Vol. 90, SM6.

INDEX

Q

R

S